Handbook of Membrane Separations

The third edition of the *Handbook of Membrane Separations: Chemical, Pharmaceutical, Food, and Biotechnological Applications* provides a comprehensive discussion of membrane applications. Fully updated to include the latest advancements in membrane science and technology, it is a one-of-its-kind overview of the existing literature. This fully illustrated handbook is written by experts and professionals in membrane applications from around the world.

Key Features:

- Includes entirely new chapters on organic solvent-resistant nanofiltration, membrane condensers, membrane-reactors in hydrogen production, membrane materials for haemodialysis, and integrated membrane distillation
- Covers the full spectrum of membrane technology and its advancements
- Explores membrane applications in a range of fields, from biotechnological and food processing to industrial waste management and environmental engineering

This book will appeal to both newcomers to membrane science as well as engineers and scientists looking to expand their knowledge on upcoming advancements in the field.

Anil Kumar Pabby served with one of the pioneer research centres of India, Bhabha Atomic Research Centre, as senior scientist and was associated with Department of Atomic Energy activities including research and development works on membrane applications under various programmes.

S. Ranil Wickramasinghe joined the Department of Chemical Engineering at the University of Arkansas in 2011 and presently serves as distinguished professor. He holds the Ross E Martin Chair in Emerging Technologies and is an Arkansas Research Alliance Scholar. His research focuses on synthetic membrane-based separation processes for purification of pharmaceuticals and biopharmaceuticals, treatment and reuse of water and the production of biofuels.

Ana-Maria Sastre is a professor of Chemical Engineering at the Universitat Politècnica de Catalunya (Barcelona, Spain), where she has been teaching Chemistry for more than 40 years. She continued working for many years in the field of solvent extraction, solvent-impregnated resins, adsorption and membrane technology processes.

Handbook of Membrane Separations

Chemical, Pharmaceutical, Food, and Biotechnological Applications

Third Edition

Edited by
Anil Kumar Pabby, S. Ranil Wickramasinghe,
and Ana-Maria Sastre

CRC Press
Taylor & Francis Group
Boca Raton London New York

CRC Press is an imprint of the
Taylor & Francis Group, an **informa** business

Designed cover image: Shutterstock

Third edition published 2024
by CRC Press
2385 NW Executive Center Drive, Suite 320, Boca Raton FL 33431

and by CRC Press
4 Park Square, Milton Park, Abingdon, Oxon, OX14 4RN

CRC Press is an imprint of Taylor & Francis Group, LLC

ISBN: 978-1-032-25918-5 (hbk)
ISBN: 978-1-032-26105-8 (pbk)
ISBN: 978-1-003-28565-6 (ebk)

DOI: 10.1201/9781003285656

Typeset in Times
by Newgen Publishing UK

Contents

About the Editors

Anil Kumar Pabby served with one of the pioneer research centres of India, Bhabha Atomic Research Centre (BARC), Tarapur, Mumbai, Maharashtra, till 2021 as Scientific Officer (Senior Scientist) and associated with Department of Atomic Energy activities including research and development work under the DAE programme. He obtained his Ph.D. from University of Mumbai, India, and subsequently carried out his postdoctoral work at Technical University of Catalunya, Barcelona, Spain. Dr. Pabby has more than 190 publications to his credit including 20 chapters and two patents on non-dispersive membrane technology. He was invited to join as an associate editor of the international *Journal of Radioanalytical and Nuclear Chemistry* during 2002–2005. He has also served as consultant to IAEA, Vienna, Austria, for developing a technical book volume on *Application of Membrane Technologies for Liquid Radioactive Waste Processing*. Dr. Pabby has taken a leading role in publishing *Handbook of Membrane Separation: Chemical, Pharmaceutical, Food and Biotechnological Applications* in July 2008 by CRC Press, New York, USA. The second edition of this book was published in April 2015.

Dr. Pabby has been a regular reviewer for several national and international journals and also served on the editorial boards of a number of reputed international journals. His research interests include membrane-based solvent extraction, liquid membranes and their modeling aspects, pressure-driven membrane processes, macrocyclic compounds, etc. Dr Pabby was elected as a Fellow of the Maharashtra Academy of Sciences (FMASc) in 2003 for his outstanding contribution in membrane science and technology. Also, he has been awarded with the prestigious Tarun Datta Memorial award (instituted by Indian Association Nuclear Chemist and Allied Scientist) for his contribution in Nuclear and Radiochemistry for the year 2005. He has been appointed as Associate Professor and Ph.D. guide and also M.Tech guide, Homi Bhabha National Institute, Mumbai. He served as Secretary, Indian Association Nuclear Chemist and Allied Scientist, Tarapur chapter, during 2008–2021.

He has delivered several keynote/plenary/invited talks in national and international conferences. He has also served as a chairman of the different sessions in conferences in India and abroad.

Ana-Maria Sastre is a professor of Chemical Engineering at the Universitat Politècnica de Catalunya (Barcelona, Spain), where she has been teaching Chemistry for more than 40 years. She received her Ph.D. from Autonomous University of Barcelona in 1982 and has been working for many years in the field of solvent extraction, solvent-impregnated resins, and membrane technology and adsorption processes.

She was a visiting fellow at Department of Inorganic Chemistry, The Royal Institute of Technology, Sweden, during 1980–1981 and carried out postdoctoral research work during October 1986–April 1987 at Laboratoire de Chimie Minerale, de l'Ecole Europeenne des Hautes Etudes des Industries Chimiques d'Estrasbourg, France.

During 2015 she was Fulbright visiting scholar at Chemical and Biomolecular Engineering Department of the University of California, Berkeley.

Prof. Sastre has published more than 200 journal publications and more than 80 papers in international conferences. She holds seven patent applications. She has advised 18 Ph.D. students and 16 master theses, and is a reviewer of many international journals. She was awarded the "Narcis Monturiol medal for scientific and technological merits" given by the Generalitat de Catalunya for outstanding contribution in science and technology in 2003.

Prof. Sastre was the head of the Chemical Engineering Department from 1999 till 2005, and from 2006 till 2013 was Vicepresident for Academic Policy of the Universitat Politècnica de Catalunya.

S. Ranil Wickramasinghe has published over 200 peer-reviewed journal articles, several book chapters, and is co-editor of a book on responsive membrane and materials. He is active in the American Institute of Chemical Engineers and the North American Membrane Society. He is executive editor of *Separation Science and Technology*. Prof. Wickramasinghe's research interests are in membrane science and technology. His research focuses on synthetic membrane-based separation processes for purification of pharmaceuticals and biopharmaceuticals, treatment and reuse of water and for the production of biofuels. Typical unit operations include: microfiltration, ultrafiltration, virus filtration, nanofiltration, membrane extraction, etc. A current research focus is surface modification of membranes in order to impart unique surface properties. His group is actively developing responsive membranes. These membranes change their physical properties in response to changed environmental conditions. A second research focus is the development of catalytic membranes for biomass hydrolysis by grafting catalytic groups to the membrane surface. In 2020 he helped found a start-up company to commercialize this technology. Prof. Wickramasinghe obtained his Bachelor's and Master's degrees from the University of Melbourne, Australia, in Chemical Engineering. He obtained his Ph.D. from the University of Minnesota, also in Chemical Engineering. He worked for 5 years in the biotechnology/biomedical industry in the Boston area before joining the faculty of Chemical Engineering at Colorado State University. He joined the Department of Chemical Engineering at the University of Arkansas in 2011 where he is a distinguished professor. He holds the Ross E Martin Chair in Emerging Technologies and is an Arkansas Research Alliance Scholar.

Contributors

Tallam Aarti
Membrane Separations Group,
Process Engineering and Technology Transfer Division,
CSIR-Indian Institute of Chemical Technology, Telangana, Hyderabad 500 007, India

Francesca Alessandro, Ph.D.
1Institute on Membrane Technology (CNR-ITM), Via P. Bucci, 17/C, Rende (CS), Italy

M. Inês G. S. Almeida, Ph.D.
School of Chemistry, The University of Melbourne, Victoria 3010, Australia

Dhaval Bhandari, Ph.D.
ExxonMobil Technology & Engineering Company
4500 Bayway Drive
Baytown, Texas, 77520, USA

Eugenio Bringas, Ph.D.
Departamento de Ingenierías Química y Biomolecular,
ETSIIyT. Universidad de Cantabria
Avda. de los Castros
39005 Santander, Spain

Robert W. Cattrall, D.Sc.
School of Chemistry,
The University of Melbourne,
Victoria 3010, Australia

S. Chang, Ph.D.
School of Engineering
The University of Guelph,
Guelph, Ontario, Canada N1G 2W1

George Q. Chen
Department of Chemical Engineering,
The University of Melbourne,
Victoria 3010, Australia.

Chidambaram Thamaraiselvan
Interdisciplinary Centre for Energy Research
Indian Institute of Science (IISc)
Bangalore-560012, INDIA

J.L. Cortina, Ph.D.
Chemical Engineering Department,
Escola d'Enginyeria de Barcelona Est (EEBE),
Universitat Politècnica de Catalunya (UPC)-BarcelonaTECH,
Campus diagonal Besòs, C/ Eduard Maristany 10-14,
08930 Barcelona, Spain.

Zhongde Dai, Ph.D.
College of Carbon Neutral Future Technology,
Sichuan University,
610065, Chengdu, China

A.P. de los Ríos, Ph.D.
Department of Chemical Engineering,
Faculty of Chemistry, University of Murcia (UMU),
P.O. Box 4021, Campus de Espinardo,
E-30100, Murcia, Spain

Liyuan Deng, Ph.D.
Department of Chemical Engineering,
Norwegian University of Science and technology (NTNU),
7491-Trondheim, Norway

Enrico Drioli, Emeritus Professor
Institute on Membrane Technology (CNR- ITM),
Via P. Bucci, 17/C, Rende (CS), Italy and
Department of Environmental Engineering,
University of Calabria, Via P. Bucci, 44/A, Rende (CS), Italy

A.G. Fane, Ph.D.
UNESCO Centre for Membrane Science & Technology
School of Chemical Engineering,
The University of New South Wales,
Sydney, NSW 2052, Australia

Ramin Farnood
Department of Chemical Engineering and Applied Chemistry,
University of Toronto, Canada

Mirko Frappa, Ph.D.
Institute on Membrane Technology (CNR- ITM),
Via P. Bucci, 17/C, Rende (CS), Italy

Vikrant Gaikwad, Ph.D.
Department of Chemical Engineering,
Dr. Vishwanath Karad MIT World Peace University,
Pune-411038, Maharashtra, India.

Olviya S Gonsalves
Department of Chemical Engineering,
Institute of Chemical Technology,
Mumbai, 400 019, India

Mattheus F. A. Goosen
Office of Research & Graduate Studies
Alfaisal University
Alfaisal University, Riyadh, Saudi Arabia

May-Britt Hägg, Dr. Techn.
Department of Chemical Engineering,
Norwegian University of Science and technology (NTNU),
N-7491 Trondheim, Norway
PO Box 50927, Riyadh 11533
Saudi Arabia,

A. Hernández-Fernández
Department of Chemical Engineering,
Faculty of Chemistry, University of Murcia (UMU),
P.O. Box 4021, Campus de Espinardo,
E-30100, Murcia, Spain

F.J. Hernández-Fernández, Ph.D.
Department of Chemical and Environmental Engineering,
Technical University of Cartagena, Campus La Muralla,
C/ Doctor Fleming S/N, E-30202 Cartagena,
Murcia, Spain

Fausto Gallucci, Ph.D.
Sustainable Process Engineering,
Chemical Engineering and Chemistry,
Eindhoven University of Technology,
De Rondom 70, 5612, Eindhoven,
the Netherlands

Arun M. Isloor, Ph.D.
Membrane & Separation Technology Laboratory,
Department of Chemistry,
National Institute of Technology Karnataka,
Surathkal, Mangalore 575 025, India

A.F. Ismail, Ph.D.
Advanced Membrane Technology
Research Centre (AMTEC),
Universiti Teknologi Malaysia,
81310 Skudai, Johor, Malaysia

Noel J. Kaleekkal
Membrane Separation Group,
Department of Chemical Engineering,
National Institute of Technology Calicut (NITC),
Kerala- 673601, India

Mohd Azraai Kassim, Ph.D.
Advanced Membrane Technology
Research Centre (AMTEC),
Universiti Teknologi Malaysia,
81310 Skudai, Johor, Malaysia

Sandra E. Kentish
Department of Chemical Engineering,
The University of Melbourne,
Victoria 3010, Australia.

Spas D. Kolev, Ph.D.
School of Chemistry,
The University of Melbourne,
Victoria 3010, Australia

Sowmya M. Kumar
Department of Prosthodontics,
Nitte Deemed to be University,
A.B Shetty Memorial Institute
of Dental Sciences (ABSMIDS),
Deralakatte, Mangalore 575 022, India.

J. López, Ph.D.
Chemical Engineering Department,
Escola d'Enginyeria de Barcelona Est (EEBE),
Universitat Politècnica de Catalunya (UPC)-BarcelonaTECH,
Campus diagonal Besòs, C/ Eduard Maristany 10-14,
08930 Barcelona, Spain.

Francesca Macedonio, Ph.D.
Institute on Membrane Technology (CNR- ITM),
Via P. Bucci, 17/C, Rende (CS), Italy

M. Madhumala, Ph.D.
Membrane Separations Group,
Process Engineering and Technology Transfer Division,
CSIR-Indian Institute of Chemical Technology,
Hyderabad, Telangana, 500 007, India

Takeshi. Matsuura, Ph.D.
Department of Chemical and
Biological Engineering, University of Ottawa,
161 Louis Pasteur St, Ottawa, K1N 6N5, Canada.

Ch Sn Murthy
Department of Mining Engineering,
National Institute of Technology Karnataka,
Surathkal, Mangalore, Karnataka – 575025, India

Nitish Mittal, Ph.D.
ExxonMobil Technology & Engineering Company
4500 Bayway Drive
Baytown, Texas, 77520, USA

Abdul Wahab Mohammad
Chemical and Process Engineering,
Universiti Kebangsaan Malaysia,
43600, Bangi, Selangor, Malaysia

Parag R. Nemade
Department of Chemical Engineering,
Institute of Chemical Technology,
Mumbai, 400 019, India

Nethravathi
Membrane and Separation Technology Laboratory,
Laboratory, Department of Chemistry,
National Institute of Technology Karnataka,
Surathkal, Mangalore, Karnataka-575025, India

Inmaculada Ortiz, Ph.D.
Departamento de Ingenierías Química y Biomolecular
ETSIIyT. Universidad de Cantabria
Avda. de los Castros
39005 Santander. SPAIN

V.M. Ortiz-Martínez
Department of Chemical and Environmental
Engineering, Technical University of Cartagena,
Campus La Muralla,
C/ Doctor Fleming S/N,
E-30202 Cartagena, Murcia, Spain

Anil Kumar Pabby, Ph.D.
Formerly associated with Nuclear Recycle Board,
Bhabha Atomic Research Centre,
Tarapur, 401502, India

K.C. Pallavi
Membrane and Separation Technology Laboratory,
Department of Chemistry,
National Institute of Technology Karnataka,
Surathkal, Mangalore, Karnataka-575025, India

Pallavi Mahajan-Tatpate
Department of Chemical Engineering,
Dr. Vishwanath Karad MIT World Peace University,
Pune-411038, Maharashtra, India

B. Prabhu
Department of Chemical Engineering,
Manipal Institute of Technology,
Manipal University, Manipal,
Karnataka – 576104, India

Dipak Rana
Department of Chemical and
Biological Engineering, University of Ottawa,
161 Louis Pasteur St., Ottawa, ON, K1N 6N5, Canada

M. Reig, Ph.D.
Chemical Engineering Department,
Escola d'Enginyeria de Barcelona Est (EEBE),
Universitat Politècnica de Catalunya
(UPC)-BarcelonaTECH,
Campus diagonal Besòs, C/ Eduard Maristany 10-14,
08930 Barcelona, Spain.

Ma.- Fresnedo San Román, Ph.D.
Departamento de Ingenierías Química y Biomolecular,
ETSIIyT. Universidad de Cantabria
Avda. de los Castros
39005 Santander. SPAIN

S. Sánchez-Segado, Ph.D.
Department of Chemical and Environmental
Engineering, Technical University of Cartagena,
Campus La Muralla, C/ Doctor Fleming S/N, E-30202
Cartagena,
Murcia, Spain

V. Sangeetha
Membrane Separation Group,
Department of Chemical Engineering,
National Institute of Technology Calicut (NITC),
Kerala- 673601, India.

Ana-Maria Sastre, Ph.D.
Chemical Engineering Department
Universitat Politècnica de Catalunya
ETSEIB, Diagonal 647
E-08028 Barcelona (Spain)

P. Satishkumar
Membrane & Separation Technology Laboratory,
Department of Chemistry,
National Institute of Technology Karnataka,
Surathkal, Mangalore 575 025, India

B. Shivarama
Membrane & Separation Technology Laboratory,
Department of Chemistry,
National Institute of Technology Karnataka,
Surathkal, Mangalore 575 025, India

S. Sridhar, Ph.D.
Membrane Separations Group,
Process Engineering and
Technology Transfer Division,
CSIR-Indian Institute of Chemical Technology,
Telangana, Hyderabad 500 007, India

David. A. Pacheco Tanaka, Ph.D.
TECNALIA, Basque Research and Technology Alliance
(BRTA),
Mikeletegi Pasealekua 2, 20009,
Donostia-San Sebastian, Spain

Elena Tocci, Ph.D.
Institute on Membrane Technology (CNR- ITM),
Via P. Bucci, 17/C, Rende (CS), Italy

Ana M. Urtiaga, Ph.D.
Departamento de Ingenierías Química y Biomolecular,
ETSIIyT. Universidad de Cantabria
Avda. de los Castros
39005 Santander, Spain

P.S.V. Vaishnavi
Department of Chemical Engineering,
Institute of Chemical Technology,
Mumbai, 400 019, India

S. Ranil Wickramasinghe, Ph.D.
Ralph E. Martin Department of
Chemical Engineering, University of
Arkansas, Fayetteville,
AR 72701-1201, USA

Grażyna Zakrzewska-Koltuniewicz, Ph.D.
Institute of Nuclear Chemistry and Technology,
03-195 Warszawa, Dorodna 16,
Poland

Foreword

A membrane is an interphase between two adjacent phases acting as a selective barrier, at the same time organizing a system into compartments and regulating the transport between the two compartments. This well-known principle from biology has in the recent decades been translated into numerous industrial separation processes using synthetic membranes in suited modules, units, and plants. The main advantages of membrane technology as compared to other unit operations in (bio)chemical engineering are related to the unique separation principle, i.e., the transport selectivity of the membrane. Separations with membranes do not require additives, and they can be performed isothermally and with very competitive energy consumption. Upscaling of membrane processes as well as their integration into other separation or reaction processes are easy. The success of membrane technology had been impressively demonstrated for various important industrial applications with very large numbers or/and sizes of membrane modules or units, for example, blood detoxification by dialysis ("artificial kidney") or water desalination by reverse osmosis. In parallel, the implementation of membranes in many other industries has been realized and is steadily expanding. It is recognized that membranes play a key role in achieving the Sustainable Development Goals (SDGs) as proclaimed by the United Nations (UN) for 2030, because many of the major challenges our societies are facing can be very effectively addressed by using membrane technologies.

From the perspective of research, development, and implementation, the field is very exciting because knowledge and expertise in different disciplines of fundamental sciences and engineering must be integrated. In order to realize the different established or emerging concepts, a critical analysis of the state-of-the-art as well as guidelines for advances in membrane materials and membrane process engineering are very important.

The *Handbook of Membrane Separations: Chemical, Pharmaceutical, Food, and Biotechnological Applications* is devoted to a wide range of industrial membrane separation processes. The previous two editions of the Handbook have been received with great interest by researchers and professionals in industry because the book provides an overview on relevant fundamentals of synthetic membranes and comprehensive insights into a large number of different applications with significant impacts on society. This new edition builds on the success of previous editions and reflects also the very dynamic development of membrane science and technology. Internationally recognized experts with proven experience in the field have updated or newly written chapters for the Handbook. The first part of the book is devoted to established applications of membranes for critical separations in various industries. It is an over-arching aspect in all chapters with their focus on a concrete topic, that the above-mentioned advantages of membranes toward more sustainable process engineering are also illustrated. The second part of the book covers waste management, environmental aspects, and future trends of membrane technologies. Hence, the content of the chapters is not only devoted to separation and purification of industrial products, but also to topics such as processes with "zero waste" or for "resource recovery" as important aspects of a future circular economy.

Overall, the Handbook provides a comprehensive overview of the current state-of-the-art and future trends of membrane science and technology and the implementation of membrane separations for many important applications. It is very well suited for readers that are newcomers to the field as well as for engineers and scientists that already have knowledge and experience in membrane science and technology, but wish to gain specific insights into specialized established and future applications of membranes. With such scope, the Handbook can also serve as a valuable reference book for graduate courses on separation engineering and related topics at universities. It is expected that the Handbook will have a significant impact on the further development of the exciting and relevant field of membrane separations.

Prof. Dr. Mathias Ulbricht
Universität Duisburg-Essen
Germany

Preface

Researchers are looking for more attractive methods to overcome some of the inherent shortcomings of conventional technologies. New strategies should be environmentally benign, cost-effective, and easily implementable. Membrane technology has been rated very highly and is widely applied successfully as an effective separation technology for various applications. This new, revised third edition of the *Handbook of Membrane Separations: Chemical, Pharmaceutical, Food, and Biotechnological Applications* addresses the latest developments and important milestones that have appeared in the literature since the second edition was published. Newer technologies that have emerged and novel applications or trends for future applications along with relevant references are the hallmark of this new edition. An expanded and updated version of its well-received predecessor, this Handbook includes new chapters in the field of membrane science and technology dealing with chitosan as a new material for membrane preparation, organic solvent nanofiltration, air purification and dehydration via membranes, metal-organic framework contained polymeric membranes, membrane material for haemodialysis, membrane bioreactors, dye removal using concentration and pressure-driven membrane techniques, and the role of membranes in reducing emissions in the petrochemical industry.

The Handbook is divided into two main sections: the first section deals with membrane applications in the chemical, biochemical, food processing and pharmaceutical industries. In the second section, membrane applications in industrial waste management (including nuclear), environmental engineering, and future trends in membrane science are presented. Each section is divided into chapters that deal with the subject matter in depth and focus on cutting-edge advancements in the field. Several well-respected authors have also written new chapters under the supervision of the editors, and each chapter was peer-reviewed for content and style before it was accepted for publication. Many of the existing chapters dealing with emerging areas have been updated in order to record the latest advances in that fields. The aim has been to maintain the perspective of a practical handbook rather than merely a collection of review chapters.

The editors would like to acknowledge the contributions of a number of authors and institutions that have played a major role in drafting the Handbook, from conception to publication. The Handbook would not have been possible without their input. These contributors are leading experts in their fields and bring a great wealth of experience to this book. The editors would also like to acknowledge the efforts of the reviewers who devoted their valuable time in critically evaluating the chapters before the set deadlines and suggested improvements to maintain the high standard of the Handbook. Finally, we would like to acknowledge the support of our home institutions at every stage in the Handbook's conception: the Ex-Bhabha Atomic Research Centre, Mumbai, India, University of Arkansas, Fayetteville, USA, and the Technical University of Catalonia, Barcelona, Spain.

Anil Kumar Pabby
S. Ranil Wickramasinghe
Ana-Maria Sastre

Acknowledgements

I am thankful to my colleagues and research group working on membranes in Bhabha Atomic Research Centre and Homi Bhabha National Institute who inspired me to initiate this project and helped make this happen. I especially want to thank my family members, my loving wife, Anju, my son Anubhav, daughter Akanksha and son-in-law, Arjun, and my cute grandson, Avyan. It is well understood that this book would not have been possible without their unconditional support and love during the different stages of this document.

Anil Kumar Pabby, India

I am very appreciative of my research group for their dedication to the pursuit of knowledge. Finally, this book would not have been possible without the love, dedication, and support of my wife Xianghong and son Aroshe which has enabled me to pursue my scientific interests.

S. Ranil Wickramasinghe, USA

I would like to acknowledge my research group and colleagues of my department for their valuable discussions and help. I would like to acknowledge, as well, my husband Antonio and my sons Carles and Jordi for their support and patience during all these times as without their support this project would not have been possible.

Ana-Maria Sastre, Spain

1 Membrane Applications in Chemical, Biochemical, and Food Processing and Pharmaceutical Industries

Introduction

Anil Kumar Pabby[1], S. Ranil Wickramasinghe[2], and Ana-Maria Sastre[3]*
[1]Formerly associated with Nuclear Recycle Board, BARC, Tarapur, Maharashtra, India
[2]Ralph E. Martin, Department of Chemical Engineering, University of Arkansas, Fayetteville, AR, USA
[3]Department of Chemical Engineering, Universitat Politècnica de Catalunya
Barcelona, Spain
*E-mail: dranilpabby@gmail.com

Over the last two decades, membrane separation has evolved into a mature process domain that is widely used in industrial applications such as water and wastewater treatment, desalination, bioseparation, and food and beverage processing [1]. The successful evolution of membrane separations has resulted from a high degree of interdisciplinary research and development, particularly between polymer chemists, material scientists, and mechanical and process engineers responsible for fabricating, manufacturing, and evaluating new membranes [2,3]. On the other hand, owing to rapid economic development globally, the climate and environment have continued to deteriorate.

Natural phenomena such as global warming and acid rain are closely related to emission of greenhouse gases, industrial production, and energy use [4–8]. According to data from the National Oceanic and Atmospheric Administration/Earth System Research Laboratory, global CO_2 emissions are steadily increasing, and the CO_2 concentration in the atmosphere has reached 411 ppm, a record high [9]. Further, the concentrations of atmospheric and other gases have increased, resulting in the greenhouse effect and the acidification of ecosystems, which have caused significant losses to the construction industry, agriculture, and public facilities. Studies have shown that fossil fuels, including coal, oil, and natural gas, will continue to dominate the energy structure over the short term [10]. Therefore, the Organization for Economic Cooperation and Development (OECD) countries and India have formulated increasingly strict regulations and laws to limit gas emissions to ensure the sustainable consumption of fossil energy [11,12]. However, typical challenges in membrane processes such as permeance-selectivity trade-off [13–15] and membrane fouling [16–18] have pushed membrane researchers to seek the next generation of membranes with better separation performance, less environmental impact, and multi-functional characteristics [19–22].

Over the last decade, process intensification has been gaining paramount importance and has slowly become part of the already established and newly developed technological processes. Drioli et al. [23] define process intensification as the strategy to bring drastic improvements in manufacturing and processing by decreasing capital cost, equipment size, energy consumption, waste production, environmental impact, etc. In this context, membrane operations have the potential to replace conventional energy-intensive techniques, accomplish the selective and efficient transport of specific components, improve the performance of reactive processes, and, ultimately, provide reliable options for sustainable growth [24]. In addition, membrane processes can be beneficially integrated at different levels because of their several advantages over conventional processes, including: compactness, ease of scale-up, and automation [25,26].

Pressure-driven processes such as ultrafiltration, nanofiltration, and microfiltration have been already established, and various applications have been commercialized in the food, pharmaceutical, and biotechnology fields. The development of a means of characterizing, controlling, and preventing membrane fouling has been shown to be critical for all membrane processes. Engineering tailored membranes, fouling prevention, and optimization of chemical cleaning will ensure a high level of membrane process performance. Over the last 10 years, developments of new techniques for membrane characterization and improvements in existing

DOI: 10.1201/9781003285656-1

techniques have increased our knowledge of the mechanisms involved in membrane fouling. The advanced techniques used for membrane fouling detection will not only provide useful insights into the fouling mechanism but also augment our understanding of the factors that affect membrane fouling.

Membrane materials can be primarily classified into organic polymer membranes, inorganic membranes, and organic–inorganic composite membranes. Polypropylene (PP) [27], polyvinylidenefluoride (PVDF) [28], polytetrafluoroethylene (PTFE) [29], polysulfone (PS) [30], polyethersulfone (PES) [31], and polyimide (PI) [32] are some of the most widely used membrane materials. In recent years, ceramic membranes have become a hot research topic owing to their high temperature resistance, strong thermal and chemical stability, and low energy consumption [33], however, their cost is relatively high. In addition, researchers have recently discovered that membrane modification treatment is an effective and economical method to improve membrane performance, and matrix and surface modifications are the two main methods to modify membranes. Surface hydrophobic modification has proved to be an effective approach to improve membrane wetting resistance. Ahmad and co-workers [34] fabricated a layer of porous superhydrophobic low-density polyethylene on a PVDF hollow-fiber membrane. The hydrophobic modified membrane displayed higher CO_2 absorption flux than that of an unmodified membrane. Zhang et al. [35] developed a highly hydrophobic organic–inorganic composite hollow-fiber membrane by incorporating a fluorinated silica inorganic layer on a polyetherimide (PEI) organic substrate. The inorganic layer provided high hydrophobicity and protected the polymeric substrate from chemical absorbents.

Sirkar et al. [36], in a novel approach, recently demonstrated that the technique of solid hollow-fiber membrane-based cooling crystallization can be employed to achieve continuous polymer precipitation from a solution and subsequent polymer coating/encapsulation of sub-micrometer particles/drug crystals as well as nanoparticles in suspension. In this technique, there is no need for high pressure as in supercritical solvent-based antisolvent crystallization; further, there is easy scale-up and production of free-flowing polymer-coated particles.

Another fascinating area is the use of ionic liquid coupled with a hollow-fiber contactor. In this direction, Irabien et al. [37] demonstrated the potential of coupling ionic liquid 1-ethyl-3-methylimidazolium ethylsulfate ([emim][$EtSO_4$]) and hollow-fiber membrane contactors for post-combustion CO_2 capture. CO_2 absorption experiments in counter-current configuration were carried out, followed by comprehensive two-dimensional dynamic modeling based on steady-state and pseudo-steady-state operating modes. This model considers the level of wetting of porous hollow fibers. The model predicted the effects of membrane wetting, porosity, tortuosity, module length, fiber inner diameter, and gas and absorbent flow rates. Membrane wetting has a noteworthy effect on CO_2 capture efficiency. A smaller amount of wetting can cause huge resistance in CO_2 transport through the membrane. The separation efficiency was enhanced by using membranes with high porosity and low tortuosity and decreased by enhancing the gas flow rate and absorbent flow rate reduction. Therefore, CO_2 capture is enhanced.

Petukhov et al. [38] proposed a combined process for gas stream dehumidification using a porous polypropylene membrane contactor heat-exchanger with a liquid coolant absorbent and reported the results on its operation performance. This technique enables the effective removal of condensable components of gas mixtures to ultimate residual pressures by a synergetic contribution of two effects—suppression of the equilibrium water vapor pressure at low temperatures and water vapor removal by hygroscopic absorbent by increasing the module length and a reduction of the inner diameter of fibers.

The use of membrane contactors for ozone diffusion in water treatment recently emerged as a very interesting option. Indeed, by using a bubbleless operation, membrane contactors can overcome these challenges [39]. As supported by the literature, membrane contactors have been pointed out as a good alternative for the transfer of gas to the liquid phase [40] and to control the dosage of ozone during ozonation processes [39]. Recently, water contamination caused by ammonia effluents has become a critical environmental issue. In this study, a coupled system of water splitting and hollow-fiber extraction was proposed for highly efficient ammonia capture from wastewater that has a high ammonia nitrogen concentration [41].

In the last few decades, the presence of active pharmaceutical materials (endocrine-disrupting compounds) in aqueous environments has attracted increased attention due to their potential to change the function of hormones and physiological conditions of humans and animals. The existence of pharmaceutical contaminants in wastewater streams is related to hospital effluents, therapeutic drugs, pharmaceutical industry effluents, and personal hygiene products. The concentration of pharmaceutical pollutants in effluent systems and environmental waters is almost between nano- and micrograms per liter [42]. Membrane contactor-based technology has been applied to remove pharmaceutical compounds from aqueous streams [43,44]. In another study [45], a hollow-fiber membrane filled with n-octanol was immersed into a sample containing trace amounts of pharmaceutical compounds and an enriched sample of contaminants was obtained using this method.

Porous membranes having a particular wetting characteristic, hydrophobic or hydrophilic, are used for nondispersive membrane solvent extraction (MSX) where two immiscible phases flow on the two sides of the membrane. The aqueous–organic phase interface across which solvent extraction/back extraction occurs remains immobilized on one surface of the membrane. This process requires the pressure of the phase not present in membrane pores to be either equal to or higher than that of the phase present in the membrane pores; the excess phase pressure should not exceed a breakthrough pressure. In countercurrent MSX with significant flow pressure drop in each phase, this often poses a problem. To overcome this problem, flat porous Janus membranes were developed by Sirkar et al. [46] using either a base polypropylene (PP), polyvinylidene fluoride (PVDF), or polyamide (Nylon) membrane, one side of which is hydrophobic with the other being hydrophilic. Such membranes were characterized using the contact angle, liquid entry pressure (LEP), and the droplet breakthrough pressure from each side of the membrane along with characterizations via scanning electron microscopy (SEM) and Fourier transform infrared spectroscopy (FTIR). Nondispersive solvent extractions were carried out successfully for two systems, octanol–phenol (solute)–water, toluene–acetone (solute)–water, with either flowing phase at a pressure higher than that of the other phase. The phenol extraction system had a high solute distribution coefficient, whereas acetone prefers both phases to be almost identical. The potential practical utility of the MSX technique will be substantially enhanced via Janus MSX membranes.

The area of food processing is growing very rapidly, with a good industrial market. The global fruit and vegetable processing market was valued at USD 230.96 billion in 2016 and was projected to grow at a CAGR of 7.1% from 2017, to reach USD 346.05 billion by 2022 [47]. The fruit juice industry is one of the food sectors that has invested the most in the implementation of new technologies, such as non-thermal technologies. Among them, membrane processes are considered today well-established separation techniques to support the production and marketing of innovative fruit juices designed to exploit the sensory characteristics and nutritional peculiarities of fresh fruits. Pressure-driven membrane operations, membrane distillation, osmotic distillation, and pervaporation have been widely investigated in the last few decades to replace the conventional technologies used in the fruit juice processing industry (i.e., clarification, stabilization, concentration, and recovery of aroma compounds) [48].

Another study investigated the potential of hybrid membrane processes including microfiltration (MF), ultrafiltration (UF), and forward osmosis (FO) for non-thermal concentration of apple juice. The process performance and characteristics (physicochemical properties, nutritional and aroma components, and microbiological quality) of apple juice were studied [49]. The clarity of apple juice was significantly promoted as the pore size of the membrane was reduced. MF and UF can also ensure microbiological safety in pre-treated apple juice. According to its filtration efficiency as well as its simultaneous clarification and cold-sterilization performance, 0.22 μm MF membrane was identified as the optimal membrane for the pre-treatment. The pre-treated apple juice can be concentrated up to 65°Brix by subsequent single-stage FO. FO retained nutritional and volatile compounds of apple juice while significant reductions were found in the juice concentrated by vacuum evaporation. Hybrid MF–FO could be a promising non-thermal technology to produce apple juice concentrate of high quality.

In this emerging area, porous metal membranes have recently received increasing attention, and significant progress has been made in their preparation and characterization. This progress has stimulated research in their applications in a number of key industries, including wastewater treatment, dairy processing, wineries, and biofuel purification. So far, porous metal membranes have been mainly used for the filtration of liquids to remove solid particles. For porous metal membranes to be more widely used across a number of separation applications, particularly for water applications, further work needs to focus on the development of smaller pore (e.g., sub-micron) metal membranes and the significant reduction of capital and maintenance costs [50].

This section addresses new developments in membrane science and technology. Included in this edition are newer practices and technologies and their applications or trends for future applications with relevant references that have appeared in the literature since the second edition was published. Several new chapters on emerging areas such as membrane separation in chitosan as new material for membrane preparation, organic solvent nanofiltration, air purification and dehydration via membranes, metal–organic framework-contained polymeric membranes, membrane materials for hemodialysis, and the role of membrane in reducing emissions in the petrochemical industry have been added to the third edition.

Chapter 1 (this chapter) presents an overview of different membrane processes and a description of all of the chapters presented in this edition. Chapter 2 presents a comprehensive review of the reverse osmosis membrane, the latest developments in the field, important installations demonstrating this technology, and the future scope of RO processes. Chapter 3 presents the latest advances in the organic solvent nanofiltration process. This chapter describes the important applications, current status of technology, and future perspectives. Chapter 4 presents new developments in air purification and dehydration via membrane-based systems. It first provides a brief introduction to the theory of the current technique and then discusses the model, the criteria for selection, and the most important latest applications. Chapter 5 focuses on the important aspects of membrane applications in gas separation. It deals with the subject comprehensively, providing an introduction and discussing transport mechanisms, different membrane materials for gas separation, module design, current and potential applications, and novel developments in the field. Chapter 6 presents the various methodologies or techniques for improving the membrane performance of liquid-phase membrane processes by improved control of concentration polarization. Chapter 7 focuses on membrane fouling and the strategies used to reduce it relative to pressure-driven processes. This chapter highlights recent strategies for minimizing membrane fouling. In particular, it discusses the latest literature on the fouling phenomena in pressure-driven membrane systems, the analytical techniques employed to quantify fouling, preventive methods, and membrane-cleaning strategies. Specific recommendations are also made on how membrane users may find it helpful to improve the performance of these systems by minimizing the membrane-fouling phenomena. Chapter 8 provides an introduction to membrane applications and recent advances in the biochemical and pharmaceutical industries, its current status, and future potential in this very important area. Chapter 9 presents studies in an emerging area dealing with new materials for developing membranes, i.e. chitosan and its derivatives as potential materials. This chapter also elaborates on the latest applications, current status, and future challenges in this field. Chapter 10 presents a review of diverse membrane materials for hemodialysis. Chapter 11 describes a new emerging area of membranes where the membrane plays an important role in reducing emissions in the petrochemical industry and its future perspectives. Chapter 12 focuses on the use of ionic liquid technology for the selective separation of organic compounds and metal ions. The current state of research and development in the sprawling industrial use of membranes in the dairy industry for various processing purposes is presented in Chapter 13. Chapter 14 presents the metal–organic framework-contained polymeric membranes for fuel cells. Finally, Chapter 15 describes recent developments on supported Pd membranes for hydrogen separation and membrane reactors.

REFERENCES

1. Baker, R.W. (2012) *Membrane Technology and Applications*. John Wiley & Sons.
2. Lonsdale, H.K. (1982) The growth of membrane technology. *Journal of Membrane Science* 10: 81–181.
3. Li, N.N., Fane, A.G., Ho, W.W., Matsuura, T. (2011) *Advanced Membrane Technology and Applications*. John Wiley & Sons.
4. Zhang, N., Pan, Z., Zhang, Z., Zhang, W., Zhang, L. (2020) CO_2 capture from coalbed methane using membranes: A review. *Environmental Chemistry Letters* 18: 79–96.
5. Dagan-Jaldety, C., Fridman-Bishop, N., Gendel, Y. (2020) Nitrate hydrogenation by microtubular CNT-made catalytic membrane contactor. *Chemical Engineering Journal* 401: 126142.
6. Nieminen, H., Järvinen, L., Ruuskanen, V., Laari, A., Ahola, J. (2020) Mass transfer characteristics of a continuously operated hollow-fiber membrane contactor and stripper unit for CO_2 capture. *International Journal of Greenhouse Gas Control* 98: 103063.
7. Cabezas, R., Prieto, V., Plaza, A., Merlet, G., Quijada-Maldonado, E., Torres, A., Yáñez-S, M., Romero, J. (2020) Extraction of vanillin from aqueous matrices by membrane-based supercritical fluid extraction: Effect of operational conditions on its performance. *Industrial & Engineering Chemistry Research* 59: 14064–14074.
8. Pan, Z., Zhang, N., Zhang, W., Zhang, Z. (2020) Simultaneous removal of CO_2 and H_2S from coalbed methane in a membrane contactor. *Journal of Cleaner Production* 273: 123107.
9. Zhang, Z., Pan, S.-Y., Li, H., Cai, J., Olabi, A.G., Anthony, E.J., Manovic, V. (2020) Recent advances in carbon dioxide utilization. *Renewable and Sustainable Energy Reviews* 125: 109799.
10. Algarini, A., Ozturk, I. (2020) The relationship among GDP, carbon dioxide emissions, energy consumption, and energy production from oil and gas in Saudi Arabia. *International Journal of Energy Economics and Policy* 10: 280–285.
11. Wang, Z.S., Pan, L.-B. (2014) Implementation results of emission standards of air pollutants for thermal power plants: A numerical simulation. *Environmental Sciences* 35: 853–863.
12. Wang, B., Lau, Y.S., Huang, Y., Organ, B., Ho, K.F. (2019) Investigation of factors affecting the gaseous and particulate matter emissions from diesel vehicles. *Air Quality, Atmosphere and Health* 12: 1113–1126.
13. Geise, G.M., Park, H.B., Sagle, A.C., Freeman, B.D., McGrath, J.E. (2011) Water permeability and water/salt selectivity tradeoff in polymers for desalination. *Journal of Membrane Science* 369: 130–138.
14. Park, H.B., Kamcev, J., Robeson, L.M., Elimelech, M., Freeman, B.D. (2017) Maximizing the right stuff: The trade-off between membrane permeability and selectivity. *Science* 356: 6343.
15. Yang, Z., Guo, H., Tang, C.Y. (2019) The upper bound of thin-film composite (TFC) polyamide membranes for desalination. *Journal of Membrane Science* 590: 117297.
16. Greenlee, L.F., Lawler, D.F., Freeman, B.D., Marrot, B., Moulin, P. (2009) Reverse osmosis desalination: Water sources, technology, and today's challenges. *Water Research* 43: 2317–2348.
17. Gao, W., Liang, H., Ma, J., Han, M., Chen, Z.-l. Han, Z.-s., Li, G.-b. (2011) Membrane fouling control in ultrafiltration technology for drinking water production: A review. *Desalination* 272: 1–8.
18. Tijing, L.D., Woo, Y.C., Choi, J.-S., Lee, S., Kim, S.-H., Shon, H.K. (2015) Fouling and its control in membrane distillation—A review. *Journal of Membrane Science* 475: 215–244.
19 Yang, Z., Ma, X.-H., Tang, C.Y. (2018) Recent development of novel membranes for desalination. *Desalination* 434: 37–59.
20. Elimelech, M., Phillip, W.A. (2011) The future of seawater desalination: Energy, technology, and the environment. *Science* 333: 712–717.
21. Werber, J.R., Osuji, C.O., Elimelech, M. (2016) Materials for next-generation desalination and water purification membranes. *Nature Reviews Materials* 1: 1–15.
22. Nunes, S.P., Culfaz-Emecen, P.Z., Ramon, G.Z., Visser, T., Koops, G.H., Jin, W., Ulbricht, M. (2020) Thinking the future of membranes: perspectives for advanced and new membrane materials and manufacturing processes. *Journal of Membrane Science* 598: 117761.
23. Drioli, E., Stankiewicz, A.I., Macedonio, F. (2011) Membrane engineering in process intensification—An overview. *Journal of Membrane Science* 380: 1–8.
24. Drioli, E., Curcio, E., Di Profio, G. (2005) State of the art and recent progresses in membrane contactors. Transactions of IChemE, *Part A: Chemical Engineering Research and Design* 83(A3): 223–233.
25. Bernardo, P., Clarizia, G. (2011) Potential of membrane operations in redesigning industrial processes. The ethylene oxide manufacture. *Chemical Engineering Transactions* 25: 617–622.
26. Pabby, A.K., Sastre, A.M. (2013) State-of-the-art review on hollow fibre contactor technology and membrane-based extraction processes. *Journal of Membrane Science* 430: 263–303.
27. Wang, D., Teo, W., Li, K. (2002) Removal of H_2S to ultra-low concentrations using an asymmetric hollow fibre membrane module. *Separation and Purification Technology* 27: 33–40.

28. Xian, Y., Shui, Y., Li, M., Pei, C., Zhang, Q., Yao, Y. (2020) pH-Dependent thermoresponsive poly[2-(diethylamino) ethyl acrylamide]-grafted PVDF membranes with switchable wettability for efficient emulsion separation. *Journal of Applied Polymer Science* 137: 49032.
29. Zhu, Z., Kang, G., Sun, Y., Yu, S., Li, M., Xu, J., Cao, Y. (2019) An experimental study on synthesis of glycolic acid via carbonylation of formaldehyde using PTFE membrane contactor. *Journal of Membrane Science* 586: 259–266.
30. Koseoglu-Imer, D.Y., Kose, B., Altinbas, M., Koyuncu, I. (2013) The production of polysulfone (PS) membrane with silver nanoparticles (AgNP): Physical properties, filtration performances, and biofouling resistances of membranes. *Journal of Membrane Science* 428: 620–628.
31. Han, L., Young, H., Hwi, R., Jin, H. (2019) Preparation of an anion exchange membrane using the blending polymer of poly(ether sulfone) (PES) and poly(phenylene sulfide sulfone) (PPSS). *Membrane Journal* 29: 155–163.
32. Zheng, P., Liu, K., Wang, J., Dai, Y., Yu, B., Zhou, X., Hao, H., Luo, Y. (2012) Surface modification of polyimide (PI) film using water cathode atmospheric pressure glow discharge plasma. *Applied Surface Science* 259: 494–500.
33. Wang, Q., Yu, L., Nagasawa, H., Kanezashi, M., Tsuru, T. (2020) High-performance molecular-separation ceramic membranes derived from oxidative cross-linked polytitanocarbosilane. *Journal of the American Ceramics Society* 103: 4473–4488.
34. Ahmad, A.L., Mohammed, H.N., Ooi, B.S., Leo, C.P. (2013) Deposition of a polymeric porous superhydrophobic thin layer on the surface of poly(vinylidenefluoride) hollow fiber membrane. *Polish Journal of Chemical Technology* 15: 1–6.
35. Zhang, Y., Wang, R. (2013) Fabrication of novel polyetherimide-fluorinated silica organic-inorganic composite hollow fiber membranes intended for membrane contactor application. *Journal of Membrane Science* 443: 170–180.
36. Jin, C., Chen, D., Sirkar, K.K., Pfeffer, R. (2020) An extended duration operation for solid hollow fiber membrane-based cooling crystallization. *Powder Technology* 365: 106–114.
37. Qazi, S., Gómez-Coma, L., Albob, J., Druon-Bocquet, S., Irabien, A., Sanchez-Marcano, J. (2020) CO_2 capture in a hollow fiber membrane contactor coupled with ionic liquid: Influence of membrane wetting and process parameters. *Separation and Purification Technology* 233: 115986.
38. Petukhov, D.I., Komkova, M.A., Brotsman, V.A., Poyarkov, A.A., Eliseeva, An.A. (2020) Membrane condenser heat exchanger for conditioning of humid gases. *Separation and Purification Technology* 241: 116697.
39. Schmitt, A., Mendret, J., Roustan, M., Indeed, S.B. (2020) Ozonation using hollow fiber contactor technology and its perspectives for micropollutants removal in water: A review. *Science of the Total Environment* 729: 138664.
40. Pabby, A.K., Swain, B., Sastre, A.M. (2017) Recent advances in smart integrated membrane assisted liquid extraction technology. *Chemical Engineering and Processing – Process Intensification* 120: 27–56.
41. Yan, H., Wu, L., Wang, Y., Irfan, M., Jiang, C., Xu, T. (2020) Ammonia capture from wastewater with a high ammonia nitrogen concentration by water splitting and hollow fiber extraction. *Chemical Engineering Science* 227: 115934.
42. Deriss, A.Q., Langari, S., Taherian, M. (2021) Computational fluid dynamics modeling of ibuprofen removal using a hollow fiber membrane contactor. *Environmental Progress & Sustainable Energy* 40: e13490.
43. Williams, N.S., Gomaa, H.G., Ray, M.B. (2013) Effect of solvent immobilization on membrane separation of ibuprofen metabolite: Agreen and organic solvent analysis. *Separation and Purification Technology* 115: 57–65.
44. Davarnejad, R., Soofi, B., Farghadani, F., Behfar, R. (2018) Ibuprofen removal from a medicinal effluent: A review on the various techniques for medicinal effluents treatment. *Environmental Technology & Innovation* 11: 308–320.
45. Müller, S., Möder, M., Schrader, S., Popp, P. (2003) Semi-automated hollow-fibre membrane extraction, a novel enrichment technique for the determination of biologically active compounds in water samples. *Journal of Chromatography A* 985: 99–106.
46. Rodrigues, L.N., Sirkar, K.K., Weisbrod, K.R., Ahern, J.C. (2021) Janus hydrophobic-hydrophilic Janus membranes for nondispersive membrane solvent extraction. *Journal of Membrane Science* 637: 119633.
47. Beuscher, U. (2022) *PoroFruit & Vegetable Processing Market by Product Type, Equipment, Operation, and Region-Global Forecast to 2022*. Available online: www.marketsandmarkets.com/Market-Reports/fruit-vegetable-processingmarket- 140232885.html (accessed on 19 June 2022).
48. Conidi, C., Castro-Muñoz, R., Cassano, A. (2020) Membrane-based operations in the fruit juice processing industry: A review. *Beverages* 6: 18.
49. Zhao, S., Li, S., Pei, J., Meng, H., Wang, H., Li, Z. (2022) Evaluation of hybrid pressure-driven and osmotically-driven membrane process for non-thermal production of apple juice concentrate. *Innovative Food Science and Emerging Technologies* 75: 102895.
50. Zhu, B., Duke, M., Dumée, L.F., Merenda, A., des Ligneris, E., Kong, L., Hodgson, P.D., Gray, S. (2018) Short review on porous metal membranes—Fabrication, commercial products, and applications. *Membranes* 8: 83.

2 Reverse Osmosis Membrane

Dipak Rana[*1], *Takeshi Matsuura*[1], *Mohd Azraai Kassim*[2], *and Ahmad Fauzi Ismail*[2]

[1]Department of Chemical and Biological Engineering, University of Ottawa, ON, Canada

[2]Advanced Membrane Technology Research Centre, Universiti Teknologi Malaysia, Johor, Malaysia

*E-mail: rana@uottawa.ca

2.1 INTRODUCTION

The fastest-growing desalination process is a membrane separation process called reverse osmosis (RO). The most remarkable advantage of RO is that it consumes little energy since no phase change is involved in the process. RO employs hydraulic pressure to overcome the osmotic pressure of the salt solution, causing water-selective permeation from the saline side of a membrane to the freshwater side since the membrane barrier rejects salts [1–4]. Polymeric membranes are usually fabricated from materials such as cellulose acetate (CA), cellulose triacetate (CTA), polyamide (PA), etc. by the dry-wet phase inversion technique, or by coating aromatic PA via interfacial polymerization (IFP) [5].

In the early 1960s, Loeb and Sourirajan developed asymmetric CA membranes by the phase inversion method for seawater desalination [6]. The fundamental principle underlying their RO membrane development is the preferential sorption-capillary flow mechanism. According to the Gibbs adsorption isotherm, a thin (about 0.5 nm) layer of pure water exists at the surface of seawater. By analogy, a pure water layer should also exist at the membrane/saltwater interface, which can be collected through the membrane pore by applying a pressure gradient, as depicted in Figure 2.1 [7]. Later, it was found that the Loeb–Sourirajan membrane had an asymmetric structure with a top thin layer that governs the salt rejection and water flux, and a porous sublayer that provides the mechanical strength (see Figure 2.2) [7]. In 1968, Cadotte was able to prepare a new type of RO membrane called a thin-film-composite (TFC) membrane that consisted of a thin layer of PA formed by an in situ polycondensation reaction of branched poly(ethylene imine) and 2,4-diisocyanate on a porous polysulfone (PS) membrane [8]. Since then, the focus has shifted to the development of TFC membranes with a thin skin layer, mostly made of aromatic PA material, and it has provided the major platform to fabricate novel RO membranes for seawater desalination. Nowadays, TFC membranes are manufactured by IFP of trimesoyl-chloride (TMC) and *m*-phenylene diamine (MPD) at the surface of a porous PS support. Although the first practical TFC PA membrane appeared nearly 40 years ago, there remains considerable uncertainty regarding the molecular mechanisms of their function, selectivity, and molecular structure–performance relationship. The main feature of the TFC membrane is that the material of the top thin layer (PA) is different from that of the porous support layer (PS), while, for the CA membrane, the material for both layers is the same. Therefore, the membrane of the latter type is called an integrally skinned asymmetric membrane to be distinguished from the TFC membrane which also has an asymmetric structure.

Another noticeable development in RO membrane fabrication was that of the PA membrane with an integrally skinned asymmetric structure. It was made by DuPont in 1971 both in flat-sheet and hollow-fiber configurations [9,10], which established a novel membrane fabrication technology referred to as hollow-fiber spinning. According to the patent disclosure, approximately 10% of the linear aromatic polyamide rings are substituted by sulfonic acid groups. As the aromatic ring is easily attacked by chlorine in water, the presence of the sulfonate group is necessary to render the material chlorine-resistant.

Compared with CA or PA membranes of integrally skinned asymmetric structure, the TFC PA membrane enjoys superior permeation flux. However, the PA TFC membranes are more susceptible to halogens compared to the CA membranes which are, in general, resistant to halogens. However, according to the data reported, the permeate flux, anti-fouling, and anti-chlorine ability of the RO membrane are still not as effective as desired, for either the CA membrane and its integrally skinned structure or the TFC PA membrane, and efforts to improve membrane performance are continuing. One such example is the asymmetric copolyamide membrane with piperazine moiety, which was developed by Toyobo in hollow-fiber configurations [11]. The rejection of the TFC flat sheet membrane starts to decrease at around 4000 h and drops almost completely at around 8000 h, whereas, the rejection retention of the Toyobo hollow-fiber membrane is quite stable, up to an 8000-h test period. The representative chemical structures of the conventional TFC of Dow, DuPont, Toray, and Toyobo membranes are represented in Figure 2.3. It is noted that the m and n values of the Dow TFC co-polyamide are approximately 0.7 and 0.3, respectively. The chemical components of the DuPont and Toyobo materials are copolyamidohydrazide and copolyamide containing piperazine as well as diphenylsulfone, respectively. It is worth mentioning that the commercial

DOI: 10.1201/9781003285656-2

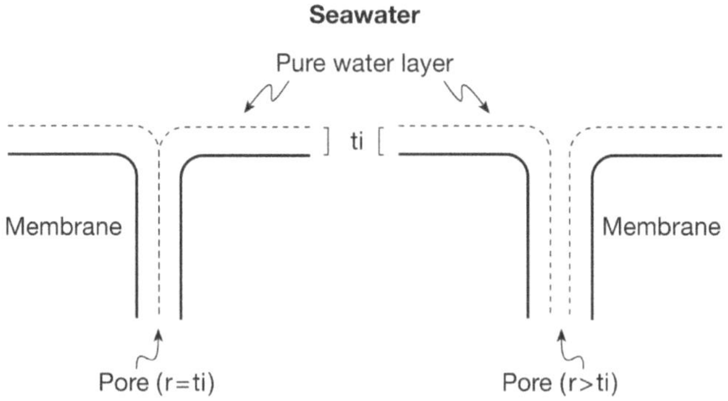

FIGURE 2.1 Preferential sorption capillary flow mechanism of RO. (Courtesy of Matsuura, T. [1994] *Synthetic Membranes and Membrane Separation Processes*. CRC Press, Boca Raton, FL.)

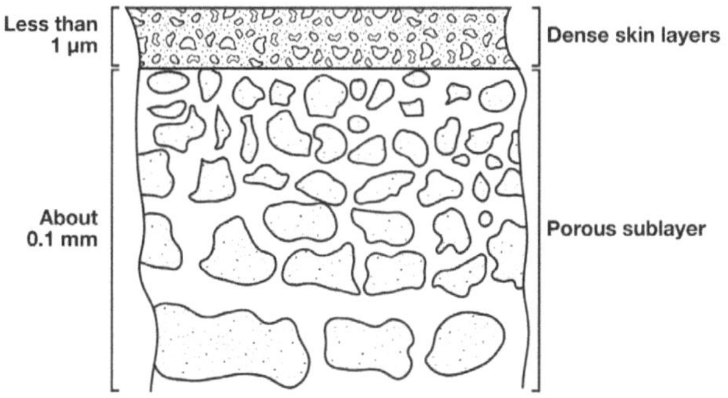

FIGURE 2.2 Asymmetric structure of the membrane. (Courtesy of Matsuura, T. [1994] *Synthetic Membranes and Membrane Separation Processes*. CRC Press, Boca Raton, FL.)

TFC membranes manufactured by Dow, DuPont, Toray, and Toyobo membranes are called, for example, FT-30 (aromatic polyamide), PPPH8273 (aromatic copolyamidohydrazide), UTC-70 (cross-linked aromatic polyamide), and 4T-PIP(30) (chlorine-resistant copolyamide), respectively.

2.2 MEMBRANE PREPARATION

2.2.1 Integrally Skinned Asymmetric Membrane and Thin-Film-Composite Membrane

The phase inversion technique is used to prepare integrally skinned asymmetric membranes [12]. This is a process in which a polymer is transformed from a liquid to a solid state. The dry–wet phase inversion technique and the temperature-induced phase separation (TIPS) are most commonly used in industrial membrane manufacturing. Loeb and Sourirajan applied the dry–wet phase inversion technique in their development of the first CA membrane for seawater desalination, which is often called the Loeb–Sourirajan method. According to the Loeb–Sourirajan method, polymer solutions are prepared by mixing polymer, solvent, and sometimes even nonsolvent. The solution is then cast on a suitable surface with a doctor blade to a thickness of about 250 µm. After partial evaporation of the solvent, the cast polymer solution film is immersed in a bath of the nonsolvent medium, often called coagulation (gelation) medium. Solidification of the polymer film takes place in two steps, i.e. evaporation of the solvent and the solvent–nonsolvent exchange in the gelation bath. It is often preferable to choose a solvent of strong dissolving power with high volatility. During the first step of desolvation by solvent evaporation, a thin skin layer of solid polymer is formed instantly at the top of the cast film due to the loss of solvent. In the next solvent–nonsolvent exchange step, nonsolvent diffuses into the polymer solution film, while solvent diffuses out through the thin solid layer. The composition of the polymer solution changes during the solvent–nonsolvent exchange process.

As already mentioned, TFC membranes are fabricated by in situ polymerization. This method, developed by Cadotte and coworkers of Filmtec in the 1970s, is currently most widely used to prepare high-performance reverse osmosis and nanofiltration membranes [8]. A thin selective layer is deposited on top of a porous substrate membrane by interfacial in situ polycondensation. There are a number of modifications of this method primarily based on the choice of monomers [13]. However, for a matter of simplicity, the polycondensation procedure is described by a pair of diamine and diacid chloride monomers.

A diamine solution in water and a diacid chloride solution in hexane are prepared. A porous substrate membrane is then dipped into the aqueous solution of diamine. The pores at the top of the porous substrate membrane are filled with the aqueous solution in this process. The membrane is then immersed in the diacid chloride solution in hexane. Since water and hexane are not miscible, an interface is formed at the boundary of the two phases. Polycondensation of diamine and diacid chloride will take place at the interface, resulting in a very thin layer of polyamide. The preparation of composite membranes by the interfacial in situ polycondensation is schematically presented in Figure 2.4.

There are a number of combinations for the choice of diamine and acid chloride monomers. For example, if trimesoyl chloride, which has three -COCl groups in an aromatic ring, is mixed with phthaloyl chloride, which has two -COCl groups, cross-linking will form between two main chains. Unreacted -COCl will become -COOH upon contact with water and the membrane will become negatively charged. Monomers with reactive groups other than amine and acid chloride can also be used.

2.2.2 Membrane Surface Modification

Since both the surface chemistry and morphology of the membrane play a crucial role in determining the membrane performance, its enhancement has been attempted through modification of the membrane surface [14,15]. In fact, the latest research on TFC membranes is mostly on membrane surface modification as summarized in Ref. [16]. To the authors' knowledge, most of these remain as research efforts only and are yet to be commercialized.

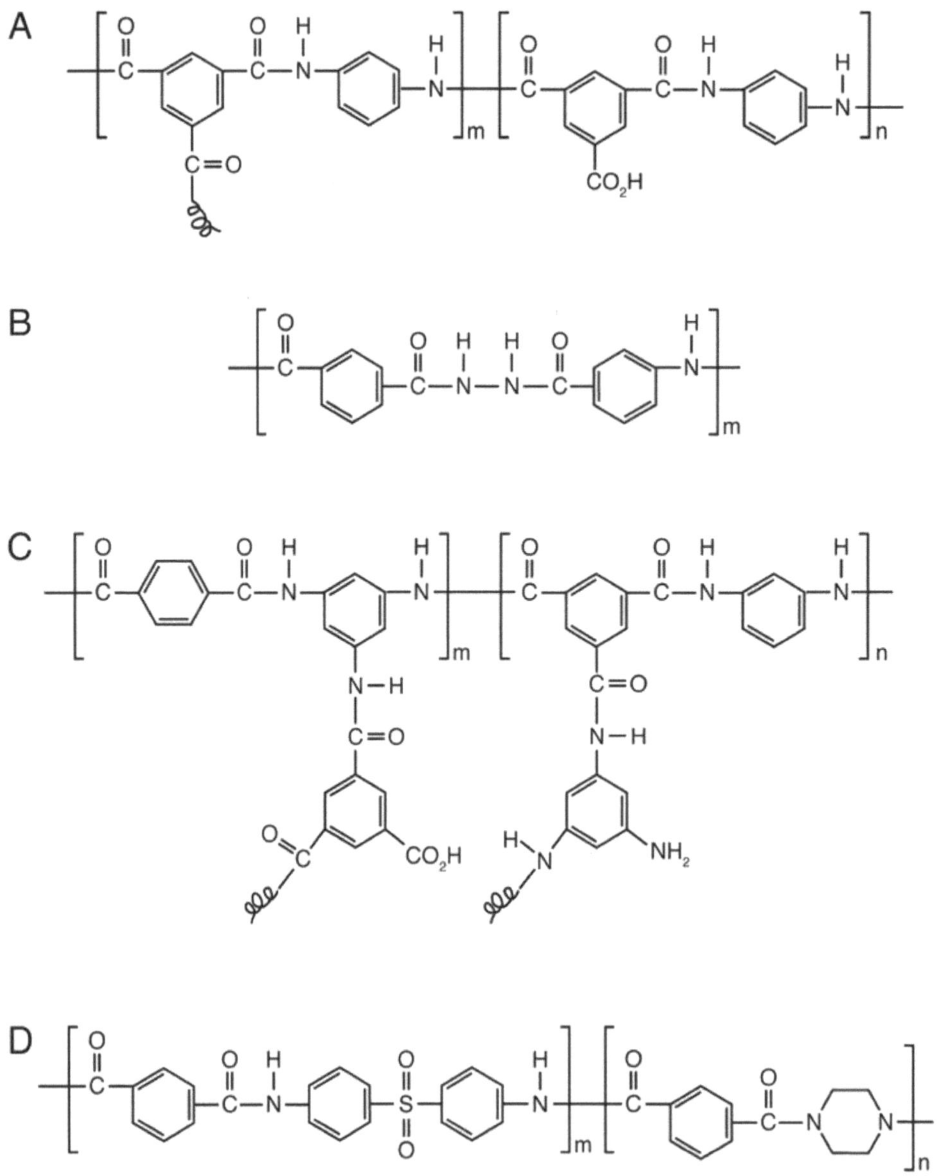

FIGURE 2.3 Representative chemical structures of the conventional TFC of (A) Dow, (B) DuPont, (C) Toray, and (D) Toyobo membranes.

Soaking the freshly prepared TFC membranes in solutions containing various organic species, including glycerol, sodium lauryl sulfate (SLS), salt of triethylamine, and camphorsulfonic acid (TEACSA) can increase the membrane flux in RO applications by 30–70% [17]. The physical properties of TFC (abrasion resistance) and flux stability can also be improved by applying an aqueous solution composed of polyvinyl alcohol (PVA) and a buffer solution as a post-treatment step during the preparation of the TFC membranes [18]. On the other hand, using polyvinyl alcohol-based amine compounds having a side-chain amino group at the aqueous phase monomer instead of MPD can produce high flux TFC membranes for low-pressure applications [19,20]. The addition of alcohols, ethers, sulfur-containing compounds, monohydric aromatic compounds, and more specifically dimethyl sulfoxide (DMSO), in the aqueous phase, can produce TFC membranes with excellent performance [21–24].

It was also determined that the water flux of TFC-RO membranes could be doubled without affecting the salt rejection by incorporating zeolite nanoparticles in the thin layer of the TFC-RO membranes [25]. To prevent microbial fouling, a new type of anti-fouling membrane was developed by introducing TiO_2 nanoparticles to TFC membranes in order to reduce the loss of reverse osmosis permeability [26,27].

The water permeability of PA-TFC-RO membranes could be increased using an oxygen plasma treatment by introducing carboxylic groups which increase the hydrophilicity of the treated membrane. On the other hand, the chlorine resistance of the TFC-RO membrane can be enhanced using an argon plasma treatment which causes cross-linking to take place at the nitrogen sites on the membrane surface [28].

Since hydrophilic surfaces are believed to attract more water than organic contaminants of high hydrophobicity, many attempts have been made to increase membrane surface

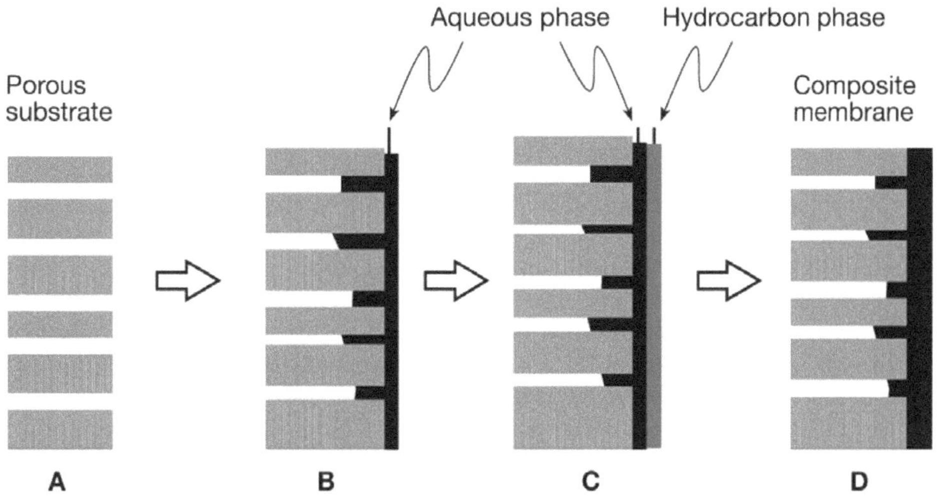

FIGURE 2.4 Schematic of in situ interfacial polymerization.

hydrophilicity by surface modification techniques, each having its own advantages and disadvantages. Hydrophilization by treating the membrane surface with water-soluble solvents (acids, alcohols, and mixtures of acids, alcohols, and water) is one of the surface modification techniques, as mentioned earlier. This method increases the flux without changing the chemical structure, but one of its disadvantages is that the water flux decreases with time because of the leaching of the hydrophilizing agent by water permeation [29]. Using a mixture of acid and alcohol in water for the surface treatment can improve the surface properties since acid and alcohol in water cause partial hydrolysis and skin modification, which produces a membrane with higher flux and higher rejection. It has been suggested that the presence of hydrogen bonding on the membrane surface encourages the acid and water to react on these sites, producing more charges [29,30]. Kulkarni et al. hydrophilized a TFC-RO membrane by using ethanol, 2-propanol, hydrofluoric acid, and hydrochloric acid [29]. They found that there was an increase in hydrophilicity, which led to a remarkable increase in water flux with no loss in ion rejection.

A hydrophilic charged TFC could be produced by using radical grafting of two monomers, methacrylic acid and poly(ethylene glycol) methacrylate, onto a commercial PA-TFC-RO membrane [31]. It was found that the use of amine-containing ethylene glycol blocks enhanced the performance of the membrane and highly improved membrane water permeability by increasing hydrophilicity [32]. Poly(ethylene glycol) (PEG) and its derivatives have been used for surface modification. TFC membrane resistance to fouling could be improved by grafting PEG chains onto the TFC-RO membranes [33,34]. Louie et al. coated the surface of commercial TFC-RO membranes with a solution of polyether–polyamide (PEBAX 1657) to produce anti-fouling membranes [35].

Another alternative and less common approach for membrane surface modification is the introduction of an active additive. The basis of this technique is the idea that those additives can move toward the top film surface during membrane formation and alter the membrane surface chemistry while keeping bulk properties unchanged. According to this method, only a very small quantity of additives is enough to change the surface chemistry of the membranes [36]. Blending is a conventional technique used for membrane surface modification, and recently much attention has been given to utilizing this technique, in which hydrophobic surface-modifying macromolecules (SMMs) are blended into a base polymer for membrane surface modification [37]. Preparation of thin-film-composite polyamide membranes for desalination using novel hydrophilic surface-modifying macromolecules improved the flux stability of the PA-TFC-RO membrane by using hydrophilic surface-modifying macromolecules formed simultaneously by an in situ polymerization reaction as the polycondensation reaction takes place within the organic solvent of the TFC system [38,39].

2.2.3 Development of Membranes for Boron Removal, Chlorine Tolerance, and Antibiofouling

Boron occurs in seawater at an average concentration of 4–6 mg/L. Toxicological effects from the exposure of the human body to boron, primarily by consumption of desalinated seawater also have been reported [40]. A provisional guideline of the WHO suggests a maximum boron concentration of 2.4 mg/L in drinking water (see, for example, Guidelines for drinking-water quality, 4_{th} edition, ISBN, 978 92 4 154815 1, World Health Organization, Geneva, Switzerland, 2011). The California Department of Public Health (CDPH) set a notification level of 1.0 mg/L for drinking water. It has been found that the RO membrane is not very effective in boron removal. According to the study for the two-stage RO desalination of Atlantic Ocean water under the conditions of feed pressures of 802–966 psi, product recovery of 38–42%, for the boron concentrations of 4.2–4.7 mg/L, and with pH 7.0–8.2, was achieved. The boron removal in the first stage was 80.0–83.4%, while in the second stage, it was 64.4–66.0%.

These low values of boron rejection cannot satisfy the WHO standard for drinking water. The difficulty of boron removal arises from the fact that boron exists in seawater either as boric acid (H_3BO_3) or borate ion ($H_2BO_3^-$) with their respective concentrations depending on the pH. Based on the first dissociation constant, $pK_{a1} = 8.7$, boron is primarily present as boric acid in neutral seawater. It was found that the charged $H_2BO_3^-$ is rejected at 97.3–99.6% by commercial spiral wound modules while the rejection of H_3BO_3 is much lower (85.7–96.7%) [41]. A recent paper published by Koseoglu et al. reported boron rejection of 85–90% from seawater of pH = 8.2 (average seawater pH) and >98% at a pH of 10.5 [42]. The experiments were carried out at 700 psig at ambient temperature using two commercial SWRO modules, i.e. Toray™ UTC-80-AB and Filmtec™ SW30HR (Dow Chemical). SWRO membranes with better boron rejection are still being sought. Further improvements in boron removal by RO membranes can be found in the literature [43,44].

There remain drawbacks in the commercial integrally skinned cellulose acetate and thin-film-composite polyamide RO membranes. CA membranes are susceptible to microbiological attack, undergo compaction at high temperatures and pressures, and are workable only within a narrow range of pH (3–7). PA membranes, on the other hand, can tolerate a wider range of pH but suffer from poor resistance to continual exposure to oxidizing agents such as chlorine that is most widely used as an oxidizing biocide in water treatment. As such, the performance of PA membranes deteriorates at continuous exposure to water containing more than a few ppb (parts per billion) of chlorine. A number of attempts have been made to change the macromolecular structure of PA, as already mentioned in the surface modification section, but the problem has not yet been completely solved.

Recently, an attempt was made to synthesize a sulfonated polysulfone using a disulfonated monomer. In this case, a random co-polymer is synthesized by direct polymerization of (3,3'-disulfonato-4,4'-dichlorodiphenyl sulfone [SDCDPS]), 4,4'-dichlorodiphenyl sulfone (DCDPS), and 4,4'-biphenol (BP) or another copolymer is synthesized from SDCDPS, BP, and 4,4'-difluorotriphenylphosphine oxide (DFIPPO) [45]. The reaction scheme is presented in Figure 2.5 [45]. In contrast to the sulfonation of polysulfone, sulfonated polysulfone of a given ion exchange capacity (IEC) is obtained by this method in a more controlled manner. Using these polymers, RO membranes were prepared and their chlorine resistance was examined. One laboratory-made membrane (BPS-40H, based on the second type of synthesized copolymer with 40% sulfonic acid component in H form) was compared with the

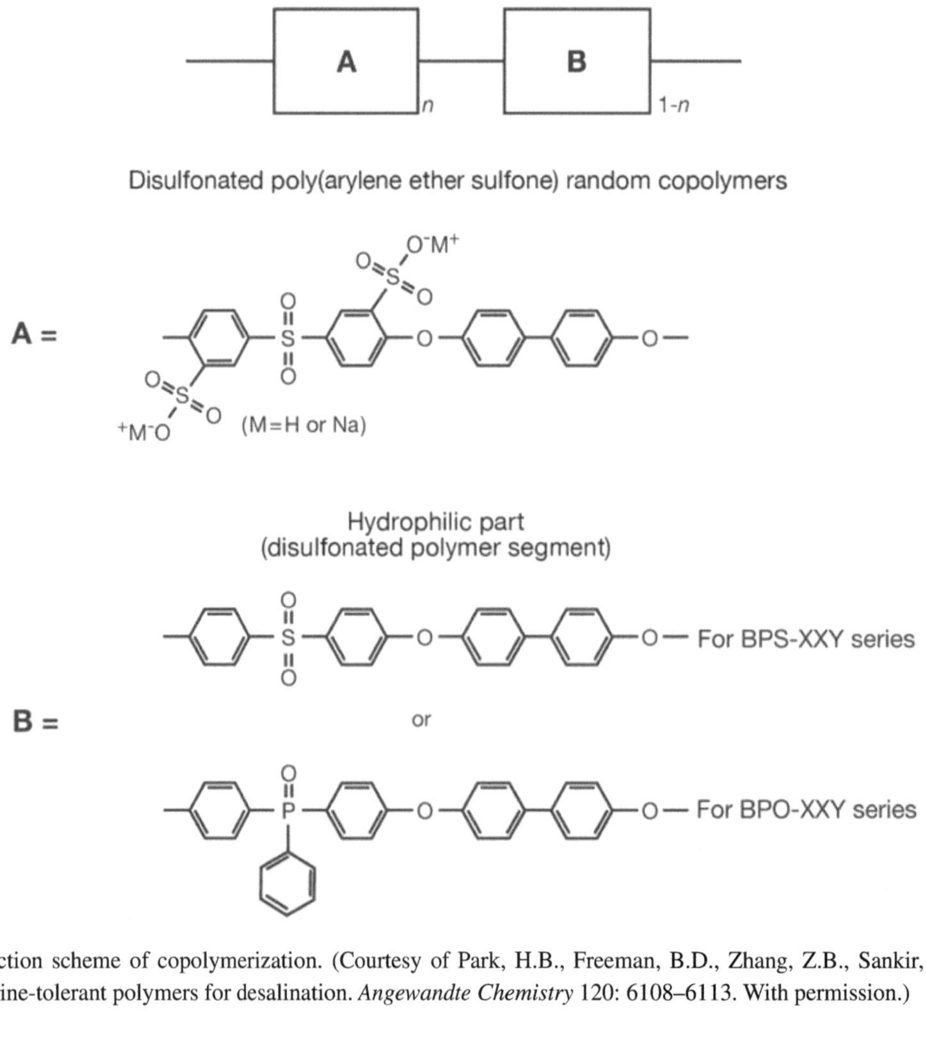

FIGURE 2.5 Reaction scheme of copolymerization. (Courtesy of Park, H.B., Freeman, B.D., Zhang, Z.B., Sankir, M., McGrath, J.M. [2008] Highly chlorine-tolerant polymers for desalination. *Angewandte Chemistry* 120: 6108–6113. With permission.)

TABLE 2.1
Advances in Dow RO membrane (Courtesy of Micklos, M. [2008] *The Development of Reverse Osmosis and Nanofiltration through Modern Times.* Plenary Lecture III, International Congress on Membranes and Membrane Processes (ICOM), Honolulu, HI, July 12–18.)

Year	Production capacity (gpd)	Salt rejection (%)
	Brackish water	
1990	8,000	98
1998	10,000	99.2
2007	11,000	99.8
	Seawater	
1990	4,000	99.4
1998	5,500	99.5
2007	7,500-8,000	99.8

SW30HR SWRO membrane from Dow with a feed solution containing 2000 ppm NaCl and 500 ppm chlorine at pH = 9.5. While the solute rejection of commercial RO membranes decreased by 20% within 10,000 ppm-hours of chlorine exposure, there was hardly any change in salt rejection in the laboratory-made membrane. This paper indicates that there are still many things to be done in RO membrane material development since, as a research topic, this has been neglected for many years.

2.2.4 Membrane Development with Low Pressure and High Flux

Kurihara summarized the improvement in RO membrane performance up to 1999 [46]. The membrane productivity in terms of water flux normalized by operating pressure has increased by an order of magnitude. This allowed the operation of RO for brackish water desalination and ultrapure-water production at pressures lower than 1 MPa, resulting in significant savings in energy costs. Kurihara attributed the high productivity of the newly developed membrane to an increase in surface roughness. Currently, there seems to be nothing in sight limiting further improvements in membrane productivity. Table 2.1 also shows similar advancements made for the Dow RO membranes [47].

2.2.5 Membrane Development for High Operating Pressure

Current seawater desalination technology aims to increase pure water recovery by a membrane module from the conventional 40% to 60%. Since the osmotic pressure of the retentate will increase from 4.5 to 7.0 MPa when the water recovery increases from 40% to 60%, the development of a high-pressure vessel as well as the development of a membrane that will show little compaction under high pressure are necessary. Kawada reported the development of an RO membrane that was suitable for operation at 9 MPa [48]. The compaction of the membrane usually takes place at the porous sub-layer rather than at the skin layer. An attempt was therefore made to reduce the compaction by making a large number of uniform pores of small sizes at the surface of the porous sub-layer on which an aromatic PA skin layer was coated by in situ interfacial polymerization. The stability of the membrane module productivity increased significantly as compared to the conventional seawater desalination membrane.

2.3 REVERSE OSMOSIS TRANSPORT

2.3.1 Single-Component System

Currently, the mainstream RO membrane transport theory is the solution-diffusion model [49]. According to this model, mass transfer occurs in three steps: absorption into the membrane, diffusion through the membrane, and desorption from the membrane. The chemical potential gradient from the feed side of the membrane to the permeate side of the membrane is the driving force for the mass transfer. When the difference in hydrostatic pressure is greater than the difference in osmotic pressure between the upstream and downstream sides of the membrane, the chemical potential difference of water across the membrane drives water against the natural direction of water flow.

Thus, the water transport through the membrane can be described by

$$N_A = L(\Delta p - \Delta\pi) \tag{2.1}$$

where N_A is the water flux through the membrane (subscript A denotes water), L is the water permeability coefficient, Δp is the transmembrane pressure difference, and $\Delta\pi$ is the difference in osmotic pressure between the upstream and downstream sides of the membrane. The osmotic pressure can be given by an equation similar to the van't Hoff equation:

$$\pi = i\varphi CRT \tag{2.2}$$

where i is the dissociation parameter which is the number of ions produced by the dissociation of the salt, φ represents a correction factor, C is the salt concentration, R is the universal gas constant, and T is the absolute temperature. The permeability coefficient can be given by

$$L = \frac{DSV}{RTl} \tag{2.3}$$

where D is the diffusivity of water in the membrane, S is water solubility in the membrane, V is the partial molar volume of water, and l is the thickness of the skin layer of the membrane.

The osmotic pressure of seawater is typically 2300–2600 kPa and can be as high as 3500 kPa [50]. The osmotic pressures of brackish water are much lower than those of seawater.

Corresponding to the concentration range of 2000–5000 mg/L, the osmotic pressure ranges from 100 to 300 kPa [50]. To overcome these osmotic pressures, the pressures applied to RO seawater desalination are typically 6000–8000 kPa and for the desalination of brackish water, they are 600–3000 kPa.

Recovery of product water is an important parameter of RO operation. It is defined as

$$R_w = \frac{Q_p}{Q_f} \tag{2.4}$$

where Q_p and Q_f are volumetric flow rates of permeate and feed stream, respectively. Typically, the recovery of RO seawater desalination is 40%. When the recovery is increased, the salt concentration on the upstream side increases and more pressure should be applied to overcome the osmotic pressure. Special designs of both membrane and membrane module are required to apply such high operating pressures. The problem of salt precipitation will arise also. On the other hand, the recovery of RO brackish water desalination can reach as high as 85%.

Assuming the passage of water and salt through the membrane are independent of each other, salt transport through the membrane occurs by the concentration difference between the upstream and downstream sides of the membrane as the driving force. Salt transport through the membrane can be given by:

$$N_B = B(C_{fs} - C_p) \tag{2.5}$$

where N_B is the salt flux (subscript B denotes salt), B is the salt transport parameter, and C_{fs} and C_p are salt concentration at the membrane surface on the feed side and in the permeate, respectively. B is further given by:

$$B = \frac{D_B K_B}{l} \tag{2.6}$$

where D_B is the diffusivity of salt through the membrane and K_B is the partition coefficient of salt between the jsolution and the membrane.

A more important parameter to express the salt passage through the membrane is salt rejection given by:

$$R = \frac{(C_{fb} - C_p)}{C_{fb}} \times 100\% \tag{2.7}$$

where C_{fb} is the salt concentration in the bulk of the solution on the feed side. $R = 100\%$ means perfect salt rejection and $R = 0$ means no salt rejection by the membrane.

It should be noted that C_{fs} is the salt concentration at the membrane surface on the feed side which was used to calculate N_B, while C_{fb} is the salt concentration in the bulk of the solution on the feed side of the membrane. C_{fs} and C_{fb} may become different due to a phenomenon called concentration polarization. This occurs due to the rejection of the salt and its accumulation at the membrane surface. The concentration polarization is schematically illustrated in Figure 2.6.

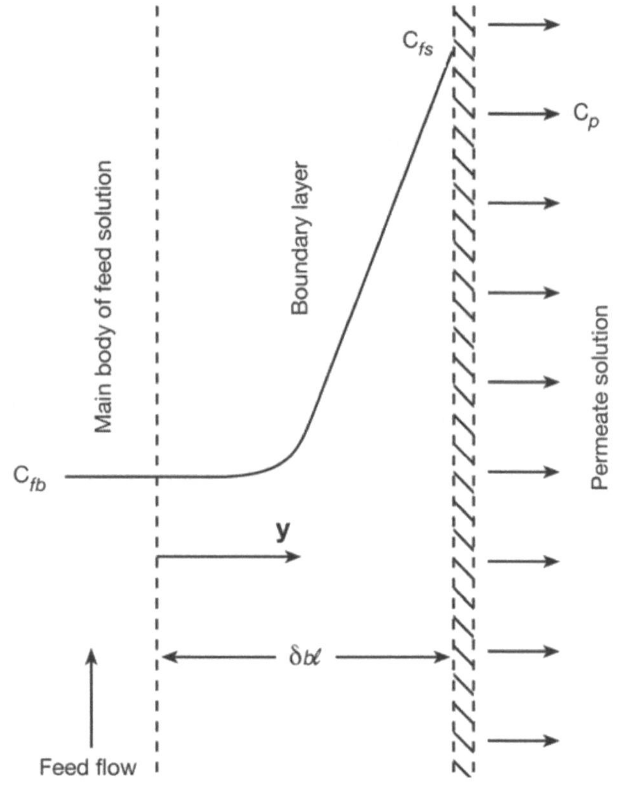

FIGURE 2.6 Schematic illustration of concentration polarization. (Courtesy of Matsuura, T. [1994] *Synthetic Membranes and Membrane Separation Processes*. CRC Press, Boca Raton, FL.)

By solving simple transport equations in the boundary layer at the membrane surface, the above two concentrations can be related to each other by the following equation.

$$\frac{(C_{fs} - C_p)}{(C_{fb} - C_p)} = \exp(\frac{N_A}{c_A k}) \tag{2.8}$$

where c_A is the molar concentration of water and k is the mass transfer coefficient of the salt.

To minimize the concentration polarization, i.e. to make C_{fs} as close to C_{fb} as possible, k should be increased, which is enabled by increasing the turbulence of the feed solution. All the reverse osmosis modules include devices to achieve this goal.

2.3.2 Multicomponent System

Seawater is a complicated mixture of many components. Hence, it is difficult to predict the membrane performance under different operating conditions based on transport theories. Most of the theories treating the separation of multicomponent electrolytic systems are based on the Debye–Hückel theory, Donnan effect, and Nernst–Planck equation [51,52]. Although they are applicable to the mixture of any number of ions involved in the feed, there are only a few works in which agreement between experimental results and theoretical predictions were tested for mixtures of more than four ions. One of the noted exceptions is the work of Rangarajan et al. [53]. These authors have expanded the transport theory to treat a multicomponent system involving nine ions of primary importance present in seawater. The separations of individual ions were predicted theoretically and the prediction was examined by experiments.

2.4 COMMERCIAL MODULES

The commercial RO modules that have appeared in the literature since 1977 are summarized in Table 2.2, together with their typical operating conditions and performance. Some of them may no longer exist in the market. From this table, it is clear that the majority of high salt rejection RO membranes are spiral wound aromatic polyamide (PA) TFC membranes, with the noted exception of Toyobo, which is based on cellulose triacetate hollow fiber. The reason for this could be easier cleaning of the spiral wound module, on one hand, and chlorine resistance of cellulose acetate, on the other.

2.5 RECENT OPERATIONAL EXAMPLES OF LARGE-SCALE RO MODULES

Table 2.3 summarizes recent operational examples of large-scale RO modules.

A brief outline of each paper mentioned in Table 2.3 follows.

This study deals with the optimization of membrane filtration performance in the treatment of AMD (acid mine drainage) using two nanofiltration (NF) membranes (NF99 from Alfa Laval and DK from GE-Osmonics) and one reverse osmosis membrane (RO HR98PP from Alfa Laval) [54]. RO experiments were performed with a laboratory-scale membrane area of 63.6 cm^2 with pressure, pH, temperature, and flow rate changed.

In 2007, two of the world's largest wastewater reclamation plants using similar energy-saving RO membrane technology were commissioned [56]. One plant, Orange County Water District's (OCWD) Groundwater Replenishment System (GWRS) located in Southern California, was expanding to 265,000 m^3/d (70 MGD), while the other, Ulu Pandan located in Singapore, was producing 148,000 m^3/d (39 MGD). This paper discusses the evolution of the design of these large-scale wastewater reclamation plants using energy-saving RO membranes (polyamide TFC membranes from Koch, Dow, and Hydranautics). Lessons learned from years of the pilot, demonstration plants, and full-scale plant experience at both OCWD and UP were presented.

A case study of a natural gas production site covering various technical issues related to the selection of an appropriate Reverse Osmosis (RO) system is presented in Ref. [58]. As part of the pretreatment selection, two types of UF membrane modules, viz., spiral wound and hollow fiber, with MWCO of 8000 and 50,000 Dalton, respectively, were tested in parallel with NF membranes of the spiral wound type with MWCO 200 Dalton. The NF plant with 50% capacity gave a recovery rate of 75% and the RO plant gave a recovery rate of 60% in comparison to the expected 92–95%. The long-term tests have indicated that the remainder of the membranes could be installed to achieve the full capacity of the plant. This study also demonstrates the importance of the selection of a proper pretreatment set-up for the RO system design.

The following three case studies of wastewater desalination are described [59]. (1) Effluent from a poultry and meat processing industry in southern Europe was treated by RO (FILMTEC polyamide TFC membrane). Pretreatment by UF and or CMF (continuous MF) was made before NF treatment followed by RO. (2) Secondary municipal and agricultural effluent was treated by FILMTEC BW30-365FR2 with BW30-330RO. (3) High-salinity brackish well water was treated by (SWRO) FILMTEC SW30-8040. In each case, cost estimation of the product water was made.

The evaluation of longer time-frame fouling in RO membranes, after 4 years of operation, was made through extensive membrane autopsy employing various techniques of analysis [60]. Inorganic foulants are comprised primarily of hydrogen aluminosilicate and halite, whereas the main organic substances emerging from this investigation were proteins, polysaccharides, and humic compounds which are attributed to biofouling. Predominant factors for the foulant deposition on the RO membranes are the increase of membrane selectivity due to biofouling, the large size of the cartridge filters, and high operating pressure.

A fouled RO membrane for the municipal wastewater reclamation (MWR) plant was autopsied after use, and the deposit on the membrane was analyzed for the organics which occupy 75% of the deposit, and also for the inorganic components involved [61]. The inorganic scaling of the RO membrane was primarily from the elements Fe, Ca, and Si. Hydrophobic acids and hydrophilic neutrals were the most significant fractions seen in the deposit, while the hydrophilic neutrals fraction in the deposit resulted from the microorganisms on the membrane rather than depositing on the membranes.

The performance and characteristics of new RO membranes operating at ultralow pressure (7–10 bar) are presented in Ref. [62]. The total cost comparison based on a 4 km^3/day plant revealed that TFC-ULP (ultralow pressure) provides about 10% savings over the traditional TFC-HR (traditional brackish water RO membrane). Generally, the use of TFC-ULP contributes to energy saving but the capital cost increases due to the higher membrane cost.

The study outlined in Ref. [63] focuses on the behavior of ultra-low pressure RO membranes in a full-scale system. The ultra-low pressure membranes have salt rejections comparable to conventional RO membranes but the hydraulic characteristics are significantly different. Hence, modifications to conventional membrane system design must

TABLE 2.2
Typical commercial modules' performance at various operating conditions

Manufacturer	Membrane code	Module, area (m^2)	Surface material	Test conditions: Operating pressure (MPa)	Test conditions: Temp (°C)	Test conditions: pH	Test conditions: Feed conc. (mg/l)	Performances: PWP (m^3/d)	Performances: Salt rejection (%)	References
Alfa Laval, Sweden	HR98PP		TFC on polypropylene	16-40 bar	25	2–10	200 NaCl	39 L/m²h at 20 bar	>96	[54]
Dow Chemical	XFS-4167.08	Hollow fiber, 85/8″ x 40″ L, 335	Aromatic polyamide	56 atm	35	4–7.5	290-470 TDS	9.4 (2500 gpd)	98.5	[55]
	BW30-400FR	400 ft²	Polyamide RO	225 psi		6	2000 NaCl	39.7 m³/h (10500 gpd)	99.5	[56]
	XLE-440	440 ft²	Polyamide RO	100 psi		6	500 NaCl	48.1 m³/h (12700 gpd)	99.0	[56]
DuPont *de Nemours*	Permsep B-10	Hollow fiber,5 1/2″ OD x 4 5/8 ″ IDx 47 ″ L, 139	Aromatic polyamide	56 atm	35	5–9	290-470 TDS	5.5 (1500 gpd)	98.5	[55]
	Permasep B-10	Hollow fiber, 4 inch dia 4 ft long	aramid	800 psig	25		30,000 NaCl	1500 gpd		[57]
	Permasep B-10	Hollow fiber, 8 inch dia 5 ft long	aramid	800 psig	25		30,000 NaCl	5000 gpd		[57]
FilmTec Corp.	BW30-4040	Spiral-wound	TFC	18 bar	30	6.81		0.89 L/m²h		[58]
	SW30-2540-2	Spiral-wound	TFC	60 bar	30	6.81		0.5 L/m²h		[58]
	SW30-2540-2	Spiral-wound	TFC	62 bar	30	6.81		0.14 L/m²h		[58]
	SW30-380			61.0 bar	19.9		33030 TDS	5.8 m³/h (8000 gpd)	99.3	[59]
FilmTec Corp.	SW30HR-380	Spiral-wound	Polyamide TFC	83 bar	45	2–11	37143 TDS	23 m³/d	99.7	[60]
FilmTec Corp.	SW30XHR-440i	Spiral-wound 8 in. elements (41 m²)	Polyamide TFC	Summer:50bar; Winter: 65 bar	2-19	7.7	32906 TDS	25 m³/d	99.82	[61]
Fluid Systems	TFC-ULP		Polyamide	100–150 psi		4–11	1300 TDS	105.8 m³/h	90.2	[62]
	TFC-HR		Polyamide	100–150 psi		4–11	1300 TDS	103.3 m³/h	96.2	[62]
	TFC 8822 HR	30.6 (330 ft²)	Polyamide TFC	1.55 (225 psi)			2000 NaCl	32.2 (8500 gpd)	99.5	[63]
	TFC 8821 ULP	30.6 (330 ft²)	Polyamide TFC	1.05 (150 psi)			500 NaCl	32.2 (8500 gpd)	99.0	[63]
Hydranautics	8040-LSY-CPA2	33.8 (365 ft²)	Polyamide TFC	1.55 (225 psi)			1500 NaCl	32.2 (8500 gpd)	99.5	[63]
	8040-UHY-ESPA	37.1 (400 ft²)	Polyamide TFC	1.05 (1500 psi)			1500 NaCl	32.2 (8500 gpd)	99.0	[63]
	LFC	37.2, 3 stage 28:14:8 array with 6 elements per vessel	Polyamide	15.5 bar	25	6.8–7.0	1500 NaCl	(2380 gpd)	98.7–99.5	[64]

Hydranautics	LFC1	400 ft^2	Polyamide RO	225 psi		6	1500 NaCl	41.6 m^3/h (11000 gpd)	99.5	[56]
	LFC3	400 ft^2	Polyamide RO	225 psi		6	1500 NaCl	36.0 m^3/h (9500 gpd)	99.7	[56]
	ESPA2	400 ft^2, 3 stage 6:4:2 array with 7 elements per vessel	Polyamide RO	150 psi	30	6	1500 NaCl	34.1 m^3/h (9000 gpd)	99.6	[56]
	ESPA2+	440 ft^2	Polyamide RO	150 psi	30	6	1500 NaCl	45.4 m^3/h (12000 gpd)	99.6	[56]
	LFC1-4040	37.2 per element	Polyamide RO	7.5–8.2 bar	30		633-987 TDS$^{\perp}$	9.4 gfd	97.1–98.1	[65]
	ESPA2-4040	37.2 per element	Polyamide RO	7.5–8.2 bar	30		633-987 TDS	11.3 gfd	97.1–98.1	[65]
	LFC3-4040	37.2 per element	Polyamide RO	7.5–8.2 bar	30		1500 NaCl	14.1 gfd	97.1–98.1	[65]
Hydranautics	Proc10	100 cm^2, 2 stage RO	Aromatic polyamide composite	1.1–1.3 MPa		7.0±0.2	4.2 TSS$^{\perp}$	16 L/m^{2}h	99.0–99.6	[66]
Hydranautics	Proc10	2 stage RO	Aromatic polyamide composite	1.55 MPa		7.0±0.2	Anti-scalant dosage 5 mg/L	0.0445 m^3/m^{2}h	99.6	[67]
Koch Fluid Systems	TFC 2822-SS	27.9 m^2, 6 spiral wound module	TFC	27-65 bar			32237 TDS$^{\perp}$	18.9 m^3/day	99.8	[68]
Koch Membrane	TFC-HR	400 ft^2	Polyamide RO	225 psi		6	2000 NaCl	42.4 m^3/h (11200 gpd)	99.5	[56]
NanoH_2O	QuantumFlux (Qfx) SW 365ES	Spiral-wound 8 in. elements (34 m^2)	Polyamide TFC	Summer:50bar; Winter: 65 bar	2-19	7.7	32906 TDS		99.8	[61]
Nitto Electric Ind. Co. / Kobo Steel	NRO-A (1st stage)	Tubular, 190 mmϕ x 2674 mmL, 1.75 (Tube) (12.5 mm ID x 18 pcs)	Cellulose di-acetate, FPR$^{\perp}$ (Tube)	50 kg/cm^2	25	5.5–6.0	35000 NaCl	0.30 m^3/d	89	[69]
	NRO-B (2nd stage)	Tubular, 190 mmϕ x 2674 mmL, 1.75 (Tube) (12.5 mm ID x 18 pcs)	Blend of cellulose di-acetate and cellulose tri-acetate, FPR (Tube)	40 kg/cm^2	25	5.5–6.0	5000 NaCl	0.56	97.5	[69]
Nitto Electric Ind. Co.	NRO A, B (1st stage)	Tubular, 12: 4 parallel 2-1 x 3 series	Blend of cellulose di-acetate and cellulose tri-acetate	50 kg/cm^2	35		3800 TDS	2.75		[55]
	NRO A, B (2nd stage)	Tubular, 3 number	Blend of cellulose di-acetate and cellulose tri-acetate	40 kg/cm^2	35		250 TDS	1.7		[55]
Toray Co.	SC-5000A (1st stage)	Spiral-wound, 12: X-mas tree series 6-4-2	Cellulose acetate	56 kg/cm^2	35		3000 TDS	12.5		[55]

(Continued)

TABLE 2.2 (Continued)
Typical commercial modules' performance at various operating conditions

Manufacturer	Membrane code	Module, area (m^2)	Surface material	Test conditions: Operating pressure (MPa)	Temp (℃)	pH	Feed conc. (mg/l)	Performances: PWP (m^3/d)	Salt rejection (%)	References
	SC-5000B (2nd stage)	Spiral-wound, 4 number	Cellulose acetate	40 kg/cm^2	35		200 TDS	10		[55]
	SC-5100A (1st stage)	Spiral-wound, 100 mmϕ x1.0 mL, 6.8	Cellulose acetate	56 kg/cm^2	25	5.5–6.0	35000 NaCl	2.0	95	[69]
	SC-5100B (2nd stage)	Spiral-wound, 100 mmϕ x1.0 mL, 6.8	Cellulose acetate	30 kg/cm^2	25	5.5–6.0	2000 NaCl	3.2	98	[69]
	PEC-1000	Spiral wound		53.4–59.5 kg/cm^2	11.8–23.8	6.4–6.9	18,400–19,800 Cl^-	706–812	99	[70]
	SU-720	Spiral wound, low pressure	Polyamide composite	1.5	25		1500 NaCl	26.0	99.4	[71]
	SU-720L	Spiral wound, low pressure	Polyamide composite	1.0	25		1500 NaCl	22.0	99.0	[71]
	SUL-G20	Spiral wound, ultra-low pressure	Polyamide composite	0.75	25		1500 NaCl	26.0	99.4	[71]
	SUL-H20	Spiral wound, super ultra-low pressure	Polyamide composite	0.5	25		1500 NaCl	26.0	99.4	[71]
Toray Membrane Inc., US	TMG20D-400	37.2 m^2, spiral wound	Polyamide TFC	40.46 atm	45	7.45–7.59	1099 TDS	29.1–29.6 m^3/h	99.5	[72]
Toyobo Co.	HR-5350 (1st stage)	Hollow fiber, 5 parallel	Cellulose acetate	38 kg/cm^2	35		2000 TDS	14		[55]
	HR-5350S (2nd stage)	Hollow fiber, 1 number	Cellulose acetate	70 kg/cm^2	35		200 TDS	10		[55]
	Hollosep HR 5350 (1st stage)	Hollow fiber, 125 mmϕ x 1.1 mL, 80	Cellulose acetate	55 kg/cm^2	25	5.5–6.0	35,000 NaCl	2.5	94	[69]
	Hollosep HR 5350S (2nd stage)	Hollow fiber, 125 mmϕ x 1.1 mL, 80	Cellulose acetate	22 kg/cm^2	25	5.5–6.0	2000 NaCl	10	97	[69]
	Hollosep HR 8350	Hollow fiber, 305 mmϕ x 1330 mmL,	Cellulose tri-acetate	55 kg/cm^2G	25	3-8	35,000 NaCl	> 10	99.7	[73]

	Hollosep High Pressure Temp. (HPT)	Hollow fiber,	Cellulose acetate	75 kg/cm²G	40		4200 NaCl	67.5 l/m²d	99.8	[74]
	JM-12	Hollow fiber, 5 element (HR1235ON-5)	Cellulose acetate	70 kg/cm²G	25		3500 NaCl	167–132.5	99.1–99.4	[74]
	Hollosep	Hollow fiber, 8″ module		54.8–58.3 kg/cm²	11.7–28.1	6.15–6.94	47,550–51,000 µs/cm	366–439	99	[70]
	Hollosep	Hollow fiber, 12″ module		53.2–57.4 kg/cm²	11.7–28.1	6.15–6.94	47,550–51,000 µs/cm	340–440	99	[70]
	HJ9155	Hollow fiber, 870	Cellulose tri-acetate	5.5	25		3500 NaCl	34	99.6	[75]
Universal Oil Products	ROGA-2B-TFC	Spiral wound, 8″ϕx10 3/4″, 0.36	Cellulose tri-acetate TFC	60 atm	35	4–7.5	290–470 TDS	120 l/d (32 gpd)	98.5	[55]
Ventron Membrane Tech., Beijing	HOR21-8040	33.9 (365 ft²)	Aromatic polyamide	4.14	25	7.5	2000 NaCl	34.1 (9000 gpd)	99.2	[71]
	HOR21-4040	7.9 (85 ft²)	Aromatic polyamide	4.14	25	7.5	2000 NaCl	8.3 (2200 gpd)	99.2	[71]
	HOR-2012	0.46 (5.0 ft²)	Aromatic polyamide	2.07	25	7.5	250 NaCl	0.19 (50 gpd)	97.5	[71]
Woongjin Chemical, Seoul	RE8040-FL	Spiral-wound, 37.2	Polyamide TFC	10–11.5 Kgf/cm²		7.5	3.85 mg/L TSS ⊥	0.5–0.8 m³/m²d	Not detected	[76]
Woongjin Chemical, Seoul	CSM BE, CSM FE and CSM BLR	Spiral-wound	Polyamide TFC	55 bar	18.6	6.5–7.0	2–1.5 g/L NaCl	105–118 L/m²h	99.4–98.5	[77]

⊥ TDS: Total dissolved solid; TSS: Total suspended solid; FRP: Fibre-reinforced plastic

TABLE 2.3
Examples of large-scale RO applications

Membrane and module	Plant[a] (element[b]) capacity	Water type	Location	Other remarks	References
			Pretreatment and fouling		
Polyamide TFC membrane	30 m^3/d	Waste water generated from natural gas production	Thailand	Selection of pretreatment (pretreatment)	[58]
Polyamide TFC	[a]2 km^3/d, 1.05 km^3/d, 14 km^3/d	Meet processing effluent, municipal and agricultural effluent, agricultural runoff	Southern Europe, South America, Southern Spain	Pretreatment study and cost evaluation (pretreatment)	[59]
Polyamide TFC membrane	[a]10 m^3/h	Municipal waste water	Bedok and Kranji, Singapore	MF and UF pretreatment (Pretreatment)	[64]
PA TFC	37.2 m^2	Municipal water	Gwangju, Korea	Characterization of organic foulant (fouling)	[76]
PA spiral wound	[a]36 m^3/h RO	River water	Steel industry Bremen, Germany	UF pretreatment of RO feed water (Pretreatment)	[78]
Film Tec BW 30 LE 440	[a]1.56 km^3/d	Surface water	Geleen (NL)	Pretreatment by UF (Pretreatment)	[79]
Filmtec SW30HR-380 spiral wound	15000 m^3/d	Sea water	Gran Canaria, Canary Islands, Spain	Deposition of foulants on the RO membranes increases membrane selectivity due to biofouling	[60]
Hydranautics Proc10 Aromatic PA	4800 m^3/d	Municipal wastewater reclamation (MWR)	Zibo, Shandong Province, China	Autopsy from MWR plant (Pretreatment by UF)	[66]
Hydranautics Proc10 Aromatic PA	11000-12000 m^3/d	Secondary effluent from wastewater treatment plants (WWTPs)	Tianjin, China	Autopsy from WWTPs (Pretreatment by UF)	[67]
Dow Filmtec, SW30HR LE-400	144 MLD, 17% of Perth's drinking water	Sea water	Perth, Western Australia	Organic and biological fouling parameters (Pretreatment by dual media filtration polishing with cartridge filter)	[80]
Polyamide TFC	9092 m^3/d	Seawater	Gijang-gun, Pusan, Korea	Role of MOM in the fouling formation of the RO membranes	[81, 82]
Not specified	Various	Various	Various	Review, pretreatment (pretreatment)	[83]
			New membranes and modules		
PA spiral wound	[a]24000 m^3/d	Brackish water	Hypothetical	Ultra low pressure (100-150 psi) (Membrane)	[62]
Fluid Systems TFC and Hydranautics TFC	[a]1.9 km^3/d	TDS 1000 and 5000 mg/L	Collier County, FA	System design for ultra-low pressure RO membrane (Membrane)	[63]
CTA hollow fiber	[b]34 m^3/d	Seawater	Arabian Gulf	Open ended hollow fiber (Membrane)	[75]
16" and 8" spiral wound element (Koch membrane)	[a]180 mgd	Colorado river water	Metropolitan Water District of California	Cost evaluation of 16" and 8" reverse osmosis element (cost)	[84]
LG Chem SW400R, SW400GR, BW400R	(218-23)x10^3 m^3/d	Seawater	Arabian Gulf	Two-stage SWRO systems perform better than single-stage SWRO to desalinate extreme seawater of high salinity and high temperature	[85]

TABLE 2.3 (Continued)
Examples of large-scale RO applications

Membrane and module	Plant[a] (element[b]) capacity	Water type	Location	Other remarks	References
Not specified, 16-inch RO membranes	45000 m^3/d	Seawater	Gijang-gun, Busan, Korea	Optimization of design factors and operating conditions of the SWRO plant	[86]
Energy					
Polyamide TFC membrane (Koch, Dow and Hydranautics	[a]265 km^3/d and 148 km^3/d	Ground water and wastewater	Orange county district, CA, and UluPandan, Singapore	Use of energy saving RO membranes (energy)	[56]
PA TFC membrane	[b]0.18 L/min (maximum)	Brackish water	Aqaba Gulf, Deep well	Photovoltaic powered (energy)	[87]
TFC membrane	158.53 hm^3/yr	Seawater	Gran Canaria, Spain	Optimum performance power input is 16-18 kWh with pressures ranging from 57-67 bar	[68]
Polyamide TFC, ultra-low pressure BWRO	1200 m^3/d	Groundwater salt wells	Arab Potash Company (APC), Jordan	Mathematical modeling of APC brackish water RO plant for energy saving	[72]
Toray Industries, SWRO and BWRO trains	40000 m^3/d	Seawater	Red Sea coast, Jeddah, Saudi Arabia	Autopsy of the membrane modules and cartridge filter operated for long-term revealed the presence of a heterogeneous fouling layer	[88]
Filmtec™ BW30-400 spiral wound	416 m^3/d	Groundwater well	Gran Canaria, Canary Islands, Spain	Renewable energy sources (RES) to power RO desalination plants for remote communities	[89]
Not specified	Various	Various	Various	Review, Solar powered RO (energy)	[90]
Cost evaluation					
Polyamide TFC and zeolite membranes	[a]50 m^3/d	Wastewater produced from various sources	San Juan basin	Cost evaluation (cost)	[91]
Not specified	Not specified	Not specified	Not specified	NF/RO/MD hybrid system cost evaluation (hybrid)	[92]
Miscellaneous					
PA TFC membrane (Alfa Laval)	Small lab scale	Acid mine drainage	-	Synthetic AMD (Miscellaneous)	[54]
-	[a]120 k + 30 k m^3/d	Seawater	Gaza strip	Model development based on Bayesian belief networks (miscellaneous)	[93]
Polyamide composite, Dow BW30FR	1 km^3/d	2000 ppm NaCl solution, fouled by BSA solution	-	Removal of the top skin layer for reuse (miscellaneous)	[94]
Dow BW 30-4040 2-stage RO	3760 m^3/d	TDS 2308 mg/L	ILVA Steelworks, Taranto, Italy	The chloride and fluoride contents, decrease the alkalinity and the total dissolved solids.	[95]
FILMTEC SW30-4040, 3-stage RO		Chloride 1064 mg/L, Fluoride 19 mg/L, and Sulphate 137.3 mg/L	ArcelorMittal Steelworks, Aviles, Spain	RO pilot plant to use Blast Furnace gas cleaning water blow-down	[95]

be considered to optimize the use of the ultra-low-pressure RO membrane. Several design options are evaluated for function, effectiveness, and cost impact.

The growing success of membranes in municipal and industrial wastewater is related to improved process designs and improved membrane products [64]. Key factors which have been determined to result from the successful operation of large-scale plants are discussed in this paper. Factors which play a key role in the use of RO membranes include ultra- or microfiltration pretreatment, low fouling membranes, flux rate, recovery, and control of fouling and scaling. In particular, high flux rates can be used when UF or MF pretreatment is used. These technologies remove most of the suspended particles that would normally cause heavy fouling of lead elements. This was demonstrated by the 2.5-year operation of the 10,000 m^3/d Bedok, Singapore, demonstration plant.

RO membranes in a full-scale municipal wastewater reclamation plant were autopsied, and an analysis of the foulant compositions on the RO membranes was made with respect to organic and inorganic matter [66]. The hydrophilic acid fraction was found to be the largest fraction among the six fractions in the RO influent but little of the fraction was deposited on the membranes due to its tendency to remain in the water. However, hydrophobic fractions, especially hydrophobic neutral and hydrophobic acid, had a tendency to gather on the membranes in much higher quantities.

An RO desalination plant powered by wind is technologically feasible even when the plant operates under fluctuating and intermittent loads resulting from variations in the supply of energy which is characteristic of a wind turbine [67]. Direct integration of energy from renewable sources into an RO plant is described in this research.

A mathematical model was developed for a spiral wound RO process based on the solution–diffusion mechanism and applied for the operation of a low-salinity Arab Potash Company (APC) medium-sized RO desalination plant to remove salt from brackish water located in Jordan [68]. The impacts of several operational parameters including feed salinity, feed water flow rate, operating pressure, and temperature on the BWRO plant's performance were discussed.

The performance of pretreatment processes including coagulation, dual media filtration, and polishing with a cartridge filter coupled with antiscalant used at the Perth Seawater Desalination Plant, Western Australia, has been studied in terms of parameters that affect organic and biological fouling [72]. These analyses were conducted employing liquid chromatography with an organic carbon detector (LC-OCD), three-dimensional fluorescence excitation-emission matrix (3D-FEEM), and assimilable organic carbon (AOC). Inorganic scalants found on the fouled RO membranes came from the feed water as well as chemicals used in the pretreatment processes and materials present in the plant. It was revealed that humic substances and low-molecular-weight neutrals were the major organic foulants on the fouled RO membranes.

The conventional hollow-fiber-type RO element is a single open-ended (SOE) structure, in which the pressure drop of the permeated water in the bore side of the hollow fibers prevents the development of a large-sized (longer) RO element [75]. In this work, a both open-ended (BOE) element was devised to reduce the permeate pressure drop. A medium-size Toyobo RO module HJ9155 (cellulose triacetate) was used for the test. It has been confirmed that the permeate flow rate of BOE was greater by about 30% than that of SOE.

Organic foulants obtained from ultrafiltration (UF, Kolon Membrane, Gwacheon, Korea) and RO (four-stage spiral-wound polyamide type thin-film composite RO membrane system, Woongjin Chemical, Seoul, Korea) membranes of a large-scale municipal water reclamation plant were rigorously characterized using conventional and advanced characterization analyses (e.g. pyrolysis and mass spectrometry). This was done in order to identify major constituents of the organic foulants and investigate fouling characteristics in a large-scale application of the UF and RO membranes [76]. Hydrophobic fractions comprising carboxylic acids and aldehydes strongly contributed to the fouling formation of the UF membrane, whereas the RO membrane foulants mainly consisted of hydrophilic fractions comprising amides and alcohols, indicating that the membrane characteristics could play an important role.

Ultrafiltration was used as a single-stage process to pretreat surface water as an alternative to conventional multistage treatment processes (e.g. ozonization–precipitation–flocculation–coagulation–chlorination–gravel filtration). In this process, the use of different chemicals necessitates special safety measures and careful harmonization and control of water chemistry in view of the requirements of downstream reverse osmosis [78]. By contrast, processes based on membrane technology enable a simply designed plant to be used with several advantages. The pretreated water was supplied to the final stage of RO treatment to produce drinking water. Two UF modules MOLPURE FW50-Technology, a four-end module produced by Daicen (cross-flow), and type S-225-X PVCUFC M5 from X-Flow (dead-end) were used for comparison. The total operating costs were 0.25–0.35 EURO/m^3, which was lower than the 0.5 EURO/m^3 of the conventional pretreatment.

The water demineralization plant in Geleen in the Netherlands for the production of boiler feed water from surface water with a capacity of 1560 m^3/h uses RO and UF for the pretreatment [79]. In this study, a series of pilot-scale studies were made using different UF membranes and a suitable UF membrane was selected. As for the RO membrane, Film Tec BW 30 LE 440 was chosen primarily due to its high flux and high salt rejection capacity.

The impacts of marine organic matter (MOM) characteristics on the fouling layer composition of seven parallel arranged RO membranes (16-inch spiral wound modules, area 148.6 m^2) were evaluated in a real-scale seawater desalination plant in South Korea to offer valuable insights into the fouling behavior of MOM [80,81]. Hydrophobic MOM fractions complexed with multivalent metal ions (i.e., Al, Ca, Cu, Fe, and Mg) were found to be major contributors to the irreversible fouling of the RO membranes. Hydrophilic MOM fractions were deposited in a preferential manner onto the membrane surfaces in the

RO_{1st} module, while residual hydrophobic MOM fractions strongly influenced the fouling formation of the RO membranes in the RO_{7th} module. The strategic pairing of three different cleaning agents could control the desorption efficiencies of the organic and inorganic foulants along with the restoration of membrane surface features associated with the differing performances of the RO membranes.

The researchers have provided a theoretical basis for the design and operation of seawater desalination plants to process extreme seawater with high salinity and high temperatures that are often encountered in the Middle East, particularly in the Arabian Gulf region [82]. The two-stage SWRO system was energy-efficient, with high levels of recovery. However, under the extreme conditions encountered, the quality of permeate resembled that of a single-stage SWRO.

This article reviews the recent representative research that is related to SWRO antifouling strategies and answers the most crucial questions about the design and operating parameters of SWRO and its pretreatment processes [83]. Also, the economic evaluation of the SWRO system in regard to antifouling strategies is discussed. The pretreatment technologies to prevent membrane fouling and to extend the lifetime of the RO membrane are commonly grouped into two categories: conventional and non-conventional (membrane). After demonstrating many case studies, the authors concluded that the conventional pretreatment system applied in various SWRO plants in various parts of the world proved to be able to provide the feed water quality required for the RO lines. However, the system was very difficult to control. On the other hand, membrane pretreatment provides a more stable and reliable system that can tolerate feed water quality variations.

Three 16" diameter and 60"-long RO-ULP elements from Koch Membrane Systems were evaluated in parallel with two commercially available 8"-diameter elements [84]. The overall specific flux for 16" was comparable to that of the 8" element. Slightly higher fouling was observed for the former element. Cost evaluation revealed that for a 185 million gpd plant, a 16" element can save as much as 12.4% in total cost. The capital cost saving alone is 27%.

To optimize the design factors and operating conditions of the plant, pretreatment investigations, split partial simulations, and evaluation of the final water quality were conducted. The production capacity and plant performance were then compared with other SWRO plants utilizing normalized specific energy consumption (SEC) [85]. The plant used in this study was the large-scale Gijang SWRO plant which was manufactured with a capacity of 10 MIGD with two different pretreatments (dual media filtration-based 8 MIGD and UF-based 2 MIGD trains) with first-pass SWRO/second-pass BWRO and a remineralization process.

To determine the nature and fate of the foulants of a full-scale SWRO desalination plant located on the Red Sea, a membrane autopsy was performed on the module that was operated longer term [86]. The foulants found on the membrane and cartridge filter were inorganic deposits which mainly consisting of aluminum, iron, and magnesium silicate.

An experimental study was conducted to investigate the potential of the development of water desalination using the photovoltaic-powered system in Jordan [87]. A testing rig was built, where the reverse osmosis (RO, polyamide TFC) desalination system was driven by photovoltaic power by directly coupling the photovoltaic power system to a DC motor, which was coupled to a pump that was capable of providing sufficient torque to run the RO system. Analysis of the results showed that gain of 25% and 15% of electrical power and pure water flow, respectively, could be achieved using the east–west one-axis tracking system compared with a fixed flat plate.

Renewable energy sources (RES) to power BWRO desalination plants have been functioning under intermittent operating conditions for 14 years (~9 h/d) in some remote communities [88]. Daily shut-downs and start-ups did not result in additional issues in the desalination plant, indicating that intermittent operation of BWRO desalination plants can be feasible over a longer time period.

Two water circuits belonging to two integrated steelworks, where high salt concentrations will cause problems, were investigated [89]. In the first circuit, the high chloride and carbonate concentrations in the cooling water of the hot strip mill (HSM) impacted the quality of the strips, due to the salt deposited on the strip surfaces, resulting in the corrosion of equipment. In the second circuit, the high content of chlorides and fluorides in the process waters of a blast furnace (BF) gas cleaning system resulted in the corrosion of various components in the system. RO has been incorporated for the treatment of make-up water before being introduced into the first circuit. In the second circuit, RO was applied for treating blow-down water with the expectation of reusing this water in the circuit.

The project described below aims to provide a comprehensive review of all the indirect solar desalination technologies along with plant-specific technical details [90]. Solar desalination plants are summarized since the year 1978 when solar RO desalination started. The factors affecting PV-RO water cost are the capital cost of the PV array and battery, the inclusion of the energy recovery device, the type of feed water, and the type of RO unit. A water cost of around $7/m^3 for 44 m^3/d is estimated, which is very high. However, for large-scale plants having a capacity of greater than 1000 m^3/d, the specific plant cost is in the range of $4500–6200 m^3/d with an estimated water cost of < $2/m^3 owing to the reduction of high-efficiency PV module cost.

Reverse osmosis membranes including polymeric (polyamide TFC membrane) and molecular sieve zeolite membranes were investigated for ion removal from the water produced at the oil field and coal bed methane sites by a cross-flow RO process [91]. Pretreatments including nanofiltration (NF) and adsorption by active carbon were implemented. The study revealed that (1) most of the permeation tests lasted only 3 months due to severe fouling, (2) multistage pretreatment is crucial to extending membrane life, and (3) only NF treatment could extend the membrane life to 6 months.

In a comparative study of an integrated hybrid membrane-based system with an earlier locally designed RO unit, such a system is comprised of nanofiltration (NF), reverse osmosis (RO), and membrane distillation (MD) subsystems [92]. The comparison was essentially based on using the NF technique in the pretreatment section, while the MD contributed to concentrating the two brine streams from both NF and RO. Thus, the objective was for the high recovery rate of product water. The proposed system was economically evaluated and compared with the RO unit. It was concluded that 76.2% water recovery was possible with a water production cost of $0.92/m^3$.

Characterization of uncertainties in the operation and economies of the proposed seawater desalination plant in the Gaza Strip was made using a Bayesian belief network (BBN) approach [93]. In particular, the model was used to (1) characterize the different uncertainties involved in the RO process; (2) optimize the RO process reliability and cost; and (3) study how uncertainty in unit capital cost, unit operation and maintenance (O&M) cost, and permeate quality were related to different input variables. The minimum specific capital cost was found to be 0.224 ± 0.064 US$/m^3$ and the minimum O&M cost was found to be 0.59 ± 0.11 US$/m^3$. This unit cost was for a production capacity of 140,000 m^3/d.

This study aimed at assessing the technical feasibility of removing the dense polyamide (PA) active layer of RO membranes, with the intent to reuse degraded RO in porous low-pressure membranes [94]. The study assessed the ability of three degrading solutions (NaOH, $KMnO_4$, and NaOCl) to remove the active layer. The most promising results were found using NaOCl. Membranes treated with at least 300,000 ppm of NaOCl presented an increased permeability to 175 ± 4 L m^{-2} h^{-1} bar^{-1} with less than 4% salt rejection. The fouling behavior of the degraded RO membranes was also compared to commercially available ultrafiltration (UF) membranes, which displayed similar fouling characteristics, and by using LC-OCD, the molecular-weight cut-off (MWCO) of the reused membranes was estimated to be 5–10 kDa.

From this review, it is clear that the most serious concern over large-scale plant operations is pretreatment and fouling prevention. Membrane pretreatment seems to be more effective and economical than the conventional pretreatment. The recent membrane module development is focused on ultralow-pressure RO membranes and the construction of a 16"-diameter module, both generating economic benefits. However, care must be taken for the large-diameter module operation. As energy costs keep increasing, the search for alternative energies continues. In the arid regions of the Middle East, the use of solar power has been attempted for quite some time to power medium-sized desalination plants. However, the production of water is quite costly due to the high cost of photovoltaic power. Moreover, the price may go down as the size of the plant increases. Last but not least in importance, most of the articles deal with the cost of water production.

2.6 RECENT DEVELOPMENTS OF RO MEMBRANES FOR DESALINATION: LITERATURE REVIEW

In this section, fundamental basic concepts related to the development of RO membranes are outlined, along with a couple of examples.

Smooth sub-8 nm polyamide nanofilms at a free aqueous–organic interface were made for the purpose of salt removal using reverse osmosis, maintaining chemical homogeneity at both aqueous and organic-facing surfaces [96]. Due to a fast and somewhat unrestrained interfacial polymerization reaction, these nanofilms were crumpled together forming the overall separation layer with a thickness of 50–200 nm. The authors were successful in reducing the separating layer thickness to 6–15 nm by interfacial polymerization at the free aqueous–organic solution interface. Reducing the separating layer thickness from 15 to 8 nm showed that water permeance increased in proportion to the decrease in thickness, but when the thickness was decreased further to 6 nm, the permeance increased irregularly to 2.7 L m^{-2} h^{-1} bar^{-1} [97]. These nanofilms were found to be of sufficient rigidity that the crumpled textures with the increased permeable area could withstand pressurized filtration. Composite membranes consisting of these crumpled nanofilms on alumina materials provided good stability with high retention of solutes.

An easy route founded upon the principles of interfacial polymerization (IFP) was reported, by which Turing-type polyamide membranes for purifying water could be produced under appropriate initial conditions of IFP. Tan et al. manufactured polyamide membranes via IFP by the addition of polyvinyl alcohol to the aqueous phase, which lowered the diffusion of the monomer [98]. By controlling the reaction conditions, membranes with bubbles or tubes with a spatial distribution of voids were formed as a result of small heterogeneous perturbations. The membranes exhibited relatively higher water permeability owing to the Turing structures at the nanoscale level. These unexpected nanostructured membranes, featuring more bumps, voids, and islands, are produced by diffusion-driven instability, enabling exceptional transport properties for both water permeability and water-salt selectivity that exceed the maximum expected level normally seen in traditional desalination membranes.

The additive approach to the fabrication of TFC membranes can control the thickness and roughness of the membrane independently with high resolution or precision, while conventional TFC membrane formation methodologies do not have the capacity to control these properties [99]. In this method, a thin polyamide layer is electro-splayed on top of the substrate at high voltage. The small droplet size coupled with low monomer concentrations creates polyamide films that are smoother and thinner (as low as 15 nm with as low as 4 nm resolution in thickness), while still demonstrating good permselectivity in comparison to a commercial benchmarking membrane. The structure of the membrane can be determined by changing the proportion of the two components or even by using polymers dissolved in solvents.

The atomic-scale transport features and energetics of water and salt ions at high pressure in 3D-printed PA RO membranes, with the different degrees of cross-linking and *m*-phenylenediamine/trimesoyl chloride ratios, were examined through non-equilibrium molecular dynamics (NEMD) simulations by He et al. [100]. With a 3D-printed PA RO membrane, water permeability is always related to the percolated water-accessible free volume providing a predictable path that links the opposite membrane surfaces. The unconnected percolated free volume in the 3D-printed membrane allows water molecules to jump from one pore to another via the route of temporary on-and-off channels. However, salt ions pass through the membrane via the continuous water channels in the membrane.

In another investigation, a series of polyamide desalination membranes were produced in an industrial-scale manufacturing line with varying process conditions. It was demonstrated that when the processing conditions are changed while retaining the chemical compositions, water permeability and active layer thickness increase at a constant sodium chloride selectivity [101]. The authors demonstrated that the energy-filtered TEM allowed obtaining nanoscale 3D images to enable the prediction of water transport by knowing the membrane morphology without any adjustable parameters. Systematic control over nanoscale polyamide inhomogeneities is a key route to maximizing water permeability without sacrificing selectivity for salt in desalination membranes.

2.7 PROGRESS IN THE DEVELOPMENT OF ULTRAHIGH PERMEATION RO MEMBRANES AND FUTURE PROSPECTS

Currently, the construction of new desalination plants is dominated by RO. Nevertheless, RO requires high pressure to overcome the osmotic pressure of the salty feed water, which results in high energy consumption. To reduce the pressure, the RO membrane should be highly permeable to water. However, due to the trade-off effect of the polymeric membrane, the increase in permeability lowers the salt rejection. Hence, it has always been a keen interest of researchers to find new materials other than polymers. Carbon nanotubes, aquaporin, graphene, and, recently, fluorous oligoamide nanorings have appeared to break this dilemma of extremely high permeability to water. Tremendous efforts have been made toward the fabrication of membranes from these materials during the last 15–16 years. These efforts and future prospects for the development of RO membranes of ultrafast water permeation are highlighted in this section.

2.7.1 Carbon Nanotubes

Hummer et al. and Kalra et al. made molecular dynamics simulations (MDS) of water transport in single-walled carbon nanotubes (SWCNTs) [102,103]. The simulation of the osmotically driven transport of water resulted in 5.8 water molecules per ns and CNT [103]. This fast water flow was ascribed to the frictionless flow of a single file of water formed in the narrow SWCNT pore with a diameter of 0.81 nm. Inspired by the MDS results, Holt et al. in 2006 made micro-fabricated membranes in which double-walled carbon nanotubes (DWCNTs) of sub-2-nm diameter were vertically aligned [104]. These researchers have grown DWCNTs vertically on the surface of a silicon tip via catalytic chemical vapor deposition, followed by low-pressure chemical deposition of silicon nitride matrix which filled the spaces between the DWCNTs. After removing excess silicon nitride by ion milling, the ends of the nanotubes were opened by reactive ion etching. The pore density was less than 0.25 α 10^{12} per cm^2 and the pore length was 2.0–3.0 μm. Water permeation experiments carried out under the feed side pressure of ca. 0.8 bar revealed that the water permeation rate was three orders of magnitude higher than that predicted by the Hagen–Poiseuille equation. The measurement of the water flow rate through the cylindrical carbon nanotube was made via the stopped-flow technique much later by Li et al. in 2020 [105]. They embedded short CNT segments of 0.8 nm diameter in lipid membranes and measured the water flow rate via self-quenching of fluorescence dye, carboxyfluorescein (CF). The obtained flow rate was 2.3 ± 0.1 cm^3/s, which was slightly higher than the value obtained earlier by Tunuguntla et al. [106].

Since the *Science* paper by Holt et al. was published, many attempts have been made to fabricate industrially viable carbon nanotube-based membranes [104]. Many of them are thin-film nanocomposite (TNC) membranes in which carbon nanotubes (CNTs) are embedded in aromatic polyamide thin-film composite (TFC) membranes. Some examples are (1) the work of Cruz-Silva et al. with MWCNTs. The membrane was not only an improvement in RO performance but also fouling and chlorine resistance were enhanced [107]; (2) Marita et al. reported in their review article that higher salt rejection (97.69% as compared with 96.1% [without CNTs]) and a near doubling of water flux (44 L/m^2 day bar as compared with 26 L/m^2 day bar [without CNTs]) were achieved by incorporation of CNTs [108]; (3) Kim et al. found CNTs were acidified and embedded in the TFC membrane. Water flux was significantly increased. Also, the durability and chemical resistance against NaCl solutions were increased [109]; (4) Zhang et al. also acidified MWCNTs and embedded them in the TFC membrane. Water flux was significantly improved. The membrane may have a potential application in the separation of an organic aqueous solution [110]; (5) Inukai et al. embedded 1 wt.% of anionic surfactant stabilized MWCNT in the TFC membrane. Membrane performance in terms of flow and antifouling was improved and chlorine degradation was inhibited [111]; (6) Kim et al. embedded a mixture of acidified CNTs and graphene oxide (GO) in a TFC membrane [112]. Mechanical strength, antimicrobial and antifouling properties, selectivity, water flux, and thermal properties were significantly improved after incorporation of the fillers; and (7) Zhao et al. pretreated MWCNTs with mixed acids before being modified with diisobutyryl peroxide to enhance their dispersity and chemical activity [113]. The modified MWCNTs were then incorporated into TFC membranes. MWCNT dispersion improved. A flux of 28 kg/m^2 s with NaCl separation of 90% was achieved at 16 bar. The more recent development of CNT-related membranes is summarized in the review paper of Kumar et al. [114].

2.7.2 Biomimetic Membrane

MDS of aquaporins revealed that the water permeability through an aquaporin pore is 10^8–10^9 water molecules/s [115]. The water permeability of aquaporin was then measured by the stop-flow method, in which the suspension of aquaporin-containing vehicles was mixed with the solution of the solutes (sorbitol, sucrose, or mannitol) that could not permeate through the membrane. Thus, the water flows from inside to outside of the aquaporin-containing vehicles. The consequent reduction in the vehicle volume was measured by the increase in the scattered light intensity. By this method, the water permeability was estimated to be $2–10 \times 10^{-14}$ cm^3/s, which was in reasonable agreement with the MDS results of $2.2 \pm 0.3 \times 10^{-14}$ cm^3/s [116]. Kumar et al., based on the water permeability of Aquaporin Z (AqpZ)-containing proteoliposomes, postulated that AqpZ-based biomimetic membranes can potentially achieve a water permeability as high as 601 LMH/bar, which is two orders of magnitude higher than the currently available commercial membranes [117,118]. Zhao et al. [119] fabricated an aquaporin-based biomimetic membrane via the interfacial polymerization method. The membrane with an area greater than 200 cm^2 had good mechanical stability up to 10 bar under RO conditions. High water permeability of 4.0 LMH/bar and NaCl rejection of 97% were obtained at 5 bar. Rejection was further improved at higher pressures. The performance of the membrane was far better than the commercial RO membranes (BW30 and SW30HR). Attempts were made to make the membranes as stable, robust, scalable, and cost-effective as existing RO membranes [120].

As for the commercial production of aquaporin membranes [121], a Danish company called Aquaporin is manufacturing flat-sheet and hollow-fiber membranes based on their technology called Aquaporin Inside. In this technology, aquaporin is incorporated into the dense layer of a thin-film polyamide membrane and the membrane can be used either for RO or forward osmosis (FO). According to Aquaporin's brochure, the hollow-fiber module shows 11 ± 1.5 LMH at the cross-membrane osmotic pressure difference of 0.2 bar (water permeability of 55 LMH/bar) at 25°C [122]. Although the aim of the fabrication of high flux RO, NF, and FO membranes is to reduce the energy consumption and water production cost of the desalination process, another advantage of the aquaporin membrane is its high efficiency in the removal of trace organic pollutants in wastewater treatment and boron removal in the seawater desalination process. For example, the aquaporin membrane could remove 97% of some typical trace organic pollutants with higher water permeability than the cellulose acetate-based membrane in FO [121,123].

2.7.3 Graphene Membrane

Cohen-Tanugi and Grossman made an MDS of the water and sodium chloride transport through nanopores of graphene [124]. Two materials of the pore wall were considered, hydrogenated, representing the hydrophobic pores, and hydroxylated, representing the hydrophilic pores. The pore sizes varied in the range of 1.2–62 Å^2. The results showed nearly 100% salt rejection for the smallest and medium-sized hydrogenated pores and also for the smallest hydroxylated pores. Water permeability increased from 0 L/cm^2 day MPa of 1.2 Å^2 to 60 L/cm^2 day MPa (2502 LMH/bar) of 52 Å^2 for the hydrogenated pores and from 50 L/cm^2 day MPa (2085 LMH/bar) of 16 Å^2 to 129 L/cm^2 day MPa (5379 LMH/bar) of 62 Å^2 for the hydroxylated pores. Thus, the graphene membrane allows permeability of as high as 2000 MLH/bar with perfect NaCl rejection, which is several orders of magnitude higher than the conventional RO membrane. Membranes with a graphene domain of 5 μm diameter were fabricated by oxygen plasma etching of graphene grown by chemical vapor deposition [125]. The pore size and pore density depended on the oxygen plasma etching time. The membrane with the lowest defect exhibited 100% salt rejection with 10^6 g/m^2 s (3.6×10^6 LMH) by a hydraulic pressure difference and 70 g/m^2 s bar (252 LMH/bar) by an osmotic pressure difference. The area of single-layer graphene was increased to cm-scale by O'Hern et al. [126]. The defective pores of the membrane were sealed by the two-step process, first with atomic layer deposition of hafnia (HfO_2), followed by interfacial polymerization of nylon. The membrane showed a nanofiltration capacity of 0.5×10^{-5} m^3/m^2 s water flux at 10 bar (1.75 LMH/bar) with 70% $MgSO_4$ rejection by FO experiments. According to the latest News [127], Clean TeQ Water (ASX:CNQ) subsidiary, NematiQ, has achieved a major milestone by producing over 1000 m of 1000 mm wide flat sheet graphene membrane by roll-to-roll coating. The membrane will be loaded into a spiral wound module and tested in a demonstration unit. The membrane is called a nanofiltration membrane but is said to remove bacteria, viruses, and dissolved organic compounds without removing the salts. Therefore, the membrane seems more like an ultrafiltration membrane. The experimental data on water flux and solute rejection were not disclosed. Some graphene-based membranes consist of multilayered sheets, often of graphene oxide (GO). Akbari et al. [128] cast discotic and nematic liquid crystals of GO on a nylon sheet using a casting blade. The membrane area was as large as 13×14 cm^2. The membrane exhibited the capacity of a nanofiltration membrane with more than 90% rejection of charged and uncharged organic probe molecules of hydrated radius above 0.5 nm. The rejection rate of mono- and divalent salts was 30–40%. The water permeability is 71 ± 5 LMH/bar for 150 ± 15 nm thick membranes. Compared with the commercial NF 270 membrane, the flux was 10 times higher. Zhang et al. exfoliated the GO nanosheet by sonication, which was then functionalized by sulfonation and mild reduction [129]. It was then cross-linked to stabilize the membrane in water. The membrane showed a water permeability of 1.8 LMH/bar and 80.5% of Na_2SO_4 rejection and 52% of NaCl rejection (data for high-R).

2.7.4 Fluorous Oligoamide Nanorings ($^{FM}NR_{N}s$)

Itoh et al. synthesized fluorous oligoamide nanorings ($^{Fm}NR_{n}s$) [130] and calculated their sizes and shapes by MDS. The calculated sizes were 5.1–6.4 nm and 0.9–1.9 nm for the external and internal diameters, respectively. The external diameters were further confirmed by transmission electron microscopy. The water permeation rate was also calculated by MDS and experimentally determined by the stopped-flow fluorescence method using 1,2-dipalmitoyl-sn-glycerol-3-phosphocholine (DPPC) for the vehicle and carboxyfluorescein (CF) to observe the fluorescence decay. The result was a flow rate of 5.5×10^{-10} cm^3/s per channel, which is much higher than those of 1.2×10^{-13} cm^3/s per channel and 2.3×10^{-13} cm^3/s per channel of aquaporin and carbon nanotube, respectively. The flow velocity was calculated by dividing by the channel area and the result was 8.7×10^{-10} $cm^3/s\ nm^2$. These values obtained for $^{F12}NR_4$ of the smallest pore size were the highest among all the synthesized $^{Fm}NR_{n}s$. These results were ascribed to the breakage of the water cluster near the fluorinated pore wall, which reduced considerably the shear force working on the water at the pore wall. This small pore also stopped the permeation of chlorine ions, thus resulting in the cessation of salt permeation since the electroneutrality of ions should be maintained.

2.7.5 Summary

When you look at the progress in membrane development from the four materials mentioned above, we realize that there is a common pattern among these, i.e., all of them follow the following steps (Scheme 2.1).

All of these, except for fluorous oligoamide nanorings, start with the discovery of the material by a Nobel laureate. First, MDS of water transport is performed, followed by the water flow rate measurement through the nano-sized cylinder by the stopped-flow method and then by the miniature-sized membrane. The results from these experiments are reported mostly in very high-impact journals. Then, to make the membranes more practical, membranes of the cm-sized area are fabricated for the filtration experiments. If the experiment turns out to be successful, a module is constructed for the pilot-scale experiments before commercialization. Although there are some exceptions to the sequence, the above steps are followed in most cases. Until now, only aquaporin and graphene have reached step 5. Aquaporin membrane has reached step 6 but its application is mainly for forward osmosis. Considering the time that has passed since the announcement of CNT-based membrane in 2006, progress has been very slow.

In contrast, not even a year passed after the first announcement of the cellulose acetate RO membrane by Loeb and Sourirajan to the construction of the plate and frame module by their own hands. When the scale-up ratio of 10^{18} (from nm^2 to m^2) for the membranes of ultra-high water permeation is compared with 10^4 (from cm^2 to m^2) for the cellulose acetate RO membrane, the slow progress is understandable. No engineering principles have been established until now for the scale-up to such a high ratio.

Giwaa et al. further commented on the following challenges for the development of large-scale aquaporin membranes [121], which seem also to be applicable to all other ultra-high flux membranes.

1. Production of a large quantity of aquaporin
2. Long-term stability of the aquaporin membrane
3. Good compatibility between aquaporin and the host membrane
4. Chemical cleaning of the membrane

In addition, there may be some other challenges, such as:

1. Vertical alignment of nano-sized cylinders in the membrane
2. Reproducibility of filtration performance
3. The environmental issue of the release of nanoparticles into drinking water, even very small amounts

Hence, there are many risks to take into account before ultra-high flux membranes enter the commercial market.

2.8 FUTURE DIRECTIONS OF THE RESEARCH AGENDA

The improvement of membrane performance to achieve higher flux, higher salt rejection, and low energy consumption simultaneously remains the subject of greatest interest. As is

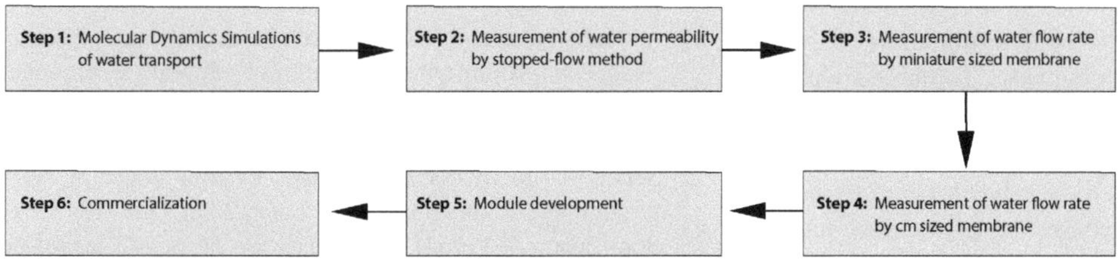

SCHEME 2.1 Schematic of the progress in membrane development

clear from this review, enhancement of membrane flux, while maintaining high salt rejection, has been achieved primarily by the major industrial RO membrane manufacturers by decreasing the skin layer of the PA TFC membrane. On the other hand, many attempts have been made, mostly in academia, to increase the flux by incorporating nano-scale hydrophilic particles. Another interesting attempt has been made to replace the conventional support membrane for in situ polymerization with a highly porous material such as a nanofibrous membrane [131]. In all of these attempts, the approach is based on the improvement of the PA TFC membrane due to the latter membrane's enormous commercial success. However, there may be polymeric materials other than PA to form the thin surface layer in situ. Moreover, attempts are continuing to fabricate membranes with overwhelmingly high fluxes [132]. The presence of boron, arsenic, and residual micro-organics in drinking water is a serious health issue. New membranes should be developed to remove these compounds effectively. Inorganic membranes with NF pore sizes are now available but the emergence of inorganic RO membranes is still awaited. Similarly, the treatment of organic solvent is now possible by polymeric NF membranes. However, inorganic material seems more appropriate for RO membranes since polymeric membrane swells in an organic environment.

Fouling continues to be of great concern for the practical application of RO membranes and the development of intrinsically fouling-free membranes is the goal of many researchers [16]. Surface modification is the most popular approach to reducing fouling. In this regard, one of the unconventional approaches of blending surface-modifying macromolecules (SMMs) into the host polymer matrix should be explored further [36–39]. Despite the simplicity of its process, i.e. the requirement for only a small amount of SMM in one step to modify the membrane surface, the method has not yet fully taken off, since amphiphilic macromolecules of rather exotic structures have so far been used. SMMs should be found among the bulk chemicals that are easily accessible to many laboratories at low cost.

As the modern characterization method advances, details of the membrane morphology are revealed. The size and size distribution of the water channels can be measured precisely by modern instruments, which makes the water and solute transport theories based on the black box approach look more obsolete. A new transport model that matches the advancement of the modern characterization method is called for. Also, precise control of the water channel size and its distribution, as well as the surface roughness, is required for future membrane design. In this respect, 3D imaging of the skin layer of the RO membrane will contribute to the precise determination of the size and size distribution of water channels. Combined with molecular dynamics simulation, prediction of water flux and salt rejection will become possible [99].

Regarding module development, the module size continues to become larger along with increases in plant size, and currently, there seems no limit to this trend. As the module size increases, so also do the sizes of pumps, pipes, etc., and so a more precise system control will be required. Also, as the construction of RO plants increases, the amount of discarded RO modules increases. The reuse of these RO modules will become increasingly desirable.

Hybrid processes are also being considered more seriously than ever [133]. There are hybrid processes, such as pretreatment by UF and NF to provide RO feed water, RO treatment of boiler water for the distillation process, etc., and these are already in use. Combining RO with emerging membrane separation processes such as membrane distillation (MD) and forward osmosis (FO) will become more important to decrease the amount of effluent from RO desalination plants. As drinking water production by the desalination of sea- and brackish water increases, an increase in the quantity of salty RO effluent will become more of an environmental concern. The RO hybrid process with MD and FO may become the answer to this problem. Also, it will increase the production of drinking water.

2.9 CONCLUSIONS

Many new polymers have been synthesized and tested for their permeation properties, specifically fouling resistance, chlorine resistance, thermal and mechanical stabilities, etc., aiming at the improvement of membrane performance for RO applications. These efforts seem to continue with insightful vision and a strong commitment to the future. However, only a handful of polymers are currently being used as materials for commercial applications, and they are not necessarily the polymers with the best performance properties. This is mainly due to the cost factor that governs the current membrane market. The RO performance of membranes, on the other hand, is known to be primarily determined by the membrane's active surface. Surface modification and surface coating research and development will focus on new membrane development techniques in the future, as only a small amount of polymer is required for the surface coating. This new knowledge of the effects of these factors on RO performance will allow for full utilization of the growing potential for polymer research applications for membrane materials, which will open up promising new avenues for further research and development in these areas. Also, attempts are being made to develop RO membranes with orders of magnitude higher fluxes than the currently available commercial membranes.

REFERENCES

[1] Sourirajan, S. (1970) *Reverse Osmosis*. Academic Press, New York, NY.

[2] (a) Sourirajan, S. (1963) The mechanism of demineralization of aqueous sodium chloride solutions by flow, under pressure, through porous membranes. *Industrial and Engineering Chemistry Fundamentals* 2: 51–55; (b) Sourirajan, S. (1964) Separation of hydrocarbon liquids by flow under pressure through porous membranes. *Nature* 203: 1348–1349; (c) Sourirajan, S. (1964) Characteristics of porous cellulose acetate membranes for the separation of some inorganic salts in aqueous solution. *Journal of Applied Chemistry* 14: 506–513.

[3] (a) Kimura, S., Sourirajan, S. (1967) Analysis of data in reverse osmosis with porous cellulose acetate membranes used. *American Institute of Chemical Engineers Journal* 13: 497–503; (b) Matsuura, T., Sourirajan, S. (1972) Reverse osmosis for water pollution control. *Water Research* 6: 1073–1086; (c) Matsuura, T., Sourirajan, S. (1978) A fundamental approach to application of reverse osmosis for food processing. *American Institute of Chemical Engineers Symposium Series* 74: 196–208.

[4] (a) Sourirajan, S., Matsuura, T. (1981) Reverse osmosis transport through capillary pores under the influence of surface forces. *Industrial Engineering Chemistry Process Design Development* 20: 273–282; (b) Sourirajan, S., Matsuura, T. (1982) Science of reverse osmosis – An essential tool for the chemical engineer. *Chemical Engineer* 385: 359–368; (c) Matsuura, T., Sourirajan, S. (1985) Materials science of reverse osmosis-ultrafiltration membranes. *American Chemical Society Symposium Series* 281: 1–19; (d) Zhu, Z., Matsuura, T. (1991) Discussion on the formation mechanism of surface pores in reverse osmosis, ultrafiltration, and microfiltration membranes prepared by phase inversion process. *Journal of Colloid and Interface Science* 147: 307–315; (e) Matsuura, T., Tan, J. (1999) Membrane separation technology and surface science. *Hyomen* 37: 255–263; (f) Rana, D., Matsuura, T. (2010) Membrane transport models. In: *Encyclopedia of Agriculture, Food and Biological Engineering: Second Edition*, Heldman, D.R., Moraru, C.I., (Eds.). Taylor & Francis, New York, NY, Vol. 2, pp. 1041–1047; (f) Rana, D., Matsuura, T., Sourirajan, S. (2014) Physicochemical and engineering properties of food in membrane separation processes. In: *Engineering Properties of Foods: Fourth Edition*, Rao, M.A., Rizvi, S.S.H., Datta, A.K., Ahmed, J. (Eds.). Taylor & Francis/CRC Press, Boca Raton, FL, Ch. 12, pp 437–525.

[5] Sourirajan, S., Matsuura, T. (1985) *Reverse Osmosis/ Ultrafiltration Process Principles*, NRCC 24188. National Research Council of Canada, Ottawa, ON, Canada.

[6] (a) Loeb, S., Sourirajan, S. (1961) *Sea water demineralization by means of a semipermeable membrane.* Department of Engineering, University of California at Los Angeles, Report No. 1961, pp. 60–60; (b) Loeb, S., Sourirajan, S. (1963) Sea water demineralization by means of an osmotic membrane. *Advances in Chemistry Series* 38: 117–132; (c) Loeb, S., Sourirajan, S. (1964) *The preparation of high-flow semi-permeable membranes for separation of water from saline solutions.* The Regents of the University California, Berkeley, CA, US Patent 3 133 132, 12 May 1964; (d) Loeb, S., Sourirajan, S., Weaver, D.E. (1964) *Porous membranes for separating water from saline.* The Regents of the University California, Berkeley, CA, US Patent 3 133 137, 12 May 1964.

[7] Matsuura, T. (1994) *Synthetic Membranes and Membrane Separation Processes.* CRC Press, Boca Raton, FL.

[8] (a) Rozelle, L.T., Cadotte, J.E., Corneliussen, R.D., Erickson, E.E. (1968) *Development of new reverse osmosis membranes for desalination.* Office of Saline Water Research and Development, Washington, DC, USA, Progress Report No. 359, June 1968; (b) Rozelle, L.T., Cadotte, J.E., Cobian, K.E., Kopp Jr., C.V. (1977) Nonpolysaccharide membranes for reverse osmosis: NS-100 membranes. In: *Reverse Osmosis and Synthetic Membranes: Theory-Technology-Engineering*, Sourirajan, S., (Ed.). National Research Council of Canada, Ottawa, ON, pp 249–261; (c) Cadotte, J.E. (1981) *Interfacially synthesized reverse osmosis membranes.* FilmTec Corporation, Minetonka, MN, US Patent 4 277 344, 7 Jul 1981.

[9] (a) Richter, J.W., Square, K., Hoehn, H.H. (1971) *Perm-selective, aromatic, nitrogen-containing polymeric membranes.* E. I. du Pont de Nemours and Company, Wilmington, DE, US Patent 3 567 632, 2 Mar 1971; (b) Pana, M., Hoehn, H.H., Heber, R.R. (1973) The nature of asymmetry in reverse osmosis membranes. *Macromolecules* 6: 777–780.

[10] (a) Sundet S.A. (1985) *Production of composite membranes.* E. I. du Pont de Nemours and Company, Wilmington, DE, US Patent 4 529 646, 16 Jul 1985; (b) Sundet, S.A. (1987) *Multilayer reverse osmosis membrane in which one layer is poly-meta-phenylene cyclohexane-1,3,5-tricarboxamide.* E. I. du Pont de Nemours and Company, Wilmington, DE, US Patent 4 643 829, 17 Feb 1987; (c) Arthur, S.D. (1991) *Multilayer reverse osmosis membrane of polyamide-urea.* E. I. du Pont de Nemours and Company, Wilmington, DE, US Patent 5 019 264, 28 May 1991; (d) Arthur, S.D. (1992) *Method of producing multilayer reverse osmosis membrane of polyamide-urea.* E. I. du Pont de Nemours and Company, Wilmington, DE, US Patent 5 173 335, 22 Dec 1992.

[11] (a) Uemura, T., Kurihara, M. (1985) *High performance semipermeable composite membrane and process for producing the same.* Toray Industries Inc., Tokyo, Japan, US Patent 4 559 139, 17 Dec 1985; (b) Kurihara, M., Himeshima, Y. (1991) The major developments of the evolving reverse osmosis membranes and ultrafiltration membranes. *Polymer Journal* 23: 513–520; (c) Fusaoka, Y. (1999) Super ultra low pressure composite reverse osmosis membrane elements for brackish water desalination and ultrapure water production. *Membrane* 24: 319–323; (d) Konagaya, S., Nita, K., Matsui, Y., Miyagi, M. (2001) New chlorine-resistant polyamide reverse osmosis membrane with hollow fiber configuration. *Journal of Applied Polymer Science* 79: 517–527.

[12] (a) Kesting, R.E. (1977) Asymmetric cellulose acetate membranes. In: *Reverse Osmosis and Synthetic Membranes: Theory-Technology-Engineering*, Sourirajan, S. (Ed.). National Research Council of Canada, Ottawa, ON, pp 89–109; (b) Kesting, R.E. (1985) Phase inversion membranes. In: *Materials Science of Synthetic Membranes*, Lloyd, D.R. (Ed.). ACS Symposium Series 269, American Chemical Society, Washington, DC, pp 131–164; (c) Kesting, R.E. (1989) The nature of pores in integrally-skinned phase inversion membranes. In: *Advances in Reverse Osmosis and Ultrafiltration*, Matsuura, T., Sourirajan, S. (Eds.). National Research Council of Canada, Ottawa, ON, pp. 3–13.

[13] Petersen, R.J. (1993) Composite reverse osmosis and nanofiltration membranes. *Journal of Membrane Science* 83: 81–150.

[14] Baker, R.W. (2004) *Membrane Technology and Applications.* John Wiley & Sons, Chichester, UK.

[15] Wavhal, D.S., Fisher, E.R. (2003) Membrane surface modification by plasma-induced polymerization of acrylamide for improved surface properties and reduced protein fouling. *Langmuir* 19: 79–85.

[16] Rana, D., Matsuura, T. (2010) Surface modifications for antifouling membrane. *Chemical Reviews* 110: 2448–2471.

[17] (a) Kuehne, M.A., Song, R.Q., Li, N.N., Petersen, R.J. (2001) Flux enhancement in TFC RO membranes. *Environmental Progress* 20: 23–26; (b) Li, N.N., Kuehne, M.A., Petersen, R.J. (2000) *High flux reverse osmosis membrane*. NL Chemicals Technologies Inc., Mount Prospect, IL, US Patent 6 162 358, 19 Dec 2000.

[18] (a) Tran, C.N., Maldonado, A.C., Somanathan, R. (1993) *Thin-film composite membrane*. Allied-Signal Inc., Morris Township, NJ, US Patent 5 234 598, 10 Aug 1993; (b) Tran, C.N., Maldonado, A.C., Somanathan, R. (1994) *Method of making thin-film composite membranes*. Fluid Systems Corporation, San Diego, CA, US Patent 5 358 745, 25 Oct 1994.

[19] Hachisuka, H., Ikeda, K. (2001) *Composite reverse osmosis membrane having a separation layer with polyvinyl alcohol coating and method of reverse osmosis treatment of water using the same*. Nitto Denko Corporation, Osaka, Japan, US Patent 6 177 011, 23 Jan 2001.

[20] Hirose, M. (2005) *Highly permeable composite reverse osmosis membrane and method of producing the same*. Nitto Denko Corporation, Osaka, Japan, US Patent 6 837 381, 4 Jan 2005.

[21] Kwak, S.-Y., Jung, S.G., Kim, S.H. (2001) Structure-motion-performance relationship of flux-enhanced reverse osmosis (RO) membranes composed of aromatic polyamide thin films. *Environmental Science and Technology* 35: 4334–4340.

[22] (a) Hirose, M., Ikeda, K. (1996) *Method of producing high permeable composite reverse osmosis membrane*. Nitto Denko Corporation, Osaka, Japan, US Patent 5 576 057, 19 Nov 1996; (b) Hirose, M., Ito, H., Kamiyama, Y. (1996) Effect of skin layer surface structures on the flux behaviour of RO membranes. *Journal of Membrane Science* 121: 209–215; (c) Gerard, R., Hachisuka, H., Hirose, M. (1998) New membrane developments expanding the horizon for the application of reverse osmosis technology. *Desalination* 119: 47–55.

[23] Hirose, M., Ito, H., Maeda, M., Tanaka, K. (1997) *Highly permeable composite reverse osmosis membrane, method of producing the same, and method of using the same*. Nitto Denko Corporation, Japan, US Patent 5 614 099, 1997.

[24] Kim, S.H., Kwak, S.Y., Suzuki, T. (2005) Positron annihilation spectroscopic evidence to demonstrate the flux-enhancement mechanism in morphology-controlled thin-film-composite (TFC) membrane. *Environmental Science and Technology* 39: 1764–1770.

[25] Jeong, B.H., Hoek, E.M.V., Yan, Y., Subramani, A., Huang, X., Hurwitz, G., Ghosh, A.K., Jawor, A. (2007) Interfacial polymerization of thin film nanocomposites: A new concept for reverse osmosis membranes. *Journal of Membrane Science* 294; 1–7.

[26] Kim, S.H., Kwak, S.Y., Sohn, B., Park, T.H. Design of TiO_2 nanoparticles self- assembled aromatic polyamide thin-film-composite (TFC) membrane as an approach to solve biofouling problem. *Journal of Membrane Science* 211: 157–165.

[27] Kwak, S.Y., Kim, S.H., Kim, S.S. (2001) Hybrid organic/inorganic reverse osmosis (RO) membrane for bactericidal anti-fouling. 1. Preparation and characterization of TiO_2 nanoparticles self-assembled aromatic polyamide thin-film composite (TFC) membrane. *Environmental Science and Technology* 35: 2388–2394.

[28] Wu, S., Xing, J., Zheng, C., Xu, G., Zheng, G., Xu, J. (1997) Plasma modification of aromatic polyamide reverse osmosis composite membrane surface. *Journal of Applied Polymer Science* 64: 1923–1926.

[29] Kulkarni, A., Mukherjee, D., Gill, W. (1996) Flux enhancement by hydrophilization of thin film composite reverse osmosis membranes. *Journal of Membrane Science* 114: 39–50.

[30] Mukherjee, D., Kulkarni, A., Gill, W.N. (1994) Flux enhancement of reverse osmosis membranes by chemical surface modification. *Journal of Membrane Science* 97: 231–249.

[31] Belfer, S., Purinson, Y., Fainshtein, R., Radchenko, Y., Kedem, O. (1998) Surface modification of commercial composite polyamide reverse osmosis membranes. *Journal of Membrane Science* 139; 175–181.

[32] Sforca, M., Nunes, S.P., Peinemann, K.V. (1997) Composite nanofiltration membranes prepared by in-situ polycondensation of amines in a poly(ethylene oxide-b-amide) layer. *Journal of Membrane Science* 135: 179–186.

[33] Kang, G., Liu, M., Lin, B., Cao, Y., Yuan, Q. (2007) A novel method of surface modification on thin-film composite reverse osmosis membrane by grafting poly(ethylene glycol). *Polymer* 48; 1165–1170.

[34] Freger, V., Gilron, J., Belfer, S. (2002) TFC polyamide membranes modified by grafting of hydrophilic polymers: An FT-IR/AFM/TEM study. *Journal of Membrane Science* 209: 283–292.

[35] Louie, J.S., Pinnau, I., Ciobanu, I., Ishida, K.P., Ng, A., Reinhard, M. (2006) Effects of polyether–polyamide block copolymer coating on performance and fouling of reverse osmosis membranes. *Journal of Membrane Science* 280: 762–770.

[36] (a) Hamza, A., Pham, V.A., Matsuura, T., Santerre, J.P. (1997) Development of membranes with low surface energy to reduce the fouling in ultrafiltration applications. *Journal of Membrane Science* 131: 217–227; (b) Matsuura, T., Santerre, P., Narbaitz, R.M., Pham, V.A., Fang, Y., Mahmud, H., Baig, F. (1999) *Membrane composition and method of preparation*. University of Ottawa, Ottawa, ON and Fielding Chemicals, Mississauga, ON, US Patent 5 954 966, 21 Sep 1999; (c) Khayet, M., Suk, D.E., Narbaitz, R.M., Santerre, J.P., Matsuura, T. (2003) Study on surface modification by surface-modifying macromolecules and its applications in membrane separation processes. *Journal of Applied Polymer Science* 89: 2902–2916.

[37] (a) Rana, D., Matsuura, T., Narbaitz, R.M., Feng, C. (2005) Development and characterization of novel hydrophilic surface modifying macromolecule for polymeric membranes. *Journal of Membrane Science* 249: 103–112; (b) Rana, D., Matsuura, T., Narbaitz, R.M. (2006) Novel hydrophilic surface modifying macromolecules for polymeric membranes: Polyurethane ends capped by hydroxy group. *Journal of Membrane Science* 282: 205–216.

[38] Abu Tarboush, B.J., Rana, D., Matsuura, T., Arafat, H.A., Narbaitz, R.M. (2008) Preparation of thin-film-composite polyamide membranes for desalination using novel hydrophilic surface modifying macromolecules. *Journal of Membrane Science* 325: 166–175.

[39] Rana, D., Kim, Y., Matsuura, T., Arafat, H.A. (2011) Development of antifouling thin-film-composite membranes for seawater desalination. *Journal of Membrane Science* 367: 110–118.

[40] Hyung, H., Kim, J.H. (2006) A mechanistic study on boron rejection by sea water reverse osmosis membranes. *Journal of Membrane Science* 286: 269–278.
[41] Mane, P.P., Park, P.K., Hyung, H., Brown, J.C., Kim, J.H. (2009) Modeling boron rejection in pilot- and full scale-reverse osmosis desalination processes. *Journal of Membrane Science* 338: 119–127.
[42] Koseoglu, H., Kabay, N., Yüksel, M., Sarp, S., Arar, Ö., Kitis, M. (2008) Boron removal from seawater using high rejection SWRO membranes – Impact of pH, feed concentration, pressure and cross-flow velocity. *Journal of Membrane Science* 227: 253–263.
[43] Güler, E., Kabay, N., Yüksel, M., Yiğit, N.O., Kitiş, M., Bryjak, M. (2011) Integrated solution for boron removal from seawater using RO process and sorption-membrane filtration hybrid method. *Journal of Membrane Science* 375: 249–257.
[44] Güler, E., Kabay, N., Yüksel, M., Yavuz, E., Yüksel, Ü. (2011) A comparative study for boron removal from seawater by two types of polyamide thin film composite SWRO membranes. *Desalination* 273: 81–84.
[45] Park, H.B., Freeman, B.D., Zhang, Z.B., Sankir, M., McGrath, J.M. (2008) Highly chlorine-tolerant polymers for desalination. *Angewandte Chemie* 120: 6108–6113.
[46] Kurihara, M., Fusaoka, Y. (1999) Recent progress on separation membrane and outlook for future. *Membrane* 24: 247–255.
[47] Micklos, M. (2008) *The Development of Reverse Osmosis and Nanofiltration through Modern Times.* Plenary Lecture III, International Congress on Membranes and Membrane Processes (ICOM), Honolulu, HI, 12–18 July 2008.
[48] Kawada, I. (1999) Development of high efficiency sea water desalination RO membrane. *Membrane* 24: 336–341.
[49] Lonsdale, H.K., Merten, U., Riley, R.L. (1965) Transport properties of cellulose acetate osmotic membranes. *Journal of Applied Polymer Science* 9: 1341–1362.
[50] Greenlee, L.F., Lawler, D.F., Freeman, B.D., Marrot, B., Moulin, P. (2009) Reverse osmosis desalination: Water sources, technology and today's challenges. *Water Research* 43: 2317–2348.
[51] (a) Chevalier, S. (1999) *Mathematical Modeling of Separation by Nanofiltration Mechanisms.* Ph.D. thesis, University of Bordeaux I, Bordeaux, France; (b) Donnan, F.G. (1924) The theory of membrane equilibria. *Chemical Reviews* 1: 73–90; (c) Planck, M. (1890) By the potential difference between two dilute solutions of binary electrolyte. *Annals of Physic*s 40: 561–576; (d) Nemst, W. (1888) Kinetic for the Befindlichen in solution body. 1. First theory diffusion. *Journal of Physical Chemistry* 2: 613–637.
[52] (a) Matsuura, T. (2001) Progress in membrane science and technology for seawater desalination – A review. *Desalination* 134, 47–54; (b) Matsuura, T., Rana, D., Qtaishat, M.R., Singh, G. (2013) Recent advances in membrane science and technology in seawater desalination – with technology development in the Middle East and Singapore. In: *Desalination and Water Resources: Membrane Processes*, Kotchetkov, V. (Ed.). Encyclopedia of Life Support Systems (EOLSS) Publ. Co. Ltd., Ramsey, Isle of Man, Vol. III, pp. 330–381.
[53] (a) Rangarajan, R., Matsuura, T., Goodhue, E.C., Sourirajan, S. (1979) Predictability of membrane performance for mixed solute reverse osmosis systems, Part II: System cellulose acetate membrane 1-1 and 2-1 electrolytes-water. *Industrial Engineering Chemistry Process Design Development* 18: 278–287; (b) Rangarajan, R., Majid, M.A., Matsuura, T., Sourirajan, S. (1985) Predictability of membrane performance for mixed-solute reverse osmosis systems. 4. System: cellulose acetate-nine seawater ions-water. *Industrial Engineering Chemistry Process Design Development* 24; 977–985.
[54] Al-Zoubi, H., Rieger, A., Steinberger, P., Pelz, W., Haseneder, R., Härtel, G. (2010) Optimization study for treatment of acid mine drainage using membrane technology. *Separation Science and Technology* 45: 2004–2016.
[55] Kunisada, Y., Murayama, Y., Hirai, M. (1977) The development of seawater desalination by reverse osmosis process in Japan. *Desalination* 22: 243–252.
[56] Franks, R., Bartels, C.R., Andes, K., Patel, M. (2007) *Implementing energy saving RO technology in large scale wastewater treatment plants.* IDA World Congress-Maspalomas, Gran Canaria-Spain October 21–26, 2007, REF: IDAWC/MP07-148.
[57] Pohland, H.W. (1980) Seawater desalination and reverse osmosis plant design. *Desalination* 32: 157–167.
[58] Visvanathan, C., Svenstrup, P., Ariyamethee, P. (2000) Volume reduction of produced water generated from natural gas production process using membrane technology. *Water Science and Technology* 41(10–11): 117–123.
[59] Redondo, J.A. (2001) Brackish-, sea- and wastewater desalination. *Desalination* 138: 29–40.
[60] Melián-Martel, N., Sadhwani, J.J., Malamis, S., Ochsenkühn-Petropoulou, M. (2012) Structural and chemical characterization of long-term reverse osmosis membrane fouling in a full scale desalination plant. *Desalination* 300: 44–53.
[61] Hofs, B., Schurer, R., Harmsen, D.J.H., Ceccarelli, C., Beerendonk, E.F., Cornelissen, E.R. (2013) Characterization and performance of a commercial thin film nanocomposite seawater reverse osmosis membrane and comparison with a thin film composite. *Journal of Membrane Science* 446: 68–78.
[62] Filteau, G., Moss, P. (1997) Ultra-low pressure RO membranes: An analysis of performance and cost. *Desalination* 113: 147–152.
[63] Nemeth, J.E. (1998) Innovative system designs to optimize performance of ultra-low pressure reverse osmosis membranes. *Desalination* 118: 63–71.
[64] Bartels, C.R., Wilf, M., Andes, K., Iong, J. (2005) Design considerations for wastewater treatment by reverse osmosis. *Water Science and Technology* 51(6–7): 473–482.
[65] Bartels, C.R. *Reverse osmosis membranes for wastewater reclamation.* www.membranes.com/.../...Retrieved on January 13, 2023.
[66] Tang, F., Hu, H.-Y., Sun, J.-L., Wu, Q.-Y., Jiang, Y.-M., Guan, Y.-T., Huang, J.-J. (2014) Fouling of reverse osmosis membrane for municipal wastewater reclamation: Autopsy results from a full-scale plant. *Desalination* 349: 73–79.
[67] Tang, F., Hu, H.-Y., Sun, J.-L., Sun, Y.-X., Shi, N., Crittenden, J.C. Fouling characteristics of reverse osmosis membranes at different positions of a full-scale plant for municipal wastewater reclamation. *Water Research* 90: 329–336.
[68] García Latorre, F.J., Pérez Báez, S.O., Gotor, A.G. (2015) Energy performance of a reverse osmosis desalination plant

operating with variable pressure and flow. *Desalination* 366: 146–153.
[69] Kunisada, Y., Murayama, Y. (1978) Long term experiments on sea water desalination at the Chigasaki Laboratory. *Desalination* 27: 333–344.
[70] Kimura, S., Ohya, H., Murayama, Y., Kikuchi, K., Hirai, M., Toyoda, M., Sonoda, T., Setogawa, S. (1985) Five years operating experience of a 800 cubic meters per day R.O. seawater desalination plant. *Desalination* 54: 45–54.
[71] Drioli, E., Macedonio, F. *Membrane research, membrane production and membrane application in China.* Report: Institute on Membrane Technology, Italy, pp. 1–126 (www.itm.cnr.it/data/section/CHINA_Report.pdf.).
[72] Al-Obaidi, M.A., Alsarayreh, A.A., Al-Hroub, A.M., Alsadaie, S., Mujtaba, I.M. (2018) Performance analysis of a medium-sized industrial reverse osmosis brackish water desalination plant. *Desalination* 443: 272–284.
[73] Ukai, T., Nimura, Y., Hamada, K., Matsui, H. (1980) Development of one pass sea water reverse osmosis module "Hollosep". *Desalination* 32: 169–178.
[74] Matsui, H., Matsumoto, H., Kuzumoto, H., Sekino, M., Nimura, Y., Ukai, T. (1981) Single-pass desalination of high salinity sea water with reverse osmosis. *Desalination* 38: 441–447.
[75] Fujiwara, N., Matsuyama, H. (2008) The analysis and design of a both open ended hollow fiber type RO module. *Journal of Applied Polymer Science* 110: 2267–2277.
[76] Chon, K., Cho, J., Shon, H.K., Chon, K. (2012) Advanced characterization of organic foulants of ultrafiltration and reverse osmosis from water reclamation. *Desalination* 301; 59–66.
[77] Widjaya, A., Hoang, T., Stevens, G.W., Kentish, S.E. (2012) A comparison of commercial reverse osmosis membrane characteristics and performance under alginate fouling conditions. *Separation and Purification Technology* 89; 270–281.
[78] Clever, M., Jordt, F., Knauf, R., Räbiger, N., Rüdebusch, M., Hilker-Scheibel, R. (2000) Process water production from river water by ultrafiltration and reverse osmosis. *Desalination* 131: 325–336.
[79] Manth, M., Frenzel, J., van Vlerken, A. (1998) Large-scale application of UF and RO in the production of demineralized water. *Desalination* 118; 255–262.
[80] Jeong, S., Naidu, G., Vollprecht, R., Leiknes, T., Vigneswaran, S. (2016) In-depth analyses of organic matters in a full-scale seawater desalination plant and an autopsy of reverse osmosis membrane. *Separation and Purification Technology* 162: 171–179.
[81] Lee, Y.-G., Kim, S., Shin, J., Rho, H., Lee, Y., Kim, Y.M., Park, Y., Oh, S.-E., Cho, J., Chon, K. (2020) Fouling behavior of marine organic matter in reverse osmosis membranes of a real-scale seawater desalination plant in South Korea. *Desalination* 485; 114305.
[82] Lee, Y.-G., Kim, S., Shin, J., Rho, H., Kim, Y.M., Cho, K.H., Eom, H., Oh, S.-E., Cho, J., Chon, K. (2021) Sequential effects of cleaning protocols on desorption of reverse osmosis membrane foulants: Autopsy results from a full-scale desalination plant. *Desalination* 500: 114830.
[83] Prihasto, N., Liu, Q.F., Kim, S.H. (2009) Pre-treatment strategies for seawater desalination by reverse osmosis system. *Desalination* 249: 308–316.
[84] Yun, T.I., Gabelich, C.J., Cox, M.R., Mofidi, A.A., Lesan, R. (2006) Reducing costs for large-scale desalination plants using large-diameter, reverse osmosis membranes. *Desalination* 189; 141–154.
[85] Kim, J., Park, K., Hong, S. (2020) Application of two-stage reverse osmosis system for desalination of high salinity and high-temperature seawater with improved stability and performance. *Desalination* 492: 114645.
[86] Chu, K.H., Lim, J., Kim, S.-J., Jeong, T.-U., Hwang, M.-H. (2020) Determination of optimal design factors and operating conditions in a large-scale seawater reverse osmosis desalination plant. *Journal of Cleaner Production* 244, 118918.
[87] Abdallah, S., Abu-Hilal, M., Mohsen, M.S. (2005) Performance of photovoltaic powered reverse osmosis system under local climate conditions. *Desalination* 183; 95–104.
[88] Fortunato, L., Alshahri, A.H., Farinha, A.S.F., Islam Zakzouk, I., Jeong, S., Leiknes, T. (2020) Fouling investigation of a full-scale seawater reverse osmosis desalination (SWRO) plant on the Red Sea: Membrane autopsy and pretreatment efficiency. *Desalination* 496: 114536.
[89] Ruiz-García, A., Nuez, I. (2020) Long-term intermittent operation of a full-scale BWRO desalination plant. *Desalination* 489: 114526.
[90] Ali, M.T., Fath, H.E.S., Armstrong, P.R. (2011) A comprehensive techno-economical review of indirect solar desalination. *Renewable and Sustainable Energy Reviews* 15: 4187–4199.
[91] Muraleedaaran, S., Li, X., Li, L., Lee, R. (2009) *Is reverse osmosis effective for produced water purification? Viability and economic analysis.* Paper presented at the 2009 SPE Western Regional Meeting held in San Jose, CA, 24–26 March 2009.
[92] El-Zanati, E., El-Khatib, K.M. (2007) Integrated membrane-based desalination system. *Desalination* 205; 15–25.
[93] Ghabayen, S., McKee, M., Kemblowski, M. (2004) Characterization of uncertainties in the operation and economics of the proposed seawater desalination plant in the Gaza Strip. *Desalination* 161: 191–201.
[94] Lawler, W., Wijaya, T., Antony, A., Leslie, G., Le-Clech, P. (2011) *Reuse of reverse osmosis desalination membranes.* IDA World Congress, Perth, Western Australia, September 4–9, 2011.
[95] Colla, V., Branca, T.A., Rosito, F., Lucca, C., Vivas, B.P., Delmiro, V.M. (2016) Sustainable reverse osmosis application for wastewater treatment in the steel industry. *Journal of Cleaner Production* 130; 103–115.
[96] Jiang, Z., Karan, S., Livingston, A.G. (2018) Water transport through ultrathin polyamide nanofilms used for reverse osmosis. *Advanced Materials* 30: 1705973.
[97] Karan, S., Jiang, Z., Livingston, A.G. (2015) Sub–10 nm polyamide nanofilms with ultrafast solvent transport for molecular separation. *Science* 348: 1347–1351.
[98] Tan, Z., Chen, S., Peng, X., Zhang, L., Gao, C. (2018) Polyamide membranes with nanoscale Turing structures for water purification. *Science* 360: 518–521.
[99] Chowdhury, M.R., Steffes, J., Huey, B.D., McCutcheon, J.R. (2018) 3D printed polyamide membranes for desalination. *Science* 361: 682–686.

[100] He, J., Yang, J., McCutcheon, J.R., Li, Y. (2022) Molecular insights into the structure-property relationships of 3D printed polyamide reverse-osmosis membrane for desalination. *Journal of Membrane Science* 658: 120731.

[101] Culp, T.E., Khara, B., Brickey, K.P., Geitner, M., Zimudzi, T.J., Wilbur, J.D., Jons, S.D., Roy, A., Paul, M., Ganapathysubramanian, B., Zydney, A.L., Kumar, M., Gomez, E.D. (2021) Nanoscale control of internal inhomogeneity enhances water transport in desalination membranes. *Science* 371: 72–75.

[102] Hummer, G., Rasaiah, J.C., Noworyta, J.P. (2001) Water conduction through the hydrophobic channel of a carbon nanotube. *Nature* 414: 188–190.

[103] Kalra, A., Garde, S., Hummer, G. (2003) Osmotic water transport through carbon nanotube membranes *Proceedings of the National Academy of Sciences U.S.A.* 100: 10175–10180.

[104] Holt, J.K., Park, H.G., Wang, Y., Stadermann, M., Artyukhin, A.B., Grigoropoulos, C.P., Noy, A., Bakajin, O. (2006) Fast mass transport through sub-2-nanometer carbon nanotubes. *Science* 312; 1034–1037.

[105] Li, Y., Li, Z., Aydin, F., Quan, J., Chen, X., Yao, Y.-C., Zhan, C., Chen, Y., Pham, T.A., Noy, A. (2020) Water-ion permselectivity of narrow-diameter carbon nanotubes. *Science Advances* 6; eaba9966.

[106] Tunuguntla, R., Henley, R.Y., Yao, Y.-C., Pham, T.A., Wanunu, M., Noy, A. (2017) Enhanced water permeability and tunable ion selectivity in subnanometer carbon nanotube porins. *Science* 25; 792–796.

[107] Cruz-Silva, R., Inukai, S., Araki, T., Morelos-Gomez, A., Ortiz-Medina, J., Takeuchi, K., Hayashi, T., Tanioka, A., Tejima, S., Noguchi, T., Terrones, M., Endo, M. (2016) High performance and chlorine resistant carbon nanotube/aromatic polyamide reverse osmosis nanocomposite membrane. *MRS Advances* 1: 1469–1476.

[108] Mattia, D. (2011) A review of reverse osmosis membrane materials for desalination – Development to date and future potential. *Journal of Membrane Science* 370; 1–22.

[109] Kim, H.J., Choi, K., Baek, Y., Kim, D.-G., Shim, J., Yoon, J., Lee, J.-C. (2014) High-performance reverse osmosis CNT/polyamide nanocomposite membrane by controlled interfacial interactions. *ACS Applied Materials & Interfaces* 6: 2819–2829.

[110] Zhang, L., Shi, G.-Z., Qiu, S., Cheng, L.-H., Chen, H.-L. (2011) Preparation of high-flux thin film nanocomposite reverse osmosis membranes by incorporating functionalized multi-walled carbon nanotubes. *Desalination and Water Treatment* 34: 19–24.

[111] Inukai, S., Cruz-Silva, R., Ortiz-Medina, J., Morelos-Gomez, A., Takeuchi, K., Hayashi, T., Tanioka, A., Araki, T., Tejima, S., Noguchi, T., Terrones, M., Endo, M. (2015) High-performance multi-functional reverse osmosis membranes obtained by carbon nanotube-polyamide nanocomposite. *Science Reports* 5: 13562.

[112] Kim, H.J., Lim, M.-Y., Jung, K.H., Kim, D.-G., Lee, J.-C. (2015) High-performance reverse osmosis nanocomposite membranes containing the mixture of carbon nanotubes and graphene oxides. *Journal of Materials Chemistry A* 3: 6798–6809.

[113] Zhao, H., Qiu, S., Wu, L., Zhang, L., Chen, H., Gao, C. (2014) Improving the performance of polyamide reverse osmosis membrane by incorporation of modified multi-walled carbon nanotubes. *Journal of Membrane Science* 450: 249–256.

[114] Kumar, M., Khan, M.A., Arafat, H.A. (2020) Recent developments in the rational fabrication of thin film nanocomposite membranes for water purification and desalination. *ACS Omega* 5: 3792–3800.

[115] Jensen, M.O., Mouritsen, O.G. (2006) Single-channel water permeabilities of Escherichia coli aquaporins AqpZ and GlpF. *Biophysics Journal* 90: 2270–2284.

[116] Hovijitra, N.T., Wuu, J.J., Peaker, B., Swartz, J.R. (2009) Cell-free synthesis of functional aquaporin Z in synthetic liposomes. *Biotechnology and Bioengineering* 104: 10.

[117] Kumar, M., Grzelakowski, M., Zilles, J., Clark, M., Meier, W. (2007) Highly permeable polymeric membranes based on the incorporation of the functional water channel protein Aquaporin Z. *Proceedings of the National Academy of Sciences U.S.A.* 104: 20719–20724.

[118] Tang, C.Y., Zhao, Y., Wang, R., Hélix-Nielsen, C., Fane, A.G. (2013) Desalination by biomimetic aquaporin membranes: Review of status and prospects. *Desalination* 308: 34–40.

[119] Zhao, Y., Qiu, C., Li, X., Vararattanavech, A., Shen, W., Torres, J., Hélix-Nielsen, C., Wang, R., Hu, X., Fane, A.G. (2012) Synthesis of robust and high-performance aquaporin-based biomimetic membranes by interfacial polymerization-membrane preparation and RO performance characterization. *Journal of Membrane Science* 423–424: 422–428.

[120] Tang, C.Y., Kwon, Y.-N., Leckie, J.O. (2009) Effect of membrane chemistry and coating layer on physiochemical properties of thin film composite polyamide RO and NF membranes II. Membrane physiochemical properties and their dependence on polyamide and coating layers. *Desalination* 242: 168–182.

[121] Giwa, A., Hasan, S.W., Yousuf, A., Chakraborty, S., Johnson, D.J., Hilal, N. (2017) Biomimetic membranes: A critical review of recent progress. *Desalination* 420: 403–424.

[122] HFFO®2 module – *Aquaporin*, https://aquaporin.com/wp-content/uploads/2021/10/Aquaporin-HFFO2-Datasheet-web.pdf. Retrieved on January 13, 2023.

[123] Madsen, N.T., Bajraktari, N., Hélix-Nielsen, C., Van der Bruggen, B., Søgaard, E.G. (2015) Use of biomimetic forward osmosis membrane for trace organics removal. *Journal of Membrane Science* 476: 469–474.

[124] Cohen-Tanugi, D., Grossman, J.C. (2012) Water desalination across nanoporous graphene. *Nano Letters* 12: 3602–3608.

[125] Surwade, S.P., Smirnov, S.N., Vlassiouk, I.V., Unocic, R.R., Veith, G.M., Dai, S., Mahurin, S.M. (2015) Water desalination using nanoporous single-layer graphene. *Nature Nanotechnology* 10: 459–464.

[126] O'Hern, S.C., Jang, D., Bose, S., Idrobo, J.-C., Song, Y., Laoui, T., Kong, J., Karnik, R. (2015) Nanofiltration across defect-sealed nanoporous monolayer graphene. *Nano Letters* 15; 3254–3260.

[127] *Groundbreaking Graphene Membrane Manufactured at Commercial Scale.* www.globenewswire.com/en/news-release/2022/03/16/2404098/0/en/Groundbreaking-Graphene-Membrane-Manufactured-at-Commercial-Scale.html. Retrieved on January 13, 2023.

[128] Akbari, A., Sheath, P., Martin, S.T., Shinde, D.B., Shaibani, M., Banerjee, P.C., Tkacz, R., Bhattacharyya, D., Majumder, M. (2016) Large-area graphene-based nanofiltration membranes by shear alignment of discotic nematic liquid crystals of graphene oxide. *Nature Communications* 7: 10891.

[129] Zhang, Z., Zou, L., Aubry, C., Jouiad, M., Hao, Z. (2016) Chemically crosslinked rGO laminate film as an ion selective barrier of composite membrane. *Journal of Membrane Science* 515: 204–211.

[130] Itoh, Y., Chen, S., Hirahara, R., Konda, T., Aoki, T., Ueda, T., Shimada, I., Cannon, J.J., Shao, C., Shiomi, J., Tabata, K.V., Noji, H., Sato, K., Aida, T. (2022) Ultrafast water permeation through nanochannels with a densely fluorous interior surface. *Science* 376: 738–743.

[131] (a) Ramakrishna, S., Fujihara, K., Teo, W.E., Yong, T., Ma, Z., Ramaseshan, R. (2006) Electrospun nanofibers: Solving global issues. *Materials Today* 9: 40–50; (b) Kaur, S., Gopal, R., Ng, W.J., Ramakrishna, S., Matsuura, T. (2008) Next-generation fibrous media for water treatment. *MRS Bulletin* 33: 21–26; (c) Kaur, S., Rana, D., Matsuura, T., Sundarrajan, S., Ramakrishna, S. (2012) Preparation and characterization of surface modified electrospun membranes for higher filtration flux. *Journal of Membrane Science* 390–391: 235–242; (d) Kaur, S., Sundarrajan, S., Rana, D., Matsuura, T., Ramakrishna, S. (2012) Influence of electrospun fiber size on the separation efficiency of thin film nanofiltration composite membrane. *Journal of Membrane Science* 392–393: 101–111; (e) Kaur, S., Sundarrajan, S., Rana, D., Sridhar, R., Gopal, R., Matsuura, T., Ramakrishna, S. (2014) Review: the characterization of electrospun nanofibrous liquid filtration membranes. *Journal of Materials Science* 49; 6143–6159.

[132] (a) Karla, A., Garde, S., Hummer, G. (2003) Osmotic water transport through carbon nanotube membranes. *Proceedings of the National Academy of Sciences U.S.A.* 100: 10175–10180; (b) Srivastava, A., Srivastava, O.N., Talapatra, S., Vajtai, R., Ajayan, P.M. (2004) Carbon nanotube filters. *Nature Materials* 3; 610–614; (c) Hinds, B.J., Chopra, N., Rantell, R., Andrews, R., Gavalas, V., Bachas, L.G. (2004) Aligned multiwalled carbon nanotube membranes. *Science* 303: 62–65; (d) Majumdar, M., Chopra, N., Andrews, R., Hinds, B.J. (2005) Nanoscale hydrodynamics: Enhanced flow in carbon nanotubes. *Nature* 438: 44; (e) Holt, J.K., Park, H.G., Wang, Y., Stadermann, M., Artyukhin, A.B., Grigoropoulos, C.P., Noy, A., Bakajin, O. (2006) Fast mass transport through sub-2-nanometer carbon nanotubes. *Science* 312: 1034–1037; (f) Hinds, B.J. (2009) *Aligned nanotubule membranes*. University of Kentucky Research Foundation, Lexington, KY, US Patent 7 611 628 B1, 3 Nov 2009; (g) Bakajin, O., Holt, J., Noy, A., Park, H.G. (2011) *Membranes for nanometer-scale mass fast transport*. Lawrence Livermore National Security LLC, Livermore, CA, US Patent 8 038 887 B2, 18 Oct 2011; (h) Ratto, T.V., Holt, J.K., Szmodis, A.W. (2012) *Asymmetric nanotube containing membranes*. NanOasis Technologies Inc., Richmond, CA, US Patent 8 177 979 B2, 12 May 2012.

[133] (a) Curcio, E., Drioli, E. (2005) Membrane distillation and related operations – a review. *Separation and Purification Reviews* 34: 35–86; (b) Cath, T.Y., Childress, A.E., Elimelech, M. (2006) Forward osmosis: Principles, applications, and recent developments. *Journal of Membrane Science* 281; 70–87; (c) Suk, D.E., Matsuura, T. (2006) Membrane-based hybrid processes: A review. *Separation Science and Technology* 41: 595–626; (d) Vetter, T.A., Perdue, E.M., Ingall, E., Koprivnjak, J.F., Pfromm, P.H. (2007) Combining reverse osmosis and electrodialysis for more complete recovery of dissolved organic matter from seawater. *Separation and Purification Technology* 56: 383–387; (e) Cath, T.Y., Drewes, J.E., Lundin, C.D., Hancock, N.T. (2009) *Forward osmosis–reverse osmosis process offers a novel hybrid solution for water purification and reuse*. AWWA Membrane Technology Conference, March 15–18, 2009 at Memphis, TN, *IDA Journal*, 2010, December issue, 16–20; (f) Voutchkov, N., Semaat, R. (2008) Seawater desalination. In: *Advanced Membrane Technology and Applications*, Li, N.N., Fane, A.G., Ho, S.W.W., Matsuura, T. (Eds.). John Wiley & Sons, Inc., New York, NY, Ch. 3, pp. 47–86; (g) Choi, Y.J., Choi, J.C., Oh, H.J., Lee, S., Yang, D.R., Kim, J.H. (2009) Toward a combined system of forward osmosis and reverse osmosis for seawater desalination. *Desalination* 249: 239–246; (h) Ge, Q., Wang, P., Wan, C., Chung, T.S. (2012) Polyelectrolyte-promoted forward osmosis-membrane distillation (FO-MD) hybrid process for dye wastewater treatment. *Environmental Science and Technology* 46: 6236–6243; (i) Im, S.-J., Jeong, S., Jang, A. (2021) Forward osmosis (FO)-reverse osmosis (RO) hybrid process incorporated with hollow fiber FO. *npj Clean Water* 4: 51; (j) Skuse, C., Gallego-Schmid, A., Azapagic, A., Gorgojo, P. (2021) Can emerging membrane-based desalination technologies replace reverse osmosis? *Desalination* 500: 114844; (k) Wang, H., Gao, Y., Gao, B., Guo, K., Shon, H.K., Yue, Q., Wang, Z. (2021) Comprehensive analysis of a hybrid FO-NF-RO process for seawater desalination: With an NF-like FO membrane, *Desalination* 515; 115203; (l) Du, C., Zhao, X., Du, J.R., Feng, X., Yang, H., Cheng, F., Ali, M.E.A. (2022) A field study of desalination of high-salinity surface brackish water via an RO-NF hybrid system. *Chemical Engineering Research and Design* 182: 133–144.

3 Organic Solvent-Resistant Nanofiltration

Parag R. Nemade[1,2,], Olviya S. Gonsalves[1], and P.S.V. Vaishnavi[1]*
[1]Department of Chemical Engineering, Institute of Chemical Technology, Mumbai, India
[2]Institute of Chemical Technology Mumbai, Marathwada Jalna Campus, Jalna, India
*Corresponding Author: Parag R. Nemade (pr.nemade@ictmumbai.edu.in)

3.1 INTRODUCTION

Separating chemicals consumes an astounding amount of energy: 10–15 percent of the world's total energy consumption. Chemicals are ubiquitous in today's environment, which contributes to this scenario. Each individual on Earth, for example, consumes around two liters of crude oil per day and 30 kg of ethene and propene per year. However, it is not simply the magnitude of these operations that necessitates the use of energy. Distillation separations are used in these instances, as well as almost all bulk chemical separations of organic liquids and gases. Distillation and other phase-change-based processes, such as evaporation, use more than 80% of the energy involved in chemical separation [1].

Separation processes account for about 40–70% of both capital and operating costs in pharmaceutical industries. In chemical and pharmaceutical industries, high added-value products are synthesized in, and separated from, organic solvents. All these products are solvent-intensive. Organic solvents are used as raw materials or cleaning agents are collected at the end of the process, are discarded, or recovered. Process energy efficiency and mass efficiency have gained attention in production plants. Solvent recovery and recycling improves the overall mass efficiency of the process.

Organic solvent nanofiltration (OSN) or solvent-resistant nanofiltration (SRNF) is a promising membrane technique that permits separation of organic mixtures down to the molecular level by applying a pressure gradient across a membrane and recovering the organic solvents with the possibility of reuse. As the pressure difference across the membrane is the driving force for the OSN, the compounds that are not stable at higher temperature can be separated. Importantly, the benefits of membrane processes apply to OSN as well, such as simple design, linear scale-up, and easy integration with other unit operations to form hybrid processes. Membrane technology also offers advantages in terms of economy, environment, and safety, but is largely limited to aqueous applications.

SRNF is a relatively new technique that broke through around the beginning of the century, despite the fact that pressure-driven solvent separations had been reported in 1965. However, although it is yet not considered a proven technology, it has huge potential. In molecular-level separation, mutual interactions between solute, solvent, and membrane play an important role in addition to mere molecular size. This makes SRNF less accessible to non-specialists and makes choosing the right membrane type for a specific separation more complex. It is even more complicated for commercially available membranes whose the physio-chemical properties are not known [2].

According to Marchetti et al. [3], OSN has three process operations based on its operating modes: concentration, solvent exchange, and purification. A concentration process involves separation of solute from solvent, where the solute is concentrated at the high-pressure side of the membrane and the solvent is permeated through the membrane. This process can be applied for the recovery of high-value solute from dilute solution (solute enrichment) or to recover the solvent by separating the solute dissolved in it (solvent recovery). In a solvent exchange process, one solvent is replaced by another without the application of thermal energy. The second solvent is added to the solution of dissolved solute in the first solvent. In the batch process, as a function of time, the second solvent is enriched in the solution replacing the first solvent. In the purification process, the two or more solutes in a solution are separated. For purification processes, highly selective membranes are required that are capable of distinguishing between molecular size and/or physicochemical features of the solutes.

3.2 MODELING OF ONF

It is found in the literature that membrane–solvent complex interactions significantly affect the permeation of the solvent and are critical in understanding the solvent behavior. There have been many efforts at predicting the transport phenomena of solute and solvent through the dense or porous SRNF membranes. Several transport models for SRNF are proposed, such as (a) the pore-flow model, (b) the Spiegler–Kedem model, and (c) the solution-diffusion model. The Spiegler–Kedem model originates from irreversible thermodynamics, treating the membrane as a black box, while the pore-flow model and the solution-diffusion model consider the membrane properties (Figure 3.1).

3.2.1 Pore-Flow Model

In a pore-flow model, it is assumed that the membrane has a porous structure and the mass transfer occurs solely through the pores by pressure-driven convective flow. Darcy's law

DOI: 10.1201/9781003285656-3

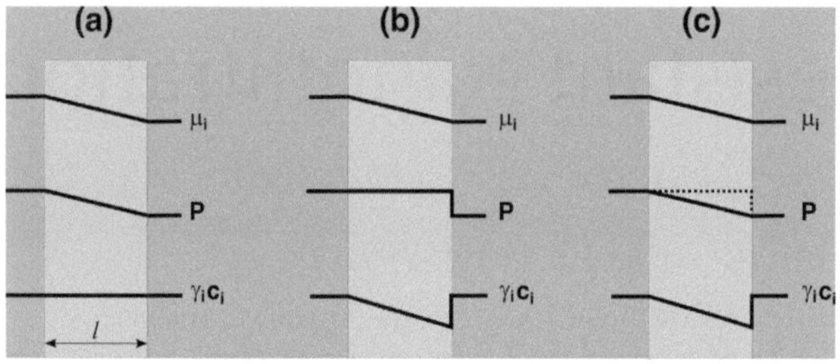

FIGURE 3.1 Pressure-driven permeation process through a membrane, according to: (a) pore–flow transport model, (b) solution–diffusion model, and (c) solution–diffusion model without constant pressure consideration (adapted from[19]).

may be used to explain the transport through the pores in the absence of a concentration gradient, where J is the flux, k is the permeability coefficient, which includes structural characteristics such as membrane pore size, surface porosity, and tortuosity, and (p_0-p_1) is the pressure gradient across the membrane of thickness l.

$$J_1 = k\frac{(p_0 - p_l)}{l} \tag{3.1}$$

In case of solvents where there is no concentration gradient and pores have uniform cylindrical shape, Darcy's law gives rise to the Hagen–Poiseuille equation for solvent flux:

$$J_1 = \frac{\varepsilon r_p^{\ 2}}{8\eta\tau} \cdot \frac{\Delta p}{l} \tag{3.2}$$

where, ε is the porosity of the membrane, r_p is the radius of the pores, η is the solvent viscosity, and τ is the tortuosity of the membrane. However, in SRNF, pore size may vary depending on the kind of organic solvent utilized, due to differences in membrane polymer swelling.

There are several empirical pore-flow models for solute flux. Bowen and Welfoot [4], Bowen [5], and Mukthar [6] developed the Donnan Steric Pore-flow model, which is based on the extended Nernst–Planck equation. For the generic solute, this model combines steric hindrance and Donnan exclusion effects. The solute flux is

$$J_i = -K_{i,d} D_i \frac{dc_i}{dx} - \frac{z_i c_i K_{i,d} D_i}{RT} F \frac{d\psi}{dx} + K_{i,c} c_i J_v \tag{3.3}$$

where $K_{i,d}$ and $K_{i,c}$ are diffusion and convective hindrance factors, respectively, z_i is the solute valence, and ψ is the electrical potential. Other than steric interactions, this model does not account for specific solute–solvent–membrane molecular interactions, which has a significant impact on membrane performance in some cases [7]. The surface force pore-flow (SFPF) model considers solute–solvent–membrane interactions, including the type of solvent. In this model, distribution of solute at the membrane surface is taken into account along with interfacial forces between the solute and the membrane material. The solute flux is given as [8]:

$$J_2 = \frac{RT}{\varsigma_{1,2} b} \cdot \frac{dC_2^m}{dx} + \frac{C_2^m v_1}{b} \ with\ b = \frac{\varsigma_{1,2} + \varsigma_{2,3}}{\varsigma_{1,2}} \tag{3.4}$$

where b is the friction parameter, a function of friction coefficients between solute and solvent $\varsigma_{1,2}$ and between solute and membrane $\varsigma_{2,3}$. The solute flux is composed of two parts: a diffusion flux due to concentration gradient and convection of solutes with total volume flux [9,10].

Robinson et al. [11] studied the transport of alkanes and aromatic solvent through PDMS/PAN composite and found that the data followed a Hagen–Poiseuille type pore-flow model. For a feed pressure above 30 bar, the transport was ruled by viscous flow. There was no change in the composition of the binary mixture observed while passing through the membrane. The hydraulic transport mechanism was justified by non-separation of binary mixtures and the dependency of flux on viscosity and membrane thickness.

3.2.2 Irreversible Thermodynamic Model

The Kedem–Katchalsky [12] and Spiegler–Kedem [13] models are the most commonly used models based on irreversible thermodynamics. In the irreversible thermodynamic model, the membrane is considered as a black box and the transport process is an irreversible process during which free energy is dissipated continuously and entropy is produced. Hence, it is not possible to describe the structural and electrical properties of the membrane. Even if the initial ideas of the two models are the same, the definitions of the driving forces are different, respectively, as differences or differentials across the membranes. According to the Kedem–Katchalsky model:

$$J_v = L_p(\Delta P - \sigma\Delta\pi) \tag{3.5}$$

$$J_2 = \omega\text{"}\ \pi + (1-\sigma)\bar{C}J_v \tag{3.6}$$

where J_v is the volume flux, J_s is the solute flux, and L_p, ω, and σ represent the solvent permeability, solute permeability, and reflection coefficient, respectively. $\bar{C}$ is defined as the logarithmic average of concentration on both sides of the membrane. Spiegler and Kedem gave the differential form of the equation:

$$J_2 = -D_{eff}\frac{d}{dx}C + (1-\sigma)CJ_v \tag{3.7}$$

where D_{eff} is the effective diffusivity inside a membrane and is reported to be different from that in the bulk solution because of hindered diffusion [14]. The first and second terms show the contribution of the diffusion and convection to solute flux, respectively.

3.2.3 Solution–Diffusion Model

The classic solution–diffusion model was developed by Lonsdale in 1965 and was revised by Wijmans and Bakers in 1995. The solution–diffusion model implies that the pressure inside a membrane is constant and that the chemical potential gradient across the membrane can only be described as a concentration gradient. As a result, the key assumption in model development is that the solute and solvent fluxes are independent of one another. According to diffusion theory:

$$J_i = \frac{D_i K_i}{l}\left[c_{if} - c_{ip}\exp\left(\frac{-v_i(p_f - p_p)}{RT}\right)\right] \quad (3.8)$$

where D_i is the diffusion coefficient of i through the membrane, K_i is the partition coefficient, l is the membrane thickness, c_{if}, c_{ip} are the feed and the permeate concentration of species i, respectively, v_i is the partial molar volume of species i, p_f, p_p are the feed and permeate pressures, respectively, R is the gas constant, and T is the temperature. In terms of the pressure and concentration difference across the membrane, Equation (3.8) is applicable for both solute and solvent fluxes over the membrane. The classical solution–diffusion model has a limitation in case of nanofiltration: it does not account for any kind of coupling between solute and solvent fluxes [15].

Silva et al. observed that for STARMEM™ 122 [16] the solvent fluxes for methanol/toluene and ethyl acetate/toluene mixtures were described by the solution–diffusion model. These authors predicted the fluxes for the solvent mixture using the pure solvent data. It was found that membrane–solvent interactions are an important phenomenon that controls the solvent transport in the specific system, yet they could not explain the difference in the flux between toluene and ethyl acetate. Han et al. found that for MPF-50, membrane–solute interactions are an important parameter in the mass transport of toluene.

3.2.4 Solution–Diffusion with Imperfection

This model is an extension of the solution–diffusion model, which takes into account the viscous flow and the interactions between the partial fluxes of the permeating species. In this model, there is a parallel connection of a matrix where the transport is by solution–diffusion other than for imperfection in which the solute is convectively transported without change of concentration. The model by Sherwood et al. [17] provides:

$$J_v = L_m(\Delta p - \Delta\pi) + L_i\Delta p \quad (3.9)$$

$$J_s = P_m(c_0 - c_p) + L_i c_0 \Delta p \quad (3.10)$$

where c_o, c_p are the solute concentrations in the feed and permeate; L_m, L_i are the partial mechanical permeabilities of the matrix and imperfections; J_v, J_s are the volume and solute fluxes; and P_m is the partial diffusional permeability of the matrix.

The SD with imperfection was further revised by Yaroshchuk [18] where the solute concentration at the end of the imperfection is decreased as the solute diffuses along the membrane towards the surrounding perfect regions. The solute flux is given as:

$$J_2 = -P_{diff}\frac{dC}{dx} + J_v C \quad (3.11)$$

where P_{diff} is the partial diffusional permeability of imperfection, C is the solute concentration inside the imperfection, J_v is the volume flux through imperfections, and x is the space variable scaled on the membrane thickness.

Fierro et al. [19] investigated the permeation of mixtures of ketones and glycols through dense polydimethylsiloxane composite membranes using a solution–diffusion with imperfections model. The modeled permeation fluxes and experimental data showed a good correlation for the ketone/glycol system in terms of concentration and diffusion coefficient. However, in many cases, the experimental data verification for calculated data has worked in pure solvent fluxes but not for solute fluxes. The solvent and solute fluxes are characterized to specific material systems and therefore a general prediction of flux and rejection is not possible. Along with solvent and membrane, solute properties also have an impact on the performance of specific OSN membrane. As a result, there are almost unlimited numbers of potential membrane, solvent, and solute combinations, as well as a huge number of known and unknown interactions. Solubility of the solute plays an important role in membrane performance [20]. Other than solvent and solute, the membrane material also has a specific solubility. To define solubility, the Hildebrand or Hansen solubility parameters are employed. The total Hansen solubility parameter (δ_{tot}) involves material-specific parameters including intermolecular forces that drive molecular affinity (dispersion δ_D, dipole–dipole forces δ_P and hydrogen bonding δ_H). The Hansen solubility parameter is three-dimensional and the difference in polymer and solvent solubility is expressed by three-dimensional spatial distance.

$$R_{3D} = \sqrt{(\delta_{D,mem-}\delta_{D,solvent})^2 + \left(\delta_{P,mem} - \delta_{P,solvent}\right)^2 + \left(\delta_{H,mem} - \delta_{H,solvent}\right)^2} \quad (3.12)$$

A projection can simplify this three-dimensional representation into two dimensions by summarizing dispersive δ_D and polar δ_P parameters, where

$$\delta_V^2 = \delta_D^2 + \delta_P^2 \quad (3.13)$$

TABLE 3.1
Application of the transport model for different membranes

S. no.	Membrane	Solute/solvent	Model	Year	References
1	PI, PA, PDMS on PAN	Quaternary alkyl ammonium bromide salts/methanol	Pore flow model	2002	[77]
2	PDMS, PA, TiO2	Different reference components/methanol, ethanol, acetone, ethyl acetate, and n-hexane	Pore flow model (lognormal model)	2005	[78]
3	PI, PDMS on PAN	TOABr/methanol, toluene, ethyl acetate	Solution–diffusion model	2005	[16]
4	PDMS	Dyes/alcohol	Solution–diffusion Kedem–Katchalsky model	2006	[79]
5	PDMS	Dimethyl methyl succinate	Solution–diffusion model with film theory	2006	[80]
6	PDMS, TiO2	Six primary alcohols	Solution diffusion with imperfection model	2009	[81]
7	TiO2/ZrO2	Polystyrene/THF	Pore-flow model	2016	[82]
8	PDMS grafted alumina	Dyes/isopropanol, toluene	Spiegler–Kedem–Katchalsky (SKK) model	2017	[83]
9	PDMS	Surfactants/1-dodecene	Solution–diffusion model	2017	[84]
10	PDMS	Oil/terpene	Maxwell–Stefan ternary mixture diffusion model	2019	[85]
11	–	Palm oil/acetone	Solution–diffusion model	2021	[86]
12	PDMS	Oil/water in dodecene	Solution–diffusion model	2021	[87]

$$R_{2D} = \sqrt{(\delta_{V,mem-}\delta_{V,solvent})^2 + \left(\delta_{H,mem} - \delta_{H,solvent}\right)^2} \quad (3.14)$$

R_{3D} and R_{2D} describe the affinity between the membrane polymer and solvent [21].

Applications of these transport models for various SRNF membranes in various solvent systems are summarized in Table 3.1.

3.3 MEMBRANE MATERIALS, PREPARATION, AND FABRICATION

3.3.1 Membrane Materials

Polymeric, ceramic, and composite materials are used to prepare SRNF membranes. For the membranes to be of good quality, the material has to be mechanically, chemically, and thermally stable. Polymeric membranes lack mechanical stability, affecting the performance by flux reduction due to aging or compaction. Ceramic membranes have good mechanical, chemical, and thermal stability; also, they do not suffer from compaction, do not swell in organic solvents, and are easy to clean. Ceramic membranes have very little variety in terms of membrane material and are inherently hydrophilic, restricting their use in aprotic solvents. Composite organic–inorganic membranes (mixed-matrix membranes, MMM) have low polymer swelling, compaction, and as well as better mechanical stability. The mixed-matrix membranes have the best properties of polymeric and ceramic membranes and can be tailor-made with desirable properties.

3.3.2 Polymeric Membranes

Polymeric membranes are the most commonly used for SRNF because of the wide variety of materials, easy processing, and reproducibility. Polymeric membranes are divided into two types: integrally skinned asymmetric (ISA) membranes and thin-film composite (TFC) membranes. They are made-up of an asymmetric sponge-like substructure for mechanical stability and a dense skin-top later where separation occurs. These membranes are prepared on a non-woven support material to improve the mechanical stability. The non-woven support must be stable to solvent as polymeric membranes. Lim et al. wrote a review on polymeric membrane for solvent-resistant nanofiltration [22].

3.3.2.1 Integrally Skinned Asymmetric Membranes (ISA)

ISA consists of a dense skin layer on the top and a porous layer underneath it of the same composition. Both layers are formed simultaneously and the top layer is responsible for membrane performance, such as rejection and permeation. Loeb and Sourirajan [23], using a phase inversion technique, prepared the first ISA. The membrane is prepared by immersing the casting solution in a coagulation bath consisting of non-solvent. The casting solution is prepared by dissolving the polymer into solvent. The solvent used has to be immiscible with the solvent used in the coagulation bath i.e., the non-solvent. The single phase of the casting solution separates into two parts: one is the polymer-rich membrane and the other is the mixture of solvent and anti-solvent in the coagulation bath. The system's thermodynamic parameters, as well as the kinetics of solvent and

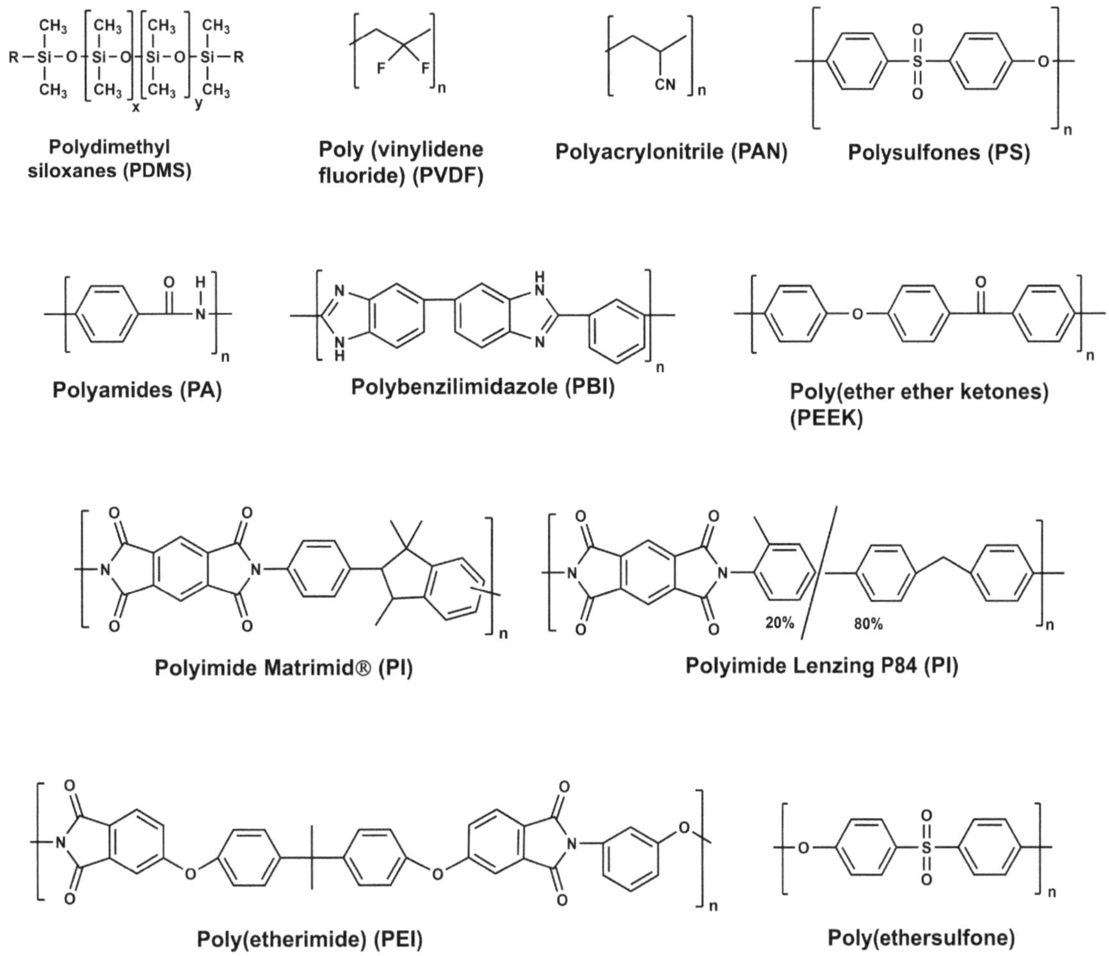

FIGURE 3.2 Structure of the different polymers used in organic solvent resistant nanofiltration.

nonsolvent exchange, have a significant influence on membrane morphology, since they affect permeance and solute rejection [24]. Due to poor membrane stability toward the solvent, the membranes undergo a post-modification, such as cross-linking, to enhance the stability of membranes in organic solvents. Various polymeric materials are used for SRNF preparation by phase inversion, as illustrated in Figure 3.2.

3.3.2.2 Thin-Film Composite Membrane

Another type of polymeric membrane is thin-film composite membranes. These membranes consist of a porous support and a chemically different thin separation layer on its top. As these membranes have a characteristic layered structure, the performance of both layers can be optimized separately to give maximum overall performance. Solvent-stable polymers are used as porous UF support such as Psf, poly(ether sulfone) (PES), PAN, PVDF, PP, PI, and PBI. The support UF membrane provides mechanical stability to the TFC membrane and allows a defect-free top layer. TFC is prepared through various techniques, of which interfacial polymerization (IP) and dip coating on porous support are mostly used. The barrier thin film is formed at the interface of two immiscible phases. The reactive monomer (diamine) is incorporated in the UF support, and on contact with the organic phase (acyl halides) forms a thin film via interfacial polymerization. Various techniques mentioned in literature for preparing TFC include:

- Interfacial polymerization
- Dip-coating/solvent casting: The dip-coating solution consists of a polymer solution or a reactive monomer solution
- Layer-by-layer: Self-arrangement of different materials in an orderly fashioned via electrostatic forces of attraction
- Deposition of the selective layer directly from gaseous-phase monomer.

3.3.3 Ceramic Membranes

Polymeric membranes are resistant to specific classes of solvents, which is not the case in terms of ceramic membranes [25]. Unlike polymeric membranes, there is no leaching of chemicals observed due to progressive degradation of membranes or by a conditioning agent. Ceramic membranes are mostly made up of metal oxides such as alumina (Al_2O_3), zirconia (ZrO_2), titania (TiO_2), or mixed oxides.

Ceramic membranes are asymmetric structures made of at least two porous layers. The porous layers are responsible for providing mechanical stability and outline the external shape of the membranes. A suspension coating with pore size of around 30 nm can be prepared with the finest available powders. An additional defect-free layer is applied to prepare NF membranes with lower pore size by the sol–gel process. A gel is prepared from polymeric or colloidal solution by hydrolysis and condensation of alkoxides or salts dissolved in water or organic solvents. The final gelation takes place on the support that is deposited by dip or spin coating prior to which viscosity modifiers or binders are added. The NF ceramic membrane is obtained after the gel is dried followed by controlled calcinations and/or sintering.

3.3.4 Mixed-Matrix Membranes

Mixed-matrix membranes (MMMs) combine the properties of both inorganic and polymeric materials to give a superior performing membrane. They can be ISA or TFC membranes. The addition of various nanostructures such as nanotubes, zeolite nanoparticles, clay, and fullerenes in the polymer matrix has been extensively studied and applied for organic solvent nanofiltration. Nanoparticles are incorporated into the membranes by three different methods: (a) addition to the casting solution before phase inversion; (b) nanoparticle deposition on the membrane surface; and (c) pore-filling of the polymeric membrane by nanoparticles. The addition of metal oxides to the casting solution before phase-inversion has an effect on the structure and properties of the membrane, in result affecting its performance. Incorporation of nanoparticles into the polymer matrix has an effect on its mechanical and thermal properties, and the crystallinity and hydrophilicity of the membrane [26].

3.4 POST-TREATMENT

In order to improve the membrane performance and life-span of the membrane, various post-treatments or conditioning processes can be used, such as annealing, cross-linking, drying, by solvent exchange and treatment with conditioning agent.

3.4.1 Annealing

Exceptional solvent resistance and high permeabilities were obtained after a thermal annealing procedure on asymmetric PI-based membrane. The membrane has potential as a support for SRNF applications [27]. A porous PI ultrafiltration membrane was utilized as a precursor for thermal annealing, and the temperature was kept around the PI glass transition temperature (T_g). The polymer chains were cross-linked and relaxed at the same time to allow in situ healing of the pores on the surface, resulting in a solvent-resistant, thin, and defect-free separation layer without changing the bulk structure. The resultant membrane displayed rejection of around 300 Da with reasonable permeabilities [28].

3.4.2 Cross-Linking

The cross-linking process is generally introduced to enhance the chemical stability and rejection properties. There are various means of cross-linking, i.e. chemical, plasma, or photo-induced cross-linking. Vanherck et al. [29] discussed various cross-linking methods for polyimide membranes. Cross-linking of PI membranes suppresses the plasticization of polymer, increases the hydrophilicity, and enhances the solvent stability at the expense of permeance. PI is chemically cross-linked by using diols or diamines as linkers. Polysulfone-based membranes were UV-cross-linked, displaying superior chemical resistance [30]. Membrane in the presence of a cross-linker and photo-initiator showed a macrovoid-like morphology. PDMS membrane was treated by $Ar-O_2$ plasma under different conditions. The cross-linking at the surface led to a decrease in permeability and changes in hydrophilicity [31].

3.4.3 Drying by Solvent Exchange

In this process, the non-solvent that remains in the membrane after immersion is replaced by the first solvent that is miscible with the non-solvent. This solvent is then replaced with a second, more volatile solvent that can be quickly evaporated to produce a dry membrane. Drying of membrane in this manner reduces the risk of pore collapse. Asymmetric polyimide membranes are dried by first replacing the non-solvent with low-molecular-weight alcohols or ketone. Further, it is exchanged with hydrocarbons to maintain its pore structure in the dry state with improved flexibility and handling characteristics [32].

3.5 MEMBRANE MODULES FOR ORGANIC SOLVENT NANOFILTRATION

The module geometry depends on the configuration of the membrane produced, such as a flat sheet or cylindrical shape. Plate-and-frame and spiral-wound modules are made from flat sheet membranes and tubular, capillary, and hollow-fiber modules involve cylindrical membrane configurations. Some important features of membrane plant are packing density, energy usage, fluid management and fouling control, ease of operation, compactness of the system, ease of manufacture, ease of cleaning and maintenance, and membrane replacement. In an SRNF system, adequate pressure resistance and polarization control along with chemical durability are important parameters. Some of the main factors to be considered while fabricating an efficient membrane module are: (a) selection of membrane material; (b) a defect-free, robust membrane; and (c) packing of the membrane into a compact, high-surface-area module. The packing densities of different membrane modules are listed in Table 3.2.

For a plate-and-frame module, the membrane, feed spacers, and permeate spacers are arranged in a sandwich-like fashion between two end-plates. There are alternating compartments for feed and permeate sealed with gaskets. A tortuous path is

TABLE 3.2
Packing density of different membrane modules

Configuration	Packing Density (m^2/m^3)
Plate and frame	100–400
Spiral wound	300–1000
Capillary	600–1200
Hollow fiber	Up to 30,000

incorporated in the gasket to minimize channeling, increase residence time, improve mass transfer coefficient to decrease fouling, and reduce concentration polarization. The plate-and-frame modules are labor-intensive during membrane replacement and have a modest surface-area-to-volume ratio.

At an industrial scale, the spiral-wound module is the most dominant configuration for polymeric membranes. In a spiral-wound module, the flat sheet membranes are wound around a central collection pipe. Membrane sheets are attached on three sides and fastened to the permeate channel along the leaf's unsealed edge. A permeate spacer is positioned between the interior edges of the leaves to provide mechanical resistance, inhibit collapse due to pressure, and channel the fluid to the permeation tube. A feed channel spacer separates the upper layers of the membrane. The pressurized feed runs parallel to the permeation channel, while the permeate flows radially toward the central collecting pipe via the spiral-wound permeate spacer. This module has a higher packing density with respect to the plate-and-frame module that depends on the height of the channel and is determined by permeate and feed channel spacers. GMT has developed an envelope-type module that has some advantages over spiral-wound modules. It has a very short permeate distance, and avoids the use of feed spacer and a glue. It has some disadvantages such as it operates at low pressure with a maximum of 40 bar and has a smaller membrane area compared to spiral wound. With poly(m-phenylene isophthalamide) (PMIA) as support, a spiral-wound module was fabricated in an industrial-scale manufacturing line showing promising results for large-scale and continuous application [33].

Ceramic membranes in the market are generally tubular shaped. In the tubular module, the active membrane surface is inside the tubes, and not self-supporting. The walls are not strong enough to withstand the pressure, and collapse or burst. Hence, they are placed inside a porous stainless steel, ceramic, or plastic tube with the diameter of the tube around 10 nm. The feed flows from the inside of the tubes, the permeate is filtered in a radial direction through the porous support and the active layer, and is collected at the outside.

Hollow-fiber modules are made-up of a bundle of fibers packed in a pressurized vessel. In hollow-fiber modules, the membranes used are self-supporting. Depending on the side pressurized, they are available in two arrangements: shell-side feed design or bore-side feed design. The most significant benefit of this module is that it can fit a large membrane area into a single module.

For SRNF, all module elements as well as the membrane and the entire separation unit have to meet the stability criteria—chemically, thermally, and mechanically. Moving from lab-scale coupon testing to larger stage-cut trials and pilot plant tests, then to a site demo plant, and finally to large-scale commercialization, is undertaken in the process of developing a viable industrial membrane process. Each phase requires more membrane area, equipment, necessary feedstock amount, execution time, analytical facilities (monitoring), technical challenges, and operational personnel. Process development also necessitates the selection of an appropriate operational concept, which is frequently determined by the feed stream composition and the desired product quality.

3.6 APPLICATION OF SRNF

Solvent-resistant nanofiltration (SRNF) is used in a variety of industries, including food, chemical processing, petrochemicals, and pharmaceuticals. As compared to other processes like crystallization and distillation, SRNF requires less energy, does not involve any additives, and does not comprise any phase transition. As this process is operated at low temperatures compared to other processes like distillation, it reduces any thermal damage which could include degradation and undesired side reactions. SRNF can be easily installed and can be either used as a continuous process or as an integrated or hybrid process along with prevailing processes (Figure 3.3) [2]. This section focuses on the applications of SRNF in different industries and how it is being applied for various operations. Different commercial membranes along with their manufacturing companies and their solvent stabilities are outlined in Table 3.3.

3.6.1 Food Industry

3.6.1.1 Edible Oil Processing

Crude vegetable oils are synthesized by pressing seeds mechanically, followed by solvent extraction, mostly with hexane. Triglycerides constitute over 95% of the crude oil fraction. The remaining components include phospholipids, free fatty acids (FFAs), pigments, sterols, proteins, carbohydrates, and their degradation products. These substances impart undesirable flavor and color to the oil, and shorten its shelf life. Crude oil needs to be refined to achieve the desirable quality [34]. The conventional method of refining vegetable oil includes the following sequence of steps [35]:

- Preparation of seed, which includes pressing of oil seeds.
- Extraction of oil from seeds using hexane as solvent.
- Removal of the residual hexane through evaporation from the extract called miscella.
- Removal of phospholipids through de-gumming.
- Removal of FFAs and traces of phospholipids through deacidification.
- Removal of color by bleaching, and deodorization by high-vacuum steam stripping.

TABLE 3.3
Properties of different commercial membranes

Tradename	Manufacturer	Series	Materials, Affinity	MWCO(Da)	Stable in	Module Configuration	Refs
Sel RO®	Koch, USA	MPF	PDMS, Hydrophilic	250 (glucose in water)	Aqueous mixtures of lower alcohols, hydrocarbons, chlorinated solvents, aromatics, diethyl ether, ketones, cyclohexane, ethyl acetate	Flat sheet or spiral wound	[88–90]
Stramem™	W.R. Grace Davidson; Membrane extraction Technology, UK	Starmem™ 120	Polyimide, hydrophobic	200 (n-alkanes in toluene)	Alcohols, alkanes, aromatics, ethers, ketones, esters, ethyl acetate	Flatsheets pre-cut discs or spiral wounds	[89]
Puramem™	Evonik, Germany	Puramem™ 280	Polyimide, hydrophobic	280 (styrene oligomers in toluene)	Ethyl acetate	Flat sheet or spiral wound	[91,92]
SolSep	SolSep BV, Netherlands	SolSep 030705	Hydrophobic (unknown)	Not specified	Methanol, ethanol, propanol, acetone, ethyl acetate, hexane, toluene, ketones, esters, dimethyl formamide		[89]
DuraMem™	Evonik, Germany	Duramem™ 150	Modified Polyimide, hydrophobic	150	Isopropanol, methanol, ethyl acetate, ethanol, heptanes		[91]
GMT-oNF-2	Borsig Membrane Technology, GmbH, Germany	GMT-oNF-2	Silicone polymer-based composite	327 (93%)	Alkanes, aromatics, alcohols, ethers, ketones		[93]
Desal	GE, USA	Desal-5	Polyamide, hydrophilic	350	Methanol, ethanol, ethyl acetate, tetrahydrofuran		[94]

The method described above is energy-intensive, generates solid wastes and polluted effluents, and results in oil loss. The energy usage can be lowered by using membrane technology to replace the degumming, deacidification, and bleaching processes. As a result, oil losses are reduced, and bleaching clays will no longer be required. Membrane technology also improves the oil quality by decreasing thermal damage. Waste streams can be minimized as there will not be the requirement for additional chemicals [36]. The general scheme for edible oil processing with opportunities for SRNF membranes is depicted in Figure 3.3. The major applications of NF reported in the edible oil industry are:

- De-gumming
- Recovery of extraction solvents
- Deacidification.

3.6.1.2 De-gumming

De-gumming is the process of the removal of phospholipids from crude oil. The conventional method of de-gumming involves hydrating the phospholipids to an insoluble form and centrifuging them. Because triglycerols and phospholipids have comparable molecular weights, membrane separation is challenging. Phospholipids are amphiphilic in nature and form reverse micelles in a non-aqueous environment, trapping the more polar components inside. These phospholipids have a molecular weight of around 20 kDa. Prior to solvent stripping, miscella is obtained from oil extraction, which contains 20–25% crude oil and 70–75% hexane. Using suitable membranes, phospholipids can be isolated from triglycerols in the miscella stage. A permeate and retentate fraction comprising triglycerols and phospholipids are produced by membrane-based crude oil degumming. Silicone-based SRNF was shown to be the most effective method at removing phospholipids. The phospholipid rejection of Desalination DS-7, a silicone-based SRNF membrane, was greater than 99%. These figures compare favorably to the traditional degumming procedure, which only removes 85% of the phospholipids [34].

Membrane fouling caused by residual micelles is one of the most difficult aspects of the de-gumming process. Phospholipid retention values above 98% were achieved with poly(vinylidene) fluoride membrane. Poly(vinylidene) fluoride membranes have a stronger fouling resistance than polyethersulfone and polysulfone membranes [37]. Several operational parameters for the de-gumming filtering process have been studied using ceramic UF membranes. Despite the minimal rejection with an alumina anodisc membrane (0.02 μm pore diameter and 1 μm thick), the accumulation of phospholipids on the membrane surface caused concentration polarization and the creation of gel layers [38]. Gel layer formation was also reported in polyimide-based ultrafiltration membranes [39].

Major challenges in SRNF include the solvent stability of the membranes which can be evaded by direct de-gumming of oil as obtained from pressed seeds. The major bottleneck of this de-gumming process is the low permeability of membrane for triglycerides, which is one order of magnitude lower than for solvent-based de-gumming. However, the absence of solvent and related energy, and safety and environmental issues could make the process competitive [2].

3.6.1.3 Recovery of Extraction Solvents

To enhance the oil content to roughly 90%, the solvents employed in the extraction are normally recovered by distillation. As a result, many tens of thousands of tonnes of explosive vapors are released into the environment each year. To improve the environmental, safety, and economical aspects of edible oil processing, SRNF is an attractive alternate to retain the oil fraction by removing the solvent partially and concentrating the oil solution before final purification by distillation [40].

Oil separation from various solvents has been tested using commercial reverse osmosis and nanofiltration membranes. In i-propanol and ethanol, cellulose acetate, poly(vinylidene) fluoride, and polyamide membranes performed well, but in n-hexane, they were either destroyed or had low permeabilities [35]. Polyamide membranes hydrophobized with polyacrylonitrile and poly(dimethylsiloxane) showed a good rejection rate of oleic acid dissolved in methanol and acetone. Good hexane permeability has also been demonstrated by the membrane [41]. A tailor-made membrane having poly(dimethylsiloxane) as an active layer and polyacrylonitrile as support was used for the separation of sunflower oil from n-hexane and showed a rejection rate of >90% [42]. Two types of nanofiltration membranes having poly(dimethylsiloxane) and zeolite-filled poly(dimethylsiloxane) as active layer with microporous poly(vinylidene) fluoride support synthesized by a phase inversion process were used for the recovery of hexane from soybean oil. Zeolite-filled PDMS gave enhanced separation with soybean oil rejection > 96.1% [43]. Because there is no phase transition, new solvents such as EtOH or acetone [44,45] can be used for oil extraction, making the process more ecologically friendly and safer.

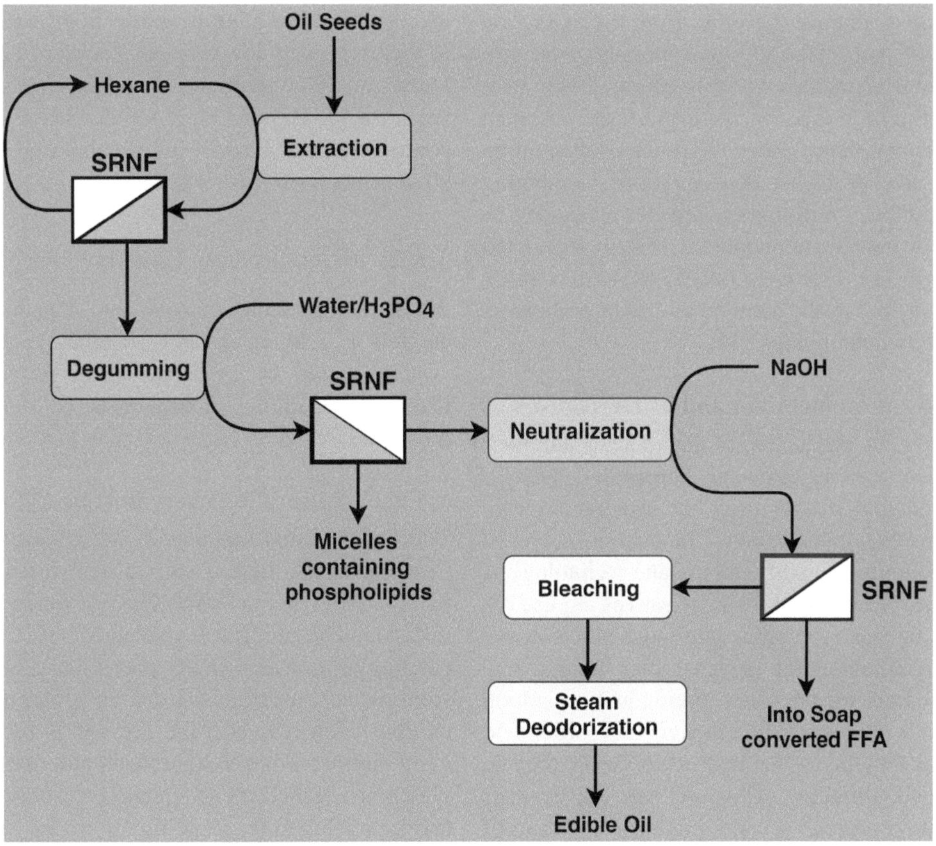

FIGURE 3.3 General scheme for edible oil processing with possible opportunities to implement SRNF (adapted from[2]).

3.6.1.4 Deacidification

Deacidification is the removal of free fatty acids from vegetable oil. Alkali refining is the conventional method for accomplishing this. Many methods are now being researched, as this process involves a lot of energy and additional chemicals, resulting in the loss of neutral oils [46]. The degumming process does not show selective removal of free fatty acids from triglycerides. A hydrophobic polymeric composite membrane, NTGS-2200, with silicon as an active layer and polyimide as support was employed for the separation of oleic acid from sunflower oil [36]. Using ethanol as a solvent, a study was conducted to remove FFAs from a groundnut oil/triglycerol mixture. Three membranes were considered: polyamide, cellulose acetate, and polysulfone. In comparison to the other two membranes, the polyamide membrane provided good separation and a moderate flux [47]. A study was carried out to assess the deacidification of model vegetable oils with and without the addition of solvents using polymeric hydrophobic non-porous membrane NTGS 2200 and hydrophilic nanofiltration membrane. For undiluted systems, the selectivity attained with nonporous hydrophobic membranes was quite low (<2). With hexane-diluted systems, selectivity between triacylglycerols and oleic acid was entirely lost. The NF membrane utilized in the study had high oleic acid selectivity and great solvent stability for acetone, but the oil throughput needs to be significantly improved before it can be used in industry [48]. Another method for deacidifying vegetable oil is to use an organic solvent with high selectivity for FFAs, such as methanol, followed by membrane filtration to remove the FFAs from the extraction solvent [46]. Desal-5 and NTR-759 RO membranes were able to extract oleic acid from methanol with rejection rates of over 90%. A significant drop in flux was the major issue observed in higher rejection membranes, especially when concentration solutions are passed. For higher concentrations, combining higher rejection and lower rejection membranes appears to be a good approach. A multistage membrane system was opted to increase the overall FFA recovery [49]. Long-term stability will need to be improved as the membrane's hydrophilicity is lost owing to polarity conditioning [46].

3.6.1.5 Synthesis of Amino Acids and Its Derivatives

The role of SRNF in the production of amino acids and their derivatives has also been investigated. Dipeptides, such as aspartyl-L-phenylalanine methyl ester, or aspartame, were synthesized via an enzymatic method in organic solvents. As enzymes have limited stability, membrane technology is a viable option. Amino acids and their derivatives are used to make dipeptides. To make the procedure more efficient, the unreacted amino acids should be recycled after the reaction. A study was carried out for the concentration and separation of the amino acids N-benzyloxycarbonyl L-aspartic acid and L-phenylalanine methyl ester hydrochloride in organic solvents using two different types of reverse osmosis membranes: cellulose acetate, a nanofiltration membrane of polyamide–polyphenylene sulfone (PA-PPSO) composite, and a gas separation membrane of polyimide composite. The highest rejection of amino acids (~0.94) was obtained with the PA-PPSO membrane using methanol as a solvent. The cellulose acetate membranes gave reasonable rejection and fluxes but the membrane stability was very poor. The performance of the polyimide composite membrane was good with ethanol but poor with other solvents [50]. PBI 17DBX membrane was also employed for liquid-phase oligonucleotide synthesis (LPOS) where the oligonucleotidyl homostar was purified using OSN technology [51].

3.6.1.6 Concentration and Purification of Bio-Active Compounds

SRNF processes are also widely applicable for the concentration and purification of bio-active compounds. Xanthophylls such as lutein and zeaxanthin are naturally occurring yellow–orange oxygenated pigments of the carotenoid family valuable as natural colorants or as neutraceuticals. Ultrafiltration (UF) was used to separate ethanol-soluble protein (zein) and other large solutes from the extract. The xanthophyll-containing stream was concentrated and separated from the solvent using nanofiltration (NF). Of all the membranes screened, Desal-DK gave the best flux, rejection, and stability [52]. Desalination membranes were also used to concentrate catechins, commonly known as polyphenols, by eliminating caffeine from a combination of bio-active components extracted from green tea using an ethanol extraction method [53]. Propolis is a natural product, rich in biologically active compounds. The removal of biologically active compounds from propolis by an ethanol–water mixture results in a very diluted extract due to their relatively low content. Starmem 122 (polyimide) and Duramem 200 (modified polyimide) were used to concentrate the extract [54]. In addition to this, SRNF was also applied to concentrate γ-oryzanol from rice bran oil [55] and fatty acid ethyl esters from fish oil [56].

3.6.2 Petrochemical Industry

As the refining sector requires a lot of energy and separations, as well as a lot of organic solvents, large-scale membrane systems could be beneficial. The petrochemical industry involves separating compounds with similar molecular properties.

3.6.2.1 Solvent Recovery in Lube Oil Dewaxing

The conventional method of the solvent dewaxing process involves the addition of a mixture of volatile solvents, usually methyl ethyl ketone (MEK) and toluene to the waxy oil raffinate. The mixture is cooled to -10^0C and brought over rotating drum filters where precipitated waxes are separated from the solvent stream. Lube oil filtrate still contains some amount of waxes and the solvent is recovered through a combination of multi-stage flash and distillation operations which are highly energy-intensive. The solvent needs to be refrigerated before recycling it to the dewaxing process. SRNF can be used to recover the solvent by replacing

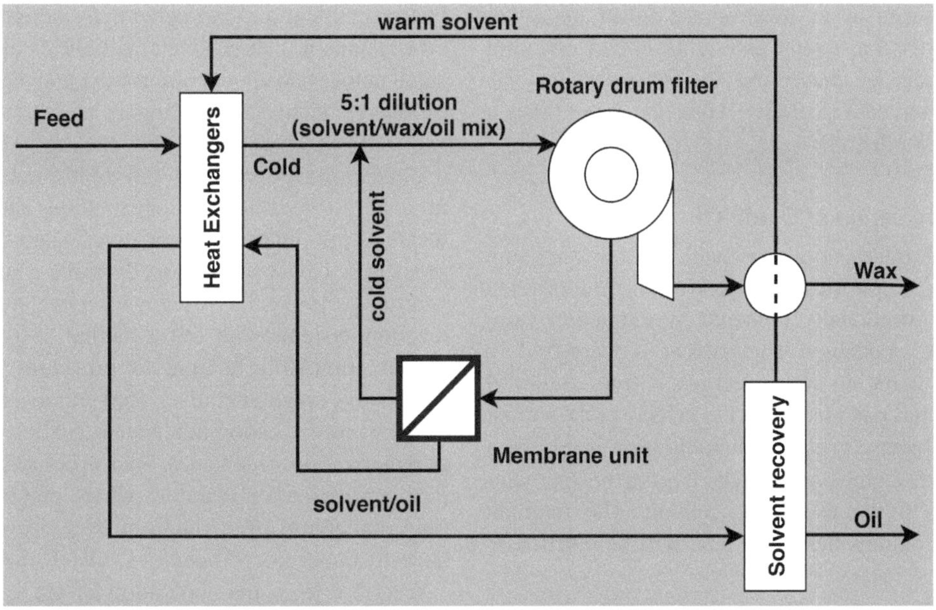

FIGURE 3.4 Schematic representation of an SRNF-assisted lube oil (adapted from [95]).

the evaporation step. Through SRNF, 99% of the solvent was recovered at refrigeration temperature, which can be directly recycled to the chilled feed stream. ExxonMobil's Beaumont (Texas) refinery established the first large-scale industrial SRNF plant in 1998, using asymmetric Matrimid 5218 PI membranes in spiral modules to recover dewaxing solvents from lubricating oil filtrates (Figure 3.4). The process trademarked as Max-Dewax™ was integrated with the existing processes and enabled for the recovery of a 99% pure solvent mixture, typically MEK and toluene, at refrigeration temperature and direct recycling to the chilled feed stream (Figure 3.4) [57,58].

The membrane method lowered energy consumption per product unit by 20%, increased the average base oil production by 25%, and resulted in significant reductions in volatile organic compound emissions and the usage of cooling water with only a third of the capital investment. For recovering cooled solvents from lube oil filtrates, a new nanofiltration membrane was made from polyimide based on 2,2-bis(3,4-dicarboxyphenyl) hexafluoropropane dianhydride (6FDA). The rejection of oil was around 96% at –18^{0}C, which was comparable to the Max–Dewax membrane [59]. Thin-film nanocomposite (TFN) membrane containing amino-functionalized UZM-5 nanoparticles was synthesized by interfacial polymerization on polyetherimide (PEI)/modified SiO_2 asymmetric substrate tailored for the SRNF process. The polyamide layer was prepared by the interfacial polymerization of m-phenylene diamine and trimesoyl chloride. UZM-5 nanoparticles (~ 73 nm) were synthesized and then functionalized by aminopropylediethoxymethyls ilane (APDEMS) and incorporated into a polyamide (PA) selective layer at concentrations ranging from 0 to 0.2 w/v%. TFN 0.2 gave an oil rejection rate of 96.27% [60].

3.6.2.2 Separation and Enrichment of Hydrocarbons

Researchers have investigated the potential of SRNF in aromatics enrichment of refinery streams [58]. Aromatic isomerizations, hydrogenations, disproportionations, and alkylations are examples of such reactions. SRNF would replace conventional liquid/liquid extraction techniques by selectively permeating aromatic hydrocarbons from a mixture of non-aromatic hydrocarbons [36], saving investment and energy expenditures. The retentate stream could be put into a higher-efficiency hydrocarbon separation or conversion process.

In the toluene disproportionation process, which mainly involves the conversion of toluene to higher value products like p-xylene and benzene, a purge stream is used to remove non-aromatic impurities from the toluene feedstock before recycling it back to the processing unit. A large amount of toluene can be recovered and recycled to the reactor loop by inserting a Lenzing P84 polyimide membrane into the purge stream. The membrane selectively permeates toluene while rejecting the non-aromatics [61]. The membrane system can be combined with the reformer or distillation units in the reforming process to enhance product yield or quality.

3.6.2.3 Desulfurization of Gasoline

The amount of sulfur in gasoline is being limited due to environmental regulations. Automobile exhaust catalysts are also poisoned by sulfur-containing components in gasoline. Sulfur-containing components are conventionally reduced by hydrotreating. Hydrotreating increases the capital and operational costs and lowers the octane number of gasoline. SRNF can be used in integration with the prevailing processes by bypassing a large fraction of feed around the hydrotreating unit. Polyimide, polyurea-urethane, and polysiloxane membranes have been tested to eliminate sulfur-containing components from naptha streams and fluid catalytic cracking, yielding clean fuels by preserving the octane value. The technology trademarked as S-Brane™ uses the membranes in the pervaporation mode [58].

3.6.2.4 Deacidification of Crude Oil

Crude oil contains traces of organic acids such as napthenic acid. These acidic impurities cause corrosion. These acidic impurities are conventionally removed by extraction using polar solvents like methanol. The solvent is recovered by distillation after separating the extract phase from crude oil. British Petroleum has patented a process to recover the solvent from the extract phase using a nanofiltration membrane. The permeate of membrane which is mostly methanol will be recycled to the extraction column. The retentate which primarily contains napthenic acid will be purified by distillation.

3.6.3 Catalytic Applications

Homogeneous catalytic reaction systems, known for their better reaction rates with a high degree of mixing with superior selectivity, have a major disadvantage with their product–catalyst separation and purification on a commercial scale. Most of the industrial processes separate the inactive catalyst and send it to the manufacturer for regeneration. Pharmaceutical and petrochemical industries allow a trade-off of some of their products with the catalyst present in very small amounts without compromising the product purity [62]. Moreover, catalyst recycling is an extensive phenomenon and has adverse effects on the economy. Conventional methods follow distillation for volatile catalysts or extraction for catalyst separation. Such methods are not found to be feasible for thermally sensitive catalysts [63]. Efficient processes can be developed with continuous or semi-continuous [64] catalyst recovery using the SRNF membrane reaction systems with lower capital cost and viable energy consumption [62]. In catalytic processes such as enzymatic catalysis, organometallic [65] and non-organometallic catalysis for oxidation of alkenes, Ziegler–Natta catalysis for polymerization of alkenes, hydroformylation in metathesis [66], and oxidation reactions where precious metals such as gold, palladium [67], rhodium, etc., are used, the catalysts have to undergo reprocessing. SRNF membranes can play a crucial role in the separation of the catalyst, which can be activated and used multiple times while purifying the reaction products without any thermal stress on the whole process (Figure 3.5). Membranes made of polymers such as polyaniline (PANI), poly(ether ether ketone) (PEEK), polybenzimidazole (PBI), polyimide (PI), polyamide-imide (PAI), mounted on the porous or fabric support, PDMS, polyurethanes, and ceramic materials such as TiO_2, ZrO_2 mounted on alumina Al_2O_3 [68], etc. are being employed because of their solvent-resistant activity. The polymeric pore

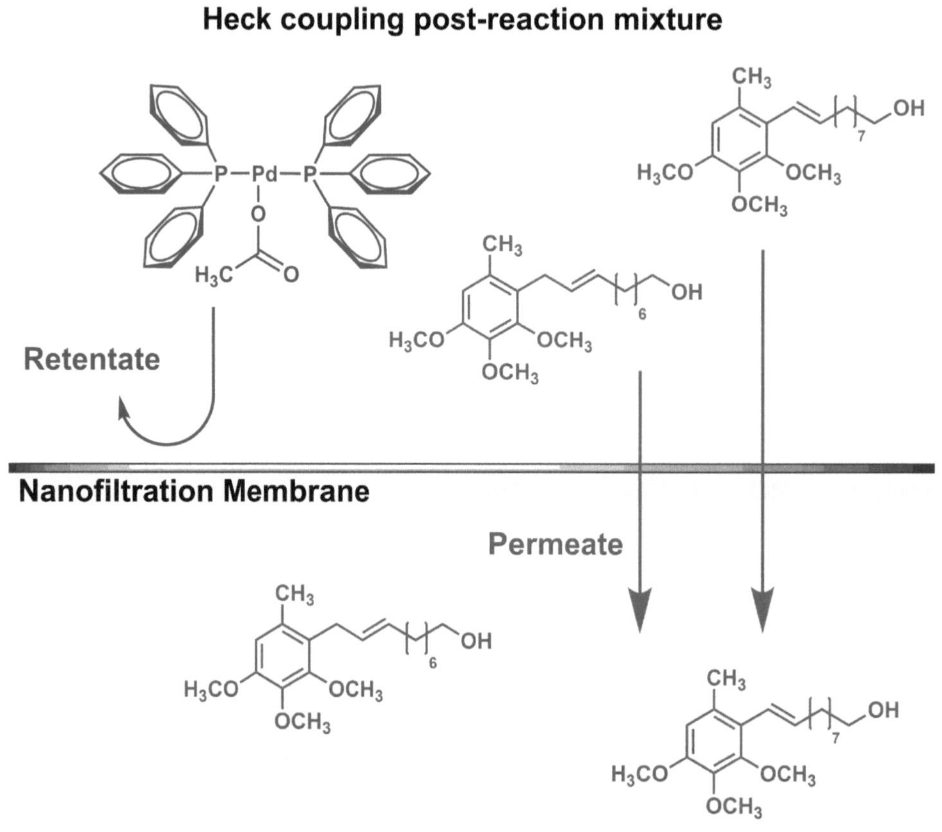

FIGURE 3.5 Separation of [Pd(PPh3)OAc]– from the products found in a Heck coupling postreaction mixture (adapted from [67]).

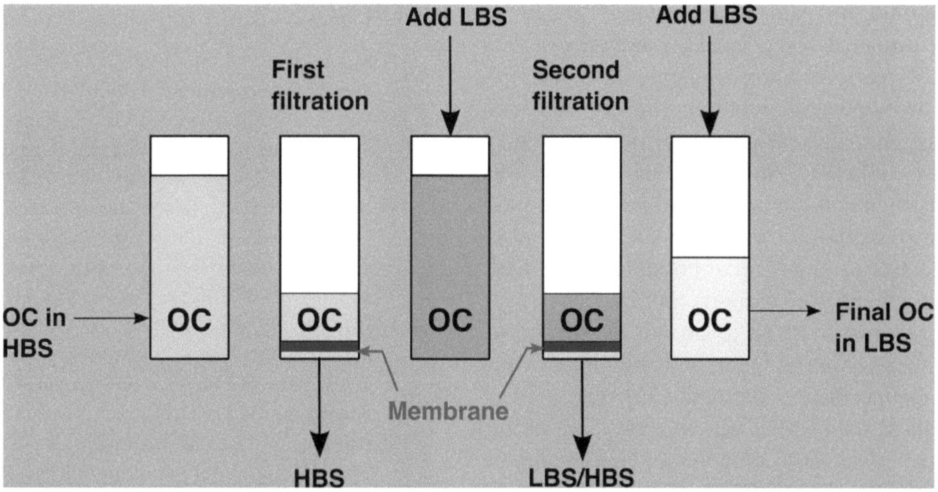

FIGURE 3.6 Operating principle of membrane solvent exchange: OC, organic compound; HBS, high boiling point solvent; LBS, low boiling point solvent (adapted from [72]).

sizes can be altered to produce fine nanopores to retain low-molecular-weight solutes between 200–1000 Da [69]. A high rejection rate of > 96% was obtained for aquinine-based organocatalyst using the commercial DuraMem membranes in Henry post-reaction mixtures.

Despite this, SRNF membrane applications in catalyst recovery of any chemical industry are only limited to the laboratory scale since the reactor function and development of the reaction kinetics is a complex phenomenon. The design of an efficient SRNF membrane catalyst recovery unit with high rejection and long-term catalyst stability under high pressure and temperatures is still being researched and has to be debated since there is direct scope for scale up of such technologies.

3.6.4 Pharmaceutical Applications

SRNF can be applied in drug synthesis either between the reactions or in the downstream processing (DSP). One of the prominent applications of SRNF in the pharmaceutical industry is the isolation and concentration of antibiotics, intermediates, or peptides from organic solvents or aqueous solutions containing organic solvents. MPS-44 membrane (Koch) was employed to recover 6-aminopencillianic acid (6-Apa, 216 Da), an intermediate in the enzymatic manufacturing of synthetic penicillin from its bioconversion solution [70]. Polyimide membrane was employed to concentrate the spiramycin extract obtained from bacterial broths with butyl acetate as a solvent, which is recovered by evaporation. The membrane has shown long-term stability with excellent solvent resistance and rejection rates of around 99% [71].

the synthesis of pharmaceuticals is a multi-step process which involves different solvents in each step. The isolation of the product is carried out in a specific solvent. Solvent exchange is one of the key unit-operations in the synthesis of active pharmaceutical intermediates (APIs). Membrane-based solvent exchange is an interesting alternative to separate mixtures forming an azeotrope where distillation is no longer be an operation (Figure 3.6). Membrane based solvent exchange can take place at room temperature, which is advantageous for thermally liable compounds. Starmem 122 polyimide-based membrane was opted for the solvent exchange of tetraoctylammonium bromide, a phase transfer catalyst exchanged from toluene to methanol [72]. MPF-50 and MPF-60 (Koch) were employed for the solvent exchange of erythromycin, an active pharmaceutical intermediate from ethyl acetate to methanol. MPF 60 gave a higher rejection rate of around 96% for erythromycin. The NF membranes, MPF-50 and MPF-60, showed no selectivity when exposed to a mixed-solvent feed, allowing the solvent mixture and low-molecular-weight impurities to permeate [73].

3.7 FUTURE SCOPE

In lab-scale efforts, the feasibility of SRNF technology in many applications has been shown, providing a strong starting point for larger-scale process development. One issue in incorporating SRNF technology into existing processes and reaping its benefits is to have more thorough demonstrations of the technology in relevant pilot testing to encourage potential end-users [74]. Furthermore, in order to boost the adoption rate of SRNF membranes and lower the market entry barrier, the environmental impact of their manufacture must be taken into account. The discharge of toxic chemicals as effluent during SRNF membrane manufacture eventually denies some of SRNF's environmental benefits. By tailoring the dope composition or employing milder membrane post-treatment settings, new approaches for producing SRNF membranes without affecting membrane performance have been successfully created. Dimethyl formamide, a solvent in the synthesis of PI membrane, was replaced with ecofriendly solvents like DMSO and acetone. Also, isopropyl alcohol in the cross-linking medium was replaced by water [75]. Meanwhile, more attention should be paid to improving membrane properties and performance. It is critical to continue research to develop new polymeric composite membranes and TFN membranes that are stable in polar aprotic solvents and extreme pH values. High solvent permeabilities would help increase the energy efficiency in solvent recovery in the chemical and petrochemical sectors. The objective is to use materials with a strong affinity for organic solvents to make the membrane selective layer as thin as possible. Skin layers constructed of diamond-like carbon (DLC) nanosheets with thicknesses ranging from 35 to 50 nm, for example, showed extraordinarily high solvent permeabilities while retaining significant mechanical strength [76]. The solvent permeabilities of these membranes can be up to three orders of magnitude higher than those of commercially available membranes. Synthesizing ultrathin PA layers on an industrial scale remains a challenge. As a result, more research on nanoscale engineering methodologies to build ultrathin separation layers with long-term stability under actual membrane operations is required to provide permeability and selectivity beyond any existing SRNF membranes.

3.8 CONCLUSION

Organic solvent-resistant nanofiltration membrane displays a potential for replacement of established separation techniques as a versatile, energy-, waste-, and cost-efficient separation technique. Despite significant studies on OSN, industries remain resistant to adapt since change is capital-intensive and risky. The investigation of the permeation mechanism through nanofiltration membranes has been done extensively. There are initiatives for the study of transport models for solutes and solvents for commercially available membranes. Considerable efforts and improvements for understanding the fundamentals of molecular transport have been made, although the most suitable mathematical model for the transport mechanism of nanofiltration membranes is not yet available. Modeling should assist the process development by minimizing the amount of time and effort required to obtain flux and rejection experimentally in order to choose the optimum membrane for a certain application.

There has been continuous development for novel methodologies for membrane fabrication and characterization. The synthesis and development of new polymers with high selectivity, high flux, and high stability in organic solvents and harsh chemicals trading off between selectivity and permeability is a major challenge. There have been new polymers, ceramics, and mixed matrix materials developed that are chemically and thermally stable. OSN has shown a promising development in terms of its application in various industries. It is used in the food industry for refining edible oil, and concentrating and purifying bio-active compounds. Processes such as concentration, purification, and solvent exchange by OSN are applied in the petrochemical and pharmaceutical industries.

REFERENCES

[1] Lively, R. P., Sholl, D. S. (2017) From water to organics in membrane separations. *Nature Materials* 16: 276–279.

[2] Vandezande, P., Gevers, L. E. M., Vankelecom, I. F. J. (2008) Solvent resistant nanofiltration: Separating on a molecular level. *Chemical Society Reviews* 37: 365–405.

[3] Marchetti, P., Jimenez Solomon, M. F., Szekely, G., Livingston, A. G. (2014) Molecular separation with organic solvent nanofiltration: A critical review. *Chemical Review* 114: 10735–10806.

[4] Bowen, W. R., Welfoot, J. S. (2002) Modelling the performance of membrane nanofiltration—critical assessment and model development. *Chemical Engineering Science* 57: 1121–1137.

[5] Bowen, W. R., Mohammad, A. W., Hilal, N. (1997) Characterisation of nanofiltration membranes for predictive purposes—use of salts, uncharged solutes and atomic force microscopy. *Journal of Membrane Science* 126: 91–105.

[6] Bowen, W. R., Mukhtar, H. (1996) Characterisation and prediction of separation performance of nanofiltration membranes. *Journal of Membrane Science* 112: 263–274.

[7] Darvishmanesh, S., Degrève, J., der Bruggen, B. V. (2010) Mechanisms of solute rejection in solvent resistant nanofiltration: the effect of solvent on solute rejection. *Physical Chemistry Chemical Physics* 12: 13333–13342.

[8] Matsuura, T., Sourirajan, S. (1981) Reverse osmosis transport through capillary pores under the influence of surface forces. *Industrial & Engineering Chemistry Process Design and Development* 20: 273–282.

[9] Mehdizadeh, H., Dickson, J. M. (1991) Evaluation of surface force-pore flow and modified surface force-pore flow models for reverse osmosis transport. *Chemical Engineering Communications* 103: 65–82.

[10] Mehdizadeh, H., Dickson, J. M. (1989) Theoretical modification of the surface force-pore flow model for reverse osmosis transport. *Journal of Membrane Science* 42: 119–145.

[11] Robinson, J. P., Tarleton, E. S., Millington, C. R., Nijmeijer, A. (2004) Solvent flux through dense polymeric nanofiltration membranes. *Journal of Membrane Science* 230: 29–37.

[12] Kedem, O., Katchalsky, A. (1958) Thermodynamic analysis of the permeability of biological membranes to non-electrolytes. *Biochimica et Biophysica Acta* 27: 229–246.

[13] Spiegler, K. S., Kedem, O. (1966) Thermodynamics of hyperfiltration (reverse osmosis): criteria for efficient membranes. *Desalination* 4: 311–326.

[14] Tsuru, T., Izumi, S., Yoshioka, T., Asaeda, M. (2000) Temperature effect on transport performance by inorganic nanofiltration membranes *AIChE Journal* 46: 565–574.

[15] Bhanushali, D., Kloos, S., Bhattacharyya, D. (2002) Solute transport in solvent-resistant nanofiltration membranes for non-aqueous systems: experimental results and the role of solute–solvent coupling. *Journal of Membrane Science* 208: 343–359.

[16] Silva, P., Han, S., Livingston, A. G. (2005) Solvent transport in organic solvent nanofiltration membranes. *Journal of Membrane Science* 262: 49–59.

[17] Sherwood, T. K., Brian, P. L. T., Fisher, R. E. (1967) Desalination by Reverse Osmosis. *Industrial Engineering Chemistry Fundamentals* 6: 2–12.

[18] Yaroshchuk, A. E. (1995) Solution-diffusion-imperfection model revised. *Journal of Membrane Science* 101: 83–87.

[19] Fierro, D., Boschetti-de-Fierro, A., Abetz, V. (2012) The solution-diffusion with imperfections model as a method to understand organic solvent nanofiltration of multicomponent systems. *Journal of Membrane Science* 413–414: 91–101.

[20] Thiermeyer, Y., Blumenschein, S., Skiborowski, M. (2018) Solvent dependent membrane-solute sensitivity of OSN membranes. *Journal of Membrane Science* 567: 7–17.

[21] Thiermeyer, Y., Blumenschein, S., Skiborowski, M. (2021) Fundamental insights into the rejection behavior of polyimide-based OSN membranes. *Separation Purification Technology* 265: 118492.

[22] Lim, S. K., Goh, K., Bae, T.-H., Wang, R. (2017) Polymer-based membranes for solvent-resistant nanofiltration: A review. *Chinese Journal Chemical Engineering* 25: 1653–1675.

[23] Loeb, S., Sourirajan, S. (1963) Sea water demineralization by means of an osmotic membrane. #In: *Saline Water Conversion—II*, pp. 117–132. Advances in Chemistry, American Chemical Society.

[24] Park, J.-S., Kim, S.-K., Lee, K.-H. 2000. Effect of $ZnCl_2$ on formation of asymmetric PEI membrane by phase inversion process. *Journal of Industrial and Engineering Chemistry* 6: 93–99.

[25] Merlet, R. B., Pizzoccaro-Zilamy, M.-A., Nijmeijer A., Winnubst, L. (2020) Hybrid ceramic membranes for organic solvent nanofiltration: State-of-the-art and challenges. *Journal of Membrane Science* 599: 117839.

[26] Guizard, C., Ayral, A., Julbe, A. (2002) Potentiality of organic solvents filtration with ceramic membranes. A comparison with polymer membranes. *Desalination* 147: 275–280.

[27] Friesen, D. T., Mccray, S. B., Miller, W. K. (1997) *Solvent resistant microporous polyimide membranes*. European patent EP0753336A2.

[28] Feng, W., Li, J., Fang, C., Zhang, L., Zhu, L. (2022) Controllable thermal annealing of polyimide membranes for highly-precise organic solvent nanofiltration. *Journal of Membrane Science* 643: 120013.

[29] Vanherck, K., Koeckelberghs, G., Vankelecom, I. F. J. (2013) Crosslinking polyimides for membrane applications: A review. *Progress Polymer Science* 38: 874–896.

[30] Strużyńska-Piron, I., Bilad, M. R., Loccufier, J., Vanmaele, L., Vankelecom, I. F.J. (2014) Influence of UV curing on morphology and performance of polysulfone membranes containing acrylates. *Journal of Membrane Science* 463: 17–27.

[31] Aerts, S., Vanhulsel, A., Buekenhoudt, A., Weyten, H., Kuypers, S., Chen, H., Bryjak, M., Gevers, L. E. M., Vankelecom, I. F. J., Jacobs, P. A. (2006) Plasma-treated PDMS-membranes in solvent resistant nanofiltration: Characterization and study of transport mechanism. *Journal of Membrane Science* 275: 212–219.

[32] White, L. S. (2001) *Polyimide membranes for hyperfiltration recovery of aromatic solvents*. US patent US6180008B1.

[33] Jin, L., Hu, L., Liang, S., Wang, Z., Xu, G., Yang, X. (2022) A novel organic solvent nanofiltration (OSN) membrane fabricated by poly(m-phenylene isophthalamide) (PMIA) under large-scale and continuous process. *Journal of Membrane Science* 647: 120259.

[34] Lin, L., Rhee, K. C., Koseoglu, S. S. (1997) Bench-scale membrane degumming of crude vegetable oil: Process optimization. *Journal of Membrane Science* 134: 101–108.

[35] Köseoglu, S. S., Engelgau, D. E. (1990) Membrane applications and research in the edible oil industry: An assessment. *Journal of American Chemists' Society* 67: 239–249.

[36] Subramanian, R., Raghavarao, K. S. M. S., Nabetani, H., Nakajima, M., Kimura, T., Maekawa, T. (2001) Differential permeation of oil constituents in nonporous denser polymeric membranes. *Journal of Membrane Science* 187: 57–69.

[37] Ochoa, N., Pagliero, C., Marchese, J., Mattea, M. (2001) Ultrafiltration of vegetable oils degumming by polymeric membranes. *Separation and Purification Technology* 22–23: 417–422.

[38] Chi-Sheng Wu, J., Lee, E. H. (1999) Ultrafiltration of soybean oil/hexane extract by porous ceramic membranes. *Journal of Membrane Science* 154: 251–259.

[39] Kim, I. C., Kim, J. H., Lee, K. H., Tak, T. M. (2002) Phospholipids separation (degumming) from crude vegetable oil by polyimide ultrafiltration membrane. *Journal of Membrane Science* 205: 113–123.

[40] Cheryan, M. (2005) Membrane technology in the vegetable oil industry. *Membrane Technology* 2005: 5–7.

[41] Kim, I. C., Lee, K. H. (2002) Preparation of interfacially synthesized and silicone-coated composite polyamide nanofiltration membranes with high performance. *Industrial & Engineering Chemistry Research* 41: 5523–5528.

[42] Stafie, N., Stamatialis, D. F., Wessling, M. (2004) Insight into the transport of hexane – solute systems through tailor-made composite membranes. *Journal of Membrane Science* 228: 103–116.

[43] Cai, W., Sun, Y., Piao, X., Li, J., Zhu, S. (2011) Solvent recovery from soybean oil/hexane miscella by PDMS composite membrane. *Chinese Journal of Chemical Engineering* 19: 575–580.

[44] Kwiatkowski, J. R., Cheryan, M. (2005) Recovery of corn oil from ethanol extracts of ground corn using membrane technology. *Journal of the American Oils Chemists' Society* 82: 221–227.

[45] Kuk, M. S., Tetlow, R., Dowd, M. K. (2005) Cottonseed extraction with mixtures of acetone and hexane. *Journal of the American Oils Chemists' Society* 82: 609–612.

[46] Bhosle, B. M., Subramanian, R. (2005) New approaches in deacidification of edible oils — a review. *Journal of Food Engineering* 69: 481–494.

[47] Krishna Kumar, N. S., Bhowmick, D. N. (1996) Separation of fatty acids/triacylglycerol by membranes. *Journal of the American Oils Chemists' Society* 73: 399–401.

[48] Bhosle, B. M., Subramanian, R., Ebert, K. (2005) Deacidification of model vegetable oils using polymeric membranes. *European Journal of Lipid Science and Technology* 107: 746–753.

[49] Raman, L. P., Cheryan, M., Rajagopalan, N. (1996) Deacidification of soybean oil by membrane technology. *Journal of the American Oils Chemists' Society* 73: 219–224.

[50] Reddy, K. K., Kawakatsu, T., Snape, J. B., Nakajima, M. (1996) Membrane concentration and separation of L-aspartic acid and L-phenylalanine derivatives in organic solvents. *Separation Science and Technology* 31: 1161–1178.

[51] Kim, J. F., Gaffney, P. R. J., Valtcheva, I. B., Williams, G., Buswell, A. M., Anson, M. S., Livingston, A. G. (2016) Organic solvent nanofiltration (OSN): A new technology platform for liquid-phase oligonucleotide synthesis (LPOS). *Organic Process Research and Development* 20: 1439–1452.

[52] Tsui, E. M., Cheryan, M. (2007) Membrane processing of xanthophylls in ethanol extracts of corn. *Journal of Food Engineering* 83: 590–595.

[53] Nwuha, V. (2000) Novel studies on membrane extraction of bioactive components of green tea in organic solvents: Part I. *Journal of Food Engineering* 44: 233–238.

[54] Tylkowski, B., Trusheva, B., Bankova, V., Giamberini, M., Peev, G., Nikolova, A. (2010) Extraction of biologically active compounds from propolis and concentration of extract by nanofiltration. *Journal of Membrane Science* 348: 124–130.

[55] Sereewatthanawut, I., Baptista, I. I. R., Boam, A. T., Hodgson, A., Livingston, A. G. (2011) Nanofiltration process for the nutritional enrichment and refining of rice bran oil. *Journal of Food Engineering* 102: 16–24.

[56] Gilmer, C. M., Bowden, N. B. (2016) Highly cross-linked epoxy nanofiltration membranes for the separation of organic chemicals and fish oil ethyl esters. *ACS Applied Materials and Interfaces* 8: 24104–24111.

[57] White, L. S., Nitsch, A. R. (2000) Solvent recovery from lube oil filtrates with a polyimide membrane. *Journal of Membrane Science* 179: 267–274.

[58] White, L. S. (2006) Development of large-scale applications in organic solvent nanofiltration and pervaporation for chemical and refining processes. *Journal of Membrane Science* 286: 26–35.

[59] Kong Y., Shi, D., Yu, H., Wang, Y., Yang, J., Zhang, Y. (2006) Separation performance of polyimide nanofiltration membranes for solvent recovery from dewaxed lube oil filtrates. *Desalination* 191: 254–261.

[60] Namvar-Mahboub, M., Pakizeh, M., Davari, S. (2014) Preparation and characterization of UZM-5/polyamide thin film nanocomposite membrane for dewaxing solvent recovery. *Journal of Membrane Science* 459: 22–32.

[61] White, L. S. (2002) Transport properties of a polyimide solvent resistant nanofiltration membrane. *Journal of Membrane Science* 205: 191–202.

[62] Vandezande, P., Gevers, L. E. M., Vankelecom, I. F. J. (2008) Solvent resistant nanofiltration: Separating on a molecular level. *Chemical Society Reviews* 37: 365–405.

[63] Cole-Hamilton, D. J. (2003) Homogeneous catalysis – New approaches to catalyst separation, recovery, and recycling. *Science* 299: 1702–1706.

[64] Kajetanowicz, A., Czaban, J., Krishnan, G. R., Malińska, M., Woźniak, K., Siddique, H., Peeva, L. G., Livingston, A. G., Grela, K. (2013) Batchwise and continuous nanofiltration of POSS-tagged Grubbs-Hoveyda-type olefin metathesis catalysts. *ChemSusChem* 6: 182–192.

[65] Großeheilmann, J., Büttner, H., Kohrt, C., Kragl, U., Werner, T. (2015) Recycling of phosphorus-based organocatalysts by organic solvent nanofiltration. *ACS Sustainable Chemistry and Engineering* 3: 2817–2822.

[66] Haibach, M. C., Kundu, S., Brookhart, M., Goldman, A. S. (2012) Alkane metathesis by tandem alkane-dehydrogenation-olefin-metathesis catalysis and related chemistry. *Accounts of Chemical Research* 45: 947–958.

[67] Tsoukala, A., Peeva, L., Livingston, A. G., Bjørsvik, H. R. (2012) Separation of reaction product and palladium catalyst after a heck coupling reaction by means of organic solvent nanofiltration. *ChemSusChem* 5: 188–193.

[68] Priske, M., Lazar, M., Schnitzer, C., Baumgarten, G. (2016) Recent Applications of Organic Solvent Nanofiltration. *Chemie Ingenieur Technik* 88: 39–49.

[69] Li, Y., Guo, Z., Li, S., Van der Bruggen, B. (2021) Interfacially polymerized thin-film composite membranes for organic solvent nanofiltration. *Advanced Materials Interfaces* 8: 1–27.

[70] Cao, X., Wu, X. Y., Wu, T., Jin, K., Hur, B. K. (2001) Concentration of 6-aminopenicillanic acid from penicillin bioconversion solution and its mother liquor by nanofiltration membrane. *Biotechnology and Bioprocess Engineering* 6: 200–204.

[71] Shi, D., Kong, Y., Yu, J., Wang, Y., Yang, J. (2006) Separation performance of polyimide nanofiltration membranes for concentrating spiramycin extract. *Desalination* 191: 309–317.

[72] Livingston, A., Peeva, L., Han, S., Nair, D., Luthra, S. S., White, L. S., Freitas Dos Santos, L.M. (2003) Membrane separation in green chemical processing. *Annals of the New York Academy of Sciences*, 984: 123–141.

[73] Sheth, J. P., Qin, Y., Sirkar, K. K., Baltzis, B. C. (2003) Nanofiltration-based diafiltration process for solvent exchange in pharmaceutical manufacturing. *Journal of Membrane Science* 211: 251–261.

[74] Buekenhoudt, A., Beckers, H., Ormerod, D., Bulut, M., Vandezande, P., Vleeschouwers, R. (2013) Solvent based membrane nanofiltration for process intensification. *Chemie Ingenieur Technik* 85: 1243–1247.

[75] Soroko, I., Bhole, Y., Livingston, A. G. (2011) Environmentally friendly route for the preparation of solvent resistant polyimide nanofiltration membranes. *Green Chemistry* 13: 162–168.

[76] Karan, S., Samitsu, S., Peng, X., Kurashima, K., Ichinose, I. (2012) Ultrafast viscous permeation of organic solvents through diamond-like carbon nanosheets. *Science* 335: 444–447.

[77] Gibbins, E., D' Antonio, M., Nair, D., White, L. S., Freitas dos Santos, L. M., Vankelecom, I. F.J., Livingston, A. G. (2002) Observations on solvent flux and solute rejection across solvent resistant nanofiltration membranes. *Desalination* 147: 307–313.

[78] Geens, J., Boussu, K., Vandecasteele, C., Van der Bruggen, B. (2006) Modelling of solute transport in non-aqueous nanofiltration. *Journal of Membrane Science* 281: 139–148.

[79] Gevers, L. E. M., Meyen, G., De Smet, K., Van De Velde, P., Du Prez, F., Vankelecom, I. F. J., Jacobs, P. A. (2006) Physico-chemical interpretation of the SRNF transport mechanism for solutes through dense silicone membranes. *Journal of Membrane Science* 274: 173–182.

[80] Silva, P., Livingston, A. G. (2006) Effect of solute concentration and mass transfer limitations on transport in organic solvent nanofiltration — partially rejected solute. *Journal of Membrane Science* 280: 889–898.

[81] Darvishmanesh, S., Buekenhoudt, A., Degrève, J., Van der Bruggen, B. (2009) General model for prediction of solvent permeation through organic and inorganic solvent resistant nanofiltration membranes. *Journal of Membrane Science* 334: 43–49.

[82] Blumenschein, S., Böcking, A., Kätzel, U., Postel, S., Wessling, M. (2016) Rejection modeling of ceramic membranes in organic solvent nanofiltration. *Journal of Membrane Science* 510: 191–200.
[83] Merlet, R. B., Tanardi, C. R., Vankelecom, I. F. J., Nijmeijer, A., Winnubst, L. (2017) Interpreting rejection in SRNF across grafted ceramic membranes through the Spiegler-Kedem model. *Journal of Membrane Science* 525: 359–367.
[84] Zedel, D., Kraume, M., Drews, A. (2017) Modelling and prediction of organic solvent flux and retention of surfactants by organic solvent nanofiltration. *Journal of Membrane Science* 544: 323–332.
[85] Abdellah, M. H., Liu, L., Scholes, C. A., Freeman, B. D., Kentish, S. E. (2019) Organic solvent nanofiltration of binary vegetable oil/terpene mixtures: Experiments and modelling. *Journal of Membrane Science* 573: 694–703.
[86] Lim, K. M., Ghazali, N. F. (2021) Nanofiltration of binary palm oil/solvent mixtures: Experimental and modeling. *Materials Today: Proceedings* 39: 1010–1014.
[87] Kempin, M. V., Schroeder, H., Hohl, L., Kraume, M., Drews, A. (2021) Modeling of water-in-oil Pickering emulsion nanofiltration – Influence of temperature. *Journal of Membrane Science* 636: 119547.
[88] Koncsag, C. I., Kirwan, K. (2012) A membrane screening for the separation/concentration of dilignols and trilignols from solvent extracts. *Separation and Purification Technology* 94: 54–60.
[89] Othman, R., Mohammad, A. W., Ismail, M., Salimon, J. (2010) Application of polymeric solvent resistant nanofiltration membranes for biodiesel production *Journal of Membrane Science* 348: 287–297.
[90] Sheth, J. P., Qin, Y., Sirkar, K. K., Baltzis, B. C. (2003) Nanofiltration-based diafiltration process for solvent exchange in pharmaceutical manufacturing *Journal of Membrane Science* 211: 251–261.
[91] Rundquist, E. M., Pink, C. J., Livingston, A. G. (2012) Organic solvent nanofiltration: a potential alternative to distillation for solvent recovery from crystallisation mother liquors. *Green Chemistry* 14: 2197–2205.
[92] Rundquist, E., Pink, C., Vilminot, E., Livingston, A. (2012) Facilitating the use of counter-current chromatography in pharmaceutical purification through use of organic solvent nanofiltration. *Journal of Chromatography A* 1229: 156–163.
[93] Székely, G., Bandarra, J., Heggie, W., Sellergren, B., Ferreira, F. C. (2011) Organic solvent nanofiltration: A platform for removal of genotoxins from active pharmaceutical ingredients. *Journal of Membrane Science* 381: 21–33.
[94] Luthra, S. S., Yang, X., Freitas dos Santos, L. M., White, L. S., Livingston, A. G. (2002) Homogeneous phase transfer catalyst recovery and re-use using solvent resistant membranes. *Journal of Membrane Science* 201: 65–75.
[95] Gould, R. M., White, L. S., Wildemuth, C. R. (2001) Membrane separation in solvent lube dewaxing. *Environmental Progress* 20: 12–16.

4 Membrane Condensers for Air Purification and Water Recovery

Mirko Frappa, Francesca Macedonio, and Enrico Drioli*
Institute on Membrane Technology, CNR-ITM, Rende, Italy
*Email: m.frappa@itm.cnr.it

4.1 INTRODUCTION

The scarcity of water is pushing the search for the exploitation of unconventional water sources. Among these, the water vapor content in exhaust gas streams could become a new source of water. However, these gases often contain pollutants and/or particulates that are harmful to the environment. The clouds of condensation flowing from cooling towers, when properly treated, can be used to remove the acid and/or bio-acid present and to capture the water contained therein. The gaseous streams that come out of industrial processes often contain heat and water. Typically, these are considered waste products and are released into the atmosphere. The possibility of recovering and reusing them could potentially reduce a process's water consumption and increase its efficiency. Air filtration technology provides a solution to clean indoor environments. Conventional air purifiers use different membranes, depending on the pollutant present.

Conventional air purifiers remove pollutants (such as suspended particles with a diameter of 0.01–100 μm) and microorganisms (such as bacteria with a diameter of 0.2–10 μm) [1,2]. However, these technologies are not still designed to recover the water molecules contained in the exhaust gases [3].

In this regard, the membrane condenser (MCo) was recently introduced as an innovative "tailor-made" membrane process for the selective recovery of the wastewater evaporated from industrial gases. According to the operating principle of MCo, the humidity contained in a gas stream is retained by a microporous hydrophobic membrane at a relatively low temperature so that the droplets of condensed water vapor join together on the retentate side. The dehydrated gas and any non-condensable compounds instead pass through the pores of the membrane and leave the MCo unit on the permeate side. In addition to water, MCo technology also allows contaminants to be retained, which in principle can be recovered and reused.

This chapter describes the use of the innovative MCo for the selective recovery of evaporated wastewater from industrial gaseous streams and the control of the composition of the outgoing liquid stream. In addition, the possibilities of MCo operation as a pretreatment unit operation are described.

4.2 AIR POLLUTION AND DEHYDRATION

4.2.1 Air Pollution

Industrialization and globalization have brought numerous advantages but also the rapid spread of dangerous toxic substances into the atmosphere. The consequences of air pollution have a negative impact on both the environment and human health. According to the World Health Organization (WHO), in 2019 air pollution and climate change topped the list as the greatest health threats. Strict environmental regulations and standards for the safety of global public health are therefore a must. Researchers advised national and international regulatory bodies to redefine European Union (EU) air quality standards to those of the WHO guidelines [3].

The chemical and medical industries require a large amount of high-purity oxygen and nitrogen [4]. Oxygen-enriched air is widely used for processes such as fuel combustion and wastewater treatment, while high-purity nitrogen has been adopted for food storage, coal mining, cryogenic storage, synthesis laboratories, and chemical transport. Air treatment and nitrogen production are therefore fundamental procurement processes. Furthermore, another factor that should not be underestimated is the cleanliness of the air to be used in the various processes. Air filtration technology is the primary solution to cleaning the indoor environment. There are two common and critical parameters adopted to evaluate filtration performance: efficiency and pressure drop [5]. While filtration efficiency is mainly determined by the interaction between particulate matter and the characteristics of the membranes (mainly pore size and distribution), pressure drop is mainly related to the interaction between gas molecules and membranes. In general, filtration occurs through a common particle capture mechanism. There are several methods to do this: interception effect, inertial effect, Brownian diffusion, gravity effect, and electrostatic effect. Generally, the smaller the pore size, the smaller the particles that can be retained and the higher the filtration efficiency, but a high packing density of the fibers also leads to a high pressure drop.

Air pollution can be due to six different types of pollutants (according to the United States Environmental Protection Agency [EPA]): particulate matter (PM), ozone (O_3), carbon monoxide (CO), sulfur dioxide (SO_2), nitric oxide (NOx), and lead (Pb). In general, these types of pollutants are non-biological in nature, and derive from natural (e.g., forest fires)

 DOI: 10.1201/9781003285656-4

and anthropogenic (e.g., industrial gases) sources. Among them, PM has attracted the greatest attention of researchers [6]. The dangerous effects of PMs depend on their size. PMs with particle sizes <10 μm are associated with lung cancer risk, while those with a size <2.5 μm can penetrate deep into the human body and cause serious problems. PM removal is a major challenge for air purification today as they are found both industrially and commercially. The main sources of these pollutants are construction materials, consumer products, and combustion products.

Air cleaning techniques include filtration, adsorption, ionization, ultraviolet germicidal irradiation, and photocatalytic oxidation [7,8]. Particle filtration efficiency classifies air filters into four categories, namely pre-filters, medium filters, high-efficiency particulate air (HEPA) filters, and ultra-low particulate air (ULPA) filters [9]. The glass filters used as pre-filters, patented in the 1940s, are used to remove particles of 10 μm with an efficiency of 99%. Medium filters remove 0.3 μm particles with an efficiency of 60–90%. HEPA filters, developed in the 1970s, were made with superfine fiberglass paper to remove particles of 0.3 μm or larger with an efficiency of 99.99%. ULPA filters, manufactured by the US Company Lydair in the early 1990s, were made with superfine glass fiber paper to remove 0.1 μm particles with an efficiency of 99.99% [3]. Conventional filters are inexpensive and convenient to install. Conventional filters are used in various sectors of industry and healthcare. HEPA filters are mainly used to reduce diesel combustion particles [10], leaching of heavy metals and radionuclides in the heat treatment of nuclear power plants [11], and in recirculating cabin air in airplanes [12]. The main drawbacks of conventional air purifiers are that they cannot remove volatile organic compounds (formaldehyde, benzene, ammonia, etc.) with a diameter of 0.1–1 nm. They also fail to remove virus particles with a diameter of approximately 0.01–0.3 μm. In the case of HEPA filters, the development of resistances linked to purification efficiency is the main drawback which varies according to the filtration technique adopted and the nature of the membrane used [10].

Nitric oxide compounds are collectively referred to as NO_x and include nitric oxide (NO), nitrogen dioxide (NO_2), nitrogen trioxide (NO_3), dinitrogen pentoxide (N_2O_5), as well as nitrogen compound acids such as nitrous acid. The sources that emit nitrogen oxides are industrial waste and the combustion of fossil fuels where first the N_2 is converted into NO and finally into NO_2 oxidized. Nitrogen dioxide is the compound of greatest concern among the other nitrogen oxides as it is the most toxic [13]. Compounds of carbon oxides include carbon monoxide (CO) and carbon dioxide (CO_2). Carbon monoxide is one of the major pollutants and, in the case of CO_2, if its concentration increase, it becomes toxic with consequent serious repercussions on the environment such as global warming [14–16]. Both can be removed from the atmosphere through the use of special plants and released by certain biological reactions. Carbon monoxide concentrations are normally 0.2 ppm in the atmosphere, and the concentration is highest during the fall. Due to the emission of CO from anthropogenic activities, their concentration reaches up to 50–100 ppm. This is due to the lack of soil, trees, and plants in urban areas where CO emissions are greatest. CO_2 is known as the thermal absorber, it is the main greenhouse gas which protects the Earth from numerous harmful rays. However, the increase in CO_2 concentration in the atmosphere causes global warming, the so-called greenhouse effect [17]. Ozone is one of the components present in the Earth's atmosphere, and it is necessary to protect the Earth from harmful radiation such as UV rays [3]. The increase of this is mainly due to the photolysis of NO_2 into NO. The atmospheric oxygen reacts with the oxygen that is released by the photolysis reaction, thus increasing the O_3 concentration in the environment. Ozone is also considered to be mutagenic and carcinogenic since a small amount of exposure to it can cause tissue damage and other damage at the molecular level [18]. Sulfur dioxide (SO_2) is a toxic gas that has a greater affinity for water, and can hydrate to sulfuric acid (H_2SO_4) [19]. About 60% of SO_2 emissions come from man-made activities in different countries. SO_2 in the air can react with water to form dilute sulfuric acid, leading to acid rain [20].

4.2.2 Dehydration and Dehumidification

The recovery of water from the atmosphere and, in particular, from the gases produced in many industrial production processes could be a way to increase the availability of water in many regions of the planet. Consequently, efficient technologies for the recovery of evaporated "waste" water and their reuse represent a critical industrial need. The technologies proposed so far for the capture of evaporated water from gas streams are: cooling with condensation [21], cryogenic separation, absorption of liquids and solids [22], and dense or hydrophilic porous membranes [22,23]. In Figure 4.1 each of these is illustrated with the main advantages and disadvantages.

Dehydration of gaseous streams is useful not only to recover water, it is also a critical aspect for the management of natural and industrial gas streams. In fact, natural gas contains many contaminants, of which the most common undesirable impurity is water. Dehydration of natural gas, which is a critical component of the natural gas conditioning process, is the removal of the water that is associated with natural gases in vapor form. The presence of water associated with the gas in the form of vapor could be harmful to all those processes that work on gaseous flows [24]. Removal of water from the gas stream reduces the potential for corrosion, hydrate formation, and freezing in the pipeline. It stops sluggish flow conditions that may be caused by condensation of water vapor in natural gas. Moreover, corrosion has become a regular problem in the oil and gas industry, and was responsible for more than 29% of global pipeline corrosion accidents from 1998 to 2008. In the event of a pipeline rupture, the release of gas into the environment can result in further undesirable consequences such as a fire outbreak, reduced air quality, and gas exposure to the surrounding flora and fauna [25]. Therefore, the water content in natural gas has been strictly regulated; the limit of which varies from country to country. For example, in

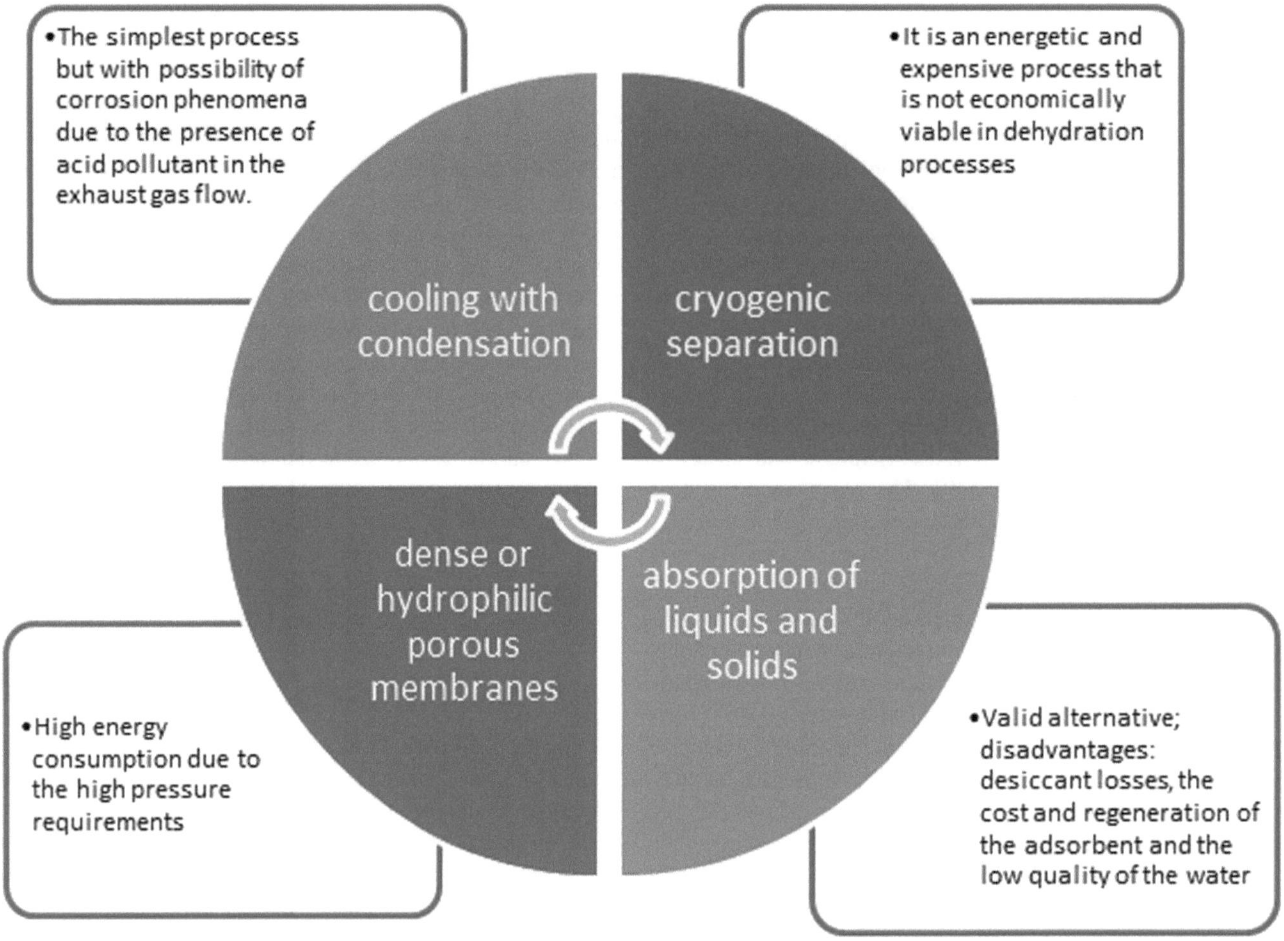

FIGURE 4.1 Traditional technologies for dehydration of gaseous streams. Modified from [22].

the United States and Canada, the water content in natural gas must not exceed 7 pounds of water per million standard cubic feet of gas (pounds MMSCF-1) and 4 pounds MMSCF-1, respectively [26]. Countries with colder climates such as Alaska often have lower limits, which vary between 1 and 2 pounds MMSCF-1. In fact, a reduction in temperature or an increase in pressure leads to water condensation.

4.3 MEMBRANE SYSTEMS FOR WASTE GASEOUS STREAMS TREATMENT

Membrane technology is of great importance for air purification, as it is more energy efficient than conventional processes. As mentioned in Section 4.2, filtration is the simplest technique for removing suspended particles, and mechanical filters are crucial components in air conditioning systems. One problem with using mechanical filtration is that this technique cannot remove gases [27]. The selective separation of some gases and the interception of solid contaminants in the air can be achieved by membrane filtration. This technique has advantages such as high efficiency, energy saving, and easy application on a large scale [28]. In order to effectively block contaminants and particulates in the air, a proper membrane pore size is required. At the same time, to have a high separation efficiency it is necessary to use membranes with high porosity [29]. A study conducted by Bortolassi et al. [30] demonstrated that an electrospun silver nanofiber/polyacrylonitrile membrane (with porosity of 96%) has high filtration efficiency for solid particles (9–300 nm) in the air and excellent antibacterial activity against *Escherichia coli* (*E. coli*).

Regarding the presence of water particles in industrial waste processes, there is currently no commercial technology for water recovery. The membrane technologies currently proposed are those based on the use of porous or dense membranes. The transport of water vapor across a membrane, whether porous or dense, follows one or a combination of the following four mechanisms: Knudsen diffusion, surface diffusion, capillary condensation, and molecular sieving. In Figure 4.2 the main transport mechanisms are reported. A particular mechanism can be dominant over another depending on the pressure, temperature, size, and interactions between the membrane surface and the feed [31].

The transport membrane condenser (TMC) process uses hydrophilic porous membranes for condensing the water and recovering the heat from the fumes. The employed membranes have a high porosity, with pore size normally in the sub-micron range (nanoporous). To achieve a good separation

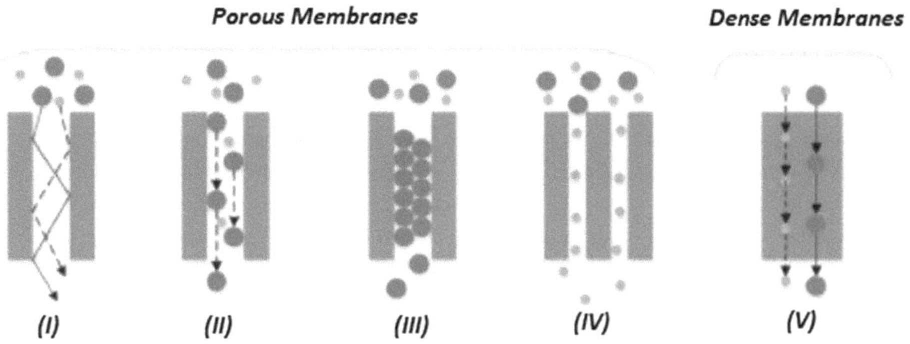

FIGURE 4.2 Transport mechanism for porous membranes [(I) Knudson diffusion, (II) surface diffusion, (III) capillary condensation, and (IV) molecular sieving] and dense membranes ((V) solution–diffusion). From [32], open access.

ratio with a porous membrane for gaseous species, including the separation of water vapor from flue gas, the typical pore size must be less than 50 nm so that the TMC can work based on the capillary condensation. The tubes are grouped in a shell to essentially form a nanoporous heat exchanger. The exhaust gas flows through the duct and comes into contact with the TMC tubes [31,33]. In the TMC, the water vapor of the fumes on the discharge side condenses inside the nanoporous membrane and passes through it by direct contact with the low-temperature water at the permeate side. In this way the transported water is recovered together with almost all of its latent heat. The conditioned fumes leave the TMC at a reduced temperature and with relative humidity below saturation. Non-condensable gases such as CO_2, O_2, NO_x, and SO_2 are inhibited from passing through the membrane by condensed water which clogs the pores of the pipe membrane. The low pressure difference across the membrane tube wall and its nanoporous membrane coating on the tubes inhibit the particulate matter from clogging the pores. The recovered water is of high quality and free of minerals, so it can be used directly as boiler make-up water, as well as for other processes. Recent studies have stated that water recovery via TMC varies between 20–60%, while the heat recovery rate is 33–85% (depending on gas flow rate, refrigerant flow rate, and temperature difference). The corresponding recoveries may decrease as the coolant temperature increases.

Not only porous membranes but also dense membranes can be used to recover and/or remove water vapor from gas streams. The gas is dehumidified by diffusion through the dense membranes in order to selectively permeate the water vapor with respect to other non-condensable gases. Vapor permeation membranes have already been applied for drying natural gas [34] and compressed air [35]. Water vapor generally has a much higher solubility and diffusivity than other gases, and therefore its permeability and selectivity are in the order of 107 (H_2O/N_2) and 105 barrer, respectively [36]. Although the perm-selectivity of the vapor separation membranes is impressive, it cannot be used fully in real processes because the saturation vapor pressure of the water in the flue-gas state (50–80°C) is only about 0.1 bar and 0.5 bar. Therefore, thermodynamically, only a small differential pressure can be applied and the membrane performance is completely dependent on the vacuum capacity of the permeate or the sweeping rate. Another very important point that must be considered for the dehydration membrane is the inevitable condensation of water within the membrane pores. Unlike dense films, permeating water will condense in the support layer of hollow-fiber membranes due to capillary condensation, potentially reducing membrane permeability [37].

4.3.1 Membrane Condenser

The innovative membrane condenser (MCo) technology was developed and used for the first time in FP7project “CAPWA” (Capture of evaporated water with novel membranes) [38]. Successively, in the context of the EU H2020 project “MATChING” (Materials &Technologies for Performance Improvement of Cooling Systems performance in Power Plants) the MCo was further improved so as to optimize the water recovery from the plume of cooling towers [39].

The membrane-assisted condenser belongs to the class of membrane contactors [40]. The feed of MCo is a gaseous stream, at a certain temperature and, in most of the cases, water saturated. As shown in Figure 4.3, it is fed to the membrane condenser module whose temperature is equal or lower than that of the feed. In doing this, water droplets already present in the feed together with the amount that condenses onto the membrane surface (and in the membrane module) are blocked by the hydrophobic nature of the membrane. Therefore, the liquid water is recovered at the retentate side, whereas the other gases are at the permeate side of the membrane unit. Moreover, in a membrane condenser unit, the modulation of contact time between saturated stream and membrane, as well as the control of temperature and/or pressure difference between membrane sides, allow controlling the fraction of components present in the feed gaseous stream that will be retained in the condensed water.

The membrane condenser can also be considered as a cleaning unit for the removal of pollutants and particulates. In this case, the humidified gaseous stream (consisting of saturated or super-saturated air with ammonia or particles) is fed to the upper part of the module. The membrane allows the liquid water to be retained in the retentate side and collected at the bottom of the module, opportunely designed with a conical shape to avoid stagnant zones, whereas the dehydrated stream is recovered in the permeate side. Depending on the operating conditions and on the membranes used during the experiments, the retentate could contain a certain concentration of ammonia and a certain number of particles [41].

The main advantages of MCo over competing technologies are the low energy consumption and the absence of corrosion phenomena. In fact, traditional condensers represent the easiest process but are very limited due to corrosion phenomena caused by the presence of acid pollutants in the waste gaseous stream with respect to liquid or solid desiccant. MCo does not suffer from desiccant losses, high costs associated with the regeneration of the absorbent, or reduced quality of recovered water [42].

The other important benefit of MCo is the ability to remove pollutants often contained in waste gas streams. These include SO_x, NO_x, VOCs, H_2S, NH_3, siloxanes, halides, hashes, particulates, and organic pollutants [16]. The need for the removal of such contaminants is related either to the processing of the gaseous streams (because the contaminants could severely damage subsequent separation units [43]) or to environmental constraints (as in the case of cooling tower plumes). Two parameters affecting the removal of chemicals in MCo systems are the temperature difference between the feed and the membrane module (ΔT), and the ratio between the feed flow rate (Q) and membrane area (A). In fact, Macedonio et al. [40] observed that the concentration of contaminants is, in fact, a decreasing function Q/A.

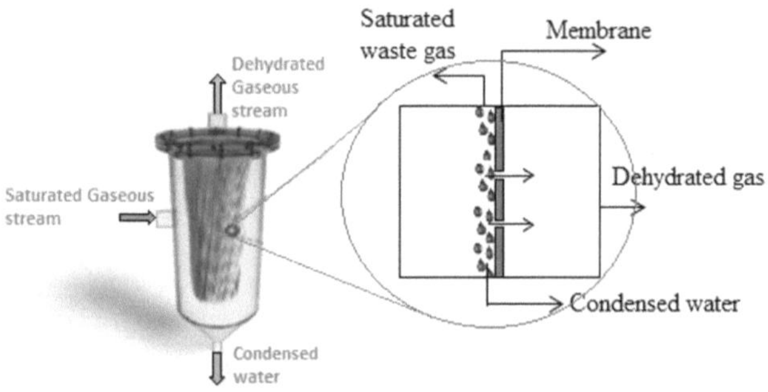

FIGURE 4.3 Scheme of the membrane condenser process. From [16], open access.

On the other hand, the concentration of contaminants in the liquid stream recovered on the retentate side increases with ΔT. In effect, considering the same feed inlet temperature, the more the membrane temperature is lowered, the higher is the condensation of water vapor. In Figure 4.4 the membrane condenser concept for the recovery of clean water and for pollutants removal is reported.

The dehydrated and decontaminated gas streams exiting MCo units can be further processed through gas separation systems, as in the case of CO_2 capture and biogas production [44–48]. In this case, MCo can be considered as an efficient pretreatment unit operation of subsequent gas separation stages.

MCo membranes are required to exhibit excellent mechanical resistance and chemical stability toward contaminants.

The main materials used for MCo applications are PP and PVDF due to their relatively low cost, high porosity, and easy fabrication procedure [49,50]. Besides PVDF, other fluoropolymer materials can be used in MCo systems, such as PTFE [51], Hyflon [52], and Teflon. Since they are quite expensive, new composite membranes in which the hydrophobic layer is deposited on cheap supports are now being developed.

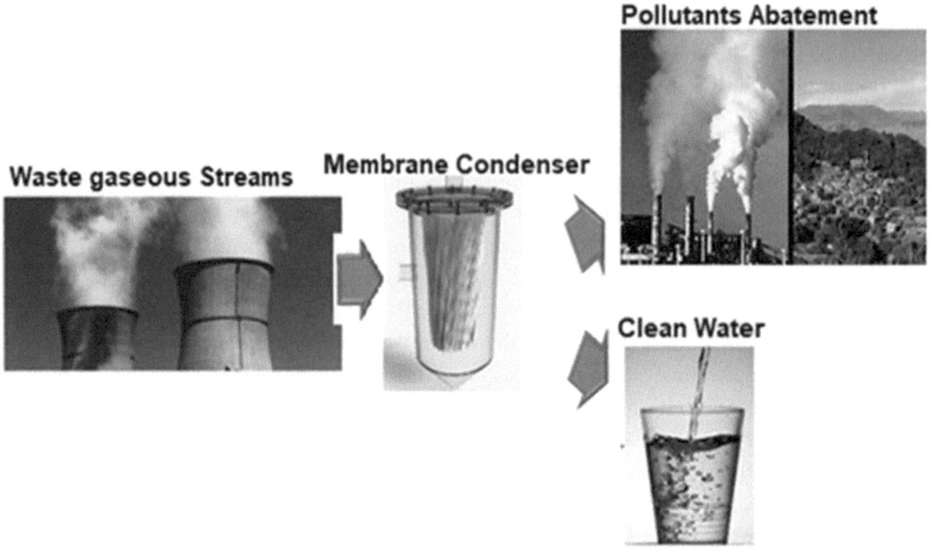

FIGURE 4.4 Membrane condenser concept: from waste gaseous stream to clean water production and pollutant abatement.

TABLE 4.1
Comparison between a membrane condenser and traditional technologies [22]

	Water recovery (%)	Main issues	Economic viability (Euro m^{-3})	Investments costs
Liquid and solid sorption [55]	22–62	Corrosion and salt crystals formation	\$4.4 m^{-3}	5.8*10^6 \$ (2006) þ200.000 \$/year (2006) as operational costs
Cooling with condensation [21]	<70	Corrosion owing to the diluted acids and fly ashes forming deposits	1.5–2	6.4*10^6 Euro (2011)
Dense membranes [61]	20–40	Flue gas desulfuration (FGD) necessary to avoid damage	1.5 for wet regions; 10 for dry regions	To be determined
Membrane condensers [22]	>70	Ashes removal to avoid membrane damage	1.5–2.5[a]	To be determined

[a] Considering only costs related to energy requirements and membrane modules.

4.4 MCO FUNDAMENTALS

The amount of water that can be recovered via a membrane condenser can vary depending on the operating conditions, such as: relative humidity and temperature of the feed, temperature of the membrane module, and the ratio between the feed flow rate and the membrane area (Q/A). Water recovery can be calculated through the following relation:

$$\text{Water Recovery}\,\% = \frac{\text{Liquid water retained in the retentate side}}{\text{Total water contained in the feed stream}}$$

Mass intensity and energy intensity are two parameters that allow evaluation of the performance of the membrane condenser. The first takes into account the water recovered on the retentated side with respect to the total mass of water contained in the feed, the second considers the power required by the system with respect to the recovered water. Lower values of these indicators are linked to a more intense process. In the ideal situation, mass intensity and energy intensity tend to the lowest possible value.

$$\text{Mass Intensity} = \frac{\text{Total inlet mass}}{H_2O\text{ mass ecovered}}$$

$$\textbf{Energy Intensity} = \frac{\textbf{Energy duty for condensation}}{\mathbf{H_2O}\textbf{ mass recovered}}$$

As reported in Table 4.1, compared to the traditional technologies, a membrane condenser offers higher water recovery with respect to all the other technologies. Additionally, adsorption units may suffer from desiccant leakage problems. Furthermore, membrane condensers do not undergo corrosion phenomena that normally occur in traditional condensers or adsorption units. However, regardless of whether the gas stream contains hash or particles, a pretreatment step may be required before feeding the membrane condenser.

Compared to dense membrane technology, the main differences are related to the greater pressure difference necessary to favor the permeation of water vapor through dense membranes. This aspect involves investment and operating costs linked to the presence of compressors or vacuum pumps.

On the contrary, the purity of the water recovered in the membrane condensers can be affected by the possible condensation of contaminants—if present in the gaseous flow—but it is sufficient as industrial water make-up. However, further purification would be required to make it drinkable.

4.5 MCO PERFORMANCES

As mentioned above, the membrane condenser was initially developed for the recovery of water from waste gaseous streams, and at a later time for the removal of contaminants. The various applications are briefly described and analyzed below.

4.5.1 MCo for Removal of Ammonia from Gaseous Streams

Ammonia is a toxic, reactive, and corrosive gas which is very harmful to human health and is mainly produced by farms where decomposing of animals, waste, and plants is the main source. Secondary sources of ammonia are integrated systems for combined cycle power generation, coal gasification, landfills for waste disposal, cremation, chemical and manufacturing industries, and wastewater treatment plants [53]. In the case of ammonia, an aspect to consider is its high solubility in water. In water, most of the ammonia turns into ammonium, which is not a gas and has no smell. Traditional strategies to reduce ammonia and its ammonium ion emissions from animal facilities and other sources include preventing ammonia formation and volatilization and controlling ammonia transmission. These strategies include the use of filtration (and/or bio-filtration) systems, impermeable and semi-permeable barriers, and dietary manipulation. A membrane condenser has been tested for the removal of ammonia contained in a gaseous waste stream

[41,54]. Experimental measurements were carried out at various operating conditions, as reported in Table 4.2.

In general, the amount of water recovered increases with the growth of ΔT. In particular, it was found that considering a gaseous stream at 25°C and using a Q/A = 1.2 m h^{-1}, the water recovery varies from 36.24% (ΔT = 5.84°C and RH = 104.9%) to over 55% (ΔT ≈ 10°C, T_{plume} ≈ 25°C and RH > 140%). At the same feed temperature (≈25°C) but higher Q/A ratio (2.7 m h^{-1}) and relative humidity of 142%, the amount of water recovered increases from 46.04% (ΔT = 6°C) to more than 70% (ΔT ≈ 13.9°C).

The ammonia concentration in the recovered liquid water increases with an increase in temperature difference ΔT between the feed and membrane module, and with the rise of NH_3 concentration in the feed or with the increase of the relative humidity of the feed, as shown in Figure 4.5.
Therefore, in addition to the recovery of water, another important advantage of the membrane condenser is the possibility of recovering the necessary chemicals in the system and, at the same time, of reducing their emissions. These are fundamental aspects to be taken into account in highly industrialized realities where the problems linked to water scarcity and air pollution (which induce health problems and climate concerns) are becoming increasingly urgent.

TABLE 4.2
Operative conditions utilized in the experimental tests [41,54]

Relative humidity, %	From 98.0 to 150
Feed temperature, °C	From 25 to 45
Feed flow rate/membrane area (Q/A), m h^{-1}	1.2; 2.7
ΔT, °C	From 2.5 to 11.6
NH_3 concentration in the feed, ppm	From 100 to 650

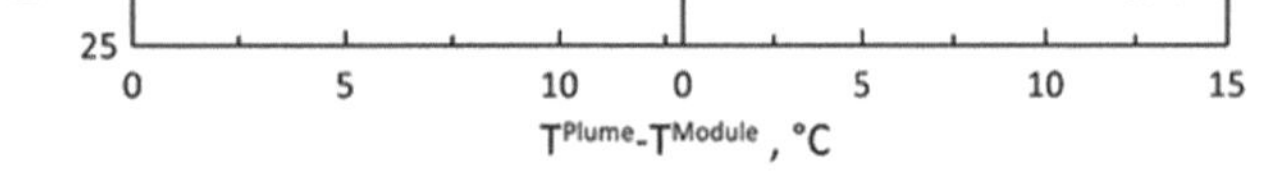

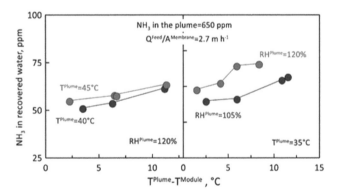

FIGURE 4.5 NH_3 in recovered water as a function of temperature difference between the plume and the membrane module, at different T_{Plume} and RH_{Plume}. From [41], open access.

4.5.2 MCo for Abatement of Particulate Contained in Gaseous Streams

Other experiments performed with MCo units were aimed at reducing the concentration of particles of gaseous streams [20]. Polystyrene particles with a nominal size of 0.5 μm were dispersed in a gaseous stream at a concentration of 30 μg m^{-3} and various membranes differing in nominal pore size were tested. Experiments were performed for 203 h and 72.5 h on PP and PVDF/Hyflon membranes, respectively. Both types of membranes offered an active surface area of around 0.25 m^2. Table 4.3 reports the main characteristics of the microparticles utilized and the Q/A ratio tested.

Analytical measurements of particle concentration in the collected samples allowed the correlation of their concentration on the retentate side with the size of the pores of the tested membranes. In fact, the two membranes showed different particle retentions, mainly due to their pore sizes. Despite the nominal pore size of PP commercial membrane being 0.2 μm, SEM analysis revealed the presence of larger pores (up to 10 μm [55]) which were not able to retain polystyrene particles. In contrast, the reduced particle concentration of the water recovered from the MCo system working with PVDF/Hyflon membranes, as shown in Figure 4.6, was attributed to the narrower pore distribution of these membranes, with a pore size of 0.15 ± 0.018 μm.

4.5.3 MCo for the Regulation of CO_2 Emission

Current constraints and regulations on CO_2 emissions from power plants have forced manufacturers and researchers to focus on the separation of CO_2 from flue gas streams and to develop specific CO_2 capture technologies that can be retrofitted to existing power plants as well as being designed into new plants with the goal of achieving 90% CO_2 capture while limiting the increase in cost of electricity to no more than 35%. Today, membrane technology for the separation and recovery of CO_2 is a well-consolidated technique, mainly based on the use of polymeric membranes. Waste gaseous streams usually contain impurities such as water vapor, acid

TABLE 4.3
Characteristics of the particles utilized in the experiments

Particle composition	Particle specific gravity, g/cm³	Nominal particle diameter, microns	Standard deviation, micron	CV, %	Particle concentration in the feed, μg/m³	Feed flow rate/ membrane area, m/h
Polystyrene	1.05	0.5	0.013	2.6	30	1.3

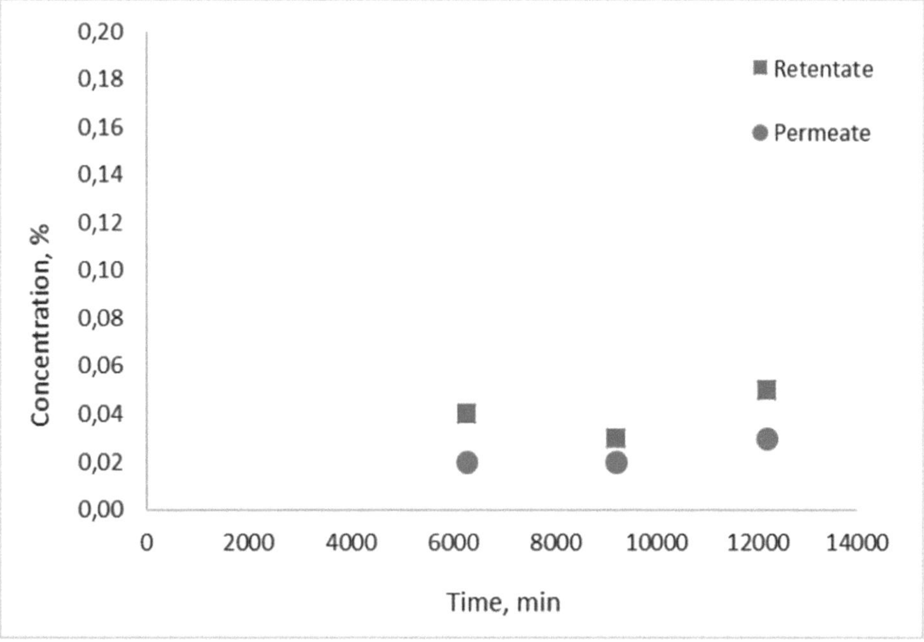

FIGURE 4.6 Particle concentration in the retentate and permeate achieved with PP hollow-fiber membranes at various times of experimentation. From [20], open access.

gases, olefins, aromatics, and other organics. At relatively low concentrations, these impurities can cause membrane plasticization and loss of selectivity, while at higher concentrations they can condense on the membrane surface, which could be damaged as a result.

Mass transport properties of polymeric membranes often used for waste gases treatment, such as CO_2 separation, are significantly affected by the presence of water vapor, which usually competes with CO_2 in permeation, reducing its permeability as well as inducing the formation of water clusters, swelling, etc., with consequent variation of permeabilities and selectivity. The solution for successful operation of polymeric modules is, also, a careful selection of feed pretreatment. In this scenario, the possibility to use integrated membrane systems as alternatives to traditional operations is becoming increasingly attractive for gas purification. Analogous to what was done in the water purification field, where integrated membrane systems are today one of the leading technologies [56–58], also in the gas field, the necessity for pretreatment stages is a fundamental step both to fit required targets of contaminants content as well as to prolong the lifetime of the other downstream separation units [16,59,60]. The reduction of water use would significantly reduce the aforementioned phenomena, better exploiting the membrane area available and, thus, reducing the footprint of membrane gas separation (GS) units currently foreseen for the specific separation.

A membrane condenser can be a suitable pretreatment stage placed before other separation technologies. The latter will be significantly reduced in size and their lifetime prolonged with respect to that without using a condenser.

4.6 CONCLUSION AND PERSPECTIVE

MCo units are used in the treatment of gaseous streams containing humidity and/or pollutants. The hydrophobic membranes used in the MCo allow concentration of the water vapor and condensable contaminants on the feed side. Earlier works on membrane condenser applications are summarized in Table 4.4.

The quantity of recovered water obtainable through MCo increases with ΔT (i.e. the temperature difference between the inlet flow and the membrane module). Furthermore, the amount of recovered water and retained pollutants increases at higher Q/A ratios [16,40].

MCo technology could also be considered as a pretreatment phase for the production of dried and depolluted gas streams to be fed to other separation units. In particular, MCo might be utilized to regulate the emission of CO_2 capture and in the field of biogas production. In this perspective, the membrane condenser can be a suitable pretreatment stage placed before other separation technologies. By modulating the contact time between saturated flow and membrane condenser, it is possible to control the fraction of contaminants that can be retained in the condensed water. Although MCo is not always sufficient to reduce the contaminant content below the regulation limit (it mainly depends on the ΔT that can be imposed), its use increases the operating time of the other pretreatment phases such as activated carbon, scrubbers, etc. and it can be used as a pretreatment unit not only for membranes but also for other separation technologies such as PSA, cryogenic, adsorption, etc. Furthermore, the appropriate use of these systems would lead not only to reduced emissions of contaminants and the

TABLE 4.4
Relevant past works on the membrane condenser

Title	Objective	Membrane material	Ref.
Application of a membrane condenser system for ammonia recovery from humid waste gaseous streams at a minimum energy consumption	MCo used for the recovery of ammonia at minimum energy consumption	Polypropylene (PP)	[54]
Preparation of ECTFE porous membrane for dehumidification of gaseous streams through membrane condenser	ECTFE membranes prepared by thermally induced phase separation (TIPS) for MCo	Poly(ethylene trifluoroethylene) (ECTFE)	Pan, J., Chen, K., Cui, Z., Bamaga, O., Albeirutty, M., Alsaiari, A. O., ... & Drioli, E. (2022). Preparation of ECTFE Porous Membrane for Dehumidification of Gaseous Streams through Membrane Condenser. Membranes, 12(1), 65. Drioli, E., Santoro, S., Simone, S., Barbieri, G., Brunetti, A., Macedonio, F., & Figoli, A. (2014). ECTFE membrane preparation for recovery of humidified gas streams using membrane condenser. Reactive and Functional Polymers, 79, 1–7
Membrane condenser for particulate abatement from waste-gaseous streams	Application of membrane condenser for water recovery and microparticles removal from gaseous streams	Lab-made Hyflon AD/polyvinylidene fluoride (PVDF) composite membranes Commercial polypropylene membranes	Frappa, M., Brunetti, A., Drioli, E., Cui, Z., Pan, J., & Macedonio, F. (2020). Membrane condenser for particulate abatement from waste-gaseous streams. Journal of Membrane Science and Research, 6(1), 81–89.
Recovery of water and contaminants from cooling tower plume	Application of membrane condenser for water and ammonia recovery from synthetic streams simulating the plume of cooling tower	Polypropylene (PP)	Macedonio, F., Frappa, M., Brunetti, A., Barbieri, G., & Drioli, E. (2020). Recovery of water and contaminants from cooling tower plume. Environmental Engineering Research, 25(2), 222–229
Membrane condenser as a new technology for water recovery from humidified "waste" gaseous streams	Application of membrane condenser for water recovery from synthetic streams simulating flue gas	Polypropylene (PP)	Macedonio, F., Brunetti, A., Barbieri, G., & Drioli, E. (2013). Membrane condenser as a new technology for water recovery from humidified "waste" gaseous streams. Industrial & Engineering Chemistry Research, 52(3), 1160–1167

recovery of water vapor contained in the gas flow, but also to the recovery of compounds with high added value such as VOC/VFA and/or to retain pollutants, which can be suspended in the waste gas stream.

The membrane condenser could therefore be a fundamental technology integrated in a system comprising other separation units for the treatment of the liquid and the downstream gas. In the case of flue gas treatment, for example, the condensate water will contain contaminants, such as SOx and possibly particulate matter. Therefore, depending on their quantity, the condensed water can be reused directly in the system as a reintegration or further treated with other technology purifiers. MCo also could play an interesting role in biogas treatment. Biomethane is currently separated from CO_2 using various technologies including membranes. However, various pretreatment steps before membrane gas separation units are necessary to preserve the membranes and meet grid injection specifications. These pretreatments are usually based on chillers, scrubbers for the absorption of H_2S with Fe_2O_3, activated carbon for siloxanes, etc. The membrane condenser will allow the separation of water vapor and contaminants (H_2S, NH_3, siloxanes and halides, VFA, VOC, etc.), from the gas stream; the condensation water can undergo a post-treatment to obtain high-quality water, while the VOC and VFA fraction can be recovered and further separated in other unit operations.

As already mentioned, the use of the membrane condenser would lead to a reduction in pretreatment loads, which translates into a lower use of solvents, a lower environmental impact, and an improvement in air quality. In addition, the integrated membrane condenser with another separation unit for VFA/VOC recovery could represent a new, highly selective and energy-efficient separation (pretreatment) technology that will improve the recovery of additional resources from the downstream bio-digester, which can be further purified and reused.

REFERENCES

[1] Quan, Z., Zu, Y., Wang, Y., Zhou, M., Qin, X., Yu, J. (2021) Slip effect based bimodal nanofibrous membrane for high-efficiency and low-resistance air purification. *Separation and Purification Technology* 275: 119258.

[2] Brigagão, G.V., de O. Arinelli, L., de Medeiros, J.L., Araújo, de Q.F. (2020) Low-pressure supersonic separator with finishing adsorption: Higher exergy efficiency in air pre-purification for cryogenic fractionation. *Separation and Purification Technology* 248: 116969.

[3] Rathna, R., Jeno, J.G.A., Sivagami, N., Bharathi, V.P., Nakkeeran, E. (2020) Invisible membrane revolution: shaping the future of air purification In: *Nanomaterials for Air Remediation*, pp. 343–358. Elsevier.

[4] Fernández-Barquín, A., Casado-Coterillo, C., Valencia, S., Irabien, A. (2016) Mixed matrix membranes for O_2/N_2 separation: The influence of temperature. *Membranes (Basel)* 6: 28.

[5] Li, Q., Wang, Z., Shao, S., Niu, Z., Xin, Y., Zhao, D., Hou, Y., Xu, Z. (2022) Experimental study on the synthetic dust loading characteristics of air filters. *Separation and Purification Technology* 284: 120209.

[6] Beesley, L., Inneh, O.S., Norton, G.J., Moreno-Jimenez, R., Pardo, T., Clemente, R., Dawson, J.J.C. (2014) Assessing the influence of compost and biochar amendments on the mobility and toxicity of metals and arsenic in a naturally contaminated mine soil. *Environmental Pollution* 186: 195–202.

[7] Siegel, J.A. (2016) Primary and secondary consequences of indoor air cleaners. *Indoor Air* 26: 88–96.

[8] Tang, X., Misztal, P.K., Nazaroff, W.W., Goldstein, A.H. (2016) Volatile organic compound emissions from humans indoors. *Environmental Science & Technology* 50: 12686–12694.

[9] Rafique, M.S., Tahir, M.B., Rafique, M., Shakil, M. (2020) Photocatalytic nanomaterials for air purification and self-cleaning. In: *Nanotechnology and Photocatalysis for Environmental Applications*, pp. 203–219. Elsevier.

[10] Yoon, I.-H., Choi, W.-K., Lee, S.-C., Min, B.-Y., Yang, H.-C., Lee, K.-W. (2012) Volatility and leachability of heavy metals and radionuclides in thermally treated HEPA filter media generated from nuclear facilities. *Journal of Hazardous Materials* 219–220: 240–246.

[11] Rabinowitz, O. (2016) Nuclear energy and desalination in Israel. *Bulletin of the Atomic Scientists* 72: 32–38.

[12] Eckels, S.J., Jones, B., Mann, G., Mohan, K.R., Weisel, C.P. (2014) Aircraft recirculation filter for air-quality and incident assessment. *Journal of Aircraft* 51: 320–326.

[13] Swamy, G.S.N.V.K.S.N. (2021) Development of an indoor air purification system to improve ventilation and air quality. *Heliyon* 7: e08153.

[14] Lazzerini, G., Lucchetti, S., Nicese, F.P. (2016) Green house gases (GHG) emissions from the ornamental plant nursery industry: a life cycle assessment (LCA) approach in a nursery district in central Italy. *Journal of Cleaner Production* 112: 4022–4030.

[15] Li, L., Yan, K., Chen, J., Feng, T., Wang, F., Wang, J., Song, Z., Ma, C. (2019) Fe-rich biomass derived char for microwave-assisted methane reforming with carbon dioxide. *Science of the Total Environment* 657: 1357–1367.

[16] Brunetti, A., Macedonio, F., Barbieri, G., Drioli, E. (2019) Membrane condenser as emerging technology for water recovery and gas pre-treatment: Current status and perspectives. *BMC Chemical Engineering*. doi:10.1186/s42480-019-0020-x.

[17] Boributh, S., Assabumrungrat, S., Laosiripojana, N., Jiraratananon, R. (2011) A modeling study on the effects of membrane characteristics and operating parameters on physical absorption of CO_2 by hollow fiber membrane contactor. *Journal of Membrane Science*. doi:10.1016/j.memsci.2011.06.029.

[18] Godish, T. (2016) *Indoor Environmental Quality*. CRC Press.

[19] Platt, S., Nyström, M., Bottino, A., Capannelli, G. (2004) Stability of NF membranes under extreme acidic conditions. *Journal of Membrane Science*. doi:10.1016/j.memsci.2003.09.030

[20] Frappa, M., Brunetti, A., Drioli, E., Cui, Z., Pan, J., Macedonio, F. (2020) Membrane condenser for particulate abatement from waste-gaseous streams. *Journal of Membrane Science Research*. doi:10.22079/JMSR.2019.112686.1282.

[21] Folkedahl, B.C., Weber, G.F., Collings, M.E. (2006) *Water Extraction from Coal-Fired Power Plant Flue Gas*. Pittsburgh, PA, and Morgantown, WV.

[22] Brunetti, A., Santoro, S., Macedonio, F., Figoli, A., Drioli, E., Barbieri, G. (2014) Waste gaseous streams: From environmental issue to source of water by using membrane condensers. *CLEAN – Soil, Air, Water* 42: 1145–1153.

[23] Sijbesma, H., Nymeijer, K., van Marwijk, R., Heijboer, R., Potreck, J., Wessling, M. (2008) Flue gas dehydration using polymer membranes. *Journal of Membrane Science* 313: 263–276.

[24] Gandhidasan, P. (2001) Dehydration of natural gas using solid desiccants. *Energy* 26: 855–868.

[25] Farag, H.A.A., Ezzat, M.M., Amer, H., Nashed, A.W. (2011) Natural gas dehydration by desiccant materials. *Alexandria Engineering Journal* 50: 431–439.

[26] Kong, Z.Y., Mahmoud, A., Liu, S., Sunarso, J. (2018) Revamping existing glycol technologies in natural gas dehydration to improve the purity and absorption efficiency: Available methods and recent developments. *Journal of Natural Gas Science and Engineering* 56: 486–503.

[27] Bui, V.K.H., Nguyen, T.N., Van Tran, V., Hur, J., Kim, I.T., Park, D., Lee, Y.-C. (2021) Photocatalytic materials for indoor air purification systems: An updated mini-review. *Environmental Technology & Innovation* 22: 101471.

[28] Zhong, C., Xiong, X. (2021) Preparation of a composite coating film via vapor induced phase separation for air purification and real-time bacteria photocatalytic inactivation. *Progress in Organic Coatings* 161: 106486.

[29] Wang, B., Wang, Q., Wang, Y., Di, J., Miao, S., Yu, J. (2019) Flexible multifunctional porous nanofibrous membranes for high-efficiency air filtration. *ACS Applied Materials and Interfaces* 11: 43409–43415.

[30] Bortolassi, A.C.C., Nagarajan, S., de Araújo Lima, B., Guerra, V.G., Aguiar, M.L., Huon, V., Soussan, L., Cornu, D., Miele, P., Bechelany, M. (2019) Efficient nanoparticles removal and bactericidal action of electrospun nanofibers membranes for air filtration. *Materials Science and Engineering C* 102: 718–729.

[31] Ratnakar, R.R., Dindoruk, B. (2022) The role of diffusivity in oil and gas industries: Fundamentals, measurement, and correlative techniques. *Processes* 10: 1194.

[32] Vermaak, L., Neomagus, H.W.J.P., Bessarabov, D.G. (2021) Recent advances in membrane-based electrochemical hydrogen separation: A review. *Membranes (Basel)* 11: 127.

[33] Wang, T., Yue, M., Qi, H., Feron, P.H.M., Zhao, S. (2015) Transport membrane condenser for water and heat recovery from gaseous streams: Performance evaluation. *Journal of Membrane Science*. doi:10.1016/j.memsci.2015.03.007.

[34] Lin, H., Thompson, S.M., Serbanescu-Martin, A., Wijmans, J.G., Amo, K.D., Lokhandwala, K.A., Merkel, T.C. (2012) Dehydration of natural gas using membranes. Part I: Composite membranes. *Journal of Membrane Science* 413–414: 70–81.

[35] Li, G.M., Feng, C., Li, J.F., Liu, J.Z., Wu, Y.L. (2008) Water vapor permeation and compressed air dehydration performances of modified polyimide membrane. *Separation and Purification Technology* 60: 330–334.

[36] Kim, J.F., Drioli, E. (2020) Transport membrane condenser heat exchangers to break the water-energy nexus—A critical review. *Membranes (Basel)* 11: 12.

[37] Scholes, C.A., Kentish, S.E., Stevens, G.W., DeMontigny, D. (2015) Comparison of thin film composite and microporous membrane contactors for CO_2 absorption into monoethanolamine. *International Journal of Greenhouse Gas Control* 42: 66–74.

[38] *Capture of evaporated water with novel membranes*, GA n° 246074, www.watercapture.eu/ (last accessed: 8 July, 2019), (n.d.).

[39] *Materials & Technologies for Performance Improvement of Cooling Systems performance in Power Plants*, GA n° 686031, http://matching-project.eu/ (last accessed: 8 July, 2019), (n.d.).

[40] Macedonio, F., Brunetti, A., Barbieri, G., Drioli, E. (2012) Water recovery from waste gaseous streams: An application of hydrophobic membranes. *Procedia Engineering* 44: 202–203.

[41] Macedonio, F., Frappa, M., Brunetti, A., Barbieri, G., Drioli, E. (2020) Recovery of water and contaminants from cooling tower plume. *Environmental Engineering Research*. doi:10.4491/eer.2018.192.

[42] Macedonio, F., Cersosimo, M., Brunetti, A., Barbieri, G., Drioli, E. (2014) Water recovery from humidified waste gas streams: Quality control using membrane condenser technology. *Chemical Engineering and Processing – Process Intensification* 86: 196–203.

[43] Park, K., Hong, S.Y., Lee, J.W., Kang, K.C., Lee, Y.C., Ha, M.-G., Lee, J.D. (2011) A new apparatus for seawater desalination by gas hydrate process and removal characteristics of dissolved minerals (Na^+, Mg^{2+}, Ca^{2+}, K^+, B^{3+}). *Desalination* 274: 91–96.

[44] Choi, W., Ingole, P.G., Park, J.-S., Lee, D.-W., Kim, J.-H., Lee, H.-K. (2015) H_2/CO mixture gas separation using composite hollow fiber membranes prepared by interfacial polymerization method. *Chemical Engineering Research and Design* 102: 297–306.

[45] Naidu, G., Tijing, L., Johir, M.A.H., Shon, H., Vigneswaran, S. (2020) Hybrid membrane distillation: Resource, nutrient and energy recovery. Journal of Membrane Science 599: 117832.

[46] Gugliuzza, A., Basile, A. (2013) *Membranes for Clean and Renewable Power Applications*. Woodhead Publishing.

[47] Gao, Y., Nan, X., Yang, Y., Sun, B., Xu, W., Dasilva, W.D.L., Li, X., Li, Y., Zhang, Q. (2021) Non-layered transition metal carbides for energy storage and conversion. *New Carbon Materials* 36: 751–778.

[48] Li, Y., Zhang, X., Wu, T., Tang, J., Deng, L., Li, W., Wang, L., Deng, H., Hu, W. (2021) First-principles study on the dissolution and diffusion behavior of hydrogen in carbide precipitates. *International Journal of Hydrogen Energy* 46: 22030–22039.

[49] Francis, L., Ghaffour, N., Al-Saadi, A.S., Amy, G. (2015) Performance of different hollow fiber membranes for seawater desalination using membrane distillation. *Desalination and Water Treatment*. doi:10.1080/19443994.2014.946723.

[50] Martínez, L., Rodríguez-Maroto, J.M. (2006) Characterization of membrane distillation modules and analysis of mass flux enhancement by channel spacers. *Journal of Membrane Science* 274: 123–137.

[51] Eykens, L., De Sitter, K., Dotremont, C., Pinoy, L., Van der Bruggen, B. (2017) Membrane synthesis for membrane distillation: A review. *Separation and Purification Technology* 182: 36–51.

[52] Xu, K., Cai, Y., Hassankiadeh, N.T., Cheng, Y., Li, X., Wang, X., Wang, Z., Drioli, E., Cui, Z. (2019) ECTFE membrane fabrication via TIPS method using ATBC diluent for vacuum membrane distillation. *Desalination*. doi:10.1016/j.desal.2019.01.004

[53] Xia, L., Huang, L., Shu, X., Zhang, R., Dong, W., Hou, H. (2008) Removal of ammonia from gas streams with dielectric barrier discharge plasmas. *Journal of Hazardous Materials* 152: 113–119.

[54] Macedonio, F., Frappa, M., Bamaga, O., Abulkhair, H., Almatrafi, E., Albeirutty, M., Tocci, E., Drioli, E. (2022) Application of a membrane condenser system for ammonia recovery from humid waste gaseous streams at a minimum energy consumption. *Applied Water Science* 12: 90.

[55] Ito, A. (2000) Dehumidification of air by a hygroscopic liquid membrane supported on surface of a hydrophobic microporous membrane. *Journal of Membrane Science* 175: 35–42.

[56] Okampo, E.J., Nwulu, N. (2021) Optimisation of renewable energy powered reverse osmosis desalination systems: A state-of-the-art review. *Renewable and Sustainable Energy Review* 140. doi:10.1016/j.rser.2021.110712

[57] Tang, C.Y., Yang, Z., Guo, H., Wen, J.J., Nghiem, L.D., Cornelissen, E. (2018) Potable water reuse through advanced membrane technology. *Environmental Science & Technology* 52: 10215–10223.

[58] Semblante, G.U., Lee, J.Z., Lee, L.Y., Ong, S.L., Ng, H.Y. (2018) Brine pre-treatment technologies for zero liquid discharge systems. *Desalination* 441: 96–111.

[59] Husnain, T., Liu, Y., Riffat, R., Mi, B. (2015) Integration of forward osmosis and membrane distillation for sustainable wastewater reuse. *Separation and Purification Technology*. doi:10.1016/j.seppur.2015.10.031

[60] Frappa, M., Xue, L., Enrico, D., Francesca, M. (2020) Membrane crystallization and membrane condenser: Two membrane contactor applications. *Journal of Chemical Science and Chemical Engineering* 1: 7–17.

[61] Isetti, C., Nannei, E., Magrini, A. (1997) On the application of a membrane air—liquid contactor for air dehumidification. *Energy and Buildings* 25: 185–193.

5 Membranes in Gas Separation

May-Britt Hägg[1], Liyuan Deng[1], and Zhongde Dai[2*]*
[1]Department of Chemical Engineering, Norwegian University of Science and Technology (NTNU), Trondheim, Norway
[2]College of Carbon Neutral Future Technology, Sichuan University, Chengdu, China
[*]E-mail: liyuan.deng@ntnu.no; zhongde.dai@scu.edu.

5.1 INTRODUCTION

The application of membranes for gas separation is a relatively young technology compared to using membranes for liquid separation. Although the basic theoretical principles were partly understood dating back to the early 19th and 20th centuries with Fick's law (1855), osmotic pressure (Van t'Hoff, 1887 and Einstein 1905), and membrane equilibrium (Donnan 1911), it was not until around 1950 that theories for gas transport through a membrane were presented and later further developed (the pore model by Schmid in 1950 and Meares in 1956, and the solution–diffusion model by Lonsdale in 1965) [1]. The breakthrough for industrial membrane applications came with the asymmetric membranes developed by Loeb and Sourirajan around 1960 [2]. These membranes were developed for reverse osmosis and consisted of a very thin dense top layer (thickness< 0.5 μm) supported by a thicker, porous sublayer; hence, the flux, which is inversely proportional to the selective membrane thickness, could be dramatically increased. The work of Loeb and Surirajan resulted in the commercialization of the reverse osmosis process for desalting water and had also a major impact on the further development of ultrafiltration and microfiltration processes. The development of gas separation membranes is based on their achievements, and about 20 years later (~1980), the work of Henis and Tripodi made industrial gas separation economically feasible. They developed further the technique of depositing a very thin homogeneous layer of a highly gas-permeable polymer on top of a porous asymmetric membrane, ensuring that pores were filled so that a leak-free composite membrane for gas separation was obtained. The first major development was the Monsanto Prism® membrane for hydrogen recovery from a gas stream at a petrochemical plant [3]. Within a few years, the Dow Chemical Company was producing systems to separate nitrogen from air and Cynara NATCO Group and Separex™ UOP LLC systems for the separation of carbon dioxide (CO_2) from natural gas. These first membranes were all composite membranes where a very thin nonporous layer with a high gas permeation rate (usually polysulfone or cellulose acetate) was placed on a support structure for mechanical strength—later other techniques for membrane formation were developed also (i.e., interfacial polymerization, multilayer casting, and coating). The interest from industry in membrane gas separation has virtually exploded over the last 10–15 years, and the potential is enormous. The most attractive feature to the industry of membrane gas separation is the simplicity of the separation process. The conventional units for gas separation usually involve large towers for absorption or stripping, adsorption beds, cryogenic distillation, large compressors, and recovery and recycling of chemicals, all resulting in expensive and energy-demanding processes—not always without harmful effects to the environment. This does not mean that a membrane process will be more economical or as efficient as the traditional separation process. However, as the trend grows, the development of tailor-made membranes for specific gas applications will most likely continue to bring the technology into focus as an economical and environment-friendly alternative for gas separation. In theory, the limitations are few for membrane applications, but in practice, the challenges to succeed are numerous. For each application, process conditions must be taken into account (volume and composition of gas stream, pressures and temperatures, durability of the material), as the membrane separation properties may vary dramatically depending on these variables. This means that the focus also must be on utilities and pretreatment of the gas in order to evaluate the economics and performance of the membrane process. A hybrid process combining membranes and standard unit operations may be the best solution.

As a rule, the driving force of membrane gas separation is the difference in partial pressures (concentrations) between the feed side and permeate side. It is, however, more correct to say that the driving force is the difference in chemical potential thereby, including the effect of temperature. An additional driving force may be an electric potential or a carrier effect for certain types of membranes. New membrane materials may combine different transport mechanisms and thereby increase the flux and selectivities. A more detailed discussion of these issues is given in Section 5.2.

The need for optimized membrane separation properties for specific gas mixtures kicked off an explosive development with respect to tailor-made polymeric membrane materials in the mid-1990s. A search using "membranes in gas separation" as the keywords yields a large number of publications from the Web of Science, denoting that membranes for gas separation have attracted significant attention from researchers in the past few decades, and great progress has been made (Figure 5.1). Until then, the approach had been to look at existing polymeric

DOI: 10.1201/9781003285656-5

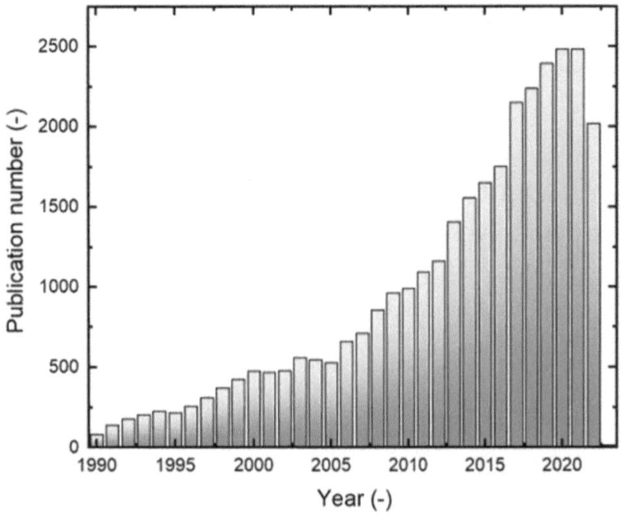

FIGURE 5.1 Publications found on the Web of Science on 22 November 2022.

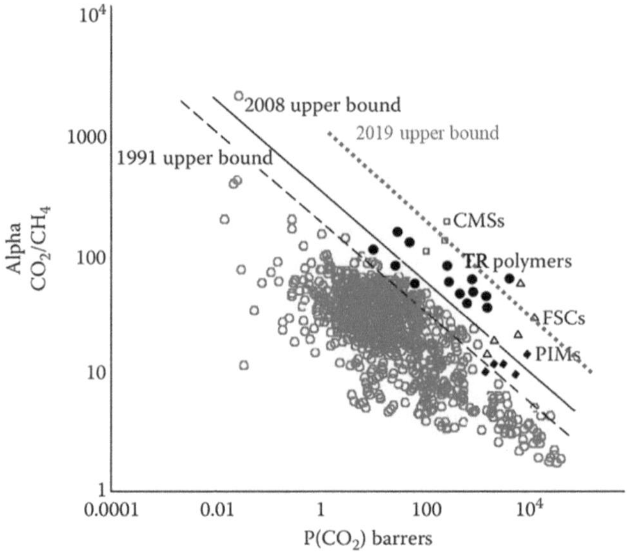

FIGURE 5.2 Selectivity for the gas pair CO_2/CH_4 as a function of CO_2 permeability and the Robeson plots (squares for CMS membranes, solid circles for TR polymers, triangles for FSC membranes, and solid diamonds for PIMs). (Robeson plots from Ref. [5] updated with literature data.)

materials and try to tailor separation properties by making moderate changes in the materials. This could be done by synthesizing families of polymers (for instance, polyimides [PIs], polycarbonates [PCs]) with different fractional free volume (FFV) and glass transition temperature (T_g), by using various methods for cross-linking or combining polymers like block copolymers.

In 1991, the now well-known Robeson plots for polymeric membranes were published [4], an upper performance limit (known as the Robeson upper bound or trade-off between gas permeability and selectivity), which later challenged membrane scientists all over the world. New types of membrane materials started to emerge, challenging the separation properties shown in this plot. In 2008, Robeson plots were revisited and the new membrane materials that emerged during these years had pushed the upper bounds significantly up and forward. The 1991 Robeson plot and its updated 2008 plot for CO_2/CH_4 separation are presented in Figure 5.2 [5]. Further up-shifts above the 2008 upper bound have been reported [6–8], such as the 2019 upper bound (included in Figure 5.2). A database of polymer gas separation membranes supported by the Membrane Society of Australasia (MSA) to provide real-time updates on membrane performances can be found online [9].

A group of highly gas permeable polymers, such as 1,2-substituted polyacetylenes (poly(1-trimethylsilyl1-propyne) [PTMSP], poly(4-metyl-2-pentyne) [PMP], poly(1-trimethylgermyl-1-propyne) [PTMGP]) and additive polynorbornenes, are particularly interesting for gas separation [10,11]. Their high free-volume structures contribute to the extremely high gas permeability. However, the open structure also brings in a very low selectivity for gas separation. The separation performance of this group of polymers is generally on the right side of the 2008 Robeson upper bound.

Some new classes of membrane materials for gas separation with separation performances above the 2008 upper bound are indicated in Figure 5.2 [12–17]: thermally rearranged (TR) polymeric membranes, polymers of intrinsic microporosity (PIM), fixed-site carrier (FSC) membranes, and carbon molecular sieve (CMS) membranes. These new materials have further elevated the Robeson plots. Thermally rearranged polymeric membranes were reported by Park et al. in 2007 [12], showing excellent gas separation performance and no detectable physical aging. The TR polymers with hourglass-shaped micropores are reported as specific gas separation membranes, for example, CO_2 separation, exhibiting molecular sieving characters with very high CO_2 permeability and selectivity [12,13]. TR polymers have rigid insoluble polymeric networks. However, they can be readily prepared into the desired membrane morphologies using soluble precursors. PIMs have received the most attention as CO_2 separation membranes since they were reported by Budd et al. in 2004 [16]. A number of PIM membranes have been published, and their excellent gas separation performance was claimed to lie above the 2008 Robeson upper bounds. The gas transport in PIMs follows the molecular sieving mechanism due to their high excess free volume (FFV, ~0.26), which enables a combination of the gas separation properties as molecular sieving inorganic membranes and the processibility as polymeric membranes [17,18]. The facilitated transport membrane is another type of membrane that can surmount the Robeson upper bounds. The concept of facilitated transport was first advocated by Scholander with the report of a supported liquid membrane (SLM) for oxygen transport in 1960 [19]. Those reported for CO_2 capture are of special interest, either as liquid membranes, supported liquid membranes (SLM), or FSC membranes [20,21]. As an approach to overcome the limitations of SLMs, FSC polymeric membranes were developed and widely studied, in which the functional groups are bound chemically (covalently) or physically to a solid polymer matrix as the CO_2-facilitated transport carriers, enabling better stability than SLMs. Among all reported FSC membrane materials, polyvinyl amine (PVAm) has the highest carriers content (over 30% amino groups), showing excellent CO_2 separation properties as well as long-term stability and durability in real gas streams [14,22–24]. Recently, Sandru et al. reported a membrane with integrated mechanisms by combining highly permeable polymers as substrates (Polydimethylsiloxane or AF 2400) for fast CO_2 transport and a grafted ultrathin amine-rich layer to enrich the concentration of CO_2 towards the surface—the separation performance is far above the upper bound [25].

The concept of carbon membranes was already known around 1970, but it was not until the late 1980s, with the published results of Koresh and Soffer [26], that these membranes caught general interest. Several scientists later reported impressive gas separation properties with CMS membranes pyrolyzed from different precursors [27–29]. In 1996, Singh and Koros presented a revised Robeson plot that included the potential of the CMS membranes [30]. Later development has shown that despite excellent separation properties, also at high temperatures, these membranes are brittle and expensive to process, so there is a need to find cheap precursors and secure mechanical strength in order to make them an economically good choice. Instead of the pure pyrolyzed carbon membranes, researchers are trying to combine the excellent molecule-sieving nano-channels of carbon membranes with the more robust polymeric matrices to make mixed-matrix membranes (MMMs), which are today in focus as a group of hybrid membranes with very high expectations. The fillers now include a broader number of highly porous, ordered nanostructured materials, such as various types of metal-organic frameworks (MOFs) [31–34]. In parallel with this development, polymeric functionalized membranes, nanocomposite materials with non-porous nanofillers, as well as new block copolymers are being reported with intriguing gas separation properties—this is elaborated on in Section 5.3.

Inorganic materials usually possess superior chemical and thermal stability compared to polymers. However, their use as gas membrane materials has been limited. Today, a growing interest is being observed and new materials are being developed for gas separation. Zeolite membranes have a very narrow pore size distribution and are usually prepared by sintering or sol–gel processes. Combined with organic surface treatment for pore tailoring, acceptable gas separation properties have been reported. A good overview of inorganic membranes for gas separation is given by Burggraaf [35].

Last but not least, special membranes being 100% selective for a specific gas component should be mentioned. One of the most interesting materials is the metal palladium-based membranes for the transport of hydrogen as protons through the membrane. Ceramic oxygen-conducting membranes producing high-purity oxygen have also been reported by several authors [35]. All these inorganic membranes are suitable for high-temperature applications. The separation properties of these materials are excellent, and a significant breakthrough may be seen in the near future. The challenge will, however, still be the module-making due to the brittleness and price of the materials.

All the materials mentioned above are discussed in this chapter. Many good review papers and chapters in books on membranes for gas applications have been published in recent years; only a few are referred to here [36–38]. A rich source of information on membrane materials and gas separation may also be found on the websites of membrane producers and research institutes [9,39].

5.2 TRANSPORT MECHANISMS FOR GAS THROUGH MEMBRANES

The most common types of membranes for gas separation in use today are still the dense polymeric materials where transport takes place according to a solution–diffusion mechanism with flux based on Fick's law (Eq. 5.1). The permeability of polymeric membranes in relation to a whole set of permanent gases, acid (toxic) gases, noble gases, and lower hydrocarbons can be found in Refs. [40,41]. For the microporous membranes (inorganic or hybrid), the transport mechanisms may be according to one of the following mechanisms or combinations of these: Knudsen diffusion, selective surface flow, or molecular sieving. The average pore size and pore size distribution are important since they will give an indication of which transport mechanism can be expected to be dominant for a given gas mixture in a defined material and at given process conditions.

Fick's law gives the mass flux through an area perpendicular to the flow direction:

$$J_i = -D_{ij}\frac{dc_i}{dx} \tag{5.1}$$

where J_i is the flux of component i [mol/(m² s)], D_{ij} is the diffusion coefficient [m²/s], and dc_i/dx is the concentration gradient for component i over the length x [mol/(m³·m)].

Fick's law integrated and applied for a membrane yields $dx = l$ (membrane thickness), and dc_i = concentration difference (i.e. partial pressures for gases) over the membrane. D_{ij} will vary according to the dominating transport mechanism.

The permeance P/l [mol/(m² Pa s)] (SI-units) for a given gas (i) is defined by:

$$\frac{P_i}{l} = \frac{J_i}{\Delta p_i} \tag{5.2}$$

P/l is also referred to as the permeability flux and expressed as [m³ (STP) /(m² bar h)] and often GPU. Δp_i is the partial pressure difference of "i" across the membrane [Pa] or [bar]. This equation shows that the flux through the membrane is proportional to the pressure difference across the membrane and inversely proportional to the membrane thickness. Table 5.1

TABLE 5.1
Conversion of common gas membrane permeance units

	GPU	10^{-7}cm³ (STP) cm⁻² s⁻¹ kPa⁻¹	10^{-10} mol m⁻² s⁻¹ Pa⁻¹	10^{-3} m³(STP) m⁻² h⁻¹ bar⁻¹
GPU*	1	7.50	3.35	2.70
10^{-7} cm³(STP) cm⁻² s⁻¹ kPa⁻¹	0.133	1	0.447	0.360
10^{-10} mol m⁻² s⁻¹ Pa⁻¹	0.299	2.24	1	0.806
10^{-3} m³(STP) m⁻² h⁻¹ bar⁻¹	0.365	2.78	1.24	1

* 1 GPU= 1×10^{-6} cm³(STP) cm⁻² s⁻¹ cmHg⁻¹.

summarizes the conversion of permeance units for a better comparison of the membrane performance from various sources.

For selectivity between gas components, Equations 5.3 and 5.4 are referred to. The "ideal" selectivity, α*, may be expressed by the ratio of the pure gas permeabilities for the individual components *i* and *j*

$$\alpha_{ij}^* = \frac{P_i}{P_j} \tag{5.3}$$

The separation factor for gases in the mixture α_{ij} (Eq. 5.4) is expressed by the mole fractions of the components in the feed (*x*) and the permeate (*y*), respectively:

$$\alpha_{ij} = \frac{y_{i/}y_j}{x_{i/}x_j} \tag{5.4}$$

The permeability, *P*, can be expressed as the product of diffusion (*D*) and solubility (*S*) of the gas through the membrane (Eq. 5.5):

$$P = D \cdot S \tag{5.5}$$

Depending on the type of gas and type of membrane material, the importance of these to the variables, *D* and *S*, will vary. Both the diffusion and the solubility coefficient for the gas are temperature dependent, while a pressure dependency is only observed for certain gases and materials.

Dense inorganic or metallic membranes for gas separation are usually ion-conducting materials, while membranes with carriers are polymers or SLM. For transport through these materials, different flux equations should be applied. Figure 5.3 sums up and generalizes the various types of transport which may take place in gas separation membranes [42]. The properties of the gases of interest are summarized in Table 5.2 [43].

5.2.1 Solution Diffusion

The transport of gas (permeability) through a dense, polymeric membrane can be described in terms of a solution–diffusion mechanism with permeability expressed as in Eq. 5.5.

TABLE 5.2
Kinetic diameter (d_k), critical volume (V_c), critical temperature (T_c), and polarizability (α) for the relevant gases [43]

Gas	d_k (Å)	V_c (cm³/mol)	T_c (K)	$\alpha \times 10^{24}$ (cm³) [27]
He	2.6	57.4	5.2	0.21
H_2	2.89	65.1	33.2	0.80
CO_2	3.3	93.9	304.2	2.91
N_2	3.64	89.8	126.2	1.74
CH_4	3.8	99.2	191.0	2.59
C_2H_4	4.16 [39]	130.4	282.4	4.25
C_2H_6	4.44 [39]	148.3	305.4	4.47
C_3H_6	4.68 [40]	181.0	364.9	6.26
C_3H_8	5.06 [40]	203.0	369.8	6.29

The thermodynamic parameter, *S*, gives the pressure normalized amount of gas sorbed in the membrane under equilibrium conditions. This parameter is usually very low for gases in polymers but will vary depending on the physical properties of the gas (ideal, nonideal) and the state of the polymer (glassy or rubbery). The state of the polymer is characterized by the material's glass transition temperature, T_g. Above T_g the polymer is in its rubbery state, while below T_g it is glassy. In glassy polymers, the segmental motions of the chains are more restricted, and these materials are, therefore, able to discriminate more effectively between small differences in molecular dimensions. They exhibit an enhanced mobility selectivity compared to rubbery polymers. Table 5.3 shows a comparison of permeabilities and selectivities for both a rubbery (PDMS) and a glassy (PC) polymer. Crystallinity in a polymer will also restrict gas transport [44].

For ideal systems (usually as in elastomers), the solubility is independent of concentration and the sorption curve follows Henry's law (Eq. 5.6); i.e., gas concentration within the polymer is proportional to the applied pressure. For non-ideal systems (usually as in glassy polymers), the sorption isotherm is generally curved and highly non-linear. Such behavior can be described by free volume models and Flory–Huggins thermodynamics—comprehensive discussions on this may be found elsewhere [1,45].

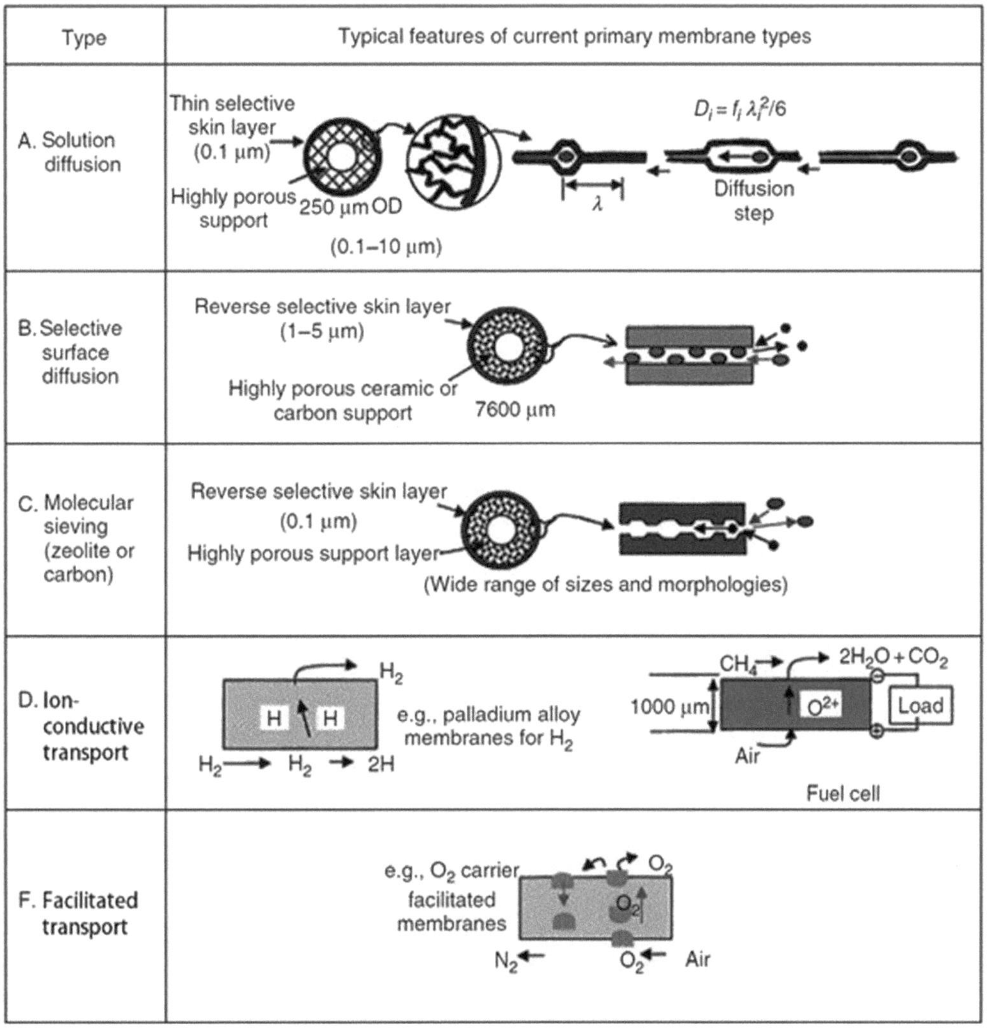

FIGURE 5.3 Illustration of transport mechanisms in gas separation membranes. (Adapted from Ref. [42].)

$$c_i = S_i p \tag{5.6}$$

The diffusivity, D (see Eq. 5.5), is a kinetic parameter and is dependent on the geometry of the polymer as well as its state (e.g., glassy, rubbery, swollen). A small molecule will more easily diffuse through a polymer than a larger one. However, large (organic or non-ideal) molecules may have the ability to swell the polymer, hence large diffusion coefficients result and "reverse selectivity" may be observed. It is quite clear that the interdependency of molecular size, ideal or non-ideal gases in mixtures, structure and state of the polymer, must be carefully evaluated in order to fully understand the transport through polymers. This fundamental understanding will also govern how a membrane material may best be tailored for a given separation. For illustration of the complexity, Figure 5.4 shows the diffusion and sorption coefficients for some gases through natural rubber [46].

Transport through a dense polymer may be considered as an activated process which can be represented by an Arrhenius type of equation. This implies that temperature may have a large effect on the transport rate. Equations 5.7 and 5.8 express the temperature dependence of the diffusion coefficient and solubility coefficient in Eq. 2.5:

$$D = D_0 \exp\left(-E_d / RT\right) \tag{5.7}$$

$$S = S_0 \exp\left(-\Delta H_s / RT\right) \tag{5.8}$$

where E_d is the activation energy for diffusion, ΔH_s is the heat of the solution, and D_0 and S_0 are temperature-independent constants.

By inserting Henry's law (Eq. 5.6) into Fick's law (Eq. 5.1), integrating across the membrane and remembering the

TABLE 5.3
Permeabilities (P) and selectivities (α) of various gas pairs in silicone rubber (PDMS) and polycarbonate (PC)

Polymer	T (°C)	P_{He} (Barrer)	α_{He/CH_4}	α_{He/C_2H_4}	P_{CO2} (Barrer)	α_{CO_2/CH_4}	α_{CO_2/C_2H_4}	P_{O2} (Barrer)	α_{O_2/N_2}
PDMS	35	561	0.41	0.15	4550	3.37	1.19	933	2.12
PC	35	14	50	33.7	6.5	23.2	14.6	1.48	5.12

[1] Barrer = 7.52 x 10^{-15} m^3(STP) m/m^2 s kPa.
Sources: [40, 41, 44, 48].

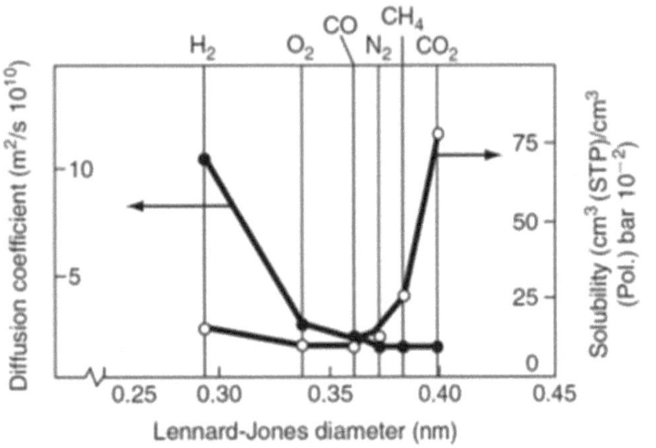

FIGURE 5.4 Solubility and diffusivity of various gases in natural rubber. (From Ref. [46].)

definition of the permeability coefficient (Eq. 5.5), Equation 5.2 was developed as the standard equation for transport through a dense polymeric membrane.

For gases in a mixture, Eq. 5.2 may be detailed out as in Eq. 5.9:

$$J_i = \frac{P_i}{l}\left(p_h x_{i0} - p_l y_p\right) \tag{5.9}$$

where p_h and p_l denote pressure on the feed side and permeate side, respectively. The fraction x_{i0} is the fraction of gas i on the feed side, and depending on the design of the membrane module and flow regime for the gas, different calculation methods are adapted for this variable—details on this may be found elsewhere [1,47]. Parameters influencing the separation efficiency of polymers are elaborated on in Section 5.3.1.

5.2.2 Knudsen Diffusion

Knudsen diffusion may take place in a microporous inorganic membrane or through pinholes in dense polymeric membranes. It may also take place in a mixed-matrix membrane with insufficient adhesion between the phases.

Knudsen flow is characterized by the mean free path (λ) of the molecules being larger than the pore size. Hence collisions between the molecules and the pore walls are more frequent than intermolecular collisions. A lower limit for the significance of the Knudsen mechanism has usually been set at $d_p > 20$ Å [48]. The classical Knudsen equation for diffusion of gas is:

$$D_{Kn} = \frac{d_p}{3} v_A = \frac{d_p}{3}\sqrt{\frac{8RT}{\pi M_A}} = 48.5 \cdot d_p \sqrt{\frac{T}{M_A}} \tag{5.10}$$

where d_p = average pore diameter [m], α_{He/CH_4} = average molecular velocity [m/s], M_A = molecular weight of gas component A [g/mol], and T = temperature [K].

Hence, for Knudsen diffusion, the square root of the inverse ratio of the molecular weights will give the selectivity. However, Gilron and Soffer [49] indicate that the Knudsen mechanism can be significant for pore sizes as small as d_p ~5Å. The Knudsen flow in this region takes on a slightly different form as indicated in the following expression derived as transport through a series of constrictions using resistance in series model:

$$D_{act,Kn} = g_d d_p \sqrt{\frac{8RT}{\pi M_A}} \exp\left(-\frac{\Delta E}{RT}\right) \tag{5.11}$$

Here g_d is the probability that a molecule can make a jump in the right direction given the jump length is d_p and the velocity is α_{He/CH_4}.

5.2.3 Selective Surface Flow

Selective surface flow is, as Knudsen diffusion, associated with transport through microporous membranes, usually inorganic materials.

The mechanism of surface diffusion is disputed and several different approaches have been proposed in the literature. Theories ranging from viewing the low surface coverage adsorbed gas as a 2D gas model, through a hopping model and into a more liquid-like sliding layer, exist. The mechanism dominating the surface diffusion coefficient will be influenced by a number of factors such as homogeneity of the surface, the temperature vs. the adsorption enthalpy, and the surface concentration, c_s [50]. All three regimes can be described by a 2D analog of Fick's law (Eq. 5.1, given for a single component, a). The flux, J_a, is now evaluated as molecules crossing a hypothetical line in the surface perpendicular to the direction *x*. D_s is the surface diffusion coefficient and dc_s/dx is the surface concentration gradient in the x-direction. The following expression may be used to determine if the surface transport is dominated by the 2D-gas model [50]:

$$\frac{q}{RT} < \frac{1}{a} \tag{5.12}$$

where q is the adsorption enthalpy [J/mole] and a is an energy fraction factor. The energy barrier for surface migration, E, is then defined as:

$$E = aq \tag{5.13}$$

The 2D-gas model is characterized by a surface mean free path, λ_s, inversely proportional to the surface concentration, c_s, and this λ_s value can be much larger than the spacing between adjacent surface sites.

If the q/RT part of Eq. 5.12 is increased, then λ_s will no longer be controlled by collisions between adsorbed molecules. As q/RT increases, λ_s decreases and is approaching the spacing between adjacent sites, and a hopping mechanism is observed.

If c_s is low, then a random walk diffusion of independent molecules can be expected, and D_s would be given as:

$$D_s = \frac{1}{4} \nu \lambda_s^2 \tag{5.14}$$

where ν is a jump frequency factor, a factor which has a temperature dependence according to Arrhenius' law, $\nu = \nu_0 \cdot \exp(-aq/(RT))$ [1/s].

When c_s increases, the chance of a molecule hitting another molecule increases, and this interaction will bear some similarity to diffusion in liquid. Thus, the region of the sliding layer prevails. A more comprehensive discussion of this theory may also be found in Ref. [51].

Selective surface diffusion is governed by selective adsorption of the larger (non-ideal) components on the pore surface. The critical temperature, T_c, of a gas will thus indicate which component in a mixture is more easily condensable. The gas with the highest T_c will most likely be the fastest permeating component where a selective surface flow can take place. For a mixed gas, an additional increase in selectivity may be achieved if the adsorbed layer now covering the internal pore walls restricts the free pore entrance so that the smaller non-adsorbed molecules cannot pass through.

5.2.4 Molecular Sieving

Molecular sieving is the dominating transport mechanism when the pore size is comparable to the molecular dimensions, 3–5 Å, hence the smallest molecule will permeate, and the larger is retained. The dimensions of a molecule are usually described with either the Lennard–Jones radii or the van der Waal's radii. For separation by molecular sieving, this is not a satisfactory way of stating the molecular size; a shape factor should also be included [52].

The sorption selectivity has little influence on the separation when molecular sieving is considered. An Arrhenius type of equation is still valid for the activated transport, but attention should be drawn to the pre-exponential term, D_0 (see Eq. 5.7). From the transition state theory, this factor may be expressed as shown in Eq. 5.15 [53]:

$$D_0 = e\lambda^2 \frac{kT}{h} \exp\left(\frac{S_{a,d}}{R}\right) \tag{5.15}$$

where k and h are Boltzmann's and Planck's constants, respectively, $S_{a,d}$ is the activation entropy for diffusion, and $e = g_d \cdot d_p$ in Eq. 5.15. A change in entropy will thus have a significant effect on the selectivity when molecular sieving is considered. This is thoroughly discussed by Singh and Koros [30]. The flux may be described as in Equation 5.16, where $E_{a,MS}$ is the activation energy for diffusion in the molecular sieving media.

$$J_a = \frac{\Delta p}{RT \cdot l} D_0 \cdot \exp\left(\frac{-E_{a,MS}}{RT}\right) \tag{5.16}$$

The selectivity for separation will normally decrease with increasing temperature because of the increased diffusion rate for permeating components, and the sorption will be of minor importance.

5.2.5 Ion Conductive Transport

There are two important types of ion-conducting membranes for gas separation: (1) the proton (H^+) conducting palladium membranes which are of great interest in combination with fuel cells, and (2) the oxygen ion (O^{2-}) conducting inorganic membranes, usually perovskite-type of oxides. Both are suitable for high-temperature/high-pressure applications, and an interesting feature is the 100% selectivity in favor of H_2 and O_2, respectively.

5.2.5.1 Proton-Conducting Membranes

Palladium and its alloys are recognized as very efficient proton-conducting membranes, which may be used for hydrogen separation and membrane reactor applications. The alloys are less susceptible to hydrogen embrittlement than pure Pd, and alloys with silver or copper represent the least expensive alternative alloys. These alloys also seem to produce membranes with enhanced chemical resistance (for instance, toward H_2S). Additionally, Pd-Ag alloys have a relatively higher H_2 permeability than pure Pd. Efforts to produce economically viable Pd membranes have focused on preparing supported composite membranes with a thin dense

Pd or Pd alloy layer. Forming this thin layer from two or more metals is quite challenging. The advantages of palladium membranes are especially the ability to separate out high-purity H_2, and that they may be used at high temperatures (300°C and above).

Hydrogen is present in many gas streams, a product of hydrocarbons' dehydrogenation, a component in syngas, or a byproduct in bioprocesses. The basic flux equation for hydrogen, J_{H2} (mol/m²s), is given in Eq. 5.17 [54]. The flux for hydrogen atoms will be twice that of J_{H2}:

$$J_{H_2} = -\frac{D_M}{2} \cdot \frac{K_S\left(p^n_{H2,ret} - p^n_{H2,perm}\right)}{l} \quad (5.17)$$

where D_M is the diffusion coefficient of a hydrogen atom in the metal (m²/s) and K_S is the Sievert constant (mole/(m³Pa$^{0.5}$)). For bulk transport of hydrogen, $n = 0.5$ but approaches 1 for transport limited by surface kinetics. The exponent of 0.5 reflects the dissociation of the gaseous hydrogen molecule into two hydrogen atoms (protons) diffusing into the metal where an ideal solution of hydrogen in palladium is formed; then association again as H_2 on the other side of the membrane. The hydrogen permeability of the palladium, here denoted as k, corresponds to the constants in Eq. 5.17, expressed as in Eq. 5.18:

$$k = \frac{1}{2} D_M K_S \text{ (mole/m s Pa0.5)} \quad (5.18)$$

Among the proton-conducting membranes, Nafion or Nafion-like sulfonated perfluorinated polymers should also be mentioned. These materials are used for polymer electrolyte membrane (PEM) fuel cells, and in addition to being chemically very stable, they exhibit high proton conductivity at temperatures lower than 100°C. It is believed that permeability and thermal stability may be increased if tailor-made lamellar nanoparticles are added to a proton-conducting polymer. The sulfonated poly ether ether ketones (S-PEEK) type of polymers is also widely reported as an alternative to fluorinated polymers such as Nafion or Hyflon [55].

5.2.5.2 Oxygen-Conducting Membranes

The zirconia and perovskite membranes may be considered solid electrolyte membranes containing an oxygen ion conductor (various oxides). Depending on the type of materials used, the oxygen separation may take place according to direct excitation of the oxygen at several hundred degrees (gas separation controlled by electric current) or a mixed conductor method where the gas separation is proportional to log (p_1/p_2); the ratio of the partial pressures. (Reactions taking place at the electrodes are: ½ $O_2 + 2e^- \leftrightarrow O^{2-}$; at the positive electrode, the reaction is shifted to the left, at the negative electrode, it is shifted to the right.)

Detailed equations for transport can be found elsewhere [56,57].

5.2.6 Facilitated Transport

Facilitated transport indicates that a carrier is introduced into the membrane matrix, such as a polymer matrix. This carrier will be selective for a certain gas component and enhance the transport of this component through the membrane.

The use of facilitated transport membranes for gas separation was first introduced by Ward and Robb [58] by impregnating the pores of microporous support with a carrier solution, and a separation factor of 1500 was reported for CO_2/O_2. These membranes, or SLM, are discussed by several, and initially, very good separation properties are observed [59–62]. They are, however, known to have serious degradation problems such as loss of carrier solution due to evaporation or entrainment with the gas stream, and the complexing agent (carrier) can be deactivated. These problems have restricted the further development of SLMs. The use of ion exchange membranes as supports was proposed as an approach to overcome the problems of SLMs, and the application of ion exchange membranes for the facilitated transport of CO_2 and C_2H_4 was first reported by LeBlanc et al. [63]. Since then, a number of papers have been published on this type of membrane. Along with the use of ion exchange membranes as supports, another approach to overcome the above-mentioned limitations was developed by introducing carriers directly into solid polymer membranes, as illustrated in Refs. [64,65]. The fixed-site-carrier membranes have carriers covalently bonded to the polymer backbone. Hence the carriers have restricted mobility, but in this way, the membranes are more stable compared to mobile carrier membranes. It is obvious that the diffusivity (and thus permeability) in an FSC membrane is lower than that of a mobile carrier membrane. The diffusivity of a swollen FSC membrane should, however, show diffusivities between that of a mobile and a fixed carrier [66]. Various ways of enhancing carrier transport have later been suggested [67,68]. It is suggested by many that CO_2 will be transported as carbonate or bicarbonate anions in anion exchange membranes and as anions of various amines in cation exchange membranes.

The characteristic of a facilitated or carrier-mediated transport is the occurrence of a reversible chemical reaction or complexation process in combination with a diffusion process. This implies that either the diffusion or the reaction is rate-limiting: the total flux of a permeant A will thus be the sum of both the Fickian diffusion and the carrier-mediated diffusion, as illustrated in Eq. 5.19 [69]:

$$J_A = \frac{D_A}{l}\left(c_{A,0} - c_{A,l}\right) + \frac{D_{AC}}{l}\left(c_{AC,0} - c_{AC,l}\right) \quad (5.19)$$

where the first term on the right-hand side of the equation is the Fickian diffusion (D_A), and the second term represents the carrier-mediated diffusion (D_{AC}). l is the thickness of the membrane, while c (as defined by Henry's law, Eq. 5.6) is the concentration of component A and its complex AC at the interfaces of the membrane, and 0 and l indicate the feed side

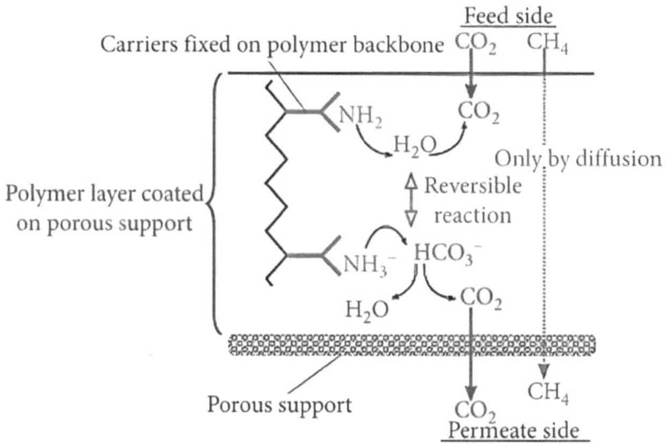

FIGURE 5.5 A proposed mechanism of facilitated transport for CO_2 in a fixed-site-carrier membrane. (From Ref. [21].)

and permeate side, respectively. The concentration difference of the complex AC in Eq. (5.19) must be further expressed by an equilibrium constant of the complexing reaction and a distribution coefficient. This is given in detail by Cussler [70]. Non-polar gases in a gas mixture will exclusively be transported through the membrane by Fickian diffusion, while by using partial pressures (p_A) instead of concentration c_A (inserting Eq. (5.6) into Eq. (5.19)), it can easily be seen that the driving force through the membrane is the difference in partial pressures for the Fickian diffusion and that transport also depends on the solubility coefficient, S_A, for the gas in the polymer. For carrier-mediated transport (second term in Eq. 5.19), the driving force is the concentration difference of the complex AC through the membrane. The permeation of the non-polar gases may additionally be hindered by polar sites introduced into the membrane matrix [21,67,68]. This should then lead to increased permeance of the carrier-transported gas compared to ideal gases in the mixture (like CH_4, N_2, and O_2), giving high selectivities in favor of the complexed gas (e.g., CO_2). A proposed mechanism for the facilitated transport of CO_2 in a fixed-site-carrier membrane is illustrated in Figure 5.5 [21].

5.3 MEMBRANE MATERIALS USED FOR GAS SEPARATION

The properties of various materials available for gas separation membranes are discussed in this section. Selecting the right membrane for a given gas separation is very challenging as the criteria are quite complex. The first choice is usually based on favorable flux and selectivity for a given gas mixture. Membrane performance will, however, have to be evaluated with respect to operating conditions as well as mechanical strength and durability. Finally, separation efficiency should be balanced against the cost for each case evaluated. The choice of "the right membrane" may therefore have more than one answer.

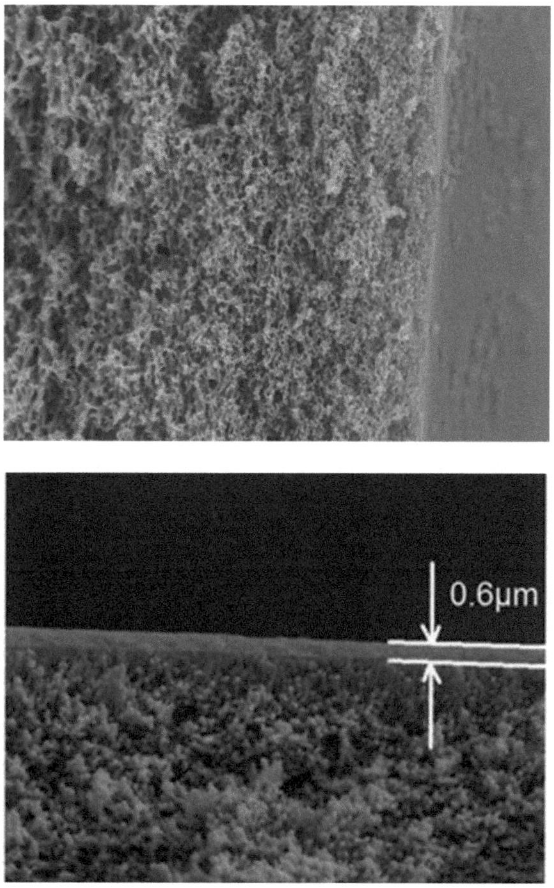

FIGURE 5.6 SEM images of (a) an asymmetric polysulfone membrane and (b) a composite membrane comprising of polysulfone porous support structure and a selective layer of PVAm.

As gas separation membranes must have sufficient flux to make it possible for their use in industrial applications, membrane materials are usually made into an asymmetric structure with an integrated dense skin layer on the porous substrate (as seen in Figure 5.6(a), a polysulfone membrane) or a composite structure consisting of a dense, thin coating layer of a selective material on a porous substrate of a different material (as shown in Figure 5.6(b), a PVAm selective layer on polysulfone porous support). It is worth noting that when membrane materials are discussed in this section, the reported data are either intrinsic material properties (i.e., ideal selectivity and permeability) obtained from single gas permeation testing of thick membranes or membrane properties (i.e., selectivity/separation factor and permeance) obtained from mixed gas permeation testing of composite membranes. It isn't easy to calculate the permeability of individual layers in a composite membrane.

5.3.1 Polymeric Membranes

Polymeric materials remain the most widely used membranes for gas separation, and for specific applications, the separation technology is well established (see Section 5.5). The polymeric membrane development today is clearly into more carefully tailored membranes for specific applications. The important material properties defining the separation performance are molecular structure, glass transition temperature (T_g), crystallinity, degree of cross-linking, and as a function of these variables; durability with respect to possible degradation or loss of performance.

The basic transport mechanism through a polymeric membrane is the solution diffusion as explained in Section 5.2.1. As noted, there is a fundamental difference in the sorption process of a rubbery polymer and a glassy polymer. Whereas sorption in a rubbery polymer follows Henry's law and is similar to penetrant sorption in low-molecular-weight liquids, the sorption in glassy polymers may be described by complex sorption isotherms related to unrelaxed volume locked into these materials when they are quenched below the glass transition temperature, *Tg*. The various sorption isotherms are illustrated in Figure 5.7 [71].

The solubility in glassy polymers is usually described by the so-called dual-mode model, which implies that there is a need for a more detailed definition of the sorption, *c*, in the flux Eq. 7.1. Equations 7.20 and 7.21 illustrate this and can be referred to Figure 5.7. The term c_D accounts for Henry's law, while c_H is the Langmuir term with *b* being the hole affinity constant (bar^{-1}) and $c`_{Hb}$ is the saturation constant (cm^3(STP)/cm^3):

$$c = c_D + c_H \quad (5.20)$$

$$c = k_D p + \frac{c'_H bp}{1+bp} \quad (5.21)$$

The dual-mode model has been extensively covered by several authors [71–73].

Figure 5.8 illustrates how the available free volume for transport increases with increasing temperature ($V_f = V_T - V_0$), and the remarkable change when passing the T_g of a polymer. According to the free volume diffusion model, the diffusion of molecules depends on the available free volume as well as sufficient energy to overcome polymer–polymer attractive forces. The specific volume at a particular temperature (V_T) can be obtained from the polymer density, whereas the volume occupied at 0 Kelvin (V_0) can be estimated from group contributions. Details on this theory may be found in relevant handbooks, textbooks, and numerous publications [72–75].

Non-ideal gases dissolve more easily in polymers; hence the separation factor may easily be in favor of a larger, non-ideal gas component compared to a small ideal gas. At the same time, the non-ideal component may swell the membrane; hence the net result is a decrease in selectivity. For a polymeric membrane, flux and selectivity are inversely related; thus, a high flux usually means low selectivity. Elastomers have higher flux and lower selectivity for a given gas pair than a glassy polymer. This problem can be addressed by various methods: controlled cross-linking, opening the matrix by inserting carefully designed side groups into the main polymer chain, and/or functionalizing the polymer.

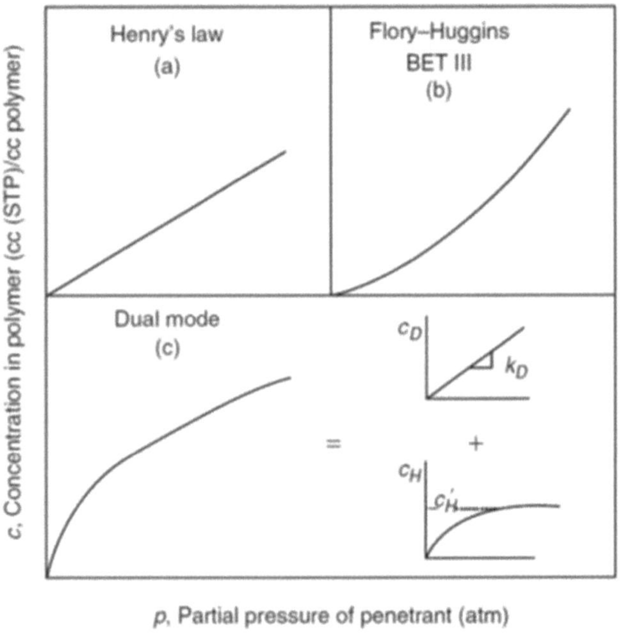

FIGURE 5.7 Typical gas sorption isotherms for polymers: (a) Henry's law illustrating ideal sorption as in a rubbery polymer where solubility is independent of concentration, (b) illustrating a highly non-linear behavior according to Flory-Huggins, as can be expected for interactions between organic liquids or liquids with polymers (swelling results), (c) illustrating the dual-mode sorption (Langmuir) typical for a glassy polymer. (From Ref. [71].)

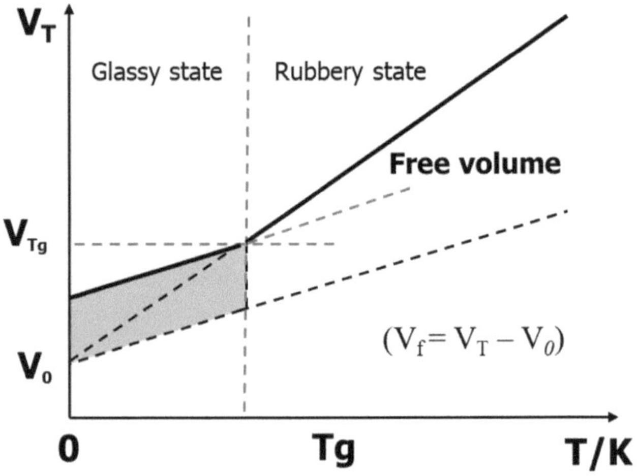

FIGURE 5.8 Polymeric specific volume as a function of temperature.

Tables 5.4 and 5.5 illustrate how separation properties may be changed within two families of polymers by the change of side groups; polycarbonates [42,76,77] and polyimides [78].

Glassy polymers may swell in the presence of plasticizing agents, hence selectivities will be reduced and flux will be increased—the membrane loses performance. One way of avoiding this problem may be to incorporate cross-linkable functional groups in the polymer backbone. The cross-linking

TABLE 5.4
Structures of representative polycarbonates[a] and polyimides[b]

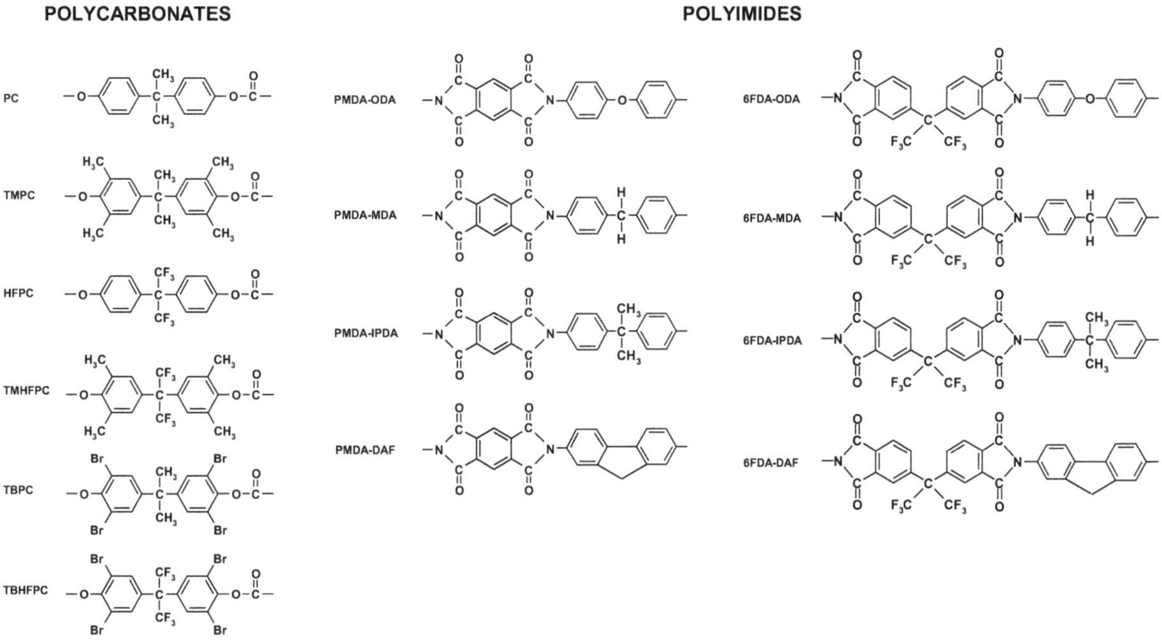

Sources: (a) Refs. [42,72,76,77]; (b) Ref. [78].

will be performed in the post-treatment process and will hinder uncontrolled swelling. Regions of crystallinity and chemical cross-linking have somewhat similar effects on the transport properties of rubber by causing restrictions of swelling and suppressing long-chain segmental motion. Transport in semicrystalline polymers is more complex due to tortuosity caused by the presence of the typically impermeable crystalline regions. Studies of gas sorption and transport strongly support the notion of the impenetrability of crystalline domains by even tiny gas molecules. The sorption coefficient seems to be essentially proportional to the volume fraction of amorphous material, while the effect on diffusion is more complex.

5.3.1.1 Robeson Upper Bounds

Robeson first revealed a general trade-off relation between gas permeability and selectivity (in log-log plots) of polymeric membranes in 1991 and updated the same in 2008 [4,5]. The more permeable polymers are found to be generally less selective, and vice versa, as shown in Figure 5.2. A line, the "upper bound" for the permeability and selectivity of known polymer membrane materials for a particular gas pair, is defined based on an empirical relationship using published permeability and selectivity data for the gas pair, which can be described by Eq. 5.22 [79].

$$\alpha_{A/B} = \beta_{A/B} / P_A^{\lambda_{A/B}} \tag{5.22}$$

Robeson reported values for slopes and intercepts ($\lambda_{A/B}$ and $\beta_{A/B}$) of upper bounds for many common gas pairs. According to Eq. 5.3, when the permeability of a polymer at the upper bound to gas A (P_A,) increases, its selectivity over gas B ($\alpha_{A/B}$) will decrease. Later, a theoretical model that fits permeability/selectivity trade-off for all gas pairs was proposed by Freeman [79], indicating that the slope of the upper-bound for a gas pair, $\lambda_{A/B}$, depends on the ratio of gas molecule diameters of the gas pair, $\lambda_{A/B} = (d_B/d_A)^2 - 1$, where d_B and d_A are the kinetic diameters of the larger and smaller gases, respectively. The front factor, $\beta_{A/B}$, depends on gas properties (e.g., $\lambda_{A/B}$), gas–polymer interaction (solubility, diffusivity), and polymer properties (e.g., free volume). When comparing the 1991 and 2008 upper-bounds, for example, the CO_2/CH_4 gas pair as shown in Figure 5.2, a modest upward shifts can be observed. These shifts are in agreement with the upper-bound model: The slopes of the upper bound, $\lambda_{A/B}$, cannot change, while the interception position of the upper bound, $\beta_{A/B}$, may move up. Thus, efforts to move up $\beta_{A/B}$ have been through improving gas–polymer interaction and membrane matrix properties, for example, synthesizing high free-volume membranes (PIMs, TR) and perfluoropolymers, introducing facilitated transport, surface modification, developing hybrid membranes (e.g., nanocomposite, mixed-matrix membranes), and phase-separated block copolymers or polymer blends.

For gas separation, however, high permeability is only one component. For industrial applications, membrane

TABLE 5.5
Permeabilities and selectivities of polycarbonates[a] and polyimides[b]

	Permeabilities at 35°C (Barrer)			Ideal selectivities at 35°C		
	He	O_2	CO_2	He/CH_4	O_2/N_2	CO_2/CH_4
Polymer	10 atm	2 atm	10 atm	10 atm	2 atm	10 atm
			Polycarbonates			
PC	13	1.6	6.8	35	4.8	19
TMPC	46	5.6	18.6	50	5.1	21
HFPC	60	6.9	24	57	4.1	23
TMHFPC	206	32	111	44	4.1	24
TBPC	18	1.4	4.2	140	7.5	34
TBHFPC	100	9.7	32	112	5.4	36
TB/TBHF-co-PC	49	4.9	16	110	6.2	34
			Polyimides			
PMDA-ODA	8.0	0.61	2.71	134.9	6.1	45.9
PMDA-MDA	9.4	0.98	4.03	94	4.9	42.9
PMDA-IPDA	37.1	7.1	26.8	41.1	4.7	29.7
PMDA-DAF	1.9	–	0.15	921	–	71.6
6FDA-ODA	51.5	4.34	23	135.4	5.2	60.5
6FDA-MDA	50	4.6	19.3	117.1	5.7	44.9
6FDA-IPDA	71.2	7.53	30	102.1	5.6	42.9
6FDA-DAF	98.5	7.85	32.2	156.3	6.2	51.1

[1] Barrer = 7.52 × 10^{-15} m^3(STP) m m^{-2} s^{-1} kPa^{-1}.
Sources: (a) Refs. [42,76,77; (b) Refs. [42,78].

performance is assessed by its gas permeation rate, i.e., permeance, and its ability to separate the gas mixture of interest. It is thus critical to coat high-permeability materials as thin membranes (e.g., <100 nm) on porous substrates. However, it is challenging to fabricate thin, defect-free membranes on a large scale. In 2017, Park et al. [80] studied the effect of membrane support on separation characteristics of gas separation membranes and extended the upper bound to examine the permeance–selectivity relationship. They "translated" the upper-bound data for CO_2/N_2 separation into selectivity and permeance (i.e., pressure-normalized flux) by assuming a thin (100 nm) membrane onto a slow (10^3 GPU), medium (10^4 GPU), or fast (10^5 GPU) porous support membrane, showing that a high flux support is needed to reach the desirable performance in a composite membrane.

5.3.1.2 Polymers Receiving Special Interest

The block copolymers form an interesting group of materials with promising separation properties for selected gas mixtures. These membranes usually combine the flexible phase of an elastomer with a dispersed phase of a glassy or crystalline polymer. The hard domains will act as physical cross-links, and the temperature should not be raised above the T_g of the glassy polymer. The morphology and properties of block-copolymers are mainly determined by the size and ratio of the blocks. The separation mechanism in these membranes is typically based on sorption–diffusion, but if correctly tailored, the sorption selectivity will govern the separation. Examples that can be mentioned are copolymers of ethylene and vinyl alcohol. These two components are able to co-crystallize in the same crystal lattice, and the material can be tailored by varying the amount of the highly polar, diffusion-inhibiting vinyl alcohol component without strongly affecting the crystallinity. Another example reported is polyether block amide or PEBA, well known under the tradename of PEBAX®, block copolymers composed of polyamide (PA) and polyethylene glycol (PEG). In this material, the semicrystalline PA blocks will ensure structural integrity while high-molecular-weight PEG will control the separation [81]. The permselectivity will typically be reversed, and these membranes may have great potential for CO_2 capture or VOC removal from gas streams. Interesting results for CO_2–H_2 separation (reverse selective) have been published [82,83]. Pebax is also used as the polymeric matrix to study the effects of different fillers in hybrid membranes [32,84].

Nexar and Nafion are two other examples of block copolymers that have been studied for gas separation. Initially, both Nexar and Nafion membranes are considered poorly gas permeable. However, by introducing CO_2-philic agents into the polymeric matrix or using non-solvent to induce proper micro-phase separation, both Nafion and Nexar demonstrated promising results in CO_2/N_2 separation, especially under humid conditions [85–87].

Perfluorinated polymers are also materials of special interest due to their exceptionally good chemical and thermal stability compared to other polymers. The challenge has, however, for many years been to prepare these materials with suitable gas

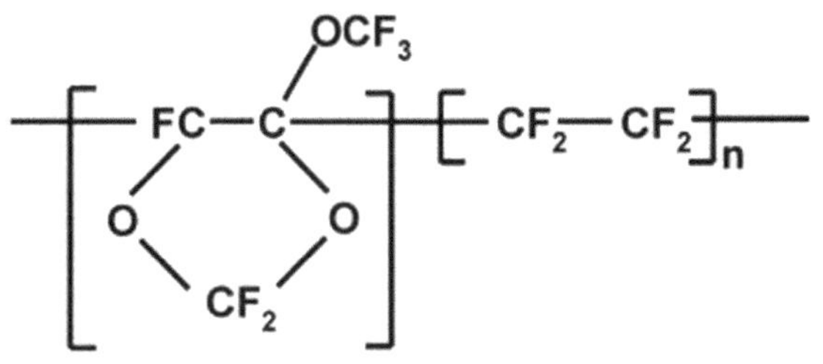

FIGURE 5.9 Structure of Hyflon AD.

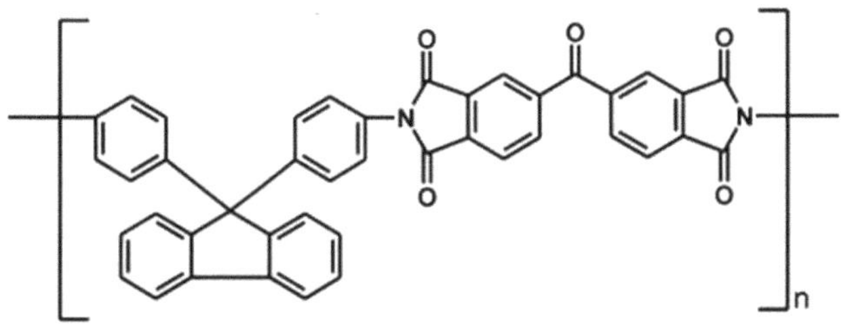

FIGURE 5.10 Example of the structure of a cardo polymer based on polyimide.

separation properties. They used to be either very crystalline or too porous; hence selectivities were low. They may, however, be prepared as high-flux–low-selectivity membranes, which is acceptable for certain applications. Materials prepared from tetrafluoroethylene (TFE) and cyclic dioxole (TTD) are highly hydrophobic and have especially good potential for use in gas–liquid membrane contactors. The copolymers known commercially as Hyflon® AD (Figure 5.9) are made from TFE and TTD, and are amorphous perfluoropolymers with a glass transition temperature (T_g) higher than room temperature. Hyflon® AD 60 shows values of permeability and selectivity for gases which make the material interesting for separation—this was documented by Arcella et al. [88].

Kharitonov et al. [89] have shown that direct fluorination of the polyimide Matrimid® is possible, hence the resulting membrane should have good potential for use in harsh environments. Perfluorinated materials were also studied by Hägg [90] for chlorine gas purification, and were shown to be exceptionally stable in these harsh environments, but the selectivity was too low. In a later publication on chlorine purification [51] it was suggested to use perfluorinated monomers as surface-modifying compounds for pore tailoring of glass membranes for this specific gas separation.

Hydrophilic perfluoromembranes may be prepared from TFE and copolymerized with perfluorosulfonylfluoridevinylet her (SFVE), making a so-called Hyflon Ion Polymer which is a rubbery polymer at room temperature. This polymer contains the group $-SO_2F$, where F can be exchanged for a metal or hydrogen atom. This makes the material suitable for a wide variety of fields ranging from electrochemical electrolyzers (chloralkali and HCl), proton exchange fuel cells, energy storage, electrodialysis, to membrane catalytic reactors and many more applications [88].

Cardo polymers are polymers containing very bulky aromatic structures in the main chain. This structure can be coupled to a polyimide, polyamide, or polysulfone. An example of a cardo polymer based on polyimide is shown in Figure 5.10. The polymer may be further modified by substituting methyl groups or halogens into the aromatic rings. The bulky structure gives the polymers high gas permeability and high solubility for non-ideal gases (like CO_2 and hydrocarbons); they can be easily processed and are fairly heat stable. These materials have been extensively studied [91–93]. This makes the cardo polymers, especially the polyimide-based ones, interesting for recovery of CO_2 in flue gas. Takeuchi et al. [94] evaluated the costs of a global process of CO_2 fixation and utilization using catalytic hydrogenation reaction and converting CO_2 to methanol. CO_2 recovery was then based on a membrane process using a cardo polymer.

The acetylene-based high free-volume polymers have received renewed interest. The well-known PTMSP has the highest gas permeability of all known polymers [36,37]. This polymer and other acetylene-based polymers are amorphous and characterized by a very high glass transition temperature (typically >200°C), high free volume, and very high gas permeabilities. These polymers show higher permeabilities to large condensable organic vapors than small permanent gases [95], and PTMSP has the highest C_{2+}/CH_4 and C_{2+}/H_2 selectivities of any known polymer [96]. The selectivity of these gases is typically reversed compared to what is expected in a polymeric membrane. This can be understood by the large free volume in these polymers and the high solubility of the hydrocarbons in the material. The transport may be described in the same way as the selective surface flow through a microporous membrane or a mixed-matrix material. Pinnau et al. [97], among others, have investigated the effects of the side-chain structure of substituted polyacetylenes on their gas permeation properties (see Figure 5.11).

However, fast physical aging limits the practical application of PTMSP membranes. One solution is the cross-linking of PTMSP, which stabilizes the large excess free-volume elements and hence improves physical stability [98–100]. Cross-linking generally reduces gas permeability due to free-volume reduction, while the polymer network becomes more size-selective and gas selectivity increases.

As another representative of high free volume, addition-polymerized polynorbornenes (APNs) were reported by Finkelshtein in 2006 [11]. Bearing $Si(CH_3)_3$ groups in the 5-position, APNs have extremely high glass transition temperatures (T_g > 340°C). The completely saturated homopolymer containing $Si(CH_3)_3$ group (APNSi) and a copolymer containing both $Si(CH_3)_3$ and n-hexyl side groups (ACPNHSi) were prepared. APNSi reveals unusually high gas permeation parameters, e.g., $P(O_2)$ of about 800 Barrer for different samples. Very large sizes of free-volume elements are characterized.

TR polymers, as a new class of high free-volume polymers with "engineered" microporous structures, have drawn great attention as membrane materials for their very high gas

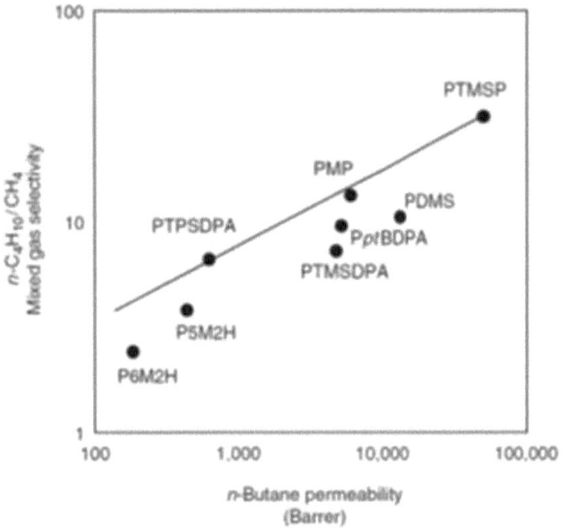

FIGURE 5.11 Relationship between mixed gas n-butane permeability and n-butane/methane selectivity for a series of glassy polyacetylenes and rubbery polydimethylsiloxane (PDMS). Feed pressure: 10 bar, permeate pressure 1 bar; temperature 25°C. (From Ref. [97].)

permeability and selectivity due to the molecular sieving effect. The microporous structures and pore size distributions can be systematically tailored. The TR polymers can be prepared from highly soluble precursors through irreversible thermal rearrangement of aromatic polyimides containing ortho-positioned functional groups between 350°C and 450°C, using different thermal protocols. The micropores are formed by the changes of chain conformations and the spatial location of rigid moieties [12,13]. Unlike zeolite or carbon membranes, TR polymers are tough, ductile, and robust. Although completely insoluble, TR polymers can be readily prepared into the desired membrane morphologies (i.e., hollow fibers and composite membranes) since the precursor polymers are soluble in common solvents. These membranes do not show permeability decay with time due to physical aging [12], which is the major drawback of acetylene-based high free-volume polymer membranes. These features markedly enhance their potential applications as membranes. Suitable TR membranes with "engineered" cavity (i.e. hourglass shape) can be produced for specific gas separation applications or applied as fuel cell membranes. Research shows that TR polymer membranes exhibit excellent CO_2 separation performance with high resistance to plasticization [101]. Moreover, the TR polymers have high thermal and chemical stability, which enables separation under harsh conditions (e.g., pre-combustion CO_2 capture in IGCC plants).

Polymers of intrinsic microporosity (PIMs) are another new class of microporous high free-volume polymers first reported by McKeown and Budd et al. [12,102–104]. PIMs are ladder polymers with rigid, contorted molecular structures that pack inefficiently. They are amorphous polymers, remaining glassy up to their decomposition temperatures (>350ºC), but they behave like inorganic molecular sieves, which may be due to their rigid microstructures without possessing a network of covalent bonds. PIMs can be synthesized as soluble or insoluble. The soluble PIMs are of most interest for membrane applications to prepare membranes with the desired morphologies. Two examples of soluble PIMs are PIM-1 and PIM-7. These polymers have no single bond in the backbone and hence no conformational freedom, but the spiro-centers incorporated in the structure introduce sharp bends into the chain. The permeability order in PIMs is $CO_2 > H_2 > He > O_2 > Ar > CH_4 > N_2 > Xe$, other than many conventional glassy polymers (where $He > CO_2$). The separation performances of CO_2/CH_4 in PIMs are located between the 1991 and the 2008 Robeson's upper bounds [104] (see Figure 5.1). Later, many PIMs with contorted and inflexible units were reported as gas separation membrane materials [105,106].

Troger's base-based polymers were reported in 2013 as PIMs containing bridged bicyclic amine, also known as the Troger's base (TB) group. The TB-based polymer membrane exhibited particularly high gas permeability with high gas selectivity exceeding or close to the upper bounds for several gas pairs [106] due to their more rigid chains, less efficient packing, and higher gas permeability compared with PIMs. As the synthesis of TB units is simple and the resultant TB-based polymers are soluble in many organic solvents, this type of membrane has the potential for large-scale fabrication. Recent studies have enlarged the family of TB-based polymers to polyimide (PI) copolymer or polymers containing only TB units [107–110]. These TB-based polymers generally have high gas permeability for small molecules because of the contorted chains. In addition, due to the narrow distribution of free-volume elements, their selectivity for various gas pairs is usually higher compared with other highly permeable polymers, for example, with a selectivity of H_2/N_2 up to 50.

5.3.1.3 Microporous Organic Materials

Microporous organic materials are hydrocarbons containing pores or voids in the microporous regime, such as PIMs, in which the backbones composed of organic moieties are connected by strong covalent bonds, forming ordered and rigid structures. A summary of common microporous organic materials is shown in Figure 5.12 [111], including PIMs, covalent organic frameworks (COFs), conjugated microporous polymers (CMPs), covalent triazine-based frameworks (CTFs), hyper cross-linked polymers (HCPs), porous cages (PCs), and porous aromatic frameworks (PAFs). Most microporous organic materials have tunable pore sizes with high surface areas, exhibit exceptionally high thermal stability, and may be made with multi-functionality. In recent years, the membrane application of microporous organic materials has drawn increasing attention, such as being a new class of molecular sieves, thanks to their high porosity and molecular range pore sizes. High gas-separation performance would be expected if microporous organics are processed into membranes for the separation of gas mixtures containing hydrogen, carbon dioxide, oxygen, and some hydrocarbons.

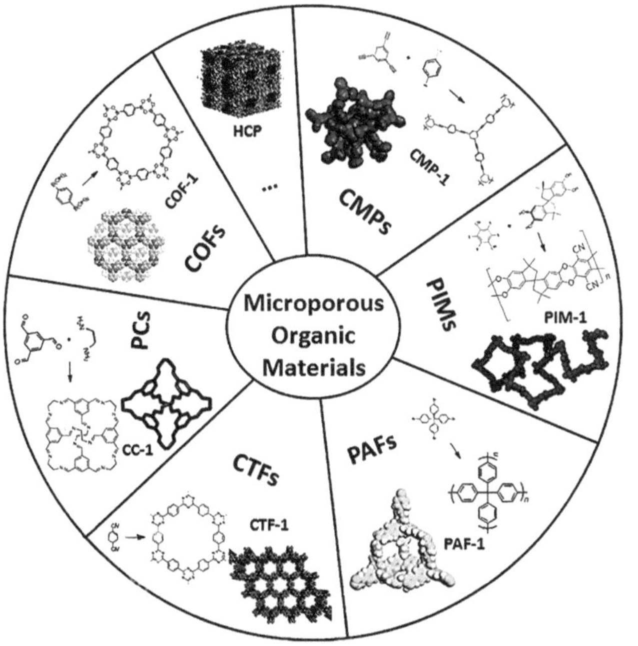

FIGURE 5.12 Chemical structures of different microporous organic materials [111].

A review of newly emerged microporous organic membranes with defined porous structures was reported by Zou et al. [111]. The correlation between the pore size, dimensionality, configuration, and surface functionality with the membrane structure and separation properties was proposed. They concluded that the highly interconnected pores provide gas-transport channels through the membranes, and the small pores and surface functionalities contribute to the selectivities. Today, except for the challenges for PIMs, the main limitation in membrane application of other microporous organic materials (COFs, CTFs, HCPs, CMPs, and PAFs) is their poor membrane-formation ability. Therefore, incorporating these materials into a polymer matrix for MMMs is considered an approach to take advantage of this material. Moreover, most microporous organic membranes suffer physical aging in gas separations, especially when CO_2 and hydrocarbon are involved. Molecularly engineering the backbone at a particular site may improve the robustness and preserve the chemical structure.

5.3.1.4 Fixed-Site-Carrier Polymers

As an alternative to conventional polymeric membranes, facilitated transport FSC membranes have attracted attention due to the potential to achieve both high permeabilities and high selectivities. Facilitated transport membranes may, for instance, selectively permeate CO_2 by means of a reversible reaction of CO_2 with an incorporated complexing agent (carrier) in the membrane, whereas gases such as H_2, N_2, and CH_4 will permeate exclusively by the solution–diffusion mechanism. As pointed out in Section 5.2.6, the ion exchange membranes were introduced as an approach to overcome the problems of SLMs, and the application of ion exchange membranes for the facilitated transport of CO_2 and C_2H_4 was first reported by LeBlanc et al. [63]. Since then, a number of papers have been published on this type of membrane [112–115]. Along with the use of ion exchange membranes as supports, another approach to overcome the above limitations was developed by introducing carriers directly into polymer membranes [64,116,117]. These FSC membranes have carriers covalently bonded to the polymer backbone; hence the carriers have restricted mobility but are favorable when stability is considered. It is obvious that the diffusivity (and thus permeability) in an FSC membrane is lower than that of a mobile carrier membrane. The diffusivity of a swollen FSC membrane should, however, show diffusivities between that of a mobile and a fixed carrier.

The aminated polymeric membranes for facilitated transport of CO_2 have been investigated extensively in recent years [21,114,118–122]. The anticipated mechanism for transport through the membrane is described in Section 5.2.6 and illustrated in Figure 5.5. The findings of several of these investigators were that the aminated polymeric membranes showed higher permselectivity in water-swollen conditions than in dry conditions. The reversible reaction between CO_2 and amines follows the zwitterion mechanism; CO_2 is converted into carbamate ions by consuming 2 mols of amines for 1 mol of CO_2. For a sterically hindered amine, such as a primary amine (the amino group is attached to a tertiary carbon) or a secondary amine (the amino group is attached to at least one secondary or tertiary carbon), the attached bulky group causes a "steric hindrance" effect, resulting in a stoichiometric CO_2 loading up to 1 mol per mole of hindered amine, as shown by Eq. 5.23.

$$CO_2+R_1\text{-}NH\text{-}R_2+H_2O—R_1R_2\text{-}NH_2+HCO_3\text{-} \quad (5.23)$$

The enhanced CO_2 loading capacity significantly improves CO_2 separation performance in membranes based on sterically hindered polyamines. Zhao and Ho [123] reported that, at 107–140°C, CO_2 permeability of greater than 3000 Barrers with a CO_2/H_2 selectivity of > 90 and CO_2/N_2 selectivity of >150 in an FSC membrane containing sterically hindered amines.

The biomimetic membranes represent a special group of carriers-facilitated transport membranes. These are artificial membranes based on biomembrane mimicking, i.e., imitation of the essential features biomembranes use for separation. Nitrocellulose filters impregnated with fatty acids, their esters, and other lipid-like substances may be used—in other words, an imitation of many nonspecific barrier properties of biomembranes. The transport of gas through these membranes will essentially be according to facilitated transport. Biomimetic membranes for CO_2 capture will transport the gas as HCO_3^-. One example is the use of immobilized carbonic anhydrase (CA) in membranes to mimic the mechanism of the mammalian respiratory system, which allows a dramatic increase in CO_2 selectivity over other gases. Enzyme CA can speed up the converting of CO_2 to bicarbonate hydration and reverse bicarbonate dehydration; CA has the ability to catalyze the hydration of 600,000 molecules of CO_2 per molecule of CA per second, 4000 times faster than MEA in terms of catalytic activity [124]. Interestingly, very high selectivities at very low CO_2 concentrations (1–0.1%) were documented, suggesting the possibility of capturing CO_2 from air and life support systems or enclosed spaces [125–127]. Yang et al. reported that the CA biomimetic membrane process can afford a 17-fold increase in membrane area or 17 times lower permeance value and still be competitive in cost with MEA technology [128]. The immobilization of CA has been extensively investigated for the preparation of effective membranes [127]. One method is the use of bifunctional reagents (e.g., glutaraldehyde) to cross-link the enzyme and form a gel. CA can also be covalently attached or encapsulated within polymeric membranes.

The major drawbacks of CA membranes are the loss of enzyme activity and long-term stability. Inorganic zinc complexes resembling the active site of CA enzymes are optimized that mimic the biocatalytic process of CA [129]. These inorganic catalysts are stable at high temperatures, can operate over a wide pH range, and show fast rates of catalysis that are comparable with CA enzymes on a weight basis, providing promising and affordable substitutes for CA enzymes. Saeed and Deng [130] reported a mimic enzyme (Zn–cyclen) that promoted and facilitated transport membrane for CO_2 capture. A thin polyvinyl alcohol (PVA) selective layer containing only 5 μmol/g of mimic enzyme at a pH value of 9 and in a fully humidified gas stream showed a CO_2 permeance of 0.69 [m^3(STP)/(m^2 bar h)] and a CO_2/N_2 selectivity of 107, which is significantly higher than that of a PVA membrane without mimic enzyme. 1.0 wt.% carbon nanotubes (CNTs) were added to the membrane to enhance CO_2 hydration [131] and resulted in 30% higher CO_2 permeance and 15% higher CO_2/N_2 selectivity. Zn-cyclen was also introduced to a CO_2 membrane absorption process using K_2CO_3 solution as a solvent in a membrane contactor. The kinetic rate constant for absorption of CO_2 in the K_2CO_3 solvent was promoted 10-fold by mimic enzyme [132].

5.3.1.5 Proton-Conducting Polymeric Membranes

Polymer electrolyte membranes (PEMs) are polymeric proton-conducting membranes suitable for low-temperature operations (<100°C) in fuel cell applications and have the advantage of low weight [133]. In a fuel cell, electricity is produced by an electrochemical reaction. The proton is produced at the anode by oxidation of the fuel and will diffuse through the "proton-conducting membrane" to the cathode, where water is formed (see illustration in Figure 5.13) [134]. The fuel may be hydrogen or hydrogenated molecules. When hydrogen is used as fuel, the PEM fuel cells will use Nafion or Nafion-like sulfonated perfluorinated polymers. These materials have high proton conductivity combined with high stability.

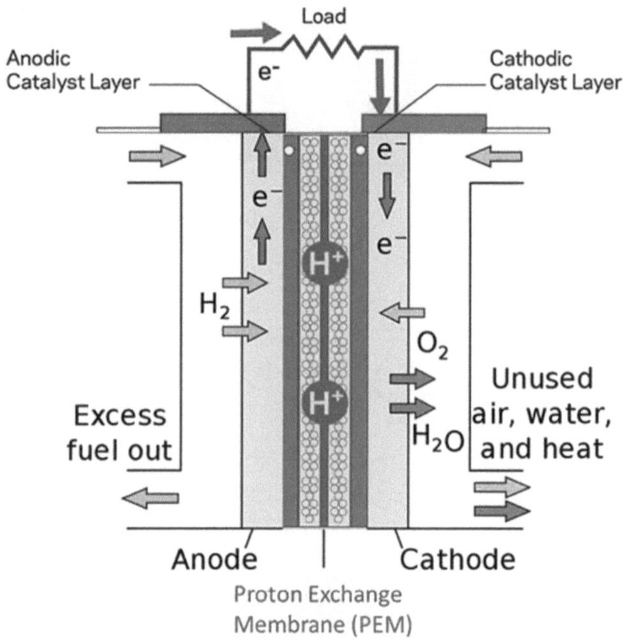

FIGURE 5.13 Transport through a fuel cell. (From Ref. [134].)

5.3.2 Inorganic Membranes

The inorganic membranes had, up until the late 1990s, received little attention for gas separation. This has mainly been due to the porous structure of most inorganic membranes that lack the ability to separate gas molecules. Within the group of inorganic membranes, however, there are materials of molecular sieving nanostructures (i.e., CMS) and dense metallic membranes and solid oxide electrolytes; these will be discussed separately.

The increasing interest in inorganic membranes for gas applications is undoubtedly due to their excellent high-temperature resistance. Inorganic membrane reactors (including carbon membranes) may thus have very good potential for industrial applications. The various configurations of membrane rectors however are not discussed in this chapter. Their separation properties may be understood on the basis of the materials used, kinetics, and process conditions.

Porous materials have very complex structures and morphology, and parameters like porosity, pore size distribution, and pore shape are extremely important variables affecting the gas separation properties. A schematic picture of different pores is given in Figure 5.14 [135]. As can be seen, pore constrictions, dead-end pores, and interconnection between pores will contribute to the characterization of the membrane; hence the tortuosity (τ) plays an important role. The tortuosity will have a value equal to unity (=1) for a cylindrical pore. The inorganic membranes may be symmetric or asymmetric. Symmetric membranes are systems with a homogeneous structure throughout the membrane. Capillary glass membranes and anodized alumina membranes are examples within this group.

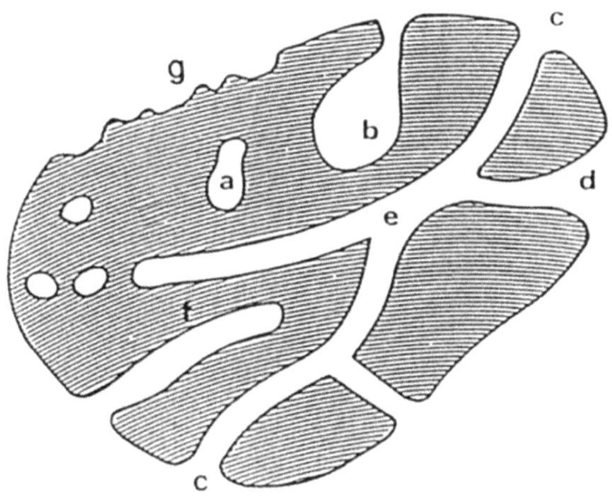

FIGURE 5.14 Schematic illustration of pore types in porous solid with open pores (c, d), locked-in (a) and dead-end pores (b, e, f.) (From Ref. [135].)

In most cases, however, the inorganic membranes are asymmetric with a composite structure consisting of several layers with a gradual decrease in pore size to the feed side. Examples are ceramic aluminas synthesized by the sol–gel technique or carbon-zirconia membranes. For gas separation, surface-modified inorganic membranes have become increasingly important. An introduction to the synthesis and properties of inorganic membranes may be found in comprehensive textbooks [35,136,137]

5.3.2.1 Carbon Molecular Sieving (CMS) Membranes

The CMS membranes are microporous carbon (as fibers, on tubes or flat sheets) prepared from the carbonization of polymeric precursors under controlled conditions. The precursors which are mostly used are cellulose or polyimides. Depending on the membrane pore size and the process conditions, the separation may take place according to (a) molecular sieving ($d_p < 5$ Å), (b) selective surface flow (5Å $< d_p < 12$ Å), (c) Knudsen diffusion ($d_p > 20$Å), or combinations of these [27,49] (see Sections 5.2.2–5.2.4). The membranes for gas separation are prepared as self-supported hollow fibers or carbonized polymeric precursors deposited on different substrates. An example is dip-coating α-alumina tubes with polyimide and pyrolyze/dip-coating in several steps, as done by Hayashi et al. [138]. CMS membranes as flat sheets are in general too brittle if self-supported and should be prepared on support for scaling up. The hollow fibers may have the greatest potential for becoming a successful separation unit on an industrial scale due to the possibility of making modules with a high packing density (m^2/m^3). The production process of these membrane modules is, however, challenging and expensive, and costs need to be reduced to make it of interest for larger gas volume applications. The use of cheap polymeric precursors is favorable. Properties that should place carbon membranes among the most promising membrane materials are their high temperature resistance and excellent chemical resistance to acids, hot organic solvents, and alkaline baths. The carbon membranes are fairly easy to produce as much is known about how carbonization conditions affect separation properties [27–30]. A carbon membrane can thus be tailored with a pore size giving excellent separation properties for a given gas mixture (high flux for permeating component, high selectivity for gas pairs).

The CMS membranes may be prepared in two ways; in both cases, the pore tailoring is the focus for the final membrane:

1. By careful control of carbonization conditions; this is done by controlling the heating rate, heating temperature, and choice of inert gas or vacuum during the process [139–141]
2. By treating microporous cellulose fibers with CVD, and pores are tailored by post-oxidation [142].

CMS membranes may also easily be functionalized; i.e. metals (like $AgNO_3$, MgO, Fe_2O_3, or others) are embedded in the structure of the precursor and will enhance the separation for certain gas pairs [143].

5.3.2.1.1 Separation Properties for CMS Membranes

The ability of microporous carbon fibers to separate gases depends on the pore size of the membrane, the physiochemical properties of the gases, and the surface properties of the membrane pore. The pore size of a carbon fiber for gas separation is usually within the range of 3.5–10 Å; depending on the conditions for preparation of the membrane during the carbonization or treatment afterward (post oxidation or chemical vapor deposition) [27–30,35]. With reference to the typical range indicated above for the transport mechanisms, one would expect that the dominating mechanisms will be either molecular sieving or selective surface flow.

Figure 5.15 illustrates a carbon membrane with pores in the range suitable for molecular sieving [143]. As expected, there is a clear and indisputable correlation between flux and molecular size. The gas pair reported is CO_2 and CH_4, and the selectivity is clearly in favor of CO_2, indicating a selective surface flow. The critical temperatures, T_c, and Lennard-Jones diameters, $d_{L\text{-}J}$, for the two gases are:

$$CO_2\text{: } d_{L\text{-}J} = 3.94 \text{ Å}, T_c = 304 \text{ K}$$

$$CH_4\text{: } d_{L\text{-}J} = 3.74 \text{ Å}, T_c = 190.4 \text{ K}.$$

The easily condensable CO_2 molecule will follow an SSF mechanism and seriously hinder CH_4 from permeating; hence high selectivities are obtained.

Due to the high selectivities of molecular sieving, CMS membranes have found applications in hydrogen purification [144], natural gas sweetening [145], and biogas upgrading [146].

The fabrication of CMS membranes is challenging, and most CMS membranes were fabricated as self-supported asymmetric flat sheets or hollow fibers. Recently, CMS membranes have been reported as composite membranes

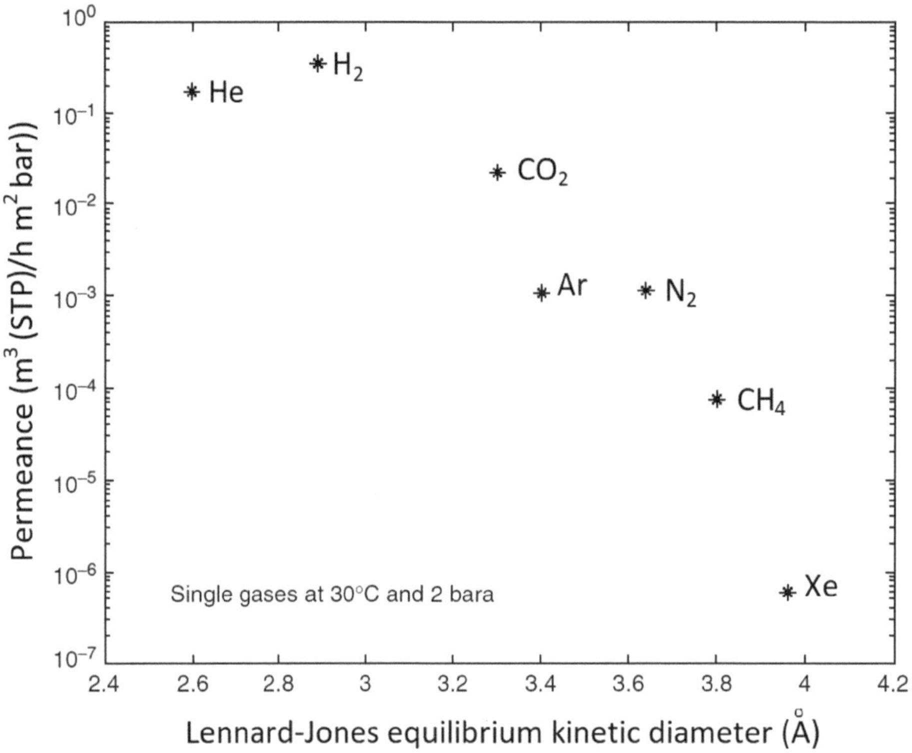

FIGURE 5.15 Permeance as a function of L-J diameter for a sieving carbon membrane. (From Ref. [143].) 1[Å] = Ångstrom = 10^{-10} m.

with a thin selective layer on porous substrates of different materials, both organic and inorganic [147,148]. The ultra-thin selective layer can be made through a controlled coating process and the subsequent carbonization process.

5.3.2.1.2 Regeneration

A CMS membrane will typically have a flux decline over time and regeneration will be necessary at intervals. Oxygen is one of the most detrimental species for a CMS membrane. When carbon materials are exposed to air at room temperature, irreversible chemisorption of oxygen may take place and C–O surface groups are formed [149]. These groups also provide sites for the adsorption of H_2O. Both phenomena will slightly reduce the effective size of micropores. The chemically bonded oxygen is only completely removed (as CO and CO_2) by heating the sample to temperatures as high as 700–800°C with an inert gas.

Adsorption of water also may result in flux decline. At low relative humidity, only active polar sites seem to be involved, and this sorption is so weak that the negative effect can easily be managed. With the relative humidity is high (> 25%), the negative effect may be substantial, caused by hydrogen bonding by neighboring water molecules forming clusters of adsorbed water [150]. It should be noted however, that water uptake from various gases with the same humidity level may differ greatly, and will, therefore, also be more or less easily removed.

Adsorption of organics may cause the same type of flux decline.

To recover a decreased membrane flux, three main approaches are reported:

1. The membrane may be treated at elevated temperatures, at least 200°C, under a vacuum or inert atmosphere. If the flux is only partially restored after regeneration, this could be the result of incomplete removal of C–O surface groups.
2. If exposed to organics, treatment of the membrane with propene may be a good solution. Jones and Koros [151] found propene to be very effective in removing sorbed organics. In some cases, the flux was completely restored—this was also confirmed by Hägg et al. [152].
3. The use of electrothermal regeneration (low voltage direct current) has been tested successfully online [153].

5.3.2.2 Modified Inorganic Membranes

As discussed in Section 53.2, the pore size of a microporous inorganic membrane has to be reduced in order to separate gases. Only for pore sizes in the range below Knudsen flow, will the separation be efficient and follow either selective surface flow or capillary condensation. Separation according to configure diffusion may take place if the pore size is sufficiently small (< 1 nm). The surface may also be modified to change its chemical nature and thus separation properties. Several methods of modifying the surface structure

TABLE 5.6
Some modified nanoscale ceramic microstructures within membranes with pore diameters of 3–5 nm

Membrane material	Modification by	Modified structure	Size (nm)	Loading (wt.%)
γ-Al_2O_3	Fe or V-oxide	Monolayer	≈ 0.3	5–10
γ-Al_2O_3	$MgO/Mg(OH)_2$	Particles		2–20
γ-Al_2O_3	$Al_2O_3/Al(OH)_3$	Particles		5–20
γ-Al_2O_3	Ag	Particles	5–20	5–65
γ-Al_2O_3	CuCl/KCl	Multilayer	> 20	
γ-Al_2O_3	ZrO_2	Surface layer	< 1	2–25
γ-Al_2O_3	SiO_2 (amorphous)	20 nm layer + porous plugs	< 1.5	5–100
a-TiO_2*	V_2O_5	Monolayer	≈ 0.3	2 – 10
Al_2O_3/TiO_2	V_2O_5 or Ag	As for a-TiO_2 or Al_2O_3		
θ/α-Al_2O_3	ZrO_2/Y_2O_3	Multilayer/porous plugs	Few nm pore size	1–100

* a-TiO_2: anatase titania.
Source: Ref. [136].

of ceramic membranes have been suggested by Burggraaf and Keizer [136], and pore sizes < 1 nm have been obtained (see Table 5.6).

The main questions related to the economy of production and brittleness of these membranes remain to be solved, although the materials may show excellent separation properties for selected gas mixtures, especially in the range of selective surface flow.

The effects of temperature and pressure will play an important role in the resulting separation. This has been discussed for different MFI zeolite membranes by Posthusta et al. [154].

A usually less expensive way for surface modification is using organic compounds. The range for application will then be determined by the decomposition temperature of the organic compound. Several papers have been published reporting promising results for separation [155,156]. The glass membranes (silicate membranes) should be specially mentioned as they have proved to be exceptionally resistant to aggressive gases like chlorine and hydrochloric acid [90,157], and they have a good potential for the separation of such gases when membranes are surface modified with chemical-resistant perfluoro compounds [51].

5.3.2.3 Ion-Conducting Membranes

5.3.2.3.1 Proton-Conducting Pd Membranes

For the transport of hydrogen through a palladium membrane, refer to Section 5.3.2. The membranes may be prepared as pure palladium membranes, but the trend has moved in the direction of preparing composites and using Pd alloys. There seems to be a number of advantages to using composite palladium membranes supported on porous substrates over palladium foils and tubes—one important aspect is the stability of the system as stability problems increase with reduced thickness [158]. Stability is also alloy-dependent. The porous supports used comprise porous alumina or glass, and porous metals, including porous Ni and porous stainless steel. However, from an industrial perspective, alumina supports have disadvantages in terms of insufficient sealing and difficult fabrication of large modules. (This is often a general problem with inorganic membranes for gas separation.)

There are various techniques for the deposition of Pd or its alloys on supports; details on this may be found elsewhere [158,159]. Among these methods, electroless plating is quite attractive due to the possibility of uniform deposition on complex shapes and large substrate areas, the hardness of the deposited film, and very simple equipment required. The main advantage of Pd alloys compared to pure Pd for use in hydrogen separation and membrane reactor applications is that Pd alloys have a reduced critical temperature for the α–β transition and may therefore be operated in the presence of hydrogen at temperatures below 300°C without risking the hydrogen embrittlement observed for pure Pd membranes [160]. The components most often used in the Pd alloys are silver (Ag) or copper (Cu) added in a weight % of up to around 30.

The palladium–silver alloys have attractively enhanced permeability compared to pure Pd, while the palladium–copper alloys are more resistant to sulfur. (The hydrogen flux through a 20 μm thick palladium membrane was measured by Mardlovich et al. [159] at 350°C to be between 2–2.5 $m^3/m^2 \cdot h$), while the performance of palladium–copper alloy membranes over a wide range of temperatures and pressures has been presented by Howard et al. [161].) Co-existing hydrocarbons in the gas stream may influence the hydrogen permeation through palladium membranes. It has been documented that especially propylene in the gas mixture may seriously affect the hydrogen permeation [162]. The propylene decomposed and the carbonaceous matter forming would chemisorb to the membrane surface. Regeneration with pure hydrogen at high temperature (600°C) may however restore the flux. Other components of concern that may cause deactivation and poisoning of a palladium membrane are CO, H_2O, H_2S, and Cl_2.

The very attractive features of palladium membranes and their alloys include the favorable selectivity for hydrogen permeating at high temperatures. This makes the membranes attractive for use in fuel cells.

5.3.2.3.2 Oxygen-Conductive Membranes

Industry is continuously in search for suitable membrane materials which may produce high-purity oxygen at low cost and preferably at low temperatures. This type of material is yet to be developed. Presently, zirconia and perovskite membranes may be used to produce pure oxygen at high temperatures as these materials (solid electrolytes) contain an oxygen ion conductor. When used as a membrane reactor, the permeating oxygen may be used for the oxidation of ethane on the permeate side, hence producing syngas [158]. (For details on the transport, see Refs. [56,57].)

Air Products and Chemicals Inc. recently presented a ceramic oxygen-transporting membrane operating at high temperature (transporting O^{2-}). This material is a "mixed-conductor" where both oxygen ions and electrodes are highly mobile within the solid. The technology makes use of an oxygen partial pressure gradient across the ceramic membrane to drive the oxygen flux [163]. The technology is suited for advanced power generation, which requires oxygen for combustion or gasification.

5.3.3 Hybrid Membranes

Organic–inorganic hybrid membranes consist of a continuous polymeric phase wherein nanoporous or dense inorganic materials such as silica, zeolites, or carbon particles/nanotubes are dispersed. The separation efficiency can be "tuned" by judiciously selecting the constituent materials (and their intrinsic properties), as well as the composition of the resultant hybrid materials. Molecular separation through the polymer occurs according to the solution–diffusion model and is combined with surface diffusion or molecular sieving through the inorganic phase when a microporous filler is added. When dense fillers are added (silica, carbon nanoparticles), the membrane may become reverse selective but still separate according to solution diffusion. The choice of polymer governs the application temperature range, and fabrication requires excellent adhesion between the filler and the polymer to avoid the formation of voids and, consequently, undesirable Knudsen diffusion [153]. Methods under current investigation to improve such adhesion rely on polymer softening (at temperatures close to T_g or through the use of plasticizing agents) or reactive coupling with or without surface treatment of the filler.

The field of hybrid membranes is developing rapidly due to the wide range of new properties that can be addressed within this unique class of materials. The resulting properties of these hybrid organic/inorganic materials (which differ significantly in property behavior), rely to a large extent on successful blending. Both enhanced thermomechanical properties and separation properties can be expected. The choice of polymers and inorganic fillers must be based on fundamental knowledge about the separation properties as well as the miscibility of the materials. For an illustration of a mixed-matrix membrane with microporous fillers, see Figure 5.16 [164]. Figure 5.17 presents commonly used porous and nonporous nanofillers in MMMs [165].

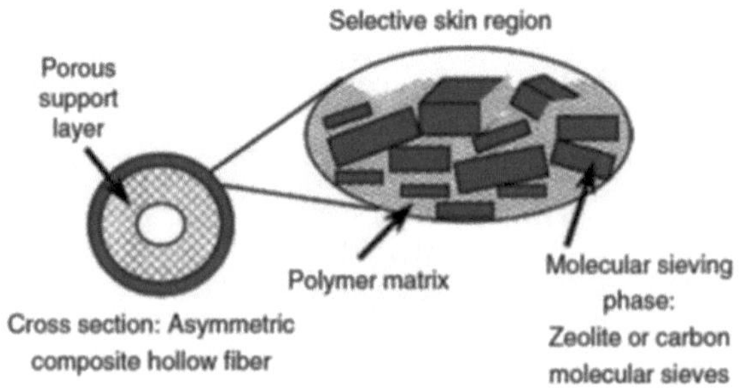

FIGURE 5.16 Illustration of a mixed-matrix membrane. (From Ref. [164].)

5.3.3.1 Mixed-Matrix Membranes: Polymer Matrix with Microporous Fillers

When microporous fillers are incorporated in a polymer matrix and gas transport involves a second mechanism (e.g., molecular sieving or selective surface diffusion), the hybrid membrane is often referred as a MMMs, which combines the high separation capacity of molecular sieving materials with the desirable mechanical properties and economical processing attributes of polymeric materials. The microporous fillers must be selected so that their pores can separate the gas molecules of interest according to size. In the case of CMSs, varying the carbonization conditions during fabrication controls the pore size. This is an advantage of CMS membranes over zeolitic molecular sieves, in which the pore size is fixed for a given zeolite type.

Vu et al. [166] incorporated carbon molecular sieve (CMS) materials into polymers to form mixed-matrix membrane films for selective gas separations. The CMS, formed by pyrolysis of a polyimide precursor and exhibiting an intrinsic CO_2/CH_4 selectivity of 200, was dispersed into a polymer matrix. Pure-gas permeation tests of such MMMs revealed the CO_2/CH_4 selectivity was enhanced by as much as 40–45% relative to that of the pure polymer. The effective permeabilities of fast-gas penetrants (e.g., O_2 and CO_2) through these MMMs are also improved relative to the intrinsic permeabilities of the unmodified polymer matrices. For a CO_2/H_2 gas mixture, the CO_2 will be the fastest permeating component, and H_2 will be retained on the feed side to avoid repressurization, in which case the polymer matrix dictates the minimum membrane performance. Properly selected molecular sieves can only improve membrane performance in the absence of defects. The polymer matrix must be chosen so that comparable permeation occurs in the two phases (to avoid starving the

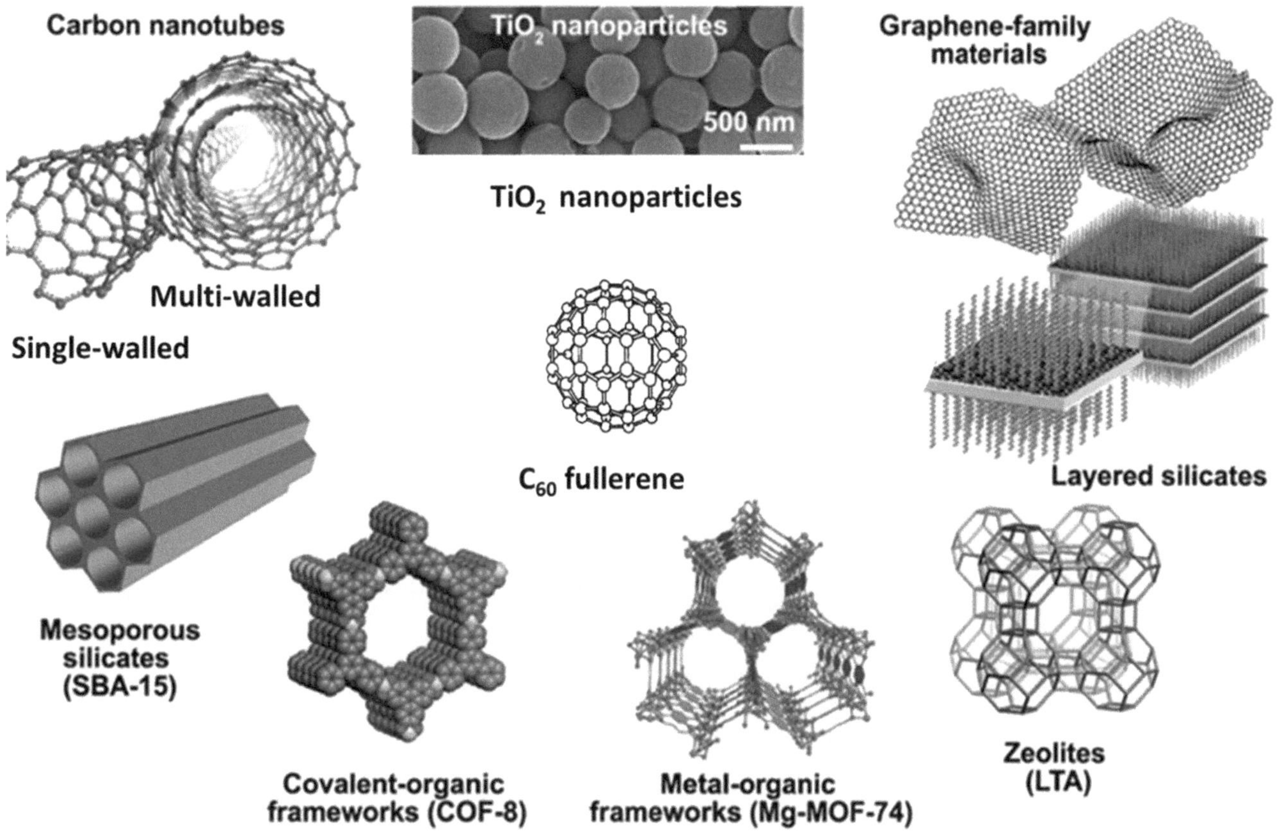

FIGURE 5.17 Commonly used porous and nonporous nanofiller in MMMs. (Reproduced from Ref. [165].)

sieves) and so the permeating molecules are directed toward (not around) the dispersed sieve particulates.

Recently, mixed-matrix membranes with more ordered microporous 3D materials as fillers (e.g., MOFs, ZIFs, POFs, zeolites) for gas separation have been reported [166–168], and recently also 2D materials (e.g., COFs, MOFs, and MXenes) [169].

MOFs are a class of porous crystalline materials formed by the self-assembly of complex subunits comprising transition metal centers connected by various polyfunctional organic ligands. By varying the chemical combination, MOFs can regulate the frameworks to have large surface areas, high porosities, and low densities, and also tune the pore size and pore structure.

The mixed-matrix MOFs-polymer membranes exhibit enhanced gas permeabilities as well as selectivities. The improved separation performance is attributed to the precisely tunable MOF porosity. Separation performances of MOFs-based MMMs comparable to the best polymer membranes have been reported. MOFs-based MMMs can also reach similar performance as pure MOF membranes for the same gas separation. Compared to traditional inorganic fillers, the MOF particles have better affinity and compatibility with polymer chains due to the presence of organic linkers in the MOF structure. The adhesion of MOF fillers to organic polymers is not a problem, as MMMs use inorganic fillers. Polyimide (PI) and polysulfone (Psf) polymers are widely used as polymer matrices for MOF fillers [32]. Due to their size and shape, ultrathin MOF-based membranes are difficult to fabricate, especially on a large scale.

Recently, porous organic frameworks (POFs) have been reported as fillers in MMMs for gas separations. Different from MOFs, POFs have an entirely organic structure, which ensures good chemical compatibility with the polymer phase [170]. In addition, POFs possess characteristics such as high surface areas, tunable pore sizes, and excellent thermal and chemical stabilities that contribute to improved gas separation when POFs are embedded in MMMs [168]. COFs are a class of crystalline POFs and can form both 2D and 3D structures, but mostly 2D structures when used in membranes, which is advantageous when making ultrathin continuous COF-based membranes. Moreover, COFs have a fully organic nature with good dispersibility in polymer solutions and excellent compatibility with polymer matrices for MMM synthesis. Therefore, a relatively high loading of COFs of up to 50 wt.% in the membrane can be easily achieved [171]. The higher content of COFs in a membrane yields additional transport pathways for the gas, water, and organic solvent molecules to go through, thus resulting in a relatively high permeability. In addition, the organic linkers used in synthesizing COFs allow for rational design and functionalization by decorating them with functional groups. The pore size of most reported COFs

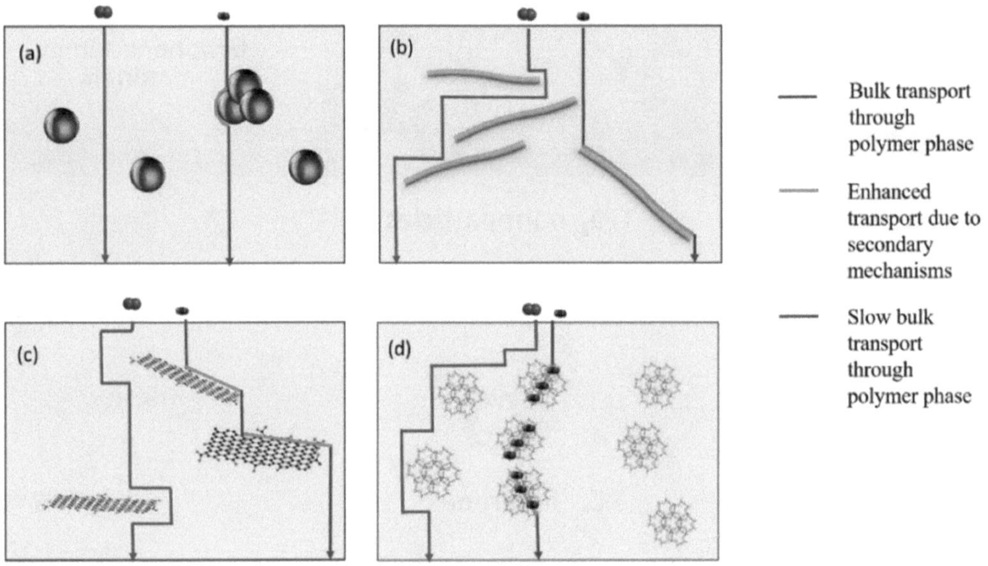

FIGURE 5.18 Schematic illustration of hybrid membranes with (a) 0D (b) 1D (c) 2D (d) 3D fillers. (From Ref. [173].)

ranges from 1 nm to 4 nm. Special attention was given to pore size, stability, hydrophilicity/hydrophobicity, and the surface charge of COFs in view of determining COFs for proper membrane applications.

5.3.3.2 Nanocomposite Membranes: Polymer Matrix with Non-Porous Nanoparticles

When the dispersed fillers in a hybrid membrane consist of non-porous nano-dimensional materials (e.g., carbon nanotubes, fumed or colloidal silica), the membrane is considered a nanocomposite membrane. The polymer may be a rigid-chain polymer or a self-organized block copolymer. In a nanocomposite membrane, the nanoscale fillers may disrupt the packing of the polymer chains if they are rigid and increase the accessible free volume in the polymer matrix and hence, the molecular transport of permeating species. This increased free volume augments molecular diffusion and weakens the size-sieving nature of the polymer, thereby increasing gas permeability and reversing the selectivity. Since permeability depends on diffusivity and solubility, these nanocomposite membranes will favor the permeation of the larger (and more soluble) components through the membrane. The introduced nanoparticles may also alter the mechanism by which a copolymer self-organizes, hence influencing the equilibrium morphology and polymer thermodynamics (and then properties).

Merkel et al. [172] mixed a substituted polyacetylene, poly(4-metyl-2-pentyne) (PMP), with fumed silica particles possessing hydrophobic trimethylsilyl surface groups. Dispersion of the particles was achieved by matching the polarity of the polymer and the particle surface groups, as well as by controlling film-drying conditions. The addition of fumed silica (up to 30 vol%) (size ~10 nm) promoted a considerable increase in the permeability of CH_4 relative to that of pure PMP.

Zero-dimensional (0D), one-dimensional (1D), and two-dimensional (2D) nanomaterials have been introduced as nanofillers in nanocomposite membranes. Figure 5.18 shows gas transport pathways in hybrid membranes with 0D–3D fillers [173]. CNTs, graphene, and graphene oxide (GO) have been widely reported as 1D and 2D nanofillers in membranes and were found significantly to influence the mechanical and transport properties [174,175]. Challenges using 2D fillers include incompatibility between the filler and polymer, upscaling complexity to form defect-free thin composite membranes, and morphology/chemical property tuning of fillers [174].

Janakiram et al. reported a hybrid-facilitated transport membrane using GO or porous GO at a very low filler loading of 0.2 wt.% to enhance CO_2 separation. The presence of hydroxyl groups on the GO surface results in hydrophilicity and increased surface interaction with CO_2 [176], while chemical modification of GO with PEG groups was found to further increase the selectivity of the membranes but reduce the CO_2 permeance.

Nanofillers in a nanocomposite membrane are not necessarily inorganic. In recent years, as a replacement for CNTs, nanocellulose fibers have been used to enhance gas separation [177]. Due to the high specific surface area, high mechanical strength, and broad possibility of surface modification, nanocellulose has obtained much attention in a wide variety of applications [177]. Uniformly dispersed nanocomposites with added nanocellulose crystals (CNCs) and nanocellulose fibers (CNFs) were made to improve the water retention and enhance CO_2 facilitated transport effects [178]. A TFC in a hollow-fiber membrane module was made into a pre-pilot scale and tested on-site in a cement process to capture CO_2. The PVA/CNC nanocomposite membrane showed high CO_2 permeance and CO_2/N_2 separation factor with good stability in a real flue gas stream.

In some cases, a third phase is used to improve the dispersion and wrapping of the fillers in a membrane matrix, e.g., ILs, or to provide additional transport enhancement, such as adding mobile carriers to a hybrid facilitated transport membrane. Janakiram et al. [179] developed a TFC hollow-fiber membrane with an ultrathin layer consisting of three phases, namely the host PAA polymer matrix, the GO nanofillers, and ILs as the added CO_2-philic mobile carriers, resulting in a hybrid membrane with promoted CO_2 permeation. CO_2 permeance of 825 GPU and a CO_2/N_2 separation factor of 30 were reported.

5.3.4 Supported Liquid Membranes (SLMs)

Supported liquid membrane (SLM) is a practical design to utilize liquid membranes with mobile carriers for the facilitated transport of certain gases, e.g., CO_2. Essentially, this is a porous membrane with pores filled with solvent and carriers by capillary force.

Compared with liquid membranes, the relatively small volume of solvent in the porous membrane offers advantages of possible usage of expensive carriers with high separation factors. The advantages of SLMs also include easy scale-up, low energy requirements, low capital and operating costs, simpler configuration and process, known interface area, and predictable separation performance.

Instability is the most common problem with SLMs, including the loss of solvents and carriers through evaporation and decomposition or the breaking through of the solvent by high pressure differences across the membranes. SLM stability can be affected by the type of polymeric support and its pore radius, solvents used in the liquid membrane, the interfacial tension between the liquid and membrane phase, the flow velocity of the aqueous phases, and the method of preparation [1].

The maximum transmembrane pressure to avoid breakthrough of the impregnating phase out of the largest pores can be calculated using the Laplace equation, as shown in Equation 5.24.

$$r_p = \frac{2\gamma}{\Delta P}\cos\theta \tag{5.24}$$

where γ is the interfacial tension between the strip or feed solution and SLM phase, θ is the contact angle between the membrane pores and the impregnating liquid, and r_p is the pore radius.

Several immobilization techniques have been developed to overcome the instability problems of SLMs, such as by gelation or microencapsulation of the liquid phase, covering surface layers (contained liquid membranes). An overview of stabilization techniques for SLMs developed over the last 10 years can be found in Ref. [180].

Supported ionic liquid membranes (SILMs) have been brought into focus due to their unique properties that can prevent the loss of solvent by evaporation. ILs are organic salts with negligible vapor pressures. They are thermally robust, with a wide temperature range in the liquid state, e.g., up to 300°C, compared to 100°C for water, and their polarity and hydrophilicity/hydrophobicity can be tuned by a suitable combination of cations and anions. Noble et al. have extensively studied SILMs for CO_2 separation, including the gelation of SILMs. Excellent CO_2 separation performance has been reported [181,182].

Block copolymers were found to be good host matrices to prepare IL-based hybrid membranes, a special type of SILM. A series of hybrid membranes containing IL nanochannels were developed by combining IL and block copolymer for CO_2 separation. Dai et al. [183] mixed 1-butyl-3-methylimidazolium tetrafluoroborate ([Bmim][BF_4]) with Nafion to prepare a SILM membrane. By testing under humid conditions, it was found that when the IL concentration was as high as 40%, the CO_2 permeability and corresponding CO_2/N_2 selectivity were 390 Barrer and 30, respectively. [Bmim][BF_4] was also introduced into a mid-block sulfonated five-block polymer (Nexar) with thermoplastic elastomer properties to prepare a SILM membrane containing 40 wt.% of ILs. The addition of IL to Nexar systematically improved the CO_2 permeability already in the dry state. The presence of water vapor further facilitated the CO_2 transport, showing CO_2 permeability and CO_2/N_2 selectivity of 194 Barrer and 128, respectively [87]. The Nexar-based SILM also exhibited very high NH_3/N_2 separation performance when ILs in the membrane were tuned to be acidic [184].

It was confirmed that, instead of simply swelling the polymer matrix, the addition of IL induced the formation of IL nanochannels inside the block copolymer matrix. The molecular transport pathway is, therefore, further enhanced, enabling membrane materials to obtain higher gas fluxes and superior separation performance.

5.4 MODULE DESIGN

Depending on the type of materials to be used and the operating conditions for the membrane separation, the module may have different configurations. The footprint of the membrane separation unit may be an important issue where it is going to be placed, and the packing density of the module (m^2/m^3) will then have to be considered. Some modules may be suitable for large-volume applications, and some for smaller ones. In most cases, the investment cost and lifetime of the membrane will decide which one should be chosen. If specific process conditions are necessary for the optimum performance of the membrane, e.g., pressure, temperature, filtering, and drying of gas, required utilities must be included in the cost estimation.

For the calculation of the required membrane permeation area, flow patterns for the various module designs must be considered. The cross-flow, counter-current, co-current, or complete mixing flow will result in different degrees of purity for the same stage cut, θ (= q_p/q_f) [1,47]. Inserting the relevant flux equation for J_i given there into Equation 5.25, the required membrane permeation area, A_m, may be calculated:

$$A_m = q_{p,i} / J_i \tag{5.25}$$

where $q_{p,i}$ is the permeation rate [m^3(STP)/h] of component i and J_i is the flux [m^3(STP)/m^2·h].

In a real case there are several additional variables to be taken into account: the possibility of concentration polarization, pressure drop, heat transfer, and Joule–Thomson effect across the membrane. The J-T effect may be significant when there is a large ΔP across the membrane and with non-ideal gases permeating. Frugality should then be used in the calculations. These effects have been discussed by several authors [185,186].

The standard module configurations are presented below. With the development of new membrane materials for various applications, new configurations for optimum gas separation may be expected on the market in the future.

5.4.1 Flat Sheet—Plate and Frame/Envelope Type

Inorganic or metallic membranes for gas separation are usually prepared as discs or flat sheets. These thin sheets or discs may be quite vulnerable to breakage, and assembling the membranes in modules can be quite challenging. These membranes are usually intended for high-temperature gas applications, and the sealing technology may be complicated. Carbon membranes also face this challenge at high temperatures.

Polymeric flat sheet membranes are easy to prepare, handle, and mount. For gas separation, the flat sheet membranes are composites with a selective polymer coated on a support. A commercial configuration that has been quite successful for hydrocarbon vapor recovery is the Borsig envelope-type module [187]. Packing densities for flat sheet membranes may be in the range of 100–400 m^3/m^2 [1].

5.4.2 Spiral-Wound Membrane

The typical spiral-wound membrane, as shown in Figure 5.19, consists of four layers wrapped around a central collection pipe: membrane, spacer (providing a permeate channel), membrane, and a new spacer (providing feed channel) [217]. The feed-side spacer acts as a turbulence promotor, whereas on the permeate side the flow is directed toward the central pipe. The spiral-wound membrane will typically be a polymeric composite material and is also often used for liquid separation. The packing density of this type of module will depend on the channel height but is usually within the range of 300–1000 m^2/m^3 [1]. Several modules may be assembled in one pressure vessel.

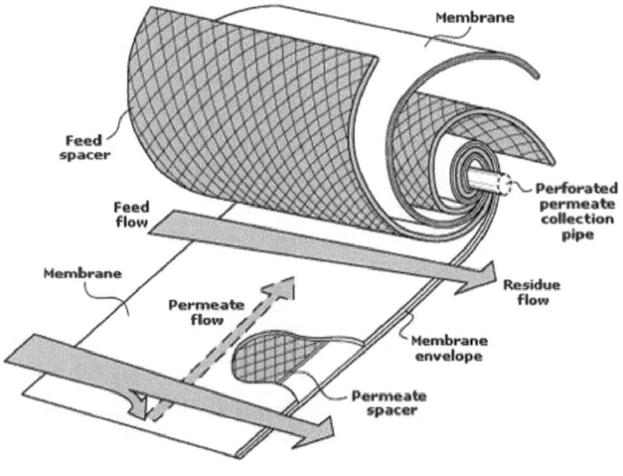

FIGURE 5.19 A spiral wound membrane module. (From Ref. [217] with permission.)

5.4.3 Hollow-Fiber Membranes

The hollow-fiber membranes are the optimum choice for gas separation modules due to their very high packing density (up to 30,000 m^2/m^3 may be attained [1]). Figure 5.20 shows alternative configurations for such modules [188]. Modifications of this configuration exist, where the possibility for the introduction of sweep gas on the permeate side is included, or fibers may be arranged transversal to the flow in order to minimize concentration polarization [189,190]. The hollow-fiber membranes are usually asymmetric polymers, but composites also exist. Carbon molecular sieve membranes may easily be prepared as hollow fibers by pyrolysis.

5.4.4 Membrane Contactors

A membrane contactor may be considered a new unit operation where the membrane offers an effective gas–liquid interface without the dispersion of one phase within another. The most important advantages of a membrane contactor compared to a traditional absorber are: (a) reduction in size and weight (important when used offshore), (b) with the gas and liquid separated by a barrier, the liquid and gas flow rate may be adjusted independently, (c) no entrainment, flooding, or channeling and (d) a reduction in solvent loss. A typical example of using a membrane contactor for gas separation is the removal of CO_2 from a gas stream using absorbent on the liquid side, known as the "membrane absorption" process [191–201]. The membrane contactor may have different module configurations. Figure 5.21 shows an illustration of how CO_2 is absorbed from a gas stream through a membrane contactor (a), and the Aker Kvaerner membrane contactor (b) [191]. A different design is presented by Liqui-Cel, as shown in Figure 5.21(c) [202].

The instability of the system is a serious challenge in a gas–liquid membrane contactor. For instance, wetting and bubbling problems occur when the pressure difference across a porous membrane is too high. Factors such as pore size, pore size distribution, hydrophobicity, and hydrophilicity of the membrane will play a major role in determining the break-through of gas or liquid across the membrane [196].

The governing equation to avoid break-through is the Laplace equation, as shown in Equation 5.24. In order to prevent wetting and bubbling, hydrophobic membranes (for aqueous solvents) or membranes with a smaller surface pore size (r_p in Equation 5.24) should be used. In addition, a liquid solvent with high surface tension (γ) should be selected, and

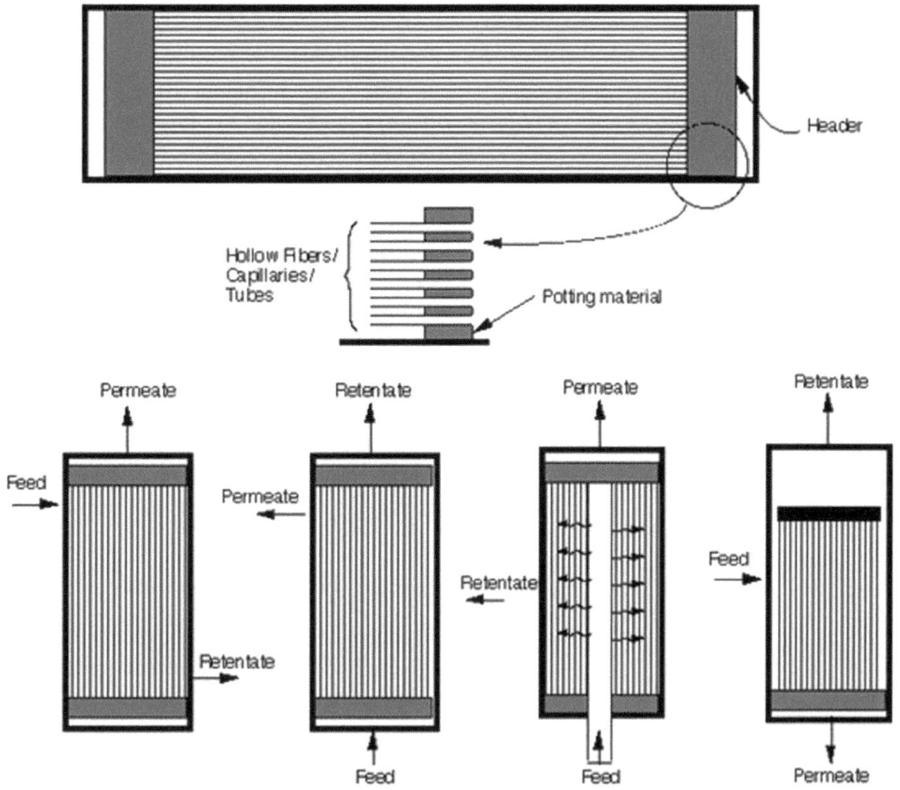

FIGURE 5.20 Hollow-fiber membrane modules with different configurations. (From Ref. [188].)

operation conditions (Δp) should be optimized. The use of composite membranes with nonporous layers is also a solution. Taplyakov et al. [203] simulated a nonporous membrane for biogas upgrading using water as an absorbent and reported promising results.

The selection of the liquid absorbents in membrane contactors is critical. The commonly used absorbents for CO_2 removal are amine-based (i.e., MEA, DEA, and TEA) [197–200]. Recently, membrane contactors using ionic liquids (ILs) as alternative absorbents for capturing acid gases have been reported [204–208]. The unique properties of ILs (non-volatile with high affinity for the acid gas component, thermally and chemically stable [209]) make them very promising as CO_2 absorbents in membrane contactors, especially for applications in harsh conditions, such as CO_2 separation from pre-combustion syngas at elevated temperatures and pressures [207,208,210].

As the wetting problem and the consequent increase in mass transfer resistance in a microporous membrane contactor are almost unavoidable, using a non-porous membrane contactor by coating a thin layer in composite membranes has been studied. The thin dense layer may also bring in additional selectivity and prevent the loss of solvent. The dense skin layer should be very thin to have negligible mass transfer resistance, and should have high gas permeability and good chemical and thermal stabilities [211]. The most studied membrane materials for the dense skin layer are highly permeable membranes (e.g., PDMS, PTMSP, and Teflon-AF 2400). Kreulen et al. [212] reported on a membrane module coated with a thin (0.7 μm) permeable silicone rubber. Even though the wetting issue may be effectively solved, the dense layer added significant mass transfer resistance and thus decreased absorption flux. Scholes et al. [213] observed that using a non-porous PDMS membrane in a membrane contactor has an overall mass transfer coefficient two or three orders of magnitude less than the porous polypropylene (PP) or polytetrafluoroethylene (PTFE) membranes, respectively. Later, Nguyen et al. fabricated a dense membrane contactor by coating two highly CO_2 permeable glassy polymers (PTMSP and Teflon-AF 2400) on porous PP supports [214]. The resultant membrane contactor has improved wetting resistance without sacrificing the overall mass transfer coefficient. More recently, Dai et al. [215] and Ansaloni et al. [216] reported non-porous membrane contactors using Teflon-AF 2400 on PP porous support for a high-pressure application and highly aggressive solvent, respectively. The highly permeable dense skin coating offers new opportunities for cheap supports (e.g., PP) to be used stably in membrane contactors for long-term operation.

5.5 CURRENT APPLICATIONS AND NOVEL DEVELOPMENTS

As already mentioned, most of the membranes used in gas applications today are still polymeric solution–diffusion materials. Among these, the glassy materials separating

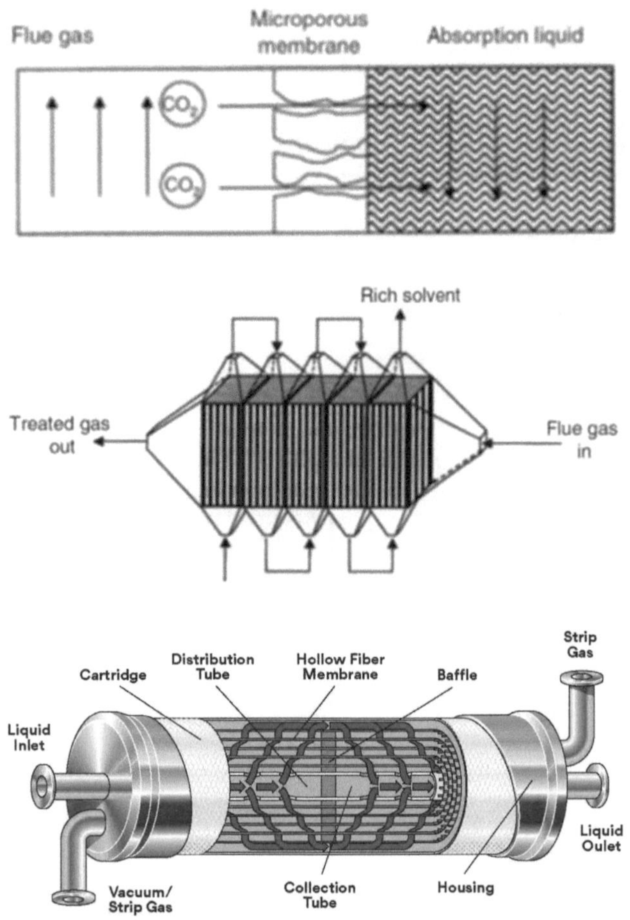

FIGURE 5.21 Illustration of the principle for removal of CO_2 from a gas stream (a) using a membrane contactor in cross flow with hollow fibers (b), and a Liqui-Cel membrane contactor (c). (From Refs. [191,202].)

according to molecular size dominate the market. This will probably change when new tailormade materials are commercialized. For all membrane applications, the gas mixture and process conditions (volumes, pressures, temperature, specifications of product for purity) will govern the choice of membrane material and module design.

The dominating processes for industrial membrane applications remain the production of high-purity nitrogen, recovery of hydrogen from gas streams, and recovery of carbon dioxide. With respect to both hydrogen and CO_2, a major increase in membrane applications may be expected with the development of tailor-made materials and with the world-wide focus on renewable energy and reduction of greenhouse gas emissions. The removal of volatile organic compounds (VOCs) is also a major potential area for membrane applications due to environmental concerns and international agreements on the reduction of emissions. There is only a selection of about 8–9 materials used for 90% of the total gas separation membranes today [217]. An overview of commercial-scale membrane suppliers is presented in Table 5.7, adapted and updated from Ref. [218].

5.5.1 Process Design Considerations

For industrial applications, a single-stage membrane unit is often not sufficient to give the desired product purity. The product may be either permeate or retentate. A cascade solution may be a viable option where a fraction of permeate streams are being recompressed and recycled, membrane units are placed in combinations of series or parallel, or permeate streams are being fed to a second stage for further permeation. These calculations will very quickly become quite complicated and should preferably be performed with an integrated membrane simulation tool hooked up to a standard simulation program (like Hysys, ProII, or Aspen).

Figure 5.22 illustrates a three-stage separation unit typically developed for the separation of CO_2 from CH_4 (natural gas) [219]. More discussion in process design is provided in the following sections in connection with specific applications.

5.5.2 Hydrogen Recovery

Hydrogen is typically recovered from gas streams at refineries (from hydrocrackers), petrochemical plants (adjustment of syngas ratio, dehydrogenation), biohydrogen production (via dark fermentation), and from other streams where hydrogen is present, for example, at ammonia plants. Syngas contains, in addition to H_2 and CO, impurities like N_2, CO_2, CH_4, and water. It is produced mainly from hydrocarbons by (a) stream reforming and (b) partial oxidation of heavy oils or gasification of coke or coal. There may be a significant variation in the stochiometric ratio of H_2/CO for the various chemical syntheses, and the adjustments of the syngas ratio may easily be performed by the application of membrane units. Current commercial units are usually either based on polysulfone hollow fibers or spiral-wound cellulose acetate membranes. Membranes compete with pressure swing adsorption (PSA) and cryogenic systems in hydrogen recovery applications over a wide range of operating conditions. Membrane systems have the advantage of low capital cost and ease of operation. The competing systems, however, usually deliver the purified hydrogen at almost the same pressure as the feed gas, which results in lower compression costs than those of the membrane system, where the hydrogen product always is at a pressure lower than the original feed when using these conventional membranes. The typical performance of membranes for hydrogen recovery in refining applications is shown in Table 5.8 [220].

5.5.2.1 Novel Applications for Hydrogen and Fuel Cells

With the new scenario of hydrogen as a future energy carrier for use in fuel cells, the hydrogen source will no longer only be fossil fuels but will also be produced from biomass and water electrolysis, and possibly from water splitting by algae. There are two ways of producing hydrogen from biomass; either via bio-oil followed by catalytic steam reforming (as for natural gas), or by carefully controlled anaerobic digestion where biogas is produced in mixtures of CH_4, CO_2, N_2, and

TABLE 5.7
Commercial-scale membrane suppliers for gas separation

Company	CO_2	H_2	O_2	N_2	Other*	Website address
AGA (Linde)			X	X		www.linde-gas.com
Aquilo (Whatman)				X		www.aquilo.nl
Asahi Glass (HISEP)			X	X		www.agc.co.jp
Borsig					X	www.borsig.de
Cynara (Dow)	X				X	www.dow.com
Generon (Dow)			X	X	X	www.generon.com
Medal (Du Pont/Air Liquide)	X	X		X		www.medal.airliquide.com
MTR	X			X	X	www.mtrinc.com
Nitto Denko	X				X	www.nitto.com
Osaka Gas			X		X	www.osakagas.co.jp
Air Products	X	X	X	X	X	www.airproducts.com
Linde (Praxair)			X	X		www.lindeus.com/
Toyobo			X		X	www.toyobo.co.jp
Ube Industries	X	X	X	X	X	www.ube.com
Union Carbide (Dow)	X	X	X			www.unioncarbide.com
UOP (Separex)	X	X			X	www.uop.com
Fujifilm	X		X		X	www.fujifilmmembranes.com

* Includes solvent recovery, dehumidification, pervaporation, and helium recovery membranes.

Source: partly adapted from Ref. [218].

TABLE 5.8
Typical hydrogen membrane performance in refining applications

	Hydrogen membrane recovery			
Process stream	**Primary separation**	**Feed purity (%)**	**Permeate purity (%)**	**Recovery (%)**
Catalytic reformer offgas	H_2/CH_4	70–80	90–97	75–95+
Catalytic cracker offgas	H_2/CH_4	15–20	80–90	70–80
Hydroprocessing unit purge gas	H_2/CH_4	60–80	85–95	80–95
Pressure swing adsorption offgas	H_2/CH_4	50–60	80–90	65–85
Butamer process	H_2/CH_4	70	90	

Source: Ref. [220].

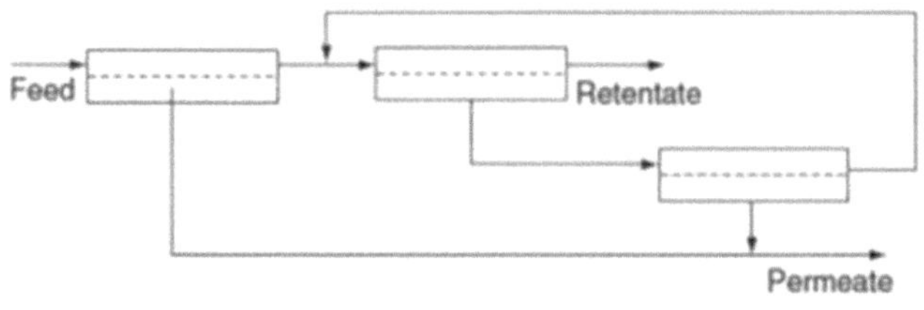

FIGURE 5.22 A three-stage membrane separation process—the recycled gas from stage three needs to be recompressed. This configuration could typically be used for the removal of CO_2 from gas streams where the retentate is the product. (From Ref. [219].)

H_2. A production based on algae and bacteria is a potentially large resource, but the hydrogen production rate is slow, and large membrane areas are needed. The most appropriate organisms have not yet been found, and the utilization of this resource is still far into the future. Depending on whether the hydrogen is going to be stored, transported, or fed directly to a fuel cell, the specifications for product purity may vary. There is a wide variety of fuel cells under development; some are already commercial and in operation, especially within the transport sector. Table 5.9 presents an overview of the various types of fuel cells, their capacities, and their potential applications [221].

A schematic illustration of how two different fuel cells work is shown in Figure 5.23: (a) with pure hydrogen as fuel and (b) with methanol as fuel [222].

When pure hydrogen is to be used as fuel, the gas must first be recovered from the process stream where it is produced. Membranes may be used for this purpose, and several of the new materials under development may be suitable depending on the gas volume to be handled and process conditions like temperature and pressure. Along with the introduction of fuel

TABLE 5.9
Overview of fuel cells (FCs); their capacity and potential application areas

Type of fuel cell	Acronym	Cell output	Temperature range (°C)	Field of application	Special features
Alkaline FC	AFC	0.5–5 kW	50–100°C	Micropower, domestic	Very pure fuel needed
Direct methanol FC	DMFC	Depending on power density	80–100°C	Domestic, residential	Easily stacked
Proton exchange FC	PEM	50–250 kW	50–100°C	Residential, transport	Pure fuel, Pt catalyst
Solid Oxide FC	SOFC	100 kW	850–1000°C	Residential	Large units
Phosphoric acid FC	PAFC	200 kW; 11 MW units	190–210°C	Power station	1.5% CO tolerance
Melted carbonate FC	MCFC	2 MW; 100 MW units	600–1000°C	Power station	

Source: adapted from Ref. [221].

cells, the development of small-scale gas processing units will follow. Several small, integrated production units with direct conversion of fuel to hydrogen gas combined with a membrane unit for hydrogen recovery have been patented [223,224].

With reference to Section 5.3 on materials, the proton conductive palladium (alloys) and carbon membranes will most likely be among the new suitable membranes for hydrogen recovery. Palladium will be a good choice when the highest purity of hydrogen is needed, while for this, an adsorption unit may be needed in combination with carbon molecular sieves. Both can operate at high temperatures, with palladium at the highest. For low-temperature applications, mixed-matrix membranes have very promising potential; these materials will be easier to produce on a large scale and, most likely, at a lower price.

The purified hydrogen needs to be stored as liquid (at –253°C) or compressed gas at around 200 bar—this is due to the low energy density of hydrogen (0.003 kWh/l at 1 bar and ambient temperature, and 0.5 kWh/l at 200 bar). When used in the transport sector, thick steel cylinders for the compressed gas are needed, and stacks of the cylinders must be carried under or on top of the vehicle (bus, truck, ferry). For private cars, the "direct methanol fuel cell" (DMFC) is more attractive. For DMFC, methanol (CH_3OH) is carried as fuel and converted directly to hydrogen. CO_2 will, however, then be produced and, depending on the source of the fuel, may add to the emissions of greenhouse gases. Impurities like CO and H_2S in the feed will poison fuel cells.

Another approach to solving the problem of transport of hydrogen over long distances is to introduce hydrogen gas into the existing domestic natural gas net. This is seriously being looked into also by gas suppliers. At the user end, the gas will be decompressed (to approximately 7 bar), and the hydrogen may be separated from the natural gas and fed to fuel cells where applicable. The ratio of H_2/natural gas can also be adjusted by membranes to specified mixtures of natural gas and hydrogen ("Hythan"; a mixture of hydrogen and methane); mixtures which are currently being tested out as an alternative fuel in the transport sector to reduce emissions while waiting for the fuel cells. The percentage of hydrogen mixed into the transport pipe system must be carefully evaluated for safety reasons, leakage, and material fatigue of the steel pipes. The preferred solution for the production of H_2 in the future would be to use renewable energy (solar, wind, waves) for the electrolysis of water, or direct production by algae.

The market predictions for fuel cell systems worldwide, as presented by Hagler Bailey, are shown in Table 5.10 [225].

5.5.3 CO_2 Removal

Separation of CO_2 from gas streams is required in four areas: (1) purification of natural gas (gas sweetening), (2) separation of CO_2 from enhanced oil recovery (EOR) gas streams, (3) removal of CO_2 from flue gas, and (4) removal of CO_2 from biogas. A fifth area vital for the "space age" should be mentioned: the removal of CO_2 from life support systems onboard spaceships and also in submarines. All these applications have different specifications for the purified gas or for the recovered CO_2, and future membrane applications will most likely be based on "tailormade" materials.

5.5.3.1 CO_2 Removal from Natural Gas

The application of membranes today for CO_2 recovery and natural gas processing is mainly used for gas streams of a moderate volume. For large-volume gas streams, membrane separation today cannot yet compete with the standard amine absorption—the flux and selectivity of the membranes are too low for processing large gas volumes. The membrane separation units found at sites today will often be hybrid solutions with membranes combined with traditional technology. The CO_2 (and H_2S) must be removed from crude natural gas in order to increase heating value and reduce corrosion during transport and distribution. The amount of CO_2 in natural gas is typically in the range of 10% by volume or less, and the gas is at very high operating pressures (35–80 bar or even higher). This will, however, vary greatly throughout the world, depending on the gas fields. Shale gas is the next generation of natural

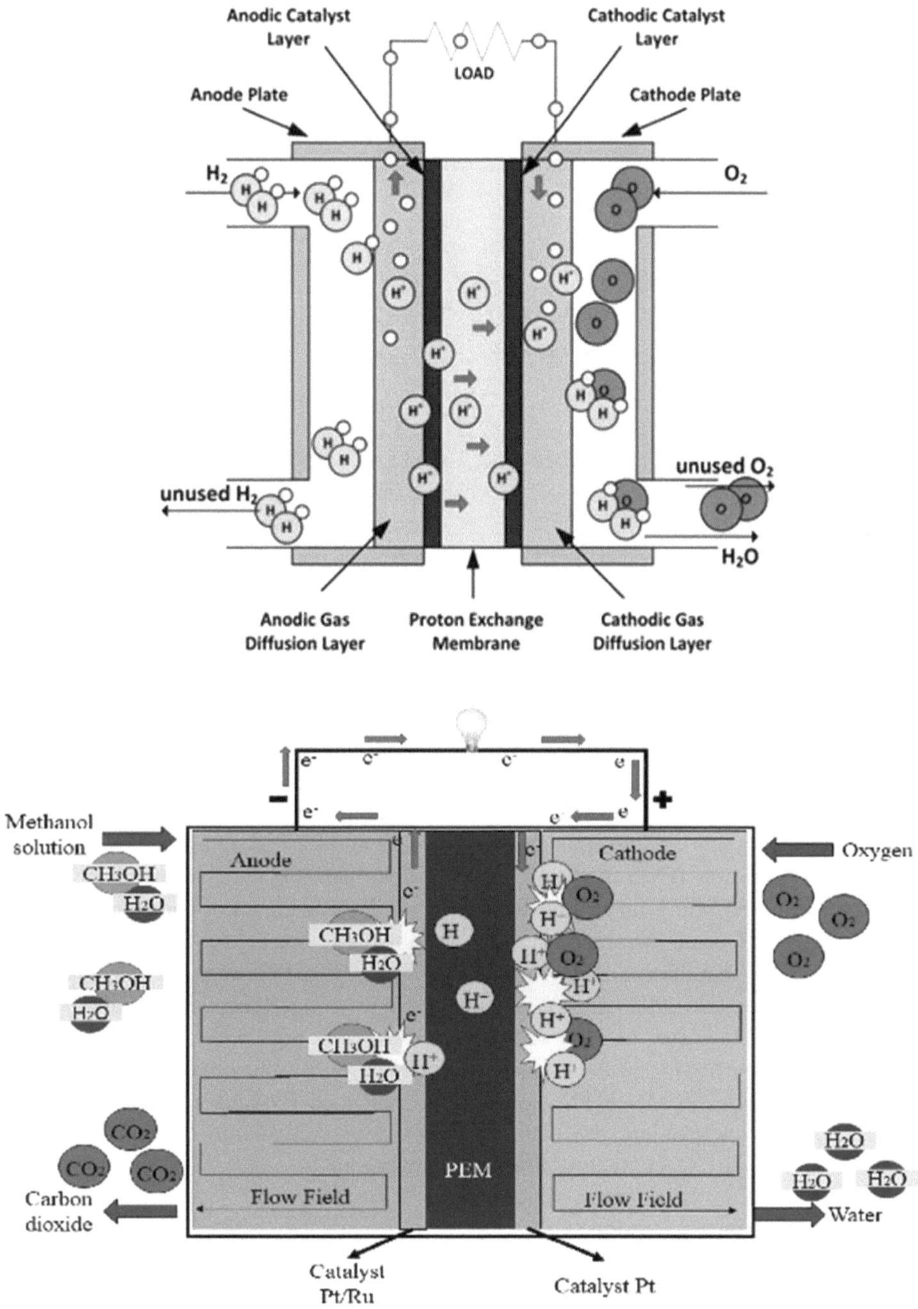

FIGURE 5.23 Schematic illustration of how two different fuel cells work; with pure hydrogen as fuel (a), and with methanol as fuel (b). (From Ref. [222].)

gas which has not started to be exploited—the pressure is generally lower for this gas. The specifications for the removal of sour gases are very strict, and the content of CO_2 should typically be brought down to < 2 vol.% for sales gas. The membranes in operation are typically made from polyimides (PIs) as hollow fibers, or asymmetric cellulose acetates (CAs) as spiral-wound modules. The PI membranes have a higher flux and CO_2/CH_4 selectivity (~20) compared to the CA membranes, are basically hydrophobic, and therefore less vulnerable to water. The CA are, however, more resistant to heavier hydrocarbons. The hollow-fiber configurations will be able to handle large gas volumes with relatively few modules due to high packing density (m^2/m^3).

Recovery of CO_2 in oil and gas production is of major importance to promote enhanced oil recovery (EOR) from depleted fields: high-pressure CO_2 is then pumped back into the reservoir at the periphery of the field and diffuses through the formation to drive residual oil towards the wells. The recycled gas generally needs to have a purity of at least 95 vol.% CO_2.

The main companies producing membranes for CO_2 removal are listed in Table 5.7.

TABLE 5.10
Market predictions for fuel cell systems

Technology application	Projected average annual shipment	of which in Europe	...in the US	... .in Japan	... .in the rest of the world
Distributed generation; low-temperature FC	370 MW (starting 2001)	50 MW	100 MW	200 MW	20 MW
Distributed generation; high-temperature FC	1400 MW (starting 2005)	300 MW	500 MW	500 MW	100 MW
Vehicle FC	200 000 engines (starting 2003)	30 000	100 000	60 000	10 000

Source: Ref. [225].

Some examples of installed membrane units for gas processing are given below:

Example 1: One of the earliest (and largest) membrane plants for EOR was at the SACROC unit in West Texas, which started up in 1984. (The hollow-fiber membrane units are owned and operated by Cynara, a subsidiary of Dow.) In this process, the purified CO_2 stream from the membranes is further treated with hot potassium carbonate prior to reinjection into the oil field. A single membrane stage is used, followed by multiple banks of membrane permeators in parallel; thus, plant performance can be optimized under varying feed conditions by adjusting the number of permeators in operation. (Over the years, the CO_2 content increased from 0.5 vol.% up to a level of approximately 60 vol.%). The Cynara membrane system would successfully process 70 million SCFD of gas containing 40–70 vol.% CO_2 around 1990 [218].

Example 2: In Qadirpur, Pakistan, the world's largest membrane-based natural gas processing plant is situated (1999). This plant is processing 265 MMSCFD natural gas at 59 bar, with plans for expanding the plant to handle 400 MMSCFD [226]. The CO_2 content is reduced from 6.5 mol.% to less than 2 mol.% using a cellulose acetate membrane. The plant is also designed for gas dehydration with membranes.

Examples 3: Other operating membrane plants around the world are: at Kadanwari, Pakistan, a two-stage unit for treatment of 210 MMSCFD gas at 90 bar is operating. In Mexico, an EOR facility is installed to process 120 MMSCFD of gas containing 70 mol.% CO_2; in Egypt, a three-stage unit is operating at Slalm & Tarek, with each unit treating 100 MMSCFD natural gas at 65 bar.

All the plants mentioned are operating with membranes based on hollow-fiber polyimides or spiral-wound cellulose acetate—which is considered to be "proven technology." The environmental aspect related to CO_2 as a greenhouse gas has triggered the development of better membranes for CO_2 removal—this is discussed in more detail in Section 5.3.3.

New materials for natural gas sweetening are basically based on further developments of existing polymeric materials, which can easily be made into hollow-fiber or spiral-wound membrane modules. In addition, new membrane materials used in the natural gas sweetening process must have good plasticization resistance and long-term stability. It is important that the sales gas specifications can be met when membranes are being used for CO_2 removal and dew point control. Several promising materials are under development and being tested out—the market potential is huge. This development is typically done in cooperation with oil and gas companies, and little information is released until the membrane is ready for commercialization. Even though it has been reported that MMMs may have much better CO_2 separation performances and lower physical aging rate, MMMs with high selectivities in favor of CO_2 will still need many years of development to compete with tailormade polymeric membranes for this application. The PIMs, TB polymers, and cardopolymers, as well as FSC membranes, are interesting candidates for natural gas sweetening. Recent investigations by Uddin et al. [24, 227] performed at high pressures (up to 80 bar) have given useful information on the effects of H_2S and higher hydrocarbons on a blended PVAm/PVA membrane over time and also the effects of compaction and plasticization of the polymeric membrane.

5.5.3.2 CO_2 Removal from Biogas

Biogas is a valuable energy resource that can be processed in small plants all over the world. Biogas will have different compositions depending on the source with which it is being produced (see Table 5.11) [228]. The gas is easily collected from dumpsites or anaerobic digesters and may be processed at moderate pressures and ambient temperature. Membrane systems are excellent for this purpose, operating at pressures in the range 5–7 bar. The choice of membrane materials is, therefore, also quite flexible, and small-scale processing units are likely to be developed—a few already exist. Highly CO_2-selective polymeric membranes (cardo polymers, fixed-site-carriers), carbon molecular sieves, mixed-matrix, or biomimetic materials are potential membrane materials for this application. The purified methane can then be compressed to 300 bar, and stored in tanks for fuel in the transport sector or for conversion to methanol used for fuel cells.

TABLE 5.11
Composition of biogas from different sources

Component	Municipal wastewater treatment plants (vol.%)	Dedicated reactors (organic waste) (vol.%)	Landfills (vol.%)
CH_4	55–75	50–90	40–55
CO_2	25–45	10–50	30–40
N_2	Traces	Traces	2–25
O_2	Traces	Traces	0–5
H_2	< 1	< 1	< 1
H_2S	< 1	< 1	< 1
H_2O	4–7	4–7	4–7

Source: Ref. [228].

If the methane is going to be used for high-energy fuel, it must be purified to contain ~95 vol% CH_4. Processing of biogas may, in some cases, be handled by using the existing CO_2-selective membranes. However, the presence of nitrogen in biogas is a major challenge, especially when the biogas is produced from a landfill (see Table 5.11). Nitrogen will remain on the feed side with the methane, and fuel specifications may be difficult to reach. About 3200 m^3(STP)/h of biogas can be collected from a medium-sized dumpsite equivalent to 1700 l/h of fuel oil [229]. The standard way of utilizing biogas (if at all considered) is by burning the gas on-site in combination with a turbine, thus producing energy. This may be sensible if an energy distribution network is easily available. A more environmentally friendly solution is to upgrade the gas to fuel specifications using membranes.

5.5.3.3 CO_2 Capture from Flue Gas

Due to the environmental focus on CO_2 emissions around the world, there are numerous CO_2-selective materials under development—several hundred polymers have been reported (in articles and patents). The main challenge for bringing these membranes into commercialization is to upscale the membranes and membrane modules for large production and document durability over time (maintaining separation properties) during real operating conditions.

The removal of CO_2 from the combustion exhaust gas is a major driver for material development for CO_2 separation membranes. Integrated membrane solutions to power cycles (gas turbines) are in focus. In the exhaust flue gas of a typical post-combustion CO_2 capture process for a coal-fired power plant, CO_2 will be in a mixture with H_2O, N_2, NO_X, O_2, possibly SO_2, and some hydrocarbons, and at low pressure and high temperature. In addition, there will also be dust particles in the gas. The membrane separation will, in this case, meet challenges different from those for natural gas sweetening. Integrated membrane solutions are often considered when combustion/energy production is discussed, both in pre-combustion and post-combustion.

Considering flue gas treatment, as membranes do not tolerate particles, the feed gas has to be filtered before separation. For several membrane materials, water vapor may also be a problem, causing plugging (inorganic microporous materials) or swelling (polymeric materials). Hence, for these applications, the gas must be dried. Depending on the specifications for the CO_2, the co-permeation of gases may be a problem.

The development of membrane materials for CO_2 capture from flue gas has received a great deal of attention during the last decade, clearly as a function of the concern about climate change and the need for carbon capture and sequestration (CCS). Many papers have been published, with only a few mentioned here [14,230–233], but the major challenges for this membrane application are the durability of the material over time as there will be exposure to SOx and NOx, and very high separation performance is needed (flux and selectivity) in order to decrease the required membrane area for the huge volume of gas streams. Very few pilots have been tested around the world, mostly in flat sheet configurations [234–237], and only a few tested with hollow-fiber membranes [238]. The number of pilots is expected to increase substantially in the coming years thanks to increasing public awareness and capital to reduce CO_2 emissions over the last decade.

Among the new membrane materials developed for CO_2 capture from flue gases, facilitated transport membranes and membranes with high CO_2 sorption based on solution-diffusion mechanism have received the most attention, which are effective in obtaining high CO_2 flux at a relatively low CO_2 partial pressure in the feed gas, and tolerate the presence of water vapor, such as cross-linked PEG membranes [239], polynorbornenes-based membranes [240], ionic liquids, or poly-ionic liquids-based membranes [37], and various types of facilitated transport membranes.

Membrane contactors for CO_2 removal deserve special attention. They can be used for natural gas treatment and dehydration, and removal of CO_2 from flue gas. A contactor patented and developed for this purpose by Aker Kvaerner-pilots have been installed and tested both in Norway (at Kårstø) and at a gas terminal in Scotland. This module was based on PTFE membranes. A different commercial contactor based on polyimide membranes was installed at Santos Gas Plant in Queensland, Australia (December 2003). Santos is the largest gas producer in Australia. The membrane contactors are receiving increased interest for CO_2 removal, both for natural gas and flue gas.

5.5.3.4 CO_2 Removal in Life Support Systems

CO_2 removal life support systems are becoming increasingly important as mankind is going into space for longer journeys. The standard method of removing CO_2 from breathing air is to adsorb the gas on molecular sieves, mainly using carbonates. This is a safe way as long as there is enough adsorbent. The preferred way would be to separate CO_2 from the air without using adsorbents but rather highly CO_2-selective membranes, then dispose of it or bring it back into the life cycle using water (from urine) and green plants grown artificially in space. Nanostructured materials, fixed-site carriers, or biomimetic membranes would most likely be suitable for this application.

5.5.4 Air Separation

Membranes for air separation have become increasingly important over the years, and the development of continuously better materials has been rapid since Dow and Permea presented their first membrane systems in 1985. The first membranes used for air separation were based on poly(4-methyl-1-pentene) (TPX) and ethyl cellulose, and had selectivities for O_2/N_2 of around 4–5 [241]. Today, the second-generation air separation membranes are on the market, usually as hollow-fiber modules with very high packing density, and based on polymers such as polysulfones (PS) and poly(phenylene)oxide (PPO). Remembering the very strong inverse relationship between selectivity and permeability for oxygen and nitrogen, as illustrated by Robeson in Figure 5.1 [4], it may easily be understood that there is a major difference in producing high-purity nitrogen compared to high-purity oxygen. Air separation using membranes is today considered proven technology, and there are several producers in addition to Dow (now upgraded system Generon II) and Permea (Prism Alpha; now Air Products). Detailed information on membrane solutions for air separation can be found in many textbooks, handbooks, and references [218,241].

5.5.4.1 Production of High-Purity Nitrogen

High-purity nitrogen is used for many applications: as blanketing gas in the oil and gas industry, as purge and blanketing in the chemical industry, in the food industry for packaging, and various others. When air is separated by polymeric membranes, oxygen will be the faster-permeating component with a selectivity in the range of 2–9 over nitrogen. Selectivity ~15 has been measured with carbon membranes. This means that when compressed air is fed to the membrane, nitrogen is retained at high pressure on the feed side, which is usually an advantage. The competing separation technologies are typically cryogenic distillation and PSA. Baker recommended a production range for applying the different technologies [241]. Even with low-selectivity membranes, high-purity nitrogen may be obtained (99.9%), but the cost of the system increases significantly in the range of 95–99% nitrogen purity. Various process designs may be considered for cost reduction. A process may be optimized to reduce the membrane area and compressor cost for a required product of 99% purity nitrogen. The three-step process will, however, be limited to large systems where energy and membrane area savings compensate for the extra complexity and higher maintenance cost of a second compressor. A comprehensive discussion of factors affecting the design of nitrogen plants is given by Prasad et al. [226]

5.5.4.2 Oxygen-Enriched Air

Oxygen-enriched air will be produced on the low-pressure permeate side of the air separation membranes. The oxygen-enriched permeate stream is usually vented, but there is increasing interest in using this gas for combustion. High-purity oxygen cannot be produced with polymeric air separation membranes. Calculations easily show that with a selectivity of 8 for O_2/N_2, an infinite pressure ratio, and the zero stage cut, the permeate can only reach, as a "best case," a 68% purity for O_2. These constraints explain why oxygen-enriched air in these systems usually is in the range of 30–50 % purity for O_2.

5.5.4.3 Developments for High-Purity Oxygen

Oxygen-conductive membranes are under development. These membranes are suitable for advanced power generation requiring oxygen for combustion or gasification, and are based on zirconia and perovskite, where oxygen is transported through the material as O^{2-}. The materials are stable at very high temperatures (> 500°C). A schematic illustration of the ITM (Ion Transport Membrane) developed by Air Products can be found in Ref. [163]. Details on this technology can be found in Ref. [242]. The Danish company, Haldor Topsoe A/S, has presented a patented solution with an integrated ion-conducting membrane for auto-thermal steam reforming; several other patents also have been announced [243].

5.5.5 Recovery of Volatile Organic Compounds (VOCs)

With the increased focus on climate change and greenhouse gases, the need for reducing VOC emissions has been brought into focus. Major sources for these emissions are evaporation from oil and gas tankers during transport and bulk handling of oil at buoys and terminals, refineries, and petrochemical plants. Based on international agreements on "Long Range Transboundary Air Pollution" (Geneva Convention of 1979 with later protocols) and "Clean Air Act Amendments" of 1990 in the US, various technologies for VOC recovery have been developed.

The main technologies are based on the recirculation of the VOC-rich stream for absorption in oil. For large gas volumes, this demands a complicated system of cooling and/or recompressions of the VOC and venting of the purified air. Process units for installation on tankers have been developed by several companies (Aker Kvaerner Process Systems, Hamworthy, and others). For moderate-volume air streams, membrane solutions exist. A rubbery membrane allowing the VOCs to permeate (and inerts to be retained) is used. A hybrid solution with PSA may be necessary to remove traces of VOC before the gas is vented [244]. The challenge remains to develop efficient process schemes for the filling and unloading tankers. The volume % VOC in the inert gas stream may vary between 8–60% from start to end of the filling operation—this complicates the recovery process. Combinations with membranes for concentrating the VOC before absorption are being looked into.

Even though significant progress has been made in membrane-based VOCs separation/recovery, the following aspects still need attention: (1) specific membrane materials for the treatment and recovery of different types of VOCs to meet the needs of industrial applications; (2) the stability issues of membranes under operation conditions; (3) compatibility and dispersibility of polymers and porous materials when developing MMMs, the most promising VOCs separation membranes; (4) process optimization, such as new module designs to maximize the performance of the membrane; (5) regeneration of membrane, such as developing multifunctional composite membranes with self-cleaning effects; and (6) using hybrid processes to couple membrane separation and other technologies to achieve maximum purification and recovery efficiency [245].

5.5.6 Separation of Hydrocarbons

Successful separation of alkanes and alkenes has been documented when microporous membranes have been used [246]. The physiochemical properties, size, and shape of the molecules will play an important role in the separation; hence critical temperatures and gas molecule configurations should be carefully evaluated for the gases in the mixture. Table 5.12 lists the properties of ethane, ethylene, propane, and propylene.

Based on the gas properties and process conditions, the separation may be performed according to selective surface flow or molecular sieving. The transport may also be enhanced by having an Ag compound in the membrane. The Ag^+ ion will form a reversible complex with the alkene and facilitate transport results. Selectivities in the range of 200–300 have been reported for the separation of ethene– ethane and propene–propane [246]. Successful separation of alkanes and alkenes will be important for the petrochemical industry. Today the surplus hydrocarbons in purge gas are usually flared. Membranes suitable for this application are the carbon molecular sieves and nanostructured materials.

An overview on the development of membranes and processes for hydrocarbon separation and removal can be found in a few recent reviews [247–251]. Different types of membranes and membrane processes have been investigated for olefin/paraffin separation, including conventional polymeric membranes [252,253], mixed-matrix membranes [254,255], facilitated membranes [256,257], and membrane contactors [247,258]. According to the literature, separation performances of various membranes for both C2 and C3 separation have been greatly improved during the past 20 years of research and development.

TABLE 5.12
Physical properties of ethane, ethylene, propane, and propylene

	C_2H_4	C_2H_6	C_3H_6	C_3H_8
Boiling point (°C)	−103.9	−88.6	−48	−42.2
σ_{LJ} (Å)	3.7	4.1	4.0	4.3

5.5.7 Other Applications

5.5.7.1 Water Vapor Removal from Air

Membranes are used to dehydrate process air streams as a replacement for desiccant dryers or adsorption systems. Such membrane units have been on the market for many years, but they are mainly for small gas streams. The membranes being used have very high water-to-air selectivity. The dehumidification units are usually connected to a compressed air line, and loss of pressurized air through the membrane may be a major cost.

5.5.7.2 Dehydration of Natural Gas

Natural gas has to be dried in order to prevent water from freezing or hydrates from forming in pipelines for gas distribution. The potential application of membranes for gas dehydration is very large. There are already numerous polymers available with very high selectivities for water vapor; 500–2000 are beneficial, and a breakthrough for the commercialization of a few materials is expected. The water permeance should preferably be at least 30 m^3 (STP)/m^2h bar [259]. The problem is that with increasing selectivity at a given water vapor flux, the necessary permeation area also increases. If the gas on the permeate side could be used for low-pressure fuel at the site, a membrane solution for dehydration would be economical and competitive to glycol dehydration. If the permeate gas has to be recompressed, the costs will probably be too high. Permea Maritime Protection (a division of Air Products) is one of the companies having presented a commercial unit already installed in the North Sea.

Zeolites have been used for many years as an adsorbent in a wide range of industrial applications (also for smaller volumes of natural gas) because of the regular and controllable pore size in their crystalline structure. The highly hydrophilic nature of these materials means that water is always preferentially adsorbed. Regeneration is, however, necessary when the adsorbent is fully loaded.

5.5.7.3 Helium Recovery

Helium is a non-renewable inert gas produced from natural gas, which is present in very low concentrations. Helium is a valuable gas that has a wide range of applications in science, medicine, and industry. It is also used as a diluent for deep-sea divers breathing gas mixtures [260]. Helium-rich natural gas is currently the only source of helium production [261]. Typical gas compositions in natural gas include methane (70–90%), C_2–C_3 hydrocarbons (0–20%), heavy hydrocarbons (1–3%), CO_2 (0–10%), nitrogen (0–10%) [261], and small amounts of hydrogen sulfide and helium. In a typical natural gas processing process, after the removal of H_2S, CO_2, H_2O heavy hydrocarbons, and most of the CH_4, the helium gas recovery unit can be used to concentrate helium from 1–30% to 50–70% (aka crude helium), and then the helium-enriched product is sent to the helium upgrading device for further purification. Since the kinetic diameters of helium (2.6 Å) are quite different from other components of natural gas (such as 3.8 Å for CH_4 and 3.64 Å for N_2), membrane technology is considered a promising alternative to conventional helium recovery and purification technology. As early as the 1960s, membrane technology emerged as a viable alternative for separating helium from natural gas [262]. However, in recent years, breakthroughs in new materials, especially new polymers and inorganic nanostructured materials, have led to the exponential growth of new helium separation membranes. In addition, a number of membrane-based helium purification processes, including hybrid processes where membranes are combined with other technologies, have been reported for process optimization through modeling and simulation [263].

5.5.7.4 Recovery of Aggressive Gases: Cl_2 and HCl

Membranes for the purification or recovery of aggressive gases have been under development for many years, and are expected to be commercialized within a few years. The main challenge for membrane separation of gases like Cl_2 and HCl is the durability of the material—this has been thoroughly documented in several works [51,90,264,265]. The only polymers that can withstand the aggressive process environment are perfluorinated materials. These membranes do not (yet) have satisfactory separation properties for Cl_2 or HCl in mixtures with more inert gases. There are two ways for membrane development for this application: (a) using perfluorinated compounds to surface modify microporous glass membranes, or (b) possibly making a mixed-matrix membrane with the perfluorinated polymer as the continuous phase and molecular sieving glass fibers as the sieving phase. The potential for membrane application within this field is very large, as chlorine is a widely used chemical in numerous industries worldwide. The competing purification methods today are based on chemical reactions with additives or cryogenic distillation.

5.6 CONCLUSIONS AND FUTURE PERSPECTIVES

There is potential for membrane separation of almost any gas from a mixture of gases if the physical and chemical properties are carefully considered as well as suitable material properties and durability, possible transport mechanisms, and optimum process conditions evaluated. Membranes for gas separation applications have attracted considerable attention in both academia and industry over the past few decades. Great R&D effort is dedicated not only to developing membrane materials with higher gas separation performance, better long-term stability, and durability under harsh operating conditions but, more importantly, also to module design, upscaling, and industrial demonstrations. Since the first industrial application of hydrogen recovery, gas separation membranes have been expanded into many other fields, such as air separation, VOCs recovery, and CO_2 capture/separation from various mixtures. There is no doubt that gas separation membranes will find more applications in the coming years. Creative reflection and advanced research will enable the development of this environmentally friendly separation technology for applications within many more areas in the future and hopefully be able to replace older, energy-intensive (and not so clean) technologies or combine them with as hybrid process solutions. However, no matter how strongly motivated, the economics of the final solution will always be the major consideration for commercialization.

ACKNOWLEDGMENTS

The authors thank colleagues and researchers in the membrane research group at NTNU for valuable discussions and patience during the writing of this chapter. Special thanks are also due to the following members and former members of the authors' research groups: Dr. Jon Arvid Lie, Dr. TaekJoong Kim, Dr. Arne Lindbrathen, and Jing Wei for valuable help with references, figures, and proofreading.

REFERENCES

1. Mulder, M., Mulder, J. (1996) *Basic Principles of Membrane Technology*. Springer Science & Business Media.
2. Loeb, S., Sourirajan, S. (1962) *Sea Water Demineralization by Means of an Osmotic Membrane*. ACS Publications.
3. Henis, J.M., Tripodi, M.K. (1980) A novel approach to gas separations using composite hollow fiber membranes. *Separation Science and Technology* 15(4): 1059–1068.
4. Robeson, L.M. (1991) Correlation of separation factor versus permeability for polymeric membranes. *Journal of Membrane Science* 62(2): 165–185.
5. Robeson, L.M. (2008) The upper bound revisited. *Journal of Membrane Science* 320(1): 390–400.
6. Swaidan, R., Ghanem, B., Pinnau, I. (2015) *Fine-tuned Intrinsically Ultramicroporous Polymers Redefine the Permeability/Selectivity Upper Bounds of Membrane-based Air and Hydrogen Separations*. ACS Publications.
7. Comesaña-Gándara, B., et al. (2019) Redefining the Robeson upper bounds for CO_2/CH_4 and CO_2/N_2 separations using a series of ultrapermeable benzotriptycene-based polymers of intrinsic microporosity. *Energy & Environmental Science* 12(9): 2733–2740.
8. Wu, A.X., Drayton, J.A., Smith, Z.P. (2019) The perfluoropolymer upper bound. *AIChE Journal* 65(12): e16700.
9. Australasia, M.S.A.; Available from: https://membrane-australasia.org/msa-activities/polymer-gas-separation-membrane-database/ (last accessed 2021).
10. Grinevich, Y., et al. (2011) Solubility controlled permeation of hydrocarbons in novel highly permeable polymers. *Journal of Membrane Science* 378(1–2): 250–256.
11. Finkelshtein, E.S., et al. (2006) Addition-type polynorbornenes with $Si(CH_3)_3$ side groups: synthesis,

gas permeability, and free volume. *Macromolecules* 39(20): 7022–7029.
12. Park, H.B., et al. (2007) Polymers with cavities tuned for fast selective transport of small molecules and ions. *Science* 318(5848): 254–258.
13. Park, H.B., et al. (2010) Thermally rearranged (TR) polymer membranes for CO_2 separation. *Journal of Membrane Science* 359(1–2): 11–24.
14. Kim, T.-J., et al., (2013) Separation performance of PVAm composite membrane for CO2 capture at various pH levels. *Journal of Membrane Science* 428: 218–224.
15. He, X., Hägg, M.-B. (2012) Structural, kinetic and performance characterization of hollow fiber carbon membranes. *Journal of Membrane Science* 390: 23–31.
16. Budd, P.M., et al., *Polymers of intrinsic microporosity (PIMs): robust, solution-processable, organic nanoporous materials.* Chemical communications, 2004(2): p. 230–231.
17. McKeown, N.B. (2012) *Polymers of Intrinsic Microporosity.* International Scholarly Research Notices.
18. Budd, P.M., et al. (2005) Gas separation membranes from polymers of intrinsic microporosity. *Journal of Membrane Science* 251(1–2): 263–269.
19. Scholander, P. (1960) Oxygen transport through hemoglobin solutions: How does the presence of hemoglobin in wet membrane mediate an eightfold increase in oxygen passage? *Science* 131(3400): 585–590.
20. Ward, I.W.J., Robb, W.L. (1968) *Liquid Membranes for Use in the Separation of Gases*. Google Patents.
21. Kim, T.J., Li, B., Hägg, M.B. (2004) Novel fixed-site–carrier polyvinylamine membrane for carbon dioxide capture. *Journal of Polymer Science Part B: Polymer Physics* 42(23): 4326–4336.
22. Deng, L., Kim, T.-J., Hägg, M.-B. (2009) Facilitated transport of CO_2 in novel PVAm/PVA blend membrane. *Journal of Membrane Science* 340(1–2): 154–163.
23. Sandru, M., Haukebø, S.H., Hägg, M.-B. (2010) Composite hollow fiber membranes for CO_2 capture. *Journal of Membrane Science* 346(1): 172–186.
24. Uddin, M.W., Hägg, M.-B. (2012) Natural gas sweetening—the effect on CO_2–CH_4 separation after exposing a facilitated transport membrane to hydrogen sulfide and higher hydrocarbons. *Journal of Membrane Science* 423: 143–149.
25. Sandru, M., et al. (2022) An integrated materials approach to ultrapermeable and ultraselective CO_2 polymer membranes. *Science* 376(6588): 90–94.
26. Koresh, J.E., Soffer, A. (1987) The carbon molecular sieve membranes. General properties and the permeability of CH_4/H_2 mixture. *Separation Science and Technology* 22(2–3): 973–982.
27. Rao, M., Sircar, S. (1993) Nanoporous carbon membranes for separation of gas mixtures by selective surface flow. *Journal of Membrane Science* 85(3): 253–264.
28. Jones, C.W., Koros, W.J. (1994) Carbon molecular sieve gas separation membranes—I. Preparation and characterization based on polyimide precursors. *Carbon* 32(8): 1419–1425.
29. Fuertes, A.B. (2000) Adsorption-selective carbon membrane for gas separation. *Journal of Membrane Science* 177(1–2): 9–16.
30. Singh, A., Koros, W. (1996) Significance of entropic selectivity for advanced gas separation membranes. *Industrial & Engineering Chemistry Research* 35(4): 1231–1234.
31. Bloch, E.D., et al. (2012) Hydrocarbon separations in a metal-organic framework with open iron (II) coordination sites. *Science* 335(6076): 1606–1610.
32. Deng, J., et al. (2020) Morphologically tunable MOF nanosheets in mixed matrix membranes for CO_2 separation. *Chemistry of Materials* 32(10): 4174–4184.
33. Li, L., et al. (2018) Ethane/ethylene separation in a metal-organic framework with iron-peroxo sites. *Science* 362(6413): 443–446.
34. Qian, Q., et al. (2020) MOF-based membranes for gas separations. *Chemical Reviews* 120(16): 8161–8266.
35. Burggraaf, A. (1996) Transport and separation properties of membranes with gases and vapours. In *Membrane Science and Technology*, pp. 331–433. Elsevier.
36. Dai, Z., Ansaloni, L., Deng, L. (2016) Recent advances in multi-layer composite polymeric membranes for CO_2 separation: A review. *Green Energy & Environment* 1(2): 102–128.
37. Dai, Z., et al. (2016) Combination of ionic liquids with membrane technology: A new approach for CO_2 separation. *Journal of Membrane Science* 497: 1–20.
38. Xie, K., et al. (2019) Recent progress on fabrication methods of polymeric thin film gas separation membranes for CO_2 capture. *Journal of Membrane Science* 572: 38–60.
39. MTR Inc. (2022) www.mtrinc.com/ (last accessed 2022).
40. Malykh, O., Golub, A.Y., Teplyakov, V. (2011) Polymeric membrane materials: New aspects of empirical approaches to prediction of gas permeability parameters in relation to permanent gases, linear lower hydrocarbons and some toxic gases. *Advances in Colloid and Interface Science* 164(1–2): 89–99.
41. Teplyakov, V., Meares, P. (1990) Correlation aspects of the selective gas permeabilities of polymeric materials and membranes. *Gas Separation & Purification* 4(2): 66–74.
42. Koros, W.J. (2002) Gas separation membranes: needs for combined materials science and processing approaches. In: *Macromolecular Symposia*. Wiley Online Library.
43. Omidvar, M., et al. (2018) Sorption-enhanced membrane materials for gas separation: a road less traveled. *Current Opinion in Chemical Engineering* 20: 50–59.
44. Jordan, S. (1988) *The Effects of Carbon Dioxide Exposure on the Permeability Behavior of Silicone Rubber and Glassy Polycarbonates*. Ph. D. diss., University of Texas at Austin.
45. Koros, W.J., Fleming, G. (1993) Membrane-based gas separation. *Journal of Membrane Science* 83(1): 1–80.
46. Baker, R., Blume, I. (1986) Permselective membranes separate gases. *Chemtech* 16(4): 232–238.
47. Geankoplis, C.J. (2003) *Transport Processes and Separation Process Principles*, 4th ed. Upper Saddle River, NJ: Pearson.
48. Burggraaf, A.J., Cot, L. (1996) *Fundamentals of Inorganic Membrane Science and Technology*. Elsevier.
49. Gilron, J., Soffer, A. (2002) Knudsen diffusion in microporous carbon membranes with molecular sieving character. *Journal of Membrane Science* 209(2): 339–352.
50. Gilliland, E.R., et al. (1974) Diffusion on surfaces. I. Effect of concentration on the diffusivity of physically adsorbed gases. *Industrial & Engineering Chemistry Fundamentals* 13(2): 95–100.
51. Lindbråthen, A., Hägg, M.-B. (2005) Glass membranes for purification of aggressive gases: Part I: Permeability and stability. *Journal of Membrane Science* 259(1–2): 145–153.

52. Stern, S., Shah, V., Hardy, B. (1987) Structure-permeability relationships in silicone polymers. *Journal of Polymer Science Part B: Polymer Physics* 25(6): 1263–1298.
53. Glasstone, S., Laidler, K.J., Eyring, H. (1941) *The Theory of Rate Processes; The Kinetics of Chemical Reactions, Viscosity, Diffusion and Electrochemical Phenomena.* McGraw-Hill Book Company.
54. Buxbaum, R.E., Marker, T.L. (1993) Hydrogen transport through non-porous membranes of palladium-coated niobium, tantalum and vanadium. *Journal of Membrane Science* 85(1): 29–38.
55. Wang, F., et al. (1999) Synthesis of poly (ether ether ketone) with high content of sodium sulfonate groups and its membrane characteristics. *Polymer* 40(3): 795–799.
56. Kondo, T. (1992) *New Developments in Gas Separation Technology: From Basic Principles to Frontier Technologies, Applications and Future Trends.* Toray Research Center.
57. Wang, H., Cong, Y., Yang, W. (2002) Oxygen permeation study in a tubular Ba0.5Sr0.5Co0.8Fe0.2O_3-δ oxygen permeable membrane. *Journal of Membrane Science* 210(2): 259–271.
58. Ward III, W.J., Robb, W.L. (1967) Carbon dioxide-oxygen separation: Facilitated transport of carbon dioxide across a liquid film. *Science* 156(3781): 1481–1484.
59. Majumdar, S., Sirkar, K.K., Sengupta, A. (1992) Hollow-fiber contained liquid membrane. In *Membrane Handbook*, pp. 764–808. Springer..
60. Majumdar, S., et al. (1994) Simultaneous SO_2/NO separation from flue gas in a contained liquid membrane permeator. *Industrial & Engineering Chemistry Research* 33(3): 667–675.
61. Sirkar, K. (1996) *Hollow Fiber-Contained Liquid Membranes for Separations: An Overview.* ACS Symposium Series, Vol. 642. American Chemical Society.
62. Jeong, S.-H., Lee, K.-H. (1999) Separation of CO_2 from CO_2/N_2 mixture using supported polymeric liquid membranes at elevated temperatures. *Separation Science and Technology* 34(12): 2383–2394.
63. LeBlanc Jr, O.H., et al. (1980) Facilitated transport in ion-exchange membranes. *Journal of Membrane Science* 6: 339–343.
64. Yoshikawa, M., et al. (1988) Selective permeation of carbon dioxide through synthetic polymer membranes having pyridine moiety as a fixed carrier. *Journal of Applied Polymer Science* 35(1): 145–154.
65. Noble, R.D. (1990) Analysis of facilitated transport with fixed site carrier membranes. *Journal of Membrane Science* 50(2): 207–214.
66. Deng, L., Hägg, M.-B. (2010) Swelling behavior and gas permeation performance of PVAm/PVA blend FSC membrane. *Journal of Membrane Science* 363(1–2): 295–301.
67. Quinn, R., Appleby, J., Pez, G. (1995) New facilitated transport membranes for the separation of carbon dioxide from hydrogen and methane. *Journal of Membrane Science* 104(1–2): 139–146.
68. Quinn, R., Laciak, D. (1997) Polyelectrolyte membranes for acid gas separations. *Journal of Membrane Science* 131(1–2): 49–60.
69. Cussler, E. (2018) Facilitated and active transport. In: *Polymeric Gas Separation Membranes*, pp. 273–300. CRC Press.
70. Cussler, E.L. (1994) Facilitated and active transport. In: E.L. Cussler (Ed.) *Polymeric Gas Separation Membranes*, 1st Edition. CRC Press.
71. Koros, W.J., Chern, R.T. (1987) Separation of gaseous mixtures using polymer membranes. *Handbook of Separation Process Technology*, 862–953.
72. Chern, R., et al. (1983) *Implications of the Dual-Dode Sorption and Transport Models for Mixed Gas Permeation.* ACS Publications.
73. Kesting, R.E., Fritzsche, A. (1993) *Polymeric Gas Separation Membranes.* Wiley-Interscience.
74. Vieth, W.R.W.R., (1991) *Diffusion In and Through Polymers.* Defect & Diffusion Forum.
75. Vrentas, J., Duda, J. (1977) Solvent and temperature effects on diffusion in polymer–solvent systems. *Journal of Applied Polymer Science* 21(6): 1715–1728.
76. Morisato, A., et al. (1993) The influence of chain configuration and, in turn, chain packing on the sorption and transport properties of poly (tert-butyl acetylene). *Journal of Applied Polymer Science* 49(12): 2065–2074.
77. Koros, W., Hellums, M. (1989) Gas separation membrane material selection criteria: Differences for weakly and strongly interacting feed components. *Fluid Phase Equilibria* 53: 339–354.
78. Kim, T., et al. (1988) Relationship between gas separation properties and chemical structure in a series of aromatic polyimides. *Journal of Membrane Science* 37(1): 45–62.
79. Freeman, B.D. (1999) Basis of permeability/selectivity tradeoff relations in polymeric gas separation membranes. *Macromolecules* 32(2): 375–380.
80. Park, H.B., et al. (2017) Maximizing the right stuff: The trade-off between membrane permeability and selectivity. *Science* 356(6343): eaab0530.
81. Lin, H., Freeman, B.D. (2004) Gas solubility, diffusivity and permeability in poly (ethylene oxide). *Journal of Membrane Science* 239(1): 105–117.
82. Patel, N.P., Miller, A.C., Spontak, R.J. (2003) Highly CO_2-permeable and selective polymer nanocomposite membranes. *Advanced Materials* 15(9): 729–733.
83. Hirayama, Y., et al. (1999) Permeation properties to CO_2 and N_2 of poly (ethylene oxide)-containing and crosslinked polymer films. *Journal of Membrane Science* 160(1): 87–99.
84. Deng, J., Dai, Z., Deng, L. (2020) Effects of the morphology of the ZIF on the CO_2 separation performance of MMMs. *Industrial & Engineering Chemistry Research* 59(32): 14458–14466.
85. Dai, Z., et al. (2019) Highly CO_2-permeable membranes derived from a midblock-sulfonated multiblock polymer after submersion in water. *NPG Asia Materials* 11(1): 53.
86. Dai, Z., et al. (2019) Nafion/PEG hybrid membrane for CO_2 separation: Effect of PEG on membrane micro-structure and performance. *Separation and Purification Technology* 214: 67–77.
87. Dai, Z., et al. (2019) Incorporation of an ionic liquid into a midblock-sulfonated multiblock polymer for CO_2 capture. *Journal of Membrane Science* 588: 117193.
88. Arcella, V., Ghielmi, A., Tommasi, G. (2003) *High performance perfluoropolymer films and membranes. Annals of the New York Academy of Sciences* 984(1): 226–244.
89. Kharitonov, A., et al. (2004) Direct fluorination of the polyimide matrimid® 5218: The formation kinetics and

physicochemical properties of the fluorinated layers. *Journal of Applied Polymer Science* 92(1): 6–17.
90. Hägg, M.-B. (2000) Membrane purification of Cl_2 gas: II. Permeabilities as function of temperature for Cl_2, O_2, N_2, H_2 and HCl in perfluorinated, glass and carbon molecular sieve membranes. *Journal of Membrane Science* 177(1–2): 109–128.
91. Hiarayama, Y., et al. (1995) Novel membranes for carbon dioxide separation. *Energy Conversion and Management* 6(36): 435–438.
92. Alghannam, A.A., et al. (2018) High pressure pure- and mixed sour gas transport properties of Cardo-type block co-polyimide membranes. *Journal of Membrane Science* ***553***: 32–42.
93. Yahaya, G.O., et al. (2020) Development of thin-film composite membranes from aromatic cardo-type co-polyimide for mixed and sour gas separations from natural gas. *Global Challenges* 4(7): 1900107.
94. Takeuchi, M., Sakamoto, Y., Niwa, S. (1999) Study on RITE CO_2 global recycling system. In: *Proceedings of the5th International Conference on CO_2 Utilization (C02 ICCDU V)*. Karlsruhe, Germany, September, 1999.
95. Schultz, J., Peinemann, K.-V. (1996) Membranes for separation of higher hydrocarbons from methane. *Journal of Membrane Science* 110(1): 37–45.
96. Pinnau, I., et al. (1997) Long-term permeation properties of poly (1-trimethylsilyl-1-propyne) membranes in hydrocarbon—Vapor environment. *Journal of Polymer Science Part B: Polymer Physics* 35(10): 1483–1490.
97. Pinnau, I., He, Z., Morisato, A. (2004) Synthesis and gas permeation properties of poly (dialkylacetylenes) containing isopropyl-terminated side-chains. *Journal of Membrane Science* 241(2): 363–369.
98. Kelman, S.D., et al. (2008) Crosslinking poly [1-(trimethylsilyl)-1-propyne] and its effect on physical stability. *Journal of Membrane Science* 320(1–2): 123–134.
99. Hill, A.J., et al. (2004) Influence of methanol conditioning and physical aging on carbon spin-lattice relaxation times of poly (1-trimethylsilyl-1-propyne). *Journal of Membrane Science* 243(1–2): 37–44.
100. Kocherlakota, L.S., et al. (2012) Enhanced gas transport properties and molecular mobilities in nano-constrained poly [1-(trimethylsilyl)-1-propyne] membranes. *Polymer* 53(12): 2394–2401.
101. Choi, J.I., et al. (2010) Thermally rearranged (TR) poly (benzoxazole-co-pyrrolone) membranes tuned for high gas permeability and selectivity. *Journal of Membrane Science* 349(1–2): 358–368.
102. Emmler, T., et al. (2010) Free volume investigation of polymers of intrinsic microporosity (PIMs): PIM-1 and PIM1 copolymers incorporating ethanoanthracene units. *Macromolecules* 43(14): 6075–6084.
103. McKeown, N.B., et al. (2005) Polymers of intrinsic microporosity (PIMs): bridging the void between microporous and polymeric materials. *Chemistry–A European Journal* 11(9): 2610–2620.
104. Budd, P.M., et al. (2008) Gas permeation parameters and other physicochemical properties of a polymer of intrinsic microporosity: Polybenzodioxane PIM-1. *Journal of Membrane Science* 325(2): 851–860.
105. Low, Z.-X., et al. (2018) Gas permeation properties, physical aging, and its mitigation in high free volume glassy polymers. *Chemical Reviews* 118(12): 5871–5911.
106. Carta, M., et al. (2013) An efficient polymer molecular sieve for membrane gas separations. *Science* 339(6117): 303–307.
107. Zhuang, Y., et al. (2016) Soluble, microporous, Tröger's base copolyimides with tunable membrane performance for gas separation. *Chemical Communications* 52(19): 3817–3820.
108. Wang, Z.G., et al. (2014) Tröger's base-based copolymers with intrinsic microporosity for CO_2 separation and effect of Tröger's base on separation performance. *Polymer Chemistry* 5(8): 2793–2800.
109. Fan, Y., et al. (2019) Tröger's base mixed matrix membranes for gas separation incorporating NH2-MIL-53 (Al) nanocrystals. *Journal of Membrane Science* 573: 359–369.
110. Deng, J., Dai, Z., Deng, L. (2020) H_2-selective Troger's base polymer based mixed matrix membranes enhanced by 2D MOFs. *Journal of Membrane Science* 610: 118262.
111. Zou, X., Zhu, G. (2018) Microporous organic materials for membrane-based gas separation. *Advanced Materials* 30(3): 1700750.
112. Way, J., et al. (1987) Facilitated transport of CO_2 in ion exchange membranes. *AIChE Journal* 33(3): 480–487.
113. Pellegrino, J., et al. (1993) Gas transport properties of solution-cast perfluorosulfonic acid ionomer films containing ionic surfactants. *Journal of Membrane Science* 84(1–2): 161–169.
114. Matsuyama, H., Teramoto, M., Iwai, H. (1994) Development of a new functional cation-exchange membrane and its application to facilitated transport of CO_2. *Journal of Membrane Science* 93(3): 237–244.
115. Noble, R.D. (1991) Analysis of ion transport with fixed site carrier membranes. *Journal of Membrane Science* 56(2): 229–234.
116. Noble, R.D. (1991) Facilitated transport mechanism in fixed site carrier membranes. *Journal of Membrane Science* 60(2–3): 297–306.
117. Matsuyama, H., et al. (1999) Effects of membrane thickness and membrane preparation condition on facilitated transport of CO_2 through ionomer membrane. *Separation and Purification Technology* 17(3): 235–241.
118. Kwon, S.H., Rhim, J.W. (2015) Facilitated transport separation of carbon dioxide using aminated polyetherimide membranes. *Membrane Journal* 25(3): 248–255.
119. Xin, Q., et al. (2015) Incorporating one-dimensional aminated titania nanotubes into sulfonated poly (ether ether ketone) membrane to construct CO_2-facilitated transport pathways for enhanced CO_2 separation. *Journal of Membrane Science* 488: 13–29.
120. Shin, D.H., Rhim, J.W. (2015) Gas permeation properties of aminated polyphenylene oxide membranes. *Membrane Journal* 25(6): 488–495.
121. Yamaguchi, T., et al. (1996) Transport mechanism of carbon dioxide through perfluorosulfonate ionomer membranes containing an amine carrier. *Chemical Engineering Science* 51(21): 4781–4789.
122. Matsuyama, H., et al. (1999) Facilitated transport of CO_2 through polyethylenimine/poly (vinyl alcohol) blend membrane. *Journal of Membrane Science* 163(2): 221–227.
123. Zhao, Y., Ho, W.W. (2013) CO_2-selective membranes containing sterically hindered amines for CO_2/H_2

separation. *Industrial & Engineering Chemistry Research* 52(26): 8774–8782.
124. Trachtenberg, M.C., et al. (1999) Carbon dioxide transport by proteic and facilitated transport membranes. *Life Support & Biosphere Science* 6(4): 293–302.
125. Cheng, L.-H., et al. (2008) Hollow fiber contained hydrogel-CA membrane contactor for carbon dioxide removal from the enclosed spaces. *Journal of Membrane Science* 324(1–2): 33–43.
126. Zhang, Y.-T., et al. (2010) Selective separation of low concentration CO_2 using hydrogel immobilized CA enzyme based hollow fiber membrane reactors. *Chemical Engineering Science* 65(10): 3199–3207.
127. Tran, D.N., Balkus Jr., K.J. (2011) Perspective of recent progress in immobilization of enzymes. ACS Catalysis 1(8): 956–968.
128. Yang, W., Ciferno, J. (2006) Assessment of carbozyme enzyme-based membrane technology for CO_2 capture from flue gas. *DOE/NETL* 401: 072606.
129. Davy, R. (2009) Development of catalysts for fast, energy efficient post combustion capture of CO_2 into water; an alternative to monoethanolamine (MEA) solvents. *Energy Procedia* 1(1): 885–892.
130. Saeed, M., Deng, L.(2015) CO_2 facilitated transport membrane promoted by mimic enzyme. *Journal of Membrane Science* 494: 196–204.
131. Saeed, M., Deng, L. (2016) Carbon nanotube enhanced PVA-mimic enzyme membrane for post-combustion CO_2 capture. *International Journal of Greenhouse Gas Control* 53: 254–262.
132. Saeed, M., Deng, L. (2016) Post-combustion CO_2 membrane absorption promoted by mimic enzyme. *Journal of Membrane Science* 499: 36–46.
133. Chen, Q., et al. (2021) Thermal management of polymer electrolyte membrane fuel cells: A review of cooling methods, material properties, and durability. *Applied Energy* 286: 116496.
134. Wikipedia (2022) *Proton-Exchange Membrane Fuel Cell.* Accessed 2022 November 28; available from: https://en.wikipedia.org/wiki/Proton-exchange_membrane_fuel_cell.
135. Rouquerol, J., et al. (1994) Recommendations for the characterization of porous solids (Technical Report). Pure and Applied Chemistry 66(8): 1739–1758.
136. Burggraaf, A., Keizer, K. (1991) Synthesis of inorganic membranes. In: *Inorganic Membranes Synthesis, Characteristics and Applications*, pp. 10–63. Springer.
137. Burggraaf, A.J., Keizer, K. (1991) Synthesis of inorganic membranes. In: *Inorganic Membranes Synthesis, Characteristics and Applications*, Bhave, R. (Ed.). Springer: Dordrecht.
138. Hayashi, J.-i., et al. (1996) Separation of ethane/ethylene and propane/propylene systems with a carbonized BPDA–pp 'ODA polyimide membrane. *Industrial & Engineering Chemistry Research* 35(11): 4176–4181.
139. Geiszler, V.C., Koros, W.J. (1996) Effects of polyimide pyrolysis conditions on carbon molecular sieve membrane properties. *Industrial & Engineering Chemistry Research* 35(9): 2999–3003.
140. Salleh, W.N.W., et al. (2011) Precursor selection and process conditions in the preparation of carbon membrane for gas separation: A review. *Separation & Purification Reviews* 40(4): 261–311.
141. Kiyono, M., Williams, P.J., Koros, W.J. (2010) Effect of pyrolysis atmosphere on separation performance of carbon molecular sieve membranes. *Journal of Membrane Science* 359(1–2): 2–10.
142. Yoshimune, M., Haraya, K. (2019) Simple control of the pore structures and gas separation performances of carbon hollow fiber membranes by chemical vapor deposition of propylene. *Separation and Purification Technology* 223: 162–167.
143. Lie, J.A. (2005) *Synthesis, performance and regeneration of carbon membranes for biogas upgrading – a future energy carrier*. PhD thesis. Norwegian University of Technology.
144. Xu, R., et al. (2020) Ultraselective carbon molecular sieve membrane for hydrogen purification. *Journal of Energy Chemistry* 50: 16–24.
145. Hazazi, K., et al. (2019) Ultra-selective carbon molecular sieve membranes for natural gas separations based on a carbon-rich intrinsically microporous polyimide precursor. *Journal of Membrane Science* 585: 1–9.
146. He, X., et al. (2018) Carbon molecular sieve membranes for biogas upgrading: Techno-economic feasibility analysis. *Journal of Cleaner Production* 194: 584–593.
147. Cao, Y., et al. (2019) Carbon molecular sieve membrane preparation by economical coating and pyrolysis of porous polymer hollow fibers.*Angewandte Chemie International Edition* 58(35): 12149–12153.
148. Zhang, C., Kumar, R., Koros, W.J. (2019) Ultra-thin skin carbon hollow fiber membranes for sustainable molecular separations. *AIChE Journal* 65(8): e16611.
149. Wu, Q., et al. (2018) Carbon defect-induced reversible carbon–oxygen interfaces for efficient oxygen reduction. *ACS Applied Materials & Interfaces* 10(46): 39735–39744.
150. Gawryś, M., et al. (2001) Prevention of water vapour adsorption by carbon molecular sieves in sampling humid gases. *Journal of Chromatography A* 933(1–2): 107–116.
151. Jones, C.W., Koros, W.J. (1994) Carbon molecular sieve gas separation membranes – II. Regeneration following organic exposure. *Carbon* 32(8): 1427–1432.
152. Hägg, M.B., Lie, J.A., Lindbråthen, A. (2003) Carbon molecular sieve membranes: a promising alternative for selected industrial applications. *Annals of the New York Academy of Sciences* 984(1): 329–345.
153. Mahajan, R., Koros, W.J. (2000) Factors controlling successful formation of mixed-matrix gas separation materials. *Industrial & Engineering Chemistry Research* 39(8): 2692–2696.
154. Poshusta, J.C., Noble, R.D., Falconer, J.L. (1999) Temperature and pressure effects on CO_2 and CH_4 permeation through MFI zeolite membranes. *Journal of Membrane Science* 160(1): 115–125.
155. Javaid, A., Ford, D.M. (2003) Solubility-based gas separation with oligomer-modified inorganic membranes: Part II. Mixed gas permeation of 5 nm alumina membranes modified with octadecyltrichlorosilane. *Journal of Membrane Science* 215(1–2): 157–168.
156. Hwang, G.-J., et al. (2003) Stability of a silica membrane prepared by CVD using γ-and α-alumina tube as the support tube in the HI–H_2O gaseous mixture. *Journal of Membrane Science* 215(1–2): 293–302.
157. Kuraoka, K., Chujo, Y., Yazawa, T. (2001) Hydrocarbon separation via porous glass membranes surface-modified

using organosilane compounds. *Journal of Membrane Science* 182(1–2): 139–149.
158. Ma, Y.H., Mardilovich, I.P., Engwall, E.E. (2003) Thin composite palladium and palladium/alloy membranes for hydrogen separation. *Annals of the New York Academy of Sciences* 984(1): 346–360.
159. Mardilovich, P.P., et al. (1998) Defect-free palladium membranes on porous stainless-steel support. *AIChE Journal* 44(2): 310–322.
160. Lewis, F.A. (1967) *The Palladium/Hydrogen System*. Academic Press: London.
161. Howard, B., et al. (2004) Hydrogen permeance of palladium–copper alloy membranes over a wide range of temperatures and pressures. *Journal of Membrane Science* 241(2): 207–218.
162. Jung, S.H., et al. (2000) Effects of co-existing hydrocarbons on hydrogen permeation through a palladium membrane. *Journal of Membrane Science* 170(1): 53–60.
163. Armstrong, P.A., et al. (2000) Ceramic membrane development for oxygen supply to gasification applications. In: *Proceedings of the Gasification Technologies Conference*, San Francisco, CA, USA.
164. Koros, W.J., Mahajan, R. (2000) Pushing the limits on possibilities for large scale gas separation: which strategies? *Journal of Membrane Science* 175(2): 181–196.
165. Chuah, C.Y., et al. (2018) Harnessing filler materials for enhancing biogas separation membranes. *Chemical Reviews* 118(18): 8655–8769.
166. Vu, D.Q., Koros, W.J., Miller, S.J. (2003) Mixed matrix membranes using carbon molecular sieves: I. Preparation and experimental results. *Journal of Membrane Science* 211(2): 311–334.
167. Jeazet, H.B.T., Staudt, C., Janiak, C. (2012) Metal–organic frameworks in mixed-matrix membranes for gas separation. *Dalton Transactions* 41(46): 14003–14027.
168. Yu, G., et al. (2019) Mixed matrix membranes derived from nanoscale porous organic frameworks for permeable and selective CO_2 separation. *Journal of Membrane Science* 591: 117343.
169. Kamble, A.R., Patel, C.M., Murthy, Z.V.P. (2021) A review on the recent advances in mixed matrix membranes for gas separation processes. *Renewable and Sustainable Energy Reviews* 145: 111062.
170. Dechnik, J., et al. (2017) Mixed-matrix membranes. *Angewandte Chemie International Edition* 56(32): 9292–9310.
171. Yuan, S., et al. (2019) Covalent organic frameworks for membrane separation. *Chemical Society Reviews* 48(10): 2665–2681.
172. Merkel, T., et al. (2002) Ultrapermeable, reverse-selective nanocomposite membranes. *Science* 296(5567): 519–522.
173. Ansaloni, L., Deng, L. (2017) Advances in polymer-inorganic hybrids as membrane materials. In: *Recent Developments in Polymer Macro, Micro and Nano Blends*, pp. 163–206. Elsevier.
174. Janakiram, S., et al. (2018) Performance of nanocomposite membranes containing 0D to 2D nanofillers for CO_2 separation: A review. *Membranes* 8(2): 24.
175. Li, X., et al. (2015) Synergistic effect of combining carbon nanotubes and graphene oxide in mixed matrix membranes for efficient CO_2 separation. *Journal of Membrane Science* 479: 1–10.
176. Janakiram, S., et al. (2020) Facilitated transport membranes containing graphene oxide-based nanoplatelets for CO_2 separation: Effect of 2D filler properties. *Journal of Membrane Science* 616: 118626.
177. Dai, Z., et al. (2019) A brief review of nanocellulose based hybrid membranes for CO_2 separation. *Fibers* 7(5): 40.
178. Dai, Z., et al. (2019) Fabrication and evaluation of bio-based nanocomposite TFC hollow fiber membranes for enhanced CO_2 capture. *ACS Applied Materials & Interfaces* 11(11): 10874–10882.
179. Janakiram, S., et al. (2020) Three-phase hybrid facilitated transport hollow fiber membranes for enhanced CO_2 separation. *Applied Materials Today* 21: 100801.
180. Figoli, A., Sager, W., Mulder, M. (2001) Facilitated oxygen transport in liquid membranes: review and new concepts. *Journal of Membrane Science* 181(1): 97–110.
181. Gin, D.L., Noble, R.D. (2011) Designing the next generation of chemical separation membranes. *Science* 332(6030): 674–676.
182. Noble, R.D., Gin, D.L. (2011) Perspective on ionic liquids and ionic liquid membranes. *Journal of Membrane Science* 369(1–2): 1–4.
183. Dai, Z., et al. (2018) Nafion/IL hybrid membranes with tuned nanostructure for enhanced CO_2 separation: Effects of ionic liquid and water vapor. *Green Chemistry* 20(6): 1391–1404.
184. Ansaloni, L., et al. (2017) Solvent-templated block ionomers for base- and acid-gas separations: Effect of humidity on ammonia and carbon dioxide permeation. *Advanced Materials Interfaces* 4(22): 1700854.
185. Ohlrogge, K., Brinkmann, T. (2003) Natural gas cleanup by means of membranes. *Annals of the New York Academy of Sciences* 984(1): 306–317.
186. Keil B., Wind J., Ohlrogge K. (2000) *Membrane Simulation Tools for Flowsheeting programmes*. AIDIC Milan: Proceedings IcheaP-5, 2.
187. Ohlrogge, K., Keil, B., Wind, J. (2001) Dehydration and hydrocarbon dewpointing of natural gas by membrane technology. *American Chemical Society, Division of Petroleum Chemistry, Preprints* 46(2): 138–141.
188. Sikdar, S. (1998). *Fundamentals and Applications of Bioremediation: Principles*, Volume I (1st ed.). Routledge.
189. Baudet, J., et al. (1976) *Hollow Fiber Assembly for Use in Fluid Treatment Apparatus*. Google Patents.
190. Nichols, R.W. (1990) *Hollow Fiber Separation Module and Method for the Use Thereof*. Google Patents.
191. Hoff, K.A. (2002) *Modeling and Experimental Study of Carbon Dioxide Absorption in a Membrane Contactor*. Fakultet for Naturvitenskap og Teknologi.
192. Nishikawa, N., et al. (1995) CO_2 removal by hollow-fiber gas-liquid contactor. *Energy Conversion and Management* 36(6–9): 415–418.
193. Lee, Y., et al. (2001) Analysis of CO_2 removal by hollow fiber membrane contactors. *Journal of Membrane Science* 194(1): 57–67.
194. Ivory, J., Feng, X., Kovacik, G. (2002) Development of hollow fiber membrane systems for nitrogen generation from combustion exhaust gas: Part II: Full-scale module test and membrane stability. *Journal of Membrane Science* 202(1–2): 195–204.
195. Klaassen, R. (2003) Achieving flue gas desulphurization with membrane gas absorption. *Filtration & Separation* 40(10): 26–28.

196. Mavroudi, M., Kaldis, S., Sakellaropoulos, G. (2003) Reduction of CO_2 emissions by a membrane contacting process. *Fuel* 82(15–17): 2153–2159.
197. Simons, K., Nijmeijer, K., Wessling, M. (2009) Gas–liquid membrane contactors for CO_2 removal. *Journal of Membrane Science* 340(1–2): 214–220.
198. Mansourizadeh, A., Ismail, A. (2009) Hollow fiber gas–liquid membrane contactors for acid gas capture: A review. *Journal of Hazardous Materials* 171(1–3): 38–53.
199. Atchariyawut, S., Jiraratananon, R., Wang, R. (2007) Separation of CO_2 from CH_4 by using gas–liquid membrane contacting process. *Journal of Membrane Science* 304(1–2): 163–172.
200. Bessarabov, D., et al. (1996) Use of nonporous polymeric flat-sheet gas-separation membranes in a membrane-liquid contactor: Experimental studies. *Journal of Membrane Science* 113(2): 275–284.
201. Zhang, H.-Y., et al. (2008) Theoretical and experimental studies of membrane wetting in the membrane gas–liquid contacting process for CO_2 absorption. *Journal of Membrane Science* 308(1–2): 162–170.
202. Schlosser, Š., Kertész, R., Marták, J. (2005) Recovery and separation of organic acids by membrane-based solvent extraction and pertraction: An overview with a case study on recovery of MPCA. *Separation and Purification Technology* 41(3): 237–266.
203. Teplyakov, V., Okunev, A.Y., Laguntsov, N. (2007) Computer design of recycle membrane contactor systems for gas separation. *Separation and Purification Technology* 57(3): 450–454.
204. Alhadidi, A., et al. (2011) The influence of membrane properties on the Silt Density Index. *Journal of Membrane Science* 384(1–2): 205–218.
205. Luis, P., Garea, A., Irabien, A. (2009) Zero solvent emission process for sulfur dioxide recovery using a membrane contactor and ionic liquids. *Journal of Membrane Science* 330(1–2): 80–89.
206. Albo, J., Luis, P., Irabien, A. (2010) Carbon dioxide capture from flue gases using a cross-flow membrane contactor and the ionic liquid 1-ethyl-3-methylimidazolium ethylsulfate. *Industrial & Engineering Chemistry Research* 49(21): 11045–11051.
207. Deng, L. (2012) Facilitated transport membranes and membrane contactors using ionic liquids for CO_2 capture. In: *Proceedings of the 3rd Asian-Pacific Conference on Ionic Liquids and Green Processes*, Beijing, China.
208. Sirkar, K., et al. (2013), *Pressure Swing Absorption Device and Process for Separating CO_2 from Shifted Syngas and Its Capture for Subsequent Storage*. New Jersey Institute of Technology, Newark, NJ (United States).
209. Blanchard, L.A., et al. (1999) Green processing using ionic liquids and CO_2. *Nature* 399(6731): 28–29.
210. Dai, Z., et al. (2016) Modelling of a tubular membrane contactor for pre-combustion CO_2 capture using ionic liquids: Influence of the membrane configuration, absorbent properties and operation parameters. *Green Energy & Environment* 1(3): 266–275.
211. Zhao, S., et al. (2016) Status and progress of membrane contactors in post-combustion carbon capture: A state-of-the-art review of new developments. *Journal of Membrane Science* 511: 180–206.
212. Kreulen, H., et al. (1993) Microporous hollow fibre membrane modules as gas-liquid contactors Part 2. Mass transfer with chemical reaction. *Journal of Membrane Science* 78(3): 217–238.
213. Scholes, C.A., et al. (2014) Membrane gas-solvent contactor pilot plant trials of CO_2 absorption from flue gas. *Separation Science and Technology* 49(16): 2449–2458.
214. Nguyen, P., et al. (2011) A dense membrane contactor for intensified CO_2 gas/liquid absorption in post-combustion capture. *Journal of Membrane Science* 377(1–2): 261–272.
215. Dai, Z., Ansaloni, L., Deng, L. (2016) Precombustion CO_2 capture in polymeric hollow fiber membrane contactors using ionic liquids: Porous membrane versus nonporous composite membrane. *Industrial & Engineering Chemistry Research* 55(20): 5983–5992.
216. Ansaloni, L., et al. (2017) Development of membrane contactors using volatile amine-based absorbents for CO_2 capture: Amine permeation through the membrane. *Journal of Membrane Science* 537: 272–282.
217. Baker, R.W. (2002) Future directions of membrane gas separation technology. *Industrial & Engineering Chemistry Research* 41(6): 1393–1411.
218. Zolands RR, (2001) Applications. In: Ho, W.S.W., Sirkar, W.S. (Eds.) *Membrane Handbook*, pp. 78–94. London, U.K. Kluwer Academic Publishers.
219. Spillman, R.W. (1989) Economics of gas separation membranes. *Chemical Engineering Progress* 85(1): 41–62.
220. Scott K. (1988) *Handbook of Industrial Membranes: Gas Separations*, pp. 271–305. Elsevier Science.
221. Nunes, S. (2003) *Review on Fuel Cells*. Presentation at EMS Summer School, NTNU, Trondheim, Norway.
222. Junoh, H., et al. (2020) Performance of polymer electrolyte membrane for direct methanol fuel cell application: Perspective on morphological structure. *Membranes* 10(3): 34.
223. Edlund, D.J., Pledger, W.A., Studebaker, T. (2002) *Hydrogen-Selective Metal Membrane Modules and Method of Forming the Same*. Google Patents.
224. Franz, A.J., et al. (2003) *Integrated Palladium-Based Micromembranes for Hydrogen Separation and Hydrogenation/Dehydrogenation reactions*. Google Patents.
225. Nunes, S.P. (ed.) (2003) *Review on Fuel Cells*. Presentation at EMS Summer School, NTNU, Trondheim, Norway.
226. Prasad, R., Shaner, R., Doshi, K. (2018) Comparison of membranes with other gas separation technologies. In: *Polymeric gas separation membranes*, pp. 513–614. CRC Press.
227. Uddin, M.W., Hägg, M.-B. (2012) Effect of monoethylene glycol and triethylene glycol contamination on CO_2/CH_4 separation of a facilitated transport membrane for natural gas sweetening. *Journal of Membrane Science* 423: 150–158.
228. Maltesson, H.Å. (1997) *Biogas för fordonsdrift*. KFB, Stockholm, Sweden.
229. May-Britt, H. (1998) Membranes in chemical processing a review of applications and novel developments. *Separation and Purification Methods* **27**(1): p. 51–168.
230. Merkel, T.C., et al. (2010) Power plant post-combustion carbon dioxide capture: An opportunity for membranes. *Journal of Membrane Science* 359(1–2): 126–139.
231. Favre, E. (2007) Carbon dioxide recovery from post-combustion processes: can gas permeation membranes

compete with absorption? *Journal of Membrane Science* 294(1–2): 50–59.
232. Shao, P., et al. (2013) Simulation of membrane-based CO_2 capture in a coal-fired power plant. *Journal of Membrane Science* 427: 451–459.
233. Hussain, A., Hägg, M.-B. (2010) A feasibility study of CO_2 capture from flue gas by a facilitated transport membrane. *Journal of Membrane Science* 359(1–2): 140–148.
234. Sandru, M., et al. (2013) Pilot scale testing of polymeric membranes for CO_2 capture from coal fired power plants. *Energy Procedia* 37: 6473–6480.
235. Han, Y., et al. (2019) Field trial of spiral-wound facilitated transport membrane module for CO_2 capture from flue gas. *Journal of Membrane Science* 575: 242–251.
236. Wu, H., et al. (2021) Membrane technology for CO_2 capture: From pilot-scale investigation of two-stage plant to actual system design. *Journal of Membrane Science* 624: 119137.
237. White, L.S., et al. (2015) Extended flue gas trials with a membrane-based pilot plant at a one-ton-per-day carbon capture rate. *Journal of Membrane Science* 496: 48–57.
238. He, X., et al. (2017) Pilot testing on fixed-site-carrier membranes for CO_2 capture from flue gas. *International Journal of Greenhouse Gas Control* 64: 323–332.
239. Liu, J., et al. (2019) Highly polar but amorphous polymers with robust membrane CO_2/N_2 separation performance. *Joule* 3(8): 1881–1894.
240. Gmernicki, K.R., et al. (2016) Accessing siloxane functionalized polynorbornenes via vinyl-addition polymerization for CO_2 separation membranes. *ACS Macro Letters* 5(7): 879–883.
241. Baker, R.W. (2000) *Membrane Technology and Applications*. New York: McGraw-Hill.
242. Dyer, P.N., et al. (2000) Ion transport membrane technology for oxygen separation and syngas production. *Solid State Ionics* 134(1–2): 21–33.
243. Aasberg-Petersen, K. (2002) *Autothermal Steam Reforming with Modified Catalyst*. U.S. Patent.
244. Ohlrogge, K., Wind, J., Behling, R. (1993) Membrane technique in the chemical and petrochemical industry. *Erdol & Kohle Erdgas Petrochemie* 46(9): 326–332.
245. Gan, G., et al. (2022) Adsorption and membrane separation for removal and recovery of volatile organic compounds. *Journal of Environmental Sciences* 123: 96–115.
246. Ryu, J.H., et al. (2001) Facilitated olefin transport by reversible olefin coordination to silver ions in a dry cellulose acetate membrane. *Chemistry–A European Journal* 7(7): 1525–1529.
247. Faiz, R., et al. (2013) Separation of olefin/paraffin gas mixtures using ceramic hollow fiber membrane contactors. *Industrial & Engineering Chemistry Research* 52(23): 7918–7929.
248. Semenova, S.I. (2004) Polymer membranes for hydrocarbon separation and removal. *Journal of Membrane Science* 231(1–2): 189–207.
249. Faiz, R., Li, K. (2012) Polymeric membranes for light olefin/paraffin separation. *Desalination* 287: 82–97.
250. Faiz, R., Li, K. (2012) Olefin/paraffin separation using membrane based facilitated transport/chemical absorption techniques. *Chemical Engineering Science* 73: 261–284.
251. Ren, Y., et al. (2020) Membrane-based olefin/paraffin separations. *Advanced Science* 7(19): 2001398.
252. Yoshino, M., et al. (2003), Olefin/paraffin separation performance of asymmetric hollow fiber membrane of 6FDA/BPDA–DDBT copolyimide. *Journal of Membrane Science* 212(1–2): 13–27.
253. Tanaka, K., et al. (1996) Permeation and separation properties of polyimide membranes to olefins and paraffins. *Journal of Membrane Science* 121(2): 197–207.
254. Askari, M., Chung, T.-S. (2013) Natural gas purification and olefin/paraffin separation using thermal cross-linkable co-polyimide/ZIF-8 mixed matrix membranes. *Journal of Membrane Science* 444: 173–183.
255. Zhang, C., et al. (2012) High performance ZIF-8/6FDA-DAM mixed matrix membrane for propylene/propane separations. *Journal of Membrane Science* 389: 34–42.
256. Merkel, T.C., et al. (2003) Olefin/paraffin solubility in a solid polymer electrolyte membrane. *Chemical Communications* 13: 1596–1597.
257. Kim, J.H., et al. (2003) Revelation of facilitated olefin transport through silver-polymer complex membranes using anion complexation. *Macromolecules* 36(12): 4577–4581.
258. Nymeijer, D.C., et al. (2004) Composite hollow fiber gas–liquid membrane contactors for olefin/paraffin separation. *Separation and Purification Technology* 37(3): 209–220.
259. Peinemann, K.V. (2017) Membrane technology in industrial processes. In: *The Forum Seminar*, Oslo, Norway.
260. Harris, P., Barnes, R. (2008) The uses of helium and xenon in current clinical practice. *Anaesthesia* 63(3): 284–293.
261. Rufford, T.E., et al. (2014) A review of conventional and emerging process technologies for the recovery of helium from natural gas. *Adsorption Science & Technology* 32(1): 49–72.
262. Stern, S., et al. (1965) Helium recovery by permeation. *Industrial & Engineering Chemistry* 57(2): 49–60.
263. Dai, Z., et al. (2021) Helium separation using membrane technology: Recent advances and perspectives. *Separation and Purification Technology* 274: 119044.
264. Hägg, M.-B. (2000) Membrane purification of Cl2 gas: I. Permeabilities as a function of temperature for Cl_2, O_2, N_2, H_2 in two types of PDMS membranes. *Journal of Membrane Science* 170(2): 173–190.
265. Eikeland, M., et al. (2002) Durability of poly (dimethylsiloxane) when exposed to chlorine gas. *Journal of Applied Polymer Science* **85**(11): 2458–2470.

6 Techniques to Enhance Performance of Liquid-Phase Membrane Processes by Improved Control of Concentration Polarization

S. Chang[1,*] *and A.G. Fane*[2]
[1]School of Engineering, University of Guelph, Ontario, Canada
[2]UNESCO Centre for Membrane Science & Technology, University of New South Wales, Sydney, NSW, Australia
*Email: schang01@uoguelph.ca

6.1 PROCESS LIMITATIONS AND ENHANCEMENT

A range of membrane processes are used to separate fine particles and colloids, macromolecules such as proteins, low-molecular-weight organics, and dissolved salts. These processes include the pressure-driven liquid-phase processes, microfiltration (MF), ultrafiltration (UF), nanofiltration (NF), and reverse osmosis (RO), and the thermal processes, pervaporation (PV) and membrane distillation (MD), all of which operate with solvent (usually water) transmission. Processes which involve "solute transport" are electrodialysis (ED) and dialysis (D), as well as applications of PV where the trace species is transmitted. In all of these applications the conditions in the liquid boundary layer have a strong influence on membrane performance. For example, for the pressure-driven processes, the separation of solutes takes place at the membrane surface where the solvent passes through the membrane and the retained solutes cause the local concentration to increase. Membrane performance is usually compromised by concentration polarization and fouling. This section discusses the process limitations caused by concentration polarization and the mitigation strategies available. Although the applications focus on RO-MF, the strategies for CP control are universal.

6.1.1 Limitations due to Concentration Polarization and Fouling

Concentration polarization is the reversible build-up of dissolved or suspended species in the solution phase, as depicted in Figure 6.1. The concentration profile of the retained solutes depends on the balance between the convective drag toward and through the membrane (resulting from the permeation flux) and back-transport away from the membrane. The properties of this deposited layer could be reversible or irreversible. The influence of concentration polarization on performance varies with different membrane processes. For RO and NF, concentration polarization can result in a significant increase in osmotic pressure. As a result, higher delivery pressure is required to provide the driving force to achieve reasonable flux values. For UF, the macromolecular solutes and colloidal species could have modest osmotic pressures and also the concentration at the membrane surface can rise to the point of incipient gel formation, which is typically in the range of 20–60% solute by volume [1]. This can lead to a transition from concentration polarization to membrane fouling. For MF membranes without a fouling layer formed on the surface or in the pores, concentration polarization of macromolecules should be negligible because large pore membranes are essentially not retentive to most macromolecules. However, MF membranes can experience particle polarization due to the retention of colloids and particulates.

Concentration polarization and particle polarization are related to the permeation-induced build-up of the concentration profile on the membrane surface without changing the characteristics of the permeability and selectivity of the membranes. Membrane fouling is related to irreversible changes in membrane permeability due to retained species deposition on the surface or the internals of the membrane. Figure 6.2 shows typical forms of membrane fouling applicable to MF and UF, including pore constriction, pore plugging, and gel/cake formation. Considering a particle with diameter d_p and a pore with d_{pore}, when $d_p << d_{pore,}$ the particle can enter the membrane pore and pore constriction may occur. When $d_p \geq d_{pore,}$ deposition of the particle onto the membrane surface may cause pore plugging and cake formation on the membrane surface, depending on the flux and crossflow conditions. In a membrane filtration process, fouling could be a combination of different mechanisms due to the distribution of species sizes in real feeds. Usually, in the initial stages, membrane fouling could be dominated by pore constriction and pore plugging. Once the first cake layer has formed on the membrane surface, gel/cake formation may become the

DOI: 10.1201/9781003285656-6

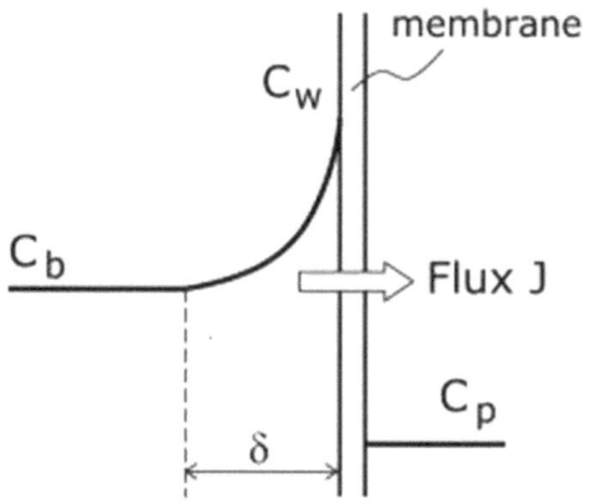

FIGURE 6.1 Schematics of the concentration–polarization boundary layer for membrane filtration.

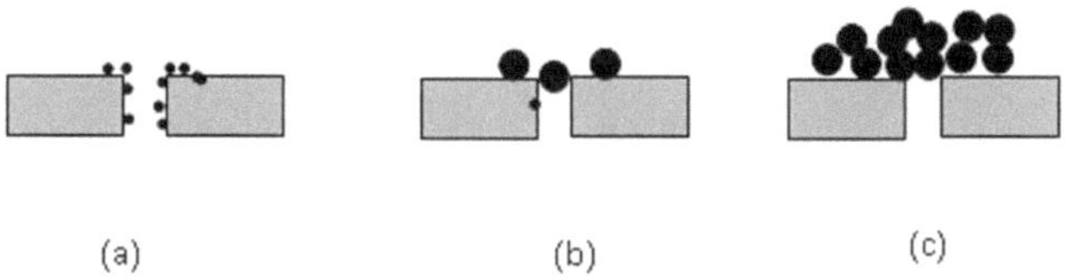

FIGURE 6.2 Fouling schematics for different mechanisms: (a) pore construction, (b) pore blocking, and (c) cake formation.

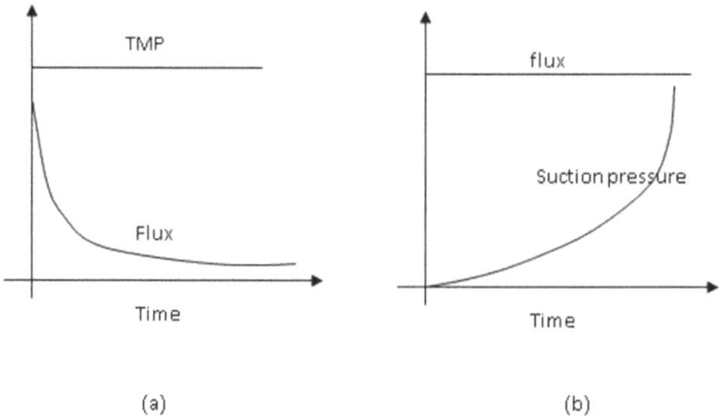

FIGURE 6.3 TMP and flux profiles for constant pressure and constant flux filtration.

dominant fouling mechanism. The filtration behaviour of the cake layer, including compressibility and permeability, is a function of the properties of the colloids deposited. For highly compressible cakes, usually formed by particles with strong interparticle interactions, the cake resistance may be dominated by a very thin layer adjacent to the membrane surface. Analogous phenomena occur with macrosolutes and sparingly soluble solutes because concentration polarization frequently leads to fouling.

Membrane fouling may result in a significant increase in filtration resistance, leading to unstable filtration behaviour. The pressure-driven membrane processes can be operated either with constant feed pressure or in constant flux mode. For constant pressure operation where the transmembrane pressure (TMP) is maintained at a constant value during the filtration, the flux will decline with time due to the increased filtration resistance as shown in Figure 6.3a. It is observed that a quasi-steady state can be reached for crossflow membrane filtration when the flux drops to a value called the "limiting flux", where the particle deposition caused by the flux is balanced by the crossflow-induced particle back-transport. The limiting flux is independent of the TMP and the filtration resistance, but depends on the hydrodynamic condition-induced back-transport and is a function of surface shear, particle size, and the physicochemical characteristics of the solution [2]. The limiting flux can be experimentally determined or estimated based on the mechanisms of back-transport, as discussed below. Membrane filtration can also be operated in the constant flux mode by extracting permeate at a constant flow rate, for example by using a suction pump. With fixed flux, the suction pressure and TMP could theoretically be maintained steady if the flux is below a "critical" flux (see below) [3]. More typically, suction pressure will increase with time due to membrane fouling, as depicted in Figure 6.3b. Constant flux operation has the advantage of avoiding rapid development of membrane fouling caused by a high initial flux as is typically found with constant pressure filtration. This observation leads to the "critical" or "sustainable flux" strategy to enhance the performance of membrane processes.

6.1.2 Critical Flux and Sustainable Flux

For crossflow filtration, the tangential flow tends to sweep the particles away from the membrane surface or promote particle back-transport. It can be assumed that the particle will not deposit onto the membrane surface if the permeate flux is controlled below a value such that the drag force exerted on the particle by the permeate flow is lower than the lift force caused by the crossflow. Based on experimental observation, Field et al. [4] first suggested that there may be a critical flux below which species in the feed have negligible interactions with the membrane and the TMP is similar to that of pure water at the same flux (the strong form of critical flux). If some interaction (such as adsorption) occurs, the flux at which the TMP-flux line becomes non-linear is the weak form of the critical flux. Using a technique called direct observation through the membrane (DOTM), Li et al. [5] were able to visually confirm that a clean membrane without deposition was attainable during the microfiltration of latex particles if the flux was controlled below a certain "critical" value that depended on the crossflow velocity and particle size.

A number of mechanisms have been related to the particle back-transport induced by crossflow. Among them, Brownian diffusion, shear-induce diffusion, and inertial lift mechanisms are most widely accepted [2]. The Brownian diffusion model suggests that molecular diffusion in the boundary layer is responsible for the particle back-transport, implying that the concentration polarization is inevitable in membrane filtration. Shear-induced diffusion arises from multiparticle interactions induced by shear flow. The essential requirement for these mechanisms is that the particle concentration be sufficiently high to result in a significant particle interaction

[6]. Inertial lift was first studied by Segre and Silberberg [7], who identified a "tubular pinch effect" in which lateral migration occurs both from the centreline and from the tube wall towards an eccentric equilibrium position for a range of particle size and flow conditions. Inertial lift was regarded arising from non-linear interactions of a particle with the surrounding flow field under conditions where the Reynolds number based on the particle size is not negligible and so the nonlinear inertial terms in the Navier–Stokes equations play a role. The critic flux for different back-transport mechanisms can be expressed in following general form:

$$J_{cr} = c\gamma^n r_p^m \phi_b^p L_f^q \eta^s \tag{6.1}$$

where γ is the shear rate at the membrane surface, r_p is particle radius, ϕ_b is the solid volume fraction in the feed, L_f is the filter length, η is the viscosity, and c, n, m, q, and s are coefficients. Equation 6.1 highlights the importance of shear rate in raising critical flux and illustrates why the majority of performance-enhancing techniques involve methods of increasing surface shear phenomena. Table 6.1 shows the coefficients in Equation 6.1 for different back-transport mechanisms.

For a given crossflow filtration, the dominant particle back-transport mechanism may depend on the shear rate and the particle size [2]. Brownian diffusion is only important for particles smaller than only a few tenths of a micron in diameter with relative low shear, whereas inertial lift is important for particles larger than several tens of microns with higher shear rates. Shear-induced back-transport appears to be important for intermediate particle sizes and shear rates. Li et al. [8] reported that the shear-induced mechanism was able to predict fluxes comparable with the critical fluxes identified by the DOTM.

For mixed feed, the limiting critical flux will be that of the component with lowest critical flux; in practice, this may be impractically low and "threshold flux" or a "sustainable" flux, requiring infrequent cleaning, may be adopted. For soluble species and fine colloids, the critical flux can be considered as the flux below which the wall concentration does not initiate fouling.

One of the earliest approaches to "sustainable" operation was the Bactocatch process applied in the dairy industry [9] where in effect the TMP was maintained at a low controlled value to limit flux and fouling. In this process, the axial feed flow in a ceramic tubular module was matched by the recirculation of permeate in the shell. It is now recognized that constant flux operation applying critical or sustainable flux concepts is attractive. A prime example is the large-scale membrane bioreactor (MBR) used for wastewater treatment where the long-term stable operation with minimum membrane chemical cleaning is desirable. In this application, controlling flux below the "sustainable flux" has become a common strategy to fight membrane fouling. The critical flux can be experimentally identified through constant flux filtration experiments by incrementing the flux until the TMP (or suction pressure) is no longer steady, as depicted in Figure 6.4. For a real feed, the suction pressure may slightly

TABLE 6.1
Coefficient in Equation 6.1 for different mechanisms

Coefficients	Brownian diffusion	Shear-induced diffusion	Inertial lift
c	1.31	0.072	0.036
m (r_p)	−0.67	1.33	3
n (γ)	0.33	1	2
p (ϕ_b)	−0.33	−0.33	0
q (L_f)	−0.33	−0.33	0
s (η)	−1	−0.33	0

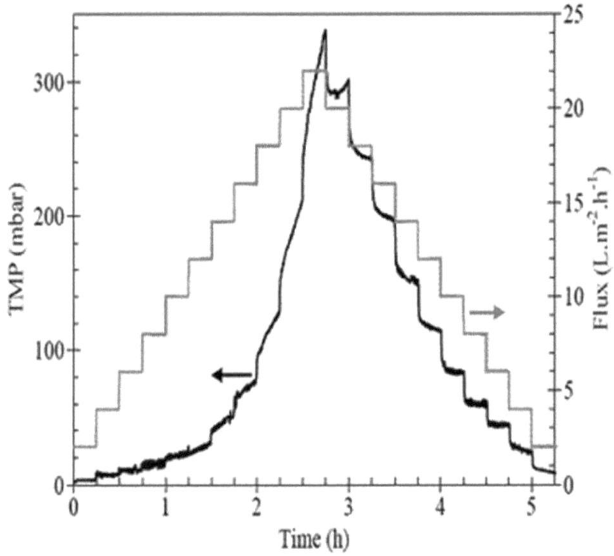

FIGURE 6.4 Determination of the critical flux by stepped TMP pressure increase protocols (Reprinted from Ref. [10], with permission from the copyright holders, Elsevier.)

increase even at a very low flux. Figure 6.4 shows a typical transmembrane pressure profile with stepped flux increase at 2 L/m²/h obtained in filtration of synthetic wastewater [10]. From Figure 6.4, it can be seen that the increase in the rate of TMP became significant when the flux was higher than 10 L/m²/h. However, closer observation indicates that the TMP started to increase even at a flux as low as 2 L/m²/h [10]. For cases like this (i.e., a membrane bioreactor), a specific rate of increase in TMP (dTMP/dt) can be used for the determination of the critical flux or "sustainable flux" [10].

6.1.3 Ultrapermeable Membranes and CP Control

To combat climate change, decarbonization of the membrane process has become a major driver for the innovation of membrane technologies. However, more water with less carbon is a grand challenge for water treatment membrane processes, and is particularly difficult for membrane desalination because of the intensive energy consumption of the RO operation [11]. One of the major approaches to decarbonizing membrane desalination is the development of ultrapermeable membranes (UPMs). UPMs are being developed and could involve incorporation of aquaporins [12,13], carbon nanotubes [14], or graphene materials [15]. The development of UPMs represents a paradigm shift for the membrane water transport mechanism from the "water-diffusion" to "nano-channel water flow" [11]. UPMs promise significant energy reduction in desalination [15]. However, one of the process limitations with UPMs is that as water permeability (A) increases, the attainable water flux becomes limited by CP. Figure 6.5 shows the relationship between the water permeability (A), the attainable water flux (J), and the boundary layer mass transfer coefficient (k) [11].

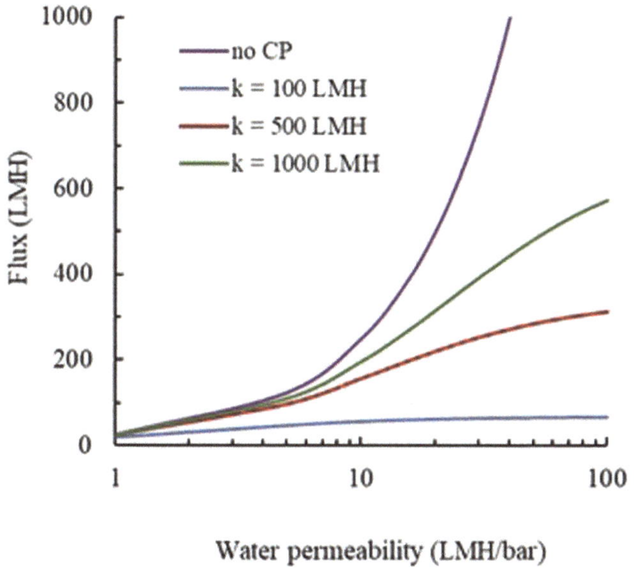

FIGURE 6.5 Potential water flux versus membrane permeability A (L/m^2/hr/bar) for range of mass transfer coefficients *k*. Basis: Flux $J = A^*(\Delta P - \exp(J/k)^*\Delta\Pi)$, with ΔP= 50 bar, $\Delta\Pi$= 25 bar, complete retention, no fouling (LMH: L/m^2/h). (Reprinted from Ref. [11] with permission from the copyright holders, Elsevier.)

The typical value of k is 100 L/m^2 h and is determined by the type of flow channel spacer used. In Figure 6.5, the "k = 100" line represents the attainable fluxes with an increase in A from 1 LMH/bar (typical for current RO membranes) to 100 L/m^{2}h/bar (potential for UPMs), while the "no CP" line indicates the attainable flux with A without CP limitation. Clearly, to exploit UPMs it would be necessary to increase the boundary mass transfer coefficient (k) up to 5–10 times the current k values without a significant increase in energy input [11].

6.1.4 Methods to Control Concentration Polarization and Membrane Fouling

Figure 6.6 summarizes the range of methods used to control concentration polarization and membrane fouling. These include chemical, hydrodynamic, and physical methods. The most direct chemical approach to control membrane fouling is modification of the surface properties of the membranes, such as surface charge and hydrophilicity. Some methods used to modify membrane properties include heterogeneous chemical modification [16], adsorption of hydrophilic polymers [17], irradiation methods [18], and low-temperature plasma activation [19]. The chemical methods also include the application of coagulants to change the particle size distribution in the treated feeds (note the importance of particle size in Equation 6.1). The addition of inorganic salts and polyelectrolytes as well as pH shifts in the feeds can destabilize colloids and result in the formation of larger flocs, reducing the concentration of the colloids in the feeds that are mainly responsible for deposition and high foulant resistance. Chemical methods can also alter solute–membrane interactions and reduce fouling [20].

Most of the hydrodynamic methods have focused on increasing the particle and solute back-transport from the membrane–liquid interface by increasing the shear rate and the flow instability in the boundary layer. These techniques include secondary flows (see Section 6.2), spacers and inserts (Section 6.3), pulsed flow (section 6.4), high shear rate devices

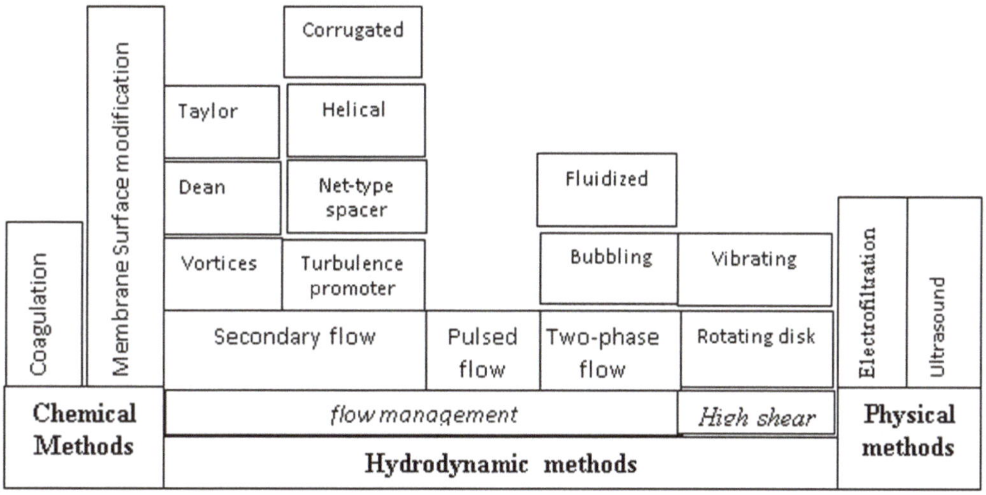

FIGURE 6.6 Methods used for performance enhancement of membrane filtration.

TABLE 6.2
Potential enhancements and estimated specific power for unsteady-state shear techniques. (Reprinted from Ref. [11], with permission from the copyright holders, Elsevier). FS: flat sheet membrane; HF: hollow-fibre membrane

Technique	Potential enhancement factor	Specific power (W/m^2)[a]	Preferred module	Challenges
Two-phase flow (particles)	3.5×	To 0.25	FS or HF	Spacer deign, membrane integrity
Two-phase flow (bubbles)	2–5×	To 5	FS or HF	Spacer design
Vibrations (axial/lateral)	2–10×	To 0.6	HF	FS difficult, spacer design
VSEP	2–20×	To >100	FS	High energy, scale up

[a] Benchmark: ca 5 W/m^2 (assume 10% of specific energy 50 W/m^2: 20 L/m^2 h @ 2.5 kWh/m^3).

(Section 6.5), and vibrations and two-phase flow (Section 6.6). The physical methods that are currently being tested to enhance the filtration performance of membranes include the application of electric fields and ultrasound (Section 6.7).

The performance enhancement by a specific hydrodynamic technique is assessed by comparing the steady flux obtained under the same filtration conditions with and without imposition of the hydrodynamic technique. However, since most of the hydrodynamic techniques used for flux enhancement also result in an increase in the energy consumption, there is a trade-off between capital cost (related to flux enhancement) and operating cost (related to energy consumption). In filtration processes, the energy will be dissipated by crossflow and driving the permeate going through the membrane. Thus, the overall power expenditure within the module (P_d) can be approximated by:

$$P_d = \left| Q_f \Delta P_L + Q_p P_{tm} \right| \quad (6.2)$$

where Q_f is the crossflow flow rate, Q_p is the permeate flow rate, P_{tm} is the transmembrane pressure, and ΔP_L is the pressure loss in the module (a part of which is used to generate surface shear). Then the specific energy expenditure per unit volume of permeate in kwh/m^3 is:

$$E_d = \frac{1}{3.6x10^6}\left(\frac{Q_f \Delta P_L}{Q_p} + P_{tm}\right) \quad (6.3)$$

The costs of the filtration processes include module and energy cost. The module cost depends on the membrane area (inversely related to flux), so the module cost per unit filtrate can be calculated by:

$$module\ cost\left(\frac{\$}{m^3}\right) = \frac{1}{J_d}\left(\frac{s}{m}\right) \times C_{mem}\left(\frac{\$}{m^2 s}\right) \quad (6.4)$$

where J_d is the design flux of the filtration process and C_{mem} is the cost of capital related to membrane area per unit time. The critical flux or sustainable flux can be used as the design flux for the processes, which is to be enhanced by the hydrodynamic techniques applied.

The energy cost can be estimated by

$$\text{Energy (\$/m}^3) = E_d\ (\text{kwh/m}^3) \times \text{cost (\$/kwh)} \quad (6.5)$$

Most of the hydrodynamic techniques will reduce the module/membrane cost due to enhancement of the design flux but increase the energy cost due to increased pressure drop through the module. The optimal process design should result in a minimum total cost or the sum of the energy and module cost. An example of this trade-off is described in Section 6.3.1.2. The important point to note is that the performance-enhancing techniques involve a cost that must provide a benefit and "return on investment".

To decarbonize membrane filtration, mitigation of CP and membrane fouling without significant energy input becomes a key criterion to assess membrane performance enhancement. Fane et al. [11] and Zamani et al. [21] indicated that it may be feasible to achieve two- to five-fold increases in mass transfer in membrane filtration using unsteady-shear strategies for module design and operation (Table 6.2).

6.2 SECONDARY FLOWS

Secondary flow is a flow region where the flow velocity and direction are significantly different from the primary flow region. The secondary flow could be generated by curved channel, coiled tubes, and rotating movement. The secondary flow generated close to the membrane surface can reduce the concentration polarization and enhance membrane filtration. This section discusses the secondary flows, including Dean vortices, Taylor flow, and helical membrane modules for the membrane performance enhancement.

6.2.1 Dean Vortices

6.2.1.1 Characteristics of Dean Flow

Dean vortices are the secondary flows that occur in the cross-section of a curved channel or helically coiled tubes. Figure 6.7 shows the secondary flow developed in a helically coiled tube. When fluid flows through the helically curved tube, the faster

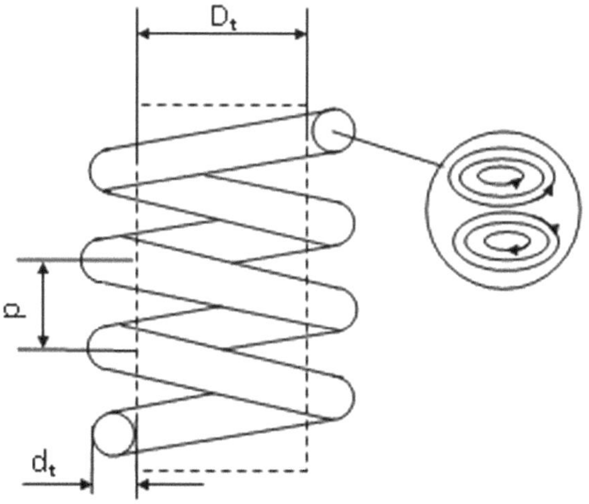

FIGURE 6.7 Dean flow formed in a coiled helical tube.

elements of the fluid in the centre of the tube tend to be moved outwards by centrifugal force, while the slower elements of the fluid are forced inwards to maintain mass balance, resulting in counter-rotating vortices in the cross-section of the channel, as shown in Figure 6.7. The intensity of the secondary flow depends on the fluid flow in the tube and the geometric features of the curved channel, characterizing by the so-called Dean number (*De*):

$$De = \text{Re}\sqrt{\frac{d_t}{D_t}} \tag{6.6}$$

where D_t is the coil diameter and the Reynolds number $Re = d_t V_m / v$, d_t is the tube diameter, V_m is the mean azimuthal velocity, and v is the kinematic viscosity.
A modified curvature diameter D_t', which takes into account the torsion effect and varies with the pitch b, is formulated as follows for helically coiled tubes:

$$D_t' = D_t\left(1+\left(\frac{\pi}{bD_t}\right)^2\right) \tag{6.7}$$

It is suggested that the appearance of the Dean vortices depends on the magnitude of the Dean number. Although it is suggested that the secondary flow can only appear above a critical Dean number [22], flow simulation indicates that a secondary flow field could be developed even at very low Dean number [23]. However, the effect of the secondary flow on mass and heat transfer is difficult to detect experimentally at Dean numbers lower than about 20.

The following empirical correlation from Mishra and Gupta [24] can be used to calculate the friction factor (f_c) in a torus, coil, or helical tube under the conditions: *Laminar,* $1 < De < 3000$; $3 \times 10^{-3} < d_t/D_t < 0.15$; $0 < (b/D_t) < 25.4$).

$$f_c = f_s(1+0.033(\log De)^4 \tag{6.8}$$

where f_s is the friction factor in a similar straight tube. For Newtonian laminar flow, f_s can be calculated by:

$$f_s = \frac{64}{\text{Re}} \tag{6.9}$$

The axial pressure drop due to friction for a helical coil tube can then be calculated by:

$$\Delta P_L = f_c \frac{L_f}{d_t}\frac{1}{2}\rho V_m^2 \tag{6.10}$$

Equations 6.9 and 6.10 indicate that the pressure drops will increase with the Dean number. As indicated by Equation 6.3, the overall efficiency of the Dean flow filtration enhancement will depend on the balance of the TMP reduction and the flow resistance increase caused by the Dean secondary flow.

6.2.1.2 Dean Flow Enhanced Membrane Process

Experimental evidence has demonstrated that Dean vortices can be effective for the enhancement of membrane performance under laminar conditions [25]. As flow conditions approach the transition and turbulent flow regimes, straight membranes have a better mass transfer and higher wall shear rate than in flows with curved membrane channels. The effects of Dean vortices on the performance of membrane filtration have been studied experimentally and theoretically by Belfort and co-workers [26–29]. Mallubholta and Belfort [28] assessed the filtration of suspensions of polydispersed polystyrene particles (mean diameter 25 μm) and silica particles (mean diameter 20 μm) with and without the presence of Dean flow using a 180° U-bend channel with a membrane attached on the lower surface of the upper linear part of the U channel after the curve. With such a test cell, the filtration can be conducted with and without the appearance of the Dean vortices in the crossflow over the membrane by using the flow setting shown in Figure 6.8. Figure 6.9 shows the permeation flux time profiles for filtration of the polydispersed polystyrene particles with and without the Dean vortices. It can be seen from this figure that a significantly higher pseudo-steady state can be reached at an earlier stage in the filtration with Dean vortices than that without Dean vortices, indicating an improvement in filtration performance.

Although some of the earlier work on Dean flow enhanced membrane filtration was conducted with modules designed by placing a half spiral tube onto a flat sheet membrane [26], most recent studies have been carried out with helical hollow-fibre membrane modules. Moulin et al. [30] compared the limiting fluxes obtained in filtration of bentonite and Dextran solutions using a coiled and a straight hollow-fibre membranes (fibre inner diameters 0.93 and 0.7 mm) and reported that the secondary flow induced by the coiled geometry could enhance the limiting flux by up to a factor of 2 and 3 for the bentonite

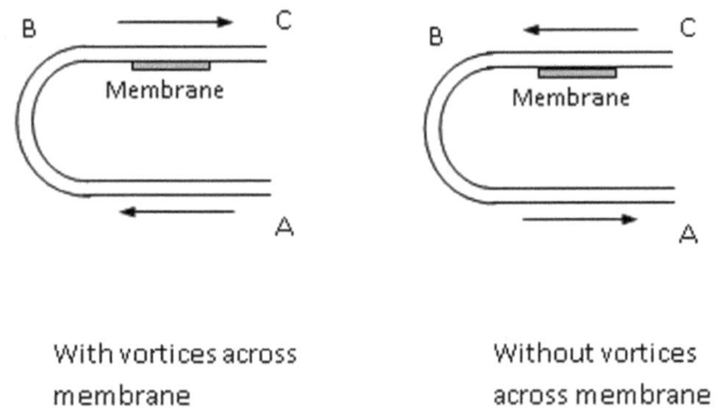

FIGURE 6.8 Configuration of Dean generator test cell: (a) with vortices across membrane (flow direction from A to B to C) and (b) without vortices across membrane (flow direction from C to B to A). (Reprinted from Ref. [28], with permission from the copyright holders Elsevier.)

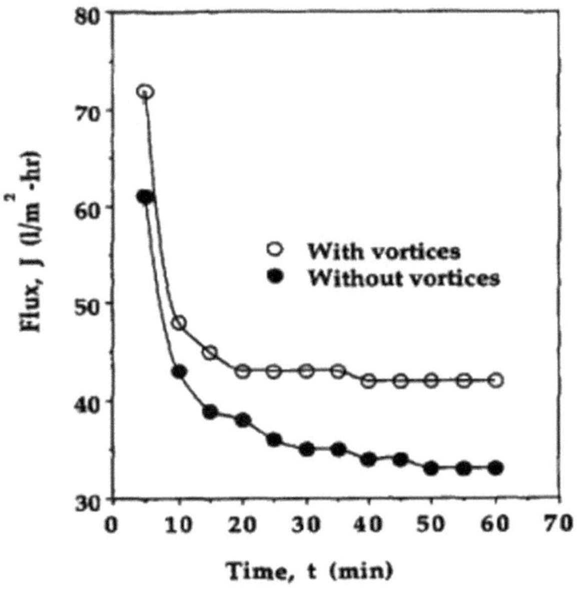

FIGURE 6.9 Flux-time profile with and without Dean flow for filtration of styrene-divinyl-benzene (S/DVB) particles. (Reprinted from Ref. [28], with permission from the copyright holders Elsevier.)

suspension and the macromolecular solution, respectively. The flux enhancement increased when the coil diameter was reduced from 11 cm to 4.1 cm, implying increased enhancement with an increase in Dean number (Equation 6.6). Mallubhotla et al. [29] studied the effect of Dean vortices on nanofiltration of inorganic and amino acid solutions using helical hollow-fibre membrane modules that were made by wrapping a 0.27 mm fibre (inner diameter) around a steel rod of 3.18 mm diameter. The results showed the flux of the helical module was about 16% higher for 0.02 mol KCl and K_2SO_4 solution and about 32% higher for K_3PO_4 than those with the straight module. For filtration of different amino acid solutions, the flux enhancement by Dean flow was found to be a strong function of the pH of the solution, ranging from less

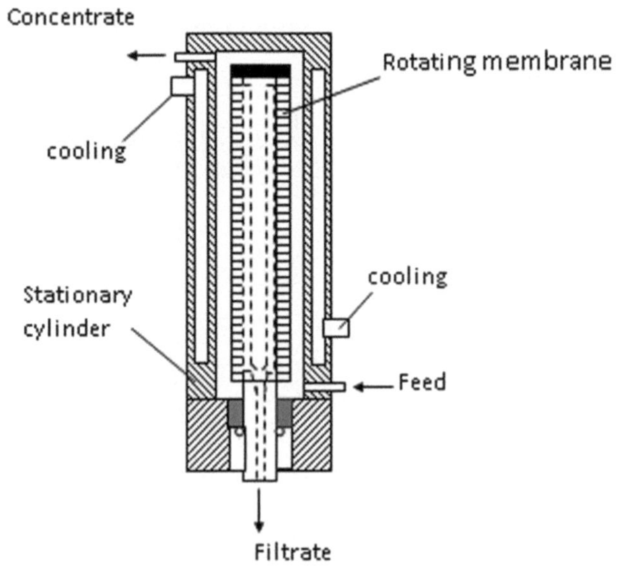

FIGURE 6.10 Schematics of axial rotatory membrane module.

than 5% to 35% in the pH range of 3–10 with an increased enhancement at high pH. Although helically coiled hollow fibres have proved to be effective for flux enhancement, no large-scale commercial hollow-fibre membrane module has adopted the design to date, probably due to the difficulty in fabrication.

6.2.2 Taylor Flow

6.2.2.1 Characteristics of Taylor Flow

Another method that has been used to promote secondary flow is the generation of Taylor vortices in the annulus of two concentric cylinders with a rotating cylindrical filter. Figure 6.10 shows a typical rotary membrane module that consists of two coaxial cylinders with the inner driven by an electric motor and the outer fixed. The membrane is usually attached to the inner rotating cylinder. The feed flows along the z-axis in the annular gap between the two concentric cylinders with the permeate collected by a duct along the axis of rotation. When the inner cylinder rotates, a hydrodynamic regime is generated with annual counter rotating vortices (Taylor vortices), as shown in Figure 6.11. The Taylor flow can be characterized by the Taylor number:

$$Ta = \frac{\omega R_1 \Delta R}{\nu}\left(\frac{\Delta R}{R_1}\right)^{0.5} \tag{6.11}$$

where
ω is angular velocity
R_1 is radius of the rotating cylinder
$\Delta R = (R_2 - R_1)$, i.e., the width of the annular gap
R_2 is the inner radius of the fixed cylinder

When the Taylor number exceeds a critical value (Ta_c), a transition from stable Couette flow to vortical Taylor–Couette

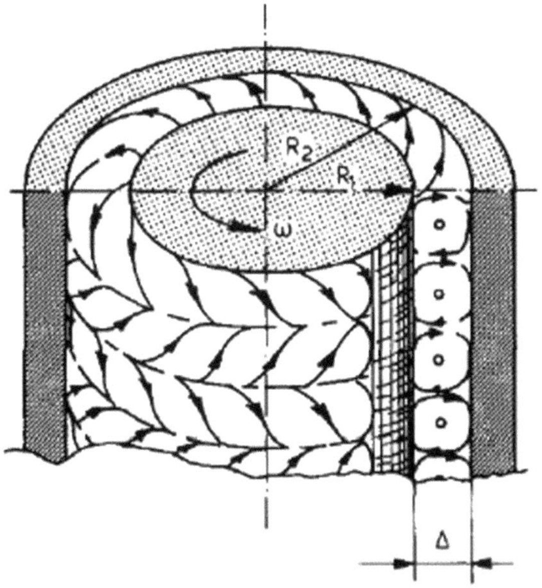

FIGURE 6.11 Schematic illustration of the Taylor vortices in the annulus of an axial rotating dynamic membrane filter. (Reprinted from Ref. [32], with permission from the copyright holders, Elsevier.)

flow occurs. The critical Taylor number can be calculated by [31]:

$$Ta_c = 20.1 + 13.6\left(\frac{\Delta R}{R_i}\right) + 1.4\left(\frac{\Delta R}{R_i}\right)^2 \quad (6.12)$$

As shown in Equation 6.12, the critical Taylor number is a function of the ratio of the annular gap width to the radius of the rotating cylinder. Combining Equations 6.11 and 6.12, a critical rotating speed to attain the flow transition can be developed as:

$$\omega_c = \frac{v}{R_1^2}\left(13.6\left(\frac{\Delta R}{R_1}\right)^{-1} + 20.1\left(\frac{\Delta R}{R_1}\right)^{-1..5} + 1.4\left(\frac{\Delta R}{R_1}\right)^{-2.5}\right) \quad (6.13)$$

The critical rotating speed will depend on the radius and the ratio of the annual gap to the radius. A smaller radius and a larger gap result in a lower flow transition rotating speed.

6.2.2.2 Membrane Filtration with Taylor Flow

The rotating filter has been tested for a wide range of applications, including concentration of biological suspensions [32], skim milk separation [33], separating plasma from whole blood [34], and oil–water separation [35]. The flux obtained with a rotating filter will depend on the rotating speed, operating pressure, and the structure of the filter such as the gap size and the roughness of the rotating surface. It was reported that a flux from 50 to 150 L/m²/h has been achieved for UF of 20% cutting oil emulsion under conditions of 300 kPa and rotating speeds from 210 to 2550 rpm [35]. Kroner and Nissinen [32] reported fluxes of 60–140 L/m²/h for MF of 3% Baker's yeast suspensions at 25 kPa and a flux range of 30–75 L/m²/h for MF of 2% *E. coli* broth at 30 kPa in the rotating speed range 1000–3000 rpm. For cell concentration, a final batch concentration of 30–70% for baker's yeast and 9–53% for bacterial suspensions, accompanied by a high protein transmission, was reported with the application of the rotating filter [32].

Figure 6.12 shows the effect of rotation speed for filtration of baker's yeast suspension and *E. coli* fermentation broth [32]. In both cases, the increase in rotation speed in the range of 1000–3000 rpm, which corresponds to a range of Ta from about 2000 to 6000 for the test device, resulted in a substantial increase in flux with a transition point observed at a rotation speed of 2000 rpm—this corresponds approximately to a Taylor number of 3500, where turbulent Couette flow can be assumed.

An empirical correlation which expresses the Sherwood number as a function of the Taylor number was proposed by Holeschovsky and Cooney [36] as:

$$Sh = 1.26Ta^{0.5}\left(\frac{2\Delta R}{R}\right)^{0.17} Sc^{0.33} \quad (6.14)$$

The main limitation of the axially rotating filter is difficulty of scale-up. Most of the reported experiments have been carried out with lab-scale units with the radius of the rotating cylinder < 50 mm and the total filtration area < 0.05 m².

6.2.3 Helical Membrane Module

A recent development of the secondary flow enhanced membrane processes is the helical membrane module, which is characterized by a DNA helical-like spacer enveloped by polyester filter cloth with a pore size around 22 μm [37–39]. Figure 6.13 shows the schematics of the helical membrane spacers with different twisted helical angles. The main parameters of the helical membrane include the angles of twist of the spacer or membrane and the ratio of the membrane width to the length.

The concept of the helical membrane module has been tested in a submerged membrane filtration mode with bubbling used for the membrane fouling control. Liu et al. [40] showed that the helical membrane with a twisted angle of 180° could achieve a 1.46–1.69 flux enhancement, compared to the membrane modules with 0° twisted angle, in the filtration of 500 mg/L kaolin suspension under constant transmembrane pressure of 2.8 and 3.2 kPa. The particle image velocimetry (PIV) analysis [37] showed that the tortuous membrane surface of the helical membrane could generate rotational flow near the membrane surface and increase the wall shear rate. Filtration experiments conducted with the helical membrane modules installed in a membrane bioreactor showed that the filtration performance of the helical membrane could be significantly affected by the

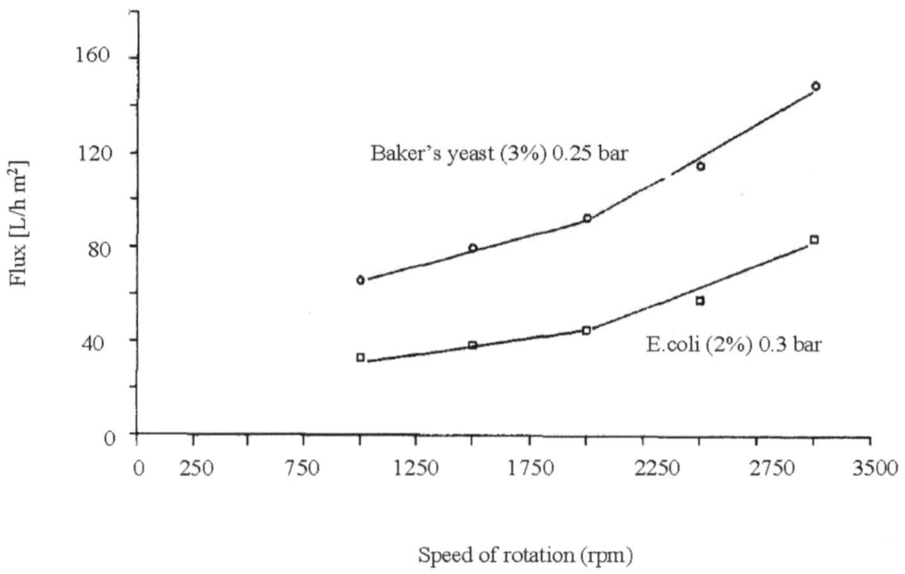

FIGURE 6.12 Effect of rotating speed on permeate flux for filtration of baker's yeast and *E. coli* suspensions. (Reprinted from Ref. [32], with permission from the copyright holders, Elsevier.)

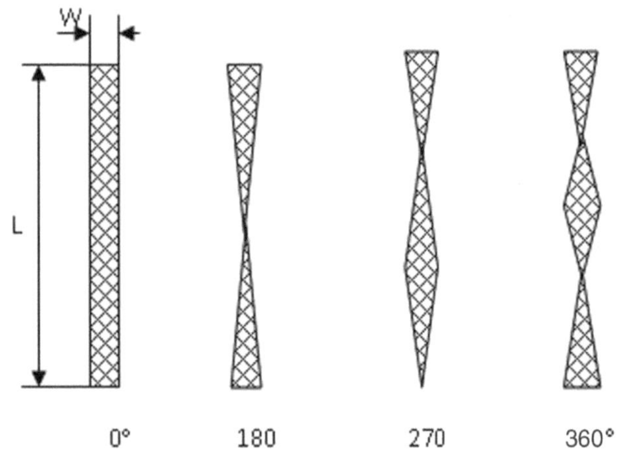

FIGURE 6.13 Helical membranes with different twisted helical angles.

twisted angle and the ratio of width to length of the helical membrane. The average flux enhancement under the constant pressure condition for the filtration in the MBR was reported in the range of 17–35.5% [37]. The slender membrane module (smaller ratio of width to length) exhibited better performance than the wider modules (larger ratio of width to length). It was also found that the module made by multi-pieces of helical membranes performed better than the module consisting of only a single piece of helical membrane.

The helical membrane has also been tested in a rotating mode. Liu et al. [38] reported that rotating a 360° twisted helical membrane module at 160 rpm can lead to a 27% flux enhancement in the filtration of 5 g/L yeast suspension under a constant pressure of 5.3 kPa. For the rotating helical membrane module, the twisted angle of the membrane was found to exert a significant effect on the membrane filtration performance. Liu et al. [38] reported the flux enhancement in the filtration of 5 g/L yeast suspension was 16% for the 270° helical angle and 27% for the 360° helical angle, while it was only 0.3% for the 180° helical angle and 5% for the 450° helical angle. Liu et al. [38] also reported that a better performance could be obtained by rotating the membrane in a confined tube. The estimated energy consumption of the rotating helical membrane was reported at around 0.069 kWh/m^3 [38].

6.3 FLOW CHANNEL SPACERS

For tubular and flat sheet membrane modules, one of the commonly used hydrodynamic techniques for the concentration polarization control is turbulence promoters, such as spacers used in spiral-wound membrane modules, helical inserts used in tubular membrane modules, and the corrugated membrane for the flat sheet membrane. Research has been conducted to assess the effect of the membrane flow channel spacers and inserts on the membrane filtration and to optimize their design.

6.3.1 Net-Type Spacers

6.3.1.1 Geometrical Characteristics of Net-Type Spacers

The use of net-type spacers, including woven and non-woven type spacers, is a key feature of the spiral-wound modules (SWMs). SWMs are widely used in RO (desalination), NF (water treatment), and some large-scale applications of UF (dairy and food). The spacers in the spiral-wound module have the dual functions of keeping adjacent membranes apart to form a feed channel and of promoting eddies in the feed channel. As shown in Figure 6.14, the net-type spacer usually consists of two layers of cylindrical filaments joined together to form a screen-like mesh. The typical mesh geometries include square, rhomboid, and parallelograms, with cell sizes of the order 4 mm and mesh heights of 1–2 mm. The diameter of the filaments in the top and the bottom layers can be identical (symmetric spacers) or different (asymmetric spacer). The main characteristic parameters of the spacers include filament diameter, mesh size, and the angle between the filaments, and the angle between the flow and the filament (Figure 6.14). Other defined spacer parameters include the specific area, S_{vsp} [39], and spacer voidage ε [41].

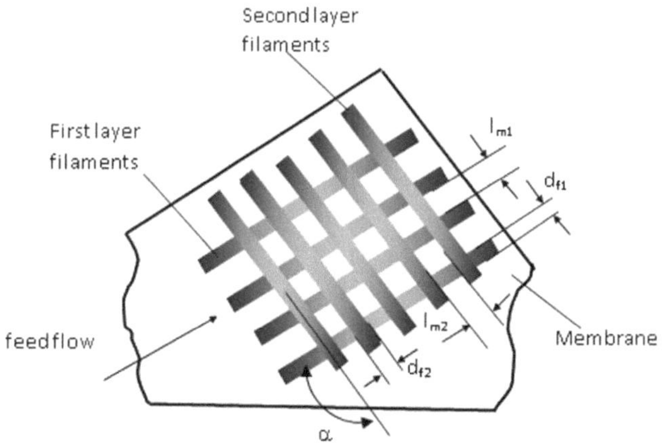

FIGURE 6.14 Schematics of net-type spacers.

6.3.1.2 Effect of Spacer Geometry

In the SWM, flow through the spacer-filled channel induces unsteady flows and increases local shear rates and local mixing, which reduce the boundary layer thickness and the concentration polarization. The mass transfer and pressure drop in a spacer filled channel can be significantly affected by the geometric characteristics of the spacer. Figure 6.15 shows some typical commercial net-type spacers studied by Da Costa et al. [42]. Table 6.3 shows their characteristic parameters. Figure 6.16 shows the steady flux at different transmembrane pressures for filtration of 1 g/L dextran (average molecular weight 500,000 Da) using a Koch Systems HFK-131 polysulphone 5 kDa UF membrane in a test cell with channel dimensions of 25 mm width and 285 mm long. Comparing the flux obtained with and without spacers, the flux was significantly higher (3- to 5-fold) for the filtration with spacers. Inspection of Figure 6.16 shows that the flux enhancement was significantly influenced by the spacer geometry. The spacers also increased the channel pressure loss from about 1 kPa (empty channel) up to as high as 167 kPa.

Da Costa et al. [42] showed that the mass transfer coefficient for various net-type spacers can be represented by a Sherwood correlation:

$$Sh = c\,\mathrm{Re}^m\, Sc^p (d_h / L_s)^q \qquad (6.15)$$

where L_s is channel length and d_h is the hydraulic diameter. The hydraulic diameter can be calculated by:

$$d_h = \frac{4 \times cross \quad section}{wetted \quad perimeter} = \frac{4\varepsilon}{2/h_{sp} + (1-\varepsilon)S_{vsp}} \qquad (6.16)$$

where h_{sp} is the thickness of the spacer. This is equal to the height of the spacer-filled channel, but it is less than $(d_{f1}+d_{f2})$ because the filaments are slightly embedded in each other. The other dimensionless numbers in Equation 6.15 can be calculated as:

$$Sh = \frac{kd_h}{D} \quad \mathrm{Re} = \frac{Vd_h}{\nu} \quad Sc = \frac{\nu}{D} \qquad (6.17–6.19)$$

where V is the superficial channel axial velocity.

TABLE 6.3
Geometrical characteristics of spacers

Spacer	Hp (mm)	d_f (mm)	l_{m*}	Angle (deg) [a]	Flow angle (deg)[b]
80 MIL-1	2.1	1.15	1.85	80	40
80 MIL-2	2.1	1.15	1.85	80	50
80 MIL-1E	2.1	1.15	4.85	80	40
80 MIL-2E	2.1	1.15	4.85	80	50
UF3	1.7	0.76 (1.07)[c]	4.06 (5.3) [d]	45	60
UF4	1.7	0.76 (1.07)[c]	4.06 (5.3) [d]	45	20

[a] Angle between the filaments facing the main flow direction.
[b] Angle between the filament and channel axis (main flow direction).
[c] Thin and thick filaments.
[d] Short and long filaments.

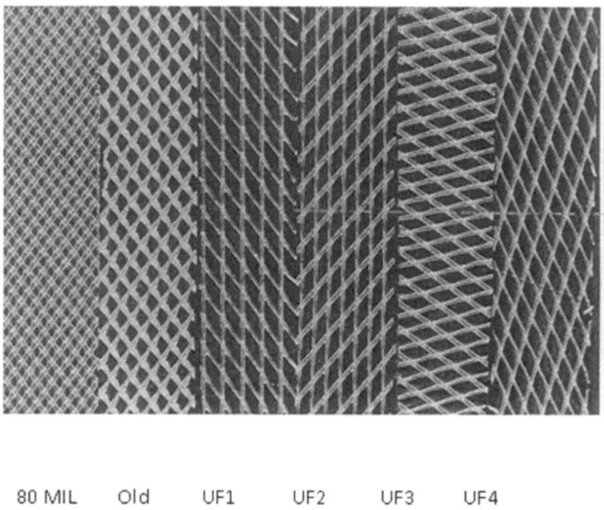

FIGURE 6.15 Typical commercial net-type spacers. (Reprinted from Ref. [42], with permission from the copyright holders, Elsevier.)

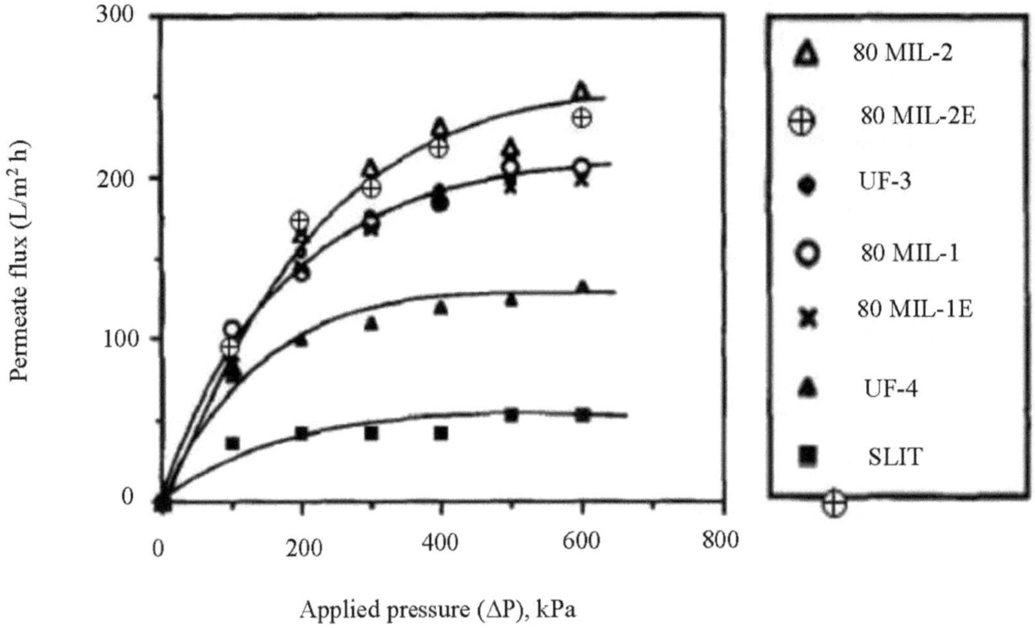

FIGURE 6.16 Flux versus transmembrane pressure for different spacers. (Reprinted from Ref. [42], with permission from the copyright holders, Elsevier.)

TABLE 6.4
Summary of mass transfer coefficients, coefficients in Equations 6.15 and 6.20, and cost estimation for filtration with spacers tested in Figure 6.17 [41]

Spacer	k (× 10⁶) m/s	c 6.15	m 6. 15	p 6.15	q 6. 15	A 6. 20	n 6. 20	Total cost $/m³
Slit	1.0 (0.24*)	1.86	0.43	0.32	0.33	–	–	1.98
80 MIL-1	6.0 (0.44)	0.0096	0.62	0.58	–	0.54	0.26	1.09
80 MIL-2	6.0 (0.44)	0.0096	0.66	0.58	–	0.91	0.23	1.73
80 MIL-1E	6.0 (0.25)	–	–	–	–	2.16	0.17	1.02
80 MIL-2E	6.0 (0.25)	–	–	–	–	4.51	0.15	1.71
UF3	5.0 (0.27)	0.0096	0.59	0.60	–	3.3	0.17	1.87
UF4	3.0 (0.27)	0.0096	0.50	0.59	–	0.61	0.30	0.80

* Flow rate (L/min).

The friction factor for channel pressure loss can be related to the Reynolds number by the following empirical correlation:

$$f = \frac{A}{\mathrm{Re}^n} \tag{6.20}$$

According to Da Costa and Fane [41], a lower value of n is indicative of a higher degree of turbulence in the fluid flow.

Table 6.4 shows the experimentally determined coefficients for the mass transfer and the friction coefficient correlations for the spacers tested in Figure 6.17 [42]. The coefficients in Equations 6.15 and 6.20 are highly spacer geometry dependent and different for various arrangements of the spacer in the channel. Da Costa et al. [42] also carried out an illustrative economic analysis on membrane (capital) and energy (operating) costs for the spacers tested based on hypothetical cost parameters. Table 6.4 shows the total costs of different spacer options under conditions of a flow rate of 3 L/min. Under such a flow condition, it seems that the UF-4 was the most efficient spacer compared with the others. However, the total cost of the membrane process would be a function of the crossflow rate. A high crossflow rate can result in a high energy cost but a reduced membrane area because an increased flux should be achieved under high crossflow conditions. This implies that there may be an optimal crossflow rate at which the minimum total cost (i.e., the sum of the operating and capital costs) can be achieved. Figure 6.17 shows the estimated costs versus crossflow for the spacers tested by Da Costa et al. [42]. Given that typical spiral-wound elements are usually operated with a crossflow velocity less than 0.6 m/s due to the limitation of the maximum pressure drop across

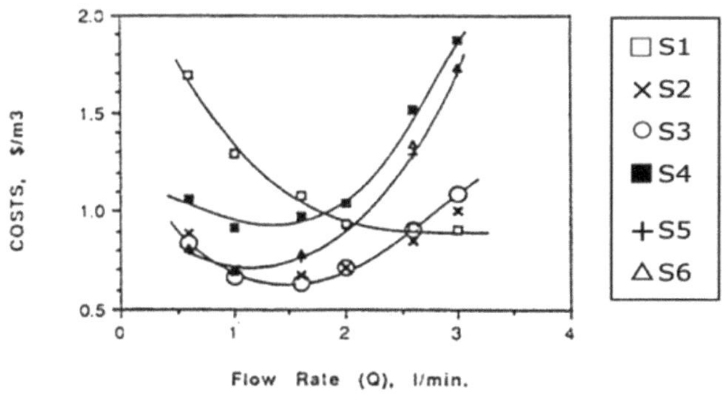

FIGURE 6.17 Unit processing costs for the spacers shown in Tables 6.3 and 6.4 under different flow conditions, where S1 = UF4, S2 = 80 MIL-1E, S3 = 80 MIL-1, S4 = UF3, S5 = 80 MIL-2E, and S6 = 80MIL-2. (Reprinted from Ref. [42], with permission from the copyright holders, Elsevier.)

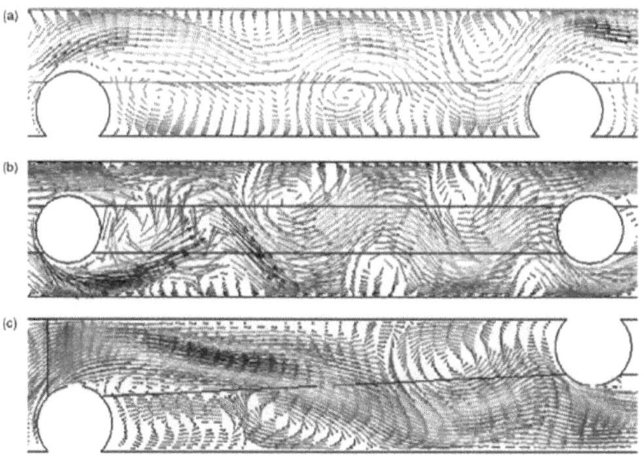

FIGURE 6.18 CFD simulation for unsteady flow for Re = 1200 for the (a) cavity, (b) submerged, and (c) zigzag spacer at a fixed mesh length and filament diameter. (Reprinted from Ref. [46], with permission from the copyright holders, Elsevier.)

the elements to avoid telescoping, the 80 MIL design could be the optimal spacer in this set of spacer designs for a practical crossflow range.

The effect of the spacer on the filtration performance can be affected by the crossflow velocity, flux conditions, and feed water characteristics. Suwarno et al. [43] investigated the effects of spacers on the TMP development and biofilm growth on the reverse osmosis membranes in the treatment of a 2000 mg/L NaCl solution under constant flux conditions. Their results showed that the effect of the spacer on the TMP increase rate was insignificant at the crossflow velocity of 0.1 m/s but evident at the higher velocity of 0.34 m/s with the beneficial effect of spacers more evident at higher feed salt concentrations, i.e. C_w in Figure 6.1. On the other hand, although the spacer was effective to enhance filtration flux, it was observed that a biofilm was still growing in the spacer-filled channel [43]. The confocal laser scanning microscopy examination revealed that the biofilm development in the spacer-filled channel was initiated on the membrane surface near the spacer filaments and subsequently grew to cover the central area. Li et al. [44] tested flexible hairy spacers made from nylon fibres for mass transfer enhancement in FO filtration and reported that the vibrations of flexible nylon fibres in the flow channel increased the FO flux by around 30% [44].

While the mass transfer enhancement by spacers is closely related to the hydrodynamic conditions in a space-filled channel, computational fluid dynamics (CFD) simulations have been demonstrated to be an effective tool to study the hydrodynamic conditions in the spacer-filled channel, including revealing transversal and longitudinal vortices, vortex shedding, and unstable flow behaviour [45]. Figure 6.18 shows the results of CFD simulations of the local flow patterns in a spacer-filled channel with the "cavity", "zigzag", and "submerged" spacer arrangements [46]. The simulations indicate that large recirculation regions may be formed behind the filaments and the shed vortices can scour the channel wall and enhance the shear stress at the membrane surface. For the cavity geometry spacer, the recirculation regions between sequential filaments influence each other and merge to form one large recirculation region between sequential filaments above a critical Reynolds number or below a critical mesh length. In contrast, the "zigzag" spacers forced the channel flow into an up and down zigzag pattern, which causes the recirculation region to reattach to the wall. Li et al. [45] suggested, based on the CFD simulation, that a moderated ratio of the length of the filament (l_m) to the thickness of the spacer (h_{sp}), e.g., $l_m/h_{sp} = 4$, is optimal for the enhancement of mass transfer due to the effective shedding of vortices between neighbouring filaments. A lower or higher l_m to h_{sp} ratio will be less effective for mass transfer enhancement because the former causes one dead eddy between neighbouring filaments, while the latter will only have shedding of vortices in the region close to the filaments [45].

To study the impact of spacers on particle deposition and membrane fouling, techniques, such as direct observation through membrane (DOTM) and 3D optical coherence tomography (OCT) imaging, have been used to visualize the cake formation in the spacer-filled channel. Figure 6.19 shows DOTM recorded images of the typical particle deposition pattern in a spacer mesh-filled channel [46], where one filament of the mesh was orientated perpendicular to the main flow direction and the other filament parallel with the main flow. This observation shows that there is a clear area with negligible deposition on the membrane behind and in front of the transverse filaments due to the effect of the inertial force and the scouring of the eddies. The clear area was observed to increase with an increase in the crossflow velocity. Using the DOTM technique, Neal et al. [47] measured the effect of spacers on the critical fluxes of micron and supra-micron-sized particles and found that critical fluxes were enhanced by up to two times those obtained with an empty channel. The 3D OCT imaging was used for in situ analyses of the variation of cake morphology in a spacer-filled channel. OCT imaging can detect the "active fouling region" through sensing the

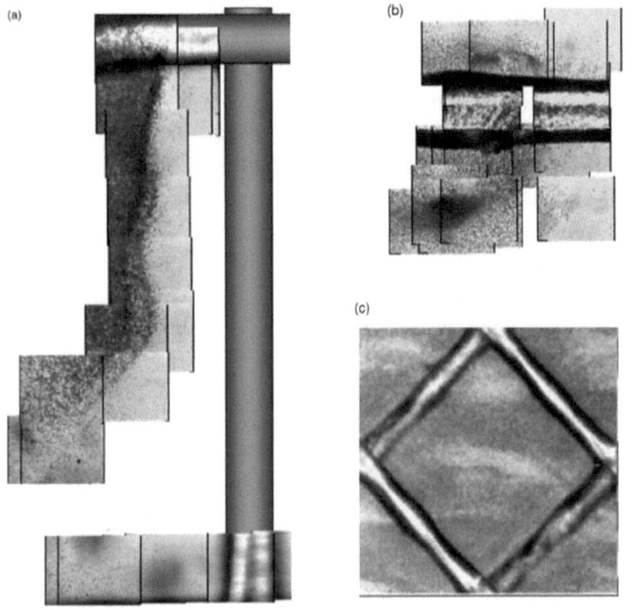

FIGURE 6.19 DOTM revealed particle deposition patterns for different spacer orientations: the attached filament was arranged at (a) 90°, (b) 0°, and (c) 45° to the direction of flow. (Reprinted from Ref. [46], with permission from the copyright holders, Elsevier.)

instantaneous changes in local cake thickness caused by deposition or shear erosion. This technique was proven to be an effective tool for the development of novel spacers [48].

6.3.1.3 3D-Printing Spacers

One of the main penalties of spacers is the increased pressure loss and consequent energy consumption. The ongoing work on spacer development has aimed to achieve mass transfer enhancement with a minimum additional resistance. A recent tool used for novel spacer design and development is 3D printing, which allows fabricating spacers with different filament structures and geometries for laboratory testing. Some novel 3D-printed spacers tested include helical filament spacers [49], 3D column-type (pillar) spacers [50], spacers with perforated filaments [51], hole-pillar spacers [52], multi-layer net spacers [53], spacers based on triply periodic minimal surfaces (TPMS) mathematical architecture [54], Turbospacers [55], and static mixing spacers [56]. Table 6.5 shows the main characteristics and performances of these spacers. From the performances of these 3D-printed spacers of different designs, it can be seen that optimization of the configurations of the spacer filaments, e.g., spacer filaments with the helical, pillar, and perforated structures [49,51,52,55] could achieve flux enhancement with reduced pressure drops. Comparing the different designs of the 3D-printed spacers shown in Table 6.5, the hole-pillar spacers could be the most efficient design based on their performance in flux enhancement and pressure drop reduction compared to those with the conventional non-woven spacers and 3D-printed pillar and perforated filament spacers [52]. As shown by the CFD simulation [46] and DOTM observations [47], the mass transfer efficiency of the conventional spacers can be limited by the dead zones formed behind the filaments. In the hole-pillar spacers, the dead zone in the spacer cell can be largely eliminated by the vortex shedding effect induced by the cylindrical nodes at the intersections of spacer elements [50] and by the micro-jet caused by the perforated filaments [51]. In the meantime, perforation on the filaments could effectively lower the pressure drop along the spacer-filled channel, which could reduce the uneven pressure distribution along the flow channel and allow a lower operation pressure. However, while various designs of 3D-printed spacers showed the feasibility to enhance mass transfer with a reduced pressure drop in spacer-assisted membrane filtration, most of these 3D-printed spacers are still in the proof-of-concept stage. Moreover, not all design features of 3D-printed spacers can be additively manufactured with accuracy and the differences in the accuracy and surface finish associated with the fabrication methods could result in surfaces and geometry deviations from the intended design [57]. Different 3D-printing methods, including solid-based fused deposition modelling (FDM), powder-based selective laser sintering (SLS), and liquid-based polyjet, have been used to fabricate 3D-printed spacers. Tan et al. [57] found that the liquid-based polyjet can achieve the most accurate representation of the intended design while spacers fabricated by the FDM method could have the greatest deviation from the specifications.

6.3.2 Other Turbulence Promoters

6.3.2.1 Helical Inserts and Corrugated Membranes

Figure 6.20 shows a typical design of a winding helical insert that is centrically located in a tubular membrane. This type of insert can be easily fabricated by winding a metal or plastic wire onto a rod. The main structural parameters of this insert include the rod diameter, pitch, and the clearance distance between the tubular membrane and the insert. Flow visualization shows that the presence of the helical baffle in the tubular membrane induces a rotational flow within the baffle and the angle of rotation depends on the pitch or number of turns over the length of the baffle. Three flow components were identified, which are the tangential flow in the clearance between the membrane and the insert, the rotational helical flow following the shape of the helical, and a reverse flow generated by the secondary flows on the cylindrical rod surface [58]. The flow near the membrane surface is dominated by tangential and rotational flows. It was reported that no obvious vortices were observed in the tube with helical inserts, even at relatively high flow rates.

Figure 6.21 shows the flux time profiles obtained in filtration of a 5% yeast cell suspension using a tubular membrane of 6 mm inner diameter and 0.14 μm pore size with a helical baffle (HB), a rod baffle (RB), and the tubular membrane without baffle (NB) [58]. A comparison has been made at the same hydraulic dissipated power which is defined as the product of the flow rate and the pressure drop along the tubular membrane, or the energy consumed to generate

TABLE 6.5
The characteristics and performance of 3D-printed spacers

	Characteristics	Performance
Helical filaments spacers [49]	1–3 helices along the spacer filaments	Compared to the standard spacers without helices (UF at the crossflow velocity of 0.182 m/s): - Increased specific flux by 291% - Decreased the pressure drop by 65% - Less fouling
3D column-type (pillar) spacers [50]	Addition of column nodes at the intersection of the filaments	Compared to standard non-woven symmetric spacer (UF of seawater spiked by yeast, alginate, and xanthan gum): - Reduced the pressure drop by 3 times - Doubled the specific water flux - Lower biomass accumulation
Spacers with perforated filaments [51]	Filaments with 1, 2, or 3 holes Perforated filament generated micro-jet in the spacer cell	One hole-spacers: - Increased flux by 75% - Reduced pressure drops by 15% 3-hole-spacers - Increased flux by 17% - Reduced pressure drops by 54%
Hole-pillar spacers [52]	Perforations at the pillar of the pillar spacers	Compared with the commercial and pillar spacers - Lower pressure drops - High permeate flux - Lower biofouling
Multi-layer net spacers [53]	Middle layer: Flow attack angle 47.7°, filament angle 85° Filament length 7 mm Filament diameter 1.1 mm Top and bottom layers: Flow attack angle 60°, filament angle 60° Filament length 4 mm Filament diameter 0.5 mm	Compared to a standard non-woven spacer: - 20% higher mass transfer - Crossflow energy consumption was reduced by 30-fold by changing the thickness of the middle spacer from 1.1 mm to 0.5 mm
TPMS-based spacers [54]	3D-printed spacers based on triply periodic minimal surfaces (TPMS) mathematical architecture	Compared to net-type commercial spacer in brackish RO and UF of alginate solution - 15.5% higher flux with RO and 38% higher flux with UF - Significant biofouling reduction - Reduced pressure drops at the higher crossflow velocity
Turbospacers [55]	Ladder-type filaments with microturbine within filament cells The rotation of the microturbines promotes turbulence for fouling mitigation.	Compared to standard non-woven spacers in UF of seawater spiked by yeast, alginate, and xanthan gum: - Reduced membrane fouling - Reduced pressure drop - 2.5 times lower in energy consumption (0.5 kwh/m^3)
Static mixing spacers [56]	Planar flow channel elements were designed for static mixing A spacer frame to fix the planar static mixer elements	Compared to the conventional spacers, the static mixer had - Higher mass transfer coefficient - Higher pressure drops

the crossflow through the tubular membrane. Using the hydraulic dissipated power rather than the crossflow rate as a control parameter for the comparison of the tubular membrane with and without inserts eliminates the effect of the reduced crossflow section by introduction of the inserts. From Figure 6.21, it can be seen that the presence of the helical baffle resulted in a higher flux but the flux with the RB mode was slightly lower than with the NB mode. The rod-type baffle (RB mode) has a similar parallel flow pattern to the NB mode but the value of the tangential velocity is lower than that with the NB mode for a given hydraulic dissipated power. This explains the slightly lower flux with the RB mode. The enhancement of filtration performance by helical baffles was also reported in the filtration of different "difficult" feeds using tubular membranes, including crude oil emulsions and mixtures of crude oil with biological solids from activated sludge processes [59], dextran T500 [60], and municipal wastewater [61].

One of the important characteristic parameters of the helical inserts is the pitch, or the turns per unit length. A large pitch means fewer turns of the helices over a certain length of the baffle. However, the limit of indefinitely increasing the number of turns or reducing the pitch will be a geometry approaching a rod-type baffle with negligible rotational flow. Thus, there should be an optimal number of helices for flux enhancement. Gupta et al. [58] experimentally determined that a helical baffle made up of four turns per 25 mm length was optimal, which is similar to the optimal 5 mm pitch determined by Xu et al. [61]. Gupta et al. [58] also suggested that a gap of about 1 mm between the membrane inner surface and the baffle height is appropriate for good filtration performance using this type of baffle.

Corrugated membranes have been tested and were formed by spreading flat sheet membranes over a corrugated supported plate. The corrugated structure of the plate was made by half-cylindrical bars attached to or grooves machined into the plate [62]. The depth of the semi-circular bar or groove could be

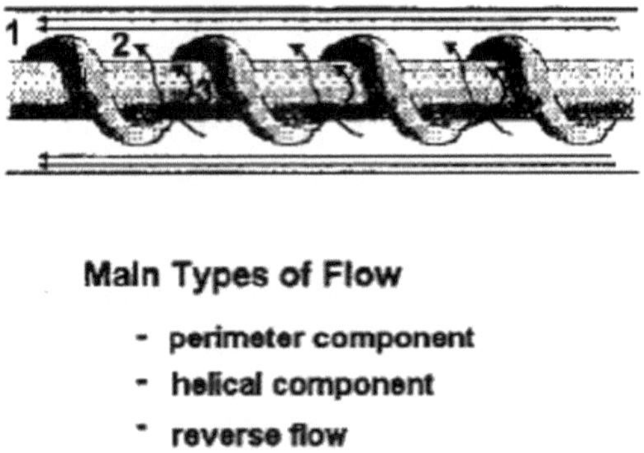

FIGURE 6.20 Schematics of a helical baffle. (Reprinted from Ref. [58], with permission from the copyright holders, Elsevier.)

from less than 1 mm to several mm. van der Waal and Racz [62] studied the effect of corrugations on filtration with 3 mm circular bars glued onto the support plate. Flow visualization revealed that the membrane corrugations resulted in a transition from laminar to turbulent flow at a lower velocity than with a flat membrane. Circulation eddies formed behind the corrugation over a downstream distance up to about 10–15 mm. The filtration experiments indicated that corrugations could result in an enhancement in mass transfer coefficient provided the separation distance between corrugations was in the range 15–40 mm. However, no significant effects were observed with larger or smaller mutual distances, such as 80 mm and 10 mm, respectively.

Stairmand and Bellhouse [63] tested the combined effect of corrugated membranes and pulsated flow on mass transfer in a blood oxygenator. Figure 6.22 shows the schematics of the device used in their experiments. The corrugated structure tested by them was formed by semi-circular grooves of 0.75 mm depth and 1.25 mm separation, much shallower and more closely spaced than that tested by van der Waal and Racz [62]. The pulsated flow was produced by a rotating ½" ball that intermittently closed the line and a section of flexible tube which could be clamped to varying degrees to alter the stroke of the pulsations. During the experiments, the oxygen was convectively transferred from the high-pressure chamber (1 bar) to the low-pressure chamber (0 bar) through the liquid flowing through the channel between these two chambers. The rate of convective transfer across the channel would depend on the intensity of mixing of the fluid flowing through the channel. Figure 6.23 [63], shows the effect of pulsations on oxygen transfer between the two corrugated membranes

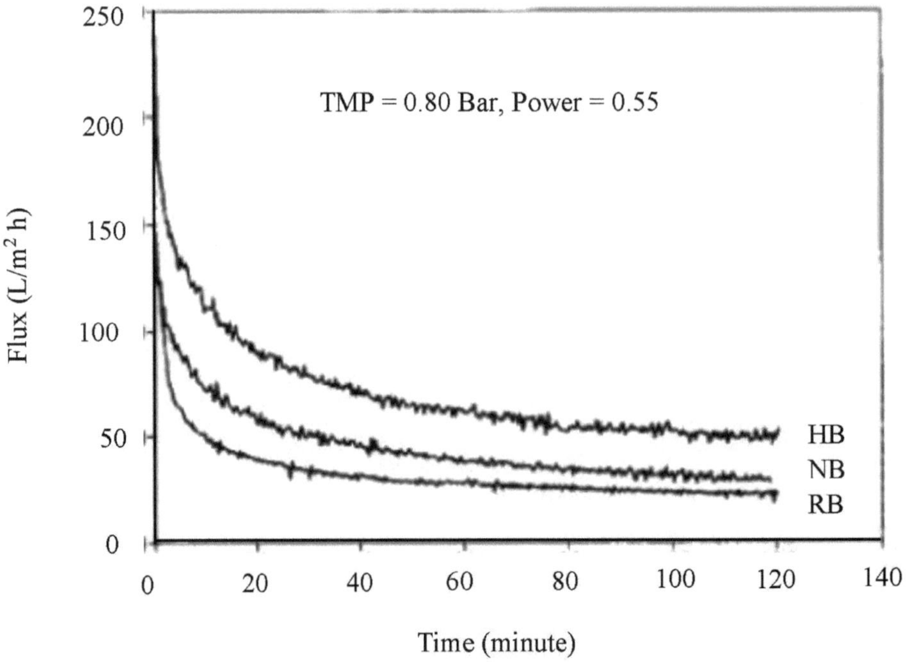

FIGURE 6.21 Flux-time profiles for filtration with HB, RB, and NB modes. (Reprinted from Ref. [58], with permission from the copyright holders, Elsevier.)

with water- and milk-filled channels under conditions of different Reynolds number. The experimental data indicate that for non-fouling liquid medium (distilled water), where the Reynolds number exerted a dominant effect on oxygen transfer rate applying an 8-Hz pulsation to the water-filled channel, could enhance the Sherwood number by about 20%. For fouling medium (milk-filled channel), the effect of Reynolds number on oxygen transfer could be considerably limited by membrane fouling. However, this phenomenon was found to be partially reversed by the application of pulsations. For Reynolds numbers larger than 10,000, the Sherwood number was enhanced by a factor of about 2 by imposition of pulsations, confirming the effect of pulsations on reducing concentration polarization with corrugated membranes.

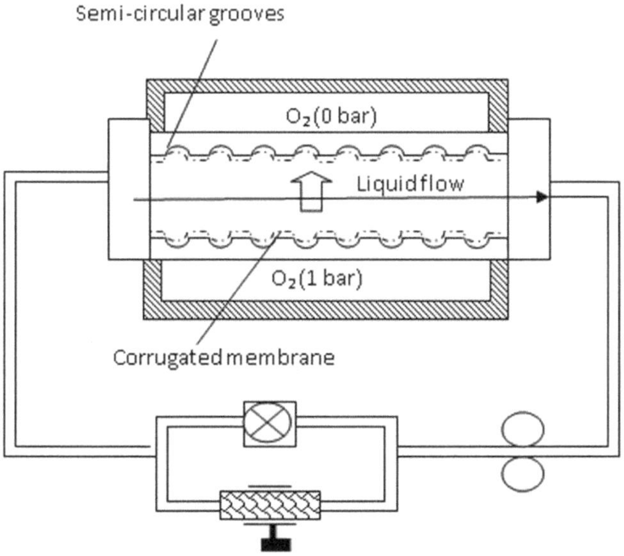

FIGURE 6.22 Schematics of simulated oxygenator employed with corrugated membrane and pulsatile flow. (Adapted from Ref. [63], with permission from the copyright holders, Elsevier.)

Instead of using a furrowed or dimpled membrane support plate, Sobey [64] observed that a single-flow deflector in a flat membrane channel could produce a number of vortices under oscillatory flow conditions, an effect named the "vortex wave". An important feature of the vortex wave is that it could occur under low crossflow velocity conditions or with laminar flow, so that it can be used for shear-sensitive fluids. Millward et al. [65] tested the effect of vortex waves on plasma filtration with waves produced by flow deflectors with a cross-sectional area of 1 × 1 mm^2 in a 2.25 mm high channel, as shown in Figure 6.24. The aim was to improve membrane applications for the separation of plasma from cellular blood components for both donor plasmapheresis and the treatment of autoimmune disease. For this application, it is important to achieve a high plasma filtration rate with negligible damage to the blood components under limited blood flow rate conditions, so there would be a need to employ secondary flows to control concentration polarization and membrane fouling. Figure 6.25 shows the plasma flux obtained with different spacings of

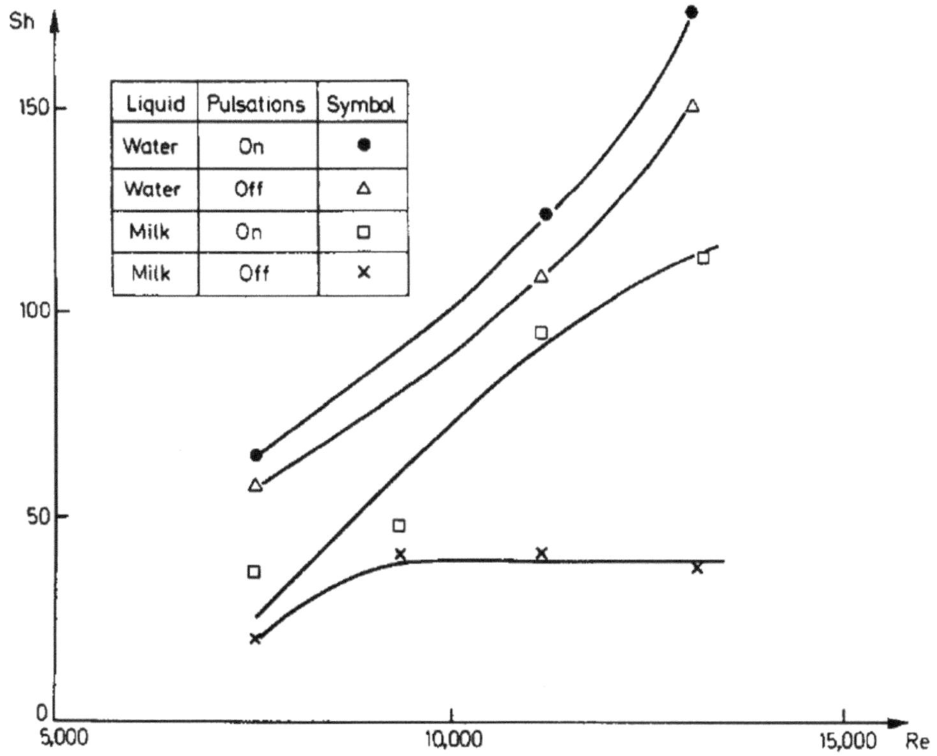

FIGURE 6.23 Effect of pulsatile flow on oxygen transfer with water and milk medium. (Reprinted from Ref. [63], with permission from the copyright holders, Elsevier.)

flow deflector under relatively low Reynolds numbers, in the range of 50–200, using 0.2 µm flat sheet membranes. The results indicate that the addition of narrowly spaced (4 mm) flow deflectors to the membrane channel enhanced the flux significantly by a factor of 3.5 compared to an empty channel at a Reynolds number of Re = 123. More importantly, it was found that the deflectors did not increase the channel pressure drop significantly in such a low Reynolds range. Millward et al. [65] also demonstrated that the oscillatory flow component was an essential feature of the vortex wave design. They found that the plasma flux dropped from 0.1 cm/min to 0.01 cm/min by removing the oscillations during the filtration with a channel with 4 mm spaced deflectors. These results demonstrate that significant flux enhancement could be obtained by combining a "turbulence" promoter and pulsation design even with laminar flow conditions.

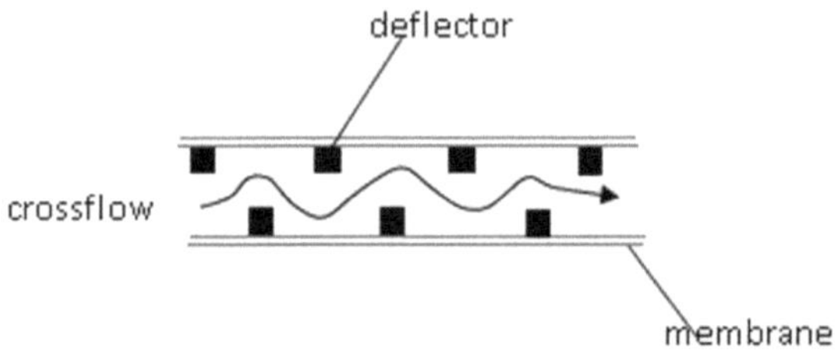

FIGURE 6.24 Schematics of membrane flow channel-equipped flow deflector.

The recent evolution of corrugated membranes is to directly alter the membrane topography through solution-based micromoulding (solution mould casting and phase separation), embossing, and 3D printing [66]. These methods either direct cast membranes with micron or submicron surface patterns (solution-based micromoulding methods and printing) or impart the patterns at the top of membranes (embossing methods). Surface patterning has been tested to fabricate polymer microfiltration, ultrafiltration, and thin-film composite membrane, which have regular, periodic corrugated surface patterns with feature sizes ranging from sub-10 nm to hundreds of µm [67]. Performance enhancement by patterned membranes was observed in reverse osmosis desalination, membrane distillation desalination, nanofiltration, particle filtration, water and wastewater treatment, forward osmosis, gas–liquid contractor, gas permeation pervaporation, proton exchange fuel cell, direct methanol fuel cells, and anion exchange membranes [68]. Ding et al. [67] showed that surface patterning increased the critical fluxes by 19–45% in the filtration of solutions containing silica particles of different sizes and significantly reduced flux decline and irreversible fouling in the filtration of bovine serum albumin (BSA) solution. In general, the degree of enhancements by patterned membranes can be affected by the crossflow velocity, the direction of feed flow, pattern geometry, and properties of the feed solution [67]. Thus, optimization is needed for different applications.

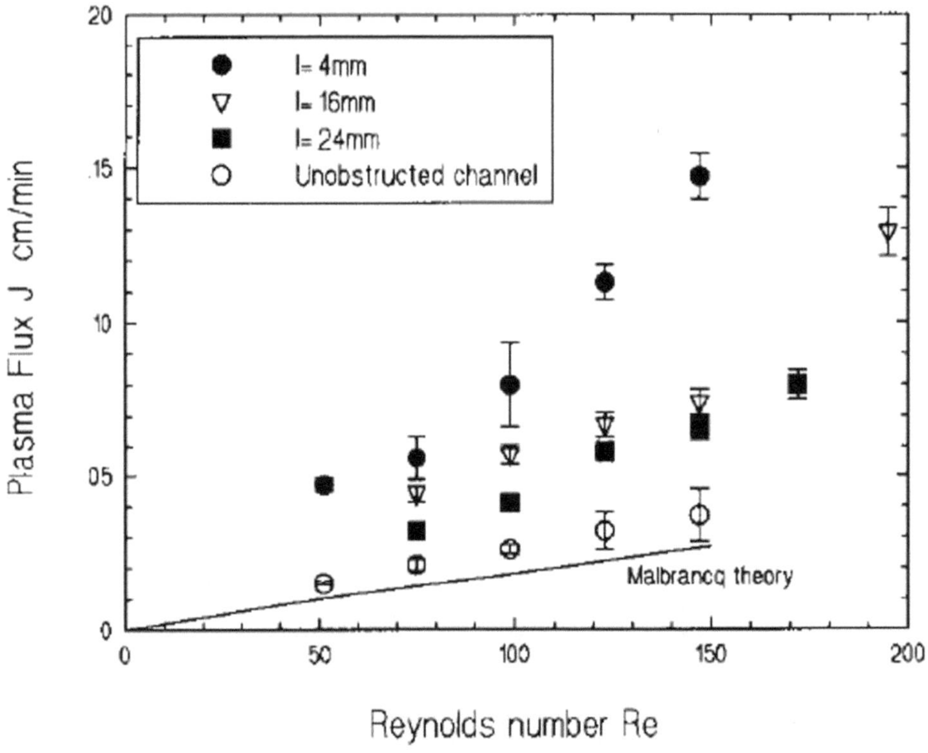

FIGURE 6.25 Plasma flux obtained with flow deflector equipped channel with and without pulsation. (Reprinted from Ref. [65], with permission from the copyright holders, Elsevier.)

6.3.2.2 Flow-Field Mitigation of Membrane Fouling

Flow-field mitigation of membrane fouling (FMMF) is a membrane fouling control technique based on the so-called field-flow fractionation (FFF) principle [69]. The FFF is induced by segregation of the components in a stream resulting from the different physicochemical properties of the components. The FMMF manipulates the flow-field by inclining the membrane flow channel to generate a transversal lift effect to counter the permeate drag experienced by particulate foulants. The experiment and simulation showed that a significant increase in critical flux with a reduced energy requirement can be achieved by applying a slight channel inclination angle of 1.15° [69]. The FMMF can be readily applied to membrane modules, e.g., tubular, SWM, and flat sheet submerged membrane modules, by inclining either the feed channel wall or two permeable membranes to divert foulant particles away from the membrane [69].

6.4 PULSED FLOW

Different from the secondary flows and flow channel spacer techniques, the pulsed flow method generates pressure fluctuation waves either in the feed or permeate flow channel using certain oscillators. The fluctuating pressure wave can enhance the membrane filtration by reducing the boundary layer or by induced instant local backflushing flow, as discussed in Section 6.4.1.

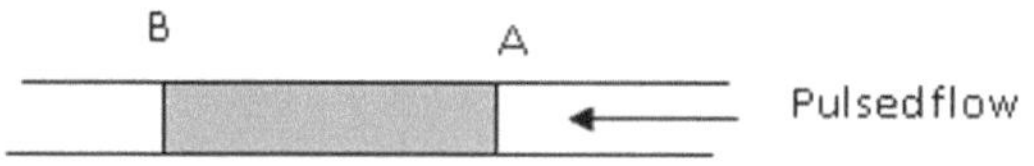

FIGURE 6.26 Pulsed flow between two points a short distance apart along a channel.

6.4.1 Hydrodynamic Characteristics of Pulsatile Flow

Pulsatile flow can be defined as flow with a periodic pressure fluctuation wave travelling along the flow path. As in a steady Poiseuille's flow, it is the pressure gradient along the flow path that determines the instantaneous pulsatile flow rate. For the flow path shown in Figure 6.26, the pressure gradient is determined by the pressure difference between the pressure recorded at upstream point A and that recorded at the downstream point B. Owing to the pressure pulse transfer from the original source of the pulsation down along the flow path, the crest of the wave reaches the first point A a short time before it reaches the downstream point B, at this time the pressure at the point A is higher than that at point B. A short time later, as the crest reaches the point B, the pressure at point B will be higher than that at point A. Thus, the pressure pulsation will cause a rapid forward-and-reverse pressure gradient oscillation over the flow path AB. Figure 6.27 shows how a travelling sinusoidal pressure waveform creates an oscillatory pressure difference. Figure 6.27(a) shows two simulated waveforms recorded at two points over a short distance of a flow path with a 10° interval and Figure 6.27(b) shows the instantaneous pressure gradient between the upstream and the downstream points. From this figure it can be seen that during one period of the pressure fluctuation the pressure gradient over the short flow path also experiences change from a negative to positive value. The motion of the fluid that is driven by an oscillating pressure gradient is complicated. The following momentum equation has been developed for an incompressible laminar pulsatile flow in a circular rigid tube:

$$\frac{\partial^2 w}{\partial r^2}+\frac{1}{r}\frac{\partial w}{\partial r}+\frac{1}{\eta}\frac{\partial P}{\partial z}=\frac{\rho}{\eta}\frac{\partial w}{\partial t} \tag{6.21}$$

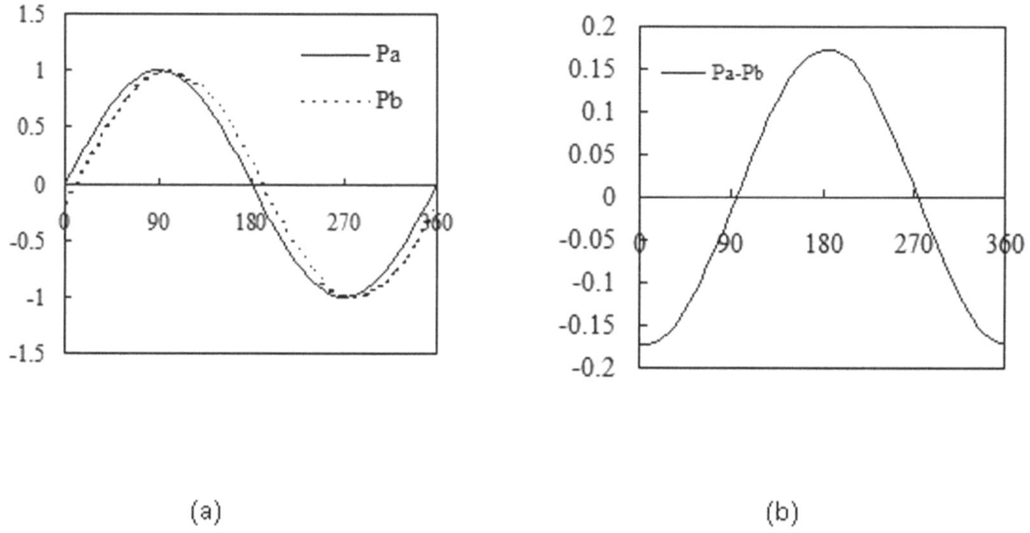

FIGURE 6.27 (a) Two simulated waveforms recorded at two points over a short distance of a flow path with 10° interval; (b) the instantaneous pressure gradient between the upstream and the downstream points.

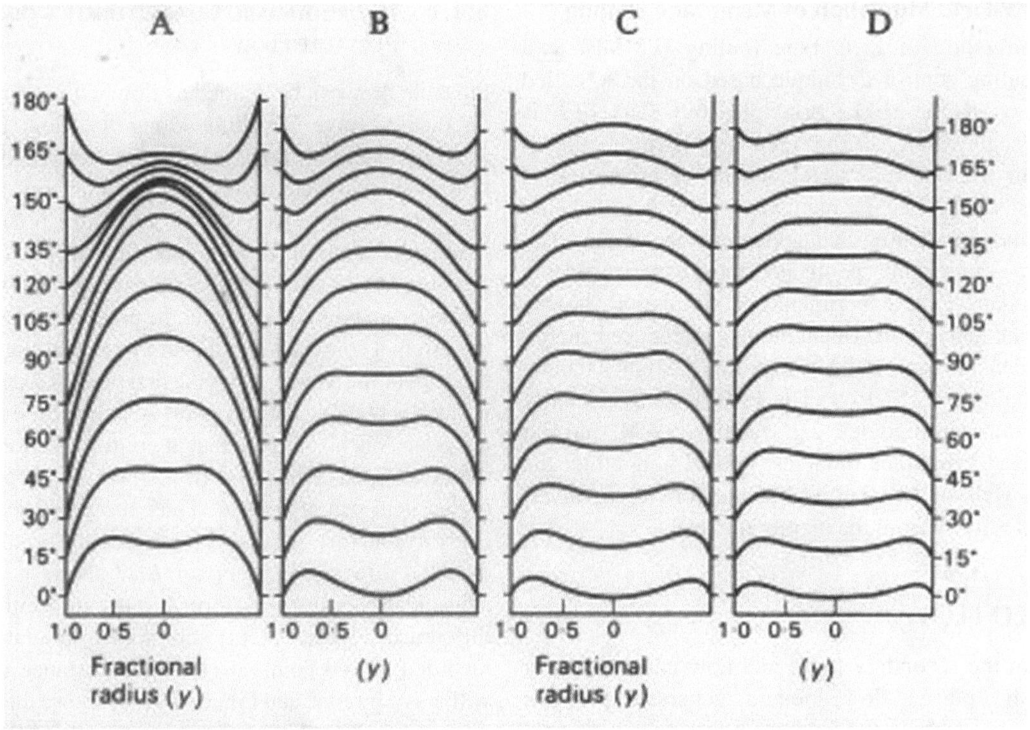

FIGURE 6.28 The velocity profiles of the flow resulting from a sinusoidal pressure gradient (cosωt) in a tube: (A) α = 3.34, (B) α = 4.72, (C) α = 5.78, (D) α = 6.67. (Reprinted from Ref. [71], with permission from the copyright holders, John Wiley & Sons.)

where P is the pressure, and w is the velocity in the axial direction.

For steady flow the velocity does not vary with time, so that $\partial w/\partial t = 0$. The pressure gradient $\partial P/\partial z$ can be measured over a finite distance and written as:

$$\frac{\partial P}{\partial z} = \frac{P_1 - P_2}{L} \tag{6.22}$$

When the pressure gradient is in the form of a single harmonic with complex components:

$$-\frac{\partial P}{\partial z} = A^* e^{i\omega t} \tag{6.23}$$

where A^* is the amplitude of pressure gradient in complex form, ω is the angular velocity. For symmetric flow with non-slip conditions applied, the solution to this equation is [70]:

$$w = \frac{A^* R^2}{i\eta\alpha^2}\left\{1 - \frac{J_0(\alpha y i^{3/2})}{J_0(\alpha i^{3/2})}\right\} e^{i\omega t} \tag{6.24}$$

where $J_0(\alpha y i^{3/2})$ is a Bessel function of the first kind of order zero and complex argument, $y = r/R$. The parameter α is a dimensionless number that characterizes the kinematic similarities in the liquid motion, known as the Womersley number. It is written as:

$$\alpha = R(\omega / \nu)^{\frac{1}{2}} \tag{6.25}$$

The volume flow can be obtained by integrating the velocity across the lumen of the tube:

$$Q = \frac{\pi A^* R^2}{i\omega\rho}\left\{1 - \frac{2J_1(\alpha i^{3/2})}{\alpha i^{3/2} J_0(\alpha i^{3/2})}\right\} e^{i\omega t} \tag{6.26}$$

where J_1 is Bessel function of order one.

The velocity profile across the tube lumen with pulsatile flow is not of the same parabolic form as that found in a steady laminar flow. The velocity profiles oscillate sinusoidally, as discussed in detail by Hale et al. [71]. For example, Figure 6.28 shows the velocity profiles, at intervals of 15°, resulting from a simple sinusoidal pressure-gradient (cos(ωt)) during the half cycle (0° to 180°); as for a simple harmonic motion, the second half is the same.

It can be seen from Figure 6.28, for a pulsed pressure gradient, that the characteristic parabolic velocity profile observed in steady laminar flow does not appear at any time during the cycle. There is a phase lag between the pressure gradient and the liquid movement, and being a cosine function, the maximum amplitude of the pressure gradient occurs at 0°,

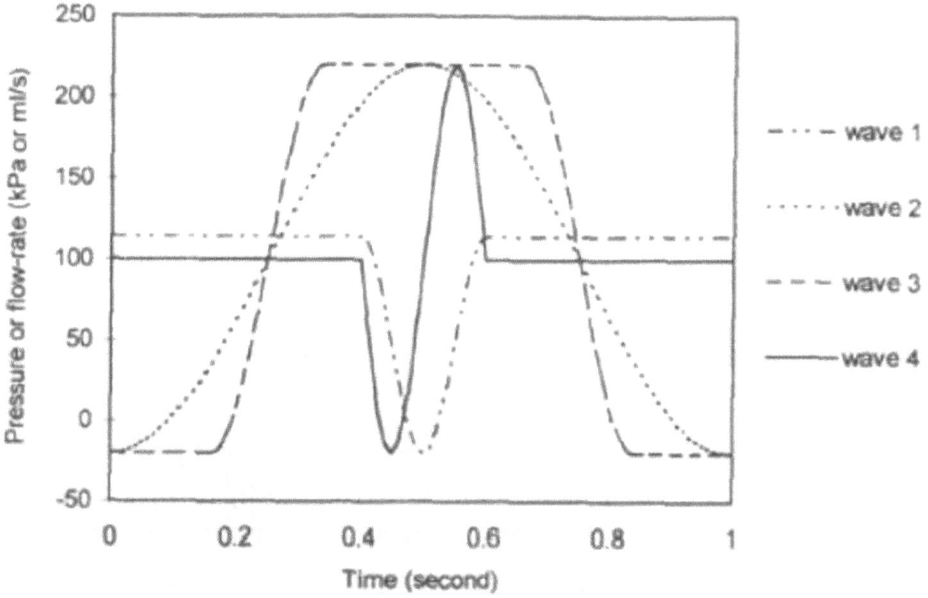

FIGURE 6.29 Design waveform for pressure and flow rate. (Reprinted from Ref. [72], with permission from the copyright holders, John Wiley & Sons.)

while the maximum for the total flow is at 60° in Figure 6.27(a) and at about 70° in Figure 6.27(b). Significant shear is only found in the region near the wall with the liquid in the central portion of the tube virtually unsheared. Thus, the liquid behaviour for pulsatile flow is rather like a solid mass sliding inside a thin layer of viscous liquid surrounding it. The higher the Wormserly number (the higher the frequency or the larger the tube diameter), the flatter is the velocity profile, implying a thinner boundary layer at a higher Wormserly number.

6.4.2 Pulsatile Flow Enhanced Membrane Processes

Two mechanisms, shear-related and oscillated backflushing, have been suggested for pulsatile flow enhanced membrane processes by Li et al. [72]. The shear-related mechanisms contribute to the filtration enhancement by a reduced boundary layer and enhanced particle back-transport. Since the shear scouring effect is not direction-dependent, the maximum of the absolute value of the shear may be used to estimate the limiting or critical flux for filtration under certain conditions. For laminar flow, the wall shear can be calculated based on the Womersley equation. For turbulent flow, there is no simple solution for the wall shear next to the membrane surface. Li [73] suggested estimating the wall shear based on the measured pressure gradient with the assumption that the same pressure gradient should also apply to the boundary layer. Thus, the maximum shear can be regarded as being proportional to the maximum absolute pressure gradient along the membrane module [73].

The oscillated backflushing mechanism accounts for the effect of the pressure waveform on transmembrane pressure. When the minimum pressure of the pulsatile pressure waveform results in a negative transmembrane pressure, a reverse permeate flow may occur that acts as a backwash flow that drives particles deposited on or near the membrane surface back to the bulk flow. In this way, concentration polarization and cake formation caused by the filtration operation could be limited, depending on the properties of the particles and the magnitude of the reverse pressure.

In order to obtain oscillated backflushing and enhanced membrane performance, the design of the pressure waveform is crucial. The pressure waveform can be characterized by the minimum and maximum pressure, the duration of the negative pressure and the positive pressure, and the pulsation frequency. Figure 6.29 shows four typical pressure waveforms studied by Li [72]. Wave-type 1 has a long steady pressure phase and a short negative pulsation. Wave-type 2 has an approximate sinusoidal form with smooth variation of the pressure. Wave-type 3 is approximately square shaped, and wave-type 4 had a shortened sinusoidal disturbance. Figure 6.30 shows the cake-resistance reduction results as a function of the minimum transmembrane pressure obtained by Li et al. [72] or pulsatile flow filtration with the four pressure waveforms shown in Figure 6.29 for two different amplitudes. It can be seen from Figure 6.30 that a significant reduction in cake resistance in filtration of 0.5 g/L silica suspension using a ceramic membrane of 0.2 μm was be obtained by using waveform types 1 and 3 with relatively large negative transmembrane pressures. Waveform type 1 had a long steady-pressure phase and a short, negative low-pressure pulse which may result in a short period of backflushing. For this waveform type about an 80% reduction in cake resistance was achieved when the minimum transmembrane pressure was reduced to about –60 kPa. Waveform type 3, which had a longer negative pressure period than waveform type 1, resulted in more than a 90% reduction in filtration resistance

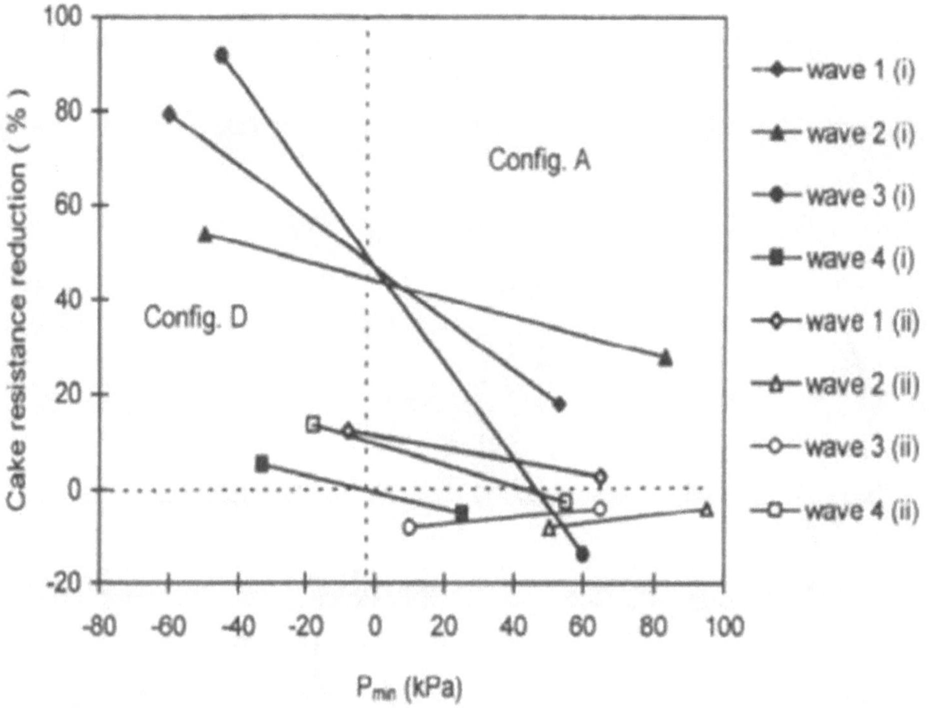

FIGURE 6.30 Cake resistance reduction with different waveform-type design. (Reprinted from Ref. [72], with permission from the copyright holders, John Wiley & Sons.)

with a minimum transmembrane pressure of about –40 kPa, implying the effect of duration of the backflushing. Waveform types 3 and 4 were long and short sinusoidal forms, providing a continuous variation of pressure. There seemed to be less effective reduction in cake resistance with the shorter negative pressure period (waveform type 4).

Pulsated flow can be achieved by various flow control methods. Figure 6.31 shows a lab-scale membrane filtration set-up with the so-called collapsible-tube oscillator used by Bertram et al. [74] for producing pulsated flow in membrane filtration. This system included a supply vessel, a membrane module, a receiving tank, a holding tank, and the collapsible-tube oscillator. The collapsible-tube oscillator consisted of a horizontally mounted 365 mm length of silicone rubber tube with unstressed inside diameter 13.2 mm and wall thickness 3.2 mm, clamped at each end by stainless steel fittings. The pulsations with this device are produced by periodic collapse and reopening of the tube induced by pressure applied to the chamber surrounding the tube. With this arrangement the frequency of the pulse flow can be controlled in a range of 7–12 Hz, depending on the volume of fluid downstream being oscillated, tube parameters, and pressure manipulation.

A number of researchers have assessed the effects of pulsatile flow on different membrane processes with a wide range of feeds. One of the first studies was by Kennedy et al. [75], who showed that flux in the RO of sucrose solution could be increase by 70% by pulsatile flow at 1 Hz. Gupta et al. [76] reported a 45% enhancement of flux in MF of raw apple juice with a pressure wave form provided by a fast piston return followed by a fast forward stroke at 1 Hz frequency. Jaffrin et al. [77], using hollow-fibre filters, demonstrated a 45% enhancement in flux in plasma filtration. Using the collapsible-tube oscillation generator described above, Bertram et al. [74] demonstrated that pulsation resulted in a 60% increase in permeate flux in the filtration of silica suspensions.

6.4.3 Transmembrane Pressure Backshock Technique

Instead of exerting pressure pulsing on the feed side, Rodgers and Sparks [78] tested periodic pressurization of the filtrate from 5 to 30 kPa above the respective operating pressure to obtain pulsatile negative transmembrane pressures. With a short period of negative pressure or backshock (0.01–0.38 second) presented at a frequency of 0–5 Hz, their experiments showed an average increase in permeate flux of 62–174% for ultrafiltration of 1% BSA solutions with laminar crossflow. However, no improvement in flux was observed with turbulent flow as a result of transmembrane pressure pulsing. Guerra et al. [79] assessed the effect of backshock on the filtration of skim milk using so-called normal and reverse asymmetric membranes. The normal asymmetric membrane refers to the asymmetric membrane with the tight skin of the membrane facing the feed (as is usual practice). The reverse asymmetric membrane was with the porous support layer facing the feed. Guerra et al. [79] reported that for filtration with the normal asymmetric membrane, the permeate flux could be increased by 100% by applying a backshock of 0.022 second at a frequency of 0.33 Hz for the filtration of skim milk. For filtration with

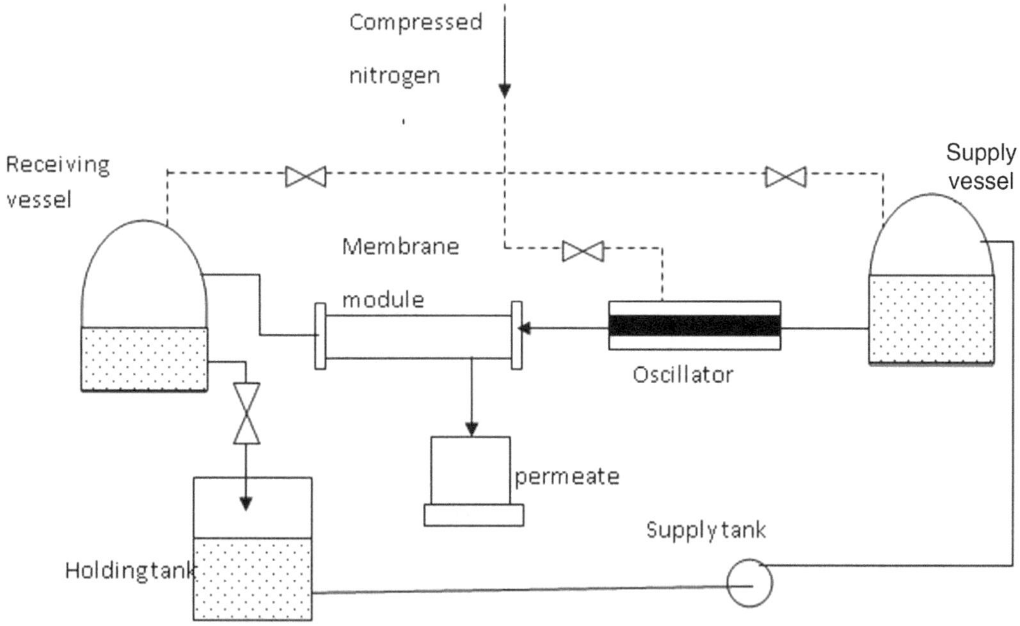

FIGURE 6.31 Schematics of the filtration set-up with collapse-tube pulsation generator. (Adapted from Ref. [74], with permission from the copyright holders, Elsevier.)

the reverse asymmetric membrane without backshock, it was found that concentration polarization formed within the porous support layer and it was impossible to filter skim milk with such a flow arrangement. However, in the filtration of skim milk using reverse asymmetric membrane with backshock at a crossflow velocity of 1 m/s, the imposed flux was able to step up to higher than 350 L/m^2/h without causing a significant increase in the transmembrane pressure, while stopping the backshock provoked an immediate decrease in the filtration flux [79]. The combination of reverse asymmetric membranes and backshock was patented by Wenten et al. [80].

Backshock was also applied in filtration using membranes called microsieves, which are very thin and smooth membranes with uniform pores made by silicon micromachining [81]. The thinness (ca. 1.2 μm) and high porosity of microsieves can result in a high flux in the order of 10^4 L/m^2/h even under very low transmembrane pressure conditions. However, such a high flux could result in the rapid formation of a cake layer on the membrane surface within a fraction of a second. In order to control the cake formation caused by high flux, transmembrane pressure pulsation is usually applied in filtration by microsieves. Kuiper et al. [82] reported that the cake layer formed by yeast particles in microsieve filtration of lager beer could be largely prevented by applying a backflush of –0.05 bar with a pulse duration of 0.05 s at a frequency of the order of seconds. Moreover, direct observation revealed that the backpulse could lift particles off the surface and these loosely attached particles or flocs were found to be eventually removed by a sudden increase in the crossflow velocity for a short period.

Lab-scale tests have been carried out to assess backshock for tubular, flat sheet, hollow-fibre, and spiral-wound membranes with the application fields covering wastewater treatment, food industry, and biotechnology applications [83]. Most of these tests have shown that backshock significantly increased filtration flux and reduced membrane fouling under optimal backshock conditions. Several analytical and empirical models have been developed to investigate the effects of various factors on backshock efficiency and to predict the net fluxes of filtration with backshock [83,84]. The net flux takes account of the flux loss caused by backshock and is defined as the difference between the forward filtration flux and backshock flux. The net flux models of backshock filtration contain a variety of parameters, including the filtration duration (t_f), backshock duration (t_b), forward flux during filtration (J_f), reverse flux during backshock (J_b), clean membrane flux (J_0), steady-state flux (J_s), the ratio between the cleaned membrane surface area and the total filtration (β), the delay of cake formation at the beginning of the filtration cycle (t^f_{crit}), the delay of cake removal at the beginning of backshock (t^b_{crit}), the ratio of the reverse and forward transmembrane pressures (α), and the time constant for permeate flux decline due to cake formation (τ_1). Gao et al. [83] assessed six backshock filtration models based on 21 experiments. They found the following flux model [85] was the best for predicting the net flux (J):

$$\text{when } t_b \geq t_b^{crit}, t_f \leq t_f^{crit}$$

$$J = \left[\beta J_0 + (1-\beta) J_s\right] \frac{t_f - \alpha t_b}{t_f + t_b}$$

$$\text{when } t_b \geq t_b^{crit}, t_f > t_f^{crit}$$

$$J = \beta J_0 \frac{t_f^{crit} + 2\tau_1\left[\left(1 + \left(t_f - t_f^{crit}\right)/\tau_1\right)^{1/2} - 1\right] - \alpha\left(t_b - t_b^{crit}\right)}{t_f + t_b} + \left(1 - \right) J_s \frac{t_f - \alpha t_b}{t_f + t_b} - \beta J_s \frac{\alpha t_b^{crit}}{t_f + t_b} \quad (6.27)$$

The modelling analysis showed that the net fluxes were 6–28 times the steady-state flux for $\beta \rightarrow 1$ and between 2–9 times the steady-state flux for $0 < \beta < 1$ for the filtration duration between 0.4–50 s and backshock durations between 0.05–2 s [83]. Given that β is the ratio between the cleaned membrane surface area and the total filtration, these results indicate that the enhancement of net flux by backshock is highly dependent on the irreversibility of the fouling layer formed during the filtration period. For hollow-fibre membrane filtration, Vinther and Jönsson, [84] showed that the average net flux over the length of the membrane with backshock was 1.36 times the flux without backshock. Therefore, the efficiency of backshock for the membrane performance enhancement is a function of backflush frequency, the period of backflushing, the operation conditions, the properties of the foulant materials, and the membrane configurations. A high operational flux usually needs a strong backflush to maintain stable performance. However, it may cause an increased loss of permeate volume and even mechanical failures of membranes [83].

6.5 ENHANCED SHEAR DEVICES

In conventional crossflow filtration, the shear on the membrane surface is usually produced by the feed flow driven by a pump. The shear rate can get up to the order of 10^4 s^{-1}. High liquid flow results in high pressure drop along the membrane module, leading to considerable energy consumption and non-uniform distribution of transmembrane pressure over the flow path of the feed channel. Such a close dependence between high crossflow velocity and pressure drop limits the use of high crossflow in various commercial membrane modules. However, in high shear membrane filtration devices, the relative movement between the liquid and the membrane surface is induced by an independent device rather than by the liquid transport pump. With shear rate independent of the feed bulk flow, the high shear membrane filtration devices can be operated at a relatively homogeneous and controllable transmembrane pressure. The typical high shear membrane filtration devices include rotating disks and vibrating dynamic membrane filtration systems, as described below.

6.5.1 Rotating Systems

Figure 6.32 shows the typical structure of a rotating dynamic membrane system which consists of a stationary flat membrane, a rotating disk, a hollow shaft, and a cylindrical housing with feed inlet and permeate outlet. During filtration, the feed is pumped into the closed housing, flowing through the membrane and evacuating through the hollow shaft. The permeate flow comes out of the system from the permeate outlet. Driven by an electric motor, the rotating disk rotates at high speed (100s to 1000s rpm) to produce shear on the membrane surface. The transmembrane pressure can be controlled by a valve in the retentate outlet line.

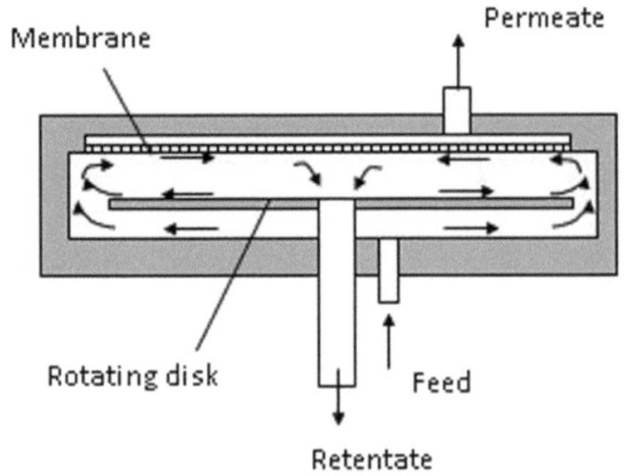

FIGURE 6.32 Schematics of rotating disk dynamic membrane filter.

The centrifugal forces generated by the disk induce an outward radial flow near the rotating disk and inward radial flow towards the axis near the membrane surface. If there is a relatively large gap (several mm) between the membrane and the rotating disk, the flow in the gap can be approximately treated as a flow core that rotates at an angular velocity of $k\omega$ between two boundary layers on the membrane and the rotating disk surface, where ω is the rotating velocity of the disk, and k is a coefficient between 0.3–0.44 [86]: the azimuthal Reynolds number can be defined as $Re = k\omega R/\nu$. The shear on the membrane surface at radius r can be calculated based on the expressions for different flow regimes [87]:

laminar boundary layer regime, $Re < 2.5 \times 10^5$, $e/R > 0.05$

$$\gamma = \frac{0.77(k\omega)^{3/2} r}{\nu^{1/2}} \quad (6.28)$$

In the turbulent regimes ($Re > 2.5 \times 10^5$, $e/R > 0.05$),

$$\gamma = \frac{0.0296(k\omega)^{9/5} r^{8/5}}{\nu^{4/5}} \quad (6.29)$$

Typically, for a disk with diameter 150 mm rotating at speeds of 500 or 2000 rpm, the calculated shear rates at the edge of the disk based on Equation 6.29 with $k = 3.7$ are 6.4×10^3 and 7.7×10^4 (1/s), respectively.

The pressure in the gap is also a function of the radius distance. The radial gradient in the boundary layer is equal to that in the inviscid core as given by [86]:

$$\frac{\partial P}{\partial r} = \rho r(k\omega)^2 \qquad (6.30)$$

The pressure distribution can be obtained by integrating Equation 6.30 for r from 0 to r, assuming k is independent of r.

$$P = \frac{1}{2}\rho(k\omega r)^2 + p_0 \qquad (6.31)$$

For a rotating disk dynamic membrane filter, the transmembrane pressure and the shear rate on the membrane surface vary radically as indicated by the above equations. Bouzerar et al. [86] studied the averaged filtration resistance at different radial zones for different rotating speeds using calcium carbonate particle (mean diameter 4.7 μm) suspension as the model fluid. In the experiments, several Nylon membranes were partially sealed with epoxy to obtain membranes with different centric permeable zones: $r < 3$ cm, $3 < r < 4.5$ cm, $4.5 < r < 6$ cm, and $6 < r < 7.5$ cm. The results indicated that for the filtration zones $r > 3$ cm, the resistance is equal to the membrane resistance for all the rotating speeds ≥ 200 rpm, implying that the particle deposition can be effectively arrested by a rotating speed as low as 200 rpm in this membrane zone. For the centre filtration zone ($r < 3$ cm), the filtration resistance was a strong function of the rotating speed, and the filtration resistance became equal to the membrane resistance only at 2100 rpm, while at lower rpm the filtration resistance was greater. Jaffrin et al. [87] reported that the permeate flux of the rotating disk can be significantly enhanced by equipping the rotating disk with radial vanes of a certain height, which can increase the shear rate on the membrane surface considerably.

The following empirical relationship has been suggested to correlate the filtration flux to the shear rate for filtration with a rotating disk dynamic membrane filter:

$$J = A\gamma_m^n \qquad (6.32)$$

$A = 1$ and $n = 0.5$, and $A = 0.30$ and $n = 0.55$, were suggested for filtration of silica suspension [88] and skim milk [89], respectively.

A number of commercially available rotating disk dynamic membrane systems have been tested for different applications. The DMF module (Pall Corp. NY) consists of several disks mounted on the same shaft, such that each can rotate between two annular membranes with a maximum rotation speed of 3450 rpm, corresponding to an azimuthal velocity of 20 m/s at the disk tip. Pall's lab-scale system has been tested for protein separation [90] and filtration of recombinant yeast cells [91]. The studies of the application of the rotating disk dynamic membrane indicated that high shear-enhanced filtration is much less sensitive to the solids concentration. High concentration factors achieved at significantly higher fluxes than the conventional processes have been reported with quite a range of feeds, including ferric hydroxide, yeast suspension, and skim milk, etc. [86,89,91–93]. The Optifilter CR (cross-rotational) system (Metso Paper, Raisio, Finland) is a commercial dynamic rotating membrane filter where the high shear on the stationary membranes is generated by a rotating blade with a tip azimuthal speed of 10–15 m/s [92]. The commercial CR system consists of 40–100 membrane disks and can provide a total membrane area from 15 m^2 up to 140 m^2. Nuortula-Jokinen and Nyström [92] compared the Optifilter CR with conventional crossflow tubular membrane for concentrating paper-mill effluents and reported that the quasi-steady flux obtained with CR at 470 rpm (a rotor-tip azimuthal velocity of 12 m/s) was four to five times higher than that obtained with the tubular membrane modules at a crossflow velocity of 2.1–2.5 m/s. Dal-Cin et al. [93] tested Spintek ST II module, a rotating membrane disk module produced by Spintek, Huntington Beach, CA, for the filtration of oil–water microemulsions. They reported that the permeate flux obtained with the Spintek ST II was proportional to the rotating speed at a power ranging from 0.75 to 1.2 until a maximum flux was reached [93]. Another commercial rotating membrane system is the Multi Shaft Disks (MSD) commercialized by Westfalia Separation Filtration Gmbh, Aalen, Germany. This system consists of ceramic disks mounted on parallel shafts with the membrane disks overlapping each other (Figure 6.33). Ding et al. [95] tested this concept using a pilot system which consists of 12 of the 9.0 cm membrane disks mounted on two parallel shafts which rotated in the same direction and at the same speed. When filtering $CaCO_3$ suspensions, it was found that the MSD system can produce very high flux provided a sufficient feed side pressure was applied to compensate for the permeate side pressure induced by the centrifugal forces. Ding et al. [95] also compared the MSD system with a rotating disk stationary membrane system and found that the MSD module could produce similar flux at a much lower disk rim velocity than the rotating disk stationary membrane filtration system, implying that the MSD system might be more energy efficient than the rotating disk stationary membrane system.

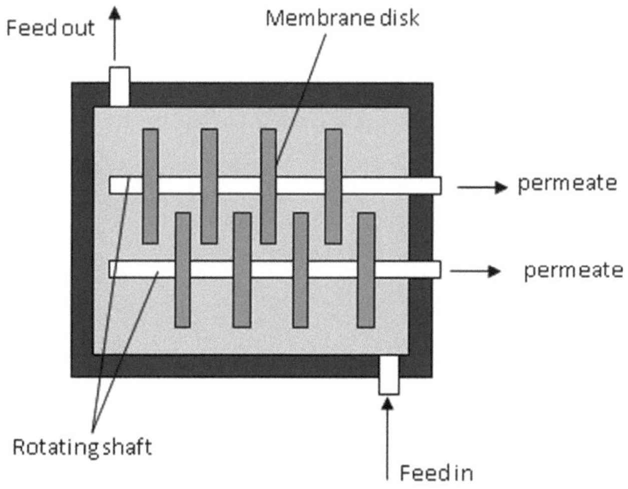

FIGURE 6.33 Schematics of multi-shaft disks (MSD).

6.5.2 Vibratory Systems

Vibratory membrane techniques are characterized by vibrating the membrane at a certain frequency and amplitude. The vibration techniques used in the membrane filtration enhancement include the high-frequency circular vibration, e.g., VSEP, and lower frequency vibrations for hollow-fibre membrane filtration,

6.5.2.1 Vibratory Shear-Enhanced Process (VSEP)

The concept of the VSEP (New Logic International, Emeryville, CA, USA), proposed by Armando et al. [96], came with the idea to combat concentration polarization and membrane fouling by directly moving the membranes rather than by moving the liquid. The VSEP system consists of a stack of parallel circular membranes mounted in a cylindrical housing which is spun in torsional oscillation at a resonant frequency of about 60 Hz, as shown in Figure 6.34.

The shear rate at the VSEP membrane is created by the inertial-induced relative motion of the fluid and can be of the order 10^5 s^{-1}. The shear rate varies sinusoidally and increases with local membrane azimuthal displacement proportionally to radius. The maximum shear rate at the periphery can be related to the vibrating frequency (F) and the membrane displacement at the periphery (d) by the following equation [97]:

$$\gamma_{max} = 1.414 d(\pi F)^{3/2} v^{-1/2} \tag{6.33}$$

In the case of VSEP, $F = 60.75$ Hz, $d = 3$ cm, so for water-like liquids the value of γ_{max} is about 1.1×10^5 (1/s). It should be noted that the displacement d is a function of the frequency, which decreases with the decrease in frequency.

The mean shear rate obtained by averaging the absolute value of shear rate over a period and over the membrane area

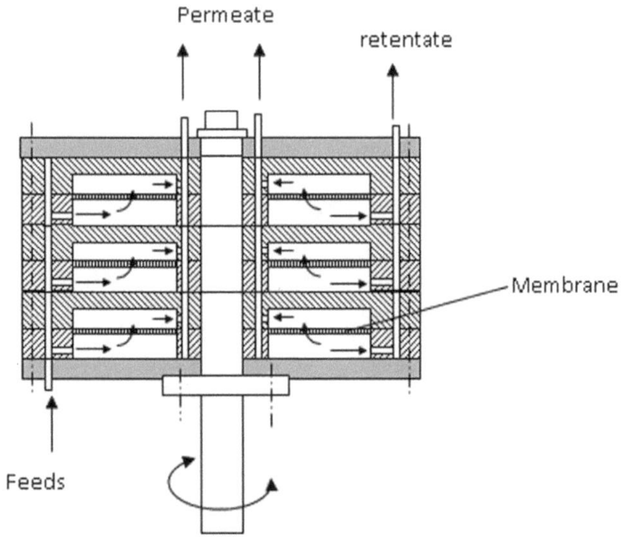

FIGURE 6.34 Schematics of vibrating flat sheet membrane module.

which, for VSEP, is an annular region with radii R_1 and R_2, can be formulated as:

$$\bar{\gamma} = \frac{2.828(R_2^3 - R_1^3)}{3\pi R_2(R_2^2 - R_1^2)}\gamma_{max} \tag{6.34}$$

Jaffrin et al. [87] made a hydrodynamic comparison between the rotating disk and the VSEP system based on the flux achieved under similar maximum shear rates for the microfiltration of baker's yeast suspensions and skim milk using 0.2 μm MF and 50 kDa UF membrane, respectively. They found that the flux variation with time in these two modules was almost identical when they were operated at the same maximum shear rate, suggesting the dominant effect of shear rate on the filtration performance.

For the VSEP module, the permeate flux can be described by $J = a\gamma^x$ [97]. The value of x reflects the effect of the shear rate on the membrane filtration, varying with applications. For example, it has been reported that for pulp and paper $x = 0.58$ [92,98]; for MF of yeast suspension, $x = 0.19$ with frequency < 59 Hz and $x = 0.50$ with frequency > 59 Hz [97]; for UF of bovine albumin solutions, $x = 0.456$ [97]; for filtration of NOM-containing surface water, $x = 0.21$ (UF) and 0.31 (NF) [99]; and for UF of powder and skim milk, $x = 0.471$ [100] and 0.533 [101], respectively. In addition, VSEP has also been tested for landfill leachate [102], arsenic removal [103], and desalination of brackish water and brine [104].

6.5.2.2 Lower-Frequency Vibratory Membranes

While VSEP demonstrated that vibrations at a relatively high frequency (55–60 Hz) can significantly improve the membrane filtration performance, vibration at frequencies lower than ~ 20 Hz has also been tested to enhance the filtration performance of hollow-fibre and flat sheet membrane filtration. Krantz et al. [105] designed the system shown in Figure 6.35(a) to assess the effect of axial membrane vibrations on mass transfer in a hollow-fibre oxygenator. In this system, a shell and tube silicon fibre module were mounted vertically and connected to the liquid flow inlet and outlet lines via flexible bellows couplings. The membrane module was rigidly attached to the plate of an electromechanical shaker which could axially vibrate the hollow-fibre module over a frequency range of 6–18 Hz with an amplitude range of 0.05–1.2 cm. Surge tanks were employed on the liquid inlet and outlet lines to eliminate liquid pressure pulsation in the fibre lumen. The experimental results showed that a maximum enhancement of 1.58 in the Sherwood number ratio was achieved for oxygen transfer to water when the surge tanks were employed to suppress liquid pulsations in the fibre lumen. It was found that the vibrations without the surge tanks could induce a secondary flow in the fibre lumen, and the combined effect of the vibrations and the secondary flow resulted in the enhancement factor increasing to 2.65.

Low-frequency vibrations have been applied to submerged hollow fibres and flat sheet membranes for flux enhancement

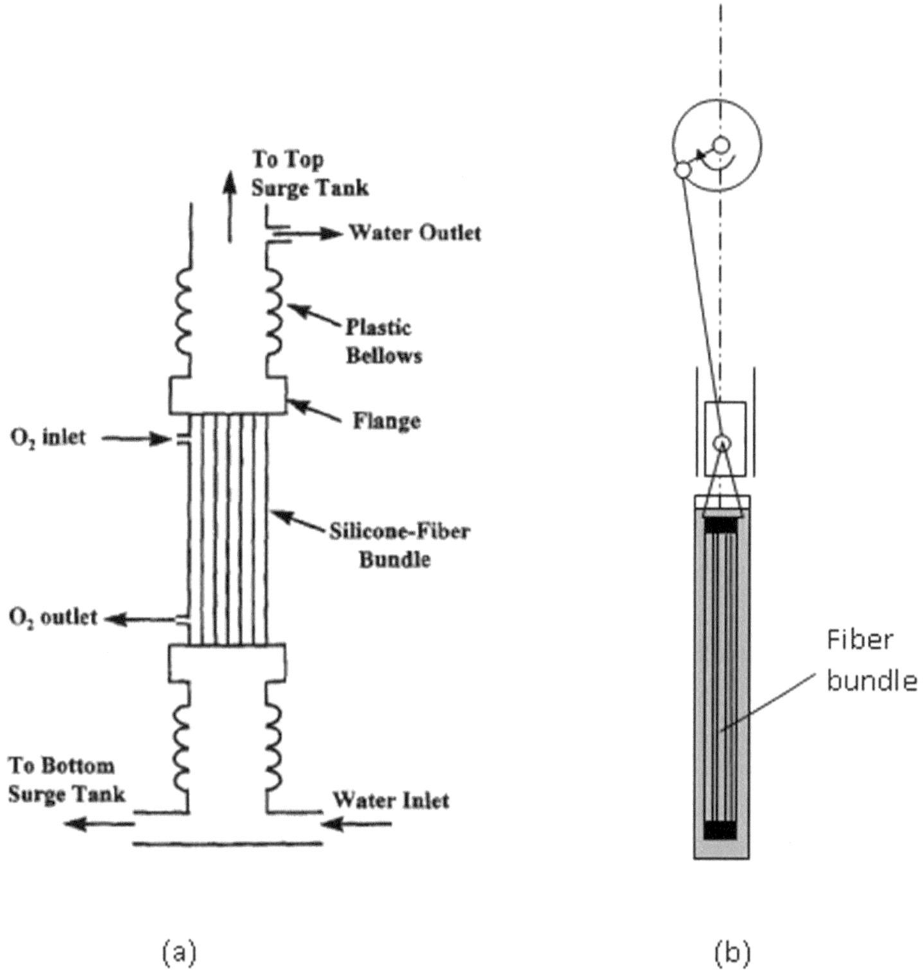

FIGURE 6.35 (a) Vibrating hollow-fibre oxygenator; (b) vibrating submerged hollow-fibre membranes. (Reprinted from Ref. [105], with permission from the copyright holders, Elsevier.)

and fouling mitigation. The mechanisms of flux enhancement by vibration of submerged membrane modules could include back diffusion caused by pulsation and unsteady-state shear on the membrane surface [21]. Equation 6.35, derived from vibration of a flat plate, can be used to calculate the shear rate at the surface (γ_{memb}) of vibrating flat sheet membranes, and the hollow-fibre membranes with the radius much larger than the thickness of the concentration boundary layer [21,106,107]:

$$\gamma_{memb} = \alpha\omega^{1.5}v^{-0.5}\cos\left(\omega t - \frac{3\pi}{4}\right) \quad (6.35)$$

and the maximum shear rate and the average of the absolute value of the shear rate can be calculated by:

$$\gamma_{max} = \alpha\omega^{1.5}v^{-0.5} \quad (6.36)$$

$$\gamma_{ave} = \frac{2}{\pi}\gamma_{max} \quad (6.37)$$

where α is the amplitude of vibration, ω is the angular frequency (=2πf) in which *f* is the linear frequency of the vibration, and ν is the kinematic viscosity.

The following equation can be used to estimate the specific power requirement of vibrating membranes (power per membrane area) [21,108]:

$$P_{vib} = -\frac{\mu\alpha^2\omega^{2.5}}{2v^{0.5}}\cos\frac{3\pi}{4} \quad (6.38)$$

Figure 6.35(b) shows a vibrating submerged hollow-fibre membrane system for filtration applications [109]. In this system, the submerged membrane is vibrated by a mechanical device which converts the rotating motion of the electric motor to vertical oscillations of the vertical fibre bundles. The system can be operated in a frequency range of 1–10 Hz with a maximum displacement of 4 cm. Figure 6.36 shows the experimentally determined relationship between critical flux and the vibration frequency for filtration of 5 g/L baker's yeast suspensions, indicating a nearly monotonic increase in critical flux with increase in the vibration frequency over the frequency range tested. Li et al. [110] showed that vibration of vertical submerged hollow-fibre membranes with a 1–2% looseness at 8 Hz and 8 mm amplitude resulted in a 90% reduction in the membrane fouling rate in the filtration of 4 g/L Bentonite solution. Gomaa et al. [111] studied the filtration performance enhancement of submerged flat sheet membranes by vertical vibration. Their experiments showed that the vibration could enhance the flux up to three-fold with the flux increasing with the shear rate to the power of around 0.22. The model developed based on a modified film theory predicted that an effective filtration performance enhancement of the flat sheet membrane can be achieved by vibration at frequencies < 25 Hz and amplitude < 15 mm [112]. Table 6.6 shows Equations 6.39–6.42 proposed to calculate the critical flux of vertical vibration membranes in the filtration of yeast solutions [21].

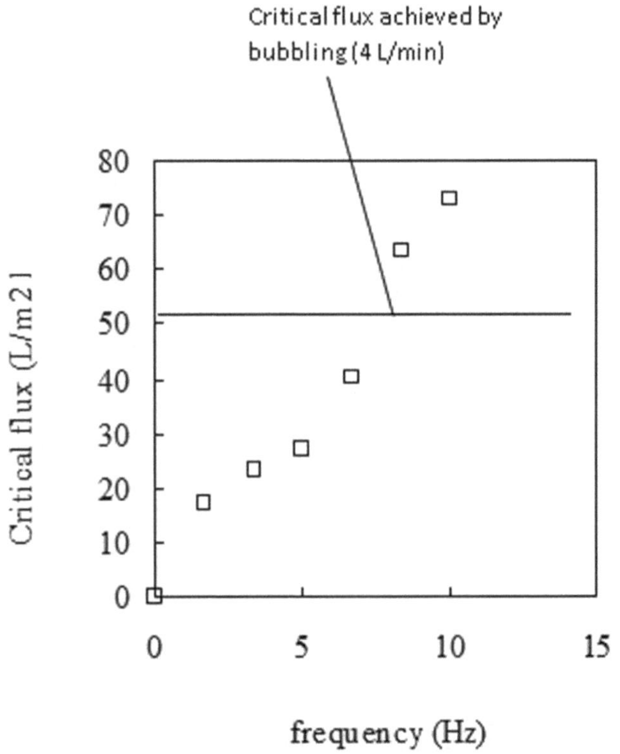

FIGURE 6.36 Experimentally determined critical flux for filtration of 5 g/L yeast using vibrating hollow-fibre membrane.

Compared to longitudinal vibrations, studies showed that continuous transverse vibration with low frequencies was more effective for hollow-fibre membrane fouling control [113–115]. Li et al. [116] reported that the membrane fouling mitigation by transverse vibration could be further enhanced by using slightly loose fibres and combining the vibration with a low flow rate aeration. The vibration energy consumption could be significantly reduced by using an intermittent vibration–continuous permeation strategy, where the non-vibration period should be controlled to less than 120 s to avoid irreversible membrane fouling [116].

In addition to membrane vibration, liquid oscillation has also been tested for the filtration performance enhancement of submerged hollow-fibre membranes (Figure 6.37). Kola et al. [114] reported that the filtration performance of submerged hollow-fibre membranes could be enhanced by "liquid oscillation" or applying a small "rotationally orbital motion" to the membrane tank in which the hollow-fibre membranes were immersed. Their experiments showed that oscillating the liquid at a frequency lower than 10 Hz resulted in a significant increase in the critical flux and the improvement of the long-term performance of the submerged hollow-fibre membrane in the filtration of yeast suspension. Kola et al. [114] also compared the strategy of oscillating the liquid with the hollow-fibre membrane transverse vibration (Figure 6.37). Under conditions of 10 Hz and 2.5 mm displacement, both strategies showed a similar effect on the critical flux enhancement. For long-term performance under conditions of 6.8 Hz and 2.5 mm displacement, the oscillating liquid exhibited a slightly better performance.

While vibration is often tested for hollow-fibre and flat sheet membranes, vibrating spacers with sinusoidal geometry (3-D), flat plates with grooves (2-D), and flat plate (1-D) [117] were also assessed for fouling control in a submerged flat sheet membrane filtration system. The results showed that effective fouling control can be achieved by vibrating the 3-D spacer. The parameters of the 3-D vibration spacers included a distance of around 0.1 mm to the membrane surface, vibration

TABLE 6.6
Correlations between permeate flux (*J*) and the maximum wall shear rate (γ_{max}) for submerged vibrating systems. (MF: microfiltration; HF: hollow-fibre membrane; FS: Flat sheet membrane. (Reprinted from Ref. [21], with permission from the copyright holders, Elsevier)

Membrane (pore size)	Feed	Correlations		References
MF-HF (0.45 µm)	Yeast	$J = 7.29\gamma_{max}^{0.26}$	(6.39)	[106]
MF-HF (0.2 µm)	Yeast	$J = K\gamma_{max}^{0.21}$ for $f < 5$ Hz	(6.40)	[109]
		$J = K\gamma_{max}^{0.95}$ for $f > 5$ Hz	(6.41)	
MF-FS (0.22 µm)	Yeast	$J = 22.39\,\gamma_{max}^{0.22}$ for $f < 5$ Hz	(6.42)	[112]

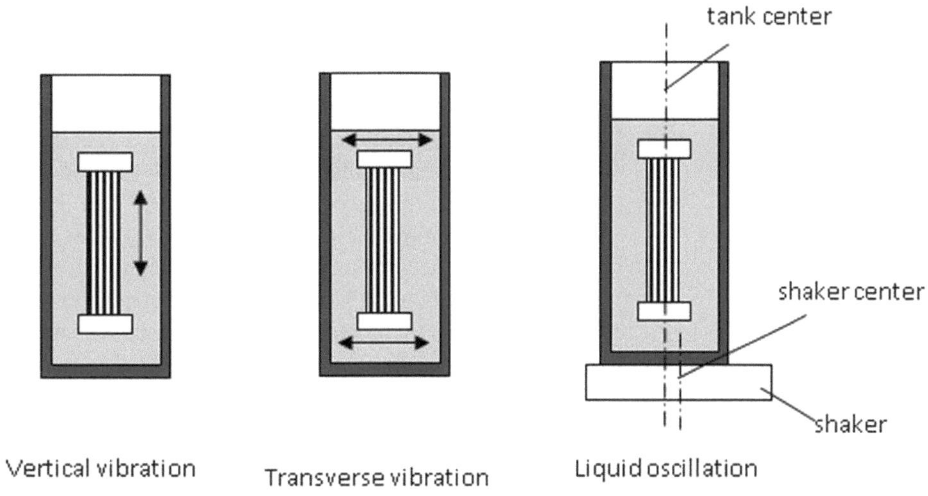

FIGURE 6.37 Schematics of different vibration strategies for submerged hollow-fibre membranes.

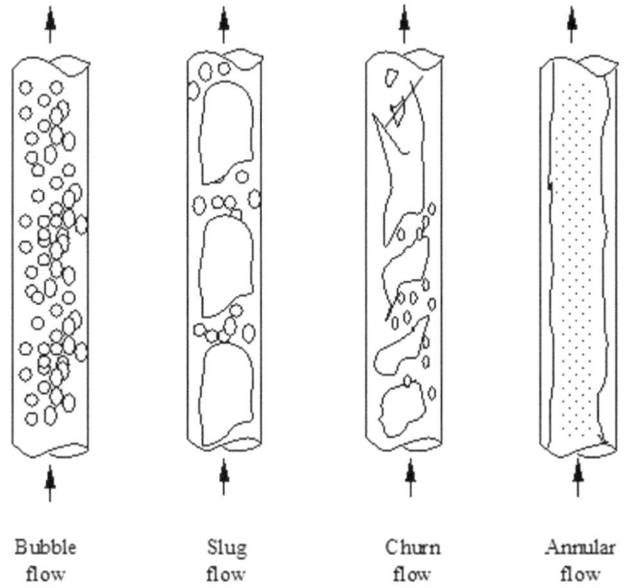

FIGURE 6.38 Flow pattern of gas–liquid two-phase flow in a tubular duct.

frequencies between 1–2.5 Hz, and amplitudes between 0.8–2 cm [117]. The simulation showed that a vibrating spacer between fixed flat sheet membranes enhanced the back-transport of foulants and the shear rate on the membrane surface.

6.6 TWO-PHASE FLOW

Another strategy to provide additional shear and/or flow instability in the boundary layer is the use of two-phase flow. Two approaches have been applied to membrane processes, bubble addition for gas–liquid two-phase flow and particle addition for solid–liquid two-phase flow.

6.6.1 Gas–Liquid Two-Phase Flow

Over the past decade, there has been an upsurge of interest in the use of gas bubbles to enhance membrane processes. The typical applications include two-phase flow filtration with tubular membranes and submerged membrane systems. A major stimulus for the latter has been the development of membrane bioreactors.

6.6.1.1 Two-Phase Flow Filtration with Tubular Membranes

As depicted in Figure 6.38, the flow patterns formed in a vertical tube follow a trend with increasing gas flow from bubble flow, slug flow, churn flow, to annular flow. These regimes are described briefly below.

- Bubble flow: the gas phase is approximately uniformly distributed in the form of discrete bubbles in a continuous liquid phase.
- Slug flow: most of the gas is located in large bullet-shaped bubbles which have diameters almost equal to the tube diameter and are sometimes designated as "Taylor bubbles". They move uniformly upward and are separated by liquid slugs which may contain small gas bubbles. Around the Taylor bubbles, there is a thin liquid falling film which causes turbulence in the wake of the Taylor bubble.
- Churn flow: somewhat similar to slug flow but is more chaotic, frothy, and disordered.
- Annular flow: characterized by the continuity of the gas phase along the core of the tube. The liquid phase moves upwards partly as a wavy liquid film and partially in the form of drops entrained in the gas flow.

The flow pattern depends on the gas and liquid flow rate as well as the diameter of the duct [118]. For two-phase flow filtration, the flow path for the gas–liquid mixture is relatively narrow and the liquid velocity is usually set at a low value, so

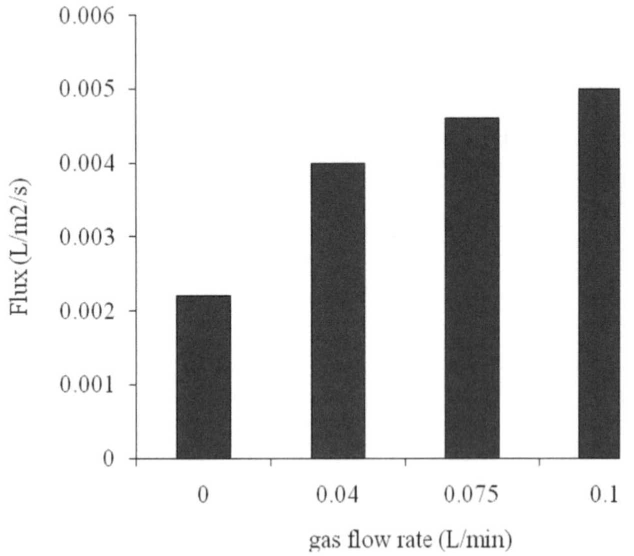

FIGURE 6.39 Bubbling enhanced flux for filtration of dextran solution.

the prevalent flow pattern for two-phase flow filtration would be slug flow. When the bubble size reaches about 60% of the tube diameter, the bubble can be categorized as a slug with its characteristic bullet-shaped nose. The rising velocity of the slug depends on the tube size. For tubes of above 10 mm diameter (d_b) with a slug length $1.5d_b$ or above, the rising velocity of the bubble can be calculated by:

$$U_b = 0.35(gd_b)^{1/2} \tag{6.43}$$

If the tube diameter is less than 10 mm, the liquid surface tension will affect the slug velocity and simulation predicts that for diameters less than about 5 mm with water the surface tension can stop the upward motion of the slug causing an air lock, implying that careful assessment is required before applying bubbling to the lumen of hollow-fibre membranes.

A number of studies have been carried out to evaluate the effect of injection of air into the lumen of tubular and hollow-fibre membranes on the performance of membrane filtration. Figure 6.39 shows the flux enhancement by injecting air bubbles into a vertically installed tubular membrane module (12.7 mm i.d., PVDF, MWCO: 100 kDa) in the filtration of dextran solution (MW: 87 kDa) [119]. As shown in the figure, the flux improvement was about 2.0× for the minimum gas rate and only increased to 2.4× for the maximum gas rate. Cabassud et al. [120] and Mercier et al. [121] studied the effect of slug flow on particle fouling in the ultrafiltration of bentonite suspensions with organic hollow fibres (0.01 µm, $d_i = 0.93$ mm, $L = 1.2$ m) and a mineral tubular membrane (0.02 µm, $d_i = 15$ mm, $L = 0.75$ m). Although an initial flux decline also occurred for filtration with air injection, an enhancement of steady-state flux of 110% for hollow fibres and of 300% for tubular membranes was observed in the experiments. Gas sparging inside tubes has also been shown to be efficient for improvement of the performance of filtration of biomass. Mercier et al. [122] combined two-phase flow ultrafiltration (tubular: 50 kDa, $d_i = 6$ mm, $L = 1.2$ m) with a continuous alcohol fermentation process to investigate the feasibility of its long-term application in a membrane bioreactor. Their results indicated that with air injection, a stable flux of 58 L/m^2/h was maintained over 100 h of fermentation until a final biomass concentration of 150 g (dry)/L was achieved, while the filtration without bubbling was prematurely stopped at about 30 hours with a final concentration of 50 g (dry/L) due to membrane fouling. Performance enhancement for biomass filtration by gas injection was also reported by Imasaka et al. [123] for filtration of methane fermentation broth and by Vera et al. [124] for activated sludge.

Two-phase gas–liquid flow clearly reduces concentration polarization, and this can improve membrane separation. For example, Ghosh et al. [125] assessed the effect of gas sparging on the protein fraction with BSA (MW 67,000) and Lysozyme (MW 14,100) as model solutes. They reported that nearly complete separation of these two model proteins was achieved with two-phase flow ultrafiltration (MWCO 100 kDa), indicating an 18-fold increase in selectivity compared to that without air injection. The enhancement in selectivity was believed to be caused by disruption of the concentration polarization so that solute retentions were closer to the intrinsic values. Although the depolarization decreased transmission for both BSA and Lysozyme, the theoretical analysis suggested that air injection affected more the transmission of the more rejected component (BSA) so that high separation efficiency was achieved. Chen et al. [126,127] studied the heat transfer enhancement and scaling mitigation in membrane distillation utilizing gas–liquid two-phase flow in a specially designed direct contact MD (DCMD) module. They reported that compared to MD without bubbling, DCMD with bubbling achieved 2.30- and 2.13-fold enhancements in the heat-transfer coefficient and temperature polarization coefficient, respectively, at an optimal gas flow rate of 0.2 L min^{-1}.

An assessment of the literature over a wide range of operating conditions and feeds indicates that flux enhancement for tubular membrane filtration can be affected by module configuration and operation conditions such as feed concentration, liquid velocity, bubble size, and transmembrane pressure. The trends have been summarized by Cui et al. [128], as follows:

1. The benefit of bubbling becomes more significant when polarization is more severe, for example, at high transmembrane pressure (or flux), a low liquid velocity, and a high feed concentration, due to the disruption effect of bubbling on concentration polarization.
2. The orientation of the tubular membrane is important. Cui et al. [129,130] indicated that greater flux enhancement can be obtained with vertical tubular membranes with increasing bubbling coupled with a moderate downwards liquid flow than with a horizontal tube. Cheng et al. [131,132] suggested that there is an optimal mounting angle for tubular membranes and

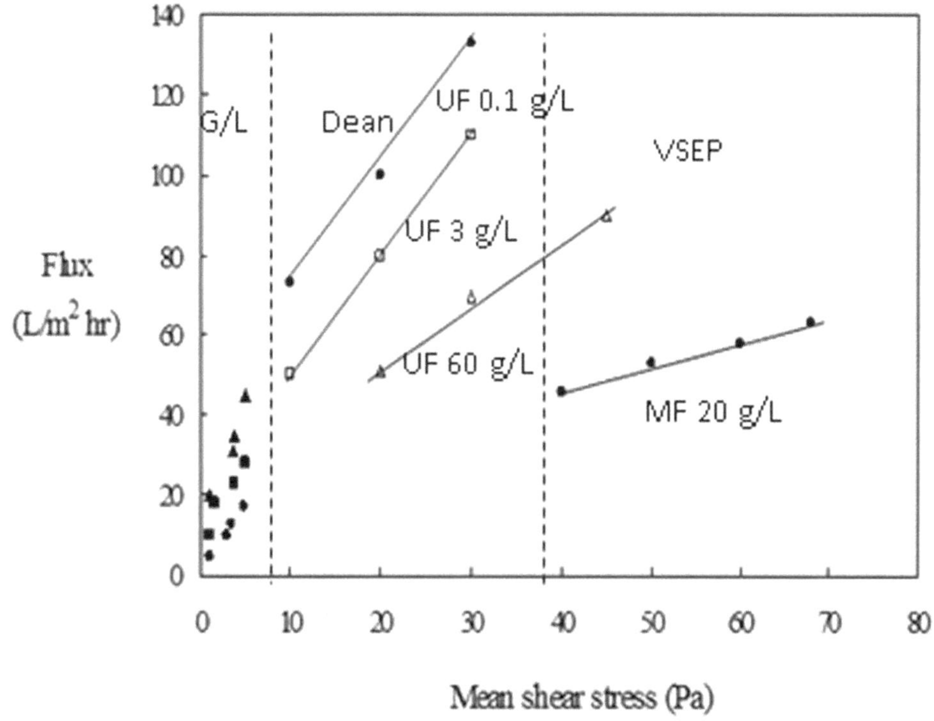

FIGURE 6.40 Comparison of the relationship of permeate flux and mean wall shear stress for two-phase flow, Dean, and VSEP systems. (Reprinted from Ref. [97], with permission from the copyright holders, Elsevier.)

they concluded that around 50° inclination gave the maximum enhancement because gas slugs on an angle move faster than in a vertical tube.

3. Flux enhancement by bubbling is significant in the laminar regime of liquid flow and becomes insignificant as the liquid flow Reynolds number approaches 2500–3000. The enhancement is relatively insensitive to the actual liquid flow over much of the laminar region. This is because the filtration is dominated by the secondary flows induced by the bubbles.

For bubbling-enhanced tubular membrane filtration, Cui et al. [128] have attributed the main mechanisms to:

1. Bubble-induced secondary flow: moving bubbles generate secondary flows and wakes which promote local mixing near the membrane surface. Slug flow also results in an annular falling film as displaced liquid flows downwards between the slug and the tube wall.
2. Physical displacement of the concentration polarization layer. Gas slugs can penetrate into the concentration polarization layer and displace the upper part it.
3. Pressure pulsing caused by passing slugs. A moving slug causes pressure pulsing in the liquid around it, with a higher pressure at its nose and lower pressure at its tail. This is similar to imparting a local pulsatile flow during the filtration. In addition, injection of bubbles may also result in an increase in mean transmembrane pressure. Both of these factors could contribute to increased flux.

Al-Akoum et al. [97] compared the bubbling, Dean flow, and vibrating enhanced membrane processes in terms of the shear stress and the permeate fluxes obtained in filtration of yeast suspension. Filtration with two-phase flow was carried out using 15 mm ceramic monotubular UF (permeability 250 L/m^2/h/bar) and MF (1500 L/m^2/h/bar) membranes with transmembrane pressures of 100 kPa and 25 kPa for UF and MF, respectively. The yeast concentrations used in the two-phase experiments were 1 and 20 g/L and different ratios of gas flow rate to liquid flow rate were tested with a superficial liquid velocity of 0.45 m/s and 0.38 m/s for UF and MF, respectively. The Dean flow experiments were conducted using yeast concentrations of 0.5–60 g/L with 0.93 mm (inner diameter) cellulose hollow-fibre membranes (permeability 270 L/m^2/h/bar) in helical form under a transmembrane pressure of 110 kPa over a liquid velocity range of 0.81–1.64 m/s. A VSEP device was used for the vibrating-enhanced process and the yeast concentration used in the experiments was 20 g/L. Figure 6.40 shows the permeate flux and mean shear stress relationship for the three different systems. The shear stress range for the three systems was determined as < 10 Pa for two-phase flow, 10–40 Pa for the Dean flow system, and > 40 Pa for the VSEP systems. It was observed that the flux with the three systems tested obeys the empirical law: $J = A\tau_w^n$, where the τ_w is the averaged wall shear stress. The coefficients A and n varied with the filtration systems and suspension conditions

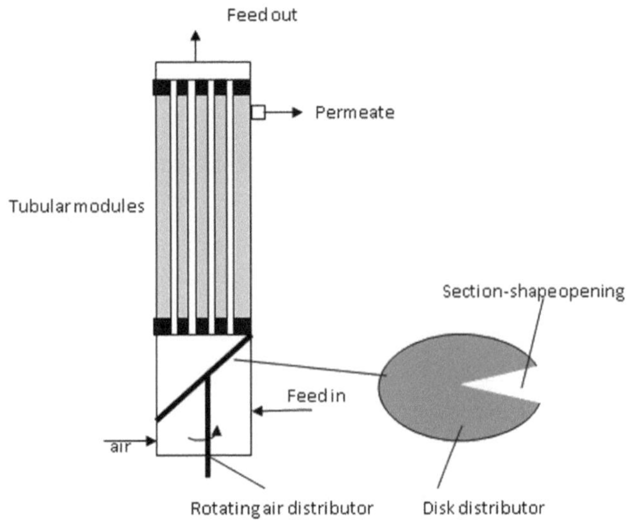

FIGURE 6.41 Schematic of a rotating air distributor.

[97]. However, although the shear stress with the VSEP system was higher than those with the other two systems, a lower flux with VSEP was observed, indicating that other factors, such as membrane properties (materials, pore size, and surface morphology, etc.) and hydrodynamic instability behaviour of the fluid, may affect membrane fouling and performance.

When injecting air into a multi-membrane tube module, the distribution of the air over the individual parallel tubular membranes can exert an important effect on the overall performance enhancement. Phattaranawik et al. [133] developed an optical sensor system to study the air slug distribution among the individual membrane tubes of a multi-membrane tube system and to assess the average bubbling parameters of the modules, which include the bubble velocity, slug length, void fraction, and bubble frequency. With the application of this optical sensor system, Phattaranawik et al. [133] revealed that for a multi-membrane tube module, the percentage of membrane tubes having bubbles was in the range of 30–50% with a reduced percentage at the high air-to-water ratio. Compared with a single tubular membrane module, the average void faction, the slug length, and the bubbling frequency with a multi-tube module could be 40–50% lower than those with a single tube module at the same superficial air velocity. In order to improve the air distribution in a multi-tube membrane module, Phattaranawik [133] developed a rotating air distributor, as shown in Figure 6.41, which is characterized by a sector-shape opening area to direct the air to a small section of the module cross-section. When the distributor is rotated at a speed around 4 rpm, the air can be uniformly delivered over the whole cross-section of the module. The experimental results showed that the rotating distributor can improve the air distribution and significantly reduce the membrane fouling of the multi-membrane tube module.

Recently, Guha et al. [134] demonstrated a chemical reaction-based in situ micromixing method to control organic membrane fouling and CP. In this technique, oxygen molecules and hydroxyl radicals were generated through anchoring nanocatalyst (CuO or MnO_2) at the surfaces of nanofiltration or RO membranes and injecting hydrogen peroxide (2–7 mM) into the feed water. The chemical reactions arrested the growth of biofilm at the membrane surface and the reaction induced micromixing which enhanced the solute back-transport. This technique could potentially be applied to UPM RO membrane applications.

6.6.1.2 Bubbling with Submerged Membrane Systems

Bubbling seems to be an obvious strategy to induce flow and produce shear at the membrane surface in submerged systems to control concentration polarization and fouling. This is particular attractive in membrane bioreactors for wastewater treatment where bubbling is already required as an oxygen supply. Figures 6.42(a) and (b) show two different submerged membrane filtration modules: the submerged flat sheet system and the submerged hollow-fibre membrane system. The flat sheet concept has been developed successfully by Kubota as a feature of their MBR system. As shown in Figure 6.42(a), the Kubota flat sheet submerged membrane unit may consist of multiple independent module panels that are arranged vertically at a distance of about 5–8 mm from each other. Each panel has its own support plate, membranes, and permeate collector. Air is injected from a distributor below the plates generating a well-defined flow in the channel. A potential advantage of the flat sheet arrangement is that the membranes are precisely located and more accessible to well-directed bubbles. The disadvantages are that membrane packing density is relatively low and vigorous backwashing is not feasible.

The submerged hollow-fibre membrane was introduced in Japan in the mid-1980s. A patent by Tajima and Yamamoto [135] described a filter for nuclear power plants which incorporates U-shaped hollow fibres (0.1 μm) in a vessel with intermittent air bubbling around the fibres for "vibrating the hollow fibres to remove solid particles trapped thereby". Yamamoto et al. [136] are believed to be the first to have reported the use of submerged hollow fibres in a wastewater membrane bioreactor (MBR); the role of the gas was described as for aeration, mixing, and inducing liquid flow. They also emphasized the importance of low imposed flux and the use of intermittent suction for long-term stable operation of the submerged hollow fibres. This concept has become the generally accepted approach for submerged membranes in MBRs. In the early 1990s, Zenon developed their commercial submerged hollow-fibre system with bubbling for wastewater and water treatment. Figure 6.42(b) shows the Zenon concept of the submerged hollow-fibre membrane module which involves use of a vertical loose fibre curtain supported at the top and bottom and a gas bubble distribution system along the edge of the fibre bundle. Vigorous intermittent coarse bubble flow is applied to prevent particle deposition on the fibre surface.

Figure 6.43 depicts the likely mechanisms of "depolarization" and flux enhancement with vertical hollow fibres [128]. For bubble flow outside hollow fibres, the vertical

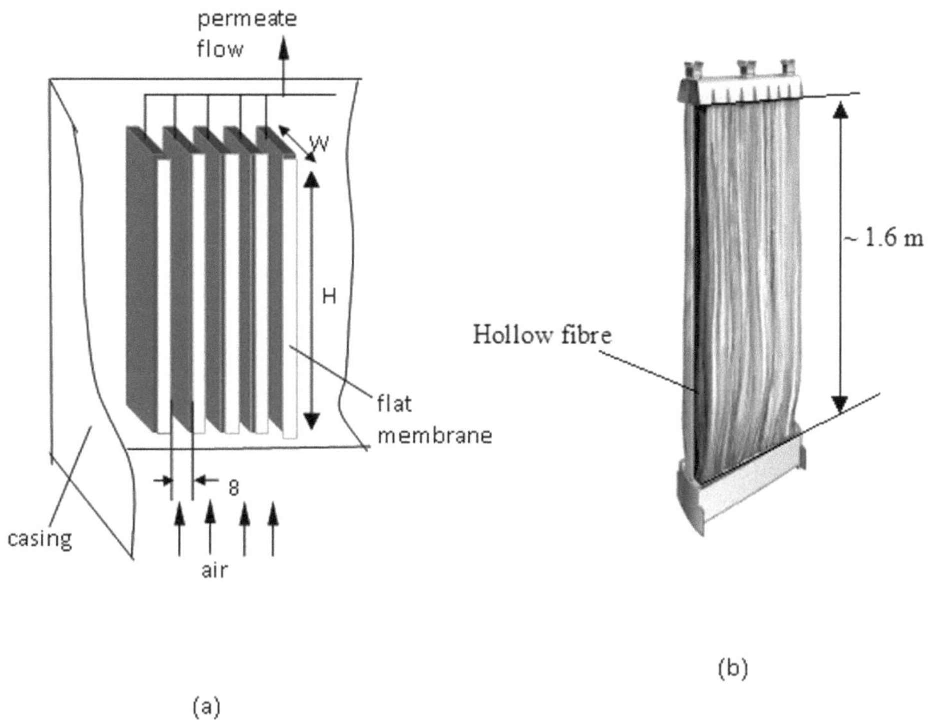

FIGURE 6.42 Submerged hollow-fibre membrane module: (a) flat sheet, (b) hollow fibre. (Reprinted from Ref. [128], with permission from the copyright holders, Elsevier.)

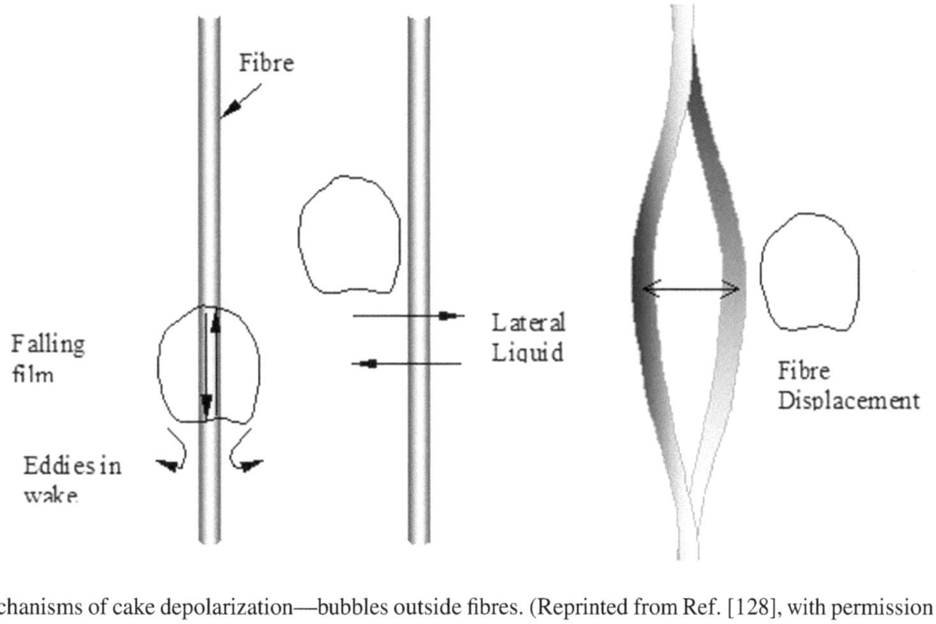

FIGURE 6.43 Mechanisms of cake depolarization—bubbles outside fibres. (Reprinted from Ref. [128], with permission from the copyright holders, Elsevier.)

flow "channels" are less distinct and comprise the outer edge of fibr2s rather than being randomly positioned. Surface shear on the hollow-fibre membrane could be produced by large-scale liquid circulation induced by bubble flow. The local bubble behaviour, as bubbles or slugs move upwards, including eddies in the wake and falling film around the slug, will enhance the mixing of the local liquid. In addition, the bubbles induce fluctuating liquid flow that transverses the fibres and causes lateral fibre movement, depending on the looseness of the fibres. Using the DOTM technique combined with a high-speed video camera, Wang et al. [137] studied the effect of bubbling on the critical flux of submerged membranes and the results revealed that: (i) the critical flux varies with the height along the membrane surface with the local critical flux values

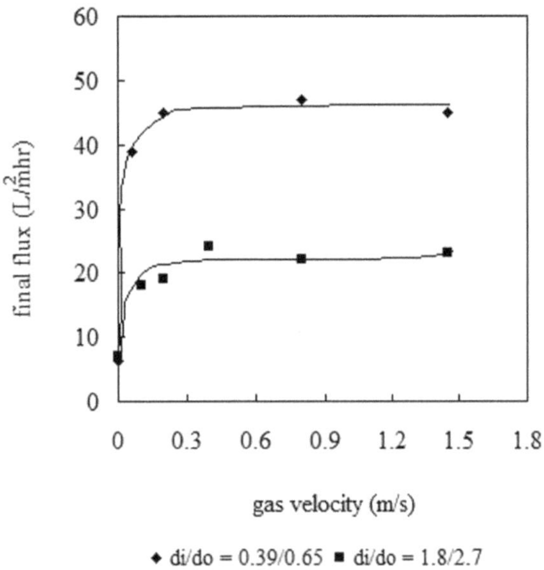

FIGURE 6.44 Effect of gas velocity on final flux in filtration with submerged membranes. (Reprinted from Ref. [138], with permission from copyright holders Elsevier.)

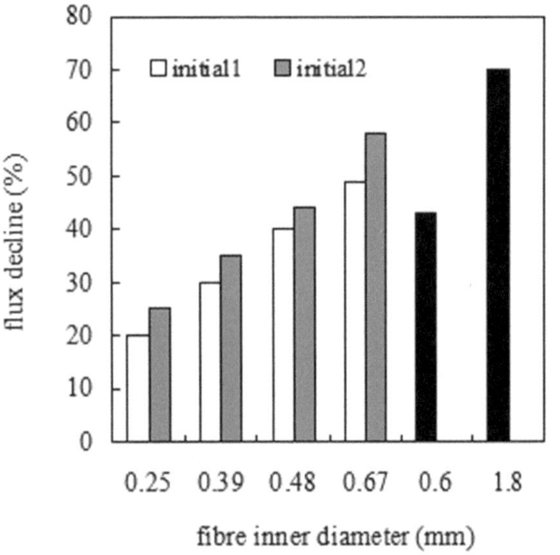

FIGURE 6.45 Effect of fibre diameter on flux decline in filtration with submerged hollow-fibre. (Reprinted from Ref. [138], with permission from copyright holders, Elsevier.)

increased with the height and the gas flow rate; (ii) the mean bubble velocity and mean bubble area per bubble increased and the number of bubbles per unit membrane area decreased with the height; (iii) the local critical fluxes were positively correlated to the bubble momentum and mean area per bubble; and (iv) the average of the local critical fluxes over the membrane length increased linearly, steeply then gradually, with respect to the ratio of gas to total flow rates. This result may suggest that in the filtration of submerged hollow-fibre membrane with bubbling, the distribution of the local critical flux could, to some extent, offset the nonuniform deposition of particles on the membrane surface caused by the distribution of the permeate flux along the fibre length.

Figure 6.44 shows the results of tests at fixed TMP and plots "the steady-state flux" versus gas flow rate using a test system with fibres vertically immersed in the stagnant feed with well-controlled spacing [138]. This study shows that bubbling could significantly enhance the performance of the filtration with a moderate gas rate but the enhancement did not increase much with further increase in the gas flow rate. The experimental results also indicate that the effect of bubbling on filtration could be affected by fibre size. From Figure 6.44, it can be seen that the final flux obtained in the filtration with the d_i/d_o =0.39/0.65 fibre was significantly higher than that with the d_i/d_o =1.8/2.7 mm fibre. Flux decline, defined as 100(initial flux-final flux)/(initial flux), is shown in Figure 6.45 for a wide range of fibre diameters; a clear advantage for the smaller-diameter fibres is evident [138]. An explanation for the better response of the smaller fibres is that they are more flexible and able to move laterally as bubbles pass. This explanation is supported by the observation that the benefit of the small fibres increased with higher crossflow velocity or the injection of air. However, it should be noted that there is a practical constraint on using very small fibre diameters due to the increased pressure drop on the lumen side that causes a significant flux distribution along the fibre and adds to the energy cost.

The flexibility of the submerged hollow fibres under bubbling conditions can be promoted by having the fibres held loosely rather than tightly. Chang and Fane [139] indicated that there are significant differences in suction pressure profiles for filtration with tightly and loosely fixed fibre modules. Similarly, Wicaksana et al. [140] showed that the rate of suction pressure rise with a tight fibre was 50% faster than for a fibre with 96% tightness, for a 5g/L yeast suspension and the same airflow rate. In addition, the "depolarization" due to fibre movement was shown to be up to 40% of the overall bubbling effect. Commercial hollow-fibre systems have recognized the importance of a certain degree of fibre looseness and flexibility [141–143].

Zhang et al. [144] investigated the effects of bubble size and frequency on the mass transfer in submerged flat sheet membrane modules. It was found that the effect of the bubble size and frequency on the mass transfer can be significant in a certain range. For the system tested in their studies, where air was injected into a 20-mm gap between two 1000 mm × 300 mm plates with control of a solenoid valve, the mass transfer was found to be sensitive to the bubble size up to a bubble volume of 60 mL and the threshold bubbling frequency for the mass transfer enhancement was 0.4 Hz. Phattaranawik et al. [145] developed a device named the bubble-size transformer (BST) to transfer fine bubbles produced by the fine bubble aerators into coarse bubbles for the membrane performance enhancement. In biological wastewater treatment processes, fine bubble aerators are widely used because fine bubbles, whose diameters are usually smaller than 1 mm, are beneficial for the oxygen transfer. As shown in Figure 6.46, the design of

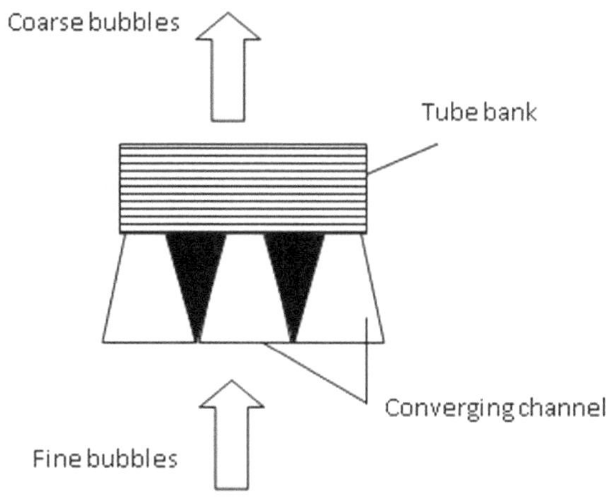

FIGURE 6.46 Schematic of the bubble size-transformer.

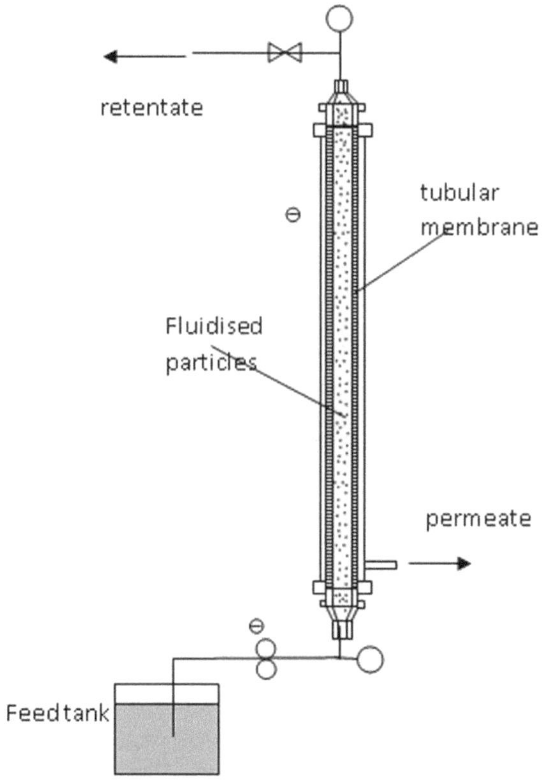

FIGURE 6.47 Schematic of tubular membrane with fluidized particles.

BST includes a converged channel and a tube bank with tubes arranged with 2 mm gaps [145]. The converging channel can sweep and collect the fine bubbles, while the tube bank can promote the bubble coalescences. The BST is designed to be located between the submerged flat sheet membrane and the fine bubble aerators. Phattaranawik et al. [145] showed that the BST can increase the size of the fine bubbles produced by a fine bubble aerator by 10–14-fold. A significant difference in the long-term filtration performance was observed for the filtration of the mixed liquor with and without BST when the fine bubble aerator was used [145].

Reducing energy consumption of aeration has always been a focus of the innovation of the submerged membrane filtration because the membrane process aeration typically counts for 30–40% of the total power consumption of MBR plants [146]. For the operation of full-scale submerged membrane modules, the air injection methods and the distributions of air bubbles within the membrane tanks are important factors for effective membrane fouling mitigation and energy saving. One of the important strategies to reduce the aeration energy consumption of submerged membrane filtration is the intermittent aeration method. The typical intermittent aeration operations include 10/10 cyclic aeration and 10/10 sequential aeration. The former is characterized by injecting air in 10-second on 10-second cyclic mode, while the latter is characterized by switching one-half of the aerators in the membrane tank alternatively on and off every 10 seconds. Compared to continuous aeration, the 10/10 cyclic or sequential aeration can save 50% aeration energy because only half of the operation times (cyclic aeration) or half of the aerators (sequential aeration) in the aeration tank are working during the operation. Furthermore, the cyclic or sequential on–off aeration can cause strong transition flows, resulting in a higher filtration performance enhancement than the continuous aeration [146]. A latter development in the MBR aeration is pulsed aeration, which is more energy efficient than the cyclic or sequential aeration. For pulsed aeration, the air is continuously injected into the cavities of the diffusers and released only after the air pressures reach some critical values. This allows an abrupt release of large air bubbles from the diffusers to generate strong turbulence in the membrane tank. Compared to the conventional cyclic aeration, pulsed aeration can achieve around a 50% reduction in aeration equipment and a 30% reduction in aeration energy in the full-scale MBR operation [147].

6.6.2 Fluidized Particles

An early study by Bixler and Rappe [148] showed that glass beads (up to 100 μm size) added to a stirred cell UF for the filtration of a macrosolute solution were able to significantly enhance flux. The mechanism was probably eddy formation and thinning of the concentration boundary layer by particle interaction. Similar effects were reported by Fane [149] who noted that enhancement required significantly supramicron particles and that smaller particles could in fact add to the deposit resistance.

The added-particle effect is also evident in the application of fluidized beds to provide turbulence promotion. Figure 6.47 is a schematic diagram of a combined tubular membrane-fluidized beds system. It consists of a tubular membrane module, fluidized particles, a feed pump, and the feed storage tank. During the operation, the particles are fluidized by the pumped liquid flow. The drag forces on the solid particles are a function of the liquid velocity and the porosity of the bed. Usually, the liquid velocity will be adjusted to a value that results in a balance between the gravity and drag forces so that the particle will not be hydraulically transported. The

TABLE 6.7
Mass transfer and flux for filtration with different fluidized particles [150]

Material	d_p (mm)	ρ_s (kg/m^3)	k(μm/s)	Flux (LMH)	Flux enhancement
Empty tube	–	–	0.54	6.26	1
Glass	1	2900	4.3	49.8	8.0
Glass	1	2900	6.4	74.2	11.8
Glass	0.46	2900	5.9	68.4	6.9
Stainless steel	1	7800	7.6	88.1	14.07
Stainless steel	2	7800	8.1	93.9	15

Note: d_p: particle diameter, ρ_s: particle density, k: mass transfer coefficient.

TABLE 6.8
Properties of scouring particles for membrane fouling mitigation [159]

Particle property	Conditions
General preferable properties	Inert, porous, hydrophilic, and moderately elastic Inert property can avoid release of fine particles Porous particle can provide large specific surface area for biofilm growth and adsorption
Particle size	1–3 mm, smaller particles are less effective for scouring but have high adsorption capacity as adsorbent
Particle density	< 1.2 g/cm^3 for gas–solid–liquid system High-density particles (e.g., 2.5 g/cm^2) for solid–liquid system
Particle dosage	Particle fill fraction 10–50% 50–100 g/L GAC
Particle shape	Particles with greater sphericity showed higher fouling mitigation than particles with lower sphericity

enhanced mass transfer is caused by the irregular flow of the liquid between the particles and the erosive action of the randomly moving particles at the membrane surface.

Noordman et al. [150] assessed the performance enhancement of ultrafiltration of 10 g/L BSA solution by fluidized particles of different materials and size using a polysulphone tubular membrane with inner diameter of 14.4 mm and 1.75 m length and a MWCO of 10 kDa. Table 6.7 summarizes the fluxes and mass transfer coefficients obtained under different experimental conditions. The mass transfer coefficients have been calculated using the film model based on the experimentally determined fluxes with the assumption of a constant "gel concentration" of 120 g/L. From Table 6.7 it can be seen the fluidized particles resulted in an enhancement in mass transfer of up to 15. The particle size and density seem to only have a limited effect on the process. Given that increasing the particle size and density may increase the risk of membrane damage by the fluidized particles [151] and require greater energy consumption, Noordman et al. [150] suggested that light and relatively small particle should be used in fluidized particle-enhanced membrane processes.

A relatively recent successful application of particle scouring in membrane filtration is the fluidized bed anaerobic MBR (FBAnMBR). In the FBAnMBR, granular activated carbon (GAC) particles are liquid fluidized to generate shear scouring at the surfaces of membranes immersed in the fluidized bed for fouling mitigation [152]. The first FBAnMBR reported by Kim et al. [152] was characterized by adding 450 g of 10 × 30 mesh GAC into a 2-litre mesophilic anaerobic membrane bioreactor equipped with a submerged hollow-fibre membrane module. Without gas scouring, the FBAnMBR achieved stable operation at a constant flux of 7 L/m^2/h in the treatment of effluent from a fluidized bed anaerobic reactor when the GAC in the FBAnMBR is fluidized by a superficial liquid velocity of 0.012 m/s [152]. Aslam et al. [153] studied the performance of an anaerobic fluidized ceramic membrane bioreactor (AFCMBR), which contained 10 × 30 mesh GAC at 50% of the packing ratio and a submerged flat-tubular ceramic membrane module with a filtration area of 0.1 m^2 and a nominal pore size of 0.5 μm. The AFCMBR achieved a net permeate flux of 22 L/m^2/h in the treatment of effluent from an anaerobic fluidized bed reactor (AFBR) treating low-strength wastewater. The energy requirement to operate the AFBR-AFCMBR was 0.024 kWh/m^3, which was only 10% of the electrical energy converted from the methane gas produced by the reactor [153]. A series of papers on the fundamentals of fluidized bed fouling control include those by Wang et al. [154–157] and Cahyadi et al. [158]. In a recent review, Wang et al. [159] suggested that filtration-enhancing particles could have several roles, acting as mechanical scouring, adsorbents of organic foulants, coagulants to form porous flocs, and dynamic membranes to protect the primary membrane matrix. For particle scouring, as shown in Table 6.8, the scouring efficiency could be affected by the type of scouring particle, particle size, particle density, particle dosage, and sphericity [159].

6.7 OTHER TECHNIQUES

In addition to the hydrodynamic techniques, electric field techniques and ultrasonic methods have also been assessed for membrane performance enhancement.

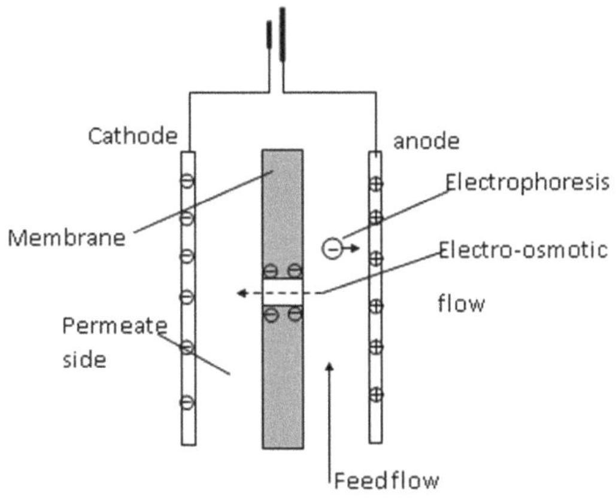

FIGURE 6.48 Schematic of electrofiltration.

6.7.1 Electrofiltration

Electrofiltration is related to the application of an electric field to improve the efficiency of pressure-driven membrane filtration [160]. Figure 6.48 shows the basic configuration of electrofiltration, where an electric field is applied across micro- or ultrafiltration membranes in flat sheet, tubular, and spiral-wound modules. The electrode is installed on either side of the membrane with the cathode in the permeate side and the anode in the feed side. Usually, the membrane support is made of stainless steel or the membrane itself is made of conductive materials, to form the cathode. Titanium coated with a thin layer of a noble metal such as platinum could, according to Bowen [161], be one of the best anode materials. Wakeman and Tarleton [162] analysed the particle trajectory in a combined fluid flow and electric field and suggested that a tubular configuration should be more effective in use of electric power than flat and multi-tubular modules.

Filtration enhancement by an electric field results from the influence of the applied electric field on particle deposition via electrophoresis. Most particles acquire a surface charge when in contact with a polar (e.g., aqueous) medium due to ionization, ion adsorption, and ion dissolution. This surface charge influences the distribution of nearby ions of opposite charge and leads to the formation of an electrical double layer at the interface between the particle and the dispersion medium (water). An imaginary boundary, the surface of shear, separates the double layer into two parts, the bound and mobile parts. If a voltage gradient is applied, the charged particle (plus bound ions) tends to move in the appropriate direction, whilst the ions in the mobile part of the double layer show a net migration in the opposite direction carrying solvent along with them. Electrophoresis is the movement of the charged particle relative to stationary liquid under the effect of an applied electric field. This movement depends on the potential at the surface of shear, known as the zeta potential, and the strength of the electric field applied. The electrophoretic mobility (u_E), or the electrophoretic velocity (v_E) produced by unit electric field strength (E), can be related to the zeta potential by:

$$u_E = \frac{v_E}{\mathbf{E}} = \frac{\varepsilon_r \xi}{\eta} \tag{6.44}$$

where ε_r is the electrolyte permittivity, ξ is the zeta potential, and η is the viscosity.

Surfaces in contact with the aqueous media are more usually negatively charged (negative zeta potential in a liquid) than positively charged; the effect is pH dependent with negative charge increasing with pH. When an external electric field is applied across the membrane with its anode in the feed side, the electrophoretic movement of negatively charged particles in the feed will be away from the membrane and augment diffusive back-transport.

This means that the conventional film model relationship can be modified to show the augmentation due to electrophoretic velocity (v_E) [160], so

$$J = \ln\left({C_W}/{C_B} \right) + (v_E) \tag{6.45}$$

where C_W and C_B are the wall and bulk concentration, respectively.

The critical electric field strength at which the net migration of the particles towards the membrane is zero can be calculated by combining Equations 6.44 and 6.45 with $C_W = C_B$.

$$\mathbf{E}_{crit} = \frac{J\eta}{\varepsilon\xi} \tag{6.46}$$

When the electric strength is equal to or larger than the critical value the particle deposition can be arrested by the electrophoretic movement or the particles will even move in the opposite direction to the permeate flow. It should be noted that electrofiltration requires charged particles in the feed, and that the zeta potential will depend on the pH and the ionic environment. This means it will not be effective close to the isoelectric point or with raised levels of salts.

Another electrokinetic phenomenon that may occur during electrofiltration is electroosmosis or the movement of liquid relative to a stationary charged surface. If the membrane itself is charged, the electrical field applied across the membrane will try to shear off the mobile part of the double layer of the pore wall, leading to an electroosmotic flow in the pores. For a negatively charged membrane, the electroosmosis flow will be in the same direction as the permeate flow if the applied electric field is as shown in Figure 6.48. It has been reported the electroosmosis could result in up to 15% enhancement in permeate flux [163], although the main process enhancement mechanism has been attributed to the electrophoretic movement of the charged particles that limits cake formation. Visvanathan and Ben Aim [164] describe the combined application of electrofiltration and electroosmotic backwashing, which involves polarity reversal to keep both the membrane and the electrode free from deposits.

Apart from electrokinetic phenomena, the electrofiltration process may also be affected by the electrochemical reactions that occur at the electrodes. A typical cathodic and anodic process in aqueous systems is the formation of hydrogen gas at the cathode and oxygen at the anode:

$$2H_2O + 2e \rightarrow H_2 + 2(OH)^- \text{ (cathode, –0.83 V)} \quad (6.47)$$

$$2H_2O \rightarrow O_2 + 2H^+ + 4e^- \text{ (anode, +0.4 V)} \quad (6.48)$$

Electrochemical reactions can be harmful to the filtration processes by causing foaming or pore obstruction due to the production of gas. However, anodic oxidation can also be used as a method of removing organic molecules from the effluent. It seems further studies are needed to justify the role of electrode reactions in electrofiltration.

The flux enhancement by an electric field has been reported by a number of researchers. Huotari et al. [165] reported a five-fold increase in flux by applying 2.4 kV/m external electric field in filtration of oily wastewater using a tubular carbon fibre–carbon composite membrane with a membrane pore size of 0.05 μm and an inner diameter of 5.7 mm with the membrane itself as the cathode and stainless steel of 2 mm in diameter located in the centre of the tubular membrane as the anode. For this enhancement, the zeta potentials of the particle and the membrane were –67 mV and –50 m, respectively. Akay and Wakeman [166] obtained 10-fold enhancement of the flux using an electric field strength in the microfiltration of an anionic, hydrophobically modified water-soluble polymer (HMWSP). Rios et al. [167] reported that the flux in filtration of gelatine solution can be increased from 4 to 12 L/m^2/h with an electric field strength of 2.4 kV/m. To achieve this effect, it was necessary to operate at a pH above the isoelectric point of gelatine to give it a net negative charge.

Although it has been reported that an external DC electric field can induce electrophoretic back-transport to enhance flux in crossflow membrane filtration, its commercial implementation appears to be restricted by factors including lack of suitably inexpensive corrosion-resistant electrode materials, concerns about energy consumption, and the complexity of module manufacture.

6.7.2 Piezoelectric Membrane Vibration

The application of an external AC field effect has also been reported to enhance the performance of RO membranes. The electromagnetic field is said to reduce fouling by scale formers and to reduce particulate fouling and biofilm development. Coster et al. [168] reported that the effect of the AC field on the membrane filtration performance could be significantly affected by the membrane properties. The results demonstrated that piezo-electric properties could be imparted to polyvinylidene difluoride (PVDF) membranes through placing the membrane in an intense electric field for electrical "poling". The typical poling conditions could include temperatures of 80–100°C, poling times of 2–4 hours, and a constant field strength of 16.3 × 10^6 V/m [169]. Analyses of the properties of piezoelectric membranes confirmed that poling can cause the formation of β-phase crystals in PVDF membranes and changes in the microstructure of the membranes [169], resulting in the formation of aligned electric dipoles in the membrane material. The laser Doppler vibrometer measurement showed that placing the "poling"-treated PVDF membrane in an AC field could cause membrane vibration with amplitudes ranging from nano- to micrometres, depending on the frequency and amplitude of the AC signals. The optimal operation strategies to operate the piezoelectric membranes for membrane fouling control include a 5–10 V AC signal at frequencies of 500–1000 Hz [170]. In this frequency range, a significant portion of the applied AC signal appears across the membrane, which is desired for membrane fouling control. The energy consumption for the filtration using 0.22 μm piezoelectric membrane excited at 10 V and 500 Hz was around 110 W/m^2 membrane area or 1.2 W h/L of the permeate [170]. Coster et al. [168] showed that the flux improvement by piezoelectric vibration in the filtration of 1% PEG solution increased from 60% to 400% when the crossflow velocity increased from 7 mL/min to 30 mL/min. Cao et al. [171] demonstrated that the combination of crossflow filtration with piezoelectric vibration at 20 V and 10 kHz achieved a 72.6% enhancement in the steady-state flux of the membrane filtration in an anaerobic membrane bioreactor application. Su et al. [172] demonstrated that incorporating piezoelectric $BaTiO_3$ nanoparticles into PVDF membrane improved the dielectric strength, piezoelectric d_{33} coefficient, and mechanical properties of PVDF membrane. They demonstrated that combination of the poled $BaTiO_3$-PVDF membranes at 10 V and 500 Hz with a crossflow at 0.1 m/s can achieve a critical flux around 60–70 LMH in the filtration of 100 mg/L silica solution [172]. Compared to the unpoled and non-AC-applied membrane, the $BaTiO_3$-PVDF membrane showed a 51% increase in critical flux in short filtration experiments and it increased the filtration duration up to a factor of 2–4 in multi-cycle filtration. The piezoelectric vibration could introduce fluid instability and enhance the crossflow-induced molecule back-transport [168], resulting in enhanced fouling mitigation [170].

6.7.3 Ultrasound-Enhanced Filtration

Ultrasound occurs at a frequency above 16 kHz and is typically associated with the frequency range of 20 kHz–500 MHz. The frequency level is inversely proportional to the power output. The high-intensity, low-frequency ultrasound can alter the state of the medium chemically or physically to be used for cleaning, emulsification, crystallization, sonochemistry, etc. [173]. The chemical and physical effects of ultrasound are related to the cavitation phenomenon induced by rarefaction and compression of the sound wave. When ultrasound is irradiated through a liquid medium, an alternating adiabatic compression and rarefaction cycle of the medium occurs. During the rarefaction period, negative pressures occur and microbubbles can be formed and grow at sites where there is some gaseous impurity. The formed bubbles may suddenly collapse during the compression period after a few acoustic

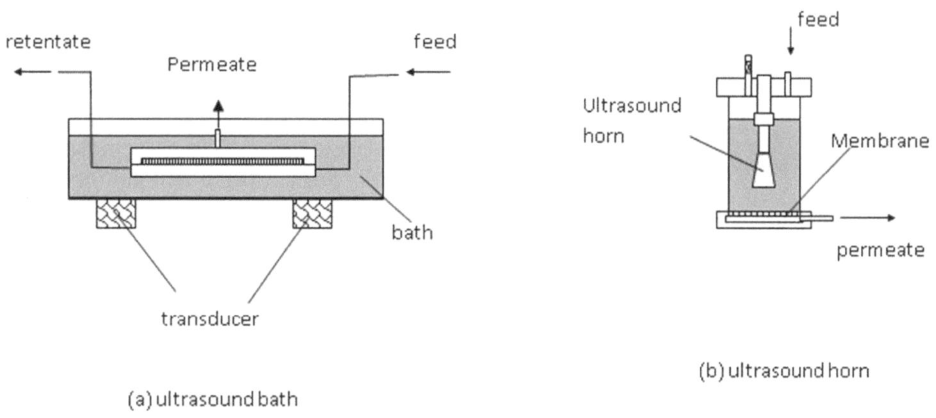

FIGURE 6.49 Schematics of ultrasound membrane filtration systems.

cycles with a release of energy. This helps to generate micromixing in the liquid, or form liquid microjets near a solid surface, or cause chemical changes in reactants in the cavitation bubbles. The power dissipated (P_{diss}) in the liquid medium can be determined calorimetrically in terms of the change in temperature of the medium with the assumption that all the energy delivered to the system is dissipated as heat, as shown by:

$$P_{diss} = (\frac{\Delta T}{\Delta t})C_p M \tag{6.49}$$

where ΔT is the change of temperature during time period Δt, C_p is the heat capacity of water, and M is the water mass.

Figure 6.49 depicts two typical configurations used to assess the effect of ultrasound on membrane filtration and cleaning. Figure 6.49(a) shows a bath configuration where the crossflow cell is immersed in a water bath with ultrasonic transducers attached to the bottom of the bath. Figure 6.49b shows a horn configuration where an ultrasonic horn is installed in a conventional cell.

Juang and Lin [174] assessed the effect of ultrasound on flux recovery in the ultrafiltration (YM 10, regenerated cellulose) of Cu^{2+}-polyethylenimine (PEI) solution and W/O/W solutions using a filtration cell equipped with an ultrasound horn [Misonix Sonicator 3000 (600 w, 20 kHz), horn type 200]. Their results indicated that the ultrasound can even increase the pure water flux due to the cavitation-induced transmembrane pressure increase. For ultrafiltration of Cu^{2+} -PEI solution at 69 kPa, a 70–80% flux recovery was obtained for an ultrasonic power > 30 W at the optimal tip height of 65 mm. For ultrafiltration of the W/O/W emulsions, a 30–60% flux recovery was achieved at 145 W, depending on the volume ratio of the water to oil in the emulsions. Analysis, based on the resistances in series model, revealed that the resistances caused by pore blocking and cake compressibility were reduced by increased ultrasound power. They also reported that the structure of the regenerated cellulose and polyethersulphone membranes tested could be affected by ultrasonics at powers greater than 80 W, but no obvious change was observed with the polyvinylidene fluoride and polyacrylonitrite membranes. Simon et al. [175] defined an imaginary "ultrasonic stirring" speed at which the mass transfer coefficient will be similar to that obtained by a traditional stirrer at the same stirring speed. They suggested that the "ultrasonic stirring" speed is proportional to the ultrasound power by comparing the mass transfer coefficients obtained in ultrafiltration of Dextran (500 kDa) using a classical stirred cell and an ultrasound-assisted cell. Kobayashi et al. [176] assessed the effect of ultrasonic frequency on ultrasound membrane cleaning for Dextran-fouled ultrafiltration membranes (10 kDa) and milk-fouled cellulose microfiltration membranes using a crossflow cell immersed in an ultrasound bath. For the tested frequencies of 28, 45, and 100 kHz, they reported that cleaning can be significantly affected by the frequency, with 28 kHz identified as optimal. Muthukumaran et al. [177] examined the effect of low-power ultrasonics on the UF of dairy whey. They found flux enhancements of 1.2–1.7× and that this could be improved when used in combination with flow channel spacers. Recently, Ghasemi et al. [178] studied the mechanisms of ultrasonic detachment of biofilm using an integrated biofilm growth–cavitation–biofilm viscoplasticity model and the simulation results showed that increasing the pressure amplitude and decreasing the frequency of the transducer results in more effective biofilm detachment, but a pressure amplitude over 500 kPa can have a negative and destructive impact on the membrane structure.

All the reported studies indicate that low-frequency ultrasound can enhance the filtration process and improve membrane cleaning efficiency to some extent. However, reducing energy losses, increasing the efficiency of the ultrasound, and system scale-up are challenges that need to be overcome before serious commercial application can be considered. In an attempt to improve cleaning efficiency, Gonzalez-Avila et al. [179] have shown the benefit of "tandem frequency excitation". This involves the generation of small bubbles on the membrane surface using high-frequency ultrasound, followed by a switch to low-frequency ultrasound that converts the bubbles into cleaning bubbles.

6.8 CONCLUSIONS AND PERSPECTIVES

Membrane separation plays an important role in many industrial processes, including water and wastewater treatment, desalination, biofuel production, and bio-pharmaceutical separation and purification. Controlling membrane fouling is critical for the successful use of membrane separation in all industrial processes. Over the years, continuous effort has been made by researchers to develop various strategies for the enhancement of performance of membrane processes. Table 6.9 summarizes the potential applications of the hydrodynamic membrane performance enhancement techniques discussed in this chapter. All of these techniques aim to limit the effect of concentration polarization at the membrane surface. The majority of techniques are hydrodynamically driven and involve additional energy input or some extra design feature. Thus, the practical applications of these techniques will depend on the membrane module configuration, the operation scale, and the process characteristics. The most prevalent strategies for enhancement are the flow channel spacers used in the spiral-wound module for RO, NF, and some UF, and two-phase flow by air sparging used in the increasingly ubiquitous submerged membranes applied to water and wastewater treatment. High-shear devices, with rotations and vibrations, also demonstrate significant ability for enhanced performance, particularly for difficult feed materials in niche applications.

TABLE 6.9
Application potential of the hydrodynamic membrane performance enhancement techniques

Hydrodynamic properties	Potential applications
Dean vortices	Tubular membranes For small-scale separation and purification
Taylor flow	Flat sheet membrane filters for small-scale separation and purification
Helical membrane modules	Modular submerged membrane modules for medium- and large-scale filtration
Net-type spacer	Widely used in spiral-wound membrane modules
Helical inserter	Tubular membrane modules
Corrugated membrane	Modulated membrane module
Pulsed flow	Applied to crossflow filtration with tubular and flat sheet membrane modules
Rotating disk	For small-scale separation and purification
VSEP	For small- and medium-scale separation and purification
Low-frequency vibration	For small and medium scale submerged membrane modules
Air–liquid flow	Large-scale membrane filtration with hollow-fibre, flat sheet, and tubular membrane modules
Solid–liquid flow	Large-scale membrane filtration with hollow-fibre, flat sheet, and tubular membrane modules

However, in each case there will be a trade-off between operating and capital cost that provides an optimum condition that minimizes total production cost to obtain a payback for the investment for the enhancement in performance. For the future development, as improvements are made in membranes (higher permeability), it will put focus on needs for enhanced CP control. For example, as shown in Figure 6.5, it would be necessary to increase the boundary mass transfer coefficient up to 5–10 times the current K values to exploit the high permeability of emerging UPMs. Meanwhile, minimizing the energy consumption of hydrodynamically driven membrane performance enhancement techniques will be a focus of innovation due to the need to decarbonize membrane processes. Recently developed novel manufacturing methods could provide access to novel technique and module designs. For example, 3D printing methods could fabricate novel 3D-printed spacers which could achieve flux enhancement with reduced pressure drops (Section 6.3.1.3) and surface patterning could be used to optimize the membrane topography to enhance membrane filtration mass transfer (Section 6.3.2.2). In addition, improved computing power will allow testing novel hydrodynamic strategies by CFD etc., to enhance boundary layer disturbance. The advanced visual observation techniques, such as DOTM and 3D OCT imaging, can also be used to evaluate different membrane/module enhancement methods. Overall, the important point of optimization of the membrane performance enhancement techniques is the energy versus productivity trade-off. In future, access to lower-cost energy (via renewables) could shift the optimum.

NOMENCLATURE

B	Pitch of helical coil tube
C	Coefficient (Equation 6.1)
CFD	Computational fluid dynamics
d_b	Bubble diameter
d_f	Filament diameter of spacer
d_p	Particle diameter
d_{pore}	Membrane pore diameter
d_{ti}	Tube internal diameter
d_v	Displacement
De	Dean number
D_t	Coil diameter
D_t'	Modified coil diameter
D	Dialysis
DOTM	Direct observation through membrane
E_t	Energy consumed per unit volume of filtrate
ED	Electrodialysis
F	Frequency
f_c	Friction factor in a coil tube
f_s	Friction factor in a straight tube
J_{cr}	Critical flux
J_d	Design flux
J_b	Reverse flux during backshocking period of backshock filtration
J_f	Forward flux during filtration period of backshock filtration
J_0	Clean membrane flux
Js	Steady state flux

K	Vibration critical flux correlation coefficient
l_m	Length of filament of spacer
L	Length
L_f	Filter length
L_s	Length of space channel
M	Coefficient
MBR	Membrane bioreactor
MF	Microfiltration
MWCO	Molecular weight cut off
N	Coefficient
NF	Nanofiltration
P	Coefficient
P	Pressure
P_d	Overall energy consumed
ΔP_L	Pressure drop along the filter
ΔP_{tm}	Transmembrane pressure
P_{vib}	Vibration power consumption per membrane area
PV	Pervaporation
Q	Coefficient
Q_f	Feed flow rate
Q_p	Permeate flow rate
R	Radial coordinate
r_p	Particle radius
R	Radius
Re	Reynolds number
RO	Reverse osmosis
S	Coefficient
Sc	Schmidt number
Sh	Sherwood number
S_{sp}	Filament surface area of spacer
S_{vsp}	Specific surface area of spacer
SWM	Spiral-wound module
U_b	Bubble rising velocity
V_m	Mean azimuthal velocity
T	Time
t_b	Backshocking duration of backshock filtration
t_f	Filtration duration of backshock filtration
t^b_{crit}	Delay of cake removal in the beginning of backshocking
t^f_{crit}	Delay of cake formation in the beginning of the filtration cycle of backshock filtration
TMP	Transmembrane pressure
UF	Ultrafiltration
u_E	Electrophoretic mobility
v_E	Electrophoretic velocity
V_{sp}	Filament volume of space filament
v	Velocity
V_{total}	Channel volume of space-filled channel
w	Axial velocity of pulsatile flow
Z	Coordinate

GREEK LETTERS

α	Ratio of the reverse and forward transmembrane pressure in backshock filtration
β	Backshocking clean efficiency defined by the ratio of cleaned membrane surface area to the total filtration area
ε	Voidage of spacer
ε_r	Electrolyte permittivity
ϕ_b	Solid volume fraction in the solution
γ	Shear rate
τ_1	Time constant for permeate flux decline due to cake formation in backshock filtration
η	Viscosity
ν	Kinematical viscosity
ρ	Liquid density
ω	Angular velocity
ξ	Zeta potential

REFERENCES

1. Porter, M.C. (1979) Membrane filtration, Chapter 2.1. In: *Handbook of Separation Techniques for Chemical Engineers*, Schweitzer, P.A. (Ed.) McGraw-Hill Book Co., New York.
2. Belfort, G., Davis, R.H., Zydney, A.L. (1994) Review: The behaviour of suspensions and macromolecular solutions in crossflow microfiltration. *Journal of Membrane Science* 96: 1–58.
3. Fane, A.G., Chong, T.H., Le-Clech, P. (2009) Fouling in membrane processes. In: Drioli, E., Giorno, L., (Eds.) *Membrane Operations*, pp. 121–138. Wiley-CH.
4. Field, R.W., Wu, D., Howell, J.A., Gupta, B.B. (1995) Critical flux concept for microfiltration fouling. *Journal of Membrane Science* 100: 259–272.
5. Li, H., Fane, A.G., Coster, H.G.L., Vigneswaran, S. (1998) Direct observation of particle deposition on the membrane surface during crossflow microfiltration. *Journal of Membrane Science* 149: 83–97.
6. Eckstein, E.C., Bailey, D.G., Shapiro, A.H. (1997) Self-diffusion of particles in shear flow of a suspension. *Journal of Fluid Mechanics* 79: 191.
7. Segre, G., Silberberg, A. (1962) Behaviour of microscopic rigid spheres in Poiseuille flow, Part 1 and 2. *Fluid Mechanics* 14: 115–157.
8. Li, H., Fane, A.G., Coster, H.G.L., Vigneswaran, S. (2000) An assessment of depolarisation models of crossflow microfiltration by direct observation through the membrane. *Journal of Membrane Science* 172: 135–147.
9. Holm, S., Maimberg, R., Svensson, K. (1986) *Method and plant for producing milk with a low bacterial content*. World Patent, WO 86/01687.
10. Le-Clech, P., Jefferson, B., Chang, I.S., Judd, S.J. (2003) Critical flux determination by the flux-step method in a submerged membrane bioreactor. *Journal of Membrane Science* 227: 81–93.
11. Fane A.G. (2018) A grand challenge for membrane desalination: More water, less carbon. *Desalination* 426: 155–163.
12. Zhao, Y., Qiu, C.Q., Li, X.S., Vararattanavech, A., Shen, W.M., Torres, J., Nielsen, C.H., Wang, R., Hu, X., Fane, A.G., Tang, C.Y. (2012) Synthesis of robust and high-performance aquaporin-based biomimetic membranes by interfacial polymerization-membrane preparation and RO performance characterization. *Journal of Membrane Science* 423–424: 422–428.
13. Li, X.S., Chou, S.R., Wang, R., Shi, L., Fang, W.X., Chaitra, G., Tang, C.Y., Torres, J., Hu, X., Fane, A.G. (2015) Nature gives the best solution for desalination: Aquaporin-based

hollow fiber composite membrane with superior performance. *Journal of Membrane Science* 494: 68–77.
14. Goh, P.S., Ismail, A.F., Ng, B.C. (2013) Carbon nanotubes for desalination: Performance evolution and current hurdles. *Desalination* 308: 2–14.
15. Cohen-Tanugi, D., Grossman, J.C. (2012) Water desalination across nanoporous graphene. *Nano Letters* 12 (7): 3602–3608.
16. Steuck, M.J., Reading, N. (1988) *Porous membrane having hydrophilic surface and process*. US Patent 4,618,533.
17. Kim, K.J., Fane, A.G., Fell, C.J.D. (1988) The performance of ultrafiltration membrane pre-treated by polymers. *Desalination* 70: 229–249.
18. Nyström, M., Jarvinen, P. (1991) Modification of polysulfone ultrafiltration membrane with UV irritation and hydrophilicity increasing agent. *Journal of Membrane Science* 60: 275–296.
19. Kramer, P.W., Yeh, Y.S., Yasuda, H. (1989) Low temperature plasma for the preparation of separation membranes. *Journal of Membrane Science* 46: 1–28.
20. Fane, A.G. (1986) Ultrafiltration: Factors influencing flux and rejection. In: Wakeman R.J. (Ed.) *Progress in Filtration and Separation*, Vol. 4, pp. 101–179. Elsevier Science, Amsterdam.
21. Zamani, F., Chew, J.W., Akhondi, E., Krantz, W.B., Fane A.G. (2015) Unsteady-state shear strategies to enhance mass transfer for use of ultrapermeable membranes in RO: a review. *Desalination* 356: 328–348.
22. Reid, W.H. (1928) On the stability of viscosity flow in a curved channel. *Proceedings of the Royal Society A* 121: 402.
23. Moulin, P.H., Veyret, D., Charbit, F. (2001) Dean vortices: Comparison of numerical simulation of shear stress and improvement of mass transfer in membrane processes at low permeation fluxes. *Journal of Membrane Science* 183: 148–162.
24. Mishra, P., Gupta, S.N. (1979) Momentum transfer in curved pipes, 1. Newtonian fluids. *Industrial & Engineering Chemistry Process Design and Development* 18: 130.
25. Bubolz, M., Wille, M., Langer, G., Werner, U. (2002) The use of Dean vortices for crossflow microfiltration: Basic principles and further investigation. *Separation and Purification Technology* 26: 81–89.
26. Winzeler, H.B., Belfort, G. (1993) Enhanced performance for pressure-driven membrane processes: The argument for fluid instabilities. *Journal of Membrane Science* 80: 35–47.
27. Brewster, M.E., Chung, K.Y., Belfort, G. (1993) Dean vortices with wall flux in a curved channel membrane system, 1. A new approach to membrane module design. *Journal of Membrane Science* 81: 127–137.
28. Mallubhotla, H., Belfort, G. (1997) Flux enhancement during Dean vortex microfiltration. 8. Further diagnostics. *Journal of Membrane Science* 125: 75–91.
29. Mallubhotla, H., Schmidt, M., Lee, K.H., Belfort, G. (1999) Flux enhancement during Dean vortex tubular membrane nanofiltration: 13. Effect of concentration and solute type. *Journal of Membrane Science* 153: 259–269.
30. Moulin, P.H., Manno, P., Rouch, J.C., Serra, C., Clifton, M.J., Aptel, P. (1999) Mass transfer improvement by Dean vortices: Ultrafiltration of colloid suspensions and macromolecular solutions. *Journal of Membrane Science* 156: 109–130.
31. Murase, T., Iritani, E., Chidphong, P., Kano K., Atsumi, K., Shirato, M. (1991) High-speed microfiltration using a rotating, cylindrical, ceramic membrane. *International Chemical Engineering* 31(2): 370–378.
32. Kroner, K.H., Nissinen, V. (1988) Dynamic filtration of microbial suspensions using an axially rotating filter. *Journal of Membrane Science* 36: 85–100.
33. Hallström, B., Lopez-Levia, M. (19780 Description of a rotating ultrafiltration module. *Desalination* 24: 273–279.
34. Ohashi, K., Tashiro, K., Kushiya, F., Matsumoto, T., Youshida, Endo, M., Horio, T., Ozawa, K., Sakai, K. (1988) Rotation-induced Taylor vortex enhances filtrate flux in plasma separation. *ASAIO Transactions* 34(3): 300–307.
35. Vigo, F., Uliana, C. (1986) Influence of the velocity at the membrane surface on the performance of the ultrafiltration rotating module. *Separation Science & Technology* 21(4): 367–381.
36. Holeschovsky, U.B., Cooney, C.L. (1991) Quantitative description of ultrafiltration in a rotating filtration device. *AIChE Journal* 37(8): 219 – 1227.
37. Jie, L., Liu, L., Yang, F., Liu, F., Lui, Z. (2012) The configuration and application of helical membrane modules in MBR. *Journal of Membrane Science* 392–393: 112–121.
38. Liu, L., Gao, B., Liu, J., Yang, F. (2012) Rotating a helical membrane for turbulence enhancement and fouling reduction. *Journal of Membrane Science* 181–182: 486–493.
39. Shock, G., Miquel, A. (1978) Mass transfer and pressure loss in spiral wound modules. *Desalination* 64: 339–352.
40. Liu, L., Xu, X., Zhao, C., Yang, F. (2010) A new helical membrane module for increasing permeate flux. *Journal of Membrane Science* 360: 142–148.
41. DaCosta, A.R., Fane, A.G. (1994) Net-type spacers: Effect of configuration on fluid flow path and ultrafiltration flux. *Industrial & Engineering Chemistry Research* 33: 1845–1851.
42. DaCosta, A.R., Fane, A.G., Fell, C.J.D., Franken, A.C.M. (1991) Optimal channel spacer design for ultrafiltration. *Journal of Membrane Science* 62: 275–291.
43. Suwarno, S.R., Chen, X., Chong, T.H., Puspitasari, V.L., McDougald, D., Cohen, Y., Rice, S.A., Fane, A.G. (2012) The impact of flux and spacers on biofilm development on reverse osmosis membranes. *Journal of Membrane Science* 405–406: 219–232.
44. Li, W., Che, K., Wang, Y.N., Krantz, W.B., Fane, A.G., Tang, C. (2016) A conceptual design of spacers with hairy structures for membrane processes. *Journal of Membrane Science* 510: 314–325
45. Li, F., Meindersma, W., de Haan A.B., Reith, T. (2002) Optimization of commercial net spacers in spiral wound membrane modules. *Journal of Membrane Science* 208: 289–302.
46. Schwinge, J., Neal, P.R., Wiley, D.E., Fletcher, D.E., Fane, A.G. (2004) Spiral wound modules and spacers: Review and analysis. *Journal of Membrane Science* 242: 129–153.
47. Neal, P.R., Li, H., Fane, A.G., Wiley, D.E. (2003) The effect of filament orientation on critical flux and particle deposition in spacer-filled channels. *Journal of Membrane Science* 214: 165–178.
48. Liu, X., Li, W.Y., Chong, T.H., Fane, A.G. (2017) Effects of spacer orientations on the cake formation during membrane fouling: Quantitative analysis based on 3D OCT imaging. *Water Research* 110: 1–14.

49. Kerdi, S., Qamar, A., Alpatova, A., Vrouwenvelder, J., Ghaffour, N. (2020) Membrane filtration performance enhancement and biofouling mitigation using symmetric spacers with helical filaments. *Desalination* 484: 114454.
50. Ali, S.M., Qamar, A., Kerdi, S., Phuntsho, S. Vrouwenvelder, J.S.,Ghaffour, N., Shon, H.K. (2019) Energy efficient 3D printed column type feed spacer for membrane filtration. *Water Research* 164: 114961.
51. Kerdi, S., Qamar, A., Vrouwenvelder, J.S., Ghaffour, N. (2018) Fouling resilient perforated feed spacers for membrane filtration. *Water Research* 140: 211–219.
52. Qamar, A., Kerdi, S., Ali, S.M., Shon, H.K., Vrouwenvelder, J.S., Ghaffour, N. (2021) Novel hole-pillar spacer design for improved hydrodynamics and biofouling mitigation in membrane filtration. *Scientific Reports* 11: 6979.
53. Balster, J., Pünt, I., Stamatialis, D.F., Wessling, M. (2006) Multi-layer spacer geometries with improved mass transport. *Journal of Membrane Science* 282: 351–361.
54. Sreedhar, N., Thomas, N., Al-Ketan, O., Rowshan, R., Hernandez, H., Abu Al-Rub, R.K., Arafata, H.A. (2018) 3D printed feed spacers based on triply periodic minimal surfaces for flux enhancement and biofouling mitigation in RO and UF. *Desalination* 425: 12–21.
55. Ali, S.M., Qamar, A., Phuntsho, S., Ghaffour, N., Vrouwenvelder, J.S., Shon, H.K. (2020) Conceptual design of a dynamic turbospacer for efficient low pressure membrane filtration. *Desalination* 496: 114712.
56. Liu, J., Iranshahi, A., Lou, Y., Lipscomb, G. (2013) Static mixing spacers for spiral wound modules. *Journal of Membrane Science* 442: 140–148.
57. Tan, W.S., Suwarno, S.R., An, J., Chua, C.K., Fane, A.G., Chong, T.H. (2017) Comparison of solid, liquid and powder forms of 3D printing techniques in membrane spacer fabrication. *Journal of Membrane Science* 537: 283–296.
58. Gupta, B.B., Howell, J.A., Wu, D., Field, R.W. (1995) A helical baffle for cross-flow microfiltration. *Journal of Membrane Science* 102: 31–42.
59. Elmaleh, S., Ghaffor, N. (1996) Cross-flow ultrafiltration of hydrocarbon and biological solid mixed suspensions. *Journal of Membrane Science* 118: 111–120.
60. Yeh, H.M., Chen, K.T. (2000) Improvement of ultrafiltration performance in tubular membrane using a twisted wire-rode assembly. *Journal of Membrane Science* 178: 43–53.
61. Xu, N., Xing, W., Xu, N., Shi, J. (2002) Application of turbulence promoters in ceramic membrane bioreactor used for municipal wastewater reclamation. *Journal of Membrane Science* 210: 307–313.
62. van der Waal, M.J., Racz, I.G. (1989) Mass transfer in corrugated-plate membrane module. 1. Hyperfiltration experiments. *Journal of Membrane Science* 40: 243–260.
63. Stairmand, J.W., Bellhouse, B.J. (1985) Mass transfer in a pulsating turbulent flow with deposition onto furrowed walls. *International Journal of Heat and Mass Transfer* 27(7): 1405–1408.
64. Sobey, I.J. (1985) Observation of waves during oscillator channel flow. *Journal of Fluid Mechanics* 151: 247.
65. Millward, H.R., Bellhouse, B.J., Sobey, I.J., Lewis, R.W.H. (1995) Enhancement of plasma filtration using the concept of the vortex wave. *Journal of Membrane Science* 100: 121–129.
66. Barambu, N.U., Bilad, M.R., Wibisono, Y., Jaafar, J., Mahlia, T.M.I., Khan, A.L. (2019) Membrane surface patterning as a fouling mitigation strategy in liquid filtration: A review. *Polymers* 11: 1687.
67. Ding, Y., Maruf, S., Aghajani, M., Greenberg, A. R. (2017) Surface patterning of polymeric membranes and its effect on antifouling characteristics. *Separation Science and Technology* 52(2): 240–257.
68. Heinz, O., Aghajani, M., Greenberg, A.R., Ding, Y. (2018) Surface-patterning of polymeric membranes: Fabrication and performance. *Current Opinion in Chemical Engineering* 20: 1–12.
69. Zamani, F., Tanudjaja, H.J., Akhondi, E., Krantz, W.B., Fane, A.G., Chew, J.W. (2017) Flow-field mitigation of membrane fouling (FMMF) via manipulation of the convective flow in cross-flow membrane applications. *Journal of Membrane Science* 526: 377–386.
70. Womersley, J.R. (1955) Method for the calculation of velocity, rate of flow and viscous drag in arteries when the pressure gradient is known. *Journal of Physiology* 127: 553–563.
71. Hale, J.F., McDonald, D.A., Womersley, J.R. (1955) Velocity profiles of oscillating arterial flow, with some calculations of viscous drag and the Reynolds number. *Journal of Physiology* 128: 629–640.
72. Li, H., Bertram, C.D. (1998) Mechanisms by which pulsatile low affects cross flow microfiltration. *AIChE Journal* 44(9): 1950 – 1961.
73. Li, H. (1995) *Mechanism study for crossflow microfiltration with pulsatile flow*. PhD thesis, The University of New South Wales, Australia.
74. Bertram, C.D., Hoogland, M.R., Li, H., Odell, R.A., Fane, A.G. (1993) Flux enhancement in crossflow microfiltration using a collapsible-tube pulsation generator. *Journal of Membrane Science* 84: 279–292.
75. Kennedy, T.J., Merson, R.L., McCoy, B.J. (1974) Improving permeation flux by pulsed reverse osmosis. *Chemical Engineering Science* 29: 1927–1931.
76. Gupta, B.B., Blanpain, P., Jaffrin, M.Y. (1992) Permeate flux enhancement by pressure and flow pulsations in microfiltration with mineral membranes. *Journal of Membrane Science* 70: 257 – 266.
77. Jaffrin, M.Y., Ding, L.H., Gupta, B.B. (1987) Rationale of filtration enhancement in membrane plasmapheresis by pulsatile blood flow. *Life Support Systems: Journal of the European Society for Artificial Organs* 5: 267–271.
78. Rodgers, V.G.J., Sparks, R.E. (1992) Effect of transmembrane pressure pulsing on concentration polarization. *Journal of Membrane Science* 68: 149–168.
79. Guerra, A., Jonsson, G., Rasmussen, A., Waagner, Nielsen, E., Edelsten, D., (1997) Low cross-flow velocity microfiltration of skim milk for removal of bacterial spores. *International Dairy Journal* 7: 849–861.
80. Wenten, I.G., Koenhen, D.M., Roesink, H.D.W., Rasmussen, A., Jonsson, G. (1996) *Method for the removal of components causing turbidity, from fluid, by means of microfiltration.* U.S. Patent 5560828.
81. Kuiper, S., van Rijn, C.J.M., Nijdam, W., Elwenspoek, M.C. (1998) Development and applications of very high flux microfiltration membranes. *Journal of Membrane Science* 150:1–8.
82. Kuiper, S., van Rijn, C., Nijdam, W., Raspe, O., van Wolferen, H., Krijnen, G., Elwenspoek, M. (2002) Filtration of lager beer with microsieves: Flux, permeate haze and

in-line microscope observations. *Journal of Membrane Science* 196: 159–170.
83. Gao Y., Qin, J., Wang, Z., Østerhus, S.W. (2019) Backpulsing technology applied in MF and UF processes for membrane fouling mitigation: A review. *Journal of Membrane Science* 587: 117136.
84. Vinther, F., Jönsson, A.S. (2016) Modelling of optimal back-shock frequency in hollow fibre ultrafiltration membranes I: Computational fluid dynamics. *Journal of Membrane Science* 506: 130–136.
85. Kuberkar, V., Czekaj, P., Davis, R. (1998) Flux enhancement for membrane filtration of bacterial suspensions using high-frequency backpulsing. *Biotechnology & Bioengineering* 60: 77–87.
86. Bouzerar, R., Jaffrin, M.Y., Lefevre, A., Paullier, P. (2000) Concentration of ferric hydroxide suspensions insaline medium by dynamic cross-flow filtration. *Journal of Membrane Science* 165: 111–123.
87. Jaffrin, M.Y., Ding, L.H., Akoum, O., Brou, A. (2004) A hydrodynamic comparison between rotating disk and vibratory dynamic filtration systems. *Journal of Membrane Science* 242: 155–167.
88. Chang, S., Li, H., Fane, A.G. (1988) Characteristics of operation of a rotating disk membrane filter. In: Schäfer, A.I., Basson, L., Richards, B.S. (Eds.) *Proceedings of the Environmental Engineering Research Event, "Environmental Engineering in Australia: Opportunities and Challenges", EERE'98*, pp. 153–158. Avoca Beach, December 1998.
89. Ding, L.H., Akoum, O., Abraham, A., Jaffrin, M.Y. (2003) High shear skim ultrafiltration using rotating disk filtration systems. *AIChE Journal* 49(9): 2433–2441.
90. Frenander, U., Jönsson, A S. (1996) Cell harvesting by crossflow microfiltration using a shear-enhanced module. *Biotechnology & Bioengineering* 52: 397.
91. Lee, S., Burt, A., Russoti, G., Buckland, B. (1995) Microfiltration of recombinant yeast cells using a rotating disk dynamic filtration system. *Biotechnology & Bioengineering* 48: 386.
92. Nuortula-Jokinen, J., Nyström, M. (1996) Comparison of membrane separation processes in the internal purification of paper mill water. *Journal of Membrane Science* 119: 99.
93. Brou, A., Ding, L., Boulnois, P., Jaffrin, M. (2002) Dynamic microfiltration of yeast suspensions using rotating disks equipped with vanes. *Journal of Membrane Science* 197: 269–282.
94. Dal-Cin, M.M., Lick, C.N., Kumar, A., Lealess, S. (1998) Dispersed phase back transport during ultrafiltration of cutting oil emulsions with a spinning disc geometry. *Journal of Membrane Science* 141: 166–181.
95. Ding, L.H., Jaffrin, M.Y., Mellal, M., He, G. (20060 Investigation of performance of a multi-shaft disk (MSD) system with overlapping ceramic membranes in microfiltration of mineral suspensions. *Journal of Membrane Science* 276: 232–240.
96. Armando, A.D., Culkin, B., Purchas, D.B. (1992) New separation system extends the use of membranes. In: *Proceedings of the Eurmembrane'92*, Vol. 6, pp. 459–462. Lavoisier, Paris.
97. Al-Akoum, O., Mercier-Bonin, M., Ding, L, Fonade, C., Aptel, P., Jaffrin, M. (2002) Comparison of three different systems used for flux enhancement: application to crossflow filtration of yeast suspensions. *Desalination* 147: 31–36.
98. Jaffrin, M.Y. (2008) Dynamic shear-enhanced membrane filtration: A review of rotating disks, rotating membranes and vibrating system. *Journal of Membrane Science* 324: 7–25.
99. Petala, M.D., Zouboulis, A.I. (2006) Viboratory shear enhanced processing membrane filtration applied for the removal of natural organic matter from surface waters. *Journal of Membrane Science* 269: 1–14.
100. Al-Akoum, O., Jaffrin, M.Y., Ding, L.H. (2005) Concentration of total milk proteins by high shear ultrafiltration in a vibrating membrane module. *Journal of Membrane Science* 247: 211–220.
101. Al-Akoum, O., Ding, L.H., Jaffrin, M.Y. (2002) Microfiltration and ultrafiltration of UHT skim milk with a vibrating module. *Separation and Purification Technology* 28: 219–234.
102. Zouboulis, I., Petala, M.D. (2008) Performance of VSEP viboratory membrane filtration system during the treatment of landfill leachates. *Desalination* 222: 165–175.
103. Ahmed, S., Rasul, M. G., Hasib, M. A., Watanabe, Y. (2010) Performance of nanofiltration membrane in a vibrating module (VSEP-NF) for arsenic removal. *Desalination* 252: 127–134.
104. Shi, W., Benjamin, M.M. (2009) Fouling of RO membranes in a vibratory enhanced membrane process (VSEP). *Journal of Membrane Science* 331: 11–20.
105. Krantz, W.B., Bilodeau, R.R., Voorhees, M.E., Elgas, R.J. (1997) Use of axial membrane vibrations to enhance mass transfer in a hollow tube oxygenator. *Journal of Membrane Science* 124: 283–299.
106. Beier, S.P., Jonsson, G. (2006) Dynamic microfiltration with a vibrating hollow fiber membrane module. *Desalination* 199: 499–500.
107. Gomaa, H.G., Rao, S., Al-Taweel, A.M. (2011) Intensification of membrane microfiltration using oscillatory motion. *Separation and Purification Technology* 78: 336–344.
108. Zamani, F., Law, W.W.K., Fane, A.G. (2013) Hydrodynamic analysis of vibrating hollow fibre membranes. *Journal of Membrane Science* 429: 304–312.
109. Genkin, G., Waite, T.D., Fane, A.G., Chang, S. (2006) The effect of vibration and coagulant addition on the filtration performance of submerged hollow fiber membranes. *Journal of Membrane Science* 281: 726–734.
110. Li, T., Law, A.W.K., Cetin, M., Fane A.G. (2013) Fouling control of submerged hollow fibre membranes by vibrations. *Journal of Membrane Science* 427: 230–239.
111. Gomaa, H.G., Rao, S., Al-Taweel, A.M. (2011) Intensification of membrane microfiltration using oscillatory motion. *Separation and Purification Technology* 78: 336–344.
112. Gomaa, H.G., Rao, S. (2011) Analysis of flux enhancement at oscillating flat surface membranes. *Journal of Membrane Science* 374: 59–66.
113. Li, T., Law, A.W.K., Fane, A.G. (2014) Submerged hollow fibre membrane filtration with transverse and longitudinal vibrations. *Journal of Membrane Science* 455: 83–91.
114. Kola, A., Ye, Y., Ho, A., Le-Clech, P., Chen, V. (2012) Application of low frequency transverse vibration on fouling limitation in submerged hollow fibre membranes. *Journal of Membrane Science* 409–410: 54–65.
115. Kola, A., Ye, Y., Le-Clech, P., Chen, V. (2014) Transverse vibration as novel membrane fouling mitigation strategy in anaerobic membrane bioreactor applications. *Journal of Membrane Science* 455: 320–329.

116. Li, T., Law, A.W.K., Jiang, Y., Harijanto, A.K., Fane, A.G. (2016b) Fouling control of submerged hollow fibre membrane bioreactor with transverse vibration. *Journal of Membrane Science* 505: 216–224.
117. Wu, B., Zhang. Y., Mao, Z., Tan, W.S., Tan, Y.Z., Chew, J.W., Chong, T.H., Fane, A.G. (2019) Spacer vibration for fouling control of submerged flat sheet membranes. *Separation and Purification Technology* 210: 719–728.
118. Hewitt, G.F. (1990) Multiphase fluid flow and pressure drop: Introduction and fundamentals. In: Hewitt, G.F. (Ed.) *Hemisphere Handbook of Heat Exchanger Design*, 2.3.2–1. Hemisphere Publishing Corporation, New York.
119. Cui, Z.F. (1993) Experimental investigation on enhancement of crossflow ultrafiltration with air sparging. In: R.P. Aterson (Ed.) *Effective Membrane Processes-New Perspective*, pp. 237–245. Mechanical Engineering Publications Ltd., London.
120. Cabassud, C., Laborie, S., Lainé, J.M. (1997) How slug flow can improve ultrafiltration flux in organic hollow fibers. *Journal of Membrane Science* 128(1): 93–101.
121. Mercier, M., Fonade, C., Lafforgue-Delorme, C. (1997) How slug flow can enhance the ultrafiltration flux in mineral tubular membranes. *Journal of Membrane Science* 128: 103–113.
122. Mercier, M., Maranges, C., Fonade, C., Lafforgue-Delorme C. (1998) Flux enhancement using upward gas liquid slug flow: Application to continuous alcoholic fermentation with cell recycle. *Biotechnology and Bioengineering* 58: 47.
123. Imasaka, T., So, H., Matsushita, H.K., Furukawa, T., Kanekuni, N. (1993) Application of gas-liquid two-phase crossflow filtration to pilot scale methane fermentation. *Drying Technology* 11: 769.
124. Vera, L., Delgado S., Elmaleh S. (2000) Gas sparged cross-flow microfiltration of biologically treatment wastewater. *Water Science & Technology* 41(10–11): 173–180.
125. Ghosh, R., Li, Q., Cui, Z.F. (1998) Fraction of BSA and Lysozyme using ultrafiltration: Effect of gas sparging. *AIChE Journal* 44(1); 61–67.
126. Chen, G., Yang, X., Wang, R., Fane, A.G. (2013) Performance enhancement and scaling control with gas bubbling in direct contact membrane distillation. *Desalination* 308: 47–55.
127. Chen, G., Yang, X., Lu, Y., Wang, R., Fane, A.G. (2014) Heat transfer intensification and scaling mitigation in bubbling-enhanced membrane distillation for brine concentration. *Journal of Membrane Science* 470: 60–69.
128. Cui, Z.F., Chang, S., Fane, A.G. (2003) The use of gas bubble to enhance membrane processes. *Journal of Membrane Science* 221: 1–35.
129. Cui, Z.F., Wright, K.I.T. (1996) Flux enhancement with gas sparging in downwards crossflow ultrafiltration: performance and mechanisms. *Journal of Membrane Science*, 117: 109–116.
130. Cui, Z.F., Wright, K.I.T. (1993) Gas-liquid two-phase crossflow ultrafiltration of dextran and BSA solutions. *Journal of Membrane Science* 90: 183–189.
131. Cheng, T.W., Yeh, H.M.G., Wu, J.H. (1999) Effects of gas slugs and inclination angle on the ultrafiltration flux in tubular membrane module. *Journal of Membrane Science* 158: 223–234.
132. Cheng, T.W. (2002) Influence of inclination on gas-sparged cross-flow ultrafiltration through an inorganic tubular membrane. *Journal of Membrane Science* 196(1): 103–110.
133. Phattaranawik, J., Fane, A.G., Pasquier, A.C.S. (2009) Studies of air slug distributions and preliminary membrane fouling by optical monitoring in a side-stream membrane module. *Separation Science and Technology* 44: 3793–3813.
134. Guha, R., Xiong, B., Geitner, M., Moore, T., Wood, T.K., Velegol, D., Kumar, M. (2017) Reactive micromixing eliminates fouling and concentration polarization in reverse osmosis membranes. *Journal of Membrane Science* 542: 8–17.
135. Tajima, F., Yamamoto, T. (1988) *Apparatus for filtering water containing radioactive substances in nuclear power plants*. Toshiba, US patent 4756875.
136. Yamamoto, K., Hiasa, M., Mahmood, T., Matsuo, T. (1989) Direct solid-liquid separation using hollow fiber membrane in an activated sludge aeration tank. *Water Science and Technology* 21: 43–54.
137. Wang, J., Fane, A.G., Chew, J.W. (2017) Effect of bubble characteristics on critical flux in the microfiltration of particulate foulants. *Journal of Membrane Science* 535: 279–293.
138. Chang, S., Fane, A.G. (2001) The effect of fiber diameter on filtration and flux distribution – relevance to submerged hollow fiber modules. *Journal of Membrane Science* 184: 221–231.
139. Chang, S., Fane, A.G. (2002) Filtration of biomass with lab-scale submerged hollow fiber membrane module: Effect of operational conditions and module configuration. *Journal of Chemical Technology & Biotechnology* 77: 1030–2212.
140. Wicaksana, F., Fane, A.G., Chen, V. (2006) Fibre movement induced by bubbling using submerged hollow fibre membranes. *Journal of Membrane Science* 271: 186–195.
141. Cote, P.L., Smith, B.M., Deutschmann, A.A., Rodrigues, C.F., Pedersen, S.K. (1994) *Frameless array of hollow fiber membranes and method of maintaining clean fiber surfaces while filtering a substrate to withdraw a permeate*. PCT WO 94/11094.
142. Mahendran, M., Pedersen, S.K., Henshaw, W.J., Behmann, H., Rodrigues, C.F. (1997) *Vertical skein of hollow fiber membranes and method of maintaining clean fiber surfaces*. PCT WO. 97/06880.
143. Ohkubo, K., Hayashi, T., Nagai, H. (1988) *Hollow fiber filter device*. Ebara, US Patent 4 876 006.
144. Zhang, K., Cui, Z., Field, R.W. (2009) Effect of bubble size and frequency on mass transfer in flat sheet MBR. *Journal of Membrane Science* 332: 30–37.
145. Phattaranawik, J., Fane, A.G., Pasquier, A.C.S., Bing, W. (2007) Membrane bioreactor with bubble-size transformer: Design and fouling control. *AIChE Journal* 53(1): 243–248.
146. Chang, S. (2011) Application of submerged hollow fiber membrane in membrane bioreactors: Filtration principles, operation, and membrane fouling. *Desalination* 283: 31–39.
147. Courtis, B. (2013) *The next generation of ZeeWeed MBR technology: Leap MBR, 7th European Wastewater Management Conference & Exhibition*. The Point, Lancashire County Cricket Club, Manchester, UK.
148. Bixler, H.J., Rappe, G.C. (1970) *Ultrafiltration process*. U.S. Patent 3,541,006.
149. Fane, A.G. (1984) Ultrafiltration of suspensions. *Journal of Membrane Science* 20: 249–260.
150. Noordman, T.R., De Jonge, A., Wesselingh, J.A., Bel, W., Dekke, M., Ter Vorde, E., Grijpma, S.D. (2002)

Application of fluidised particles as turbulence promoters in ultrafiltration improvement of flux and rejection. *Journal of Membrane Science* 208: 157–169.

151. van der Waal, M.J., van der Velden, Koning, J., Smolders, C.A., van Swaay, W.P.M. (1977) Use of fluidised beds as turbulence promoters in tubular membrane systems. *Desalination* 22: 465–483.
152. Kim, J., Kim, K., Young, H., Ye, H., Lee, E., Shin, C., Mccarty, P., Bae, J. (2011) Anaerobic fluidized bed membrane bioreactor for wastewater treatment. *Environmental Science & Technology* 45: 576–581.
153. Aslam, M., Yang, P., Lee, P.-H., Kim, J. (2018) Novel staged anaerobic fluidized bed ceramic membrane bioreactor: Energy reduction, fouling control and microbial characterization. *Journal of Membrane Science* 553: 200–208.
154. Wang, J.W., Wu, B., Liu, Y., Fane, A.G., Chew, J.W. (2016) Characterizing the scouring efficiency of granular activated carbon (GAC) particles in membrane fouling mitigation via wavelet decomposition of accelerometer signals. *Journal of Membrane Science* 498: 105–115.
155. Wang, J.W., Wu, B., Liu, Y., Fane, A.G., Chew, J.W. (2017) Effect of fluidized granular activated carbon (GAC) on critical flux in the microfiltration of particulate foulants. Journal of Membrane Science 523: 409–417.
156. Wang, J., Fane, A.G., Chew, J.W. (2018a) Relationship between scouring efficiency and overall concentration of fluidized granular activated carbon (GAC) in microfiltration. *Chemical Engineering Research and Design* 132: 28–39.
157. Wang, J., Fane, A.G., Chew, J.W. (2018) Characteristics of non-spherical fluidized media in a fluidized bed–membrane reactor: Effect of particle sphericity on critical flux. *Separation and Purification Technology* 202: 185–199.
158. Cahyadi, A., Fane, A.G., Chew, J.W. (2018) Correlating the hydrodynamics of fluidized media with the extent of membrane fouling mitigation: Effect of bidisperse GAC mixtures. *Separation and Purification Technology* 192: 309–321.
159. Wang, J.W., Cahyadi, A., Wu, B., Pee, W., Fane, A.G., Chew, J.W. (2020) Roles of particles in enhancing membrane filtration: A review. *Journal of Membrane Science* 595: 117570.
160. Henry, J.D., Lawler L.F., Kuo, C.H.A, (1977) A solid/liquid separation process based on crossflow and electrofiltration. *AIChE Journal* 23(6): 851–859.
161. Bowen, W.R. (1993) Electrochemical aspects of microfiltration and ultrafiltration. In: Howell, J.A., Scanchez, V., Field, R.W. (Eds.) *Membranes in Bioprocessing*, pp. 265–292. Chapman & Hall, Cambridge.
162. Wakeman, R.J., Tarleton, E.S. (1987) Membrane fouling prevention in crossflow microfiltration. *Chemical Engineering Science* 42(4): 829–842.
163. Radovich, J.M. Behnam, B. (1983) Concentration ultrafiltration and diafiltration of albumin with an electric field. *Separation Science and Technology* 18(3): 215–222.
164. Visvanathan, C., Ben Aim, R. (1990) Enhancing electrofiltration with the aid of an electro-osmotic backwashing arrangement. *Filtration and Separation* 27(1): 42–44.
165. Huotari, H.M., Huisman, I.M., Trägårdh, G. (1991) Electrically enhanced crossflow membrane filtration of oily waste water using the membrane as a cathode. *Journal of Membrane Science* 156: 49-60.
166. Akay, G., Wakeman, R.J. (1997) Electric field enhanced crossflow microfiltration of hydrophobically modified water soluble polymers. *Journal of Membrane Science* 131: 229–236.
167. Rios, G.M., Rakotoririsoa, H., Tarodo de la Fuente, B. (1988) Basic transport mechanisms of ultrafiltration in the presence of an electric field. *Journal of Membrane Science* 38: 147–159.
168. Coster, H.G.L., Farahani, T.D., Chilcott, T.C. (2011) Production of piezo-electric membranes. *Desalination* 283: 52–57.
169. Darestani, M.T., Coster, H.G.L., Chilcott, T.C., Fleming, S., Nagarajan, V., An, H. (2013) Piezoelectric membranes for separation processes: Fabrication and piezoelectric properties. *Journal of Membrane Science* 434: 184–192.
170. Darestani, M.T., Coster, H.G.L., Chilcott, T.C. (2013) Piezoelectric membranes for separation processes: Operating conditions and filtration performance. *Journal of Membrane Science* 435: 226–232.
171. Cao, P., Shi, J., Zhang, J., Wang, X., Jung, J.T., Wang, Z., Cui, Z., Lee, Y.M. (2020) Piezoelectric PVDF membranes for use in anaerobic membrane bioreactor (AnMBR) and their antifouling performance. *Journal of Membrane Science* 603: 118037.
172. Su, Y.P., Sim, L.N., Coster, H.G.L., Chong, T.H. (2021) Incorporation of barium titanate nanoparticles in piezoelectric PVDF membrane. *Journal of Membrane Science* 640:119861.
173. Thompson, L.H., Doraiswamy, L.K. (1999) Sonochemistry: Science and engineering. *Industrial & Engineering Chemistry Research* 38: 1215–1249.
174. Juang, R.S., Lin, K.H. (2004) Flux recovery in the ultrafiltration of suspended solution with ultrasound. *Journal of Membrane Science* 243: 115–124.
175. Simon, A., Penpenic, L., Gondrexon, N., Taha, S., Dorange, G. (2000) A comparative study between classical stirred and ultrasonically-assisted dead-end ultrafiltration. *Ultrasonics Sonochemistry* 7: 183–186.
176. Kobayshi, T., Kobayashi, T., Hosaka, Y., Fujii, N. (2003) Ultrasound-enhanced membrane-cleaning processes applied water treatments: Influence of sonic frequency on filtration treatments. *Ultrasonic* 41: 185–190.
177. Muthukumaran, S., Kentish, S.E., Ashokkumar, M., Stevens, G.W. (2005) Mechanisms for the ultrasonic enhancement of dairy whey ultrafiltration. *Journal of Membrane Science* 205(258): 106 – 114.
178. Ghasemi, M., Chang, S., Sivaloganathan, S. (2021) Development of an integrated ultrasonic biofilm detachment model for biofilm thickness control in membrane aerated bioreactors. *Applied Mathematical Modelling* 100: 596–611.
179. Gonzalez-Avila, S.R., Prabowo, F., Kumar, A., Ohl, C.D. (2012) Improved ultrasonic cleaning of membranes with tandem frequency excitation. *Journal of Membrane Science* 415–416: 776–783.

7 Current Challenges in Reducing Membrane Fouling

Mattheus F.A. Goosen
Alfaisal University, Riyadh, Saudi Arabia
E-mail: mgoosen@alfaisal.edu mgoosen31@hotmail.com

7.1 INTRODUCTION

Membrane bioreactors have been widely used for municipal and industrial wastewater treatment around the world due to their advantages, such as higher efficiency, smaller footprint, and lower sludge production compared with conventional activated sludge processes [1]. However, membrane fouling resulting from physicochemical interactions between the membrane and feed components such as mixed liquor remains a most challenging issue preventing the broad application of this technology. Despite the development of low-fouling membrane systems, more research and engineering activities focusing on surface modification, wastewater specifications, pretreatment and treatment conditions, and efficient fouling control strategies are still needed to minimize fouling. Pichardo-Romero [2], in a review of advances in biofouling mitigation in membranes for water treatment, noted that it was possible to obtain high-performing membranes in terms of permeation and rejection with better antifouling resistance by embedding, for example, nanomaterials including clays, zeolites, metal oxides, graphene-based materials, carbon nanotubes, and metal-organic frameworks, into the polymer matrix.

Industries such as food, petrochemicals, and fossil fuel refining generate oily wastewater that contaminates soil and water and has adverse effects on human health [3,4]. The Earth's overall oily wastewater production, for example, has already reached well over 15 billion m^3, and this figure is expected to grow dramatically [5]. It can be argued that direct disposal of this effluent needs to be restricted by government regulations, as it could result in severe water and soil contamination. Furthermore, Zulkefli et al. [4], in assessing mitigation strategies on membrane fouling for oily wastewater treatment, recommended the application of a pretreatment system before the filtration process. To save space and expense, membrane-based pretreatments such as microfiltration and ultrafiltration were preferable for high removal rates of contaminants by significantly reducing suspended particles and microorganisms from contaminated water with low energy consumption. According to Zulkefli et al. [4], most researchers used hydrophilic materials to prevent foulants from attaching to the modified membrane surface. However, the long-term stability of the modified layer of the membrane was unclear and still needs to be addressed. Optimizing the operating parameters such as backpulsing time, temperature, and transmembrane pressure was also effective in preventing fouling. However, optimizing various parameters consumed considerable time and cost. Jepsen et al. [5], in an excellent review of membrane fouling for produced water treatment from a process control perspective, argued that with the benefits of membrane filtration, it is predicted that membrane technology will be incorporated in produced water treatment if zero-discharge policies are enforced globally.

The sustainable economic development of society is determined not only by technical progress but also by environmental management [6–10]. Zakari et al. [11] endeavored to connect the *United Nations Sustainable Development Goals (SDGs)* with energy efficiency for 20 Asian and Pacific countries using data envelopment to empirically measure the productive efficiency of decision-making units. They found that sustainable economic development and energy efficiency were positively related, suggesting that sustainable economic development is associated with increased energy efficiency. In further analyses, the authors confirmed a positive impact of green innovation on energy efficiency. Policies were recommended by the authors that tended to promote sustainable economic and financial development.

The application of renewable energy sources such as solar, wind, wave, and geothermal, for example, is now considered by many as a good option for maintainable water desalination [10,12–14]. Elimelech and Philip [15], for instance, reported on the possible reductions in energy demand by desalination technologies and the role of innovative technologies in improving the sustainability of desalination as a technological solution to global water shortages (Figure 7.1). They argued, for example, that future research to improve the energy efficiency of desalination should focus on, for example, the pretreatment and post-treatment stages of seawater reverse osmosis plants. However, this argument is questionable as pretreatment and post-treatment processes represent less than 10% of the total energy consumption. Better opportunities for reducing seawater reverse osmosis energy consumption exist in the actual desalination process and optimization techniques.

It can be reasoned that the best way to approach the world's limited water resources problem lies in a coordinated approach involving water management, water purification, and water conservation [16–22]. Thermal and

DOI: 10.1201/9781003285656-7

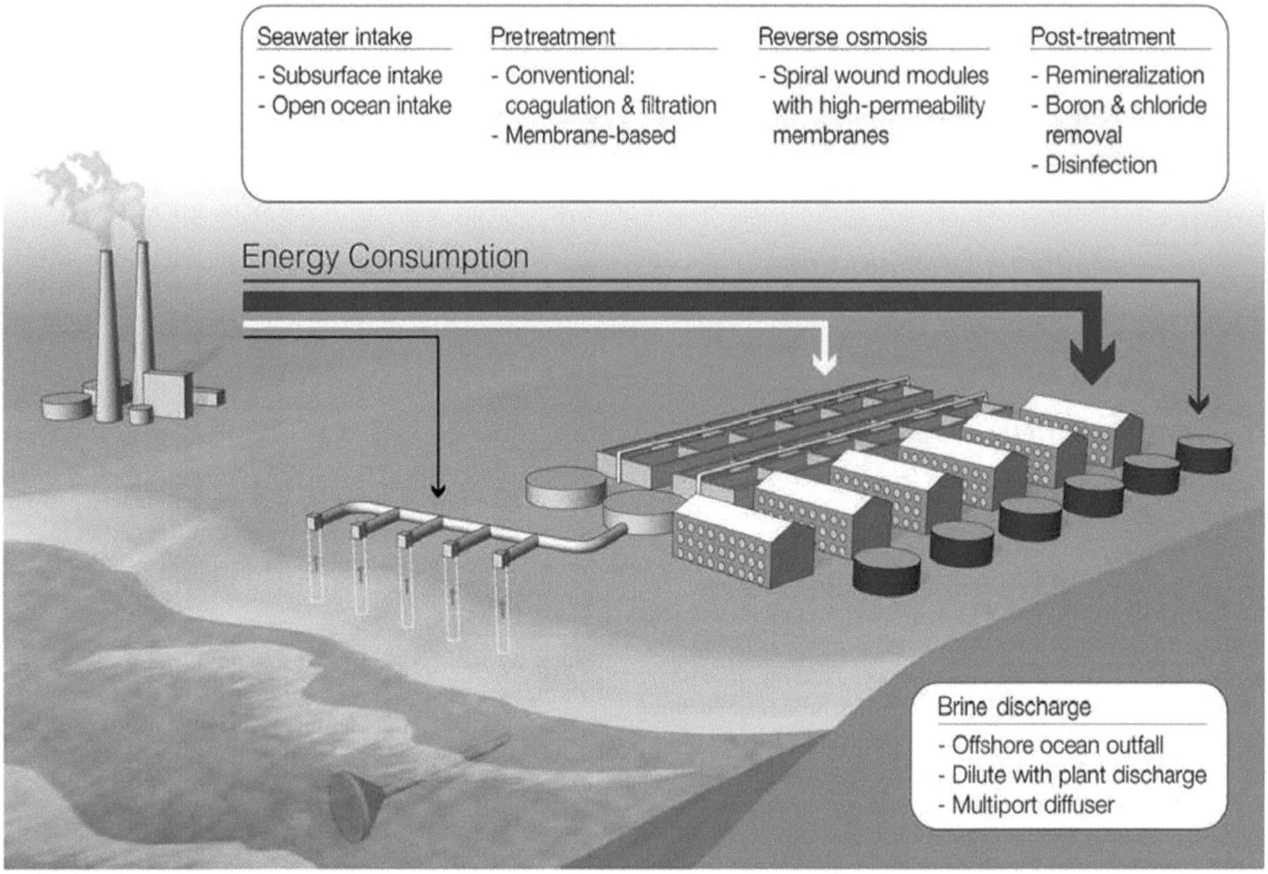

FIGURE 7.1 Energy consumption in a seawater reverse-osmosis (SWRO) desalination plant. Conceptual drawing showing the various stages: seawater intake (lower black line), pretreatment (white line), reverse osmosis (thick dark gray line), post-treatment (upper black line), and brine discharge—and their interactions with the environment. The thickness of the arrows for the energy consumption represents the relative amount of energy consumed at the various stages [15].

membrane systems are the two most successful commercial water purification techniques. Large-scale desalination plants built several decades ago, typically in the arid Arabian Gulf nations, were based on thermal desalination, where the seawater is heated and the evaporated water is condensed to yield fresh water. However, such plants consume considerable amounts of fossil fuels for producing thermal and electric energy, resulting in high greenhouse gas emissions. On the other hand, most desalination plants constructed in the past two decades are based on reverse osmosis technology, where seawater is pressurized against a semipermeable membrane that lets water pass through but retains salt. Over the past 40 years, reverse osmosis (RO) technology has advanced so that it controls well over 44% of the market share in the world's desalting production capacity [23]. Furthermore, RO holds an 80% share in the total number of desalination plants installed worldwide. This technique has low capital and operating costs compared to alternative processes like multistage flash [24]. In addition, ultrafiltration may be used prior to reverse osmosis for feed water pretreatment [25]. Membrane separation processes are also widely used in biochemical processing, industrial wastewater treatment, food and beverage production, and pharmaceutical applications [3,4,26].

Huang et al. [3,27] explained that based on the operating transmembrane pressure, membranes for water treatment use could be broadly classified as high pressure and low pressure. The latter are operated at relatively low transmembrane pressures (less than 1–2 bar, typically) and include microfiltration (MF) membranes and ultrafiltration (UF) membranes. With pore sizes ranging from approximately 10 to 100 nm, low-pressure membranes effectively remove suspended solids and particulates to reduce turbidity and pathogens. Still, they are not effective for substances such as organic micropollutants.

Membrane lifetime and permeate (i.e., pure water) fluxes are primarily affected by the phenomena of concentration polarization (i.e., solute build-up) and fouling (e.g. microbial adhesion, gel layer formation, and solute adhesion) at the membrane surface (Figure 7.2) [20,28,29]. Koltuniewicz and Noworyta [30], in a highly recommended paper, summarized the phenomena responsible for limiting the permeate flux during cyclic operation (i.e., permeation followed by cleaning). During the initial period of operation within a cycle,

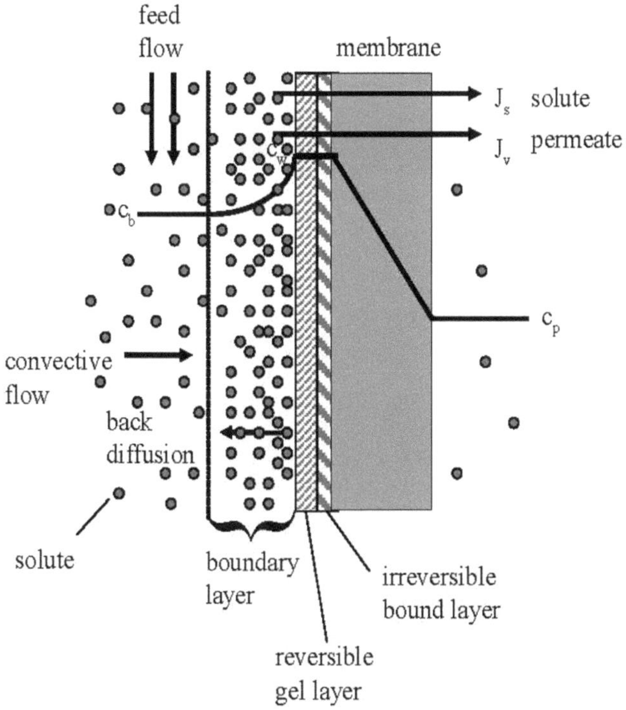

FIGURE 7.2 A schematic representation of concentration polarization and fouling at the membrane surface [20].

concentration polarization is one of the primary reasons for flux decline, J_a, (Figure 7.3). Large-scale membrane systems operate in a cyclic mode, where clean-in-place operation alternates with the normal run. The figure shows a decrease in the flux for pure water from cycle to cycle, $J_o(t)$, due to fouling, the flux decline within a cycle due to concentration polarization, $J(t_p)$, and the average flux under steady-state concentration, J_a. The latter, also decreasing from cycle to cycle, suggests irreversible solute adsorption or fouling. Accumulation of the solute retained on a membrane surface leads to increasing permeate flow resistance at the membrane wall region.

Membrane fouling remains a major challenge in the development and long-term operation of microfiltration, ultrafiltration, and reverse osmosis membranes [3,27]. As a pretreatment, coagulation, oxidation, and adsorption have been studied and established to reduce membrane fouling effectively. Huang et al. [27] demonstrated, for instance, that coagulation is the most successful pretreatment for fouling reduction. A detailed comparison between pretreatment methods is shown in Table 7.1. Additionally, the combination of coagulation and microfiltration (MF) can remove pathogenic microorganisms, including viruses, from the raw water [31].

The most frequently applied adsorbent for removing organic pollutants in wastewaters is currently activated

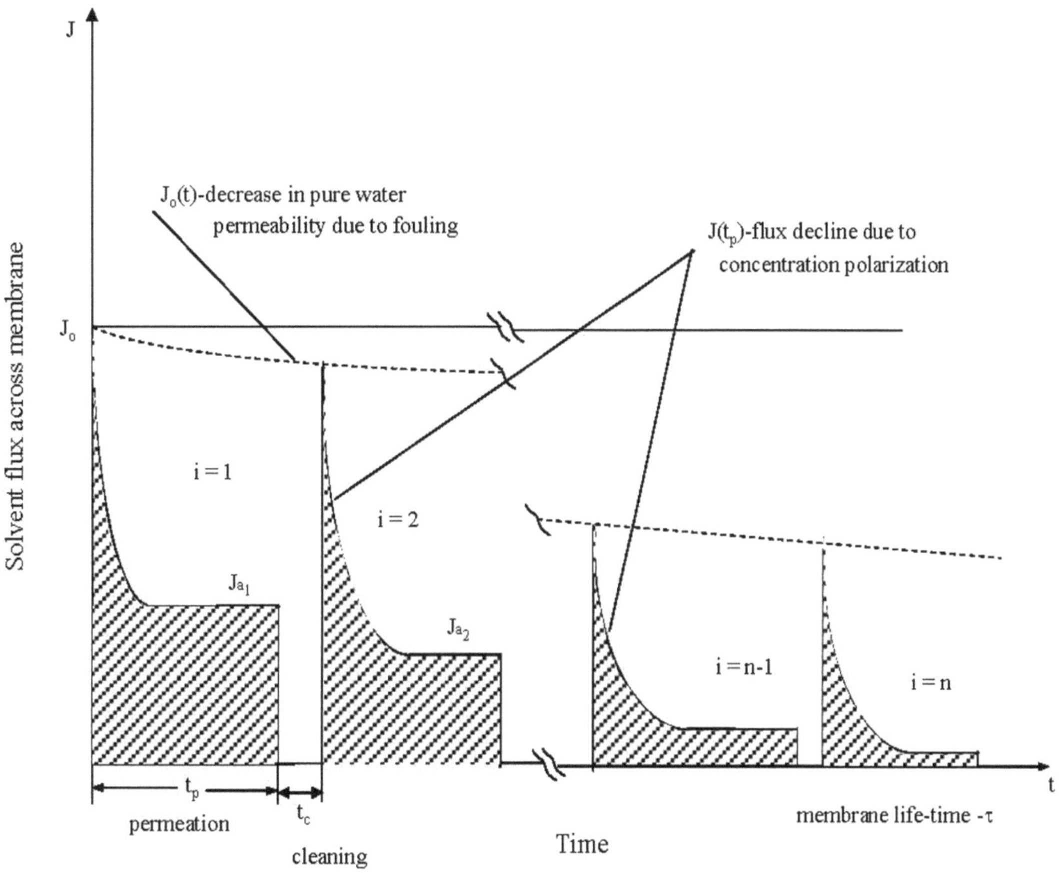

FIGURE 7.3 Diagram of typical flux–time dependency during cyclic operation in large-scale ultrafiltration systems. Adapted from Koltuniewicz and Noworyta [30].

TABLE 7.1
List of the mechanisms, effects, and applications of major pre-treatments for reducing fouling in membrane filtration [3,27]

Pretreatment	Coagulation	Adsorption	Preoxidation	Prefiltration
Chemicals applied	Coagulants (or flocculants) at proper dose	Porous or nonporous adsorbents in suspension or fixed contactor	Gaseous or liquid oxidants	Granular media with/without coagulants, membranes
Dose effects	Under-, optimal, or overdose (optimal for enhanced coagulation)	Minimal effective dose if used as suspended particles	Minimal effective dose	None
Physical mechanisms	Increases the size of aquatic contaminants to filterable level	Binds small contaminants to adsorbents much larger than membrane pores	May cause dissociation of organic colloids into smaller sizes or the release of EPS by aquatic organisms	Removes coarse materials that may cause cake/gel layer formation on downstream membranes
Chemical mechanisms	Destabilizes contaminants to cause aggregation or adsorption on coagulant precipitates or membrane surfaces	Provides new interfaces to adsorb/accumulate substances detrimental to membrane performance	Oxidizes and/or partially decomposes NOM, possible mineralization if VUV used	Selectively removes contaminants or other particles that are sticky to filter media and downstream membranes
Biological mechanisms	Partially removes autochthonous NOM and hinders bacterial growth in feedwater or on membrane	May adsorb organic contaminants relevant to biofouling	Suppresses microbial growth	Partially removes microorganisms that can cause biofouling
Targeted contaminants	Viruses, humic/fulvic acids, proteins, polysaccharides with acidic groups, colloids smaller than membrane pores	Humic/fulvic acids, small natural organic acids, some DBPs, pesticides, and other synthetic organic compounds	Viruses and organic contaminants with ozonation	Particulate and colloidal organic/inorganic substances, microbiota
Effects on membrane fouling	Reduces colloidal fouling and NOM fouling	May increase or decrease membrane fouling	May reduce biofouling and NOM fouling	May reduce fouling to different extents
Advantages	Significantly improves LPM performance (less fouling and greater rejection)	Increases the removal of DBPs and DBP precursors	Reduces the occurrence of biofouling; increases organic removal (ozonation)	May reduce biofouling, colloidal fouling, and/or solids loading
Disadvantages	(1) Requires proper dose that can be difficult to meet if feedwater quality varies rapidly/significantly, (2) may exacerbate fouling, (3) produce solid wastes, (4) ineffective in mitigating the fouling by hydrophilic neutral organics	(1) Possible exacerbation of LPM fouling, (2) difficulty in removing PAC powders from treatment facilities	(1) Formation of DBPs; (2) may damage membranes incompatible with oxidants; (3) may be ineffective in suppressing the growth of some microbiota resistant to oxidation	(1) Performance of prefilters may deteriorate and be difficult to recover, (2) may require pretreatment (e.g. coagulation or preoxidation) to enhance the efficacy

carbon, an expensive material. Mounting apprehension about environmental issues has prompted the textile industry to investigate appropriate and environmentally friendly treatment technologies for aqueous waste effluent containing color and heavy metals. Dubey and Rao [32] evaluated several alternatives and cheaper adsorbents. The results indicated that wood was the most economic adsorbent, followed by peat, carbon, pith, and Fuller's earth. They argued that even though activated carbons may exhibit the highest adsorption capacity for dyes in most of the systems studied, some natural and cheaper materials such as peat and lignite also have high capacities. We can speculate that the practical information obtained from these studies can be applied to reducing membrane fouling during wastewater pretreatment using inexpensive adsorbents.

One of the most serious forms of RO membrane fouling is bacterial adhesion and growth [33]. Once they form, biofilms can be very difficult to remove, either through disinfection or chemical cleaning. This waste energy degrades salt rejection and leads to shortened membrane life. This is one area, for example, where further research is required. A vital aspect of controlling fouling in RO membranes is the need to improve resistance to oxidation. The main reason biofilms prevail in RO membranes compared to MF/UF is the susceptibility of thin-film composite (TFC) membranes to oxidation damage. Li and Wang [38] have reported progress in developing TFC chlorine-resistant RO membranes. Their article focused on the modification of current polymeric membrane materials and synthesis and separation performance of new polymer membranes, inorganic membranes, and mixed-matrix membranes.

A variety of liquids have been treated with reverse osmosis and ultrafiltration membranes ranging from seawater to wastewater to milk and yeast suspensions. Each liquid varies in composition and the type and fraction of the solute to be retained by the membrane. Complicating factors include the presence of substances such as, for example, oil in seawater and wastewater [4,34–37]. The presence of the oil normally necessitates an additional pretreatment step and further complicates the fouling process. The presence of humic acids in surface water and wastewater also needs special attention [39,40]. The fouling phenomena, the preventive means (i.e., pretreatment), and the frequency and type of membrane cleaning cycle are all dependent on the type of liquid being treated.

Membrane materials for reverse osmosis and ultrafiltration applications range from polysulfone and polyethersulfone to cellulose acetate and cellulose diacetate [41–47]. Commercially available polyamide composite membranes for desalination of seawater, for example, are available from a variety of companies in the US, Europe, and Japan [48]; the specific choice of which membrane material to use will depend on the process (e.g., type of liquid to be treated, operating conditions) and economic factors (e.g., cost of replacement membranes, cost of cleaning chemicals). In addition, the exact chemical composition and physical morphology of the membranes may vary from manufacturer to manufacturer. Since the liquids to be treated and the operating conditions also vary from application to application, it becomes difficult to draw general conclusions on which materials are the best to use to inhibit membrane fouling.

The main scope of this chapter is to review recent studies on fouling phenomena in reverse osmosis and ultrafiltration membrane systems, characterization of the foulant layer, challenges in fouling mitigation and feed water pretreatment, and the economic aspects of membrane fouling. Explicit recommendations have also been made on how scientists, engineers, and technical staff can assist in improving the performance of these systems through minimization of membrane fouling.

7.2 MEMBRANE FOULING PHENOMENA

The foremost mechanisms of membrane fouling are adsorption of feed components, clogging of pores, chemical interaction between solutes and membrane material, gel formation, and bacterial growth (Table 7.2). The fouling mechanism is dependent on the type of membrane being utilized; the mechanism for low-pressure membranes (i.e., MF/UF) is different from high-pressure membranes (i.e., NF/RO). This difference is critical for understanding how to mitigate fouling. Let us first consider algal and bacterial growth on membranes.

7.2.1 Algal and Microbiological Fouling

Microbiological fouling of reverse osmosis membranes is one of the main factors in flux decline and loss of salt rejection [50,51,55,56,69,74,106,140] (Table 7.2). Good reviews of membrane fouling due to algae-containing waters were provided by Du et al. [57], Ly et al. [56], Asif and Zhang [58], and Stork [54]. Bacterial fouling of a surface (i.e., formation of a biofilm) can be divided into three phases: transport of the organisms to the surface, attachment to the substratum, and growth at the surface. Fleming et al. [49] have shown that it takes about three days to cover the reverse osmosis membrane with a biofilm completely. Ghayeni et al. [50,140] studied the initial adhesion of sewage bacteria belonging to the genus *Pseudomonas* to reverse osmosis membranes. It was found that bacteria would sometimes aggregate upon adhering. While minimal bacterial attachment occurred in a very low ionic strength solution, significantly higher numbers of attached microbes occurred when using salt concentrations corresponding to wastewater. Understanding the mechanism of bacterial attachment may assist in developing antifouling technologies for membrane systems.

After a few minutes of contact between a membrane and raw water, Flemming and Schaule [44] demonstrated that the first irreversible attachment of cells occurs. *Pseudomonas* was identified as a fast-adhering species out of a tap water microflora. If non-starving cells were utilized (i.e., sufficient nutrients and dissolved oxygen in the raw water), the adhesion process improved with an increase in the number of cells in suspension. When starving cells were employed, incomplete coverage of the surface occurred. This was like the surface aggregate formations observed for membranes by Ghayeni et al. [50,140]. Flemming and Schaule [44] also detected a biological affinity of different membrane materials toward bacteria. Polyetherurea, for example, had a significantly lower biological affinity than polyamide, polysulfone and polyethersulfone.

In a similar but more thorough study than that performed by Ghayeni et al. [50], Ridgway et al. [50,52], in two excellent papers, have reported on the biofouling of reverse-osmosis membranes with wastewater. Cellulose diacetate membranes became uniformly coated with a fouling layer that was primarily organic in composition. The major inorganic constituents detected were calcium, phosphorus, sulfur, and chlorine. Protein and carbohydrate represented as much as 30% and 17%, respectively, of the dry weight of the biofilm. Electron microscopy revealed that the biofilm on the feed water side surface of the membrane was 10–20 μm thick and was composed of several layers of compacted bacterial cells,

TABLE 7.2
Summary of membrane fouling studies reported in literature. Specific papers are (•) recommended and (• •) highly recommended

Fouling studies	References
Membrane Fouling Phenomena	
Algal and microbial cell attachment	49, 50, ••51, ••52, 53, •54, •55, •56,• 57, •58
Humic acids and morphology of fouling layer	39, 59, 60, 40, •61, 62, ••52
Inorganics, interface characteristics	63, 64
Proteins and colloids	65, 66, 59, 67, 68
Reversible adsorbed layer and transition from reversible to irreversible fouling	••69, ••30, ••70, 71
Gel layer thickness, evaluation of layering	••72, 73
Pore blockage and cake formation	74, 75
Temperature and pressure effects on membrane	76
Fouling Layer Characterization	
Fouling layer morphology and growth, optical tomography	77, 37, 7, 78, •79
Adhesion kinetics	42, 43
Hydrodynamics	80, •81, 82
Passage of bacteria through membrane	•83
Analysis of deposits: ATR, FTIR, measuring fouling in real time	34, 84, 46, 85, 86, ••47
Measuring concentration polarization	•87, •88, 36, 28
Mathematical modeling of flux decline	89, ••30, •90, 91
Fouling Mitigation and Pretreatment	
Feed water pretreatment	92, 93, 50, •83, 94, 95, 96, 22, 97, 98, 99, ••100, 35, 101, •103, 32,
Microfiltration and ultrafiltration	27, 31, 104, 23, 105, •106, 107, 108, 102
Coagulation and flocculation	••109, 110, 25, 47,111, 112, 113
Ozone oxidation	34, 37, •58, •114, 115
Spacers	•116, 44, 43, 48, 117, 41, • 118, ••119, 165
Corrugated, ceramic, and silicon carbide membranes	40, •61, 82, 3
Surface chemistry and modification, nanoparticles	•120, ••70, ••30, 121, 122, 123, 19, 76, •124, 125, 126, 57, 127, •128, ••5
Hydrophobic and hydrophilic membranes	
Control of operating parameters, critical flux, membrane bioreactors, retention time	
Rinsing water quality	129, •34
Cleaning agents and chemical cleaning	130, 131
Back pulsing	132, 133, 134
Membrane wear and degradation	135, •45, 43
Economic Aspects	136, 137, 23, 105, •138, 139

many of which were partially or completely autolyzed. The bacteria were firmly attached to the membrane surface by an extensive network of extracellular polymeric fibrils. They showed that mycobacteria adhered to the cellulose acetate membrane surface 25-fold more effectively than a wild-type strain of *Escherichia coli*. In a key finding, the ability of *Mycobacterium* and *E. coli* to adhere to the membrane was correlated with their relative surface hydrophobicities as determined by their affinities for *n*-hexadecane [52]. The results suggested that hydrophobic interaction between bacterial cell surface components and the cellulose membrane surface plays an important role in the initial stages of bacterial adhesion and biofilm formation. A key question that arises is whether the importance of this hydrophobic interaction

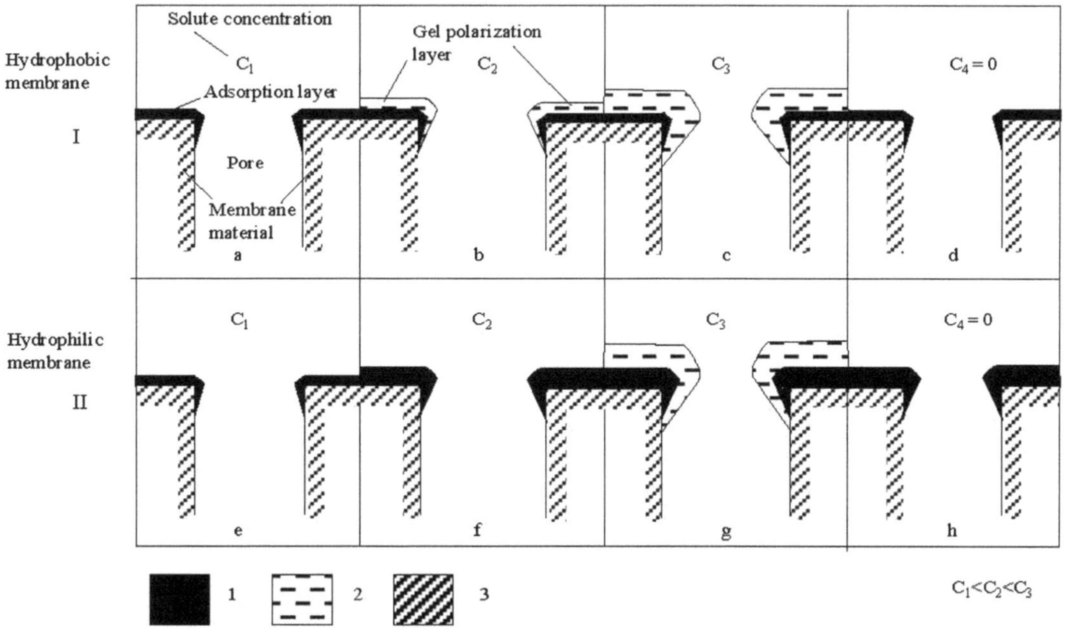

FIGURE 7.4 Gel layer formation on surface of an ultrafiltration membrane made from (I) hydrophobic and (II) hydrophilic material. C, solute concentration; $C_1 < C_2 < C_3$, 1 adsorption layer, 2 gel-polarization layer, 3 membrane material. Adapted from Cherkasov et al. [82].

between the cell and the membrane also holds true for other polymers. This work is similar to that reported by Cherkasov et al. [82] on fouling resistance of hydrophilic and hydrophobic membranes (Figure 7.4). A later research study by Ridgway [53] confirmed these results and conclusions. For cleaning mechanisms for biofilms, please see Section 7.4.4.

7.2.2 Effect of Humic Acids, Inorganics, Proteins, and Colloids on the Fouling Layer

The breakdown of organic material, such as plants, in the soil produces a mixture of complex macromolecules called humic acids. These complex molecules have polymeric phenolic structures with the ability to chelate metals, especially iron. It is recommended that humic acids be removed from process water before filtration by complexation (i.e., flocculation/coagulation; see Section 7.4.1). Humic acids give surface water a yellowish to brownish color and often cause fouling problems in membrane filtration [39,62,68,141,142]. The fouling tendency of humic acids appears to be due to their ability to bind to multivalent salts. Nystrom et al. [39], for example, showed that humic acids were most harmful in positively charged membranes [i.e., containing alumina (Al) and silica (Si)]. This is because humic acids formed chelates with the metals (i.e., multivalent ions) and could be seen as a gel-like layer on the filter surface.

Schafer et al. [59] studied the role of concentration polarization and solution chemistry on the morphology of the humic acid fouling layer. Irreversible fouling occurred with all membranes at high calcium concentrations. Interestingly, it was found that the hydrophobic fraction of the humic acids was deposited preferentially on the membrane surface. This result is similar to the work of Ridgeway et al. [52]. They showed that the hydrophobic interaction between a bacterial cell surface and a membrane surface plays a key role in biofilm formation. The formation of two layers, one on top of the other, was also observed by Khatib et al. [60]. The formation of a Fe–Si gel layer directly on the membrane surface was mainly responsible for the fouling; reducing the electrostatic repulsion between the ferric gel and the membrane surface encouraged adhesion. Tu et al. [61] also demonstrated that membranes with a higher negative surface charge and greater hydrophilicity were less prone to fouling due to fewer interactions between the chemical groups in the organic solute and the polar groups on the membrane surface. These studies tell us that to reduce fouling due to humic acids, it is best to employ hydrophilic membranes, to have feed water with a low mineral salts content (e.g., calcium), and to work at low pH.

Sahachaiyunta et al. [63] conducted dynamic tests to investigate the effect of silica fouling on RO membranes in the presence of minute amounts of various inorganic cations such as iron, manganese, nickel, and barium that are present in industrial and mineral processing wastewaters. Experimental results showed that the presence of iron greatly affected the scale structure on the membrane surface compared to the other metal species. Fouling due to silica normally occurs only in the tail elements of a RO process. It is important to remember that scaling due to inorganics such as calcium sulfate and calcium carbonate also occurs.

A dual-mode fouling process, like that observed for humic acids [59], was found for protein [i.e., bovine serum albumin (BSA)] fouling of microfiltration membranes. Protein aggregates first formed on the membrane surface, followed by native (i.e., non-aggregated) protein. The native protein is attached to an existing protein via the formation of intermolecular disulfide linkages.

Stable colloidal suspensions can cause less fouling. Yiantsios and Karabelas [65], in a very interesting paper, found that apart from particle size and concentration, colloid stability plays a major role in RO and UF membrane fouling. They demonstrated that standard fouling tests and most well-known fouling models are inadequate. A key finding was that the use of acid, a common practice to avoid scaling in desalination, might promote colloidal fouling. Lowering the pH reduces the negative charge on particles, causing aggregate formations that deposit on the membrane surface. Jarusutthirak et al. [66] previously mentioned that organic matter of isolated wastewater effluent could be divided into different fractions. Each isolate exhibited different characteristics in fouling of nanofiltration (NF) and UF membranes. Polysaccharides and amino sugars were found to play an important role in fouling. The colloidal fractions gave a high flux decline due to pore blockage, and hydrophobic interactions were very important for hydrophobic membranes causing a reduction in permeate flux.

7.2.3 Membrane Bioreactors and Transition from Reversible Adsorption to Irreversible Fouling

Membrane bioreactors (MBRs) are gaining acceptance as an effective method for wastewater treatment and water reclamation [(71,75,126–128,143,144]. The performance of the different configurations of anaerobic MBRs is affected by operating parameters, including solid retention time (SRT), hydraulic retention time (HRT), organic loading rate (OLR), sludge recycle rate, temperature, and wastewater characteristics [125]. Generally, the optimum operating condition is characterized by a longer SRT and HRT, and higher OLR at a mesophilic temperature range. These parameters also affect the tendency of membrane fouling which is considered the major challenge, limiting its sustainable use. The common fouling control strategies, described in a thorough review by De Vela [125], include sub-critical flux operation, membrane relaxation, backwashing, biogas sparging, sludge recycling, and use of coagulants and turbulence promoters. The authors reported that emerging technologies like enzymatic or bacterial degradation through quorum quenching technology are rapidly being investigated. Additionally, integration of rotary disks/membrane, attached growth system, dynamic membrane, electrochemical membrane, and baffled reactor configurations have been introduced as modifications to the conventional anaerobic MBR. Similarly, Lei et al. [79] provided further understanding of fouling on anaerobic membrane bioreactors treating sewage through physicochemical and biological characterization of the cake and gel layers. Likewise, Sabalanvand et al. [145] performed an interesting investigation on how Ag and magnetite nanoparticles affected fouling in membrane bioreactors. It was observed that overall, the application of nanoparticles resulted in better removal of organic matter in the system. The authors concluded that applying nanoparticles in MBR systems leads to significantly improved performance and reduced membrane fouling.

Hermanowicz [146] has shown that although an MBR can be operated at biomass concentrations 5–10 times higher than activated sludge, the standard wastewater treatment method, these concentrations are limited in practice by increasing biomass suspension viscosity that in turn increases "reversible" membrane fouling and decreases oxygen transfer rates. "Irreversible" fouling also occurs at the same time. This is a major operational challenge since it depends on complex interactions of membranes with various fractions of soluble microbial products resulting from microbial metabolism.

"Reversible" fouling is typically caused by a high solids concentration that increases the viscosity of the biomass (sludge). High viscosity also affects oxygen transfer rates, thus limiting biomass concentrations in practical applications for economic reasons. "Irreversible" fouling requires chemical membrane cleaning to remove fouling molecules and particles that obstruct filtration through the membrane. Fouling mechanisms are not well understood, according to Hermanowicz [146] and Wang et al. [147]. Interactions between the membrane and high-molecular-weight carbohydrates have been implicated in "irreversible" fouling.

In a key study, Nikolova and Islam (69) reported that adsorption resistance was the decisive factor in flux decline. With the development of a concentration polarization layer, the adsorbed layer resistance at the membrane wall increased linearly as a function of the solute concentration at the wall. They described the flux by the following relationship:

$$J = \frac{\Delta P - \Delta\pi(w)}{\mu\left(R_m + kC_w\right)} \tag{7.1}$$

where ΔP is the hydraulic pressure difference across the membrane, C_w is the concentration at the membrane surface and $\Delta\pi(w)$ is the corresponding osmotic pressure, R_m is the membrane resistance, kC_w is the adsorbed layer resistance, and μ is the fluid viscosity. The key finding showed that the adsorption resistance was of the same order of magnitude as the membrane resistance. Surprisingly, the osmotic pressure was negligible compared to the applied transmembrane pressure. The significance of this study showed that the reversibly adsorbed solute layer at the membrane surface is the primary cause of flux decline and not the higher osmotic pressure at the membrane surface. This is supported by the work of Koltuniewicz and Noworyta [30] (Figure 7.3).

The transition between the reversible adsorption described by Nikolova and Islam [69] and irreversible fouling is crucial in determining the strategy for improved membrane performance and understanding the threshold values for which optimal flux and rejection can be maintained. In a very thorough study, Chen et al. [70] reported on the dynamic transition from concentration polarization to cake (i.e., gel-layer) formation for membrane filtration of colloidal silica. Once a critical flux, J_{crit}, was exceeded, the colloids in the polarized layer formed a consolidated cake structure that was slow to depolarize and reduced the flux. This paper is a very valuable source of information for membrane plant operators. By operating just below J_{crit}, they can maximize the flux while at the same time reducing the frequency of membrane cleaning. The study by Chen et al. [70] showed that by controlling the flux below J_{crit}, the polarization layer might form, and solute adsorption may occur, but it is reversible and responds quickly to any changes in convection. For additional reading on the topics of membrane bioreactors, control of operating parameters and concentration polarization, readers may refer to the works by May et al. [28], Banti et al. [124], De Vela [125], Di Bella et al. [126], Du et al. [57], Szabo-Corbacho et al. [127], Yuliwati et al. [128], and an excellent paper by Jepsen et al. [5] (Table 7.2).

Control of mixed-liquor suspended solids concentration in a membrane bioreactor is critical for controlling fouling on the membrane surface. Membrane fouling is also attributed to the larger soluble molecules plugging and narrowing the pores of membranes or the particles depositing on the membrane surfaces to form a cake layer [148]. For instance, mixed liquors contain two fractions: microbial floc and supernatant, including colloids and solutes. Attempts have been made to quantify the fouling caused by each fraction of the mixed liquor, such as the suspended solids (i.e., bacterial flocs), colloids, and solutes, although the results are inconsistent [148]. Some researchers have concluded that the resistance of the cake layer formed by microbial floc appeared to determine the overall resistance. For example, a study by Wu et al. [148] showed that more serious cake-fouling happened in an MBR, which correlated with the activated sludge characteristics such as smaller floc size and greater amounts of extracellular polymeric substances, which are excreted or autolyzed by microorganisms. See also a related study by Stumme et al. [73]. They demonstrated that when layer-dominated film growth prevailed from the early stages of the coating process, permeability values were higher at similar solute rejection rates compared to an initial pore-dominated and then layer-dominated film growth.

7.2.4 Temperature and Pressure Effects on Membrane Permeation Properties

Changes in membrane permeation due to variations in temperature and pressure have also been reported by numerous researchers [117,147,149]. Goosen et al. [76] assessed the effects of cyclic changes in temperature and pressure of feed water on permeate flux, solute rejection, and compaction in spiral-wound composite polyamide seawater reverse osmosis membranes using pure water and 4% NaCl solutions. Although these studies did not directly involve membrane fouling phenomena, they are important as they provide a better understanding of what physically occurs to the membrane during changes in temperature and pressure and thus have applications in membrane fouling phenomena and mitigation. In their studies, Goosen et al. [76] observed a membrane permeability hysteresis or memory effect due to the up and down temperature and pressure sequences which were only seen with the saline water studies. However, the observed changes appeared to be reversible and were consistent with the Spiegler-Kedem/Film Theory and the Kimura-Sourirajan Analysis/Film Theory models. The overall effects of cyclic changes in operating temperature and pressure on permeation properties of composite polyamide RO membranes are represented in Figure 7.5. Increasing feed/operating temperatures increase the movement of polymer chains and hence increase membrane porosity. In addition, the higher temperatures of the feed solution decrease the water/solute cluster size, which in turn lowers viscosity. Both of these increase permeance and decrease rejection. Increasing transmembrane pressure, on the other hand, may cause membrane compaction, resulting in lower permeance. The net effect on permeance and retention is a combination of these factors. The permeance, for example, is greatest and solute rejection lowest under operating conditions of high temperature and low pressure (case B). The outcomes of this investigation imply or suggest that cyclic changes in the operating temperature and pressure should not damage the spiral-wound composite polyamide seawater reverse osmosis, thus allowing for longer membrane lifetimes before replacement is needed. We can speculate that reversible and irreversible membrane fouling may be affected by the physical changes occurring in the membrane due to changes in temperature and pressure. It would be interesting to repeat these experiments with fouled membranes as well as fouled membranes that have been chemically cleaned. This would simulate the real operating conditions normally found in, for example, the Arabian Gulf, and the information gained would help improve operating times.

7.2.5 Membrane Fouling in Gas Separation Processes

Fouling in gas separation processes is less severe than in microfiltration, nanofiltration, and reverse osmosis, where it is the main cause of permanent flux decline and loss of product quality [117,150]. Many industrial activities such as gas production [150–152], catalysis [153], and fuel cells [152] require gas separation. Saracco and Specchia [153] noted that inorganic membranes have great potential in gas separation, catalytic reactors, gasification of coal, water decomposition, and solid electrolyte fuel cells [150–158]. Inorganic membranes are usually made from alumina, silica, carbon, and zeolites [154]. We will only assess porous membranes since fouling is virtually absent in dense membranes [150]. Roque-Malherbe et al. [159,160] have studied the transport of hydrogen and carbon dioxide through porous ceramic membranes. The flux was found to decline because membrane fouling can be attributed to pore blocking due to the deposition of particles flowing with the gas to be cleaned. In addition, a decrease in the membrane porosity and, consequently, flux decay during the gas separation process could be produced in ceramic membranes because of the sinterization of the particles which constitute the membrane if the gas separation

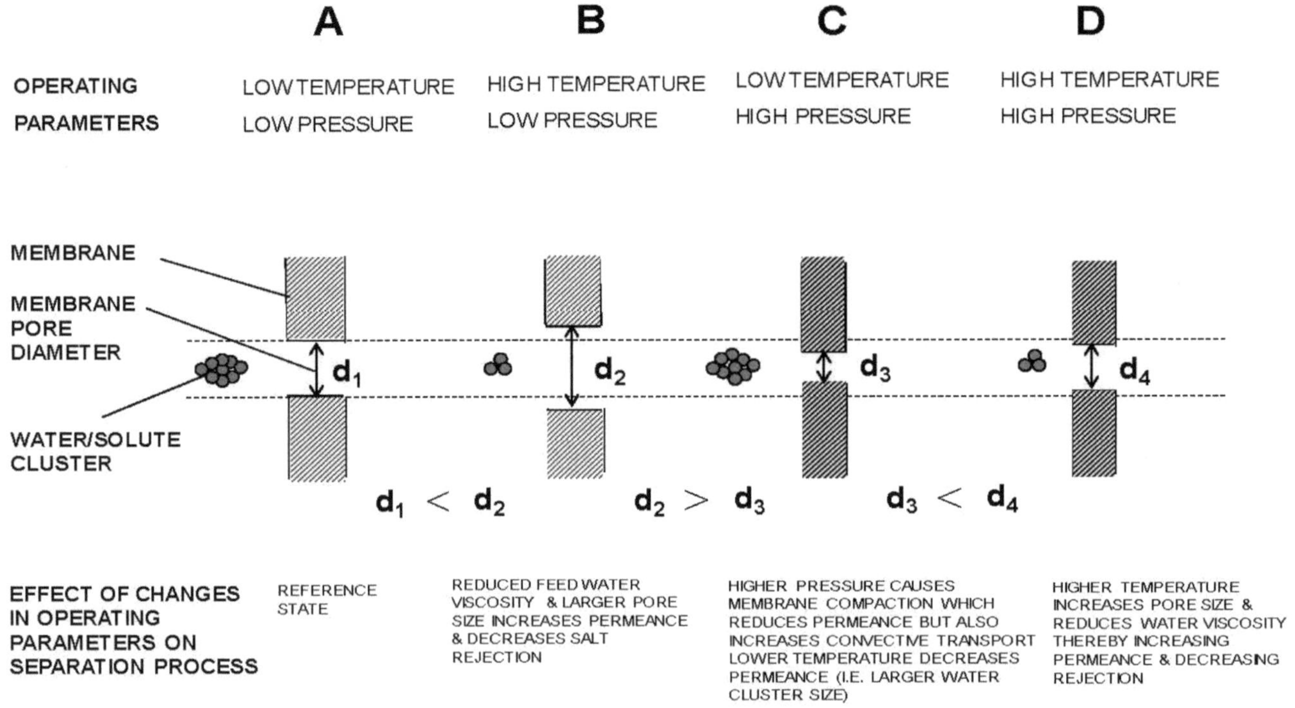

FIGURE 7.5 Graphical representation of the effects of cyclic changes in operating temperature and pressure on permeation properties of composite polyamide RO membranes [76].

process is carried out at high temperatures. This means physical changes in the membrane can occur because of the operating parameters. Pore blockage and membrane compaction are two factors that contribute to the fouling of membranes during gas separations.

Dilshad et al. [117] assessed the effect of silica nanoparticles on carbon dioxide separation performances of PVA/PEG (i.e., polyvinyl alcohol/polyethylene glycol) cross-linked membranes. PVA/PEG cross-linked membranes were prepared with (0–20 wt.%) of silica nanoparticles. It was found that the permeability of all the gases increased with an increase in silica loading. In contrast, the ideal selectivity of carbon dioxide with respect to nitrogen and methane increased up to 10 wt.% loading and then became nearly constant with further loading. A 20 wt.% silica-loaded membrane was found to give the best performance. The gas permeability of CO_2 was also compared with different gas permeation models and was in close agreement with the Maxwell Model. The authors also reported that optimum performances were achieved at 65°C.

7.3 FOULING LAYER CHARACTERIZATION

7.3.1 Measuring Fouling Layer Morphology and Passage of Bacteria through Membranes

The fouling mechanisms and interface characteristics were investigated by Liu et al. [64] and Guo et al. [78]. The results of Liu et al. [64] showed that ClO_2 pre-oxidation at low doses (1–2 mg/L) could alleviate membrane flux decline caused by humus, polysaccharides, and simulated natural water, but had a limited alleviating effect on the irreversible resistance of the membrane. Interfacial free energy analysis disclosed that the interaction force between the membrane and the simulated natural water was also repulsive after the pre-oxidation, indicating that ClO_2 pre-oxidation was an effective way to alleviate cake layer fouling by reducing the interaction between the foulant and the membrane. Guo et al. [78] investigated the fouling mechanism in membrane distillation using in situ optical coherence tomography with green regeneration of the fouled membrane. Two commercial membranes, polyvinylidene fluoride (C-PVDF) and polytetrafluoroethylene (C-PTFE), were applied in dye wastewater treatment by membrane distillation, where the C-PTFE membrane achieved 99.9% dye removal. 3D-OCT images indicated that the foulant attachment on the C-PTFE membrane was much looser than on the C-PVDF membrane, owing to its pattern-shaped surface morphology and higher hydrophobicity as explained through mathematic modeling.

Riedl et al. [77] employed an atomic force microscopy technique to measure membrane surface roughness and scanning electron microscopy to assess the fouling layer. Based on the findings, the membrane surface's smoothness can influence the fouling layer's morphology. It was shown that smooth membranes produced a dense surface fouling layer, whereas this same layer or biofilm on rough membranes was much more open. The primary conclusion of Riedl et al.'s study was that the fluxes through rough membranes are less affected by fouling formation than fluxes through smooth membranes.

The kinetics of adhesion of *Mycobacterium* sp. to cellulose diacetate reverse-osmosis membranes have been described by Ridgway et al. [43]. Adhesion of the cells to the membrane surface occurred within 1–2 h. It exhibited saturation-type kinetics, which conformed closely to the Langmuir adsorption isotherm, a mathematical expression describing the partitioning of substances between a solution and a solid–liquid interface. This suggested that cellulose diacetate membrane surfaces may possess a finite number of available binding sites to which the mycobacteria can adhere. Treatment of the attached mycobacteria with different enzymes suggested that cell surface polypeptides, 4- or α-1.6-linked glucan polymers and carboxyl ester bond-containing substances (possibly peptiglycolipids) may be involved in the adhesion process. However, the exact molecular mechanisms of adhesion have not yet been clearly defined. This is one area where further research is needed.

Altena and Belfort [80] and Drew et al. [81] performed fundamental studies of the membrane fouling process based on the movement of rigid neutrally buoyant spherical particles (i.e., a model bacterial foulant) towards a membrane surface. While these researchers did not work directly with microbial cells, their hydrodynamic studies provide useful information on how the particle size and fluid flow affect microbial adhesion. Their studies were an attempt to give a clearer insight into the hydrodynamics behind the mechanism of microbial adhesion in RO systems. Under typical laminar flow conditions, particles with a radius smaller than 1 μm were captured by a porous membrane surface (i.e., the microbial adhesion step), resulting in cake formation. Due to convective flow into the membrane wall, particles moved laterally towards the membrane. As a result, the particle concentration near the membrane surface increased significantly over that in the bulk solution, resulting in a fouling layer. In their cross-flow membrane filtration experiments, there appeared to be two major causes of lateral migration: a drag force exerted by the fluid on the particle due to the convective flow into the membrane wall (i.e., wall suction effect or permeation drag force) that carried particles towards the membrane, and an inertial lift force which carried particles near the membrane away from the porous wall. For the small particles (<1 μm), the permeation drag forces dominated. An expression was developed from first principles to predict conditions under which a membrane module exposed to dilute suspensions of spherical particles will not foul.

In a recommended paper, Ghayeni et al. [83] studied the passage of bacteria (0.5 μm diameter) through microfiltration membranes in wastewater applications. Membranes with a pore size smaller than 0.2 μm still transmitted secondary effluent cells. This interesting study showed that based on total cell counts (DAPI), up to 1% of the bacteria in the feed could pass to the permeate side. While a significant portion of the cells (e.g., 50%) in the permeate showed biological (CTC) activity, none of the cells were able to be reproduced (i.e., culture on agar or in suspension). This is a good quantitative method for measuring cell injury. We can speculate that smaller cells, or membranes with larger pores, would allow for the passage of viable bacteria which would be able to reproduce. This could occur at some critical cell/pore ratio (Figure 7.6). For further reading on algal and microbial cell attachment, see several good papers by Aydin et al. [55], Ly et al. [56], Du et al. [57], and Asif and Zhang [58] (Table 7.2).

7.3.2 Pore Blockage, Cake Formation, and Analysis of Deposits on the Membrane Surface

Cake formation, shear forces, and the kinetics of the boundary layer are described in studies by Hermia [161], Stumme et al. [73], and Suárez et al. [162]. To further understand the effect of membrane fouling on system capacity, the V_{max} test is often used to accelerate testing. This test assumes that fouling occurs by uniform constriction of the cylindrical membrane

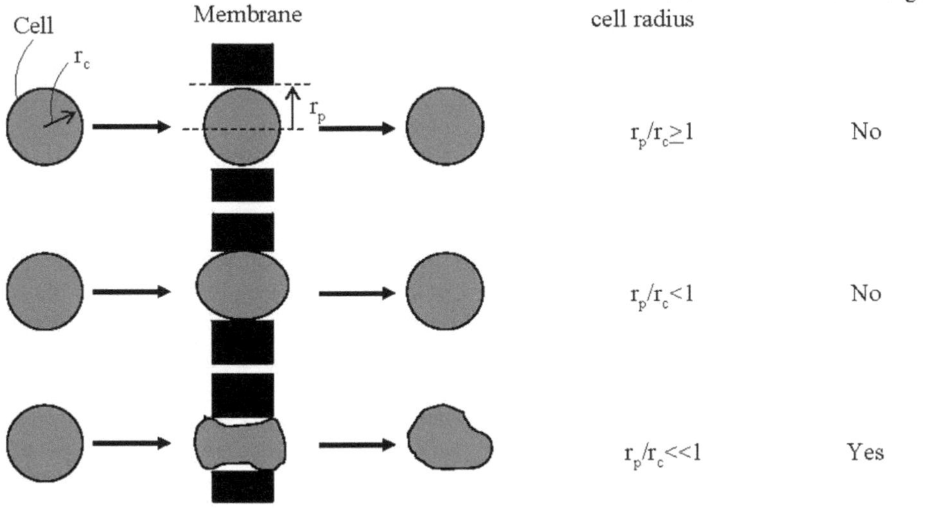

FIGURE 7.6 Passage of bacterial cells through membrane pores. Cell damage occurs at the critical pore radius/cell radius ratio [20].

pores, which does not happen in practice. Zydney and Ho [74] examined the validity of the V_{max} model and compared the results with predictions from a new model that accounts for fouling due to pore blockage and cake formation. It was found that the V_{max} analysis significantly overestimates the system capacity for proteins that foul primarily by pore blockage. However, it underestimates the capacity for compounds that foul primarily by cake formation. In contrast, the pore blockage–cake filtration model provides a much better description of membrane fouling, leading to more accurate sizing and scale-up of normal flow filtration devices. In a related study, Stumme et al. [73] performed a theoretical evaluation of polyelectrolyte layering during layer-by-layer coating of ultrafiltration hollow-fiber membranes. Their approach gave a new insight into layering conformation and helped assess the interaction between the membrane surface and the polyelectrolyte film. Their data confirmed that a rejection for $MgSO_4$ could only be achieved in the regime of layer-dominated film growth. Additionally, when layer-dominated film growth prevailed from the early stages of the coating process, permeability values were higher at similar $MgSO_4$ rejection rates compared to an initial pore-dominated and then layer-dominated film growth. The authors concluded that the interaction between the membrane pore size and molecular weight of the polyelectrolytes in the coating solutions plays an important role during layer-by-layer coating. Furthermore, in another relevant investigation, Suarez et al. [162] presented a mathematical model that aimed to predict distillate fluxes in direct-contact membrane distillation when fouling occurs as salts are deposited onto the membrane surface, forming an inorganic fouling layer. Their mathematical model used the heat and mass-transfer formulation to predict the distillate flux under steady-state conditions and was combined with the cake-filtration theory to represent the distillate fluxes after the onset of membrane fouling. Model results agreed well with the experimental observation of distillate fluxes, both before and after the onset of membrane fouling. Attenuated total reflection (ATR) Fourier transform infrared (FTIR) spectroscopy can provide insight into the chemical nature of deposits on membranes [84]. The spectra of the foulants can be easily distinguished from the spectra of the membrane material. ATR-FTIR can also indicate the presence of inorganic foulants as well as the ratio of inorganic to organic foulants.

The surface deposits on UF polyethersulfone (PES) membranes fouled by skimmed milk have been studied using ATR-FTIR to detect the functional groups of the fouling species [46]. Some milk components (lactose and salts) were eliminated by water rinsing, whereas proteins were only partially removed by chemical cleaning at basic pH. For dynamic conditions, the cleanliness of the membrane was evaluated through two criteria: hydraulic (i.e., recovery of initial flux) and chemical (i.e., no more contaminants detected). The hydraulic cleanliness of the membrane was achieved, whereas the membrane's initial surface state was not restored. ATR-FTIR is also a useful tool for evaluating other fouling species such as oil and humic acids.

Deposits on a membrane surface, before and after cleaning, can also be analyzed using scanning electron microscopy (SEM) in combination with energy dispersive X-ray (EDX) combined with a micro-analysis system permitting quantitative determination of elements [34]. Furthermore, identifying specific species deposited onto membrane surfaces can be carried out using matrix-assisted laser desorption ionization mass spectroscopy (MALDI-MS). Chan et al. [85] employed this technique to differentiate between the desorption of proteins from the membrane surface, inside pores, and the membrane substrate. It has the potential for quantitative measurement of protein fouling on membrane surfaces. It was shown that the technique is a powerful tool for distinguishing between different proteins in fouling deposits.

Atomic force microscopy (AFM) has proved to be a rapid method for assessing membrane–solute interactions (fouling) of membranes under process conditions [(86]. Given the good agreement between the correlations using AFM and operating performance, it should be possible, in the future, to use these techniques to allow prior assessment of the fouling propensity of process streams.

The non-destructive, real-time observation techniques to detect and monitor fouling during liquid separation processes are of great importance in developing strategies to improve operating conditions. As recommended by a paper from Li et al. [47], ultrasonic time-domain reflectometry (UTDR) was used to measure organic fouling, in real-time, during ultrafiltration with polysulfone (PS) membranes. The feed solution was a paper-mill effluent, which contained breakdown products of lignin or lignosulfonate, from a wastewater treatment plant. Experimental results showed that the ultrasonic signal response could be used to monitor fouling-layer formation and growth on the membrane in real time. In addition, the differential signal developed indicated the state and progress of the fouling layer and warned of advanced fouling during operation.

7.3.3 Measurement of Concentration Polarization and Variation in Gel Layer Thickness along the Flow Channel

Concentration polarization results from the emergence of gradients at a membrane–solution interface due to selective transfer through the membrane. According to Field and Wu [163], it is distinguishable from fouling in at least two ways: the state of the molecules involved (in solution for concentration polarization, although no longer in solution for fouling); and by the timescale, normally less than a minute for concentration polarization, although generally at least two or more orders of magnitude more for fouling. Thus, the phenomenon of flux decline occurring over a timescale of tens of minutes should not be attributed to concentration polarization establishing itself. This distinction and several questions surrounding modeling were addressed and clarified by Field and Wu [163].

The authors reported on two approaches for modeling flux, one used the overall driving force (in which case allowance for osmotic effects was expressed as additional resistances), and the other used the net driving force across the separating layer or fouled separating layer, although often the two are unfortunately comingled. In the discussion of flux decline, models' robust approaches for the determination of flux–time relationships, including the integral method of fouling analysis, were described, and various concepts clarified. The researchers emphasized that for design purposes, pilot plant data are vital.

May et al. [28] investigated the bottleneck challenge of membrane fouling by establishing a scalable concentration polarization-enabled and surface-selective hydrogel coating using zwitterionic cross-linkable macromolecules as building blocks. Their results disclosed that zwitterionic hydrogel-coated membranes exhibited lower surface charge and higher flux during protein filtration compared to original unspoiled membranes. However, the salt rejection was found to remain unchanged. The results further revealed that the hydrogel coating thickness and consequently the reduction in membrane permeance due to the coating could be adjusted by variation of the filtration time and polymer feed concentration, illustrating the novel modification method's promising potential for scale-up to real applications.

Gowman and Ethier [164,165] established an automated laser-based refractometric technique to measure the solute concentration gradient during dead-end filtration of a biopolymer solution. These are good papers that attempt to reconcile theory with experimental data. The refractometric technique may be useful to other researchers working on quantifying membrane fouling. Similarly, Pope et al. [36] employed a nuclear magnetic resonance technique to quantitatively measure the concentration polarization layer thickness during crossflow filtration of an oil–water emulsion. This method will help to clarify the relative quantitative contributions to flux decline of the adsorbed layer resistance and the concentration polarization layer gradient and thickness. In addition, it can help to explain the flux declines due to different resistances, as shown in Figure 7.3. The technique, which measured layer thickness using chemical shift selective micro-imaging, may be useful in studying other membrane fouling situations that occur in food processing and desalination.

In the case of cross-flow filtration, one can expect that the gel-layer thickness and/or the surface concentration of the solute will vary with distance from the channel entrance. Consequently, the local permeate flux will also vary with longitudinal position. In a highly recommended article, Denisov [72] presented a mathematically rigorous theory of concentration polarization in crossflow ultrafiltration, which takes into account the non-uniformity of the local permeate membrane flux. He derived equations describing the pressure–flux curve.

In the case of the gel-layer model, the theory led to a simple analytical formula for a limiting or critical flux, J_{lim}. The flux turned out to be proportional to the cube root of the ratio of the gel concentration to the feed solution concentration, rather than to the logarithm of this ratio, as the simplified Michaels–Blatt theory predicted:

$$J_{\lim} = (3/2)^{(2/3)} KP_g = 1.31\left(\frac{C_g}{C_o}\right)\frac{m^{1/3}D^{2/3}U^{1/3}}{L^{1/3}h^{1/3}} \tag{7.2}$$

where

$$P_g = \left(\frac{C_g mD^2U_o}{C_oK^3Lh}\right) \tag{7.3}$$

where K is the hydraulic permeability of the membrane to pure solvent (m^3/Ns), C_g is the gel concentration (kmol/m^3), C_o is the solute concentration in feed solution (kmol/m^3), m is the channel parameter, D is the solute diffusion coefficient (m^2/s), U_o is the longitude component of fluid velocity averaged over the channel cross-section (m/s), L is the channel length (m), and h is the transversal dimension of the channel (m).

In the case of the osmotic-pressure model, the rigorous theory allowed the conclusion that at high applied transmembrane pressure, the permeate flux increased as a cube root of the pressure, thus that the limiting flux was never reached:

$$\bar{J} \approx (3/2)^{2/3} K\bar{P}^{1/3}P_o^{2/3} \approx 1.31\left(\frac{\bar{P}}{RTC_o}\right)^{1/3}\frac{m^{1/3}D^{2/3}U_o^{1/3}}{L^{1/3}h^{1/3}} \tag{7.4}$$

where

$$P_o = \left(\frac{mD^2U_o}{RTC_oK^3Lh}\right) \tag{7.5}$$

where J is the average flux over the channel (m/s), P is the transmembrane pressure (N/m^2), R is the gas constant (J/kmolK), and T is the temperature (K). However, one minor weakness of this study was that the analysis ignored the viscosity concentration dependence and the solute's partial transmission through the membrane.

7.3.4 Mathematical Models for Flux Decline

Suarez et al. [162] reported on a mathematical model that aimed to predict distillate fluxes in direct-contact membrane distillation when fouling occurs as salts are deposited onto the membrane surface, forming an inorganic fouling layer. Their mathematical model used the heat and mass-transfer formulation to predict the distillate flux under steady-state conditions and was combined with the cake-filtration theory to represent the distillate fluxes after the onset of membrane fouling. The model results agreed well with the experimental observation of distillate fluxes, both before and after the onset of membrane fouling, and suggested that the cake-filtration theory can be used to represent water flux decline in distillation membranes prone to inorganic fouling. From their experiments and modeling, they found that the onset of membrane failure was relatively constant; the precipitation reaction constant was conditioned by the physicochemical interaction between the feed solution and the membrane, and the rate of flux decline after membrane fouling depended on flow conditions as well as on the precipitation compound. However, the authors argued that the model had limitations that must be addressed in future investigations to validate it under a wider range of operating conditions, for membranes composed of other materials and with different feed solutions. In addition, organic, biological, and/or colloidal fouling, which typically occur under real conditions, need to be addressed. AlSawaftah et al. [90], in an excellent comprehensive review on membrane fouling, focused on mathematical modeling, prediction, diagnosis, and mitigation. The authors noted that mathematical fouling prediction models are valuable because they facilitate the optimization of fouling removal and prevention methods and help establish interactions and relationships between different filtration variables. Most equations in this paper were aimed at relating the time-dependent decline of the permeate flux with the water permeability coefficient.

In any membrane filtration, the prediction of permeate flux is critical to calculating the membrane surface required, which is an essential parameter for scaling-up, equipment sizing, and cost determination. Modeling approaches were reviewed by Quezada et al. [91] based on phenomenological or theoretical derivation (such as gel-polarization, osmotic pressure, resistance-in-series, and fouling models) and non-phenomenological models. The aims were to better describe the limiting phenomena as well as to predict the permeate flux. The results obtained by the authors showed that phenomenological models presented a high variability of prediction among the investigated juices. The non-phenomenological models demonstrated a great capacity to predict permeate flux with R-squares higher than 97%. The findings suggested that non-phenomenological models are a useful tool from a practical point of view for predicting the permeate flux, under defined operating conditions, in membrane separation processes. However, the phenomenological models are still a proper tool for scaling up and understanding the UF process.

Dal-Cin et al. [89] developed a series resistance model to quantify the relative contributions of adsorption, pore plugging, and concentration polarization to flux decline during UF of a pulp mill effluent. They proposed a relative flux loss ratio as an alternative measure to the conventional resistance model, which was found to be a misleading indicator of flux loss. Using experimental and simulated flux data, the series resistance model was shown to underpredict fouling due to adsorption and overpredict concentration polarization. This appears to be a disadvantage and would make the model of limited use in its current form. As mentioned in the introduction, Koltuniewicz and Noworyta [30] modeled the flux decline because of the growth of a concentration polarization layer based on the surface renewal theory developed by Danckwerts [166]. The surface renewal model is more realistic than the commonly used film model since mass transfer at the membrane boundary layer is random in nature due to membrane roughness. Specifically, the membrane is not covered by a uniform concentration polarization layer, as was assumed in the film model, but rather by a mosaic of small surface elements with different ages and, therefore, different permeate flow resistance. Any element can be swept away randomly by a hydrodynamic impulse. Then a new element starts building up a layer of retained solute at the same place on the membrane surface. They showed that the decrease in flux with respect to time, $J(t_p)$, due to the development of the concentration polarization layer, is given by the following equation, which also considers the rate of membrane surface renewal, s (area/unit time):

$$\bar{J}\left(t_p\right) = \left(J_o - J^*\right)\frac{s}{s+A}\,\frac{1-e^{-(s+A)t_p}}{1-e^{-st_p}} + J^* \tag{7.6}$$

where A is the rate of loss of membrane surface area as a function of time, J_o is the initial value of the flux, J^* is the flux observed after infinite time, and t_p is the time of permeation.

$$s = A\frac{J_{\text{lim}} - J^*}{J_o - J_{\text{lim}}} \tag{7.7}$$

where J_{lim} is the limiting flux, which is like critical flux, J_{crit}. The former can be obtained from literature data. The average flux under steady-state conditions, J_a, can be calculated directly from Equation (7.2) as a limit, giving Equation (7.8):

$$\bar{J}_a = \lim_{t_p \to \infty} \bar{J}\left(t_p\right) = \left(J_o - J^*\right)\frac{s}{A+s} + J^* \tag{7.8}$$

In support of this model, calculated flux values using Equations (7.6) and (7.7) agreed well with experimental data. The two equations describe a permeation cycle of duration, t_p, as shown in Figure 7.3. This is a highly recommended paper for those operating large-scale continuous ultrafiltration plants and, to a certain extent, RO plants. The model developed describes not only the dynamic behavior of a plant but also allows for optimization of operating conditions (i.e., permeation time, cleaning time, cleaning strategy). For further reading about the mathematical modeling of flux decline, please see studies by AlSawaftah et al. [90] and Quezada et al. [91] (Table 7.2).

7.4 FOULING MITIGATION AND PRETREATMENT

7.4.1 Feed Water Pretreatment Using Filtration, Flocculation, and Ozone Oxidation

In a paper by Vatanpour et al. [108], the effect of UV/ H_2O_2 advanced oxidation as a pretreatment process on the performance of a TiO_2/ PVDF nanocomposite membrane was investigated in the degradation and mineralization of Reactive Green 19 dye solution as an organic contaminant model. In the first step, the UV/H_2O_2 process was optimized by response surface methodology. The results showed that the number of UV lamps, the H_2O_2 concentration, and the temperature were the important factors for fouling reduction of PVDF/ TiO_2 nanocomposite ultrafiltration membranes. The PVDF membrane was fabricated by a nonsolvent-induced separation process and modified by embedding TiO_2 nanoparticles. The morphology of the modified membrane was assessed by SEM images and EDX mapping analysis. Moreover, the performance of bare and modified membranes was evaluated by measuring pure water flux, dye rejection, and flux decline. The results disclosed that by TiO_2 introduction, membrane performance significantly improved. Finally, the effect of UV/H_2O_2 pretreatment on the performance of membranes was evaluated. The membrane antifouling feature after the pretreatment of the dye solution considerably improved. Additionally, the COD removal efficiency of the UV/H_2O_2-membrane process was measured in batch and continuous regimes. The best performance was obtained for the continuous (UV/H_2O_2) + (PVDF+TiO_2) process, with a 64% removal yield. In a related investigation, Gong et al. [107] employed electrocoagulation pretreatment of pulp and paper wastewater for low-pressure reverse osmosis membrane fouling control. Similarly, Yu et al. [167] reported that a combined process of coagulation–ultrafiltration as a pretreatment of nanofiltration or reverse osmosis should be an ideal way to prevent membrane fouling caused by residual aluminum.

Huang et al. [27] established that coagulation is the best pretreatment for fouling reduction by removing around 30% of dissolved organic carbon (DOC), which is an important membrane fouling factor. A comprehensive assessment amongst pretreatment methods is shown in Table 7.1. In addition, Huang et al. [27] argued that even though periodic backwashing has been considered as a solution for fouling alleviation, frequent backwashing increases maintenance and discharging costs, leading to low recovery and high energy consumption. A crucial observation was the deposition of a cake layer from the coagulated flocs on the membrane surface. This acted as protection for the membrane; instead of removing the flocs, the fouling formed by the accumulated cake layer was reversible since it could easily be removed. Furthermore, the combination of coagulation and microfiltration (MF) can remove pathogenic microorganisms, including viruses, from the raw water [31]. For further related reading on backpulsing for fouling mitigation, please see Gao et al. [133,134] (Table 7.2).

Reverse osmosis seawater systems that operate on surface feed water normally require an extensive pretreatment process to control membrane fouling. Over the past two decades, effective water microfiltration technologies have been introduced commercially. Wilf and Klinko [92] and Glueckstern et al. [136] noted that these developments could improve the quality of surface seawater feed to a level comparable to or better than the water quality from well water sources. The utilization of capillary ultrafiltration as a pretreatment step enabled the operation of the reverse osmosis system at a high recovery (15%) and permeate flux rate. In a similar study utilizing micro- and ultrafiltration as seawater pretreatment steps for reverse osmosis, Glueckstern and Priel [93] showed that such technology can dramatically improve the feed water quality. This is especially important if cooling water from existing power stations is used as feed water for desalination plants.

Reusing municipal wastewater requires treatment to an acceptable quality level that satisfies regulatory guidelines. Ghayani et al. [50] employed hollow-fiber microfiltration (MF) as a pretreatment for wastewater for RO in producing high-quality water. Organisms present in MF-treated secondary effluent were able to attach to RO membranes and proliferate to form a biofilm. Total cell counts in this treated effluent (i.e., permeate from the MF unit) were several orders of magnitude higher than viable cell counts. This was confirmed in a later study Ghayani et al. [83,140]. These results indicate that microfiltration membranes will not be totally effective in removing bacteria from the feed water stream. In addition, the result showed that most cells were severely damaged by passage through the membrane (Figure 7.6). However, we can speculate that this damaging effect may be cell strain-specific and/or dependent on the cell/pore diameter.

In a study by Chapman et al. [95], a flocculator was used to remove suspended solids, organics, and phosphorus from wastewater. The flocculator produced uniform microflocs, which were removed by cross-flow microfiltration. Flocculated particles can form a highly porous filtration cake on a membrane surface. This will help inhibit fouling on the membrane by preventing the deposition of particles and reducing the number of membrane cleaning cycles [96].

Another commonly used method is coagulation which was reported by Gong et al. [107] and Zhao et al. [5] (Table 7.2). This technique removes turbidity from water by the addition of cationic compounds. The usefulness of coagulation as a pretreatment to remove microparticles in aqueous suspension before a membrane filtration was shown by Choksuchart et al. [97]. There are several types of coagulation systems. Comparisons were made by Park et al. [98] between coagulation with only rapid mixing in a separate tank (i.e., ordinary coagulation) and coagulation with no mixing tank (i.e., in-line coagulation) prior to an ultrafiltration process. The former was superior. The in-line coagulation (without settling) UF process was also employed by Guigui et al. [99]. Floc cake resistance was found to be lower than resistance due to the unsettled floc and the uncoagulated organics.

A reduction in coagulant dose induced an increase in the mass transfer resistance.

Combining flocculation and coagulation in a pretreatment process has also been studied. In a key paper by Lopez-Ramirez et al. [100] the secondary effluent from an activated sludge unit was pretreated, prior to RO, with three levels: intense (coagulation-flocculation with ferric chloride and polyelectrolite and high pH sedimentation), moderate (coagulation-flocculation with ferric chloride and polyelectrolite and sedimentation), and minimum (only sedimentation). The optimum for membrane protection, in terms of calcium, conductivity, and bicarbonates reduction was the intense treatment. Membrane performance varied with pretreatment but not reclaimed water quality. The study recommended intense pretreatment to protect the membrane.

A modular pilot-size plant involving coagulation/flocculation, centrifugation, ultrafiltration, and sorption processes was designed and constructed by Benito et al. [35] for the treatment of oily wastewaters. Empirical equations developed by Shaalan [101] predicted the impact of water contaminants on flux decline and will aid decision-makers in choosing a suitable water pretreatment scheme and in the selection of the most appropriate cleaning cycle.

Van Geluwe et al. [103] reported on the use of ozone, O_3, for reducing membrane fouling by natural organic matter (NOM). Membrane fouling by natural organic matter is one of the main problems that slow the application of membrane technology in water treatment. In their review, Van Geluwe et al. [103] explained that O_3 could efficiently change the physicochemical characteristics of natural organic matter in order to reduce membrane fouling. The electrophilic character of O_3 accounts for the fast reaction of O_3 molecules with unsaturated bonds. The direct reaction of unsaturated bonds in NOM with O_3 can lead to the consumption of O_3 or the production of an ozonide ion radical, which decomposes upon protonation into a hydroxyl (OH) radical. The OH radical is a strong oxidizing agent that can react with NOM molecules. This is referred to as the indirect oxidation pathway. Although the OH radical is considered an unselective oxidant, it can be regarded as an electrophilic oxidant. Van Geluwe et al. [103] recounted that several researchers proved that the application of O_3 oxidation of the feed water prior to membrane filtration resulted in a significant decrease in membrane fouling. However, only minor dissolved organic carbon (DOC) removal (10–20%) could be achieved. This is explained by the fact that O_3 causes substantial structural changes to the NOM present in the feed water.

Significant developments in organic removal have also been achieved using magnetic ion exchange (MIEX). Zhang et al. [168] showed that magnetic ion-exchange resin could remove a majority of hydrophilic compounds and a significant amount of hydrophobic compounds from biologically treated secondary effluent within a short contact time of 20 min. In addition, it removed a majority of lower-molecular-weight organic compounds from the wastewater. The resin could easily be regenerated, and even after several regenerations it gave almost the same organic removal efficiency. The process, when used as pretreatment to a submerged membrane reactor, resulted in very high organic removal while significantly reducing the membrane fouling. The membrane fouling was further reduced by the addition of a small dose of powdered activated carbon in the submerged membrane reactor.

7.4.2 Control of Operating Parameters, Spacers, and Critical Flux

A promising solution for membrane fouling reduction in membrane bioreactors is the adjustment of operating parameters, such as hydraulic retention time, food/microorganisms loading, and dissolved oxygen concentration, with the aim to modify, for example, the sludge morphology to allow for improvement in membrane filtration. In a study by Banti et al. [124], these parameters were investigated in a step-aerating pilot membrane bioreactor that treated municipal wastewater to control the filamentous population. When food/microorganism loading in the first aeration tank was $\leq$0.65 g COD/g MLSS/d at 20^{0}C, DO = 2.5 mg/L, and HRT = 1.6 h, the filamentous bacteria were controlled effectively at a moderate filament index of 1.5–3. The moderate population of filamentous bacteria improved the membrane performance, leading to low transmembrane pressure at values $\leq$ 2 kPa for a long period. In contrast, with the control membrane bioreactor, the transmembrane pressure gradually increased, reaching 14 kPa.

Muro et al. [104] showed the beneficial effects of hybrid processes in wastewater treatment in the food industry. These hybrid systems included traditional techniques such as centrifugation, cartridge filtration, disinfection, and different membrane techniques resulting in a cascade design which could be applied in various applications. They argued that the risk of membrane damage due to contact with particles, salt conglomerates, chemicals, or other substances must be minimized to prevent short membrane life. Operation parameters must also be carefully selected to obtain good results, especially not to overpass the maximum temperature and transmembrane pressures recommended by membrane manufacturers. From the point of view of each process, it is necessary to work at permeate flow rates below critical flux to improve processes in wastewater treatment. This will ensure longer operating runs before membrane cleaning is required due to fouling.

A comprehensive difference model was developed by Madireddi et al. [121] to predict membrane fouling in commercial spiral wound membranes with various spacers. This is a useful paper for experimental studies on the effect of flow channel thickness on flux and fouling. Avlonitis et al. [123] presented an analytical solution for the performance of spiral-wound modules with seawater as the feed. A key finding showed that it was necessary to incorporate the concentration and pressure of the feed into the correlation for the mass transfer coefficient. In a similar study, Boudinar et al. [169]

developed the following relationship for calculating mass transfer coefficients in channels equipped with a spacer:

$$k = 0.753\left(\frac{K}{2-K}\right)^{1/2}\frac{D_s}{h_B}Sc^{-1/6}\left(\frac{Peh_B}{M}\right) \tag{7.9}$$

where Pe is Peclet number, $K = 0.5$, and $M = 0.6$ (cm).

Controlled centrifugal instabilities (called Dean vortices), resulting from flow around a curved channel, were used by Mallubhotla and Belfort [122] to reduce concentration polarization and the tendency toward membrane fouling. These vortices enhanced back-migration through convective flow away from the membrane–solution interface and allowed for increased membrane permeation rates.

Goosen et al. [19,76] demonstrated that the polymer membrane could be very sensitive to changes in the feed temperature. There was up to a 100% difference in the permeate flux between feed temperatures of 30°C and 40°C. Another study showed that the improved flux was due primarily, though not completely, to viscosity effects on the water [18]. Reversible physical changes in the membrane may also have occurred.

The transition from concentration polarization to fouling is a key phase in membrane separation processes. This occurs at a critical flux. Song [120] indicated that in most theories developed, the limiting or critical flux is based on semi-empirical knowledge rather than being predicted from fundamental principles. To overcome this shortcoming, he developed a mechanistic model, based on first principles, for predicting the limiting flux. Similar to the critical flux results of Chen et al. [70] and the limiting flux of Koltuniewicz and Noworyta [30], Song [120] showed that there is a critical pressure for a given suspension. When the applied pressure is below the critical pressure, only a concentration polarization layer exists over the membrane surface. A fouling layer, however, will form between the polarization and the membrane surface when the applied pressure exceeds the critical pressure. The limiting or critical flux values predicted by the mechanistic model compared well with the integral model for a low concentration feed. Operators of RO/UF plants/units should therefore operate their systems just below the critical flux in order to maximize productivity while minimizing membrane fouling. How will the operator know about critical flux? Critical flux is never mentioned in any membrane manufacturers' specification sheets or design models. It depends on the feed water quality and operating conditions (i.e., flux and recovery). The concept and application of critical flux is one area where further education is needed for both operators and manufacturers. For further reading, please see papers by Banti et al. [124], De Vela [125], Di Bella et al. [126], Du et al. [57], Szabo-Corbacho et al. [127], Yuliwati et al. [128], and Jepsen et al. [5] (Table 7.2).

The influences of spacer thickness in spiral-wound membrane units on permeate flow and its salinity were studied by Sablani et al. [25]. Membrane parameters were also estimated using an analytical osmotic pressure model for high-salinity applications. The effects of spacer thickness on permeate flux showed that the observed flux decreases by up to 50% in going from a spacer thickness of 0.1168 to 0.0508 cm. The authors commented that the different geometry/configurations of the spacer influenced turbulence at the membrane surface and that, in turn, affected concentration polarization. This suggested less turbulence with the smaller spacer thickness and is opposite to what is normally expected. A membrane module with an intermediate spacer thickness of 0.0711 cm was the best economically since it gave the highest water production rate (L/h).

Geraldes et al. [110] assessed the effect of a ladder-type spacer configuration in NF spiral-wound modules on concentration boundary layer disruption. The results showed that the average concentration polarization for the membrane wall was independent of the distance to the channel inlet. In contrast, for the membrane wall without adjacent filaments, the average concentration polarization increased with the channel length. This was since, in the first case, the transverse filaments periodically disrupted the concentration boundary layer. In contrast, in the second case, the concentration boundary layer grew continuously along the channel length. The experimental results of the apparent rejection coefficients were compared to model predictions, the agreement being good. Their results clearly established how crucial the spacer configuration is in optimizing the spiral-wound module efficiency.

The unexpected results of Sablani et al. [25] (i.e., less turbulence with smaller spacer thickness) may be best explained by an excellent paper by Schwinge et al. [109]. The latter employed computational fluid dynamics (CFD) in a study of unsteady flow in narrow spacer-filled channels for spiral-wound membrane modules. The flow patterns were visualized for different filament configurations incorporating variations in mesh length, filament diameter, and for channel Reynolds numbers, Re_{ch}, up to 1000. The simulated flow patterns revealed the dependence of the formation of recirculation regions on the filament configuration, mesh length, filament diameter, and the Reynolds number. When the channel Reynolds number was increased above 300, the flow became super-critical, showing time-dependent movements for a filament located in the center of a narrow channel; when the channel Reynolds number was increased above 500, the flow became super-critical for a filament adjacent to the membrane wall. For multiple filament configurations, flow transition can occur at channel Reynolds numbers as low as 80 for the submerged spacer at a very small mesh length [mesh length/channel height (l_m/h_{ch}) = 1] and at a slightly larger Reynolds number at a larger mesh length (l_m/h_{ch} = 4). The transition occurred above Re_{ch} of 300 for a cavity spacer and above Re_{ch} of 400 for a zigzag spacer. We can speculate that the conclusion of Sablani et al. [25], less turbulence with smaller spacer thickness, was due to fewer recirculating regions because of smaller mesh length and filament diameter.

CFD simulations were used by Li et al. [111] to determine mass transfer coefficients and power consumption in channels filled with non-woven net spacers. The geometric parameters of a non-woven spacer were found to greatly influence the performance of a spacer in terms of mass transfer enhancement and power consumption. The results from the CFD simulations indicated that an optimal spacer geometry exists. For further reading on mathematical modeling studies related to the prediction, diagnosis, and mitigation of membrane fouling, please see reports by AlSawaftah et al. [90] and Quezada et al. [91] (Table 7.2).

7.4.3 Membrane Surface Modification and Fouling Resistance of Hydrophilic and Hydrophobic Membranes

Nanomaterial aggregation within the polyamide layer of a thin-film nanocomposite membrane can negatively affect membrane filtration performance. Khoo et al. [41] reported on an approach to develop thin-film nanocomposite membranes with improved desalination and antifouling properties by employing post-treatment of the membrane surface to solve the aggregation problem. In the study by Khoo et al. [41], an eco-friendly technique based on plasma-enhanced chemical vapor deposition was employed to deposit hydrophilic acrylic acid onto the polyamide surface of the membrane with the aims of simultaneously minimizing the polyamide surface defects caused by nanomaterial incorporation and improving the membrane surface hydrophilicity for reverse osmosis application. The results showed that the sodium chloride rejection of the modified plasma membrane was improved, with salt passage being reduced without significantly altering the pure water flux. In addition, the modified membrane also exhibited a remarkable antifouling property with a higher flux recovery rate than the unmodified membrane. This was attributed to enhanced membrane hydrophilicity and a smoother surface. Furthermore, the modified membrane also showed higher performance stability throughout a 12-h filtration period. The deposition of hydrophilic material on the membrane surface has the potential for developing a defect-free thin-film nanocomposite membrane with enhanced fouling resistance for improved desalination. In a related investigation, Shen et al. (118) reported on polymeric membranes incorporated with ZnO nanoparticles for membrane fouling mitigation. Likewise, Usman et al. [115] reviewed recent innovations in superhydrophobic methods for the modification of ceramic membranes for oil–water recovery. Different types of hydrophobic ceramic membrane modification using chemical agents and consequential effects on oil–water separation were reviewed in detail. The technical challenges and issues associated with applications of superhydrophobic–superoleophilic ceramic membranes for oil–water separation were discussed as well as future directions in the research on cost-efficient approaches.

Improving membrane chlorine resistances and anti-biofouling properties are the main challenges for the widespread applications of aromatic polyamide reverse osmosis membranes. To enhance membrane chlorine resistance and anti-biofouling property, a commercial aromatic polyamide (RO) membrane was modified by Zhang et al. [170] by free-radical graft polymerization of 3-allyl-5,5-dimethylhydantoin (ADMH) and further cross-linked with N,N′-methylenebis(acrylamide) (MBA). Graft polymerization increased the nitrogen atom content and surface hydrophilicity of the raw membrane. As a result, the salt rejections of the modified membranes were higher than those of the raw membranes, but the water fluxes decreased. Besides that, the repeatable chlorine resistances and anti-biofouling properties of the membranes modified by ADMH and MBA were also significantly enhanced.

In a related study, Belfer et al. [48] described a simple method for surface modification of commercial composite polyamide reverse-osmosis membranes. The procedure involved radial grafting with a redox system consisting of potassium persulfate/sodium methabisulfite. ATR-FTIR provided valuable information about the degree of grafting and the microstructure of the grafted chain on the membrane surface. Both acrylic and sulfo-acidic monomers and neutral monomers such as polyethylene glycol methacrylate were used to demonstrate the wide possibilities of the method in terms of grafting of different monomers and initiators. It was shown that some of the modified membranes conserved their previous operating characteristics, flux, or rejection, but exhibited higher resistance to humic acid. Additional work needs to be done to find out what happens to the fouling resistance of such membranes over the long term (i.e., after initial biofilm formation).

A fouling-resistant reverse-osmosis membrane that reduces microbial adhesion was reported by Jenkins and Tanner [116]. In this interesting study that confirmed the results of Flemming and Schaule [44], they compared two types of thin-film composite membranes with different chemistries. One type was classified as a polyamide; the other utilized new chemistry that formed a polyamide–urea barrier (i.e., surface) layer. The latter composite membrane proved superior in reverse-osmosis operation like that of the polyether–urea membrane of Flemming and Schaule [44], including rejection of certain dissolved species and fouling resistance. These results suggest that the presence of urea groups in the membrane reduces microbial adhesion, perhaps through charge repulsion. In addition, the results from work done by Ridgeway et al. [43] on the kinetics of adhesion of *Mycobacterium* sp. to cellulose diacetate reverse-osmosis membranes have similar implications. Scientists should therefore be able to minimize microbial adhesion by controlling the surface chemistry of polymer membranes through, for example, the inclusion of urea groups.

Chemical modification of a membrane surface can be used in combination with spacers and periodic applications of bioacids [171]. The paper by Redondo [171], however, is short on specifics (e.g., details of chemical modification of aromatic polyamides membrane surface) and therefore not very useful to those looking for insights into membrane fouling

Cherkasov et al. [82] presented an analysis of membrane selectivity from the standpoint of concentration polarization and adsorption phenomena. The results of their study showed that hydrophobic membranes attracted a thicker irreversible adsorption layer than hydrophilic membranes. The layer thickness was determined by the intensity of concentration polarization (Figure 7.4). This may be due to the stronger attraction of water to hydrophilic membranes. Kabsch-Korbutowicz et al. [40] also demonstrated that the most hydrophilic of the membranes tested (i.e., regenerated cellulose) had the lowest proneness to fouling by organic colloids (i.e., humic acids). These conclusions were further supported by the thorough work of Tu et al. [61]. They showed that membranes with a higher negative surface charge and greater hydrophilicity were less prone to fouling due to fewer interactions between the chemical groups in the organic solute and the polar groups on the membrane surface.

A review by Eray et al. [114] focused on developing silicon carbide membranes from an industrial standpoint. In addition to describing the significant steps in the fabrication of silicon carbide membranes, the advantages, disadvantages, and key challenges related to these approaches were emphasized. The wide range of silicon carbide membrane applications in water and wastewater treatment and other applications were reviewed. Cleaning methods for silicon carbide membranes are also described to address fouling issues during filtration processes along with the commercialization of these membranes. Their review paper aimed to provide a roadmap for potential applications and further development of silicon carbide membranes in liquid filtration. In an excellent related study, Wang et al. [119] gave microscopic new insights into the effects of surface chemistry and roughness on membrane fouling. The authors reported a molecular simulation analysis for fouling on alumina and graphene membrane surfaces during water treatment. For two foulants (sucralose and bisphenol A), the fouling on alumina surfaces was reduced with increasing surface roughness; however, the fouling on graphene surfaces was enhanced by roughness. This was unexpected. The authors argued that the foulant–surface interaction became weaker in the ridge region of a rough alumina surface, thus allowing foulant to leave the surface and reducing fouling. Such behavior is not observed on a rough graphene surface because of the strong foulant–graphene interaction. Moreover, with increasing roughness, the hydrogen bonds formed between water and alumina surfaces were found to increase in number as well as stability. This exciting simulation study revealed that surface chemistry and roughness play a crucial role in membrane fouling and that microscopic insights are useful for designing new membranes for high-performance water treatment.

7.4.4 Membrane Cleaning Using Chemical Agents and Backpulsing

Membrane cleaning is crucial to maintain the permeability and selectivity of membranes. Physical cleaning can mitigate membrane fouling, lower the frequency of chemical cleaning, thus prolonging membrane lifetime, and reduce operational costs. Backpulsing, for example, is a promising physical cleaning method which can effectively mitigate external and non-adhesive fouling and has been used in many industrial fields [133]. Backpulsing is effective in removing hydraulically reversible fouling and reducing concentration polarization. This includes external fouling and nonadhesive fouling. However, a comprehensive understanding of backpulsing and the optimization of this technology is still lacking. Gao et al. [133] critically reviewed the development of backpulsing techniques in microfiltration and ultrafiltration processes. Results of the pilot- and commercial-scale operations were presented. Factors influencing backpulsing efficiency included feed and membrane properties and operating parameters. In a related investigation, Gao et al. [134] reported on a multivariate study of backpulsing for membrane fouling mitigation in produced water treatment. The results showed that backpulsing was efficient in mitigating membrane fouling. However, the cleaning efficiency varied between different backpulsing conditions. The effect of backpulsing parameters (amplitude, duration, and frequency) and their interactions on membrane performance were studied by a 23 full factorial design. The authors reported that amplitude was the most crucial variable for fouling removal and final specific flux, while the frequency was the most significant for membrane net yield.

Al-Obeidani et al. [131] described the development of more effective and economical procedures for cleaning polyethylene hollow-fiber microfiltration membranes that had been used for removing oil from contaminated seawater. In their study, alkaline cleaning showed higher recovery of operating cycle time but lower permeate flux recovery than acid cleaning. The combination of alkaline and acid cleaning agents gave the best operating cycle time and flux recoveries (e.g., 96% and 94%, respectively) (Figure 7.7). As the cleaning agent soaking time was reduced, the actual operating cycle time was reduced. However, the ratio of operating time/chemical cleaning time increased as the soaking time was reduced. Therefore, the soaking time was recommended to be as short as possible (8–10 hours) in the design of small-capacity plants and 30 hours or higher in the case of large-capacity plants. SEM analysis showed that in the case of alkaline cleaning, most of the pores remained covered with a fouling layer, resulting in low flux recovery. The SEM results of acid-cleaned membranes showed more complete removal of the fouling layer from the pores, resulting in better flux recovery. Surface analysis of membranes cleaned with combined acid/base agents showed the best results. A membrane surface like the original one was obtained.

Zulkefli et al. [4], in assessing mitigation strategies on membrane fouling for oily wastewater treatment, recommended the application of a pretreatment system before the filtration process. To save space and expense, membrane-based pretreatments such as microfiltration and ultrafiltration were preferable for high removal rates of contaminants by significantly reducing suspended particles and microorganisms from contaminated water with low energy consumption. In addition, optimizing the operating parameters such as backpulsing time, temperature, and transmembrane pressure was also effective in preventing fouling. Jepsen et al. [5], in an

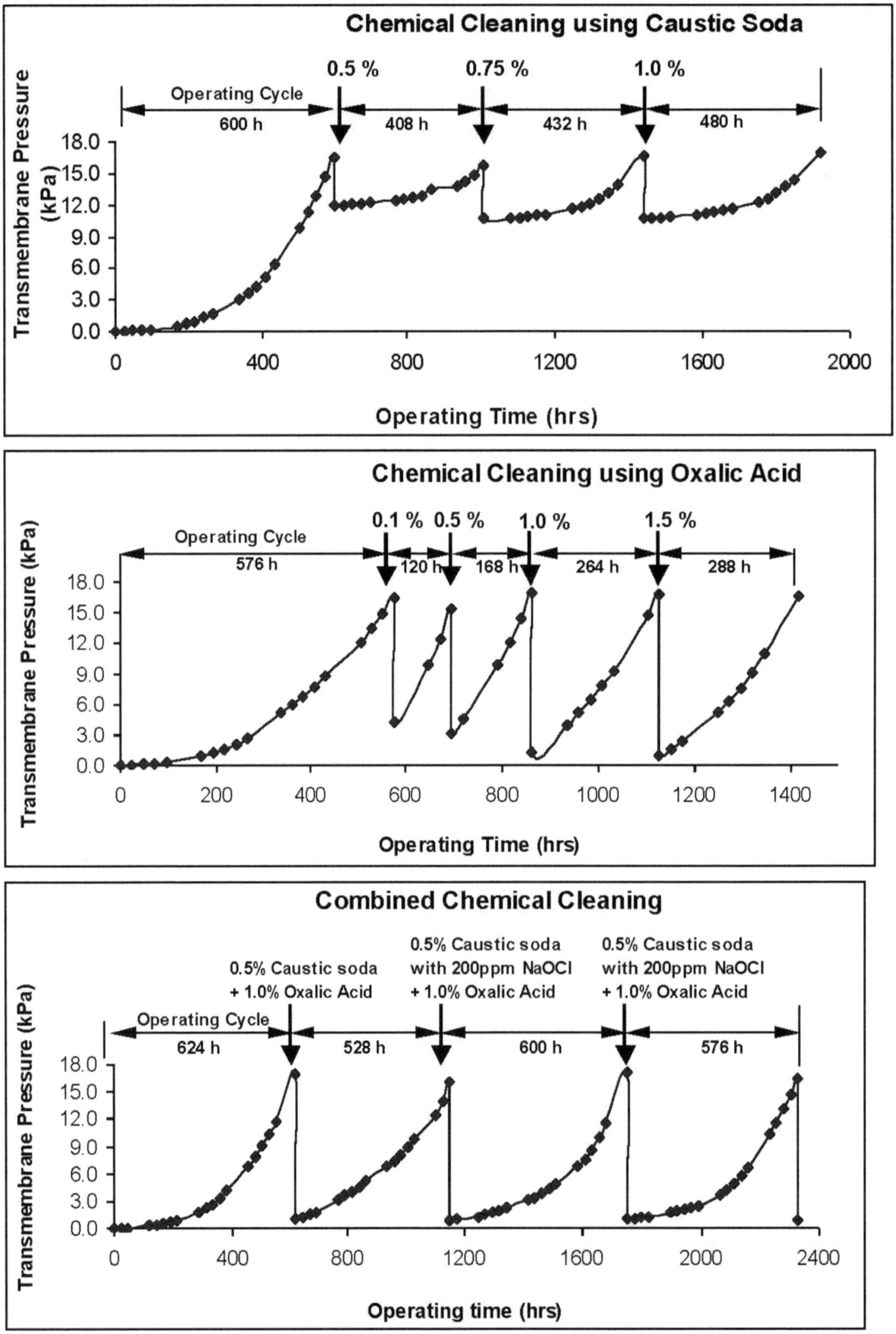

FIGURE 7.7 Top: Alkaline cleaning agent effects. Flux recoveries using three concentrations of caustic soda, 0.5%, 0.75%, and 1.0% by volume. Middle: Acid cleaning agent effects. Bottom: Effect of combination of both cleaning agents on flux recovery and operating time [131].

excellent review of membrane fouling for produced water (i.e., oily salt water) treatment from a process control perspective, argued that with the benefits of membrane filtration, it was predicted that membrane technology would be incorporated in produced water treatment, if zero-discharge policies are enforced globally.

Membranes used in the food industry for ultrafiltration of milk or whey are cleaned regularly with water and various aqueous solutions to ensure hygienic operation and maintain membrane performance. Water quality, therefore, is of special importance in the rinsing and cleaning process as impurities present in the water could affect cleaning efficiency and, in the long term, contribute to a reduction in the performance and life of the membrane [129]. Membrane manufacturers generally recommend using high-quality water such as filtered and demineralized water. Installing and running water purification systems, however, is expensive. Alternatively, water treatment chemicals such as sequestering agents (e.g., EDTA, polyphosphates) can be added to low-quality water to increase the solubility of metal ions such as calcium, magnesium, manganese, and iron. Reverse osmosis permeate may also be of suitable quality for use in cleaning.

In a study by Tran-Ha and Wiley [129], it was shown that impurities such as particulate and dissolved salts present in the water could affect the cleaning efficiency of a polysulfone ultrafiltration membrane. The water used for cleaning was doped with a known amount of specific ions (i.e., calcium, sodium, chloride, nitrate, and sulfate). The presence of calcium in the water, at the usual concentrations found in tap water, did not greatly affect cleaning efficiency, while chloride was found to reduce it. Sodium, nitrate, and sulfate appeared to improve the flux recovery during membrane cleaning. The cleaning efficiency was also improved at higher ionic strengths. For further reading, please see the studies by Lindau and Jonsson [34], Fortunato et al. [106], Gong et al. [107], Vatanpour et al. [108], and Zhao et al. [102] (Table 7.2).

The effect of different cleaning agents on the recovery of the fouled membrane was investigated by Mohammadi et al. [130]. The results showed that a combination of sodium dodecyl sulfate and sodium hydroxide could be used as a cleaning material to reach the optimum recovery of the polysulfone membranes used in milk concentration industries. Likewise, a mixture of sodium hypochlorite and sodium hydroxide showed acceptable results where washing with acidic solutions was ineffective.

Mores and Davis [132], to view membrane surfaces at different times in crossflow microfiltration, employed direct visual observation (DVO) of yeast suspensions with rapid backpulsing at varied backpulsing durations and pressures. The DVO photos showed that the membranes were more effectively cleaned by longer backpulse durations and higher backpulse pressures. However, tradeoffs existed between longer and stronger backpulses and permeate loss during the backpulse. Shorter, stronger backpulses resulted in higher net fluxes than longer, weaker backpulses.

Roth et al. [135] proposed a method to determine the state of membrane wear by analyzing sodium chloride stimulus–response experiments. The shape of sodium chloride's distribution in the membrane's permeate flow revealed the solute permeation mechanisms for used membranes. For new membranes, the distribution of sodium chloride collected in the permeate side as well in the rejection side was unimodal. For fouled membranes, they noted the presence of several modes. The existence of a salt leakage peak and earlier detection of salt for all the fouled membranes gave evidence of membrane structure modification. The intensive use of the membranes might have enlarged the pore sizes. Salt and solvent permeability increased also. While this is a difficult paper to follow, it may be useful to those who want to develop new methods for measuring membrane degradation.

Ammerlaan et al. [45] reported on membrane degradation resulting in a premature loss of salt rejection by cellulose acetate membranes. Tests were initiated to find a solution to the problem and to gain a better understanding of the mechanisms involved. It was found that the removal of all free chlorine solved the problem. This was accomplished by injecting ammonia in the feed water, presumably forming ammonium chloride. Membrane damage by chlorine was also reported by Ridgway et al. [42]. They studied membrane fouling at a wastewater treatment plant under low- and high-chlorine conditions. High-chlorine residuals damaged the membrane structure and reduced mineral rejection capacity.

7.5 ECONOMIC ASPECTS OF MEMBRANE SEPARATIONS

Vinardell et al. [139] investigated the impact of permeate flux and gas sparging rate on membrane performance and process costs of granular anaerobic membrane bioreactors. Economic analysis showed that operating the membranes at moderate fluxes was the best alternative. A sensitivity analysis demonstrated that electricity and membrane cost were the most sensitive economic parameters, highlighting the importance of reducing specific gas demand requirements and improving membrane permeability to reduce costs. In a related report, Tanudjaja et al. [172] evaluated membrane-based separation for oily wastewater from a practical perspective. Commercially available membranes, membrane modules, operation modes, and hybrids were assessed, and their economics were discussed. The authors noted that cost inevitably depended on the source of the oily wastewater since each industry has unique blends of oil and grease as well as other foulants specific to the process. For example, the total cost (including costs of membranes, labor, electricity, cleaning, and maintenance) of treating oily wastewater from the fatty acid industry was \$2.65/m^3 [173], that from the railroad industry was \$1.03–1.48/m^3, while that for metalworking wastewater using ultrafiltration was \$2.8/m^3 [164]. Clearly, the cost of membrane-based treatment of oily wastewater is quite varied but is nevertheless lower than that for conventional technologies.

Harclerode et al. [138] performed a life cycle assessment and economic analysis of anaerobic membrane bioreactor whole plant configurations for resource recovery from domestic wastewater. The study determined that two process

TABLE 7.3
Cost analysis comparison of conventional and UF/MF pretreatments [105]

Parameter	Conventional pretreatment	UF/MF pretreatment	Benefits
Capital costs	Cost competitive with MF/UF	Slightly higher than conventional pretreatment. Costs continue to decline as developments are made	Capital costs of MF/UF could be 0–25% higher, whereas life cycle costs using either of the treatment schemes are comparable
Energy requirements	Calls for larger footprint	Significantly smaller footprint, higher than conventional	Footprint of MF/UF could be 30–50% of conventional filters
Footprint	Less than MF/UF as in requirements	Higher than conventional	MF/UF requires pumping of water through the membranes. This can vary depending on the type of membrane and water quality
Chemical costs	High due to coagulant and process chemicals needed for optimization	Chemical use is low, dependent on raw water quality	Less chemicals
RO capital cost	Higher than MF/UF since RO operates at lower flux	Higher flux is logically possible resulting in lower capital cost	Due to lower SDI values, RO can be operated at 20% higher flux if feasible, reducing RO capital costs
RO operating costs	Higher costs as fouling potential of RO feed water is high, resulting in higher operating pressure. One experiences frequent cleaning of RO membrane	Lower RO operating costs are expected due to less fouling potential and longer membrane life	The NDP (net driving pressure) is likely to be lower if the feed water is pretreated by MF/UF. Membrane cleaning frequency is reduced by 10–100%, reducing system downtime and prolonging element life

subcomponents (sulfide and phosphorus removal and sludge management) drove up chemical use and residuals generation, and in turn increased the environmental impacts and overall cost. Integrating primary sedimentation and a vacuum degassing tank for dissolved methane removal maximized net energy recovery. In addition, sustainability impacts were mitigated by operating at a higher flux and temperature, as well as by substituting biological sulfide removal for chemical coagulation.

Depending on the feed water quality, it may be necessary to consider the proper integration of multiple pretreatments and combine the benefits of each separate pretreatment [23]. Raw water, with its severe and changeable chemistry quality, presents a challenge for the appropriate pretreatment technology for fending off fouling. In economic terms, although mixing multiple pretreatments may increase the system's capital costs, operational expenditures can decline if membrane fouling can be effectively reduced by the integration. Vedavyasan [105] performed a cost analysis between the application of conventional and UF/MF pretreatments (Table 7.3). Their overview indicated that combining UF/MF with RO is feasible. In addition, lower chemical cleaning frequency and RO membrane replacement are expected due to improved feed water quality as a result of UF/MF pretreatment.

New separation techniques must, at a minimum, be comparable in overall cost and preferably be lower in cost than traditional technology. Unfortunately, scientists often forget that the successful commercialization of new technology depends on economic factors. Just because a novel separation technique works in the laboratory, for example, does not mean it will replace current methods.

The competitiveness of UF pretreatment in comparison to conventional pretreatment (i.e., coagulation and media filtration) was assessed by Brehant et al. [137] by looking at the impact on RO hydraulic performances. The study showed that ultrafiltration provided permeate water with high and constant quality, resulting in higher reliability of the RO process than with conventional pretreatment. The combination of UF with a precoagulation at a low dose helped control UF membrane fouling. The authors concluded that the combined effect of a higher recovery and a higher flux rate promised to reduce the RO plant costs significantly. The conclusions reached were the opposite of those reported in the paper by Glueckstern et al. [136] above and demonstrated the complexity of the overall economics of a membrane separation process.

Field evaluation of a hybrid membrane system consisting of a UF membrane pretreatment unit and a RO seawater unit was conducted by Glueckstern et al. [136]. For comparison, a second pilot system consisting of conventional pretreatment and an RO unit was operated in parallel. The conventional pretreatment unit included in-line flocculation followed by media filtration. The study showed that UF provided a very reliable pretreatment for the RO system independent of the raw water quality fluctuations. However, the cost of membrane pretreatment was higher than conventional pretreatment. Thus, this suggested that membrane pretreatment for RO desalting systems is only economical for sites that require extensive, conventional pretreatment or where wide fluctuations in the raw water quality are expected.

7.6 CONCLUDING REMARKS

Membrane fouling resulting from physicochemical interactions between the membrane and components of mixed liquor remains the most challenging issue preventing the broad application of this technology. Despite the development of low-fouling membrane systems, more research and engineering activities with a focus on surface modification, wastewater specifications, pretreatment and treatment conditions, and efficient fouling control and remedy strategies are still needed to minimize the occurrence of fouling. In addition, depending on the feed water quality, it may be necessary to consider the proper integration of multiple pretreatments and combine the benefits of each separate pretreatment.

In economic terms, although mixing multiple pretreatments may increase the system's capital costs, operational expenditures can decline if membrane fouling can be effectively reduced by the integration. The anaerobic membrane bioreactor, for example, is an emerging technology for municipal sewage treatment that combines anaerobic digestion and membrane separation. Economic analysis showed that operating the membranes at moderate fluxes could be the most favorable alternative. Furthermore, a sensitivity analysis illustrated that electricity and membrane cost were the most sensitive economic parameters, highlighting the importance of reducing specific gas demand requirements and improving membrane permeability to reduce costs.

To reduce the tendency to irreversible fouling, it is essential to operate the plant/unit below the critical flux. This must go hand in hand with reliable feed water pretreatment schemes. Membrane fouling is a complex process where the physicochemical properties of the membrane, the type of cells, the quality of the feed water, the type of solute molecules, and the operating conditions all play a role. The result of most membrane separations is a fouled surface that the operator will not be able to clean to its original state. Therefore, studies are required on effectively removing biofilms without damaging the membrane. Furthermore, it was possible to obtain high-performing membranes in terms of permeation and rejection with better antifouling resistance by embedding nanomaterials including clays, zeolites, metal oxides, graphene-based materials, carbon nanotubes, and metal-organic frameworks into the polymer matrix.

Additional work needs to be done to find out what happens to the fouling resistance of chemically modified membranes over the long term (i.e., after initial biofilm formation). Membrane resistance to humic acids is another area for further study. Additionally, the molecular tools needed for exploring the biochemical details of the microbial adhesion process to membranes are now available.

There is a need to accelerate the development and scale-up of novel water production systems, which will help to minimize environmental concerns. Renewable energy technologies, for instance, along with improved membranes, are rapidly emerging with the promise of economic and environmental viability for desalination. We can stipulate that in future, serious conflicts will arise not because of a lack of oil but due to water shortages. As scientists and engineers continue to improve the technical and economic efficiency of membrane desalination systems, it is imperative that we do not lose sight of the bigger water resources picture. A three-pronged approach, therefore, needs to be taken by society: water needs to be effectively managed; it needs to be economically purified; and it needs to be conserved. It can be argued that part of the solution to the world's water scarcity is not only to produce more water but to do this in an environmentally sustainable way, which remains a challenge for us.

ACKNOWLEDGMENTS

Special thanks are extended to Dr. S. Sablani from the Department of Biological Systems Engineering, Washington State University, Pullman, WA, USA, and Dr. R. Roque-Malherbe from the Universidad del Turabo, Gurabo, Puerto Rico, for their contribution to a previous version of this chapter which was updated for the current edition of this Handbook.

NOMENCLATURE

A	rate of loss of membrane surface area as function of time (m^2/s)
A	effective membrane area (m^2)
AFM	atomic force microscopy
ATR	attenuated total reflection
B	Permeability (mol/m s Pa)
$C = 4.8 \pm 0.3$	Carman–Kozeny constant
c_b	bulk solute concentration ($moles/cm^3$)
C_g	gel concentration ($kmol/m^3$)
C_o	solute concentration in feed solution ($kmol/m^3$)
c_p	permeate solute concentration ($moles/cm^3$)
C_w	concentration at membrane surface ($moles/cm^3$)
d_p	average grain diameter (mm)
d_v	membrane pore diameter
D	solute diffusion coefficient (m^2/s)
FTIR	Fourier transform infrared
G	geometrical factor (m)
h	transversal dimension of channel (m)
i	cycle number
J	solvent flux across membrane (m^3/m^2 s)
J^*	flux at infinite time (m^3/m^2 s)
J_a	average flux under steady state conditions (m^3/m^2 s)
J_{ai}	solvent flux at time a and in cycle i (m^3/m^2 s)
J_{crit}	limiting or critical flux (m^3/m^2 s)
J_{lim}	limiting or critical flux (m^3/m^2 s)
J_o	solvent flux at beginning of cycle (m^3/m^2 s)
J_s	solute flux ($moles/cm^2$ s)
$J(t_p)$	solvent flux as function of permeation time (m^3/m^2 s)
J_v	permeate flux ($moles/cm^2$ s)
J	molar gas flow (mol/m^2 s)
k	permeation factor (m^2)

K	hydraulic permeability of the membrane to pure solvent (m^3/Ns)
k	mass transfer coefficient
kC_w	adsorbed layer resistance
L	channel length (m)
l	membrane thickness (m)
M	molecular weight of the gaseous permeating species (kg/mol)
m	channel parameter
ΔP	hydraulic pressure difference across membrane (cm/s)
P	transmembrane pressure (N/m^2)
Pe	Peclet number
ΔP	$= P_1 - P_2$ transmembrane pressure (Pa)
Q	gas filtrate flux (m^3/s)
RO	reverse osmosis
R_m	membrane resistance
R	gas constant (J/kmol K)
Sc	Schmidt number
T	temperature (K)
t_p	permeation time (h)
t_c	cleaning time (h)
UF	ultrafiltration
UTDR	ultrasonic time domain reflectometry
U_o	longitudinal component of fluid velocity averaged over channel cross section (m/s)
Vm	molar volume of the flowing gas

GREEK SYMBOLS

ε	membrane porosity
Π	gas permeance (mol/m^2 s Pa)
ρ_A	apparent membrane density (g/cm^3)
ρ_R	real membrane density (g/cm^3)
μ:	dynamic viscosity of the gas (Pa.s)
$\Delta\pi$(w)	osmotic pressure at membrane surface (cm/s)
μ	fluid viscosity
τ	membrane lifetime (y)

REFERENCES

1. Hamedi, H., Ehteshami, M., Mirbagheri, S.A., Rasouli, S.A., Zendehboudi, S. (2019) Current status and future prospects of membrane bioreactors (MBRs) and fouling phenomena: A systematic review. *The Canadian Journal of Chemical Engineering* 97(1): 32–58.
2. Pichardo-Romero, D., Garcia-Arce, Z.P., Zavala-Ramírez, A., Castro-Muñoz, R. (2020) Current advances in biofouling mitigation in membranes for water treatment: An overview. *Processes* 8(2): 182.
3. Huang, S., Ras, R.H., Tian, X. (2018) Antifouling membranes for oily wastewater treatment: Interplay between wetting and membrane fouling. *Current Opinion in Colloid & Interface Science* 36: 90–109.
4. Zulkefli, N.F., Alias, N.H., Jamaluddin, N.S., Abdullah, N., Abdul Manaf, S.F., Othman, N.H., Marpani, F., Mat-Shayuti, M.S., Kusworo, T.D. (2022) Recent mitigation strategies on membrane fouling for oily wastewater treatment. *Membranes* 12(1): 26.
5. Jepsen, K.L., Bram, M.V., Pedersen, S., Yang, Z. (2018) Membrane fouling for produced water treatment: A review study from a process control perspective. *Water* 10(7): 847.
6. Abdulla, F.A., Abu-Dieyeh, M.H., Qnais, E. (2008) Human activities and ecosystem health. In: *Environmental Management, Sustainable Development and Human Health*, E.N. Laboy-Nieves, F. Schaffner, A. Abdelhadi, M.F.A. Goosen (Eds.), pp. 341–352. London: Taylor & Francis.
7. Goosen, M.F.A., Sablani, S.S., Roque, R. (2008) Membrane fouling: Recent strategies and methodologies for its minimization. In: *Handbook of Membrane Separations: Chemical, Pharmaceutical, Food, and Biotechnological Applications*, A.K. Pabby, S.S.H. Rizvi, and A.M. Sastre (Eds.), pp. 325–341. CRC Press Taylor and Francis.
8. Gottinger, H., Goosen, M.F.A. (Eds.) (2012) *Strategies of Economic Growth and Catch-up: Industrial Policies and Management*. Hauppauge, NY: Nova Science Publ Inc.
9. Laboy, E., Goosen, M.F.A., Emmanuel, E. (Eds.) (2010) *Environmental and Human Health: Risk Management in Developing Countries*. London: Taylor & Frances.
10. Mahmoudi, H., Ghaffour, N., Goosen, M.F., Bundschuh, J. (Eds.) (2017) *Renewable Energy Technologies for Water Desalination*. CRC Press.
11. Zakari, A., Khan, I., Tan, D., Alvarado, R. and Dagar, V. (2022) Energy efficiency and sustainable development goals (SDGs). *Energy* 239: 122365.
12. Goosen, M.F.A., Laboy-Nieves, E.N., Schaffner, F., Abdelhadi, A. (2009) The environment, sustainable development and human wellbeing: An overview. In: *Environmental Management, Sustainable Development and Human Health*, E.N. Laboy-Nieves, F. Schaffner, A. Abdelhadi, M.F.A. Goosen (Eds.), pp. 3–12. London: Taylor & Frances Publ.
13. Goosen, M.F.A., Al-Obeidani, S.K.S., Al-Hinai, H., Sablani, S., Taniguchi, Y., Okamura, H. (2009) Membrane fouling and cleaning in treatment of contaminated water. In: *Environmental Management, Sustainable Development and Human Health*, E.N. Laboy-Nieves, F. Schaffner, A. Abdelhadi, M.F.A. Goosen (Eds.), Chapter 36. London: Taylor & Frances Publ.
14. Misra, B. (2000) New economic policy and economic development. *IASSI Quarterly* 18 (4): 20.
15. Elimelech, M., Phillip, W.A. (2011) The future of seawater desalination: Energy, technology, and the environment. *Science* 333: 712–717.
16. Yi, E., Kang, H.S., Lim, S.M., Heo, H.J., Han, D., Kim, J.F., Park, A., Park, Y.I., Park, H., Cho, Y.H., Sohn, E.H., (2022) Superamphiphobic blood-repellent surface modification of porous fluoropolymer membranes for blood oxygenation applications. *Journal of Membrane Science* 648: 120363.
17. Goosen, M.F.A., Shayya, W.H. (1999) In: *Water Management, Purification and Conservation in Arid Climates; Vol. 1, Water Management*, Goosen, M. F.A., Shayya, W.H. (Eds.), pp. 1–6. Lancaster, PA: Technomic.
18. Goosen, M.F.A., Al-Hinai, H., Sablani, S. (2001) Capacity-building strategies for desalination: Activities, facilities and educational programs in Oman. *Desalination* 141: 181–189.
19. Goosen, M.F.A., Sablani, S.S., Al-Maskari, S.S., Al-Belushi, R.H., Wilf, M. (2002) Effect of feed temperature on

permeate flux and mass transfer coefficient in spiral-wound reverse osmosis systems. *Desalination* 144: 367–372.
20. Goosen, M.F.A., Sablani, S.S., Al-Hinai, H., Al-Obeidani, S., Al-Belushi, R., Jackson, D. (2004) Fouling of reverse osmosis and ultrafiltration membranes: A critical review. *Separation Science and Technology* 39: 2261–2298.
21. Ahmed, M., Arakel, A., Hoey, D., Thumarukudy, M.R., Goosen, M.F.A., Al-Haddabi, M., Al-Belushi, A. (2003) Feasibility of salt production from inland RO desalination plant reject brine: A case study. *Desalination* 158: 109–117.
22. Han, B., Runnels, T., Zimbron, J., Wickramasinghe, R. (2002) Arsenic removal from drinking water by flocculation and microfiltration. *Desalination* 145: 293–298.
23. Valavala, R., Sohn, J., Han, J., Her, N., Yoon, Y. (2011) Pre-treatment in reverse osmosis seawater desalination: A short review. *Environmental Engineering Research* 16(4): 205–212.
24. Voros, N.G., Maroulis, Z.B., Marinos-Kouris, D. (1996) Salt and water permeability in reverse osmosis membranes. *Desalination* 104: 141–154.
25. Sablani, S.S., Goosen, M.F.A., Al-Belushi, R., Gerardos, V. (2002) Influence of spacer thickness on permeate flux in spiral-wound seawater reverse osmosis systems. *Desalination* 146: 225–230.
26. Singh, R., Tembrock, J. (1999) Effectively controlled reverse osmosis systems. *Chemical Engineering Progress* 95: 57–64.
27. Huang, H., Schwab, K., Jacangelo, J.G. (2009) Pretreatment for low pressure membranes in water treatment: A review. *Environmental Science & Technology* 43: 3011–3019.
28. May, P., Laghmari, S., Ulbricht, M. (2021) Concentration polarization enabled reactive coating of nanofiltration membranes with zwitterionic hydrogel. *Membranes* 11(3): 187.
29. Sablani, S.S., Goosen, M.F.A., Al-Belushi, R., Wilf, M. (2001) Concentration polarization in ultrafiltration and reverse osmosis: A critical review. *Desalination* 141: 269–289.
30. Koltuniewicz, A., Noworyta, A. (1994) Dynamic properties of ultrafiltration systems in light of the surface renewal theory. *Industrial & Engineering Chemistry Research* 33: 1771–1779.
31. Lafi, W.K., Al-Anber, M., Al-Anber, Z.A., Al-Shannag, M., Khalil, A. (2010) Coagulation and advanced oxidation processes in the treatment of olive mill wastewater (OMW). *Desalination and Water Treatment* 24: 251–256.
32. Dubey, S.S., Rao, B.S. (2012) Removal of dyes and heavy metals by using low cost adsorbents – a review. *Journal of Pharmacy Research* 5: 461–470.
33. Upen, J., Barwada, S.J., Coker, S.D., Terry, A.R. (2000) Winning the battle against biofouling of reverse osmosis membranes. *Desalination Water Reuse* 10(2): 53–54.
34. Lindau, J., Jonsson, A.-S. (1994) Cleaning of ultrafiltration membranes after treatment of oily wastewater. *Journal of Membrane Science* 87: 71–78.
35. Benito, J.M., Rios, G., Ortea, E., Fernandez, E., Cambiella, A., Pazos, C., Coca, J. (2002) Design and construction of a modular pilot plant for the treatment of oil-containing wastewaters. *Desalination* 147: 5–10.
36. Pope, J.M., Yao, S., Fane, A.G. (1996) Quantitative measurements of the concentration polarization layer thickness in membrane filtration of oil-water emulsions using NMR micro-imaging. *Journal of Membrane Science* 118: 247: 257.
37. Scott, K., Mahood, A.J., Jachuck, R.J., Hu, B. (2000) Intensified membrane filtration with corrugated membranes. *Journal of Membrane Science* 173: 1–16.
38. Li, D., Wang, H. (2010) Recent developments in reverse osmosis desalination membranes. *Journal of Materials Chemistry* 20: 4551–4566.
39. Nystrom, M., Ruohomaki, K., Kaipa, L. (1996) Humic acid as a fouling agent in filtration. *Desalination* 106: 78–86.
40. Kabsch-Korbutowicz, M., Majewska-Nowak, K., Winnicki, T. (1999) Analysis of membrane fouling in the treatment of water solutions containing humic acids and mineral salts. *Desalination* 126: 179–185.
41. Khoo, Y.S., Lau, W.J., Liang, Y.Y., Karaman, M., Gürsoy, M., Ismail, A.F. (2022) Eco-friendly surface modification approach to develop thin film nanocomposite membrane with improved desalination and antifouling properties. *Journal of Advanced Research* 36: 39–49.
42. Ridgway, H.F., Justice, C.A., Whittaker, C., Argo, D.G., Olson, B.H. (1984) Biofilm fouling of RO membranes – its nature and effect on treatment of water for reuse. *Journal AWWA* 76: 94–101.
43. Ridgway, H.F., Rigby, M.G., Argo, D.G. (1984) Adhesion of a *Mycobacterium* sp. to cellulose diacetate membranes used in reverse osmosis. *Applied and Environmental Microbiology* 47: 61–67.
44. Flemming, H.-C., Schaule, G. (1988) Biofouling of membranes – a microbiological approach. *Desalination* 70: 95–119.
45. Amerlaan, A.C.F., Franklin, J.C., Moody, C.D. (1992) Yuma desalting plant membrane degradation during test operations. *Desalination* 88: 33–49.
46. Rabiller-Baudry, M., Le Maux M., Chaufer B., Begoin, L. (2002) Characterization of cleaned and fouled membranes by ATR-FTIR and EDX analysis coupled with SEM: Application to UF of skimmed milk with a PES membrane. *Desalination* 146: 123–128.
47. Li, J., Sanderson, R.D., Hallbauer, D.K., Hallbauer-Zadorozhnaya, V.Y. (2002) Measurement and modeling of organic deposition in ultrafiltration by ultrasonic transfers signals and reflections. *Desalination* 146: 177–185.
48. Belfer, S., Purinson, Y., Kedem, O. (1998) Reducing fouling of RO membranes by redox-initiated graft polymerization. *Desalination* 119: 189–195.
49. Flemming, H.-C., Schaule, G., McDonough, R. (1993) How do performance parameters respond to initial biofouling on separation membranes? *Vom Wasser* 80: 177–186.
50. Ghayeni, S.B.S., Beatson, P.J., Schncider, R.P., Fane, A.G. (1998) Adhesion of wastewater bacteria to reverse osmosis membranes. *Journal of Membrane Science* 138: 29–42.
51. Ridgway, H.F., Kelly, A., Justice, C., Olson, B.H. (1983) Microbial fouling of reverse osmosis membranes used in advanced wastewater treatment technology: Chemical bacteriological and ultra-structural analyses. *Applied and Environmental Microbiology* 46: 1066–1084.
52. Ridgway, H.F., Rigby, M.G., Argo, D.G. (1985) Bacterial adhesion and fouling of reverse osmosis membranes. *Journal AWWA* 77: 97–106.
53. Ridgway, H.F. (1991) Bacteria and membranes: Ending a bad relationship. *Desalination* 83: 53.

54. Stork, D. (2008) *An Investigation into Membrane Fouling from Algae Containing Waters.* Master's Thesis RMIT University: Australia
55. Aydin, S., Ünlü, İ.D., Arabacı, D.N., Duru, Ö.A. (2022) Evaluating the effect of microalga *Haematococcus pluvialis* bioaugmentation on aerobic membrane bioreactor in terms of performance, membrane fouling and microbial community structure. *Science of the Total Environment* 807: 149908.
56. Ly, Q.V., Maqbool, T., Hur, J. (2017) Unique characteristics of algal dissolved organic matter and their association with membrane fouling behavior: A review. *Environmental Science and Pollution Research* 24(12): 11192–11205.
57. Du, X., Shi, Y., Jegatheesan, V., Haq, I.U. (2020) A review on the mechanism, impacts and control methods of membrane fouling in MBR system. *Membranes* 10(2): 24.
58. Asif, M.B., Zhang, Z. (2021) Ceramic membrane technology for water and wastewater treatment: A critical review of performance, full-scale applications, membrane fouling and prospects. *Chemical Engineering Journal* 418: 129481.
59. Schafer, A.I., Martrup, M., Lund Jensen, R. (2002) Particle interactions and removal of trace contaminants from water and wastewaters. *Desalination* 147: 243–250.
60. Khatib, K., Rose, J., Barres, O., Stone, W., Bottero, J-Y., Anselme, C. (1997) Physico-chemical study of fouling mechanisms of ultrafiltration membrane on Biwa Lake (Japan). *Journal of Membrane Science* 130: 53–62.
61. Tu S.-C., Ravindran V., Den W., Pirbazari, M. (2001) Predictive membrane transport model for nanofiltration processes in water treatment. *AIChE Journal* 47: 1346–1362.
62. Domany, Z., Galambos, I., Vatai, G., Bekassy-Molnar, E. (2002) Humic substances removal from drinking water by membrane filtration. *Desalination* 145: 333–337.
63. Sahachaiyunta, P., Koo, T., Sheikholeslami, R. (2002) Effect of several inorganic species on silica fouling in RO membranes. *Desalination* 144: 373–378.
64. Liu, B., Wang, M., Yang, K., Li, G., Shi, Z. (2022) Alleviation of ultrafiltration membrane fouling by ClO_2 pre-oxidation: Fouling mechanism and interface characteristics. *Membranes* 12(1): 78.
65. Yiantsios, S.G., Karabelas, S. (1998) The effect of colloid stability on membrane fouling. *Desalination* 118: 143–152.
66. Jarusutthirak, C., Amy, G., Croue, J.-P. (2002) Fouling characteristics of wastewater effluent organic matter (EfOM) isolates on NF and UF membranes. *Desalination* 145: 247–255.
67. Bacchin, P., Aimar, P., Sanches, V. (1995) Model of colloidal fouling of membranes. *AIChE Journal* 41: 368–376.
68. Qin, W., Zhang, J., Xie, Z., Ng, D., Ye, Y., Gray, S.R., Xie, M. (2017) Synergistic effect of combined colloidal and organic fouling in membrane distillation: Measurements and mechanisms. *Environmental Science: Water Research & Technology* 3(1): 119–127.
69. Nikolova, J.D., Islam, M.A. (1998) Contribution of adsorbed layer resistance to the flux-decline in an ultrafiltration process. *Journal of Membrane Science* 146: 105–111.
70. Chen, V., Fane, A.G., Madaeni, S., Wenten, I.G. (1997) Particle deposition during membrane filtration of colloids: transition between concentration polarization and cake formation. *Journal of Membrane Science* 125: 109–122.
71. Poojamnong, K., Tungsudjawong, K., Khongnakorn, W., Jutaporn, P. (2020) Characterization of reversible and irreversible foulants in membrane bioreactor (MBR) for eucalyptus pulp and paper mill wastewater treatment using fluorescence regional integration. *Journal of Environmental Chemical Engineering* 8(5): 104231.
72. Denisov, G.A. (1994) Theory of concentration polarization in cross-flow ultrafiltration: Gel-layer model and osmotic-pressure model. *Journal of Membrane Science* 91: 173–187.
73. Stumme, J., Ashokkumar, O., Dillmann, S., Niestroj-Pahl, R., Ernst, M. (2021) Theoretical evaluation of polyelectrolyte layering during layer-by-layer coating of ultrafiltration hollow fiber membranes. *Membranes* 11(2): 106.
74. Zydney, A.L., Ho, C.C. (2002) Scale-up of microfiltration systems: Fouling phenomena and Vmax analysis. *Desalination* 146: 75–81.
75. Choi, J., Kim, W.S., Kim, H.K., Yang, S.C., Han, J.H., Jeung, Y.C., Jeong, N.J. (2022) Fouling behavior of wavy-patterned pore-filling membranes in reverse electrodialysis under natural seawater and sewage effluents. *npj Clean Water* 5(1): 1–12.
76. Goosen, M.F.A., Sablani, S., Del-Cin, M., Wilf, M. (2011) Effect of cyclic changes in temperature and pressure on permeation properties of composite polyamide spiral wound reverse osmosis membranes. *Separation Science and Technology* 46: 14–26.
77. Riedl, K., Girard, B., Lencki, W. (1998) Influence of membrane structure on fouling layer morphology during apple juice clarification. *Journal of Membrane Science* 139: 155–166.
78. Guo, J., Wong, P.W., Deka, B.J., Zhang, B., Jeong, S., An, A.K. (2022) Investigation of fouling mechanism in membrane distillation using in-situ optical coherence tomography with green regeneration of fouled membrane. *Journal of Membrane Science* 641: 119894.
79. Lei, Z., Wang, J., Leng, L., Yang, S., Dzakpasu, M., Li, Q., Li, Y.Y., Wang, X.C., Chen, R. (2021) New insight into the membrane fouling of anaerobic membrane bioreactors treating sewage: Physicochemical and biological characterization of cake and gel layers. *Journal of Membrane Science* 632: 119383.
80. Altena, F.W., Belfort, G. (1984) Lateral migration of spherical particles in porous flow channels: Application to membrane filtration. *Chemical Engineering Science* 19: 343–355.
81. Drew, D.A., Schonberg, J.A., Belfort, G. (1991) Lateral inertial migration of small sphere in fast laminar flow through a membrane duct. *Chemical Engineering Science* 46: 3219–3224.
82. Cherkasov, A.N., Tsareva, S.V., Polotsky, A.E. (1995) Selective properties of ultrafiltration membranes from the standpoint of concentration polarization and adsorption phenomena. *Journal of Membrane Science* 104: 157–164.
83. Ghayeni, S.B.S., Beatson, P.J., Fane, A.G., Schneider, R.P. (1999) Bacterial passage through micro filtration membranes in wastewater applications. *Journal of Membrane Science* 153: 71–82.
84. Howe, K.J., Ishida, K.P., Clark, M.M. (2002) Use of ATR/FTIR spectrometry to study fouling of microfiltration membranes by natural waters. *Desalination* 147: 251–255.
85. Chan, R., Chen, V., Bucknall, M.P. (2002) Ultrafiltration of protein mixtures: measurement of apparent critical flux, rejection performance, and identification of protein deposition. *Desalination* 146: 83–90.

86. Bowen, W.R., Doneva, T.A., Yin, H.B. (2002) Atomic force microscopy studies of membrane-solute interactions (fouling). *Desalination* 146: 97–102.
87. Gowman, L.M., Ethier, C.R. (1997) Concentration and concentration gradient measurements in an ultrafiltration concentration polarization layer. Part I: A laser-based refractometric experimental technique. *Journal of Membrane Science* 131: 95–105.
88. Gowman, L.M., Ethier, C.R. (1997) Concentration and concentration gradient measurements in an ultrafiltration concentration polarization layer. Part II: Application to hyaluronan. *Journal of Membrane Science* 131: 107–123.
89. Dal-Cin, M.M., McLellan, F., Striez, C.N., Tam, C.M., Tweddle, T.A., Kumar, A. (1996) Membrane performance with a pulp mill effluent: Relative contributions of fouling mechanisms. *Journal of Membrane Science* 120: 273–285.
90. AlSawaftah, N., Abuwatfa, W., Darwish, N., Husseini, G. (2021) A comprehensive review on membrane fouling: Mathematical modelling, prediction, diagnosis, and mitigation. *Water* 13(9): 1327.
91. Quezada, C., Estay, H., Cassano, A., Troncoso, E., Ruby-Figueroa, R. (2021) Prediction of permeate flux in ultrafiltration processes: A review of modeling approaches. *Membranes* 11(5): 368.
92. Wilf M., Klinko, K. (1998) Effective new pretreatment for seawater reverse osmosis systems. *Desalination* 117: 323–331.
93. Glueckstern, P., Priel, M. (1998) Advanced concept of large seawater desalination systems for Israel. *Desalination* 119: 33–45.
94. Karakulski, K., Gryta, M., Morawski, A. (2002) Membrane processes used for potable water quality improvement. *Desalination* 145: 315–319.
95. Chapman, H., Vigneswaran, S., Ngo, H. H., Dyer, S., Ben Aim, R. (2002) Pre-flocculation of secondary treated wastewater in enhancing the performance of microfiltration. *Desalination* 146: 367–372.
96. Nguyen M.T., Ripperger, S. (2002) Investigation on the effect of flocculants on the filtration behavior in microfiltration of fine particles. *Desalination* 147: 37–42.
97. Choksuchart, P., Heran, M., Grasmick, A. (2002) Ultrafiltration enhanced by coagulation in an immersed membrane system. *Desalination* 145: 265–272.
98. Park, P.K., Lee, C.H., Choi, S.J., Choo, K.H., Kim, S.H., Yoon, C.H. (2002) Effect of the removal of DOMs on the performance of a coagulation –UF membrane system for drinking water production. *Desalination* 145: 237–245.
99. Guigui, C., Rouch, J.C., Durand-Bourlier, L., Bonnelye, V., Aptel, P. (2002) Impact of coagulation conditions on the in-line coagulation/UF process for drinking water production. *Desalination* 147: 95–100.
100. Lopez-Ramirez, J.A., Marquez, D.S., Alonso, J.M.Q. (2002) Comparison studies of feedwater pre-treatment in reverse osmosis pilot plant. *Desalination* 144: 347–352.
101. Shaalan, H.F. (2003) Development of fouling control strategies pertinent to nanofiltration membranes. *Desalination* 153(1–3): 125–131.
102. Zhao, C., Zhou, J., Yan, Y., Yang, L., Xing, G., Li, H., Wu, P., Wang, M., Zheng, H. (2021) Application of coagulation/flocculation in oily wastewater treatment: A review. *Science of the Total Environment* 765: 142795.
103. Van Geluwe, S., Braeken, L., Van der Bruggen, B. (2011) Ozone oxidation for the alleviation of membrane fouling by natural organic matter: A review. *Water Research* 45: 355–370.
104. Muro, C., Riera, F., Diaz, M.C. (2012) Membrane separation process in wastewater treatment of food industry. *Intech* 253–280. http://cdn.intechweb.org/pdfs/29163.pdf (accessed 6 May 2012).
105. Vedavyasan, C.V. (2007) Pretreatment trends – an overview. *Desalination* 203:296–299.
106. Fortunato, L., Lamprea, A.F., Leiknes, T. (2020) Evaluation of membrane fouling mitigation strategies in an algal membrane photobioreactor (AMPBR) treating secondary wastewater effluent. *Science of the Total Environment* 708: 134548.
107. Gong, C., Ren, X., Zhang, Z., Sun, Y., Huang, H. (2022) Electrocoagulation pretreatment of pulp and paper wastewater for low pressure reverse osmosis membrane fouling control. *Environmental Science and Pollution Research* 29(24): 36897–36910.
108. Vatanpour, V., Hazrati, M., Sheydaei, M., Dehqan, A. (2022) Investigation of using UV/H_2O_2 pre-treatment process on filterability and fouling reduction of PVDF/TiO_2 nanocomposite ultrafiltration membrane. *Chemical Engineering and Processing-Process Intensification* 170: 108677.
109. Schwinge, J., Wiley, D.E., Fletcher, D.F. (2002) A CFD study of unsteady flow in narrow spacer-filled channels for spiral-wound membrane modules. *Desalination* 146: 195–201.
110. Geraldes, V., Semiao, V., de Pinho, M.N. (2002) The effect of the ladder-type spacers configuration in NF spiral wound modules on the concentration boundary layers disruption. *Desalination* 146: 187–194.
111. Li, F., Meindersma, G.W., de Haan, A.B., Reith, T. (2002) Optimization of non-woven spacers by CFD and validation by experiments. *Desalination* 146: 209–212.
112. Lipnizki, J., Jonsson, G. (2002) Flow dynamics and concentration polarization in spacer-filled channels. *Desalination* 146: 213–217.
113. Abid, H.S., Johnson, D.J., Hashaikeh, R., Hilal, N. (2017) A review of efforts to reduce membrane fouling by control of feed spacer characteristics. *Desalination* 420: 384–402.
114. Eray, E., Candelario, V.M., Boffa, V., Safafar, H., Østedgaard-Munck, D.N., Zahrtmann, N., Kadrispahic, H., Jørgensen, M.K. (2021) A roadmap for the development and applications of silicon carbide membranes for liquid filtration: Recent advancements, challenges, and perspectives. *Chemical Engineering Journal* 414: 128826.
115. Usman, J., Othman, M.H.D., Ismail, A.F., Rahman, M.A., Jaafar, J., Raji, Y.O., Gbadamosi, A.O., El Badawy, T.H., Said, K.A.M. (2021) An overview of superhydrophobic ceramic membrane surface modification for oil-water separation. *Journal of Materials Research and Technology* 12: 643–667.
116. Jenkins, M., Tanner, M.B. (1998) Operational experience with a new fouling resistant reverse osmosis membrane. *Desalination* 119: 243–250.
117. Dilshad, M.R., Islam, A., Haider, B., Sajid, M., Ijaz, A., Khan, R.U., Khan, W.G. (2021) Effect of silica nanoparticles on carbon dioxide separation performances of PVA/PEG cross-linked membranes. *Chemical Papers* 75(7): 3131–3153.

118. Shen, L., Huang, Z., Liu, Y., Li, R., Xu, Y., Jakaj, G., Lin, H. (2020) Polymeric membranes incorporated with ZnO nanoparticles for membrane fouling mitigation: A brief review. *Frontiers in Chemistry* 8: 224.
119. Wang, M., Wang, J., Jiang, J. (2022) Membrane fouling: microscopic insights into the effects of surface chemistry and roughness. *Advanced Theory and Simulations* 5(1): 2100395.
120. Song, L. (1998) A new model for the calculation of the limiting flux in ultrafiltration. *Journal of Membrane Science* 144: 173–185.
121. Madireddi, K., Babcock, R. B., Levine, B., Kim, J.H., Stenstrom, M.K. (1999) *Journal of Membrane Science* 157: 13–22.
122. Mallubhotla, H., Belfort, G. (1988) Flux enhancement during Dean vortex micro filtration. 8. Further diagnostics. *Journal of Membrane Science* 125: 75–91.
123. Avlonitis, S., Hanbury, W.T., Boudinar, M.B. (1993) Spiral wound modules performance: an analytical solution Part II. *Desalination* 89: 227–246.
124. Banti, D.C., Mitrakas, M., Samaras, P. (2021) Membrane fouling controlled by adjustment of biological treatment parameters in step aerating MBR. *Membranes* 11(8): 553.
125. De Vela, R.J. (2021) A review of the factors affecting the performance of anaerobic membrane bioreactor and strategies to control membrane fouling. *Reviews in Environmental Science and Bio/Technology* 20(3): 607–644.
126. Di Bella, G., Di Trapani, D., Judd, S. (2018) Fouling mechanism elucidation in membrane bioreactors by bespoke physical cleaning. *Separation and Purification Technology* 199: 124–133.
127. Szabo-Corbacho, M.A., Pacheco-Ruiz, S., Míguez, D., Hooijmans, C.M., Brdjanovic, D., García, H.A., van Lier, J.B. (2022) Influence of the sludge retention time on membrane fouling in an anaerobic membrane bioreactor (AnMBR) treating lipid-rich dairy wastewater. *Membranes* 12(3): 262.
128. Yuliwati, E., Ismail, A.F., Othman, M.H.D., Shirazi, M.M.A. (2022) Critical flux and fouling analysis of PVDF-mixed matrix membranes for reclamation of refinery-produced wastewater: Effect of mixed liquor suspended solids concentration and aeration. *Membranes* 12(2): 161.
129. Tran-Ha, M.H., Wiley, D.E. (1998) The relationship between membrane cleaning efficiency and water quality. *Journal of Membrane Science* 145: 99–110.
130. Mohammadi, T., Madaeni, S.S., and Moghadam, M.K. (2002) Investigation of membrane fouling. *Euromed 2002 Conference Proceedings* 1: 14–16. Sharm El-Sheikh, Egypt.
131. Al-Obeidani, S., Al-Hinai, H., Goosen, M.F.A., Sablani, S., Taniguchi, Y., Okamura, H. (2008) Chemical cleaning of oil contaminated polyethylene hollow fiber microfiltration membranes. *Journal of Membrane Science* 307: 299–308.
132. Mores, W.D., Davis, R.H. (2002) Direct observation of membrane cleaning via rapid backpulsing. *Desalination* 146: 135–140.
133. Gao, Y., Qin, J., Wang, Z., Østerhus, S.W. (2019) Backpulsing technology applied in MF and UF processes for membrane fouling mitigation: A review. *Journal of Membrane Science* 587: 117136.
134. Gao, Y., Zhang, Y., Dudek, M., Qin, J., Øye, G., Østerhus, S.W. (2021) A multivariate study of backpulsing for membrane fouling mitigation in produced water treatment. *Journal of Environmental Chemical Engineering* 9(2): 104839.
135. Roth, E., Kessler, M., Fabre, B., Accary, A. (1999) Sodium chloride stimulus-response experiments in spiral wound reverse osmosis membranes: A new method to detect fouling. *Desalination* 121: 183–193.
136. Glueckstern, P., Priel, M., Wilf, M. (2002) Field evaluation of capillary UF technology as a pretreatment for large seawater RO systems. *Desalination* 147: 55–62.
137. Brehant, A., Bonnelye, V., Perez, M. (2002) Comparison of MF/UF pretreatment with conventional filtration prior to RO membranes for surface seawater desalination. *Desalination* 144: 353–360.
138. Harclerode, M., Doody, A., Brower, A., Vila, P., Ho, J., Evans, P.J. (2020) Life cycle assessment and economic analysis of anaerobic membrane bioreactor whole-plant configurations for resource recovery from domestic wastewater. *Journal of Environmental Management* 269: 110720.
139. Vinardell, S., Sanchez, L., Astals, S., Mata-Alvarez, J., Dosta, J., Heran, M., Lesage, G. (2022) Impact of permeate flux and gas sparging rate on membrane performance and process economics of granular anaerobic membrane bioreactors. *Science of the Total Environment* 825: 153907.
140. Ghayeni, S.B.S., Beatson, P.J., Schneider, R.P., Fane, A.G. (1998) Water reclamation from municipal wastewater using combined micro filtration-reverse osmosis (ME-RO): Preliminary performance data and microbiological aspects of system operation. *Desalination* 116: 65–80.
141. Anis, S.F., Hashaikeh, R., Hilal, N. (2019) Reverse osmosis pretreatment technologies and future trends: A comprehensive review. *Desalination* 452: 159–195.
142. Jagaba, A.H., Kutty, S.R.M., Lawal, I.M., Abubakar, S., Hassan, I., Zubairu, I., Umaru, I., Abdurrasheed, A.S., Adam, A.A., Ghaleb, A.A.S., Almahbashi, N.M.Y. (2021) Sequencing batch reactor technology for landfill leachate treatment: A state-of-the-art review. *Journal of Environmental Management* 282: 111946.
143. Hedayati, M., Krapf, D., Kipper, M.J. (2021) Dynamics of long-term protein aggregation on low-fouling surfaces. *Journal of Colloid and Interface Science* 589: 356–366.
144. Torre-Celeizabal, A., Garea, A., Casado-Coterillo, C. (2022) Chitosan: Polyvinyl alcohol based mixed matrix sustainable coatings for reusing composite membranes in water treatment: Fouling characterization. *Chemical Engineering Journal Advances* 9: 100236.
145. Sabalanvand, S., Hazrati, H., Jafarzadeh, Y., Jafarizad, A., Gharibian, S. (2021) Investigation of Ag and magnetite nanoparticle effect on the membrane fouling in membrane bioreactor. *International Journal of Environmental Science and Technology* 18(11): 3407–3418.
146. Hermanowicz, S.W. (2011) *Membrane Bioreactors: Past, Present and Future?* Working Papers, Water Resources Collections and Archives, University of California Water Resources Centre, UC Berkeley.
147. Wang, X., Wang, B., Wang, M., Liu, Q., Wang, H. (2021) Cyclohexane dehydrogenation in solar-driven hydrogen permeation membrane reactor for efficient solar energy conversion and storage. *Journal of Thermal Science* 30(5): 1548–1558.
148. Wu, B., Yi, S., Fane, A.G. (2011) Microbial behaviors involved in cake fouling in membrane bioreactors under

different solids retention times. *Bioresource Technology* 102(3): 2511–2516.
149. Gao, C., Liao, J., Lu, J., Ma, J., Kianfar, E. (2021) The effect of nanoparticles on gas permeability with polyimide membranes and network hybrid membranes: A review. *Reviews in Inorganic Chemistry* 41(1): 1–20.
150. Mulder, M. (1996). *Basic Principles of Membrane Technology*. Dordrecht: Kluwer Academic Publ.
151. Baker, R.W. (2004) *Membrane Technology and Applications*. New York: J. Wiley & Sons.
152. Ogden, J.M. (2002) Hydrogen: The fuel of the future. *Physics Today* April: 69–75.
153. Saracco, G., Specchia, V. (1994) Catalytic inorganic-membrane-reactors: Present experience and future opportunities. *Catalysis Reviews – S&E* 36: 305–384.
154. Morooka, S., Kusakabe, K. (1999) Microporous inorganic membranes for gas separation. *MRS Bulletin* March: 25–29.
155. Saracco, G., Neomagus, H.W.J.P., Versteeg, G.F., Swaaij, W.P.M. (1999) High-temperature membrane reactors: potential and problems. *Chemical Engineering Science* 54: 1997–2017.
156. Lai, Z., Bonilla, G., Diaz, I., Nery, J.G., Sujaoti, K., Amat, M., Kokkoki, E., Terakasi, O.,.Thompson, R.W., Tsapatis M., Vlachos, D.G. (2003) Microstructural optimization of a zeolite membrane for organic vapor separation. *Science* 300: 456–460.
157. Mauran, S., Rigaud, L., Coudeville, O. (2001) Application of the Carman-Kozeny correlation to a high-porosity anisotropic consolidated medium: The compressed expanded natural graphite. *Transport in Porous Media* 43: 355–376.
158. Burggraaf, A.J. (1999) Single gas permeation of thin zeolite (MFI) membranes: Theory and analysis of experimental observations. *Journal of Membrane Science* 155: 45–65.
159. Roque-Malherbe, R., Wendelbo, R., Mifsud, A., Corma, A. (1995) Diffusion of aromatic hydrocarbons in H-ZSM-5, H-Beta and H-MCM-22 zeolites. *The Journal of Physical Chemistry* 99: 14064–14071.
160. Roque-Malherbe, R. (2001) Applications of natural zeolites in pollution abatement and industry, In: *Handbook of Surfaces and Interfaces of Materials*, Vol. 5, Nalwa, H.S. (Ed.), pp. 495–522. New York: Academic Press.
161. Hermia, J. (1982) Constant pressure blocking filtration laws: Application to power-law non-Newtonian fluids. *Transactions of the Institution of Chemical Engineers* 60: 183–187.
162. Suárez, F., Del Río, M.B., Aravena, J.E. (2022) Water flux prediction in direct contact membrane distillation subject to inorganic fouling. *Membranes* 12(2): 157.
163. Field, R.W., Wu, J.J. (2022) Permeate flux in ultrafiltration processes—Understandings and misunderstandings. *Membranes* 12(2): 187.
164. Cheryan, M. (1998) *Ultrafiltration and Microfiltration Handbook*. CRC Press.
165. Saffarimiandoab, F., Gul, B.Y., Tasdemir, R.S., Ilter, S.E., Unal, S., Tunaboylu, B., Menceloglu, Y.Z., Koyuncu, İ. (2021) A review on membrane fouling: Membrane modification. *Desalination and Water Treatment* 216: 47–70.
166. Danckwerts, P.V. (1951) Significance of liquid film coefficients in gas absorption. *Industrial & Engineering Chemistry* 43: 460–1470.
167. Yu, H.B., Zhang, X.Q., Han, X., Yang, Z.Z., Zhou, Y.W., Ding, W., Du, M.X. (2022) Nanofiltration membrane fouling and control caused by residual aluminum in feed water. *Water, Air, & Soil Pollution* 233(1): 1–9.
168. Zhang, R., Vigneswaran, S., Ngo, H.H., Nguyen, H. (2006) Magnetic ion exchange (MIEX®) resin as a pre-treatment to a submerged membrane system in the treatment of biologically treated wastewater. *Desalination* 192: 296–302.
169. Boudinar, M.B., Hanbury, W.T., Avlonitis, S. (1992) Numerical simulation and optimization of spiral-wound modules. *Desalination* 86: 273–290.
170. Zhang, Z., Wang, Z., Wang, J., Wang, S. (2013) Enhancing chlorine resistances and anti-biofouling properties of commercial aromatic polyamide reverse osmosis membranes by grafting 3-allyl-5,5-dimethylhydantoin and N,N′-Methylenebis (acrylamide). *Desalination* 309: 187–196.
171. Redondo, J.A. (1999) Improve RO system performance and reduce operating cost with FILMTEC fouling resistant (FR) elements. *Desalination* 126: 249–259.
172. Tanudjaja, H.J., Hejase, C.A., Tarabara, V.V., Fane, A.G., Chew, J.W. (2019) Membrane-based separation for oily wastewater: A practical perspective. *Water Research* 156: 347–365.
173. Dangel, R.A., Astraukis, D., Palmateer, J. (1995) Fatty acid separation from hydrolyzer wastewater by ultrafiltration. *Environmental Progress* 14(1): 65–68.

8 Membrane Processes and Developments for Biochemical and Pharmaceutical Separations

Chidambaram Thamaraiselvan[1,2,*] *and S. Ranil Wickramasinghe*[2]
[1]Interdisciplinary Centre for Energy Research, Indian Institute of Science, Bangalore, India
[2]Ralph E. Martin Department of Chemical Engineering, University of Arkansas, Fayetteville, AR, USA
*Email: chidambaramt@iisc.ac.in

8.1 INTRODUCTION

Old textbooks discuss various separation techniques such as syringe filtration, centricon centrifugal filtration, dialysis, gel filtration, and solid-phase extraction in today's modern biochemistry applications [1]. Membrane processes are increasingly being used for biochemical separation in various industries, including the biotechnology [2], pharmaceuticals [3], food processing, and dairy industries [4]. These processes utilize a semipermeable membrane that selectively separates desired molecules from a mixture based on size exclusion, Donnan exclusion, and mass transfer [5,6]. Membrane processes have several advantages over conventional separation techniques, including their ability to operate at lower temperatures and pressures, their cost-effectiveness, and their ability to achieve high selectivity without the use of chemicals or solvents [7]. Moreover, membrane processes can be easily integrated into existing processes, making them a popular choice for industrial applications [8]. In biochemical separation and pharmaceutical applications, the most commonly used membrane processes are microfiltration (MF), ultrafiltration (UF), nanofiltration (NF), and reverse osmosis (RO). The removal characteristics of the membranes are shown in Figure 8.1. Each of these processes has its unique features and is suitable for different types of biochemical and pharmaceutical separations. So far, novel membranes, modules, and processes have been developed to meet the criteria of the biochemical and pharmaceutical industries. The objective of this chapter is to provide an overview of these developments. The chapter focuses mainly on the recent developments of pressure-driven membrane processes for the sustainable separation of biochemicals and pharmaceuticals. In addition, membrane performance results and fabrication techniques are discussed.

8.2 BIOCHEMICAL SEPARATION

Biomolecules are essential to our life; carbohydrates, proteins, peptides, amino acids, and nucleic acids are some of the most essential biomolecules that play crucial roles in maintaining the structure and function of living organisms. These are very important in many biological processes, including catalyzing chemical reactions [9,10]. The recent advances in polymeric membranes for biochemical separation could potentially improve the efficiency and selectivity of separation processes in various applications, including biotechnology, pharmaceuticals, and food processing. This section mainly focuses on these biochemicals' separations using membrane technology and its recent advancements.

8.2.1 Progress in Novel Membrane Fabrication

8.2.1.1 Polymeric Membranes

Polymeric membranes for biochemical separation use stimuli-responsive materials that can change their properties in response to environmental changes such as temperature, pH, and ionic strength. Many polymeric membranes were developed for biochemical applications. Membrane based on poly(n-isopropylacrylamide) (PNIPAAm) is a thermoresponsive polymer that exhibits a lower critical solution temperature (LCST) at around 32°C, above which it becomes hydrophobic and can be used for the separation of biomolecules based on size and charge [11]. In addition, there has been research on the development of novel membrane materials based on natural polymers such as chitosan and cellulose, which offer advantages such as biodegradability and biocompatibility. For example, cellulose diacetate nitrate membranes have been developed by the electrospinning method to separate protein with a higher adsorption capacity of 300.11 mg/g [12].

Low-cost chemical and thermal stable membrane fabrication are vital for industrial applications. Notably, resource recovery is significant for the industrial circular economy. Thus, chemical engineering research focuses on recovering valuable sources from waste. For example, fatty acids are among the valuable biochemicals recovered from synthetic organic waste streams by supported liquid membranes (SLMs) [13]. Cellulose acetate membranes used in the dialysis process

DOI: 10.1201/9781003285656-8

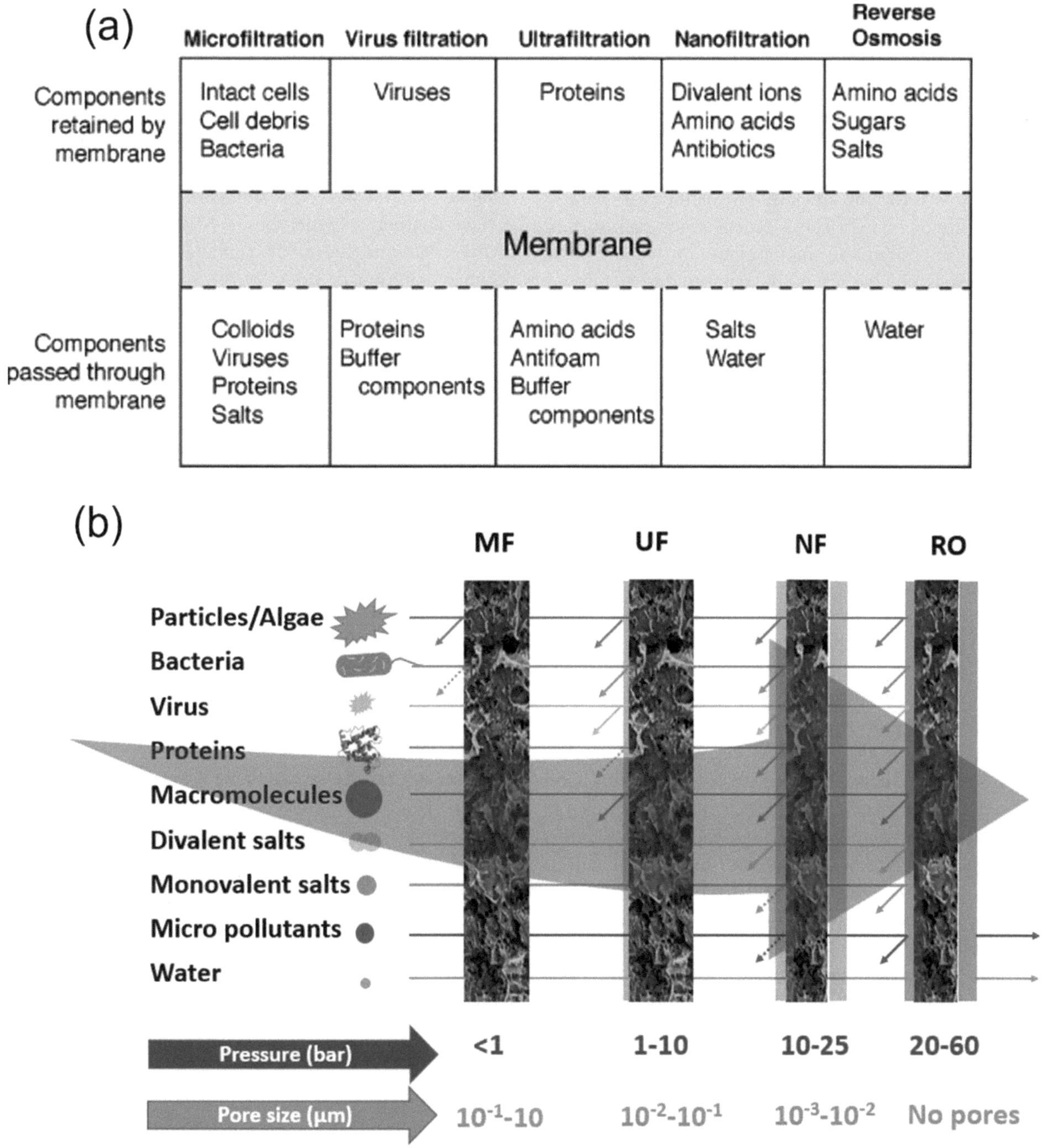

FIGURE 8.1 Removal characteristics of different pressure-driven membrane processes (a,b) (adapted with permission from Refs. [6,83]).

remove waste products and excess fluids from the blood of patients with kidney failure. These membranes are designed to allow the passage of small molecules like ions and urea while preventing the passage of larger molecules like proteins, DNA, and polysaccharides [14].

Surface modifications of polymeric membranes can enhance their performance in biochemical separation processes by altering their surface properties, such as hydrophobicity, charge, and roughness. Several recent works have been carried out on surface modifications of polymeric membranes for biochemical separation. Grafting of the membrane involves the covalent attachment of polymer chains to the surface of the membrane to modify its surface properties. For example, polyethylene glycol (PEG) can be grafted onto the surface of the membrane to improve its biocompatibility and reduce fouling. Polyacrylonitrile (PAN) UF membranes were fabricated with a dense coating of the copolymer polyacrylonitrile-graft-poly(ethylene oxide). These grafted membranes show high performances in terms of flux and antifouling properties [15]. Zwitterionic polymers such as poly(sulfobetaine methacrylate) (pSBMA)-catechol have been used to modify the surfaces of polymeric membranes for improved fouling resistance and antifouling properties [16]. In this study, pSBMA-catechol was grafted onto a polyethersulfone (PES) membrane, resulting in a membrane with reduced fouling and enhanced permeability for protein separation [16]. Thus, these surface modifications enhance the performance of polymeric membranes for industrial applications.

Interestingly, temperature-responsive microgel coatings on porous support for the modern biochemical process is a promising strategy [17]. These recent works on surface modifications of polymeric membranes for biochemical separation demonstrate the potential for improving membrane performance and selectivity through surface engineering [15–18]. Plasma treatment can also modify the surface of polymeric membranes by introducing functional groups such as carboxyl and hydroxyl groups, which can enhance their wettability and protein-binding properties. In a recent study, a poly(vinylidene fluoride) (PVDF) membrane was treated with plasma to introduce carboxyl groups, resulting in a membrane with improved antibiofouling [18].

Hollow-fiber membranes have found remarkable applications in the biomedical field, showcasing their versatility and potential. One notable use is in the development of bioartificial liver systems. These membranes play a crucial role in mimicking the functions of a natural liver, allowing for the removal of toxins and the maintenance of metabolic processes in patients with liver failure. The unique structure of the hollow-fiber membranes enables efficient exchange of molecules between the patient's blood and the device, facilitating effective detoxification and metabolic support [19,20]. Furthermore, hollow-fiber membranes are employed in wound-healing applications. Another exciting application is in drug-delivery systems. Hollow-fiber membranes can be utilized as carriers for controlled and targeted drug release. Their porous structure and selective permeability enable the precise modulation of drug diffusion, allowing for sustained and localized delivery to specific target sites within the body [19,20]. Also, it is used in tissue engineering applications. The utilization of hollow-fiber membranes in these biomedical applications highlights their ability to address critical challenges and improve patient outcomes. As research and development in this field continue to progress, it is expected that further advancements will be made, leading to even more sophisticated and tailored applications in bioartificial organs, wound healing, drug delivery, and tissue engineering [19,20].

In general, polymeric membranes can be used for hemodialysis applications, particularly for individuals with end-stage renal disease. This remarkable technology enables the replacement of impaired kidney function by employing flat membranes or empty fiber dialyzers to facilitate the removal of excess water, salts, and metabolic waste products. Beyond hemodialysis, polymeric membranes find wide application in various other biomedical fields. They play a pivotal role in the development of artificial organs, including the artificial liver, oxygenator, and artificial pancreas. Moreover, membranes offer tremendous potential in drug-delivery systems, serving as efficient carriers for targeted medication release. Additionally, they find application in diverse areas such as oxygenators, osteosynthesis membranes, and biosensors, further expanding their utility in the medical realm [21].

8.2.1.2 Composite Membranes

Within the realm of membrane fabrication, there are several methods for fabricating nanocomposite membranes. Two notable approaches are mixed-matrix membranes and surface-coating membranes. Mixed-matrix membranes involve the integration of nanomaterials directly into the polymer solution or dope before casting the membrane or employing electrospinning techniques. By incorporating these nanomaterials, the resulting membranes gain enhanced properties and functionalities. Surface-coating membranes, on the other hand, follow a different route. First, a polymer support is created, and then nanomaterials are introduced during the interfacial polymerization process. This strategic incorporation of nanomaterials at the interface further enhances the performance and characteristics of the resulting membranes. In some cases, vacuum filtration is employed in combination with cross-linking to achieve the desired membrane structure and functionality. Another fascinating approach involves the creation of free-standing membranes using direct spinning chemical vapor deposition (CVD) techniques or electrospinning techniques. Through these methods, membranes are formed directly from the vapor phase or via the controlled electrostatic deposition of nanomaterial-infused polymer solutions.

There is a rapid increase in the utilization of nanomaterials (NMs) in the field of environmental applications. Nanoparticles such as silica and titanium dioxide have been grafted onto polymeric membranes to enhance their surface roughness and hydrophilicity. A recent study incorporated silica nanoparticles into a PES membrane, improving water permeability, hydrophilicity, and antifouling properties for separating proteins [22]. High chemical and thermal stable membranes are needed for harsh environments. Membranes made of UiO-66 metal-organic frameworks (MOFs) are successfully created on prestructured yttria-stabilized zirconia hollow fibers, offering outstanding separation capabilities and impressive flux rates. Specifically, the flux rates achieved are 4.81, 5.95, and 4.06 $kg.m^{-2}\ h^{-1}$ for the separation of water from different organic compounds, namely i-butanol, furfural, and tetrahydrofuran, respectively. Moreover, these membranes exhibit rejection rates of over 99.98% for all of these compounds. These membranes have a high separation factor and flux rate, outperforming commercially available polymeric and silica membranes [23]. The manufactured HF membrane exhibits sponge-like layers that enhance the mechanical reinforcement of the membrane, as illustrated in Figure 8.2. The hollow fiber has an approximate outer diameter of 1.1 mm, a wall thickness of 200 μm, and a pore size of approximately 80 nm. It consists of three sponge-like layers interleaved with two layers of microchannels.

However, when nanoparticles are incorporated into membranes, there is a possibility that these particles may be released or leached into the surrounding solution or fluid

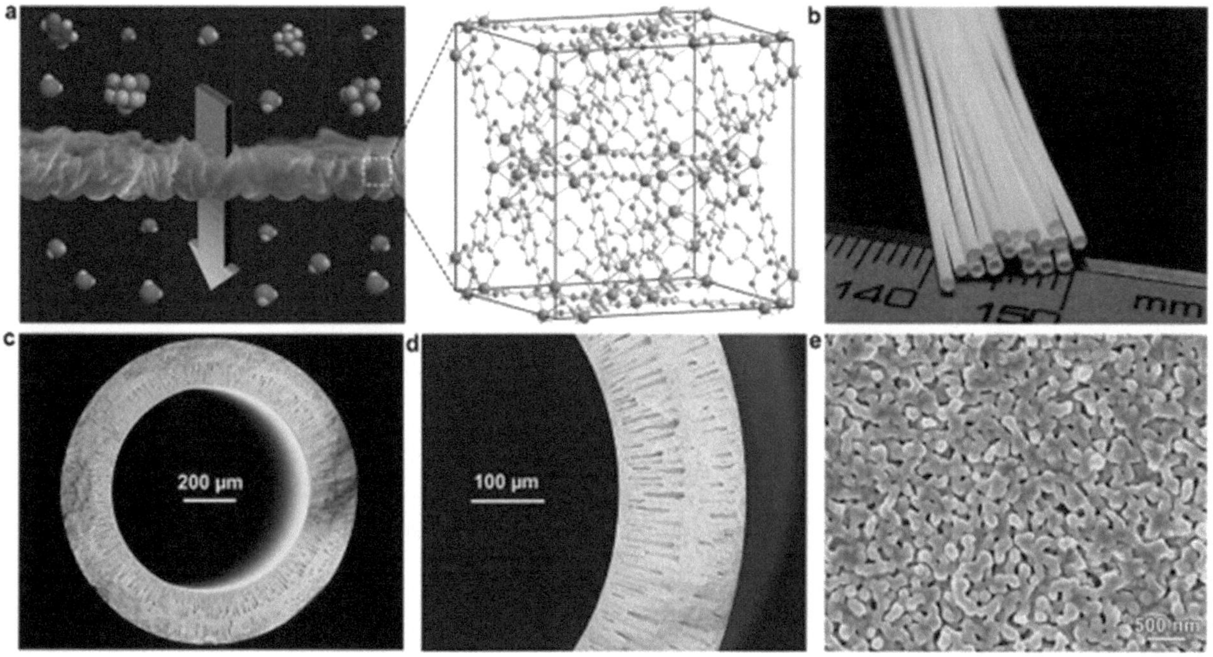

FIGURE 8.2 (a) Illustration of water separation from organics using UiO-66. (b) Photo and SEM images of YSZ hollow fibers: (c,d) cross-section; (e) outer surface [23].

during the membrane's operation. This leaching phenomenon can potentially impact the biochemical processes or samples being treated or analyzed. Thus, leaching is a significant concern that warrants careful consideration.

Leached nanoparticles can contaminate the product (solution), affecting the purity and integrity of the biochemical samples. Moreover, if the nanoparticles have the potential to interact with biological systems, their leaching can lead to unintended interactions with cells, proteins, enzymes, or other biological components present in the system. This can result in altered biological activity or even toxicity. In addition, the presence of leached nanoparticles may modify the performance characteristics of the membrane itself. This can include changes in permeability, selectivity, or fouling behavior, ultimately affecting the overall efficiency and reliability of the biochemical process. Thus, to mitigate the potential negative effects of nanoparticle leaching, it is crucial to carefully select and design membranes that minimize or prevent the release of nanoparticles.

8.2.1.3 Carbon-Based Membranes

One of the advantages of carbon-based membranes is their high selectivity toward specific biomolecules. This selectivity is due to the specific surface chemistry of the membrane material, which can interact with the target molecules through hydrogen bonding, van der Waals forces, and electrostatic interactions [24]. Hence, carbon-based membranes for biochemical separation have recently gained much attention [25,26]. Carbon-based membranes comprise carbonaceous materials such as graphite, graphene oxide, carbon nanotubes, and activated carbon. They can be prepared by different methods, such as carbonization [27,28], pyrolysis [29], and chemical vapor deposition [30]. These membranes can be used for various separation processes, such as gas separation, liquid separation, and ion separation. Moreover, researchers use nanocomposite membranes that incorporate nanoparticles to enhance their properties. For example, graphene oxide (GO) has been incorporated into polymeric membranes to improve their mechanical strength and water flux while maintaining selectivity for biomolecules [31].

For desalting applications, graphene-based large-area nanoporous atomically thin membranes (NATMs) were fabricated by scalable chemical vapor deposition (CVD) [14]. These membranes also have high permeability, which is essential in biochemical separation as it allows for a high flux of the desired biomolecule. The high permeability of carbon-based membranes is due to their unique structural properties such as high porosity and pore size distribution, which allow for easy transport of molecules through the membrane [14]. Another advantage of carbon-based membranes is their stability in harsh environments. The membrane materials are resistant to chemical and thermal degradation [32], which is essential for their long-term use in biochemical separation processes.

8.2.2 Progress in Process-Based Techniques

8.2.2.1 Microfiltration

Microfiltration (MF) is a bio-separation technique that uses a semipermeable membrane to separate particles based on size. MF has been widely used in various industries, including biotechnology, food, and pharmaceuticals [6]. The semipermeable membrane used in MF has a pore size

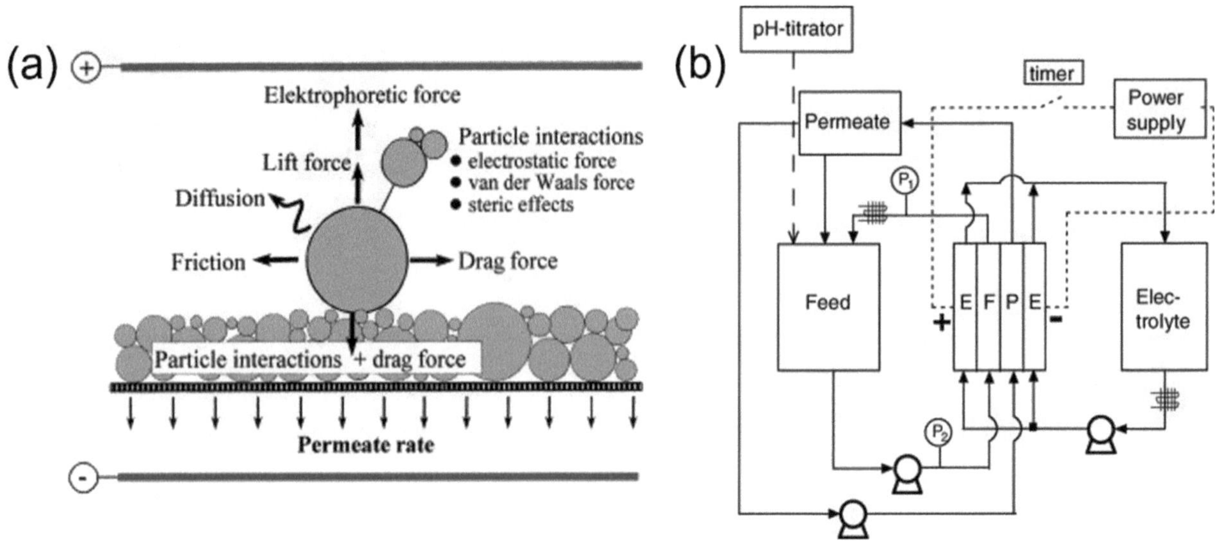

FIGURE 8.3 (a) Phenomenon of electro-UF; (b) electro-UF set up (adapted with permission from Refs. [37,91]).

ranging from 0.1 to 10 μm, allowing particles smaller than the pore size to pass through the membrane. The membrane is typically made of materials such as polyethersulfone, polysulfone, and ceramic. MF offers several advantages over other bio-separation techniques, including gentle separation, continuous operation, cost-effectiveness, and particle removal without altering the liquid composition. However, MF has some limitations, including membrane fouling and limited selectivity. Despite these limitations, MF remains a popular and promising technique for separating and purifying biological particles [33].

The MF membrane separated an antivenom antibody precipitate from a contaminated protein molecule. The utilization of microfiltration (MF) as a replacement for the centrifugation technique is recommended for antibody purification and serum purification processes in industry [34]. In the existing procedure, immunoglobulins (IgG) are precipitated from sheep serum using Na_2SO_4 at 30°C, followed by separation using a continuous-flow disk-stack centrifuge to isolate the precipitated IgG from the mixture containing contaminants. A proposed approach suggests the use of MF instead of the disk-stack centrifuge. This MF-based approach offers a significant advantage by significantly reducing albumin contamination [34].

Poly(l-lactic acid) microfiltration membranes were prepared with Tween 80 surfactant as an additive using nonsolvent-induced phase inversion (NIPS) method. This membrane is stable at 25°C but deteriorates in wet conditions at 60°C. Thus, this membrane is suitable as a compostable microfiltration membrane in the food and biochemical industries [35]. Moreover, hollow-fiber membranes have various biomedical applications such as blood oxygenation, hemodialysis, drug delivery, and tissue engineering due to their ability to provide a large surface area for mass transfer in a compact configuration. A hollow-fiber membrane was used for immunoglobulin G separation and clinical uses [36].

Notably, the development of electrically aided membranes has emerged as an intriguing approach for protein purification. The application of an electrically aided ultrafiltration (UF) process offers distinct advantages in protein purification, as it combines the effects of electroosmosis and electrophoresis to effectively purify colloidal molecules under the influence of an electric field [37]. By utilizing this process, the formation of colloidal aggregates or gel layers on the membrane surface is reduced, leading to improved permeate flux. A schematic representation of the electro-UF process is depicted in Figure 8.3 [37].

The MF process offers numerous advantages while also presenting certain limitations. One of its key strengths lies in its simplicity and ease of operation, as it operates under low-pressure conditions. This makes it a versatile solution applicable to a wide range of applications, as we explore in detail throughout this chapter. However, it is important to acknowledge that the MF process may not be suitable for low-molecular-weight compounds. Due to this limitation, it may not achieve significant rejection rates for such substances. Nonetheless, the MF process remains a valuable and effective method for various other applications, showcasing its versatility and practicality in many scenarios.

8.2.2.2 Ultrafiltration (UF)

On the other hand, UF is very similar to the MF process, separating molecules based on size and charge. UF is often used for the concentration of proteins, enzymes, and other biomolecules from complex mixtures such as cell lysates and fermentation broths [38,39]. UF can also be used for the purification of proteins by removing impurities such as salts, small molecules, and endotoxins [40]. Additionally, UF is applicable for fractionating proteins based on their molecular weight. Geoffroy et al. studied electrodialysis using a 20-kDa UF membrane to separate bioactive cationic peptides. The study primarily examined the influence of the initial concentration of whey protein hydrolysate (WPH) and investigated the effects of varying current intensity [41].

Some interesting works have reported using hollow-fiber (HF) membranes also. For example, one study showed the successful synthesis of carboxylated graphene oxide nanosheets decorated with iron oxide nanoparticles (Fe_3O_4/cGO nanohybrid) and then incorporated into polyethersulfone (PES) hollow-fiber ultrafiltration membranes (HFMs) to create modified membranes. The performance of these modified membranes was evaluated for the separation of various proteins, including lysozyme, trypsin, pepsin, human serum albumin, γ-globulin, and fibrinogen. The modified HFMs exhibited favorable physicochemical properties, such as improved mechanical strength, hydrophilicity, porosity, pore size, and surface roughness. These properties played a significant role in enabling the composite membranes, specifically those modified with 0.1 wt.% Fe_3O_4/cGO nanohybrid, to achieve a remarkably high pure water flux of 110.0 ± 3.8 LMH, along with a high flux recovery rate of up to 97.8%. The rejection rates for the tested proteins were as follows: lysozyme (92.9 ± 1.3%), trypsin (94.5 ± 1.1%), pepsin (96.9 ± 1.2%), human serum albumin (99.5 ± 0.5%), human γ-globulin (100%), and human fibrinogen (100%) [42]. Furthermore, the effective elimination of bovine serum albumin (BSA) has been effectively showcased using alumina HF membranes [43]. It is crucial to select the appropriate size of alumina nanoparticles (NPs) for optimal protein separation. Smaller-sized NPs have been found to improve the BSA removal rate [43].

UF is a versatile bio-separation technique widely used in various industries, including biotechnology, food, and pharmaceuticals. For example, the process of ultrafiltration of soy (*Glycine max*) proteins isolate (SPI) hydrolysates has been proved to be a convenient method for retrieving a substantial quantity of bioactive peptides [44]. The integration of ultrafiltration (UF) and diafiltration (DF) systems has gained widespread use in whey production [45]. However, a drawback of this approach is that the UF permeates retains certain low-molecular-weight peptides and amino acids, leading to incomplete recovery. Peptides and amino acids with molecular weights below 10 kDa can pass through the membrane pores. This occurrence not only diminishes protein recovery rates but also has implications for lactose recovery and purification. [45]. Thus tight UF membranes are required for high recovery of protein.

UF offers several advantages over other bio-separation techniques, including gentle separation, continuous operation, cost-effectiveness, and selective separation. However, UF has some limitations, including membrane fouling and limited selectivity. Despite these limitations, UF remains a popular and promising technique for the separation and purification of biological molecules.

8.2.2.3 Nanofiltration (NF)

NF is a semipermeable membrane that separates molecules based on size and charge. Due to its higher selectivity, NF is a promising bio-separation technique that has been widely used in various industries, including biotechnology, food, and pharmaceuticals [46]. The NF membrane is typically made of polyamide [47], and some ceramic- [48] and graphene-based membranes [49] have also been studied. Tuning the polyamide selective layer is very important during membrane fabrication for specific applications. The polyamide layer forms by interfacial polymerization reaction with piperazine and tannic acid on PES support and subsequently post-etching treatment was done with NaOH [50]. NF involves the concentration and purification of proteins, separation of sugars and amino acids, and removal of unwanted ions and impurities [31].

Carbon-based membranes exhibit greater potential for industrial applications. For instance, high flux nanofiltration was enabled utilizing graphene oxide quantum dots and employed for the removal of BSA protein. The rejection rate was almost 100% [51]. Similarly, a nitrogen-doped GO-doped PES membrane demonstrated over 95% rejection of BSA [52]. Carbon-based membranes show higher potential for industrial applications. For example, high flux nanofiltration was fabricated with graphene oxide quantum dots and used for BSA protein removal. The rejection rate was almost 100%. Similarly, nitrogen-doped GO-doped PES membrane had a >95% rate of BSA rejection. NF offers several advantages over other bio-separation techniques, including high selectivity and purity, mild conditions, small ions and impurities removal, and continuous operation [46]. However, NF has some limitations, including membrane fouling and limited selectivity for some target molecules [53]. Despite these limitations, NF remains a popular and promising technique for the separation and purification of biological molecules.

8.2.2.4 Reverse Osmosis (RO)

RO is based on the principle of size and charge exclusion. The semipermeable membrane used in RO has a pore size of less than 1 nanometer, allowing the separation of molecules based on their size and charge. The polyamide-based RO membranes are well-known in water treatment industries. In addition, RO can be used in other industries, including biotechnology, food, and pharmaceuticals. RO offers several advantages over other biochemical separation techniques, including high selectivity and purity, mild conditions, removal of small ions and impurities, and continuous operation [54,55]. However, RO has limitations, including membrane fouling and high energy requirements compared with UF and NF. Thus, RO membranes are not generally used in biomolecule separation applications, rather it is used for desalination.

8.2.2.5 Electro-Membrane Processes

Electrodialysis (ED) is a membrane-based separation technology that uses an electric field to selectively transport ions across ion exchange membranes for the separation and purification of charged species such as proteins, amino acids, and organic acids. ED has several advantages over traditional separation techniques, such as high selectivity, low energy consumption, and continuous operation. It was reported in earlier studies that bioconversion of fumaric acid to L-malic acid is feasible with ED [56]. Recent research on ED for biochemical separation has focused on improving the process's selectivity, efficiency, and scale-up. Separation of pharmaceutical products from the highly concentrated aqueous solution is challenging. Thus, Garcia et al. reported that electrodialysis is one of the feasible techniques to achieve this effect. They demonstrated that more than 70% salt removal (ammonium sulfate and sodium dihydrogen phosphate) from a mixed solution of amino acids so that isolation of amino acids is feasible with high purity [57].

Electro deionization (EDI) is an emerging technology for biochemical processes. The laminated ion exchange resin (porous ionomer-binder resin) wafer was prepared for p-coumaric acid separation. The prepared resin wafer shows an extraordinary performance, almost sevenfold higher than for the existing EDI modules [58]. Moreover, electrochemical treatments have demonstrated a promising approach for protein (BSA) removal using laser-induced graphene membranes as electrodes [59].

8.3 PHARMACEUTICAL APPLICATIONS

Membrane processes have emerged as a promising technology for pharmaceutical separation and purification [53–55]. These processes utilize a semipermeable membrane that separates desired molecules from a mixture based on their size, shape, and chemical properties. Membrane processes have several advantages over traditional separation techniques, including their ability to operate at low temperatures and pressures, their cost-effectiveness, and their ability to achieve high levels of purity without using chemicals or solvents. This section explores the different membranes studied in pharmaceutical applications and their benefits.

8.3.1 Polymeric Membranes

Pharmaceutical purification and separation performance can be improved by tuning the membrane properties. Thus, the focus of this section is mainly on the development of novel separation membranes which are used in the pharmaceutical industry. Several novel membranes have been developed for pharmaceutical applications, with unique properties and potential advantages over traditional membranes [60–62]. Membrane performances can be improved in multiple ways; surface modification is one of the most promising methods. Surface modification includes surface functionalization with carboxyl, amine, and sulfonate groups, graft polymerization with polyethylene glycol (PEG), and surface coating with biopolymer chitosan can improve their selectivity, stability, and biocompatibility, making them more suitable for pharmaceutical applications.

For example, novel high-performance NF membranes were developed using amino functional polyethylene glycol (PEG) and trimesoyl chloride (TMC), which exhibited strong resistance to fouling and chlorine [62]. A novel TFC membrane was also developed based on polyester and cellulose support for the separation of dyes from organic solvent [63]. High-performance ultrafiltration membranes were fabricated with PES and cellulose nanocrystals (CNCs) using the NIPS method. These membranes have a hydrophilic nature and exhibit good rejection for BSA and antifouling resistance for pharmaceutical and biotechnological industries [64].

The ultrahigh specific surface areas, well-defined apertures, and variable topological characteristics of zeolitic imidazolate frameworks (ZIFs) have made them a popular choice in the field of membrane separation. However, due to their limited aperture sizes, ZIFs face challenges when separating organic solvents. To overcome this, a methodology based on thermal annealing has been developed for defect engineering on the ZIF-67 membrane by Sun et al. [65].

A regenerated cellulose-based ultrafiltration membrane was developed by Zheng et al. [66]. The results indicated that RC membranes have superior selectivity compared to BSA rejection, with over 98% rejection and efficient removal of endotoxins below 0.25 EU/mL. Furthermore, these membranes demonstrated excellent antifouling properties, with a flux recovery rate (FRR) of over 95% and, in some cases, as high as 100%. Notably, the separation of many proteins including lysozyme and BSA was successfully demonstrated using a novel approach (Figure 8.4) with a thin-film nanocomposite (TFN) membrane integrated with Arg-MMT (arginine-montmorillonite) clay nanoparticles [67]. The fabricated membrane exhibited impressive rejection rates of approximately 98% for cephalexin antibiotics and 99% for BSA, while maintaining a permeability of 26 LMH at a pressure of 5 bar. Furthermore, these membranes achieved the highest flux recovery ratio of 86.48% [67].

Overall, the pressure-driven membrane fails due to significant fouling on the membrane surface during the separation of solutes. To overcome these, NF membranes were developed using a novel zwitterionic amine monomer for antifouling in the pharmaceutical and chemical industry [68]. Another important requirement for successful industrial applications is selective removal of target compounds. A modified interfacial polymerization was used to fabricate the NF membrane. 3,5-diaminobenzoic acid (BA) was additionally used along with piperazine (PIP) as a diamine monomer with trimesoyl chloride (TMC) [69] for selective removal of organic compounds. To achieve the desired pore size in membranes, the ratio between BA and PIP was adjusted, and it was discovered that a concentration ratio of 50:50 yielded superior results in removing pharmaceutical compounds. The rejection rate for six specific pharmaceutical compounds (trimethoprim, sulpiride, primidone, nalidixic acid, sulfamethoxazole, and indometacin) was found to be above 80%, while still allowing the passage of inorganic salts like $CaCl_2$ and $MgCl_2$. A schematic of the typical and modified interfacial polymerization reaction is shown in Figure 8.5.

NF membranes had a significant role in separating antibiotics from wastewater due to their ability to selectively remove small molecules. Recent studies have shown some achievements in using the NF process for antibiotic separation. Many research works have demonstrated the successful

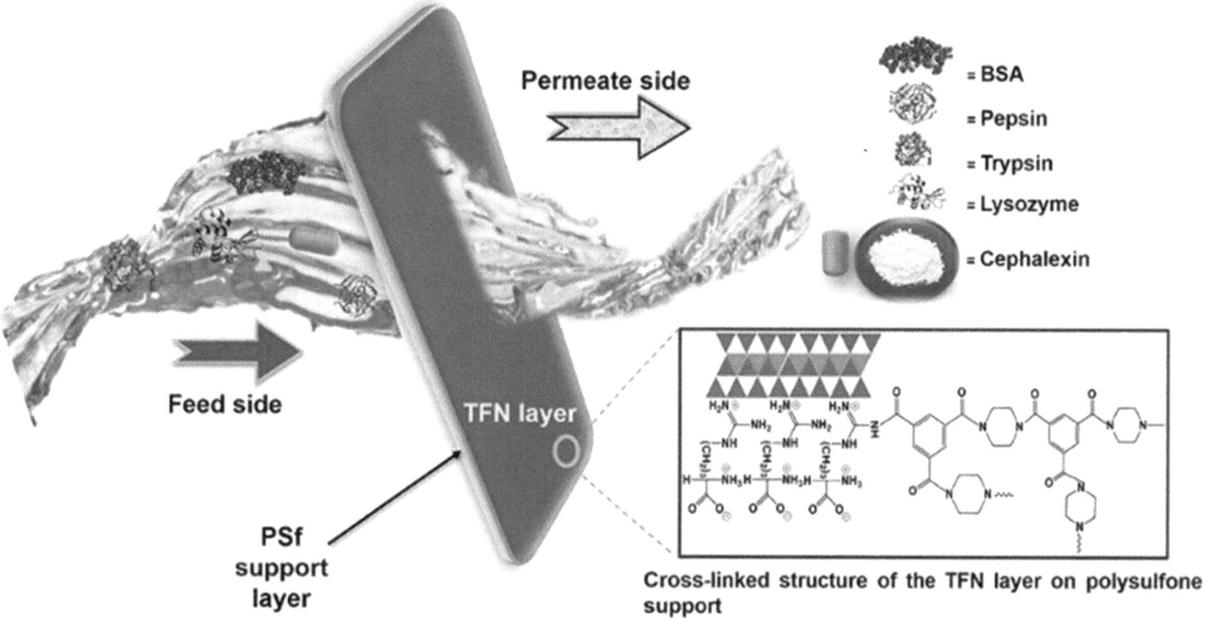

FIGURE 8.4 (a) Schematic of the novel approach for biomacromolecule separation using low-cost polymeric thin-film nanocomposite membranes [67].

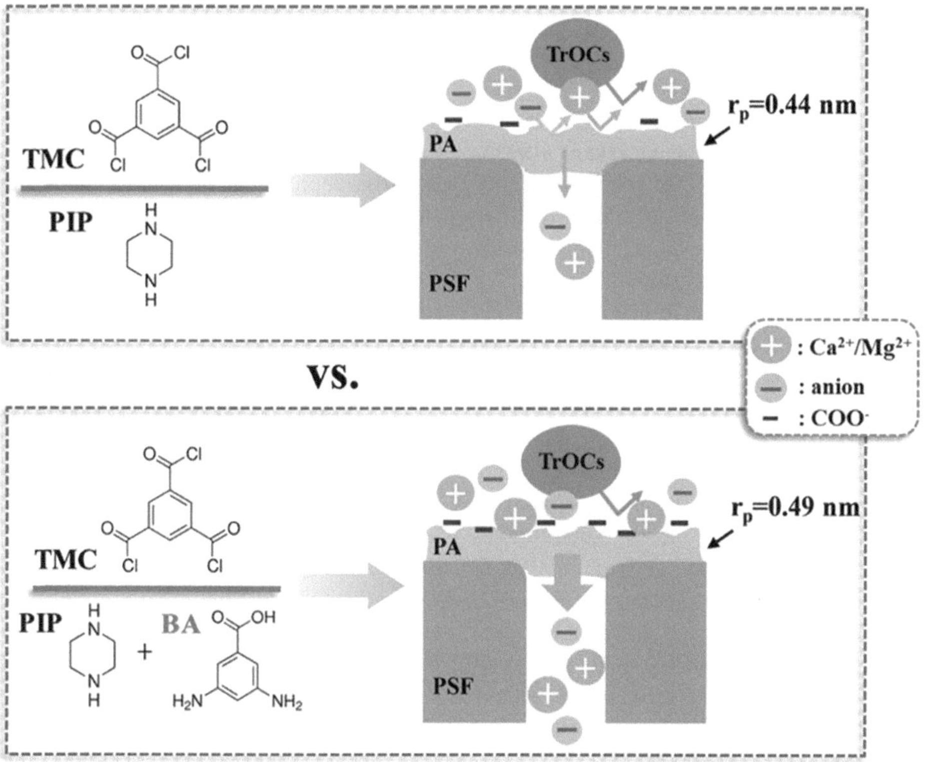

FIGURE 8.5 A schematic of the organic selective membranes (adapted with permission from Ref. [69]).

removal of commonly used antibiotics such as amoxicillin [70], cephalexin [71], sulfamethoxazole [72], and diclofenac [72] from aqueous solutions using a novel TFC NF membrane. These studies showed high removal efficiency (>90–95%) and the potential for the membrane to be used in wastewater treatment. To increase the removal of pharmaceutical compounds, an integrated process of NF with the Fenton reaction has been recommended [73]. Another study reported

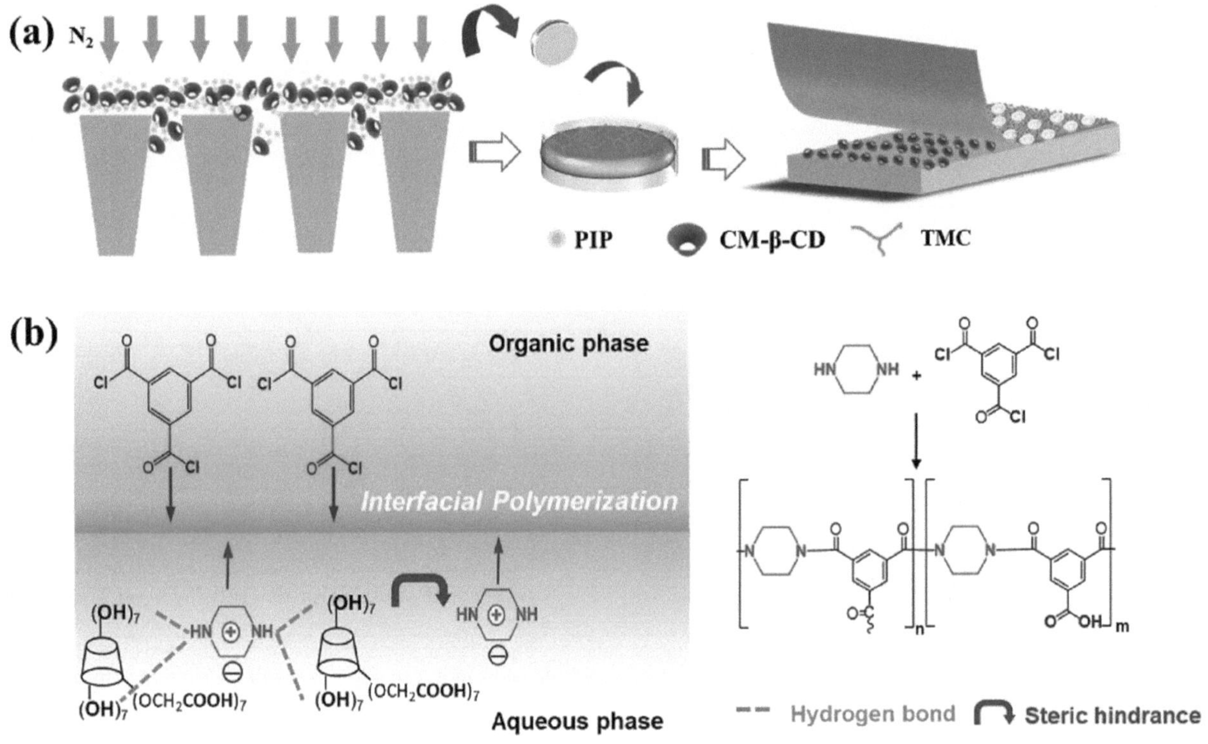

FIGURE 8.6 (a) Scheme of the fabrication process of the modified TFN NF membranes; (b) scheme of the role of CM-β-CD for regulating the diffusion of PIP monomers through a hydrogen bond, electrostatic attraction, steric hindrance (adapted with permission from Ref. [76]).

on the use of a novel composite NF membrane fabricated by polydopamine/poly(sulfobetaine methacrylate) (PDA/PSBMA) as the interlayer to enhance the water permeance of polyamide (PA) NF membrane for the separation of three antibiotics: erythromycin (ERY), tetracycline hydrochloride (TC), and cephalexin (CA). This study found that the NF membrane achieved a high removal efficiency (>97%) [74]. A similar study conducted by Mahdavi et al. reported that 98.2% ceftriaxone and 97.6% amoxicillin rejections were achieved with a novel NF membrane [75]. To improve the mechanical properties of PES membranes, a blend of polyethersulfone and thermoplastic polyurethane was used to prepare the membranes. The vapor-induced phase separation (VIPS)-NIPS method was employed to fabricate membranes to prevent the formation of large voids and improve their mechanical stability. Following the formation of PES-TPU membranes, a layer of Sabja seed mucilage (SSM), a low-cost and eco-friendly polysaccharide, was applied to the surface of the membranes. To enhance the rejection performance of the prepared membranes, the SSM layer was chelated with Cu^{2+} ions [75].

An interfacial polymerization (IP) process was used to fabricate a nanofiltration membrane modified with carboxymethyl-β-cyclodextrin (CM-β-CD). A solution of 0.1% w/v TMC in n-heptane and 0.2% w/v PIP was employed for the process of interfacial polymerization (IP). A varied concentration of CM-β-CD was mixed with an aqueous solution of PIP for IP. Subsequently, the resulting membrane was heated at 60°C for 15 minutes [76]. A schematic of the fabrication process including the mechanism and role of CM-β-CD is depicted in Figure 8.6. The presence of numerous carboxyl groups in CM-β-CD led to steric hindrance and electrostatic repulsion, which contributed to the high permeability (122.4 LMH) of the optimized membrane while still maintaining a competitive rejection rate for a concentrated cephalexin solution (94.7%) [76].

A successful membrane separation process heavily relies on a significant factor, which is the high flux. To improve the permeability, novel NF membranes such as amino-functional polyethylene glycol (PEG)-based TFC membrane have been studied [77]. Polysulfone-based NF membrane also exhibited higher flux during amoxicillin separation [70]. Understanding the mechanism of nanofiltration is very important during solute separation. Most of the separation is due to size exclusion. In addition to size exclusion, the role of electrostatic interactions in separating pharmaceuticals is an important mechanism [78]. Moreover, Yadav et al. developed a low-cost polymeric thin-film composite (TFC) membrane with polysulfone and Arg-MMT (arginine-montmorillonite) clay nanoparticles for cephalexin antibiotic removal. The rejection rate was >98%, with a permeation flux of 26.14 LMH [67].

8.3.2 Composite Membranes

Metal-organic frameworks (MOFs) are porous materials that consist of metal ions or clusters coordinated with organic ligands. They have a high surface area and tunable pore sizes, which make them suitable for separation applications. MOF membranes have been developed for the separation and purification of small molecules, such as drug molecules and chiral compounds. Highly water-stable MOF-based membranes have been demonstrated for dyes and antibiotics [79]. The rejection rate of the MOF membranes was above 94%. Many commercial polymeric membranes lack higher permeability for organic solvents, and so, recently, ultrathin polyamide membranes have been developed to solve the issue of permeability–rejection tradeoff using MOFs [60]. These composite membranes showed excellent improvements in water permeance (increased by 84%), and a high rate of rejection of pharmaceuticals, such as rose bengal (100%) and azithromycin (97.6%), while showing excellent separation performance in ethyl acetate [60].

Moreover, MXene membrane displays added advantages for membrane separation applications. For example, titanium carbide membrane displays extraordinary permeance, one order magnitude higher than polymeric NF due to its large aspect ratio of nanosheets while retaining the same degree of antibiotic rejection [80].

Using the phase inversion technique, researchers fabricated composite hollow-fiber membranes of sulfonated polyphenylsulfone (SPPSu) that contain titanium oxide (TiO_2). These membranes demonstrated a 25.5% increase in flux compared to the control membrane, as well as enhanced thermal and mechanical properties [81]. Bhattacharya et al. fabricated a composite ceramic UF membrane using CuO NPs and TiO_2 NPs on a clay-alumina-based macroporous support to remove antibiotics (ciprofloxacin) from an aqueous solution [82]. Some of the composite NF membranes showed extraordinary removal efficiency for pharmaceutical compounds such as erythromycin (100%) and tetracycline hydrochloride (99.4%) [74]. The advantages of these composite membranes include high stability and good rejection properties.

8.3.3 Carbon-Based Membranes

Carbon-based membranes have shown promise for various pharmaceutical applications due to their unique properties, including high selectivity and resistance to fouling. There are different classes of carbon allotrope and only carbon nanotube (CNT) and graphene membranes are focused on due to their standalone properties. CNT membranes are made from aligned arrays of carbon nanotubes and can provide high selectivity for small molecules, such as drug molecules. CNT membranes have been explored for drug delivery and water purification applications [32,83]. A solvent-resistant hollow-fiber membrane fabricated with robust polyamide P84 and NH_2-MWCNT/P84 is an alternative membrane for organic solvent nanofiltration. The developed membranes are suitable for selectively separating tetracycline/IPA, L-alpha-lecithin/hexane, and BINAP-Ru(II)/methanol solution [61].

Graphene oxide (GO) membranes are two-dimensional materials with high mechanical strength and high water permeability. GO membranes have been developed for various applications, including desalination, gas separation, and water purification. In the pharmaceutical industry, GO membranes have the potential for drug-delivery and protein-separation applications due to their biocompatibility and selectivity. For example, GO membranes were used to remove chemicals present in consumer products, such as triclosan and triclocarban [84]. High-permeable ultrathin GO laminar composite membrane was developed for active pharmaceutical ingredients and food additives, e.g., tetracycline, rifampicin, roxithromycin, spiramycin, vitamin B12, and lecithin. Generally, Donnan exclusion is significant in aqueous solutions; the rejection and flux are significantly altered if there is a charged molecule or membrane. However, Donnan exclusion is less effective in organic solvent nanofiltration [85]. Moreover, electrochemical treatments have demonstrated a promising approach for pharmaceutical compounds removal using laser-induced graphene membranes as electrodes [86,87] and carbon nanotube membranes [26].

These membranes can provide high selectivity and fouling resistance. For harsh environments, and in the pharmaceutical applications industry, chemical-resistant or solvent-resistant membranes are needed. Carbon-based laminated membranes are promising if they overcome the issue with high-viscosity solvents. Nie et al. designed hyper-looping channels using multiwalled carbon nanotubes (MWCNTs) intercalated within lanthanum (III) toward high-viscosity solvents. GO nanochannel and CNTs' hyper-looping pathways are shown in Figure 8.7. The carbon-based membranes that have been optimized to exhibit a very high ethanol flux of 138 LMH, and they demonstrate a separation efficiency of nearly 99% for organic dyes from ethanol [88]. The enhanced separation efficiency resulting in a higher flux was accomplished by creating a nano-architecture within the structure of the MWCNTs, thereby improving the overall performance. Generally, graphene-based membranes enhance the performance of molecular-based separation processes. Preparing large-scale single-layer porous graphene membranes poses a significant challenge. However, the development of multilayer graphene offers a promising approach to address this challenge. To fabricate a self-supporting multilayer GO membrane, optimization is required for various factors such as the concentration of GO suspension, the GO preparation process, and the filtration technique [49].

A recently discovered and developed graphene material called "laser-induced graphene" (LIG) is a 3D layer formed on polymeric surfaces (membranes) during a 10.6 μm CO_2 laser exposure. These membranes are electrically conductive, demonstrating effective removal of pharmaceutical compounds. An electrically conductive membrane was used as electrodes and, in the presence of electric potential, pollutants undergo electrochemical oxidation and reduction and electro sorption [86,87]. Self-supporting CNT-based membranes showed excellent removal of pharmaceutical compounds during the filtration process under an electric field [26]. Gao et al. presented an adjustable arch-bridged reduced graphene

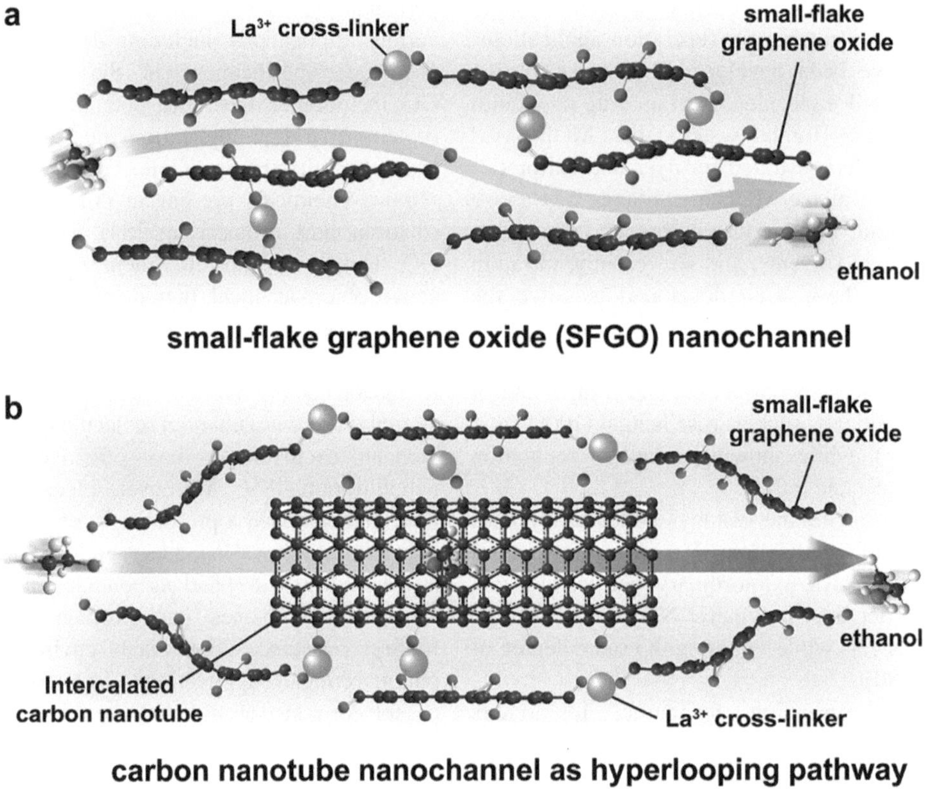

FIGURE 8.7 Organic solvent nanofiltration strategy: (a) small-flake graphene oxide (SFGO) nanochannel cross-linked by La^{3+} exhibits low permeance toward high viscosity solvent, such as ethanol. (b) MWCNT intercalated SFGO cross-linked membrane with MWCNTs as hyper-looping nanochannels can better facilitate fast permeance for high-viscosity solvents (adapted with permission from Ref. [88]).

oxide (rGO) nanofiltration membrane that can be aided with an applied voltage. At 5 V, the rGO membrane entirely rejects organic/anionic molecules. The voltage-assisted positive-charge-modified rGO membrane could reject both cationic and anionic dyes, with the ability to modulate harsh organic solvents [89].

Carbon-based membranes are generally biocompatible and can be modified to provide additional biocompatibility for specific applications. Overall, carbon-based membranes have the potential to enable new applications and improve existing processes in the pharmaceutical industry due to their unique properties and potential advantages for fouling control and higher selectivity. Overall, these electrically conductive membranes have the potential to address some of the limitations of traditional membranes, such as low selectivity, fouling, and poor stability, and to enable new applications in the pharmaceutical industry.

Overall, to fabricate high-performing membranes for bio-separation and pharmaceutical applications, the following key points should be considered.

- Selecting membrane materials with precise pore sizes, optimal surface charges, and excellent chemical compatibility according to the target of compound separation is crucial to ensure efficient separation while minimizing fouling.
- The careful selection of pore size distribution and membrane thickness is essential to achieving the desired levels of selectivity and flux rates, thereby enhancing membrane performance.
- Implementing suitable surface modification techniques, such as appropriate functionalization or coating, can significantly enhance membrane performance by reducing fouling and boosting separation efficiency.

8.4 CHALLENGES AND SOLUTIONS TO THE MEMBRANE PROCESS FOR PHARMACEUTICAL AND BIOCHEMICAL APPLICATIONS

Several challenges are associated with membrane separation processes for biochemical separation and pharmaceutical applications. These challenges include fouling, membrane degradation, limited selectivity, scale-up, and cost. The factors which affect the challenges and their impact on the process are given in Table 8.1. However, these challenges can be overcome by the proper implementation and development of high-performance membranes as discussed in the previous sections. In addition, membrane separation processes used in pharmaceutical applications must meet stringent regulatory requirements to ensure product purity, safety, and efficacy.

TABLE 8.1
Membrane process challenges and their impacts on the membrane separation process

Membrane challenges	Causes	Impacts on process
Membrane fouling	Feed properties, membrane surface properties, and operating conditions	Particle or molecule accumulation on the membrane surface or within the membrane pores, leading to reduced separation efficiency and increased energy consumption
Membrane degradation	Chemical, thermal, or mechanical stresses	Leading to reduced separation efficiency and shorter membrane lifetimes
Limited selectivity	Membrane selection, membrane properties	Loss of valuable products or contamination of the final product
Scaling up	Membrane fouling, fluid dynamics, and energy consumption	Limited product quantity
Cost	Specialized equipment, maintenance, and membrane replacement	Decreased industrial profit margin

TABLE 8.2
Strategies to overcome the membrane process challenges

Membrane challenges	Solutions	Outcomes
Fouling control	Membrane surface modification, pretreatment and integrated processes	Reduced fouling and extend membrane life
Membrane durability	Improving mechanical, chemical, thermal stability	Increased lifespan of the membranes
limited selectivity	new surface modifications and functionalization	Selectivity improvements
Scaling up	Developing optimized scaling-up strategies	Limiting the product quantity
Cost	More cost-effective membrane materials and process designs	Can reduce capital and operating costs

Addressing these challenges will require continued research and the development of new membrane materials, surface modifications, and process designs, as well as advances in monitoring and control technologies.

There are several strategies that can be employed to overcome the challenges associated with membrane separation processes for biochemical separation and pharmaceutical applications (Table 8.2). In addition, developing optimized scaling-up strategies for membrane separation processes, considering factors such as feed properties, operating conditions, and fluid dynamics, can lead to efficient and effective large-scale production. Transport modeling toward scale-up for pharmaceutical applications is essential [90]. Moreover, adhering to regulatory guidelines and standards and utilizing advanced monitoring and control technologies can help ensure product purity, safety, and efficacy.

Addressing these challenges will require a multidisciplinary approach involving materials science, chemistry, engineering, and process optimization. Continued research and development efforts can help to overcome these challenges and improve the efficiency, efficacy, and affordability of membrane separation processes for biochemical separation and pharmaceutical applications.

8.5 FUTURE PERSPECTIVES OF THE MEMBRANE PROCESS FOR BIOCHEMICAL AND PHARMACEUTICAL APPLICATIONS

Membrane processes have become increasingly important in bio-separation and pharmaceutical applications due to their high efficiency, low energy consumption, and minimal environmental impact. As the demand for biopharmaceuticals and other high-value products continues to grow, the use of membrane processes is expected to become even more widespread.

1. Modeling and simulation: Modeling and simulation of membrane processes significantly influence the development and optimization of membrane separation processes for pharmaceutical and biochemical industries. Some of the key benefits of modeling and simulation include process optimization by membrane process design and operation parameters, cost reduction by optimizing the use of resources such as energy, materials, and labor; additionally, modeling and simulation can facilitate the scale-up of membrane separation processes by providing insights into the effects of changes in operating conditions, membrane

properties, and feedstock properties on separation efficiency.

2. Production of biologics: One of the key areas where membrane processes are expected to have a significant impact is in the production of biologics, such as monoclonal antibodies, vaccines, and gene therapies. These products require high-purity and high-quality separation processes to ensure safety and efficacy.
3. Production of medicinal products: Another area where membrane processes are expected to make a significant impact is in the production of personalized medicine. Personalized medicine requires the ability to manufacture small batches of products with high precision and quality.
4. Membrane processes are also expected to have a significant impact on the development of continuous manufacturing processes in the pharmaceutical industry. This offers reduced costs and increased flexibility. Membrane processes, such as chromatography and filtration, are key components of continuous manufacturing processes and are expected to play an increasingly important role in the pharmaceutical industry.

8.6 CONCLUSION

In this chapter, our goal was to provide a comprehensive overview of membrane processes in biochemical and pharmaceutical applications, highlighting recent advancements. Each application requires specific membrane characteristics, tailored to its needs. For macromolecular separation, low-pressure UF membranes suffice, unlike high-pressure options such as NF or RO. Membranes find broad use in biomedical applications like artificial kidneys and livers, and drug delivery. Their versatility underscores their significance in advancing the healthcare and pharmaceutical fields.

In conclusion, membrane processes are poised to play an increasingly important role in bio-separation and pharmaceutical applications in the future. Their ability to provide high-purity and high-quality products with low energy consumption and minimal environmental impact makes them an attractive option for the pharmaceutical industry. As new technologies and materials continue to be developed, the use of membrane processes is expected to become even more widespread in the years to come. Advancements in membrane processes have revolutionized biochemical separation and pharmaceutical applications. These processes offer many advantages, including improved selectivity, efficiency, and cost-effectiveness compared to traditional separation techniques. The future looks promising for membrane processes as they continue to be optimized and utilized in various industries. The development of new membrane materials and processes is also expected to drive the future growth of membrane processes in bio-separation and pharmaceutical applications. Advances in membrane technology, such as the development of high-performance nanofiltration and reverse osmosis membranes, are expected further to increase the efficiency and effectiveness of membrane processes.

In conclusion, membrane processes are poised to play a vital role in bio-separation and pharmaceutical applications. They offer high-purity products, low energy consumption, and minimal environmental impact. Continued development of technologies and materials will further propel their widespread adoption. Membrane processes revolutionize biochemical separation and pharmaceutical applications, boasting improved selectivity, efficiency, and cost-effectiveness. Future growth is anticipated with the optimization of membrane materials and processes, including advancements in nanofiltration and reverse osmosis membranes, enhancing efficiency and effectiveness.

REFERENCES

[1] Nilsson, M.R. (2007) Survey of biochemical separation techniques. *Journal of Chemical Education* 84: 112–114.
[2] Saxena, A., Tripathi, B.P., Kumar, M., Shahi, V.K. (2009) Membrane-based techniques for the separation and purification of proteins: An overview. *Advances in Colloid and Interface Science* 145: 1–22.
[3] Shojaee Nasirabadi, P., Saljoughi, E., Mousavi, S.M. (2016) Membrane processes used for removal of pharmaceuticals, hormones, endocrine disruptors and their metabolites from wastewaters: A review. *New Pub Balaban* 57: 24146–24175.
[4] Daufin, G., Escudier, J.P., Carrère, H., Bérot, S., Fillaudeau, L., Decloux, M. (2001) Recent and emerging applications of membrane processes in the food and dairy industry. Food and Bioproducts Processing: *Transactions of the Institution of Chemical Engineers, Part C* 79: 89–102. https://doi.org/10.1205/096030801750286131.
[5] Michaels, A.S., Matson, S.L. (1985) Membranes in biotechnology: State of the art. *Desalination* 53: 231–258.
[6] van Reis, R., Zydney, A. (2007) Bioprocess membrane technology. *Journal of Membrane Science* 297: 16–50.
[7] Razdan, U., Joshi, S.V., Shah, V.J. (2003) Novel membrane processes for separation of organics. *Current Science* 85: 761–771.
[8] Ghosh, R. (2016) Bioseparations using integrated membrane processes. In: *Integrated Membrane System Process*, pp. 23–34. John Wiley & Sons, Ltd.
[9] Bao, G. (2002) Mechanics of biomolecules. *Journal of the Mechanics and Physics of Solids* 50: 2237–2274.
[10] Karplus, M., McCammon, J.A. (2002) Molecular dynamics simulations of biomolecules. *Nature Structural & Molecular Biology* 9: 646–652.
[11] Nagase, K. (2021)Thermoresponsive interfaces obtained using poly(N-isopropylacrylamide)-based copolymer for bioseparation and tissue engineering applications. *Advances in Colloid Interface Science* 295: 102487.
[12] Lan, T., Shao, Z.Q., Gu, M.J., Zhou, Z.W., Wang, Y.L., Wang, W.J., Wang, F.J., Wang, J.Q. (2015) Electrospun nanofibrous cellulose diacetate nitrate membrane for protein separation. *Journal of Membrane Science* 489: 204–211.
[13] Fukuda, H., Lee, J. (2022) Medium-chain fatty acids recovery from synthetic organic waste streams using supported liquid membranes. *ACS Sustainable Chemistry & Engineering* 10: 8370–8379.

[14] Kidambi, P.R., Jang, D., Idrobo, J.C., Boutilier, M.S.H., Wang, L., Kong, J., Karnik, R. (2017) Nanoporous atomically thin graphene membranes for desalting and dialysis applications. *Advanced Materials*https://doi.org/10.1002/adma.201700277.

[15] Asatekin, A., Olivetti, E.A., Mayes, A.M. (2009) Fouling resistant, high flux nanofiltration membranes from polyacrylonitrile-graft-poly(ethylene oxide). *Journal of Membrane Science* 332: 6–12.

[16] Yang, W., Sundaram, H.S., Ella, J.R., He, N., Jiang, S. (2016) Low-fouling electrospun PLLA films modified with zwitterionic poly(sulfobetaine methacrylate)-catechol conjugates. *Acta Biomaterialia* 40: 92–99.

[17] Bell, D.J., Ludwanowski, S., Luken, A., Sarikaya, B., Walther, A., Wessling, M. (2021) Hydrogel membranes made from crosslinked microgel multilayers with tunable density. *Journal of Membrane Science* 620.

[18] Venault, A., Wei, T.C., Shih, H.L., Yeh, C.C., Chinnathambi, A., Alharbi, S.A., Carretier, A., Aimar, P., Lai, J.Y., Chang, Y. (2016) Antifouling pseudo-zwitterionic poly(vinylidene fluoride) membranes with efficient mixed-charge surface grafting via glow dielectric barrier discharge plasma-induced copolymerization. *Journal of Membrane Science* 516: 13–25.

[19] Maleki, H., Azimi, B., Ismaeilimoghadam, S., Danti, S. (2022) Poly(lactic acid)-based electrospun fibrous structures for biomedical applications. *Applied Science* 12: 3192.

[20] Morelli, S., Piscioneri, A., Salerno, S., de Bartolo, L. (2022) Hollow fiber and nanofiber membranes in bioartificial liver and neuronal tissue engineering. *Cells Tissues Organs* 211: 447–476.

[21] Radu, E.R., Voicu, S.I., Thakur, V.K. (2023) Polymeric membranes for biomedical applications. *Polymers* 15: 619.

[22] Li, X., Janke, A., Formanek, P., Fery, A., Stamm, M., Tripathi, B.P. (2020) High permeation and antifouling polysulfone ultrafiltration membranes with in situ synthesized silica nanoparticles. *Materials Today Communications* 22: 100784.

[23] Liu, X., Wang, C., Wang, B., Li, K. (2017) Novel organic-dehydration membranes prepared from zirconium metal-organic frameworks. *Advanced Functional Materials* 27.

[24] Su, S., Wu, W., Gao, J., Lu, J., Fan, C. (2012) Nanomaterials-based sensors for applications in environmental monitoring. *Journal of Materials Chemistry* 22: 18101.

[25] Thamaraiselvan, C., Ronen, A., Lerman, S., Balaish, M., Ein-Eli, Y., Dosoretz, C.G. (2018) Low voltage electric potential as a driving force to hinder biofouling in self-supporting carbon nanotube membranes. *Water Research* 129: 143–153.

[26] Thamaraiselvan, C., Lau, W.J., Dosoretz, C.G. (2022) Coupled electrochemical transformation and filtration of water pollutants by cathodic-carbon nanotube membranes. *Journal of Environmental Chemical Engineering* 10: 107670.

[27] Ye, R., James, D.K., Tour, J.M. (2019) Laser-induced graphene: From discovery to translation. *Advanced Materials* 31: 1803621.

[28] Lin, J., Peng, Z., Liu, Y., Ruiz-Zepeda, F., Ye, R., Samuel, E.L.G., Yacaman, M.J., Yakobson, B.I., Tour, J.M. (2014) Laser-induced porous graphene films from commercial polymers. *Nature Communications* 5: 5714.

[29] Zhao, S., Cheng, Y., Veder, J.P., Johannessen, B., Saunders, M., Zhang, L., Liu, C., Chisholm, M.F., De Marco, R., Liu, J., Yang, S.Z., Jiang, A.P. (2018) One-pot pyrolysis method to fabricate carbon nanotube supported Ni single-atom catalysts with ultrahigh loading. *ACS Applied Energy Materials* 1: 5286–5297.

[30] Zhang, Y., Zhang, L., Zhou, C. (2013) Review of chemical vapor deposition of graphene and related applications. *Accounts of Chemical Research* 46: 2329–2339.

[31] Cao, Y., Chen, G., Wan, Y., Luo, J. (2021) Nanofiltration membrane for bio-separation: Process-oriented materials innovation. *Engineering in Life Sciences* 21: 405–416.

[32] Thamaraiselvan, C., Lerman, S., Weinfeld-Cohen, K., Dosoretz, C.G. (2018) Characterization of a support-free carbon nanotube-microporous membrane for water and wastewater filtration. *Separation and Purification Technology* 202: 1–8.

[33] Anis, S.F., Hashaikeh, R., Hilal, N. (2019) Microfiltration membrane processes: A review of research trends over the past decade. *Journal of Water Process Engineering* 32: 100941.

[34] Neal, G., Francis, R., Shamlou, P.A., Keshavarz-Moore, E. (2004) Separation of immunoglobulin G precipitate from contaminating proteins using microfiltration. *Biotechnology and Applied Biochemistry* 39: 241.

[35] Minbu, H., Ochiai, A., Kawase, T., Taniguchi, M., Lloyd, D.R., Tanaka, T. (2015) Preparation of poly(L-lactic acid) microfiltration membranes by a nonsolvent-induced phase separation method with the aid of surfactants. *Journal of Membrane Science* 479: 85–94.

[36] Pitiot, O., Legallais, C., Darnige, L., Vijayalakshmi, M.A. (2000) A potential set up based on histidine hollow fiber membranes for the extracorporeal removal of human antibodies. *Journal of Membrane Science* 166: 221–227.

[37] Weigert, T., Altmann, J., Ripperger, S. (1999) Crossflow electrofiltration in pilot scale. *Journal of Membrane Science* 159: 253–262.

[38] Van Reis, R., Zydney, A.L. (2010) Protein ultrafiltration. In: *Encyclopedia of Industrial Biotechnology*, pp. 1–25. John Wiley & Sons, Ltd.

[39] Flickinger, M.C. (2009) *Encyclopedia of Industrial Biotechnology*. Wiley.

[40] Li, Q., Yan Bi, Q., Lin, H.H., Bian, L.X., Wang, X.L. (2013) A novel ultrafiltration (UF) membrane with controllable selectivity for protein separation., *Journal of Membrane Science* 427: 155–167.

[41] Geoffroy, T.R., Thibodeau, J., Faucher, M., Langevin, M.E., Lutin, F., Bazinet, L. (2022) Relationship between feed concentration and bioactive cationic peptide recovery: Impact on ecoefficiency of EDUF at semi-industrial scale. *Separation and Purification Technology* 286: 120403.

[42] Modi, A., Bellare, J. (2019) Efficient separation of biological macromolecular proteins by polyethersulfone hollow fiber ultrafiltration membranes modified with Fe3O4 nanoparticles-decorated carboxylated graphene oxide nanosheets. *International Journal of Biological Macromolecules* 135: 798–807.

[43] Awang Chee, D.N., Ismail, A.F., Aziz, F., Mohamed Amin, M.A., Abdullah, N. (2020) The influence of alumina particle size on the properties and performance of alumina hollow fiber as support membrane for protein separation. *Separation and Purification Technology* 250: 117147.

[44] Roblet, C., Amiot, J., Lavigne, C., Marette, A., Lessard, M., Jean, J., Ramassamy, C., Moresoli, C., Bazinet, L. (2012) Screening of in vitro bioactivities of a soy protein hydrolysate separated by hollow fiber and spiral-wound ultrafiltration membranes. *Food Research International* 46: 237–249.

[45] Wen-qiong, W., Yun-chao, W., Xiao-feng, Z., Rui-xia, G., Mao-lin, L. (2019) Whey protein membrane processing methods and membrane fouling mechanism analysis. *Food Chemistry* 289: 468–481.

[46] Mallakpour, S., Azadi, E. (2022) Nanofiltration membranes for food and pharmaceutical industries. *Emergent Materials* 5: 1329–1343.

[47] Mohammad, A.W., Teow, Y.H., Ang, W.L., Chung, Y.T., Oatley-Radcliffe, D.L., Hilal, N. (2015) Nanofiltration membranes review: Recent advances and future prospects. *Desalination* 356: 226–254.

[48] Weber, R., Chmiel, H., Mavrov, V. (2003) Characteristics and application of new ceramic nanofiltration membranes. *Desalination* 157: 113–125.

[49] Nie, L., Chuah, C.Y., Bae, T.H., Lee, J.M. (2021) Graphene-based advanced membrane applications in organic solvent nanofiltration. *Advanced Functional Materials* 31: 2006949.

[50] Guo, S., Zhang, H., Chen, X., Feng, S., Wan, Y., Luo, J. (2021) Fabrication of antiswelling loose nanofiltration membranes via a "selective-etching-induced reinforcing" strategy for bioseparation. *ACS Applied Materials Interfaces* 13: 19312–19323.

[51] Zhao, G., Hu, R., Zhao, X., He, Y., Zhu, H. (2019) High flux nanofiltration membranes prepared with a graphene oxide homo-structure. *Journal of Membrane Science* 585: 29–37.

[52] Vatanpour, V., Mousavi Khadem, S.S., Dehqan, A., Al-Naqshabandi, M.A., Ganjali, M.R., Sadegh Hassani, S., Rashid, M.R., Saeb, M.R., Dizge, N. (2021) Efficient removal of dyes and proteins by nitrogen-doped porous graphene blended polyethersulfone nanocomposite membranes. *Chemosphere* 263: 127892.

[53] Van der Bruggen, B., Mänttäri, M., Nyström, M. (2008) Drawbacks of applying nanofiltration and how to avoid them: A review. *Separation and Purification Technology* 63: 251–263.

[54] Jiang, S., Li, Y., Ladewig, B.P. (2017) A review of reverse osmosis membrane fouling and control strategies. *Science of the Total Environment* 595: 567–583.

[55] Kang, G., Cao, Y. (2012) Development of antifouling reverse osmosis membranes for water treatment: A review. *Water Research* 46: 584–600.

[56] Bélafi-Bakó, K., Nemestóthy, N., Gubicza, L. (2004) A study on applications of membrane techniques in bioconversion of fumaric acid to L-malic acid. *Desalination* 162: 301–306.

[57] García-García, V., Montiel, V., González-García, J., Expósito, E., Iniesta, J., Bonete, P., Inglés, M. (2000) The application of electrodialysis to desalting an amino acid solution. *Journal of Chemical Education* 77: 1477–1479.

[58] Jordan, M.L., Kokoszka, G., Dona, H.K.G., Senadheera, D.I., Kumar, R., Lin, Y.J., Arges, C.G. (n.d.) Integrated ion-exchange membrane resin wafer assemblies for aromatic organic acid separations using electrodeionization. *ACS Sustainable Chemistry & Engineering*. https://doi.org/10.1021/acssuschemeng.2c05255.

[59] Thakur, A.K., Singh, S.P., Thamaraiselvan, C., Kleinberg, M.N., Arnusch, C.J. (2019) Graphene oxide on laser-induced graphene filters for antifouling, electrically conductive ultrafiltration membranes. *Journal of Membrane Science* 591: 117322.

[60] Cheng, X.Q., Jiang, X., Zhang, Y.Q., Lau, C.H., Xie, Z.L., Ng, D., Smith, S.J.D., Hill, M.R., Shao, L. (2017) Building additional passageways in polyamide membranes with hydrostable metal organic frameworks to recycle and remove organic solutes from various solvents. *ACS Applied Materials Interfaces* 9: 38877–38886.

[61] Farahani, M., Chung, T.S. (2018) Solvent resistant hollow fiber membranes comprising P84 polyimide and amine-functionalized carbon nanotubes with potential applications in pharmaceutical, food, and petrochemical industries. *Chemical Engineering Journal* 345: 174–185.

[62] Cheng, X.Q., Liu, Y.Y., Guo, Z.H., Shao, L. (2015) Nanofiltration membrane achieving dual resistance to fouling and chlorine for "green" separation of antibiotics. *Journal of Membrane Science* 493: 156–166.

[63] Abdellah, M.H., Perez-Manriquez, L., Puspasari, T., Scholes, C.A., Kentish, S.E., Peinemann, K.V. (2018) A catechin/cellulose composite membrane for organic solvent nanofiltration. *Journal of Membrane Science* 567: 139–145.

[64] Zheng, S., Yang, S., Ouyang, Z., Zhang, Y. (2023) Robust and highly hydrophilic ultrafiltration membrane with multi-branched cellulose nanocrystals for permeability-selectivity anti-trade-off property. *Applied Surface Science* 614: 156157.

[65] Sun, H., Li, X., Wang, N., An, Q.F. (2023) Defect engineering on zeolitic imidazolate framework membrane via thermal annealing for organic solvent nanofiltration. *Separation and Purification Technology* 310: 123220.

[66] Zheng, S., Yang, S., Ouyang, Z., Chen, T., Kuang, Y., Shen, H., Yang, F., Zhang, Y. (2023) Performance investigation of hydrophilic regenerated cellulose ultrafiltration membranes with excellent anti-fouling property via hydrolysis technology. *Journal of Environmental Chemical Engineering* 11: 109041.

[67] Yadav, D., Borpatra Gohain, M., Karki, S., Ingole, P.G. (2022) A novel approach for the development of low-cost polymeric thin-film nanocomposite membranes for the biomacromolecule separation. *ACS Omega* 7: 47967–47985.

[68] Ma, T.Y., Su, Y.L., Li, Y.F., Zhang, R.N., Liu, Y.N., He, M.R., Li, Y.F., Dong, N.X., Wu, H., Jiang, Z.Y. (2016) Fabrication of electro-neutral nanofiltration membranes at neutral pH with antifouling surface via interfacial polymerization from a novel zwitterionic amine monomer. *Journal of Membrane Science* 503: 101–109.

[69] Liu, Y.L., Wang, X.M., Yang, H.W., Xie, Y.F.F., Huang, X. (2019) Preparation of nanofiltration membranes for high rejection of organic micropollutants and low rejection of divalent cations. *Journal of Membrane Science* 572: 152–160.

[70] Derakhsheshpoor, R., Homayoonfal, M., Akbari, A., Mehrnia, M.R. (2013) Amoxicillin separation from pharmaceutical wastewater by high permeability polysulfone nanofiltration membrane. *Journal of Environmental Health Science and Engineering* 11: 9.

[71] Zhen, H., Wu, M., Yuan, Z., Qi, Z., Meng, Y., Zu, X., Liu, D., He, G., Jiang, X. (2023) Nanofiltration membrane with CM-β-CD tailored polyamide layer for high concentration cephalexin solution separation. *Journal of Membrane Science* 672: 121445.

[72] Gomes, D., Cardoso, M., Martins, R.C., Quinta-Ferreira, R.M., Gando-Ferreira, L.M. (2020) Removal of a mixture of pharmaceuticals sulfamethoxazole and diclofenac from water streams by a polyamide nanofiltration membrane. *Water Science and Technology* 81: 732–743.

[73] Karimnezhad, H., Navarchian, A.H., Tavakoli Gheinani, T., Zinadini, S. (2020) Amoxicillin removal by Fe-based nanoparticles immobilized on polyacrylonitrile membrane: Individual nanofiltration or Fenton reaction, vs. engineered combined process. *Chemical Engineering Research and Design* 153: 187–200.

[74] Lei Wang, X., Wang, Q., Xian Xue, Y., Zhang, B., Rui Han, S., Zhang, H., Yin Zhao, K., Wang, W., Fu Wei, J. (2023) Preparation of composite nanofiltration membrane with interlayer for pharmaceutical rejection. *Separation and Purification Technology* 312: 123411.

[75] Mahdavi, H., Hosseini, F., Heidari, A.A., Karami, M. (2023) Polyethersulfone-TPU blend membrane coated with an environmentally friendly sabja seed mucilage-Cu^{2+} cross-linked layer with outstanding separation performance and superior antifouling. *Journal of Industrial and Engineering Chemistry* 121: 421–433.

[76] Zhen, H., Wu, M., Yuan, Z., Qi, Z., Meng, Y., Zu, X., Liu, D., He, G., Jiang, X. (2023) Nanofiltration membrane with CM-β-CD tailored polyamide layer for high concentration cephalexin solution separation. *Journal of Membrane Science* 672: 121445.

[77] Cheng, X.Q., Shao, L., Lau, C.H. (2015) High flux polyethylene glycol based nanofiltration membranes for water environmental remediation. *Journal of Membrane Science* 476: 95–104.

[78] Nghiem, L.D., Schäfer, A.I., Elimelech, M. (2006) Role of electrostatic interactions in the retention of pharmaceutically active contaminants by a loose nanofiltration membrane. *Journal of Membrane Science* 286: 52–59.

[79] Fang, S.Y., Zhang, P., Gong, J.L., Tang, L., Zeng, G.M., Song, B., Cao, W.C., Li, J., Ye, J. (2020) Construction of highly water-stable metal-organic framework UiO-66 thin-film composite membrane for dyes and antibiotics separation. *Chemical Engineering Journal* 385.

[80] Li, Z.K., Wei, Y.Y., Gao, X., Ding, L., Lu, Z., Deng, J.J., Yang, X.F., Caro, J., Wang, H.H. (2020) Antibiotics separation with MXene membranes based on regularly stacked high-aspect-ratio nanosheets. *Angew Chemie-International Ed.* 59: 9751–9756.

[81] Dass, L.A. Alhoshan, M., Alam, J., Muthumareeswaran, M.R., Figoli, A., Shukla, A.K. (2017) Separation of proteins and antifouling properties of polyphenylsulfone based mixed matrix hollow fiber membranes. *Separation and Purification Technology* 174: 529–543.

[82] Bhattacharya, P., Mukherjee, D., Dey, S., Ghosh, S., Banerjee, S. (2019) Development and performance evaluation of a novel CuO/TiO_2 ceramic ultrafiltration membrane for ciprofloxacin removal. *Materials Chemistry and Physics* 229: 106–116.

[83] Thamaraiselvan, C., Wang, J., James, D.K., Narkhede, P., Singh, S.P., Jassby, D., Tour, J.M., Arnusch, C.J. (2020) Laser-induced graphene and carbon nanotubes as conductive carbon-based materials in environmental technologydoi.org/10.1016/j.mattod.2019.08.014.

[84] Oh, Y., Armstrong, D.L., Finnerty, C., Zheng, S.X., Hu, M., Torrents, A., Mi, B.X. (2017) Understanding the pH-responsive behavior of graphene oxide membrane in removing ions and organic micropollulants. *Journal of Membrane Science* 541: 235–243.

[85] Li, B.F., Cui, Y., Japip, S., Thong, Z.W., Chung, T.S. (2018) Graphene oxide (GO) laminar membranes for concentrating pharmaceuticals and food additives in organic solvents. *Carbon* 130: 503–514.

[86] Thamaraiselvan, C., Bandyopadhyay, D., Powell, C.D., Arnusch, C.J. (2021) Electrochemical degradation of emerging pollutants via laser-induced graphene electrodes. *Chemical Engineering Journal Advances* 8: 100195.

[87] Thamaraiselvan, C., Thakur, A.K., Gupta, A., Arnusch, C.J. (2021) Electrochemical removal of organic and inorganic pollutants using robust laser-induced graphene membranes. *ACS Applied Materials Interfaces* 13: 1452–1462.

[88] Nie, L., Goh, K., Wang, Y., Velioğlu, S., Huang, Y., Dou, S., Wan, Y., Zhou, K., Bae, T.H., Lee, J.M. (2023) Hyperlooping carbon nanotube-graphene oxide nanoarchitectonics as membranes for ultrafast organic solvent nanofiltration. *ACS Materials Letters* 5: 357–369.

[89] Gao, T., Wen, Y., Li, C., Cheng, H., Jin, X.R., Ai, X., Yang, Y., Zhou, K.G., Qu, L. (2022) Electrically modulated nanofiltration membrane based on an arch-bridged graphene structure for multicomponent molecular separation. *ACS Nano* 17: 17.

[90] Chakrabortty, S., Nayak, J., Pal, P., Kumar, R., Chakraborty, P. (2020) Separation of COD, sulphate and chloride from pharmaceutical wastewater using membrane integrated system: Transport modeling towards scale-up. *Journal of Environmental Chemical Engineering* 8: 104275.

[91] Enevoldsen, A.D., Hansen, E.B., Jonsson, G. (2007) Electro-ultrafiltration of industrial enzyme solutions. *Journal of Membrane Science* 299: 28–37.

9 Expansive Applications of Chitosan and Its Derivatives in Membrane Technology

P. Satishkumar[1], *Arun M. Isloor*[1*], *and Ramin Farnood*[2]
[1]Membrane and Separation Technology Laboratory, Department of Chemistry, National Institute of Technology Karnataka, Surathkal, Mangalore, India
[2]Department of Chemical Engineering and Applied Chemistry, University of Toronto, Canada
*Corresponding author: isloor@yahoo.com

9.1 INTRODUCTION

9.1.1 Membrane Technology

Over the past few decades, membrane technology has earned great importance, especially in the separation, fuel cells, biomedical, and water purification fields. The reasons for this include low energy demand, ease of preparation, molecular level tunability, precise selectivity, and the ability to scale up. However, membrane technology is a young offshoot of chemical research when compared to other traditional separation techniques such as distillation, sedimentation, centrifugation, crystallization, extraction, etc. A membrane is an interface that functions as a selective hurdle and controls the movement of chemical species through it [1]. Studies on membranes were started in 1748 by Nollet with the effect of osmotic pressure [2]. In 1861, Graham used a synthetic membrane for his dialysis experiment. Synthetic membranes were extensively used in numerous applications on a commercial scale. A vast array of polymers have been utilized in developing synthetic membranes, which include polysulfone (PSF) [3], polyvinylidene fluoride (PVDF) [4], polyphenyl sulfone (PPSU) [5], polyether sulfone (PES) [6], polyimide (PI) [7], polyamide (PA) [8], etc. Membranes are segregated into several categories, among which membranes based on pore size and morphology are of significant importance. Pressure as well as pore size-dependent membranes such as ultrafiltration (UF) [9], nanofiltration (NF) [10], and reverse osmosis (RO) [11] membranes have undergone many modifications due to research and utilitarian interest. These modifications include the addition of inorganic additives and surface functionalization. This has helped to overcome certain membrane impediments such as fouling and hydrophobicity [12]. Membrane properties can be altered by preparing a precise pore size of interest to a particular application. Based on morphology, membranes are categorized into asymmetric, symmetric, and composite membranes. Composite membranes have gained remarkable attention from scientists due to their bountiful advantages involving better selectivity from the active layer and robust mechanical strength from microporous support. In an endeavor to make the membrane process more eco-friendly, natural polymers have been significantly employed in the last few years. Many biopolymers are profusely available in nature and accessible through renewable resources. The advantage of natural polymers over synthetic polymers is that they can considerably lower the dependence on exhaustible, non-renewable fossil fuels and cause less harm to the environment since most of them are biodegradable in nature. Many of the natural polymers, for instance, polylactic acid (PLA), chitosan (CS), and cellulose acetate (CA) [13], lack robust mechanical strength. However, the development of composite membranes has resolved this issue and boosted the usage of natural polymers in membrane technology [14].

9.1.2 Chitosan

Crustaceans are widespread marine life forms, including more than 52,000 species globally [15]. Among these, crabs, shrimp, and lobsters are most popular. These crustacean species contain chitin in their shell, and chitin has been found to be one of the most abundant biopolymers in the world. Other than crustaceans, chitin is also found in the cell walls of microorganisms, cartilage of mollusks, and skin of insects [16]. N-deacetylation of chitin yields chitosan [poly-β-(1→4)-2-amino-2-deoxy-D-glucose]. Chitosan is a polysaccharide formed by the D-glucosamine monomer units and is shown in Figure 9.1 [17].

Overriding aspects of chitosan include that it is nontoxic and biodegradable. Chitosan has also been revealed to have antimicrobial activity, immunological activity, biocompatibility, bioactivity, and renewability [18]. The film-forming ability of chitosan enabled its extensive usage in the field of membrane technology. The presence of hydroxyl and primary amine groups in chitosan effectively augmented membrane hydrophilicity and eventually reduced the fouling tendency [19]. However, chitosan is not devoid of limitations such as low mechanical strength and poor solubility in many the organic solvents. To overcome these drawbacks and for the special applications, chitosan has been modified in many different forms.

DOI: 10.1201/9781003285656-9

9.1.3 Chitosan Membranes and Their Modifications

The presence of the primary amine group plays a key role in developing a variety of modified biopolymers by reacting with numerous substituents. Padaki and co-workers treated chitosan with phthalic anhydride in dimethylformamide to obtain N-phthaloylchitosan as displayed in Figure 9.2 [20]. The formation of N-phthaloylchitosan (NPCS) was confirmed by IR spectroscopic data in which carbonyl peaks appeared at 1664 cm^{-1} and 1679 cm^{-1}. This N-phthaloylchitosan-PSF composite membrane showed 95% rejection of $MgSO_4$, 75% of NaCl, and 78% of Na_2SO_4, indicating its superior salt rejection capability.

Shenvi et al. cross-linked poly(1,4-phenylene ether ether sulfone) (PPEES)-supported chitosan membrane with glutaraldehyde at different concentrations and checked for its salt rejection property [21]. It showed a 34% rejection rate for NaCl and 53% for $MgSO_4$. Cross-linking plays a crucial role in preventing partial dissolution of chitosan in an acidic medium owing to the protonation of primary amine moiety. In another study, they ionically cross-linked the chitosan layer with sodium tripolyphosphate, which is supported by PPEES [22]. The prepared composite was successful in rejecting 55% $MgSO_4$ and 21% NaCl. Blending is one of the predominant techniques in membrane processing in addition to grafting and coating [23]. For the first time, Kumar and co-workers blended the chitosan membrane with PSF and incorporated titanium dioxide nanotubes into it [24]. TiO_2 nanotubes were uniformly dispersed in the chitosan-PSF composite membrane and exhibited a maximum fouling resistance ratio of 76%. The difficulty in blending chitosan was due to its insolubility in many of the common organic solvents. A water-soluble chitosan derivative N-propylphosphonic chitosan (NPPCS) was synthesized by reacting chitosan with hydroxybenzotriazole (HOBt) [25]. NPCS-PSF composite membrane showed a notable flux recovery ratio of 74%. Another water-soluble modification for chitosan was in the form of N-succinyl chitosan (NSCS) [26]. Twenty percent NSCS incorporated PSF-NSCS membrane displayed improved hydrophilicity and a flux recovery ratio of 70%. Other than the polymer nature and additives incorporated, the composition of the coagulation bath also significantly affected the membrane pore formation during the phase inversion process. Finger-like macrovoids formation takes place with a high coagulation rate, whereas porous sponge-like structures dominate with a slow coagulation rate. In one of the experiments, glutaraldehyde cross-linked chitosan solution was used as a coagulation bath for immersing PSF-poly(isobutylene-alt-maleic anhydride) (PIAM) membrane [27]. Later, PIAM was hydrolyzed by treating with an NaOH solution. This PSF-PIAM-CS membrane illustrated a fouling resistance ratio of 74%, and salt rejection was found to be in the order of Na_2SO_4 > $MgSO_4$ > NaCl. Chitosan-based membranes were also extensively used for the removal of heavy metals from water sources. PSF-N-succinyl chitosan membrane exhibited excellent rejection of Cu, Ni, and Cd at the values of 98%, 95%, and 92%, respectively [28]. Organo clays, such as cloisite 15 A and 30B, were mixed with chitosan solution and a PVDF-chitosan-clay composite membrane was prepared [29]. These membranes showed

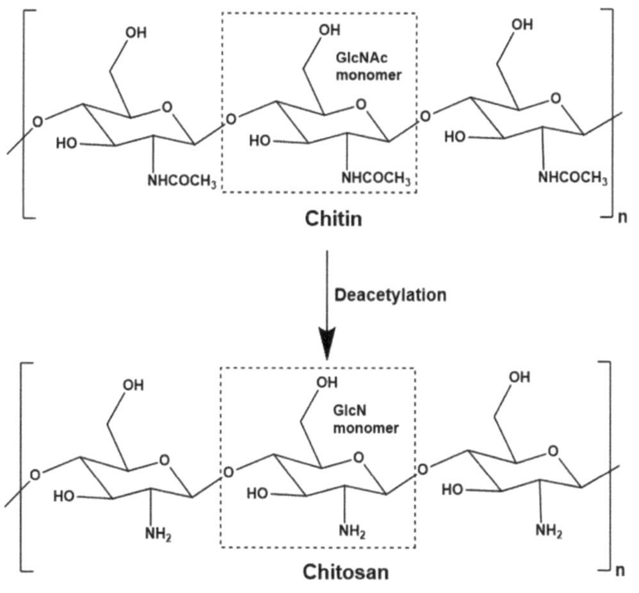

FIGURE 9.1 Structure of chitin and chitosan. Adapted from Ref. [17].

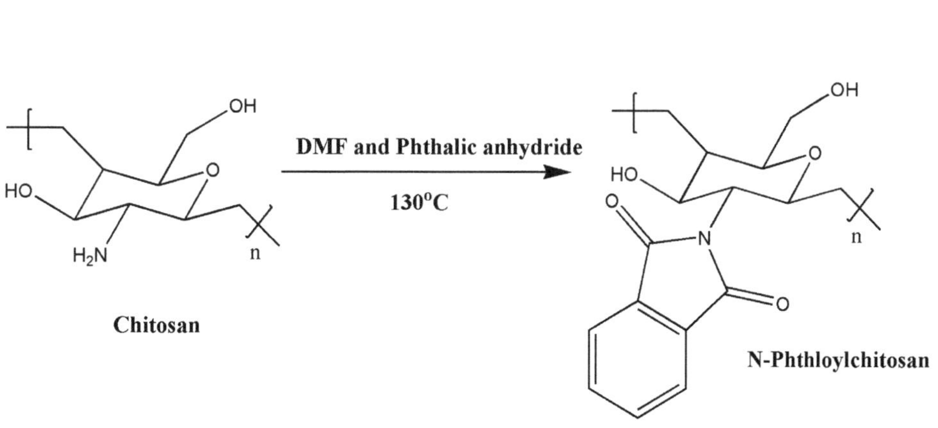

FIGURE 9.2 Modification of chitosan to N-phthloylchitosan. Adapted from Ref. [20].

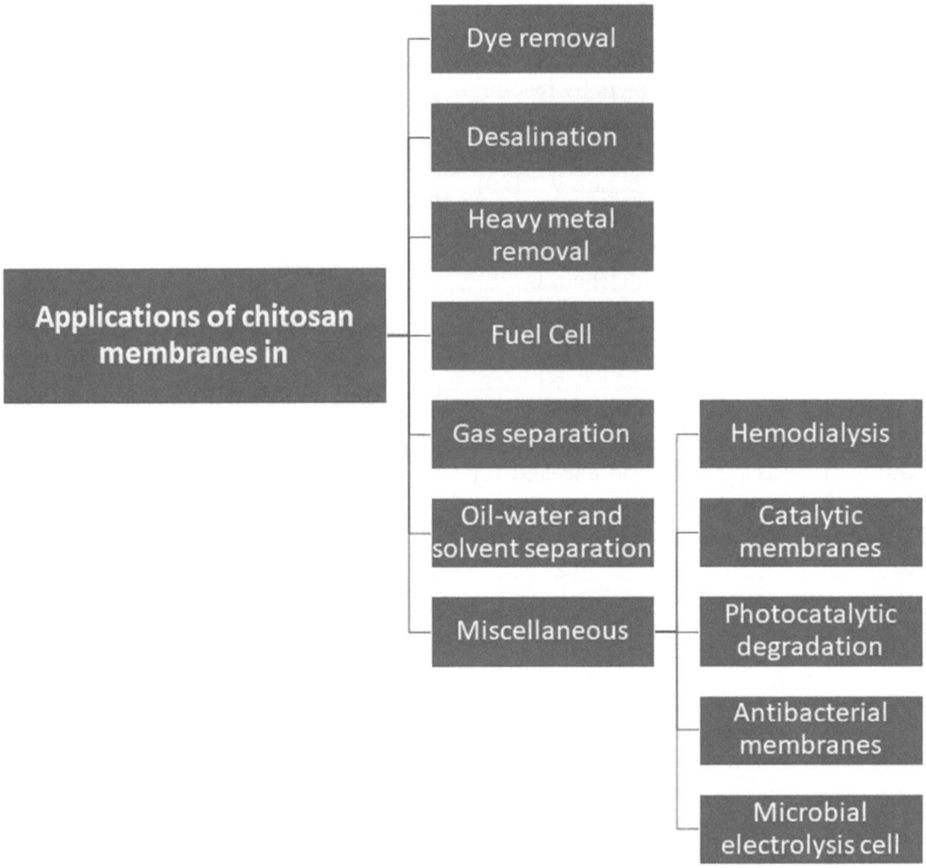

FIGURE 9.3 A brief outline of the various applications of chitosan membranes that were included in this review.

remarkable dye (methylene blue) removal efficiency. Chitosan membranes also found their application in gas separation. A water-swollen chitosan membrane effectively separated CO_2 from a ternary mixture of 10% CO_2, 10% H_2, and 80% N_2 [30]. Studies have been carried out with a wide temperature range from 20 to 150^{0}C and a pressure range from 1.5 to 5 atm. It finds valuable application in flue gas purification. Efforts have been made to replace commercial Nafion proton exchange membrane (PEM) with natural polymers like chitosan in fuel cells. Chitosan PEMs showed higher hydrophilicity and water uptake capacity than Nafion 117 [31]. However, the proton conductivity chitosan membrane was found to be 0.005 S cm^{-1}, which is less than that of the pure Nafion 117 membrane (0.08 S cm^{-1}). Numerous review papers have been published so far on chitosan applications. However, many of these review papers focused only on one or two applications of chitosan membranes among their wide range of applications in dye removal [32], heavy metal removal [33], fuel cells [34], wound healing [35], catalysis [36], desalination [37], wastewater treatment [38], gas separation [39], solvent separation, food industry [40], controlled drug delivery [41], and many more. This chapter covers diverse applications of chitosan-based membranes (Figure 9.3) and describes recent developments that have taken place in this field.

9.2 APPLICATIONS OF CHITOSAN-BASED MEMBRANES

9.2.1 Dye Removal

Dyes are contaminating water sources at an alarming rate due to their immense usage in the paper, plastic, paint, and textile industries [42]. To reduce the environmental pollution and harmful effects caused by dyes, it is of foremost importance to remove them from effluents before discharge. He et al. prepared a hybrid chitosan membrane by cross-linking oxidized starch (OS) and silica (Si) [43]. This chitosan/oxidized starch/silica (CS/OS/Si) composite membrane showed notable improvements in the thermal stability and swelling property. CS/OS/Si membrane exhibited excellent dye adsorption capacity against blue 71 and red 31. The dye adsorption efficiency of the membrane is found to be maximum at 60°C temperature and pH 9.82. Li and co-workers developed a chitosan nanofibrous membrane using the electrospinning technique to remove the dye acid blue 113 [44]. Among the various nanofibrous membranes prepared with distinct fiber diameters, fibers with 86 nm showed a remarkable adsorption capacity of 1338 mg/g. The adsorption isotherm obtained was in line with the Langmuir adsorption isotherm. They also studied the membrane efficiency in the long-term treatment of colored water. The membrane displayed appreciable regeneration of pure chitosan nanofibrous membrane even after four cycles. Chitosan and humic acid (HA)–chitosan gels were treated with polyurethane foams (PUF) to prepare PUF-CS-HA membranes by the hot pressing method [45]. Membrane filtration experiments were carried out using anionic methyl orange (MO), cationic methylene blue (MB), and neutral rhodamine B (RB) dyes. PUF-CS membrane showed retention rates of 99.7 % and 65% for MO and RB, respectively, while retention of MB was not observed. This illustrates the PUF-CS membrane ability in selectively separating MB from RB and MO solutions. PUF-CS-HA membrane with a 0.2:1 ratio concerning HA to CS was successful in retaining 62.1% MB, while retention rates for MO and RB were found to be 97.7% and 71.6%, respectively. In this experiment adsorption isotherms were in concurrence with the Freundlich adsorption isotherm. Figure 9.4 [45] clearly shows the retention capacity of the two different membranes for three different dyes along with photographs and UV-visible spectra.

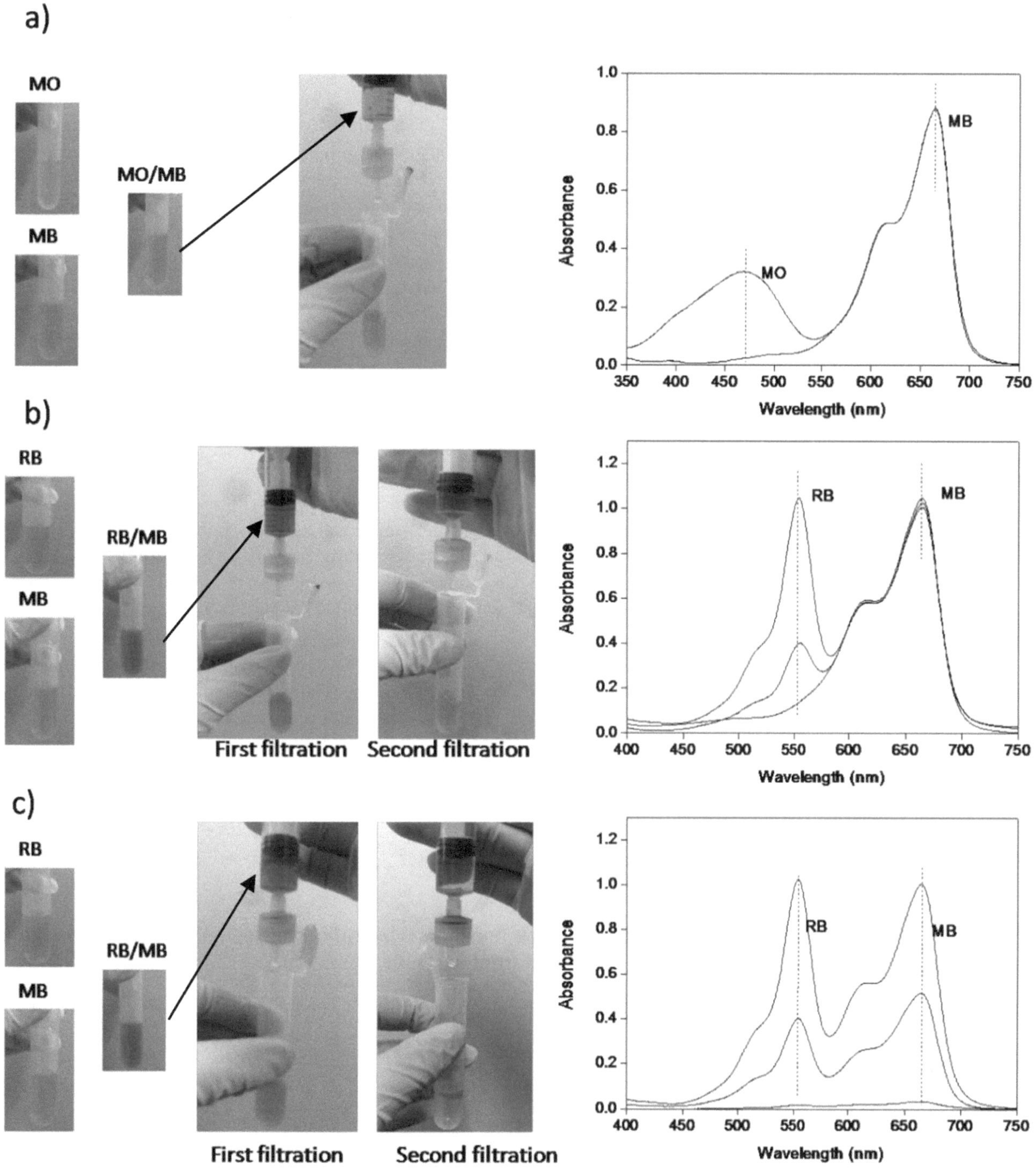

FIGURE 9.4 Photographs of dye separation of (a) MO/MB mixed solution, (b) RB/MB mixed solution by PUF-CS membrane, and (c) RB/MB mixed solution by PUF-CS-HA membrane along with the UV-visible spectra of the permeate. Adapted from Ref. [45].

Shi and co-workers utilized hydroxyapatite (HAp) to develop a chitosan/HAp porous membrane for azo dye DB15 removal [46]. Within 15 minutes, the developed membrane showed high-speed dynamic azo dye DB15 removal with an efficiency of 98% and greater. The proposed preparation route for the membrane was also found to be more economical and eco-friendly. This experiment used polyethylene glycol (PEG) as a pore-forming agent. The nonporous membrane formed without PEG showed a 2.5 times lower adsorption strength. Membranes developed with 30 wt.% HAp with CS exhibited the highest DB 15 dye adsorption rate. Even after five cycles of dynamic adsorption, the chitosan/HAp porous membrane displayed more than 80% efficiency. Weng and co-workers prepared a bamboo cellulose (BC)/chitosan nanofiltration membrane involving interfacial polymerization (IP) of trimesoyl chloride and piperazine [47]. The phase inversion technique is utilized for preparing BC/CS/IP membrane, and N-methylmorpholine-N-oxide was used as a solvent to dissolve chitosan and cellulose. The BC/CS/IP membrane showed excellent dye rejection rates of 93.65% and 98.86% for methyl orange and methyl blue, respectively, at 0.5 MPa pressure. The membrane also was successful in retaining NaCl, $MgCl_2$, $MgSO_4$, and Na_2SO_4 at rates of 40.26%, 53.28%, 62.55%, and 71.34%, respectively. Different chitosan/PEG/multiwalled carbon nanotubes (MWCNTs)/iodine membranes were developed by immersing dried CS/PEG/MWCNT membranes in 250, 500, and 750 ppm iodine solutions [48]. The incorporation of iodine significantly reduced the water contact angle of the membrane from 70.3° to 59.3°, which indicated amplification in membrane hydrophilicity. A 0.31 wt.% iodine-incorporated CS/PEG/MWCNT/I membrane exhibited direct blue 151 dye rejection with an efficiency rate of 83.7% in the dead-end filtration module. The same membrane displayed remarkable antibacterial activity, with 99.2% and 100% killing ratios of *Escherichia coli* and *Staphylococcus aureus* due to the incorporation of iodine. Mousavi and co-workers wrapped MWCNTs in chitosan, and so formed CSMWCNTs embedded in PEBAX 1657 matrix [49]. A mixed matrix developed from that was coated on polyethersulfone support to obtain a thin-film nanocomposite (TFN) membrane. The highest retention of malachite green was observed for 0.1 wt.% CSMWCNT-incorporated TFN membrane, while the peak permeates flux of 13.85 L/m^2h was observed for 1 wt.% CSMWCNTs addition. Chitosan membranes of different thicknesses were prepared by Long and co-workers using different knife gaps in the casting instrument (50, 75, and 150 μm) [50]. Additionally, membranes were repeatedly coated from one to four times, and PEG 400 was utilized as a pore former agent. In this study, eight different dyes were tested, with their details tabulated in Table 9.1 [50]. With an increase in membrane thickness, dye rejection increased while water flux decreased.

Dye retention was affected by three chief factors: size exclusion, adsorption, and electrostatic repulsion. Chitosan membranes that were cast four times with 75 and 150 μm knife gaps showed 91.3 and 99.9% dye rejection abilities for BB, OG, MnB, MB, MV, and RosB. Khajavian et al. coated metal-organic framework (ZIF-8) embedded chitosan/polyvinyl alcohol on PVDF support to remove malachite green dye from contaminated water [51]. The developed (CPZP) membrane exhibited a 90.32% rejection of malachite green. CPZP membrane showed a good fouling resistant ratio of 92% against bovine serum albumin and excellent pure water flux of 78.94 L/m^2h. Hydrogen titanate nanowire (HTN)/CS membranes were prepared by Tan and co-workers using the immersion filtration method [52]. The HTN/CS membrane revealed an outstanding rejection rate of 99.5% against congo red. The adsorption strength of the membrane reached up to 374.4 mg g^{-1} for congo red. Even after 15 adsorption–desorption cycles, the HTN/CS membrane showed a removal rate of 98.7% for congo red (Figure 9.5) [52].

Amino-functionalized montmorillonite (MMT) nanoparticles were added to chitosan/PVA electrospun membranes by Hosseini et al. [53]. CS/PVA/MMT membrane with 2 wt.% MMT showed 80% removal of basic blue 14 dye after 15 minutes. The incorporation of 2 wt.% MMT also enhanced the mechanical strength of the membrane by increasing the Young's modulus from 0.9 to 2.4 MPa. A chitosan/activated carbon(AC) membrane was prepared with the addition of PEG_{10000} as a pore-forming agent and tripolyphosphate (TPP) as a cross-linking agent [54]. A 30% AC embedded membrane displayed a 91.29% dynamic

TABLE 9.1
Various dye molecules studied in an experiment with their maximum absorption wavelengths of UV–vis spectra, molecular weights, hydrated radius, and charge. Adapted from Ref. [50]

Dye molecules	Abbreviation	UV–vis λ_{max} (nm)	Molecular weight (Da)	Hydrated radius (Å)	Charged type
Methyl viologen	MV	580	257.16	4.36	+ at pH 6
Methyl red	MR	410	269.3	4.87	neutral
Methyl orange	MO	463	327.33	4.96	– at pH 6
Methylene blue	MnB	664	319.85	5.04	+ at pH 6.5
Orange G	OG	477	452.36	5.21	– at pH 6
Rose bengal	RosB	508	1015.46	5.88	– at pH 6
Methyl blue	MB	586	821.78	7.29	– at pH 6.5
Brilliant blue	BB	610	854.02	7.96	– at pH 6.5

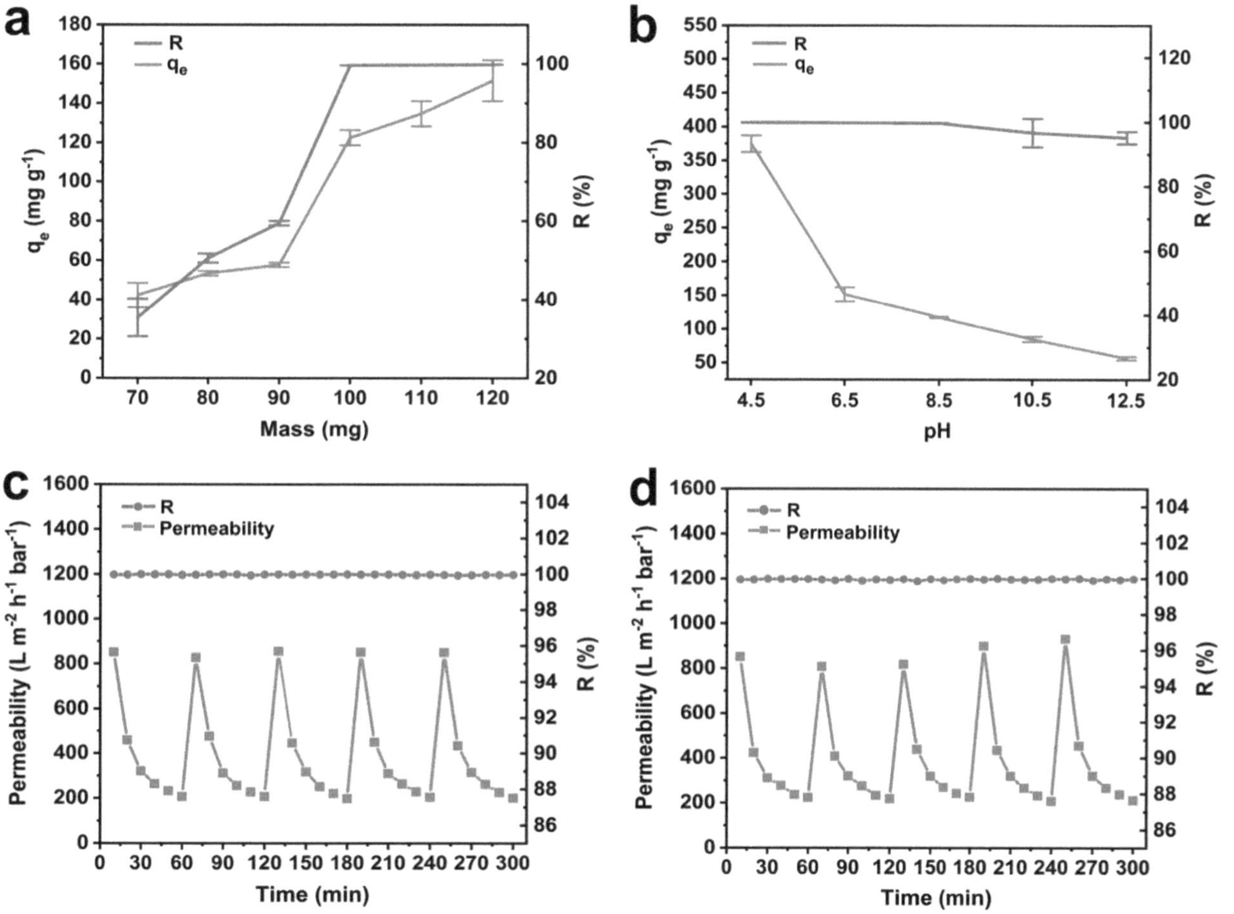

FIGURE 9.5 Adsorption and removal rates HTN/CS membranes: (a) with an increase in HTN-CS mass and (b) with a change in the pH of the solution. Permeability and removal rates of HTN/CS membranes at (c) pH 3.5 and (d) 2.5, respectively. Adapted from Ref. [52].

adsorption rate of rhodamine B dye. Guo and co-workers prepared a chitosan/piperazine(PIP)/trimesoyl chloride (TMC) composite membrane via interfacial polymerization [55]. CS/PIP/TMC membrane exhibited 99% removal of three different dyes such as coomassie brilliant blue G250, congo red, and methyl blue in the pH range of 2–12. Chitosan solutions of several concentrations were made, and in that PVDF/graphitic carbon nitride (g-C_3N_4) membrane was dipped to obtain PVDF/g-C_3N_4/CS membrane. In this study, direct blue 14 was used for dye rejection studies, and it was found that PVDF/g-C_3N_4/CS 4% w/v showed a peak dye rejection rate of 93% [56]. However, an increase in chitosan addition from 2% to 4% drastically reduced the water flux from 70.98% to 14.7%. To remove methylene blue dye via adsorption, Ulu and co-workers developed chitosan/κ-carrageenan/acid-activated bentonite composite membrane [57]. The developed membrane exhibited a notable MB dye removal rate of 98% at 50°C, and even after six adsorption–desorption the membranes maintained a dye removal efficiency of 77%. Vedula and Yadav prepared a novel chitosan lignin membrane via the solvent evaporation method [58]. CS/lignin membrane exhibited remarkable MB dye removal efficiency of 95%. The striking aspect of this membrane was that it was made from compostable and low-cost natural waste products.

9.2.2 Desalination

Water shortage is increasing at an alarming rate around the world due to the destruction of natural sources and population explosion. Seawater desalination is one of the prominent techniques to address this problem [59]. Ma and co-workers synthesized metal-organic frameworks (MOFs) such as NH_2-MIL-101(Al) and NH_2-MIL-101(Cr) and dispersed them in a chitosan polymer solution to prepare CS/MOF nanofiltration membrane [60]. A 15 wt.% NH_2-MIL-101(Al) added chitosan membrane showed the highest water flux, while a membrane with 20 wt.% NH_2-MIL-101(Al) displayed peak rejection rates for $MgCl_2$ of 93% and for $CaCl_2$ of 86.5%. Deng and co-workers tried to develop a desalination reverse-osmosis membrane from reduced graphene oxide, titania, and chitosan [61]. However, they achieved only a 30% rejection capacity, which was a threefold increment over only graphene oxide membrane. For desalination, three kinds of buckypaper membranes were prepared: chitosan, chitosan/polyethylene glycol diglycidyl ether (PEGDE), and chitosan/glycerin

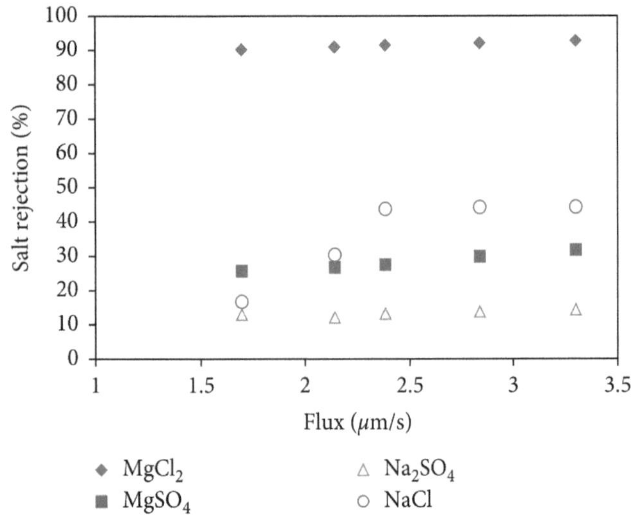

FIGURE 9.6 Salt removal efficiency of PVDF/CS/MWCNT membrane at different operating pressures. Adapted from Ref. [67].

(Gly) [62]. Gly and PEGDE escalated MWCNT dispersion in water and helped the buckypaper chitosan membrane to attain excellent tensile strength (59 MPa). Buckypaper chitosan membranes showed a salt rejection rate of $MgCl_2$ in the range of 80–95%. Tetraethylorthosilicate cross-linker utilized CS/PVA RO membrane showed 80% salt rejection [63]. In addition, it displayed a decrease in water contact angle and an increase in pure water flux. A CS/PIP composite membrane prepared via interfacial polymerization by Tang and group members enhanced the pure water flux by twofold, while the Na_2SO_4 rejection value was found to be 89% with a slight decrement [64]. In this study, 35,000 ppm seawater and 5000 ppm brackish water were tested for desalination, and CS/PIP nanofiltration membrane exhibited adequate performance. To enhance water desalination via membrane distillation, Kebria and co-workers embedded a zeolitic imidazolate framework (ZIF-8) into the chitosan layer-coated PVDF membrane [65]. For 3.5 wt.% NaCl solution, PVDF/CS/ZIF-8 membrane exhibited a 350% increment of permeate flux over neat PVDF membrane, and NaCl rejection reached above 99.5%. Real seawater utilized in antifouling studies revealed that PVDF/CS/ZIF-8 membrane exhibited a flux recovery rate of 90%, while virgin PVDF showed only 67%. Chitosan nanoparticles were incorporated into cellulose acetate (CA) RO membranes, and salt rejection studies were done using 35g/L NaCl solution [66]. The membrane containing 2% CS nanoparticles significantly amplified water flux from 6 L/m^2 h to 18 L/m^2 h and salt removal from 89 to 94 % when compared with a pristine CA membrane. PVDF/CS/MWCNT nanofiltration membranes were developed for desalination using an evaporating casting method. The prepared membrane showed a notable tensile strength of 28 MPa and Young's modulus of 1.2 GPa. Salt rejection studies of PVDF/CS/MWCNT membrane are illustrated in Figure 9.6 [67].

Polyamide-6/CS membranes were prepared and tested with 2000 ppm NaCl solution to check desalination properties [68]. A 2 wt.% chitosan blended membrane had the highest salt rejection rate, while a 4 wt.% CS blended membrane displayed peak permeate flux. Optimum salt rejection was performed by 3 wt.% CS blended polyamide-6 membrane with a salt rejection efficiency of 75%. Unugul and Nigiz prepared halloysite nanotube (HNT)-incorporated chitosan membranes for pervaporation-based desalination of water [69]. HNT loading from 0 to 20 wt.% showed remarkable enhancement in the water flux from 1.6 to 4.89 kg/m^{2}h at 30°C. An outstanding salt rejection rate of 99.90% was observed for 10 wt.% HNT-incorporated chitosan membrane.

9.2.3 Fuel Cells

Purwanto and co-workers prepared an MMT-embedded chitosan-based membrane electrolyte membrane for direct methanol fuel cell (DMFC) [70]. To make MMT organophilic (O-MMT) it was treated with a surface modifier called 3-glicidoxy propyltrimethoxysilane (GPTMS). A 5 wt.% O-MMT added CS/O-MMT membrane exhibited excellent methanol permeability of 3.03×10^{-7} and proton conductivity of 4.66 mS cm^{-1}. Equimolar CS and PVA blend anion exchange membranes were prepared for alkaline fuel cells with four different additives involving two organic ionomers (4VP and AS4) and two inorganic ionomers (layered titanosilicate AM-4 and stannosilicate UZAR-S3). The incorporation of additives intensified the thermal stability and ionic conductivity of the membranes (Table 9.2) [71]. CS/PVA/4VP membrane showed peak ionic conductivity, and CS/PVA/UZAR-S3 membrane showed the least alcohol crossover.

Liu and co-workers improved the oxidative stability of the chitosan PEM from 125 minutes to 217 minutes by inserting 5 wt.% silica-coated CNTs (SCNTs) [72]. In addition, an increase in proton conductivity was also observed from 0.015 S cm^{-1} to 0.025 S cm^{-1} when compared to a pristine CS membrane. Uniform distribution of additives enhanced the mechanical strength of PEM also. In another study, they used titania-coated CNTs (TCNT) in place of SCNTs. Similarly, TCNT-incorporated chitosan PEM also showed enhancement of its proton conductivity and oxidation stability with an increment of TCNT loading [73]. The tensile strength of the membrane was increased from 17.8 to 29.0 MPa when TCNT loading was elevated from 0 to 5 wt.%. Shirdast et al. attempted to improve the proton conductivity of the chitosan membrane by incorporating 10 wt.% sulfonated CS (SCS) and different amounts of sulfonated graphene oxide (SGO) [74]. A 5 wt.% SGO-embedded CS/SCS PEM membrane exhibited a 454% enhancement in conductivity and a 23% reduction in permeability in comparison with a neat CS membrane. Phosphorylated chitosan (CS-P) PEM membrane was developed by Holder and co-workers for microbial fuel cells [75]. CS-P PEM displayed a power density of 130.03 mW/m^2, a 1.7-fold increase in tensile strength, and a 5.9-fold enhancement in cation exchange capacity when compared to a neat CS membrane. Wu et al. prepared a novel CS/phosphotungstic acid (HPW) PEM membrane, which showed a peak proton conductivity of 2.9×10^{-2} S cm^{-1} at

TABLE 9.2
Water content (WC), water uptake (WU), ion exchange capacity (IEC), and conductivity of different compositions of ionomer-filled membranes. Adapted from Ref. [71]

Membrane	WC (%)	*WU* (%)	IEC (mmol g^{-1})	Conductivity (mS cm^{-1})
CS	28.64	88.5	0.12	0.07
PVA	18.28	157.5	0.096±0.018	–
CS/PVA	23.11	139.5	0.253±0.050	0.15–0.29
4VP /CS/PVA	21.89	134.0	0.266±0.0 04	1.15
AS4/CS/PVA	22.5	131.5	0.325±0.026	0.32
AM4/CS/PVA	26.67	128.3	0.176±0.001	0.38
UZAR-S3/CS/PVA	30.24	121.6	0.310±0.060	0.03

80°C with 40% HPW [76]. A 60% reduction in methanol permeability (4.7 × 10^{-7} cm^2 s^{-1}) was observed for 40% HPW-incorporated CS membrane. Wang and co-workers prepared a novel reactive cationic dye (RCD)-loaded chitosan alkaline exchange membrane for fuel cells [77]. This membrane displayed good OH^- conductivity of 4.59 × 10^{-3} S cm^{-1}. When CS/RCD membrane was dipped in a KOH solution, it showed enhanced OH^- conductivity of 1.057 × 10^{-2} S cm^{-1}, which was a strong validation for its stability in strong alkali solutions. N-phthaloyl chitosan (NPCS)-incorporated sulfonated polyethersulfone (SPES) membrane was developed by Muthumeenal [78]. SPES/NPCS membrane improved proton conductivity from 3.15 × 10^{-3} S cm^{-1} to 9.2 × 10^{-3} S cm^{-1} when compared to a neat SPES membrane. Glutaraldehyde cross-linked chitosan–alginate membrane was prepared by Eldin et al. as PEM [79]. CS/alginate membrane displayed a significantly lower methanol permeability value of 2.5 × 10^{-10} cm^2 s^{-1} when compared to the Nafion membrane (1.14 × 10^{-9} cm^2 s^{-1}). The proton conductivity of the chitosan membrane was greatly improved by adding anatase titania-coated CNTs (TCNTs) and sodium lignin sulfonate (SLS) [80]. A 5% TCNT and 2% SLS-incorporated CS/TCNT/SLS membrane exhibited a remarkable increase in proton conductivity with a value of 0.0647 S cm^{-1} at 60°C. Figure 9.7 [80] shows the increase in proton conductivity with TCNT and SLS loading. To overcome the poor dispersion ability of CNTs, through an ion-exchange method, CNT fluids were developed and incorporated into the chitosan matrix [81]. A 3 wt.% CNT nanofluid-added chitosan membrane showed a good proton conductivity value of 0.044 S cm^{-1} at 80°C. Polydopamine (PDA)-functionalized CNTs were prepared by sulfuric acid and embedded into chitosan matrix to prepare a PEM [82]. A 2 wt.% CS/CNT-PDA membrane displayed peak proton conductivity of 0.028 S cm^{-1}. The mechanical and oxidation stability of the PEM were also enhanced due to hydrogen bonding between PDA and CNT. A sulfonated polyether ether ketone (SPEEK) and chitosan composite PEM was prepared to reduce the methanol permeability of pure chitosan membrane to half its value from 6.02 × 10^{-6} cm^2 s^{-1} to 2.46 × 10^{-6} cm^2 s^{-1} [83]. Divya and co-workers prepared chitosan-based PEM for DMFC with excellent properties by incorporating two-dimensional exfoliated MoS_2 nanoparticles [84]. CS/MoS_2 nanocomposite membrane with 0.75% E-MoS_2 exhibited proton conductivity value of 2.92 × 10^{-3} S cm^{-1} and appreciably low methanol permeability value (3.28 × 10^{-8} cm^2 s^{-1}). For alkaline ethanol fuel cells, chitosan-based anion exchange membranes were developed by inserting some additives including $Mg(OH)_2$, GO, and benzyl trimethylammonium chloride (BTMAC) [85]. Hydroxide conductivity, power density, and ethanol permeability of CS/GO/BTMAC membranes were found to be 142.5 mS cm^{-1} at 40°C, 72.7 mW cm^{-2}, and 6.17 × 10^{-7} cm^2 s^{-1}, respectively. Ryu and co-workers modified chitosan to quaternized poly[O-(2-imidazoly-ethylene)-N-picolylchitosan] (QPIENPC) by multi-step reactions [86]. QPIENPC membrane acted as a fine anion exchange membrane in DMFC with a hydroxyl ion conductance of 10.15 × 10^{-3} S cm^{-1} at 80°C. At a current density of 28.76 mA cm^{-2} QPIENPC membrane exhibited the highest power density of 10.42 mW cm^{-2}. The ion exchange capacity of the chitosan membranes was raised by blending it with poly-2-acrylamido-2-methylpropane sulfonic acid (PAMPS) [87]. A 1.85 meq/g ion exchange capacity was observed for a CS/PAMPS membrane in which CS and PAMPS were in a ratio of 2:1. Blended membrane showed reduced methanol permeability of 9 × 10^{-7} cm^2 s^{-1}, which is less than that of Nafion 117. Surface-modified zeolite (MZ) was added to a CS/PVA proton exchange membrane to improve its conductivity [88]. CS/PVA/MZ membrane showed 0.0527 S cm^{-1} proton conductivity and 2.3 × 10^{-7} cm^2 s^{-1} methanol permeability. A 25.1% higher proton conductivity compared to neat chitosan membrane was achieved by incorporating 4 wt.% attapulgites (APG) into the chitosan membrane [89]. CS/APG-4 wt.% membrane at 100% relative humidity exhibited a proton conductance of 26.2 mS cm^{-1} at 80°C. In a PEM development, sulfonic acid functionalized chitin nanowhiskers (CW) were added to chitosan for DMFC [90]. A 7 wt.% CW added CS membrane showed the highest proton conductivity value of 0.0221 S cm^{-1}, while 5 wt.% CW added CS membrane displayed the lowest methanol permeability value of 0.93 × 10^{-6} cm^2 s^{-1}. A markedly lower methanol permeability of 2.57 × 10^{-8} cm^2 s^{-1} was observed for SCS and SGO blend proton exchange membrane in DMFC [91]. A 1 wt.% SGO added SCS/SGO membrane revealed a proton conductivity value of 4.86 × 10^{-3} S cm^{-1} at 25°C with a selectivity factor

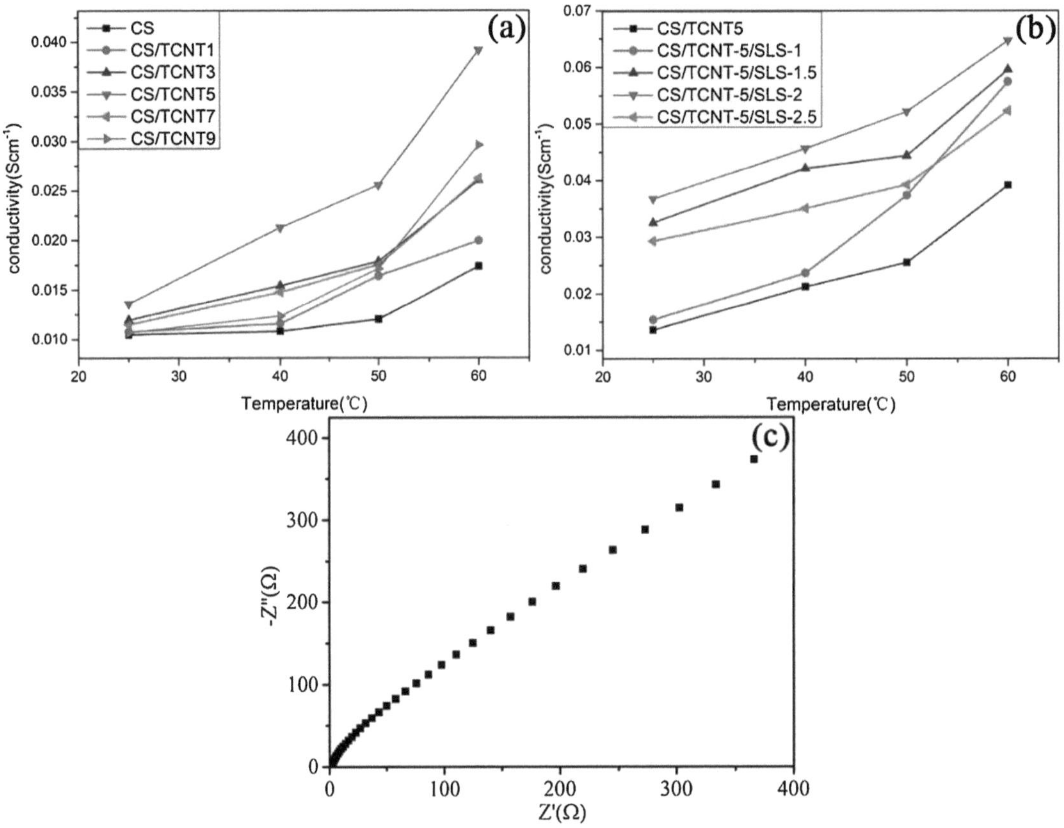

FIGURE 9.7 Effect of temperature on membrane conductivity with different (a) TCNT loading and (b) SLS loading. Adapted from Ref. [80].

of 1.89×10^5 S cm^{-3} s. Rosli and co-workers prepared SiO_2 and added it to N-methylene phosphonic chitosan/poly(vinyl alcohol) (NMPC/PVA) composite membrane to fabricate a PEM [92]. A 4 wt.% SiO_2 added NMPC/PVA membrane expressed proton conductance of 5.08×10^{-4} S cm^{-1} at 100°C. An anion exchange membrane for direct alkaline fuel cell (DAFC) was developed by embedding quaternary ammonium groups containing modified cellulose nanofibrillar fillers (CNF) into CS/$Mg(OH)_2$ composite membrane [93]. The fabricated CS/$Mg(OH)_2$/CNF membrane displayed a low ethanol permeability of 8.97×10^{-5} cm^2 s^{-1}.

9.2.4 Heavy Metal Removal

Heavy metals create serious problems for the ecosystem and human health. Many heavy metals are acutely toxic, and some are also carcinogenic. Heavy metals are non-biodegradable, and they tend to accumulate in the food chain over time. Hence the removal of heavy metals from water sources is of prime importance. Liu and co-workers fabricated glutaraldehyde cross-linked CS/MMT membranes to remove Pd(II) from water, showing an adsorption capacity of 193 mg g^{-1} at highly acidic pH 2 [94]. Chitosan nanofibers of 75 nm diameter were deposited on polyester (PE) support via the electrospinning method to fabricate CS/PE membrane for efficient removal of Cr(VI) ions [95]. The peak adsorption capacity of the membrane was found to be 16.5 mg chromium/g of chitosan for 1 mg/L Cr(VI) solution. A chitosan membrane was fabricated for the removal of copper ions from an aqueous solution utilizing silica as porogen [96]. The prepared membrane exhibited 87.5 mg/g adsorption power against Cu(II) solution and the membrane showed the ability to regenerate. Habiba and co-workers prepared CS/PVA/zeolite nanofibrous membrane by the electrospinning method. CS/PVA/zeolite nanofibrous membrane displayed excellent effectiveness in removing Ni(II), Fe(III), and Cr(VI) from contaminated water (Figure 9.8) [97].

Qin and co-workers studied various parameters such as the temperature of the coagulation bath, time of quenching, and solvent constitution effect on the preparation and performance of chitosan membrane [98]. The fabricated chitosan membrane displayed supreme adsorption of 2.57 mmol/g of Cu^{2+} ions when it was immersed for 12 hours in a coagulation bath containing 1% Na_2CO_3 in an equimolar mixture of water and ethanol maintained at –20°C. Chitosan membranes prepared by the electrospinning method were grafted with polyethyleneimine (PEI) and poly(glycidyl methacrylate) (PGMA) to produce abundant amine functional groups on the surface [99]. CS/PGMA/PEI membrane showed excellent removal efficiency of Cr(VI), Cu(II), and Co(II) with an adsorption capacity of 138.96, 69.27, and 68.31 mg g^{-1}, respectively. The adsorption isotherm can be correlated with the Langmuir model, and it also revealed appreciable regeneration capacity and stability. Efficacious removal

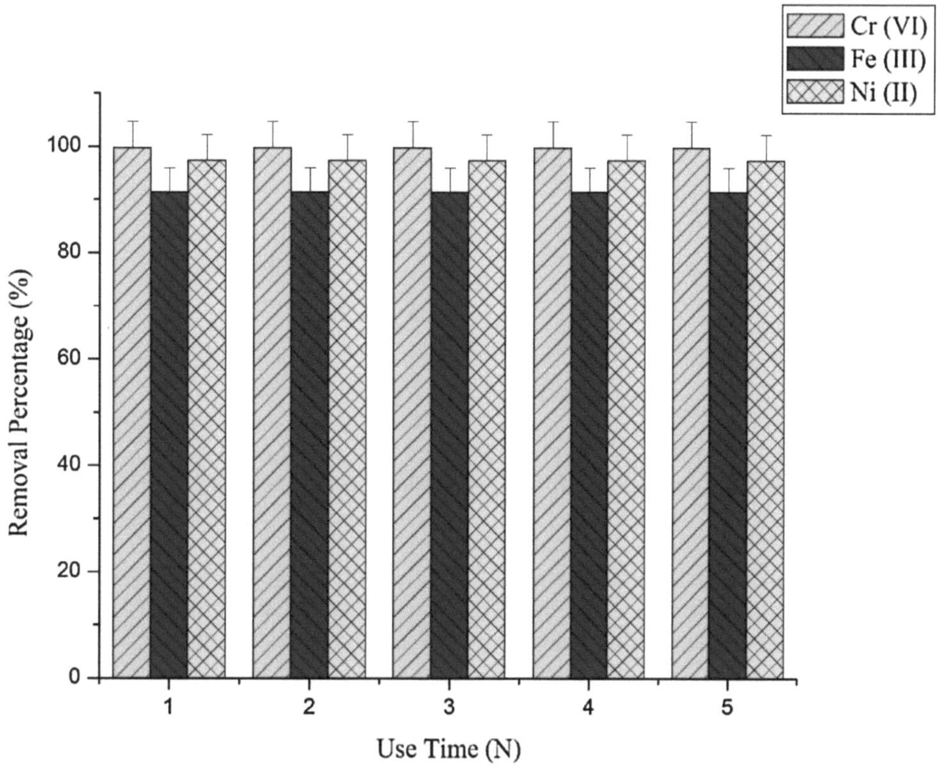

FIGURE 9.8 Heavy metal removal efficiency of CS/PVA/zeolite nanofibrous membrane in five cycle adsorption experiments. Adapted from Ref. [97].

of Pb(II) and Cu(II) from polluted water was achieved by grafting amine function groups on the chitosan nanofibers [100]. Diethylenetriamine-grafted CS nanofibers exhibited 94.34 and 166.67 mg g^{-1} adsorption of Pb(II) and Cu(II), respectively. Amine-grafted CS nanofibers showed notable stability in an aqueous solution with only 6% weight loss. Enhanced amine moieties on the surface facilitated additional chelation. Peak adsorption capacities of 266.12 and 148.79 mg g^{-1} were reported for Pb(II) and Cd(II), respectively, for PVA/CS membrane prepared via the electrospinning method [101]. PVA/CS membrane efficiency in removing Pb(II) and Cd(II) ions at different pH and time durations was also studied, and it was found that the highest adsorption was reached in the range of pH 5–9. UiO-66-NH_2 MOF with a BET surface area of 1118 m^2 g^{-1} was incorporated into polyacrylonitrile (PAN)/CS electrospun nanofiber membrane for the removal of heavy metals from contaminated water [102]. The highest adsorption strengths of 10 wt.% MOF added PAN/CS/UiO-66-NH_2 MOF membrane for Cr(VI), Cd(II), and Pb(II) were determined to be 372.6, 415.6, and 441.2 mg g^{-1}, respectively. When the temperature was gradually increased from 25 to 45°C, the metal removal strength of the membrane slightly decreased, however water flux improved remarkably. Sangeetha and co-workers fabricated a clay-incorporated CS/PVA/MMT membrane that displayed efficient removal of chromium with a notable water flux of 25.72 L m^{-2} h^{-1} [103]. With the addition of silver nanoparticles, CS/PVA/MMT/Ag membrane manifested outstanding anti-biofouling properties. A 40% increase in Cu(II) ions removal capability of the CS membranes was observed when they were cross-linked with sulfuric acid [104]. Mukhopadhyay and co-workers prepared CS/zeolite composite membranes for appreciable removal of arsenic (III) from an aqueous solution [105]. The addition of 1.25 wt.% zeolite to the CS/zeolite membrane resulted in a peak arsenic rejection rate of 94% when an As_2O_3 aqueous solution of concentration 1000 μg L^{-1} was used. CA/PEG/CS-Ag composite membrane was developed to remove iron ions from industrial wastewater through an electrodialysis system [106]. The fabricated membrane showed improved thermal stability up to 350°C and more than 60% removal of iron contents. To adsorb heavy metals such as Pb(II) and Cd (II), chitosan nanofibers were deposited over the PAN nanofiber layer embedded with ZnO or TiO_2 via the electrospinning method that resulted in a bilayer [107]. This PAN/(ZnO or TiO_2)/CS bilayer showed an exceptional increase of 405% and 102% in adsorption strength for Cd(II) and Pb(II) when compared to a single-layer PAN/ZnO or PAN/TiO_2 membrane. The bilayer formed from electrospun CS notably enhanced the tensile strength of the membrane by 68%. ZIF-8 and UiO-66-NH_2 MOF nanoparticle-incorporated PVDF/CS membrane was successful in rejecting 98.1% bovine serum albumin (BSA) protein and 95.6% Cr(VI) ions [108]. The highest water flux and adsorption capacity of 20 wt.% UiO-66-NH_2 MOF added PVDF/CS membrane was ascertained to be 470 L m^{-2} h^{-1} and 602.3 mg g^{-1}, respectively. It is noteworthy that the removal of uranium (VI) from polluted water was accomplished

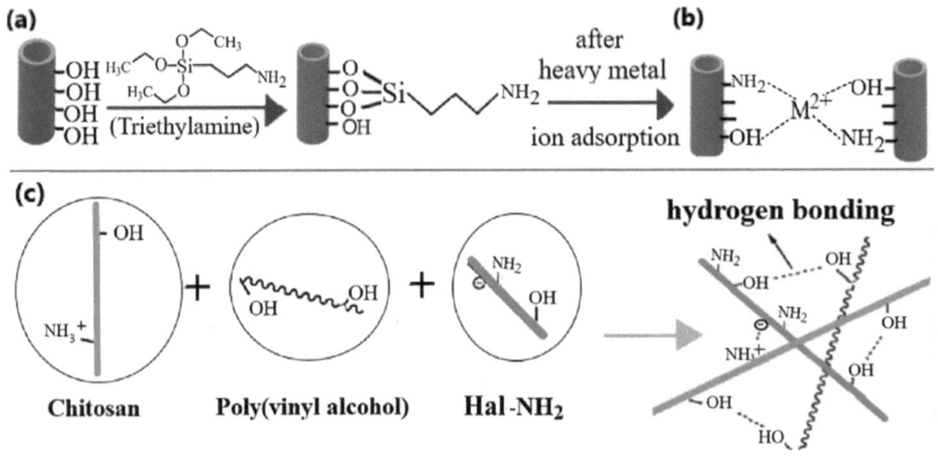

FIGURE 9.9 (a) Modification of halloysite nanotubes, and (b) their interaction with heavy metals and (c) chitosan. Adapted from Ref. [115].

by Hu et al. by fabricating graphene oxide nanoribbons/CS membrane [109]. A top adsorption capacity of 320 mg g^{-1} was noted, and the membrane possessed good reusability. Machodi and co-workers prepared PES/CS membrane for the removal of metal ions from acid mine drainage [110]. A 0.75 wt.% CS inserted PES/CS membrane displayed remarkable rejection of Fe^{2+}, Mn^{2+}, Mg^{2+}, and Ca^{2+} ions in the 90–70% range. Chromium ions can be removed from tannery effluents using PES/CS membrane [111]. A 2.5 wt.% CS-incorporated PES/CS membrane exhibited exceptional (>99%) rejection of chromium. A mechanically strong MWCNT/CS-carrageenan membrane with Young's modulus of 30.69 MPa showed 94% and 91% rejection of copper and lead, respectively, at 1 bar pressure [112]. The prepared membrane displayed an appreciable water flux of 180 L m^{-2} h^{-1}. Zia and co-workers coated polydopamine on electrospun polylactic acid nanofibers, and in the next step, chitosan was grafted on the polydopamine coating [113]. The prepared PLA/PDA/CS membrane exhibited notable rejection of Cu^{2+} ions with an adsorption capacity of 270.27 mg g^{-1}. Orabi et al. embedded chitosan-grafted p-phenylenediamine (PPD) into PSF and CA membranes to remove uranium (VI) from an aqueous solution [114]. The highest adsorption strengths of 44 mg g^{-1} and 39 mg g^{-1} were observed for PSF/PPD and CA/PPD membranes, respectively. Halloysite nanotubes were altered by reacting with triethylamine to obtain amine-functionalized halloysite nanotubes (NH_2-HAL). A 7 wt.% NH_2-HAL-incorporated CS/PVA membrane exhibited excellent adsorption capacities of 516.3 and 551.6 mg g^{-1} for Cd(II) and Pb(II) ions, respectively. CS/PVS/NH_2-HAL membrane was successful in removing 84% Cd(II) and 78% Pb(II) ions. Modification of halloysite nanotubes and their interactions with heavy metals and membranes are displayed in Figure 9.9 [115].

Lu and co-workers fabricated CS/β-cyclodextrin(CD) membrane, which showed 97% removal efficiency of Cu^{2+} ions [116]. The removal efficiency was enhanced by an increase in pH, and with an increase in temperature, the removal strength declined. The regeneration ability of the membrane was found to be above 95% after five cycles. Effective removal of arsenic (As) from the polluted water was achieved by fabricating chitosan-grafted polypropylene (PP) membrane [117]. PP/CS membrane displayed a peak adsorption strength of 0.031 mg g^{-1} of As(V) through the coordination of As(V) with polar groups of CS. The adsorption process followed pseudo-second-order kinetics, and the membrane was successful in removing 75% of As(V) ions.

9.2.5 Gas Separation

A substantial increase in global temperature was observed in the last century due to the escalation in the release of greenhouse gases like CO_2. Selective gas separation provides a promising solution to this issue. Santos and co-workers fabricated a chitosan-supported ion liquid membrane (SILM) added to a room temperature ionic liquid called 1-ethyl-3-methylimidazolium acetate (EMI) for effective CO_2 gas separation [118]. A 5 wt.% EMI-incorporated CS SILM significantly reduced the activation energy for CO_2 permeation and, in turn, lowered the effect of temperature on diffusivity and permeability. The mechanical strength and flexibility of the CS membrane were enhanced with the addition of EMI. Dubek and co-workers studied the effect of cross-linking agents such as glutaraldehyde and sulfuric acid with chitosan membranes in the separation of ethanol and water vapors [119]. A 15 w/w% iron oxide-inserted CS membrane cross-linked with glutaraldehyde exhibited an outstanding increase in water vapor permeation compared to ethanol. However, magnetite-embedded CS membrane cross-linked with sulfuric acid favored ethanol vapor transport. Chitosan was modified into N,N-diethylaminopropylcarbamate (NAPC) chitosan via multi-step reactions and then embedded into the PVA matrix for the selective separation of CO_2 gas [120]. A 20 wt.% NAPC-embedded PVA/NAPC-CS membrane showed the highest selective separation of CO_2 from a CO_2 and CH_4 mixture. Increased temperature and feed pressure reduced the membrane efficacy. Increased polyamine content encouraged CO_2 transport inside the membrane matrix. PVA/CS membrane was used as an air filtration membrane. A 30 wt.% CS-blended PVA/CS membrane restricted NaCl aerosol particles and displayed an average tensile strength of 2.82 MPa [121]. The PVA/CS membrane with 37 μm thickness showed a remarkable air filtration efficacy of 95.59%. Electrospun nanofibrous PLA/CS membranes were tested for air filtration. A 98.99% removal rate of NaCl aerosol particles was attained by PLA/CS membrane with a mass ratio of 8:2.5 concerning PLA and CS, respectively [122]. Along with this, PLA/CS membrane exhibited remarkable antibacterial properties, with 99.5% removal of *Staphylococcus aureus* and 99.4% removal of *Escherichia coli*. To separate CO_2 from a CO_2/N_2 gas concoction, Rajashree and co-workers fabricated a carboxymethyl chitosan (CMC)/MWCNTs mixed-matrix membrane [123]. The amine groups existing in CMC facilitate CO_2 transport by acting as their carrier and showed a CO_2 gas permeation unit (GPU) of 43. The CMC/MWCNT membrane displayed CO_2/N_2 selectivity of 45 at 80°C. In another work, Rajashree and Bishnupada significantly enhanced the CO_2 separation from CO_2/N_2 gas mixture by incorporating polyamidoamine (PAA) dendrimers into CMC membrane [124]. an excellent CO_2 permeance of 98 GPU was achieved with 10 wt.% PAA added CMC/PAA membrane. At 90°C, the CO_2/N_2 selectivity of the membrane was found to be 149 making it suitable for extensive CO_2 segregation applications. Piperazine (PZ)-embedded CMC composite membrane was fabricated for CO_2 partition from CO_2/N_2 mixture [125]. The prepared CMC/PZ membrane with 20 wt.% PZ exhibited a notable selectivity value of 103 for CO_2/N_2 and appreciable CO_2 permeance of 89 GPU at 80°C. CS/PVA nanofibrous membrane was incorporated with Cu-1,3,5-benzenetricarboxylic acid (BTC) MOF for selective capture of CO_2 from a CO_2/N_2 combination [126]. CS/PVA/Cu-BTC membrane revealed an 11-fold higher surface area compared to pristine PVA/CS nanofibrous membrane, and adsorption of CO_2 was observed to be 14 times greater than N_2 adsorption. Hydrotalcite (HT) and PAA dendrimer-loaded CMC mixed-matrix membrane showed effective CO_2 separation from a CO_2/N_2 gaseous mixture [127]. CMC/PAA/HT membrane with 10 wt.% PAA dendrimer and 1 wt.% HT was found to be ideal for selective CO_2 capture. It showed CO_2/N_2 selectivity of nearly 677 and 123 GPU for CO_2 at 90°C. Li and co-workers took CMC and piperazine (PIP) in aqueous and trimesoyl chloride (TMC) in the hexane phase to fabricate PIP/CMC/TMC membrane via interfacial polymerization for selective CO_2 segregation from a CO_2/N_2 blend. Figure 9.10 [128] depicts the structure of the PIP/CMC/TMC membrane along with nanostructures and smooth transportation of CO_2. At 1.1 bar pressure and room temperature, 0.2 μm PIP/CMC/TMC membrane exhibited outstanding CO_2/N_2 selectivity of 119 and CO_2 gas permeation of 1479 GPU.

9.2.6 Oil–Water and Solvent Separation

Extensive industrialization and transport of oils have led to large-scale pollution of water bodies with oils. Thus, the effective separation of oil–water emulsion is of the utmost importance. Solvent separation is also an urgent need in industries. Liu and co-workers cross-linked chitosan and silica nanoparticles on the PVDF membrane using glutaraldehyde to form a PVDF/CS/SiO_2 composite membrane that displayed outstanding hydrophilicity and superior oleophobicity inside the water body [129]. The efficacy of the PVDF/CS/SiO_2 membrane for oil–water emulsion segregation was found to be greater than 99%. Zheng and co-workers prepared homogeneous chitosan and sulfonated poly sodium vinyl sulfonate (S-PVS) by complexing between CS and PVS, followed by sulfonation [130]. The fabricated CS/S-PVS membrane successfully dehydrated a 10 wt.% water/ethanol mixture, and the permeate was found to contain 99.55 wt.% water at 70°C. Permeate flux was observed to be 1980 g m^{-2} h^{-1}. Shiva Prasad and co-workers cross-linked chitosan with tetraethyl orthosilicate (TEOS) and deposited it over poly(ether-block-amide) (PEBA-2533) porous support [131]. This PEBA/CS/TEOS membrane exhibited exceptional water/n-methyl pyrrolidone (NMP) separation ability. Notable selectivity of 225 for water against water/NMP mixture and water flux was observed to be 0.019 kg m^{-2} h^{-1} for a feed solution containing 4.6 wt.% water. A 90% NMP solution can be purified to more than 99% via distillation and pervaporation. Chitosan was cross-linked with TEOS and coated on polytetrafluoroethylene membrane and used for the separation of a methanol and toluene mixture via the pervaporation technique [132]. PTFE/CS/TEOS membrane showed notable selectivity of 58.4 for a feed concentration of 68 wt.% methanol and appreciable methanol flux of 0.13 kg m^{-2} h^{-1}. Xu co-workers fabricated titanium Mxene-incorporated chitosan pervaporation membrane for solvent dehydration [133]. At 50°C, 3 wt.% 2D titanium carbide added CS membrane exhibited exceptional separation factors of 906, 1421, and 4898 for the dehydration of dimethyl carbonate, ethanol, and ethyl acetate, respectively. An appreciable total flux of 1.4–1.5 kg m^{-2} h^{-1} was observed. Vinu et al. enhanced the selectivity factor of chitosan membrane for water–ethanol segregation from 347 to 3429 by inserting 0.15 wt.% aluminum MOF called Al(OH)(BPDC) (BPDC = biphenyl-4,40-dicarboxylate) [134]. The high surface area provided by Al MOF improved the flux of the CS/Al(OH)(BPDC) membrane to 378 g m^{-2} h^{-1}. Dubek and co-workers incorporated four different modified chitosan particles (modification with glutaraldehyde, glycidol, sulfation, and phosphorylation) into the PVA matrix and studied their effect on water–ethanol separation via pervaporation [135]. Among these, the 3 wt.% phosphorylated CS particle-embedded PVA membrane showed a good separation factor of 263.3 and peak pervaporation separation index value of 380.3 kg m^{-2} h^{-1}. The toxic cross-linking agent glutaraldehyde was replaced by naturally occurring genipin, and a fabricated genipin cross-linked CS membrane was tested for dehydration of isopropanol. Genipin cross-linked CS membrane showed enhanced total flux and increased the water content in the permeate also [136]. Through the electrospinning technique, Doan and co-workers fabricated nanofibrous membranes from chitosan and polycaprolactone (PCL). The prepared CS/PCL membrane displayed excellent oil/water separation efficiency with a more than 94.6% value. CS/PCL membrane showed

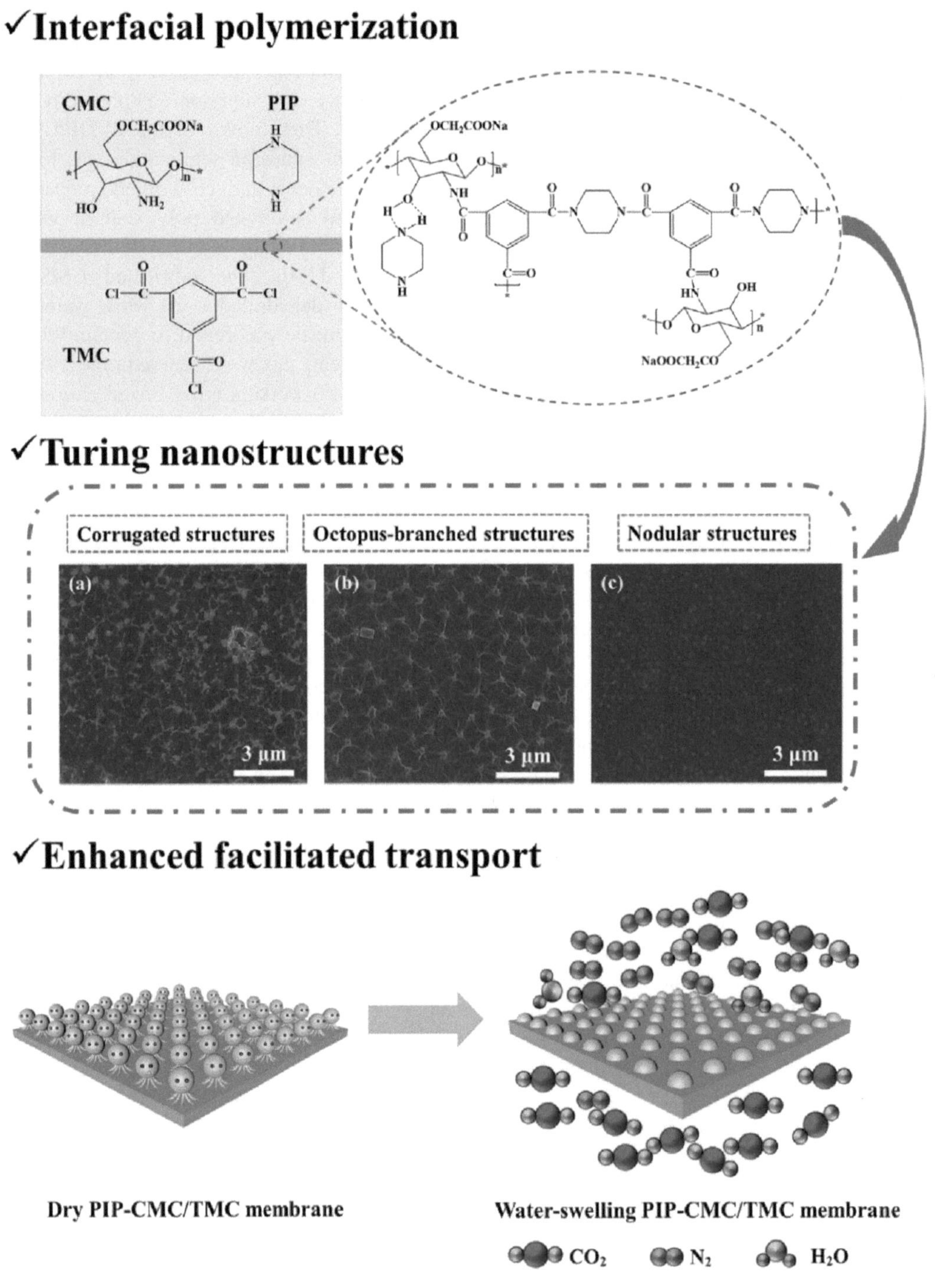

FIGURE 9.10 (a) Structure of PIP/CMC/TMC composite formed via interfacial polymerization, (b) nanostructures, and (c) selective CO_2 transport through PIP/CMC/TMC composite membrane. Adapted from Ref. [128].

superoleophobicity underwater and superhydrophobicity when prewetted with oil. Photographs of oil–water separation using CS/PCL membrane are provided in Figure 9.11 [137]. Krishnamoorthi and co-workers coated chitosan and caffeic acid (CFA) on cotton fiber (CF) substrate to get CF/CS/CFA composite membrane for oil–water emulsion partition. At a low pressure of 0.1 bar, the CF/CS/CFA membrane exhibited an excellent permeation flux of 50,050 L m^{-2} h^{-1} and separation strength of more than 99.9% [138]. Even after 10 cycles of usage, the permeation flux remained unchanged and showed superior reusability. UIO-66 MOF synthesized from $ZrCl_4$ was incorporated into the CS membrane matrix to obtain a CS/UIO-66 mixed-matrix membrane to efficiently separate the methanol and dimethyl carbonate mixture [139]. A remarkable separation factor of 337 and flux of 355 g m^{-2} h^{-1} was observed for 10 wt.% UIO-66 embedded CS/UIO-66

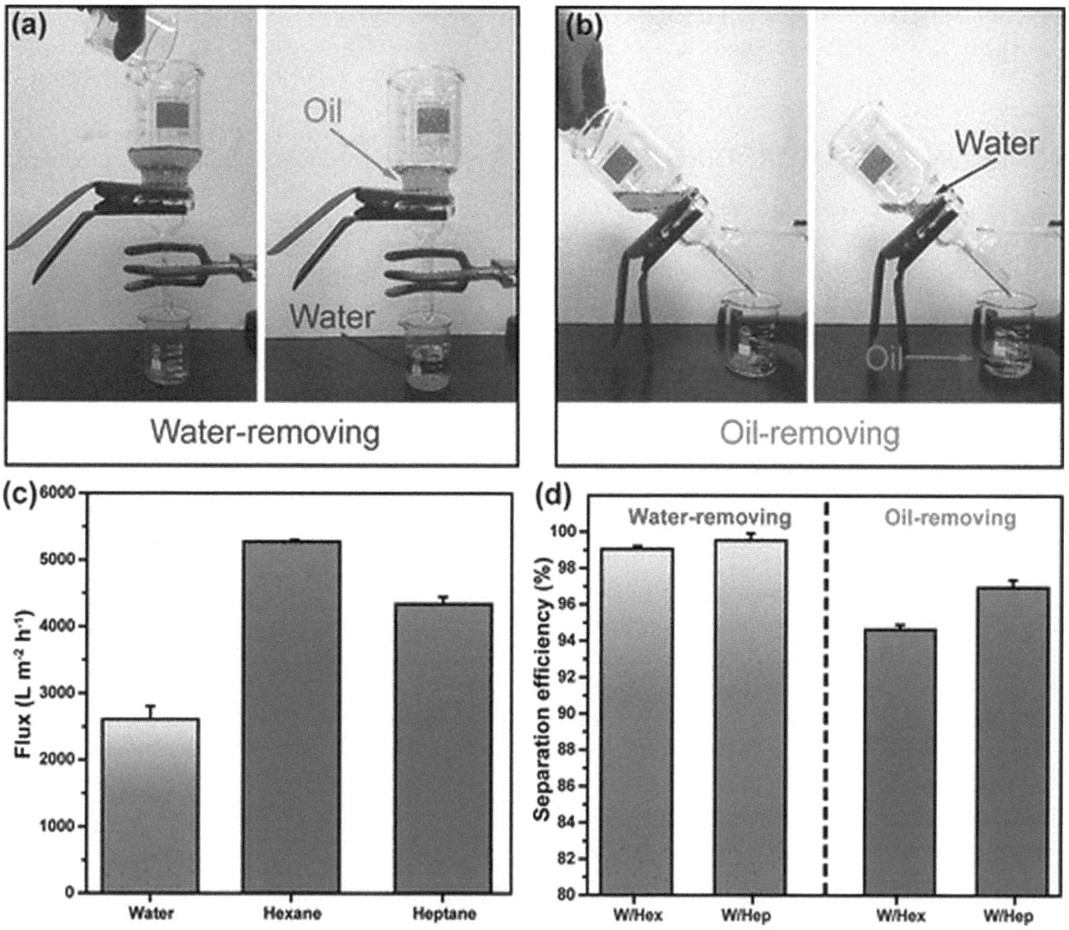

FIGURE 9.11 Photographs of removal of (a) water from oil–water mixture and (b) oil from oil–water mixture. (c) Permeate flux and (d) separation efficiency of CS/PCL membrane. Adapted from Ref. [137].

mixed-matrix membrane at 50°C. The separation efficacy of the CS/UIO-66 membrane was found to be 25 times greater than that of the pristine CS membrane.

9.2.7 Miscellaneous Applications

Lusiana and co-workers fabricated CS/PVA blend membranes cross-linked by citric acid to effectively transport the creatinine through the membrane, which is a crucial aspect of hemodialysis. CS/PVA-citric acid membrane displayed a creatinine transport value of 6.3 mg L^{-1} [140]. Unlu et al. developed a chitosan-based pervaporation catalytic membrane through the incorporation of a $Zr(SO_4)_2.4H_2O$ (ZSH) catalyst [141]. The prepared CS/ZSH catalytic membrane effectively assisted the formation of ethyl acetate from ethanol and acetic acid. The product formation rate was enhanced from 22% to 85% by simultaneous removal of water from the hydrophilic CS/ZSH membrane at 70°C. Chitosan forward osmosis membranes were fabricated for portable water filter bags by loading dimethylformamide [142]. A water flux of 4.25 J m^{-2} h^{-1} was obtained when seawater was used as feed, and 3M sucrose solution was used as the draw solution. Chitosan membranes revealed more than 90% efficacy for the photocatalytic degradation of organic pollutants called tetracycline hydrochloride (TC) when the CS membrane was irradiated with visible light for 60 minutes [143]. The photocatalytic activity of CS was compared with chitosan powders and TiO_2-P25. Both showed low photocatalytic degradation of TC compared to CS membrane. CS photocatalytic membrane showed notable reusability when washed with 2 wt.% NaOH solution. PVA/CS membranes were developed for microbial electrolysis cells, and showed effective hydrogen production of 974 mL H_2 $g^{-1}_{acetate}$ [144]. The fabricated PVA/CS membranes exhibited improved ion transport compared with the Nafion membrane and displayed a hydrogen production rate of 1277 mL H_2 L^{-1}_{cat} d^{-1}. CS/GO composite membranes were utilized for the purification of real surface water [145]. CS/GO membrane removed more than 95% BSA, humic acid, and sodium alginate. A rate of 40–50% for total organic carbon removal was reported with cross-linked CS/GO membrane for lake water. Ghasemzadeh and co-workers cross-linked chitosan with chitosan dialdehyde (CSD) and incorporated silver (Ag) nanoparticles into it. This CS/CSD/Ag composite membrane exhibited outstanding antibacterial properties against both Gram-negative and Gram-positive bacteria [146]. The ratio of minimal bacterial concentration to minimum inhibitory concentration was found to be 2.0, 1.0, and 2.0 for *E. coli*, *P. aeruginosa*, and *S. aureus*, respectively.

9.3 CONCLUSION AND FUTURE DIRECTION

The widespread application of chitosan in membrane fabrication illustrate its remarkable compatibility with a wide range of nanoparticles and polymers. It showed appreciable blending quality with polymers such as PVA, PSF, PES, SPEEK, SPES, PVDF, and PLA via different cross-linking agents. Various nanoparticles including silver, GO, TiO_2, HNT, MWCNTs, SiO_2, chitin nanowhiskers, montmorillonite, and MOFs such as ZIF-8, UIO-66 have been well incorporated in the chitosan matrix for the fabrication of membranes. Incorporating nano-additives enhanced the hydrophilicity, thermal stability, mechanical strength, and antifouling strength of chitosan membranes. Chitosan membranes have found potential applications in dye removal, heavy metal removal, desalination, fuel cells, gas separation, catalytic membranes, hemodialysis, photocatalysis, and many others. Expanding applications of chitosan membranes create room for novel composite membrane fabrication and their efficacy improvement. In the future, more work can be done on developing mechanically robust chitosan membranes without synthetic polymer membrane support. Opportunities are available for developing more efficient chitosan-based gas separation membranes. Further studies are needed to develop ultraviolet radiation- and oxidation-resistant effective chitosan-based photocatalytic membranes for dye and pollutant degradation.

REFERENCES

[1] Ulbricht, M. (2006) Advanced functional polymer membranes. *Polymer* 47(7): 2217–2262.

[2] Kumar, R., Isloor, A.M. (2015) *Chitosan and Its Derivatives as Potential Materials for Membrane Technology*. CRC Press Taylor and Francis Group.

[3] Ibrahim, G.P.S., Isloor, A.M., Inamuddin, Asiri, A.M., Ismail, A.F., Kumar, R., Ahamed, M.I. (2018) Performance intensification of the polysulfone ultrafiltration membrane by blending with copolymer encompassing novel derivative of poly(styrene-co-maleic anhydride) for heavy metal removal from wastewater. *Chemical Engineering Journal* 353: 425–435.

[4] Pereira, V.R., Isloor, A.M., Bhat, U.K., Ismail, A.F. (2014) Preparation and antifouling properties of PVDF ultrafiltration membranes with polyaniline (PANI) nanofibers and hydrolysed PSMA (H-PSMA) as additives. *Desalination* 351: 220–227.

[5] Isloor, A.M., Nayak, M.C., Inamuddin, Prabhu, B., Ismail, N., Ismail, A.F., Asiri, A.M. (2019) Novel polyphenylsulfone (PPSU)/nano tin oxide (SnO_2) mixed matrix ultrafiltration hollow fiber membranes: Fabrication, characterization and toxic dyes removal from aqueous solutions. *Reactive and Functional Polymers* 139: 170–180.

[6] Sun, M., Su, Y., Mu, C., Jiang, Z. (2010) Improved antifouling property of PES ultrafiltration membranes using additive of silica–PVP nanocomposite. *Industrial & Engineering Chemistry Research* 49(2): 790–796.

[7] Zaman, N.K., Rohani, R., Mohammad, A.W., Isloor, A.M., Jahim, J.M. (2020) Investigation of succinic acid recovery from aqueous solution and fermentation broth using polyimide nanofiltration membrane. *Journal of Environmental Chemical Engineering* 8(2): 101895.

[8] Kolangare, I.M., Isloor, A.M., Asiri, A.M., Ismail, A.F. (2019) Improved desalination by polyamide membranes containing hydrophilic glutamine and glycine. *Environmental Chemistry Letters* 17(2): 1053–1059.

[9] Jafar Mazumder, M.A., Raja, P.H., Isloor, A.M., Usman, M., Chowdhury, S.H., Ali, S.A., Inamuddin, Al-Ahmed, A. (2020) Assessment of sulfonated homo and co-polyimides incorporated polysulfone ultrafiltration blend membranes for effective removal of heavy metals and proteins. *Scientific Reports* 10(1): 7049.

[10] Ibrahim, G.S., Isloor, A.M., Moslehyani, A., Ismail, A.F. (2017) Bio-inspired, fouling resistant, tannic acid functionalized halloysite nanotube reinforced polysulfone loose nanofiltration hollow fiber membranes for efficient dye and salt separation. *Journal of Water Process Engineering* 20: 138–148.

[11] Ibrahim, G.S., Isloor, A.M., Farnood, R. (2020) Fundamentals and basics of reverse osmosis. In: *Current Trends and Future Developments on (Bio-) Membranes*, pp. 141–163. Elsevier.

[12] Vijesh, A.M., Arathi Krishnan, P.V., Isloor, A.M., Shyma, P.C. (2021) Fabrication of PPSU/PANI hollow fiber membranes for humic acid removal. *Materials Today: Proceedings* 41: 541–548.

[13] Kumar, M., Rao, T.S., Isloor, A.M., Ibrahim, G.P.S., Inamuddin, Ismail, N., Ismail, A.F., Asiri, A.M. (2019) Use of cellulose acetate/polyphenylsulfone derivatives to fabricate ultrafiltration hollow fiber membranes for the removal of arsenic from drinking water. *International Journal of Biological Macromolecules* 129: 715–727.

[14] Padaki, M., Isloor, A.M., Fernandes, J., Prabhu, K.N. (2011) New polypropylene supported chitosan NF-membrane for desalination application. *Desalination* 280(1): 419–423.

[15] Ghafor, I.M. (2020) *Crustacean*. IntechOpen.

[16] Balakrishna, P., Saidutta, M.B., Isloor, A.M., Hebbar, R. (2017) Improvement in performance of polysulfone membranes through the incorporation of chitosan-(3-phenyl-1h-pyrazole-4-carbaldehyde). *Cogent Engineering* 4(1): 1403005.

[17] Schmitz, C., González Auza, L., Koberidze, D., Rasche, S., Fischer, R., Bortesi, L. (2019) Conversion of chitin to defined chitosan oligomers: Current status and future prospects. *Marine Drugs* 17(8): 452.

[18] Divya, K., Jisha, M.S. (2018) Chitosan nanoparticles preparation and applications. *Environmental Chemistry Letters* 16(1): 101–112.

[19] Kolangare, I.M., Isloor, A.M., Karim, Z.A., Kulal, A., Ismail, A.F., Inamuddin, Asiri, A.M. (2019) Antibiofouling hollow-fiber membranes for dye rejection by embedding chitosan and silver-loaded chitosan nanoparticles. *Environmental Chemistry Letters* 17(1): 581–587.

[20] Padaki, M., Isloor, A.M., Wanichapichart, P. (2011) Polysulfone/N-phthaloylchitosan novel composite membranes for salt rejection application. *Desalination* 279(1): 409–414.

[21] Shenvi, S.S., Rashid, S.A., Ismail, A.F., Kassim, M.A., Isloor, A.M. (2013) Preparation and characterization of PPEES/chitosan composite nanofiltration membrane. *Desalination* 315: 135–141.
[22] Shenvi, S., Ismail, A.F., Isloor, A.M. (2014) Preparation and characterization study of PPEES/chitosan composite membrane crosslinked with tripolyphosphate. *Desalination* 344: 90–96.
[23] Kumar, R., Isloor, A.M., Ismail, A.F., Rashid, S.A., Matsuura, T. (2013) Polysulfone–chitosan blend ultrafiltration membranes: Preparation, characterization, permeation and antifouling properties. *RSC Advances* 3(21): 7855–7861.
[24] Kumar, R., Isloor, A.M., Ismail, A.F., Rashid, S.A., Ahmed, A.A. (2013) Permeation, antifouling and desalination performance of TiO_2 nanotube incorporated PSf/CS blend membranes. *Desalination* 316: 76–84.
[25] Kumar, R., Isloor, A.M., Ismail, A.F., Matsuura, T. (2013) Synthesis and characterization of novel water soluble derivative of chitosan as an additive for polysulfone ultrafiltration membrane. *Journal of Membrane Science* 440: 140–147.
[26] Kumar, R., Isloor, A.M., Ismail, A.F., Matsuura, T. (2013) Performance improvement of polysulfone ultrafiltration membrane using N-succinyl chitosan as additive. *Desalination* 318: 1–8.
[27] Kumar, R., Ismail, A.F., Kassim, M.A., Isloor, A.M. (2013) Modification of PSf/PIAM membrane for improved desalination applications using chitosan coagulation media. *Desalination* 317: 108–115.
[28] Kumar, R., Isloor, A.M., Ismail, A.F. (2014) Preparation and evaluation of heavy metal rejection properties of polysulfone/chitosan, polysulfone/N-succinyl chitosan and polysulfone/N-propylphosphonyl chitosan blend ultrafiltration membranes. *Desalination* 350: 102–108.
[29] Daraei, P., Madaeni, S.S., Salehi, E., Ghaemi, N., Ghari, H.S., Khadivi, M.A., Rostami, E. (2013) Novel thin film composite membrane fabricated by mixed matrix nanoclay/chitosan on PVDF microfiltration support: Preparation, characterization and performance in dye removal. *Journal of Membrane Science* 436: 97–108.
[30] El-Azzami, L.A., Grulke, E.A. (2008) Carbon dioxide separation from hydrogen and nitrogen by fixed facilitated transport in swollen chitosan membranes. *Journal of Membrane Science* 323(2): 225–234.
[31] Mukoma, P., Jooste, B.R., Vosloo, H.C.M. (2004) Synthesis and characterization of cross-linked chitosan membranes for application as alternative proton exchange membrane materials in fuel cells. *Journal of Power Sources* 136(1): 16–23.
[32] Nasrollahzadeh, M., Sajjadi, M., Iravani, S., Varma, R.S. (2021) Starch, cellulose, pectin, gum, alginate, chitin and chitosan derived (nano)materials for sustainable water treatment: A review. *Carbohydrate Polymers* 251: 116986.
[33] Salehi, E., Daraei, P., Arabi Shamsabadi, A. (2016) A review on chitosan-based adsorptive membranes. *Carbohydrate Polymers* 152: 419–432.
[34] Rosli, N.A.H., Loh, K.S., Wong, W.Y., Yunus, R.M., Lee, T.K., Ahmad, A., Chong, S.T. (2020) Review of chitosan-based polymers as proton exchange membranes and roles of chitosan-supported ionic liquids. *International Journal of Molecular Science* 21(2): 632.
[35] Miguel, S.P., Moreira, A.F., Correia, I.J. (2019) Chitosan-based asymmetric membranes for wound healing: A review. *International Journal of Biological Macromolecules* 127: 460–475.
[36] Vedula, S.S., Yadav, G.D. (2021) Chitosan-based membranes preparation and applications: Challenges and opportunities. *Journal of the Indian Chemical Society* 98(2): 100017.
[37] Yang, T., Zall, R.R. (1984) Chitosan membranes for reverse osmosis application. *Journal of Food Science* 49(1): 91–93.
[38] Thakur, V.K., Voicu, S.I. (2016) Recent advances in cellulose and chitosan based membranes for water purification: A concise review. *Carbohydrate Polymers* 146: 148–165.
[39] Borgohain, R., Pattnaik, U., Prasad, B., Mandal, B. (2021) A review on chitosan-based membranes for sustainable CO_2 separation applications: Mechanism, issues, and the way forward. *Carbohydrate Polymers* 267: 118178.
[40] Dutta, P.K., Tripathi, S., Mehrotra, G.K., Dutta, J. (2009) Perspectives for chitosan based antimicrobial films in food applications. *Food Chemistry* 114(4): 1173–1182.
[41] Ali, A., Ahmed, S. (2018) A review on chitosan and its nanocomposites in drug delivery. *International Journal of Biological Macromolecules* 109: 273–286.
[42] Ibrahim, G.P.S., Isloor, A.M., Inamuddin, Asiri, A.M., Farnood, R. (2020) Tuning the surface properties of Fe_3O_4 by zwitterionic sulfobetaine: Application to antifouling and dye removal membrane. *International Journal of Environmental Science and Technology* 17(9): 4047–4060.
[43] He, X., Du, M., Li, H., Zhou, T. (2016) Removal of direct dyes from aqueous solution by oxidized starch cross-linked chitosan/silica hybrid membrane. *International Journal of Biological Macromolecules* 82: 174–181.
[44] Li, C., Lou, T., Yan, X., Long, Y., Cui, G., Wang, X. (2018) Fabrication of pure chitosan nanofibrous membranes as effective absorbent for dye removal. *International Journal of Biological Macromolecules* 106: 768–774.
[45] Yang, H.-C., Gong, J.-L., Zeng, G.-M., Zhang, P., Zhang, J., Liu, H.-Y., Huan, S.-Y. (2017) Polyurethane foam membranes filled with humic acid-chitosan crosslinked gels for selective and simultaneous removal of dyes. *Journal of Colloid and Interface Science* 505: 67–78.
[46] Shi, C., Lv, C., Wu, L., Hou, X. (2017) Porous chitosan/hydroxyapatite composite membrane for dyes static and dynamic removal from aqueous solution. *Journal of Hazardous Materials* 338: 241–249.
[47] Weng, R., Huang, X., Liao, D., Xu, S., Peng, L., Liu, X. (2020) A novel cellulose/chitosan composite nanofiltration membrane prepared with piperazine and trimesoyl chloride by interfacial polymerization. *RSC Advances* 10(3): 1309–1318.
[48] Khoerunnisa, F., Rahmah, W., Seng Ooi, B., Dwihermiati, E., Nashrah, N., Fatimah, S., Ko, Y.G., Ng, E.-P. (2020) Chitosan/PEG/MWCNT/iodine composite membrane with enhanced antibacterial properties for dye wastewater treatment. *Journal of Environmental Chemical Engineering* 8(2): 103686.
[49] Mousavi, S.R., Asghari, M., Mahmoodi, N.M. (2020) Chitosan-wrapped multiwalled carbon nanotube as filler within PEBA thin film nanocomposite (TFN) membrane to improve dye removal. *Carbohydrate Polymers* 237: 116128.
[50] Long, Q., Zhang, Z., Qi, G., Wang, Z., Chen, Y., Liu, Z.-Q. (2020) Fabrication of chitosan nanofiltration membranes

by the film casting strategy for effective removal of dyes/salts in textile wastewater. *ACS Sustainable Chemistry & Engineering* 8(6): 2512–2522.

[51] Khajavian, M., Salehi, E., Vatanpour, V. (2020) Nanofiltration of dye solution using chitosan/poly(vinyl alcohol)/ZIF-8 thin film composite adsorptive membranes with PVDF membrane beneath as support. *Carbohydrate Polymers* 247: 116693.

[52] Tan, Y., Kang, Y., Wang, W., Lv, X., Wang, B., Zhang, Q., Cui, C., Cui, S., Jiao, S., Pang, G., et al. (2021) Chitosan modified inorganic nanowires membranes for ultra-fast and efficient removal of congo red. *Applied Surface Science* 569: 150970.

[53] Hosseini, S.A., Daneshvar e Asl, S., Vossoughi, M., Simchi, A., Sadrzadeh, M. (2021) Green electrospun membranes based on chitosan/amino-functionalized nanoclay composite fibers for cationic dye removal: Synthesis and kinetic studies. *ACS Omega* 6(16): 10816–10827.

[54] Yang, J., Han, Y., Sun, Z., Zhao, X., Chen, F., Wu, T., Jiang, Y. (2021) PEG/sodium tripolyphosphate-modified chitosan/activated carbon membrane for rhodamine B removal. *ACS Omega* 6(24): 15885–15891.

[55] Guo, Q., Wu, X., Ji, Y., Hao, Y., Liao, S., Cui, Z., Li, J., Younas, M., He, B. (2021) pH-responsive nanofiltration membrane containing chitosan for dye separation. *Journal of Membrane Science* 635: 119445.

[56] Hassanzadeh, P., Gharbani, P., Derakhshanfard, F., Memar Maher, B. (2021) Preparation and characterization of PVDF/g-C_3N_4/chitosan polymeric membrane for the removal of direct blue 14 dye. *Journal of Polymers and the Environment* 29(11): 3693–3702.

[57] Ulu, A., Alpaslan, M., Gultek, A., Ates, B. (2022) Eco-friendly chitosan/κ-carrageenan membranes reinforced with activated bentonite for adsorption of methylene blue. *Materials Chemistry and Physics* 278: 125611.

[58] Vedula, S.S., Yadav, G.D. (2022) Wastewater treatment containing methylene blue dye as pollutant using adsorption by chitosan lignin membrane: Development of membrane, characterization and kinetics of adsorption. *Journal of the Indian Chemical Society* 99(1): 100263.

[59] Shenvi, S.S., Isloor, A.M., Ismail, A.F. (2015) A review on RO membrane technology: Developments and challenges. *Desalination* 368: 10–26.

[60] Ma, X.-H., Yang, Z., Yao, Z.-K., Xu, Z.-L., Tang, C.Y. (2017) A facile preparation of novel positively charged MOF/chitosan nanofiltration membranes. *Journal of Membrane Science* 525: 269–276.

[61] Deng, H., Sun, P., Zhang, Y., Zhu, H. (2016) Reverse osmosis desalination of chitosan cross-linked graphene oxide/titania hybrid lamellar membranes. *Nanotechnology* 27(27): 274002.

[62] Alshahrani, A.A., Al-Zoubi, H., Nghiem, L.D., in het Panhuis, M. (2017) Synthesis and characterisation of MWNT/chitosan and MWNT/chitosan-crosslinked buckypaper membranes for desalination. *Desalination* 418: 60–70.

[63] Shafiq, M., Sabir, A., Islam, A., Khan, S.M., Hussain, S.N.Z., Butt, M.T.Z., Jamil, T. (2017) Development and performance characteristics of silane crosslinked poly(vinyl alcohol)/chitosan membranes for reverse osmosis. *Industrial & Engineering Chemistry Research* 48: 99–107.

[64] Tang, Y.-J., Wang, L.-J., Xu, Z.-L., Zhang, H.-Z. (2018) Novel chitosan-piperazine composite nanofiltration membranes for the desalination of brackish water and seawater. *Journal of Polymer Research* 25(5): 118.

[65] Kebria, M.R.S., Rahimpour, A., Bakeri, G., Abedini, R. (2019) Experimental and theoretical investigation of thin ZIF-8/chitosan coated layer on air gap membrane distillation performance of PVDF membrane. *Desalination* 450: 21–32.

[66] El-Ghaffar, M.A.A., Elawady, M.M., Rabie, A.M., Abdelhamid, A.E. (2020) Enhancing the RO performance of cellulose acetate membrane using chitosan nanoparticles. *Journal of Polymer Research* 27(11); 337.

[67] Alsuhybani, M., Alshahrani, A., Haidyrah, A.S. (2020) Synthesis, characterization, and evaluation of evaporated casting MWCNT/chitosan composite membranes for water desalination. *Journal of Chemistry* 2020: e5207680.

[68] Ali, H., Dilshad, M.R., Haider, B., Islam, A., Akram, M.S., Jalal, A., Hussain, S.N. (2022) Preparation and characterization of novel polyamide-6/chitosan blend dense membranes for desalination of brackish water. *Polymer Bulletin* 79(6): 4153–4169.

[69] Ünügül, T., Nigiz, F.U. (2022) Evaluation of halloysite nanotube-loaded chitosan-based nanocomposite membranes for water desalination by pervaporation. *Water, Air, & Soil Pollution* 233(2): 34.

[70] Purwanto, M., Atmaja, L., Mohamed, M.A., Salleh, M.T., Jaafar, J., Ismail, A.F., Santoso, M., Widiastuti, N. (2015) Biopolymer-based electrolyte membranes from chitosan incorporated with montmorillonite-crosslinked GPTMS for direct methanol fuel cells. *RSC Advances* 6(3): 2314–2322.

[71] García-Cruz, L., Casado-Coterillo, C., Iniesta, J., Montiel, V., Irabien, Á. (2016) Chitosan:poly (vinyl) alcohol composite alkaline membrane incorporating organic ionomers and layered silicate materials into a PEM electrochemical reactor. *Journal of Membrane Science* 498: 395–407.

[72] Liu, H., Gong, C., Wang, J., Liu, X., Liu, H., Cheng, F., Wang, G., Zheng, G., Qin, C., Wen, S. (2016) Chitosan/silica coated carbon nanotubes composite proton exchange membranes for fuel cell applications. *Carbohydrate Polymers* 136: 1379–1385.

[73] Liu, H., Wang, J., Wen, S., Gong, C., Cheng, F., Wang, G., Zheng, G., Qin, C. (2016) Composite membranes of chitosan and titania-coated carbon nanotubes as promising materials for new proton-exchange membranes. *Journal of Applied Polymer Science* 133(17):.

[74] Shirdast, A., Sharif, A., Abdollahi, M. (2016) Effect of the incorporation of sulfonated chitosan/sulfonated graphene oxide on the proton conductivity of chitosan membranes. *Journal of Power Sources* 306: 541–551.

[75] Holder, S.L., Lee, C.-H., Popuri, S.R., Zhuang, M.-X. (2016) Enhanced surface functionality and microbial fuel cell performance of chitosan membranes through phosphorylation. *Carbohydrate Polymers* 149: 251–262.

[76] Wu, Q., Wang, H., Lu, S., Xu, X., Liang, D., Xiang, Y. (2016) Novel methanol-blocking proton exchange membrane achieved via self-anchoring phosphotungstic acid into chitosan membrane with submicro-pores. *Journal of Membrane Science* 500: 203–210.

[77] Wang, B., Zhu, Y., Zhou, T., Xie, K. (2016) Synthesis and properties of chitosan membranes modified by reactive cationic dyes as a novel alkaline exchange membrane for low

temperature fuel cells. *International Journal of Hydrogen Energy* 41(40): 18166–18177.

[78] Muthumeenal, A., Neelakandan, S., Kanagaraj, P., Nagendran, A. (2016) Synthesis and properties of novel proton exchange membranes based on sulfonated polyethersulfone and N-phthaloyl chitosan blends for DMFC applications. *Renewable Energy* 86: 922–929.

[79] Mohy Eldin, M., Hashem, A., Tamer, T., Omer, A., Youssef, M.E., Sabet, M. (2017) Development of cross linked chitosan/alginate polyelectrolyte proton exchanger membranes for fuel cell applications. *International Journal of Electrochemical Science* 12: 3840–3858.

[80] Wang, W., Shan, B., Zhu, L., Xie, C., Liu, C., Cui, F. (2018) Anatase titania coated CNTs and sodium lignin sulfonate doped chitosan proton exchange membrane for DMFC application. *Carbohydrate Polymers* 187: 35–42.

[81] Wang, J., Gong, C., Wen, S., Liu, H., Qin, C., Xiong, C., Dong L. (2018) Proton exchange membrane based on chitosan and solvent-free carbon nanotube fluids for fuel cells applications. *Carbohydrate Polymers* 186: 200–207.

[82] Wang, J., Gong, C., Wen, S., Liu, H., Qin, C., Xiong, C., Dong L. (2019) A facile approach of fabricating proton exchange membranes by incorporating polydopamine-functionalized carbon nanotubes into chitosan. *International Journal of Hydrogen Energy* 44(13): 6909–6918.

[83] Hidayati, N., Harmoko, T., Mujiburohman, M., Purnama, H. (2019) Characterization of SPEEK/chitosan membrane for the direct methanol fuel cell. *AIP Conference Proceedings* 2114(1): 060008.

[84] Divya, K., Rana, D., Alwarappan, S., Abirami Saraswathi, M.S.S., Nagendran, A. (2019) Investigating the usefulness of chitosan based proton exchange membranes tailored with exfoliated molybdenum disulfide nanosheets for clean energy applications. *Carbohydrate Polymers* 208: 504–512.

[85] Kaker, B., Hribernik, S., Mohan, T., Kargl, R., Stana Kleinschek, K., Pavlica, E., Kreta, A., Bratina, G., Lue, S.J., Božič, M. (2019) Novel chitosan–$Mg(OH)_2$-based nanocomposite membranes for direct alkaline ethanol fuel cells. *ACS Sustainable Chemical Engineering* 7(24): 19356–19368.

[86] Ryu, J., Seo, J.Y., Choi, B.N., Kim, W.-J., Chung, C.-H. (2019) Quaternized chitosan-based anion exchange membrane for alkaline direct methanol fuel cells. *Industrial & Engineering Chemistry Research* 73: 254–259.

[87] Abu-Saied, M.A., Soliman, E.A., Desouki, E.A.A. (2020) Development of proton exchange membranes based on chitosan blended with poly (2-acrylamido-2-methylpropane sulfonic acid) for fuel cells applications. *Materials Today Communications* 25: 101536.

[88] Altaf, F., Batool, R., Gill, R., Shabir, M.A., Drexler, M., Alamgir, F., Abbas, G., Sabir, A., Jacob, K.I. (2020) Novel N-p-carboxy benzyl chitosan/poly (vinyl alcohol)/ functionalized zeolite mixed matrix membranes for DMFC applications. *Carbohydrate Polymers* 237: 116111.

[89] Hu, F., Li, T., Zhong, F., Wen, S., Zheng, G., Gong, C., Qin, C., Liu, H. (2020) Preparation and properties of chitosan/acidified attapulgite composite proton exchange membranes for fuel cell applications. *Journal of Applied Polymer Science* 137(36): 49079.

[90] Nasirinezhad, M., Ghaffarian, S.R., Tohidian, M. (2021) Eco-friendly polyelectrolyte nanocomposite membranes based on chitosan and sulfonated chitin nanowhiskers for fuel cell applications. *Iranian Polymer Journal* 30(4): 355–367.

[91] Divya, K., Rana, D., Rameesha, L., Sri Abirami Saraswathi, M.S., Nagendran, A. (2021) Highly selective custom-made chitosan based membranes with reduced fuel permeability for direct methanol fuel cells. *Journal of Applied Polymer Science* 138(46): 51366.

[92] Rosli, N.A.H., Loh, K.S., Wong, W.Y., Lee, T.K., Ahmad, A. (2021) Hybrid composite membrane of phosphorylated chitosan/poly (vinyl alcohol)/silica as a proton exchange membrane. *Membranes* 11(9): 675.

[93] Hren, M., Gorgieva, S. (2022) Composite chitosan and quaternary ammonium modified nanofibrillar cellulose anion exchange membranes for direct ethanol fuel cell applications. *E3S Web Conference* 334: 04001.

[94] Liu, J., Zheng, L., Li, Y., Free, M., Yang, M. (2016) Adsorptive recovery of palladium(II) from aqueous solution onto cross-linked chitosan/montmorillonite membrane. *RSC Advances* 6(57): 51757–51767.

[95] Li, L., Li, Y., Yang, C. (2016) Chemical filtration of Cr (VI) with electrospun chitosan nanofiber membranes. *Carbohydrate Polymers* 140: 299–307.

[96] Wang, X., Li, Y., Li, H., Yang, C. (2016) Chitosan membrane adsorber for low concentration copper ion removal. *Carbohydrate Polymers* 146: 274–281.

[97] Habiba, U., Afifi, A.M., Salleh, A., Ang, B.C. (2017) Chitosan/(polyvinyl alcohol)/zeolite electrospun composite nanofibrous membrane for adsorption of Cr^{6+}, Fe^{3+} and Ni^{2+}. *Journal of Hazardous Materials* 322: 182–194.

[98] Qin, W., Li, J., Tu, J., Yang, H., Chen, Q., Liu, H. (2017) Fabrication of porous chitosan membranes composed of nanofibers by low temperature thermally induced phase separation, and their adsorption behavior for Cu^{2+}. *Carbohydrate Polymers* 178: 338–346.

[99] Yang, D., Li, L., Chen, B., Shi, S., Nie, J., Ma, G. (2019) Functionalized chitosan electrospun nanofiber membranes for heavy-metal removal. *Polymer* 163: 74–85.

[100] Haider, S., Ali, F.A.A., Haider, A., Al-Masry, W.A., Al-Zeghayer, Y. (2018) Novel route for amine grafting to chitosan electrospun nanofibers membrane for the removal of copper and lead ions from aqueous medium. *Carbohydrate Polymers* 199: 406–414.

[101] Karim, M.R., Aijaz, M.O., Alharth, N.H., Alharbi, H.F., Al-Mubaddel, F.S., Awual, Md. R. (2019) Composite nanofibers membranes of poly(vinyl alcohol)/chitosan for selective lead(II) and cadmium(II) ions removal from wastewater. *Ecotoxicology and Environmental Safety* 169: 479–486.

[102] Jamshidifard, S., Koushkbaghi, S., Hosseini, S., Rezaei, S., Karamipour, A., Jafari rad, A., Irani, M. (2019) Incorporation of UiO-66-NH_2 MOF into the PAN/chitosan nanofibers for adsorption and membrane filtration of Pb(II), Cd(II) and Cr(VI) ions from aqueous solutions. *Journal of Hazardous Materials* 368: 10–20.

[103] Sangeetha, K., Vinodhini, A.P., Sudha, P.N., Faleh, A.A., Sukumaran, A (2019) Novel chitosan based thin sheet nanofiltration membrane for rejection of heavy metal chromium. *International Journal of Biological Macromolecules* 132: 939–953.

[104] Marques, J.S., Pereira, M.R., Sotto, A., Arsuaga, J.M. (2019) Removal of aqueous copper(II) by using crosslinked

chitosan films. *Reactive and Functional Polymers* 134: 31–39.

[105] Mukhopadhyay, M., Lakhotia, S.R., Ghosh, A.K., Bindal, R.C. (2019) Removal of arsenic from aqueous media using zeolite/chitosan nanocomposite membrane. *Separation Science and Technology* 54(2): 282–288.

[106] Căprărescu, S., Zgârian, R.G., Tihan, G.T., Purcar, V., Eftimie Totu, E., Modrogan, C., Chiriac, A.-L., Nicolae, C.A. (2020) Biopolymeric membrane enriched with chitosan and silver for metallic ions removal. *Polymers* 12(8): 1792.

[107] Alharbi, H.F., Haddad, M.Y., Aijaz, M.O., Assaifan, A.K., Karim, M.R. (2020) Electrospun bilayer PAN/chitosan nanofiber membranes incorporated with metal oxide nanoparticles for heavy metal ion adsorption. *Coatings* 10(3): 285.

[108] Pishnamazi, M., Koushkbaghi, S., Hosseini, S.S., Darabi, M., Yousefi, A., Irani, M. (2020) Metal organic framework nanoparticles loaded PVDF/chitosan nanofibrous ultrafiltration membranes for the removal of BSA protein and Cr(VI) ions. *Journal of Molecular Liquids* 317: 113934.

[109] Hu, X., Wang, Y., Yang, J.O., Li, Y., Wu, P., Zhang, H., Yuan, D., Liu, Y., Wu, Z., Liu, Z. (2020) Synthesis of graphene oxide nanoribbons/chitosan composite membranes for the removal of uranium from aqueous solutions. *Frontiers of Chemical Science and Engineering* 14(6): 1029–1038.

[110] Machodi, M.J., Daramola, M.O. (2020) Synthesis of PES and PES/chitosan membranes for synthetic acid mine drainage treatment. *Water SA* 46.

[111] Zakmout, A., Sadi, F., Portugal, C.A.M., Crespo, J.G., Velizarov, S. (2020) Tannery effluent treatment by nanofiltration, reverse osmosis and chitosan modified membranes. *Membranes* 10(12): 378.

[112] Alshahrani, A., Alharbi, A., Alnasser, S., Almihdar, M., Alsuhybani, M., AlOtaibi, B. (2021) Enhanced heavy metals removal by a novel carbon nanotubes buckypaper membrane containing a mixture of two biopolymers: Chitosan and i-carrageenan. *Separation and Purification Technology* 276: 119300.

[113] Zia, Q., Tabassum, M., Meng, J., Xin, Z., Gong, H., Li, J. (2021) Polydopamine-assisted grafting of chitosan on porous poly (L-lactic acid) electrospun membranes for adsorption of heavy metal ions. *International Journal of Biological Macromolecules* 167: 1479–1490.

[114] Orabi, A.H., Abdelhamid, A.E., Salem, H.M., Ismaiel, D.A. (2021) Uranium removal using composite membranes incorporated with chitosan grafted phenylenediamine from liquid waste solution. *Cellulose* 28(6): 3703–3721.

[115] Shirazi, R., Mohammadi, T., Asadi, A.A. (2022) Incorporation of amine-grafted halloysite nanotube to electrospun nanofibrous membranes of chitosan/poly (vinyl alcohol) for Cd (II) and Pb(II) removal. *Applied Clay Science* 220: 106460.

[116] Lu, Q., Li, N., Tang, Q., Zhang, X., Zhang, F., Bi, J. (2022) Knitted tube reinforced chitosan/β-cyclodextrin composite ultrafiltration membrane for removing copper ions from water. *Journal of Applied Polymer Science* 139(15): 51917.

[117] García-García, J.J., Gómez-Espinosa, R.M., Rangel, R.N., Romero, R.R., Morales, G.R. (2022) New material for arsenic (V) removal based on chitosan supported onto modified polypropylene membrane. *Environmental Science and Pollution Research* 29(2): 1909–1916.

[118] Santos, E., Rodríguez-Fernández, E., Casado-Coterillo, C., Irabien, Á. (2016) Hybrid ionic liquid-chitosan membranes for CO_2 separation: Mechanical and thermal behavior. *International Journal of Chemical Reactor Engineering* 14(3): 713–718.

[119] Dudek, G., Gnus, M., Strzelewicz, A., Turczyn, R., Krasowska, M. (2016) Permeation of ethanol and water vapors through chitosan membranes with ferroferric oxide particles cross-linked by glutaraldehyde and sulfuric(VI) acid. *Separation Science and Technology* 51: 2649–2656.

[120] Seidi, F., Haghdust, S., Saedi, S., Xu, X. (2016) Synthesis and characterization of a new amino chitosan derivative for facilitated transport of CO_2 through thin film composite membranes. *Macromolecular Research* 24(1): 1–8.

[121] Wang, Z., Yan, F., Pei, H., Li, J., Cui, Z., He, B. (2018) Antibacterial and environmentally friendly chitosan/polyvinyl alcohol blend membranes for air filtration. *Carbohydrate Polymers* 198: 241–248.

[122] Li, H., Wang, Z., Zhang, H., Pan, Z. (2018) Nanoporous PLA/(chitosan nanoparticle) composite fibrous membranes with excellent air filtration and antibacterial performance. *Polymers* 10(10): 1085.

[123] Borgohain, R., Jain, N., Prasad, B., Mandal, B., Su, B. (2019) Carboxymethyl chitosan/carbon nanotubes mixed matrix membranes for CO_2 separation. *Reactive and Functional Polymers* 143: 104331.

[124] Borgohain, R., Mandal, B. (2019) pH responsive carboxymethyl chitosan/poly(amidoamine) molecular gate membrane for CO_2/N_2 separation. *ACS Applied Materials Interfaces* 11(45): 42616–42628.

[125] Borgohain, R., Prasad, B., Mandal, B. (2019) Synthesis and characterization of water-soluble chitosan membrane blended with a mobile carrier for CO_2 separation. *Separation and Purification Technology* 222: 177–187.

[126] Jiamjirangkul, P., Inprasit, T., Intasanta, V., Pangon, A. (2020) Metal organic framework-integrated chitosan/poly(vinyl alcohol) (PVA) nanofibrous membrane hybrids from green process for selective CO_2 capture and filtration. *Chemical Engineering Science* 221: 115650.

[127] Borgohain, R., Mandal, B. (2020) Thermally stable and moisture responsive carboxymethyl chitosan/dendrimer/hydrotalcite membrane for CO_2 separation. *Journal of Membrane Science* 608: 118214.

[128] Li, N., Wang, Z., Wang, J. (2022) Water-swollen carboxymethyl chitosan (CMC)/polyamide (PA) membranes with octopus-branched nanostructures for CO_2 capture. *Journal of Membrane Science* 642: 119946.

[129] Liu, J., Li, P., Chen, L., Feng, Y., He, W., Lv, X. (2016) Modified superhydrophilic and underwater superoleophobic PVDF membrane with ultralow oil-adhesion for highly efficient oil/water emulsion separation. *Materials Letters* 185: 169–172.

[130] Zheng, P.-Y., Ye, C.-C., Wang, X.-S., Chen, K.-F., An, Q.-F., Lee, K.-R., Gao, C.-J. (2016) Poly(sodium vinylsulfonate)/chitosan membranes with sulfonate ionic cross-linking and free sulfate groups: Preparation and application in alcohol dehydration. *Journal of Membrane Science* 510: 220–228.

[131] Prasad, N.S., Moulik, S., Bohra, S., Rani, K.Y., Sridhar, S. (2016) Solvent resistant chitosan/poly(ether-block-amide) composite membranes for pervaporation of

n-methyl-2-pyrrolidone/water mixtures. *Carbohydrate Polymers* 136: 1170–1181.

[132] Moulik, S., Vani, B., Chandrasekhar, S.S., Sridhar, S. (2016) Chitosan-polytetrafluoroethylene composite membranes for separation of methanol and toluene by pervaporation. *Carbohydrate Polymers* 193: 28–38.

[133] Xu, Z., Liu, G., Ye, H., Jin, W., Cui, Z. (2018) Two-dimensional MXene incorporated chitosan mixed-matrix membranes for efficient solvent dehydration. *Journal of Membrane Science* 563: 625–632.

[134] Vinu, M., Pal, S., Chen, J.-D., Lin, Y.-F., Lai, Y.-L., Lee, C.-S., Lin, C.-H. (2019) Microporous 3D aluminum MOF doped into chitosan-based mixed matrix membranes for ethanol/water separation. *JCCS* 66(9): 1165–1171.

[135] Dudek, G., Turczyn, R., Konieczny, K. (2020) Robust poly(vinyl alcohol) membranes containing chitosan/chitosan derivatives microparticles for pervaporative dehydration of ethanol. *Separation and Purification Technology* 234: 116094.

[136] Du, J.R., Hsu, L.H., Xiao, E.S., Guo, X., Zhang, Y., Feng, X. (2020) Using genipin as a "green" crosslinker to fabricate chitosan membranes for pervaporative dehydration of isopropanol. *Separation and Purification Technology* 244: 116843.

[137] Doan, H.N., Vo, P.P., Baggio, A., Negoro, M., Kinashi, K., Fuse, Y., Sakai, W., Tsutsumi, N. (2021) Environmentally friendly chitosan-modified polycaprolactone nanofiber/nanonet membrane for controllable oil/water separation. *ACS Applied Polymer Materials* 3(8): 3891–3901.

[138] Krishnamoorthi, R., Anbazhagan, R., Tsai, H.-C., Wang, C.-F., Lai, J.-Y. (2022) Biodegradable, superwettable caffeic acid/chitosan polymer coated cotton fibers for the simultaneous removal of oils, dyes, and metal ions from water. *Chemical Engineering Journal* 427: 131920.

[139] Zhu, H., Li, R., Liu, G., Pan, Y., Li, J., Wang, Z., Guo, Y., Liu, G., Jin, W. (2022) Efficient separation of methanol/dimethyl carbonate mixtures by UiO-66 MOF incorporated chitosan mixed-matrix membrane. *Journal of Membrane Science* 652: 120473.

[140] Lusiana, R.A., Siswanta, D., Mudasir, M. (2018) Preparation of citric acid crosslinked chitosan/poly(vinyl alcohol) blend membranes for creatinine transport. *Indonesian Journal of Chemistry* 16(2): 144–150.

[141] Unlu, D., Hilmioglu, N.D. (2016) Pervaporation catalytic membrane reactor study for the production of ethyl acetate using $Zr(SO_4)2.4H_2O$ coated chitosan membrane. *Journal of Chemical Technology & Biotechnology* 91(1): 122–130.

[142] Saiful; Riana, U., Marlina, Ramli, M., Mahmud, N. (2019) Drinking water bags based on chitosan forward osmosis membranes for emergency drinking water supply. *IOP Conference Series: Earth Environmental Science* 273(1): 012047.

[143] Liang, H., Lv, C., Chen, H., Wu, L., Hou, X. (2020) Facile synthesis of chitosan membranes for visible-light-driven photocatalytic degradation of tetracycline hydrochloride. *RSC Advances* 10(73): 45171–45179.

[144] González-Pabón, M.J., Cardeña, R., Cortón, E., Buitrón, G. (2021) Hydrogen production in two-chamber MEC using a low-cost and biodegradable poly(vinyl) alcohol/chitosan membrane. *Bioresources Technology* 319: 124168.

[145] Liu, T., Graham, N., Yu, W. (2021) Evaluation of a novel composite chitosan–graphene oxide membrane for NOM removal during water treatment. *Journal of Environmental Chemical Engineering* 9(4): 105716.

[146] Ghasemzadeh, H., Sheikhahmadi, M., Nasrollah, F. (2016) Full polysaccharide crosslinked-chitosan and silver nano composites, for use as an antibacterial membrane. *Chinese Journal of Polymer Science* 34(8): 949–964.

10 A Review of Diverse Membrane Materials for Haemodialysis

K.C. Pallavi[1], Arun M. Isloor[1], Sowmya M. Kumar[2], and Abdul Wahab Mohammad[3]*

[1]Membrane and Separation Technology Laboratory, Department of Chemistry, National Institute of Technology Karnataka, Surathkal, Mangalore, India

[2]Department of Prosthodontics, Nitte Deemed to be University, A.B. Shetty Memorial Institute of Dental Sciences (ABSMIDS), Deralakatte, Mangalore, India

[3]Chemical and Process Engineering, Universiti Kebangsaan Malaysia, Bangi, Selangor, Malaysia

*Email: isloor@yahoo.com

10.1 INTRODUCTION

Kidney malfunctioning or diseases are major health concerns currently. Haemodialysis is the treatment method applied for the purification of extracorporeal blood. The dialysis market is a service for a number of patients with chronic kidney disease (CKD). It is expected to reach to USD 105.1 billion by 2026 from USD 73.55 billion in 2022. The average annual cost per individual for haemodialysis (HD) stands at around USD 41–518 [1,2].

The kidney is an essential part of the body, purifying the blood by eliminating excess water and metabolic byproducts like potassium, uric acid, urea, sodium, creatinine, waste, and toxins. Kidney well-being is a function of the glomerular filtration rate (GFR) which is given in mL/min/m^2. A healthy individual has a GFR in the range of 100–130 mL/min/1.73 m^2 for men and 90–120 mL/min/1.73 m^2 for women aged under 40. GFR declines with age by an average of 1 mL/min per year after the age of 40. Children have a GFR in the range of 100–120 mL/min/1.73 m^2, irrespective of gender. The kidney can start malfunctioning or end up non-functioning due to several renal diseases such as cystinosis, glomerulonephritis, lupus nephritis, and polycystic kidney disease. In patients with these problems, the kidney loses its normal functioning due to the affected glomerulus. Treatment methods such as renal replacement through artificial kidney transplantation, external dialysis through peritoneal dialysis, and haemodialysis are employed. Of these, the most commonly offered is haemodialysis. It involves an extracorporeal method of blood purification using a semipermeable membrane due to the concentration-driven diffusion process. It involves a membrane-driven molecular weight-dependent separation technique along with dissolution-diffusion. Metabolic wastes have moderate molecular weight and extra water in the body, and more minor metabolites such as urea and creatinine get dissolved and diffused due to the concentration. Therefore, the haemodialysis membrane is the heart of a dialyzer which in turn is at the core of haemodialysis treatment [3,4].

In a filtration process, the molecular size is crucial for the permeability of compounds. Blood cells have a size in the range of 7–8 μm in diameter and 2.5 μm in thickness. Platelets are about one-third the size of blood cells. Urea and creatinine have molecular weights of 60 Da and 113 Da, respectively. The average sizes of sodium and potassium ions are 0.36 nm and 0.133 nm, respectively. Bacteria usually have a size of about 0.2 μm in diameter and 2–8 μm in length. Viruses are in the range of 120–200 nm. Figure 10.1 depicts the basic principles involved in haemodialysis [5].

The choice of a proper dialyzer is down to the membrane efficacy, which relies on the extent of solute rejection and compatibility. Technological modifications in designing a membrane, its composition, and the sterilization technique are important for attaining greater potency. According to the membrane type, surface morphology, and characteristics, along with the sterilization technique, there is a wide variety of haemodialysis membranes. Currently, there is increased demand that dialyzers should remove endotoxins from the exterior surface because this results in safer purification, preventing blood infection through transferring from dialysate [6].

The usually employed sterilization techniques rely on ethylene oxide gas (ETO), autoclaving (steam), and gamma/beta irradiation. Dialyzers are available with varyingly sterilized forms as some patients are especially sensitive. There is at least one fault with each of these methods, such as allergic responses to ETO, and changes in hydraulic permeability or material degradation with the use of steam or γ-rays. The biocompatibility profile of a particular polymer changes as a function of sterilization-induced changes. ETO is a cost-effective and comparatively safest method. However, the performance of dialysis membranes does not depend on this, as these are mild processes. The alkylation of sulphur-possessing proteins will destroy bacteria through temperature, time, concentration, humidity, and pressure. A minimal amount of ETO is maintained in the devices to avoid allergies. Irradiation is employed for radiation sterilization, which

DOI: 10.1201/9781003285656-10

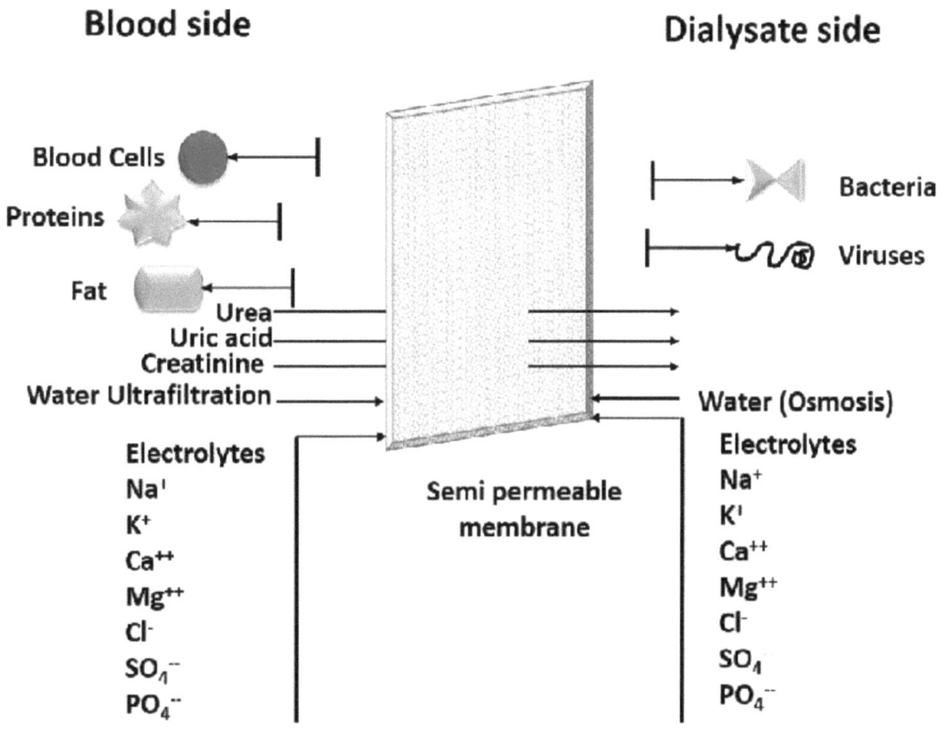

FIGURE 10.1 Schematic diagram of a typical haemodialysis treatment principle. Adapted from Ref. [5].

involves exposure to gamma/beta rays. It is also a safe and easy handled method. It proceeds through ionization and thus leads to free radical formation and DNA dimerization, with the arrest of bacterial growth. Usually, a beta source is used because it can be perfectly dosed and sterilized. Gamma irradiation is employed in the case of high-density membrane materials and delicate materials that deteriorate under heat. The extent of sterilization is a function of the amount of adsorbed radiation. The dialyzers to be sterilized are irradiated using a radiation source for the recommended duration, for example a few seconds for beta and minutes or hours for gamma sources. Deterioration of the membrane occurs as the irradiation time increases. Obstructions can form due to impairments caused by ionizing rays, which relies on cross-linking and chain breakages. Stabilizers or chain terminators are added to minimize the damaging effects of irradiation, but they can also be harmful. Therefore, the proper sterilization method is essential to enhance dialysis treatment [7].

Haemodialysis deals with the bidirectional phenomena of solute transport from the blood to the dialysate solution and vice versa. Thus, it proceeds via the interactions between the membrane surface and fluid phases. Factors such as the characteristics of the membrane, blood components, and waste content all affect dialysis. The design of the haemodialysis membrane requires the determination of various factors in terms of biocompatibility, allergic reactions associated with the materials used, efficacy in retaining the protein content and essential vitamins present in the blood, and many more [8].

This chapter enlightens the information about haemodialysis membranes, the materials employed so far, and how effective it has been.

10.2 INTRODUCTION TO MEMBRANES: DEFINITION, TYPES, AND FABRICATION

The membrane is the soul of the membrane processes, which act as a selective channel to block the passage of certain molecules or to allow the passage of certain molecules depending on the surrounding conditions. It is defined as "an interface between two adjacent phases acting as a selective barrier, regulating the transport of substances between the two compartments" [9].

Relying on the pore size and morphological characteristics, the membrane categorization is provided in Figure 10.2.

Microfiltration (MF) membrane is characterized by pore sizes in the range of 0.05–10 μm and thickness of 10–150 μm. It is symmetric with a porous nature. Phase-inversion, sintering, track-itching, anodic oxidation, sol/gel process, and stretching are the fabrication techniques. The flux out of these MF membranes is a function of the pressure employed. This is a pressure-driven process, with the highest pressure applied in the process being less than 2 bar. It retains particles according to a sieving technique [10].

Ultrafiltration (UF) membranes are asymmetric and porous, which allows characteristics in-between those of NF and MF membranes. They have a 1–100 nm pore size and 150 μm thickness. The phase inversion method is employed to prepare UF membranes. It is a pressure-driven process, with 1–10 bar being the optimum range of applied pressure. It retains particles according to a sieving mechanism [11].

The NF membranes are a comparatively newer variety of the pressure-driven membranes that have actions in-between those of RO and UF membranes. Their pore size is less than

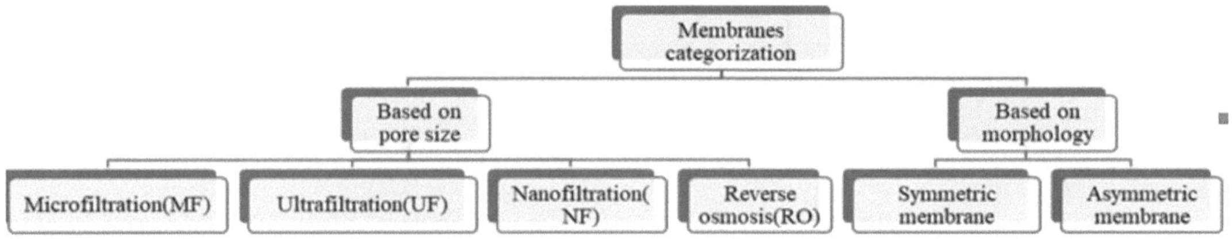

FIGURE 10.2 Categorization of membranes.

2 nm. The appearance of NF membranes is between porous and dense nonporous membranes. Accordingly, a maximum pressure in the range of 10–20 bar is needed for NF membranes, involving the solution diffusion principle [12].

RO membranes are a distinct type of membranes, which are almost non-porous. The rejection of inorganic salts is achieved by employing the RO membranes, which work based on the solution diffusion mechanism, as in NF membranes. RO membranes mainly remove both monovalent and divalent inorganic salts. Therefore, in order to overcome hydrodynamic repulsion, maximum pressure is needed. The applied pressure must be greater than that of osmotic pressure, in order to allow the water to pass through the membrane; that is, from a high solute concentration to a low solute concentration. The reverse happens when the pressure is less than the osmotic pressure. Therefore, the working pressure range for RO is 15–100 bar. It is employed for the desalination of brackish water and sea water. Likewise, forward osmosis is an interesting field of membrane technology, which is enhanced by low energy consumption, antifouling nature, and simple morphology. It employs an osmotic pressure gradient to drive the water passage through a semipermeable membrane from the side with low osmotic pressure (feed) to the side with higher osmotic pressure (draw). External energy need not be applied as it is an osmotic pressure-driven process and thus it is an excellent tool for obtaining pure water from waste water. FO exhibits a very high rejection rate for a wide range of solutes and it requires much less pressure in comparison with RO and NF [13].

Symmetric membranes have a uniform pore size distribution throughout the membrane, consequently they are called isotropic membranes [14]. Their main areas of application are microfiltration, dialysis, and electrodialysis. Their membrane thickness is 30–500 μm, and they have a porous or dense structure which can be prepared by sintering or track-itching technique. The recorded permeation is high as the mass transfer depends on membrane thickness [15].

The morphology of asymmetric membranes consists of a porous sublayer (100–300 μm) beneath the thin dense top layer (0.1–5 μm). The top skin layer acts as a selective barrier to the sublayer and enhances the membrane selectivity [16]. Consequently, the performance of asymmetric membranes depends on the pore size of the skin layer and the material used. Applications of these membranes include gas separation, reverse osmosis, and ultrafiltration, and they are also sometimes employed for microfiltration [14].

Hollow-fibre membranes with ultrafiltration capacity are preferably employed in haemodialysis.

10.3 HOLLOW-FIBRE MEMBRANE FORMATION

Hollow fibres are formed when the dope solution (a polymer solution) gets pumped from a dope reservoir through a spinneret using different methods, including wet spinning, dry spinning, dry/wet spinning, and melt spinning. A typical dry/wet spinning hollow-fibre system is depicted in Figure 10.3. The dope ejection rate and bore flow rate in terms of mL per min, air gap in cm, spinneret size in mm, external and internal coagulant temperature, and winder tank speed are the main parameters that need to be optimized in the set up [17–19].

An internal coagulant penetrates between the running dope solution in a spinneret which then reaches the external coagulant bath. A good-quality hollow-fibre membrane is collected in a winder tank. The usually employed coagulant baths contain water, ethanol, or ethanol/water mixture.

10.4 HAEMODIALYSIS MEMBRANES: THE CHALLENGES

Haemodialysis has become a relatively common procedure to solve kidney malfunctioning problems. Depending on the medical conditions, an individual with kidney malfunction requires dialysis twice or thrice weekly. A single visit can be very expensive, depending upon the medical centre. The high cost makes this treatment unaffordable for many. The cartridges available in the market are high-priced making haemodialysis treatment expensive. Therefore, there is a need to develop a cost-effective, blood-compatible, and effective membrane system for blood purification.

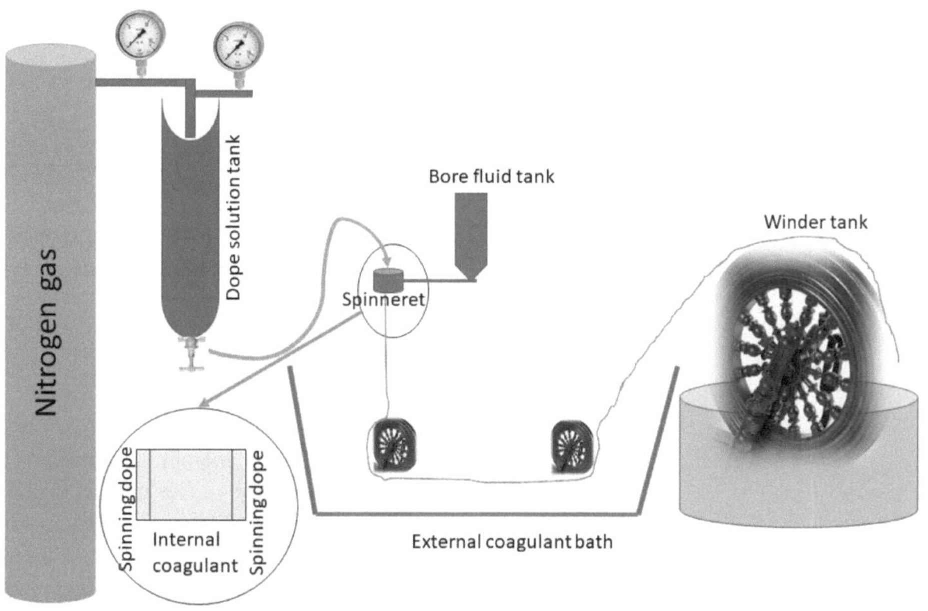

FIGURE 10.3 Schematic diagram of typical hollow-fibre membrane spinning set-up.

10.5 HAEMODIALYSIS MEMBRANES: PRINCIPLES, CRITERIA FOR ACCEPTANCE, AND CHARACTERIZATION

As already mentioned, haemodialysis proceeds via the chemical potential difference existing between feed and dialysate components. Ultrafiltration (a pressure-driven process) membranes are employed for blood purification. When a foreign particle enters the blood, it should not adsorb the plasma proteins. The activation of platelets and coagulants must not happen to avoid making the process more difficult. The most popular method for improving hemocompatibility is modification of the materials themselves into antithrombogenic materials.

10.5.1 Morphology of Modifiers and Membrane

The morphological characteristics of the modifiers or additives and also the membranes are provided by FESEM images. To perform this, the membrane substrates are dried and then immersed in liquid nitrogen to create brittleness for ease of sampling. Fracturing of the membranes is carried out, and once a gold nanoparticle coating is completed by employing sputtering, FESEM scanning is done [20].

10.5.2 Spectral and Thermal Study

To identify the functional groups and blending or immobilization, FT-IR is carried out. In order to discover the thermal stability of the modifiers and membranes, thermogravimetric analysis (TGA) and differential scanning calorimetry (DSC) are performed. The TGA result gives information on the thermal stability of the substrates and also the incorporated material through new degradation peaks. The DSC result gives the glass transition temperature value of the prepared polymer and additive.

10.5.3 Flux Study

Recording the pure water flux gives information on the water permeation characteristics of the membrane. Initially, the membranes are immersed in distilled water for a duration of about one day prior to the analysis. Then the effective membrane surface is chosen. Compaction is carried out for half an hour. The working pressure is set and recording of flux with respect to time is measured. The permeance of distinct membranes is then assessed.
The equation used is:

$$J = \frac{Q}{\Delta t * A}$$

where "J" is water flux in L/m^2 h, "Q" is the quantity of water passing through the membrane in L, "Δt" is the time recorded in hours, and "A" is the effective membrane area which is measured in m^2 [19,21–24].

10.5.4 Antifouling Nature

Initially, the pure water flux value of the membrane is determined as "Jw1" (L/m^2 h). The standard protein selected to investigate the antifouling nature is bovine serum albumin (BSA). After filtration of water for about 1 hour, the BSA solution is taken as feed and the permeance recorded. BSA filtration gives the value "J_p". The membranes are cleaned by flushing the water through instrument lines followed by recording the water flux once again, which gives the value "Jw2" (L/m^2 h) [19,21–23,25]. Finally, the membrane antifouling characteristics are expressed in terms of flux recovery ratio (FRR), total fouling ratio (R_t), reversible fouling ratio (R_r), and irreversible fouling ratio (R_{ir}) according to the following formulae:

$$FRR(\%) = \frac{J_{w2}}{J_{w1}} * 100$$

$$\mathrm{Rt}(\%) = \frac{J_{w1} - Jp}{J_{w1}} * 100$$

$$\mathrm{Rr}(\%) = \frac{J_{w2} - Jp}{J_{w1}} * 100$$

$$\mathrm{Rir}(\%) = \frac{J_{w1} - J_{w2}}{J_{w1}} * 100$$

10.5.5 Water Uptake and Porosity Studies

Small identically cut samples of a membrane were kept in demineralized water for around one day at ambient temperature. After the samples were removed from the water, extra water was removed by blotting the air. Weighting of the membrane samples provided the wet weight. The corresponding cut membranes were dried in an oven maintaining a temperature of 333K for about one day. The weights were taken again to provide the dry weights.

The water content present inside the membranes was estimated by using the equation:

$$\%\ Water\ uptake = \left(\frac{Ww - Wd}{Ww}\right) * 100$$

where "*Ww*" is the weight of the membrane in wet condition and "*Wd*" is the weight of the membrane after drying.

The percentage porosity (ϵ) is calculated by employing the equation

$$\epsilon\ (\%) = \left(\frac{Ww - Wd}{Al\rho}\right) * 100$$

where "*l*" is the thickness of the membrane in cm, "A" is the area of the membrane in cm^2, and "$\boldsymbol{\rho}$" is the density of pure water [26–29].

10.5.6 Contact Angle Values

Membrane hydrophilicity is vital in deciding its properties. It has an effect on membrane antifouling nature as well as the rejection profiles. The contact angle value decides the membrane hydrophilicity as shown in Figure 10.4. Initially, a water droplet is made to fall on a membrane surface with the help of a microsyringe. Then the contact angle is recorded by employing a sessile drop method [17,21,22,25–27,29–43].

10.5.7 Mechanical Strength Testing

The tensile strength gives an idea about the mechanical robustness of the membrane. The sample is kept in a universal testing machine (UTM) in the middle of the grips. The system traces the change in length of the sample placed and its stress–strain behaviour accordingly. The machine keeps inserting weights over the specimen. For every change in load, the machine records the compressibility of the sample and related characteristics.

$$Burst\ pressure,\ P = \left(\frac{T\left(a^2 - b^2\right)}{b^{2\left(1+\left(\frac{a^2}{b^2}\right)\right)}}\right)$$

where "*T*" is the tensile strength of the membrane, and "a" and "b" are half of the outer diameter and inner diameter of the hollow fibres, respectively.

10.5.8 Dialysis Tests

The dialysis potential is depicted in terms of the BSA rejection, and urea and creatinine clearance. Urea, creatinine, and

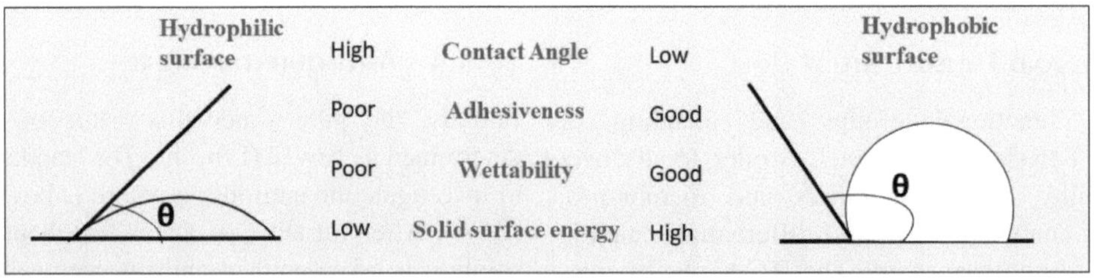

FIGURE 10.4 Diagrammatic representation of contact angle measurement and interactions during measurement.

BSA solutions are prepared using distilled water at ambient temperature. These are used as a feed stream solution, and permeances for them are recorded. To prevent concentration polarization, feed stream solutions are mixed via bead rotation at around 600 rpm. Working pressure is applied and readings are recorded. The studies are done for a set duration and the concentrations of solutes in the permeances detected using a spectrophotometer.

The rejection extent as a percent is calculated by making use of the following formula:

$$Percentage\ rejection = 1 - \frac{A_p}{A_f} * 100$$

where "A_p" and "Af" are the concentration of permeated solution and feed solution, respectively.

10.5.9 Hemocompatibility Analysis

Hemocompatibility of the developed membranes is depicted in terms of resistance to platelet and protein adherence, haemolysis ratio, extent of thrombus formation, and plasma recalcification time [3,6–8,20,44–49].

Platelet adhesion: A blood sample (around 10 mL) was taken in the centrifuge tube followed by centrifugation at high speed (around 1000 rpm) for 10 min. A platelet-rich plasma (PRP) layer formed in the supernatant. The selectively cut membranes were washed with phosphate buffer solution (pH of 7.4) followed by covering up with around 0.5 mL PRP over the surface. The analysis was conducted by maintaining the temperature of 310 K for one hour for each sample. Extra PRP which was non-settled on the membrane surface was removed by washing the membranes with phosphate buffer solution. Then the platelet uptake by the membrane was protected by adding glutaraldehyde solution (2.5 g glutaraldehyde per 100g) to the analysis media, which was kept overnight (around 18 hours). Then, dehydration of the samples was performed by employing 50%, 75%, 85%, 95%, and 100% (v/v) ethanol/water systems for about 10 minutes. The extent of platelet adhesion was examined by scanning electron microscopy (SEM).

Thrombus formation test: A small cut piece of membrane sample was immersed in blood (1.5 mL) and stabilized in carbon dioxide (5%) for about 2 hours at 310 K. Phosphate buffer solution was employed to wash out the sample and ethanol/water was utilized to dehydrate the sample, weights were recorded before and after dehydration.

$$\%\ \text{Thrombus Formation} = \left(\frac{W_b - W_d}{W_d} \right) * 100$$

where W_b and W_d denote the weight of the blood-encapsulated membrane and the weight of the dry membrane in grams, respectively.

Haemolysis ratio: The cut membrane samples were washed thrice with deionized water followed by NaCl solution (concentration of about 0.9 g in 100 mL) for about 10 min. Then the samples were submerged in an NaCl solution (concentration of about 0.9 g in 100 mL) at 310 K for half an hour by maintaining in a water bath. Around 200 μL of blood was injected into the NaCl solution and stabilized for 1 h at 310 K. After that, the same blood was centrifugated for about 10 min at a speed of 1500 rpm. The extent of absorption of the top layer was recorded using a UV spectrophotometer.

The haemolysis ratio was obtained using the formula:

$$\%\ \text{haemolysis ratio} = \left(\frac{HT - HA}{HC - HA} \right) * 100$$

where *HT* denotes the absorption value of supernatant after membrane incubation, and *HC* and *HA* designate the absorption value of positive reference and negative reference using NaCl solution (concentration of about 0.9 g in 100 mL), respectively.

Plasma recalcification time: The selected membrane cut sample was placed in a 48-well plate. A plasma poor plasma (PPP) obtained after centrifugation of anticoagulated blood for about 15 minutes at a speed of 3000 rpm was poured into it. Incubation was done for 10 min at a temperature 310 K, and then it was poured into a 100 μL aqueous calcium chloride solution (concentration of 0.025 mol/L). After stirring, fibrinogen thread generation was measured and the time taken was recorded. The average measurement was used after the procedure was repeated.

A schematic representation of the dialysis set-up is shown in Figure 10.5.

10.6 CELLULOSE ACETATE HAEMODIALYSIS MEMBRANES

Cellulose after treatment with acetic anhydride yields cellulose acetate (CA). The structure of cellulose acetate contains free acetate groups, as shown in Figure 10.6 [51].

Sufficiently moderate flux rate, very good salt removal properties, cost-friendliness, biocompatibility, ease of production, abundance in availability, and safety are the attractive benefits of using cellulose acetate as a haemodialysis membrane material. Countless research has been carried out into the modification of CA using various methods such as blending, surface functionalization, coagulation bath variation, and many more, so that the permeance and rejection characteristics are improved. The design and fabrication of multifunctional CA membranes are essential in order to find applicability in the biomedical domain. Such membranes should possess sufficiently high hemocompatibility, antioxidant, and antimicrobial characteristics. [52–54].

Peng et al. fabricated a polysaccharide-derived CA ultrafiltration membrane consisting of multiple layers. A layer-by-layer (LBL) formation of chitosan (CS) and polysaccharide

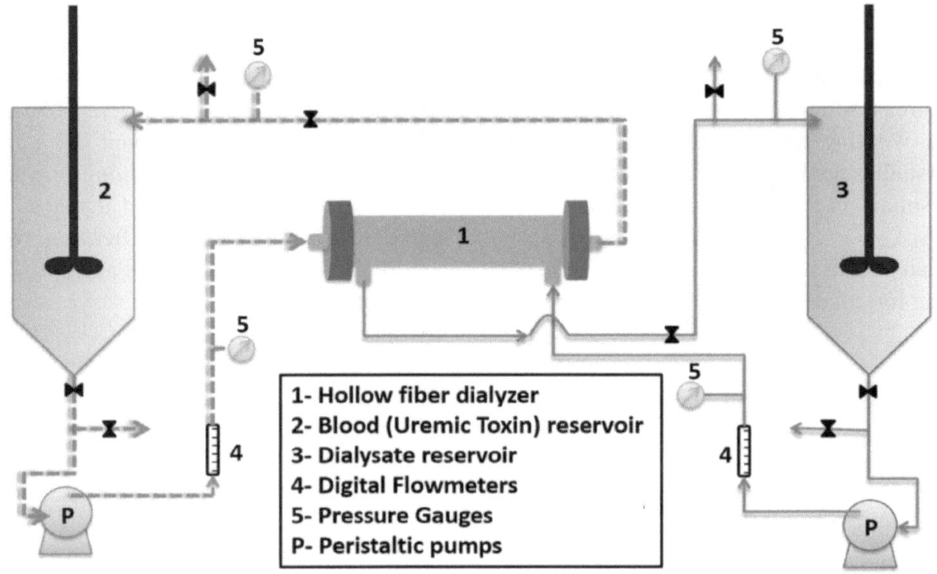

FIGURE 10.5 Depiction of dialysis experiments. Reproduced from Ref. [50].

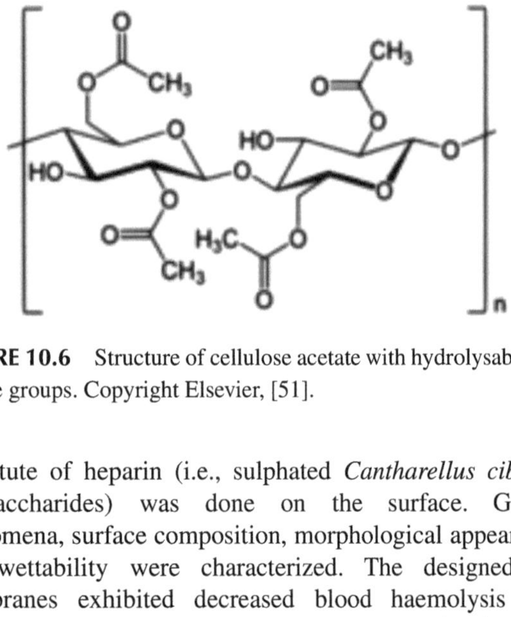

FIGURE 10.6 Structure of cellulose acetate with hydrolysable free acetate groups. Copyright Elsevier, [51].

substitute of heparin (i.e., sulphated *Cantharellus cibarius* polysaccharides) was done on the surface. Growth phenomena, surface composition, morphological appearance, and wettability were characterized. The designed CA membranes exhibited decreased blood haemolysis with enhanced biocompatibility, resistance to platelet loss and activation, protection against non-specific protein sticking, extended coagulation duration, and arrest of complement activation when the step-wise analysis of hemocompatibility was completed. Resistance to bacteria like *Escherichia coli* and *Staphylococcus aureus* proved its antibacterial behaviour. In addition to biocompatibility, antibacterial and antioxidant properties are very attractive, and could be applicable in haemodialysis [55].

Shen et al. developed CA/CS hollow-fibre membranes with the immobilization of copper ions because protein adsorption becomes high, which is non-compatible with haemodialysis [53].

Faria et al. fabricated monophasic ultrafiltration CA and silica membranes with hemocompatibility through phase inversion and the sol–gel method. The so-developed membranes exhibited an asymmetric structure in SEM images. Incorporation of silica into the CA resulted in enhancement of hydraulic permeability with excellent urea rejection. The membranes fabricated were protective against protein leaching. They also showed resistance to haemolysis and had a low thrombogenic value [56]

One of the studies focussed on both permeance and hemocompatibility. The main recipe for ultrafiltration membrane consists of cellulose acetate (CA), acetic acid, and polyvinyl alcohol (PVA). The pore size declines with the utilization of PVA, which indirectly affects the membrane potency. Due to the excellent hydrophilicity of PVA biomaterial, membrane hemocompatibility is enhanced. As a step towards casting solution preparation, 25 wt.% PVA solution was made using distilled water. A temperature of 90°C and stirring duration of 12 hours were maintained which resulted in a homogeneous polymer solution. The compositions of the materials in additive-varied casting solutions are provided in Table 10.1.

The temperature of about 343 K with a stirring duration of 24 h were maintained in order to get homogeneity. The phase inversion method was followed to form a membrane. The fabricated membranes, upon characterization, showed pure water flux, BSA removal, urea, and creatine rejection of 42 L/m^2 per hour, 95%, 93%, and 89%, respectively. The perfect blend containing homogeneously distributed additives imparted lower haemolysis and enhanced plasma recalcification duration. This increases the design and development of haemodialysis membranes with higher permeances along with blood compatibility [57].

Alterations to the characteristics of a flat sheet haemodialysis membrane by the incorporation of D-glucose monohydrate as an additive have been investigated by one research work. The

TABLE 10.1
Composition of casting solutions with varying additive concentration. Adapted from Materials Science and Engineering, 2021 [57]

Membrane	CA (wt.%)	Acetic acid (wt.%)	PVA (wt.%)	PEG (wt.%)
M-0	11	87.00	–	2
M-1	11	86.00	1.00	2
M-2	11	85.50	1.50	2
M-3	11	85.00	2.00	2
M-4	11	84.50	2.50	2

TABLE 10.2
Dope solution formulation. Data taken from Jurnal Teknologi, 2012[58]

Material	Cellulose acetate (wt.%)	Formic acid (wt.%)	D-glucose monohydrate (wt.%)
1	20	80	0
2	20	78	2
3	20	76	4
4	20	74	6
5	20	72	8
6	20	70	10

various additive-added dope solution compositions are given in Table 10.2.

Dope recipes were prepared in a polymer reaction container as per planned compositions using the microwave method. Formic acid, employed as a solvent, was placed in the vessel, followed by CA and D-glucose monohydrate addition. As an additive dosage increases, membrane efficiency for urea and creatinine rejection is enhanced noticeably. A selected membrane consisting of a D-glucose monohydrate of 10 wt.% showed urea clearance, creatinine removal, and BSA rejection rate of about 50%, 20%, and 97%, respectively. The performance was doubled compared to the membranes without additives. The whole investigation revealed that D-glucose monohydrate is applicable as an additive in haemodialysis membranes [58].

10.7 CHITOSAN HAEMODIALYSIS MEMBRANES

The biopolymer chitosan is natural and abundant in the environment. It exists as a linked N-acetyl glucosamine and glucosamine units, as in Figure 10.7 [59]. Chitin is a source of chitosan which can be isolated from prawn/crab shells (Figure 10.8) [60]. Chitosan possesses extremely beneficial physicochemical characteristics and biologically important properties [61]. In aqueous media, it is not soluble, which slightly limits its utilization in biomedical applications. Because of the hydroxyl and amine functionalization, it is susceptible to some graft modifications. It can undergo carboxylation, hydroxylation, acylation, alkylation, and esterification through which the pendant groups are formed. Such surface or graft modifications allow it to be more soluble, thereby expanding its utilization in various domains [18,22,26,27,29,32,52,53,62].

The intensive biodegradability of chitosan limits its utilization as a membrane. An investigation consisted of chitosan grafted with 2-hydroxyethylmethacrylate for blood purification. In vitro observations at 310 K for various solutes, urea, creatinine, glucose, and albumin gave very good results. A membrane with 425% grafting had high permeation to creatinine after reaching equilibrium within an hour. Compared to pristine chitosan membrane, the membranes with grafting presented enhanced permeation to glucose. When all the samples were analysed in rat fibroblast cells, they were found to be highly blood-compatible and noncytotoxic. Thus, the biodegradability of pristine chitosan could be managed with the help of HEMA groups [63].

Fe_3O_4 nanoparticles were made that are magnetic, and then membranes were fabricated by a phase conversion method through the homogeneous dispersal of nanoparticles in a solution of chitosan dissolved in ethylene glycol. The transparent homogeneous solution was obtained with a reaction. Then a magnetic natured chitosan membrane was fabricated by casting technique after proper degassing. By continuous washing with deionized water, more residual particles were cleaned. After characterization, the developed membranes exhibited excellent creatinine clearance, which is a major metabolic waste removed by the kidneys. Thus, there is an opportunity to utilize the developed membrane as an environmentally friendly material for removal of creatinine from the body [64].

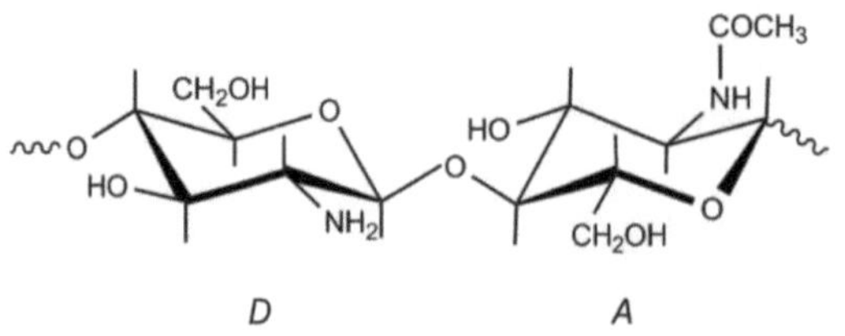

FIGURE 10.7 Structure of chitosan with N-acetyl glucosamine unit (A) and glucosamine unit (D). Copyright *Marine Drugs*, 2017 [59].

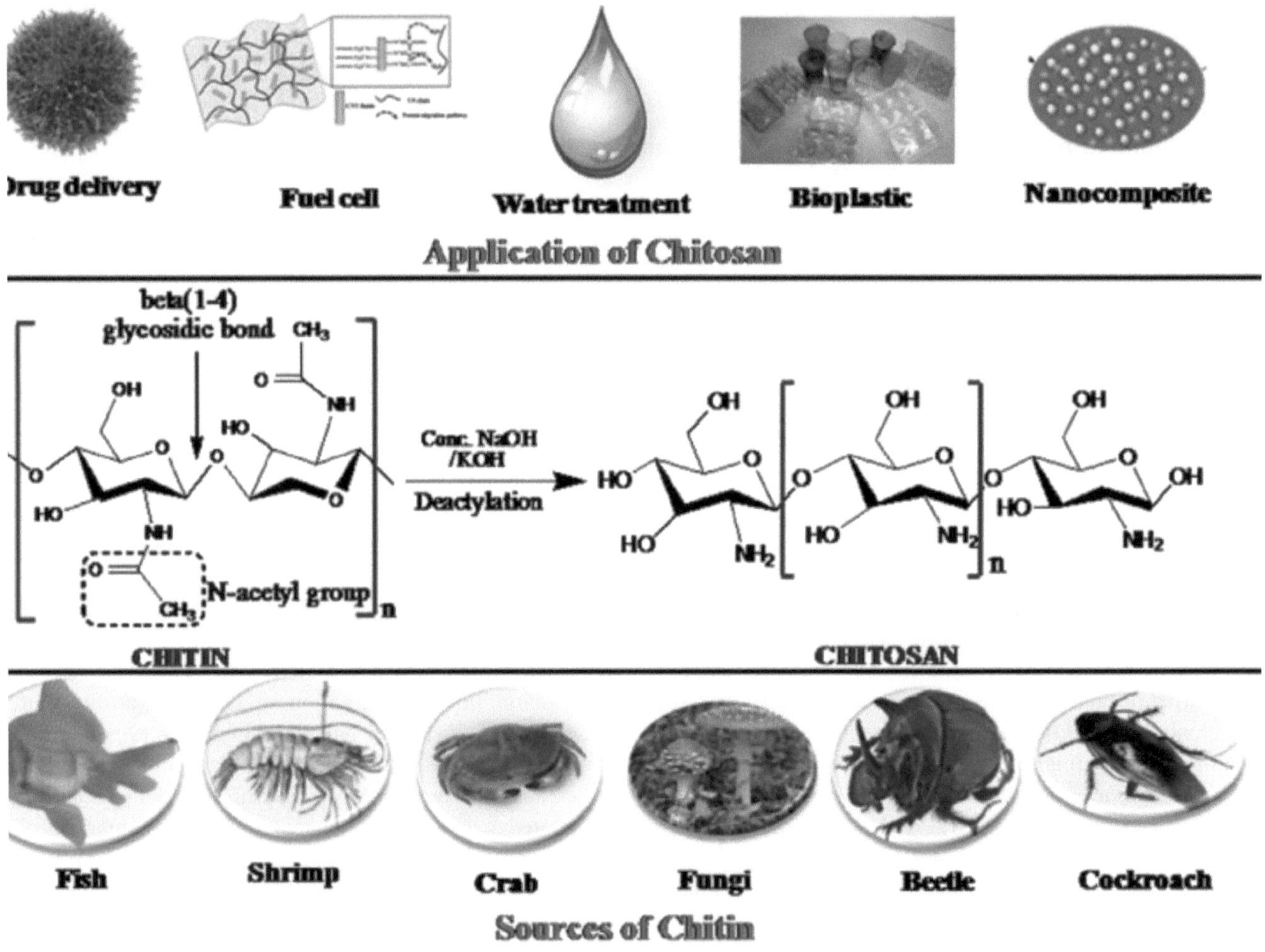

FIGURE 10.8 Sources and applications of chitosan. Adapted from Ref. [60].

10.8 POLY(LACTIC ACID) HAEMODIALYSIS MEMBRANES

Poly(lactic acid) (PLA) polymer is biodegradable and biocompatible and therefore has great utilization in many biomedical applications. Renewability, thermoplastic nature, and potential to mimic commercial petrochemical-derived polymers are among the major benefits of PLA. It has a simple structure, as shown in Figure 10.9 [65].

It has a notable thermal resistance. By employing the phase inversion technique and simple electrospinning set-ups, PLA membranes are fabricated [43].

Gao et al. improved the performance of PLA porous membranes by heparin immobilization. A conventional PLA membrane was fabricated first using the phase inversion method. In this study, a homogeneous polymer solution of 18 wt.% was prepared in N-methyl pyrrolidone under mechanical agitation, maintaining a reaction temperature of about 80°C. Polyethylene glycol is employed as a pore-former, and a water-based coagulation bath is maintained. After some hours of reaction, the formed bubbles were destroyed using ultrasonication followed by casting over a glass plate using a 150 μm thickness casting rod. In order to further proceed with the heparinization, polydopamine was used as a heparin-binder. An aqueous-mediated dopamine solution (dopamine concentration of 2 mg/mL) was taken after adjusting the pH to 8.5 using tris-buffer. An alkaline condition maintained in aqueous media causes oxidation of the pyrocatechol part of dopamine into benzoquinone through the utilization of dissolved oxygen. Pyrocatechol and benzoquinone undergo a disproportionation reaction, coupling follows as there is formation of a semiquinone radical, and finally polydopamine forms through the resultant polymerization. Then the fabricated PLA membranes were surface-coated with polydopamine by dipping in the solution for a set time at ambient temperature. Excess polydopamine was leach out by washing with deionized water. The polydopamine-coated membranes were then further surface-modified by heparinization, through immersion in a heparin solution (2.0 mg/mL heparin concentration, phosphate buffer saline as media, pH 7.4). Excess heparin was removed by washing with running deionized water. Polydopamine present in the membrane surface binds the heparin, which results in the formation of heparin-immobilized PLA membranes. After a series of characterizations, the developed membranes were improved in terms of platelet adhesion, plasma-recalcification duration, and a significant decrease in haemolysis. The dialysis performance revealed that the heparin surface-modified PLA membrane emerged with 80% urea clearance, 18% lysozyme rejection, and about 90% protein retention. The designed membranes meet the criteria for potent dialysis membranes [66].

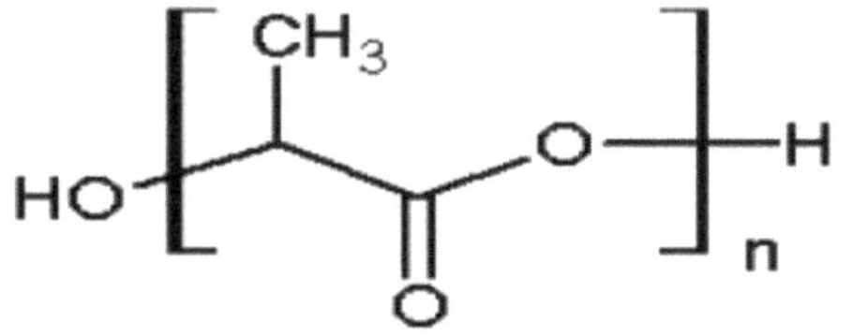

FIGURE 10.9 Structure of PLA. Reproduced from Ref. [65].

In order to impart hydrophilicity to PLA substrate, various methods have been utilized, including blending, surface immobilization, and self-assembly. Blending is the easiest method of membrane modification with efficacy through the use of polyethylene glycol (PEG), poly(vinylpyrrolidone) (PVP), and poly(ethylene oxide) (PEO) as hydrophilic additives. Along with the ease of incorporation, there is a problem of leaching out because of the water-solubility. PLA amphiphilic block copolymers, such as PLA-block-poly (N, N-dimethyl aminoethyl methacrylate) (PLA−PDMAEMA) and PLA block-poly (2-hydroxyethyl methacrylate) (PLA−PHEMA), have been found to be effective in retaining the hydrophilicity for a longer duration. Excellent cellular compatibility is the main advantage of PHEMA utilization. Aminolysed PLA bunches produce PLA−PHEMA through reversible addition−fragmentation (RAFT) polymerization. This was then used as an additive while preparing a PLA polymer casting solution in one of the investigations. Haemodialysis membranes were fabricated by a phase inversion technique and were finally analysed for their dialysis performance. The overall membrane development scheme is provided in Figure 10.10 [67].

After characterization, the membranes were found to be potent with enhanced hydrophilicity, better blood-compatibility, and increased antifouling character (reduced total fouling and increased water flux recovery ratio). They also had much less protein adsorption, extended plasma recalcification duration, and arrested platelet adhesion, which are favourable attributes. The modified membranes showed excellent urea and creatinine rejection as the PLA-PHEMA loading was increased. By retaining larger proteins, which are essential, other smaller solutes and wastes were successfully cleared from the blood sample [67].

The intrinsically brittle nature of PLA limits its application in various domains. A membrane with porous nature, less thickness, and enhanced hydrophilicity is very rare. A membrane was developed by utilizing polysulfone-graft poly(lactic acid) (PSf-g-PLA) copolymer as an additive. The specialty of brush-type PSf-g-PLA copolymer is its strong backbone and soft flexible side groups, which increase the

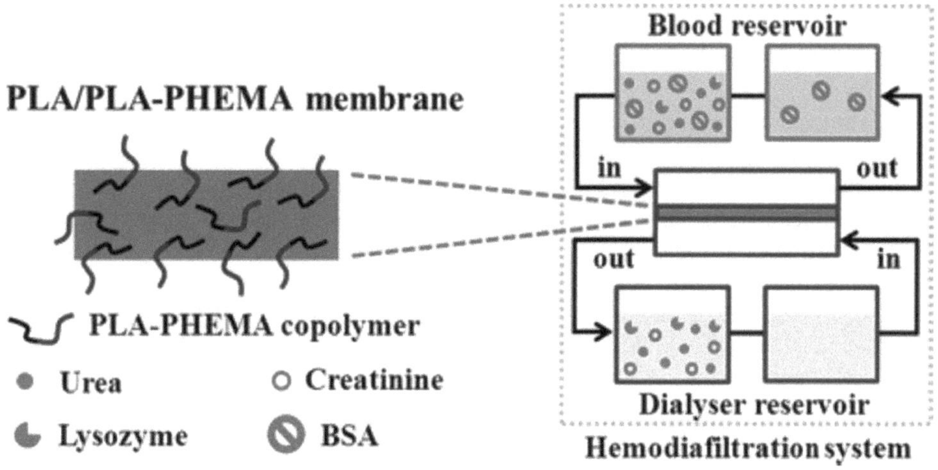

FIGURE 10.10 The overall membrane fabrication and characterization scheme. Adapted from Ref. [67].

mechanical strength of the membranes. In order to obtain PSf-g-PLA, PSf was made to undergo chloromethylation to maintain multiple reaction exposure timings to produce CMPSf with different chloromethylation extents per repeat chain of polysulfone. This was followed by amino functionalization. For that, 10 wt.% CMPSf solution was made in dimethylacetamide (DMAc) solvent with 10% ethylene diamine (EDA) dissolved in it. By maintaining room temperature, the reaction mixture drives the $-CH_2Cl$ groups present in CMPSf to interact with EDA. The resultant mixture was precipitated in water after around 4 hours in reaction conditions. Once the precipitated mixture was filtrated, the polymer solution was remade by re-dissolving it in DMAc and precipitated in water. The residue was obtained after filtration and drying in a vacuum for about one day. The formation of PSf-g-PLA occurs by the aminolysis of PLA with the PSf which is amino-functionalized (PSf-EDA). To the taken solution of PLA in DMSO, the respective PSf-EDA was incorporated. A homogeneous polymer solution resulted after the reaction was continued by maintaining around 353 K temperature and mechanical stirring for about 5 h. Finally, membranes were fabricated by casting the solution over a glass plate using a rod following the immersion of the plate in a coagulation bath (water). As an observation, it was found that an increase in the extent of chloromethylation in the first step aided in the proper binding of PLA, which improved the mechanical strength. The fabricated membranes were good not only in terms of resistance to mechanical and thermal breakage, but also in terms of dialysis performance. A denser and homogeneous soft surface natured membrane was formed due to the special attachment of brush type natured PSf-g-PLA. This characterization revealed that the modified PLA membranes emerged with the urea and creatinine clearance along with protein retention, which is a major requirement for haemodialysis membranes [43].

10.9 POLYETHER SULFONE HAEMODIALYSIS MEMBRANES

Polyether sulfone (PES) belongs to the polysulfone class, and is an example of the high-temperature engineering thermoplastics possessing a glass transition temperature of around 503 K. It has the polymeric structure shown in Figure 10.11 [68].

It is a polymer with an amorphous nature and is transparent. Under a wide temperature range, it is stable, robust, and dimensionally fit. Its resistance to flammability and chemicals, and low smoke generation are among the advantages of PES

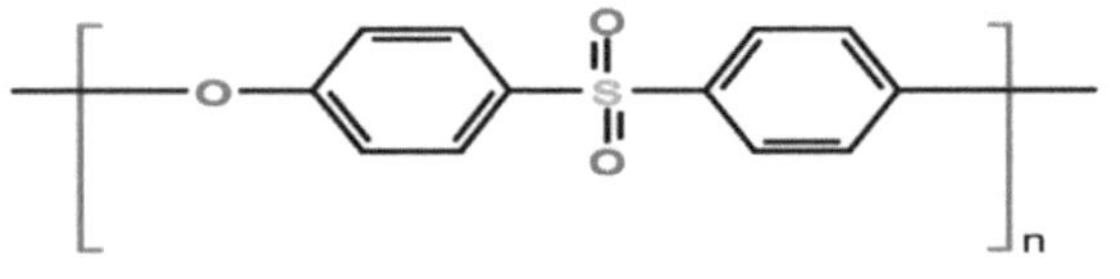

FIGURE 10.11 Polymeric structure of polyether sulfone. Copyright *Environmental Research*, 2022 [68].

utilization. PES membrane has been conventionally employed in the medical field, especially as a haemodialysis membrane. The disadvantage of applying a conventional membrane which works on diffusion is their capability to adhere uraemic toxins is much reduced [69–71].

One research work included the development of PES mixed-matrix membrane (PES/CA/IZC) blended with cellulose acetate using imprinted zeolite as an additive to aid in targeting the analyte for rejection. Zeolites are aluminosilicates with a microporous nature, uniformly distributed pores, and crystalline framework structure, which are built by GO_4 tetrahedral (G=Al, Si) with atoms holding nearer the tetrahedral position. The uniform distribution of pores allow it to absorb uremic toxins present in blood. Because of their non-toxic nature and resistance to quick degradation, zeolites enter the membranes as an additive. The micropores present in zeolite infrastructure can be modified further to transform them into more effective versions. Methods used include dealumination and desilication (demetalization type); hard and soft templating and indirect templating (assembly type); and a combined procedure to upgrade the zeolite mesopores. The molecularly imprinting method based on soft templating was used to improve the pore characteristics. In the initial step, zeolite A was produced by following ageing and hydrothermal steps. Sodium aluminate (($NaAlO_2$), having an origin of sodium oxide and alumina, silicon dioxide (SiO_2) as a source of silica, and water as a solvent were the starting materials. The chemical content of raw material mixture was silica:alumina:sodium oxide:water at a 1.8:1:4:270 ratio until colloidal dispersion of white colour was formed after ageing. It was then subjected to heat in an oven at 373 K for around two days, which is a hydrothermal process. After that, the bottom side was filled with precipitate and a clear solution was obtained on the top. The precipitate was subjected to neutrality in order to make it free of alkalinity, otherwise it can disturb the characterization. Centrifugation is continued maintaining a speed of 3000 rpm with the addition of distilled water for neutralization. The resultant thick precipitate was dried at 353 K, which yields zeolite A. The imprinting is performed prior to crystallization after zeolite synthesis. The pores are printed using a creatinine template before making zeolite to attain neutrality. For that, zeolite A was blended with creatinine for around 1 hour for proper mixing under magnetic stirring. Then the solution was kept for around 3 hours for proper imprinting and then extra creatinine was washed off using water. After incorporating the prepared IZC into a PES/CA dope solution, the hollow-fibre membranes were fabricated. The IZC successfully increased the selectivity of the hollow-fibre membrane applied in the haemodialysis process, and significantly reduced the creatinine by up to 74.99%. The morphologies for this mixed-matrix membrane (PES/CA/IZC) include an asymmetrical structure with compact spongy sublayer, selective layer, and finger-like macro-pores. Future work should explore the improvement of the selectivity for the removal of protein-bound uremic toxins such as p-cresol and indoxyl sulphate using imprinted-zeolite. Based on the results obtained in this study, the suggestion for the future modification of membrane for haemodialysis applications is focused on enhancement of the selectivity membrane. The nanoparticles such as zeolites can be further studied as selective nanoparticles mixed with a polymer [72].

Graphene is a carbon allotrope having a two-dimensionally located single layer of atoms with overall depiction of a honeycomb-type structure. It has a huge number of double bonds and special characteristics. Polyether sulfone (PES) haemodialysis membrane has been incorporated with graphene oxide (GO) in one work. GO is produced from tartaric acid via the pyrolysis method. A glass reaction vessel aired with nitrogen circulation was loaded with about 5 grams of tartaric acid followed by thermal treatment for 3 hours at different temperatures (543K, 573K, 623K, and 673K) maintaining a heating proportion of 303K. The resulting GO powder was utilized in casting solutions, from which the membranes were fabricated. The tensile stress along with strain parameter of about 5.55 MPa and 0.039 m illustrated that the mechanical robustness was enhanced through GO usage. GO is characteristically hydrophilic due to its oxygen-possessing functionalities, once the functionalities are converted from free availability of oxygen, it becomes hydrophobic. However, base graphene has hydrophilicity properties. This hydrophilic nature is in association with contact angle measurements, showing that GO incorporation instils better hydrophilicity. The creatinine clearance was very much more noticeable with the developed membranes. Hence, the haemodialysis field has been enriched with another more effective membrane system [73].

10.10 POLYACRYLONITRILE HAEMODIALYSIS MEMBRANES

Polyacrylonitrile (PAN), or polyvinyl cyanide, is a man-made polymer with a semicrystalline and thermoplastic nature. It is a linear organic polymer substrate of composition $(C_3H_3N)_n$, as shown in Figure 10.12 [74].

It is thermally a very robust material under normal conditions, and it is stable and withstands destruction to its polymeric nature. If the heating extent is maintained for a minute at 323 K then it starts melting at 573K. However, prior to melting, it starts to degrade. Clusters of monomers having acrylonitrile as the principal monomer yield different PAN substrates. This is a distinct type of polymer employed

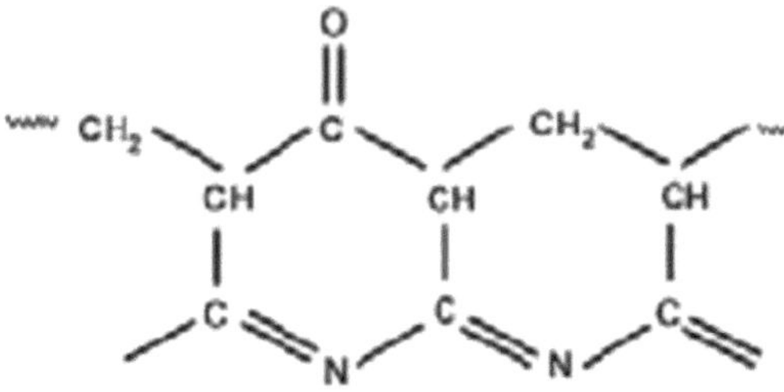

FIGURE 10.12 Structure of poly acrylonitrile. Adapted from Ref. [74].

to develop a wide variety of commercially valuable products such as hollow fibres for the reverse-osmosis process, ultra-filtration membranes, and textile fibres. PAN fibres also act as a precursor of carbon fibre. Oxidized PAN is formed once the PAN is thermally treated with air at 303K, which is then subjected to carbonization above 1273K in an inert atmosphere to make carbon fibres [75].

The blood-suitability of a PAN is not defined, which demands anticoagulant addition during haemodialysis. Antithrombogenicity-possessing materials have gained great interest in the design and establishment of artificial body organs. Heparin (HEP) is a well-known anticoagulating agent which leads to the formation of antithrombin and arrests thrombus formation along with other coagulation-causing proteases. Being a blend of distinctly sulphated polysaccharide chains with repeating strands of d-glucosamine and d-glucuronic acids, HEP has been found to be potent in anticoagulation during haemodialysis. Through immobilization, the membrane surface can easily discharge heparin, which causes continuous heparinization. However, at the same time, it may be disadvantageous also. HEP can induce spontaneous haemorrhage, which is dangerous. Chitosan has been selected to take part in the blending for various applications. It is already used as a blood oxygenator, dialysis material, skin-mimicking and wound-healing dress material, enzyme- or cell-bearable substrate to interact with fatty acids, and a drug-discharge system. One work included covalent bonding of polyelectrolyte complex (PEC) consisting of chitosan (CS)/heparin (HEP) onto the polyacrylonitrile membrane surface. Initially, the PAN membranes were fabricated by the usual phase inversion technique. The membrane formed was kept in deionized water for complete solvent–non-solvent de-mixing for around 12 hours followed by vacuum drying for around 6 hours. The PAN membranes were subjected to carboxylation by immersing the membranes in 1 molar sodium hydroxide at 313K, which converts CN groups to –COOH groups. Rinsing was performed using phosphate buffer solution in order to leach out any extra sodium hydroxide. The heparinization occurs via two different routes, as depicted in Figure 10.13. Direct heparination (route 1 in Figure 10.13) includes activation of carboxylated membranes through immersion in citric buffer with ethylene dichloride. Then the membranes are subjected to heparinization via immersion in HEP dissolved citric buffer solution maintaining a pH and temperature of 4.8 and 277K, respectively. After washing with the phosphate buffer and water, the substrates are lyophilized at 40°C employing a freeze-dryer for about one day. Two-step heparinization (route 2 in Figure 10.13) involves quite a different procedure. The EDC-activated membranes (formed as in route 1) are immersed into chitosan (CS) solution (0.25 mg per mL, 1% acetic acid as solvent) at 277K for about one day. Then the formed CS-decorated membranes are placed into glutaraldehyde (0.2 mm concentration) at 298K for about half an hour. Finally, glutaraldehyde media are replaced by a HEP bath in phosphate buffer solution at 277K for about half an hour. Extra adhered glutaraldehyde is removed with

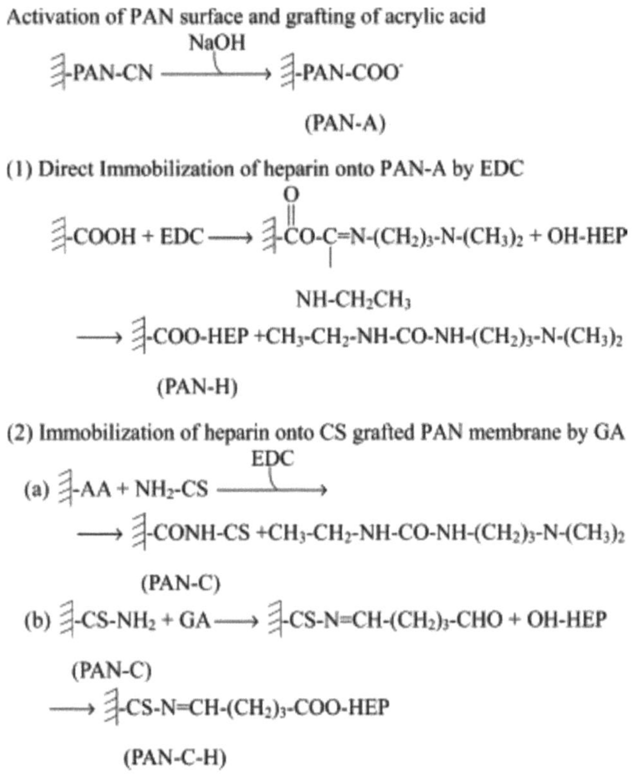

FIGURE 10.13 Heparinization of PAN membrane surface. Adapted from Ref. [76].

the help of Soxhlet extraction for one day. The substrates are freeze-dried at 313K for one day. The effect of surface immobilization was studied with respect to the metabolite's clearance, protein and platelet adherence, and anticoagulation nature. The water contact angle significantly decreased due to the PEC immobilization, which in turn indicated the enhanced hydrophilicity. In addition, the utilization of heparin instilled resistance to thrombus formation, and platelet and protein adhesion. The anti-blood-clotting nature was investigated with respect to prothrombin time (PT), fibrinogen time, activated partial thrombin time (APTT), and thrombin time (TT). The developed membrane system was found to be suitable for haemodialysis applications [76].

10.11 POLYETHERIMIDE HAEMODIALYSIS MEMBRANES

Polyetherimide (PEI), being a robust and mechanically stable material, has been selected as a membrane material many times. Figure 10.14 depicts the structure of PEI [77].

Being an amorphous polymer with amber-transparency, it has very good chemical resistance and film-forming nature [35]. The fabricated membranes are excellently hydrophobic in nature, which reduces the flux. Fouling is the main issue with hydrophobic membranes. The colloids, dispersions, microbes, and solutes deposit on the membrane surface and get stacked on the surface and pores, leading to membrane fouling. PEI is excellently flame-retardant, having stability at

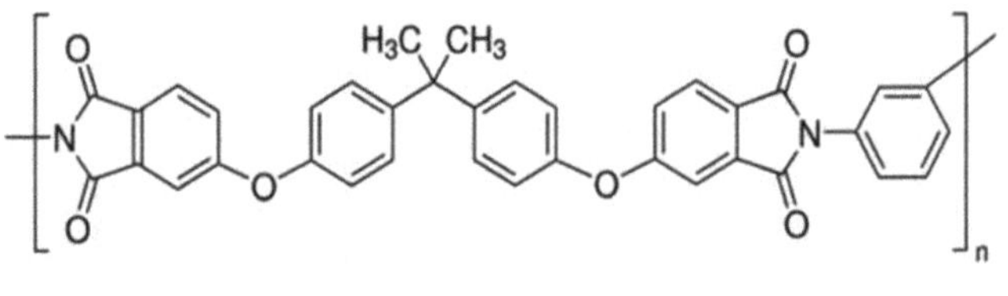

FIGURE 10.14 Structure of polyetherimide. Copyright William Andrew Publishing: Boston [77].

elevated temperatures. With all these advantages, it has been applicable in many biomedical areas [78,79].

Kaleekkal et al. fabricated PEI membranes for haemodialysis by incorporating GO-polyvinylpyrrolidone nanocomposite. Initially, graphene oxide was synthesized. For that, powdered graphite was taken in a round-bottomed flask containing 70 mL of concentrated sulphuric acid. The mixture was mixed thoroughly with a stirrer. In order to maintain a temperature of less than 288 K, 9 g of potassium permanganate was added as a stabilizer. After half an hour, the reaction flask was transferred into a 313 K bath and thoroughly mixed for 1 hour. Then 150 mL of water was added to the reaction mixture in a dropwise manner followed by temperature maintenance of 368 K for one hour. Excess water was added followed by 30% hydrogen peroxide which resulted in a colour change from dark brown to yellow. This mixture was rendered to get a solid residue in the bottom. Washing of the solid residue was performed using dilute hydrochloric acid to leach out the metal ions. Then the solution was centrifugated in hot conditions to get rid of the residue which is soluble in warm water. Ultrasonication of the yellow-brownish residue results in the formation of graphene oxide. The sample was dried in an oven for a longer time and incorporated into a PEI membrane. The water flux was increased due to the better hydrophilicity as illustrated by the contact angle and flux studies. Due to the integration of membrane through a graphene oxide/PVP nanocomposite, the dialysis needs were completely fulfilled. The developed membranes maintained the dialysis performance in terms of excellent clearance of uremic toxins, prolonged clogging time, less protein adherence, and increased plasma recalcification duration. The biocompatibility was noted so that the membranes were effective in dialysis treatment [80].

In another work of Kaleekkal et al., graphene oxide (GO) was synthesized using an altered Hummers' procedure by selecting potassium permanganate, sodium nitrate, and concentrated sulphuric acid as the starting materials. Then the graphene oxide was modified by polydopamine grafting, which resulted in PDA-g-GO formation. PDA-g-GO was further integrated with heparin decoration. Polydopamine is an excellent binder, because of which heparin was immobilized properly onto the GO sheets. The modification method resulted in very good porosity/pore-connectivity, and reduced surface roughness, which in turn opposed protein adsorption and gave an improvement in the hydrophilicity. The hemocompatibility of the developed membrane was very good when the performance was traced as a function of cell viability, lagging clotting times, reduced complement

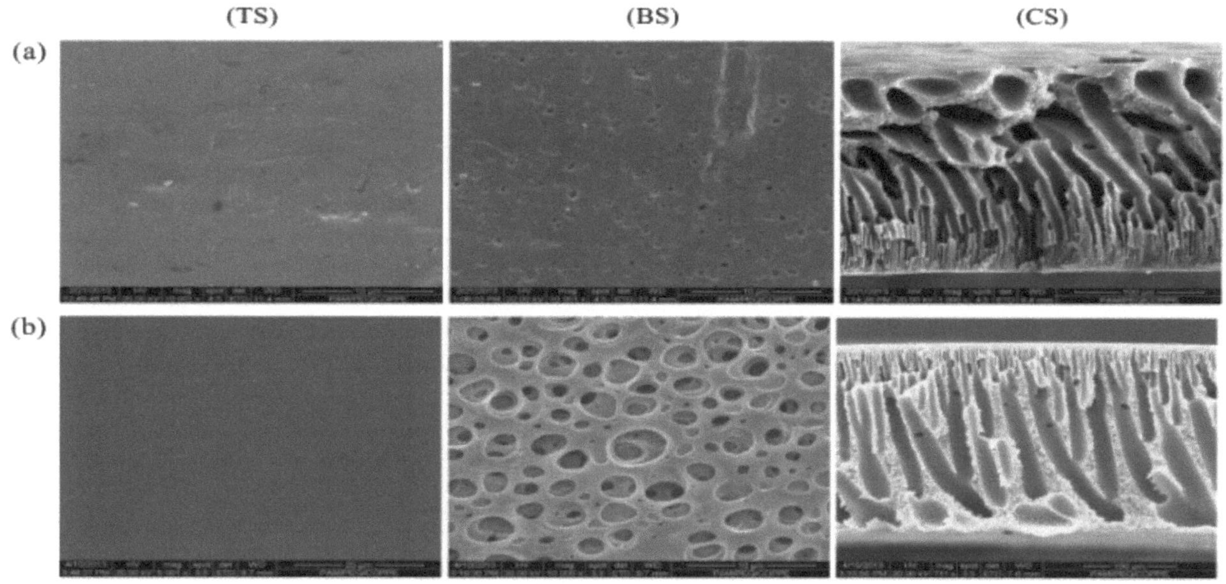

FIGURE 10.15 SEM images of (a) pristine PEI membranes and (b) nanocomposite immobilized PEI membranes. Reproduced from Ref. [81].

activation, resistance to protein adsorption, and very low platelet activation. The dialysis potency was good in terms of uremic toxin rejection and creatinine clearance [50].

Santos et al. presented a work which involved the fabrication of membranes from a PEI/PVP blend with the surface immobilization of heparin. The morphology of fabricated membranes is depicted in the SEM images shown in Figure 10.15.

The surface binding of heparin resulted in increased hydrophilicity, with very good blood compatibility. The membranes were antithrombogenic in nature which supported in efficient removal of uremic toxins [81].

10.12 POLY(VINYLIDENE FLUORIDE) HAEMODIALYSIS MEMBRANES

Poly(vinylidene fluoride) is a commonly utilized polymer in water treatment and many other applications including in the biomedical field. It is a favourable polymer because of its outstanding characteristics such as stability in terms of thermal and chemical properties, good film-forming nature, excellent blood compatibility, and many more. It has a linear polymer chain as shown in Figure 10.16 [51].

With modification, great improvements can be gained in its properties [24,82,83].

A novel tracing has been used to enhance the blood compatibility of hollow-fibre membrane made from poly(vinylidene fluoride) (PVDF) for haemodialysis. A copolymer consisting of amphiphilic side parts of polyacryloylmorpholine (PACMO) and a prime chain of PVDF (PVDF-g-PACMO) was used as an additive. By applying non-solvent-induced phase separation (NIPS), hollow-fibre membranes were fabricated by incorporating the developed co-polymer additive. These membranes were very effective in removing uremic toxins and metabolites generated in the body. The biocompatibility was recorded as good in terms of antithrombus formation and there was less platelet and serum albumin adsorption [82].

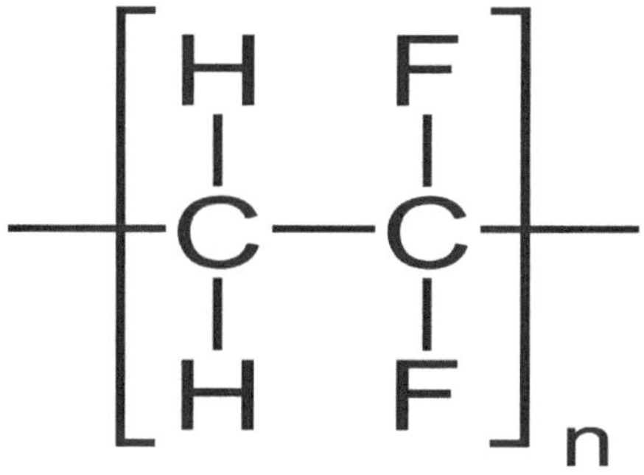

FIGURE 10.16 Linear polymeric structure of PVDF. Adapted from Ref. [51].

In another study, hollow-fibre membranes were utilized for haemodialysis. While fabricating the membranes through the NIPS method, polyethylene glycol was incorporated in the casting solution. PEG addition instilled very good hydrophilicity and pore characteristics. The dialysis performance of the developed membranes was good [83].

10.13 POLYSULFONE HAEMODIALYSIS MEMBRANES

Polysulfone (PSf) is a principal polymer utilized for fabricating membranes. It is in the class of thermoplastic materials famous for stability in terms of chemical and thermal properties with a structure containing a polymer chain as in Figure 10.17 [51].

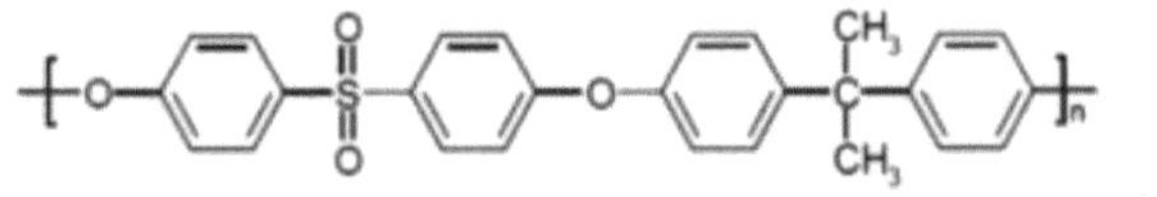

PSf

FIGURE 10.17 Polymeric structure of polysulfone. Adapted from Ref. [51].

PSf meets the biomedical requirements, i.e. high thermal and chemical resistance and a glass transition temperature of >463 K. Hence, polysulfone is employed to make many polymer recipes for fabricating membranes, but in haemodialysis it is less preferred because of low blood compatibility [17,21,22,25–27,29–43].

A sulfonated citric chitosan (SCACS) which mimics heparin in morphology and characteristics was employed to modify the PSf membranes. SCACS was prepared via acylation and sulfonation. The incorporation of a modifier improved the dialysis performance [33]. However, polysulfone still lags behind in terms of usage in haemodialysis membranes.

10.14 CONCLUSIONS AND FUTURE DIRECTION

Kidney diseases are associated with malfunctioning of the kidney in regard to removing waste products formed during metabolic reactions. Haemodialysis is a treatment technique applied for uremic toxin clearance and removal of metabolites generated in the body. A wide variety of polymer materials have been employed as dialysis materials. Chitosan, cellulose acetate, polylactic acid, and many more biologically important biodegradable and biocompatible substrates have been selected as base materials in research. Polysulfone, poly(vinylidene fluoride), polyether sulfone, and other polymers have been utilized for their robust properties. These membrane materials have the capability to be improved at the molecular level by immobilization with modifiers such as heparin, polydopamine binder, graphene oxide, co-polymers, bioactive glass, halloysite nanotubes, and numerous other materials. The selection of the most suitable developed membranes as a haemodialysis membrane is assessed as a function of blood compatibility (resistance to platelet and protein adsorption, prolonged plasma recalcification time, and low thrombus formation), dialysis potency (urea and creatinine rejection), and good membrane-related characteristics. As a result, many dialysis membranes have been created with very good results, which in turn has saved many lives. Research and development is a vast field which will keep solving problems in this area. The haemodialysis membrane stream is still finding more advanced membranes for dialysis.

Recent research into membrane materials for haemodialysis has identified several trends for improving their performance and biocompatibility. Nanotechnology is being used to enhance the mechanical and permeability properties of membranes, such as incorporating graphene oxide nanosheets into membranes. Biomimetic membranes that mimic the natural filtration process of the kidney are also being developed, resulting in improved filtration performance and biocompatibility. Researchers are also working on membranes that can selectively remove specific toxins from the blood, such as those with functional groups that bind to uric acid molecules. To combat fouling of membranes, antifouling membranes with zwitterionic surfaces are being created that resist the build-up of proteins and other materials that can clog the membrane pores. Hybrid membranes that combine different materials or processes are also being developed to improve the mechanical strength, biocompatibility, and removal of uremic toxins. Existing membrane materials are being modified with functionalized graphene oxide nanosheets to improve the water flux, biocompatibility, and removal of uremic toxins.

In addition, researchers are using 3D printing technology to produce haemodialysis membranes with precise structures and properties. Also, the advancements in artificial intelligence and machine learning are being integrated to create adaptive dialysis systems that can continuously monitor a patient's condition and adjust the filtration process in real-time. Such 'smart' membranes, integrated with automated monitoring, could lead to more precise and patient-specific treatments. These trends suggest that future research will continue to focus on improving the performance, biocompatibility, and selectivity of haemodialysis membranes to improve outcomes for patients with kidney disease.

REFERENCES

1. Bharati, J., Jha, V. (2020) Global dialysis perspective: India. *Kidney360* 1: 1143–1147.
2. Mushi, L., Marschall, P., Fleßa, S. (2015) The cost of dialysis in low and middle-income countries: a systematic review. *BMC Health Services Research* 15(1): 506.
3. Lee, G.T., Hong, Y.K. (2022) Manufacturing and separation characteristics of hemodialysis membranes to improve toxin removal rate. *Advances in Polymer Technology* 2022: 1–18.
4. Cusumano, A.M., Tzanno-Martins, C., Rosa-Diez, G.J. (2021) The glomerular filtration rate: from the diagnosis of kidney function to a public health tool. *Front Med (Lausanne)* 8: 769335
5. Jujjavarapu, S., Naik, S. (2019) A critical analysis on various technologies and functionalized materials for manufacturing dialysis membranes. *Materials Science for Energy Technologies* 3: 116–126.
6. Haroon, S., Davenport, A. (2018) Choosing a dialyzer: What clinicians need to know. *Hemodialysis International* 22(S2): S65–S74.
7. Nalesso, F., Claudio, R. (2017) Chapter 17 – Selecting a dialyzer: Technical and clinical considerations. In: *Handbook of Dialysis Therapy (Fifth Edition)*, A.R. Nissenson, R.N. Fine (Eds.), pp. 227–238.e4. Elsevier.
8. Ronco, C., Clark, W. (2018) Haemodialysis membranes. *Nature Reviews Nephrology* 14: 1.
9. Ulbricht, M. (2006) Advanced functional polymer membranes. *Polymer* 47(7): 2217–2262.

10. Charcosset, C. (2012) *3 – Microfiltration*. In: *Membrane Processes in Biotechnology and Pharmaceutics*, C. Charcosset (Ed.), pp. 101–141. Elsevier: Amsterdam.
11. Singh, R. (2015) Chapter 1 – Introduction to membrane technology. In: *Membrane Technology and Engineering for Water Purification (Second Edition)*, R. Singh (Ed.), pp. 1–80. Butterworth-Heinemann: Oxford.
12. Alaei Shahmirzadi, M.A., Kargari, A. (2018) 9 – Nanocomposite membranes. In: *Emerging Technologies for Sustainable Desalination Handbook*, V.G. Gude (Ed.) pp. 285–330. Butterworth-Heinemann.
13. Hai, F., et al. (2015) *Trace Organic Contaminants Removal by Combined Processes for Wastewater Reuse*, pp. 1–39. Research Gate.
14. Abdullah, N., et al. (2018) Chapter 2 – Membranes and membrane processes: Fundamentals. In: Current *Trends and Future Developments on (Bio-) Membranes*, A. Basile, S. Mozia, R. Molinari (Eds), pp. 45–70. Elsevier.
15. Gohil, J.M., Choudhury, R.R. (2019) Chapter 2 – Introduction to nanostructured and nano-enhanced polymeric membranes: Preparation, function, and application for water purification. In: *Nanoscale Materials in Water Purification*, S. Thomas, et al. (Eds.), pp. 25–57. Elsevier.
16. Robeson, L.M. (2012) 8.13 – Polymer membranes. In: *Polymer Science: A Comprehensive Reference*, K. Matyjaszewski, M. Möller (Eds.), pp. 325–347. Elsevier: Amsterdam.
17. Ibrahim, G.P.S., et al. (2017) Bio-inspired, fouling resistant, tannic acid functionalized halloysite nanotube reinforced polysulfone loose nanofiltration hollow fiber membranes for efficient dye and salt separation. *Journal of Water Process Engineering* **20:** 138–148.
18. Kolangare, I.M., et al. (2019) Antibiofouling hollow-fiber membranes for dye rejection by embedding chitosan and silver-loaded chitosan nanoparticles. *Environmental Chemistry Letters* 17(1): 581–587.
19. Nayak, M.C., et al. (2018) Fabrication of novel PPSU/ZSM-5 ultrafiltration hollow fiber membranes for separation of proteins and hazardous reactive dyes. *Journal of the Taiwan Institute of Chemical Engineers* 82: 342–350.
20. Shehadat, S.A., et al. (2018) Optimization of scanning electron microscope technique for amniotic membrane investigation: A preliminary study. *European Journal of Dentistry* 12(04): 574–578.
21. Padaki, M., et al. (2012) Synthesis, characterization and desalination study of novel PSAB and mPSAB blend membranes with polysulfone (PSf). *Desalination* 295: 35–42.
22. Padaki, M., et al. (2012) Preparation and characterization of sulfonated polysulfone and N-phthloyl chitosan blend composite cation-exchange membrane for desalination. *Desalination* 298: 42–48.
23. Panchami, H.R., Isloor, A.M., Ismail, A.F. (2022) Improved hydrophilic and antifouling performance of nanocomposite ultrafiltration zwitterionic polyphenylsulfone membrane for protein rejection applications. *Journal of Nanostructure in Chemistry* 12(3): 343–364.
24. Pereira, V.R., et al. (2014) Preparation and antifouling properties of PVDF ultrafiltration membranes with polyaniline (PANI) nanofibers and hydrolysed PSMA (H-PSMA) as additives. *Desalination* 351: 220–227.
25. Nair, A.K., et al. (2013) Antifouling and performance enhancement of polysulfone ultrafiltration membranes using $CaCO_3$ nanoparticles. *Desalination* 322: 69–75.
26. Kumar, R., et al. (2013) Synthesis and characterization of novel water soluble derivative of Chitosan as an additive for polysulfone ultrafiltration membrane. *Journal of Membrane Science* 440: 140–147.
27. Kumar, R., et al. (2013) Performance improvement of polysulfone ultrafiltration membrane using N-succinyl chitosan as additive. *Desalination* 318: 1–8.
28. Kumar, R., et al. (2013) Permeation, antifouling and desalination performance of TiO_2 nanotube incorporated PSf/CS blend membranes. *Desalination* 316: 76–84.
29. Kumar, R., et al. (2013) Polysulfone–chitosan blend ultrafiltration membranes: preparation, characterization, permeation and antifouling properties. *RSC Advances* 3(21): 7855.
30. Ibrahim, G.P.S., et al. (2018) Performance intensification of the polysulfone ultrafiltration membrane by blending with copolymer encompassing novel derivative of poly(styrene-co-maleic anhydride) for heavy metal removal from wastewater. *Chemical Engineering Journal* 353: 425–435.
31. Koga, Y., et al. (2018) Biocompatibility of polysulfone hemodialysis membranes and its mechanisms: Involvement of fibrinogen and its integrin receptors in activation of platelets and neutrophils. *Artificial Organs* 42(9): E246–E258.
32. Kumar, R., Isloor, A.M., Ismail, A.F. (2014) Preparation and evaluation of heavy metal rejection properties of polysulfone/chitosan, polysulfone/N-succinyl chitosan and polysulfone/N-propylphosphonyl chitosan blend ultrafiltration membranes. *Desalination* 350: 102–108.
33. Lin, B., Liu, K., Qiu, Y. (2021) Preparation of modified polysulfone material decorated by sulfonated citric chitosan for haemodialysis and its haemocompatibility. *Royal Society Open Science* 8(9): 210462.
34. Mohammadi, F., Mohammadi, F., Yavari, Z. (2021) Characterization of the cylindrical electrospun nanofibrous polysulfone membrane for hemodialysis with modelling approach. *Medical & Biological Engineering & Computing* 59(7–8): 1629–1641.
35. Mousavi, S.A., et al. (2021) Modification of porous polyetherimide hollow fiber membrane by dip-coating of Zonyl® BA for membrane distillation of dyeing wastewater. *Water Science and Technology* 83(12): 3092–3109.
36. Oshihara, W., Ueno, Y., Fujieda, H. (2017) A new polysulfone membrane dialyzer, NV, with low-fouling and antithrombotic properties. Contributions to Nephrology.
37. Ponnaiyan, P., Nammalvar, G. (2019) Effect of additives on graphene oxide incorporated polysulfone (PSF) membrane. *Polymer Bulletin* 76(8): 4003–4015.
38. Radu, E.R., Voicu, S.I. (2022) Functionalized hemodialysis polysulfone membranes with improved hemocompatibility. *Polymers* 14(6): 1130.
39. Ronco, C., et al. (1999) In vitro and in vivo evaluation of a new polysulfone membrane for Hhmodialysis. Reference methodology and clinical results: (Part 1: In vitro study). *The International Journal of Artificial Organs* 22(9): 604–615.
40. Roy, A., et al. (2015) In vitro cytocompatibility and blood compatibility of polysulfone blend, surface-modified polysulfone and polyacrylonitrile membranes for hemodialysis. *RSC Advances* 5(10): 7023–7034.

41. Tsuchida, K., et al. (2017) Effects of hydrophilic polymer-coated polysulfone membrane dialyzers on intradialytic hypotension in diabetic hemodialysis patients (ATHRITE BP Study): A pilot study. *Renal Replacement Therapy* 3(1): 1–10.
42. Wagner, S., et al. (2020) Hemocompatibility of polysulfone hemodialyzers – exploratory studies on impact of treatment modality and dialyzer characteristics. *Kidney360* 1(1): 25–35.
43. Yu, X., et al. (2015) Robust poly(lactic acid) membranes improved by polysulfone-g-poly(lactic acid) copolymers for hemodialysis. *RSC Advances* 5(95): 78306–78314.
44. Abe, M., et al. (2021) Dialyzer classification and mortality in hemodialysis patients: A 3-year nationwide cohort study. *Front Med (Lausanne)* 8: 740461.
45. Bowry, S.K., Chazot, C. (2021) The scientific principles and technological determinants of haemodialysis membranes. *Clinical Kidney Journal* 14(Supplement 4): i5–i16.
46. Hothi, D.K., Geary, D.F. (2008) Chapter 57 – Pediatric hemodialysis prescription, efficacy, and outcome. In: *Comprehensive Pediatric Nephrology*, D.F. Geary, F. Schaefer (Eds.), pp. 867–893. Mosby: Philadelphia.
47. Leypoldt, J.K. (2008) Chapter 31 – Methods and complications of dialyzer reuse. In: *Handbook of Dialysis Therapy (Fourth Edition)*, A.R. Nissenson, R.N. Fine (Eds.), pp. 469–477. W.B. Saunders: Philadelphia.
48. Smith, A.T., et al. (2019) Synthesis, properties, and applications of graphene oxide/reduced graphene oxide and their nanocomposites. *Nano Materials Science* 1(1): 31–47.
49. Waniewski, J., et al. (1994) Bidirectional solute transport in peritoneal dialysis. *Peritoneal Dialysis International* 14(4): 327–37.
50. Jacob Kaleekkal, N. (2021) Heparin immobilized graphene oxide in polyetherimide membranes for hemodialysis with enhanced hemocompatibility and removal of uremic toxins. *Journal of Membrane Science* 623: 119068.
51. He, X., Lei, L., Chu, Y. (2018) Chapter 9 – Facilitated transport membranes for CO_2 removal from natural gas. In: *Current Trends and Future Developments on (Bio-) Membranes*, A. Basile, E.P. Favvas (Eds.), pp. 261–288. Elsevier.
52. Kumar, M., et al. (2019) Use of cellulose acetate/polyphenylsulfone derivatives to fabricate ultrafiltration hollow fiber membranes for the removal of arsenic from drinking water. *International Journal of Biological Macromolecules* 129: 715–727.
53. Shen, S.S., et al. (2017) Immobilization of copper ions on chitosan/cellulose acetate blend hollow fiber membrane for protein adsorption. *RSC Advances* 7(17): 10424–10431.
54. Shenvi, S., Ismail, A.F., Isloor, A.M. (2014) Enhanced permeation performance of cellulose acetate ultrafiltration membranes by incorporation of sulfonated poly(1,4-phenylene ether ether sulfone) and poly(styrene-*co*-maleic anhydride). *Industrial & Engineering Chemistry Research* 53(35): 13820–13827.
55. Peng, L., Li, H., Meng, Y. (2017) Layer-by-layer structured polysaccharides-based multilayers on cellulose acetate membrane: Towards better hemocompatibility, antibacterial and antioxidant activities. *Applied Surface Science* 401: 25–39.
56. Faria, M., et al. (2020) Hybrid flat sheet cellulose acetate/silicon dioxide ultrafiltration membranes for uremic blood purification. *Cellulose* 27: 3847–3869.
57. Azhar, O., et al. (2021) Cellulose acetate-polyvinyl alcohol blend hemodialysis membranes integrated with dialysis performance and high biocompatibility. *Materials Science and Engineering: C* 126: 112127.
58. Idris, A., Hew, K.Y., Chan, M.K. (2012) Preparation of cellulose acetate dialysis membrane using Dglucose monohydrate as additive. *Jurnal Teknologi* 51(1): 67–76.
59. de Queiroz Antonino, R., et al. (2017) Preparation and characterization of chitosan obtained from shells of shrimp (*Litopenaeus vannamei* Boone). Marine Drugs 15(5): 141.
60. Kumari, S., Kishor, R. (2020) Chapter 1 – Chitin and chitosan: origin, properties, and applications. In: *Handbook of Chitin and Chitosan*, S. Gopi, S. Thomas, A. Pius (Eds.), pp. 1–33. Elsevier.
61. Vo, T.S., et al. (2022) Graphene oxide–chitosan network on a dialysis cellulose membrane for efficient removal of organic dyes. *ACS Applied Bio Materials* 5(6): 2795–2811.
62. Shenvi, S.S., et al. (2013) Preparation and characterization of PPEES/chitosan composite nanofiltration membrane. *Desalination* 315: 135–141.
63. Radhakumary, C., et al. (2006) *HEMA-grafted chitosan for dialysis membrane applications. Journal of Applied Polymer Science* 101: 2960–2966.
64. Siddiq, F., et al. (2022) Magnetic chitosan membrane as an effective analytical tool for adsorptive removal of creatinine from biological samples. *Journal of Taibah University for Science* 16(1): 250–258.
65. Mahapatro, A., Singh, D. (2011) Biodegradable nanoparticles are excellent vehicle for site directed in-vivo delivery of drugs and vaccines. *Journal of Nanobiotechnology* 9: 55.
66. Gao, A., Liu, F., Xue, L. (2014) Preparation and evaluation of heparin-immobilized poly (lactic acid) (PLA) membrane for hemodialysis. *Journal of Membrane Science* 452: 390–399.
67. Zhu, L., et al. (2015) Poly(lactic acid) hemodialysis membranes with poly(lactic acid)-*block*-poly(2-hydroxyethyl methacrylate) copolymer as additive: Preparation, characterization, and performance. *ACS Applied Materials & Interfaces* 7(32): 17748–17755.
68. Yi Tong, C., Derek, C.J.C. (2022) Membrane surface roughness promotes rapid initial cell adhesion and long term microalgal biofilm stability. *Environmental Research* ***206***: 112602.
69. Jia, Y., et al. (2010) [Polyethersulfone hollow fiber membrane for hemodialysis – preparation and evaluation]. *Sheng Wu Yi Xue Gong Cheng Xue Za Zhi* 27(1): 91–96.
70. Kaleekkal, N.J., et al. (2015) A functional PES membrane for hemodialysis — Preparation, characterization and biocompatibility. *Chinese Journal of Chemical Engineering* 23(7): 1236–1244.
71. Samtleben, W. (2003) Comparison of the new polyethersulfone high-flux membrane DIAPES(R) HF800 with conventional high-flux membranes during on-line haemodiafiltration. *Nephrology Dialysis Transplantation* 18(11): 2382–2386.
72. Raharjo, Y., et al. (2019) Incorporation of imprinted-zeolite to polyethersulfone/cellulose acetate membrane for creatinine removal in hemodialysis treatment. *Jurnal Teknologi* 81(3): 137–144.
73. Fahmi, M.Z., et al. (2018) Incorporation of graphene oxide in polyethersulfone mixed matrix membranes to enhance hemodialysis membrane performance. *RSC Advances* 8(2): 931–937.

74. Md Salleh, S., Abdullah, M., Wahab, A. (2014) Chemical characterization of stabilized and carbonized polyacrylonitrile (PAN) fibers treated with oleic acid. *MATEC Web of Conferences* 13: 04014.
75. Scharnagl, N., Buschatz, H. (2001) Polyacrylonitrile (PAN) membranes for ultra- and microfiltration. *Desalination* 139(1): 191–198.
76. Lin, W.-C., Liu, T.-Y., Yang, M.-C. (2004) Hemocompatibility of polyacrylonitrile dialysis membrane immobilized with chitosan and heparin conjugate. *Biomaterials* 25(10): 1947–1957.
77. Lau, K.S.Y. (2014) 10 – High-performance polyimides and high temperature resistant polymers. In: *Handbook of Thermoset Plastics (Third Edition)*, H. Dodiuk, S.H. Goodman (Eds.), pp. 297–424. William Andrew Publishing: Boston.
78. Hebbar, R.S., et al. (2016) Fabrication of polydopamine functionalized halloysite nanotube/polyetherimide membranes for heavy metal removal. *Journal of Materials Chemistry A* 4(3): 764–774.
79. Hebbar, R.S., et al. (2018) Removal of metal ions and humic acids through polyetherimide membrane with grafted bentonite clay. *Scientific Reports* 8(1): 4665.
80. Kaleekkal, N.J., et al. (2015) Graphene oxide nanocomposite incorporated poly(ether imide) mixed matrix membranes for in vitro evaluation of its efficacy in blood purification applications. *Industrial & Engineering Chemistry Research* 54(32): 7899–7913.
81. Santos, A.M.D., Habert, A.C., Ferraz, H.C. (2017) Development of functionalized polyetherimide/polyvinylpyrrolidone membranes for application in hemodialysis. *Journal of Materials Science: Materials in Medicine* 28(9): 131.
82. An, Z., et al. (2017) PVDF/PVDF-g-PACMO blend hollow fiber membranes for hemodialysis: Preparation, characterization, and performance. *RSC Advances* 7(43): 26593–26600.
83. Zhang, Q., Lu, X., Zhao, L. (2014) Preparation of polyvinylidene fluoride (PVDF) hollow fiber hemodialysis membranes. *Membranes* 4(1): 81–95.

11 Role of Membranes and Membrane Reactors in Hydrogen Production

Nitish Mittal and Dhaval A. Bhandari
Exxon Mobil Technology and Engineering Company, Baytown, TX, USA
E-mail: nitish.mittal@exxonmobil.com; dhaval.a.bhandari@exxonmobil.com

11.1 INTRODUCTION

Hydrogen has emerged as one of the promising energy carriers owing to its relatively high heating value and low-carbon combustion characteristics with existing and potential applications in automotive power supply, fuel cells, petrochemical plants and refineries, hydrogen burners, and power generation [1]. The global hydrogen market was valued at $130 billion in 2021 and is expected to rise to $1.6 trillion by 2050 [2]. The global annual demand for hydrogen was estimated at 90 Mt in 2020 [1] (Figure 11.1).

While the current major applications include ammonia production and petroleum refining, the transport and power industries are the major development areas contributing to the future growth of hydrogen demand. Other current applications include methanol synthesis, polymer production, electronics, metal/glass and food industries, pharmaceuticals, etc. (Figure 11.2) [3].

The feedstocks and technology used to produce hydrogen play a relevant role in its production profile. Currently, fossil fuel-based resources are used for almost all hydrogen production; ~59% is produced by steam methane reforming (SMR), while coal gasification and by-product generation facilities such as refineries contribute evenly for the remaining ~40% [1]. Electricity and other methods contribute only ~1% to the total production [1]. Innovative and efficient technologies can play an important role in the production of hydrogen [4–6]. In this chapter, we focus on the roles of membranes and membrane reactors.

11.2 HYDROGEN PRODUCTION USING STEAM METHANE REFORMING

Methane reforming is the most conventional method of producing hydrogen and is expected to remain the prevalent technology in the near–medium term [7]. During this process, natural gas is catalytically (majorly Ni-based catalyst) converted to hydrogen in a conventional reformer under harsh operating conditions (800–1000°C and 14–20 bar) [7,8]. The following major reactions (Rxn. 11.1 and Rxn. 11.2) take place during the reforming process.

$$CH_4 + H_2O = CO + 3H_2 \quad \Delta H^0_{298} = 206\ kJ/mol \qquad \text{(Rxn 11.1)}$$

$$CH_4 + 2H_2O = CO_2 + 4H_2 \quad \Delta H^0_{298} = 165\ kJ/mol \qquad \text{(Rxn 11.2)}$$

This is followed by high-temperature (300–450°C) and low-temperature (180–250°C) water gas shift reactions (Rxn. 11.3) that convert carbon monoxide (CO) into additional hydrogen [7,8]. The two reactors are needed because it is an exothermic reaction which is kinetically limited but thermodynamically favored at low temperature [7].

$$CO + H_2O = CO_2 + 3H_2 \quad \Delta H^0_{298} = -41\ kJ/mol \qquad \text{(Rxn 11.3)}$$

Wassie et al. [9] illustrates the SMR hydrogen production process (Figure 11.3). The resultant product goes through several purification steps to achieve high-purity hydrogen. Table 11.1 lists different commercial techniques that are used based on the desired hydrogen purity, such as chemical and physical scrubbing, methanation, cryogenic separation, and membrane separation [7,8].

Hydrogen-selective membranes can also be used in the form of a membrane reactor which provides several benefits over a conventional reformer. The conventional reforming process suffers from equilibrium limitations and requires high-temperature operation, which results in irreversible coke formation, considerable energy losses, creation of temperature profile inside the catalyst bed due to heat transfer limitation, and the higher possibility of NO_x formation in the furnace [7]. Further, the endothermic reforming reactor operates at significantly higher temperature than the exothermic water–gas shift reactors. This restricts the heat integration between the two reactions and the process suffers from a large energy penalty required to cool down the reformed effluent before entering the shift-gas reactors. The use of a hydrogen-selective membrane reactor enables the potential of greater conversion under lower temperature of operation and simultaneous purification of hydrogen [7,8,10]. A membrane reactor operating at 500°C can increase methane conversion to 0.80–0.90 as opposed to ~0.5 in a high-temperature conventional process. The membrane reactor also can help avoid catalyst fouling caused due to the high-temperature operation in a conventional process. Simultaneous hydrogen removal can also favorably shift the equilibrium of water–gas shift reactions and can allow simultaneous progression of both

DOI: 10.1201/9781003285656-11

reactions. This results in intensification of the endothermic and exothermic reactions which can reduce the energy penalty associated with the conventional process.

11.3 HYDROGEN SELECTIVE CARBON MOLECULAR SIEVES

While polymeric membranes are already commercialized for hydrogen separation, these are not stable for high-temperature (> 100°C) operation. On the other hand, carbon molecular sieves (CMS) have superior chemical and thermal stability, which makes them relevant for hydrogen production reactions [11,12]. CMS membranes are synthesized from the pyrolysis of polymeric precursors and possess a bimodal pore size distribution of micropores and ultra-micropores which provides superior sieving properties and outperforms the permeability–selectivity trade off associated with polymeric membranes [13,14]. Several factors affect the transport and separation properties of carbon membranes – the type of polymer precursor, the precursor's microstructure, and pyrolysis conditions and parameters, etc. Several polymeric precursors based on cellulose derivatives [15–17], polyimide [18–20], poly(vinylidene fluoride) (PVDF) [21], and polyacrylonitrile (PAN) [22], have been used to make carbon membranes. One polyimide, Matrimid™, has gained a lot of attention as a precursor due to its good rigidity with high temperature for both melting point and glass transition temperature while retaining its chemical and thermal stability [23].

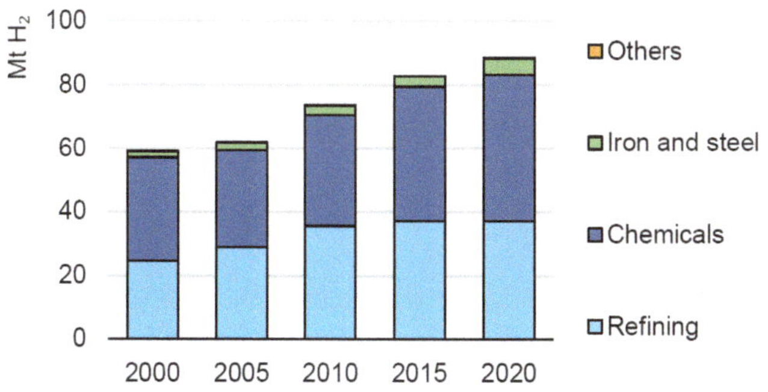

FIGURE 11.1 Global demand of hydrogen economy by technology category. Note: "Others" refers to small volumes of demand in industrial applications, transport, grid injection, and electricity generation. Reprinted from International Energy Agency, Global Hydrogen Review 2021, available from www.iea.org/reports/global-hydrogen-review-2021 [1].

Carbon membranes can further be supported on hollow fibers, suggesting their potential for fabricating high-density membrane modules for large-scale applications. While the use of carbon membranes in enriching hydrogen to 56–60% before PSA purification goes back to the early 1990s [24,25], exploring their application in reforming reactions is more recent. Tsotsis and co-workers [26–28] have studied supported carbon membranes for membrane reactors for hydrogen production. Although the membranes show good stability at temperatures and pressures of up to 250°C and 25 bar, respectively, a lower selectivity due to carbon dioxide permeation via the presence of a selective surface flow (SSF) transport mechanism was found to be a challenge. Improvement in selectivity has been observed by changing carbonization parameters, especially pyrolysis end temperature. Studies by Llosa Tanco et al. [29] and Zhang et al. [30] have demonstrated the increase in hydrogen/carbon dioxide selectivity at a temperature of 900–1000°C. However, this increment was attained at the expense of decreasing pore size and hydrogen permeability. The preparation of asymmetric CMS membranes with a dense selective layer and a lower transport-resistant porous integral support layer is a potential solution for maintaining permeance at higher carbonization temperature [18].

Process optimization studies and techno-economic analysis have been performed using a two-stage membrane configuration with recycling from the second-stage retentate to the first-stage feed for hydrogen purification post water gas shift reactor [11]. The feed pressure was optimized, and the temperature sensitivity was analyzed to high-purity hydrogen with minimal losses. A 99.5% hydrogen purity was achieved using hydrogen permeance of 111 GPU and hydrogen/carbon dioxide selectivity of 37 at 10 bar and 110°C. For ~99.9%

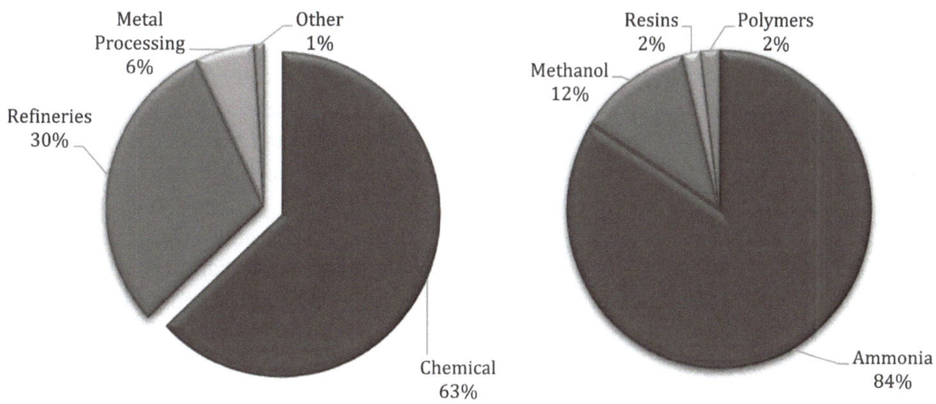

FIGURE 11.2 Breakdown of the global hydrogen market sectors (left) and further breakdown of chemical industry sector (right). Reprinted from Ref. [3] available from open access article distributed under the terms of the Creative Commons CC-BY license.

TABLE 11.1
Commercial techniques for hydrogen purification [7,8]

Technique	Description	Drawbacks	Benefits	Recovery (%)	Purity (mol%)
Preferential oxidation	Selective oxidation of carbon monoxide into carbon dioxide	Added complexity to the whole system	Carbon monoxide concentration of the final hydrogen-rich stream around 500 ppm		
PSA (pressure swing adsorption)	Selective adsorption of compounds from a gas stream	Around 20% of hydrogen is lost under operation (hydrogen recovery between 70 and 85%)	Hydrogen purity > 99.99%	70–85	99.99
Methanation	Methanation reactors make possible the carbon monoxide content reduction	Unnecessary consumption of hydrogen	Methanation reactors are simpler, and no addition of air is required		
Cryogenic distillation	Partial condensation of gas mixtures at low temperatures	Relatively low hydrogen purity, between 90 and 98%	Low operating temperature and high recovery	Up to 98	90–98
Dense palladium membrane	Selective diffusion of hydrogen through a palladium alloy membrane	High cost and low mechanical resistance	High recovery and purity	Up to 99	>99.999
Polymeric membrane	Differential rate of diffusion of gases through a permeable membrane	Relatively low hydrogen purity between 92–98	Low cost, high hydrogen recovery	>85	92–98
Metal hydride separation	Reversible reaction of hydrogen with metals to form hydrides	Hydrogen recovery between 75 and 95%	High purity	75–95	99
Solid polymer electrolyte cell	Electrolytic passage of hydrogen ions across a solid polymer membrane	Sluggish anode reaction coupled with the inefficient cathode reaction (low overall cell performance)	High recovery and purity	95	99.8

hydrogen, the membrane area and compressor duties increase significantly due to higher recycling to avoid hydrogen losses. Extrapolating the model in the range of 100–200°C shows that both the permeance and selectivity increase with temperature. Using the predicted values of hydrogen permeance of ~350 and selectivity of 40 at 200°C, the specific cost decreases only by ~20%. The results suggest obtaining >99.9% hydrogen is limited by hydrogen/carbon dioxide selectivity and can be economically challenging for the membrane process and that the development of highly selective membranes is needed for a high-purity hydrogen product.

11.4 HYDROGEN SELECTIVE METALLIC MEMBRANES

Dense metallic membranes, such as Pd-membranes, can provide high hydrogen purity and recovery, and are considered a promising technology for high-purity hydrogen production via membrane separation and/or membrane reactor operation [4,6–8]. Historically, Johnson Matthey and Tokyo Gas Company Ltd. have successfully demonstrated pilot-scale operations for membranes and membrane reactors in hydrogen-rich gas purification and methane steam reforming, respectively [8]. However, some challenges such as high membrane cost, mechanical resistance, and low hydrogen permeability limit their widespread use in commercial-scale applications [31]. The major drawback is the embrittlement phenomenon which causes cracking as the absorbed hydrogen results in a phase transition from α to β palladium hydride [8]. Thus, Pd-alloys, specifically silver and gold, are used to reduce the brittleness. Further, carbon and sulfur compounds can cause Pd deactivation and poisoning, which results in major stability concerns. On the other hand, porous membranes made of aluminum, titanium, or silica oxides exhibit completely opposite characteristics. These are chemically inert and show high hydrogen permeability but poor separation properties [8]. Thus, composite membranes with a selective Pd-layer on a porous support integrate the benefit of both and offer high hydrogen permeability and perm-selectivity, and good mechanical resistance at a moderate cost [8]. Nowadays, commercial metallic membranes for hydrogen separation are provided by several vendors, such as Tokyo Gas, HySep

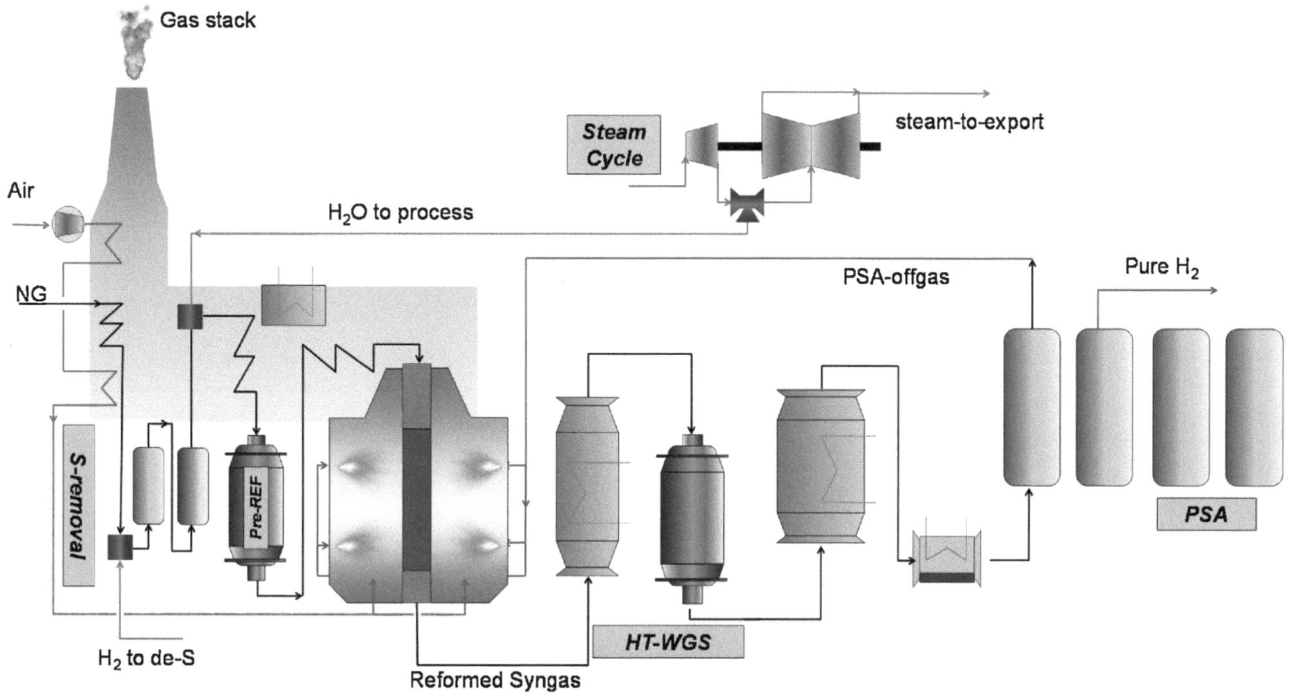

FIGURE 11.3 Schematic description of a reference fired tubular plant for hydrogen production. Reprinted from Ref. [9] with permission.

membranes by Energy Research Centre of the Netherlands (ECN), MRT, and CRI-Criterion [4,8] (Table 11.2).

Despite their commercial progress, membrane instability of the Pd-layer support interface is a challenge and an active area for research and improvement. While currently used only as stand-alone membrane, both dense self-supported and composite Pd-based membranes exhibit the potential to be used in membrane reactors in reforming and water gas shift applications.

TABLE 11.2
Commercial Pd metallic membranes for hydrogen separation [4,8]

Supplier/company	Hydrogen purity (mol%)
Tokyo Gas	99.99
ECN (HySep)	> 99.5
MRT	99.99
CRI-Criterion	> 99

11.5 MEMBRANE REACTORS FOR STEAM METHANE REFORMING

Several configurations of membrane reactor technology have been studied for steam methane reforming. Usually, they can be classified as either packed bed or fluid-bed membrane reactors [7,8]. Figure 11.4 shows a schematic of both these types of reactors. In a packed bed membrane reactor, the catalyst is confined in a tubular or planar configuration and is in direct contact with the membrane. While this is a simpler configuration, it has the same trade-offs in terms of the catalyst particle dimension, i.e., small particles increase pressure drop while large particles can limit the internal mass transfer. Further, hydrogen permeation is also influenced by mass transfer limitations from the bed to the membrane wall and the resulting temperature profile can be detrimental to the catalyst and the membrane. On the other hand, a fluidized bed consists of typically tubular bundles of membranes immersed in a fluid bed of catalyst. While the heat and mass transfer limitations can be overcome using a fluidized membrane reactor, it has the typical challenges of any fluid bed operation including maintaining minimum fluidization velocity, back-mixing, and possible bubble-to-emulsion mass transfer limitations.

Several articles have been published highlighting the application of membrane reactors equipped with palladium or palladium-alloy membranes and have been summarized elsewhere [7,8]. Some of the past relevant and more recent modeling literature is reviewed next. High methane conversion (almost 100%) and high hydrogen recovery (> 95%) have been achieved experimentally. Saric et al. [32] and Dittmar et al. [33] have also studied the long lifetime (500–1000 hours) at 600–650°C of a Pd composite membrane to demonstrate the scale-up potential of the technology (Figure 11.5). Patil et al. [34] studied the effects of fluidization velocity, temperature, and pressure to achieve 97% conversion at 650°C and 4 bar.

More recently, Vigneult et al. [35] have demonstrated 91% methane conversion and 99.99% hydrogen purity in a novel multichannel membrane reactor. The prototype consists of alternate channels for steam methane reforming with

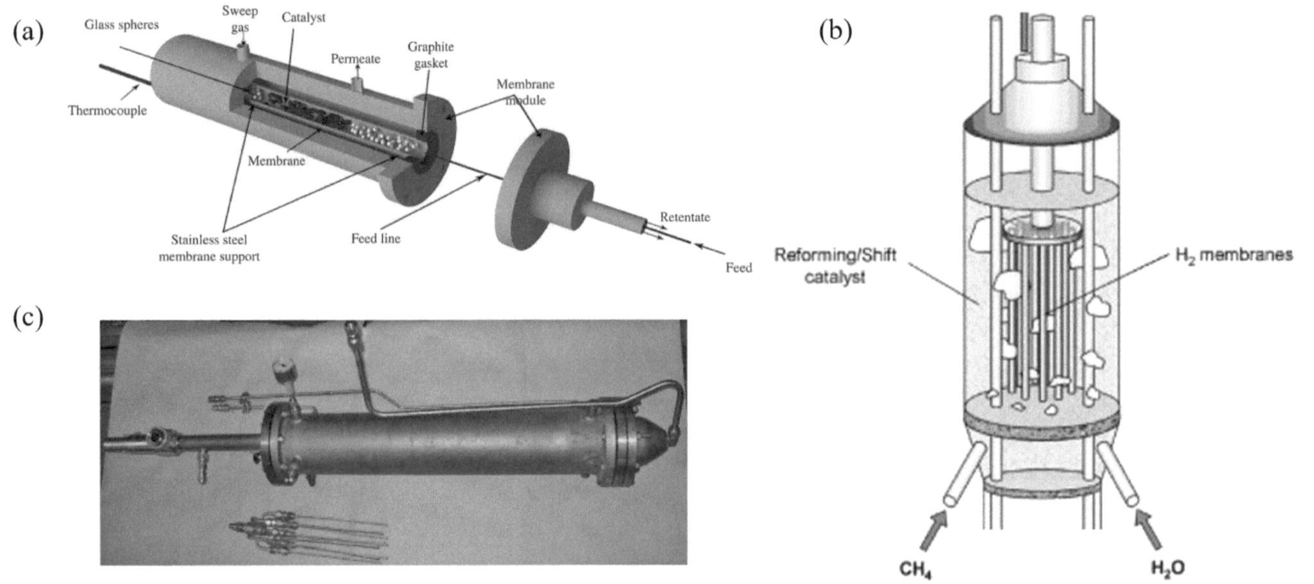

FIGURE 11.4 (a) Schematic of a packed-bed membrane reactor in tubular configuration, (b) schematic of a fluidized-bed membrane reactor in tubular configuration, and (c) picture of tubes in a fluidized bed membrane reactor. Reprinted "a" from Ref. [8], "b" and "c" from Ref. [34] with permission.

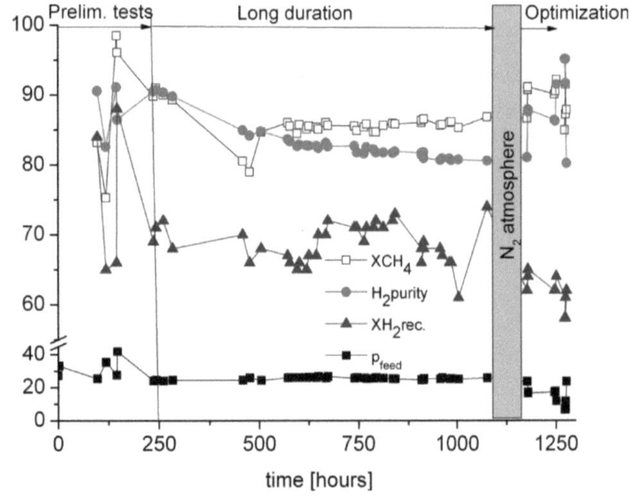

FIGURE 11.5 Long-term testing and performance of steam reforming of methane in a bench-scale packed bed Pd-based membrane reactor at realistic working conditions – measured methane conversion, hydrogen purity, and hydrogen recovery. Reprinted from Ref. [32] with permission.

palladium–silver membrane and methane combustion with heat integration. Patrascu et al. [36] obtained 90% conversion and 80% hydrogen recovery with a 175 cm^2 high-flux Pd membrane and a foam catalyst. They observed hydrogen equilibrium between retentate and permeate which restricted the performance of the membrane. An increase in driving force by increasing retentate pressure and by introducing sweep flow of N_2 in the permeate side was employed to improve hydrogen permeation. They also highlighted the possibility of permeance inhibition due to methane and carbon monoxide co-adsorption. They also investigated the effect of concentration polarization and concluded that resistances due to membrane transport and gas phase mass transfer are comparable. Hawa et al. [37] obtained > 80% methane conversion and >93% hydrogen purity by using a Pd–Ru/YSZ/PSS membrane-based catalytic reactor. Long-term testing (> 1000 h) demonstrated stable methane conversion and hydrogen flux at 580°C and 29 bar. Their results also highlight that the membrane performance was limited by concentration polarization. Kim et al. [38,39] obtained 82% methane conversion and 97.7% pure hydrogen using a Pd composite membrane on a tubular stainless-steel support. They demonstrated stable performance for > 100 h and almost nil carbon monoxide in their reformed stream, which made a water–gas shift reactor unnecessary (Figure 11.6).

Anzelmo et al. [40] carried out experiments under mild operating conditions (420°C and 3 bar) and obtained 50% methane conversion and 90% hydrogen recovery. Post-experiment scanning electron microscope (SEM) analysis of the membrane showed that the smoothness of the support delaminates the Pd-layer from the alumina support and led to the poor separation performance (Figure 11.7).

Wang et al. [41] demonstrated a novel concept by using metallic nickel hollow-fiber membranes consisting of a dense skin layer integrated on a porous nickel substrate where the porous internal surface functioned as a catalyst bed for methane steam reforming and the external dense skin layer served as the membrane. Although the operation requires temperature > 800°C, they showed high chemical stability in the reformate gases and high resistance to carbon deposition.

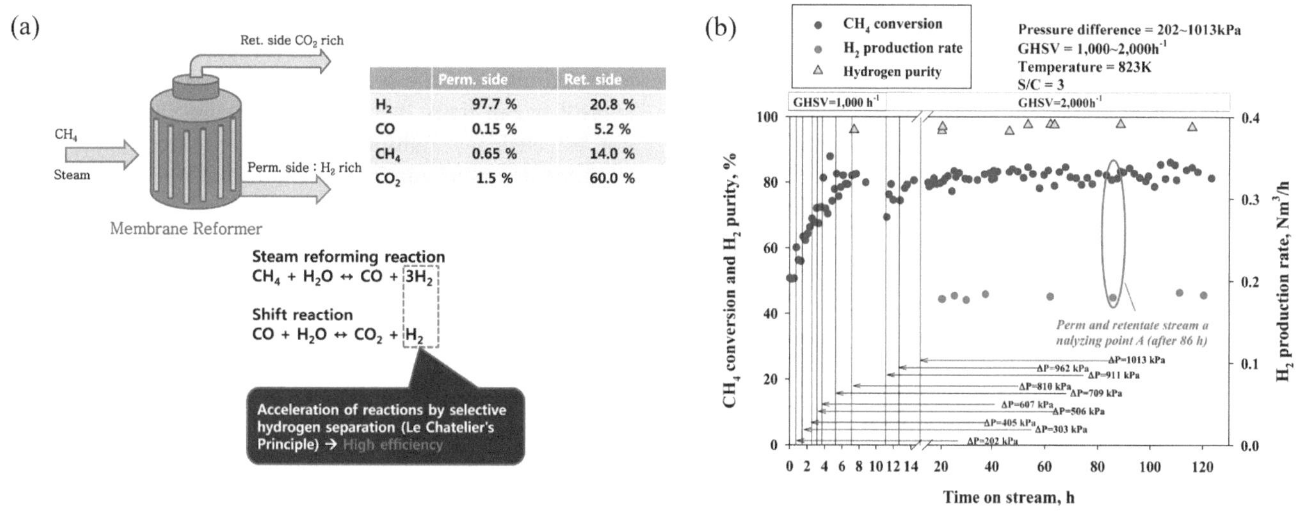

FIGURE 11.6 (a) Schematic of the steam reforming of methane in a Pd-based membrane reactor; (b) methane conversion, hydrogen purity, and hydrogen production rate for ~100 hours testing. Reprinted from Refs. [38,39] with permission.

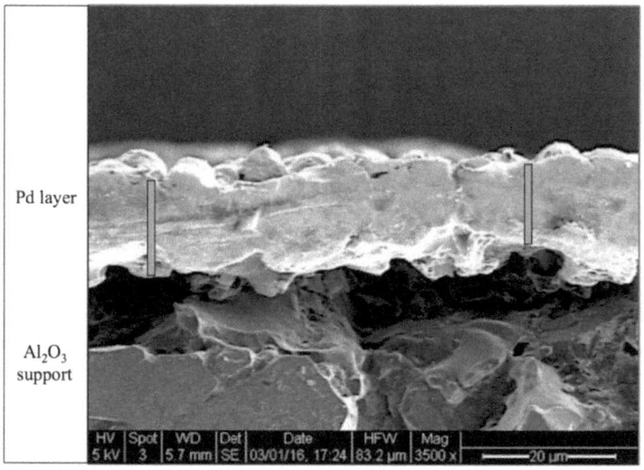

FIGURE 11.7 SEM analyses of membrane cross-section shows delamination of the Pd-layer which could be responsible for the presence of the defects in the separative layer. Reprinted from Ref. [40] with permission.

11.6 MODELING OF MEMBRANE REACTORS

The process intensification of the catalytic reactor and membrane separation into a single process requires both operations to occur at similar rates. In this context, detailed mathematical and process models are highly beneficial to understand the fundamentals and synergies of both phenomena. Several modeling approaches have been used for this purpose [7,8]. In terms of geometry, simple one-dimensional and complex two- and three-dimensional packed bed membrane reactor models have been studied. From a heat transfer perspective, both isothermal and non-isothermal modeling approaches have also been considered. For a fluid bed reactor model, pseudo-homogeneous and two-phase approaches have been proposed. Several researchers have explored the application of these models for membrane reactor modeling, which have been summarized elsewhere [7,8]. Some of the more recent modeling literature is reviewed next.

Silva et al. [42] performed a dynamics study to analyze the temperatures of gaseous and solid phases for conventional fixed-bed and fixed-bed membrane reactors for the steam reforming of methane. Medeiros et al. [43] used a non-isothermal pseudo-homogeneous dynamic model to model the reforming and a half-power pressure law to model the permeation rate of hydrogen through the membrane. They highlighted the use of a coupled integral equation approach to reduce the partial differential equations into ordinary differential equations using the boundary conditions. Cruz et al. [44] developed a two-dimensional pseudo-heterogeneous model to analyze the temperature profiles (at the surface of the solid phase and gas phase for two reactors), methane conversion at different temperatures, hydrogen production, and the product distribution. They showed that the process may be limited by hydrogen permeation as the hydrogen partial pressure difference becomes small and almost negligible halfway through the membrane area.

Murmura et al. [45–47] have developed a simple model with significantly less computational time that can be used to predict the high-level performance and identification of a limiting phenomenon based on physical parameters and operating conditions only. They used three dimensionless numbers, namely, Peclet number, Damkohler number, and permeability parameter, to study the mass transport in the packed bed, hydrogen permeation across the membrane, and reaction. They proposed a scaling law of critical efficiency versus the main dimensionless parameters to strategize the design of reactor geometry and operating conditions. They also developed a simplified model by deriving an enhanced Sherwood number to account for radial hydrogen concentration gradients due to both reaction and permeation. Leonzio et al. [48] used modeling and analysis of variance

(ANOVA) of a pilot-scale integrated membrane reactor and proposed a response surface methodology to determine more accurate optimal solutions.

Several computational fluid dynamics (CFD) studies have also been performed to analyze the performance of membrane reactors. Chompupun et al. [49] developed a three-dimensional model in COMSOL™ (a Multiphysics software platform) which includes momentum and energy balances, along with multicomponent diffusion effects and proposed scale-up strategies. They studied different geometric arrangements and suggested a square annular honeycomb monolith design with provision for simultaneous heat supply and hydrogen removal. The optimal performance was obtained when the product of Damkohler and Peclet numbers was close to unity, i.e., the rate of hydrogen production is similar to the rate of hydrogen permeation through the membrane. Their results showed a surface area-to-reactor volume requirement of 255 m^2/m^3 for an optimal geometry arrangement. Sanusi et al. [50] performed CFD simulations in the Ansys-Fluent platform. They studied the effects of reformer pressure, gas hourly space velocity, steam-to-methane ratio, and excess air factor in the combustion zone on the hydrogen yield and methane conversion efficiency. They showed that high pressure and low gas hourly space velocity favored higher hydrogen yield but resulted in a higher combustor exit temperature. A more recent CFD study by Upadhyay et al. [51] also studied the effect of various operating parameters, such as temperature, steam-to-carbon ratio, sweep gas flow configuration, and space velocity on the performance of a membrane reactor. They concluded that temperature is one of the most influencing operating parameters and suggested operation in the range 400–700°C.

Process flowsheeting and heat integration designs have also been developed and analyzed to compare different configurations for steam methane reforming. Bruni et al. [52] compared a two-step process that uses a traditional high-temperature reformer followed by a membrane reactor for water–gas shift reaction and a simple packed-bed membrane reactor. Both systems were optimized through a pinch analysis for heat recovery. They found similar performance for both plant configurations under their optimal performances. Pihcardo et al. [53] developed several flowsheet configurations depicting the use of a hybrid methane combustion-enhanced steam methane reforming and compared these to the heat-integrated flowsheets. They demonstrated a reduction in both heat load and the total work required for separation through these configurations. Franchi et al. [54] compared the membrane reactor and a two-stage reformer followed by the membrane configuration. A two-stage design helped in improving conversion and demonstrated an overall conversion of 92% and hydrogen recovery of ~85%. Although two different reactor and membrane modules may increase capital investment, a separate reactor and membrane operation provides more flexibility in the design. Kyriakides et al. [55] developed and implemented an integrated process design and control framework, considering economic and controller dynamic performance criteria simultaneously. They used the designs obtained from a multi-objective optimization study, where only the annualized equipment and operational costs are minimized as the reference. They also showed that the controller can track the imposed setpoint changes and was not affected by multiple simultaneous disturbances. The implementation of this framework also resulted in significant economic improvements.

11.7 SORPTION ENHANCED MEMBRANE REACTOR

Another advancement in the membrane reactor for steam methane reforming is carried out by integrating it with carbon dioxide adsorption, known as a sorption-enhanced membrane reactor (SEMR). This dual-enhanced reforming allows simultaneous hydrogen membrane separation and operation at lower temperature. Lee et al. [56] compared several conventional reactor configurations to membrane reactors with and without carbon dioxide sorption and concluded that simultaneous hydrogen separation and carbon dioxide sorption improves hydrogen yield (Figure 11.8).

Wu et al. [57] demonstrated this concept by using adsorbent and catalyst in one particle and thus achieving real "in situ" carbon dioxide sorption. This lowers the heat and mass transfer resistance during the reaction and sorption, which otherwise can be a challenge when using a mix of different adsorbent and catalyst particles. They exhibited a methane conversion rate of 91% at 600°C and 1.3 atm and a hydrogen purity of >98% with < 1000 ppm of carbon monoxide and showed the potential of this technique to obtain high-purity hydrogen. Anderson et al. [58,59] demonstrated this concept in a variable-volume membrane reactor at temperatures as low as 400°C. They also analyzed the time scale of different phenomena to explore the relationship between reactor component design characteristics and the rate-limiting steps of the process.

Ji et al. [60] performed computational fluid dynamic simulations to obtain dynamic and distributed information of reactants and products and compared it to a membrane reactor without sorption (Figure 11.9). The results showed higher hydrogen and provided greater insight into the reaction and separation process. These simulations can be used as a general tool to interpret reactor performance and perform parametric studies. One of the major challenges of this concept is regeneration of the sorbent which requires a batch operation or more complex moving bed or circulating fluidized bed continuous techniques. Moreover, the interaction of the sorbent particles with the catalyst and the membrane may affect the performance and stability of these operations. Further, the higher capital and operating costs due to the addition of sorbent and high-temperature (>850°C) regeneration needs to be evaluated against the advantages.

A recent techno-economic study [61] shows that the cost for a sorption-enhanced membrane reactor is 1.5× compared to a membrane reactor without sorption (Figure 11.10).

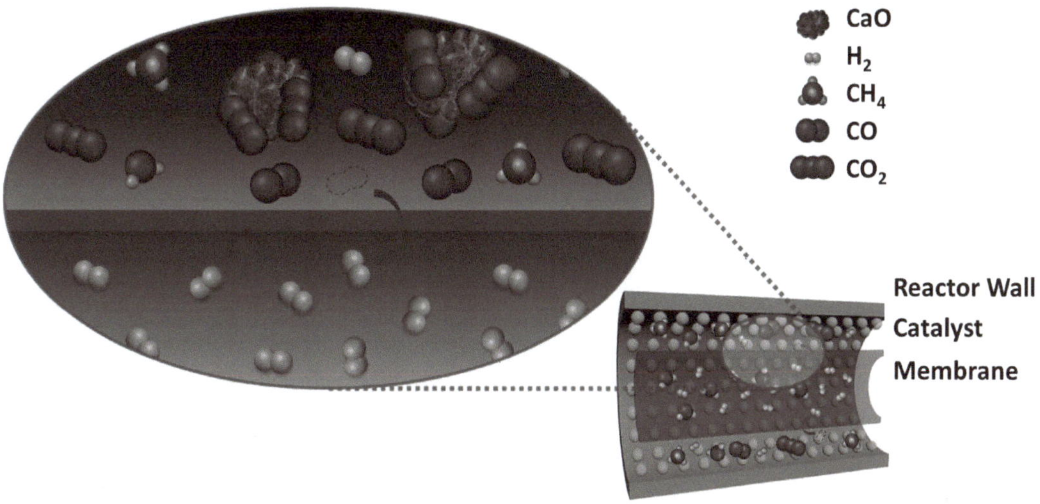

FIGURE 11.8 Illustration of phenomenon in a sorption-enhanced membrane reactor (SEMR). Reprinted from Ref. [56] with permission.

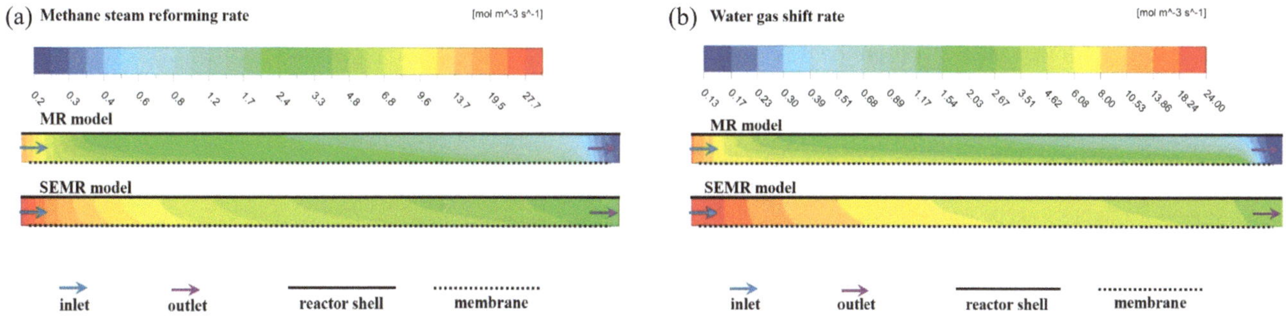

FIGURE 11.9 The distribution of (a) methane stream reforming rate, and (b) water gas shift rate in membrane reactor and sorption enhanced membrane reactor at the 100th second. Reprinted from Ref. [60] with permission.

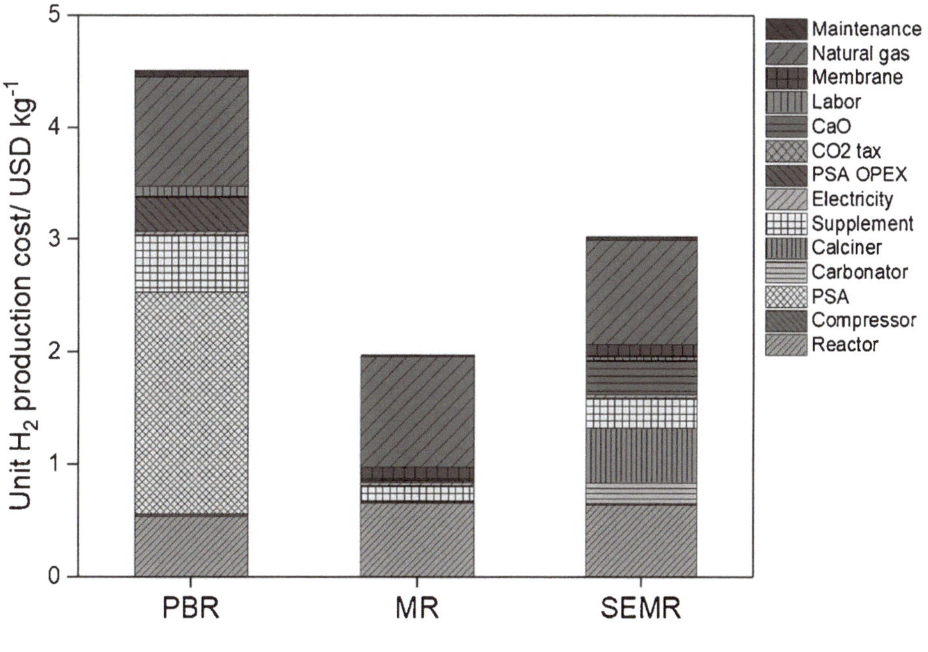

FIGURE 11.10 Itemized cost estimation results of a packed-bed reactor (PBR), a membrane reactor (MR), and a sorption enhanced-membrane reactor (SEMR). Reprinted from Ref. [61] with permission.

11.8 MEMBRANE REACTORS FOR AUTOTHERMAL REFORMING

Since steam reforming is a highly endothermic reaction, it is sometimes combined with partial oxidation of methane in the same reactor, commonly known as auto-thermal reforming (ATR). The thermal energy produced from the exothermic partial oxidation reaction is used by the endothermic steam reforming reaction. While the enthalpy change for reforming is lower, ATR requires oxygen production which adds to the investments. Although air can be used instead of O_2, the large amount of N_2 adds to the equipment cost and operation. The use of a membrane reactor can also enhance the ATR reforming.

Yan et al. [62,63] used a Pd-based membrane reactor to study hydrogen production by methane auto-thermal reforming. They analyzed the effect of temperature, air/methane feed mole ratio, steam/methane feed mole ratio, and the purge gas flow rate, and found improved performance with a membrane reactor, obtaining 99% methane conversion and 73% hydrogen yield. Further, the membrane reactor could be operated at ~100°C lower than a conventional reactor. One disadvantage of using air as an oxygen source is the large amount of N_2 that needs to be processed. In more recent articles [64,65], the application of chemical looping using oxygen carriers instead of air has been demonstrated. The authors used a novel concept where both catalytic and gas–solid reactions occur simultaneously or a single fluidized bed consisting of catalytically active oxygen-carrier particles goes through alternating exothermic and endothermic redox reaction by switching the feed between air and methane/steam and integrated with a Pd-based membranes for selective hydrogen separation. The lab-scale results showed 90% conversion with a hydrogen recovery of 30% and separation factors above 50% at temperature of ~600°C.

Techno-economic studies have also been performed to compare membrane-assisted chemical looping with traditional steam methane reforming [9]. While the novel concept achieved a higher hydrogen yield, economic analysis show ~2.2× higher cost than traditional steam methane reforming. The major challenge has been found to be a low utilization rate of membranes. Advancements in membrane performance are needed to make the overall process competitive with existing technologies.

Another novel configuration is proposed by Cloete et al. [66] where methane reforming is performed in lower parts of the fluidized reactor with hydrogen perm-selective membranes and autothermal oxidation is performed in the upper region by injecting high-purity oxygen from an air separation unit (Figure 11.11). Although this concept solves the challenge associated with achieving reliable solids circulation between large-scale looping and fuel reactors, the advantageous heat integration between reforming and autothermal oxidation cannot be realized. Nonetheless, the economic analysis shows the similar cost of this novel configuration and is an attractive alternate to membrane-assisted chemical looping reforming.

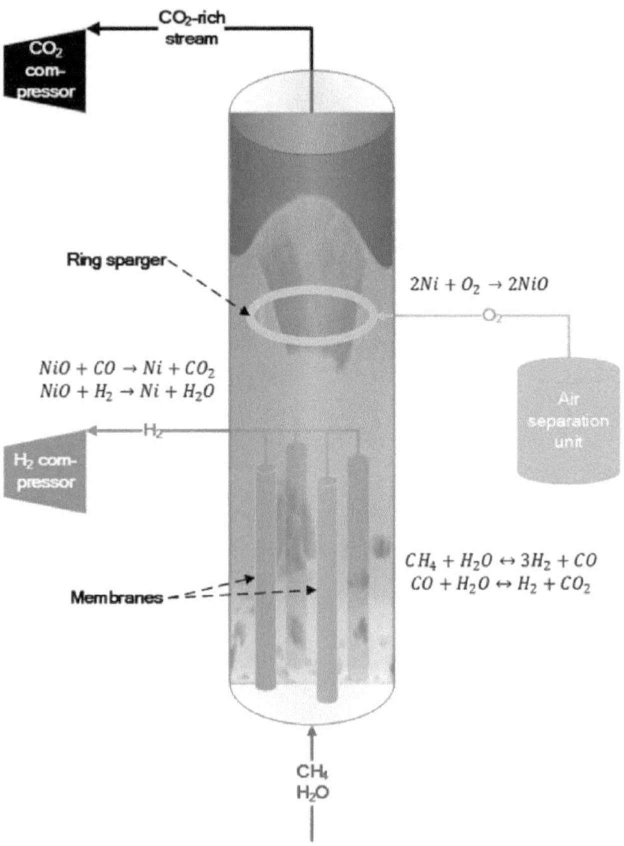

FIGURE 11.11 Illustration of the membrane-assisted ATR concept. The reactions indicate that reforming takes place in the lower regions around the membranes, oxygen carrier reduction with slipped fuel takes place directly above the membranes, and oxygen carrier oxidation with evenly injected oxygen from the ASU takes place in the upper regions. Reprinted from Ref. [66] with permission.

11.9 CARBON-DIOXIDE SELECTIVE MEMBRANES

Another way to enhance production of hydrogen is to use carbon dioxide-selective membranes. Since carbon dioxide is a larger molecule than hydrogen, separation mechanisms that favor solution/adsorption or surface diffusion are needed for high hydrogen purity [5]. Membranes of different kinds have been used, such as polymeric membranes with solution-diffusion, facilitated transport mechanism, mixed-matrix membranes, and porous inorganic membranes [67,68]. Lin and co-workers [69–71] recently developed an asymmetric ceramic-carbonate dual-phase membrane that shows great potential for high-temperature operation. A 90% hydrogen yield was achieved in a reactor demonstrating the water–gas shift conversion during methane reforming. Scanning electron microscope (SEM) and X-ray diffraction (XRD) analysis showed the good chemical stability of these membranes. Further, these membranes exhibited superior sulfur-resistant stability over a single-layered membrane and simultaneously removed traces of hydrogen. A model based on the membrane performance suggested significant benefits of operating the

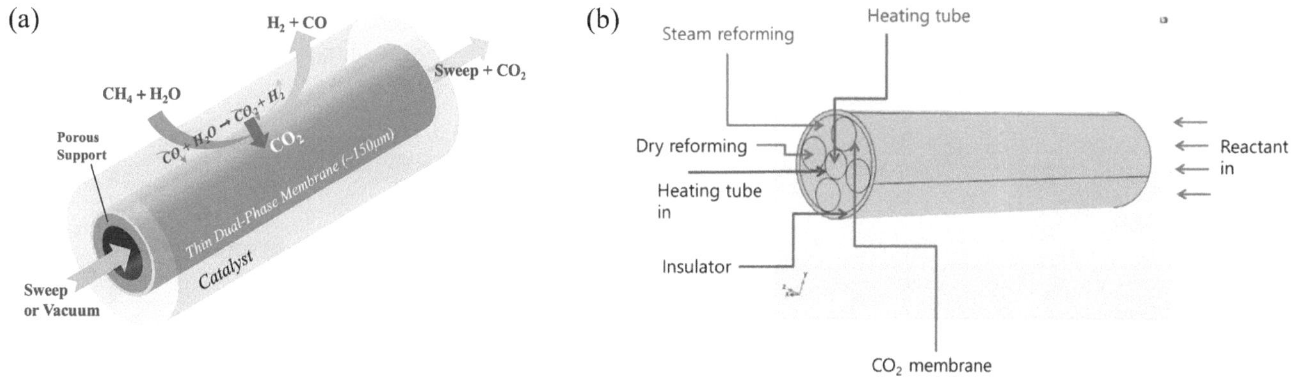

FIGURE 11.12 Illustration of (a) high-purity hydrogen production by dual-phase ceramic-carbonate membrane reactor via steam reforming of methane, (b) novel reactor concept of integrating steam and dry reforming of methane. Reprinted from Refs. [69,72] with permission.

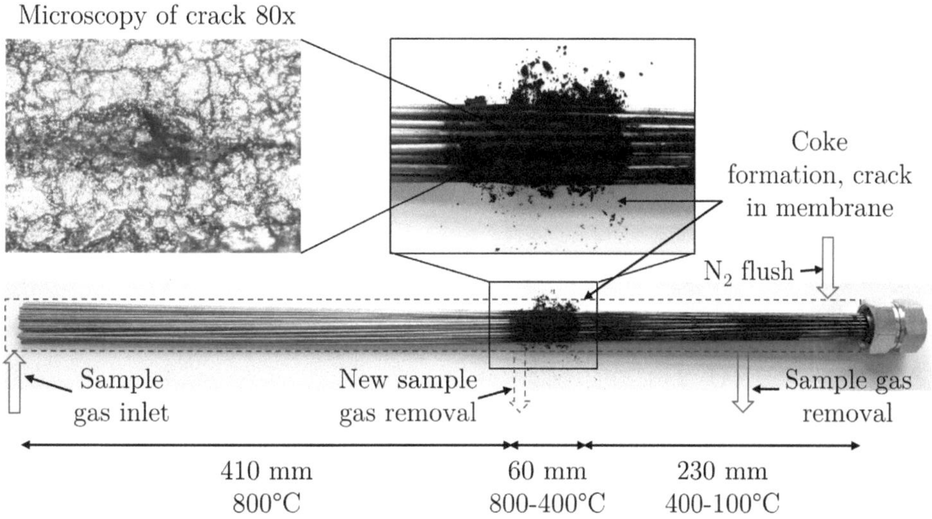

FIGURE 11.13 Photograph of coke formation on membrane bundle and microscopy of membrane crack after usage in dry reforming reaction at 800°C. Reprinted from Ref. [75], published under an open access Creative Common CC BY license.

permeate under a vacuum and predicted that essentially 100% methane conversion and hydrogen yield can be achieved in the membrane reactor at high Damkohler number. Lee et al. [72] demonstrated simultaneous steam methane reforming and dry reforming of methane during methane reforming, which was then used to drive dry reforming (see Section 11.10 and Figure 11.12). While they could only achieve low conversion, this CFD study shows the main advantage of this concept is that two different catalysts can be used. Although significant progress has been made in these membranes/membrane reactors in recent years, they are less selective and considerably less advanced than hydrogen membranes. More advancements are needed for their commercial application.

11.10 MEMBRANE REACTORS FOR DRY REFORMING OF METHANE

Dry reforming of methane has also received significant attention. A modeling study by Jokar et al. [73] compared steam reforming and dry reforming of methane using a Pd-Ag membrane reactor and concluded that dry reforming showed higher methane conversion, as well as higher hydrogen yield, even while the dry reforming membrane reactor could be operated at ~200°C lower than the steam reforming one. The modeling study projected lower operating cost and improved process reliability for the dry reforming of methane. In another modeling study, Kumar et al. [74] compared a membrane reactor to a fixed bed reactor for dry reforming. Their study also showed the membrane reactor to be superior, with 94.5% hydrogen and 98% carbon monoxide yields. Leimart et al. [75] used a nickel membrane for the generation of hydrogen from dry reforming. They achieved methane conversions of up to 70% without using an additional catalyst and 50–70% hydrogen recovery. The membrane showed coke deposit between 400–800°C and it thus requires operation at temperatures higher than 800°C for stable performance (Figure 11.13).

Lee and co-workers [76–78] performed CFD studies that compared dry reforming in membrane and packed bed reactors. These studies helped in three-dimensional

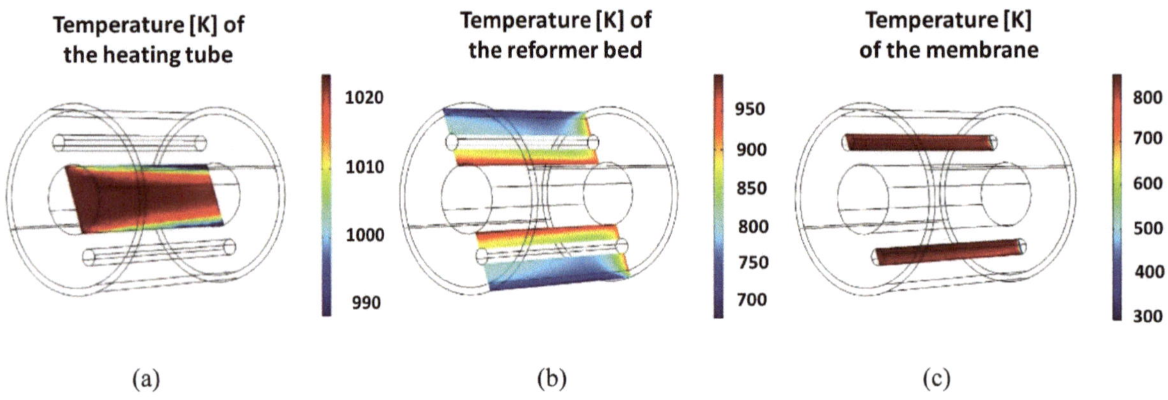

FIGURE 11.14 CFD results showing the longitudinal sectional temperatures [K] of the (a) heating tube, (b) reformer bed, and (c) membrane for hydrogen flux 5E-5 kg m^{-2} s^{-1} in two membranes. Reprinted from Ref. [78] with permission.

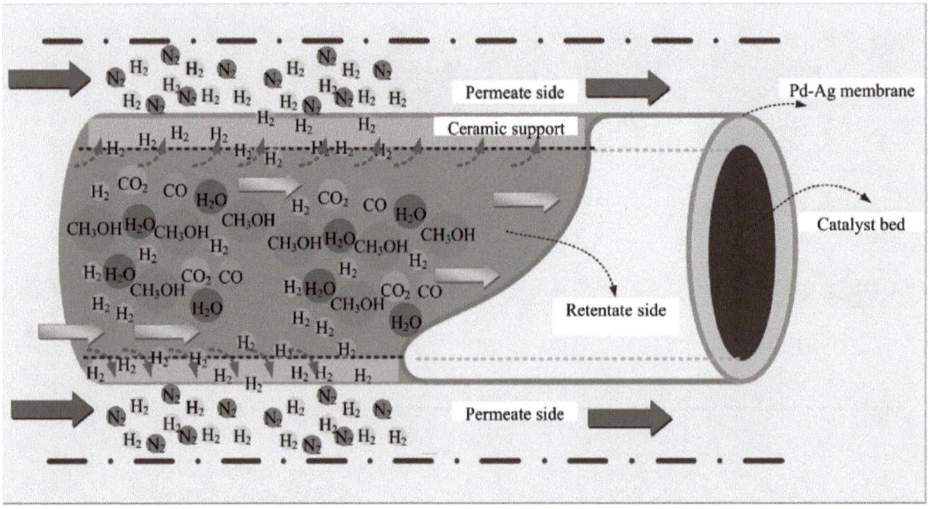

FIGURE 11.15 Illustration of a supported Pd-Ag membrane reactor for methanol steam reforming. Reprinted from Ref. [84], published under an open access Creative Common CC BY license.

visualization of temperature, component concentration, and the effect of rector geometry (Figure 11.14). The study also showed that a minimum hydrogen mass flux is required in a membrane reactor to outperform a packed-bed reactor and identified an optimum heating tube radius for maximum hydrogen yield. The computational fluid dynamics study also helps in providing guidelines for the proper dimensions of the membrane reactor and the number of membranes. Parametric studies on hydrogen permeance, sweep gas flow rate, operating temperature, and carbon dioxide/methane ratio suggested higher hydrogen permeance and a higher sweep gas flow rate enhanced hydrogen yield.

A techno-economic study of Pd membranes showed the potential for membrane reactors to outperform packed-bed reactors [79]. However, membrane reactors for dry reforming are less advanced than steam reforming, and the enhancements in conversion due to hydrogen separation are relatively minor. Typical conversion of methane and carbon dioxide did not exceed 50%. Further, other challenges such as undesired reverse water–gas shift reaction and catalyst and membrane deactivation due to extensive carbon formation also restrict the application of the membrane reactor for dry reforming.

11.11 MEMBRANE REACTORS FOR REFORMING OF LIQUID HYDROCARBONS

In recent years, other reforming and dehydrogenation reactions have been proposed for hydrogen production. Steam reforming of liquid hydrocarbon, specifically methanol, ethanol, and glycerol as feedstocks, has been studied due to several advantages, such as multiple sources of alcohol, safe and easy liquid transportation and storage, mild operating conditions, etc. [4,10].

Basile and co-workers [80–84] have extensively studied steam reforming of different alcohols. Steam reforming of methanol (Figure 11.15) performed in membrane reactors at increased temperature and pressure exhibits selective removal of hydrogen and enhanced hydrogen yield, while shifting the equilibrium toward higher methanol conversion. Steam reforming of methanol is carried out at a relatively low temperature of 200–350°C and a pressure of 1.2–25 bar. Steam reforming of ethanol is carried out at a relatively low temperature of 400–600°C and a pressure of 1–14 bar. For this system, increasing temperature enhances hydrogen production, while increasing pressure has a slight negative effect on ethanol conversion. For steam reforming of glycerol, membrane reactors can be more attractive than traditional reactors, and increasing temperature and pressure improves the hydrogen recovery and glycerol conversion.

11.12 MEMBRANE REACTORS FOR DEHYDROGENATION

Selective removal of hydrogen has attracted the attention of researchers to explore the application of membrane reactors in several dehydrogenation reactions also. Some of the widely researched reactions include dehydrogenation of cyclic and aromatic hydrocarbons, and alkanes to alkenes reactions. While the details and recent articles have been discussed in Ref. [4], a summary is presented here. Cyclohexane dehydrogenation is an endothermic reaction that produces benzene and hydrogen with equilibrium conversion of ~18.4%. However, almost complete conversion can be attained in a membrane reactor at 200°C. Both pressure and temperature have a positive effect due to the endothermic nature of the reaction and the increased permeation of hydrogen in a membrane reactor. Dehydrogenation of cyclohexane in a membrane reactor is typically conducted at 150–300°C and 1–3 bar and results in 100% conversion and 83–98% hydrogen recovery. Another commonly studied reaction is the dehydrogenation of methyl-cyclohexane to toluene and hydrogen. Like cyclohexane dehydrogenation, both temperature and pressure have a positive effect in a membrane reactor. The separation is typically carried out at a temperature of 220–400°C and pressure of 1–9 bar. Almost complete conversion of methyl-cyclohexane and > 95% hydrogen recovery have been realized with the application of a membrane reactor. Dehydrogenation of ethylbenzene is another application of membrane reactors to produce styrene and hydrogen. Both temperature and pressure have a positive effect, and the separation is typically carried out at a temperature of 500–650°C and pressure of 1–6 bar. Ethane and propane dehydrogenation are potential alternates to steam cracking to produce ethylene and propylene. While conversion is limited by equilibrium, the membrane reactor provides a promising solution to this challenge. Both pressure and temperature have positive effects on ethane and propane conversion due to the endothermic nature of dehydrogenation and increased permeation of hydrogen in a membrane reactor. Ethane dehydrogenation in a membrane reactor is carried out at 350–600°C and 1–5 bar, while dehydrogenation of propane to propylene takes place at 450–600°C and 1–6 bar. While hydrogen is a side-product of these dehydrogenation reactions, the development of such processes is mainly motivated by the chemical rather than hydrogen production. However, the application of membrane reactors would also simultaneously progress the field for other applications such as methane reforming.

11.13 CHALLENGES AND OUTLOOK

Although steam methane reforming should continue to be a dominant industrial process for hydrogen production, process intensification opportunities will become relevant for hydrogen production. The challenges will however spawn significant opportunities for the introduction of innovative membrane processes and technologies [7,8]. Membrane separation is already a mature technology for hydrogen purification and recovery from gas streams in refineries, syngas process, ammonia and methanol production, etc. Polymeric membranes are commercially sold by several vendors including Air Liquide ALaS™ membranes, Air Products Prism™ membranes, Evonik Sepuran™ membranes, MTR Vaporsep-H2™ membranes, etc. [85–88]. Several high-purity metallic membranes are now also available for commercial applications. The use of membranes in a membrane reactor is the next paradigm in the hydrogen generation. Media and Process Technology Inc. performed a field test using pilot-scale carbon molecular sieves membrane reactors for hydrogen production and successfully demonstrated the membrane stability and resistance to fouling [28]. Praxair Inc., now Linde Inc., has made great progress in the scale-up of oxygen transport membrane (OTM) reactors with applications in producing syngas for methanol and in an integrated gasification combined cycle (IGCC) [89–91]. CoorsTek Membrane Sciences have made process innovation by introducing an interconnect in proton ceramic membrane reactors to achieve complete conversion of methane with pressurized hydrogen recovery of > 99% [92]. These advances have further supported the investment and development of membrane reactor technology in hydrogen production, albeit technical challenges remain [93].

Pd-membrane-based membrane reactor technologies have shown promise for hydrogen process intensification applications, but the current success stories are limited to the lab-scale [32–41]. The membranes should be tested under actual industrial feed and process conditions in long-time operations. Efforts to improve the support layer in terms of its permeation and the integration with the selective Pd-layer are other areas of research to improve the membrane performance [40]. The cost of Pd membranes is another challenge against the industrial realization of this technology [8]. Pd alloys, especially non-precious metals, with enhanced sulfur resistance and thermal stability can decrease the cost with improved performance [8]. More computational and experimental efforts can efficiently explore a wide space and screen the relevant alloys.

Process intensification is always challenging as it often reduces the degrees of freedom and requires the two processes to occur simultaneously and synergistically. The use of a membrane reactor also requires the right selection of the most adequate and compatible catalyst with the membrane [7,8]. From a process perspective, the rates of reaction and separation should be of the same order to effectively utilize the capacity of both the catalyst and the membrane area [49]. Current studies have shown several challenges suggesting the presence of concentration polarization and pinch point in terms of hydrogen permeation [39,40]. Although the use of sweep gas has been proposed, it adds another separation step and reduces the process efficiency. Optimal module design studies and engineering analysis can be performed to tackle these challenges. Multiphysics simulations combined with representative geometries can play a significant role in predicting and validating the process hydrodynamics.

Integrated techno-environmental assessment for different hydrogen production pathways can help set realistic targets for membrane reactor applications [94]. Implementation at several scales can be evaluated – within the scope of both the large-scale and compact membrane reactors. The process designs implemented for conventional SMR would be different than that for membrane reactor SMR because of the different composition and process conditions and requires a holistic assessment [52,61]. Detailed uncertainty and sensitivity analyses can assess the favorable scenarios for membrane and membrane reactor technologies. These process systems studies could be instrumental in decision-making at several levels.

11.14 CONCLUDING REMARKS

Hydrogen has emerged as one of the promising solutions to address the growing energy demand. The current state-of-the-art methods for hydrogen production, i.e., reversible hydrocarbon reforming/dehydrogenation and water–gas shift reactions in conventional reactors are limited by equilibrium with modest opportunity for heat integration. Membranes and membrane reactor technology can provide simultaneous and enhanced product separation. At the same time, this can improve the conversion beyond equilibrium. Several novel concepts and other applications of membrane reactors for hydrogen production have been explored and further studies and pilot plant trials can pave the way for future developments.

REFERENCES

[1] IEA (2021) *Global Hydrogen Review 2021 – Analysis – IEA*. www.iea.org/reports/global-hydrogen-review-2021 (accessed May 9, 2022).

[2] Dou, Y., Sun, L., Ren, J., Dong, L. (2017) Opportunities and future challenges in hydrogen economy for sustainable development. In: *Hydrogen Economy: Supply Chain, Life Cycle Analysis and Energy Transition for Sustainability*, pp. 277–305. Academic Press.

[3] Liguori, S., Kian, K., Buggy, N., Anzelmo, B.H., Wilcox, J. (2020) Opportunities and challenges of low-carbon hydrogen via metallic membranes. *Progress in Energy and Combustion Science* 80: 100851.

[4] Mamivand, S., Binazadeh, M., Sohrabi, R. (2021) Applicability of membrane reactor technology in industrial hydrogen producing reactions: Current effort and future directions. *Journal of Industrial and Engineering Chemistry* 104: 212–230.

[5] Voldsund, M., Jordal, K., Anantharaman, R. (2016) Hydrogen production with CO_2 capture. *International Journal of Hydrogen Energy* 41(9): 4969–4992.

[6] IEA (2015) *Energy Technology Perspectives 2015 – Analysis – IEA*. www.iea.org/reports/energy-technology-perspectives-2015 (accessed May 9, 2022).

[7] Basile, A., Liguori, S., Iulianelli, A. (2015) Membrane reactors for methane steam reforming (MSR). *Membrane Reactors for Energy Applications and Basic Chemical Production*, pp. 31–59. Woodhead Publishing.

[8] Iulianelli, A., Liguori, S., Wilcox, J., Basile, A. (2016) Advances on methane steam reforming to produce hydrogen through membrane reactors technology: A review. *Catalysis Reviews*. 58(1): 1–35.

[9] Wassie, S.A., et al. (2018) Techno-economic assessment of membrane-assisted gas switching reforming for pure hydrogen production with CO_2 capture. *International Journal of Greenhouse Gas Control* 72: 163–174.

[10] Amiri, T.Y., Ghasemzageh, K., Iulianelli, A. (2020) Membrane reactors for sustainable hydrogen production through steam reforming of hydrocarbons: A review. *Chemical Engineering and Processing – Process Intensification* 157: 108148.

[11] Lei, L., Lindbråthen, A., Hillestad, M., He, X. (2021) Carbon molecular sieve membranes for hydrogen purification from a steam methane reforming process. *Journal of Membrane Science* 627: 119241.

[12] Sazali, N. (2020) A comprehensive review of carbon molecular sieve membranes for hydrogen production and purification. *The International Journal of Advanced Manufacturing Technology* 107(5): 2465–2483.

[13] Rungta, M., *et al.* (2017) Carbon molecular sieve structure development and membrane performance relationships. *Carbon N Y* 115: 237–248.

[14] Ogieglo, W., Puspasari, T., Hota, M.K., Wehbe, N., Alshareef, H.N., Pinnau, I. (2020) Nanohybrid thin-film composite carbon molecular sieve membranes. *Materials Today Nano* 9: 100065.

[15] Lei, L., et al. (2020) Preparation of carbon molecular sieve membranes with remarkable CO_2/CH_4 selectivity for high-pressure natural gas sweetening. *Journal of Membrane Science* 614: 118529.

[16] Lei, L., Lindbråthen, A., Hillestad, M., Sandru, M., Favvas, E.P., He, X. (2019) Screening cellulose spinning parameters for fabrication of novel carbon hollow fiber membranes for gas separation. *Industrial and Engineering Chemistry Research* 58(29): 13330–13339.

[17] He, X., Hägg, M.B. (2013) Hollow fiber carbon membranes: From material to application. *Chemical Engineering Journal* 215–216: 440–448.

[18] Bhuwania, N., et al. (2014) Engineering substructure morphology of asymmetric carbon molecular sieve hollow fiber membranes. *Carbon N Y* 76: 417–434.

[19] Cao, Y., Zhang, K., Sanyal, O., Koros, W.J. (2019) Carbon molecular sieve membrane preparation by economical coating and pyrolysis of porous polymer hollow fibers. *Angewandte Chemie International Edition* 58(35): 12149–12153.

[20] Fu, S., Sanders, E.S., Kulkarni, S.S., Koros, W.J. (2015) Carbon molecular sieve membrane structure–property

relationships for four novel 6FDA based polyimide precursors. *Journal of Membrane Science* 487: 60–73.

[21] Koh, D.Y., McCool, B.A., Deckman, H.W., Lively, R.P. (2016) Reverse osmosis molecular differentiation of organic liquids using carbon molecular sieve membranes. *Science* 353(6301): 804–807.

[22] David, L.I.B., Ismail, A.F. (2003) Influence of the thermastabilization process and soak time during pyrolysis process on the polyacrylonitrile carbon membranes for O_2/N_2 separation. *Journal of Membrane Science* 213(1–2): 285–291.

[23] Inagaki, M., Ohta, N., Hishiyama, Y. (2013) Aromatic polyimides as carbon precursors. *Carbon N Y* 61: 1–21.

[24] Rao, M.B., Sircar, S. (1996) Performance and pore characterization of nanoporous carbon membranes for gas separation. *Journal of Membrane Science* 110: 109–118.

[25] Rao, M.B., Sircar, S. (1993) Nanoporous carbon membranes for separation of gas mixtures by selective surface flow. *Journal of Membrane Science* 85: 253–264.

[26] Cao, M., et al. (2020) A carbon molecular sieve membrane-based reactive separation process for pre-combustion CO_2 capture. *Journal of Membrane Science* 605: 118028.

[27] Harale, A., Hwang, H.T., Liu, P.K.T., Sahimi, M., Tsotsis, T.T. (2007) Experimental studies of a hybrid adsorbent-membrane reactor (HAMR) system for hydrogen production. *Chemical Engineering Science* 62(15): 4126–4137.

[28] Parsley, D., et al. (2014) Field evaluation of carbon molecular sieve membranes for the separation and purification of hydrogen from coal- and biomass-derived syngas. *Journal of Membrane Science* 450: 81–92.

[29] Llosa Tanco, M.A., Pacheco Tanaka, D.A., Mendes, A. (2015) Composite-alumina-carbon molecular sieve membranes prepared from novolac resin and boehmite. Part II: Effect of the carbonization temperature on the gas permeation properties. *International Journal of Hydrogen Energy* 40(8): 3485–3496.

[30] Zhang, C., Koros, W.J. (2017) Ultraselective carbon molecular sieve membranes with tailored synergistic sorption selective properties. *Advanced Materials* 29(33): 1701631.

[31] Liguori, A., Morrone, S., Basile, P., Iulianelli, A. (2012) Membrane and membrane reactor technologies for COx purification of gaseous streams. Chapter 6. In: *Advances in Chemistry Research* 16.

[32] Sarić, M., van Delft, Y.C., Sumbharaju, R., Meyer, D.F., de Groot, A. (2012) Steam reforming of methane in a bench-scale membrane reactor at realistic working conditions. *Catalysis Today* 193(1): 74–80.

[33] Dittmar, B., et al. (2013) Methane steam reforming operation and thermal stability of new porous metal supported tubular palladium composite membranes. *International Journal of Hydrogen Energy* 38(21): 8759–8771.

[34] Patil, C.S., van Sint Annaland, M., Kuipers, J.A.M. (2007) Fluidised bed membrane reactor for ultrapure hydrogen production via methane steam reforming: Experimental demonstration and model validation. *Chemical Engineering Science* 62(11): 2989–3007.

[35] Vigneault, A., Grace, J.R. (2015) Hydrogen production in multi-channel membrane reactor via steam methane reforming and methane catalytic combustion. *International Journal of Hydrogen Energy* 40(1): 233–243.

[36] Patrascu, M., Sheintuch, M. (2015) On-site pure hydrogen production by methane steam reforming in high flux membrane reactor: Experimental validation, model predictions and membrane inhibition. *Chemical Engineering Journal* 262: 862–874.

[37] Abu El Hawa, H.W., Paglieri, S.N., Morris, C.C., Harale, A., Douglas Way, J. (2015) Application of a Pd–Ru composite membrane to hydrogen production in a high temperature membrane reactor. *Separation and Purification Technology* 147: 388–397.

[38] Kim, C.H., et al. (2018) Hydrogen production by steam methane reforming in a membrane reactor equipped with a Pd composite membrane deposited on a porous stainless steel. *International Journal of Hydrogen Energy* 43(15): 7684–7692.

[39] Kim, C.H., Han, J.Y., Lim, H., Lee, K.Y., Ryi, S.K. (2018) Methane steam reforming using a membrane reactor equipped with a Pd-based composite membrane for effective hydrogen production. *International Journal of Hydrogen Energy* 43(11): 5863–5872.

[40] Anzelmo, B., et al. (2018) Fabrication & performance study of a palladium on alumina supported membrane reactor: Natural gas steam reforming, a case study. *International Journal of Hydrogen Energy* 43(15): 7713–7721.

[41] Wang, M., Tan, X., Motuzas, J., Li, J., Liu, S. (2021) Hydrogen production by methane steam reforming using metallic nickel hollow fiber membranes. *Journal of Membrane Science* 620: 118909.

[42] Silva, J.D., de Abreu, C.A.M. (2016) Modelling and simulation in conventional fixed-bed and fixed-bed membrane reactors for the steam reforming of methane. *International Journal of Hydrogen Energy* 41(27): 11660–11674.

[43] de Medeiros, J.P.F., da Fonseca Dias, V., da Silva, J.M., da Silva, J.D. (2020) Thermochemical performance analysis of the steam reforming of methane in a fixed bed membrane reformer: A modelling and simulation study. *Membranes (Basel)* 11(1): 1–26.

[44] Cruz, B.M., da Silva, J.D. (2017) A two-dimensional mathematical model for the catalytic steam reforming of methane in both conventional fixed-bed and fixed-bed membrane reactors for the production of hydrogen. *International Journal of Hydrogen Energy* 42(37): 23670–23690.

[45] Murmura, M.A., Cerbelli, S., Annesini, M.C. (2017) An equilibrium theory for catalytic steam reforming in membrane reactors. *Chemical Engineering Science* 160: 291–303.

[46] Murmura, M.A., Cerbelli, S., Annesini, M.C. (2017) Transport-reaction-permeation regimes in catalytic membrane reactors for hydrogen production. The steam reforming of methane as a case study. *Chemical Engineering Science* 162: 88–103.

[47] Murmura, M.A., Cerbelli, S., Annesini, M.C., Sheintuch, M. (2021) Derivation of an enhanced Sherwood number accounting for reaction rate in membrane reactors. Steam reforming of methane as case study. *Catalysis Today* 364: 285–293.

[48] Leonzio, G. (2019) ANOVA analysis of an integrated membrane reactor for hydrogen production by methane steam reforming. *International Journal of Hydrogen Energy* 44(23): 11535–11545.

[49] Chompupun, T., Limtrakul, S., Vatanatham, T., Kanhari, C., Ramachandran, P.A. (2018) Experiments, modeling and scaling-up of membrane reactors for hydrogen production via steam methane reforming. *Chemical Engineering and Processing – Process Intensification* 134: 124–140.

[50] Sanusi, Y.S., Mokheimer, E.M.A. (2019) Performance analysis of a membrane-based reformer-combustor reactor for hydrogen generation. *International Journal of Energy Research* 43(1): 189–203.

[51] Upadhyay, M., Lee, H., Kim, A., hun Lee, S., Lim, H. (2021) CFD simulation of methane steam reforming in a membrane reactor: Performance characteristics over range of operating window. *International Journal of Hydrogen Energy* 46(59): 30402–30411.

[52] Bruni, G., Rizzello, C., Santucci, A., Alique, D., Incelli, M., Tosti, S. (2019) On the energy efficiency of hydrogen production processes via steam reforming using membrane reactors. *International Journal of Hydrogen Energy* 44(2): 988–999.

[53] Pichardo, P.A., Manousiouthakis, V.I. (2020) Intensified energetically enhanced steam methane reforming through the use of membrane reactors. *AIChE Journal* 66(2): e16827.

[54] Franchi, G., Capocelli, M., de Falco, M., Piemonte, V., Barba, D. (2020) Hydrogen production via steam reforming: A critical analysis of MR and RMM technologies. *Membranes* 10(1): 10.

[55] Kyriakides, A.S., Voutetakis, S., Papadopoulou, S., Seferlis, P. (2019) Integrated design and control of various hydrogen production flowsheet configurations via membrane based methane steam reforming. *Membranes* 9(1): 14.

[56] Lee, H., Kim, A., Lee, B., Lim, H. (2020) Comparative numerical analysis for an efficient hydrogen production via a steam methane reforming with a packed-bed reactor, a membrane reactor, and a sorption-enhanced membrane reactor. *Energy Conversion and Management* 213: 112839.

[57] Wu, X., Wu, C., Wu, S. (2015) Dual-enhanced steam methane reforming by membrane separation of hydrogen and reactive sorption of CO_2. *Chemical Engineering Research and Design* 96: 150–157.

[58] Anderson, D.M., Yun, T.M., Kottke, P.A., Fedorov, A.G. (2017) Comprehensive analysis of sorption enhanced steam methane reforming in a variable volume membrane reactor. *Industrial and Engineering Chemistry Research* 56(7): 1758–1771.

[59] Anderson, D.M., Nasr, M.H., Yun, T.M., Kottke, P.A., Fedorov, A.G. (2015) Sorption-enhanced variable-volume batch-membrane steam methane reforming at low temperature: Experimental demonstration and kinetic modeling. *Industrial and Engineering Chemistry Research* 54(34): 8422–8436.

[60] Ji, G., Zhao, M., Wang, G. (2018) Computational fluid dynamic simulation of a sorption-enhanced palladium membrane reactor for enhancing hydrogen production from methane steam reforming. *Energy* 147: 884–895.

[61] Lee, H., Lee, B., Byun, M., Lim, H. (2021) Comparative techno-economic analysis for steam methane reforming in a sorption-enhanced membrane reactor: Simultaneous hydrogen production and CO_2 capture. *Chemical Engineering Research and Design* 171: 383–394.

[62] Yan, Y., Li, H., Li, L., Zhang, L., Zhang, J. (2018) Properties of methane autothermal reforming to generate hydrogen in membrane reactor based on thermodynamic equilibrium model. *Chemical Engineering and Processing – Process Intensification* 125: 311–317.

[63] Yan, Y., et al. (2016) Experimental investigation of methane auto-thermal reforming in hydrogen-permeable membrane reactor for pure hydrogen production. *International Journal of Hydrogen Energy* 41(30): 13069–13076.

[64] Medrano, J.A., et al. (2018) The membrane-assisted chemical looping reforming concept for efficient hydrogen production with inherent CO_2 capture: Experimental demonstration and model validation. *Applied Energy* 215: 75–86.

[65] Wassie, S.A., et al. (2018) Hydrogen production with integrated CO_2 capture in a membrane assisted gas switching reforming reactor: Proof-of-concept. *International Journal of Hydrogen Energy* 43(12): 6177–6190.

[66] Cloete, S., Khan, M.N., Amini, S. (2019) Economic assessment of membrane-assisted autothermal reforming for cost effective hydrogen production with CO_2 capture. *International Journal of Hydrogen Energy* 44(7): 3492–3510.

[67] Scholes, C.A., Smith, K.H., Kentish, S.E., Stevens, G.W. (2010) CO_2 capture from pre-combustion processes—Strategies for membrane gas separation. *International Journal of Greenhouse Gas Control* 4(5): 739–755.

[68] Ramasubramanian, K., Zhao, Y., Winston Ho, W.S. (2013) CO_2 capture and hydrogen purification: Prospects for CO_2-selective membrane processes. *AIChE Journal* 59(4): 1033–1045.

[69] Wu, H.C., Rui, Z., Lin, J.Y.S. (2020) Hydrogen production with carbon dioxide capture by dual-phase ceramic-carbonate membrane reactor via steam reforming of methane. *Journal of Membrane Science* 598: 117780.

[70] Ovalle-Encinia, O., Wu, H.C., Chen, T., Lin, J.Y.S. (2022) CO_2-permselective membrane reactor for steam reforming of methane. *Journal of Membrane Science* 641: 119914.

[71] Chen, T., et al. (2019) A novel study of sulfur-resistance for CO_2 separation through asymmetric ceramic-carbonate dual-phase membrane at high temperature. *Journal of Membrane Science* 581: 72–81.

[72] Lee, S., Lim, h. (2020) Utilization of CO_2 arising from methane steam reforming reaction: Use of CO_2 membrane and heterotic reactors. *Journal of Industrial and Engineering Chemistry* 91: 201–212.

[73] Jokar, S.M., Parvasi, P., Basile, A. (2018) The evaluation of methane mixed reforming reaction in an industrial membrane reformer for hydrogen production. *International Journal of Hydrogen Energy* 43(32): 15321–15329.

[74] Kumar, S., Kumar, B., Kumar, S., Jilani, S. (2017) Comparative modeling study of catalytic membrane reactor configurations for syngas production by CO_2 reforming of methane. *Journal of CO_2 Utilization* 20: 336–346.

[75] Leimert, J.M., Karl, J., Dillig, M. (2017) Dry reforming of methane using a nickel membrane reactor. *Processes* 5(4): 82.

[76] Lee, B., Lee, S., Lim, H. (2016) Numerical modeling studies for a methane dry reforming in a membrane reactor. *Journal of Natural Gas Science and Engineering* 34: 1251–1261.

[77] Lee, B., et al. (2019) CO_2 reforming of methane for hydrogen production in a membrane reactor as CO_2 utilization: Computational fluid dynamics studies with a reactor geometry. *International Journal of Hydrogen Energy* 44(4): 2298–2311.

[78] Lee, S., Lim, H. (2020) The effect of changing the number of membranes in methane carbon dioxide reforming: A CFD

study. *Journal of Industrial and Engineering Chemistry* 87: 110–119.

[79] Kim, S., Ryi, S.K., Lim, H. (2018) Techno-economic analysis (TEA) for CO_2 reforming of methane in a membrane reactor for simultaneous CO_2 utilization and ultra-pure hydrogen production. *International Journal of Hydrogen Energy* 43(11): 5881–5893.

[80] Ghasemzadeh, K., Ghahremani, M., Amiri, T.Y., Basile, A. (2019) Performance evaluation of PdAg membrane reactor in glycerol steam reforming process: Development of the CFD model. *International Journal of Hydrogen Energy* 44(2): 1000–1009.

[81] Basile, A., et al. (2011) Ethanol steam reforming reaction in a porous stainless steel supported palladium membrane reactor. *International Journal of Hydrogen Energy* 36(3): 2029–2037.

[82] Gallucci, F., de Falco, M., Tosti, S., Marrelli, L., Basile, A. (2008) Co-current and counter-current configurations for ethanol steam reforming in a dense Pd-Ag membrane reactor. *International Journal of Hydrogen Energy* 33(21): 6165–6171.

[83] Dalena, F., Senatore, A., Basile, M., Knani, S., Basile, A., Iulianelli, A. (2018) Advances in methanol production and utilization, with particular emphasis toward hydrogen generation via membrane reactor technology. *Membranes (Basel)* 8(4): 1–27.

[84] Iulianelli, A., Ghasemzadeh, K., Basile, A. (2018) Progress in methanol steam reforming modelling via membrane reactors technology. *Membranes (Basel)* 8(3): 1–22.

[85] Air Liquide Advanced Separations (ALaS ™) www.airliquideadvancedseparations.com/our-membranes/hydrogen

[86] Air Products PRISM ™ Membrane www.airproducts.com/supply-modes/prism-membranes

[87] Evonik's SEPURAN ™ Noble membrane www.membrane-separation.com/en/hydrogen/recovery-with-sepuran-noble

[88] MTR VAPORSEP-H2™ membranes www.mtrinc.com/our-business/refinery-and-syngas/hydrogen-separations-in-syngas-processes/

[89] Li, J., et al. (2018) Advanced gasification and novel transformational coal conversion technologies development. *DOE/NETL Cooperative Agreement DE-FE0023543*, www.osti.gov/servlets/purl/1474434

[90] Li, J., et al. (2017) OTM-enhanced coal syngas for carbon capture power systems and fuel synthesis applications. *Project Review Meeting for Crosscutting Research, Gasification Systems, and Rare Earth Elements Research Portfolios*, March 20–23, https://netl.doe.gov/sites/default/files/event-proceedings/2017/crosscutting/20170320-Track-C/20170320_1300C_Presentation_FE0023543_Praxair.pdf

[91] Li, J., et al. (2015) Praxair's oxygen transport membrane technology for syngas and power applications. *Gasification Systems and Coal & Coal-Biomass to Liquids Workshop*, August 10–11, https://netl.doe.gov/sites/default/files/event-proceedings/2015/gas-ccbtl-proceedings/Gasification-and-C-CBTL-Conference-OTM-Presentation-2015.pdf

[92] Clark, D.,et al. (2022) Single-step hydrogen production from NH_3, CH_4, and biogas in stacked proton ceramic reactors. *Science* 376: 390.

[93] Schwartz, J., Lim, H., Drnevich, R. (2010) *Integrated Ceramic Membrane System for Hydrogen Production.* OSTI 984651, United States. www.osti.gov/servlets/purl/984651.

[94] Kannah, R.Y., et al. (2021) Techno-economic assessment of various hydrogen production methods – A review. *Bioresource Technology* 319: 124175.

12 Advances and Applications of Ionic Liquids for the Extraction of Organics and Metal Ions

A. Hernández-Fernandez[1], V.M. Ortiz-Martínez[2], A.P. de los Ríos[1,], F.J. Hernández-Fernández[1], and S. Sánchez-Segado[2]*

[1]Department of Chemical Engineering, Faculty of Chemistry, University of Murcia (UMU), Campus de Espinardo, Murcia, Spain

[2]Department of Chemical and Environmental Engineering, Technical University of Cartagena, Campus La Muralla, Cartagena, Murcia, Spain

*Corresponding author. E-mail address: aprios@um.es

12.1 IONIC LIQUIDS: GENERAL PROPERTIES

Ionic liquids (ILs) have recently drawn much attention in the scientific community because of their unique and highly tunable properties. ILs are ionic compounds formed by an anion and a cation with melting points below 100°C [1]. A wide range of ILs are in the liquid state at ambient temperature and are called RTILs (room temperature ionic liquids) [2]. The cation usually consists of a large organic compound, while the anion is generally smaller in volume and of inorganic nature, although organic anions can also be found in IL structures [3]. Among the organic cations are imidazolium, pyridinium, ammonium, pyrrolidinium, and phosphonium groups. Common inorganic anions include halide anions (Br^-, Cl^-, I^-), tetrafluoroborate, hexafluoroborate, dihydrogen phosphate, and others, while the most common organic anions as IL constituents are bis[(trifluoromethyl)sulfonyl]imide or trifluoromethyl sulfonate [4,5]. The chemical structures of several of these typical anions and cations are presented in Figure 12.1.

Unlike common salts, ILs display a lower tendency for crystallization as the cation structure presents an asymmetrical and bulky form [6]. Due to the high number of possible combinations between different anions and cations, ILs are called "designer solvents," and thus their properties can be suitably tailored for specific applications [7]. Other advantageous properties include non-volatility with negligible vapor pressure even at high temperatures, high chemical and thermal stability (generally > 300°C), high ionic conductivity, non-flammability, and recyclability [8,9].

The density and viscosity of ILs are higher than for conventional solvents by from one to three orders of magnitude, which is the result of electrostatic and van der Waals forces as well as hydrogen-bonding capacity [10]. The range of viscosity varies from 10 to 500 mPa/s. These features must be considered in terms of molecular transport and diffusion. As in the case of other properties, the tuning of chemical structure can lead to desirable physicochemical properties in terms of viscosity and density [9]. ILs have been shown to be efficient solvents for compounds of both inorganic and organic nature, from small and simple molecules to polymeric materials, which highlights their versatility [11,12].

Due to their volatile features, organic solvents pose higher environmental risks than ILs. The toxicity of ILs is under continuous study and several works have shown that there is a structure–toxicity relationship that can be used for the sustainable design of low-toxicity ILs. For instance, it has been observed that a rise in the length of alkyl chains and thus in lipophilicity is related to an increase in toxicity [13]. On the other hand, the addition of functional polar groups into the alkyl chains of ILs can help to decrease the toxicity effects on biological systems [14]. In addition, research strategies have been developed in order to create ILs with better biodegradability, for example, by introducing ester moieties in imidazolium compounds [15], or by designing new biodegradable ILs [16].

The number of applications of ILs has been constantly rising over the last two decades. They have been greatly studied in many chemical and engineering fields. To mention some examples, ILs have been researched as reaction media [17], in separation and extraction [18], catalysis [19] and biocatalysis [20], material synthesis [21], and energy and electrochemical systems [22,23]

Within separation technology, ILs can be used as a neat liquid phase in liquid–liquid extraction processes [24]. On the other hand, ILs can be efficiently immobilized into polymeric materials and be used to modify the solid supports in order to exploit IL capability in the adsorptive removal processes. Membranes using ILs as liquid phase offer enormous potential in a broad range of applications [25]. The possible configurations include supported IL membranes (SLIMs), in which a support is filled with ILs, polymer/IL blends to form a solid membrane by casting methods or polymer inclusion

DOI: 10.1201/9781003285656-12

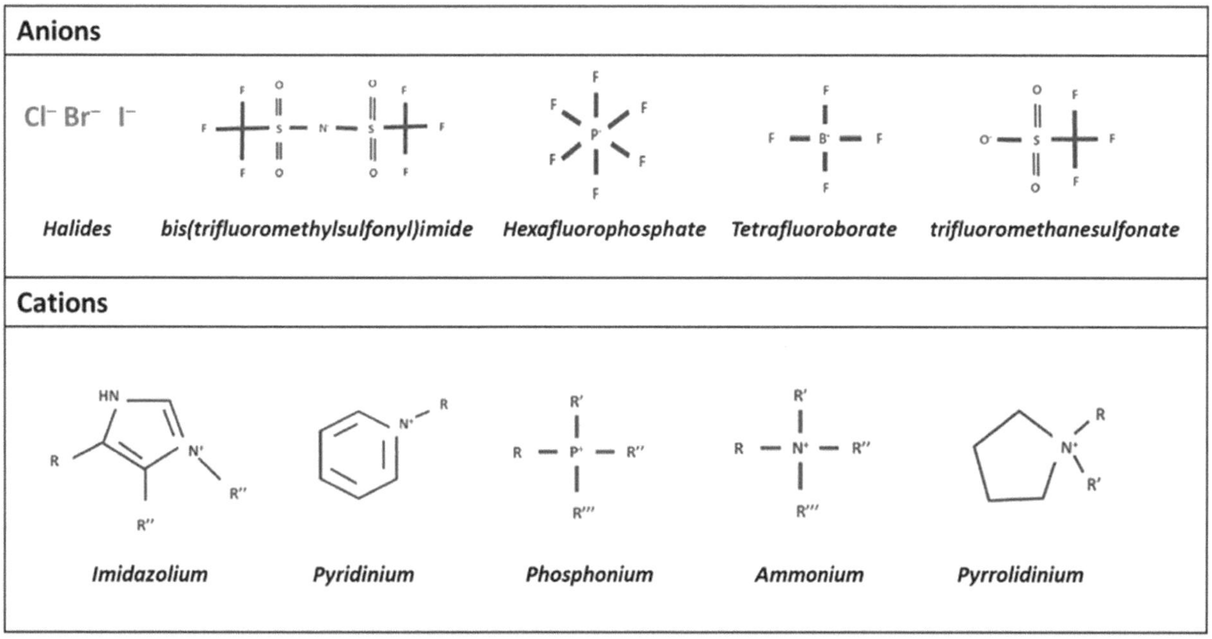

FIGURE 12.1 Common anions and cations in ionic liquids (ILs).

membranes (PIMs), and membranes prepared through the polymerization of ILs or polymeric IL membranes (PyLs) [26]. All these options have been applied for the successful and selective separation of organic compounds, including the gas state, and metal ions.

This chapter offers an overview of the recent developments achieved in the field of ILs when employed as solvents for the extraction of organic compounds and metals. The advantages and limitations of the technology are presented when ILs are used in liquid–liquid extraction and as well as immobilized in membranes. Among other aspects, this work reviews the applications of ILs in the field, their chemical composition and effects, transport mechanisms, and separation system configurations.

12.2 I ONIC LIQUIDS FOR THE SEPARATION OF ORGANIC COMPOUNDS AND METAL IONS BY LIQUID–LIQUID EXTRACTION

The use of conventional organic solvents has been a common practice in the chemical industry for the extraction of organic compounds. Despite its separation efficiency, organic solvent-based processes offer undesirable operational conditions such as significant losses of volatile solvents with associated toxicity and environmental issues. ILs have been used for the separation of a wide range of organic compounds by liquid–liquid extraction. They include phenolic compounds [27], aromatics and sulfur compounds from oil [28], amino acids from aqueous biphasic systems [29], organic acids such as carboxylic acids [30], azo dyes [31], and asphaltenes from liquefaction residues [32], among others. Also, ILs show high extraction and isolation capacity toward natural secondary metabolites from plants such as alkaloids, terpenoids, flavonoids, and polysaccharides [33,34].

The individual effect of anions and cations can be analyzed in order to study the optimal ion combinations in IL formulations for extraction purposes. For example, the individual ion contribution to phenol separation has been reported in the literature showing a clear effect of the nature of the ions. As a representative case, Fan et al. [35] studied the separation efficiencies of several organics including phenol, p-nitrophenol, guaiacol, and o-cresol from water, with functionalized ILs based on the imidazolium group, as can be seen in Figure 12.2. These cations were combined with several anions. It was found that with the hydroxyl group-functionalized ILs in combination with the anion PF_6^- and NTf_2^-, the introduction of the OH group on the cation notably increased the extraction ability. With the anion BF_4^- it was also observed that the extraction efficiency was also higher than using the dialkyl-based ILs. The authors suggested that the main driving force for resorcinol extractions is due to hydrogen bonding, while hydrophobic interactions would be the main mechanisms that facilitated the extraction of the rest of the phenolic compounds from water to ILs.

Another representative case of the applications of ILs is the recovery of volatile organic compounds (VOCs), which covers an important group of organics that are considered toxic to human health and the environment [36]. ILs have been commonly studied in ternary and higher-order systems formed by aromatic and aliphatic compounds to determine their efficiency under the perspective of real operations according to the needs of the chemical and oil industries. In this sense, the separation of aromatics from middle distillation products is of great importance for the obtention of clean fuels in accordance with legislation. ILs based on imidazolium, pyridinium, and alkyl sulfate groups have been the most widely analyzed in liquid–liquid extraction [37]. Recent examples include the IL

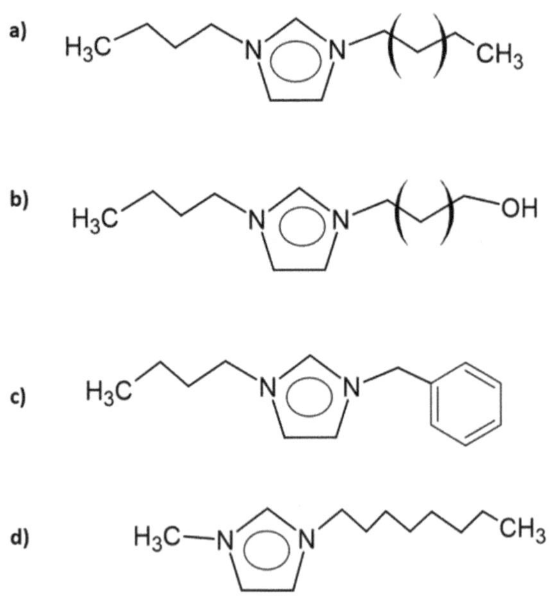

FIGURE 12.2 Ionic liquid cations tested in the work of Fan et al. [35]: (a) 1-butyl-3-n-alkylimidazolium; (b) 1-butyl-3-(hydroxy-n-alkyl)-imidazolium; (c) 1-butyl-3-benzylimidazolium; (d) 1-methyl-3-octyl-imidazolium (n = 4, 6, and 9).

1-butyl-3-methylimidazolium tricyanomethanide ([Bmim][TCM]) for the extraction of xylene from alkanes such as decane and octane [38], the use of alky methylimidazolium ILs in combination with hexafluorophosphate ([PF_6]) for the separation of benzene from cyclohexane [39] or the application of pyridinium-based ILs such as hexylpyridinium nitrate, [hpy][NO_3], and butylpyridinium nitrate, [bpy][NO_3], for the separation of benzene from alkanes [40].

In order to observe the influence of the cation, it is interesting to note from the examples mentioned above, that, in the case of the extraction of benzene from cyclohexane, it has been observed that a longer alkyl chain on the cation (1-butyl-3-methylimidazolium> 1-pentyl-3-methylimidazolium> 1-hexyl-3-methylimidazolium) increased the distribution coefficient of benzene but simultaneously reduced the solvent selectivity. Nevertheless, ILs such as [bmim][PF_6] can offer very high solvent selectivity [39]. Moreover, the advantage of using ILs in liquid–liquid extractions for the recovery of VOCs is the significant extraction efficiency achievable even at room temperature without the necessity of applying additional energy.

Among other interesting examples reported for the application of ILs in the separation of organics are the selective extraction of vinyl ester, alkyl ester, alcohol, and organic acids, this being a common problem in transesterification processes. The use of the ILs alkyl-methylimidazolium and alkyl-methylpyridinium cations in combination with several anion types such as bis(trifluoromethylsulfonyl)imide, tetrafluoroborate, hexafluorophosphate, sulfates, nitrate, dicyanamide, chloride, and acetate, has been studied for the separation of this class of compounds from hexane solutions [41]. High percentages of extraction were obtained for 1-butanol (92%) and butyric acid (99%), while the yields for other compounds such as vinyl butyrate and butyl butyrate were lower (40 and 22%, respectively). Among the most important factors in determining extraction efficiency was anion nature, which greatly affected the partitioning coefficients of these compounds between ILs and hexane. Also, cations with longer alkyl-chain lengths were shown to facilitate the recovery of the target molecules.

ILs have also been investigated for the separation of metal ions in solvent extraction as substituents of organic media. Typically, metal ions can be removed using extraction agents and ligands in an organic solvent [42,43]. Because of the versatility of ILs, they can be used for metal ion recovery as diluents and extract agents. Thus, they have been employed for the extraction of transition metals including heavy metals, noble metals, actinides, and lanthanides, among others [44].

As a representative case of diluents, Ni^{2+}, Cu^{2+}, and Pb^{2+} have been separated with the IL 1-butyl-3-methylimidazolium hexafluorophosphate, [bmim][PF_6], using 2-amino thiophenol as the extract agent, reaching equilibria extraction between 30 and 120 min. In comparison with the use of chloroform as a diluent, the IL was capable of obtaining higher efficiencies [45]. More recently, the ILs trioctylmethylammonium bis(2-ethylhexyl)phosphinate, [TOMA][D_2EHP], and 1-octyl-3-methylimidazolium bis(trifluoromethylsulfonyl)imide, [omim][NTf_2], have been studied as alternatives to kerosene as a diluent, increasing the extraction selectivity of the ions Mo(VI) and Re(VII). These ILs were prepared in a highly acidic solution [46].

Apart from imidazolium-based compounds, ILs with other cation groups are also effective methods for their application in metal extraction. For example, phosphonium ILs like triethyl(pentyl)phosphonium bis(trifluoromethylsulfonyl)imide [P_{2225}][NTf_2] have been used as diluents in the presence of trioctylamine acting as an extract for the separation of W(VI) and Co(II), offering higher selectivity and efficiency in comparison to an extraction system based on trioctylamine diluted in kerosene [47]. Similar ILs ([P_{2226}][NTf_2]) have also been proved to be effective for the recovery of platinum(IV) with chloride [48].

Other interesting examples include the utilization of ILs with extraction agents like Cyanex-923 [49], dithizone [50], and several other organics acting as ligands agents [51] and the design of tailored compounds for the recovery of highly toxic heavy metals such as mercury [52]. Other works have thoroughly analyzed the selectivity of several mixtures of metals such as Cu(II), Fe(III), Zn(II), and Cd(II) from aqueous hydrochloride media employing ILs based on imidazolium and ammonium groups [53]. Among the phases studied, the IL methyltrioctylammonium chloride, [MTOA][Cl], achieved extraction efficiencies higher than 90% for Fe(III), Cd(II), and Zn(II). On the other hand, several mechanisms have been identified for the transfer of metals in IL phases, such as solvent ion-pair extraction and ion exchange or a combined mechanism between them.

Many other works have reported the use of ILs as extract agents, mostly based on positively charged heteroatom cations like imidazolium, ammonium, and phosphonium. An interesting application is the use of butyl and octyl imidazolium compounds for the separation of M^{2+} contaminants from hydrocarbons such as diesel, with high recovery efficiencies (>99 %) of Zn(II) and Cu(II) [54]. Imidazolium-based ILs functionalized with keto groups on the cation and in combination with the bis(trifluoromethylsufonyl)imide anion have also been effective alternatives to methyl isobutyl ketone, frequently employed for the extraction of tantalum, from sulfuric acid media reaching efficiencies of about 96% in the recovery of Ta(IV) [55].

ILs with ammonium cations have been intensively studied for the extraction of metals within the 5 and 6 Groups, transition metals of Period 4, alkali-metals, rare earths, and noble metals [44]. Some representative examples are given to highlight the efficiency of this group of ILs. For instance, a mixture of tridecylmethylammonium and trioctylmethylammonium chlorides has been used for the extraction of vanadium in sulfonated kerosene [44]. A high efficiency of 99% was achieved. The same types of ILs in kerosene were effective to separate Mo(VI) from alkaline solutions in the presence of aluminum(III), copper(II), and nickel(II) with an efficiency of over 98% [57]. ILs formed by tetraoctylammonium and tetrahexylammonium in combination with bromide, thiocyanate, and dicyanamide are a good option for the extraction of noble metals, including gold and platinum [57].

Among pyridinium case studies, it is worth noting their application in the extraction of noble metals like Ag(I), Pd(II), and Au(III) from water-based solutions at ambient temperature. It has been found that pyridinium ILs including nitrile and disulfide groups in combination with the bis(trifluoromethylsulfonyl)imide anion are efficient for the extraction of these metals [58]. Octylpyridinium-based ILs can offer efficiencies of over 90% for the extraction of other metals such as vanadium. Specifically, the ILs $[C_8Py]$[Cl] (N-octylpyridinium chloride) and $[C_8Py][BF_4]$ (N-octylpyridinium tetrafluoroborate) offered recovery rates of >95% and >93%, respectively. These extraction systems could be used for the extraction of vanadium in the hydrometallurgy industry, using vanadium slag-leaching solutions. For these examples, it was also found that the extraction mechanism was based on anion exchange [59].

As can be seen, ILs offer promising applications for the extraction of metal ions, being used as extractants or diluents, or as a combination of them. Due to their near-zero vapor pressure, they can be considered safer in comparison to conventional organic diluents. In addition, the great electrical conductivity of ILs avoids the accumulation of static charge that is considered a risk due to potential ignition [44]. Due to the versatility of ILs, new functionalized compounds are expected to be synthesized and researched in this field in the coming years.

12.3 IONIC LIQUID MEMBRANES

Liquid membranes that use ILs as the liquid phase have shown enormous potential in different applications [60–62]. The transport mechanism in these membranes is a combination of three simultaneously occurring processes: molecule extraction from the feed phase to the IL membrane, diffusion through the IL membrane, and re-extraction to the received phase. This technique presents many advantages, including selectivity, reusability, the combination of the extraction and stripping steps into a single step, and the reduction of the amount of solvent required [63,64].

12.3.1 Separation System Configurations Based on Ionic Liquids

ILs have been used in various membrane configurations for the selective transport of organic compounds. Among the most commonly used configurations are supported IL membranes (SILMs), in which ILs are used as carriers in supported membrane operation. As a representative case, Kamaz et al. [62] studied the successful fractionation of an organic stream consisting of benzene, naphthalene, and phenanthrene in tetradecane, and *cis*- and *trans*-stilbene in hexane using an imidazolium-based supported ionic liquid membrane (SILM), showing that the rates of transport can be affected by the interaction between the imidazolium cation and the π electron cloud density of the aromatic solutes. On its part, Branco et al. [65] demonstrated the superiority of SILMs over bulk membranes in the selective transport of morpholine, methylmorpholine, cyclohexanol, cyclohexanone, propan-1-ol, butan-1-ol, and 1,4-dioxane, improving the selectivity in the transport of the target compounds. These membranes are easy to implement in chemical processes and require less amount of liquid phase.

In order to obtain preliminary data on the permeation of a given solute, flat-sheet supported liquid membranes (FSSLMs) are very useful. To this end, FSSLMs based on ILs have been used to study the separation of organic compounds [66–68], metal ions [69,70], and gases [71,72]. Klingberg et al. [72] reported an interesting application of a CO_2/N_2-selective FSSLM for the separation of carbon dioxide from coal-fired power plant flue gases. A polyacrylonitrile ultrafiltration membrane as a support and a selective layer of the IL 1-ethyl-3-methylimidazolium bis(trifluoromethylsulfonyl) imide ($[emin][NTf_2]$) was studied. The results showed that the volume fraction of CO_2 in the flue gas could be raised from 14% vol. (feed) to 40% vol. (permeate) by gas permeation using these SILMs.

However, even with these good results, it is required to develop membrane configurations with a higher surface area to volume ratio. With this purpose, spiral-wound and hollow-fiber modules were created, the latter being the most widespread. In hollow-fiber modules based on ILs, they are confined, by capillarity, within the pores of the hollow fibers. Normally, the feed solution would circulate on the inside of the fiber while the strip solution would circulate on the outer side. Thus, we would have two aqueous/IL interfaces with well-defined transfer areas and the solid membrane would simultaneously act as a uniform barrier between the aqueous solutions and as a support for the IL. Pang et al. [73] successfully applied this configuration to the microextraction of dichlorodiphenyltrichloroethane and its main metabolites using 1-hexadecyl-3-methylimidazolium bis(trifluoromethylsulfonyl)imide as liquid phase.

Other IL/polymer matrix membrane configurations include: (1) gelation of SILMs, which allows utilizing the functionality of the ILs providing a more solid/semi-solid structure while maintaining the mobility of the liquid phase [74]; and (2) polymer inclusion membranes (PIM), in which the solid support is formed with the carrier phase by casting methods. Most recently, IL-based PIMs membranes have been used for the selective separation of volatile fatty acids (acetic and hexanoic acids) using quaternary ammonium salt (Aliquat 336) and phosphonium salt (Cyphos IL-101) as the carrier [75]. In addition, these membranes have been successfully used as proton exchange membranes in bioelectrochemical systems such as microbial fuel cells for the purification of pollutants in wastewater [76].

There are other types of advanced SILMs that use a micro-porous solid barrier to separate the organic phase from the aqueous one. The most prominent are non-dispersive solvent extraction (NDSX) and pseudo-emulsion hollow fiber strip dispersion (PEHFSD). The NDSX operation mode, based on ILs, utilizes two hollow-fiber units, like those used for supported liquid membranes. In the first unit, the feed solution circulates on one side of the fibers, while the IL circulates on the other side. The IL then passes to the second module where the stripping operation is taking place. Finally, the IL returns to the first module and can be reused. NDSX technology has been tested in the recovery of metal ions using trihexyl(tetradecyl)phosphonium chloride (CYPHOS IL101) dissolved in cumene as the carrier phase. In particular, Comesaña et al. [77] used non-dispersive solvent extraction with strip dispersion (NDSXSD) to analyze the transport of Cd(II) from a hydrochloric acidic medium. Alternatively, NDSX technology in a ceramic hollow-fiber contactor with the ionic liquid ethyl-3-methylimidazolium ethylsulfate as a carrier was used for the recovery of SO_2 from a gas stream with a composition typical of roasting processes [78].

Then, we have the PEHFSD technology, which uses a single membrane module for both, the extraction and stripping processes. It also employs a stirred tank for homogenization of the feed phase, as well as a stirred tank for the preparation of a pseudo-emulsion. This pseudo-emulsion is formed by an organic phase, containing an IL, and a strip solution dispersed within the organic solution. Two gear pumps, with variable flow capacity, pump both phases into the module. When the operation begins, the organic phase of the pseudo-emulsion wets the micro-porous fiber wall due to its hydrophobic nature. At the end of the operation, the clarity and rapid separation of phases characterize the pseudo-emulsion once the mixing has stopped. Normally, a few minutes after the mixing of the organic and receiver phase stops, the pseudo-emulsion breaks down, and the organic and strip phases separate automatically. Then, the recovery of the extracted solute from the pseudo-emulsion becomes possible. Interesting applications of pseudo-emulsion-based hollow-fiber strip dispersion technology have been successfully reported, such as the recovery of zinc(II) from chloride solutions using two ILs based on the cation 3-[1-(hydroxyimine)undecyl]-1-propylpyridinium in combination with the anions chloride and bromide, respectively, as the carriers [79]. Another example is the permeation of chromium(III) from an alkaline medium using the IL $[RNH_3][HSO_4]$ as the carrier, dissolved in n-decane [80].

It should be emphasized that the interface between the organic and aqueous phases, in these two configurations, is maintained at the pore by applying a higher pressure to the aqueous phase than to the organic or pseudo-emulsion phases. Furthermore, both the aforementioned advanced membrane technologies can either operate in counter-current or co-current flows of the respective phases, as well as in recirculation or feed solution pass-through modes.

The following section focuses on the preparation methods of supported IL membranes since they have been widely applied in the separation of organics and metal ions.

12.3.2 Synthesis of SILMs

The selection of the method preparation of SILMs is important and has a strong influence on membrane performance. This occurs due to the fact that ILs are substances that exhibit a relatively high viscosity. The reparation of SILMs is usually carried out using three methods: direct immersion, pressure, or vacuum [81,82]. In the direct immersion method, immobilization occurs spontaneously without any external force, by bringing the supporting membrane in contact with the IL, allowing the membrane to absorb the liquid. The periodic weighting of SILMs until an equilibrium state is reached is used to control mass uptake [82]. Immobilization in the pressure method is achieved by forcing the IL to flow into the membrane pores by applying pressure using gaseous nitrogen. This requires the use of an ultrafiltration unit, in which the membrane is placed. Thus, the IL displaces the air from the membrane pores [83]. In the latter method, vacuum is applied to release all air occluded in the pores of the supporting membrane, allowing the IL, in which it is immersed, to take its place [81]. Upon completion of any of these methods, the excess IL should be removed from the membrane surface. This can be accomplished by wiping with paper tissue or leaving it to dry overnight.

Scanning electron microscopy (SEM) combined with energy dispersive X-ray (EDX) analysis is used to analyze the influence of the immobilization method on the SILM performance [81,83]. These techniques have allowed observing that after the immobilization of ILs by either method (vacuum or pressure), they were homogeneously distributed in the membrane pores, except for the largest macropores, which remained partially filled. In membranes created via the pressure method, SEM-EDX and weight studies showed that the amount of IL immobilized was independent of the IL used. However, in membranes created from the vacuum immobilization method, it was seen that the higher the viscosity of the IL used, the lower its absorption. This could be explained by the fact that the high viscosity of ILs makes it difficult for them to penetrate into the center of the deepest pores of the membrane when vacuum is applied. Based on these results, it was concluded that immobilization under pressure ensures that all membrane pores are filled with the ILs, regardless of the viscosity. In contrast, vacuum immobilization could only be considered suitable for low-viscosity ILs, with the added advantage that it is an easier preparation method.

The nature of the support membranes also plays an important role in the performance of SILMs. In this context, Grünauer et al. [84] compared two thin-film coating techniques, dip coating and spin coating, using polyacrylonitrile as supporting material. The dip-coating technique was found to perform best to create membranes that meet stability needs. Likewise, De los Rios et al. [85] studied the performance of two polymeric membranes, Nylon and Mitex, as supporting membranes. It was observed that the Mitex membranes absorbed less IL, which is explained by its high hydrophobic character and the different textural properties of these membranes, which probably restrains the interaction with the hydrophilic ILs used. The determination of wettability has also been used as a criterion in the selection of a suitable supporting membrane [86].

The use of nanofiltration membranes as support membranes has also been reported. In this regard, Xiao et al. [87] compared the resource recovery, from pigment industry wastewater, of traditional nanofiltration membranes versus functionalized interfacially polymerized nanofiltration membranes based on amino acid ILs. The results showed that the use of amino acid ILs provides nanofiltration membranes with 63% higher pure water permeability than conventional nanofiltration membranes, a more negatively charged surface, and slightly larger pore size, resulting in higher selectivity toward the studied compounds.

12.3.3 Applications of Ionic Liquid-Based Membranes in the Separation of Organics and Metal Ions

Over the last decade, the separation of organic compounds and metal ions using SILMs has received increasing attention [61,88–90]. This can be seen from the exponential increase in the number of papers being published in this field during these years. In the following sections of this chapter, we provide detailed descriptions of some representative examples of these applications. Several of these works are summarized in Table 12.1 for organics and Table 12.2 for metal ions.

12.3.3.1 Separation of Organic Compounds

The selective separation of volatile organic compounds is one of the most studied applications of SILMs. For instance, Cichowska et al. [91] studied toluene permeability and separation selectivity as a function of cation and anion type of the different ILs used. The different SILMs obtained were created using porous polypropylene support on which the different ILs, based on the pyridinium and pyrrolidinium cations ([C_4Py][NTf_2], [C_6Py][NTf_2], [C_4Pyrr][NTf_2], [C_6Pyrr][NTf_2], and [C_4Pyrr][TfO]), were absorbed, applying a combination of pressure and vacuum methods. Tests were carried out at different temperatures showing that [C_6Py][NTf_2] would be the most promising for the separation of toluene from gaseous streams, reaching permeation values of 2034 barrers at 318 K.

The possibility of using SILMs for the selective separation of substrates and products from the biosynthesis of organic esters, commonly used in the perfume and flavor industries, has also been reported [92,93]. This biosynthesis process can be carried out in non-conventional media (i.e., n-hexane [94] and ionic liquid media [95]), with low water content, by using enzymes that catalyze the transesterification of vinyl esters and alcohols.

For the immobilization of imidazolium-based ILs containing the anions [PF_6], [BF_4], and [dca] anions on polymeric membranes, substantial differences in permeability were detected among the organic compounds, depending not only on their functional groups but also on the lengths of their alkyl chains. Some authors have made efforts to establish rules for the optimal design of ILs after learning about the feasibility of their use as a liquid phase for the selective separation of transesterification reaction mixtures [92]. For this purpose, the permeability of vinyl butyrate, 1-butanol, butyl butyrate, and butyric acid through 13 supported liquid membranes based on different IL compositions with imidazolium and pyridinium cations and various anions, immobilized on Nylon membranes, was determined. According to the results reported by de Los Ríos et al. [92], selectivity highly depended on IL anion composition. For a fixed cation such as 1-butyl-3-methylimidazolium, the highest selectivity was reached with [$MeSO_4^-$], and, on the contrary, the lowest selectivity was reached with [$OcSO_4$]. Good selectivity was also obtained with anions such as [$EtSO_4$] and [$MDEGSO_4$]. On the other hand, and in regard to cations, SILMs based on dialkylimidazolium-based ILs were more efficient than dialkylpyridinium-based ILs and a decrease in terms of alkyl substituent length was shown to lead to an increase in the selectivity. From these results, it is concluded that to achieve the highest selectivity in the separation of these organic compounds, ILs combining dialkylimidazolium cations displaying a short-length alkyl chain with sulfate anions with a short alkyl substituent would be the most suitable.

The separation of aromatic hydrocarbons from aliphatic ones using SILMs has also been investigated. Kamaz et al. [62] studied the fractionation of benzene, naphthalene, and phenanthrene in tetradecane, and *cis*- and *trans*-stilbene in hexane. For this purpose, they used three ionic liquids based on the imidazolium cation: 1-allyl-3-vinylimidazolium, 1-hexyl-3-vinylimidazolium, and 1-octyl-3-vinylimidazolium, combined with the bromide anion, using a porous polypropylene membrane for their immobilization. For both

TABLE 12.1
Examples of ionic liquids used for the extraction of organics reported in the literature

Organic compounds	Ionic liquids used	Support material	Membrane type	Preparation method	References
Toluene	$[C_4Py][NTf_2]$, $[C_6Py][NTf_2]$, $[C_4Pyrr][NTf_2]$, $[C_6Pyrr][NTf_2]$, and $[C_4Pyrr][TfO]$	Polypropylene	SILM	Pressure and vacuum	[91]
rac-1-phenylethanol	$[bmim][BF_4]$	Nylon	SILM	Pressure	[94]
Vinyl acetate, vinyl propionate, vinyl butyrate, vinyl laurate, methyl acetate, methyl propionate, butyl butyrate, ethyl decanoate, methanol, 1-propanol, 1-butanol, 1-octanol, acetic acid, propionic acid, butyric acid, and lauric acid	$[bmim][PF_6]$ and $[omim][PF_6]$	Nylon	SILM	Pressure	[93]
Vinyl butyrate, 1-butanol, butyl butyrate, and butyric acid	[bmim][dca], [bmim][BF4], $[bmim][PF_6]$, $[bmim][NTf_2]$, [omim][dca], $[omim][BF_4]$, $[omim][PF_6]$, and $[omim][NTf_2]$	Nylon	SILM	Pressure	[96]
Vinyl acetate, vinyl propionate, vinyl butyrate, vinyl laurate, methyl acetate, methyl propionate, butyl butyrate, ethyl decanoate, methanol, 1-propanol, 1-butanol, 1-octanol, acetic acid, propionic acid, butyric acid, and lauric acid	$[bmim][BF_4]$ and $[omim][BF_4]$	Nylon	SILM	Pressure	[97]
Vinyl butyrate, 1-butanol, butyl butyrate, and butyric acid	[bmim][Cl], $[bmim][PF_6]$, $[bmim][BF_4]$, $[bmim][MeSO_4]$, $[bmim][EtSO_4]$, $[bmim][OcSO_4]$, $[bmim][MDEGSO_4]$, [bmim][dca], $[bmim][NO_3]$, $[omim][PF_6]$, $[omim][BF_4]$, $[emim][EtSO_4]$, $[empy][EtSO_4]$, and $[bmpy][BF_4]$	Nylon	SILM	Pressure	[92]
Benzene, naphthalene, phenantrene, *cis*-stilbene, and *trans*-stilbene	1-allyl-3-vinylimidazolium bromide, 1-hexyl-3-vinylimidazolium bromide, and 1-octyl-3-vinylimidazolium bromide	Polypropylene	SILM	Pressure	[62]
Penicillin G	$[bmim][PF_6]$, $[hmim][PF_6]$, $[omim][PF_6]$, and tri-n-octylmethylammonium chloride	Polyvinylidene fluoride	SILM	Direct immersion	[98]
Ibuprofen and alfa-pinene	$[C_{11}H_{21}N_2O][GdCl_3Br_3]$, $[bmim][FeCl_4]$, and $[omim][FeCl_4]$	Polyvinylidene fluoride	SMILM	Vacuum	[99]
Diethylstilbestrol, estrone, 17-beta-estradiol, and bisphenol A	$[omim][PF_6]$	Polypropylene	SILM (hollow fiber)	Direct immersion	[100]
Bisphenol A	$[TBP][PF_6]$, [TBTDP][BTMPP], [THTDP][Br], [THTDP][Cl], [MTONH][Cl], $[BMIM][PF_6]$, [C2DMIM][PF6], [BTNH][BTA], and [BMPYR][BTA]	Polyvinylidene fluoride	SILM	Direct immersion	[66]

2,3-DMBT, 2,4-DMBT, 2,6-DMBT, 2,3,6-TMBT, 2,3,4-TMBT, DBT, 2-prop-DBT, 1-MDBT, 2-MDBT, 3-MDBT, 4-MDBT, 4-EDBT, 4,6-DMDBT, 2,4-DMDBT, 3,6-DMDBT, 2,8-DMDBT, 1,4-DMDBT, 1,3-DMDBT, 4E,6M-DMDBT, and 2,4,8-TMDBT	N-butyl-3-methyl pyridinium bis(trifuoromethylsulfonyl)imide, 1,2-dimethyl-3-propylimidazolium bis(trifuoromethylsulfonyl)imide, 1-ethyl-3-methylimidazolium bis(trifluoromethylsulfonyl)imide, N-propyl-3-methylpyridinium bis(trifuoromethylsulfonyl)imide, and 1-ethyl-3-methylimidazolium trifluoromethanesulfonate	Polypropylene	SILM (sheet membrane)	Direct immersion	[101]
1,3-Propanediol	Tetrapropylammonium tetracyanoborate	Polydimethyl-siloxane	SILM (ceramic nanofiltration module)	Direct immersion	[102]
Butan-1-ol and acetone	1-Ethenyl-3-ethyl-imidazolium hexafluorophosphate and tetrapropylammonium tetracyano-borate	Polydimethyl-siloxane	SILM (ceramic ultrafiltration module)	Direct immersion	[103]
Benzene and cyclohexane	[bmim][PF_6], [hmim][PF_6], [omim][PF_6], diethyl (2-methoxyethyl)methylammonium, bis(trifluoromethanesulfonyl)imide, and diethyl (2-methoxyethyl) methylammonium tetrafluoroborate	Polyvinylidene fluoride	SILM	Direct immersion	[104]
Acetone, n-butanol, and ethanol	Trihexyl(tetradecyl) phosphoniumtetracyanoborate and 1-hexyl-3-methylimidazolium hexafluorophosphate	Polysulfone (+ poly(ether block amide) coating)	SILM	Direct immersion	[105]
Dimethyl carbonate and methanol	[omim][NTf2] and [C_8C_1Pyrr][NTf_2]	Polyacrylonitrile	SILM (flat sheet ultrafiltration membrane)	Vacuum	[106]
n-Hexane, n-heptane, cyclohexane, metylcyclohexane, benzene, and toluene	[emim][$B(CN)_4$], [emim][dca], [emim][BF_4], and [emim][HSO_4]	Polyvinylidene fluoride	SILM (flat sheet)	Direct immersion	[107]
Dioxins	[Aliquat][dca] and [omim][dca]	TiO_2 and Al_2O_3	SILM (inorganic ceramic membrane)	–	[108]
4-Phenoxybutyric acid, 3-phenoxypropionic acid, 2-phenylpropionic acid, 2-phenoxybutyric acid, mandelic acid, and 2-amino-2-phenylbutyric acid	[bmim][PF_6], [hmim][PF_6], [omim][PF_6], and [bmim][NTf_2]	Polypropylene	SILM	Direct immersion	[109]
(R)-2-pentanol, (S)-2-pentanol, vinyl propionate, (R)-2-pentyl propionate, (S)-2-pentyl propionate, and propionic acid	[bmim][BF_4], [bmim][PF_6], [bmim][NTf_2], [omim][BF_4], [omim][PF_6], and [omim][NTf_2]	Nylon	SILM	Pressure	[110]
(R)-2-phenyletanol, (S)-1-phenyletanol, vinyl propionate, (R)-1-phenylethyl propionate, (S)-1-phenylethyl propionate, and propionic acid	[bmim][BF_4], [bmim][PF_6], [bmim][NTf_2], [omim][BF_4], [omim][PF_6], and [omim][NTf_2]	Nylon	SILM	Pressure	[111]

fractionations, the IL 1-octyl-3-vinylimidazolium bromide presented the highest mass transfer coefficient.

Studies employing SILMs to treat organic compounds of pharmaceutical interest have also been carried out [98,99]. Matsumoto et al. [98] investigated the concentration of penicillin G using SILMs based on polyvinylidene fluoride as support and [bmim][PF_6], [hmim][PF_6], [omim][PF_6], and [Oc_3MeN][Cl] as ILs. Daniel et al. [99] studied the effects of a magnetic field on the separation of ibuprofen and α-pinene using supported magnetic ionic liquids membranes (SMILMs). In all membranes, polyvinylidene fluoride was used as the supporting material, whereas [$C_{11}H_{21}N_2O$]$_3$[GdCl_3Br_3] was used as the IL for ibuprofen transport, and [bmim][$FeCl_4$] and [omim][$FeCl_4$] were used for α-pinene transport. The results showed that upon exposure of the different membranes to a magnetic field, permeability increases by 51% for α-pinene through the [bmim][$FeCl_4$]-based SMILM, by 29% for α-pinene through the [omim][$FeCl_4$]-based SMILM, and by 59% for ibuprofen through the [$C_{11}H_{21}N_2O$]$_3$[GdCl$_3Br_3$]-based SMILM.

SILMs have also been successfully applied to the concentration of endocrine-disrupting compounds such as diethylstilbestrol, estrone, 17-β-estradiol, or bisphenol A [100] from water samples and for the selective separation of the latter compound from aqueous solutions. Yang et al. [63] used a polyvinylidene fluoride membrane as a support and different cation-based ILs such as imidazolium, pyridinium, ammonium, and phosphonium, obtaining a maximum permeation of 62% for the IL [TBP][PF_6] under pH control. Zou et al. [100] employed polypropylene hollow-fiber membranes as support and 1-octyl-3-methylimidazolium hexafluorophosphate ([omim][PF_6]) as IL for microextraction of the four targeted compounds, obtaining recovery percentages of over 80.2 ($n = 5$).

Another interesting application of SILMs was implemented by Ibrahim et al. [101], who applied a propylene sheet membrane with the IL 1-ethyl-3-methylimidazolium trifluoromethane sulfonate immobilized in its pores for the removal of sulfur compounds from crude oil in a membrane extraction flow reactor. The best removal rate achieved by this configuration was 58% of the total sulfur content in light crude oil.

The separation of various liquids or vapor mixtures by pervaporation and vapor permeation is another field in which SILM technology has been successfully applied [105–107]. The restrictive compromise between selectivity and flux is the main problem associated with solute recovery using these techniques. To overcome this problem, SLMs containing ILs have been studied, which would prevent the feed solvent from permeating through the membrane, increasing the selectivity toward the solute and ultimately allowing the process to operate under conditions that provide high fluxes [102]. In this context, the successful removal of 1,3-propanediol from aqueous solutions by pervaporation using SLMs based on a ceramic nanofiltration module and the IL [Pr_4N][B$(CN)_4$] was studied. The results showed that although the average permeation flux decreased by one order, the selectivity of the process was increased by more than two orders. Furthermore, Rdzanek et al. [105] studied the concentration of n-butanol by pervaporation from a typical acetone/butanol/ethanol (3:6:1) mixture known as ABE, which is obtained from anaerobic fermentation processes. For this purpose, they used membranes based on two ILs, trihexyl(tetradecyl)phosphoniumtetracyanoborate and 1-hexyl-3-methylimidazolium hexafluorophosphate, immobilized on a polysulfone (PS20) support using a polyether block amide (PEBA) coating. Different initial concentrations of butanol were tested at different temperatures, concluding that the configuration that showed the best results was the one created from trihexyl(tetradecyl) phosphoniumtetracyanoborate, with 1.5 wt.% of butanol at 50ºC, obtaining an enrichment factor of 14.9 and an average selectivity of butanol over water of 8.34.

Ultrafiltration membranes appear to be good support for SILMs to be used for the separation of organic compounds by pervaporation [102,106]. Izák et al. [102] compared the removal performances of a polydimethylsiloxane (PDMS) membrane versus two ultrafiltration membranes impregnated with the IL [eneim][PF_6] and [Pr_4N][B$(CN)_4$] and PDMS for the removal of butan-1-ol and acetone from aqueous solution by pervaporation. Higher average enrichment factors for the removal of both acetone and butan-1-ol were found when supported liquid membranes containing ILs were employed. More recently, Li et al. [106] studied the separation of dimethyl carbonate (DMC)/methanol mixtures employing SILMs with ILs such as [omin][NTf_2] or [OMPyrr][NTf_2] on flat sheet ultrafiltration membranes made of polyacrylonitrile. This investigation determined that the separation performance of the DMC/methanol mixture decreases at high methanol concentrations due to strong coupling effects. Despite this issue, SILM containing the IL [bmim][NTf_2] achieved a separation factor of 21 and a selectivity toward this mixture of 67.

Regarding vapor permeation using SILMs, we can find older studies on the separation of benzene and cyclohexane or more recent ones dealing with the separation of numerous aliphatic, alicyclic, and aromatic organic compounds mixtures. Matsumoto et al. [98] found that the membrane that obtained the best results was the hydrophilic liquid membrane containing the IL N,N-diethyl-N-methyl-N-(2-methoxyethyl) ammonium tetrafluoroborate, with a separation factor of 950 for 11 wt.% benzene and 185 for 53 wt.% benzene for the vapor permeation. Kamiya et al. [107] found that the SILM based on the IL [emim][HSO_4] obtained the best separation factors: 85 for the n-hexane/benzene mixture; 46 for cyclohexane/benzene; 59 for n-heptane/toluene; and 11 for methylcyclohexane/toluene. Both works showed the high efficiency of SILMs as selective membranes for the vapor permeation of organic compounds.

The removal of persistent organic pollutants, in particular dioxins, is another interesting application studied with SLMs based on ILs. Kulkarni et al. [108] specifically analyzed the removal of dioxins from high-temperature vapor sources. All membranes tested were stable at 200°C, demonstrating the

TABLE 12.2
Examples of ionic liquids used for metal ion extraction reported in the literature

Metal ions	Ionic liquid used	Supporting material	Membrane type	Method	References
Cr(III)	Trioctyl methylammonium chloride	Polypropylene	SILM	PEHFSD	[122]
Fe(III)	$[PJMTH^+]_2[SO_4]^{2-}$	Polypropylene	SILM	PEHFSD	[119]
Cr(VI)	Cyphos IL101	Polyvinylidene fluoride and polytetrafluoroethylene	SILM	PEMSD	[118]
Cr(III)	$[PJMTH^+]_2[SO_4]^{2-}$	Polypropylene	SILM	PEHFSD	[80]
Cr(VI)	Cyphos IL102 mainly and some others (DBBP, Cyanex 471X, Cyanex 921, Cyphos IL101, TBP, Primene JMT, Cyanex 923, Hostarex A 324, Aliquat 336 and Amberlite LA2)	Polyvinylidene fluoride	SILM	Direct immersion (FSSLM) and PEMSD	[117]
Hg(II), Cd(II), and Cr(III)	Isooctylmethylimidazolium bis-2-ethylhexylphosphate	Polyvinylidene fluoride	SILM	Direct immersion	[123]
Cr(III)	Aliquat 336	Cellulose triacetate, nitrophenyl octyl ether, and 2-ethylhexyl phosphoric acid or ionic liquid	PIM	Casting	[124]
Zn(II) and Fe(III)	Cyphos IL101 and Cyphos IL104	Cellulose triacetate, nitrophenyl octyl ether, and ionic liquid	PIM	Casting	[125]
Co and Ni	Tri(hexyl)tetradecyl phosphonium chloride	Polyvinylidene fluoride	SILM	Direct immersion	[69]
Fe(III), Zn(II), Cd(II), and Cu(II)	[MTOA][Cl]	Nylon	SILM	Pressure	[90]

potential of ceramic membranes with immobilized ILs for this operation.

An interesting example of the integration of supported liquid membranes and bio-catalyzed reactions was reported by Miyako et al. [109], who carried out lipase-facilitated transport of (S)-ibuprofen through SLMs based on the ILs [bmim][PF_6], [hmim][PF_6], [omim][PF_6], and [bmim][NTF_2]. In this system, enantioselective esterification of (rac)-ibuprofen occurred in the feed phase by *Candida rugosa* lipase. Then, esters selectively cross the SILM toward the receiving phase where they are eventually hydrolyzed by lipase once again. Later, other studies on kinetic resolutions using that integrated system of the enantioselective catalytic action of an enzyme and the selective permeability of SILMs were applied to the resolutions of racemic alcohols, such as rac-2-pethanol [110] and rac-1-phenylethanol [111]. For both studies, a commercial immobilized lipase B from *Candida antarctica* (CaLB) was used for transesterification with a vinyl ester. In this case, the enantio-selective esterification takes place in the feed phase and the non-reactive isomer of the alcohol diffuses through the SILM.

Finally, the use of SILMs for gas separation is a specific field of application of great interest. For the study of gas separation through this kind of membrane, the solution diffusion model of transport is usually employed and the dissolution of gases in ILs follows the regular solution theory. On the other hand, several studies have reported that the diffusivity of a gas through the active phase of the membrane is mainly linked to the viscosity of the IL and that the selectivity can be correlated with its molar volume [112]. Different gas mixtures have been studied as model mixtures to test the efficiency of SILMS in gas separation. The most studied binary gas mixtures are those formed by CO_2/CH_4, CO_2/N_2, CO_2/H_2, and H_2S/CO_2 [113–115], although other more complex mixtures are also being analyzed such as alkane–alkene separation. The separation of CO_2 from other gases such as CH_4, H_2, and N_2 with SILMs offers economic advantages as it requires lower energy consumption [112]. On the other hand, imidazolium-based ILs with anions such as BF_4 and bis(trifluoromethanesulfonyl) imide (Tf_2N^-) are among the most investigated chemical species [113,114]. Regarding membrane supports employed for the IL active phase, porous polyethersulfone (PES) and porous poly(vinylidene fluoride) (PVDF) are among the leading supports [112], but other functionalized materials can be also successfully employed, such as graphene oxide. For example, novel membranes based on graphene oxide and using the IL [BMIM][BF_4] have shown good selectivity for CO_2 from other gases [115]. Other authors such as Zhang et al. [116] reported the transport of H_2S in SILMs based on the IL 1-butyl-3-methylimidazolium acetate ([BMIM][Ac]), offering selective separation of H_2S from natural gas. Other presented highly CO_2 selective membranes (for CO_2/CH_4 mixtures) based on imidazolium ILs, using polypropylene as the support, have also been developed. It is remarkable that these separations have been reported to be stable and efficient in a lower temperatures range, from 283 to 298 K [112].

12.3.3.2 Separation of Metal Ions

The extraction of metal ions from aqueous solutions is another interesting field that has been studied in the application of SILMs. Several examples are given in Table 12.2. In this context, Alguacil et al. [80,117–119] have evaluated over several years the extraction of mainly Cr(VI), Cr(III), and Fe(III) from aqueous solutions, as well as many other metal ions and even against the presence of other metals, using different ILs. In their first assays, they employed a pseudoemulsion membrane strip dispersion (PEMSD) using the ionic liquid CYPHOS IL101 as ionophore to study the transport of Cr(VI) from hydrochloric acid medium. In their most recent assays, Alguacil et al. [117] highlighted the potential of a liquid membrane using CYPHOS IL102 [(trihexyl(tetradecyl) phosphonium bromide)] as a carrier for the facilitated transport of Cr(VI) from similar acidic media. Furthermore, in the same work, they performed an extensive comparison of this configuration against different carriers, as well as using two different types of liquid membrane operations. Similar results to those achieved in the assays discussed above for the removal of Cr(VI) anions from aqueous solutions but using organic solvents can be found in the literature. Kozlowski et al. [120] achieved them using polymeric inclusion membranes (PIMs) based on organic solvents, while Roy et al. [121] achieved them using amine-functionalized tubular ceramic membranes. Thus, we can observe the suitability of using ILs in supported liquid membranes for metal ion extraction.

As mentioned above, within the topic of metal ion recovery, the separation of chromium has been usually addressed. The recovery of Cr(III) from alkaline solutions employing a pseudoemulsion-based hollow-fiber strip dispersion (PEHFSD) has been addressed in the literature by Alguacil et al. [80,122]. For this purpose, they used the IL trioctylmethylammonium chloride obtaining a good permeation of chromium(III). A few years later, they employed the same PEHFSD configuration now with the IL [PJMTH][HSO_4] dissolved in n-decane as carrier obtaining Cr(II) recovery percentages close to 98% after 2 h of operation. More recently, assays with promising results have been reported addressing the separation of Cr(III), together with other metal ions, in acidic media using supported liquid membranes with the IL isooctylmethylimidazolium bis-2-ethylhexylphosphate as carrier [123]. Fe(III) is another metal ion that has been reported to be removed using SILMs. Alguacil et al. [119] used pseudoemulsion-based hollow-fiber strip dispersion with the ionic liquid $(PJMTH^+)_2(SO_4)^{2-}$. Another recurrent configuration for the removal of these metal ions is polymer inclusion membranes (PIMs). In previous years, PIMs containing tricaprylylmethylammonium chloride (Aliquat 336) as a plasticizer and 2-ethylhexylphosphonic acid (D2EHPA) as a metal ion carrier were studied for the removal of chromium(III) from acidic solutions [124]. In recent years, PIMs containing o-nitrophenyloctyl ether (NPOE) as a plasticizer and ionic liquids such as CYPHOS IL101 and CYPHOS IL 104 as metal ion carriers have been applied for hydrometallurgical separation of Zn(II) and Fe(III) from acidic solutions [125]

More recently, Zante et al. [69] investigated the use of SILMs for the selective separation of cobalt and nickel. In their study they focused on the recovery of cobalt by employing a membrane with tri(hexyl)tetradecyl phosphonium chloride as a carrier, an IL with hydrophobic properties, together with polyvinylidene difluoride (PVDF) as support, reaching recovery percentages of 86% under optimum conditions. Furthermore, they carried out two assays using flat-sheet supported IL membranes, hydrophobic and hydrophilic, achieving a separation factor of 218, which demonstrates the selectivity of these configurations to cobalt over nickel. This flat-sheet liquid membrane configuration was previously tested by de los Ríos et al. [90] in the selective extraction of Fe(III), Cu(III), Zn (II), and Cd(II) using the IL methyltrioctylammonium chloride. The influence of the composition of the extraction-stripping receiving phase [milliQ water or Na_2CO_3 (0.1 M) or NH_3 (6M)] was analyzed. Water milliQ enabled an extraction factor of 9 for Fe(III) at 31 h of operation, while Na_2CO_3 achieved an extraction factor of 373 for Zn(II). On the other hand, NH_3 was more selective toward Cd(II), reaching an extraction factor of 15.1 at 24 h. These results highlight the dependence of ion selectivity on stripping composition.

12.3.4 Stability of Ionic Liquid Membranes

Among the most important factors for IL membranes is their stability. In this sense, we can differentiate between supported ionic liquids membranes (SILMs) and quasi-solidified ionic liquid membranes, such as polymer inclusion IL membranes (PIMs), that include those configurations prepared by casting methods.

SILM consists of membranes in which an IL is introduced into support material pores by capillary forces. They can be prepared by several methods such as vacuum, pressure, and immersion procedures. Support materials can range from polymeric to inorganic membranes. The use of ILs as liquid phases in membranes is advantageous due to their properties like low vapor pressure and the option of reducing their solubility in the surrounding medium by tailoring the ion composition. However, the stability can be determined by the immobilization method, the compatibility between IL and support properties, the IL composition itself, as well as the interfacial tension between the aqueous and membrane, among other factors.

The method employed for the immobilization of ILs influences membrane stability. This has become evident in several research works. For instance, the differences between pressure- and vacuum-based methods on membrane stability have been reported when studying several imidazolium ILs. A comparative study of the preparation of SILMs by two different methods, under pressure and vacuum, was performed using Nylon membranes as support for ILs with 1-butyl-3-methylimidazolium ([bmin]) cation in combination with the respective anions Cl^-, BF_4^-, PF_6^-, and NTf_2^- [62]. The results showed that the pressured method was a better option for the immobilization of these ILs in comparison to vacuum. This effect was more noticeable for the ILs with higher viscosity, e.g. those based on Cl^- and PF_6^- anions, due to their difficulty in penetrating the pores.

The nature of the support and their chemical affinity are other important factors in determining the stability of SILMs. Neves et al. [126] compared the stability of SILMs formed by hydrophobic imidazolium ILs when combined with hydrophobic and hydrophilic polyvinylidene fluoride, respectively, observing that the configurations prepared with the hydrophobic support offered higher stability because of the better chemical compatibility of the liquid phases and the support. On the other hand, it has been reported that SLIMs can tolerate notable transmembrane pressure with suitable pore size. Membrane pore size within the range of 100–200 nm is usually considered adequate for these systems. The interactions of an electrostatic nature between charge functional groups present in the membrane surface and the organic cation in the IL liquid phase also affect stability [127,128].

The influence of the IL composition and the surrounding environment has also been analyzed. When studying the stability of SLIMs using ILs with the cation 1-butyl-3-methylimidazolium in combination with different anions in contact with organic phases (e.g. n-hexane, n-hexane/water, n-hexane/acetone, etc.) two important factors were observed to design stable SLIMs, the hydrophilicity of the IL and the polarity of the surrounding receiving phase in a separation system [128,129]. According to these studies, stability increases as such polarity decreases and the hydrophilicity of the IL rises. Other mechanisms that can compromise membrane stability include the detachment of ILs from the support and the possibility of dissolution and emulsification in contact with aqueous surrounding phases and the creation of water micro-environments inside the IL phase. In this last case, they can create unselective pathways for molecule transfer contributing to the constitution of new and non-selective environments for solute transport, leading to a decline in membrane performance and stability [130].

As described earlier, quasi-solidified IL membranes such as polymer inclusion membranes can be relatively simply synthesized using casting techniques by mixing an IL with a dissolved polymer in a solvent or gel which, after solvent evaporation, leads to the formation of a thin polymer film. Several studies have approached the stability of this type of membrane. These studies prove that the stability of the membranes mostly depends on the solubility of the IL in contact with the surrounding phase. For example, PIMs based on imidazolium ILs and polyvinyl chloride as polymer matrix were tested in aqueous phases [131]. Among them, [omin][PF_6] and [omin][NTF_2] showed losses of ILs lower than 10 wt.% in contact with aqueous phases since they are highly insoluble in water. Also, the stability was affected by the concentration of the IL used. Membranes with 30% IL were much more stable than membranes based on the same IL but with a concentration of 70%, which indicates that IL excess cannot be properly cast into the polymer matrix and thus IL content must be balanced. The compatibility of the IL and the polymer or gel is also crucial. For example, it has been reported that the thermal stability of IL membranes can decrease with the length of alkyl chains on imidazolium cations in VDF-HFP-based membranes, since a long alkyl chain shows less compatibility with the polymer matrix, observing a significant phase separation [132].

Other configurations based on thermostable polymeric materials such as polybenzimidazole with the IL 1-butyl-3-methylimidazolium bis(trifluoromethylsulfonyl)imide [bmim][NTf_2] can offer very high mechanical and thermal stability with transmembrane pressures between 2 and 6 bars and temperatures over 100ºC and up to 200ºC [133].

More recently, highly stable membranes were synthesized for the separation of magnesium and lithium ions. These membranes were prepared by mixing the IL 1-butyl-3-methylimidazolium bis(trifluoromethylsulfonyl)imide, [bmim][NTf_2], and tributylphosphate (as carriers) along with cellulose triacetate as polymer matrix [134]. The high stability was due to the presence of strong interactions of the cellulose triacetate with the IL and tributylphosphate by the formation of hydrogen bonding.

In any case, the new IL configurations that may be developed in the future must be studied from a stability point of view, since the stability of the synthesized membranes is crucial to maintaining efficient and long-term operation.

12.4 CONCLUSIONS AND FUTURE PERSPECTIVES

Ionic liquids (ILs) are considered task-specific solvents that can be tailored for many specific applications by the combination of suitable anions with cations, offering advantageous properties over conventional organic solvents. Through these possible combinations, the number of new ILs that can be synthesized is very high, and thus also the potential uses. As seen in this chapter, ILs have been intensively researched in separation processes both in neat form taking part in liquid–liquid extraction operations (as diluents and ligands) and in immobilized form in polymer supports to create IL-based membranes. These membranes include supported IL membranes (SILMs) and polymer inclusion membranes (PIMs), among others, which have been successfully applied for the separation of organic compounds from liquid and gas streams and the separation of metal ions. The works discussed in this chapter show that ILs can operate with high recovery yields and that high selectivity can be achieved depending on the chemical nature of the anion and cation and the characteristics of their substituents. Moreover, the versatility of ILs allows these compounds to be used in innovative separation configurations that include flat-sheet and hollow-fiber membranes, and under special conditions such as under a magnetic field for separation rate enhancement. These notable results and the designs of new functionalized ILs are expected to increase the efficiencies of these solvents in the separation of organics and metal ions opening new research pathways.

REFERENCES

[1] Singh, S.K., Savoy, A.W. (2020) Ionic liquids synthesis and applications: An overview. *Journal of Molecular Liquids* 297: 112038.

[2] Walsh, D.A., Goodwin, S. (2018) The oxygen reduction reaction in room-temperature ionic liquids. In: Wandelt, k. (Ed.) *Encyclopedia of Interfacial Chemistry: Surface Science and Electrochemistry*, pp. 898–907.

[3] Hayes, R., Warr, G.G., Atkin, R. (2015) Structure and nanostructure in ionic liquids. *Chemical Reviews* 115: 6357–6426.

[4] Khan, A.S., Ibrahim, T.H., Jabbar, N.A., Khamis, M.I., Nancarrow, P., Mjalli, F.S. (2021) Ionic liquids and deep eutectic solvents for the recovery of phenolic compounds: Effect of ionic liquids structure and process parameters. *RSC Advances* 11: 12398–12422.

[5] Silva, W., Zanatta, M., Ferreira, A.S., Corvo, M.C., Cabrita, E.J. (2020) Revisiting ionic liquid structure-property relationship: A critical analysis. *International Journal of Molecular Sciences*. 21: 7745.

[6] Sánchez-Sánchez, C.M. (2018) Electrocatalytic reduction of CO_2 in imidazolium-based ionic liquids. In: *Encyclopedia of Interfacial Chemistry: Surface Science and Electrochemistry*, pp. 539–551. Elsevier: Oxford, UK.

[7] Feldmann, C., Ruck, M. (2017) Ionic liquids – Designer solvents for the synthesis of new compounds and functional materials. *Zeitschrift Für Anorganische Und Allgemeine Chemie* 643: 2–2. https://doi.org/10.1002/ZAAC.201700001.

[8] Kianfar, E., Mafi, S. (2020) *Ionic Liquids: Properties, Application, and Synthesis, Fine Chemical Engineering*, pp. 22–31. Universal Wiser Publisher.

[9] Mallakpour, S., Dinari, M. (2012) *Ionic Liquids as Green Solvents: Progress and Prospects, Green Solvents II: Properties and Applications of Ionic Liquids*, pp. 1–32. Springer.

[10] Roland, C.M., Bair, S., Casalini, R. (2006) Thermodynamic scaling of the viscosity of van der Waals, H-bonded, and ionic liquids. *The Journal of Chemical Physics* 125: 124508.

[11] Abushammala, H., Mao, J. (2020) A review on the partial and complete dissolution and fractionation of wood and lignocelluloses using imidazolium ionic liquids. *Polymers (Basel)* 12: 194.

[12] Malolan, R., Gopinath, K.P., Vo, D.V.N., Jayaraman, R.S., Adithya, S., Ajay, P.S., Arun, J. (2021) Green ionic liquids and deep eutectic solvents for desulphurization, denitrification, biomass, biodiesel, bioethanol and hydrogen fuels: A review. *Environmental Chemistry Letters* 19: 1001–1023.

[13] Costa, S.P.F., Azevedo, A.M.O., Pinto, P.C.A.G., Saraiva, M.L.M.F.S. (2017) Environmental impact of ionic liquids: Recent advances in (eco)toxicology and (bio) degradability. *ChemSusChem* 10: 2321–2347.

[14] Vraneš, M., Tot, A., Jovanović-Šanta, S., Karaman, M., Dožić, S., Tešanović, K., Kojić, V., Gadžurić, S. (2016) Toxicity reduction of imidazolium-based ionic liquids by the oxygenation of the alkyl substituent. *RSC Advances* 6: 96289–96295.

[15] Harjani, J.R., Singer, R.D., Garcia, M.T., Scammells, P.J. (2008) The design and synthesis of biodegradable pyridinium ionic liquids. *Green Chemistry*. https://doi.org/10.1039/b800534f.

[16] Noshadi, I., Walker, B.W., Portillo-Lara, R., Sani, E.S., Gomes, N., Aziziyan, M.R., Annabi, N. (2017) Engineering biodegradable and biocompatible bio-ionic liquid conjugated hydrogels with tunable conductivity and mechanical properties. *Scientific Reports* 7: 1–18.

[17] Gupta, G.R., Girase, T.R., Kapdi, A.R. (2019) Ionic liquid as a sustainable reaction medium for Diels-Alder reaction. In: *Encyclopedia of Ionic Liquids*, pp. 1–22. Springer.

[18] Ullah, H., Wilfred, C.D., Shaharun, M.S. (2018) Ionic liquid-based extraction and separation trends of bioactive compounds from plant biomass. *Separation Science and Technology* 54: 559–579.

[19] Steinrück, H.P., Wasserscheid, P. (2015) Ionic liquids in catalysis. *Catalysis Letters* 145: 380–397.

[20] Xu, P., Liang, S., Zong, M.H., Lou, W.Y. (2021) Ionic liquids for regulating biocatalytic process: Achievements and perspectives. *Biotechnology Advances* 51: 107702.

[21] Knies, M., Groh, M.F., Pietsch, T., Lê Anh, M., Ruck, M. (2021) Metal assisted synthesis of cationic sulfidobismuth cubanes in ionic liquids. *ChemistryOpen* 10: 59–59.

[22] Watanabe, M., Thomas, M.L., Zhang, S., Ueno, K., Yasuda, T., Dokko, K. (2017) Application of ionic liquids to energy storage and conversion materials and devices. *Chemical Reviews* 117: 7190–7239.

[23] Tiago, G.A.O., Matias, I.A.S., Ribeiro, A.P.C., Martins, L.M.D.R.S. (2020) Application of ionic liquids in electrochemistry—Recent advances. *Molecules* 25: 5812.

[24] vander Hoogerstraete, T., Onghena, B., Binnemans, K. (2013) Homogeneous liquid-liquid extraction of metal ions with a functionalized ionic liquid. *Journal of Physical Chemistry Letters* 4: 1659–1663.

[25] Wang, J., Luo, J., Feng, S., Li, H., Wan, Y., Zhang, X. (2016) Recent development of ionic liquid membranes. *Green Energy & Environment* 1: 43–61.

[26] Ayati, A., Ranjbari, S., Tanhaei, B., Sillanpää, M. (2019) Ionic liquid-modified composites for the adsorptive removal of emerging water contaminants: A review. *Journal of Molecular Liquids* 275: 71–83.

[27] Sas, O.G., Domínguez, I., González, B., Domínguez, Á. (2018) Liquid-liquid extraction of phenolic compounds from water using ionic liquids: Literature review and new experimental data using [C2mim]FSI. *Journal of Environmental Management* 228: 475–482.

[28] al Kaisy, G.M.J., Abdul Mutalib, M.I., Bustam, M.A., Leveque, J.M., Muhammad, N. (2016) Liquid-liquid extraction of aromatics and sulfur compounds from base oil using ionic liquids. *Journal of Environmental Chemical Engineering* 4: 4786–4793.

[29] Domínguez-Pérez, M., Tomé, L.I.N., Freire, M.G., Marrucho, I.M., Cabeza, O., Coutinho, J.A.P. (2010) (Extraction of biomolecules using) aqueous biphasic systems formed by ionic liquids and aminoacids. *Separation and Purification Technology* 72: 85–91.
[30] Sprakel, L.M.J., Schuur, B. (2019) Solvent developments for liquid-liquid extraction of carboxylic acids in perspective. *Separation and Purification Technology* 211: 935–957.
[31] Mahajan, S., Singh, N., Kushwaha, J.P., Rajor, A. (2019) Evaluation and mechanism of cationic/anionic dyes extraction from water by ionic liquids. *Chemical Engineering Communications* 206: 697–707.
[32] Nie, Y., Bai, L., Dong, H., Zhang, X., Zhang, S. (2012) Extraction of asphaltenes from direct coal liquefaction residue by dialkylphosphate ionic liquids. *Separation Science and Technology* 47: 386–391.
[33] Ventura, S.P.M., Silva, F.A.E., Quental, M.v., Mondal, D., Freire, M.G., Coutinho, J.A.P. (2017) Ionic-liquid-mediated extraction and separation processes for bioactive compounds: Past, present, and future trends. *Chemical Reviews* 117: 6984–7052.
[34] Calla-Quispe, E., Robles, J., Areche, C., Sepulveda, B. (2020) Are ionic liquids better extracting agents than toxic volatile organic solvents? A combination of ionic liquids, microwave and LC/MS/MS, applied to the lichen *Stereocaulon glareosum*. *Frontiers in Chemistry* 8: 450.
[35] Fan, Y., Li, Y., Dong, X., Hu, G., Hua, S., Miao, J., Zhou, D. (2014) Extraction of phenols from water with functionalized ionic liquids. *Industrial and Engineering Chemistry Research* 53: 20024–20031.
[36] Berenjian, A., Chan, N., Malmiri, H.J. (2012) Volatile organic compounds removal methods: A review. *American Journal of Biochemistry and Biotechnology* 8: 220–229.
[37] Salar-García, M.j., Ortiz-Martínez, V.M., Hernández-Fernández, F.J., de los Ríos, A.P., Quesada-Medina, J. (2017) Ionic liquid technology to recover volatile organic compounds (VOCs). *Journal of Hazardous Materials* 321: 484–499.
[38] Królikowski, M. (2016) Liquid-liquid extraction of p-xylene from their mixtures with alkanes using 1-butyl-1-methylmorpholinium tricyanomethanide and 1-butyl-3-methylimidazolium tricyanomethanide ionic liquids. *Fluid Phase Equilibria* 412: 107–114.
[39] Zhou, T., Wang, Z., Chen, L., Ye, Y., Qi, Z., Freund, H., Sundmacher, K. (2012) Evaluation of the ionic liquids 1-alkyl-3-methylimidazolium hexafluorophosphate as a solvent for the extraction of benzene from cyclohexane: (liquid + liquid) equilibria. *Journal of Chemical Thermodynamics* 48: 145–149.
[40] Enayati, M., Mokhtarani, B., Sharifi, A., Mirzaei, M. (2016) Extraction of benzene from heptane with pyridinium based ionic liquid at (298.15, 308.15 and 318.15) K. *Fluid Phase Equilibria* 411: 53–58.
[41] Hernández-Fernández, F.J., de los Ríos, A.P., Gómez, D., Rubio, M., Víllora, G. (2010) Selective extraction of organic compounds from transesterification reaction mixtures by using ionic liquids. *AIChE Journal* 56: 1213–1217.
[42] Alonso, M., López-Delgado, A., Sastre, A.M., Alguacil, F.J. (2006) Kinetic modelling of the facilitated transport of cadmium (II) using Cyanex 923 as ionophore. *Chemical Engineering Journal* 118: 213–219.
[43] Harvie, A.J., Mello, J.C. (2021) Efficient extraction of metal ions using a recirculating two-stage flow reactor. *Chemistry–Methods* 1: 492–493.
[44] Vieira, M., Rodrigues, A.E., Yudaev, P.A., Chistyakov, E.M. (2022) Ionic liquids as components of systems for metal extraction. *ChemEngineering* 6: 6.
[45] Lertlapwasin, R., Bhawawet, N., Imyim, A., Fuangswasdi, S. (2010) Ionic liquid extraction of heavy metal ions by 2-aminothiophenol in 1-butyl-3-methylimidazolium hexafluorophosphate and their association constants. *Separation and Purification Technology* 72: 70–76.
[46] Quijada-Maldonado, E., Allain, A., Pérez, B., Merlet, G., Cabezas, R., Tapia, R., Romero, J. (2020) Selective liquid-liquid extraction of molybdenum (VI) and rhenium (VII) from a synthetic pregnant leach solution: Comparison between extractants and diluents. *Minerals Engineering* 145: 106060.
[47] Song, Y., Tsuchida, Y., Matsumiya, M., Uchino, Y., Yanagi, I. (2018) Separation of tungsten and cobalt from WC-Co hard metal wastes using ion-exchange and solvent extraction with ionic liquid. *Minerals Engineering* 128: 224–229.
[48] Matsumiya, M., Song, Y., Tsuchida, Y., Ota, H., Tsunashima, K. (2019) Recovery of platinum by solvent extraction and direct electrodeposition using ionic liquid. *Separation and Purification Technology* 214: 162–167.
[49] Sun, X.Q., Peng, B., Chen, J., Li, D.Q., Luo, F. (2008) An effective method for enhancing metal-ions' selectivity of ionic liquid-based extraction system: Adding water-soluble complexing agent. *Talanta* 74: 1071–1074.
[50] Wei, G.T., Yang, Z., Chen, C.J. (2003) Room temperature ionic liquid as a novel medium for liquid/liquid extraction of metal ions. *Analytica Chimica Acta* 488: 183–192.
[51] Cocalia, V.A., Holbrey, J.D., Gutowski, K.E., Bridges, N.J., Rogers, R.D. (2006) Separations of metal ions using ionic liquids: The challenges of multiple mechanisms. *Tsinghua Science and Technology* 11: 188–193.
[52] Harjani, J.R., Friščić, T., MacGillivray, L.R., Singer, R.D. (2008) Removal of metal ions from aqueous solutions using chelating task-specific ionic liquids. *Dalton Transactions* 34: 4595–4601.
[53] de Los Ríos, A.P., Hernández-Fernández, F.J., Alguacil, F.J., Lozano, L.J., Ginestá, A., García-Díaz, I., Sánchez-Segado, S., López, F.A., Godínez, C. (2012) On the use of imidazolium and ammonium-based ionic liquids as green solvents for the selective recovery of Zn(II), Cd(II), Cu(II) and Fe(III) from hydrochloride aqueous solutions. *Separation and Purification Technology* 97: 150–157.
[54] Corbett, p.J., McIntosh, A.J.S., Gee, M., Hallett, J.P. (2018) Use of ionic liquids to remove harmful M2+ contaminants from hydrocarbon streams. *Molecular Systems Design and Engineering* 3: 408–417.
[55] Turgis, R., Arrachart, G., Michel, S., Legeai, S., Lejeune, M., Draye, M., Pellet-Rostaing, S. (2018) Ketone functionalized task specific ionic liquids for selective tantalum extraction. *Separation and Purification Technology* 196: 174–182.
[56] Imam, D.M., El-Nadi, Y.A. (2018) Recovery of molybdenum from alkaline leach solution of spent hydrotreating catalyst by solvent extraction using methyl tricaprylammonium hydroxide. *Hydrometallurgy* 180.
[57] Boudesocque, S., Mohamadou, A., Conreux, A., Marin, B., Dupont, L. (2019) The recovery and selective extraction of

gold and platinum by novel ionic liquids. *Separation and Purification Technology* 210: 824–834.
[58] Lee, J.M. (2012) Extraction of noble metal ions from aqueous solution by ionic liquids. *Fluid Phase Equilibria* 319: 30–36.
[59] Zhou, C., Li, Y., Xue, X.X. (2020) Extraction mechanism of vanadium from vanadium slag leaching solution by N-octylpyridine ionic liquids. *Zhongguo Youse Jinshu Xuebao/Chinese Journal of Nonferrous Metals* 30: 172–179.
[60] Muthuraman, G., Palanivelu, K. (2006) Transport of textile dye in vegetable oils based supported liquid membrane. *Dyes and Pigments* 70: 99–104.
[61] Zante, G., Boltoeva, M., Masmoudi, A., Barillon, R., Trébouet, D. (2022) Supported ionic liquid and polymer inclusion membranes for metal separation. *Separation and Purification Reviews* 51: 100–116.
[62] Kamaz, M., Vogler, R.J., Jebur, M., Sengupta, A., Wickramasinghe, R. (2020) π Electron induced separation of organic compounds using supported ionic liquid membranes. *Separation and Purification Technology* 236: 116237.
[63] Yang, X.J., Fane, A.G., Soldenhoff, K. (2003) Comparison of liquid membrane processes for metal separations: Permeability, stability, and selectivity. *Industrial and Engineering Chemistry Research* 42: 392–403.
[64] Ren, Y., Zhang, J., Guo, J., Chen, F., Yan, F. (2017) Porous poly(ionic liquid) membranes as efficient and recyclable absorbents for heavy metal ions. *Macromolecular Rapid Communications* 38: 1700151.
[65] Branco, L.C., Crespo, J.G., Afonso, C.A.M. (2002) Studies on the selective transport of organic compounds by using ionic liquids as novel supported liquid membranes. *Chemistry – A European Journal* 8: 3865–3871.
[66] Panigrahi, A., Pilli, S.R., Mohanty, K. (2013) Selective separation of Bisphenol A from aqueous solution using supported ionic liquid membrane. *Separation and Purification Technology* 107: 70–78.
[67] Kouki, N., Tayeb, R., Dhahbi, M. (2014) Recovery of acetaminophen from aqueous solutions using a supported liquid membrane based on a quaternary ammonium salt as ionophore. *Chemical Papers* 68: 457–464.
[68] Kim, D.L., Vovusha, H., Schwingenschlögl, U., Nunes, S.P. (2017) Polyethersulfone flat sheet and hollow fiber membranes from solutions in ionic liquids. *Journal of Membrane Science* 539: 161–171.
[69] Zante, G., Boltoeva, M., Masmoudi, A., Barillon, R., Trébouet, D. (2020) Selective separation of cobalt and nickel using a stable supported ionic liquid membrane. *Separation and Purification Technology* 252: 117477.
[70] Lee, L.Y., Morad, N., Ismail, N., Rafatullah, M. (2021) Selective separation of cadmium(Ii), copper(ii) and nickel(ii) ions from electroplating wastewater using dual flat sheet supported liquid membrane. *Desalination and Water Treatment* 224: 291–301.
[71] Santos, E., Albo, J., Daniel, C.I., Portugal, C.A.M., Crespo, J.G., Irabien, A. (2013) Permeability modulation of supported magnetic ionic liquid membranes (SMILMs) by an external magnetic field. *Journal of Membrane Science* 430.
[72] Klingberg, P., Wilkner, K., Schlüter, M., Grünauer, J., Shishatskiy, S. (2019) Separation of carbon dioxide from real power plant flue gases by gas permeation using a supported ionic liquid membrane: An investigation of membrane stability. *Membranes (Basel)* 9: 35.
[73] Pang, L., Yang, P., Pang, R., Li, S. (2017) Bis(trifluoromethylsulfonyl)imide-based frozen ionic liquid for the hollow-fiber solid-phase microextraction of dichlorodiphenyltrichloroethane and its main metabolites. *Journal of Separation Science* 40: 3311–3317.
[74] Marr, P.C., Marr, A.C. (2015) Ionic liquid gel materials: Applications in green and sustainable chemistry. *Green Chemistry* 18: 105–128.
[75] Wang, B.Y., Zhang, N., Li, Z.Y., Lang, Q.L., Yan, B.H., Liu, Y., Zhang, Y. (2019) Selective separation of acetic and hexanoic acids across polymer inclusion membrane with ionic liquids as carrier. *International Journal of Molecular Sciences* 20: 3915.
[76] Hernández-Fernández, F.J., de Los Ríos, A.P., Mateo-Ramírez, F., Juarez, M.D., Lozano-Blanco, L.J., Godínez, C. (2016) New application of polymer inclusion membrane based on ionic liquids as proton exchange membrane in microbial fuel cell. *Separation and Purification Technology* 160: 51–58.
[77] Comesaña, A., Rodriguez-Monsalve, J., Cerpa, A., Alguacil, F.J. (2011) Non-dispersive solvent extraction with strip dispersion (NDSXSD) pertraction of Cd(II) in HCl medium using ionic liquid CYPHOS IL101. *Chemical Engineering Journal* 175: 228–232.
[78] Luis, P., Garea, A., Irabien, A. (2009) Zero solvent emission process for sulfur dioxide recovery using a membrane contactor and ionic liquids. *Journal of Membrane Science* 330: 80–89.
[79] Wojciechowska, A., Reis, M.T.A., Wojciechowska, I., Ismael, M.R.C., Gameiro, M.L.F., Wieszczycka, K., Carvalho, J.M.R. (2018) Application of pseudo-emulsion based hollow fiber strip dispersion with task-specific ionic liquids for recovery of zinc(II) from chloride solutions. *Journal of Molecular Liquids* 254: 369–376.
[80] Alguacil, F.J., Garcia-Diaz, I., Lopez, F.A. (2013) Modeling of facilitated transport of Cr(III) using (RNH3+HSO4-) ionic liquid and pseudo-emulsion hollow fiber strip dispersion (PEHFSD) technology. *Journal of Industrial and Engineering Chemistry* 19: 1086–1091.
[81] Pilli, S.R., Banerjee, T., Mohanty, K. (2015) Performance of different ionic liquids to remove phenol from aqueous solutions using supported liquid membrane. *Desalination and Water Treatment* 54: 3062–3072.
[82] Dahi, A., Fatyeyeva, K., Langevin, D., Chappey, C., Poncin-Epaillard, F., Marais, S. (2017) Effect of cold plasma surface treatment on the properties of supported ionic liquid membranes. *Separation and Purification Technology* 187: 127–136.
[83] Hernández-Fernández, F.J., de los Ríos, A.P., Tomás-Alonso, F., Palacios, J.M., Víllora, G. (2009) Preparation of supported ionic liquid membranes: Influence of the ionic liquid immobilization method on their operational stability. *Journal of Membrane Science* 341: 172–177.
[84] Grünauer, J., Filiz, V., Shishatskiy, S., Abetz, C., Abetz, V. (2016) Scalable application of thin film coating techniques for supported liquid membranes for gas separation made from ionic liquids. *Journal of Membrane Science* 518: 178–191.
[85] de los Ríos, A.P., Hernández-Fernández, F.J., Tomás-Alonso, F., Palacios, J.M., Gómez, D., Rubio, M., Víllora, G. (2007) A SEM-EDX study of highly stable supported liquid membranes based on ionic liquids. *Journal of Membrane Science* 300: 88–94.

[86] Cichowska-Kopczyńska, I., Joskowska, M., Debski, B., Aranowski, R., Hupka, J. (2018) Separation of toluene from gas phase using supported imidazolium ionic liquid membrane. *Journal of Membrane Science* 566: 367–373.
[87] Xiao, H.F., Chu, C.H., Xu, W.T., Chen, B.Z., Ju, X.H., Xing, W., Sun, S.P. (2019) Amphibian-inspired amino acid ionic liquid functionalized nanofiltration membranes with high water permeability and ion selectivity for pigment wastewater treatment. *Journal of Membrane Science* 586: 44–52.
[88] Fatyeyeva, K., Rogalsky, S., Tarasyuk, O., Chappey, C., Marais, S. (2018) Vapour sorption and permeation behaviour of supported ionic liquid membranes: Application for organic solvent/water separation. *Reactive and Functional Polymers* 130: 16–28.
[89] Li, J., Li, B., Sui, G., Du, L., Zhuang, Y., Zhang, Y., Zou, Y. (2021) Removal of volatile organic compounds from air using supported ionic liquid membrane containing ultraviolet-visible light-driven Nd-TiO_2 nanoparticles. *Journal of Molecular Structure* 1231: 130023.
[90] de Los Ríos, A.P., Hernández-Fernández, F.J., Lozano, L.J., Sánchez-Segado, S., Ginestá-Anzola, A., Godínez, C., Tomás-Alonso, F., Quesada-Medina, J. (2013) On the selective separation of metal ions from hydrochloride aqueous solution by pertraction through supported ionic liquid membranes. *Journal of Membrane Science* 444: 469–481.
[91] Cichowska-Kopczyńska, I., Aranowski, R. (2020) Use of pyridinium and pyrrolidinium ionic liquids for removal of toluene from gas streams. *Journal of Molecular Liquids* 298: 112091.
[92] de Los Ríos, A.P., Hernández-Fernández, F.J., Presa, H., Gómez, D., Víllora, G. (2009) Tailoring supported ionic liquid membranes for the selective separation of transesterification reaction compounds. *Journal of Membrane Science* 328: 81–85.
[93] de los Ríos, A.P., Hernández-Fernández, F.J., Rubio, M., Tomás-Alonso, F., Gómez, D., Víllora, G. (2008) Prediction of the selectivity in the recovery of transesterification reaction products using supported liquid membranes based on ionic liquids. *Journal of Membrane Science* 307: 225–232.
[94] Hernández-Fernández, F.J., de los Ríos, A.P., Tomás-Alonso, F., Gómez, D., Víllora, G. (2009) Kinetic resolution of 1-phenylethanol integrated with separation of substrates and products by a supported ionic liquid membrane. *Journal of Chemical Technology and Biotechnology* 84: 337–342.
[95] Hernández-Fernández, F.J., de Los Ríos, A.P., Lozano-Blanco, L.J., Godínez, C. (2010) Biocatalytic ester synthesis in ionic liquid media. *Journal of Chemical Technology and Biotechnology* 85: 1423–1435.
[96] Hernández-Fernández, F.J., de los Ríos, A.P., Tomás-Alonso, F., Gómez, D., Víllora, G. (2009) Improvement in the separation efficiency of transesterification reaction compounds by the use of supported ionic liquid membranes based on the dicyanamide anion. *Desalination* 244: 122–129.
[97] de los Ríos, A.P., Hernández-Fernández, F.J., Rubio, M., Gómez, D., Víllora, G. (2010) Highly selective transport of transesterification reaction compounds through supported liquid membranes containing ionic liquids based on the tetrafluoroborate anion. *Desalination* 250: 101–104.
[98] Matsumoto, M., Ohtani, T., Kondo, K. (2007) Comparison of solvent extraction and supported liquid membrane permeation using an ionic liquid for concentrating penicillin G. *Journal of Membrane Science* 289: 92–96.
[99] Daniel, C.I., Rubio, A.M., Sebastião, P.J., Afonso, C.A.M., Storch, J., Izák, P., Portugal, C.A.M., Crespo, J.G. (2016) Magnetic modulation of the transport of organophilic solutes through supported magnetic ionic liquid membranes. *Journal of Membrane Science* 505: 36–43.
[100] Zou, Y., Zhang, Z., Shao, X., Chen, Y., Wu, X., Yang, L., Zhu, J., Zhang, D. (2014) Hollow-fiber-supported liquid-phase microextraction using an ionic liquid as the extractant for the pre-concentration of bisphenol A, 17-β-estradiol, estrone and diethylstilbestrol from water samples with HPLC detection. *Water Science and Technology* 69: 1028–1035.
[101] Ibrahim, Z.M., Hendrik, M., Basheer, C., Abdulraheem, T.O., Al-Arfaj, A.R., Siddiqui, M.N. (2021) Membrane-assisted flow reactor for the extraction of sulfur compounds in petroleum crude and its fractions. *Arabian Journal for Science and Engineering* 47: 7013–7022.
[102] Izák, P., Köckerling, M., Kragl, U. (2006) Solute transport from aqueous mixture through supported ionic liquid membrane by pervaporation. *Desalination* 199: 96–98.
[103] Izák, P., Ruth, W., Fei, Z., Dyson, P.J., Kragl, U. (2008) Selective removal of acetone and butan-1-ol from water with supported ionic liquid–polydimethylsiloxane membrane by pervaporation. *Chemical Engineering Journal* 139: 318–321.
[104] Matsumoto, M., Ueba, K., Kondo, K. (2009) Vapor permeation of hydrocarbons through supported liquid membranes based on ionic liquids. *Desalination* 241: 365–371.
[105] Rdzanek, P., Marszałek, J., Kamiński, W. (2018) Biobutanol concentration by pervaporation using supported ionic liquid membranes. *Separation and Purification Technology* 196: 124–131.
[106] Li, W., Molina-Fernández, C., Estager, J., Monbaliu, J.C.M., Debecker, D.P., Luis, P. (2020) Supported ionic liquid membranes for the separation of methanol/dimethyl carbonate mixtures by pervaporation. *Journal of Membrane Science* 598: 117790.
[107] Kamiya, T., Takara, E., Ito, A. (2017) Separation of aromatic compounds from hydrocarbon mixtures by vapor permeation using liquid membranes with ionic liquids. *Journal of Chemical Engineering of Japan* 50: 684–691.
[108] Kulkami, P.S., Neves, L., Coelhoso, I., Afonso, C.A.M., Crespo, J.G. (2012) Supported ionic liquid membranes for removal of persistent organic pollutants (dioxins). *Procedia Engineering* 44: 140–142.
[109] Miyako, E., Maruyama, T., Kamiya, N., Goto, M. (2003) Use of ionic liquids in a lipase-facilitated supported liquid membrane. *Biotechnology Letters* 25: 805–808.
[110] Hernández-Fernández, F.J., de los Ríos, A.P., Tomás-Alonso, F., Gómez, D., Víllora, G. (2008) On the development of an integrated membrane process with ionic liquids for the kinetic resolution of rac-2-pentanol. *Journal of Membrane Science* 314: 238–246.
[111] Hernández-Fernández, F.J., de los Ríos, A.P., Tomás-Alonso, F., Gómez, D., Rubio, M., Víllora, G. (2007) Integrated reaction/separation processes for the kinetic resolution of rac-1-phenylethanol using supported liquid membranes based on ionic liquids. *Chemical Engineering and Processing: Process Intensification* 46: 818–824.

[112] Friess, K., Izák, P., Kárászová, M., Pasichnyk, M., Lanč, M., Nikolaeva, D., Luis, P., Jansen, J.C. (2021) A review on ionic liquid gas separation membranes. *Membranes* 11: 97.

[113] Ziobrowski, Z., Rotkegel, A. (2019) Enhanced CO_2/N_2 separation by supported ionic liquid membranes (SILMs) based on PDMS and 1-ethyl-3-methylimidazolium acetate. *Chemical Engineering Communications* 208: 137–147.

[114] Bui, T.T.L., Uong, H.T.N., v Nguyen, L., Pham, N.C., Thang, D., Tu Liem, B., Noi, H., Nam, V., Giay, C. (2018) Synthesis, characterization, and impregnation of some ionic liquids... *Chemical and Biochemical Engineering Quarterly* 32: 41–53.

[115] Ying, W., Cai, J., Zhou, K., Chen, D., Ying, Y., Guo, Y., Kong, X., Xu, Z., Peng, X. (2018) Ionic liquid selectively facilitates CO_2 transport through graphene oxide membrane. *ACS Nano* 12: 5385–5393.

[116] Zhang, X., Tu, Z., Li, H., Huang, K., Hu, X., Wu, Y., MacFarlane, D.R. (2017) Selective separation of H_2S and CO_2 from CH_4 by supported ionic liquid membranes. *Journal of Membrane Science* 543: 282–287.

[117] Alguacil, F.J. (2019) Facilitated chromium(VI) transport across an ionic liquid membrane impregnated with cyphos IL102. *Molecules* 24: 2437.

[118] Alguacil, F.J., Alonso, M., Lopez, F.A., Lopez-Delgado, A. (2010) Pseudo-emulsion membrane strip dispersion (PEMSD) pertraction of chromium(VI) using CYPHOS IL101 ionic liquid as carrier. *Environmental Science and Technology* 44: 7504–7508.

[119] Alguacil, F.J., Alonso, M., Lopez, F.A., Lopez-Delgado, A., Padilla, I., Tayibi, H. (2010) Pseudo-emulsion based hollow fiber with strip dispersion pertraction of iron(III) using (PJMTH+)2(SO42-) ionic liquid as carrier. *Chemical Engineering Journal* 157: 366–372.

[120] Kozlowski, C.A., Walkowiak, W. (2002) Removal of chromium(VI) from aqueous solutions by polymer inclusion membranes. *Water Research* 36: 4870–4876.

[121] Roy, S., Majumdar, S., Sahoo, G.C., Bhowmick, S., Kundu, A.K., Mondal, P. (2020) Removal of As(V), Cr(VI) and Cu(II) using novel amine functionalized composite nanofiltration membranes fabricated on ceramic tubular substrate. *Journal of Hazardous Materials* 399: 122841.

[122] Alguacil, F.J., Alonso, M., Lopez, F.A., Lopez-Delgado, A. (2009) Application of pseudo-emulsion based hollow fiber strip dispersion (PEHFSD) for recovery of Cr(III) from alkaline solutions. *Separation and Purification Technology* 66: 586–590.

[123] Jean, E., Villemin, D., Hlaibi, M., Lebrun, L. (2018) Heavy metal ions extraction using new supported liquid membranes containing ionic liquid as carrier. *Separation and Purification Technology* 201: 1–9.

[124] Konczyk, J., Kozlowski, C., Walkowiak, W. (2010) Removal of chromium(III) from acidic aqueous solution by polymer inclusion membranes with D2EHPA and Aliquat 336. *Desalination* 263: 211–216.

[125] Baczyńska, M., Słomka, Ż., Rzelewska, M., Waszak, M., Nowicki, M., Regel-Rosocka, M. (2018) Characterization of polymer inclusion membranes (PIM) containing phosphonium ionic liquids and their application for separation of Zn(II) from Fe(III). *Journal of Chemical Technology and Biotechnology* 93: 1767–1777.

[126] Neves, L.A., Crespo, J.G., Coelhoso, i.M. (2010) Gas permeation studies in supported ionic liquid membranes. *Journal of Membrane Science* 357: 160–170.

[127] Close, J.J., Farmer, K., Moganty, S.S., Baltus, R.E. (2012) CO_2/N_2 separations using nanoporous alumina-supported ionic liquid membranes: Effect of the support on separation performance. *Journal of Membrane Science* 390–391: 201–210.

[128] Fortunato, R., Afonso, C.A.M., Reis, M.A.M., Crespo, J.G. (2004) Supported liquid membranes using ionic liquids: study of stability and transport mechanisms. *Journal of Membrane Science* 242: 197–209.

[129] Hernández-Fernández, F.J., de Los Ríos, A.P., Tomás-Alonso, F., Palacios, J.M., Víllora, G. (2012) Understanding the influence of the ionic liquid composition and the surrounding phase nature on the stability of supported ionic liquid membranes. *AIChE Journal* 58: 583–590.

[130] Bijani, S., Fortunato, R., v. Martínez de Yuso, M., Heredia-Guerrero, F.A., Rodríguez-Castellón, E., Coehloso, I., Crespo, J., Benavente, J. (2009) Physical-chemical and electrical characterizations of membranes modified with room temperature ionic liquids: Age effect. *Vacuum* 83: 1283–1286.

[131] Tomás-Alonso, F., Rubio, A.M., Giménez, A., de los Ríos, A.P., Salar-García, M.J., Ortiz-Martínez, V.M., Hernández-Fernández, F.J. (2017) Influence of ionic liquid composition on the stability of polyvinyl chloride-based ionic liquid inclusion membranes in aqueous solution. *AIChE Journal* 63: 770–780.

[132] Jansen, J.C., Clarizia, G., Bernardo, P., Bazzarelli, F., Friess, K., Randová, A., Schauer, J., Kubicka, D., Kacirková, M., Izak, P. (2013) Gas transport properties and pervaporation performance of fluoropolymer gel membranes based on pure and mixed ionic liquids. *Separation and Purification Technology* 109: 87–97.

[133] Liang, L., Gan, Q., Nancarrow, P. (2014) Composite ionic liquid and polymer membranes for gas separation at elevated temperatures. *Journal of Membrane Science* 450: 407–417.

[134] Xu, L., Zeng, X., He, Q., Deng, T., Zhang, C., Zhang, W. (2022) Stable ionic liquid-based polymer inclusion membranes for lithium and magnesium separation. *Separation and Purification Technology* 288: 120626.

13 Membrane Applications in Dairy Science

Sandra E. Kentish and George Q. Chen*
Department of Chemical Engineering, The University of Melbourne, Victoria, Australia
*Email: sandraek@unimelb.edu.au

13.1 INTRODUCTION

Bovine milk is a complex, multiphase system. It is both a suspension of casein protein micelles and an emulsion of fat droplets, within an aqueous phase that also contains soluble proteins, lactose, and mineral salts. A typical bovine milk composition is provided in Table 13.1, with the breakdown of proteins and mineral salts in Tables 13.2 and 13.3.

The bulk of the milk proteins are held within casein micelles, which are spherical structures of around 0.1–0.5 micron in size. These structures have an internal core comprised of calcium phosphate colloids, stabilized by α_s and β casein proteins (Figure 13.1). This core is surrounded by κ-casein proteins, which have a hydrophilic tail that protrudes into the aqueous phase, forming a "hairy layer" on the surface. This "hairy layer" prevents the micelles from coagulating through both steric and electrostatic repulsion.

During cheese manufacture, the κ-casein "hairs" are cleaved from this protein molecule by the enzyme chymosin, causing the micelles to coagulate to form a gel, which after a sequence of cutting, pressing, and ripening forms the cheese. The remaining liquid is known as sweet whey and contains the whey proteins, as well as the κ-casein "hairs", which are known as casein macropeptides (CMP). Both the protein and lactose within this sweet whey can be recovered as spray-dried powders of significant commercial value.

In Greek yoghurt and cream cheese manufacture, the casein micelles coagulate due to changes in electrostatic interactions, which are induced by a reduction in the milk pH to around 4.6. In turn, this generates an acid whey, instead of sweet whey. The later brining of some cheeses can also produce a "salty whey" that is significantly higher in ionic strength (Table 13.4). These salty whey and acid whey streams are of much less commercial value than sweet whey. They can instead form waste streams that can be difficult to dispose of, due to increasing environmental regulation.

13.2 PRESSURE-DRIVEN MEMBRANE-BASED PROCESSES

The most common membrane operations used within the dairy factory are pressure-driven and are routinely categorized as microfiltration (MF), ultrafiltration (UF), nanofiltration (NF), and reverse osmosis (RO). As shown in Figure 13.2, these categories reflect the pore size within the membrane, with smaller pores requiring a larger driving force to effect separation.

MF has pore sizes of around 0.1–10 microns and so can retain casein micelles, fat globules, and micro-organisms. UF has smaller pore sizes of around 0.01–0.1 micron, or 1–500 kDa molecular weight, so are ideal for retaining whey proteins, while allowing the passage of lactose and dairy minerals. Conversely, NF retains mainly the larger divalent salts and lactose, while passing monovalent salts. Reverse osmosis membranes retain all charged species and pass only water and some very small non-charged molecules such as urea. The typical deployment of these membrane processes within the dairy factory is summarized in Figure 13.3.

In most applications, polymeric spiral-wound membrane elements are used (Figure 13.4). These are formed from two polymer membrane sheets that are glued together into an envelope. Several of these envelopes are then wound around a central permeate collection tube. One of more of these spiral-wound elements is then placed end-to-end within a tube, usually formed from polycarbonate or stainless steel, to form a single module. Polymeric hollow fibres can also be used in wastewater treatment applications. These are formed through a spinning process and resemble a bundle of drinking straws in the same configuration as a shell and tube heat exchanger. Common polymers used are polyvinylidene fluoride (PVDF), polypropylene (PP), polyethersulfone (PES), and polyamide. These membranes can be prepared by melt stretching, phase inversion, or interfacial polymerization.

Tubular ceramic membranes are also employed in specific applications (see Section 13.2.2.2). The ceramic structures are formed by sintering layers that have been cast onto supporting materials.

The membrane elements manufactured for dairy processing are not the same, however, as those used in water treatment applications. Specifically, they are referred to as "sanitary", meaning that the membrane elements are designed to fit snugly within the module with no dead spots available for microbial growth to occur. Membranes may also be referred to as "hot water sanitizable", meaning that they can be exposed to high temperatures for short periods to allow for pasteurization during cleaning cycles.

Given the scale of operation is small relative to water treatment, a single pass flow through operation is generally not viable due to the high capital cost. Rather, the membranes are often used in a "feed and bleed" mode (Figure 13.5), which allows continuous production with fewer membrane modules. The use of a recirculating flow also increases the crossflow velocity and so reduces membrane fouling.

DOI: 10.1201/9781003285656-13

TABLE 13.1
The composition of bovine milk [1]

	Weight %
Total solids	12.7
Fat	3.7
Protein	3.4
Lactose	4.8
Salts (ash)	0.7

TABLE 13.2
The concentration and molecular weight of the major proteins present in milk [2]

		Concentration in milk (g/L)	Molecular Weight (kDa, or kg/mol)
Caseins	α_{S1} Casein	10.0	24
	α_{S2} Casein	2.6	25
	β Casein	9.3	24
	$_\kappa$ Casein	3.3	19
	γ Casein	0.8	12–20
Whey proteins	β-Lactoglobulin	3.2	18*
	α-Lactalbumin	1.2	14
	Bovine serum albumin	0.4	66
	Immunoglobulins	0.7	150–900
	Lactoferrin	<0.3	76.5

* β-Lactoglobulin is found in native milk as a dimer of 36 kDa.

13.2.1 Characterization of Pressure-Driven Membrane Performance

The important parameters for characterization of dairy membrane processes are outlined in Figure 13.6 and include:

- The volumetric flux (J_v), which is the permeate flowrate through the membrane, divided by the membrane area. The SI units for this parameter are m^3/m^2.s or m/s but the industry tends to use units of L/m^2.h or LMH.
- The transmembrane pressure (TMP or ΔP), which is the difference between the feed/retentate side and permeate pressures:

$$TMP\ or\ \Delta P = \frac{P_F + P_R}{2} - P_p \tag{13.1}$$

- The rejection (R), determined as a function of the feed concentration (C_F) and permeate concentration (C_p):

$$R = \frac{C_F - C_p}{C_F} = 1 - \frac{C_p}{C_F} \tag{13.2}$$

TABLE 13.3
The distribution of salts in bovine milk and the proportion that is present as insoluble colloids [1]

Species	Concentration (g/L)	Proportion that is insoluble (% by mass)
Calcium	1–1.4	66.5
Magnesium	0.1–0.15	33
Sodium	0.35–0.6	8
Potassium	1.35–1.55	8
Phosphorus (total)	0.75–1.1	57
Chloride	0.8–1.4	0
Citrate	1.75	6

TABLE 13.4
Compositions of the three types of whey generated in dairy processing [4–11]

Type of whey	Sweet whey	Acid whey	Salty whey
pH	5.9–6.3	4.0–4.6	5.0–5.5
Conductivity (mS/cm)	3.3–6.4	7.8–8.3	39–127
Total solid (%)	6.3–7.7	5.2–6.2	12
Fat (%)	0.50	0.1–0.25	0.60
Protein (%)	0.6–1.0	0.47–0.55	0.60
Lactose (%)	4.6–5.2	4.0–4.3	Not reported
Minerals (%)	0.45–0.50	0.57–0.8	6.9
Lactic acid (%)	0.05–0.20	0.5–0.8	Not reported

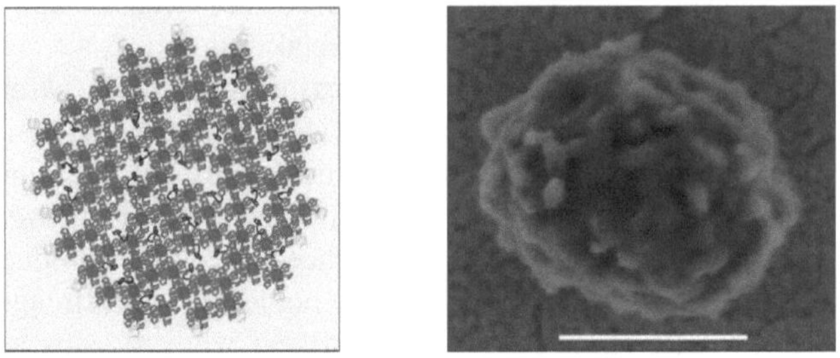

FIGURE 13.1 – Schematic representation of the casein micelle incorporating calcium phosphate nanoclusters (grey) with their attached α and β caseins and the surface-located κ-casein. Hydrophobically bound "mobile" β-casein is present within the water cavities (blank spaces) inside the micelle and (b) a field-emission scanning electron micrograph of a casein micelle. The scale bar is 200 nm. Both images reproduced from Ref. [3] with permission from Royal Society of Chemistry.

- The crossflow velocity in m/s, which is the feed flowrate divided by the cross-sectional area of the flow path.
- The volume concentration factor (VCF), determined as a function of the feed concentration(C_F) and retentate

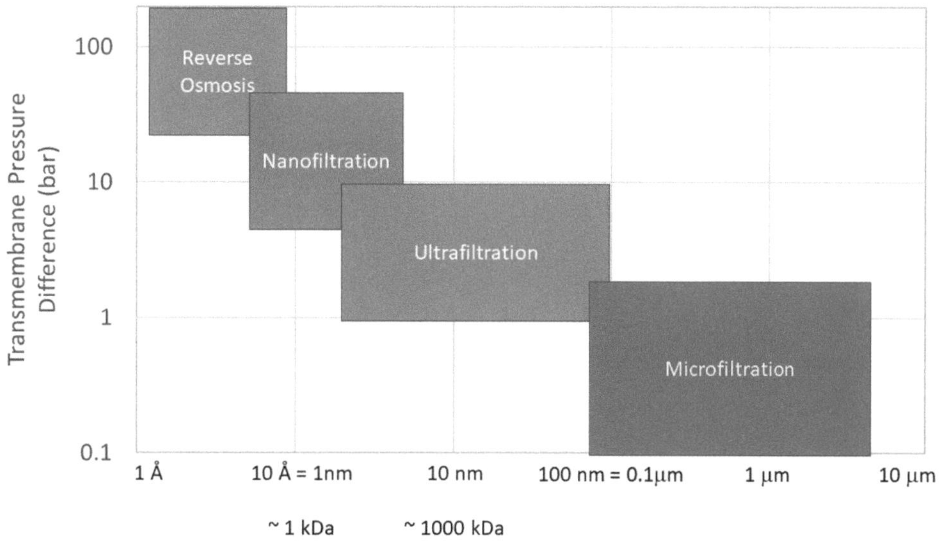

FIGURE 13.2 A schematic showing the types of membranes used in the dairy industry, with the typical pore size in both length units (nanometre, micron) and molecular weight (kiloDaltons).

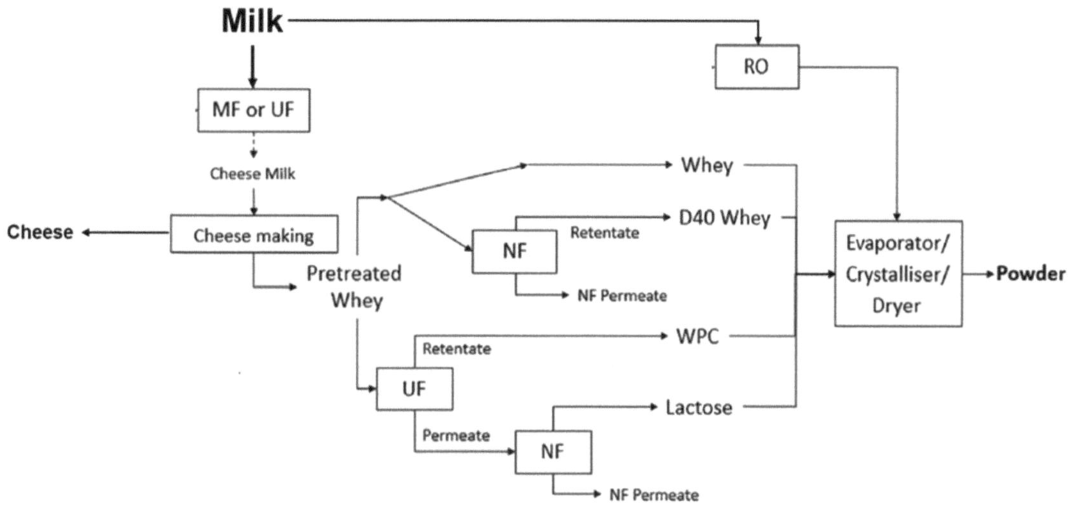

FIGURE 13.3 An overview of the dairy streams in a typical dairy factory showing where pressure-driven membrane operations can be used. D40, demineralized whey with 40% mineral removal; MF, microfiltration; NF, nanofiltration; UF ultrafiltration; WPC, whey protein concentrate. Adapted from Ref. [12].

concentration (C_R), or alternatively from the initial feed volume (V_F) and the final retentate volume (V_F):

$$VCF = \frac{C_R}{C_F} = \frac{V_F}{V_R} \tag{13.4}$$

13.2.1.1 Critical Flux and Limiting Flux

At very low transmembrane pressures, the volumetric flux through an MF or UF membrane varies linearly with the transmembrane pressure and is inversely proportional to the solution viscosity (μ). This can be written in terms of the membrane resistance (R_m) or as a function of the membrane pore size (r_p), effective thickness (Δx), and porosity(ε)[14]:

$$J_v = \frac{\Delta P}{R_m} = \frac{r_p^2 \Delta P \varepsilon}{8\mu\Delta x} \tag{13.5}$$

Microfiltration and ultrafiltration processes within the dairy industry, however, are dominated by protein fouling so that this relationship is rarely observed. At a point known as the critical flux linearity is lost, due either to this protein fouling or to concentration polarization (see Section 13.2.1.2). At even higher transmembrane pressures, the flux becomes independent of pressure (known as the limiting flux). It can even fall due to compaction of the fouling cake at very high applied pressures (Figure 13.7).

The limiting flux (J_{lim}) can be predicted based on theoretical considerations (Equation 13.6) [15].

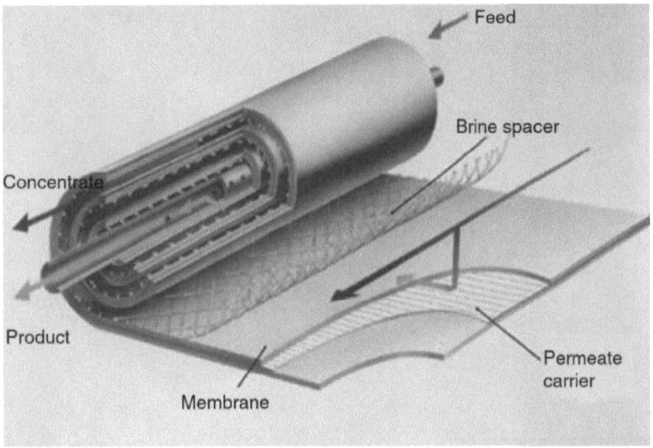

FIGURE 13.4 Diagram showing how a spiral-wound membrane is formed. Reproduced from Ref. [13] with permission from John Wiley and Sons.

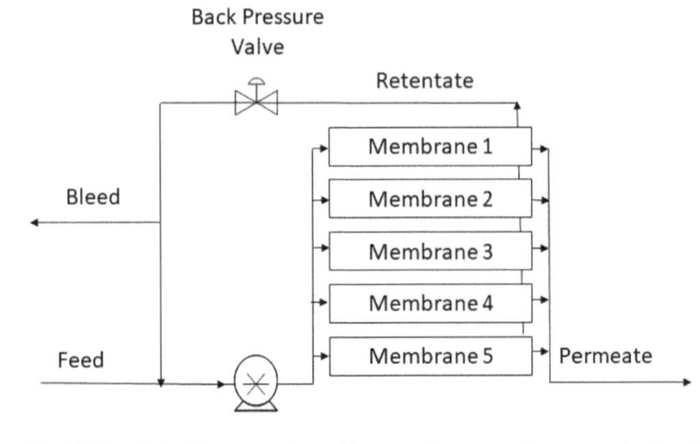

FIGURE 13.5 Process flow diagram for operation of a bank of membranes in "feed and bleed" mode. The solids concentration within the retentates is significantly higher than that in the feed, due to the use of recirculation.

$$J_{lim} = 0.072\frac{\tau}{\mu}\left(\frac{\Phi_w r_f^4}{\Phi_b L}\right)^{1/3} \tag{13.6}$$

where τ is the shear stress arising from the crossflow velocity, Φ_w and Φ_B represent the volume fraction of foulant particles in the boundary layer and bulk, respectively, r_f is the foulant particle radius, and L is the module length. While this equation may not be of practical use, it shows that the limiting flux can be increased by increasing the crossflow velocity, reducing the viscosity, or using a shorter membrane element length.

The limiting flux that can be achieved is also a function of the way this transmembrane pressure is applied. Starting a manufacturing process below the critical flux and then increasing it gradually allows the fouling cake to develop in a loose manner, with more particles in the boundary layer (Φ_w) leading to an ultimately higher flux at a given operational condition. Conversely, a rapid application of a large feed pressure results in a highly compacted cake and a lower limiting flux [16].

13.2.1.2 Concentration Polarization

Once the critical flux is exceeded, mass transfer limitations come into play. This is due to the accumulation of protein and other impermeable solutes on the surface of the membrane. The concentration of these solutes at the membrane surface (C_m) is greater than that in the bulk (C_B) and this leads to backward diffusion, according to Fick's Law ($\mathfrak{D}\frac{dC}{dx}$ where $\mathfrak{D}$ is the diffusion coefficient of the solute). This is known as concentration polarization (Figure 13.8).

Once the solute concentration on the membrane surface exceeds its solubility limit, precipitation occurs. The concentration at the membrane surface is then dictated by this solubility limit. The concentration cannot increase beyond this value, known as the gel concentration (C_G). Rather, any further increase in the magnitude of concentration polarization leads only to an increase in the rate of fouling deposition.

A mass balance around the boundary layer under these conditions gives:

$$J_v C - J_v C_P = -\mathfrak{D}\frac{dC}{dx} \tag{13.7}$$

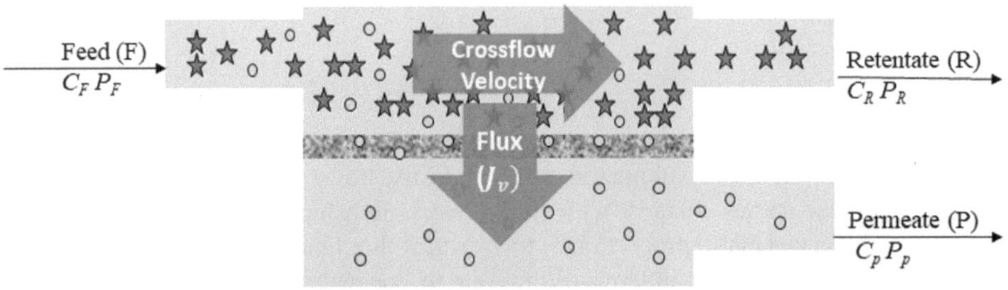

FIGURE 13.6 An overview of relevant parameters in membrane processing.

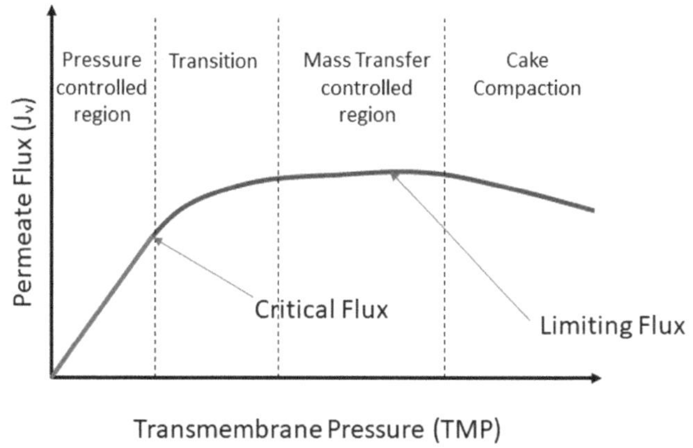

FIGURE 13.7 Change in permeate flux that occurs as a function of transmembrane pressure for dairy microfiltration or ultrafiltration.

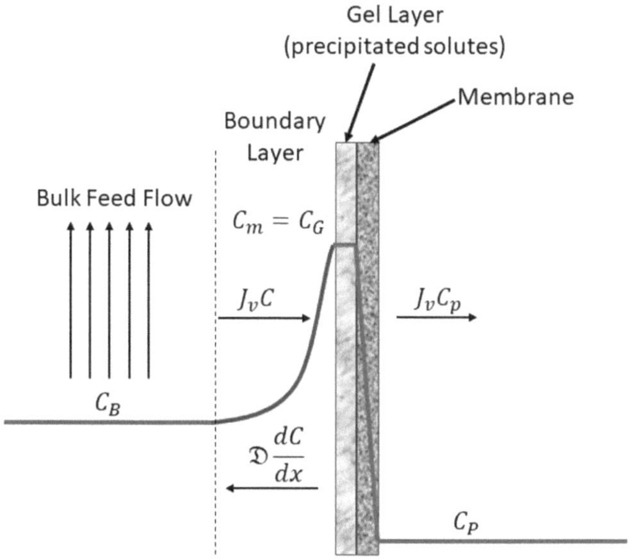

FIGURE 13.8 Schematic showing the change in concentration within the boundary layer during concentration polarization once protein fouling occurs.

Integration of this expression, noting that the mass transfer coefficient in the boundary layer (k_{CP}) can be described by the diffusion coefficient divided by the thickness of this layer, gives:

$$J_v = k_{CP} \ln\left(\frac{C_G - C_P}{C_B - C_P}\right) \tag{13.8}$$

Experimental data at varying solute concentrations can be used to identify the mass transfer coefficient and the gel concentration by a log-linear plot of flux versus ccentration. Specifically, if the permeate concentration approaches zero, the slope of a line plotting J_v versus $\ln(C_B)$ will have a slope of $-k_{CP}$ and an x-intercept of C_G. Alternatively, the mass transfer coefficient can be estimated from correlations. These correlations generally consider the Sherwood (*Sh*), Reynolds(*Re*), and Schmidt (*Sc*) numbers, as shown in Equation (13.9).

$$Sh = \frac{k_{CP} d_h}{\mathfrak{D}} = aRe^b Sc^c \left(\frac{d_h}{L}\right)^e \tag{13.9}$$

where L is the module length. The hydraulic diameter (d_h) and the parameters a, b, c, and e are available for spiral-wound membrane spacer channels and slit-shaped channels with no spacer (Table 13.5). For hollow-fibre bundles, suitable correlations are summarized in Shen et al. [17].

13.2.1.3 Osmotic Pressure

There is an added complication when processing solutions of high molar concentration. In these circumstances, there is a natural tendency for water to transfer across the membrane into the feed solution to reduce the molar concentration gradient. To go the other way and remove water thus requires an additional force to be applied. This force per unit area is referred to as the difference in osmotic pressures ($\Delta\pi$). Equation (13.5) should then be modified to Equation (13.10):

$$J_v = \frac{\Delta P - \Delta\pi}{R_m} \tag{13.10}$$

The osmotic pressure difference is determined by evaluating the values in the retentate and permeate streams respectively. For highly concentrated salt solutions, this requires knowledge of the water activity (a_w), which can be measured experimentally using an osmometer or evaluated from equations of state such as the Pitzer model [21]. In these situations, Equation (13.11) applies:

$$\pi = -\left(\frac{RT}{V_w}\right) ln(a_w) \tag{13.11}$$

where V_w is the molar volume of water (1.802×10^{-5} m^{3}mol^{-1}, 1.822×10^{-5} m^{3}mol^{-1}, and 1.85×10^{-5} m^{3}mol^{-1} at 10°C, 20°C, and 50°C, respectively [22], R is the universal gas constant (8.315 J.mol^{-1}K^{-1}), and T is the temperature in Kelvin. However, in most cases, the Van't Hoff relationship can be used to give a good approximation to this value (Equation 13.12).

$$\pi = C_S \mathrm{RT} \tag{13.12}$$

where C_S is the total solute concentration in mol/L. Note that this is the concentration of all dissolved species. Thus for 1 mol/L NaCl, C_S= 2 mol/L to account for both the sodium and chloride ions. Similarly, for 1 mol/L $MgCl_2$, the value of C_S would be 3 mol/L.

TABLE 13.5
Parameters for determining the boundary layer mass transfer coefficient from Equation (13.9), where H is the spacer channel height, μ is the spacer porosity, and s is the ratio of wetted surface of the spacer to spacer volume

	Flow	Hydraulic diameter (d_h)	a	b	c	e	
Slit	Laminar	$2H$	1.62	0.33	0.33	0.33	[18]
	Turbulent		0.023	0.80	0.33	–	[19]
Spacer channel		$\frac{4\varepsilon}{2/H+(1-\varepsilon)s}$	0.664	0.5	0.33	0.5	[20]

The high molecular weight of dairy proteins prevents a high molar concentration and so osmotic pressure effects can be ignored in MF or UF processes. However, they are important for NF or RO of salty streams, such as can occur during processing of salty whey (Section 13.2.4.1), or other waste streams.

13.2.1.4 Solute Permeation

The equations above can be used in combination to determine the volumetric flux of liquid flowing through the membrane (J_v). For membranes with larger pores (MF and UF), the solutes within this liquid move through convective forces and so their flowrate can be described simply by Equation (13.13).

$$J_s = J_v.C_{i,P} = K_{ic}C_{i,R}J_v \tag{13.13}$$

where C_i now represents the concentration of a single species such as an ion in either the retentate (R) or the permeate (P). K_{ic} is known as the convective hindrance factor and can be estimated from Equation (13.14a), based on the ratio of the species radius to the pore radius ($\lambda_i = r_i / r_p$) [23]:

$$K_{ic} = 1.0 + 0.054\lambda_i - 0.988\lambda_i^2 + 0.441\lambda_i^3 \quad 0 < \lambda_i \le 0.8 \tag{13.14}$$

$$K_{ic} = -6.83 + 19.348\lambda_i - 12.518\lambda_i^2 \quad 0.8 < \lambda_i \le 1 \tag{13.14b}$$

At the other extreme, for reverse osmosis membranes, which are essentially non-porous, the solution diffusion model implies that Fick's Law governs solute permeation, as given by Equation (13.15):

$$J_s = B\frac{dC_i}{dx} \tag{13.15}$$

where B is the salt permeability coefficient and dx represents the thickness of the membrane active layer.

Nanofiltration fits between these two extremes and is thus more complicated to describe correctly. The best representation is through the extended Nernst Planck equation (Equation 13.16), which allows for both convective flow and the solution–diffusion transport mechanism. It is, however, written in terms of the concentration within the membrane itself (c_i). Correct evaluation of this equation requires evaluation of the partition coefficient between this concentration and that in the bulk external solution (c_i / C_i), which in turn requires consideration of both the Donnan potential and dielectric exclusion effects [23].

$$J_s = K_{ic}c_iJ_v - \mathfrak{D}_{i,p}\frac{dc_i}{dx} - z_ic_iD_{i,p}\frac{F}{RT}\frac{d\psi}{dx} \tag{13.16}$$

In this equation, $\mathfrak{D}_{i,p}$ is the diffusion coefficient within the membrane pores, F is the Faraday constant, and ψ is the electrical potential.

13.2.1.5 Fouling Mechanisms and Models

In dairy systems, membrane fouling is frequent and often intense. The fouling is most commonly caused by proteins, in which case a dense gelatinous cake forms on the surface of the membrane. The initial adsorption of the proteins is related to the charge on the protein and both the charge and hydrophobicity of the membrane surface [24]. This protein initially reduces flux by blocking access to the membrane pores. As time proceeds it then builds up a dense cake on top of the membrane surface, further increasing the resistance to flow.

Classically, the development of the fouling cake over time (t) was explained through "blocking models", first developed by Hermans and Bredee [25] in 1936 (Equation 13.17). In this case, complete pore blocking corresponds to N = 2, pore constriction to N = 1.5, intermediate blocking to N = 1, and cake filtration to N = 0. The cumulative permeate volume passing through the membrane is V while k is an arbitrary constant.

$$\frac{d^2t}{dV^2} = k\left(\frac{dt}{dV}\right)^N \tag{13.17}$$

These models assume a parallel array of uniform, non-interconnected, cylindrical pores, which is quite unrealistic. Zydney [26] recently developed a pore constriction model that better accounts for the highly irregular and interconnected pore structure of MF and UF membranes (Equation 13.18).

TABLE 13.6
Typical neutral salts that can be present in dairy systems and their solubility product at 20–25°C and neutral pH [24,29–31]

Ca/P mol ratio	Compound	Formula	Solubility -log (Ksp)	pH[a]
0.5	Monocalcium phosphate monohydrate (MCPM)	$Ca(H_2PO_4)_2.H_2O$	1.14	0.0–2.0
1.0	Dicalcium phosphate (monetite)	$CaHPO_4$	6.9	
1.0	Dicalcium phosphate dihydrate (DCPD, brushite)	$CaHPO_4.2H_2O$	6.59	2.0–6.0
1.3	Octacalcium phosphate (OCP)	$Ca_8(HPO_4)_2(PO_4)_4.5H_2O$	96.6	5.5–7.0
1.2-2.2	Amorphous calcium phosphates (ACP)	$Ca_xH_y(PO_4)_z.nH_2O$ n=3–4.5	– [b]	5.0–12.0
1.5	Tricalcium phosphate	$Ca_3(PO_4)_2$	25.5(α), 28.9 (β)	
1.5-1.7	Calcium-deficient hydroxyapatite (CDHA)	$Ca_{10-x}(HPO_4)_x(PO_4)_{6-x}(OH)_{2-x}$ (0<x<1)	≈ 85	6.5–9.5
1.7	Hydroxyapatite (HAP)	$Ca_{10}(PO_4)_6(OH)_2$	116.8	9.5–12.0
	Tricalcium citrate	$Ca_3(C_6H_5O_7)_2$	17.6	
	Magnesium phosphate	$MgHPO_4.3H_20$	4	

[a] pH where the salt can exist in aqueous solution at room temperature.
[b] Cannot be precisely measured. In acidic buffer, ACP < CDHA < HAP.

$$\frac{d^2t}{dV^2} = \frac{k_b J_0^{\frac{2}{3}}(1-\beta)}{\left[\frac{2}{3} + \beta J_0\left(\frac{dt}{dV}\right)\right]}\left(\frac{dt}{dV}\right)^{5/3} \quad (13.18)$$

where $k_b = 1/V_{max}$ with V_{max} being the volume of permeate at which the pore radius reduces to zero, J_0 is the initial permeate flux through the clean membrane, and β is the ratio of the permeabilities of the fouling deposit to that of the clean membrane. If the fouling deposit is impermeable, $\beta = 0$.

These models, however, assume that only one type of fouling occurs. An alternative approach was developed by Ho and Zydney [27], who assumed that pore blockage occurred in the initial stages of fouling and was followed by cake filtration (Equation 13.19)

$$\frac{J_v}{J_0} = \left[\exp\left(-\frac{\alpha \Delta P C_B}{\mu R_m} t\right) + \frac{R_m}{R_m + R_p}\left(1 - \exp\left(-\frac{\alpha \Delta P C_B}{\mu R_m} t\right)\right)\right] \quad (13.19)$$

The first term in this expression is equivalent to classical pore blockage and dominates over shorter time frames. In this term, α is the pore blockage parameter (m²/kg), μ is the fluid viscosity, and R_m is the resistance of the clean membrane. At longer times, the second term is dominant and reflects cake filtration. In this expression, R_p is the resistance of the protein deposit, which in turn can be evaluated from Equation (13.20).

$$R_p = \left(R_m + R_{p0}\right)\sqrt{1 + \frac{2f'R'\Delta P C_B t}{\mu\left(R_m + R_{p0}\right)^2}} - R_m \quad (13.20)$$

where R_{po} is the initial resistance of the fouling deposit (m^{-1}) and the group $f'R'$ represents the rate of increase of the protein layer resistance with time (m/kg). While Equation (13.19) and Equation (13.20) seem complex, they have the advantage of mainly relying on known parameters such as the transmembrane pressure, clean membrane resistance, and fluid viscosity, with only three unknowns, which themselves have a physical meaning (α, R_{po} and $f'R'$).

Fouling can also be caused by precipitation of salts, in which case the fouling layer is referred to as "scaling". The inorganic salts in dairy systems are generally referred to as "ash" because their concentration is determined by heating the sample to a temperature where all organic matter burns away and only the inorganic ash remains. These salts can be present either as soluble free ions, dissolved neutral salts, or within insoluble colloids (see Table 13.3). Most of the insoluble calcium phosphate is associated with casein micelles and hence is separated from the whey products during cheese formation.

A variety of these salts can precipitate during membrane concentration processes such as NF and RO. Calcium phosphate is the most common precipitant, but this can exist as a number of different crystal polymorphs (Table 13.6). The specific salt that deposits on the membrane surface is a function of the solubility equilibria (Table 13.6) but also the crystallization kinetics. Indeed, it is changes in these kinetics that mean that calcium salt precipitation shows an inverse relationship with temperature. That is, most workers observe calcium precipitation to increase as the temperature increases. This is because the kinetics become more favourable. The equilibrium solubility suggest the opposite – calcium is more soluble at 50°C, relative to that at 10 °C and 30°C [28]. The extent of calcium phosphate precipitation can

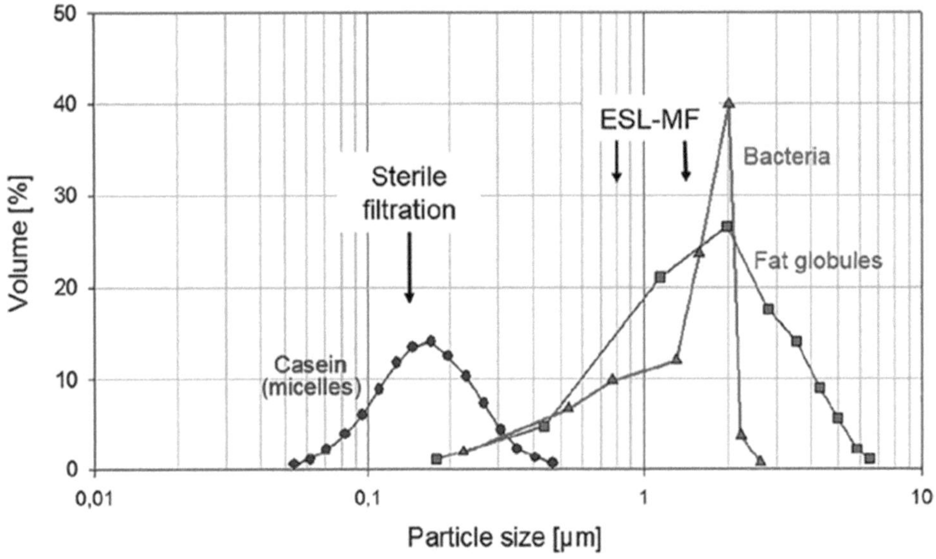

FIGURE 13.9 Particle size distributions of casein micelles, bacteria, and milk fat globules showing the range over which sterile filtration and extended shelf life (ESL) microfiltration (MF) membranes operate. Reproduced from Refs. [36,37] with permission from Elsevier.

also be influenced by the presence of residual casein, the pH, as well as the presence of citrate, lactate, and even sodium chloride [28]. The pH has the greatest effect, with the less soluble salts such as OCP and HAP tending to precipitate at higher pH.

13.2.2 Microfiltration

Microfiltration (MF) membranes have large pores that can retain particles of micron size. They were first developed for the removal of bacteria from water in the 1920s [32] and municipal wastewater treatment remains the largest application. MF is also used within the dairy industry to clean up wastewater streams. For example, in a membrane bioreactor, the organic matter in the wastewater is broken down by micro-organisms in an aerobic environment. Clean water is continuously removed by applying a vacuum to hollow-fibre membranes suspended within the wastewater tank.

13.2.2.1 Microbial Load Reduction

MF can be used to remove bacteria and spores from raw skim milk and other dairy streams. These streams must contain little fat, which would otherwise also be retained and cause fouling. It is attractive as an alternative to pasteurization because the natural flavour and heat-sensitive vitamins can be retained. It has been found however, that concurrent removal of the raw milk microflora reduces the pungency and flavour of any resulting cheese products, due to reduced proteolysis and propionic acid fermentation [33,34].

MF can provide a reduction in the spore count of raw skim milk of up to five $\log_{10}$ [35]. Complete removal of bacteria (sterile filtration), however, requires a membrane of pore size 0.1–0.2 μm (Figure 13.9), which also removes a significant fraction of the casein micelles and hence is not economically viable in most situations. Membranes of larger pore size (around 1 μm) can be used to remove the bulk of the bacteria and in this manner can be used in conjunction with pasteurization to provide milk of extended shelf life (ESL) that has better taste and stability than milk treated with ultra-high temperature (UHT). One example is the patented Bactocatch® system. In this system, the permeate from MF of skim milk is treated with a standard pasteurization cycle (e.g. 72°C for 15 seconds), while the smaller retentate stream is sterilized with a more aggressive heat treatment, such as 115–130°C for 4–6 s) (Figure 13.10) [35,36].

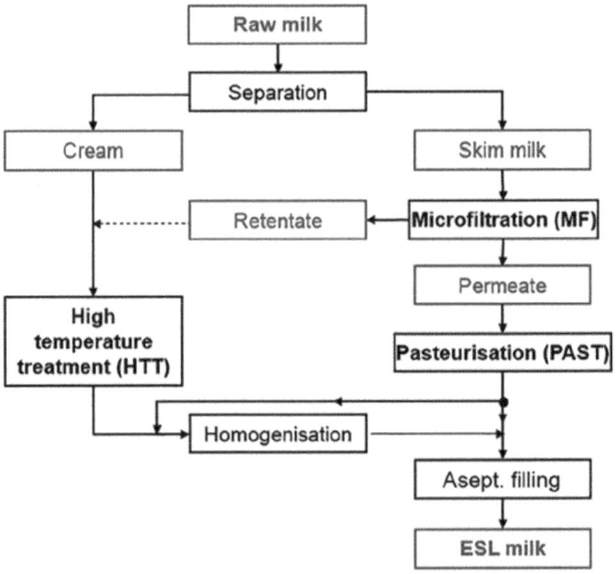

FIGURE 13.10 The process steps in preparing extended shelf life (ESL) milk using MF. Reproduced from Ref. [36] with permission from Elsevier.

13.2.2.2 Separation of Caseins from Whey

As discussed in Section 13.1, the casein proteins are predominantly held within large micelle structures of around 0.15 micron in size [2] (see also Figure 13.9), so they should be readily separable from the much smaller, soluble whey proteins during microfiltration of skim milk. Such a separation can provide multiple benefits. Firstly, the retained casein micelles can be used to prepare an "ideal cheese" as the coagulation time can be significantly reduced, the curd firming process accelerated, and the final curd firmness increased [38]. Casein and fat retention in the curd increases, leading to a higher yield. Secondly, the "native" whey protein permeate is of higher quality than traditional cheese whey, as it does not contain casein peptides, enzymes, or fat, and is less denatured. This allows it to be further separated to produce high-value ingredients, particularly for infant formula. The whey proteins are at their natural pH, whereas the classical cheese-making process produces whey at an acidic pH, leading to a reduced flavour profile. The gelation, foaming, and solubility of the proteins is also claimed to be significantly improved [38].

In reality, this separation proves much more difficult than it would suggest. This is because the proteins rapidly form a "dynamic" membrane on the true membrane surface. It is this "dynamic" membrane that controls the filtration process and can often lead to very high rejection of all proteins, inclusive of whey proteins. The thickness and density of the dynamic membrane is a function of the transverse force provided by the flow through the membrane (J_v) that drives the proteins towards the membrane surface; and the shear force (τ) which is controlled by the crossflow velocity, that sweeps these species in a direction parallel to the membrane surface. Indeed, Gesan, Daufin, and Merin [39] indicate that it is the ratio of these two forces (J_v / τ) that is the most critical parameter.

Achieving the necessarily low values of J_v / τ for effective casein/whey separation is extremely difficult using a standard polymeric membrane [40–42]. This is because a high value of τ requires a high crossflow velocity, which, in turn, implies a high pressure drop along the length of the module. In turn, this leads to a high transmembrane pressure difference at the inlet end of the module and thus a large value of J_v (see Figure 13.11(a)). Inevitably, cake compaction occurs at the inlet end and poor separation is achieved. The use of short membrane element lengths can reduce the extent of the problem but is less efficient.

The most effective approach to avoid the dynamic membrane and thus to obtain effective casein/whey protein separation is to use one of the ceramic membrane designs that are readily available specifically for this application. These ceramic membranes have a tubular design (Figure 13.12). The advantages of such membranes are a much longer life and greater tolerance of harsh chemical cleaning procedures and temperatures. The disadvantages are the high cost and the low membrane area per unit volume.

Importantly, the tubular design facilitates recycling of the permeate, which was the first approach developed to solve this issue. This was patented by Alfa Laval [43] and is referred

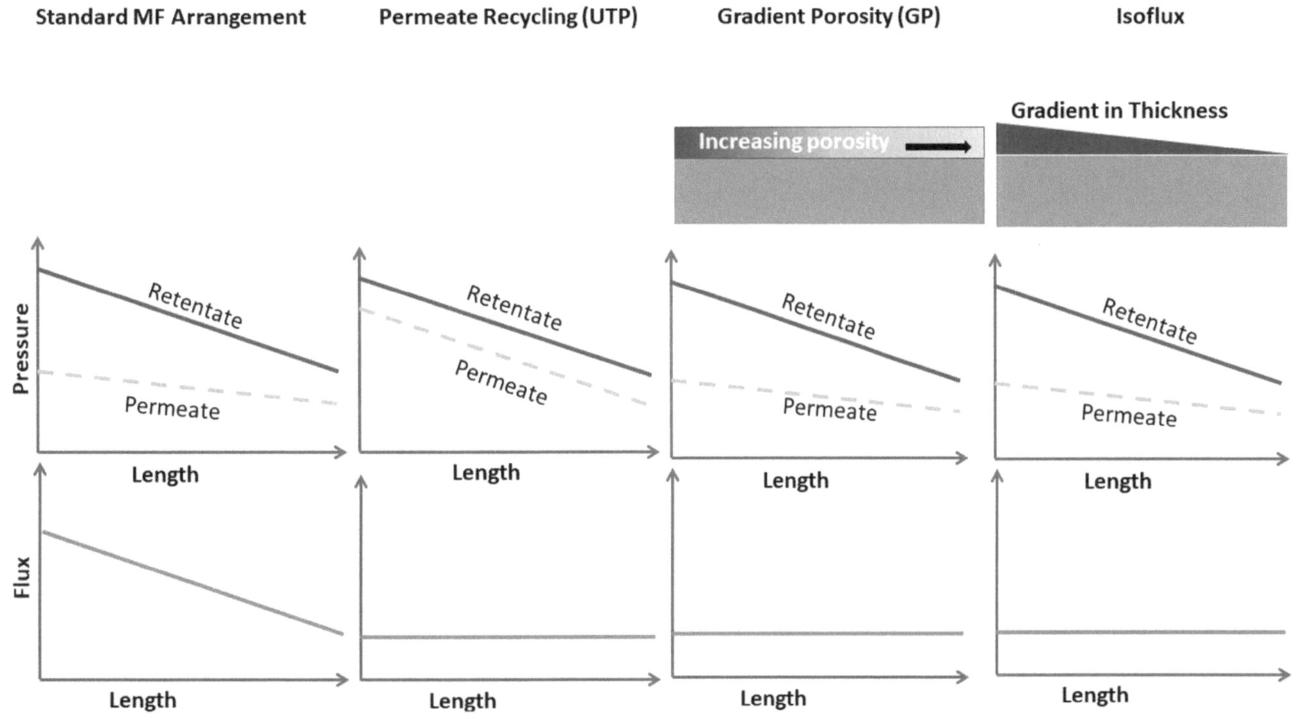

FIGURE 13.11 The different arrangements possible for casein/whey separation using MF: (a) standard arrangement which gives high flux but low selectivity; (b) permeate recycling; (c) increasing porosity in the active layer with length; and (d) reducing thickness of the active layer with length all give low flux but high selectivity.

FIGURE 13.12 Selection of ceramic membranes used for microfiltration (Koch Separation Systems; https://gp-systems.com.au/product/koch-ceramic-mf-and-uf-membranes/).

to as the uniform transmembrane pressure (UTP) approach (Figure 13.11(b)). Permeate recycling provides a longitudinal pressure drop on the permeate side which balances that on the feed side. This means that the transmembrane pressure drop maintains a constant low value across the element length. While this implies a low initial flux, the flux is much more stable with time and much greater casein/whey selectivity can be achieved [44]. As one example, Vadi and Rizvi [45] found that conventional processing led to very rapid membrane fouling that prevented operation beyond a VCF of 6 because of the resulting rapid dramatic flux decline. Conversely, when the same ceramic membrane was used in the UTP mode, the operation started with a lower flux but this could be maintained up to a VCF of 10.

More recently, companies have provided bespoke ceramic membranes that achieve similar results without permeate recycling. The gradient porosity (GP) [46,47] approach uses a membrane that has increasing porosity along the membrane length (Figure 13.11(c)). Similarly, the Isoflux approach patented by Tami Industries [48,49] employs an active membrane layer of decreasing thickness along the membrane length (Figure 13.11(d)). Both approaches are effective at achieving a stable flux and high casein/whey selectivity.

Hu et al. [50] compared the overall benefits of ceramic GP versus a polymeric membrane system. The ceramic membrane system had a much higher permeate flux and whey protein transmission than the polymeric membrane system, resulting in much less membrane area required (570 versus 3,000 m^2). However, due to the high price of the ceramic membranes, the capital cost for this approach was still about 27% higher and the membrane replacement cost was 35% higher. Cleaning costs were lower for the ceramic system due to the reduced membrane area but the energy consumption was higher. Overall, the case for either system was not clear. More recently, Schopf et al. [51] have argued that polymeric hollow-fibre membranes may represent the best choice, as they are cheaper than the ceramic membranes and can provide similar combinations of uniform transmembrane pressure and high crossflow velocities. Such membranes, while common in wastewater treatment, have yet to be commercially trialled for this application.

13.2.3 Ultrafiltration

Ultrafiltration processes are used to separate solutes of molecular weight greater than 5 kDa to 300 kDa and are the most widely used membrane processes in the dairy industry. UF separates these molecules based on their size and charge, and their affinity to the membrane. Materials with low protein-binding characteristics such as regenerated cellulose (RC), polyethersulfone (PES), and polyvinylidene fluoride (PVDF) can lead to reduced membrane fouling, and therefore are commonly used as the separation membrane for processing protein solutions [52]. PES membranes, in particular, are commonly adopted in dairy processing due to their good thermal and mechanical stabilities and low cost [53]. UF was first used in dairy processing in 1972 to separate and concentrate whey proteins from cheese whey [54,55]. In the late 1970s, UF was introduced to cheese manufacturing as a technology for preparing the milk for cheese making, improving cheese yield by adjusting and standardizing the protein level in cheese milk. Since then, these have become the main applications of UF in the dairy industry.

13.2.3.1 Cheese Milk Preparation

Over the past couple of decades, ultrafiltration has become a standard unit operation in cheese production to prepare cheese milk with a desired protein content, to enhance cheese yield, and provide better compositional and quality control [56–58]. Ultrafiltration can retain all the milk proteins and simultaneously remove lactose and minerals. This allows milk to be standardized for cheese making and yoghurt manufacturing, without being affected by any inherent variation in the raw milk caused by factors such as seasons, stage of lactation, breed, and feeding [59]. As a non-thermal process, UF can also avoid protein denaturation and any associated changes in the nutritional value and functionality of milk proteins, leading to high quality and consistent end-products [58,59].

Spiral-wound polymeric membranes are more often used in ultrafiltration of milk, compared to tubular ceramic membranes. UF is typically performed either at around 10°C or 50°C to minimize the growth of pathogenic bacteria. While the flux is greater at higher temperatures due to reduce fluid viscosity, fouling becomes more severe and the permeate flux declines at a higher rate [60]. Under a TMP of 2–4 bar, a VCF of 1.2–6 can be achieved using UF membranes with molecular weight cutoffs (MWCOs) from 10 kDa to 50 kDa [61–64]. Low VCFs (below 2) can increase the protein content by ~5%, which is suitable for making many types of cheese such as cheddar, Colby, mozzarella, cottage cheese, brick, edam, Saint Paulin, and quarg. A further increase in VCFs (between 2 and 6) can increase cheese yield by 6–8%, when special cheese-making equipment is used for making cheddar, feta, Havarti, Gouda, and blue cheese. High VCFs (between 5 and 7) are also achievable, resulting in a UF concentrate that has the same protein concentration as the final cheese. This concentrate, also called "pre-cheese", can be obtained using an MMV process, named after its inventors Maubois, Mocquot, and Vassal. The general principle behind this process is to eliminate the need for whey–curd separation, by removing the excess water and lactose. Therefore, the use of a cheese vat is not required for this process. To date, the MMV process has been used for manufacturing fresh unripened cheeses such as cream cheese, quarg, and ricotta, as well as soft and semi-hard cheeses such as mozzarella, Saint Paulin, and feta [65]. The overall cheese yield using this process has been reported to be 10–30% higher than conventional processes, mainly because of the effective concentration of whey proteins and hence the reduction in enzyme usage.

Some studies, however, have reported that ultrafiltered milk causes compromised sensory and functional properties of some semi-hard and hard cheeses, caused by the impairment of ripening. Due to the presence of more whey proteins and a decrease in chymosin and rennin action, the rate of ripening decreases, leading to a reduction in proteolysis [66,67]. Other workers report a bitter taste in cream cheese made from UF retentate and relate this to the high calcium content (265 mg 100 g^{-1}) that results from the retention of the calcium-rich micelles [68].

13.2.3.2 Whey Protein Separation

Whey is generated as a by-product when milk proteins are coagulated to make cheese, yoghurt, and casein. This by-product stream contains whey proteins, lactose, and minerals, which are recovered as whey powder and other value-added products. In whey processing, ultrafiltration is typically used to separate whey proteins from the lactose-rich permeate stream. By removing both the lactose and minerals in the UF permeate, UF membranes (MWCO: 10–20 kDa) can produce a whey protein concentrate (WPC) with 30–80% protein. This concentrate is transformed into whey powder upon evaporation and drying. The protein content in the final WPC powder is determined by the VCF of the UF process. A VCF of 5–10 can produce WPC35 powder (i.e., at least 35% protein in total solids). WPC80, however, requires a VCF of 35 followed by a diafiltration step (with a three-fold diafiltration volume) [69]. During diafiltration, water is continuously added to the feed solution so that lactose and minerals are removed simultaneously in the permeate, hence effectively enhancing the protein purity in the WPC stream. As discussed previously, the precipitation of calcium phosphate at high temperature and the accumulation of whey proteins on the membrane surface cause fouling in UF processes. Operating at a low temperature (<10°C) can avoid calcium phosphate precipitation. Since the flux is half of that obtained at ~50°C [70], this potentially allows the flux to remain below the critical flux and helps to maintain the microbiological quality of the end-product by eliminating the growth of thermophilic microbes.

The production of WPC powder has become a common pathway chosen by many dairy processors to utilize whey. However, the enrichment of certain proteins in whey powder and the isolation of individual proteins can also add value to dairy manufacturing, due to the functional and nutraceutical properties of whey proteins [71]. For example, β-lactoglobulin is used in the manufacture of protein hydrolysates for ingredient formulation, as well as in emulsifiers, forming, and gelling agents. α-Lactalbumin is used for nutraceutical foods such as infant formula, as it is high in the essential amino acid tryptophan.

Whey protein concentrate is a common starting material for whey fractionation. The most abundant whey proteins, α-lactalbumin and β-lactoglobulin, however, are very similar in size (14 kDa and 18 kDa, respectively), making their fractionation challenging using a single-step membrane filtration process. To enhance the yield and purity, the protein molecules need to be manipulated before membrane filtration can be used [72]. Heating α-lactalbumin under acidic conditions (e.g., 55°C for 30 minutes at pH 3.8 [73]), for example, can cause these molecules to unfold and precipitate, because α-lactalbumin loses its bound calcium and subsequently its stability in solution. This allows α-lactalbumin to be precipitated together with BSA and immunoglobins from whey protein concentrate, so that β-lactoglobulin can be recovered and purified using UF or MF membranes in diafiltration processes [61,62].

More recent studies on ultrafiltration of whey proteins have focused on the optimization of operational modes (e.g. batch, continuous, and diafiltration) [74,75], operating parameters (e.g. VCF [76] and TMP [77]), and membrane MWCO [75, 77], aiming to achieve better separation of α-lactalbumin and β-lactoglobulin using PVDF and PES membranes. Marella et al. [77], for example, developed a UF process to produce a WPC rich in α-lactalbumin, using a wide-pore UF membrane (50 kDa) to retain most of the β-lactoglobulin (present as dimers at 36 kDa) and a second UF step (5 kDa) with diafiltration to improve the purity of α-lactalbumin in the first UF permeate (Figure 13.13). While the first UF retentate could be converted to a β-lactoglobulin-enriched WPC powder, the yield of α-lactalbumin in the α-lactalbumin-enriched powder was only ~20%, with a purity of around 60%. This purity is among the highest reported for α-lactalbumin fractionation by ultrafiltration. Higher purity requires membranes with higher selectivity towards these proteins. Membrane modification has been attempted with polymer membranes [78,79] to enhance the electrostatic repulsion between the charged membrane and the proteins. Cowan and Richie [78] modified PES membranes to have an open pore structure with charged sulphonated grafted polymer chains, with selectivity improved five-fold compared to the untreated membrane.

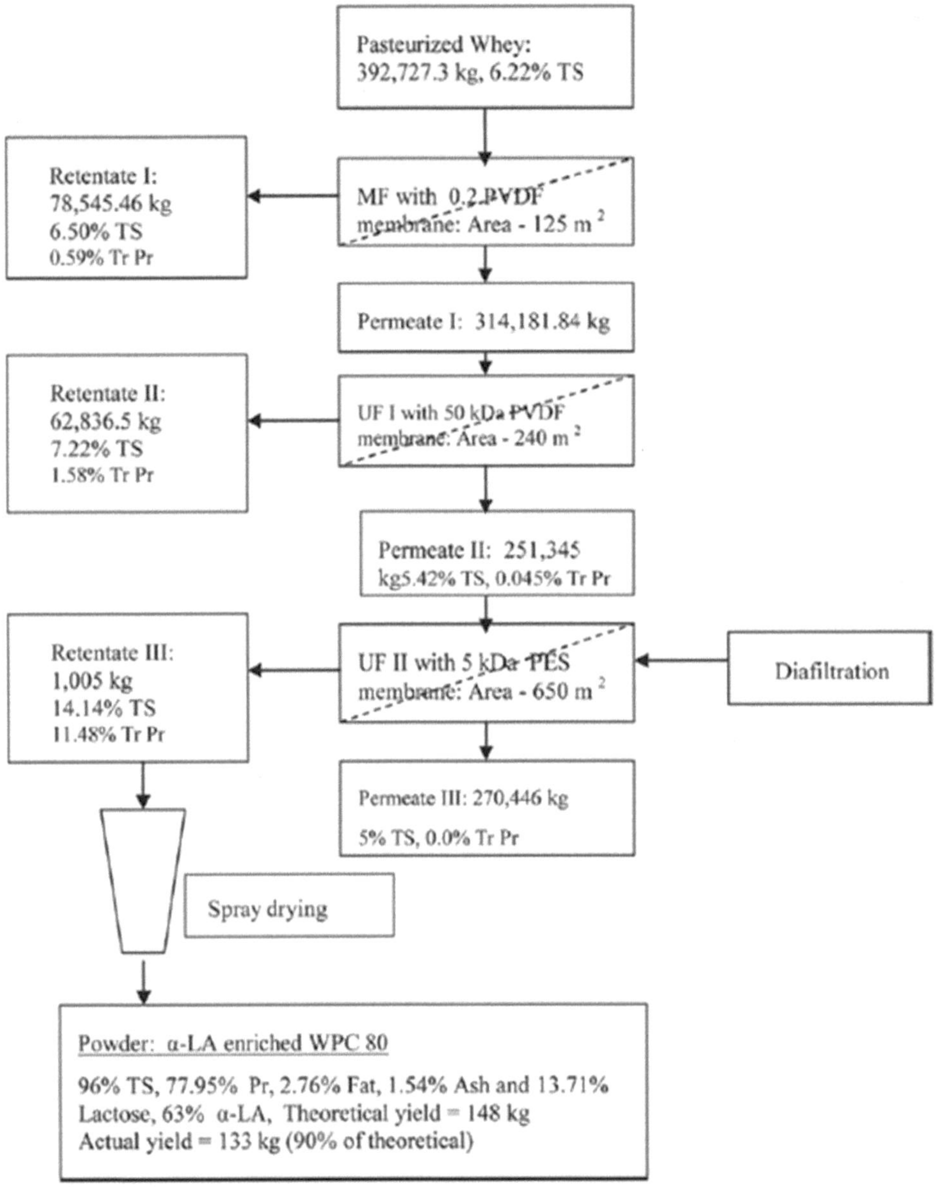

FIGURE 13.13 A typical ultrafiltration process to produce α-lactalbumin-enriched whey powder (WPC 80). Reproduced from Ref. [77] with permission from Elsevier. (Tr Pr: true protein.)

13.2.4 Nanofiltration

Nanofiltration fits between ultrafiltration and reverse osmosis with a membrane pore size ranging from 1 nm to 10 nm; and typically operates under feed pressures ranging from 5 to 20 bar. Nanofiltration membranes can retain molecules with molecular weight greater than 100–500 Da such as lactose, while smaller uncharged molecules such as organic acids (e.g., lactic acid) permeate freely. Charged species are separated by nanofiltration membranes based on subtle balances between their hydrated size and their electric charge. This means that while multivalent ions are mostly retained by nanofiltration, the removal of monovalent ions varies between 50% and 90% depending upon the membrane material and the operating parameters [83,84]. In whey demineralization, this leads to typically 35% removal of minerals from whey ultrafiltration permeate.

Salt removal or partial demineralization of dairy streams such as milk and whey is the primary application of nanofiltration in dairy processing [80–82]. Most of the commercially available NF membranes are spiral-wound membranes where a polyamide active layer (around 200 nm in thickness) is supported by a porous polysulfone interlayer and a nonwoven polyester backing. NF membranes are typically charged [85], due to the presence of carboxyl groups (–COOH) and amine groups ($-NH_2$). The charge of the membrane is often characterized by zeta potential or streaming potential which can be measured at low solute concentrations. The net surface charge of the membrane can vary with solution composition when charged solutes

TABLE 13.7
The Stokes radius and hydrated radius of ions relevant to dairy processing

	Stokes radius	Hydrated radius	References
	Å	Å	
H^+	0.28	2.82	[92]
Na^+	1.84	3.58	[92]
K^+	1.25	3.31	[92]
Ca^{2+}	3.1	4.12	[92]
Mg^{2+}	3.47	4.28	[92]
OH^-	0.46	3.00	[92]
Cl^-	1.21	3.32	[92]
Lactate ($C_3H_6O_3^-$)	2.31		[93]
Phosphate (PO_4^{3-})	7.9		[94]
Hydrogen phosphate (HPO_4^{2-})	10.4		[94]
Dihydrogen phosphate ($H_2PO_4^-$)	2.7		[94,95]
Dihydrogen citrate ($C_6H_7O_7^-$)	3.03		[96]
Citrate ($C_6H_5O_7^{3-}$)	3.94		[97]

interact with the charged functional groups, resulting in different separation performance [86]. Calcium, for example, is known to decrease the magnitude of the streaming potential as it tends to adsorb to the membrane surface and is difficult to remove during chemical cleaning [86]. The effect of anions such as citrate and phosphate on the membrane charge, however, is minimal.

The rejection of ionic compounds by nanofiltration is based on both steric exclusion and charge effects. In general, ions with hydrated radii (r_i) smaller than the membrane pore radius (r_p) are retained (see Table 13.7). However, the difference in charge between the ion and the membrane also creates a potential difference across the membrane due to charge repulsion, which is known as the Donnan effect. These two effects together determine ion partition at the membrane–solution interface, which can be related to the ratio of ion concentration within the membrane (c_i) to that in the external solution (C_i):

$$\frac{c_i}{C_i} = \phi_i exp\left(-\frac{z_i F}{RT}\Delta\psi_D\right) \quad (13.21)$$

where ϕ_i is the steric partition coefficient which can be evaluated by $\left(1-r_i/r_p\right)^2$, and $\Delta\psi_D$ is the Donnan potential difference which is a function of the fixed charge density in the membrane. Under most operating conditions (i.e., pH >5), nanofiltration membranes are negatively charged, and hence repel anions and attract cations. According to Equation (13.21), if the ions are of similar size, anions with high valency will be more strongly rejected, while strongly charged multivalent cations will permeate in preference to cations with lower valency. It was reported that sodium rejection decreased when multivalent cations such as calcium were added, which can be explained by this Donnan exclusion effect [87]. Conversely, stronger calcium rejection has been observed in other studies due to the larger size of this divalent ion compared to monovalent ions such as sodium and potassium [88–91] (see Table 13.7).

13.2.4.1 Demineralization of Whey

Demineralization is a term used in dairy processing to describe the removal of minerals in dairy systems. This process can be applied to dairy streams such as skim milk, whey, and UF permeates to reduce their ash content. The permeate from the UF process, for example, is the starting point for lactose production, an important ingredient in the food and pharmaceutical industries. To ensure a high yield and purity, nanofiltration (MWCO 100–300 Da) can be used to reduce the concentration of salts (by >30%) in the UF permeate which contains 4–8%w/w lactose and ~0.5%w/w salt [98,99]. The NF process also pre-concentrates the lactose-rich stream, which is then concentrated to 60% total solids using evaporators. This allows lactose crystallization to occur, after which lactose crystals are separated and dried into lactose powder.

The reduction in salt content in sweet whey can lead to the production of whey powder that meets the ingredient requirements in some human food applications. Partially demineralized whey concentrate is used in ice-cream and bakery products. Highly demineralized whey power (90–95% mineral removal), on the other hand, is an essential ingredient for making infant formula. Nanofiltration can concentrate sweet whey to 20–22% at a VCF of ~4, producing a D40 whey (>35% ash removal). The removal of ash content can increase to 45% when diafiltration is used [100,101]. The typical breakdown in ion concentration reduction in NF retentate is 16% for phosphate, 35% for sodium, 42% for potassium, and 71% for chloride [102]. A higher degree of demineralization requires more advanced processes such as electrodialysis (see Section 13.3.1) and ion exchange.

A special type of whey generated from cream cheese and Greek yoghurt manufacturing is called acid whey, which contains ~10 times more lactic acid than sweet whey (Table 13.4). The hygroscopic nature of the lactic acid makes drying of acid whey into whey powder difficult, due to the formation of agglomerates and sticky deposits within the processing equipment. Nanofiltration has been demonstrated to be effective in reducing the lactic acid concentration in acid whey by up to 40%, at both the laboratory [103,104] and pilot scale [105]. The concentration of the ionized lactate ions and non-ionized lactic acid molecules can be calculated using the Henderson–Hasselbalch equation:

$$pH = pK_a + \log\frac{[lactate]}{[lactic\ aicd]} \quad (13.22)$$

where, pK_a is the dissociation constant of lactic acid (3.86 at 25°C [106]).

The dissociation of lactic acid increases with increasing pH, with a higher rejection of lactic acid observed [103] using polypiperazinamide nanofiltration membranes with a MWCO of 150–300 Da. Lactic acid rejection of over 70% was observed when the pH was above 4.5. As these commercial membranes are negatively charged at high pH values (isoelectric point: 3.8–4), these results indicate that charge repulsion between the dissociated lactate anion and the membrane surface dominates the rejection. A 30% removal rate of lactic acid could be achieved at the natural pH of acid whey which ranges from 4 to 4.6 (corresponding to 50–80% lactic acid dissociation). The removal of lactic acid could be improved to greater than 50% when pH was adjusted to 3 (corresponding to 20% of lactic acid dissociation). A semi-industrial study conducted by Bédas et al. [107] demonstrated the ability of NF to remove 30% of the lactic acid and up to 60% of the monovalent ions from acid whey, resulting in a significant reduction in powder stickiness upon spray drying.

Salty whey is another type of whey generated from cheese manufacturing (Table 13.4). Sodium chloride, in the form of saturated brine or dry salt, is added to the cheese curds in the brining process when making semi-hard and hard cheeses such as cheddar and Colby. In the subsequent pressing step, more than half of the added salt ends up in a brine stream, which is commonly referred to as salty whey. However, the high salinity level of salty whey (>50 g/L Na [108]) makes its treatment and disposal challenging. It is commonly discharged to evaporation ponds, leading to potential environmental issues such as land degradation, odour, and dust.

A commercial nanofiltration spiral-wound membrane, NTR-7450-S2F (Nitto Denko), was used by Hinkova et al. [109] to remove the salts from salty whey, with less than 38% removal of monovalent ions and 50% removal of calcium. The Ultra-Osmosis® process [110] was reported to be effective in partially removing salts from salty whey, producing a retentate that has similar mineral composition as sweet whey.

13.2.4.2 Chemical Cleaning Agent Recovery

Clean-in-place (CIP) is critical to dairy processors as it can be implemented to clean and sanitize the interior surfaces of equipment, pipes, filters, and fittings without dismantling. CIP operations consume a significant amount of process water and cleaning chemicals. The spent CIP caustic soda solution, for example, can be filtered by microfiltration and polished by nanofiltration, with high chemical oxygen (COD) rejection (around 98%) and high recovery of caustic (up to 95%) [111]. The clean NF permeate is then ready to be used in subsequent CIP cycles. Membranes used in this application need to be alkali resistant. KOCH Membrane Systems has developed a caustic recovery system using pH-stable SelRO™ MPS-34 nanofiltration membranes (MWCO: 200 Da), which can be used to remove more than 90% of COD and the brown, burnt-coloured contaminants from spent acid or caustic, as well as a substantial amount of calcium and carbonates [112].

13.2.5 Reverse Osmosis

13.2.5.1 On-Farm Milk Concentration

RO or NF can be used to concentrate milk on the farm, or in regional centres, increasing farm water availability and reducing transport costs. This is an approach that is used commercially in New Zealand. RO can also be effective in concentrating skim milk in the factory, prior to incorporation into a variety of products such as yoghurts and cheeses.

The use of RO means that essentially all of the milk minerals are retained and only water permeates. While this implies maximum retention of the original milk nutrients, the concentration process alters the equilibrium of calcium phosphate between the insoluble colloids and the dissolved state. For example, Lauzin et al. [113] noted an increase in the ratio of colloidal to total calcium from 68% to 75% during a twofold concentration of skim milk. In turn, the precipitation causes a decrease in pH. Lauzin et al. [113] observed a fall from 6.64 to 6.31, while Syrios et al. [114] observed a fall from 6.64 to 6.3 in a threefold concentration. The increased ionic strength also causes some casein micelles to dissolve, with Lauzin et al. [113] noting an increase in the ratio of soluble to total protein, from 2.9% to 5.7% during a twofold concentration [113].

As with MF of skim milk, the filtration process is governed by the formation of a dense foulant cake of protein on the membrane surface. The thickness and density of the cake is again a function of the shear stress at the surface (governed by the crossflow velocity) and the transmembrane pressure difference [115].

13.2.5.2 Wastewater Treatment and Boiler Feedwater Treatment

Reverse osmosis is commonly used in dairy factories to purify wastewater streams. As one example, the feedwater required for use to make steam in factory boilers needs to be ultrapure to prevent corrosion. The steam is generated at high temperatures through the use of high pressures and as the water evaporates, the impurity concentration increases. Hence even minor salt concentrations can cause damage.

Another example occurs in the re-use of the large quantities of water generated within the evaporators used to concentrate skim milk, whey, or lactose prior to spray drying. This is known as "condensate of whey" or COW water, which is likely to contain a range of small volatile organic molecules and has a strong milk odour. RO can be used to remove such contaminants, allowing re-use in a wide range of applications [116,117].

13.3 ELECTRICAL SEPARATIONS

Electro-membrane processes are a domain of membrane separation processes where ionic species (e.g., salts and organic acids) are separated by charged polymer membranes under an electrical driving force. These membranes are called ion exchange membranes, and are generally made from dense, non-porous and charged polymers that have anionic or cationic fixed-charge groups, which only permit the passage of mobile counter-ions (ions with the opposite charge to the fixed-charge group) and reject the migration of mobile co-ions (ions with the same charge as the fixed-charge group). These fixed-charge groups are typically sulphonic acid groups ($-SO_3^-$) for cation exchange membranes (CEMs), and quaternary ammonium groups ($-NH_4^+$) for anion exchange membranes (AEMs). The most commonly used electro-membrane process is electrodialysis (ED), which consists of two electrode compartments and multiple flow channels in between, separated by alternative CEMs and AEMs (Figure 13.14). Under an electrical potential driving force, cations (e.g., Na^+) in the diluate stream migrate towards the cathode through the CEMs but are then retained by the AEMs. Similarly, anions (e.g., Cl^-) migrate towards the anode through the AEMs but are retained by the CEMs. Ultimately, the diluate stream is depleted of ionic species while the salts accumulate in the concentrate stream. Commercial ED stacks can contain up to 200 flow channels, with the current and voltage supplied to the stack up to 185 A and 400 V, respectively [118–121]. For dairy systems, neutral and large molecules such as proteins are completely rejected by ion exchange membranes, while minerals and organic acids can be removed as they are dissociated. This makes electrodialysis an effective technology for removing such species from dairy streams such as milk and whey.

The rate of transport of ionic species across an ion exchange membrane can be described by the extended Nernst–Planck equation (Equation 13.16), where J_s is replaced by J_i denoting the flux of any ionic species (i) through the membrane. The contribution of ion convection towards ion transport is often insignificant, compared to ion diffusion driven by concentration gradient and electromigration driven by an electric potential. Therefore, Equation (13.16) can be simplified as:

$$J_i = -D_{i,p}\frac{dc_i}{dx} - z_i c_i D_{i,p}\frac{F}{RT}\frac{d\Psi}{dx} \qquad (13.23)$$

Similar to nanofiltration, the concentration of an ion in the membrane phase can be related to the ion concentration in the bulk solution through the partition coefficient which dictates the equilibrium boundary conditions (see Equation 13.21). This equation suggests that Donnan exclusion is very effective if the ion concentration in the bulk solution is much lower than that in the membrane phase ($C_i \ll c_i$). Given the high concentration of fixed anionic groups within a CEM membrane, anions are readily excluded on this basis. Similarly, cations will not flow through an AEM, resulting in high permselectivity of the membrane. The transport number t_i of a particular ion i represents the fraction of the current used to transport those ions through the membrane and can be estimated by:

$$t_i = \frac{|z_i| J_i}{\sum_i^n |z_i| J_i} \qquad (13.24)$$

J_i can be related to the current density that ions carry through the IEM, according to Faraday's law:

$$\frac{I}{A} = F\sum_i^n |z_i| J_i \qquad (13.25)$$

where I is the current and A is the effective area of a single membrane. As the permselectivity for counter-ions in commercial IEMs is normally very high, particularly in dilute electrolyte solutions, the transport number of counter-ions ranges from 0.9 to 1 and that of co-ions is between 0 and 0.1 [123,124].

The utilization of current to migrate ions across the membrane is often not complete, due to the increasing mass transfer resistance, or concentration polarization established in the boundary layer adjacent to the membrane. Similarly, Fickian diffusion of ions can occur in the opposing direction due to concentration differences. The current efficiency (CE) is used to estimate the percentage of the total applied current used for effective migration of ions, and is calculated using the following equation:

$$CE = \frac{F \cdot \sum_i z_i \Delta n_i}{N_{cell}\int I dt} \times 100\% \qquad (13.26)$$

where Δn_i is the number of moles of target ions migrating from the diluate to the concentrate over time t and N_{cell} is the number of cell pairs.

Specific energy consumption (SEC) is an important parameter for assessing the viability of an electro-membrane system, and is estimated as the energy required per unit mass of the target ion removed:

$$SEC = \int_0^t \frac{U \cdot I}{\Delta n_i \times M} dt \qquad (13.27)$$

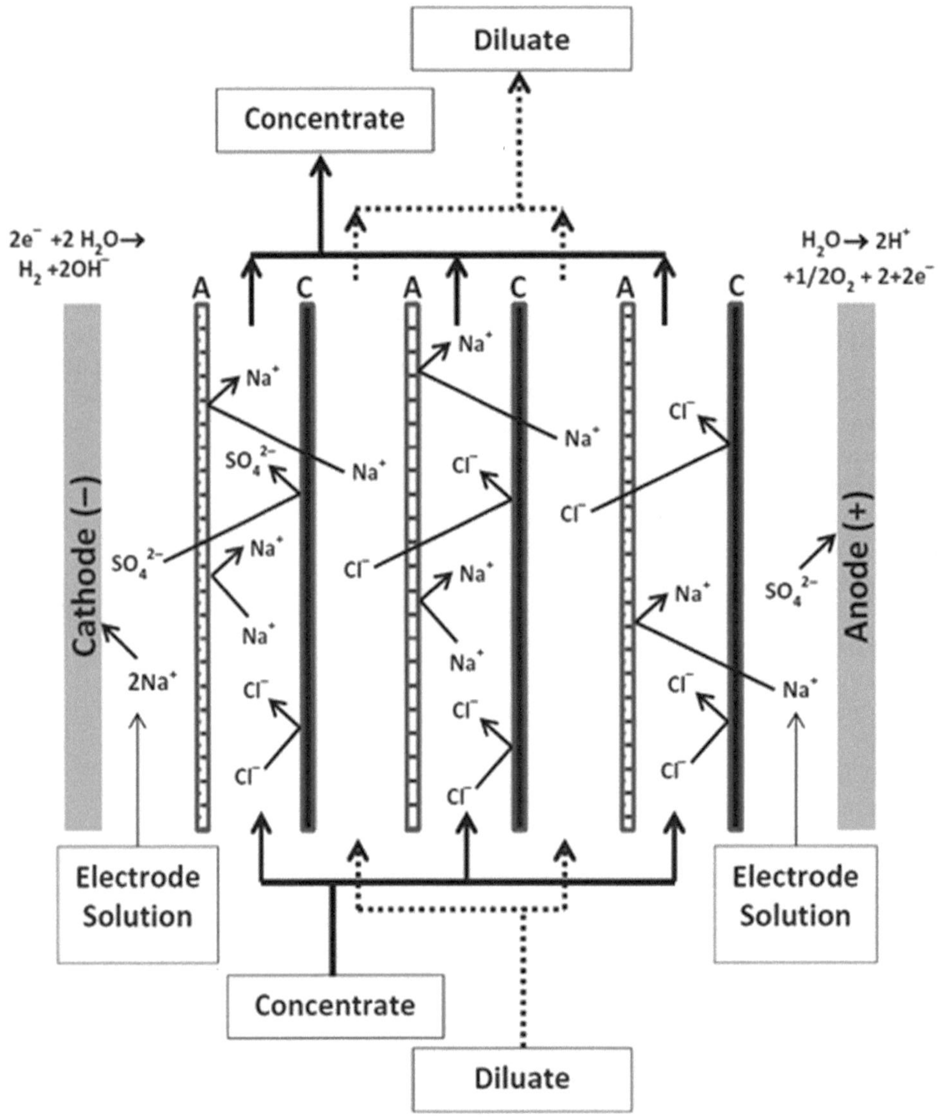

FIGURE 13.14 Illustration of an electrodialysis stack removing sodium chloride from the diluate (i.e., feed) into the concentrate. (A: Anion exchange membrane; C: Cation exchange membrane.) Reprinted from Ref. [122] with permission.

where U is the voltage applied across the ED stack and M is the molar mass of the target ion.

13.3.1 Electrodialysis for Whey Demineralization

Electrodialysis has been used in commercial operations to demineralize sweet whey. To maximize the degree of demineralization, ED should be operated at a feed pH close to the isoelectric point of the major whey protein, as uncharged proteins are not involved in ion transport, hence leading to improved current utilization [125]. Reducing the protein content in whey is effective in reducing the boundary layer thickness and fouling effects, as demonstrated by Perez et al. [126] where they compared the demineralization achieved from UF permeate (lactose rich) and UF retentate (protein rich). Monovalent ions are removed more effectively by electrodialysis compared to divalent ions, due to their smaller hydration diameter and greater mobility [127,128]. Electrodialysis is more cost-effective than ion exchange for demineralization of sweet whey to levels up to 70%, suitable for producing D70 whey. Pre-concentration to 18–24% total solids by RO or evaporation has been found to enhance the efficiency of electrodialysis [129].

Electrodialysis has also been used for lactic acid removal from acid whey, as the dissociated lactate ions and other mineral ions can be removed by the IEMs, while the whey proteins and lactose are retained. As recently demonstrated by Chen et al. [7] and Dufton et al. [130], using both ultrafiltered acid whey and raw acid whey, the ratio of lactic acid to lactose can be reduced by at least 80%, from 15 g/100 g to below 3 g/100 g (Figure 13.15). This falls within the range of the lactic acid/lactose ratio found in sweet whey. Simultaneously, the mineral content in the acid whey processed by electrodialysis decreases by greater than 80%, resulting in a demineralized

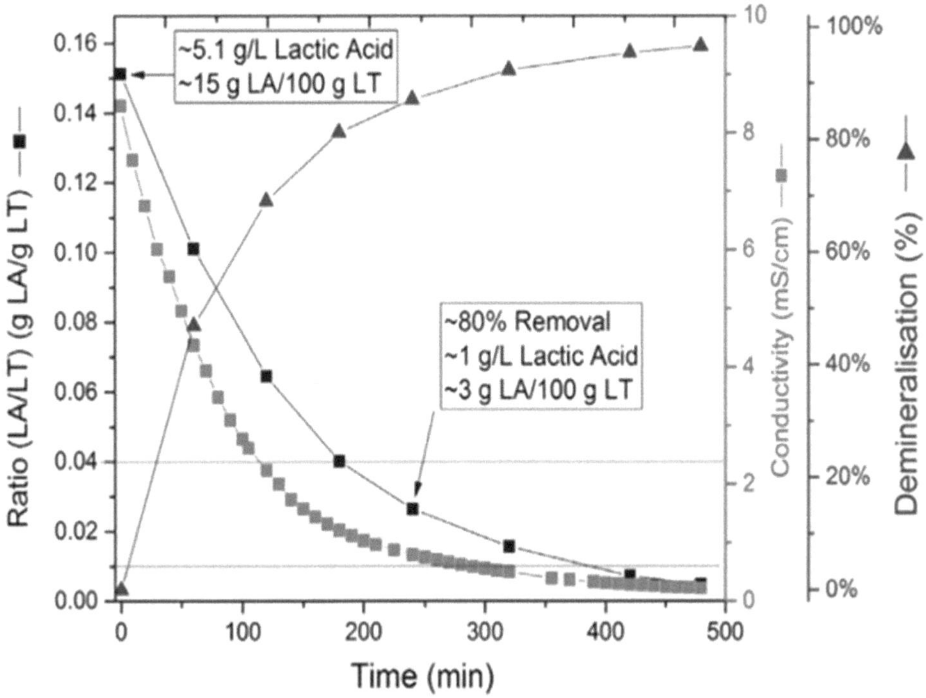

FIGURE 13.15 Performance of a batch electrodialysis process during the treatment of ultrafiltered acid whey 45°C. Reproduced from Ref. [7] with permission from Elsevier.

whey stream that can be further processed into whey powder. Similar to sweet whey demineralization, chloride is observed to be removed by the AEMs in preference to lactate ions [7]. At a demineralization degree of 70%, this process consumes less than 12 Wh per kg whey processed, for both acid whey and sweet whey.

The presence of proteins, calcium, and phosphate is also known to cause fouling in electrodialysis systems used in dairy processing. Fouling on IEMs can compromise membrane integrity and cause a reduction in ion transfer rate, leading to increased electrical resistance and energy consumption [131–134]. The type of fouling depends on the type of IEMs and the compartment in which fouling occurs. For AEMs, protein fouling on the diluate side dominates [135], as the major proteins such as β-lactoglobulin and α-lactalbumin are likely to be negatively charged at the natural pH of sweet and acid whey (pI of β-lactoglobulin and α-lactalbumin: 5.2 and 4.8). Hence, these proteins tend to migrate towards the anode but are too large to penetrate through the AEMs, hence forming deposits on the diluate side. Mineral fouling is also possible on the concentrate side of the AEMs, due to the interaction of the calcium ions that migrate through the CEMs and the phosphate ions transferred through the AEMs [133,135]. Mineral fouling involving calcium phosphate salts, however, can also be found on the CEMs facing the concentrate, particularly under alkaline concentrate pH [131,136]. Further, protein fouling on CEMs facing the diluate was observed by some studies when the concentrate was kept at acidic pH [128,137,138]. This is because the migration of protons from the concentrate to the diluate lowers the pH on the surface of the CEM facing the diluate, promoting protein precipitation as this pH approaches the isoelectric points of some whey proteins.

The pH of the concentrate is often adjusted to reduce fouling [135,138–140]. In a recent study, Talebi et al. [141] demonstrated that the use of an alkaline concentrate led to mineral deposits on IEMs during electrodialysis of sweet whey and acid whey, although these deposits appeared to detach readily. Using an acidic concentrate could enhance significantly the rate of lactic acid removal, but protein fouling took place. It was recommended that the concentrate pH should be adjusted to between 4 and 5.5 to minimize protein fouling and avoid calcium salt precipitation, while a high level of removal of monovalent ions and lactate ions can be obtained. These strategies were implemented in their pilot-scale studies on acid whey processing using electrodialysis with ultrafiltration and nanofiltration as optional pre-treatment processes [142]. They successfully demonstrated that acid whey treated with ultrafiltration followed by nanofiltration could lead to the highest lactic acid removal rate (>80%) during electrodialysis, producing a non-sticky whey powder with a moisture content of 2.5%.

The idea of separating salts from salty whey using electrodialysis was explored by Diblíková et al. [128] and Ecer et al. [143]. The increased salt content in the feed solution (such as salty whey) can improve electrodialysis performance mainly due to lower diluate resistance. The concentrate itself, however, becomes a brine stream that needs to be treated or disposed of, and often is limited to ~2–3M salt concentration due to osmotic water transport. The benefit could be the recovery of a lactose-rich stream from salty whey by removing the salts, and to produce a high-purity NaCl solution using monovalent ion exchange membranes for salt recovery [144]. In addition, the use of salty whey as the concentrate solution in an electrodialysis process for sweet whey demineralization was also recently investigated by Talebi et al. [144]. Their approach had no effect on the rate of sweet whey demineralization or energy consumption, but more calcium and less sodium were removed.

13.3.2 Electrodialysis with Bipolar Membranes

The conventional electrodialysis processes have been combined with bipolar membranes, forming an advanced electro-membrane process called electrodialysis with bipolar membranes (EDBM). A bipolar membrane (BPM) is typically composed of an anion exchange layer, a cation exchange layer, and an intermediate layer which facilitates water splitting to form H^+ and OH^- ions. This technology was initially developed for splitting a saline solution into its corresponding acid and base [145–147]. Under an electric field, the H^+ ions produced by the BPM combine with the anions (e.g., Cl^-) migrating from the feed solution through the AEM, forming the acid. Simultaneously, the OH^- ions combine with the cations (e.g., Na^+) transferred through the CEM, producing an alkaline stream. The feed solution is therefore desalted, allowing re-use or disposal as appropriate.

The application of bipolar membranes in dairy processing has only been explored in the last two decades. Merkel et al. [148] investigated the use of OH^- ions generated from the BPMs to raise the pH of the acid whey processed by ED or NF to around 6, without the need to add alkaline solutions such as potassium hydroxide or sodium hydroxide. In contrast, the H^+ ions generated from the BPMs can be used to lower the pH in milk to the isoelectric point of caseins (i.e., pH 4.6), allowing the production of caseins without chemical acidification [149]. This approach was further optimised recently by Mikhaylin et al. where they used 10 kDa ultrafiltration membranes to separate the proteins from milk permeate, directing only the protein-free permeate through the EDBM process for pH adjustment [150,151]. The pH-adjusted permeate is then mixed with the UF retentate in the milk reservoir to allow casein precipitation to occur at pH 5. Operating at this pH can minimize both protein fouling in EDBM and calcium scaling. Furthermore, the permeate is simultaneously demineralized during EDBM, leading to the complete precipitation of caseins at pH 5. The shift in the isoelectric point (from 4.6 to 5) is caused by easier destabilization of casein micelles when the ionic strength in the milk reservoir is greatly reduced [150].

EDBM was also evaluated recently as a mechanism to transform salty whey into sodium hydroxide and hydrochloric acid for reuse within the factory [108]. Using an industrial salty whey sample containing ~50 g/L Na (0.86 M), the maximum acid and base concentrations produced by EDBM were 3.6 ± 0.2 mol/L HCl and 3.0 ± 0.3 mol/L NaOH using a volume ratio of 6:1 (feed: acid and base). These concentrations are lower than the theoretical maximum (i.e., 5.1 M) because of the water transport into the acid and base compartments from the feed due to electro-osmosis and the leakage of small and highly mobile H^+ and OH^- ions from these compartments. The presence of calcium phosphate salts in salty whey has a negative effect on EDBM performance, hence pre-treatment of salty whey to precipitate calcium phosphate at pH 11 is required. In large-scale operations, this pH adjustment step can be conducted using less than 10% of the sodium hydroxide produced from the EDBM process. The NaOH solution has a purity of 97%w/w with most of the impurity being potassium, allowing re-use as CIP solutions. In general, electrically driven membrane processes such as EDBM are more capital intensive than pressure-driven filtration processes such as reverse osmosis, mainly because ion exchange membranes are significantly more expensive.

13.3.3 Electrodialysis with Filtration Membrane for Protein Separation

Ion exchange membranes are typically polymeric dense membranes which are only permeable to molecules and ions smaller than 500 Da [152]. When ion exchange membranes are replaced with filtration membranes in an electrodialysis process, it provides the option to use the electrical driving force to selectively transfer large charged molecules through the filtration membrane. This leads to the development of a special type of electro-membrane process called electrodialysis with filtration membrane (EDFM) (Figure 13.16). Ion exchange membranes can be used as "restriction membranes" in this process to prevent large molecules from entering the electrode compartments. Alternatively, filtration membranes with a MWCO smaller than the smallest organic molecules in the mixture can also be employed [153,154]. One of the key features in EDFM is that uncharged membranes are often used as the separation membrane to prevent water splitting, due to the high electrical field strength (up to 2000 V/cm [155]) needed to migrate large organic molecules such as proteins [153,155,156]. Due to the elimination of a pressure-driving force, EDFM minimizes the compressive force that leads to the formation of a highly impermeable fouling cake which often occurs in protein filtration [157]. However, because of the large pore size in the filtration membranes, small inorganic ions and organic acids are no longer discriminated by EDFM.

EDFM was first proposed by Gradipore Ltd. [155,158], who further demonstrated its application in the separation of immunoglobulin G from human plasma, proteins from egg white, and haemoglobin from a mixture with BSA [159,160]. The isolation of high-value minor whey proteins such as lactoferrin and immunoglobulins from whey has recently been investigated. Ndiaye et al. attempted the use of an EDFM process with a PES UF membrane (MWCO 500 kDa) to transfer LF from an LF-enriched whey solution at pH 3.0, with a yield of only 15% and low purity due to the migration of other whey proteins [161]. In contrast, Wang et al. [162] used a polyvinyl alcohol (PVA) membrane prepared in house to serve as the separation membrane to demonstrate high rejection of

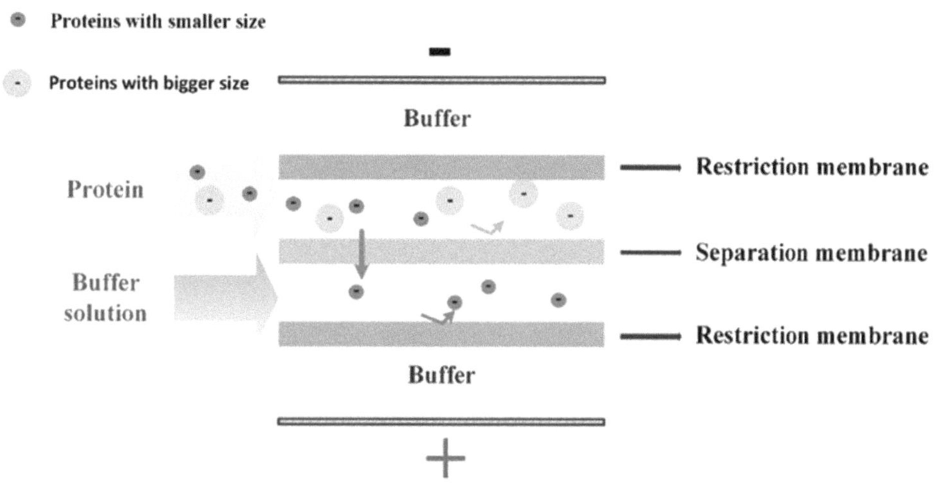

FIGURE 13.16 Illustration of a typical electrodialysis with filtration membrane (EDFM) process [162].

lactoferrin (present as tetramers) and immunoglobins in whey solutions and a high flux of other whey proteins such as BSA. Significant protein loss was observed in their study due to the adsorption of proteins within the membranes, although this could be minimized by reducing the ionic strength of the whey solutions. A washing process using a salt solution was also able to recover the protein deposits. Back diffusion of the proteins through the separation membrane needs to be avoided, by maintaining their concentration in the product stream lower than that in the feed. Overall, EDFM is a relatively less mature technology compared to conventional ED and EDBM but has the potential to facilitate the fractionation of proteins and other macro-functional molecules in dairy systems.

13.4 EMERGING MEMBRANE PROCESSES

13.4.1 Membrane Chromatography

Chromatography is an established unit operation within the dairy industry, where it can be used to recover specific proteins at high purity. The process uses resin beads formed from polymers that are functionalized with specific functional groups, or ligands. For example, lactoferrin and lactoperoxidase can be recovered from skim milk or whey using a cation exchange resin [163–165]. This resin contains negatively charged functional groups that capture the positively charged proteins, including lactoferrin. After the resins are completely loaded with proteins, salt solutions are passed through the resin beads. In the first elution, the less strongly bound proteins are released, while a later strong eluant is used to release the lactoferrin and lactoperoxidase fraction. A sequence of ultrafiltration and diafiltration is then used to remove the salt from the proteins prior to evaporation and spray drying. Similarly, immunoglobulin G can be recovered using protein A affinity chromatography, where the resins are functionalized with ligands that bind strongly to the fc portion of this antibody; or mixed-mode chromatography, where a number of different ligand groups are used.

The capital costs associated with chromatography are very high, due to the long residence times required to ensure the proteins can migrate to the interior of the polymer resin beads [166,167]. This requires large stainless-steel vessels and large volumes of resin. The viscous nature of the protein solutions also means that the pressure drop through the resin bed is high, which can cause bed compaction and channelling [168,169]. In the pharmaceutical industry, these costs are leading to a move towards membrane chromatography units. In this case, the ligands are distributed throughout the bulk of a porous polymeric membrane with a surface area of 0.6–3 m^2g^{-1} and pore sizes ranging from 2 to 4 µm [170]. As the feed flow passes throughout this structure, the mass transfer resistance is reduced and so shorter residence times are required (Figure 13.17). The pressure drop throughout the system is also significantly reduced. The devices can be formed completely from plastic, so the capital cost is reduced. Conversely, the number of ligand sites is much lower than what can be provided in packed bed chromatography. This means that a smaller volume of feed is processed in each cycle, resulting in a much greater consumption of eluant solutions and buffers.

While the use of such membrane adsorbers has yet to be adopted by the dairy industry at scale, there have been a number of preliminary studies. For example, a Sartobind® anion exchange membrane adsorber has been tested to bind β-lactoglobulin and bovine serum albumin (BSA), which were then eluted separately with salt solutions of varying strength [168,171]. In a second step, a similar cation exchange adsorber was used to separate the remaining minor proteins.

13.4.2 Forward Osmosis

Forward osmosis (FO) was originally developed to produce fresh water from brackish water or seawater and to remove

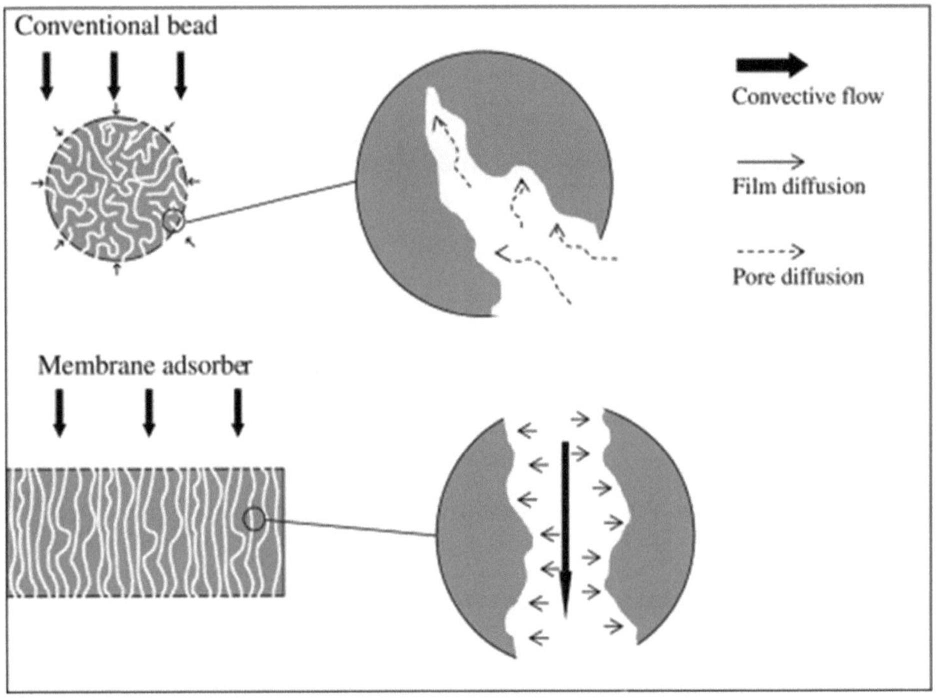

FIGURE 13.17 Mass transfer in a classical chromatography resin bead is slow, as it relies upon pore diffusion. Conversely, a membrane adsorber relies on convective flow and film diffusion [172].

water from a contaminated stream. In this process, water is "drawn" from a feed stream of low ionic strength into a concentrated salt solution (called draw solution), under the osmotic pressure difference created by the difference in concentration between the feed and draw solutions, which are separated by a semi-permeable membrane. In this case, hydraulic pressure is not required so the flux in forward osmosis can be described by an alternative form of Equation (13.10):

$$J_v = A\,\Delta\pi_{eff} = A\left(\pi_{D,m} - \pi_{F,m}\right) \tag{13.28}$$

where A is the water permeability coefficient, which is an intrinsic parameter of the active layer of the membrane, and $\Delta\pi_{eff}$ is the effective osmotic pressure difference between the draw solution side ($\pi_{D,m}$) and the feed solution side ($\pi_{F,m}$) of the membrane active layer.

Membranes used in FO are asymmetric membranes where the active polymer layer is fabricated on top of an optimised support layer that is thin, highly porous, and with little tortuosity. In these processes, concentration polarization on both sides of the membrane must be considered as the effective osmotic driving force is diminished at the active layer interface (Figure 13.18) [174]. When the active layer of an FO membrane is facing the feed solution, which is the likely orientation employed in dairy processing (to avoid any milk solids to enter the porous support layer), the osmotic pressure increases on the active layer surface as solute is concentrated. This is referred to as concentrative external concentration polarization (ECP) and can be described by manipulation of Equation (13.8) in terms of the change of solute concentration:

$$\frac{C_{F,m}}{C_{F,b}} = exp\left(\frac{J_v}{k_{cp}}\right) \tag{13.29}$$

where $C_{F,b}$ and $C_{F,m}$ are the concentration of the feed solution in the bulk and at the membrane surface, respectively. Assuming the osmotic pressure is proportional to solute concentration, the ratio of $C_{F,m}$ to $C_{F,b}$ is equal to the corresponding ratio of $\pi_{F,m}$ to $\pi_{F,b}$:

$$\frac{\pi_{F,m}}{\pi_{F,b}} = exp\left(\frac{J_v}{k_{cp}}\right) \tag{13.30}$$

In contrast, the draw solution concentration within the porous structure is diluted due to the permeation of water from the feed side, reducing the salt concentration near the active layer (Figure 13.18). This is referred to as dilutive internal concentration polarization (ICP). The solute resistivity, K, can be used to account for this effect [173]:

$$K = \left(\frac{1}{J_v}\right)\ln\left(\frac{B + A\pi_{D,b}}{B + J_v + A\pi_{F,m}}\right) \tag{13.31}$$

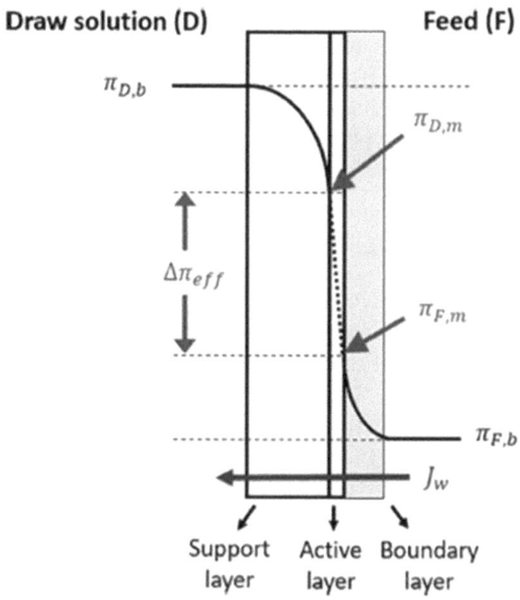

FIGURE 13.18 Illustration of concentration polarization occurring on the feed and draw solution sides of the active layer of a forward osmosis membrane [174].

where K is calculated as the ratio of the structure parameter of the FO membrane (S) to the solute diffusivity ($\mathfrak{D}$), $\pi_{D,b}$ is the osmotic pressure in the bulk of the draw solution, and B is the salt permeability coefficient of the active layer. Assuming salt permeability through the active layer is negligible (i.e., B= 0), Equation (13.28) becomes:

$$J_v = A\left(\pi_{D,b}\exp\left(-J_vK\right) - \pi_{F,m}\right) \tag{13.32}$$

An implicit expression (Equation 13.33) can be obtained by combining Equation (13.30) and Equation (13.32), and can be solved numerically to estimate the water flux through an FO membrane:

$$J_v = A\left[\pi_{D,b}\exp\left(-J_vK\right) - \pi_{F,b}\exp\left(\frac{J_v}{k_{cp}}\right)\right] \tag{13.33}$$

The main application of forward osmosis in dairy processing is the concentration of dairy streams prior to evaporation and drying. A proprietary thin-film composite membrane developed by HydrOxSys has also been commercialized to target on-farm concentration of milk [175]. Integrated systems of forward osmosis with ultrafiltration and reverse osmosis were investigated by Aydiner et al. for water recovery from whey using NaCl [176,177] and ammonia and carbon dioxide (NH_3/CO_2) salts as the draw solution [178,179]. Using a NaCl draw solution, the typical flux for FO membranes ranged from 10 to 25 LMH at low whey concentrations [177,180]. Using a 2 M NaCl draw solution, unprocessed whey (6.75%TS) could be concentrated up to 28%TS. In forward osmosis, the reverse diffusion of salts from the draw solution to the feed solution cannot be eliminated and can potentially lead to contamination of concentrated dairy products. The back diffusion of ammonium carbonate, for example, adds new ions into the natural dairy systems, meaning that final products would not meet regulatory requirements. More recently, Chen et al. [181] evaluated the use of a cellulose triacetate forward osmosis membrane to concentrate a range of dairy streams, including skim milk, whey, D40 whey, whey protein concentrate, and lactose. Due to the high osmotic pressure (150 bar) available from the 1.6 M $MgCl_2$ draw solution, the dairy streams could reach a concentration of up to 40%TS, almost doubling the concentrations which can be obtained by nanofiltration and reverse osmosis. The addition of magnesium to the dairy streams due to solute back diffusion was shown to be under 100 mg per 100 g dry powder, which is below the regulatory limit. This indicated that product quality was unlikely to be jeopardized as the presence of magnesium ions in food is known to be beneficial for human consumption [182].

Forward osmosis is claimed to be a highly energy-efficient process since the transfer of water occurs spontaneously in response to an osmotic driving force. However, the draw solution is effectively diluted by the water drawn from the feed, meaning that it loses its osmotic potential unless it is regenerated. Removing the water from a diluted draw solution is often energy intensive, involving processes such as reverse osmosis and thermal evaporation. Therefore, it is critical to identify a suitable draw solution that is readily available within the dairy processing site and that can be discharged upon dilution. To mimic the use of salty whey as the draw solution, a pilot-scale study was conducted by Chen et al. [183] to concentrate skim milk and whey using a ~50 g/L NaCl solution. This work found that a concentration factor of up to 2.5 can be achieved, without regenerating the draw solution, although the water flux declined from ~5 LMH to below 0.5 LMH. In addition to the solute back diffusion, migration of small organic molecules from the feed into the draw solution was also observed. The FO process alone requires much less energy per tonne of water removed compared to reverse osmosis, provided that a suitable brine stream such as salty whey can be used without the need to regenerate.

13.4.3 Membrane Distillation

Membrane distillation (MD) is another emerging membrane process that is driven by a vapour pressure gradient between the feed and permeate side of the membrane. The vapor partial pressure difference (ΔP_m) is typically generated by the temperature difference across a hydrophobic membrane [e.g., polytetrafluoroethylene (PTFE)], allowing only water vapour to pass through. This mass flux of water vapor (N) across the membrane can be described by Equation (13.34):

$$N = K_m\Delta P_m \tag{13.34}$$

where K_m is the mass transfer coefficient of water vapour through the membrane. The vapour pressure on the feed side can be influenced by the solute concentration. As water activity decreases with increasing solute concentration, this results in a reduction in its partial vapour pressure and hence a loss in vapour pressure driving force.

One of the key advantages of MD is the potential to use waste heat or solar energy to create the moderate temperatures required (e.g., 50°C for the feed side). Because a porous membrane (e.g., 0.22–0.7 μm in pore size) is used in MD however, pore wetting can be an issue. In this case, liquid from the feed penetrates through the pores, substantially reducing the mass transfer coefficient. The liquid entry pressure (LEP) is a property used to describe the susceptibility of a membrane toward pore wetting and can be determined experimentally or estimated by the Laplace–Kelvin equation:

$$LEP = -\frac{2\, B_g\, \gamma_L\, cos\theta}{r_p} \tag{13.35}$$

where B_g is a geometrical factor ranging between 0 to 1 (unity is used for cylindrical pores), γ_L is the liquid surface tension, θ is the liquid/membrane contact angle, and r_p is the maximum pore radius. Membrane distillation should operate below the LEP to avoid pore wetting. With increasing temperature, both liquid surface tension and contact angle are likely to reduce, leading to a reduced LEP. Based on Equation (13.35), using a hydrophobic membrane with small and uniform pores can minimize pore wetting.

Membrane distillation is an alternative for concentrating dairy streams, as all solutes can be retained in the feed stream. The concentrations of whey protein [184], milk, whey, and lactose [185,186] have been reported. The initial flux of water vapour for membrane distillation of skim milk and whey ranges from 12 kg/m^{2}h to 20 kg/m^{2}h. Salty whey can also be concentrated to a final total dissolved solids (TDS) concentration of up to 30%w/w, recovering greater than 80% of the water from the feed [8]. This leads to the potential for recovering the salt in the concentrated salty whey using crystallization processes. However, the use of a hydrophobic membrane in membrane distillation promotes protein fouling (Figure 13.19). In contrast to conventional filtration systems, fouling in membrane distillation is less likely to take place inside the membrane pores because convective flow of the solutes does not occur. Surface fouling due to organic attachment, however, can reduce membrane hydrophobicity, leading to liquid penetration and deterioration of permeate quality. In contrast, the development of mineral scaling such as calcium phosphate precipitation is often impermeable to liquid, resulting in a sharp decline in permeate flux [187].

Membrane distillation can be used to recover water from the diluted draw solution in a forward osmosis process, potentially reducing the energy required for draw solution regeneration. The hybridization of forward osmosis and membrane distillation has been demonstrated, for example, to concentrate protein solutions using NaCl as the draw solution [188] and the treatment of dye wastewater using a polyacrylic acid sodium (PAA-Na) salt as the draw solution [189]. These studies concluded that such hybrid processes were repeatable and controllable, performing better than individual FO processes.

13.4.4 Membrane Capacitive Deionization

Membrane capacitive deionization (MCDI) is an emerging electro-membrane process where the separation of ionic species is achieved through the adsorption of ions onto the electrode materials. It was first proposed by Biesheuvel and Van der Wal in 2010 [191]. As illustrated in Figure 13.20, a CEM and an AEM are placed in front of the cathode and anode, respectively. Both electrodes are typically prepared by depositing a layer of activated carbon on to a graphite sheet (as the current collector). When the feed solution is passed through the flow channel formed between the IEMs, under an electric potential, the cations will pass through the CEM and adsorb into the cathode, while the anions will flow through the AEM and adsorb into the anode. This process produces a desalinated stream. Once the adsorption capacity of the carbon electrode is reached, the voltage is reversed so that the adsorbed ions are desorbed back into the stream flowing through the cell, resulting in a salty waste stream. MCDI is an advanced version of CDI (capacitive deionization) where no IEMs are used. The use of IEMs can prevent co-ions from reaching the similarly charged electrodes during adsorption, and the desorbed ions from approaching the opposite electrodes during desorption.

The voltage required to drive ion migration in MCDI is often lower than 1.2 V, to avoid high energy consumption and loss of process efficiency caused by water electrolysis which occurs at 1.23 V and above. It has been investigated for brackish water desalination [193], wastewater treatment [194], ultrapure water production [195], and nitrate ion removal [196]. In general, salt removal efficiency decreases with increasing salinity in the feed solution, due to the limited adsorption capacity of the electrodes [193]. It has been estimated that MCDI could be more energy efficient than reverse osmosis if the feed water salinity is below 2000 ppm TDS, consuming less than 0.5 kWh per m^3 fresh water produced [197]. This demonstrates the potential for MCDI to compete with other desalting technologies in removing salts from low-salinity feed streams.

A preliminary study recently reported the use of MCDI to process three different pre-treated acid whey solutions in a lab-scale unit, including ultrafiltration permeate, nanofiltration retentate, and dia-nanofiltration retentate [198]. These three solutions are free of proteins with 4–15% total solids and 6–13 g/L lactic acid. Lower demineralization degree and lactic acid removal were achieved for the concentrated feed streams (i.e., nanofiltration retentate and dia-nanofiltration retentate), while energy consumption was higher for the ultrafiltration permeate (6.5 kWh/equivalent of cations removed from nanofiltration retentate vs. 13.9 kWh/equivalent cations removed from

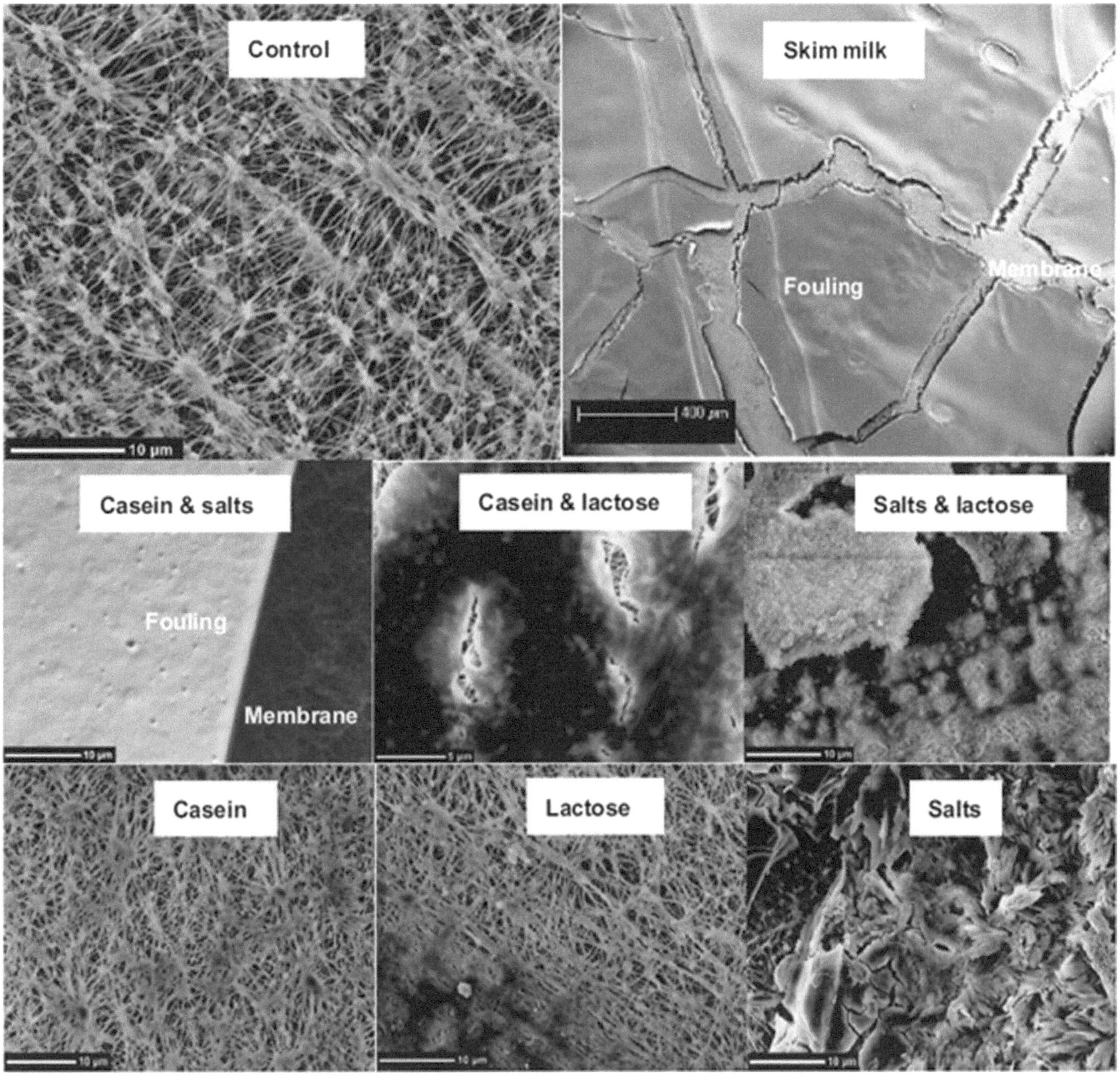

FIGURE 13.19 Scanning electron microscopy (SEM) images of different types of fouling on the surface of a hydrophobic membrane. Reproduced from Ref. [190] with permission from Elsevier.

ultrafiltration permeate). The energy consumption for the treatment of ultrafiltration permeate is also comparable to the energy required for treating the same stream using the electrodialysis process [9].

13.5 CLEANING AND SANITATION

Membrane cleaning is an essential requirement in dairy processing to maximize production efficiency, minimize energy demand, and ensure food quality. The separate, but related, step of sanitization is integral to food safety. However, every cleaning and sanitization cycle takes valuable production time and has an impact on membrane integrity. The chemicals used and the duration of application must thus be carefully chosen to ensure maximum membrane life and to ensure that any manufacturer warranty is not compromised. Most manufacturers dictate the pH range of cleaning chemicals that can be used and these recommendations should be strictly adhered to.

The cleaning chemical used should be specific to the type of foulant. Alkali cleaning is used to remove both protein foulants and dairy fats, often in combination with surfactants and sequestrants. Sodium hydroxide is widely used for this purpose, but phosphates can also be used. Sodium carbonate is the cheapest alkali available, but forms an insoluble salt with calcium, which may hinder cleaning. The optimum pH

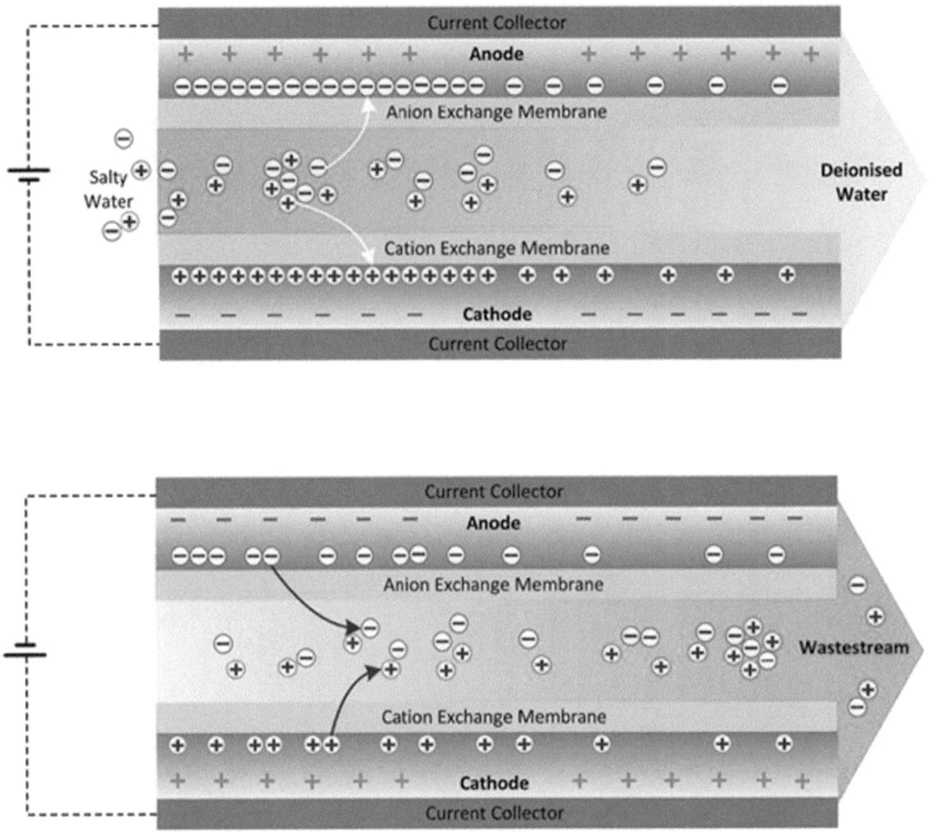

FIGURE 13.20 Schematic representation showing the adsorption and desorption process during MCDI. Reproduced from Ref. [192] under a Creative Commons Attribution (CC BY) license.

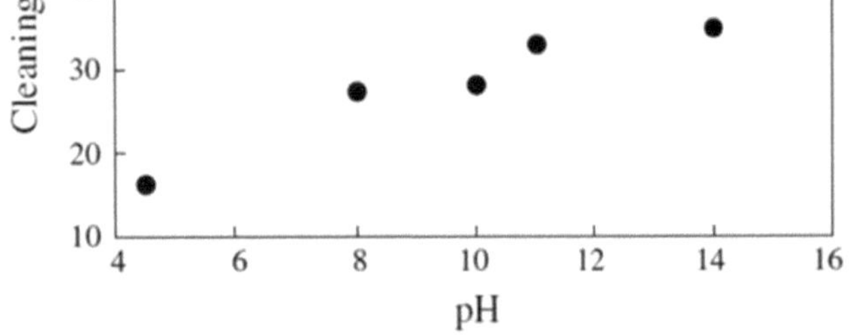

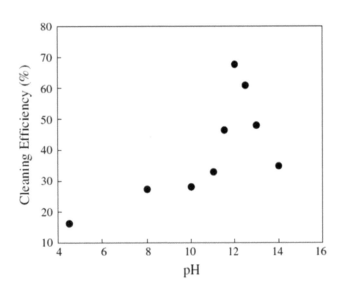

FIGURE 13.21 Effect of alkali solution pH on the cleaning efficiency of reconstituted whey powder solutions from UF membranes. Adapted from Ref. [201].

for cleaning is around 12, or 0.4 wt.% NaOH [199–201] as this is where the proteins are generally most swollen [202]. Operating at a higher pH is less effective, as the proteins tend to become more compacted with the increasing ionic strength (Figure 13.21). Importantly, PVDF membranes are not alkali tolerant and so cannot be used with alkaline cleaners. Polyamide membranes, used in NF and RO, can also suffer if frequently exposed to high pH.

Enzyme cleaners can be used as an alternative approach to alkali cleaners, particularly if the membrane is not tolerant of high pH. These typically contain a mixture of proteases, but can also include lipases. They take longer times to act, typically 60–90 min [16], which can be a deterrent to their use. Some operators can also be concerned that they may persist beyond the cleaning cycle and thus cause product damage. A short elevation in temperature following their use is usually effective in denaturing any residual activity, but a further cleaning cycle may be required to remove the denatured enzyme itself.

Surfactants are long-chain hydrocarbon molecules with a hydrophilic tail. These molecules can solubilize fats and other hydrophobic species. Anionic surfactants such as sodium dodecyl sulfonate (SDS) and sodium dodecyl benzene sulfonate (SDBS), as well as carboxylates and phosphates, are used due to their good wetting and dispersing power [203]. Their optimum pH is 10.5 [204], so a compromise with the alkali cleaners is clearly needed. Cationic surfactants are not generally recommended. Aside from a lower wetting ability [203], they can react with free carboxyl groups on the surface of polyamide membranes [16] and their accumulation in

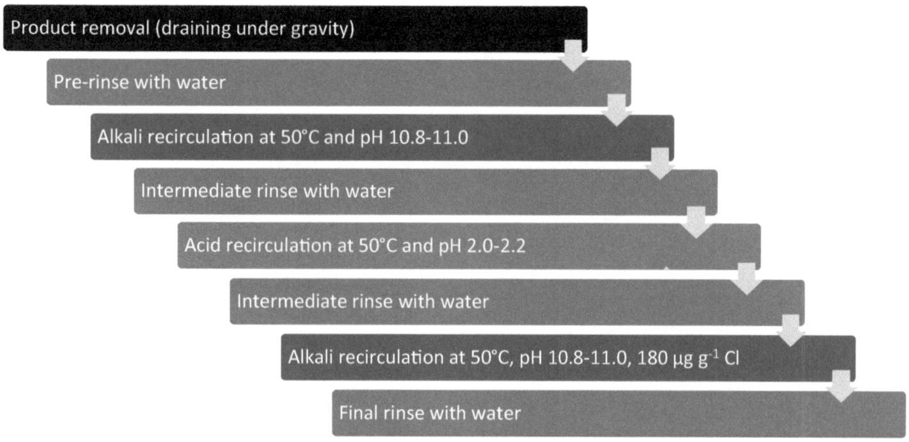

FIGURE 13.22 A typical cleaning cycle that might be used with a protein-rich dairy membrane operation.

powdered products can be associated with birth defects. Non-ionic surfactants are sometimes used. These are molecules that contain both hydrophobic regions as well as hydrophilic groups such as SO_3, NH_4, CHO, and CH_2OH [16].

Acid cleaners are generally used to remove calcium phosphate and other inorganic scales. Phosphoric acid, nitric acid, or their combination is common. Phosphoric acid is a good buffer [204], which helps to maintain the correct pH and an effective sequestering agent, which helps to remove metal cations. However, it is expensive and can cause brittleness when used with ceramic membranes. Nitric acid is a strong oxidant, which can help to remove organics, but can be a little corrosive. Organic acids, specifically citric acid, are weaker than their inorganic counterparts but do not damage membranes or ancillary equipment and have good buffering capacity. Citric acid is a good chelating agent, helping to sequester calcium. Conversely, sulphuric and hydrochloric acids are not generally used, as they are quite corrosive towards stainless steel.

Sequesterants can be added to either acid or alkali cleaners to assist in the uptake of metal ions and specifically calcium. Sodium tripolyphosphate and ethylene diamine tetraacetic acid (EDTA) are the two most commonly used.

Disinfection or sanitization is usually the final step in the cleaning cycle. The intent here is to ensure that all pathogenic micro-organisms have been removed and thus to ensure food safety. The most common disinfectant is sodium hypochlorite, which dissociates in water to provide hydrochlorous acid and sodium hydroxide [205]. At pH below 5, while having a greater bactericidal effect, it can dissociate further to generate chlorine gas, which is highly toxic and corrosive. For this reason, it should be stored at pH 8 and used at pH 6–8 by dilution [205]. Even at these pH levels, care must be taken to ensure damage to membranes does not occur. Polysulfone-based membranes can tolerate 200 μg/g Cl for short time periods and 50 μg/g Cl for longer-term exposure. Polyethersulfone membranes can provide greater chlorine tolerance. Conversely, sodium hypochlorite cannot be used with the polyamide membranes used in NF and RO operations at all. For these membrane systems, other disinfectants such as sodium metabisulfite, peracetic acid, or hydrogen peroxide must be used.

A cleaning and sanitization cycle will usually use many of the above cleaning chemicals in a sequence of steps. These cycles will operate at the highest possible crossflow velocity and the lowest transmembrane pressure to optimize the removal of deposits. A high temperature is also desirable – chemical reaction rates double with a 10°C rise in temperature [16]. This temperature, however, is limited to around 50°C for polymeric membranes and 120°C for ceramic modules to limit any damage to either the membranes themselves or the glues used to assemble them.

A typical cleaning and sanitization cycle (Figure 13.22) will commence with a purge of the dairy fluid that was being processed. This is often simply drainage under gravity, allowing the residual material to be recovered. A water rinse will then be used to further remove the dairy products. This will be followed by both alkali and acid cleans of around 20–30 min duration each, with the final alkali clean including a sanitizing agent.

13.6 CONCLUSIONS AND FUTURE DIRECTIONS

This chapter has provided an overview of the wide range of applications where membranes are used within the dairy industry. Pressure-driven membrane processes, including microfiltration ultrafiltration, nanofiltration, and reverse osmosis, have been used for many decades and their performance is well understood, as described here. Other membrane processes, such as electrodialysis, membrane distillation, and membrane chromatography, are not yet well established as modes of operation. This is partly because they are still seen as immature technologies and thus risky to implement. They can also be considerably more expensive from both capital and operating cost perspectives.

Climate change will bring an increasing perspective within the dairy industry on reducing both energy and water consumption. This will change the economics for many of the emerging membrane processes, as the costs of both energy and water increase. Electrically driven membrane processes, such as electrodialysis and electrodialysis with bipolar membranes will also benefit from the broader industry focus on processes such as electrolysis for hydrogen production. There will be a dramatic increase in the scale of production of the ion exchange membranes required for these technologies. This much greater manufacturing scale will reduce the capital cost for the corresponding dairy processes, again improving the economics. There will be a growing acceptance of such electrically driven processes, particularly when driven by cost-effective solar energy.

REFERENCES

[1] Fox, P.F., McSweeney, P.L.H. (1998) *Dairy Chemistry and Biochemistry*. Blackie Academic & Professional, London.

[2] Dupont, D., Croguennec, T., Brodkorb, A., Kouaouci, R. (2013) Quantitation of Proteins in Milk and Milk Products. In: McSweeney, P., Fox, P. (eds) *Advanced Dairy Chemistry, Volume 1A: Proteins: Basic Aspects*, pp. 87–134. Springer, Boston, MA.

[3] Dalgleish, D.G. (2011) On the structural models of bovine casein micelles—review and possible improvements. *Soft Matter* 7: 2265–2272.

[4] Blaschek, K., Wendorff, W., Rankin, S. (2007) Survey of salty and sweet whey composition from various cheese plants in Wisconsin. *Journal of Dairy Science* 90: 2029–2034.

[5] Riera Rodríguez, F.A., Fernández Martínez, A., Muro Urista, C. (2014) Cheese whey as a source of active peptides: Production, analysis, purification and animal and human trials. *Agricultural Research Updates* 8: 1–52.

[6] Jelen, P. (2011) Whey processing. *Encyclopedia of Dairy Sciences* 4: 731–737.

[7] Chen, G.Q., Eschbach, F.I., Weeks, M., Gras, S.L., Kentish, S.E. (2016) Removal of lactic acid from acid whey using electrodialysis. *Separation and Purification Technology* 158: 230–237.

[8] Kezia, K., Lee, J., Weeks, M., Kentish, S. (2015) Direct contact membrane distillation for the concentration of saline dairy effluent. *Water Research* 81: 167–177.

[9] Chen, G.Q., Eschbach, F.I.I., Weeks, M., Gras, S.L., Kentish, S.E. (2016) Removal of lactic acid from acid whey using electrodialysis. *Separation and Purification Technology* 158: 230–237.

[10] Faucher, M., Perreault, V., Ciftci, O., Gaaloul, S., Bazinet, L. (2021) Phospholipid recovery from sweet whey and whey protein concentrate: Use of electrodialysis with bipolar membrane combined with a dilution factor as an ecoefficient method. *Future Foods* 4: 100052.

[11] Cuartas-Uribe, B., Miranda, M.I., Soriano-Costa, E., Mendoza-Roca, J., Iborra-Clar, M.I., García, J. (2009) A study of separation of lactose from whey ultrafiltration permeate using nanofiltration. *Desalination* 241: 244–255.

[12] Chen, G.Q., Leong, T.S.H., Kentish, S.E., Ashokkumar, M., Martin, G.J.O. (2019) Chapter 8 – Membrane separations in the dairy industry. In: C.M. Galanakis (Ed.) *Separation of Functional Molecules in Food by Membrane Technology*, pp. 267–304. Academic Press.

[13] Wilf, M. (2014) The reverse osmosis process. In: J. Kucera (Ed.) *Desalination: Water from Water*. John Wiley & Sons, Incorporated, Somerset, United States.

[14] Nakao, S.-I., Kimura, S. (1981) Analysis of solutes rejection in ultrafiltration. *Journal of Chemical Engineering of Japan* 14: 32–37.

[15] Samuelsson, G., Huisman, I.H., Trägårdh, G., Paulsson, M. (1997) Predicting limiting flux of skim milk in crossflow microfiltration. *Journal of Membrane Science* 129: 277–281.

[16] Cheryan, M. (1998) *Ultrafiltration and Microfiltration Handbook*. Technomic Publishing AG, Basel, Switzerland.

[17] Shen, S., Kentish, S.E., Stevens, G.W. (2010) Shell-side mass-transfer performance in hollow-fiber membrane contactors. *Solvent Extraction and Ion Exchange* 28: 817–844.

[18] Leveque, M.D. (1928) Les lois de la transmission de chaleur par convection. *Ann. Mines* 13: 201.

[19] Gekas, V., Hallstrom, B. (1987) Mass transfer in the membrane concentration polarization layer under turbulent cross flow: I. Critical literature review and adaptation of existing Sherwood correlations to membrane operations. *Journal of Membrane Science* 30: 153–170.

[20] Da Costa, A.R., Fane, A.G., Wiley, D.E. (1994) Spacer characterization and pressure drop modelling in spacer-filled channels for ultrafiltration. *Journal of Membrane Science* 87: 79–98.

[21] Pitzer, K.S., Press, C. (1991) *Activity Coefficients in Electrolyte Solutions*. CRC Press: Boca Raton, FL.

[22] Weinhold, F. (1998) Quantum cluster equilibrium theory of liquids: Illustrative application to water. *The Journal of Chemical Physics* 109: 373–384.

[23] Bandini, S., Vezzani, D. (2003) Nanofiltration modeling: the role of dielectric exclusion in membrane characterization. *Chemical Engineering Science* 58: 3303–3326.

[24] Marshall, A., Daufin, G. (1995) Physico-chemical aspects of membrane fouling by dairy fluids. *International Dairy Federation (Special Issue)* S.I. 9504: 8–35.

[25] Hermans, P.H., Bredée, H.L. (1936) Principles of the mathematical treatment of constant-pressure filtration, *Journal of the Society of Chemical Industry* 55 T: 1–4.

[26] Zydney, A.L. (2022) Development of a new blocking model for membrane fouling based on a composite media model. *Journal of Membrane Science Letters* 2: 100018.

[27] Ho, C.-C., Zydney, A.L. (2000) A combined pore blockage and cake filtration model for protein fouling during microfiltration. *Journal of Colloid and Interface Science* 232: 389–399.

[28] Kezia, K., Lee, J., Zisu, B., Chen, G.Q., Gras, S.L., Kentish, S.E. (2017) Solubility of calcium phosphate in concentrated dairy effluent brines. *Journal of Agricultural and Food Chemistry* 65: 4027–4034.

[29] Dorozhkin, S. (2014) Calcium orthophosphates: Occurrence, properties and major applications. *Bioceramics Development and Applications* 4: 2.

[30] Walstra, P., Jenness, R. (1984) *Dairy Chemistry and Physics*. John Wiley & Sons, Canada.

[31] Martell, A.E., Smith, R.M., Motekaitis, R.J. (2002) NIST critically selected stability constants of metal complexes. In: *NIST Standard Reference Database 46*. National Institute of Standards and Technology, College Station, TX.

[32] Zsigmondy, S., Bachmann, R. (1922) *Filter and Method of Producing Same*. U.S.P. Office (Ed.), United States.

[33] Beuvier, E., Berthaud, K., Cegarra, S., Dasen, A., Pochet, S., Buchin, S., Duboz, G. (1997) Ripening and quality of Swiss-type cheese made from raw, pasteurized or microfiltered milk. *International Dairy Journal* 7: 311–323.
[34] Bachmann, H.P., Banks, J., Beresford, T., Bütikofer, U., Grappin, R., Lavanchy, P., Lindblad, O., McNulty, D., McSweeney, P.L., Skeie, S. (1998) Interlaboratory comparison of cheese making trials: Model cheeses made from raw, pasteurized and microfiltered milks. *LWT – Food Science and Technology* 31: 585–593.
[35] Hoffmann, W., Kiesner, C., Clawin-Rädecker, I., Martin, D., Einhoff, K., Lorenzen, P.C., Meisel, H., Hammer, P., Suhren, G., Teufel, P. (2006) Processing of extended shelf life milk using microfiltration. *International Journal of Dairy Technology* 59: 229–235.
[36] Kulozik, U. (2019) Chapter 1 – Ultra- and microfiltration in dairy technology. In: A. Basile, C. Charcosset (Eds.) *Current Trends and Future Developments on (Bio-) Membranes*, pp. 1–28. Elsevier.
[37] Kaufmann, V., Scherer, S., Kulozik, U. (2010) Verfahren zur Verlängerung der Haltbarkeit von Konsummilch und ihre stofflichen Veränderungen: ESL-Milch. *Journal of Consumer Protection and Food Safety* 5: 9–64.
[38] Maubois, J.L. (2002) Membrane microfiltration: a tool for a new approach in dairy technology. *Australian Journal of Dairy Technology* 57: 92–96.
[39] Gésan, G., Daufin, G., Merin, U. (1995) Performance of whey crossflow microfiltration during transient and stationary operating conditions. *Journal of Membrane Science* 104: 271–281.
[40] Saboyainsta, L.V., Mauboi, J.-L. (2000) Current developments of microfiltration technology in the dairy industry. *Le Lait* 80(6): 541–553.
[41] Karasu, K., Glennon, N., Lawrence, N., Stevens, G., O'Connor, A., Barber, A., Yoshikawa, S., Kentish, S. (2010) A comparison between ceramic and polymeric membrane systems for casein concentrate manufacture. *International Journal of Dairy Technology* 63: 284–289.
[42] Lawrence, N., Kentish, S., O'Connor, A., Stevens, G., Barber, A. (2006) Microfiltration of skim milk for casein concentrate manufacture. *Desalination* 200: 305–306.
[43] Sandblom, R.M. (1978) Filtering process In: *Alfa Laval*. United States.
[44] Le Berre, O., Daufin, G. (1996) Skimmilk crossflow microfiltration performance versus permeation flux to wall shear stress ratio. *Journal of Membrane Science* 117: 261–270.
[45] Vadi, P.K., Rizvi, S.S.H. (2001) Experimental evaluation of a uniform transmembrane pressure crossflow microfiltration unit for the concentration of micellar casein from skim milk. *Journal of Membrane Science* 189: 69–82.
[46] Garcera, D., Toujas, E. (2002) *Graded permeability macroporous support for crossflow filtration*. United States Patent No. 6375014.
[47] Zulewska, J., Newbold, M., Barbano, D.M. (2009) Efficiency of serum protein removal from skim milk with ceramic and polymeric membranes at 50°C. *Journal of Dairy Science* 92: 1361–1377.
[48] Grangeon, A., Lescoche, P., Fleischmann, T., Ruschel, B. (2002) *Cross-flow Filter Membrane and Method of Manufacturing It*. United States: Tami Industries.
[49] Skrzypek, M., Burger, M. (2010) Isoflux® ceramic membranes — Practical experiences in dairy industry. *Desalination* 250: 1095–1100.
[50] Hu, K., Dickson, J.M., Kentish, S. (2015) Microfiltration for casein and serum protein separation. In: K. Hu, J.M. Dickson (Eds.) *Membrane Processes for Dairy Ingredient Separation*, pp. 1–34. Wiley-Blackwell.
[51] Schopf, R., Schmidt, F., Linner, J., Kulozik, U. (2021) Comparative assessment of tubular ceramic, spiral wound, and hollow fiber membrane microfiltration module systems for milk protein fractionation. *Foods* 10: 692.
[52] Ho, C.-C. (2007) Chapter 7 – Membranes for bioseparations. In: S.-T. Yang (Ed.) *Bioprocessing for Value-Added Products from Renewable Resources*, pp. 163–183. Elsevier, Amsterdam.
[53] Aguero, R., Bringas, E., Román, S.M.F., Ortiz, I., Ibañez, R. (2017) Membrane processes for whey proteins separation and purification. A review. *Current Organic Chemistry* 21: 1740–1752.
[54] Cheryan, M., Kuo, K.P. (1984) Hollow fibers and spiral wound modules for ultrafiltration of whey: Energy consumption and performance. *Journal of Dairy Science* 67: 1406–1413.
[55] Atra, R., Vatai, G., Bekassy-Molnar, E., Balint, A. (2005) Investigation of ultra- and nanofiltration for utilization of whey protein and lactose. *Journal of Food Engineering* 67: 325–332.
[56] Guinee, T.P., Pudja, P.D., Reville, W.J., Harrington, D., Mulholland, E.O., Cotter, M., Cogan, T.M. (1995) Composition, microstructure and maturation of semi-hard cheeses from high protein ultrafiltered milk retentates with different levels of denatured whey protein. *International Dairy Journal* 5: 543–568.
[57] Guinee, T.P., O'Kennedy, B.T., Kelly, P.M. (2006) Effect of milk protein standardization using different methods on the composition and yields of cheddar cheese. *Journal of Dairy Science* 89: 468–482.
[58] Johnson, M.E., Lucey, J.A. (2006) Major technological advances and trends in cheese. *Journal of Dairy Science* 89: 1174–1178.
[59] Rattray, W., Jelen, P. (1996) Protein standardization of milk and dairy products. *Trends in Food Science & Technology* 7: 227–234.
[60] Ng, K.S.Y., Haribabu, M., Harvie, D.J.E., Dunstan, D.E., Martin, G.J.O. (2017) Mechanisms of flux decline in skim milk ultrafiltration: a review. *Journal of Membrane Science* 523: 144–162.
[61] Mistry, V.V., Maubois, J.L. (2004) Application of membrane separation technology to cheese production. In: P.F. Fox (Ed.) *Cheese: Chemistry, Physics, and Microbiology*, pp. 261-285. Elsevier.
[62] Rosenberg, M. (1995) Current and future applications for membrane processes in the dairy industry. *Trends in Food Science & Technology* 6: 12–19.
[63] Henning, D.R., Baer, R.J., Hassan, A.N., Dave, R. (2006)Major advances in concentrated and dry milk products, cheese, and milk fat-based spreads. *Journal of Dairy Science* 89: 1179–1188.
[64] van Leeuwen, H.J., Freeman, N.H., Sutherland, B.J., Jameson, G.W. (1990) *Preparation of hard cheese from concentrated milk*. Google Patents.

[65] Maubois, J.L., Mocquot, G., Vassal, L. (1969) *A method for processing milk and dairy products*. FR Patent 2052121.

[66] Mistry, V.V., Maubois, J.-L. (1993) Application of membrane separation technology to cheese production. In: P.F. Fox (Ed.) *Cheese: Chemistry, Physics and Microbiology: Volume 1 General Aspects*, pp. 493–522. Springer US, Boston, MA.

[67] Neocleous, M., Barbano, D.M., Rudan, M.A. (2006) Impact of low concentration factor microfiltration on the composition and aging of cheddar cheese. *Journal of Dairy Science* 85: 2425–2437.

[68] Salhab, H.H. (1999) *The Application of Ultrafiltration in the Manufacture of Cream Cheese* (Master's thesis). University of Alberta.

[69] Walstra, P., Jenness, R., Badings, H.Y. (1984) *Dairy Chemistry and Physics*. Wiley.

[70] Meyer, P., Kulozik, U. (2016) Impact of protein removal by an upstream ultrafiltration on the reverse osmosis of skim milk and sweet whey. *Chemie Ingerieur Technik* 88: 585–590.

[71] Gésan-Guiziou, G. (2013) 12 – Separation technologies in dairy and egg processing.I In: *Separation, Extraction and Concentration Processes in the Food*, pp. 341–380. Beverage and Nutraceutical Industries, Woodhead Publishing.

[72] Slack, A.W., Amundson, C.H., Hill, C.G. (1986) Production of enriched β-lactoglobulin and α-lactalbumin whey protein fractions. *Journal of Food Processing and Preservation* 10: 19–30.

[73] Bramaud, C., Aimar, P., Daufin, G. (1995) Thermal isoelectric precipitation of α-lactalbumin from a whey protein concentrate: Influence of protein–calcium complexation. *Biotechnology and Bioengineering* 47: 121–130.

[74] Muller, A., Daufin, G., Chaufer, B. (1999) Ultrafiltration modes of operation for the separation of α-lactalbumin from acid casein whey. *Journal of Membrane Science* 153: 9–21.

[75] Cheang, B., Zydney, A.L. (2004) A two-stage ultrafiltration process for fractionation of whey protein isolate. *Journal of Membrane Science* 231: 159-167.

[76] Espina, V., Jaffrin, M.Y., Ding, L. (2009) Extraction and separation of α-lactalbumin and β-lactoglobulin from skim milk by microfiltration and ultrafiltration at high shear rates: A feasibility study. *Separation Science and Technology* 44: 3832–3853.

[77] Marella, C., Muthukumarappan, K., Metzger, L.E. (2011) Evaluation of commercially available, wide-pore ultrafiltration membranes for production of alpha-lactalbumin enriched whey protein concentrate. *Journal of Dairy Science* 94: 1165–1175.

[78] Cowan, S., Ritchie, S. (2007) Modified polyethersulfone (PES) ultrafiltration membranes for enhanced filtration of whey proteins. *Separation Science and Technology* 42: 2405–2418.

[79] van Reis, R., Brake, J.M., Charkoudian, J., Burns, D.B., Zydney, A.L. (1999) High-performance tangential flow filtration using charged membranes. *Journal of Membrane Science* 159: 133–142.

[80] Matsui, H., Mizota, Y., Sumi, M., Ikeda, M., Iwatsuki, K. (2006) Sensory characteristics of membrane treated milk (studies on the sensory characteristics and physicochemical properties of UHT processed milk part VI). *Nippon Shokuhin Kagaku Kogaku Kaishi* 53: 644–650.

[81] Cuartas-Uribe, B., Alcaina-Miranda, M.I., Soriano-Costa, E., Mendoza-Roca, J.A., Iborra-Clar, M.I., Lora-García, J. (2009) A study of the separation of lactose from whey ultrafiltration permeate using nanofiltration. *Desalination* 241: 244–255.

[82] Gernigon, G., Schuck, P., Jeantet, R., Burling, H. (2011) Whey processing | Demineralization. In: J.W. Fuquay (Ed.) *Encyclopedia of Dairy Sciences (Second Edition)*, pp. 738–743. Academic Press, San Diego.

[83] Nath, K. (2008) *Membrane Separation Processes*. PHI Learning Pvt. Ltd, 2008.

[84] Oatley-Radcliffe, D.L., Walters, M., Ainscough, T.J., Williams, P.M., Mohammad, A.W., Hilal, N. (2017) Nanofiltration membranes and processes: A review of research trends over the past decade. *Journal of Water Process Engineering* 19: 164–171.

[85] Rice, G., Kentish, S., Vivekanand, V., O'Connor, A., Stevens, G., Barber, A. (2005) Membrane-based dairy separation: A comparison of nanofiltration and electrodialysis. *Developments in Chemical Engineering and Mineral Processing* 13: 43–54.

[86] Rice, G., Barber, A.R., O'Connor, A.J., Pihlajamaki, A., Nystrom, M., Stevens, G.W., Kentish, S.E. (2011) The influence of dairy salts on nanofiltration membrane charge. *Journal of Food Engineering* 107: 164–172.

[87] Balannec, B., Gésan-Guiziou, G., Chaufer, B., Rabiller-Baudry, M., Daufin, G. (2002) Treatment of dairy process waters by membrane operations for water reuse and milk constituents concentration. *Desalination* 147: 89–94.

[88] Schaep, J., Vandecasteele, C., Mohammad, A.W., Bowen, W.R. (1999) Analysis of the salt retention of nanofiltration membranesuUsing the Donnan–steric partitioning pore model. *Separation Science and Technology* 34: 3009–3030.

[89] Labbez, C., Fievet, P., Szymczyk, A., Vidonne, A., Foissy, A., Pagetti, P. (2003) Retention of mineral salts by a polyamide nanofiltration membrane. *Separation and Purification Technology* 30: 47–55.

[90] Garcia-Aleman, J., Dickson, J.M. (2004) Permeation of mixed-salt solutions with commercial and pore-filled nanofiltration membranes: membrane charge inversion phenomena. *Journal of Membrane Science* 239: 163–172.

[91] Rice, G., Barber, A.R., O'Connor, A.J., Stevens, G.W., Kentish, S.e. (2011) Rejection of dairy salts by a nanofiltration membrane. *Separation and Purification Technology* 79: 92–102.

[92] Nightingale, E.R. (1959) Phenomenological theory of ion solvation. Effective radii of hydrated ions. *The Journal of Physical Chemistry* 63: 1381–1387.

[93] Bouchoux, A., Balmann, H.R.-d., Lutin, F. (2005) Nanofiltration of glucose and sodium lactate solutions: Variations of retention between single- and mixed-solute solutions. *Journal of Membrane Science* 258: 123–132.

[94] Marcus, Y. (1997) *Ion Properties*. Marcel Dekker, New York.

[95] Bouchoux, A., Roux-de Balmann, H., Lutin, F. (2006) Investigation of nanofiltration as a purification step for lactic acid production processes based on conventional and bipolar electrodialysis operations. *Separation and Purification Technology* 52: 266–273.

[96] Muller, G.T.A., Stokes, R.H. (1957) The mobility of the undissociated citric acid molecule in aqueous solution. *Transactions of the Faraday Society* 53: 642–645.
[97] Buffle, J., Zhang, Z., Startchev, K. (2007) Metal flux and dynamic speciation at (bio)interfaces. Part I: Critical evaluation and compilation of physicochemical parameters for complexes with simple ligands and fulvic/humic substances. *Environmental Science & Technology* 41: 7609–7620.
[98] Suárez, E., Lobo, A., Álvarez, S., Riera, F.A., Álvarez, R. (2006) Partial demineralization of whey and milk ultrafiltration permeate by nanofiltration at pilot-plant scale. *Desalination* 198: 274–281.
[99] Cuartas-Uribe, B., Vincent-Vela, M.C., Álvarez-Blanco, S., Alcaina-Miranda, M.I., Soriano-Costa, E. (2010) Application of nanofiltration models for the prediction of lactose retention using three modes of operation. Journal of Food Engineering 99: 373–376.
[100] Kumar, P., Neelesh, S., Ranjan, R., Kumar, S., Bhat, Z.F., Jeong, D.K. (2013) Perspective of membrane technology in dairy industry: A review. *Asian-Australasian Journal of Animal Sciences* 26: 1347–1358.
[101] Salehi, F. 92014) Current and future applications for nanofiltration technology in the food processing. *Food and Bioproducts Processing* 92: 161–177.
[102] Greiter, M., Novalin, S., Wendland, M., Kulbe, K.-D., Fischer, J. (2002) Desalination of whey by electrodialysis and ion exchange resins: analysis of both processes with regard to sustainability by calculating their cumulative energy demand. *Journal of Membrane Science* 210: 91–102.
[103] Chandrapala, J., Chen, G.Q., Kezia, K., Bowman, E.G., Vasiljevic, T., Kentish, S.E. (2016) Removal of lactate from acid whey using nanofiltration. Journal of Food Engineering 177: 59–64.
[104] Chandrapala, J., Duke, M.C., Gray, S.R., Weeks, M., Palmer, M., Vasiljevic, T. (2016)Nanofiltration and nanodiafiltration of acid whey as a function of pH and temperature. *Separation and Purification Technology* 160: 18–27.
[105] Bédas, M., Tanguy, G., Dolivet, A., Méjean, S., Gaucheron, F., Garric, G., Senard, G., Jeantet, R., Schuck, P. (2017) Nanofiltration of lactic acid whey prior to spray drying: Scaling up to a semi-industrial scale. *LWT – Food Science and Technology* 79: 355–360.
[106] Budavari, S., O'Neil, M.J., Smith, A., Heckelman, P.E. (1989) *The Merck Index*. Merck Rahway, NJ.
[107] Bédas, M., Tanguy, G., Dolivet, A., Méjean, S., Gaucheron, F., Garric, G., Senard, G., Jeantet, R., Schuck, P. (2017) Nanofiltration of lactic acid whey prior to spray drying: Scaling up to a semi-industrial scale. *LWT – Food Science and Technology* 79: 355–360.
[108] Chen, X., Chen, G.Q., Wang, Q., Xu, T., Kentish, S.E. (2020) Transforming salty whey into cleaning chemicals using electrodialysis with bipolar membranes. *Desalination* 492: 114598.
[109] Hinkova, A., Zidova, P., Pour, V., Bubnik, Z., Henke, S., Salova, A., Kadlec, P. (2012) Potential of membrane separation processes in cheese whey fractionation and separation. *Procedia Engineering* 42: 1425–1436.
[110] Heldman, D.R., Lund, D.B., Sabliov, C. (2006) *Handbook of Food Engineering, Second Edition*. Taylor & Francis.
[111] Alvarez, N., Gésan-Guiziou, G., Daufin, G. (2007) The role of surface tension of re-used caustic soda on the cleaning efficiency in dairy plants. *International Dairy Journal* 17: 403–411.
[112] KOCH Membrane Systems (2013) Recovery of caustic and acids in the dairy industry. *Dairy Foods Magazine* 135
[113] Lauzin, A., Dussault-Chouinard, I., Britten, M., Pouliot, Y. (2018) Impact of membrane selectivity on the compositional characteristics and model cheese-making properties of liquid pre-cheese concentrates. *International Dairy Journal* 83: 34-42.
[114] Syrios, A., Faka, M., Grandison, A.S., Lewis, M.J. (2011) A comparison of reverse osmosis, nanofiltration and ultrafiltration as concentration processes for skim milk prior to drying. *International Journal of Dairy Technology* 64: 467–472.
[115] Kulozik, U., Kessler, H.-G. (1988) Permeation rate during reverse osmosis of milk influenced by osmotic pressure and deposit formation. *Journal of Food Science* 53: 1377–1383.
[116] Vourch, M., Balannec, B., Chaufer, B., Dorange, G. (2008) Treatment of dairy industry wastewater by reverse osmosis for water reuse. *Desalination* 219: 190–202.
[117] Suárez, A., Riera, F.A. (2015) Production of high-quality water by reverse osmosis of milk dairy condensates. *Journal of Industrial and Engineering Chemistry* 21: 1340–1349.
[118] Greiter, M., Novalin, S., Wendland, M., Kulbe, K.-D., Fischer, J. (2004) Electrodialysis versus ion exchange: comparison of the cumulative energy demand by means of two applications. *Journal of Membrane Science* 233: 11–19.
[119] Xu, T., Huang, C. (2008) Electrodialysis-based separation technologies: A critical review. *AIChE Journal* 54: 3147–3159.
[120] Strathmann, H. (2010) Electrodialysis, a mature technology with a multitude of new applications. *Desalination* 264: 268–288.
[121] Kánavová, N., Machuca, L. (2014) A novel method for limiting current calculation in electrodialysis modules. *Periodica Polytechnica. Chemical Engineering* 58: 125.
[122] Kentish, S.E., Kloester, E., Stevens, G.W., Scholes, C.A., Dumée, L.F. (2014) Electrodialysis in aqueous-organic mixtures. *Separation & Purification Reviews* 44: 269–282.
[123] Simons, R. (1984) Electric field effects on proton transfer between ionizable groups and water in ion exchange membranes. *Electrochimica Acta* 29: 151–158.
[124] Koter, S. (2001) Transport number of counterions in ion-exchange membranes. *Separation and Purification Technology* 22–23: 643–654.
[125] de Boer, R., Robbertsen, T. (1983) Electrodialysis and ion-exchange processes: The case of milk whey. *Progress in Food Engineering* 393–403.
[126] Pérez, A., Andrés, L., Alvarez, R., Coca, J., Hill Jr, C. (1994) Electrodialysis of whey permeates and retentates obtained by ultrafiltration. *Journal of Food Process Engineering* 17: 177–190.

[127] Šímová, H., Kysela, V., Černín, A. (2010) Demineralization of natural sweet whey by electrodialysis at pilot-plant scale. *Desalination and Water Treatment* 14: 170–173.

[128] Diblíková, L., Čurda, L., Kinčl, J. (2013) The effect of dry matter and salt addition on cheese whey demineralisation. *International Dairy Journal* 31: 29–33.

[129] Bylund, G., Hellman, M. (2015) *Dairy Processing Handbook*. Tetra Pak Processing Systems, Lund, Sweden.

[130] Dufton, G., Mikhaylin, S., Gaaloul, S., Bazinet, L. (2018) How electrodialysis configuration influences acid whey deacidification and membrane scaling. *Journal of Dairy Science* 101: 7833–7850.

[131] Bazinet, L., Araya-Farias, M. (2005) Effect of calcium and carbonate concentrations on cationic membrane fouling during electrodialysis. *Journal of Colloid Interface Science* 281: 188–196.

[132] Bazinet, L., Araya-Farias, M. (2005) Electrodialysis of calcium and carbonate high concentration solutions and impact on composition in cations of membrane fouling. *Journal of Colloid Interface Science* 286: 639–646.

[133] Casademont, C., Farias, M.A., Pourcelly, G., Bazinet, L. (2008) Impact of electrodialytic parameters on cation migration kinetics and fouling nature of ion-exchange membranes during treatment of solutions with different magnesium/calcium ratios. *Journal of Membrane Science* 325; 570–579.

[134] Wang, Q., Yang, P., Cong, W. (2011) Cation-exchange membrane fouling and cleaning in bipolar membrane electrodialysis of industrial glutamate production wastewater. *Separation and Purification Technology* 79: 103–113.

[135] Ayala-Bribiesca, E., Pourcelly, G., Bazinet, L. (2007) Nature identification and morphology characterization of anion-exchange membrane fouling during conventional electrodialysis. *Journal of Colloid Interface Science* 308: 182–190.

[136] Bazinet, L., Montpetit, D., Ippersiel, D., Amiot, J., Lamarche, F. (2001) Identification of skim milk electroacidification fouling: A microscopic approach. *Journal of Colloid Interface Science* 237: 62–69.

[137] Bazinet, L., Montpetit, D., Ippersiel, D., Mahdavi, B., Amiot, J., Lamarche, F. (2003) Neutralization of hydroxide generated during skim milk electroacidification and its effect on bipolar and cationic membrane integrity. *Journal of Membrane Science* 216: 229–239.

[138] Ayala-Bribiesca, E., Pourcelly, G., Bazinet, L. (2006) Nature identification and morphology characterization of cation-exchange membrane fouling during conventional electrodialysis. *Journal of Colloid Interface Science* 300: 663–672.

[139] Ayala-Bribiesca, E., Araya-Farias, M., Pourcelly, G., Bazinet, L. (2006) Effect of concentrate solution pH and mineral composition of a whey protein diluate solution on membrane fouling formation during conventional electrodialysis. *Journal of Membrane Science* 280: 790–801.

[140] Casademont, C., Pourcelly, G., Bazinet, L. (2007) Effect of magnesium/calcium ratio in solutions subjected to electrodialysis: Characterization of cation-exchange membrane fouling. *Journal of Colloid Interface Science* 315: 544–554.

[141] Talebi, S., Chen, G.Q., Freeman, B., Suarez, F., Freckleton, A., Bathurst, K., Kentish, S.E. (2019) Fouling and in-situ cleaning of ion-exchange membranes during the electrodialysis of fresh acid and sweet whey. Journal of Food Engineering 246: 192–199.

[142] Talebi, S., Suarez, F., Chen, G.Q., Chen, X., Bathurst, K., Kentish, S.E. (2020) Pilot study on the removal of lactic acid and minerals from acid whey using membrane technology. *ACS Sustainable Chemistry & Engineering* 8: 2742–2752.

[143] Ečer, J., Kinčl, J., Čurda, L. (2015) Using foil membranes for demineralization of whey by electrodialysis. *Desalination and Water Treatment* 56: 3273–3277.

[144] Talebi, S., Kee, E., Chen, G.Q., Bathurst, K., Kentish, S.E. (2019) Utilisation of salty whey ultrafiltration permeate with electrodialysis. International Dairy Journal 99: 104549.

[145] Ghyselbrecht, K., Silva, A., Van der Bruggen, B., Boussu, K., Meesschaert, B., Pinoy, L. (2014) Desalination feasibility study of an industrial NaCl stream by bipolar membrane electrodialysis. *Journal of Environmental Management* 140: 69–75.

[146] Li, Y., Shi, S., Cao, H., Wu, X., Zhao, Z., Wang, L. (2016) Bipolar membrane electrodialysis for generation of hydrochloric acid and ammonia from simulated ammonium chloride wastewater. *Water Research* 89: 201–209.

[147] Reig, M., Casas, S., Gibert, O., Valderrama, C., Cortina, J.L. (2016) Integration of nanofiltration and bipolar electrodialysis for valorization of seawater desalination brines: Production of drinking and waste water treatment chemicals. *Desalination* 382: 13–20.

[148] Merkel, A., Ashrafi, A.M., Ečer, J. (2018) Bipolar membrane electrodialysis assisted pH correction of milk whey. *Journal of Membrane Science* 555: 185–196.

[149] Mier, M.P., Ibañez, R., Ortiz, I. (2008) Influence of process variables on the production of bovine milk casein by electrodialysis with bipolar membranes. *Biochemical Engineering Journal* 40: 304–311.

[150] Mikhaylin, S., Nikonenko, V., Pourcelly, G., Bazinet, L. (2016) Hybrid bipolar membrane electrodialysis/ultrafiltration technology assisted by a pulsed electric field for casein production. *Green Chemistry* 18: 307–314.

[151] Mikhaylin, S., Patouillard, L., Margni, M., Bazinet, L. (2018) Milk protein production by a more environmentally sustainable process: bipolar membrane electrodialysis coupled with ultrafiltration. *Green Chemistry* 20: 449–456.

[152] Dlask, O., Václavíková, N., Dolezel, M. (2016) Insertion of filtration membranes into electrodialysis stack and its impact on process performance. *Periodica Polytechnica. Chemical Engineering* 60: 169.

[153] Deng, H., Chen, G.Q., Gras, S.L., Kentish, S.E. (2017) The effect of restriction membranes on mass transfer in an electrodialysis with filtration membrane process. *Journal of Membrane Science* 526: 429–436.

[154] Chen, G., Song, W., Qi, B., Li, J., Ghosh, R., Wan T. (2015) Separation of protein mixtures by an integrated electro-ultrafiltration–electrodialysis process. *Separation and Purification Technology* 147: 32–43.

[155] Ogle, D., Vigh, G., Rylatt, D. (2003) *Multi-port separation apparatus and method*. Google Patents.

[156] Bazinet, L., Amiot, J., Poulin, J.-F., Labbé, D., Tremblay, D. (2005) *Process and system for separation of organic charged compounds*. WO 2005/082495 A1.
[157] Lawrence, N., Kentish, S., O'Connor, A., Barber, A., Stevens, G. (2008) Microfiltration of skim milk using polymeric membranes for casein concentrate manufacture. *Separation and Purification Technology* 60: 237–244.
[158] Margolis, J. (1991) *Electrophoretic method for preparative separation of charged molecules in liquids*. Google Patents
[159] Ogle, D., Ho, A., Gibson, T., Rylatt, D., Shave, E., Lim, P., Vigh, G. (2002) Preparative-scale isoelectric trapping separations using a modified Gradiflow unit. *Journal of Chromatography A* 979: 155–161.
[160] Horvath, Z.S., Corthals, G.L., Wrigley, C.W., Margolis, J. (1994) Multifunctional apparatus for electrokinetic processing of proteins. *Electrophoresis* 15: 968–971.
[161] Ndiaye, N., Pouliot, Y., Saucier, L., Beaulieu, L., Bazinet, L. (2010) Electroseparation of bovine lactoferrin from model and whey solutions. *Separation and Purification Technology* 74: 93–99.
[162] Wang, Q., Chen, G.Q., Kentish, S.E. (2020) Isolation of lactoferrin and immunoglobulins from dairy whey by an electrodialysis with filtration membrane process. *Separation and Purification Technology* 233: 115987.
[163] Law, B.A., Reiter, B. (1977) The isolation and bacteriostatic properties of lactoferrin from bovine milk whey. *Journal of Dairy Research* 44: 595–599.
[164] Tomita, M., Wakabayashi, H., Yamauchi, K., Teraguchi, S., Hayasawa, H. (2002) Bovine lactoferrin and lactoferricin derived from milk: production and applications. *Biochemistry and Cell Biology* 80: 109–112.
[165] Wakabayashi, H., Yamauchi, K., Takase, M. (2006) Lactoferrin research, technology and applications. *International Dairy Journal* 16: 1241–1251.
[166] Ghosh, R. (2002) Protein separation using membrane chromatography: opportunities and challenges. *Journal of Chromatography A* 952: 13–27.
[167] Boi, C., Malavasi, A., Carbonell, R.G., Gilleskie, G. (2020) A direct comparison between membrane adsorber and packed column chromatography performance. *Journal of Chromatography A* 1612: 460629.
[168] Voswinkel, L., Kulozik, U. (2014) Fractionation of all major and minor whey proteins with radial flow membrane adsorption chromatography at lab and pilot scale. *International Dairy Journal* 39: 209–214.
[169] Brand, J., Dachmann, E., Pichler, M., Lotz, S., Kulozik, U. (2016) A novel approach for lysozyme and ovotransferrin fractionation from egg white by radial flow membrane adsorption chromatography: Impact of product and process variables. *Separation and Purification Technology* 161: 44–52.
[170] Schwellenbach, J., Kosiol, P., Sölter, B., Taft, F., Villain, L., Strube, J. (2016) Controlling the polymer-nanolayer architecture on anion-exchange membrane adsorbers via surface-initiated atom transfer radical polymerization. *Reactive and Functional Polymers* 106: 32–42.
[171] Voswinkel, L., Kulozik, U. (2011) Fractionation of whey proteins by means of membrane adsorption chromatography. *Procedia Food Science* 1: 900–907.
[172] Boi, C. (2019) Chapter 6 – Membrane chromatography for biomolecule purification. In: A. Basile, C. Charcosset (Eds.) *Current Trends and Future Developments on (Bio-) Membranes*, pp. 151–166. Elsevier.
[173] Loeb, S., Titelman, L., Korngold, E., Freiman, J. (1997) Effect of porous support fabric on osmosis through a Loeb-Sourirajan type asymmetric membrane. *Journal of Membrane Science* 129: 243–249.
[174] Artemi, A., Chen, G.Q., Kentish, S.E., Lee, J. (2020) Pilot scale concentration of cheese whey by forward osmosis: A short-cut method for evaluating the effective pressure driving force. *Separation and Purification Technology* 250: 117263.
[175] Adams, C. (2013) "Magic" to revolutionise dairy industry. *The New Zealand Herald*.
[176] Phuntsho, S., Shon, H., Hong, S., Lee, S., Vigneswaran, S., Kandasamy, J. (2021) Fertiliser drawn forward osmosis desalination: the concept, performance and limitations for fertigation. *Reviews in Environmental Science and Biotechnology* 11: 147–168.
[177] Aydiner, C., Sen, U., Topcu, S., Sesli, D., Ekinci, D., Altınay, A.D., Ozbey, B., Koseoglu-Imer, D.Y., Keskinler, B. (2014) Techno-economic investigation of water recovery and whey powder production from whey using UF/RO and FO/RO integrated membrane systems. *Desalination and Water Treatment* 52: 123–133.
[178] Seker, M., Buyuksari, E., Topcu, S., Babaoglu, D.S., Celebi, D., Keskinler, B., Aydiner, C. (2017) Effect of pretreatment and membrane orientation on fluxes for concentration of whey with high foulants by using NH_3/CO_2 in forward osmosis. *Bioresources Technology* 243: 237–246.
[179] Seker, M., Buyuksari, E., Topcu, S., Sesli, D., Celebi, D., Keskinler, B., Aydiner, C. (2017) Effect of process parameters on flux for whey concentration with NH_3/CO_2 in forward osmosis. *Food and Bioproducts Processing* 105: 64–76.
[180] Aydiner, C., Topcu, S., Tortop, C., Kuvvet, F., Ekinci, D., Dizge, N., Keskinler, B. (2013) A novel implementation of water recovery from whey: "forward–reverse osmosis" integrated membrane system. *Desalination and Water Treatment* 51: 786–799.
[181] Chen, G.Q., Gras, S.L., Kentish, S.E. (2020) The application of forward osmosis to dairy processing. *Separation and Purification Technology* 246: 116900.
[182] Ford, E.S., Mokdad, A.H. (2003) Dietary magnesium intake in a national sample of U.S. adults. *The Journal of Nutrition* 133: 2879–2882.
[183] Chen, G.Q., Artemi, A., Lee, J., Gras, S.L., Kentish, S.E. (2019) A pilot scale study on the concentration of milk and whey by forward osmosis. *Separation and Purification Technology* 215: 652–659.
[184] Christensen, K., Andresen, R., Tandskov, I., Norddahl, B., du Preez, J.H. (2006) Using direct contact membrane distillation for whey protein concentration. *Desalination* 200; 523–525.
[185] Hausmann, A., Sanciolo, P., Vasiljevic, T., Ponnampalam, E., Quispe-Chavez, N., Weeks, M., Duke, M. (2011) Direct contact membrane distillation of dairy process streams. *Membranes* 1: 48–58.
[186] Hausmann, A., Sanciolo, P., Vasiljevic, T., Kulozik, U., Duke, M. (2014) Performance assessment of membrane distillation for skim milk and whey processing. *Journal of Dairy Science* 97: 56–71.

[187] Yun, Y., Ma, R., Zhang, W., Fane, A.G., Li, J. (2006) Direct contact membrane distillation mechanism for high concentration NaCl solutions. *Desalination* 188: 251–262.
[188] Wang, K.Y., Teoh, M.M., Nugroho, A., Chung, T.-S. (2011) Integrated forward osmosis–membrane distillation (FO–MD) hybrid system for the concentration of protein solutions. *Chemical Engineering Science* 66: 2421–2430.
[189] Ge, Q., Wang, P., Wan, C., Chung, T.-S. (2012) Polyelectrolyte-promoted forward osmosis–membrane distillation (FO–MD) hybrid process for dye wastewater treatment. *Environmental Science & Technology* 46: 6236–6243.
[190] Hausmann, A., Sanciolo, P., Vasiljevic, T., Weeks, M., Schroën, K., Gray, S., Duke, M. (2013) Fouling mechanisms of dairy streams during membrane distillation. *Journal of Membrane Science* 441: 102–111.
[191] Biesheuvel, P., Van der Wal, A. (2010) Membrane capacitive deionization. *Journal of Membrane Science* 346: 256–262.
[192] Hassanvand, A., Wei, K., Talebi, S., Chen, G.Q., Kentish, S.E. (2017) The role of ion exchange membranes in membrane capacitive deionisation. *Membranes* 7: 54.
[193] Li, H., Zou, L. (2011) Ion-exchange membrane capacitive deionization: a new strategy for brackish water desalination. *Desalination* 275: 62–66.
[194] Lee, J.-B., Park, K.-K., Eum, H.-M., Lee, C.-W. (2006) Desalination of a thermal power plant wastewater by membrane capacitive deionization. *Desalination* 196: 125–134.
[195] Lee, J.-H., Choi, J.-H. (2012) The production of ultrapure water by membrane capacitive deionization (MCDI) technology. *Journal of Membrane Science* 409: 251–256.
[196] Kim, Y.-J., Kim, J.-H. Choi, J.-H. (2013) Selective removal of nitrate ions by controlling the applied current in membrane capacitive deionization (MCDI). *Journal of Membrane Science* 429: 52–57.
[197] Zhao, R., Porada, S., Biesheuvel, P.M., van der Wal, A. (2013) Energy consumption in membrane capacitive deionization for different water recoveries and flow rates, and comparison with reverse osmosis. *Desalination* 330: 35–41.
[198] Talebi, S.(2020) *Whey Management for the Dairy Industry: Acid and Salty Whey Treatment and Processing Using Membrane Technology*. PhD Thesis, University of Melbourne.
[199] Kim, K.-J., Sun, P., Chen, V., Wiley, D.E., Fane, A.G. (1993) The cleaning of ultrafiltration membranes fouled by protein. *Journal of Membrane Science* 80: 241–249.
[200] Madaeni, S.S., Mansourpanah, Y. (2004) Chemical cleaning of reverse osmosis membranes fouled by whey. *Desalination* 161: 13–24.
[201] Muthukumaran, S., Kentish, S., Lalchandani, S., Ashokkumar, M., Mawson, R., Stevens, G.W., Grieser, F. (2005) The optimisation of ultrasonic cleaning procedures for dairy fouled ultrafiltration membranes. *Ultrasonics Sonochemistry* 12: 29–35.
[202] Bartlett, M., Bird, M.R., Howell, J.A. (1995) An experimental study for the development of a qualitative membrane cleaning model. *Journal of Membrane Science* 105: 147–157.
[203] Watkinson, W.J. (2002) Chemistry of detergents and disinfectants. In: *Cleaning-in-Place: Dairy Food and Beverage Operations,*, pp. 56–80. John Wiley & Sons.
[204] Zeman, L.J., Zydney, A.L. (1996) *Microfiltration and Ultrafiltration: Principles and Applications*. Marcel Dekker, New York.
[205] Walstra, P., Wouters, J.T.M., Geurts, T.j. (2006) Cleaning and sanitising. In: *Dairy Science and Technology*. Boca Raton, Taylor and Francis.

14 Metal-Organic Framework Containing Polymeric Membranes for Fuel Cells

B. Shivarama[1], Arun M. Isloor[1], Ch. Sn. Murthy[2], Balakrishna Prabhu[3], and Ahmad Fauzi Ismail[4]*

[1]Membrane and Separation Technology Laboratory, Department of Chemistry, National Institute of Technology Karnataka, Surathkal, Mangalore, Karnataka, India

[2]Department of Mining Engineering, National Institute of Technology Karnataka, Surathkal, Mangalore, Karnataka, India

[3]Department of Chemical Engineering, Manipal Institute of Technology, Manipal University, Karnataka, India

[4]Advanced Membrane Technology Research Centre, Universiti Teknologi Malaysia, Skundai, Malaysia

*Author for correspondence: isloor@yahoo.com

14.1 INTRODUCTION

Global energy demand has undergone a tremendous increase over the last two centuries. The uninterrupted use of non-renewable resources of energy has led to the rapid development of technology around the globe. This uncontrolled usage of fossil fuels, including natural gas, coal, and oil products, has caused severe damage to the environment. At the same time, the disappearance of these resources may stop the growth of technology and lead to a scarcity of food, shelter, and energy [1]. Unlimited power generation through nuclear power and dependence on renewable energy sources seem unrealistic and pose a risk to the environment [2]. To meet the high energy demand and prevent the extinction of non-renewable energy sources and environmental pollution aspects, potential alternatives need to be rapidly developed [3].

Sir William Robert Grove made a major discovery in fuel cells in 1838 when he developed the first wet cell battery. The usage of electrical energy to break water into oxygen and hydrogen leads to a hypothesis that the opposite reaction will yield electrical power. The first gas battery was developed into a fuel cell. Later, Friedrich Wilhelm Ostwald extended the connection between the various parts of the fuel cell. Then, in the 1950s, Francis Thomas Bacon developed high-pressure fuel cells that guided the evolution of the alkali cell. This work was developed further by Pratt and Whitney who applied the fuel cells for the Apollo spacecraft [4].

Fuel cells are divided according to the electrolyte used and the operating temperature into proton exchange membrane fuel cells (PEMFCs, also called polymer electrolyte fuel cells, PEFCs, or solid electrolyte polymer fuel cells, SPEFCs) and direct methanol fuel cells (DMFCs). These two types of fuel cell depend on membranes made from polymeric sheets that serve as electrolytes. The DMFC is very similar to the PEFC, with the main difference being that methanol is used as fuel in place of hydrogen at the anode. Alkaline electrolyte fuel cells (AFCs) and phosphoric acid fuel cells (PAFCs) use concentrated potassium hydroxide in water and undiluted strong phosphoric acid. All the above types fall under fuel cells operating under reduced temperature. Solid oxide fuel cells (SOFCs) and molten carbonate fuel cells (MCFCs) use molten carbonate retained in a ceramic matrix of $LiAlO_2$ and yttrium-stabilized zircon dioxide as an electrolyte and are considered as high-temperature fuel cells [5]. Some of the physical, chemical, and operating properties of all these categories of fuel cells are provided in Table 14.1 [6].

There are two main transport mechanisms that operate with respect to transport in membrane fuel cells, namely ionic conduction and water transport mechanisms. The movement of ions especially takes place through a mechanism called conduction. The hydrophilic groups attached to the hydrophobic backbone of membranes enhance proton movement during high humid and wet conditions [7]. The Grotthuss mechanism and vehicle mechanism are the two main modes for proton transport in membranes. The former explains ionic transport at low water content, while the latter describes the transport of ions through electroosmosis diffusion. The water transport mechanism is also closely related to ion transport with the formation of hydronium ions [8]. The electrostatic interactions between hydrophilic groups of membrane and water create a hydrophilic channel for the transportation of water. The swelling of sulfonic groups through water absorption will increase ionic cluster and the diameter of the ionic cluster, which increase the water transport capacity [9]. Apart from these, the grain boundary resistance also plays a vital role in deciding the proton conductivity in polymeric membranes. Lowering of grain boundary resistance increases the proton conductivity [10]. The larger-sized

DOI: 10.1201/9781003285656-14

TABLE 14.1
Various properties of fuel cells. The table data have been taken from Ref. [6]

Fuel Cell	Temperature (°C)	Efficiency (%)	Application	Advantages	Disadvantages
Alkaline fuel cell (AFC)	50–90	50–70	Space application	High efficiency	Intolerant to CO_2 in impure H_2 and air, corrosion, expensive
Phosphoric acid fuel cell (PAFC)	175–220	40–45	Standalone & combine heat & power	Tolerant to impure H_2, commercial value	Low power, density, corrosion, and sulfur poisoning
Molten carbonate fuel cell (MCFC)	600–650	50–60	Central, standalone, combined, heat & power	High efficiency, commercial value	Electrolyte instability, corrosion, and sulfur poisoning
Solid oxide fuel cell (SOFC)	800–1000	50–60	Central, standalone, and combined heat & power	High efficiency, direct fossil fuel	High temperature, thermal stress failure, cooking and sulfur poisoning
Polymer electrolyte membrane fuel cell (PEMFC)	50–100	40–50	Vehicle and portable	High power density, low temperature	Intolerant to CO in impure H_2, expensive
Direct efficiency, methanol fuel cell (DMFC)	50–120	25–40	Vehicle and small portable	No reforming, high power density, and low temperature	Low methanol crossover and poisonous byproduct

MOFs have lower grain boundaries and hence exhibit higher conductivity [11,12]. Therefore, the size of the MOFs in membranes results in a variation in conductivity values. If the size of the MOFs incorporated in PEMs is larger, then the grain boundaries resistance is less, which results in increased proton conductivity, and vice versa [13].

The PEMFCs) and DMFCs, which use a sheet of polymer membrane as the solid electrolytic material, are discussed among the different fuel cells in this chapter. The membrane material which serves as an electrolyte will allow protons to pass through it but not electrons. Since it plays the role of both electrolyte and separator, the membrane functionality is doubled. Hence, for productive fuel cell functioning, proper movement of protons across the membrane and sufficient moisture level are considered significant, influential parameters. Conditions of dehydration and excess water affect the membrane performance [1].

The improvements shown in the field of polymer electrolyte technology directly impact the functioning of fuel cells. Varieties of membranes such as Dow membranes (Dow Chemical Co.) are perfluorinated ionomers with short-chain ionomers and commercially called DuPont FilmTec Membrane. Nafion (DuPont de Nemours) is also made up of perfluorinated ionomers, but it is referred to as a long-side-chain (LSC) ionomer. However, only Nafion has been used commercially in fuel cells [14]. Dow membranes' functioning was superior to that of Nafion, but they are much more costly. The Dow and Nafion membranes may have structural and morphological similarities but Dow membranes have shorter side chains compared to Nafion. The Dow membranes have higher conductivity and lower equivalent weights in comparison with Nafion membranes. The membranes also showed some acute limitations such as excessive methanol crossover, humidity management issues, lack of safety during production/operation, stability at smaller temperature ranges, etc. [15]. A novel type of proton-transportation solids has appeared to overcome these challenges, which is comprised of MOFs.

14.1.1 Proton Exchange Membrane Fuel Cells

These fuel cells have been widely accepted as the most promising substitute for the current sources of electricity. A simple working scheme has been employed in designing PEMFCs. It has an anode where hydrogen gas [$H_2(g)$] has been fed as fuel, and through the cathode, oxygen is fed into the fuel cell. The catalyst used will activate the hydrogen atom and convert it into a proton and electron. The membrane electrolyte allowed protons to pass through it, whereas the electrons generate current, and water is generated through the connection between protons and oxygen. The overall reactions are provided in Equations (14.1)–(14.3) [16]. A schematic representation of PEMFCs is provided in Figure 14.1 [17]

$$\text{At anode: } H_2\,(g) \rightarrow 2H^+ + 2e^- \quad (14.1)$$

$$\text{At cathode: } \frac{1}{2}O_2\,(g) + 2H^+ + 2e^- \rightarrow H_2O\,(l) \quad (14.2)$$

$$\text{Overall reaction: } \frac{1}{2}\,H_2\,(g) + O_2\,(g) \rightarrow H_2O\,(l) \quad (14.3)$$

The membrane used plays the role of both electrolyte and reactant gas separator. The problem of corrosion does not occur as the liquid is water. However, the proper balance of water inside the cell is crucial for functioning since a decline in the humidity of the membrane decreases the rate of proton transfer. If the level of water increases, then it may result in drowning of the electrodes. The combination of cathode,

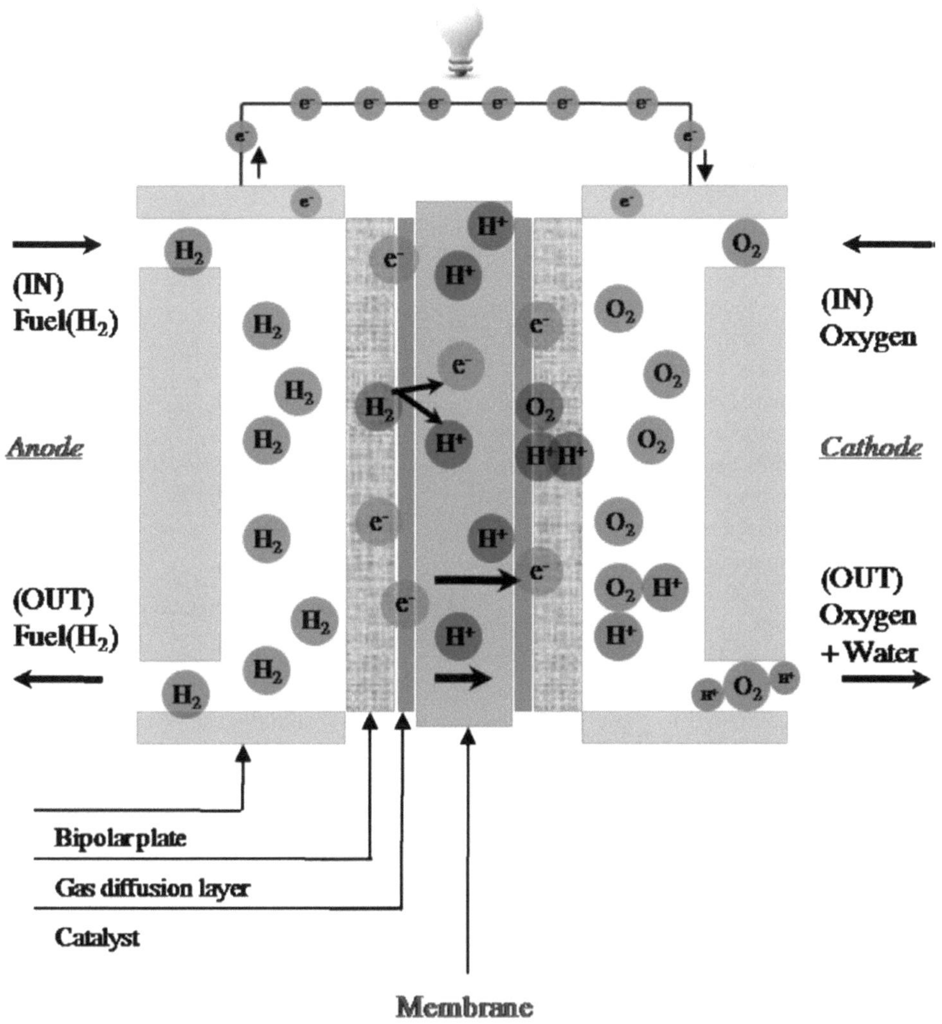

FIGURE 14.1 Working of proton exchange membrane fuel cells. Adapted from Ref. [17].

anode, and electrolyte assembly is described as a membrane electrode assembly (MEA), having thicknesses measured in microns, including the short diffusion paths of oxygen and hydrogen into the reaction sites and reduction of ionic resistance of membranes [18].

As mentioned earlier, the essential part of PEMFCs is the polymer electrolyte membrane (PEM), which acts as an electrolyte and a blockade for fuel and electron movement. Therefore, researchers have carried out intensive work to improve proton-carrying solids, which are suitable, display reasonable proton transportation rate, have less methanol crossover, excellent thermal, chemical, mechanical stability, are affordable for production, and show long-term durability [17]. In the last two decades of the 20th century, varieties of membranes were developed such as perfluoro sulfonic acid (PFSA) polymer membranes (Nafion) [19]. Along with these, sulfur-functionalized aromatic polymers like poly(ether ether ketone) (PEEK) [20], poly(ether sulfone) (PES) [21], and polybenzimidazole (PBI) [22] have also played a significant role as membranes. Figure 14.2 shows the chemical structure of perfluoro sulfonic acid (PFSA) polymer membrane [23]. Nafion is made up of aquaphobic tetrafluoroethylene (TFE).

The conductivity rate will increase with an increase in the extent of sulfonation; however, it leads to enlargement of the membrane, which in turn decreases the chemical and mechanical strength and persistence of the membrane [17]. Hybrid membranes have been invented to overcome these shortcomings, fabricated using various polymeric materials with excellent proton transportation capability and fillers with excellent chemical and thermal stability, mechanical robustness, and reduced methanol crossover. Since PEMFCs work at reduced temperatures, their performance is greatly affected by fuel passing across the membrane, carbon monoxide poisoning, reduced electrode reaction rate, optimum humidity percentage, inefficacious chilling of liberated heat, and improper waste management. Hence, the use of polymers like polybenzimidazole (PBI) or poly(ether ether ketone) (PEEK) will provide high-temperature operation and high conductivity. However, the acid doping of these polymers increases the proton mobility, which also leads to corrosion of the electrodes [24].

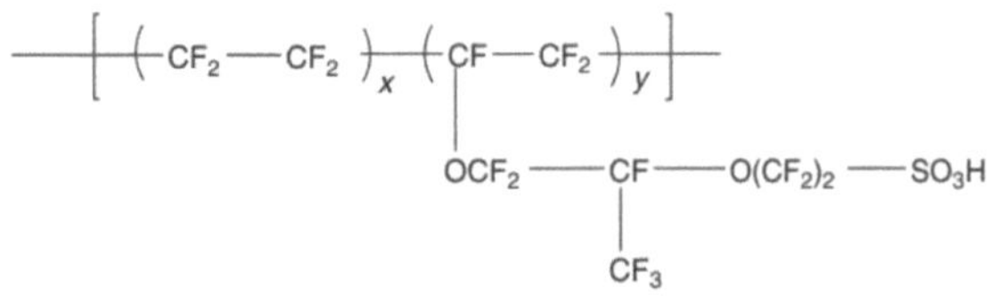

FIGURE 14.2 Chemical structure of perfluorosulfonic acid (PFSA) membrane. Adapted from Ref. [23].

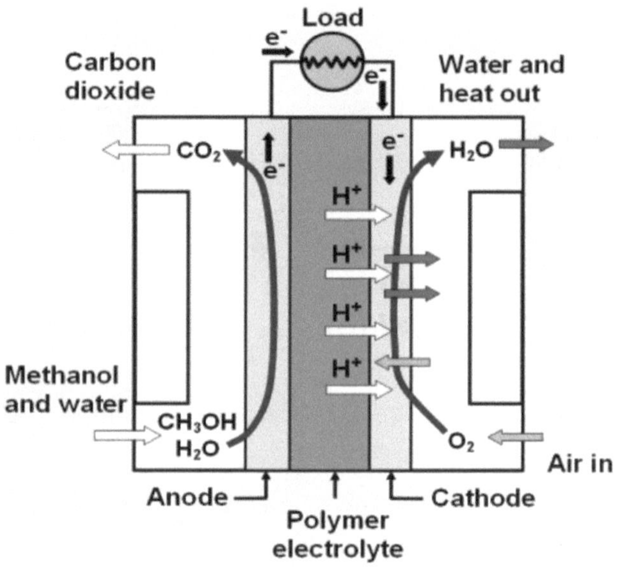

FIGURE 14.3 Diagrammatic representation of DMFC. Adapted from Ref. [38].

Many inorganic materials have been used as fillers during the fabrication of mixed-matrix membranes (MMMs). An ideal inorganic filler should possess characteristics such as the ability to absorb moisture, increased surface area, acidic nature at the surface, and best adaptability with the polymer matrix [25]. The inorganic materials used to fabricate proton-conducting mixed-matrix membranes (PC-MMMs) include hydrophilic inorganic compounds, namely silicon dioxide [26], titanium dioxide [27], aluminum oxide [28], zirconium dioxide [29], and clay [19]. These form dynamic cross-links with acidic functional groups of Nafion, and materials such as montmorillonite (MMT) or laponite (Lp) can also be used, which will decrease the hydrogen permeation, and show higher performance compared to the commercial Nafion [30,31]. The introduction of perovskites also leads to improvements in the oxidative stability and conductivity of the membranes [32]. In the manufacturing of transitional temperature and low-moisture PEMFCs, heteropolyacids (HPAs) such as phosphotungstic acid (PWA) [33] and silicotungstic acid (SiWA) [34] have been used, which exhibit high internal proton transportation with an elevation of the moisture content.

Similarly, nanoscale hollow tubes composed of carbon atoms, carbon nanotubes (CNTs), are of two principal forms, single-walled (SWCNTs) and multiwalled carbon nanotubes (MWCNTs), also have been considered to exhibit higher ionic conductivity at a temperature above 100°C [35]. Another primary inorganic filler used during the fabrication of membranes is graphene oxide (GO). Reports have revealed that the introduction of a small amount of graphene oxide (GO) in polybenzimidazole (PBI) composite membrane had an optimistic effect on characteristics such as proton transportation, acid preservation, stretching capacity, and hardiness [36]. Similarly, porous materials such as porous carbon, porous silica, and, most importantly, MOFs are newly emerging substances with exceptional quality to enhance the membrane functioning in fuel cells [25].

14.1.2 Direct Methanol Fuel Cells (DMFCs)

The DMFC class of fuel cell is considered as a fuel-cell type with membrane technology. Pure hydrogen is the primary source of fuel in the generation of electrical energy. However, the amount of risk involved in the storage, distribution, and production of hydrogen has led researchers to look for alternative sources that can indirectly produce hydrogen. Hydrogen can be extracted directly from any source of hydrocarbon, and the use of methanol has been considered as an ideal alternative [4]. In these types of fuel cells, methanol is directly fed into the anode where the oxidation of fuel takes place. The reaction usually takes place between 85–105°C. The reactions at the anode, cathode, and the overall reactions are provided in Equations (14.4)–(14.6) [5].

$$\text{At anode: } CH_3OH_{(l)} + H_2O_{(l)} \rightarrow CO_{2(g)} + 6\,H^+ + 6e^- \quad (14.4)$$

$$\text{At cathode: } {}^{3}\!/_{2}O_2(g) + 6\,H^+ + 6e^- \rightarrow 3\,H_2O(l) \quad (14.5)$$

$$\text{Overall: } CH_3OH(l) + {}^{3}\!/_{2}O_2(g) \rightarrow CO_2(g) + 2H_2O(l) \quad (14.6)$$

One of the significant drawbacks of these types of cells is methanol crossover. The proton exchange membranes have been developed for maximum proton conductivity, but they do not block methanol transport. Since methanol exhibits identical properties to water, it crosses over to the cathode and causes a decline in cell performance. Hence, a thick membrane with a methanol feed concentration slightly below 50% will give maximum power density [37]. A schematic, working illustration of a DMFC has been provided in Figure 14.3 [38].

14.2 METAL-ORGANIC FRAMEWORK-INDUCED PROTON-CONDUCTING MIXED-MATRIX MEMBRANES (PC-MMMS)

MOFs have gained tremendous attention as a promising compound in fuel cell operations. Electrochemical energy storage studies have seen considerable research, which has led to an increase in energy use from green and renewable sources. Most importantly, the proper usage and storage for future

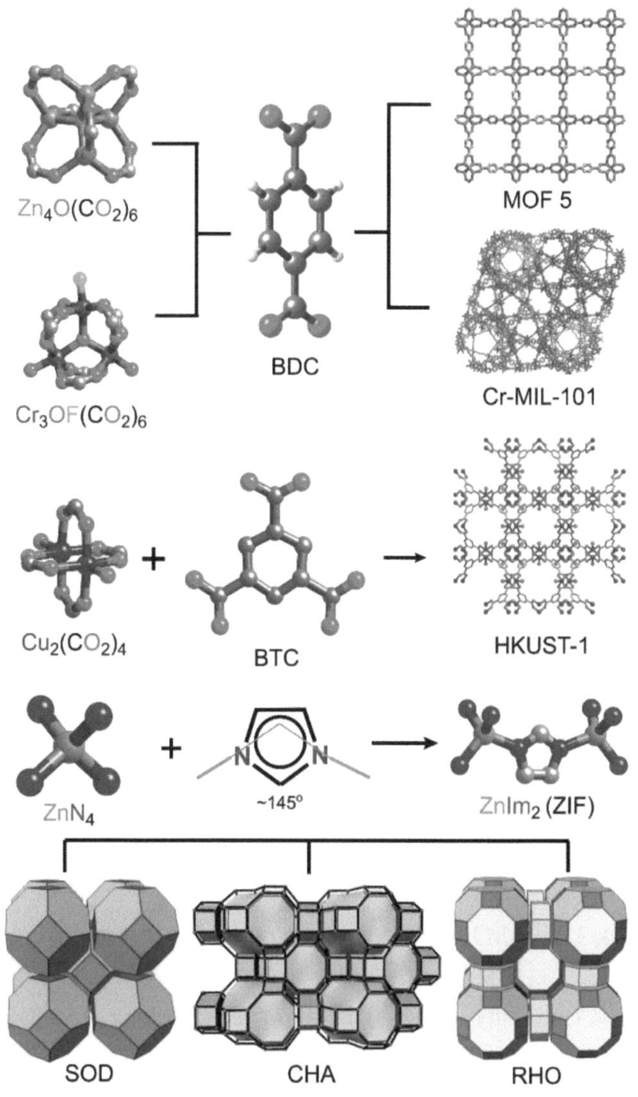

FIGURE 14.4 Illustration of SBUs and MOFs. Adapted from Ref. [40].

requirements has become a challenge. The main challenge in the construction of these PC-MMMs is to achieve the highest chemical and mechanical stability, and proton conductivity.

14.2.1 Synthesis and Structural Studies of Metal Organic Frameworks

An MOF is a class of compounds made up of metal clusters and organic compounds as linkers. Due to excellent surface area, tunable opening size, high selectivity, and capability to provide space for the carriers' transfer, they have been considered novel materials for efficient energy transformation and storing in fuel cells. MOFs can be prepared through diffusion, hydrothermal, sonochemical, microwave, mechanochemical, and electrochemical methods [39]. Metal-cantered primary and secondary building units are held together by organic molecules, which act as linkers to form the network with different dimensions. The structure of a few secondary building units (SBUs) and MOFs is provided in Figure 14.4 [40].

14.2.2 Classification and Naming of Various Metal-Organic Frameworks

In general, metal-organic frameworks are abbreviated as "MOF," and ordinal numbers always follow the individual compound. The various families of metal-organic frameworks have been classified according to symmetries, namely Isoreticular Metal-Organic Frameworks (IRMOF), Universitetet i Oslo (UiO), Materials of Institut Lavoisier (MIL), Hong Kong University of Science and Technology (HKUST), Zeolite Imidazolate Framework (ZIF) [41], Leiden Institute of Chemistry (LIC), Coordination Polymer with pillared Layer structure (CPL), Fluorinated Metal-Organic Framework (F-MOF), and Metal-Organic Polyhedra (MOP) [39].

14.2.3 Metal-Organic Framework as Filler in Proton Exchange Membranes

The efficiency of PEMFCs mainly depends on the efficiency of the PEMs. In the first polymer, Nafion, the efficiency declines rapidly as the temperature rises above 90°C or drops below –5°C [42]. Although many other proton-conducting materials have been used, all have shown a decline in conductivity value above 90°C. Four diverse methods have been designed to control proton transport in MOFs. The first approach is to produce proton carriers as counterions, such as ammonium (NH_4^+), hydronium $H(H_2O)^+$, and methylammonium ($Me_2NH_2^+$), into the pores of mesoporous MOFs. The subsequent approach enhances the proton transport and concentration through modification of the skeletal structure by incorporating acidic functional groups. The third and fourth approaches are post-synthetic modifications where charge-neutral carriers are introduced into pores, or metal–ligand substitution is included [43].

Two mechanisms operate the proton transport across the PEMs. The first is called the Grotthuss mechanism, in which the proton (H^+) is transported through water molecules in a concerted fashion through a network of hydrogen bonding [44]. The second mechanism is the vehicle mechanism in which the proton transport takes place through the diffusion mechanism. The protons (H^+) will be attached to H_2O and NH_3 to form H_3O^+ and NH_4^+ and be transported. The laden and unladen vehicles move in opposite directions to ensure continuity of the process. The activation enthalpy of the water translation process in $HUO_2AsO_4.4H_2O$ compound was found to be up to 0.8 eV [45]. The MOFs show excellent proton conductivity well above 100°C. Since these materials are porous, hydrogen fuel crossover may lead to voltage loss. This can be overcome by blocking the pores with guest molecules, preventing hydrogen fuel crossover [46].

First, research on the proton conductivity was carried out by Kitagawa's group [47] using MOFs made up of copper as

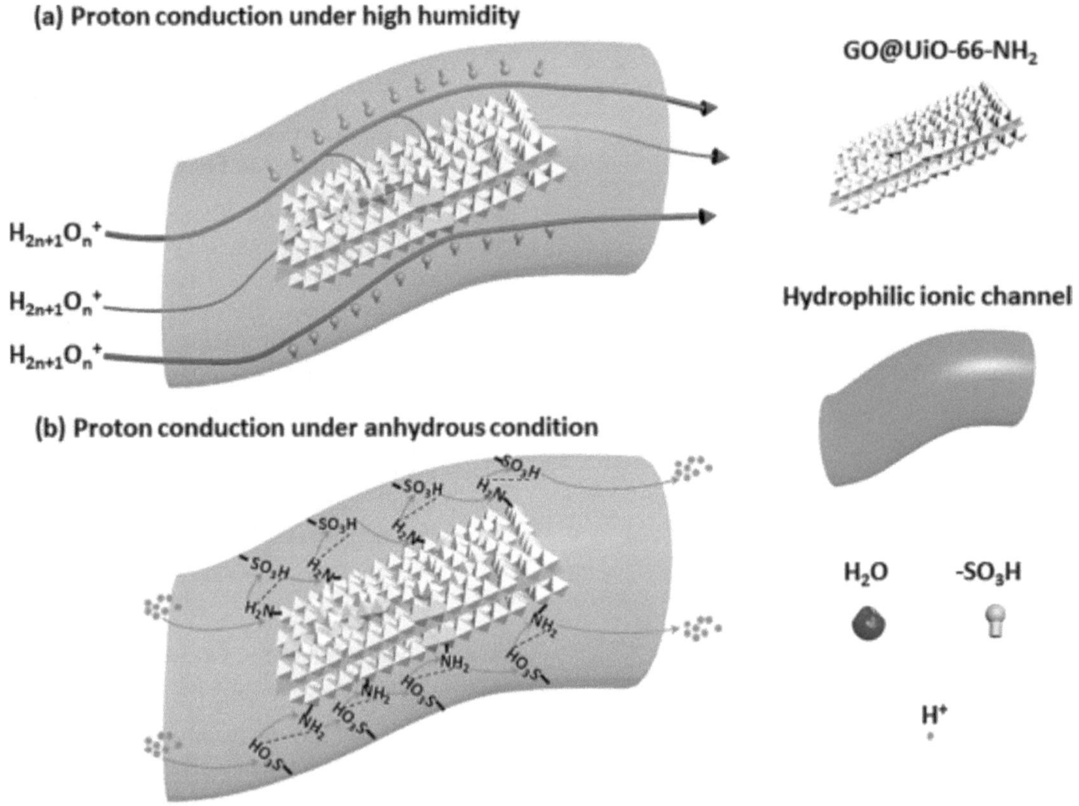

FIGURE 14.5 Proton conductivity under moisture and moisture-less conditions. Adapted from Ref. [52].

a metal atom and dithiooxamide as an organic linker. They showed that the RH will increase the amount of water present in MOFs, which increases the conductivity. Initially they have reported an increase in conductance from 3.3×10^{-9} S cm^{-1} to 2.0×10^{-6} S cm^{-1} when RH increased from 58% to 100% at 26.85°C using $(HOC_3H_6)_2$dtoaCu (dtoa = dithiooxamide) [48]. Further studies have been conducted using N,N'-diethyldithiooxamidatocopper polymer $(H_5C_2)_2$dtoaCu, which yielded a proton conductance of 4.2×10^{-6} S cm^{-1} [49]. Hydroxyl groups were introduced into the "R" moieties of MOFs and tested for the results to get a higher proton transportation rate. The polymer material N,N'-bis-(2-hydroxyethyl) dithiooxamidatocopper (II) (HO-$H_4C_2)_2$dtoaCu has given a conductance of 1.2×10^{-5} S cm^{-1} [50]. Further improvement in the conductivity was achieved with $(NH_4)_2$(adp)$\{Zn_2(ox_3)\}.3H_2O$ (adp = adipic acid) MOF, which resulted in a proton conductance of 8×10^{-3} S cm^{-1} [51]. Figure 14.5 illustrates the proton conductivity under moisture and moisture-less conditions [52].

The working model of proton exchange membranes containing MOFs works with two different mechanisms. In the first type, the openings of the MOFs will be filled with proton carriers, and in the second approach, the insertion of other functional groups into the organic linkers improves the acidic nature and affection toward water [53].

14.2.4 UiO Series Metal-Organic Framework Proton Exchange Membranes

The University of Oslo (UiO) designed these types of MOFs using zirconium as a metal-ligand. There are three categories of UiO MOFs reported, namely UiO-66, UiO-67, and UiO-68, which have identical structures and differ only by the linkers. The tailoring of UiO-66 is done using terephthalate as an organic linker for UiO-67 biphenyl dicarboxylate, and UiO-68 synthesis is done with the help of terphenyl dicarboxylate [54]. Recently, UiO-66 modified with 1,2-oxathiolane,2,2-dioxide has shown an excellent conductivity of 1.64×10^{-1} S cm^{-1}, at 80°C, maximum with systems operating below 100°C [55]. Different functional groups present in the MOFs have influenced the extent of proton conductivity. When UiO-66 is subjected to post-synthetic modification using acidic functional groups such as sulfonic acid and a carboxylic acid, the conductivity is 0.34×10^{-2} and 0.10×10^{-2}, respectively. Similarly, it was found to be 1.40×10^{-5} and 2.23×10^{-7}, respectively, at a temperature of 29.85°C and 97% RH for amine and bromide substitution. This result indicates that acidic and hydrophilic functional groups are responsible for higher proton dissociation [56].

Balanced generation and release of protons have been observed when the acidic groups are introduced through post-synthetic modification, leading to filling the acidic molecules into the crystal's openings. The impact of a post-synthetic improvement in the UiO-66 series has been studied

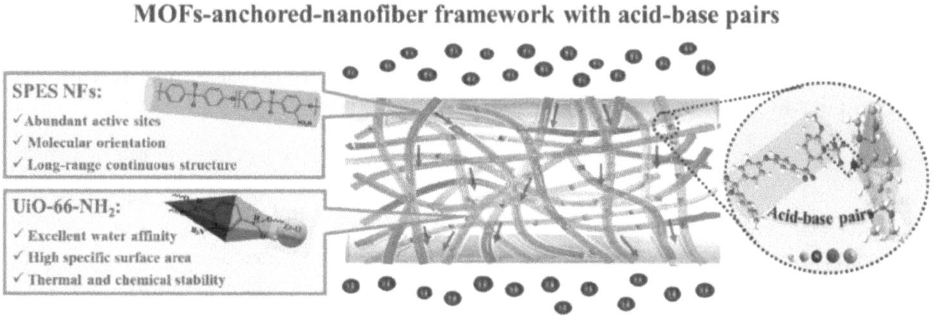

FIGURE 14.6 Construction of NFs/NH_2 – UiO-66 membrane. Adapted from Ref. [58].

using UiO-66-NH_2 and UiO-66-AS (AS stands for amino and sulfonic acid groups, respectively). Further, a Schiff base condensation of imidazole-2-carboxaldehyde with UiO-66-AS resulted in the formation of IM-UiO-66-AS metal organic frameworks, and further modification of it to a membrane displayed an excellent proton conductance value of 1.19×10^{-2} S cm^{-2} at a temperature of 80°C and RH of 98% [57]. However, when used on a large scale, the proton-conducting channels of MOFs become interrupted, resulting in poor conductivity. This limitation can be overcome by constructing uninterrupted proton-conducting channels by incorporating polymeric materials to form a composite. Further integrating the composite material into Nafion or another matrix will solve the drawbacks mentioned above. Incorporation of nanoparticles such as graphene oxide into MOFs and further combination with a polymer like sulfonated poly(ether ether ketone) (SPEEK) have resulted in elevated proton conductance of 2.68×10^{-1} at 70°C and 95% RH [54].

The MOF-induced nanofibrous membrane material (UiO-66-NH_2@NFs) was synthesized using an electrospinning technique where UiO-66-NH_2 was induced into sulfonated poly(ether sulfone) (SPES) polymer matrix, which resulted in elevated proton conductance of 2.7×10^{-1} S cm^{-1} at a temperature of 80°C and RH value of 100%. This elevation in conductance is due to the development of uninterrupted proton-transporting networks, which assist the free movement of protons. Figure 14.6 displays the structure of the UiO-66-NH_2@NFs membrane [58]. Interestingly, amine-functionalized and sulfonic acid-functionalized UiO-66 materials have been subjected to cooping and incorporated into the Nafion matrix with 0.6 wt.% of the filler. This 2:1 NH_{2-} UiO-66+ SO_3H – UiO-66/Nafion-0.6 gave an elevated conductivity result of 2.56×10^{-1} S cm^{-1} at a temperature of 90°°C and 95% RH due to the higher water-holding nature of the functional groups present [59]. In another research study, covalent-ionically cross-linked composite membrane made up of UiO-66-NH_2-5/sulfonated poly(arylene ether nitrile)s (SPENs) gave a proton conductance value of 1.35×10^{-1} S cm^{-1} at a temperature of 80°C [60]. The Nafion/GO-UiO-66-NH_2-0.6 hybrid membrane showed excellent proton conductance of 3.03×10^{-1} S cm^{-1} at a temperature of 90°C and 95% RH, which was much more compared to recast Nafion membrane, which show a conductance of 1.18×10^{-1} S cm^{-1}. This excellent stability has been stable beyond 54.17 hours, reflecting an excellent fuel cell application [52]. Another research work using a post-synthetic modification technique has been reported. In this work, UiO-66-NH_2 metal-organic framework was subjected to post-synthetic modification and named PSM 1. PSM 2, which has been incorporated into poly 2,2′-(p-oxydiphenylene)-5,5′-benzimidazole (OPBI) polymeric matrix and further doping with phosphoric acid (PA), resulted in a proton exchange membrane with a conductivity value of 2.9×10^{-1} S cm^{-1} for a PSM 1 hybrid membrane with 10% filler, and for PSM 2 hybrid membrane with 10% filler it was found to be 3.08×10^{-1} S cm^{-1} at an elevated temperature of 160°C [61]. Recently, nanofibers n fabricated using cellulose and UiO-66-NH_2 and further blending with sulfonated polysulfone polymer matrix to fabricate proton exchange membranes have been reported with a proton conductance value 1.96×10^{-1} S cm^{-1} at 80°C and RH value of 100% [62]. Table 14.2 gives the proton conductance properties of UiO-series MOF hybrid proton exchange membranes.

The restricted use of Nafion membranes in proton exchange membranes due to the polymeric material, high operating cost, low efficiency, low durability, and declined performance at higher temperatures demand an alternative material that will help to overcome these drawbacks. The excellent thermal and chemical stability, and efficiency in conductivity of the UiO-66 series MOF composite membranes with various polymeric materials have been considered one of the best proton exchange membranes. Incorporating nanoparticles, hybrid materials, implementation of functional groups, and post-synthetic modifications will drastically improve the conductivity.

14.2.5 MIL Series Metal-Organic Framework Proton Exchange Membranes

This class of metal MOFs was designed by the Materials of Institute Lavoisier (MIL) using transition metals and aromatic carboxylic derivatives as linkers [63]. These materials exhibit exceptional superiority in terms of increased porosity, large surface area, excellent surface area, tunable pore volume, high selectivity, and ability to provide space for the carriers compared to other porous materials. Chitosan has been used as a membrane material and Cr-MIL-101 MOFs with several

TABLE 14.2
Proton conductivity properties of UiO-series metal organic framework incorporated mixed-matrix membranes NH_2_ UiO-66+ SO_3H – UiO-66/Nafion-0.6

Membrane	σ (S cm^{-1})	T (°C)	RH (%)	References
S-UiO-66@GO-10/SPEEK	2.68×10^{-1}	70	95	[54]
UiO-66-NH_2@NFs/Nafion	2.7×10^{-1}	80	100	[58]
NH_2_ UiO-66+ SO_3H – UiO-66 /Nafion-0.6	2.56×10^{-1}	90	95	[59]
Recast Nafion	1.18×10^{-1}	90	95	[59]
(SPENs)/NH_2_ UiO-66- 5	1.351×10^{-1}	80	n/a	[60]
GO@UiO-66-NH_2/Nafion-0.6	3.03×10^{-1}	90	95	[52]
PSM 1 – 10%	2.9×10^{-1}	160	n/a	[61]
PSM 2 – 10%	3.08×10^{-1}	160	n/a	[61]
Cell-UiO-66-NH_2-5/SPS	1.96×10^{-1}	80	100	[62]
CBOPBI@UiO-66 40%	1.0×10^{-1}	160	n/a	[109]
OPBI/UiO-66	9.2×10^{-2}	160	Anhydrous	[110]
Chitosan/UiO-66(SO_3H)-6	1.52×10^{-3}	90	0	[119]
Chitosan/UiO-66(NH_2)-15	5.64×10^{-2}	100	98	[119]
Chitosan/SO_3H-UiO-66-6 wt.% NH_2-UiO-66-15	3.78×10^{-3}	120	0	[119]

functional groups as filler material. Chitosan/S-MIL-101-4, chitosan/H_2SO_4@MIL-101-8, chitosan/H_3PO_4@MIL-101-6, and chitosan/CF_3SO_3H@MIL-101-10 have shown proton conductances of 6.4×10^{-2} S cm^{-1}, 9.5×10^{-2} S cm^{-1}, 8.3×10^{-2} S cm^{-1}, and 9.4×10^{-2} S cm^{-1}, respectively, which is much higher than the conductivity of pure chitosan membrane at 3×10^{-2} S cm^{-1} at a temperature ranging between 25–100°C and 100% RH. This improved conductivity is due to an electrostatic interaction between the amine and hydroxyl groups of chitosan and functional groups of the MOF [64].

The impact of sulfonic acid groups on enhancing proton conductance in MOF composite membranes has been studied. Both Cr-MIL-101 MOFk and poly(ether ether ketone) (SPEEK) polymer were sulfonated, and membrane sheets were prepared by the solution casting method. The proton conductance was found to be 3.06×10^{-1} S cm^{-1} at 75°C and RH of 100% [65]. Meanwhile, a rare study on hydroxide ions' conductivity was carried out using chloromethylated MIL-101(Cr) MOF and chloromethylated poly(ether ether ketone) (PEEK) polymer. The composite membrane imidazolium PEEK/imidazolium MIL-101(Cr) was found to give a hydroxide conductance of 3.6×10^{-2} S cm^{-1} at 50°C and 100% RH [66]. Similarly, the Nafion/phytic@MIL-101 (Cr) composite membrane gives a conductivity value of 6.08×10^{-2} S cm^{-1} at a temperature range of 20–80°C and 57.4% RH, which is about 2.8-fold more than pure Nafion membrane. Nafion/phytic@MIL-12 has shown a conductivity charge of 2.28×10^{-1} S cm^{-1} at 100°C and RH of 100%, but higher loading and lowering the humidity lead to a reduction in conduction. The conductivity efficacy was 6.08×10^{-2} S cm^{-1} at 57.4% RH and 7.63×10^{-4} S cm^{-1} at 10.5% RH, which are 2.8- and 11.0-fold more than for the pure membrane [67].

In another study, a composite membrane prepared using phosphotungstic acid-induced MIL-101(Cr) MOF and sulfonated poly(ether ether ketone) (SPEEK) was found to yield a conductivity of 2.72×10^{-1} S cm^{-1} at a temperature of 65°C and 100% RH, which was 45.5% more than for pure SPEEK membrane sheet [68]. The hybrid membrane made up of sulfonated poly(ether sulfone) and amine group incorporated MIL-101 (Cr) showed a proton conduction of 4.1×10^{-2} S cm^{-1} at 160°C and 0% RH [69]. A composite membrane made up of Fe-MIL-101-NH_2 MOF and sulfonated poly(2, 6-dimethyl-1, 4-phenylene oxide) (SPPO) achieved a conductance of 1×10^{-1} S cm^{-1} at ambient temperature and 2.5×10^{-1} S cm^{-1} at 90 °C [70]. Another work with MIL-101-NH_2-SO_3H incorporated poly(arylene ether ketone) (MNS-PAEK) mixed-matrix membrane attained conduction of 1.98×10^{-1} S cm^{-1} at a temperature of 80°C and 100% RH [71]. Another remarkable work with high-temperature proton exchange membranes was carried out using a zwitterionic liquid 1-(1-ethyl-3-imidazolio) propane-3-sulfonate (EIMS) and N,N-bis(trifluoromethanesulfonyl) amide (HTFSA), which gave a conduction of 2×10^{-4} S cm^{-1} at 140°C, making it a prominent material in a high-temperature fuel cell [72]. With all the above observations, it has been proved that the conductivity efficiency and water uptake capacity of hybrid membrane sheets have been elevated by the addition of MIL MOFs and hence they have great potential for application in proton exchange membrane cells. The chitosan membranes induced with MIL-series MOFs show the highest conductivity value compared with other series. Table 14.3 summarizes the proton conduction properties of MIL-series MOF-induced proton exchange composite membranes.

14.2.6 ZIF Series Metal-Organic Framework Proton Exchange Membranes

Zeolitic imidazolate frameworks (ZIFs) are microporous MOFs made up of zinc or cobalt metal atoms, imidazolate, and their derivatives as linker materials. The cation exhibits

TABLE 14.3
Proton conductivity properties of MIL-series metal organic framework impregnated composite proton exchange membranes

Membrane	σ (S cm^{-1})	T (°C)	RH (%)	References
Chitosan	3×10^{-2}	25–100	100	[64]
Chitosan/S-MIL-101-4	6.4×10^{-2}	25–100	100	[64]
Chitosan/H_2SO_4/MIL-101 (8 wt.%)	9.5×10^{-2}	25–100	100	[64]
Chitosan/H_3PO_4/MIL-101 (6 wt.%)	8.3×10^{-2}	25–100	100	[64]
Chitosan/CF_3SO_3H@MIL-101-10	9.4×10^{-2}	25–100	100	[64]
SPEEK/S- MIL-101 (Cr)	3.06×10^{-1}	75	100	[65]
Imidazolium PEEK/imidazolium MIL-101(Cr)	3.6×10^{-2}	50	100	[66]
Phytic@MIL-101 (Cr) /Nafion	6.08×10^{-2}	20–80	57.4	[67]
Phytic@MIL-12 (Cr) /Nafion	2.28×10^{-1}	100	100	[67]
Phytic@MIL-12 (Cr) /Nafion	6.08×10^{-2}	100	57.4	[67]
Nafion/phytic@MIL-12 (Cr)	7.63×10^{-4}	100	10.5	[67]
SPEEK/phosphotungstic acid MIL-101(Cr)	2.72×10^{-1}	65	100	[68]
SPES/NH_2-MIL-101 (Cr)	4.1×10^{-2}	160	0	[69]
SPPO/NH_2-Fe-MIL-101	1×10^{-1}	RT	na	[70]
SPPO/Fe-MIL-101-NH_2	2.5×10^{-1}	90	na	[70]
MNS@SNF-PAEK-3%	1.98×10^{-1}	80	100	[71]
EIMS/HTFSA	2×10^{-4}	140	na	[72]
QCS/MIL-9	2.3×10^{-2}	80	na	[120]

an oxidation state of +2, and the metal atom tetrahedrally coordinates the linkers. The M-Im-M (M= metal, Im = imidazolate linkers) bond angle is 145° nearest to the Si-O-Si bond angle of zeolites. Due to the magnificent chemical, mechanical, and thermal stability, and solvent resistivity, they have a prominent commercial application in the membrane industry [73]. Researchers have revealed that these classes of porous materials have an excellent Langmuir surface area of 1,810 m^2/g and temperature solidity up to 550°C [74].

Several polymeric materials have been used, and ZIF MOFs have been used to prepare the hybrid membranes used in proton exchange membrane cells. Hybrid membranes made up of polyetherimide (PEI), ZIF-67/ZIF-8, or a combination of both MOFs have been studied for conductivity, revealing that the conductivity depends on the temperature. There was a six-fold increase in the conductivity as the temperature was raised from 25 to 55°C. The conductivity values were found to be 1.5×10^{-6} S cm^{-1} for PEI/ZIF-8, 3.5×10^{-6} S cm^{-1} for PEI/ZIF-67, and 2.9×10^{-6} S cm^{-1} for PEI/ZIF mix under 95% RH [75]. Another excellent work has been conducted using polybenzimidazole (PBI) as membrane material and ZIF-67, ZIF-8, and a combination of both MOFs. These membranes have shown excellent thermal stability up to 250°C. PBI/ZIF-8 has attained a conductivity of 3.1×10^{-3} S cm^{-1}, ZIF-67/PBI attained 4.1×10^{-2} S cm^{-1}, and ZIF/PBI mix resulted in 9.2×10^{-2} S cm^{-1} at 200°C and dry environments. Low conductivity has been observed with polybenzimidazole membranes at low humidity; however, this has been improved through doping with phosphoric acid (PA). The higher conductivity value for PBI/ZIF-67 compared with PBI/ZIF-8 is because phosphonate anions are less attracted by Zn (II) ions than Co (II) cations [76]. A similar type of study has been reported with sulfonated poly(ether ether ketone) (SPEEK) as the polymer medium. The ZIF-8/SPEEK, ZIF-67/SPEEK and ZIF/SPEEK mixture attained conductivities of 2.5×10^{-3} S cm^{-1}, 1.6×10^{-3} S cm^{-1}, and 8.5×10^{-3} S cm^{-1}, respectively, at 120°C. However, the SPEEK/ZIF mixture attained a value of 2.9×10^{-2} S cm^{-1} at 100°C. This variation in conductance at elevated and lower temperatures is because of the evaporation of water molecules from the hybrid membranes when the temperature exceeds 100°C [77]. A novel nanocomposite material synthesized out of ZIF-8 and graphene oxide (GO) and its fabrication in a Nafion-based membrane resulted in a proton conduction value of 2.8×10^{-2} S cm^{-1} at 120°C and 40% RH. A decrease in conduction rate has been observed when the humidity drops below 40%; however, not much change is observed between 40–100°C. This observation may be due to the binding reaction between filler and sulfonic acid functional groups of polymers, hindering the swelling and obstructing the proton path. The methanol permeation is also less as compared to recast Nafion [78].

The insertion of hybrid porous materials into the polymer matrix is very influential in increasing the conductivity. A two-dimensional (2D) CNT/ZIF-8 hybrid material was mixed with sulfonated poly(ether ether ketone) (SPEEK) matrix. The SPEEK/ZCN-2.5 was found to attain a conductivity of 5.024×10^{-2} S cm^{-1} at 120°C and 30% humidity, 11.2-fold higher than for the recast SPEEK membrane sheet. Since the methanol permeation reduces the cathodic catalyst activity, it is essential to see that only proton transportation occurs, not methanol. Here methanol permeability was found to be much lower ($1.86 \pm 0.23 \times 10^{-7}$ cm^2/s) compared to recast

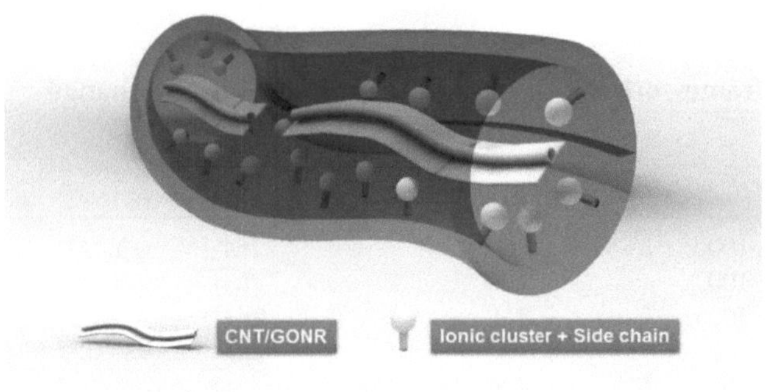

FIGURE 14.7 Schematic representation of the reaction between GONR/CNT and Nafion to generate proton transportation channels. Adapted from Ref. [81].

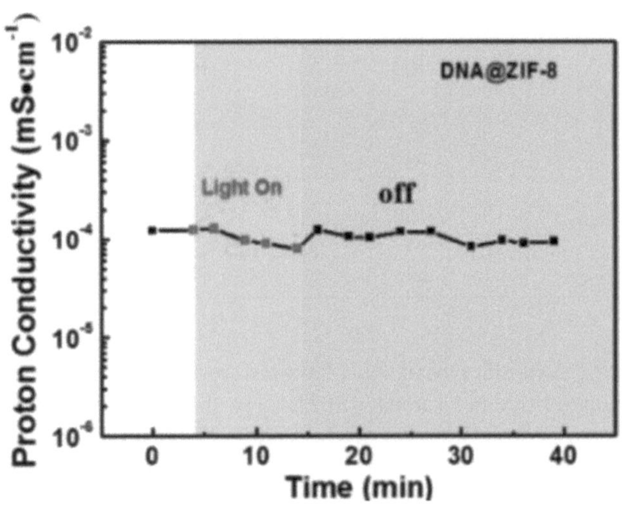

FIGURE 14.8 Display of the conduction of DNA@ZIF-8 membrane. Adapted from Ref. [85].

SPEEK membrane (20.23 ± 2.01 × 10^{-7} cm^2/s). There has been a significant observation that SPEEK/ ZCN hybrid membrane shows higher conductivity than SPEEK/ZIF hybrid membrane, indicating that the structure of MOFs alone does not increase the conductivity of hybrid membranes. Instead, a substantial interaction between fillers and polymeric materials could rearrange the hydrophilic groups of polymer to accumulate around the porous material to reduce the system's activation energy for permeation through the membrane which is measured in kJ^{-1} [79]. The thickness, surface area, temperature, pH of the medium, and the membrane material type affect the permeability. The transfer of molecules across the membrane is correlated with the number of hydrogen bonds. These hydrogen bonds should be broken to enter the membrane phase. The energy required to break these hydrogen bonds is considered as activation energy. This activation energy decides the rate of permeation across membrane [80]. A similar study with hybrid fillers of carbon nanotube/graphene oxide nanoribbon (GONR/CNT) with Nafion-based hybrid proton exchange membranes (PEMs) has been reported with a conductivity of 1.8 × 10^{-1} S cm^{-1} at an operational temperature of 120°C and 40% humidity, which is nine-fold more than recast Nafion (2 × 10^{-2} S cm^{-1}) in identical circumstances. The methanol permeability was also significantly decreased by 2.84 × 10^{-9} cm^2/s compared to recast Nafion PEM 1.03 × 10^{-8} cm^2/s at a temperature of 40°C. Figure 14.7 shows the reaction between GONR/CNT and Nafion to generate the proton transportation path [81].

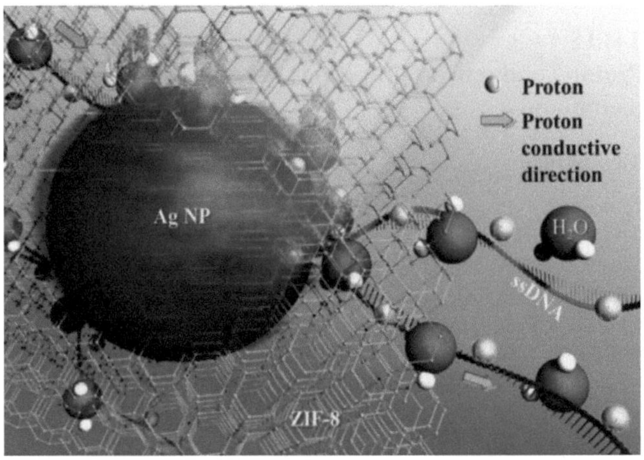

FIGURE 14.9 Mechanism of proton transportation through AgNP/DNA@ZIF-8 membrane. Adapted from Ref. [85].

A hybrid membrane fabricated using ZIF-8 and poly (vinyl phosphonic acid) (PVPA) displayed a conductivity value of 3.2 (±0.12) × 10^{-3} S cm^{-1} at 140°C under a dry environment which is much higher than PVPA membrane and pure ZIF-8 [82]. Similarly, ZIF-8 material incorporated poly(2-acrylamido-2-methylpropane sulfonic acid) (PAMPS), poly(vinyl alcohol) (PVA) hybrid material attained a conductivity value of 1.34 × 10^{-1} S cm^{-1} at 80°C and 100% humidity conditions, which is comparable with commercial Nafion-117 membrane [83]. Apart from synthetic polymers, biomolecules have also been used to construct composite membranes—introduction of a biomolecule-like single-strand DNA to ZIF-8 to form membranes through a solid-confined conversion process. The developed membranes attained a conduction value of 3.4 × 10^{-4} S cm^{-1} at 25°C and 97% RH. The methanol permittivity was found to be 1.25 × 10^{-8} cm^2 s^{-1} due to the minute pores of the membranes [84]. Figure 14.8 shows the conduction of DNA@ZIF-8 membrane, and Figure 14.9 shows the mechanism of proton transport through Ag NP/DNA@ZIF-8 membrane [85].

Introduction of ZIF-8 into the 3D network structure (3DNWS) of poly-m phenylene isophthalamide nanofibers (PMIA NFs) to construct the proton transport channels and subsequent fabrication of membrane using Nafion matrix resulted in an excellent conductivity result of 2.58 × 10^{-1} S cm^{-1} at 80°C and 100% RH. The methanol permittivity was also 7.98 × 10^{-7} cm^2 s^{-1}, which is much less than the recast Nafion permeability value of 13.8 × 10^{-7} cm^2 s^{-1}. These excellent results could be due to the reaction between the sulfonic acid groups of Nafion and the amine groups of linkers of ZIF-8, which favors the Grotthuss mechanism. Along with

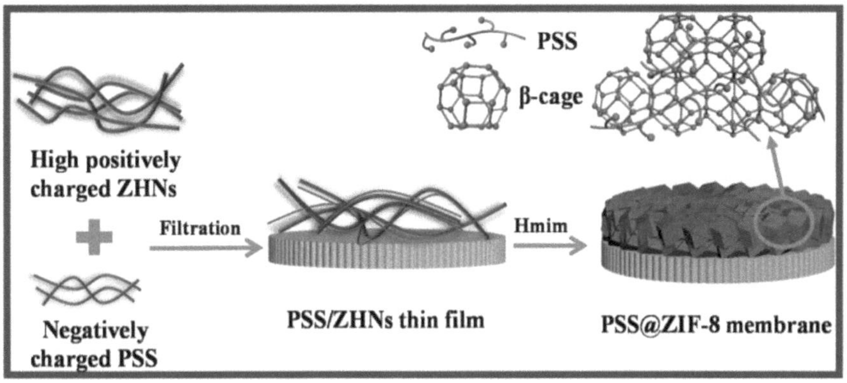

FIGURE 14.10 Fabrication of PSS/ZIF-8 hybrid membrane sheet. Adapted from Ref. [87].

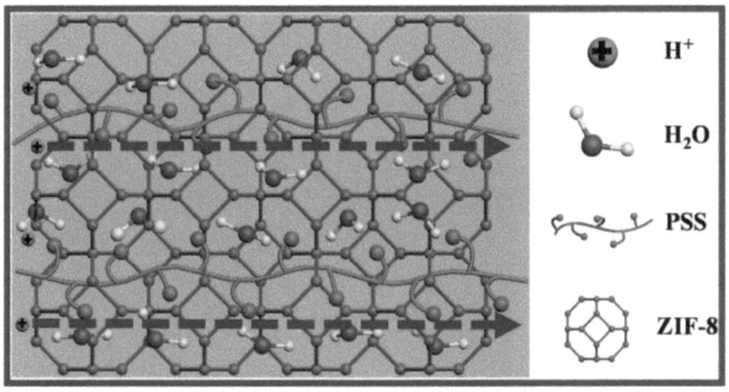

FIGURE 14.11 Mode of the proton carriage mechanism in ZIF-8/PSS. Adapted from Ref. [87].

this, the 3D structure of nanofibers enhances the connectivity, and the larger surface area of ZIF-8 and the hydrophilicity of nanofibers improve the water-holding capacity [86]. An excellent proton exchange membrane has been developed using poly(4-styrene sulfonate) (PSS) and ZIF-8, which resulted in a conduction value of 2.59×10^{-1} S cm^{-1} at a temperature of 80°C and 100% RH due to the presence of higher proton transporters in the networks at optimized conditions. The methanol permeability was found to be 2.9×10^{-9} cm^2 s^{-1} which is much less than for Nafion-117 (4.16×10^{-6} cm^2 s^{-1}) due to smaller openings and interknitted PSS molecules. Figure 14.10 illustrates the fabrication of the ZIF-8/PSS hybrid membrane sheet. Figure 14.11 shows the mode of the proton transport mechanism [87].

The carboxylic acid functional group induced ZIF-8 has been introduced into the sulfonated poly(ether ether ketone) (SPEEK) matrix to fabricate SPEEK/ZIF-COOH mixed-matrix membrane. These have shown excellent conductivity of 1.52×10^{-2} S cm^{-1} at 80°C and RH of 100% at an optimized ZIF-COOH amount of 5 wt.%, which is 46.2% more than the recast sulfonated poly(ether ether ketone) SPEEK value. This is due to more functioning sites for proton transportation in the form of imidazole and carboxylic groups and the interaction with sulfonic acid groups of SPEEK. It also exhibits a methanol penetration value of 1.1×10^{-7} cm^2 s^{-1}, about 10 times less than for pristine SPEEK. This is due to forming a dense-packed polymer matrix and the physical interference outcome of the filler [88]. A two-dimensional ZIF/GO filler was impregnated into the sulfonated poly(ether ether ketone) (SPEEK) matrix. This resulted in a proton conduction of 2.65×10^{-1} S cm^{-1} at 70°C and 100% RH. This hybrid SP-ZIF-L@GO-5 membrane has been subjected to moisture- and temperature-dependent conductivity studies. The SP-ZIF-L@GO-5 reported conduction of 1.83×10^{-1} S cm^{-1} at 70°C and 90% RH, which is 2.45-fold more than for pure SPEEK (7.46×10^{-2} S cm^{-1}). Similarly, the rate of 3.64 $\times 10^{-2}$ S cm^{-1} at 90 °C and RH value of 40% were 6.24-fold more than for pure poly(ether ether ketone) (SPEEK) [89]. Table 14.4 summarizes the proton conductivity properties of ZIF-series MOF hybrid proton exchange membranes.

14.2.7 Some Other Kinds of Metal-Organic Framework Proton Exchange Membranes

Apart from the above-mentioned MOF subclasses, a few other classes of materials are also used as fillers in the fabrication of composter membranes in fuel cells. A novel electrospun nanofiber membrane made up of Zn-aminotriazolato-oxalate compound (ZAO) and sulfonated poly(phthalazinone ether sulfone ketone) (SPPESK) has been tested in fuel cells at high temperatures and dry environments. This attained a conductance of $8.2 \pm 0.16 \times 10^{-2}$ S cm^{-1} at a high temperature of 160°C and anhydrous state, but the methanol permeability was significantly less compared to Nafion-115 [90]. In another study, phosphoric acid doped HKUST-1/Nafion® was fabricated, and reached a conductance rate of 1.8×10^{-2} S cm^{-1} at 25°C and 100% RH due to the presence of phosphate groups. However, a higher filler content leads to a reduction in conduction [91]. The development of water- and acid-stable proton exchange membranes was done using novel hexaphosphate ester-based MOF, and its incorporation into poly(vinyl alcohol) (PVA) matrix resulted in conduction of 1.25×10^{-3} S cm^{-1} at 50°C [92].

TABLE 14.4
Proton conductivity properties of ZIF-series metal organic framework induced proton exchange membranes

Membrane	σ (S cm^{-1})	T (°C)	RH (%)	References
ZIF-8/PEI	1.5×10^{-6}	55	95	[76]
ZIF-67/PEI	3.5×10^{-6}	55	95	[76]
ZIF mix/PEI	2.9×10^{-6}	55	95	[76]
ZIF-8/PBI	3.1×10^{-3}	200	0	[108]
ZIF-67/PBI	4.1×10^{-2}	200	0	[108]
ZIF mix/PBI	9.2×10^{-2}	200	0	[108]
ZIF-8/SPEEK	2.5×10^{-3}	120	na	[77]
ZIF-67/SPEEK	1.6×10^{-3}	120	na	[77]
SPEEK/ZIF mix	8.5×10^{-3}	120	na	[77]
SPEEK/ZIF mix	2.9×10^{-2}	100	na	[77]
ZIF-8@GO/Nafion	2.8×10^{-2}	120	40	[78]
SPEEK/ZIF-8/CNT	5.024×10^{-2}	120	30	[79]
CNT/GONR/Nafion	1.8×10^{-1}	120	40	[81]
Nafion	2×10^{-2}	120	40	[81]
PVPA/ZIF-8	3.2×10^{-3}	140	0	[82]
PVA/PVMPS/ZIF-8	1.34×10^{-1}	80	100	[83]
DNA@ZIF-8	3.4×10^{-4}	25	97	[84]
PMIA NFs/ZIF-8	2.58×10^{-1}	80	100	[86]
PSS/ZIF-8	2.59×10^{-1}	80	100	[87]
SPEEK/ZIF-COOH	1.52×10^{-2}	80	100	[88]
SP-ZIF-L@GO-5	2.65×10^{-1}	70	100	[89]
SP-ZIF-L@GO-5	1.83×10^{-1}	70	90	[89]
SP-ZIF-L@GO-5	3.64×10^{-2}	90	40	[89]
SPEEK	7.46×10^{-2}	70	90	[89]

14.3 HYBRID MEMBRANES WITH VARIOUS METAL ORGANIC FRAMEWORKS AS NANOMATERIALS

14.3.1 Poly(Perfluorosulfonic Acid) Nafion-Type Polymers

This variety of polymeric membranes is widely used in fuel cells, with five variations of these membranes being commercially available. Nafion membranes with thicknesses of 25.4 μm, 50.8 μm, 127 μm, 177.8 μm, and 254 μm are commercially called Nafion 111, Nafion 112, Nafion 115, Nafion 117, and Nafion 1110. These polymers have shown excellent stability towards water, heat, and chemicals. The mechanical strength is also found to be remarkable. However, their main limitation is the decrease in proton conductivity when the system temperature exceeds 90°C due to the evaporation of water molecules acting as proton carriers. This leads to the fabrication of membranes that can operate at elevated temperatures keeping all other necessary characteristics at the highest level. Therefore, recently, MOF materials with hydrophilic and proton-carrying functional groups have been introduced into Nafion membranes to make them sustainable at higher temperatures [93].

The Nafion membranes with MIL-101 (Cr) as fillers were found to exhibit 2.28×10^{-1} S cm^{-1} at 100°C and 100% RH [94]. Similarly, magnesium-induced MOF material (Mg-CPO-27) and aluminum-induced MOF material (Al-MIL-53) have been used as nanomaterials in Nafion polymer. The Nafion/CPO-27-Mg (3 wt.%) has a proton conductance of 1.10×10^{-2} S cm^{-1} at 50°C and 99.9% RH, with better water absorption and storage of water molecules. The Nafion/MIL-53(Al) (3 wt.%) gave 9.86×10^{-3} S cm^{-1} at 50°C and 99.9% RH. This decrease is due to lower water absorption capacity. These membranes were also reported to have the highest power density values of 853 mW cm^{-2} at 50°C and 568 mW cm^{-2} at 80°C and 15.0% humidity. When the humidity was raised to 99.9%, the power density values were found to be 818 mW cm^{-2} and 591 mW cm^{-2} at 50°C and 80°C, respectively [95]. The ZIF-8@GO/Nafion showed conduction of 2.8×10^{-2} S cm^{-1} at 120°C and 40% humidity (RH) [78]. A copper MOF HKUST-1 incorporated into Nafion polymeric matrix produced a proton conductance of 1.8×10^{-2} S cm^{-1} at 25°C and 100% RH [91]. The impact of amine and sulfonic functional groups on the hybrid membrane was studied using NH_2- UiO-66+ SO_3H – UiO-66/Nafion-0.6, which gave a conductance of 2.56×10^{-1} S cm^{-1} at 90°C and 95% RH [59]. Porous organic cages are crystalline, soluble materials that are gaining attention in fuel cells. The thermal stability was found to be up to 320°C; however, no research on mechanical stability has been reported. The Nafion–Cage 3 (5 mass fraction) composite membrane gave a conductivity of 2.7×10^{-1} S cm^{-1} at a temperature of 90°C and 95% RH, and the methanol permeability was found to be $5.82 \pm 0.27 \times 10^{-8}$ cm^2 s^{-1} at 40°C. Figure 14.12 shows the fabrication of Nafion–Cage 3 hybrid membrane. Figure 14.13 shows

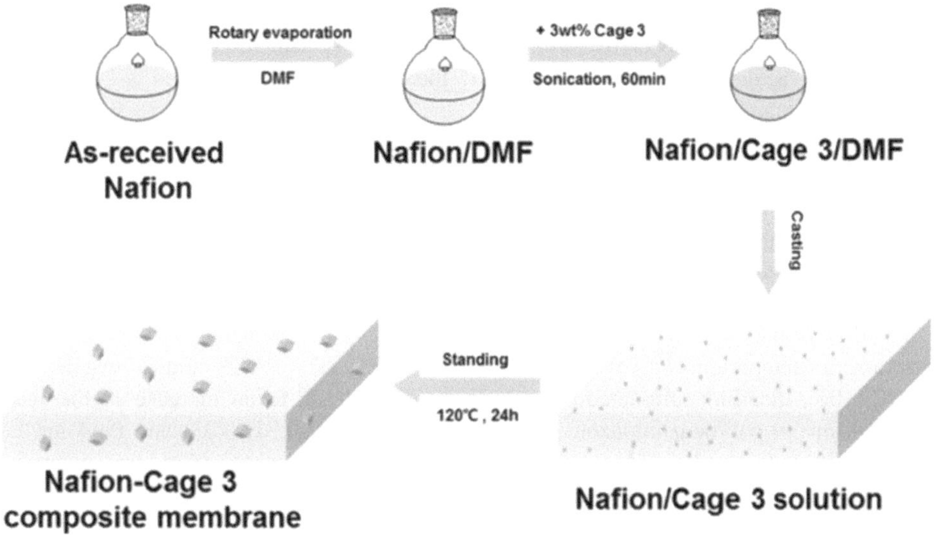

FIGURE 14.12 Fabrication of Nafion–Cage 3 hybrid membrane. Adapted from Ref. [96].

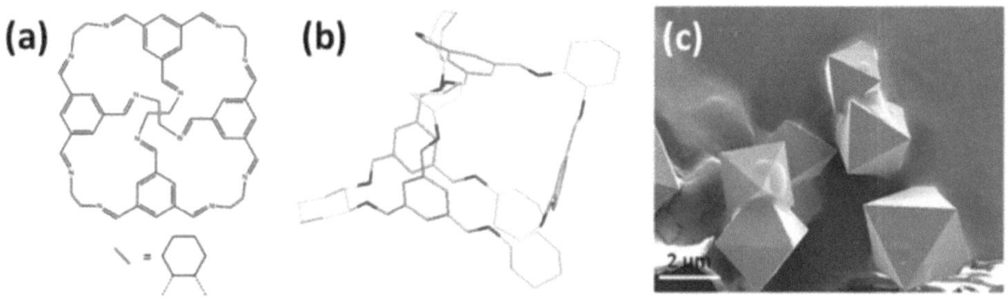

FIGURE 14.13 (a) Structural design of Cage 3, (b) 3D structural representation of Cage 3, and (c) Cage 3 field emission scanning electron microscopy (FESEM) images. Adapted from Ref. [96].

the chemical formula, structure, and field-emission scanning electron microscopy (FESEM) image of cage-3 material [96]. Similarly, Nafion/UiO-66-200 2wt.% membrane with a filler size of 200 nm resulted in a proton conductance of 2.07×10^{-1} S cm^{-1} at 110°C and 90% RH, which is around 30% more than for pristine Nafion membrane [97]. The steadiness and working of Nafion membranes at low humidity have been studied using SO_3H– Zr–MOF–808 (1 wt.%) incorporated Nafion composite membrane. This resulted in a conductivity value of 2.96×10^{-3} S cm^{-1} at a temperature of 80°C and 35% RH, suggesting that water molecules present inside the framework will have a strong connection with the acidic group and can be evaporated only at high temperature. These water molecules serve as an excellent proton carrier and enhance the proton conductivity mechanism [98].

The proton exchange membranes made up of Nafion are used commercially on a larger scale. This class of membranes heads the membrane market in the fuel cell industry. The most necessary characteristics needed for a proton exchange membrane, such as proton conductivity, water-holding capacity, mechanical, chemical, and thermal stability, have increased significantly due to incorporating simple MOFs and tailored MOFs. However, some limitations such as operation at reduced temperature due to the water-dependent transport mechanism decreased the chemical and mechanical strength at elevated temperatures, with lower resistivity toward methanol permeation, and the cost of the process needs to be addressed in the future.

14.3.2 Sulfonated Poly(Arylene Ether)-Based Membranes

Due to the drawbacks mentioned above of Nafion membranes, researchers and scientists have a primary objective of the development of membranes that can overcome all the limitations of Nafion. A polymeric membrane capable of operating at low humid conditions of 25–40% and elevated temperatures of 120–160°C with low production/operation cost and proton conductivity equaling the Nafion membrane is the primary objective of this research. Polymeric substances such as poly(arylene ether ether ketone) (PEEK), poly(arylene ether sulfone), and their derivatives are excellent alternatives to Nafion membranes due to their superb oxidative and hydrolytic stability in extreme circumstances [99,100]. Some of the most probable polymers that can be designed are shown in Figure 14.14 [100]. These polymers have magnificent thermal, chemical, and mechanical stability, and proton conductivity at less humidity and elevated temperatures, especially sulfonated polymers. Figure 14.15 shows the design of various polymers used for proton exchange membrane cells [101].

Poly(ether ether ketone) (PEEK) is a crystalline polymeric material made up of aromatic rings linked together through oxygen (-O-) atoms. These are shown to have an insufficient capacity to get dissolved in organic solvents due to their crystalline nature. Sulfonation to the aromatic foundation decreases the crystalline nature and increases solubility. The sulfonated poly(ether ether ketone) (PEEK) polymers were found to have a glass transition temperature (T_g) of 223.9°C and melting point of 263°C [102]. Blending sulfonated polyether ketones with polyethersulfone or polybenzimidazole gave an excellent proton conductivity with good membrane swelling and reduced solvent transport. Hence, they also serve as an ideal material in DMFCs due to the lower water and methanol cross over [103]. Methanol cross over is the transport of methanol from the anode to the cathode in fuel cells. This cross over is caused by the inherent permeability of the proton exchange membrane and miscibility of methanol and water.

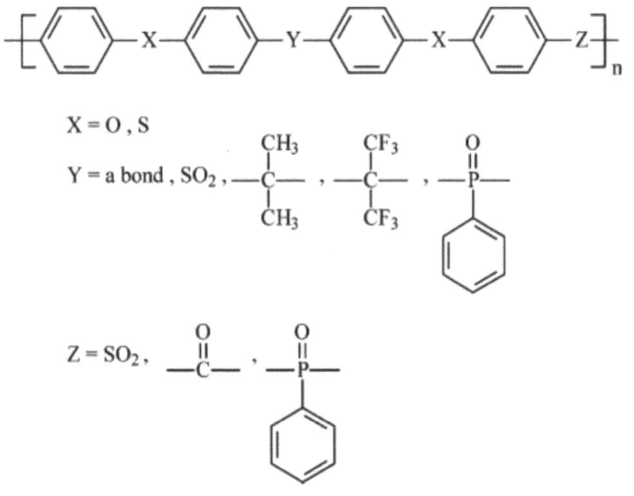

FIGURE 14.14 Probable poly(arylene ether)-based polymer structures. Adapted from Ref. [100].

Methanol at the cathode undergoes a combustion reaction with oxygen which reduces the cathode potential and lowers the cell performance [104].

The SPEEK membranes with ZIF-8/CNT give good conductivity (5.03×10^{-2} S cm^{-1} at 120°C and 30% humidity, which is 11.2-fold higher than for the recast SPEEK membrane (4.5×10^{-3} S cm^{-1}). A different wt.% of filler has been used in the polymer matrix, and 2.5 wt.% resulted in the highest conductivity because of the connection between sulfonic acid functional groups and filler material. Due to compressed channels, the methanol permeability was also found to be much less. The SAXS studies revealed that the incorporation of ZCN leads to an increase of the scattering vector from 0.28 nm^{-1} to 0.31, 0.33, and 0.34 nm^{-1} respectively, when 1, 2.5, and 5 wt.% ZCN was incorporated. The average ionic cluster dimension will decrease with an increase in the scattering vector value as they are inversely proportional [79]. A MIL-101 (Cr) on SPEEK resulted in a conductance value of 3.06×10^{-1} S cm^{-1} at 75°C and RH value of 100% [65]. Introduction of acidic phosphotungstic acid into MIL-101 (Cr) and its fabrication with SPEEK polymer gave a value of 2.72×10^{-3} S cm^{-1} at 65°C and 100% RH, which is 45.5% more than for pristine SPEEK membrane [69]. The impact of sulfone groups on conductivity has been studied using MIL-101-NH_2-SO_3H (3 wt.%) on SPEEK with a proton conductance of 1.98×10^{-1} S cm^{-1} at a temperature of 80°C and 100% RH. The methanol penetrability was less for hybrid membranes with pure MIL-101 filler, but the values were found to be much higher for membranes impregnated with sulfonated MIL-101. The excellent thermal and mechanical stability are due to filler acting as an ionic physical cross-linking agent [71]. Graphene oxide impregnated S-UiO-66 SPEEK membrane was an excellent conductor of protons with a value of 2.68×10^{-1} S cm^{-1} at 70°C and 95% RH, which is 2.6-fold more than for recast SPEEK (1.05×10^{-1} S cm^{-1}). These excellent results are due to proper filler dispersion in the polymer surface, increased number of proton carriers and enhanced water-holding capacity, and generation of proton-passing channels due to the interaction between polymer and filler [105]. SPEEK/phosphotungstic acid@MIL-101 (9

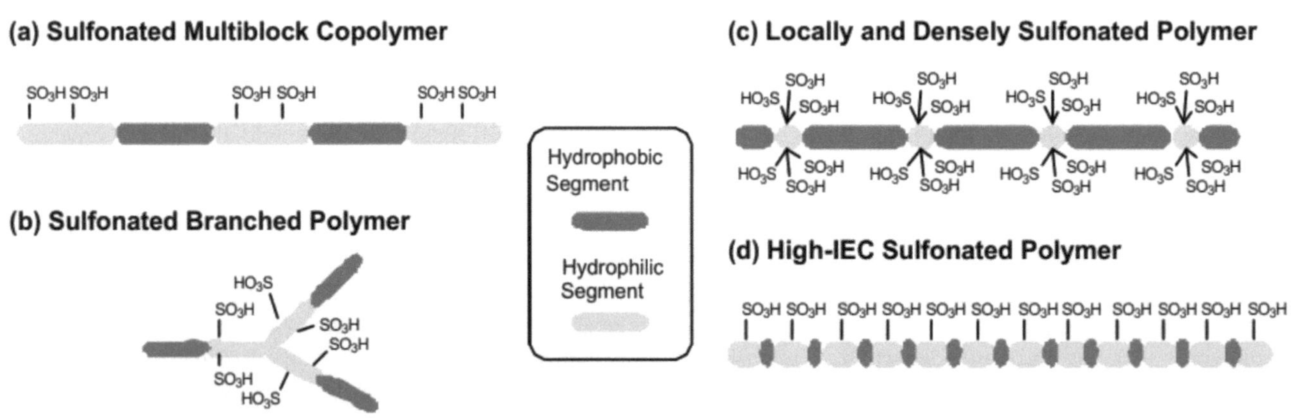

FIGURE 14.15 Design of various sulfonated polymers used for proton exchange membrane cells. Adapted from Ref. [101].

wt.%) hybrid membranes attained 2.72 × 10^{-1} S cm^{-1} at 65°C temperature and 100% RH [68].

Compared with Nafion membranes, poly(arylene ether)-based membranes exhibit some advantages in steadiness and proton conduction. The impregnation of functionalized MOF fillers has elevated the conductivity to the Nafion membrane level. The operation temperature is up to 250°C but limited to 120°C due to a drop-down in conductivity. Hence, the research focus should be on improving the longevity, performance, and stability for commercial applications.

14.3.3 Polybenzimidazole-Related Membranes

Although Nafion and SPEEK are the dominant polymeric materials in manufacturing proton exchange membrane cells for fuel cells, polybenzimidazole (PBI) has become an excellent material for high-temperature production PEMFCs. The limitation of Nafion membranes at elevated temperatures due to water dependency leads to a decline in fuel efficiency. Polybenzimidazole (PBI) has a glass transition temperature of 420°C. Proton conductivity can be increased by incorporating hydrophilic and acidic groups such as phosphoric acid, transforming the membrane to work under moisture-less conditions. These doped membranes exhibit excellent proton conductivity and high resistance to gas crossover, thermal and oxidative stability. From the cathode the nitrogen and oxygen pass through the membrane to the anode and from the anode to the cathode the hydrogen crossover takes place. The phosphoric acid doping quantity and temperature are inversely proportional to the mechanical strength and directly proportional to the molar mass of polybenzimidazole [106]. The phosphoric acid doping limitations such as acid leaching, degradation of the membrane, and reduced durability can be overcome with the help of fillers like metal oxides, graphene oxide, heteropolyacids, carbon nanotubes, etc. Most recently, imidazolate MOFs have emerged as promising materials as fillers during membrane fabrication [93]. Figure 14.16 shows the chemical structure of polybenzimidazole (PBI) [107].

In a minimal work, zinc metal ion-based porous material ZIF-8 and cobalt ion-based porous material ZIF-67 and a mixture of both have been used as fillers during the fabrication of polybenzimidazole membrane. The performance test was conducted at 200°C, and the optimum wt.% of filler is required to get the best result. The proton conductivity value for ZIF-8/PBI (5 wt.%) is 3.1 × 10^{-3} S cm^{-1}, ZIF-67/PBI reached 4.1 × 10^{-2} S cm^{-1}, and ZIF/PBI mix resulted in 9.2 × 10^{-2} S cm^{-1} [108]. Recently, the impact of acidic functional groups and cross-linked branched polybenzimidazole composite membranes on conductivity has been reported. The prepared UiO-66/cross-linked branched PBI hybrid membranes have enhanced proton conductivity with a low phosphoric acid content. The UiO-66 MOF has been incorporated into branched poly(2,2'-(p-oxydiphenylene)-5,5'-benzimidazole) (BOPBI) with different mass ratios. The hybrid membrane CBOPBI@MOF40% attained a proton conductance value of 1.0 × 10^{-1} S cm^{-1} at 160°C, which is 2.44-fold higher than for CBOPBI membranes (4.1 × 10^{-2} S cm^{-1}). As the filler percentile exceeds 30 wt.%, uninterrupted proton transportation channel formation occurs. These membranes display excellent thermal stability, and thermogravimetric analysis reveals that less than 1% weight loss has been observed up to 200°C. The incorporation of fillers into the membranes lowers the interaction among the polymer matrix. However, as The MOF content increases, mechanical stability increases due to suppression of the PBI backbone separation, which decreases the plasticizing outcome of phosphoric acid. Also, when the filler percentage is increased to 40 wt.%, the mechanical strength decreases due to the impact of MOF on the aforementioned polymer matrix [109]. Since polybenzimidazole membranes require a high percentage of phosphoric acid doping to exhibit higher proton conductivity, it leads to a decline in the membranes' mechanical properties and phosphoric acid discharge from the membranes. To overcome this drawback, UiO-66 impregnated poly 2,2'-(p-oxydiphenylene)-5,5'-benzimidazole) (OPBI) hybrid membranes have been used in fuel cells. OPBI/UiO-66 (40 wt.%) has attained a proton conductance value of 9.2 × 10^{-2} S cm^{-1} at 160°C with a low phosphoric acid uptake of 73%. The pristine OPBI membrane displayed 5.0 × 10^{-2} S cm^{-1} with an excessive phosphoric acid intake of 217.43%. The membranes had exceptional thermal stability at 250°C. The pristine OPBI membrane showed excellent tensile strength of 106.13 MPa and elongation of 63% under undoped conditions. However, the OPBI/UIO-66-40% membrane had the lowest tensile strength (79 MPa) and elongation (3.95%) without phosphoric acid doping. Phosphoric acid doping also decreased the tensile strength from 106.13 MPa to 21.37 MPa and increased the elongation from 63% to 258% for 0%UIO-66@OPBI membrane. However, for 40%UIO-66@OPBI membrane, phosphoric acid doping has improved tensile strength from 21.37 MPa to 27.02 MPa, and elongation has decreased from 257% to 239%. However, all these values are much higher than the mechanical requirements of fuel cell tests [110].

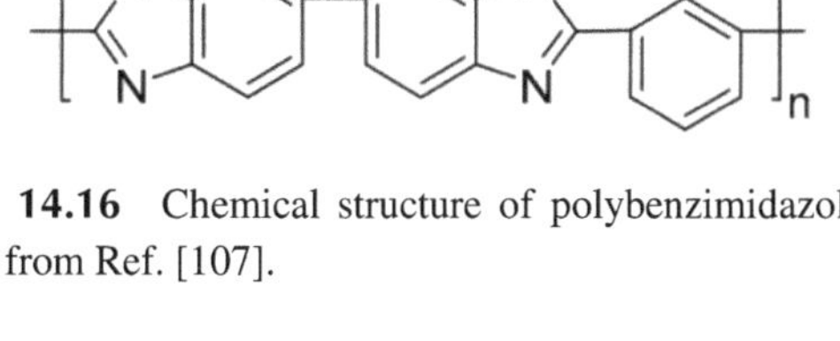

FIGURE 14.16 Chemical structure of polybenzimidazole (PBI). Adapted from Ref. [107].

Very little research has been done on the use of MOF-incorporated polybenzimidazole as a membrane material. Due to its excellent thermal stabilities and mechanical strengths, it is a promising material in high-temperature proton exchange membrane cells. The phosphoric acid doping and introduction of porous framework materials have increased the conductivity competition with Nafion and SPEEK membranes. The high openings present in the MOF make the membrane permeable to gas and methanol and reduce the mechanical properties, limiting the membrane usage in commercial applications; further research is focused on overcoming these limitations.

14.3.4 Vinyl-Based Membranes

These classes of polymeric materials have gained a lot of attention from researchers who work in the field of fuel cells. Poly(vinyl alcohol) (PVA) has been considered one of the promising alternatives to commercially available Nafion and SPEEK membranes. PVA is a decomposable membrane material, shows excellent solubility in water, and has a lower methyl alcohol permeability value. It has fewer functional groups attached to it, and the presence of hydroxyl groups exerts a restriction on the methanol permeability due to close molecular packing. However, due to the absence of proton carriers, these membranes are very poor proton conductors compared to Nafion membranes. Hence, incorporation/modification to the PVA membrane makes it suitable for use in fuel cells [111]. Figure 14.17 displays the structures of PVA and polyvinylpyrrolidone (PVP) [112].

A chiral 2-D metal organic framework dimensional MOF, $\{Ca(D\text{-}Hpmpc)(H_2O)_2.2\ HO_{0.5}\}_n$ (1, $_D$-H_3pmpc = $_D$-1-(phosphonomethyl) piperidine-3-carboxylic acid), has been incorporated into polyvinylpyrrolidone (PVP) matrix and the hybrid membranes PVP/chiral MOF (50 wt.%) produced a conduction of 3.2×10^{-4} S cm^{-1} at 25°C and 97% RH [113]. Water- and acid-stable MOF-induced hybrid membranes for fuel cells have been studied. A hexaphosphate ester-based MOF and its hybrid membrane with PVA have been studied. The phytic acid is used as an organic linker along with zinc metal atoms during the construction of MOF $\{Zn_{10}(C_6H_8P_6O_{24})_2(\text{-}\ H_2O)_{14}\cdot x(H_2O)\}$, abbreviated as JUC-200. The PVA/JUC-200 (10 wt.%) achieved a proton conductance of 1.25×10^{-3} S cm^{-1} at 50°C and 97% RH [92]. Another polymer belongs to the class of vinyl-based polymers poly(vinyl phosphonic acid) (PVPA) incorporated with ZIF-8 metal-organic framework giving a hybrid membrane with conductivity 3.2×10^{-3} S cm^{-1} at 140°C and dry condition [82]. A fluorinated polymeric material poly(vinylidene fluoride) (PVDF) has also been tested as a proton exchange membrane cell by incorporating MOF fillers. The PVDF material has good thermal, chemical, and mechanical stability, solubility in some polar solvents, and low production cost. A zirconium metal organic framework, MOF-808, has been incorporated into PVDF matrix, and the fabricated PVDF/MOF-808 (55 wt.%) attained a proton conductance of 1.56×10^{-4} S cm^{-1} at 65°C and 99% RH. The pristine PVDF membrane showed a rate of 7.38×10^{-6} S cm^{-1} at 27°C, which is less than for PVDF/MOF-808 (10 wt.%) of 3.69×10^{-5} S cm^{-1}, indicating an enhancement in proton conduction [114]. Another composite membrane fabricated using zirconium-induced metal organic framework MOF-801 (60 wt.%) and PVDF have resulted in conductance values of 1.84×10^{-3} S cm^{-1} at a temperature of 52°C and 98% RH, indicating a prominent application as commercial membrane [115].

PVA PVP

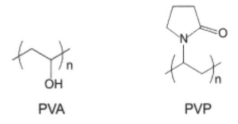

FIGURE 14.17 Structures of PVA and PVP. Adapted from Ref. [112].

Vinyl-based membranes are developing as potential PEMs for fuel cell applications. However, irrespective of tremendous work and modifications to the membranes, they have not reached the height of Nafion membranes and hence lack practical applications. Many improvements have to be brought in terms of their thermal, chemical, and mechanical properties and durability.

14.3.5 Other Polymeric Membranes

Apart from the polymers mentioned above, which are extensively used in fuel cells, some other polymers have also been used in preparing the composite membrane through the integration of MOFs. The fabrication of a novel hybrid membrane for proton exchange membrane cells has been done using sulfonated poly(2,6-dimethyl-1,4-phenylene oxide) (SPPO) as polymer and Fe-MIL-101-NH_2 MOF. These membranes attained a proton conduction value of 1.0×10^{-1} S cm^{-1} at ambient temperature and 2.5×10^{-1} S cm^{-1} at 90°C with an RH value of 98% and 6 wt.% filler materials. The iron atom enhances the acidity and helps the Grotthuss mechanism [70]. Chitosan is the second most abundant natural polymer with outstanding biodegradability, biocompatibility, and nontoxicity characteristics. Generally, heteropoly acids are doped to the chitosan membrane to enhance conductivity, strength, and stability [116]. The proton exchange membrane designed using nanohybrid chitosan and three different materials has been found to give improved conductivity, mechanical properties, and good stability compared to pristine chitosan membranes. One-dimensional (1D) sodium lignin sulfonate-induced carbon nanotubes (SCNTs), two-dimensional (2D) graphene oxide (GO), and three-dimensional (3D) zirconium-centered MOFs (UiO-66) are combined as nanomaterials and impregnated into the nanohybrid chitosan polymer matrix to get a mixed-matrix membrane. The composite membrane CS/U-S@GO-7 attained a conductivity rate of 6.4×10^{-2} S cm^{-1} at 70°C. However, the CS/U-S@GO-7 showed a conductance of 3.61×10^{-2} S cm^{-1} at 20°C, which is less when compared to pure chitosan membrane (1.6×10^{-2} S cm^{-1}). The methanol permeability of UiO-66-SCNT/Go was less than that of UiO-66-SCNT membranes due to the fillers acting as suitable methanol barriers. The methanol crossover of the U-S/CS/GO-7 membrane was 1.86×10^{-7} cm^2 s^{-1}, but for pristine chitosan membrane, it was 9.96×10^{-7} cm^2 s^{-1}, which could be due to poor compatibility [117].

Recently, a zinc and 2-amino benzene dicarboxylic acid assembled MOF, abbreviated as IRMOF-3, was introduced into quaternized chitosan (QCS) by the chemical cross-linking method. These novel hybrid membranes provided good hydroxide conductivity and exhibited good mechanical

TABLE 14.5
The proton conductivity values of some general polymeric membranes incorporated with various metal organic frameworks

Membrane	σ (S cm^{-1})	T (°C)	RH (%)	References
SPPESK/ ZCCH	8.2×10^{-2}	160	Anhydrous	[90]
PA-HKUST-1/Nafion®	1.8×10^{-2}	25	100	[91]
JUC-200/PVA (10 wt.%)	1.25×10^{-3}	50	97	[92]
Chiral MOF/PVP	3.2×10^{-4}	25	97	[113]
PVDF/MOF-808 (55 wt.%)	1.56×10^{-4}	65	99	[114]
PVDF/MOF-808 (10 wt.%)	3.69×10^{-5}	27	99	[114]
PVDF	7.38×10^{-6}	27	99	[114]
PVDF/MOF-801	1.84×10^{-3}	52	98	[115]
QCS/ IRMOF-3	2.99×10^{-2}	80	na	[118]

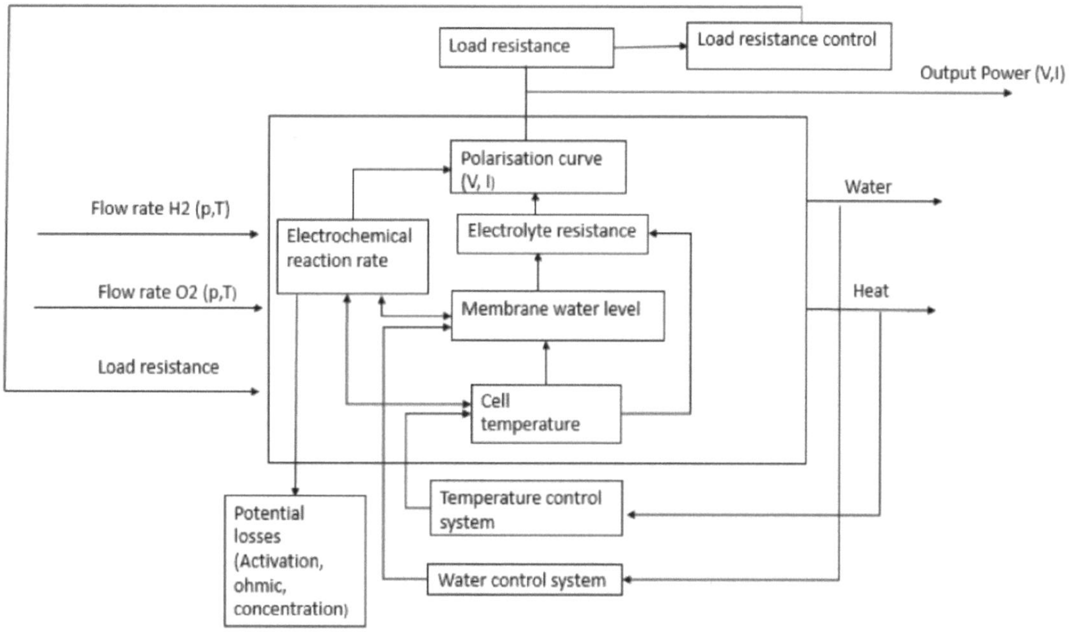

FIGURE 14.18 The different parameters and working nature of a fuel cell. Adapted from Ref. [121].

properties which were ideal for them to be used in fuel cells. The hydroxide conductivity was 2.99×10^{-2} S cm^{-1} at 80°C, a tensile strength of 22.4 MPa, and water uptake of 78.4%. The alkaline resistance test revealed that they were the most suitable as anion-conducting membranes in the alkaline polymer electrolyte fuel cells [118]. Interestingly chitosan composite membranes that can work with higher temperatures under anhydrous conditions have been designed using MOFs. An acidic functionalized MOF material, UiO-66(SO_3H), and basic functionalized MOF material, NH_2-UiO-66, have been used as fillers during the fabrication of novel chitosan composite membrane. The hybrid membrane CS/UiO-66(SO_3H) (6 wt.%) gave a proton conductance value of 1.52×10^{-3} S cm^{-1} at 90°C and 0% RH, while chitosan/SO_3H-UiO-66-(6 wt.%) + NH2-UiO-66-(15 wt.%) resulted in a proton conduction rate of 3.78×10^{-3} S cm^{-1} at 120°C and 0% RH and the CS/UiO-66(NH_2)–(15 wt.%) attained a proton conduction rate of 5.64×10^{-2} S cm^{-1} at 100°C and 98% RH. The improved conductivity could be due to amine and hydroxyl groups, which create a proton-conducting channel, and the sulfonic acid group acts as a proton carrier [119]. In designing anionic exchange membranes, chromium MOF MIL-101 (Cr) has been introduced into quaternized chitosan (QCS). The hybrid MIL-9-QCS membrane with 9 wt.% of filler attained a hydroxide conductance of 2.30×10^{-2} S cm^{-1} at a temperature of 80°C and tensile strength of 20.12 MPa, which is more than for Nafion-115 membrane (18.7 MPa). The methanol permeability was 24% less compared to the pristine quaternized chitosan membrane [120].

The polyetherimide (PEI) composite membranes were also found to be useful in fuel cells. A ZIF MOF-induced hybrid membrane has been fabricated using PEI and ZIF-8/ZIF-67. The PEI/ZIF-8 attained a rate of 1.5×10^{-6} S cm^{-1}, PEI/ZIF-67 gave 3.5×10^{-6} S cm^{-1}, and PEI/ZIF-8+ZIF-67 resulted in 2.9×10^{-6} S cm^{-1} [76]. The proton conduction in DNA/ZIF-8 has also been studied with a conductivity of 3.4×10^{-4} S cm^{-1} at

25°C and 97% RH. Methanol permittivity was found to be 1.25 × 10^{-8} cm^2 s^{-1} [84]. Table 14.5 gives the proton conductivity values of some general polymeric membranes incorporated with various metal organic frameworks. The overall working scheme and the influence of various parameters of fuel cells have been illustrated in Figure 14.18 [121]

14.4 CONCLUSION

This chapter summarizes the potential applications of MOFs as doping agents in fabricating polymeric mixed-matrix membranes. Due to the greater surface area and capacity to accommodate a high percentage of water molecules and other functional groups necessary to improve the proton conduction mechanism, MOFs are an excellent promising material for use in fuel cells. Although several types of organic and inorganic materials capable of enhancing cell performance have been utilized as fillers during the fabrication of hybrid membranes, MOFs were an excellent filler material to fabricate high-temperature proton exchange membrane cells (HT-PEMCs). The operational temperature for this class of membranes is above 100°C. This capacity of MOFs to work at high temperatures and anhydrous conditions makes them potential candidates in commercial applications.

The use of pristine MOF material crystalline membranes as a commercial membrane in fuel cells is greatly limited due to the excessive synthesis and operational costs and low-performance factors. Hence, using these filler materials to fabricate composite polymeric membranes through incorporation into a polymeric matrix has been an found to be easy and affordable method. Studies have reported that MOFs can be tuned according to requirements through various techniques such as post-synthetic modification, functionalization through hydrophilic and proton carrier groups, and co-doping with other porous materials so that the proton conductivity can be improved. The best performing membranes are listed in Table 14.6

Various classes of MOFs have been impregnated into the polymer matrix. This incorporation was found to improve the conductivity to a more considerable extent than with the pristine membrane. The UiO series MOFs have a high-density spatial structure, due to which they exhibit good chemical and mechanical stability. Incorporating these classes of MOFs will positively impact cell performance. Similarly, the MIL series MOFs have open metal sites capable of holding the hydroxyl groups, enhancing the proton conduction mechanism. These MOFs exhibit excellent chemical, mechanical, and thermal stability and have a more significant influence in reducing methanol permeability. Their advantages, disadvantages, and main uses are listed in Table 14.7.

The introduction of MOFs and other fillers into the polymeric matrix significantly improves membrane performance. These developments are due to the capacity of fillers to generate ion transportation channels inducing changes in the ionomeric

TABLE 14.6
Some important membranes showing good conductance and their properties

Membrane	σ(S cm^{-1})	T (°C)	RH (%)	References
OPBI/UiO-66	9.2 × 10^{-2}	160	Anhydrous	[110]
Chitosan/SO_3H-UiO-66-6 wt.% NH_2-UiO-66-15	3.78 × 10^{-3}	120	0	[119]
Chitosan/H_2SO_4/MIL-101 (8 wt.%)	9.5 × 10^{-2}	25–100	100	[64]
Chitosan/H_3PO_4/MIL-101 (6 wt.%)	8.3 × 10^{-2}	25–100	100	[64]
Chitosan/CF_3SO_3H@MIL-101-10	9.4 × 10^{-2}	25–100	100	[64]
Nafion/phytic@MIL-12 (Cr)	7.63 × 10^{-4}	100	10.5	[67, 94]
ZIF mix/PBI	9.2 × 10^{-2}	200	0	[108]
SPEEK/ZIF mix	8.5 × 10^{-3}	120	na	[77]
SPEEK	7.46 × 10^{-2}	70	90	[89]
SPPESK/ ZCCH	8.2 × 10^{-2}	160	Anhydrous	[90]
PVDF	7.38 × 10^{-6}	27	99	[114]
OPBI/UiO-66	9.2 × 10^{-2}	160	Anhydrous	[110]
Chitosan/SO_3H-UiO-66-6 wt.% NH_2-UiO-66-15	3.78 × 10^{-3}	120	0	[119]
Chitosan /H_2SO_4/MIL-101 (8 wt.%)	9.5 × 10^{-2}	25–100	100	[64]
Chitosan /H_3PO_4/MIL-101 (6 wt.)	8.3 × 10^{-2}	25–100	100	[64]
Chitosan/CF_3SO_3H@MIL-101-10	9.4 × 10^{-2}	25–100	100	[64]
Nafion/phytic@MIL-12 (Cr)	7.63 × 10^{-4}	100	10.5	[67,94]
ZIF mix/PBI	9.2 × 10^{-2}	200	0	[108]
SPEEK/ZIF mix	8.5 × 10^{-3}	120	na	[77]
SPEEK	7.46 × 10^{-2}	70	90	[89]
SPPESK/ ZCCH	8.2 × 10^{-2}	160	Anhydrous	[90]
PVDF	7.38 × 10^{-6}	27	99	[114]

TABLE 14.7
Advantages and disadvantages of PEMFC and DMFC

	Proton exchange membrane fuel cell	Direct methanol fuel cell
Advantages	1) Power density will be more compared to other types of fuel cells	1) High energy density
	2) Simple scaling is observed	2) The methanol fuel is cheaper
	3) The power operation range is wide	3) Methanol storage will be much easier
Disadvantages	1) High pure fuel is required	1) Expensive catalyst like Pt, Ru required
	2) Generation of CO	2) Performance of the cell depends on the fuel concentration
	3) More water in cel.	3) Possibility of cathodic electrode poisoning

chains. A comparative study between hybrid proton exchange membranes comprising MOFs and other fillers revealed that the MOF-induced membranes show better performance in membrane swelling, higher humidity, enhanced mechanical, chemical properties, and higher thermal stability. However, there are some issues to be addressed, such as the selection of relevant MOFs and polymer matrix to obtain the composite membrane with the highest performance, guarantee the proper dispersion of MOFs on the polymer matrix, and optimization of the wt.% of filler to get the highest performance from the membrane. To elude the accumulation of fillers in the membrane matrix, surface modifications can be done with MOFs. Along with thermal and chemical stability issues, the mechanical properties are also to be addressed. Since most proton exchange membranes are functionalized with hydrophilic acid groups, the design of fillers should withstand imminent hydrolysis under extreme conditions.

Apart from the many advantages of MOFs as fillers in polymeric membranes, there are some important issues to be addressed. The lack of processability, reduced mechanical strength, and weaker proton conductive nature of MOFs make them unable to be used directly as proton exchange membranes, instead the selection of a suitable polymer that can overcome these lacunae and incorporation of the MOF into it are necessary. The selection of proper-sized MOFs is essential to reduce the grain boundary resistance of PEMs. A larger size of MOF will reduce the grain boundary resistance of PEMs and a smaller-sized MOF will increase the grain boundary resistance of PEMs, thus decreasing the proton conductivity. Work has to be done on the uniformity of the membranes via uniform dispersion of MOFs. To achieve this, selecting the proper solvent for the dissolution of the polymer is necessary.

It is also imperative to pay attention to the best ratio between MOFs and polymer to get the best conductivity value as there may be differences in the correlation between the two moieties. Along with achieving high proton conductivity, scientists should also target achieving good mechanical strength for the membranes. The commercial application of polymeric membranes for fuel cells requires both performance and durability. Considering these challenges, there is a need to design next-generation MOF-incorporated membranes for fuel cells. The bulk synthesis of MOFs and stability study of the designed MOFs need to be carried out at the commercial level. The work design should be to reach the goal of industrialization of the MOF-incorporated polymeric membranes for fuel cells. With the expansion of knowledge and innovations in technology, MOF-based hybrid membranes will play a vital role in the future of fuel cell science.

Finally, designing proton exchange membranes to operate at high temperatures is the main target facing researchers. As per the review literature survey on polybenzimidazole (PBI), membranes can operate at temperatures ranging between 100–200°C. However, the proton conductivity and other necessary properties are greatly reduced when compared to commercial Nafion membranes. An MOF with a higher surface area will help in improving proton conductivity by accumulating more proton carriers. Similarly, MOFs with smaller pore sizes will hinder fuel–oxidant diffusion, resulting in enhanced selectivity. The fuel cell has been commercially used in automobile industries. They could be the best alternative to batteries that have to be recharged regularly and disposed of after degradation. However, the internal and operational parameters pose a significant challenge in their commercialization. Hence, it is a great challenge to improve these polymeric membranes to obtain the highest qualities for commercialization.

REFERENCES

1. Zaidi, S.M.J., Rauf, M.A. (2009) Fuel cell fundamentals. In: *Polymer Membranes for Fuel Cells*, S.M.J. Zaidi, T. Matsuura (Eds.), pp. 1–6. Springer US: Boston, MA6.
2. Haines, A., et al. (2007) Policies for accelerating access to clean energy, improving health, advancing development, and mitigating climate change. *The Lancet* 370(9594): 1264–1281.
3. Wen, X., Zhang, Q., Guan, J. (2020) Applications of metal–organic framework-derived materials in fuel cells and metal-air batteries. *Coordination Chemistry Reviews* **409:** 213214.
4. Rayment, C., Sherwin, S.J.D.o.A., U.o.N.D. (2003) Department of Aerospace and Mechanical Engineering, University of Notre Dame, Notre Dame, IN, *Introduction to Fuel Cell Technology.* 46556: 11–12.

5. Carrette, L., Friedrich, K.A., Stimming, U. (2000) *Fuel Cells: Principles, Types, Fuels, and Applications* Chemphyschem : a European journal of chemical physics and physical chemistry 1(4): 162–193.
6. Sin, Y.T., Najmi W.A. (2013) Industrial and academic collaboration strategies on hydrogen fuel cell technology development in Malaysia. *Procedia – Social and Behavioral Sciences* 90: 879–888.
7. Pan, M., et al. (2021) A review of membranes in proton exchange membrane fuel cells: Transport phenomena, performance and durability. *Renewable and Sustainable Energy Reviews* 141: 110771.
8. Shi, S., Weber, A.Z., Kusoglu, A. (2016) Structure-transport relationship of perfluorosulfonic-acid membranes in different cationic forms. *Electrochimica Acta* 220: 517–528.
9. Ferrara, A., Polverino, P., Pianese, C. (2018) Analytical calculation of electrolyte water content of a proton exchange membrane fuel cell for on-board modelling applications. *Journal of Power Sources* 390: 197–207.
10. Sun, Z., et al. (2011) Lowering grain boundary resistance of BaZr0.8Y0.2O3−δ with LiNO3 sintering-aid improves proton conductivity for fuel cell operation. Physical Chemistry Chemical Physics 13(17): 7692–7700.
11. Wei, M.-J., et al. (2019) Supramolecular hydrogen-bonded organic networks through acid–base pairs as efficient proton-conducting electrolytes. *CrystEngComm* 21(33): 4996–5001.
12. Bunzen, H., et al. (2018) Anisotropic water-mediated proton conductivity in large iron (II) metal–organic framework single crystals for proton-exchange membrane fuel cells. *ACS Applied Nano Materials* 2(1): 291–298.
13. Liu, Q., et al. (2020) Metal organic frameworks modified proton exchange membranes for fuel cells. *Frontiers in Chemistry* 8: 694.
14. Ghielmi, A., et al. (2005) Proton exchange membranes based on the short-side-chain perfluorinated ionomer. *Journal of Power Sources* 145(2): 108–115.
15. Zaidi, S.M.J. (2009) Research trends in polymer electrolyte membranes for PEMFC. In: *Polymer Membranes for Fuel Cells*, S.M.J. Zaidi, T. Matsuura (Eds.), pp. 7–25. Springer US: Boston, MA.
16. Lee, J.S., et al. (2006) Polymer electrolyte membranes for fuel cells. *Journal of Industrial and Engineering Chemistry* 12(2): 175–183.
17. Kim, D.J., Jo, M.J., Nam, S.Y. (2015) A review of polymer–nanocomposite electrolyte membranes for fuel cell application. *Journal of Industrial and Engineering Chemistry* 21: 36–52.
18. Ismail, A.F., Naim, R., Zubir, N.A. (2009) Fuel cell technology review. In: *Polymer Membranes for Fuel Cells*, S.M.J. Zaidi, T. Matsuura (Eds.), pp. 27–49. Springer US: Boston, MA.
19. Bébin, P., Caravanier, M., Galiano, H. (2006) Nafion®/clay-SO3H membrane for proton exchange membrane fuel cell application. *Journal of Membrane Science* 278(1): 35–42.
20. Gil, M., et al. (2004) Direct synthesis of sulfonated aromatic poly(ether ether ketone) proton exchange membranes for fuel cell applications. *Journal of Membrane Science* 234(1): 75–81.
21. Wang, L., et al. (2008) Characteristics of polyethersulfone/sulfonated polyimide blend membrane for proton exchange membrane fuel cell. *The Journal of Physical Chemistry B* 112(14): 4270–4275.
22. Xiao, L., et al. (2005) Synthesis and characterization of pyridine-based polybenzimidazoles for high temperature polymer electrolyte membrane fuel cell applications. *Fuel Cells* 5(2): 287–295.
23. Kim, Y.S., Pivovar, B.S. (2007) Chapter Four – Polymer Electrolyte Membranes for Direct Methanol Fuel Cells. In: *Advances in Fuel Cells*, T.S. Zhao, K.D. Kreuer, T. Van Nguyen (Eds.), pp. 187–234. Elsevier Science.
24. Sun, X., et al. (2019) Composite membranes for high temperature PEM fuel cells and electrolysers: A critical review. *Membranes* 9(7): 83.
25. Bakangura, E., et al. (2016) Mixed matrix proton exchange membranes for fuel cells: State of the art and perspectives. *Progress in Polymer Science* 57: 103–152.
26. Li, X., et al. (2019) Construction of high-performance, high-temperature proton exchange membranes through incorporating SiO_2 nanoparticles into novel cross-linked polybenzimidazole networks. *ACS Applied Materials & Interfaces* 11(34): 30735–30746.
27. Di Vona, M.L., et al. (2007) SPEEK-TiO_2 nanocomposite hybrid proton conductive membranes via in situ mixed sol–gel process. *Journal of Membrane Science* 296(1): 156–161.
28. Aricò, A.S., et al. (2003) Influence of the acid–base characteristics of inorganic fillers on the high temperature performance of composite membranes in direct methanol fuel cells. *Solid State Ionics* 161(3): 251–265.
29. Saccà, A., et al. (2008) Phosphotungstic acid supported on a nanopowdered ZrO_2 as a filler in Nafion-based membranes for polymer electrolyte fuel cells. *Fuel Cells* 8 (3–4): 225–235.
30. Jung, D.H., et al. (2003) Preparation and performance of a Nafion®/montmorillonite nanocomposite membrane for direct methanol fuel cell. *Journal of Power Sources* 118(1): 205–211.
31. Felice, C., Ye, S., Qu, D. (2010) Nafion−montmorillonite nanocomposite membrane for the effective reduction of fuel crossover. *Industrial & Engineering Chemistry Research* 49(4): 1514–1519.
32. Muthuraja, P., et al. (2018) Novel perovskite structured calcium titanate-PBI composite membranes for high-temperature PEM fuel cells: Synthesis and characterizations. *International Journal of Hydrogen Energy* 43(9): 4763–4772.
33. Zhou, Y., et al. (2014) Insight into proton transfer in phosphotungstic acid functionalized mesoporous silica-based proton exchange membrane fuel cells. *Journal of the American Chemical Society* 136(13): 4954–4964.
34. Shao, Z.-G., Joghee, P., Hsing, I.M. (2004) Preparation and characterization of hybrid Nafion–silica membrane doped with phosphotungstic acid for high temperature operation of proton exchange membrane fuel cells. *Journal of Membrane Science* 229(1): 43–51.
35. Kannan, R., Kakade, B.A., Pillai, V.K. (2008) Polymer electrolyte fuel cells using Nafion-based composite membranes with functionalized carbon nanotubes. *Angewandte Chemie International Edition* 47(14): 2653–2656.
36. Üregen, N., et al. (2017) Development of polybenzimidazole/graphene oxide composite membranes for high temperature PEM fuel cells. *International Journal of Hydrogen Energy* 42(4): 2636–2647.
37. Küver, A., Vielstich, W. (1998) Investigation of methanol crossover and single electrode performance during

PEMDMFC operation: A study using a solid polymer electrolyte membrane fuel cell system. *Journal of Power Sources* 74(2): 211–218.
38. Viet, N., et al. (2012) Novel Pt and Pd based core-shell catalysts with critical new issues of heat treatment, stability and durability for proton exchange membrane fuel cells and direct methanol fuel cells. In: *Heat Treatment – Conventional and Novel Applications*, 1st Edition pp. 235–268.
39. Safaei, M., et al. (2019) A review on metal-organic frameworks: Synthesis and applications. *TrAC Trends in Analytical Chemistry* 118: 401–425.
40. Dey, C., et al. (2014) Crystalline metal-organic frameworks (MOFs): synthesis, structure and function. *Acta Crystallographica Section B* 70(1): 3–10.
41. Cheong, V.F., Moh, P.Y. (2018) Recent advancement in metal–organic framework: synthesis, activation, functionalisation, and bulk production. *Materials Science and Technology* 34(9): 1025–1045.
42. Ye, Y., et al. (2015) High anhydrous proton conductivity of imidazole-loaded mesoporous polyimides over a wide range from subzero to moderate temperature. *Journal of the American Chemical Society* 137(2): 913–918.
43. Ye, Y., et al. (2020) Metal–organic frameworks as a versatile platform for proton conductors. 32(21): 1907090.
44. Agmon, N. (1995) The Grotthuss mechanism. *Chemical Physics Letters* 244(5): 456–462.
45. Kreuer, K.-D., Rabenau, A., Weppner, W. (1982) Vehicle mechanism, a new model for the interpretation of the conductivity of fast proton conductors. *Angewandte Chemie International Edition in English* 21(3): 208–209.
46. Hurd, J.A., et al. (2009) Anhydrous proton conduction at 150 °C in a crystalline metal–organic framework. *Nature Chemistry* 1(9): 705–710.
47. Nagao, Y., et al. (2003) A new proton-conductive copper coordination polymer, (HOC3H6) 2dtoaCu (dtoa= dithiooxamide). *Synthetic Metals* 135: 283–284.
48. Nagao, Y., et al. (2003) A new proton-conductive copper coordination polymer, (HOC 3 H 6) 2 dtoaCu (dtoa = dithiooxamide). *Synthetic Metals* 135-136: 283–284.
49. Nagao, Y., et al. 92005) Preparation and proton transport property of N,N′- diethyldithiooxamidatocopper coordination polymer. *Synthetic Metals* 154(1): 89–92.
50. Nagao, Y., et al. (2003) Highly proton-conductive copper coordination polymers. *Synthetic Metals* 133–134: 431–432.
51. Sadakiyo, M., Yamada, T., Kitagawa, H. (2009) Rational Designs for Highly Proton-Conductive Metal−Organic Frameworks. *Journal of the American Chemical Society* 131(29): 9906–9907.
52. Rao, Z., et al. (2017) Construction of well interconnected metal-organic framework structure for effectively promoting proton conductivity of proton exchange membrane. *Journal of Membrane Science* 533: 160–170.
53. Liu, Q., et al. (2020) Metal organic frameworks modified proton exchange membranes for fuel cells. *Frontiers in Chemistry* 8: 694.
54. Feng, L., Hou, H.-B., Zhou, H. (2020) UiO-66 derivatives and their composite membranes for effective proton conduction. *Dalton Transactions* 49(47): 17130–17139.
55. Mukhopadhyay, S., et al. (2019) Designing UiO-66-based superprotonic conductor with the highest metal–organic framework based proton conductivity. *ACS Applied Materials & Interfaces* 11(14): 13423–13432.
56. Yang, F., et al. (2015) Proton conductivities in functionalized UiO-66: Tuned properties, thermogravimetry mass, and molecular simulation analyses. *Crystal Growth & Design* 15(12): 5827–5833.
57. Li, X.-M., et al. (2019) Strategic hierarchical improvement of superprotonic conductivity in a stable metal–organic framework system. *Journal of Materials Chemistry A* 7(43): 25165–25171.
58. Wang, L., et al. (2020) Metal-organic framework anchored sulfonated poly(ether sulfone) nanofibers as highly conductive channels for hybrid proton exchange membranes. *Journal of Power Sources* 450: 227592.
59. Rao, Z., Tang, B., Wu, P. (2017) Proton conductivity of proton exchange membrane synergistically promoted by different functionalized metal–organic frameworks. *ACS Applied Materials & Interfaces* 9(27): 22597–22603.
60. Zheng, P., et al. (2020) Preparation of covalent-ionically cross-linked UiO-66-NH2/sulfonated aromatic composite proton exchange membranes with excellent performance. *Frontiers in Chemistry* 8: 56.
61. Mukhopadhyay, S., et al. (2020) Fabricating a MOF material with polybenzimidazole into an efficient proton exchange membrane. *ACS Applied Energy Materials* 3(8): 7964–7977.
62. Wang, S., et al. (2021) UiO-66-NH2 functionalized cellulose nanofibers embedded in sulfonated polysulfone as proton exchange membrane. *International Journal of Hydrogen Energy* 46(36): 19106–19115.
63. Janiak, C., Vieth, J.K.J.N.J.o.C. (2010) MOFs, MILs and more: concepts, properties and applications for porous coordination networks (PCNs). *New Journal of Chemistry* 34(11): 2366–2388.
64. Dong, X.-Y., et al. (2017) Tuning the functional substituent group and guest of metal–organic frameworks in hybrid membranes for improved interface compatibility and proton conduction. *Journal of Materials Chemistry A* 5(7): 3464–3474.
65. Li, Z., et al. (2014) Enhanced proton conductivity of proton exchange membranes by incorporating sulfonated metal-organic frameworks. *Journal of Power Sources* 262: 372–379.
66. He, X., et al. (2017) Highly conductive and robust composite anion exchange membranes by incorporating quaternized MIL-101(Cr). *Science Bulletin* 62(4): 266–276.
67. Li, Z., et al. (2014) Enhanced proton conductivity of Nafion hybrid membrane under different humidities by incorporating metal–organic frameworks with high phytic acid loading. *ACS Applied Materials & Interfaces* 6(12): 9799–9807.
68. Zhang, B., et al. (2017) Proton exchange nanohybrid membranes with high phosphotungstic acid loading within metal-organic frameworks for PEMFC applications. *Electrochimica Acta* 240: 186–194.
69. Anahidzade, N., et al. (2018) Metal-organic framework anchored sulfonated poly(ether sulfone) as a high temperature proton exchange membrane for fuel cells. *Journal of Membrane Science* 565: 281–292.
70. Wu, B., et al. (2013) A novel route for preparing highly proton conductive membrane materials with metal-organic frameworks. *Chemical Communications* 49(2): 143–145.
71. Ru, C., et al. (2018) Enhanced proton conductivity of sulfonated hybrid poly(arylene ether ketone) membranes by incorporating an amino–sulfo bifunctionalized metal–organic framework for direct methanol fuel cells. *ACS Applied Materials & Interfaces* 10(9): 7963–7973.

72. Sun, X.L., et al. (2017) Communication. *Chemistry – A European Journal* 23(6): 1248–1252.
73. Lee, Y.-R., Kim, J., Ahn, W.-S. (2013) Synthesis of metal-organic frameworks: A mini review. *Korean Journal of Chemical Engineering* 30(9): 1667–1680.
74. Park, K.S., et al. (2006) Exceptional chemical and thermal stability of zeolitic imidazolate frameworks. *Proceedings of the National Academy of Sciences of the United States of America* 103(27): 10186–10191.
75. Park, K.S., et al. (2006) Exceptional chemical and thermal stability of zeolitic imidazolate frameworks. *Proceedings of the National Academy of Sciences of the United States of America* 103(27): 10186–10191.
76. Vega, J., et al. (2017) Conductivity study of zeolitic imidazolate frameworks, tetrabutylammonium hydroxide doped with zeolitic imidazolate frameworks, and mixed matrix membranes of polyetherimide/tetrabutylammonium hydroxide doped with zeolitic imidazolate frameworks for proton conducting applications. *Electrochimica Acta* 258: 153–166.
77. Barjola, A., et al. (2018) Enhanced conductivity of composite membranes based on sulfonated poly(ether ether ketone) (SPEEK) with zeolitic imidazolate frameworks (ZIFs). *Nanomaterials* 8(12): 1042.
78. Yang, L., Tang, B., Wu, P. (2015) Metal–organic framework–graphene oxide composites: a facile method to highly improve the proton conductivity of PEMs operated under low humidity. *Journal of Materials Chemistry A* 3(31): 15838–15842.
79. Sun, H., Tang, B., Wu, P. (2017) Two-dimensional zeolitic imidazolate framework/carbon nanotube hybrid networks modified proton exchange membranes for improving transport properties. *ACS Applied Materials & Interfaces* 9(40): 35075–35085.
80. Nilam, M., et al. (2021) Membrane permeability and its activation energies in dependence on analyte, lipid, and phase type obtained by the fluorescent artificial receptor membrane assay. *ACS Sensors* 6(1): 175–182.
81. Jia, W., Tang, B., Wu, P. (2016) Novel composite PEM with long-range ionic nanochannels induced by carbon nanotube/graphene oxide nanoribbon composites. *ACS Applied Materials & Interfaces* 8(42): 28955–28963.
82. Sen, U., et al. (2016) Proton conducting self-assembled metal–organic framework/polyelectrolyte hollow hybrid nanostructures. *ACS Applied Materials & Interfaces* 8(35): 23015–23021.
83. Erkartal, M., et al. (2016) Proton conducting poly(vinyl alcohol) (PVA)/ poly(2-acrylamido-2-methylpropane sulfonic acid) (PAMPS)/ zeolitic imidazolate framework (ZIF) ternary composite membrane. *Journal of Membrane Science* 499: 156–163.
84. Guo, Y., et al. (2018) A DNA-threaded ZIF-8 membrane with high proton conductivity and low methanol permeability. *Advanced Materials* 30(2): 1705155.
85. Li, P., et al. (2020) Ag-DNA@ZIF-8 membrane: A proton conductive photoswitch. *Applied Materials Today* 20: 100761.
86. Zhao, G., et al. (2020) Zeolitic imidazolate framework decorated on 3D nanofiber network towards superior proton conduction for proton exchange membrane. *Journal of Membrane Science* 601: 117914.
87. Cai, Y.Y., et al. (2019) Achieving efficient proton conduction in a MOF-based proton exchange membrane through an encapsulation strategy. *Journal of Membrane Science* 590: 117277.
88. Hu, F., et al. (2020) Enhanced properties of sulfonated polyether ether ketone proton exchange membrane by incorporating carboxylic-contained zeolitic imidazolate frameworks. *New Journal of Chemistry* 44(32): 13788–13795.
89. Cai, Y.Y., et al. (2021) Two-dimensional metal-organic framework-graphene oxide hybrid nanocomposite proton exchange membranes with enhanced proton conduction. *Journal of Colloid and Interface Science* 594: 593–603.
90. Wu, B., et al. (2014) Oriented MOF-polymer composite nanofiber membranes for high proton conductivity at high temperature and anhydrous condition. *Scientific Reports* 4(1): 4334.
91. Kim, H.J., Talukdar, K., Choi, S.-J. (2016) Tuning of Nafion® by HKUST-1 as coordination network to enhance proton conductivity for fuel cell applications. *Journal of Nanoparticle Research* 18(2): 47.
92. Cai, K., et al. (2017) An acid-stable hexaphosphate ester based metal–organic framework and its polymer composite as proton exchange membrane. *Journal of Materials Chemistry A* 5(25): 12943–12950.
93. Escorihuela, J., et al. (2019) Proton conductivity of composite polyelectrolyte membranes with metal-organic frameworks for fuel cell applications. *Advanced Materials Interfaces* 6(2): 1801146.
94. Li, Z., et al. (2014) Enhanced proton conductivity of Nafion hybrid membrane under different humidities by incorporating metal–organic frameworks with high phytic acid loading. *ACS Applied Materials & Interfaces* 6(12): 9799–9807.
95. Tsai, C.-H., et al. (2014) Enhancing performance of Nafion®-based PEMFC by 1-D channel metal-organic frameworks as PEM filler. *International Journal of Hydrogen Energy* 39(28): 15696–15705.
96. Han, R., Wu, P. (2018) Composite proton-exchange membrane with highly improved proton conductivity prepared by in situ crystallization of porous organic cage. *ACS Applied Materials & Interfaces* 10(21): 18351–18358.
97. Donnadio, A., et al. (2017) Mixed membrane matrices based on Nafion/UiO-66/SO3H-UiO-66 nano-MOFs: Revealing the effect of crystal size, sulfonation, and filler loading on the mechanical and conductivity properties. *ACS Applied Materials & Interfaces* 9(48): 42239–42246.
98. Patel, H.A., et al. (2016) Superacidity in Nafion/MOF hybrid membranes retains water at low humidity to enhance proton conduction for fuel cells. *ACS Applied Materials & Interfaces* 8(45): 30687–30691.
99. Alberti, G., et al. (2001) Polymeric proton conducting membranes for medium temperature fuel cells (110–160°C). *Journal of Membrane Science* 185(1): 73–81.
100. Hickner, M.A., et al. (2004) Alternative polymer systems for proton exchange membranes (PEMs). *Chemical Reviews* 104(10): 4587–4612.
101. Higashihara, T., Matsumoto, K., Ueda, M. (2009) Sulfonated aromatic hydrocarbon polymers as proton exchange membranes for fuel cells. *Polymer* 50(23): 5341–5357.

102. Xing, P., et al. (2004) Synthesis and characterization of sulfonated poly(ether ether ketone) for proton exchange membranes. *Journal of Membrane Science* 229(1): 95–106.
103. Kreuer, K.D. (2001) On the development of proton conducting polymer membranes for hydrogen and methanol fuel cells. *Journal of Membrane Science* 185(1): 29–39.
104. Samimi, F., Rahimpour, M.R. (2018) Chapter 14 – Direct methanol fuel cell. In: *Methanol*, A. Basile, F. Dalena (Eds.), pp. 381–397. Elsevier.
105. Sun, H., Tang, B., Wu, P. (2017) Rational design of S-UiO-66@GO hybrid nanosheets for proton exchange membranes with significantly enhanced transport performance. *ACS Applied Materials & Interfaces* 9(31): 26077–26087.
106. Zhang, J., et al. (2007) Polybenzimidazole-membrane-based PEM fuel cell in the temperature range of 120–200°C. *Journal of Power Sources* 172(1): 163–171.
107. Escorihuela, J., et al. (2020) Recent progress in the development of composite membranes based on polybenzimidazole for high temperature proton exchange membrane (PEM) fuel cell applications. *Polymers* 12(9): 1861.
108. Escorihuela, J., et al. (2018) Phosphoric acid doped polybenzimidazole (PBI)/zeolitic imidazolate framework composite membranes with significantly enhanced proton conductivity under low humidity conditions. *Nanomaterials* 8(10): 775.
109. Wu, Y., et al. (2021) Achieving high power density and excellent durability for high temperature proton exchange membrane fuel cells based on crosslinked branched polybenzimidazole and metal-organic frameworks. *Journal of Membrane Science* 630: 119288.
110. Chen, J., Wang, L., Wang, L. (2020) Highly conductive polybenzimidazole membranes at low phosphoric acid uptake with excellent fuel cell performances by constructing long-range continuous proton transport channels using a metal–organic framework (UIO-66). *ACS Applied Materials & Interfaces* 12(37): 41350–41358.
111. Wong, C.Y., et al. (2020) Development of poly(vinyl alcohol)-based polymers as proton exchange membranes and challenges in fuel cell application: A review. *Polymer Reviews* 60(1): 171–202.
112. Lilleby Helberg, R.M., et al. (2020) PVA/PVP blend polymer matrix for hosting carriers in facilitated transport membranes: Synergistic enhancement of CO_2 separation performance. *Green Energy & Environment* 5(1): 59–68.
113. Liang, X., et al. (2013) From metal–organic framework (MOF) to MOF–polymer composite membrane: enhancement of low-humidity proton conductivity. *Chemical Science* 4(3): 983–992.
114. Luo, H.-B., et al. (2017) Proton conductance of a superior water-stable metal–organic framework and its composite membrane with poly(vinylidene fluoride). *Inorganic Chemistry* 56(7): 4169–4175.
115. Zhang, J., et al. (2018) Extra water- and acid-stable MOF-801 with high proton conductivity and its composite membrane for proton-exchange membrane. *ACS Applied Materials & Interfaces* 10(34): 28656–28663.
116. Santamaria, M., et al. (2015) Chitosan–phosphotungstic acid complex as membranes for low temperature H_2–O_2 fuel cell. *Journal of Power Source* 276: 189–194.
117. Santamaria, M., et al. (2015) Chitosan–phosphotungstic acid complex as membranes for low temperature H_2–O_2 fuel cell. *Journal of Power Sources* 276: 189–194.
118. Ma, Y., et al. (2021) Preparation of anion exchange membrane by incorporating IRMOF-3 in quaternized chitosan. *Polymer Bulletin* 78(7): 3785–3801.
119. Dong, X.-Y., et al. (2018) Synergy between isomorphous acid and basic metal–organic frameworks for anhydrous proton conduction of low-cost hybrid membranes at high temperatures. *ACS Applied Materials & Interfaces* 10(44): 38209–38216.
120. Zeng, X., et al. (2021) Construction of ordered OH− migration channels in anion exchange membrane by synergizes of cationic metal-organic framework and quaternary ammonium groups. *International Journal of Energy Research* 45(7): 10895–10911.
121. Alaswad, A., et al. (2020) Technical and commercial challenges of proton-exchange membrane (PEM) fuel cells. *Energies* 14(1): 144.

15 Recent Developments on Supported Pd Membranes for Hydrogen Separation and Membrane Reactors

David A. Pacheco Tanaka[1,2,*] *and Fausto Gallucci*[2,3]

[1]TECNALIA, Basque Research and Technology Alliance (BRTA), Donostia-San Sebastian, Spain

[2]Sustainable Process Engineering, Chemical Engineering and Chemistry, Eindhoven University of Technology, Eindhoven, The Netherlands

[3]Eindhoven Institute for Renewable Energy Systems (EIRES), Eindhoven University of Technology, Eindhoven, The Netherlands

*Email: alfredo.pacheco@tecnalia.com

15.1 INTRODUCTION

Hydrogen is expected to play a key role in the future of energy and in achieving a zero-carbon economy. More than 80% of hydrogen is produced by reforming of natural gas or water–gas shift of coal-derived syngas with subsequent separation by pressure swing adsorption. Palladium-based membranes have received growing interest for hydrogen purification and production due to their very high permeability and exclusive selectivity toward H_2 due to the unique mechanism of hydrogen permeation (Figure 15.1); thus, H_2 molecules should be transported from the bulk to the surface of the Pd membrane; then, on the surface, the H_2 molecule is chemisorbed (chemical reaction) and broken into two atoms of hydrogen (H), this reaction is exclusive for H_2 and Pd. Then, the hydrogen atoms are transferred from one side of the membrane to the other driven by the difference in the square root of partial pressure of H_2. In the other side, the hydrogen atoms recombine, catalysed by Pd, forming H_2.

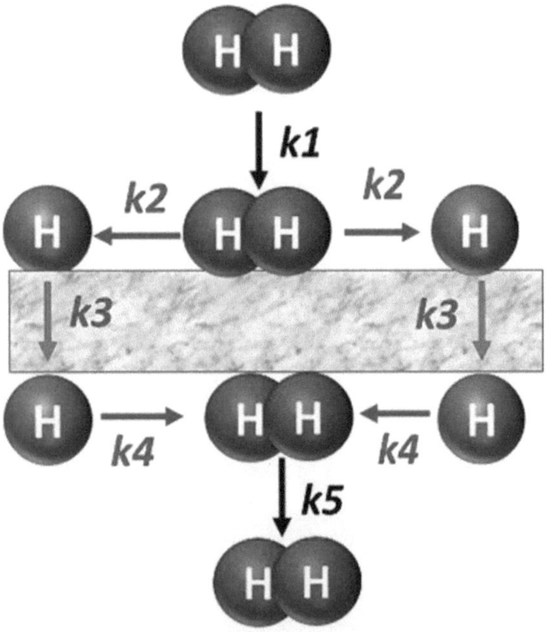

FIGURE 15.1 Hydrogen permeation through a Pd membrane.

The flux of H_2 (J_{H2}) follows Equation (15.1)

$$J_{H2} = \frac{Q_H}{L}\left(P^{0.5}_{H2\,ret} - P^{0.5}_{H2\,per}\right) \tag{15.1}$$

where Q_H is the membrane permeability, L is the thickness of the selective layer, and $P^{0.5}{}_{H2ret}$ and $P^{0.5}{}_{H2perm}$ are the H_2 partial pressures in the retentate and permeate, respectively. Pd is a precious metal, and its price has increased considerable in recent years; therefore, to reduce cost, thin (< 5 μm) supported membranes without defects are required; in this way, less Pd is required, the flux is increased, and therefore, the number of membranes are reduced. Among the various methods of deposition of the Pd layer on the support, the electroless plating (ELP) method is the most used because of its low cost and versatility. The choice of the support is very important; this should have the lowest resistance to the passage of gas, be mechanically strong, and should not interact with Pd at working temperature. The surface where the thin Pd will be deposited should have smoothness, pore size and pore size distribution (without defects) that enable good attachment, with minimum resistance to the passage of gas. Ceramic supports have good properties for the deposition of a thin Pd layer, however, mechanically they are not as strong. In contrast, metallic supports are mechanically strong and can be welded directly to metallic reactors; however, the surface properties are not suitable for thin Pd membranes. Pd suffers from embrittlement at low temperatures (< 300°C) and high pressures due to mismatch between the α and β phases of Pd-hydride which could destroy the membrane; this embrittlement can be decreased by alloying with other metals, mainly Ag. At more than 500–550°C, Pd-based membranes

DOI: 10.1201/9781003285656-15

start to sinter and, gradually, defects start to appear and consequently the H_2 selectivity is affected. Pd is very sensitive to low concentrations of H_2S (10 ppm); depending on the temperature, the H_2 permeation could be decreased by H_2 adsorption, or even destroy it through the formation of Pd_4S.

Pd membrane reactors (Pd-MR) combine reaction and H_2 separation in one device and are used in processes limited by the equilibrium where the removal of H_2 increases the yield of the process, reducing the reaction temperature and obtaining pure H_2. These systems have been studied for a long time now, starting from the pioneering work of Gryaznov and co-workers [1,2], followed by the works of Basile [3,4] and Itoh and co-workers [5].

Two types of Pd-MR are considered: packed bed (PB) [6] and fluidized bed (FB) reactors [7]; FB has several advantages, however, the continuous collision of the catalyst with the Pd layer can damage the membrane, especially at high temperatures. The addition of a protective nanoporous layer on the Pd surface (double-skin membrane) can decrease this damage. Recently, several reviews on Pd membranes and Pd membrane reactors have been published in a special issue on Pd membranes [8] and others [9–11]. In this chapter, recent advances in Pd membranes for H_2 separation and membrane reactors at Tecnalia and Eindhoven University of Technology (TUE) developed under several European projects will be presented [12]. Recently, the product of these developments, the H2Site company, was created (h2site.eu/en/) which is focused on the fabrication of Pd membranes to be used in processes involving H_2 separation.

15.2 PALLADIUM-BASED MEMBRANES

15.2.1 Preparation of Supported Thin-Film Pd-Based Membranes

Bredesen et al. at SINTEF reported the fabrication of Pd-based membranes by first manufacturing a thin Pd-based film on a polished surface by PVD magnetron sputtering, and then transferred to the porous support [13,14], however, the most studied method is the direct deposition of Pd on porous supports.

PVD magnetron sputtering. In this technique, Pd with or without the presence of other metals is removed from a target by the high energy ions from a plasma and then deposited on the pore support. The as-deposited film is not completely dense, and empty spaces are present between the grains of Pd, therefore, the hydrogen selectivity is low [15]; the selectivity was increased by deposition on top of a Pd layer by electroless plating (ELP) [16]. The advantage of PVD is that several membranes can be deposited simultaneously, several metals can be deposited at the same time, and the composition of Pd alloy membranes is controlled; however, Pd is not deposited only on the support, but also in the chamber.

Chemical vapor deposition (CVD). In a chamber, an organometallic compound containing Pd (generally Pd acetate) and the supports are placed; then, vacuum and heat are applied in such way that the Pd compound sublimates and is deposited on the porous support where the Pd^{2+} is reduced to Pd^0; the deposition continues until a film is produced [17]. With this technique, it is difficult to deposit simultaneously Pd and other metals.

Electroless plating (ELP). This technique was first developed by Uemiya in Japan [18,19]; the deposition is an autocatalytic reaction where the catalyst is Pd^0, therefore, the deposition (reaction) will only take place where Pd^0 is present. Thus, before ELP, Pd^0 should be deposited on the surface of the porous substrate (Pd seeding); this step is very important to obtain a good membrane. The most common method is by sequential immersion of the support in $SnCl_2$ (to reduce Pd^{+2}) and $[PdCl_4]^{-2}$ solutions, respectively, this cycle is repeated at least eight times. With this method, Pd and also Sn are deposited, decreasing the active area of Pd. A more efficient method for Pd seeding is by sequential immersion in solutions of Pd acetate in chloroform and hydrazine in water [20,21]. With this method, uniform nanoparticles of Pd are deposited on the surface; since the nanoparticles are close to each other, a continuous film is formed from the early stages of ELP. Pd-Ag alloy membranes are prepared by the sequential deposition of a Pd and Ag layer and subsequent annealing at high temperatures (> 600°C) for several hours; this treatment may damage the membrane, producing pinholes and peeling of the membrane by the treatment at high temperature. The simultaneous Pd-Ag co-deposition allows the formation of an alloy by annealing at lower temperatures and shorter time. The kinetics of deposition of Ag is faster than Pd, therefore, as active Pd must be present on the surface, the co-deposition must be carefully controlled since Ag can cover the Pd surface, hence stopping the reaction. The co-deposition can be controlled by the continuous addition of Ag to the ELP solution [20,22].

15.2.2 Effect of the Porous Support on the Preparation of Defect-Free Pd Membranes

The choice of the support is very important to prepare thin supported Pd-based membranes without defects; they should: (a) have the minimum resistance to the passage of gases (very high flux), (b) not suffer from Pd support strong interaction (do not react with Pd), (c) be mechanically strong, and (d) have a surface with low rugosity, sharp pore size distribution, and adequate pore size. The Pd layer does not have good attachment to surfaces with very small pores, i.e. < 40 nm; it has been reported that the minimum thickness of the Pd layer should be more than three times the diameter of the largest pores present in the support [23]. The porous supports are usually tubular; and to reduce the resistance to the passage of gases, asymmetric supports are used. In this configuration, the pore size is not uniform, it is gradually reduced to the surface where the selective layer will be deposited. In Figure 15.2, asymmetric ceramic supports with various external/internal diameters (in mm) are shown. The porous support can be classified as ceramic or metallic. Metallic supports are more robust and can be welded directly to the separators or reactors (which are mainly metallic).

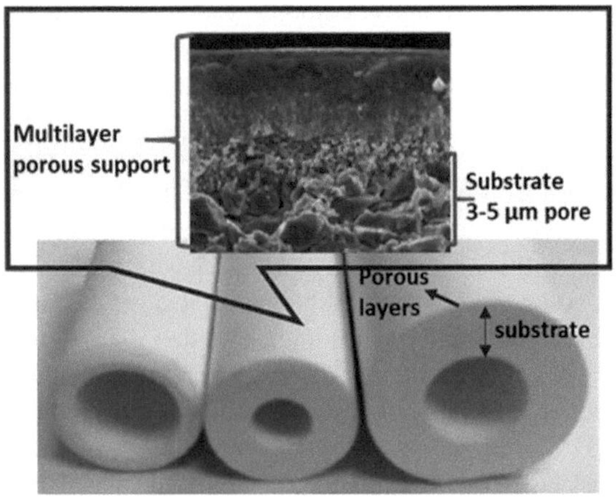

FIGURE 15.2 10/7, 10/4, and 14/7 outer and inner diameter (mm) asymmetric supports.

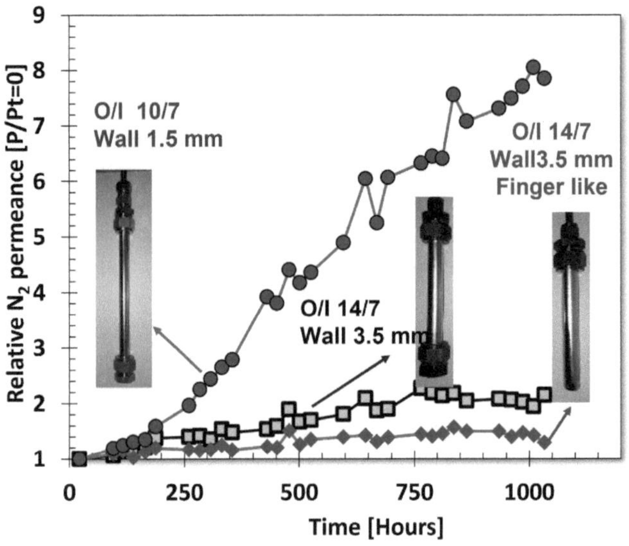

FIGURE 15.3 Relative nitrogen permeance at 500°C as a function of time [29].

Metallic supports. The commercially available metallic supports are more expensive than ceramics, they have large and non-uniform pores, and the surface is rough; therefore, they must be treated before deposition of the Pd film. At around 400–450°C in a hydrogen environment, strong interaction between the Pd layer and the metal from the support occurs, and an alloy is formed reducing considerably the H_2 permeation. To avoid the inter-metallic diffusion or strong support Pd interaction at high operating temperatures and to improve the quality of the support surface, an inter-metallic diffusion barrier layer between the metallic support and the Pd selective layer is required [24,25]. The roughness and pore size of a porous Hastelloy X tube were reduced by filling the pores with a mixture of alumina-YSZ grains and this was used as support for a thin Pd-Ag membrane (4–5 µm). This membrane was tested at 400°C for 1250 h, after that time, the hydrogen permeance was 0.9×10^{-6} mol m^{-2} s^{-1} Pa^{-1} at 1 atm pressure difference and a H_2/N_2 perm-selectivity of 150,000 [26].

Ceramic supports. Ceramic supports have good surface properties; asymmetric alumina supports are used for water treatment; however, for gas separation, the presence of defects on the surface is more critical, especially for highly selective membranes such as Pd. To facilitate the manipulation and deposition of the Pd layer, alumina supports are usually connected to a dense alumina tube using a glass paste [20,27]. N_2 permeation showed the presence of leaks in the interface of the glass with the Pd layer, the product of the difference in their expansion coefficient and poor adhesion between the smooth glass surface and Pd film. To solve this problem, the glass sections were removed, and the membrane was sealed using Swagelok fittings and graphite ferrules; sealing is obtained by applying a torque to deform the graphite. This sealing was used for supported Pd-Ag on asymmetric porous alumina (100 nm pore size in the outside) and 10/7 mm external and internal diameters (1.5 mm wall thickness) [28]. The wall of the support was only 1.5 mm and high torque could not be applied because the ceramic supports could be broken; in the project FERRET, two 25-cm-long Pd-Ag membranes were connected by Swagelok sealings to obtain an effective length of 40 cm of membrane (four sealings per membrane); these membranes were fragile in the interface between the metallic seals and the Pd membrane; especially when they are assembled in a vertical position due to the weight of the Swagelok. In the project FluidCELL, 40-cm-long membranes were prepared (only two sealings per membrane); at high temperatures of permeation (300–500°C), due to the differences in the expansion coefficient of the materials, leaks started to appear, requiring thicker supports to resist the higher torque required for sealing.

Long-term tests of H_2 and N_2 permeation at 500°C of three Pd-Ag membranes with similar Pd-Ag thickness and composition on 100 nm external layer but different supports were carried out (Figure 15.3). Two membranes with two sealings but different external and internal diameters (10/7 and 14/7) and finger-like (14/7) asymmetric supports which require only one sealing were tested. Torques of 12 Nm and 6 Nm were applied to the 14/7 and 10/7 asymmetric supports, respectively [29]. The change in the N_2 flux relative to the value at time zero with the time of permeation is shown in Figure 15.3. Since the Pd-Ag layers are similar, the N_2 leaks are due to defects in the sealings. It can be observed that in the 10/7 membrane, leaks soon start to appear and increase with time, which can be attributed to the lower force applied to the sealing. For 14/7 (two sealings), the rate of leaking is greatly diminished due to the higher force used in the sealings. Comparing 14/7 membranes with two and one sealings (finger-like), the leaks were diminished even more. These results suggest that the Pd-Ag layer is stable during the experiment and the leaks can be mainly attributed to the sealings.

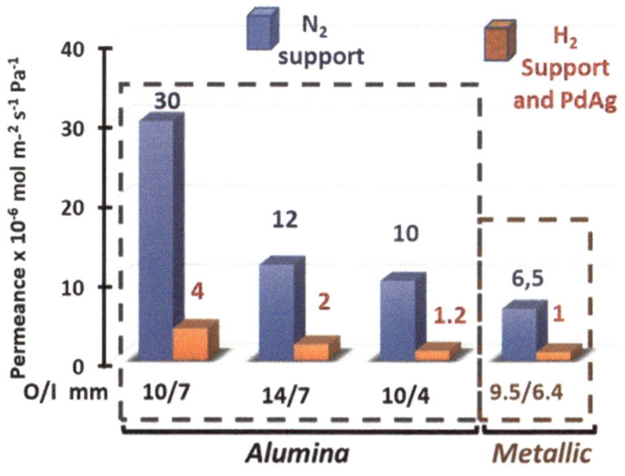

FIGURE 15.4 N_2 permeance of several supports at room temperature and H_2 permeance at 400°C with their corresponding Pd-Ag membrane.

The N_2 permeance at room temperature of several asymmetric alumina supports having 100 nm pore size on the surface and different OD and ID (10/7, 14/7, 10/7) and a metallic support, as shown in Figure 15.4. The alumina support with the thinnest wall has the higher permeance and decreases considerable when the support becomes thicker. The surface of metallic supports is usually rough, which is not suitable for the deposition of thin Pd-based membranes; thus, the surface was polished resulting in the closure of some pores in the surface, therefore, the permeance of treated Hastelloy X support [16] shows the lowest N_2 permeance. The resistance of the support to the passage of gas is important for thin supported Pd-based membranes; H_2 permeance at 400°C of 4-μm-thick Pd-Ag membranes deposited by electroless plating on these supports shows that H_2 permeance depends on the N_2 permeance of the support. Therefore, before deposition of the Pd layer, it is convenient to measure the permeance of the support to estimate the H_2 permeance of the supported Pd membrane.

One way to increase the hydrogen flux at a given feed pressure is to increase the driving force of the process by feeding a sweep gas in the permeate side. This effect can, however, be significantly reduced if mass transfer limitations in the permeate side exist. The main contribution to the mass transfer limitation is the porous support which leads to an increase in the H_2 partial pressure close to the surface of the Pd layer, decreasing hydrogen flux. The presence of stagnant sweep gas in the pores of the membrane support can reduce the hydrogen diffusion through the support with an increment of the hydrogen partial pressure at the interface between the Pd-layer and support, with a consequent decrease in the driving force of the permeation [30].

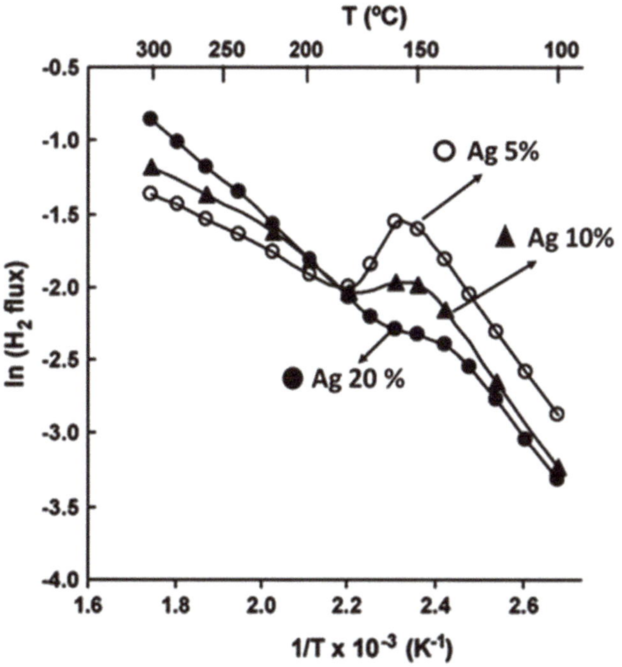

FIGURE 15.5 Arrhenius plot of H_2 flux (mol m^{-2} s^{-1}) for palladium membranes having 5, 10, and 20% silver at 200 kPa pressure difference. Reproduced from Ref. [21] with permission of Elsevier.

15.2.3 Effect of Temperature on Supported Pd-Based Membranes

Pd hydride at below 293°C can exist in two forms (α and β) which have different cell sizes (0.389 and 0.402 nm, respectively) producing distortion in the structure; the stress is accompanied of lattice expansion (3.4%) generating defects and often destruction and delamination of the membrane (hydrogen embrittlement). The embrittlement can be reduced by alloying Pd with other metals, among them, Pd-Ag alloys have been shown to be the most efficient since they can form alloys in all proportions. Supported Pd-Ag membranes having various Ag compositions (Ag = 0, 5, 10, 15, 20, and 23%) were prepared by the simultaneous ELP deposition method and the H_2 permeation studied in the range of 100–300°C, as can be seen in Figure 15.5 [21]. The membrane with 5% Ag has a peak in flux at around 150–170°C. As the Ag content increases, the intensity of the peak decreases, shifting to lower temperatures, at Ag 20% just a shoulder appears. The peak is related to the α–β transition, the membrane with 20% Ag was stable after several cycles of hydrogen and nitrogen permeation at 100 and 300°C.

Between 300 and 500°C, Pd-alloy membranes have good H_2 permeation, and the permeation is stable with time. By changing the plating time, ultra-thin (0.46–1.3μm) Pd-Ag membranes supported on asymmetric porous alumina (200 nm pore size on the surface) were prepared by Pd-Ag co-deposition ELP [31]; the 1.3-μm-thick membrane has a stable H_2 permeance (9–9.4 × 10^{-6} mol m^{-2} s^{-1} Pa^{-1}) and H_2/N_2 selectivity (1900) during 1000 h at 400°C and 1 atm pressure

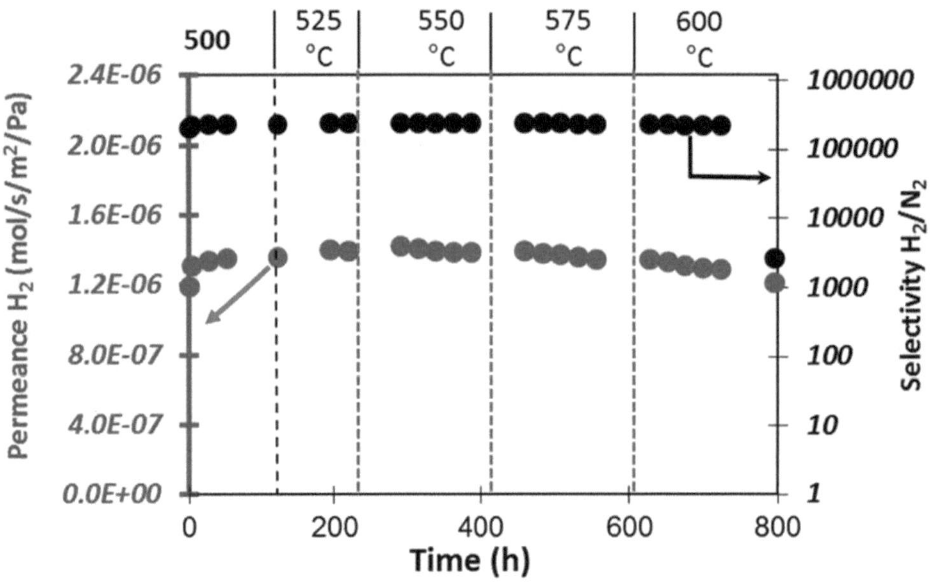

FIGURE 15.6 Long-term hydrogen permeance of a PdAg membrane supported on porous Hastelloy X. Reproduced from Ref. [26].

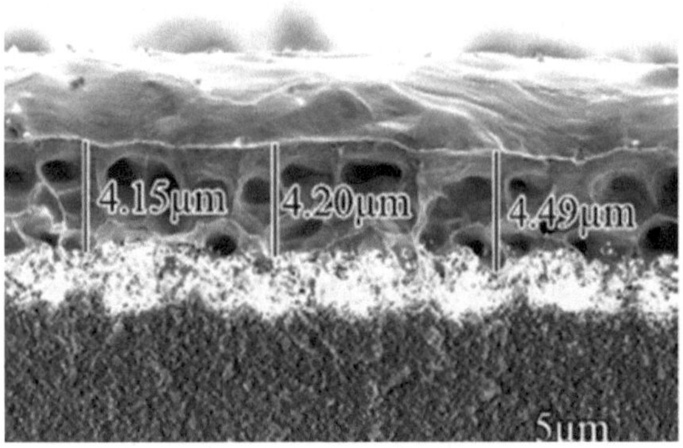

FIGURE 15.7 SEM image of a Pd-based membrane after hydrogen permeation test at 600°C. Reproduced from Ref. [33].

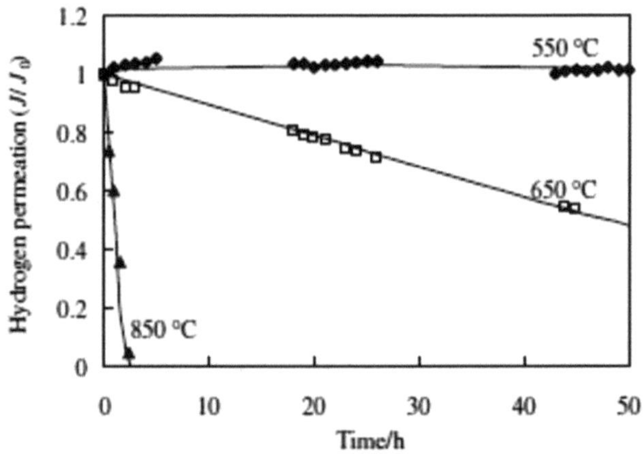

FIGURE 15.8 Reduction in the hydrogen permeation at high temperatures of a Pd membrane supported on alumina. Reproduced from Ref. [37] with permission from the Royal Society of Chemistry.

difference. A 4–5 μm Pd-Ag supported on porous Hastelloy X tube similar to that described above was tested for a long time at temperatures from 500 to 600°C with a stepwise increase of 25°C [26] (Figure 15.6). The H_2 permeance shows very good stability (~1.3×10^{-6} mol m^{-2} s^{-1} Pa^{-1}), however, from around 550°C the permeance slightly decreases with the temperature and time, probably due to densification by the sinterization of the Pd grains. The H_2/N_2 perm-selectivity was extremely high (> 200,000) for almost 800 h; the temperature was increased to 600°C and after a 795 h test, the selectivity dropped quickly to 2650.

Similar behaviour was observed in a Pd-Ag membrane on alumina support, the formation of defects was more evident at 600°C [29]. At around Tamman temperature (half of the melting temperature) of Pd (640°C), the atoms and clusters of Pd start to sinter, forming larger particles leaving behind cavities which will be filled with hydrogen; the pressure of the gas will produce plastic deformation of the PdH walls, forming spherical bubbles [32] and defects as shown in the SEM picture of a Pd-Ag membrane after hydrogen permeation at 600°C (Figure 15.7) [33]; similar behaviour at 600°C was reported in other Pd-based membranes [34,35].

Porous alumina is the most used ceramic porous support; however, at high temperatures, H_2 permeation is reduced with time due to the strong interaction between Pd and alumina [36]. At 550°C, the H_2 permeation is constant, at 650°C, it decreases gradually, and at 850°C the permeation very quickly is almost zero (Figure 15.8). At high temperatures, in the presence of Pd, Al_2O_3 reacts with H_2 producing aluminium and water; aluminium forms alloys with Pd, which have very low H_2 permeation; this strong interaction was not observed at 850°C when YSZ was used as support [37].

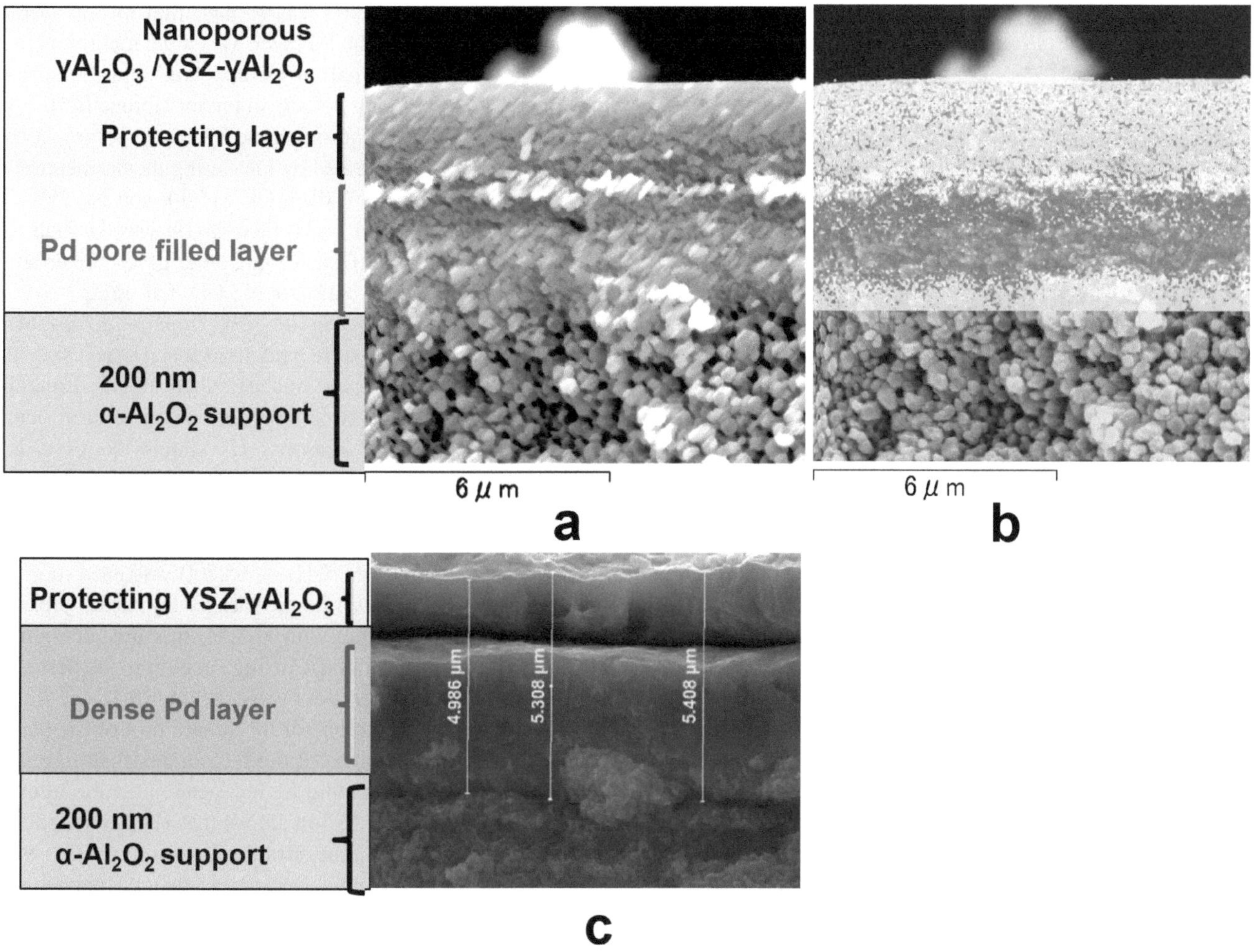

FIGURE 15.9 SEM image of (a) palladium pore-filled membrane, (b) EDX analysis of the nanoporous layer of the membrane (green Al, red Pd), (c) double-skin Pd membrane. Reproduced from Ref. [40].

15.2.4 Pd-Based Membranes with a Porous Ceramic Protective Layer

The surface of the Pd-based membrane is prone to damage during handling, strong interaction with catalyst, and abrasion with moving particles, resulting in a reduction of the permeation properties and the generation of defects or cracks. The Pd layer can be protected by deposition of a nano-porous ceramic layer on top of the Pd-containing layer. In the "Pd pore-filled" configuration, Pd particles are deposited inside the nano-porous (3–5 nm pore size) γ-alumina [38] or a mixture of γ-Al_2O_3-YSZ [39], which is placed below a nano-porous protecting layer, as shown in Figure 15.9a. EDX analysis demonstrates that Pd (red) is inside the nanopores (Figure 15.9b). As Pd particles are smaller than the pores, due to the nano-size effect, the amount of H_2 adsorbed on the surface is much higher than inside the particle, therefore, embrittlement of the α–β PdH does not occur, and these membranes can permeate H_2 at low temperatures (~100°C) without damaging the membrane. Since the expansion coefficient of Pd (11.8 K^{-1}) is not close to that of alumina (8.2 K^{-1}), when H_2 permeation is carried out at temperatures higher than 350°C, the membrane could be destroyed. The thermal expansion coefficient of YSZ is between those of Pd and Al_2O_3, and the pore-filled membranes are more stable and can work at higher than 500°C [39].

15.2.5 Pd Membranes under Fluidization Conditions

In fluidized bed reactors, small granular solid catalyst is forced to swirl around the reactor; if a Pd membrane is introduced, the catalyst will collide with the membrane with the probability of damaging or interacting with the Pd layer. De Nooijer et al. studied the effect of the fluidization of Ru/Al_2O_3 particles (180 μm particle size) on a Pd-Ag membrane [29]. The permeation of H_2 and N_2 was stable for 850 h of fluidization at 400°C. Then, the temperature was increased to 500°C and tested under fluidization for another 300 h; during this period, H_2 increased due to the temperature, then it remained constant; however, the N_2 permeation increased constantly with time.

FIGURE 15.10 SEM image showing surface defects in Pd-Ag layer after 1150 h under fluidization and temperatures up to 500°C. Reproduced from Ref. [29].

A post-mortem SEM image of the membrane (Figure 15.10) shows that the surface looks like fish scales with very small pinholes as a product of the plastic deformation of the Pd-Ag-hydride layer by the collision of particles during fluidization at 500°C.

In the "double-skin" type membrane (Figure 15.9c) [41], a nanoporous γ-Al_2O_3-YSZ protecting layer is deposited on a dense Pd membrane, in this configuration, small deffects in the Pd layer coverd by the protecting layer, increasing the selectivity of the membrane. These membranes have been succesfully used in Pd membrane reactors; it was observed that the membrane without the protective layer suffered from significant erosion, leading to an increase in nitrogen leakage under fluidization conditions, while the performance of the double-skin membrane remained stable for more than 615 h at temperatures of 400–500°C and at 4 bar of pressure difference [42]. The protecting layer prevents direct contact between catalyst particles and the Pd layer. When Ru supported on TiO_2 was used as catalyst, the hydrogen permeation decreased gradually due to the strong interaction beetwen Pd and TiO_2 (similar to that in Figure 15.8 with alumina), however, the reduction in the permeation was not observed when a Pd double-skin membrane was used [43].

15.2.6 Effect of the Presence of Other Gases in Hydrogen Permeation

In thick Pd-based membranes (> 30 μm), the limiting step in H_2 permeation is the passage of the hydrogen atoms through the membrane (Figure 15.1). For thinner membranes (> 5–10 μm), the limiting step is the adsorption of the molecule of H_2 on the surface and the splitting into hydrogen atoms, then, the blockage of the active sites of Pd on the surface of the membrane can reduce considerably the H_2 permeation. Before H_2 permeation, the surface of the membrane is usually cleaned by introducing air or oxygen for a short time [44]. In situ TEM SAED studies of the oxidation of Pd membrane showed that up to 300°C and 300 mbar O_2, PdO was not present; as the temperature increases, PdO was formed first on the surface and then on the bulk of the Pd layer. When the membrane was reduced again by H_2, small voids appeared and the surface became rough, producing defects in the membrane [45].

The effect of CO inhibition on the hydrogen permeation of ultrathin 0.78 μm was studied by measuring the permeation of a mixture containing H_2 (60%), CO (15%), and N_2 25%. At 400°C, a decrease of 15% on H_2 flux was observed which was reduced to 9% at 450°C. Considering other gases, the order of inhibition is CO >> CO_2 > H_2O > N_2, CH_4 [27,46].

Separation of H_2 present at 10% from a mixture with methane. A method to store and transport H_2 is using the existing natural gas storage and distribution infrastructure; then, at the end, the used H_2 can be recovered using membranes. For safety reasons, H_2 should be at a low concentration (~10%). From permeability and selectivity perspectives, Pd alloy membranes are the best candidates for high-purity H_2 recovery. A Pd-Ag double-skin membrane having ideal selectivity H_2/CH_4 of 65,200 was used to study the effect on the H_2 flow rate, at various H_2 partial pressure differences, when working with H_2/CH_4 mixtures containing 10, 50, and 70% of H_2 at 400°C using vacuum in the permeate (Figure 15.11a) [46]. Acording to Equation (15.1), the flux is proportional to the difference of the square root of the partial pressure of H_2 in both sides of the membrane, therefore, it is expected that the flow should be independent of the dilution of H_2. However, Figure 15.11a shows that H_2 flux is reduced considerably when the concentration of H_2 decreases; thus, at 10%, the flow is very low. Decreasing the concentration of H_2 in the mixture, the probability that H_2 is in contact with the surface of the membrane decreases. The purity of the permeate of the sample containing 10% of H_2 as a function of the total pressure difference (Figure 15.11b) shows that at 10 bar difference, the purity is > 99.9%, but it decreases with the pressure, because CH_4 passes through the defects of the membrane with the flux proportional to the pressure.

15.2.7 H_2S Poisoning

Sulphur is a very common impurity in fossil fuels; the presence of a few ppm of sulphur reduces the flux of Pd membranes drastically. Depending on the concentration and temperature, H_2S can be desorbed recovering the membrane's permeation properties (physorption) or reacting with Pd forming PdS structures ($Pd_{16}S_7$ and PdS_4). The constant lattices of Pd and Pd_4S are quite different, therefore the crystalline structure of the membrane is destroyed. by alloying Pd with Au, the resistance of the membrane to corrosive degradation by sulphur can be improved. Pd-Ag membranes were prepared by ELP [36], then, one half of the membrane was cover with a Teflon tape and Au was deposited by ELP (Figure 15.12a) [47]. After annealing to form the alloy, permeation studies were carried out. The activation energy of permeation of the Pd-Ag-Au membrane was higher than Pd-Ag (16.0 and 9.5 kJ mol^{-1}, respectively) indicating that splitting of the molecule of H_2 to atoms of H is more difficult. It was found that the addition of Au increased the resistance to H_2S poisoning in

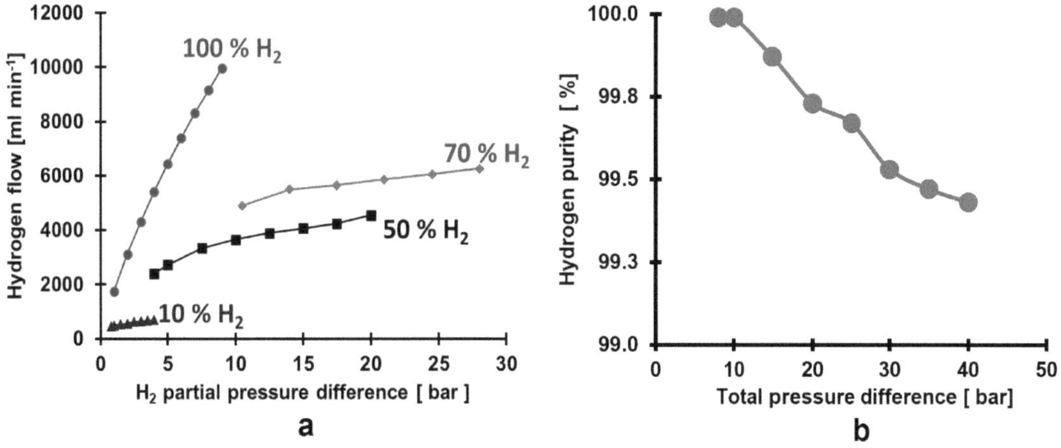

FIGURE 15.11 (a) H_2 flow rate at various partial pressure differences and various H_2 contents; (b) H_2 purity of the permeate at various pressure differences from a mixture of 90% CH_4, 10% H_2. Temperature 400°C, double-skin membrane [40].

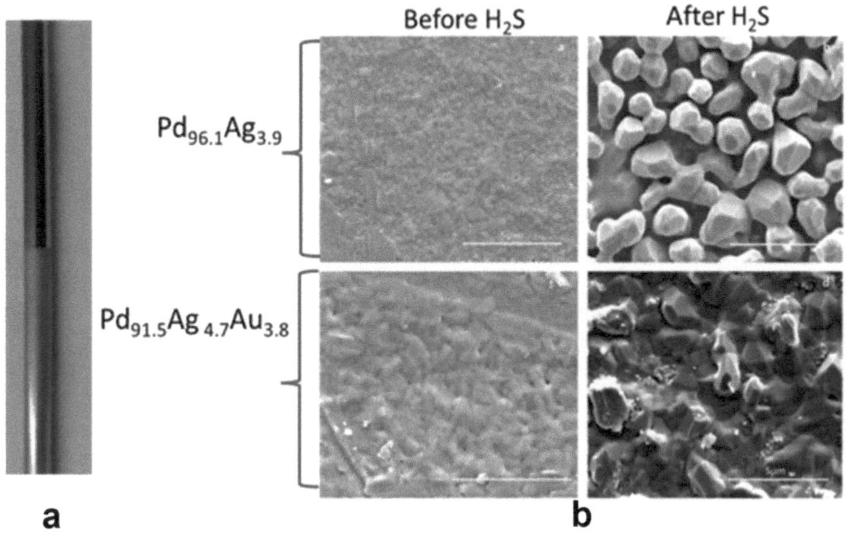

FIGURE 15.12 Pd-Ag membrane before annealing with one half deposited with Au, (b) PdAg and PdAgAu after annealing and tested with up to 17 ppm of H_2S at 550°C. Reproduced from Ref. [47].

the presence of up to 17 ppm H_2S at 550°C. Figure 15.12b shows an SEM image of the half membranes of the membrane without addition and that with Au before and after the H_2S test, the formation of polyhedral structures is clearly observed [47]. At 2 ppm of H_2S, the H_2 permeation of the Pd-Ag-Au was reduced, but the process was reversible, indicating that H2S was physisorbed [48]. In all cases, for gas mixtures which are expected to contain sulphur, this should be removed before coming in contact with the Pd membrane.

15.3 THE MEMBRANE REACTOR CONCEPT

A membrane reactor is a multifunctional system where a reaction and separation through a membrane are integrated in a single vessel, resulting in process integration. The integration of the reaction and separation in a single vessel has benefits in terms of energy efficiency and yield (thus OPEX reduction) as well as benefits in the number of process steps required (thus a decrease of CAPEX).

An example (amongst many others reported in literature) of the benefits of a membrane reactor is reported in the work of Brunetti et al. [49] for water–gas shift reaction (WGS) in terms of energy and costs. WGS is exothermic and equilibrium limited, therefore, it is generally carried out in two reactors, the first working at high temperature to make use of high reaction rates, and another (much larger) at low temperature to achieve higher conversions. If a membrane reactor is used for hydrogen separation, the equilibrium is circumvented, so WGS can be carried out with a single reactor, working at high temperatures, as reported in Figure 15.13.

Because of the integration of H_2 separation during the reaction, the equilibrium can be shifted towards the products,

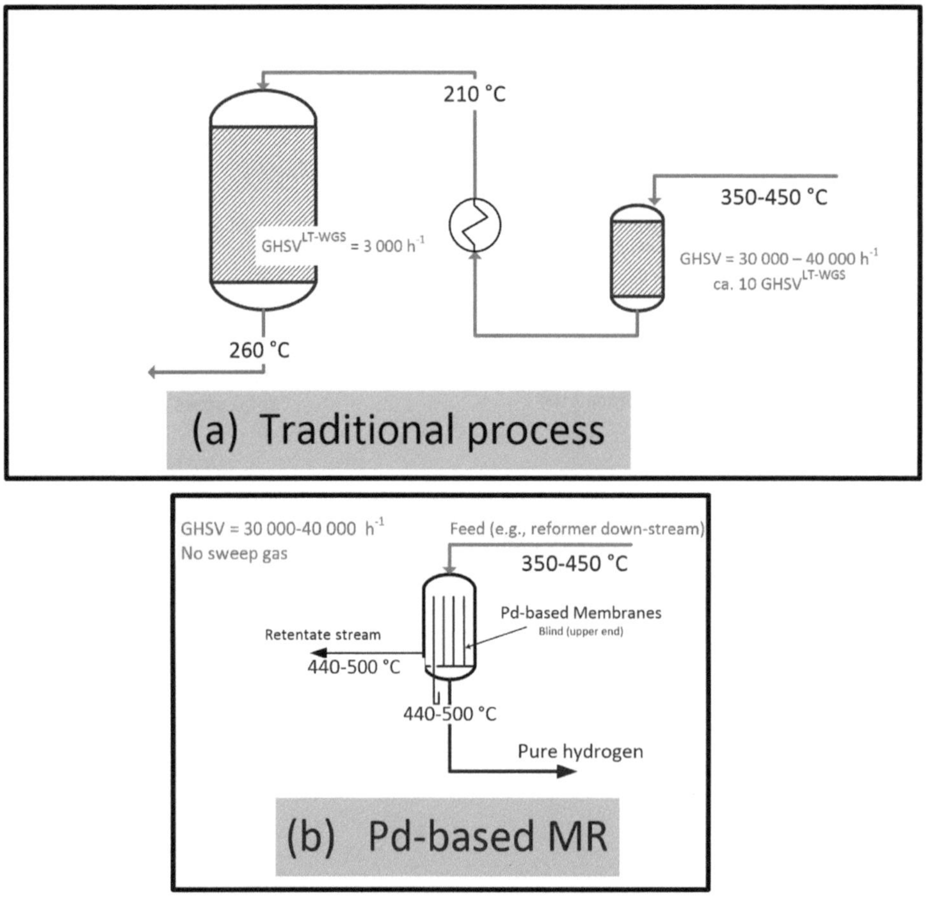

FIGURE 15.13 Comparison of a conventional reactor system and membrane reactors for WGS reaction system. Reprinted from Ref. [49] with permission of RSC.

thus circumventing the limitations of conventional reactors. This allows working at higher temperatures and makes use of faster kinetics without the conversion limitations. As a result, one can achieve a reduction of reactor volume that can go down to 20% or more compared to conventional systems by using high pressures (or more permeable membranes) as reported in Figure 15.14 [50].

The membrane reactor concept can be used for different types of reactions. The systems where membrane reactors can give clear benefits (see Figure 15.15) are reaction systems in which the conversion/yields are either limited by thermodynamic equilibrium or by consecutive/parallel reactions. In these cases, removal of the product (or feeding of a reactant) can give great benefits in terms of product yields.

Most of the membrane reactors using Pd-based membranes are applied to equilibrium reactions, so that the removal of one of the products shifts the equilibrium toward higher yields. This is discussed in more detail in the sections below.

15.3.1 Packed Bed Membrane Reactors

Packed bed membrane reactors are a combination of a catalyst bed in a fixed configuration and a membrane through which either products are removed or reagents are fed to the catalyst bed. A typical example, and a comparison with other reactors, for methane reforming has been reported in Gallucci et al. [6] and is shown in Figure 15.16.

A typical packed bed membrane reactor confguration consists of a catalytic particle bed in contact with one side of a membrane, as illustrated in Figure 15.17. The particles are generally at the high-pressure side of the reactor, and in contact with the selective membrane layer to drive the permeation driving force through the membrane, while the permeate side is in the low-pressure side and in the side of the membrane where a vaccum or carrier gas can be used to increase the pressure difference.

Although simple in terms of design, this configuration has several drawbacks. Firstly, packed beds are often affected by excessive pressure drops. This problem is more significant in membrane reactors, as permeation through the membrane is driven by pressure, thus an excessive pressure drop in the reaction zone would result in a lower driving force and lower permeation. To avoid this, larger particles need to be used which in turn result in lower catalyst loading per unit of membrane area. The larger the particle size, the lower is the possibility to use the configuration in Figure 15.17a. Indeed, for tubular membranes, the configuration in Figure 15.17b would allow only particles of few hundred microns. Larger

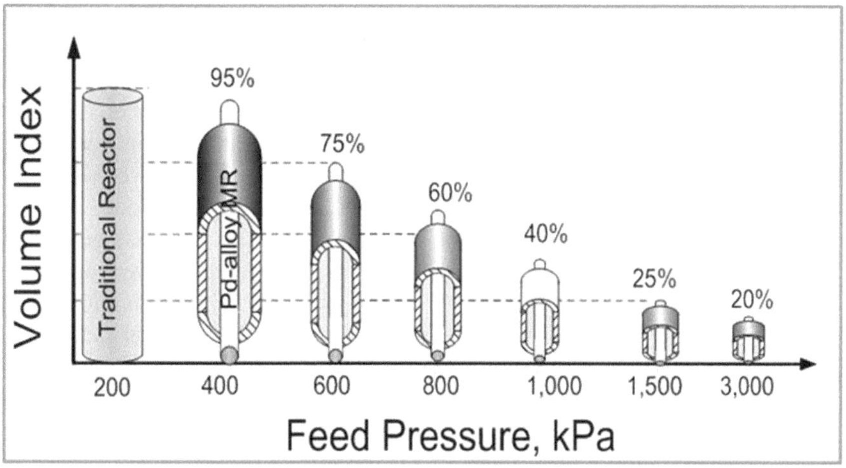

FIGURE 15.14 Volume index as a function of feed pressure at 280°C. Set CO conversion 90%. Reproduced from Ref. [50] with permission of Elsevier.

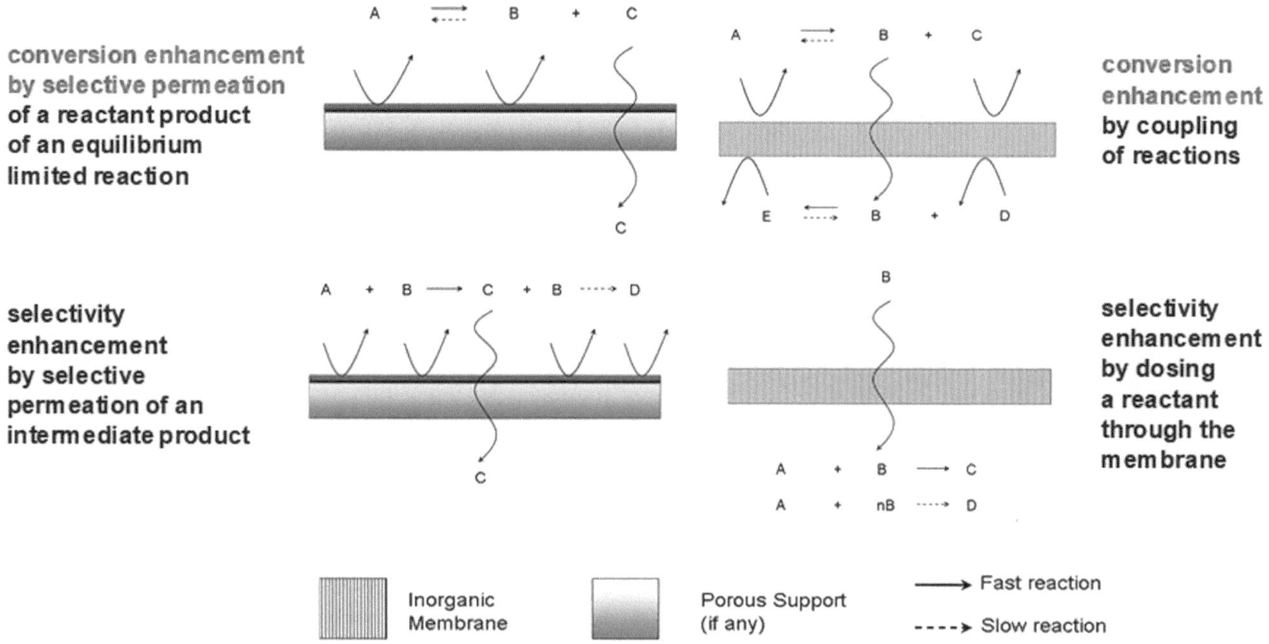

FIGURE 15.15 Main application possibilities of (inorganic) membrane reactors. Reprinted from Ref. [51].

particles can only be used in configuration b (i.e., in the shell side of the reactor). While this is not a huge problem for lab-scale reactors, for industrial-scale ones this configuration may lead to excessive heat transfer limitations and consequent sub-optimal use of the membranes. In the most extreme cases, the membrane may experience a temperature that is either too low for its correct functioning or too high compared to the maximum membrane working temperature. In this regard, the simulation work of Tiemersma et al. [52] demonstrates that for a highly exothermic reaction such as the autothermal reforming of methane, the low heat transfer limitations may lead to a high temperature peak at the start of the membrane reactor (see Figure 15.18) which will be detrimental to the stability of the membrane.

Another issue of packed bed membrane reactors is that the bed to wall mass transfer limitations affect its performance, especially for very thin and highly permeable membranes. These mass transfer limitations are often referred to as concentration polarization, a term very often used in membrane processes like desalination and reverse osmosis. While concentration polarization is frequent in membrane processes where liquid phases and solid phases are involved, for a long time it was not considered for gas-phase reactions. However, with the continuous improvements in membranes fluxes, especially with very thin supported membranes, it has become evident that this phenomenon is very detrimental even for membrane reactors dealing with gas-phase reactions. A typical example of this kind of issue has been reported for H_2

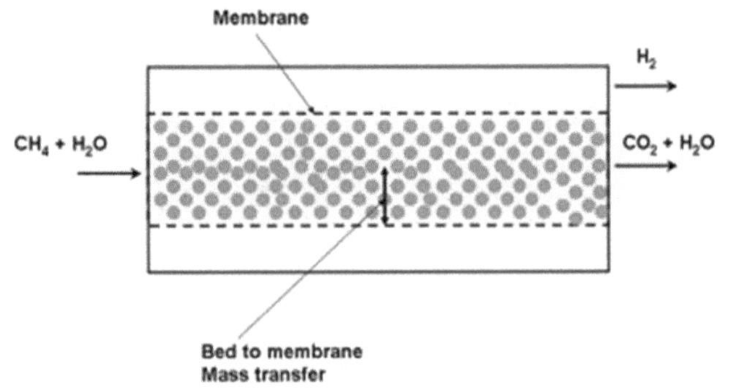

FIGURE 15.16 Scheme of the packed bed membrane reactor. Reprinted from Ref. [6].

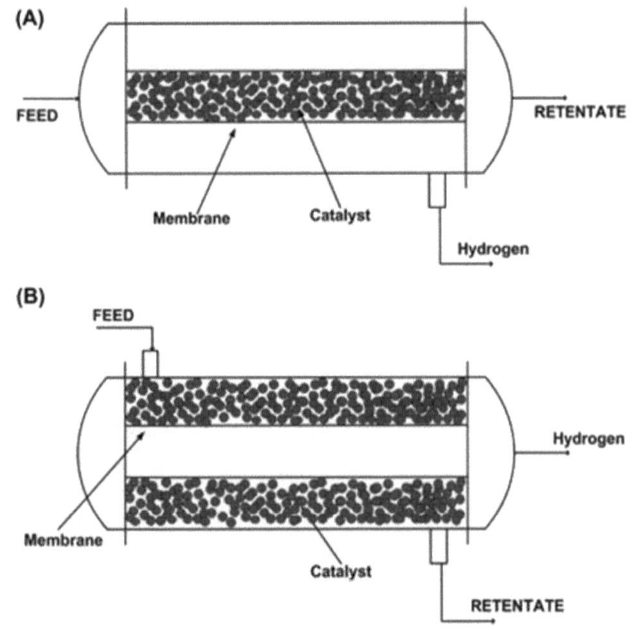

FIGURE 15.17 Membrane reactor catalyst in tube (A) and catalyst in shell (B) configurations. Reprinted from Ref. [51].

production in membrane reactors. While this system has been long studied at lab scale, the early use of thick membranes has masked the presence of concentration polarization issues. However, to make these reactors economically feasible, very thin membranes need to be used. These membranes being developed have demonstrated that at a certain point the concentration polarization become dominant in the performance of the reactor. An indicative figure is the one reported in Figure 15.19, where it is evident that the higher the flux through the membrane, the higher is the difference between the hydrogen fraction (and thus partial pressure) in the bulk ($r/R = 0$) and the hydrogen fraction at the membrane wall ($r/R = 1$). The increased mass transfer limitations result in a much larger membrane area required for the same recovery of product (in this particular case H_2).

To circumvent these limitations, other reactor concepts have been identified and studied such as the micro-structured reactor concept and the fluidized bed reactor concept, and these are discussed in more detail in the coming section.

Table 15.1 reports the main reaction systems investigated in the papers/works considered. From the table it is clear that most of the research on this kind of reactor aims for increased efficiency for chemical conversion, in particular for H_2 production (either as product itself or as an energy carrier). While a lot of research has been carried out on this type of reactors for methane conversion (which is still the case as the table shows) it is interesting to note that more effort is spent nowadays for other conversions, and in particular ammonia. The reason for this is that ammonia can be converted into hydrogen without producing CO_2 emissions, while ammonia has a very high energy content compared with hydrogen gas (Figure 15.20).

Ammonia can thus be produced with green hydrogen (or better with hydrogen produced using renewable energy) and used as transportation media. Once at the location, it can be converted again into hydrogen and used either as an industrial gas, gas for hydrogen combustion, or in fuel cells. The conversion of the ammonia to hydrogen is an endothermic equilibrium reaction which requires temperatures higher than 550°C for achieving high conversion rates. At temperatures higher than 600°C the conversion is almost complete. For use in gas turbines this conversion is more than enough, however for use of hydrogen in fuel cells, concentrations of ammonia below 0.1 ppm need to be achieved (thus a separation step is required). Integrating the ammonia conversion and hydrogen recovery in the same reactor would be beneficial for achieving high conversion rates at lower temperatures and high purities of H_2. A conceptual design of the membrane reactor for ammonia decomposition was reported by Kim et al. [54]. They reported a schematic representation of their packed bed membrane reactor as depicted in the top of Figure 15.21.

The work of Kim reports a technical feasibility study by considering ideal membranes with infinite selectivities, demonstrating that, from a model point of view, lower temperatures, high recoveries, and high efficiencies can be achieved by using membrane reactors.

High conversion rates and high purities have been demonstrated by Jo and co-workers [55] who have carried out ammonia decomposition using a Ru-based catalyst and a composite membrane made of a thin layer (0.4 micron) of palladium covering a 250 micron Ta layer. The authors reported a conversion rate of >99.5% at temperatures higher than 475°C using low pressure and a very high purity in the permeate (circa 800 ppb by Nessler methods). They have also shown that there is no degradation in the performance of a fuel cell for more than 80 h, confirming a very low ammonia content in the permeate. For the relatively short period of the operation, the authors also confirmed that there was no interdiffusion of metals between Pd and Ta, a phenomenon that may influence the ability of the membrane to split and separate hydrogen.

Zhang et al. [56] reported a catalytic membrane reactor, by depositing the catalyst directly in the support of the membrane. This can be considered as a special packed bed membrane

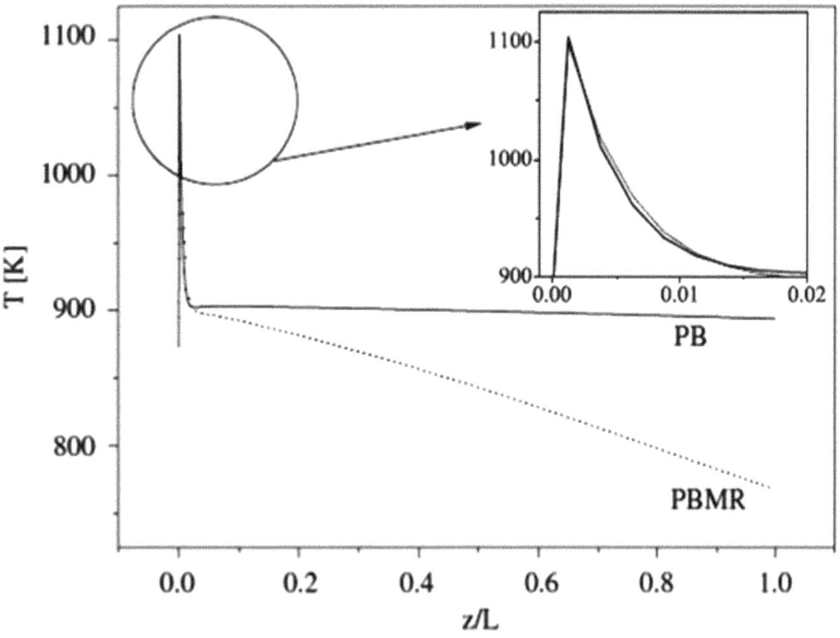

FIGURE 15.18 Axial temperature profiles in a packed bed (PB) and a packed bed membrane reactor (PBMR) for autothermal methane reforming. Reproduced from Ref. [52].

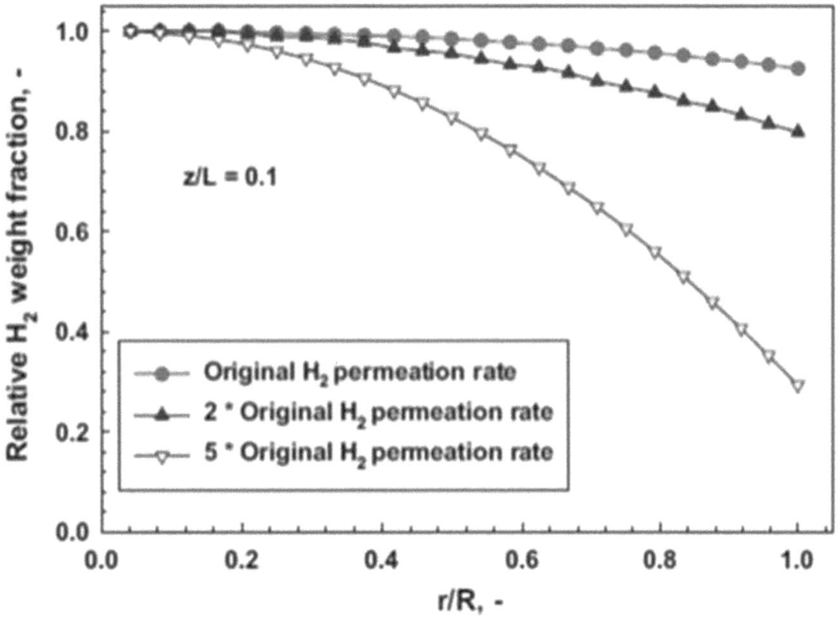

FIGURE 15.19 Relative H_2 weight fraction for the isothermal operation mode at different hydrogen permeabilities. Reproduced from Ref. [6].

reactor that may be used for decreasing the external mass transfer limitations between the catalyst and the membrane surface.

A similar configuration but with completely different materials has been also reported by Cheng and co-workers [57]. In this case the authors reported catalytic dual-layered hollow-fibre membranes made of a dense mixed protonic–electronic conducting hydrogen-selective layer over a porous Ni-activated layer of the same material as the dense one. While the mixed conducting layers need high temperature for activation, the authors reported nice enhancements of the catalytic membrane reactor compared to a conventional reactor operated in the same conditions. This work is very interesting because these kinds of membranes are foreseen

TABLE 15.1
Main reactions systems recovered in the latest works on membrane reactors

Reaction system	Product of interest	Membrane type
Ammonia decomposition/cracking	H_2	Pd-based, Ta-based, Pd/Ta,
Methane reforming, oxidative reforming, coupling	H_2 or C_2+ aromatics	Pd membranes, O_2 selective membranes, unselective porous membranes
Propane	Propylene	Pd, zeolite
Syngas	H_2	Pd based
CO_2 hydrogenation	Methanol, DME	Zeolite, carbon, polymeric

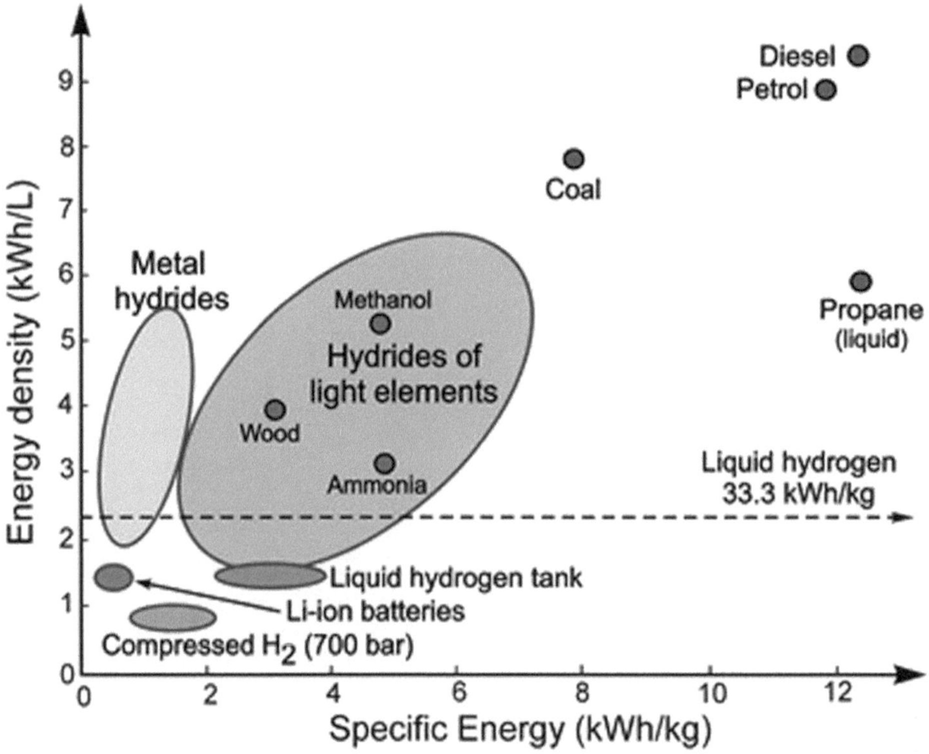

FIGURE 15.20 Energy densities of various energy storage materials and technologies, illustrating the respective volumetric and gravimetric densities. Reproduced from Ref. [53].

to be very interesting in combination with electrically driven processes.

For the same reaction system, Cechetto et al. [58] demonstrated the high conversion and high purity of hydrogen using a supported thin Pd-Ag layer membrane (around 4 microns) and a Ru-based catalyst. Especially interesting is the stability test reported for this system as the data reported are very stable for around 600 h of operation as depicted in Figure 15.22.

Table 15.1 also reports that methane conversion is a very popular reaction studied in packed bed membrane reactors. Most of the reactions of interest are either reforming of methane to produce H_2 or oxidative coupling of methane to produce ethylene and other products. While in the first case, the membranes are selective to H_2 and used to extract the product of reaction, thus increasing the yields and productivities at lower temperatures, in the case of oxidative coupling, the membranes are used for feeding of the reactant (generally pure oxygen) to improve the selectivity towards the product ethylene and reduce the full conversion of products to CO_2.

15.3.2 Fluidized Bed Membrane Reactors

Fluidized bed membrane reactors are reactor concepts in which a bundle of membranes is immersed in a bed of small catalyst particles that move because of fluid flow. Indeed, when a fluid flow passes through a bed of particles it creates a frictional force that can be measured by the pressure drop over the bed. Different regimes exist depending on the fluid flow rate for a

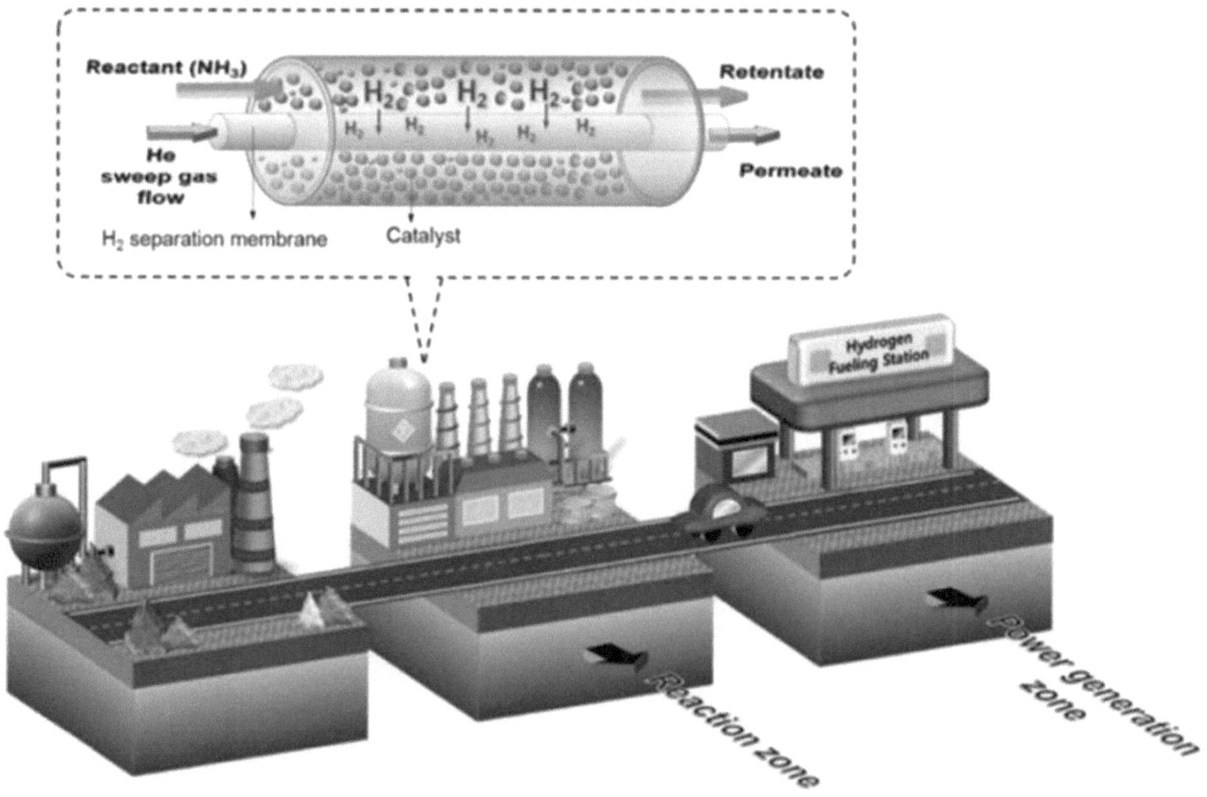

FIGURE 15.21 Schematic diagram of ammonia decomposition in a membrane reactor (MR) for PEM fuel cells (PEMFCs). Reproduced from Ref. [54].

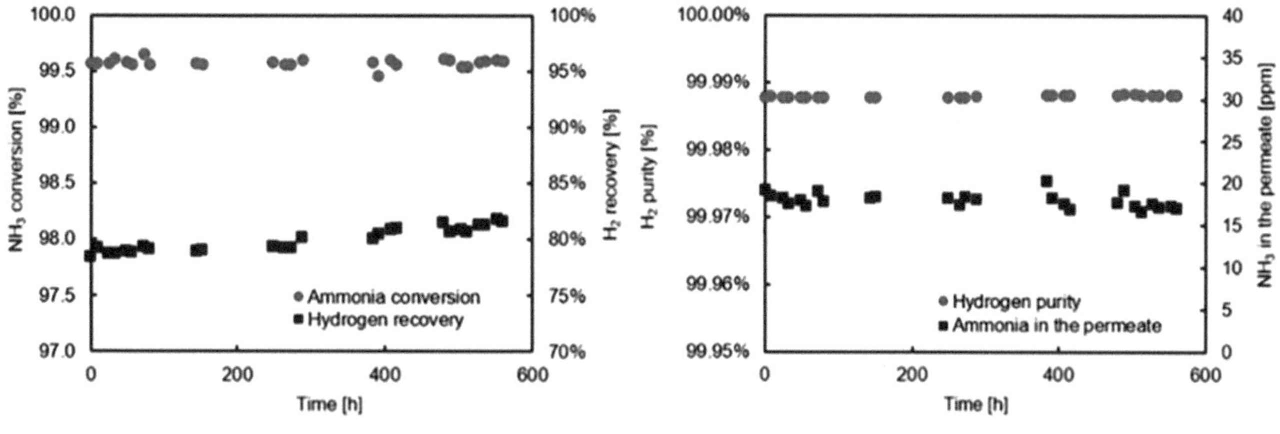

FIGURE 15.22 Stability test of ammonia cracking using a Pd-Ag membrane. Reproduced from Ref. [58].

given volume of particles. At low flow rates (fluid velocities) the bed remains stationary, in a fixed bed configuration. Here the pressure drop increases as function of the flow rate. At a certain point, the drag force created by the fluid is equal to the gravitational force. Here the bed expands and the situation is called minimum fluidization (and the corresponding velocity is the minimum fluidization velocity). At higher velocities, the gas in excess of the minimum fluidization will pass through the bed in the form of bubbles that, for certain particles called Geldart B particles, increase in size as they travel inside the bed. If the reactor size is relatively small this increase in bubble size results in a sluggish regime. At higher flow the bed behaves like a turbulent regime and when the flow is so high that the velocity exceeds the terminal velocity of the particles, then we have pneumatic transport. Typical fluidized bed (membrane) reactors are operated in either bubbling or turbulent regimes. As described above, the fluidization depends on different factors, including the type of particles. Interested readers are referred to books on fluidization for more detail.

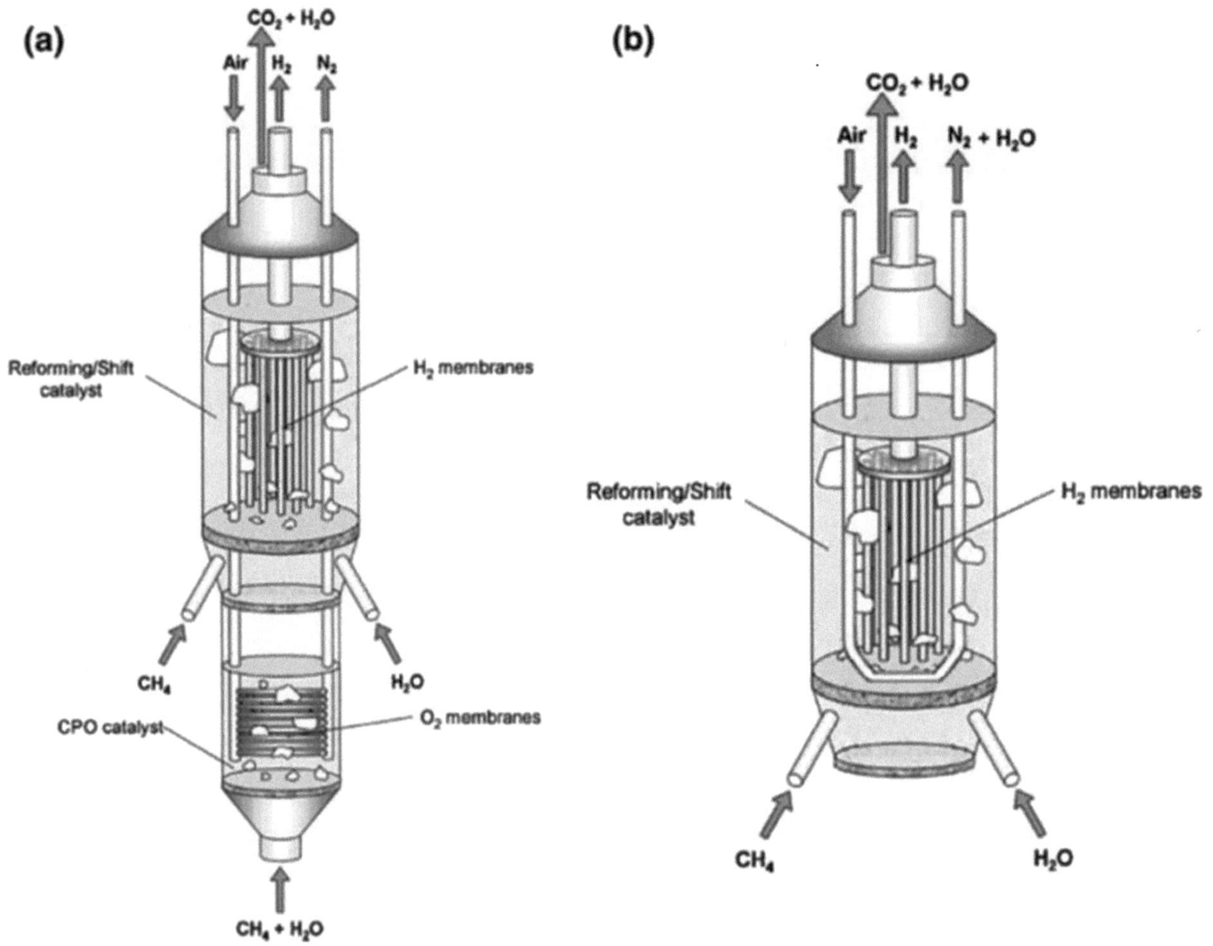

FIGURE 15.23 Schematic representation of the two fluidized membrane reactor concepts for autothermal methane reforming with integrated CO_2 capture. (a) Methane combustion configuration. (b) Hydrogen combustion configuration. Reproduced from Ref. [59].

Examples are reported in our previous work [59] and shown in Figure 15.23, where two membrane fluidized bed concepts have been introduced that use different kinds of membranes immersed in a fluidized bed for both recovering H_2 (Pd membranes) and supplying oxygen (perovskite membranes). The use of a fluidized bed of particles improves on the shortcomings of packed bed membrane reactors, as discussed above. In particular, the continuous movement of particles improves heat and mass transfer rates. Firstly, even for very exothermic or endothermic reactions, virtually uniform temperature profiles have been observed in fluidized beds. This means that, for reactions like autothermal methane reforming or steam reforming, the temperature profile in the reactor is uniform, with clear stability of membranes and catalysts. Secondly, the continuous circulation of particles also improves the mass transfer rate with much decreased external mass transfer limitations and thus less concentration polarization.

One of the first long-term experiments using ultrathin membranes in fluidized bed reactors was reported by Helmi et al. [60] who showed stable operation of the membrane reactor for up to 900 h in a water–gas shift reaction system with stable hydrogen production with CO content below the fuel cell requirements (see Figure 15.24).

Stable operation of the fluidized bed membrane reactor has also been reported by de Nooijer in his PhD thesis for biogas autothermal reforming [29].

An interesting concept that integrates hydrogen production with CO_2 capture is the membrane-assisted chemical looping reformer (MA-CLR) that has been developed and demonstrated by our group [61]. The concept has been experimentally demonstrated by Medrano et al. [62] and is depicted in Figure 15.25. In this concept, the energy required for the endothermic steam methane reforming is supplied by the oxidation of a metal to metal oxide that occurs in the air reactor which is very exothermic. The metal oxide is then transferred to the reforming reactor where it is reduced by H_2 supplying the energy for the reforming, While the hydrogen produced is separated by the Pd membranes. The proof of concept has shown that stable operation can be attained and the model was successfully validated. Medrano et al. [63] made a comparison of this concept, with several other systems proposed in literature for H_2 production including CO_2 capture. MA-CLR

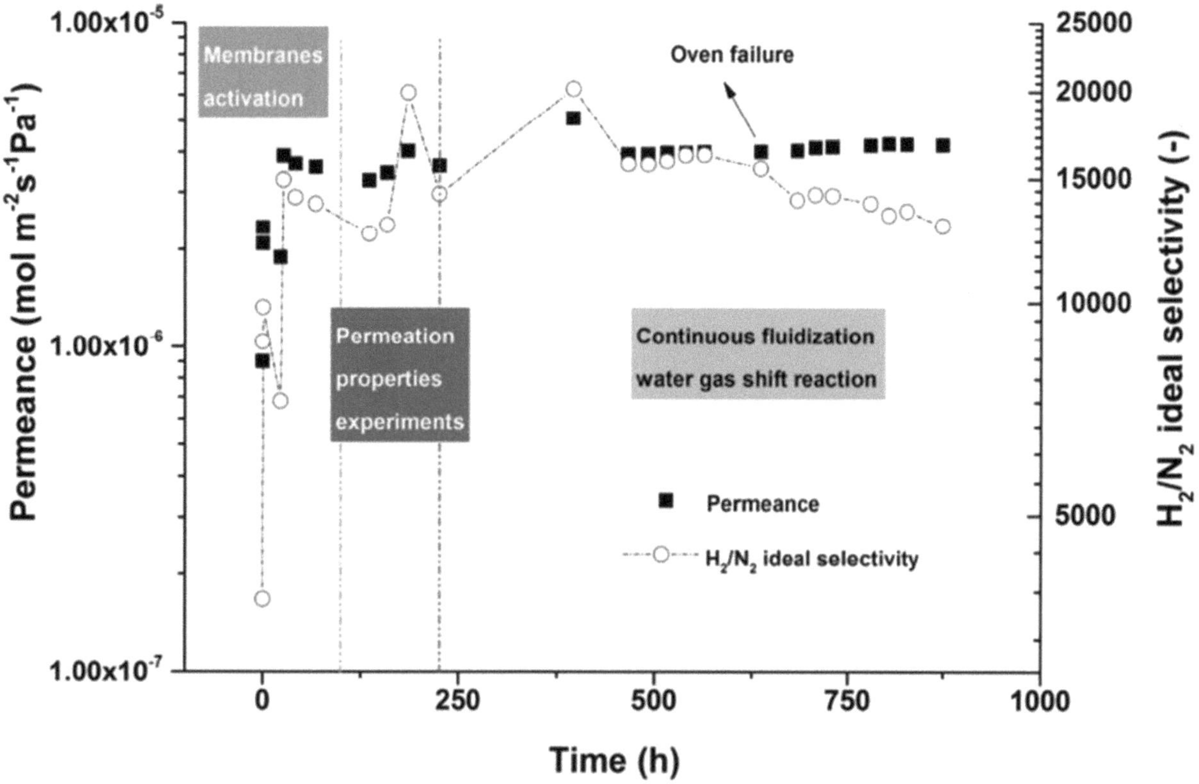

FIGURE 15.24 Long-term performance of the membrane module during 900 h of continuous operation in the bubbling fluidization regime at high-temperature WGS conditions. Reproduced from Ref. [60].

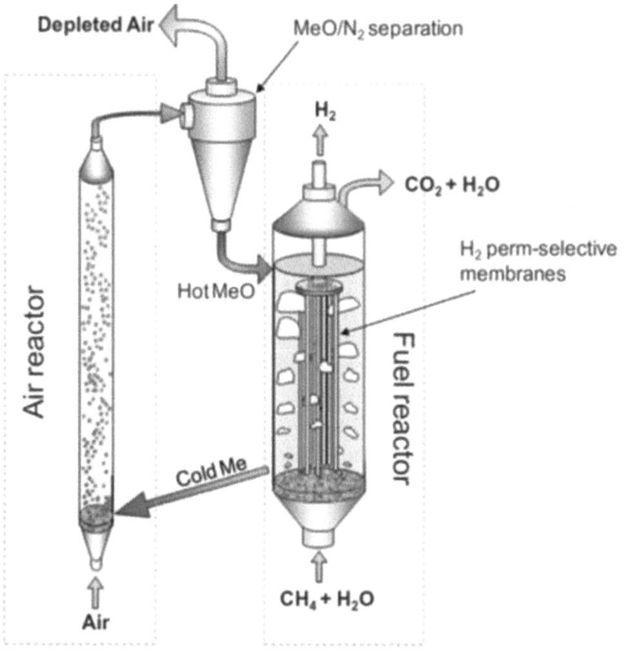

FIGURE 15.25 Schematic representation of the MA-CLR concept for pure H_2 production via Pd-based selective membranes combined with CO_2 capture via chemical looping. Reproduced from Ref. [62].

provides a high degree of process intensification and allows for much higher efficiencies compared to other systems. MA-CLR has shown very high reforming efficiencies at low temperatures because higher fuel conversion is achieved with relatively low membrane area installed and with lower steam requirements compared to other systems. With this concept, H_2 can be produced with inherent CO_2 capture at the same cost (or even cheaper) than conventional steam methane reforming without CO_2 capture.

A variation of this concept has been proposed and demonstrated by Wassie et al. [64] who demonstrated a concept called membrane-assisted gas switching reforming. Compared to the previous concept by Medrano et al. [62] (MA-CLR), gas switching uses a single fluidized bed membrane reactor and the gas is periodically switched between oxidation and reforming. This reactor concept combines ultra-pure H_2 production with integrated CO_2 capture from steam methane reforming through the use of an oxygen carrier, which acts as catalyst and heat carrier to the endothermic reforming and is periodically fed with either the fuel (methane and steam) or with air for the oxidation step. When air is fed into the reactor, the oxygen carrier is heated by the exothermic solids oxidation reaction to be utilized in the fuel stage, where endothermic reduction and catalytic reactions regenerate the oxygen carrier and produce syngas. The H_2 perm-selective membranes used in the fluidized bed directly recover pure H_2 generated during steam-methane reforming and water–gas shift reactions.

15.4 CONCLUSIONS AND FUTURE DIRECTIONS

This chapter reports the latest developments in membranes and membrane reactors for hydrogen production. Both membrane production and membrane reactors are being developed at an industrial scale. From the membrane part, H2Site has just opened a production line for Pd membranes capable of producing up to 12,000 membrane tubes per year. On the reactor side, the size has now reached TRL7 within the European project MACBETH, where two reactors are being tested in industrial conditions.

To pave the way to full exploitation of this technology, the remaining steps are the long-term testing of the system for at least one year and automatization of the system integration. This will lead to cheaper units that fulfil the demands of a fast-changing market based on hydrogen as an energy carrier.

REFERENCES

[1] Gryaznov, V. (2000) Metal containing membranes for the production of ultrapure hydrogen and the recovery of hydrogen isotopes. Separation and Purification Methods 29: 171–187.

[2] Ermilova, M.M., Orekhova, N.V., Tereshchenko, G.F., Malygin, A.A., Malkov, A.A., Basile, A., Gallucci, F. (2008) Methanol oxidative dehydrogenation on nanostructured vanadium-containing composite membranes. *Journal of Membrane Science* 317: 88–95.

[3] Gallucci, F., Tosti, S., Basile, A. (2008) Pd-Ag tubular membrane reactors for methane dry reforming: A reactive method for CO_2 consumption and H_2 production. *Journal of Membrane Science* 317: 96–105.

[4] Tosti, S., Basile, A., Bettinali, L., Borgognoni, F., Chiaravalloti, F., Gallucci, F. (2006) Long-term tests of Pd-Ag thin wall permeator tube. Journal of Membrane Science 284: 393–397.

[5] Itoh, N. (2012) Analysis of equilibrium-limited dehydrogenation and steam reforming in palladium membrane reactors. *Journal of the Japan Petroleum Institute* 55: 160–170.

[6] Gallucci, F., Van Sintannaland, M., Kuipers, J.A.M. (2010) Theoretical comparison of packed bed and fluidized bed membrane reactors for methane reforming *International Journal of Hydrogen Energy* 35: 7142–7150.

[7] Ye, G., Xie, D., Qiao, W., Grace, J.R., Lim, C.J. (2009) Modeling of fluidized bed membrane reactors for hydrogen production from steam methane reforming with Aspen Plus. *International Journal of Hydrogen Energy* 34: 4755–4762.

[8] Peters, T., Caravella, A. (2019) Pd-based membranes overview and perspectives. *Membranes MDPI* 9: 25.

[9] Jokar, S.M., Farokhnia, A., Tavakolian, M., Pejman, M., Parvasi, P., Javanmardi, J., Zare, F., Gonçalves, M.C., Basile, A. (2022) The recent areas of applicability of palladium based membrane technologies for hydrogen production from methane and natural gas: A review. *International Journal of Hydrogen Energy* 48(16): 6451–6476.

[10] Habib, M.A., Harale, A.,Paglieri, S., Alrashed, F.S., Al-Sayoud, A., Rao, M.V., Nemitallah, M.A., Hossain, S., Hussien, M., Ali, A., Haque, M.A., Abuelyamen, A., Shakeel, M.R., Mokheimer, E.M.A., Ben-Mansour, R. (2021) Palladium-alloy membrane reactors for fuel reforming and hydrogen production: A review. *Energy and Fuels* 35: 5558–5593.

[11] Amiri, T.Y., Ghasemzageh, K., Iulianelli, A. (2020) Membrane reactors for sustainable hydrogen production through steam reforming of hydrocarbons: A review. *Chemical Engineering and Processing – Process Intensification* 157: 108148.

[12] Viviente Sole, J.L., Pacheco Tanaka, D.A., Medrano, J.A., Gallucci, F. (2020) An overview of some recent European projects on metallic membranes. In: Basile, A., Gallucci, F. (Eds.) *Current Trends and Future Developments on (Bio-) Membranes*, pp. 313–379. Elsevier.

[13] Klette, B.H., Materials, R.B.S. (2005) Sputtering of very thin palladium-alloy hydrogen. *Membrane Technology* 7–9.

[14] Peters, T.A., Stange, M., Sunding, M.F., Bredesen, R. (2015) Stability investigation of micro-configured Pd-Ag membrane modules – Effect of operating temperature and pressure. *International Journal of Hydrogen Energy* 40: 3497–3505.

[15] Fernandez, E., Sanchez-Garcia, J.A., Viviente, J.L., Van Sint Annaland, M., Gallucci, F., Pacheco Tanaka, D.A. (2016) Morphology and N2 permeance of sputtered Pd-Ag ultra-thin film membranes. *Molecules* 21: 1–13.

[16] Fernandez, E., Medrano, J.A., Melendez, J., Parco, M., Viviente, J.L., Annaland, M. van S., Gallucci, F., Pacheco Tanaka, D.A. (2016) Preparation and characterization of metallic supported thin Pd–Ag membranes for hydrogen separation. *Chemical Engineering Journal* 305: 182–190.

[17] Itoh, N., Akiha, T., Sato, t. (2005) Preparation of thin palladium composite membrane tube by a CVD technique and its hydrogen permselectivity. *Catalysis Today* 104: 231–237.

[18] Uemiya, S., Sato, N., Ando, H., Kude, Y. (1991) Separation of hydrogen through palladium thin film supported on a porous glass tube. *Journal of Membrane Science* 56: 303–313.

[19] Uemiya, S., Matsuda, T., Kikuchi, E. (1991) Hydrogen permeable palladium-silver supported on porous ceramics alloy membrane. *Journal of Materials Science* 56: 315–325.

[20] Pacheco Tanaka, D.A., Llosa Tanco, M.A., Niwa, S., Wakui, Y., Mizukami, F., Namba, T., Suzuki, T. (2005) Preparation of palladium and silver alloy membrane on a porous α-alumina tube via simultaneous electroless plating. *Journal of Membrane Science* 247: 21–27.

[21] Okazaki, J., Pacheco Tanaka, D.A., Llosa Tanco, M.A., Wakui, Y., Mizukami, F., Suzuki, T.M. (2006) Hydrogen permeability study of the thin Pd-Ag alloy membranes in the temperature range across the α-β phase transition. *Journal of Membrane Science* 282: 370–374.

[22] Pacheco Tanaka, D.A., Okazaki, J., Llosa Tanco, M.A., Suzuki, T.A. (2015) Fabrication of supported palladium alloy membranes using electroless plating techniques. In: *Palladium Membrane Technology for Hydrogen Production, Carbon Capture and Other Applications Principles, Energy Production and Other Applications*, pp. 83–99. Woodhead Publishing.

[23] Mardilovich, I.P., Engwall, E., Ma, Y.H. (2002) Dependence of hydrogen flux on the pore size and plating surface topology of asymmetric Pd-porous stainless steel membranes. *Desalination* 144: 85–89.

[24] Fernandez, E., Coenen, K., Helmi, A., Melendez, J., Zuñiga, J., Pacheco Tanaka, D.A., Van Sint Annaland, M., Gallucci,

F. (2015) Preparation and characterization of thin-film Pd-Ag supported membranes for high-temperature applications. *International Journal of Hydrogen Energy* 40: 13463–13478.

[25] Agnolin, S., Melendez, J., Di Felice, L., Gallucci, F. (2022) Surface roughness improvement of Hastelloy X tubular filters for H2 selective supported Pd–Ag alloy membranes preparation. *International Journal of Hydrogen Energy* 47: 28505–28517.

[26] Medrano, J.A., Fernandez, E., Melendez, J., Parco, M., Pacheco Tanaka, D.A., Van Sint Annaland, M., Gallucci, F. (2016) Pd-based metallic supported membranes: High-temperature stability and fluidized bed reactor testing. *International Journal of Hydrogen Energy* 41: 8706–8718.

[27] Fernandez, E., Helmi, A., Coenen, K., Melendez, J., Viviente, J.L., Pacheco Tanaka, D.A., Van Sint Annaland, M., Gallucci, F. (2015) Development of thin Pd-Ag supported membranes for fluidized bed membrane reactors including WGS related gases. *International Journal of Hydrogen Energy* 40: 3506–3519.

[28] Di Marcoberardino, D., Binotti, M., Manzolini, G., Viviente, J.L., Arratibel, A., Roses, L., Gallucci, F. (2017) Achievements of European projects on membrane reactor for hydrogen production. *Journal of Cleaner Production* 161: 1442–1450.

[29] Nooijer, N., Arratibel Plazaola, A., Meléndez Rey, J., Fernandez, E., Pacheco Tanaka, D., Sint Annaland, M., Gallucci, F. (2019) Long-term stability of thin-film Pd-based supported membranes. *Processes* 7: 106.

[30] Nordio, M., Soresi, S., Manzolini, G., Melendez, J., Van Sint Annaland, M., Pacheco Tanaka, D.A., Gallucci, F. (2019) Effect of sweep gas on hydrogen permeation of supported Pd membranes: Experimental and modeling. *International Journal of Hydrogen Energy* 44: 4228–4239.

[31] Melendez, J., Fernandez, E., Gallucci, F., van Sint Annaland, M., Arias, P.L., Pacheco Tanaka, D.A. (2017) Preparation and characterization of ceramic supported ultra-thin (~1 μm) Pd-Ag membranes. *Journal of Membrane Science* 528.

[32] Pacheco Tanaka, D.A., Medrano, J.A., Viviente Sole, J.L., Gallucci, F. (2020) Metallic membranes for hydrogen separation In: *Current Trends and Future Developments on (Bio-) Membranes Recent Advances in Metallic Membranes*, pp. 1–29.

[33] Liu, J., Bellini, S., de Nooijer, N.C.A., Sun, Y., Pacheco Tanaka, D.A., Tang, C., Li, H., Gallucci, F., Caravella, A. (2020) Hydrogen permeation and stability in ultra-thin Pd–Ru supported membranes. *International Journal of Hydrogen Energy* 45: 7455–7467.

[34] Fernandez, E., Coenen, K., Helmi, A., Melendez, J., Zuñiga, J., Pacheco Tanaka, D.A., van Sint Annaland, M., Gallucci, F. (2015) Preparation and characterization of thin-film Pd–Ag supported membranes for high-temperature applications. *International Journal of Hydrogen Energy* 40: 13463–13478.

[35] Wassie, S.A., Medrano, J.A., Zaabout, A., Cloete, S., Melendez, J., Pacheco Tanaka, D.A., Amini, S., van Sint Annaland, M., Gallucci, F. (2018) Hydrogen production with integrated CO_2 capture in a membrane assisted gas switching reforming reactor: Proof-of-concept. *International Journal of Hydrogen Energy* 43: 6177–6190.

[36] Okazaki, J., Pacheco Tanaka, D.A., Llosa Tanco, M.A., Wakui, Y., Ikeda, T., Mizukami, F., Suzuki, T.M. (2008) Preparation and hydrogen permeation properties of thin Pd-Au alloy membranes supported on porous α-alumina tube. *Materials Transactions* 49: 449–452.

[37] Okazaki, J., Ikeda, T., Pacheco Tanaka, D.A., Llosa Tanco, M.A., Wakui, Y., Sato, K., Mizukami, F., Suzuki, T.M. (2009) Importance of the support material in thin palladium composite membranes for steady hydrogen permeation at elevated temperatures. *Physical Chemistry Chemical Physics* 11: 8632–8638.

[38] Pacheco Tanaka, D.A., Llosa Tanco, M.A., Nagase, T., Okazaki, J., Wakui, Y., Mizukami, F., Suzuki, t.M. (2006) Fabrication of hydrogen-permeable composite membranes packed with palladium nanoparticles. *Advanced Materials* 18: 630–632.

[39] Pacheco Tanaka, D.A., Llosa Tanco, M.A., Okazaki, J., Wakui, Y., Mizukami, F., Suzuki, T.M. (2008)Preparation of “pore-fill” type Pd-YSZ-γ-Al_2O_3 composite membrane supported on α-Al_2O_3 tube for hydrogen separation. *Journal of Membrane Science* 320: 436–441.

[40] Nordio, M., Melendez, J., van Sint Annaland, M., Pacheco Tanaka, D.A., Llosa Tanco, M., Gallucci, F. (2020) Comparison between carbon molecular sieve and Pd-Ag membranes in H2-CH4 separation at high pressure. *International Journal of Hydrogen Energy* 45: 28876–28892.

[41] Arratibel, A., Pacheco Tanaka, A., Laso, I., van Sint Annaland, M., Gallucci, F. (2018) Development of Pd-based double-skinned membranes for hydrogen production in fluidized bed membrane reactors. *Journal of Membrane Science* 550: 536–544.

[42] Arratibel, A., Medrano, J.A., Melendez, J., Pacheco Tanaka, D.A., van Sint Annaland, M., Gallucci, F. (2018) Attrition-resistant membranes for fluidized-bed membrane reactors: Double-skin membranes. *Journal of Membrane Science* 563: 419–426.

[43] Arratibel, A., Pacheco Tanaka, A., van Sint Annaland, M., Gallucci, F. (2021) On the use of double-skinned membranes to prevent chemical interaction between membranes and catalysts. *International Journal of Hydrogen Energy* 46: 20240–20244.

[44] Vicinanza, N., Svenum, I.H., Peters, T., Bredesen, R., Venvik, H. (2018) New insight to the effects of heat treatment in air on the permeation properties of thin Pd77%Ag23% membranes. *Membranes (Basel)* 8: 1–14.

[45] Okazaki, J., Ikeda, T., Pacheco Tanaka, D.A., Suzuki, T.M., Mizukami, F. (2009) In situ high-temperature X-ray diffraction study of thin palladium/α-alumina composite membranes and their hydrogen permeation properties. *Journal of Membrane Science* 335: 126–132.

[46] Fernandez, E., Sanchez-Garcia, J.A., Melendez, J., Spallina, V., Annaland, M. van S., Gallucci, F., Pacheco Tanaka, D.A., Prema, R. (2016) Development of highly permeable ultra-thin Pd-based supported membranes. *Chemical Engineering Journal* 305: 149–155.

[47] Melendez, J., de Nooijer, N., Coenen, K., Fernandez, E., Viviente, J.L., van Sint Annaland, M., Arias, P.L., Pacheco Tanaka, D.A., Gallucci, F. (2017) Effect of Au addition on hydrogen permeation and the resistance to H2S on Pd-Ag alloy membranes. *Journal of Membrane Science* 542: 329–341.

[48] de Nooijer, N., Sanchez, J.D., Melendez, J., Fernandez, E., Pacheco Tanaka, D.A., van Sint Annaland, M., Gallucci, F. (2020) Influence of H2S on the hydrogen flux of thin-film

PdAgAu membranes. *International Journal of Hydrogen Energy* 45: 7303–7312.

[49] Barbieri, G., Brunetti, A., Caravella, A., Drioli, E. (2011) Pd-based membrane reactors for one-stage process of water gas shift. *RSC Advances* 1: 651–661.

[50] Brunetti, A., Caravella, A., Barbieri, G., Drioli, E. (2007) Simulation study of water gas shift reaction in a membrane reactor. *Journal of Membrane Science* 306: 329–340.

[51] Gallucci, F., Pacheco Tanaka, D.A., Medrano, J.A., Viviente Sole, J.L. (2020) *Membrane Reactors using Metallic Membranes*. Elsevier Inc.

[52] Tiemersma, T.P., Patil, C.S., Van Sint Annaland, M., Kuipers, J.A.M. (2006) Modelling of packed bed membrane reactors for autothermal production of ultrapure hydrogen. *Chemical Engineering Science* 61: 1602–1616.

[53] Sartbaeva, A., Kuznetsov, V.L., Wells, S.A., Edwards, P.P. (2008) Hydrogen nexus in a sustainable energy future. *Energy & Environmental Science* 1: 79–85.

[54] Kim, S., Song, J., Lim, H. (2018) Conceptual feasibility studies of a COX-free hydrogen production from ammonia decomposition in a membrane reactor for PEM fuel cells, *Korean Journal of Chemical Engineering* 35: 1509–1516.

[55] Jo, Y.S., Cha, J., Lee, C.H., Jeong, H., Yoon, C.W., Nam, S.W., Han, J. (2018) A viable membrane reactor option for sustainable hydrogen production from ammonia. *Journal of Power Sources* 400: 518–526.

[56] Zhang, Z., Liguori, S., Fuerst, T.F., Way, J.D., Wolden, C.A. (2019) Efficient ammonia decomposition in a catalytic membrane reactor to enable hydrogen storage and utilization. *ACS Sustainable Chemistry & Engineering* 7: 5975–5985.

[57] Cheng, H., Meng, B., Li, C., Wang, X., Meng, X., Sunarso, J., Tan, X., Liu, S. (2020) Single-step synthesized dual-layer hollow fiber membrane reactor for on-site hydrogen production through ammonia decomposition. *International Journal of Hydrogen Energy* 45: 7423–7432.

[58] Cechetto, V., Di Felice, L., Medrano, J.A., Makhloufi, C., Zuniga, J., Gallucci, F. (2021) H2 production via ammonia decomposition in a catalytic membrane reactor. *Fuel Processing Technology* 216: 106772.

[59] Gallucci, F., Annaland, M., Kuipers, J. (2008) Autothermal reforming of methane with integrated CO2 capture in a novel fluidized bed membrane reactor. Part 1: experimental demonstration. *Topics in Catalysis* 51: 133–145.

[60] Helmi, A., Fernandez, E., Melendez, J., Pacheco Tanaka, D.A., Gallucci, F., Van Sint Annaland, M. (2016) Fluidized bed membrane reactors for ultra pure H_2 production – A step forward towards commercialization. *Molecules* 21: 376.

[61] Medrano, J.A., Spallina, V., van Sint Annaland, M., Gallucci, F. (2013) Thermodynamic analysis of a membrane-assisted chemical looping reforming reactor concept for combined H_2 production and CO_2 capture. *International Journal of Hydrogen Energy* 39: 4725–4738.

[62] Medrano, J.A., Potdar, I., Melendez, J., Spallina, V., Pacheco Tanaka, D.A., van Sint Annaland, M., Gallucci, F. (2018) The membrane-assisted chemical looping reforming concept for efficient H_2 production with inherent CO_2 capture: Experimental demonstration and model validation. *Applied Energy* 215: 75–86.

[63] Medrano, J.A., Spallina, V., Van Sint Annaland, M., Gallucci, F. (2014) Thermodynamic analysis of a membrane-assisted chemical looping reforming reactor concept for combined H_2 production and CO_2 capture. *International Journal of Hydrogen Energy* 39: 4725–4738.

[64] Wassie, S.A., Medrano, J.A., Zaabout, A., Cloete, S., Melendez, J.,Pacheco Tanaka, D.A., Amini, S., van Sint Annaland, M., Gallucci, F. (2018) Hydrogen production with integrated CO_2 capture in a membrane assisted gas switching reforming reactor: Proof-of-concept. *International Journal of Hydrogen Energy* 43(12): 6177–6190.

16 Membrane Applications in Industrial Waste Management (Including Nuclear), Environmental Engineering and Future Trends in Membrane Science

Introduction

Anil Kumar Pabby[1], S. Ranil Wickramasinghe[2], and Ana-Maria Sastre[3]*

[1]Formerly associated with BARC Complex, Nuclear Recycle Board, BARC, Tarapur, Maharashtra, India

[2]Ralph E. Martin, Deptartment of Chemical Engineering, University of Arkansas, Fayetteville, AR, USA

[3]Department of Chemical Engineering, Universitat Politècnica de Catalunya, Barcelona, Spain

*Author for communication, E-mail: dranilpabby@gmail.com

Growing population and dwindling natural resources have made it imperative to look for methods for maximization of resource utilization and to develop more sustainable manufacturing processes. The basic resources, both natural such as water, air, or soil, and synthetic such as organic chemicals like phenol, pharmaceuticals, nutrients like phosphorus, nitrogen, and trace metallic species such as heavy metals like uranium, copper, zinc, chromium, etc., are becoming costlier and scarcer. Therefore, researchers are looking to extract heavy and valuable metallic species like rubidium, caesium, uranium, etc., present in trace and ultra-trace levels from the ocean [1].

The species although it is useful can only considered as waste when it cannot be economically retrieved. Even though conventional processes, such as chemical precipitation, co-precipitation, ion exchange, etc., can be tuned to recover these low concentrations for reuse, they are economically unviable, particularly due to economy of scale. The other difficulty arises due to the mixing of all the spent streams together in a particular industrial complex, which makes it more difficult to extract the individual valuable species. Since the objective, in most instances, is to comply with the pollution control norms, wastewater streams are subjected to generalized treatment before being discharged into the environment. This situation has changed as water has become a valuable resource with increasing demand and reduced availability. However, no attention has been given to its recovery value except in a few sectors, mostly due to the non-availability of proven sustainable technologies [2].

Similarly, water resources are constantly subjected to contamination from various sources such as sewage, agricultural waste, and oil spillage. The discharge of these wastes into water bodies has caused serious chain effects that not only challenge water security but also threaten the balance of ecosystems and the health of populations. Myriads of wastewater treatment and reclamation approaches have been established to restore the quality of polluted water to meet discharge standards or to be reused for other purposes [3–5].

The three main methods of wastewater treatment are biological, physical, and chemical treatments. These methods have been conventionally used in a wide range of industries. The selection of a method depends on many factors such as the nature and characteristics of the wastewater, and the effectiveness, cost, and practicability of the treatment processes [5]. These wastewater treatment technologies have been long applied at a commercial scale across industries, however, they also suffer from some inevitable downsides.

The need for more reliable solutions has resulted in a call for alternatives to remove hazardous substances from both air and water media. The establishment of more sustainable remedial strategies and technologies is also desired to cope with the emergence of complex contaminants and evolving regulatory enforcement [6].

Among emerging technologies, there are two options, the first is the pressure-driven membrane processes such as RO, NF, UF, and MF applied for the removal of contaminants from wastewater, and the second option is membrane

DOI: 10.1201/9781003285656-16

contactor operations which are a convenient and simple tool that combines absorption by a liquid absorbent and membrane as a mass transfer facilitator. If we consider the first option, then we reveal more established and fully fledged commercial environmentally friendly technology to answer the multifarious demands of industry [7,8].

In the second option, the membrane contactor is more of an emerging technology being applied due to its several advantages as compared to other techniques. Since this is a new technology, researchers are required to understand the basics of membrane contactor operation and design. The principle of the membrane contactor lies on the mass transfer between two phases through a porous membrane. In this system, the membrane serves as a non-selective interface to enable the contact of two phases. One point of interest is that the pores of the contacting membranes, which are normally in the pore size range of 0.002–1 μm, are sufficiently small to prevent direct mixing of the phases as long as the transmembrane pressure is lower than the liquid entry pressure.

During operation, the pressure in the system is carefully controlled to avoid the intermixing of phases [2,7]. On the other hand, it has been demonstrated that pressure-driven membrane processes can be used successfully in the removal of radioactive substances, with some distinct advantages over conventional processes. Following the development of suitable membrane materials and their long-term verification in conventional water purification, membrane processes have been adopted by the nuclear industry as a viable alternative for the treatment of radioactive liquid wastes [9,10]. Among the several techniques used, the advantages offered by membrane bioreactor (MBR) technology have been recognized for some time. An MBR comprises a conventional activated sludge process coupled with membrane separation to retain the biomass. Since the effective pore size is generally below 0.1 μm, the MBR effectively produces a clarified and substantially disinfected effluent. In addition, it concentrates the biomass and, in doing so, reduces the necessary tank size and also increases the efficiency of the biotreatment process [7].

In this section, updated and detailed chapters are provided in addition to several new chapters to describe the latest developments and methodologies in this area, giving special emphasis to the use of various membrane techniques for the treatment of waste/nuclear waste generated by the chemical/nuclear industry. Chapter 16 (this chapter) is an introduction to the chapters in this section. Chapter 17 describes integrated membrane distillation approaches for sustainable water recovery and desalination. Chapter 18 details advances in membrane processes for nuclear waste processing and describes the current global situation of membrane processes for the treatment of nuclear waste. This chapter also covers the future role of membrane processes in nuclear technology and the advantages and limitations of applied membrane processes. Chapter 19 focuses on a new emerging area of membrane separation science, i.e. polymer inclusion membrane. An introduction to the principles is provided and associated mechanisms including chemistry aspects and important applications are described. Chapter 20 presents the electromembrane process and its recent advances, applications, and future perspectives.

Chapter 21 describes membrane applications for valorization routes of industrial brines and mining waters while providing examples of resource recovery schemes. Chapter 22 describes the applications of concentration-driven membrane processes for the recovery of valuable compounds from industrial waste. Chapter 23 presents on the recovery of salts from brines through membrane crystallization processes. A sustainable approach for dye removal from industrial effluent using concentration and pressure-driven membrane techniques is summarized in Chapter 24. Chapter 25 deals with membrane bioreactors for wastewater treatment, including its challenges and future perspectives. The early membranologists have always been optimistic about the possibilities of membrane operations, but the scientific and technical results reached today are even greater than their expectations. A variety of technical challenges must be overcome to permit the successful industrial application of new membrane solutions. The last chapter, Chapter 26, on the same theme, presents the future scenario of membrane processes and focuses on future progress in membrane engineering including process intensification in water treatment.

REFERENCES

[1] Bardi, U. (2010) Extracting minerals from seawater: an energy analysis. *Sustainability* 2: 980–992.

[2] Kavitha, E., Poonguzhali, E., Nanditha, D., Kapoor, A., Arthanareeswaran, G., Prabhakar, S. (2022) Current status and future prospects of membrane separation processes for value recovery from wastewater. *Chemosphere* 291: 132690.

[3] Ahmad, N.A., Goh, P.S., Karim, Z.A., Ismail, A.F. (2018) Thin film composite membrane for oily waste water treatment: recent advances and challenges. *Membranes (Basel)* 8(4): 86.

[4] Wang, P., Chung, T.S. (2015) Recent advances in membrane distillation processes: membrane development, configuration design and application exploring. Journal of Membrane Science 474: 39–56.

[5] Ahmed, M.B., Zhou, J.L., Ngo, H.H., Guo, W., Thomaidis, N.S., Xu, J. (2017)Progress in the biological and chemical treatment technologies for emerging contaminant removal from wastewater: a critical review. Journal of Hazardous Materials 323: 274–298.

[6] Goh, P.S., Naim, R., Rahbari-Sisakht, M., Ismail, A.F. (2019) Modification of membrane hydrophobicity in membrane contactors for environmental remediation. *Separation and Purification Technology* 227: 115721.

[7] Pabby, A.K., et al. (2015) *Handbook of Membrane Separations: Chemical, Pharmaceutical, Food, and Biotechnological Applications*, 2nd Edition. CRC Press: Boca Raton, FL, USA.

[8] Pabby, A.K., et al. (2021) *Industrial Experience with Fukushima Daiichi Accident Wastewater Treatment: Part I & II, Membrane Technology*, August/September 2021. Mark Allen Group: UK

[9] Pabby, A.K., Swain, B., Sonar, N.L., Mittal, V.K., Valsala, T.P., Ramsubramanian, S. (2022) Radioactive waste processing using membranes: State of the art technology, challenges and perspectives. *Separation & Purification Reviews* 51(2): 143–173.

[10] Zhang, X., Gu, P., Liu, Y. (2019) Decontamination of radioactive wastewater: State of the art and challenges forward. *Chemosphere* 215: 543.

17 Integrated Membrane Distillation Approaches for Sustainable Desalination and Water/Resource Recovery

*V. Sangeetha and Noel Jacob Kaleekkal**

Membrane Separation Group, Department of Chemical Engineering, National Institute of Technology Calicut (NITC), Kerala, India

*Corresponding Author E-mail ID: noeljacob89@gmail.com; noel@nitc.ac.in

17.1 INTRODUCTION

17.1.1 Membrane Distillation

Membrane distillation (MD) is an emerging potential technology for desalination, water recovery, and resource recovery applications. It is a thermally driven process in which the vapours are transported through a porous hydrophobic membrane from a hot feed solution [1]. The hydrophobic nature of the membrane prevents the liquid from entering the pores. The driving force of separation is the partial pressure difference across the membrane arising from the temperature gradient between the feed and the permeate side [2]. The liquid–liquid separation across the membrane pores facilitates the MD operation independent of the osmotic pressure difference between the feed and permeate solutions, thereby making it a potential technology for draw solution recovery in integrated hybrid systems [3].

17.1.2 Membrane Distillation Configurations

Major configurations adopted include direct contact membrane distillation (DCMD), wherein vapour diffused through the membrane pores is condensed by a stream of colder liquid which is in direct contact with the membrane, air gap membrane distillation (AGMD), wherein the vapours are condensed on the permeate side by a cold surface, vacuum membrane distillation (VMD), wherein the volatile compounds are removed by the application of vacuum on the permeate side, and sweep gas membrane distillation (SGMD), wherein the volatile components are removed by flowing a sweep gas on one of the membrane surfaces. Other configurations for enhanced performance used are vacuum-enhanced direct-contact membrane distillation (VE-DCMD), material gap membrane distillation (MGMD), multi-effect membrane distillation (MEMD), where "n" is the number of MD units as subsystems connected in series, vacuum-multi-effect membrane distillation (V-MEMD), permeate gap membrane distillation (PGMD), liquid gap membrane distillation (LGMD), pressure-retarded membrane distillation (PRMD), water gap membrane distillation (WGMD), solar vapour gap membrane distillation (SVGMD), and osmotic membrane distillation (OMD) [4]. However, DCMD, owing to its simplicity in operation and maintenance, is widely preferred and used for all applications [5].

17.1.3 Membrane Distillation–Advantages and Limitations

The advantages of MD include high permeate quality, energy efficiency, and the ability to use low-grade heat and treat high-salinity water [6]. Unlike pressure-driven membranes, MD is less susceptible to membrane incrustation, and membrane fouling is reversible. The MD process is less sensitive to the feed salinity as only the vapour molecules penetrate through the membrane due to partial vapour pressure, thus facilitating water recovery from highly concentrated feed solutions such as RO brines [7]. Also, adding chemicals for water treatment is eliminated, which would yield hazardous by-products that persist in the environment. However, the presence of volatile contaminants in wastewater is a significant challenge to the membrane distillation process [2].

The major limitation in MD is membrane fouling and pore wetting over long-term use [1], temperature, and concentration polarization [8]. Membrane pore wetting occurs as the liquid fills up the membranes leading to permeation of feed through the membrane, ultimately deteriorating the permeate quality. Membrane wetting depends on the membrane's intrinsic factors, such as surface free energy and surface morphology, the concentration of feed solution, operational parameters and intermittent operation, high transmembrane pressure, and membrane degradation [9]. Membrane fouling in MD hydrophobic membranes is inevitable due to the hydrophobic–hydrophobic interaction between the membrane and hydrophobic foulants leading to pore blockage and a decline in permeate flux [10]. Also, temperature polarization due to the temperature difference between the feed solution

DOI: 10.1201/9781003285656-17

and feed–membrane interface reduces the permeate flux with greater energy consumption [11]. Concentration polarization occurs when the salt concentration is higher at the feed–membrane interface than in bulk and is related to membrane scaling and fouling [12]. To overcome these challenges, various approaches have been adopted, such as surface modification of membranes, operational parameters, and integration with other desalination models have been investigated.

17.1.4 Commercial Membrane Distillation Membranes

The commercial membranes that are widely used include polytetrafluoroethylene (PTFE), polyvinylidene fluoride (PVDF), and polypropylene (PP) membranes [13] in tubular, hollow-fibre, or flat-sheet membrane configurations. PTFE is a widely used ideal material with chemical resistance and thermal stability that is prepared by complicated extrusion, rolling, and stretching or sintering procedures. PP and PVDF membranes are prepared by a molten extrusion technique followed by stretching or thermal phase separation process and phase inversion, respectively [14]. Commercial membranes are robust with high durability and reusability [5]. Table 17.1 details the most commonly used commercial membranes for MD and their properties. It can be inferred that almost all membranes depict a nominal pore size of 0.2 μm with an average porosity of 70%. However, these membranes are manufactured to fulfil the characteristics of microfiltration and the permeability reported is of liquid mass transfer rather than vapour transfer.

17.1.5 Need for Integrated Membrane Distillation Approaches

The commercialization of membrane distillation and its implementation on a large scale are a challenge due to its high thermal energy requirements and lack of suitable robust non-wetting membranes. Despite its numerous advantages, MD as a standalone technology is uneconomical, and the integrated systems could enhance the overall performance efficiency. Reverse osmosis (RO) systems with only 60–70% water recovery discharge brine with recoverable salts and water. MD could further concentrate the brine from the RO retentate, increasing the overall water recovery to 80–90% and achieving near zero liquid discharge (ZLD) [24].

FO for its continuous and stable operation requires reconcentration of its draw solute and can be achieved with the integration of MD. Integrating MD with pressure-retarded osmosis (PRO) and reverse electrodialysis (RED) makes the process more energy efficient. Through a stack of alternating cation–anion membranes, RED converts the salinity gradient into electrical energy and PRO facilitates the conversion of osmotic energy into electrical energy. Also, in the PRO–MD hybrid process, PRO requires massive volumes of highly concentrated draw solute to drive the turbines, where the MD plays a role in re-concentrating the draw solution. Also, the multi-barrier systems provide an advantage in removing microorganisms and persisting organic pollutants in wastewater treatment plants. MD integration with membrane bioreactors can address the fouling issue of the hydrophobic MD membranes and obtain high-purity water. The hybrid systems are less energy-intensive, and integrating renewable energy sources can further reduce the conventional energy requirements [25].

Thereby, integrated processes are a potential solution for water recovery, brine reclamation, and resource recovery, thus promoting ZLD, ensuring environmental protection, sustainability, and circular economy [7]. In this chapter, membrane distillation integration with other membrane processes such as forward osmosis (FO), reverse osmosis (RO), reverse electrodialysis (RED), pressure-retarded osmosis (PRO), and renewable energy sources such as solar energy, low-grade and waste heat, wind, and geothermal energy are discussed.

17.2 INTEGRATED MD APPROACHES FOR WASTEWATER TREATMENT

17.2.1 Reverse Osmosis–Membrane Distillation (RO-MD)

Reverse osmosis (RO) is the most reliable and well-established technology for seawater/brackish water desalination [26]. The RO process employs semi-permeable membranes in which water diffuses from the more concentrated to the less concentrated aqueous solution due to an applied pressure (greater than the osmotic pressure) [27].

Flue gas desulphurization (FGD) wastewater was treated with an integrated RO-MD system to achieve a near-zero liquid discharge [28]. The FGD wastewater pretreated by chemical coagulation and ultrafiltration was sent to an RO unit containing a composite polyamide spirally wound RO membrane (SWC-2540, Matrix Desalination Inc., USA). The decrease in membrane permeability from 1.73 kg/m^2.h.bar to 0.81 kg/m^2.h.bar and a fouling index of 53.1% were attributed to the increase in feed concentration and membrane fouling. The RO process exhibited a recovery rate of 63% and displayed excellent flux recovery (86% – deionized water and 96% – acid treatment). The RO retentate was further concentrated using the DCMD process for increased water recovery. An oleophobic polyethylene flat-sheet membrane (porosity: 80%; contact angle > 118°; mean pore size: 0.3 μm; membrane area: 0.05 m^2) was used in the DCMD process, which yielded an average flux of 11 kg/m^{2}h and could recover ~ 90% high-quality water (< 80 μS/cm).

17.2.2 Forward Osmosis–Membrane Distillation (FO-MD)

Forward osmosis (FO) is an emerging membrane technology for desalination and wastewater treatment applications. The osmotic pressure difference between the feed and the draw solute is the driving force for FO, thus eliminating the need for hydraulic pressure [3]. FO is operated at low transmembrane

TABLE 17.1
Commercial membranes for MD and their properties

References	Membrane Type	Manufacturer	Membrane Material	Pore size, μm	Thickness, μm	Porosity, %	Water Contact Angle, °	Membrane Area, m^2	Additional Information
[2]	Flat, Microporous, hydrophobic	Sterlitech, USA	PTFE	0.22	NR	70	NR	NR	Nil
[15]	Microporous	Porous Membrane Technology, Ningbo, China	PTFE on PP support	0.2	60	80	NR	NR	Nil
[16]	Flat sheet	Durapore®-GVHP	PVDF	0.22	125	75	131 ± 1	NR	Nil
[17]	Flat sheet	Sterlitech, USA	PTFE in nonwoven PP	0.2	76-125	NR	NR	NR	Water entry pressure > 37 psi
[18]	Hollow fibre	Econity, Republic of Korea	PVDF	0.1	Membrane Wall thickness – 250 nm	–	106 ± 2	–	18 fibres, each 0.2 m in length; LEP – 2.0–2.3 bar; ID – 1.2 mm; OD – 0.7 mm
[19]	Hollow fibre	ModuleTM Liqui-Cel®, Membrana GmbH, Germany	PP, Potting Material-PE	0.4	NR	40	112	1.4	Nil
[20]	Flat sheet	Sterlitech, USA	PTFE laminated on a PP	0.2	NR	NR	NR	NR	Nil
[21]	NR	Fluoropore® (Millipore, Burlington, NJ, USA)	PTFE	0.22	175	70	138	0.0034	Nil
[22]	NR	3M, Maplewood, Minnesota, USA	ECTFE	0.2	82 ± 15	71	130 ± 1	NR	LEP – 330 kPa
		Pall Corporation, New York, USA	PTFE	0.02	121 ± 5	76	153 ± 4	NR	LEP – 540 kPa
[23]	Flat Sheet	Membrane Solutions, China	PTFE on a PP support	0.22	PTFE – 20 ± 0.4 PP – 80 ± 1.6	39.59 ±4.7	NR	0.00096	LEP – 3.4 bar

PTFE, polytetrafluorethylene; PP, polypropylene; PVDF, polyvinylidene fluoride; PE, polyethylene; ECTFE, ethylenechlorotrifluoroethylene copolymer; NR, not reported; LEP, liquid entry pressure; ID, inner diameter; OD, outer diameter.

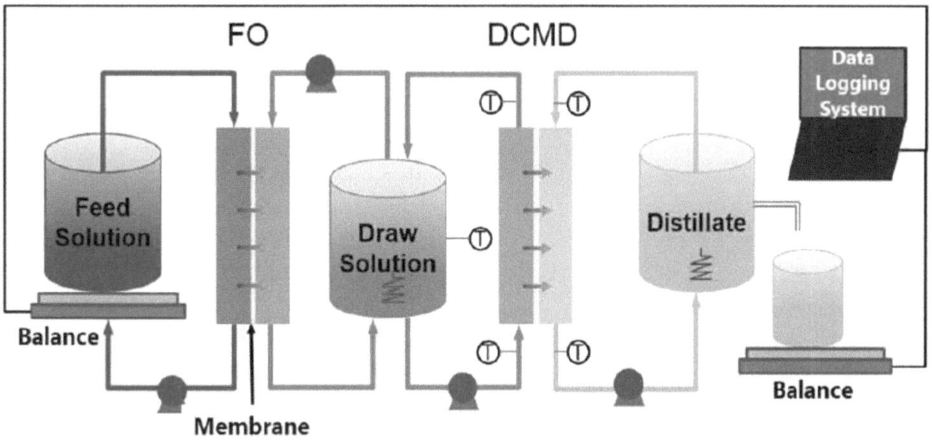

FIGURE 17.1 Schematic of FO-MD bench-scale setup. Reproduced with permission from Ref. [30].

pressure with less propensity for fouling and a reduced operational cost [20].

The feasibility of textile wastewater treatment with a forward osmosis–membrane distillation (FO-MD) hybrid system where the MD process regenerated the draw solute was evaluated [29]. The synthetic feed solution containing 1g/L Congo red dye could be concentrated ~ 10 times (with 90% water recovery) using a 1.5 M Na_2SO_4 draw solution in the FO process. The diluted DS was preheated to 55°C and sent to the MD module to recover water and concentrate the DS. A commercial PTFE membrane (PTFE0214225, Sterlitech Corporation, USA) reconcentrated the DS to produce distillate of <5 μS/cm in the DCMD mode. The initial water transfer rate (WTR) was more significant in the FO process, and the dynamic water equilibrium (equal WTR in FO and MD) was established after 12 hours of operation.

Domestic wastewater treatment was investigated with an integrated FO-MD system [30] (Figure 17.1) equipped with commercial cellulose triacetate FO membrane (HTI Albany, OR) and a PVDF hollow-fibre MD membrane (average pore size: 0.073 μm). The FO process could effectively retain 90% of the contaminants in domestic wastewater (feed), where 35 g/l NaCl solution was used as the draw solute. The negatively charged FO membrane exhibited lower removal of positively charged ammonia nitrogen (NH_4-N) and total nitrogen. The hybrid system was operated for 120 hours and provided a stable flux of 17.60 LMH with a permeate conductivity < 30 μS/cm. The fouled membranes' scanning electron microscopy (SEM) images were evaluated in detail, and confirmed deposits of complex foulants containing organic matter, calcium salts, magnesium salts, sodium salts, and silicates on the FO membrane, and solely sodium chloride (NaCl) crystals on the MD membrane. The integrated system demonstrated excellent operational stability with a yield of high-quality permeate.

A similar hybrid system was used to treat landfill leachate [31], and the system was operated continuously for 24 hours. A flat-sheet thin-film composite (TFC) membrane (Aquaporin A/S Singapore) and 4 M NaCl draw solution (DS) were used to treat the actual leachate. The diluted DS was continuously reconcentrated in the MD module using the polytetrafluoroethylene (PTFE) membrane (pore size of 0.22 μm). However, the lower retention of the foulants in the FO process resulted in fouling of the MD membrane leading to a flux decline from 9.3 to 6.8 LMH. As confirmed by the EDS analysis, the membrane surface fouling was attributed to the Ca^{2+} ions (22%) on the membrane surface. Heterogeneous crystallization is favoured as the calcium ions concentrate at the feed–membrane interface more than in bulk leading to membrane scaling. The membrane scaling was avoided by adding an antiscalant hexamethylene diamine tetra(methylene phosphonic acid) (HDTMPA) to the DS. The organic phosphine anion undergoes reverse diffusion, chelates the divalent ions and conceals the active sites for crystal growth, thus preventing scaling. This resulted in an improved permeate quality (<14 μS/cm) from the hybrid system.

A submerged FO-MD hybrid configuration with two MD membranes sandwiched between two FO membranes (Figure 17.2) was evaluated to treat wastewater containing trace organic compounds [32]. The rejection efficiencies for total organic carbon (TOC), total nitrogen (TN), and NH_4^+ were 94.9%, 93.8%, and 99.8%, respectively, in the FO process, where 1 M NaCl was used as the draw solute. The MD membrane could effectively retain > 97% of the NaCl, making this synergistically advantageous for draw solute recovery and producing high-quality freshwater.

17.2.3 Reverse Electrodialysis–Membrane Distillation (RED-MD)

Reverse electrodialysis (RED) with a stack of alternating anion and cation exchange membranes converts the osmotic energy into electrical energy. A small-scale RED-MD hybrid system for decentralized sanitation was investigated for water and energy recovery from human urine [33]. The system (Figure 17.3) consists of a VMD membrane module (G542, Mini-Module, Membrana, DE) to recover high-quality water from human urine and produce a urine concentrate rich in inorganic salts. After the VMD process, the feed solution was concentrated from 12.65 mS/cm to 24.8 mS/cm, and a permeate

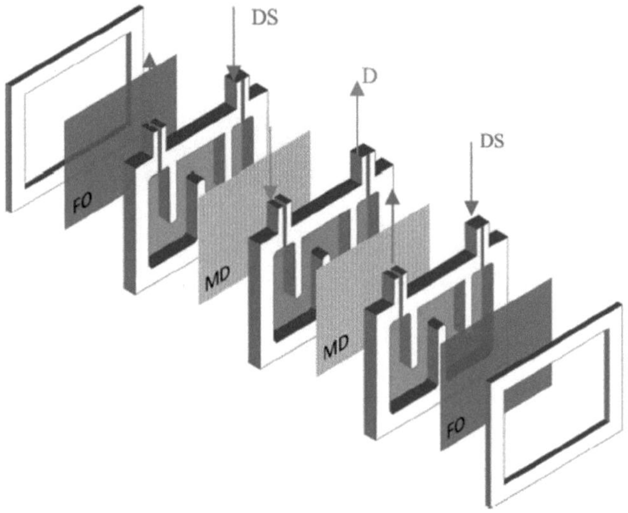

FIGURE 17.2 FO-MD hybrid module set up. Reproduced with permission from Ref. [32]. DS, draw solute of FO; D, distillate of MD.

of extremely low conductivity (0.207 mS/cm) was recovered. The permeate obtained had an NH_4-N concentration of 38.9 mg/l (feed NH_4-N concentration: 207 mg/l) and COD of 253 mg/l (feed COD concentration: 6330 mg/l). The Gibbs free energy of concentrated human urine was converted into electrical energy with a custom-made RED stack with five repeating cell units of anion and cation exchange membranes (Neosepta AMX and CMX, Eurodia, France). The retentate (concentrated feed) from VMD is fed into the RED stack, and the salinity gradient is harnessed to create an electrochemical potential across the membrane stack, and at the electrodes, a redox reaction converts the ionic potential into electrical power. A power density close to 0.2 W/m² was achieved with 40% of Gibb's free energy recovery. The recovered energy was used for low power requirements, such as operating an axial fan for sweep gas or a micro-pump for liquid transport. Therefore, membrane distillation recovers high-quality water while its integration with RED has synergistic potential for the water–energy nexus.

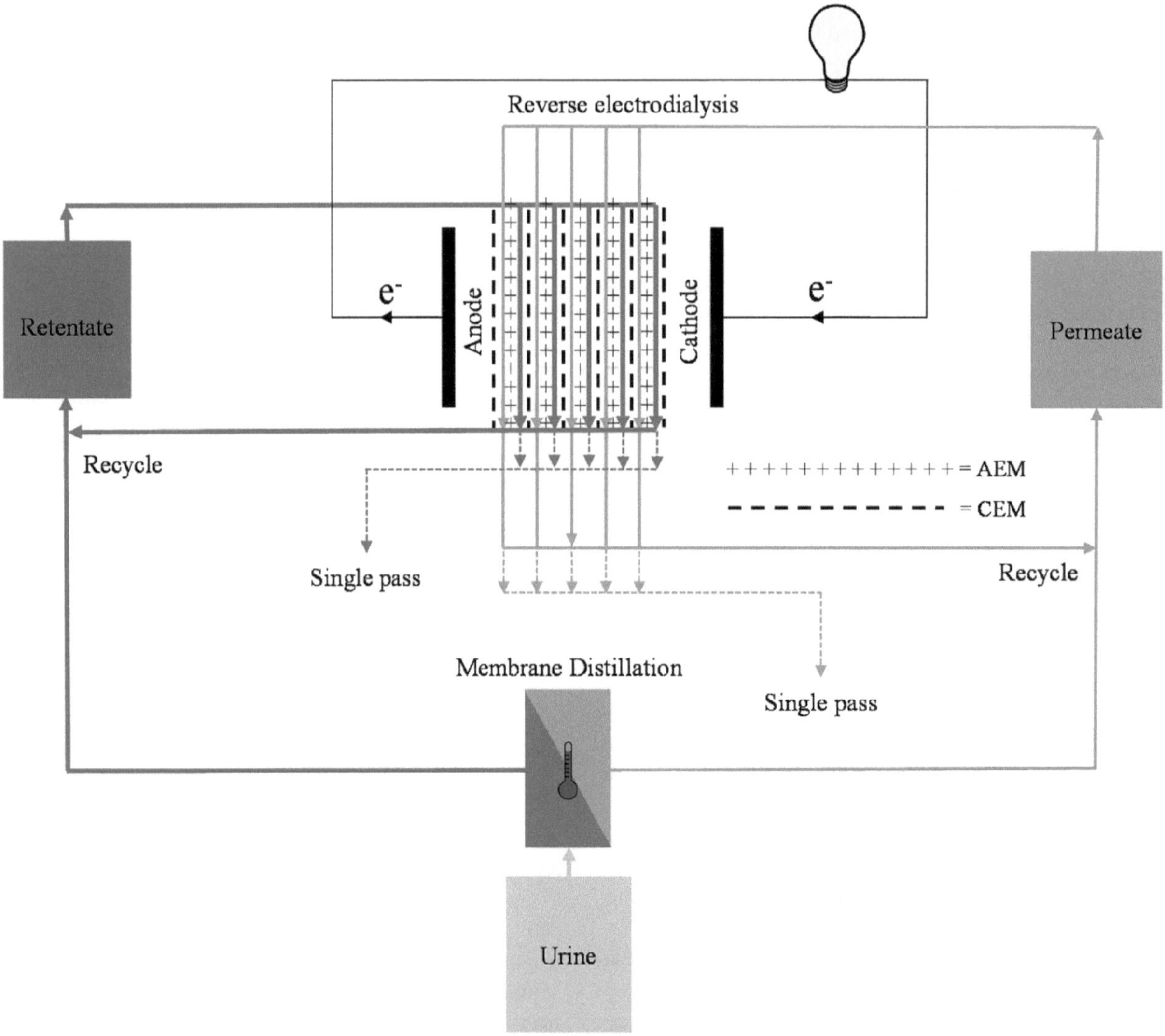

FIGURE 17.3 Reverse electrodialysis–membrane distillation. Reproduced from Ref. [33].

17.2.4 Pressure-Retarded Osmosis–Membrane Distillation (PRO-MD)

Pressure-retarded osmosis (PRO) is an emerging technique that converts osmotic energy (chemical potential energy) into mechanical energy to drive turbines for power generation. PRO utilizes the concentration difference in two solutions across a semi-permeable barrier, permitting water transport from the low solute concentration side to the higher solute concentration side that is partially pressurized (hydraulic pressure is less than the osmotic pressure, thus permitting water transport). The energy in this pressurized compartment can be harnessed to drive a turbine (any mechanical energy) to produce electricity [34]. As a large volume of concentrated draw solute is required for the PRO process, it becomes necessary to integrate with any desalination processes, such as MD, which is relatively insensitive to high salinity feedwater [35].

The retentate (wastewater brine, 0.5 bar osmotic pressure) from a low-pressure RO unit from municipal wastewater treatment was used as the feed for the integrated PRO-MD system [36]. A polyether sulphone (PES)-TFC hollow-fibre membrane (water permeability: 3.3 ± 0.3 L/[m^2.h.bar], burst pressure: >20 bar) was utilized in the counter-current flow using 0.6 M NaCl as the draw solute (DS). The PRO process achieved a power density of 9.3 W/m^2, which could be attributed to the excellent selectivity and relatively small structural parameter of the PES-TFC membrane. The DS was regenerated with a PVDF membrane (mean pore size of 168.4 ± 1.6 μm, water contact angle of 86.3 ± 1.3°, and porosity of 78.6 ± 2.3%) that could produce a permeate with conductivity <5 μS/cm (salt rejection >99.9%). The MD concentrated the draw solute for PRO from 0.6 M NaCl to 2 M NaCl. Thus, the PRO-MD hybrid unit facilitates a high-water recovery rate, osmotic power generation, and controlled membrane fouling [36] with zero carbon emissions as only the salinity gradient is used for energy production [34].

17.3 INTEGRATED MEMBRANE DISTILLATION FOR SEAWATER/BRACKISH WATER DESALINATION

17.3.1 Reverse Osmosis–Membrane Distillation (RO-MD)

The RO-MD hybrid configuration enables more than 80% water recovery with the RO reject being sent to the MD system (rejection >90%) [37]. Simulation of different design configurations concluded that a multi-stage hybrid system connected in series could improve recovery rates up to 90% with the lowest specific energy consumption of 0.9 MJ/kg [38].

An electrospun polyimide fibrous membrane coated with PDMS to recover water from RO concentrate using the DCMD process was evaluated [39]. A flux of 48 LMH and a water recovery rate of 68.09% were achieved with the conductivity of the distilled water (permeate) less than 11.72 μS.cm^{-1}. The performance was attributed to the interconnected pores, hierarchical roughness, and special wettability of the electrospun membrane. In another study, a dual-layer membrane was prepared by sequentially electrospraying oligomeric silsesquioxane onto an electrospun polyvinylidene fluoride-hexafluoropropylene (PVDF-co-HFP) fibrous membrane. This superhydrophobic membrane was investigated for DCMD application for water recovery from seawater [8]. The dual-layer superhydrophobic electrospun membrane with a bead-on-string structure showed excellent stability against fouling and scaling, exhibiting a freshwater recovery rate of 85% during five operation cycles. The conductivity of the distilled water remained below 20 μS.cm^{-1}. SEM images revealed the penetration of calcium ions into the membrane pores while sodium and magnesium ions remained on the membrane surface.

The major disadvantage of the RO-MD hybrid system is the scaling of membranes which necessitates pretreatment. To avoid membrane scaling due to the calcium sulphate present in concentrated brackish water (simulate RO concentrate), the feed was pretreated with barite to precipitate the scalant [40]. A PTFE membrane (Sterlitech, USA) in the DCMD configuration recovered nearly 90% of the water with almost no sulphate fouling. The flux and final permeate conductivity were 63 LMH and 1.5 μS.cm^{-1}, respectively. The effect of water recovery from an integrated RO-MD system with and without the use of antiscalants (AS) (before MD) was evaluated using a pilot-scale plant (Aqualstill) in the vacuum-assisted–air gap membrane distillation (VA-AGMD) mode [41].

The seawater reverse osmosis (SWRO) reject from a desalination plant with a total dissolved solids (TDS) concentration of 71,963 ppm was the feed to the MD system, and various antiscalants were chosen depending on the ions to be precipitated. The application of vacuum in the air gap reduces mass and heat resistance. When no antiscalant was added, the distillate flux decreased due to the membrane scaling, which further led to the intrusion of salt crystals through the membrane pores, diminishing the permeate quality. With AS, a maximum concentration factor of 3.24 was attained with a resulting TDS concentration of 204,719 ppm during the total test time of 1.42 hours. This configuration was best suited for maximum water recovery (~84.5%) with the lowest production costs (0.633 USD/m^3). In yet another study, a multi-step pretreatment strategy was followed to minimize fouling of the AGMD membranes. The steps included high-pH pretreatment, pH re-adjustment, and final AS addition, which could salt out various contaminants – SiO_2, Mg, and Ca – thus lowering the RO concentrate by a factor of 3.2 before being sent to the MD process [42].

17.3.2 Forward Osmosis–Membrane Distillation (FO-MD)

MD integration helps reconstitute the DS and produce high-quality water as permeate. The low fouling tendency of the hydrophobic membrane and the ability to utilize low-grade or waste heat makes this hybrid configuration economical [43]. A solar-powered FO-MD (Figure 17.4) was investigated for

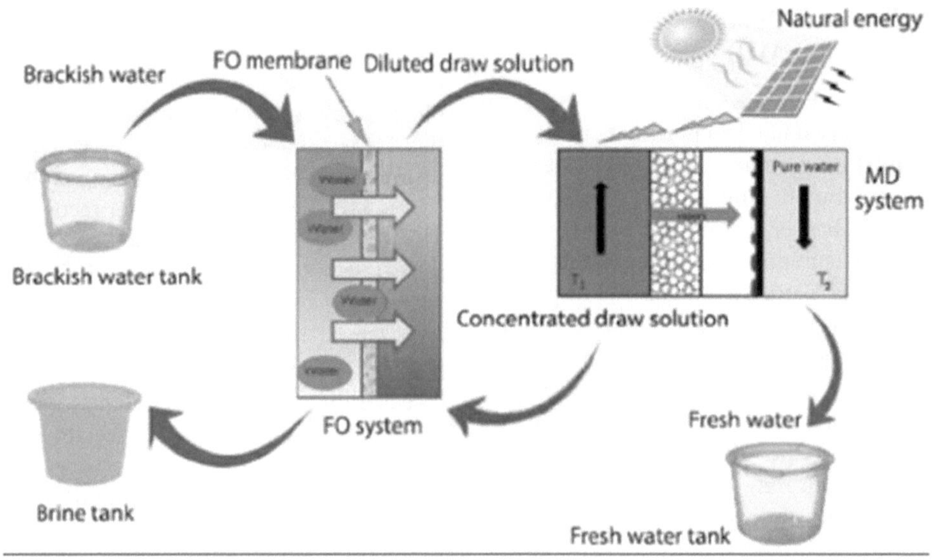

FIGURE 17.4 Illustration of the FO-MD hybrid system for brackish water treatment. Reproduced with permission from Ref. [25].

brackish water desalination [25]. The FO module consisted of a polyethersulphone (PES) support (Microdyn-Nadir GmbH, Germany), onto which a positively charged selective layer was fabricated. This membrane demonstrated a flux of ≃ 12 LMH for a brackish water feed of 20,000 mg/l using 1.5 M DS (KCl and KNO_3), resulting in a 53.3% water recovery rate. Solar energy was utilized to supply heat to the diluted DS for the AGMD process, which employed a commercial sponge-like PVDF membrane. A flux of 5.7 LMH and rejection rate of 99.5% at an MD feed temperature of 60°C were obtained. Also, a 67% reduction in energy consumption with solar power integration was achieved.

17.3.3 Reverse Electrodialysis–Membrane Distillation (RED-MD)

Integrated reverse electrodialysis and membrane distillation facilitate simultaneous water and energy production in seawater desalination, wherein the concentrated retentate from MD is used as input for power generation in RED [37]. A novel sweep gas membrane distillation (SGMD)–reverse electrodialysis (RED) method was investigated for hypersaline water desalination, as shown in Figure 17.5 [44]. Synthetic seawater (0.5 M NaCl) was used as feed water for the MD subsystem in SGMD mode with custom-made composite PVDF membranes coated with colloidal silver nanoparticles (AgNPs) to demonstrate the photothermal effect. The photothermal SGMD produced a pure water flux of 8.6 LMH (low conductivity) with a water recovery of 87.5%. Here, a UV lamp irradiated the composite membranes, which converted the adsorbed light into thermal energy, increasing the membrane interface temperature (thermoplasmonic effects). However, a 13% decline in water flux was observed as the feed concentration increased to 4 M NaCl due to the lower partial pressure difference. The hypersaline SGMD retentate or brine (4 M NaCl) was tapped for its electrochemical potential through RED, where a mixture of 0.3 M $K_3[Fe(CN)_6]$, 0.3 M $K_4[Fe(CN)_6]\cdot 3H_2O$, and 2.5 M NaCl was used as an electrolyte solution for the stack, and a maximum power density of 0.9 W/m^2 could be obtained.

In a similar configuration [45], the hot concentrate from a DCMD module employing a microporous polypropylene capillary membrane (Accurel PPS6/2) was used in the high-concentration compartment of the RED system. The DCMD membrane produced a pure water flux of 4.5 kg.m^{-2}.h^{-1} with a water recovery rate of > 80% for a feed solution of 0.5 M NaCl (synthetic seawater), and the decrease of the flux was due to the increase in the feed concentration which lowered the activity coefficient of the feed. The RED stack was used in cross-flow configuration (RED Stack B.V., The Netherlands) with ion-exchange membranes (Fujifilm Manufacturing Europe B.V., The Netherlands). The low-concentration compartment was fed with 0.5 M NaCl, and the high-concentration compartment was fed with MD brine (2–5 M). About 4500 kJ of energy was generated using 1 m^3 seawater (0.5 M NaCl) and 1 m^3 MD brine (5 M NaCl). An exergetic efficiency of 49% for the best scenario using 5 M NaCl feed at 60°C temperature was obtained, and a power density of 2.2 W/m^{2}MP was generated. It has been inferred that increasing the high salt concentration compartment (HCC) solution temperature increases the power density. A 17% reduction in energy consumption by incorporating RED in hybrid systems can be achieved.

Membrane distillation–reverse electrodialysis hybrid systems significantly enhance water recovery. Concentrated brine with high electrochemical potential for energy generation is also produced. MD-RED hybrid systems facilitate the implementation of ZLD in industries. Thereby, brine disposal and consequent environmental implications are averted. MD-RED also converts the waste steam into electrical energy. The MD-RED hybrid system promotes a circular economy with

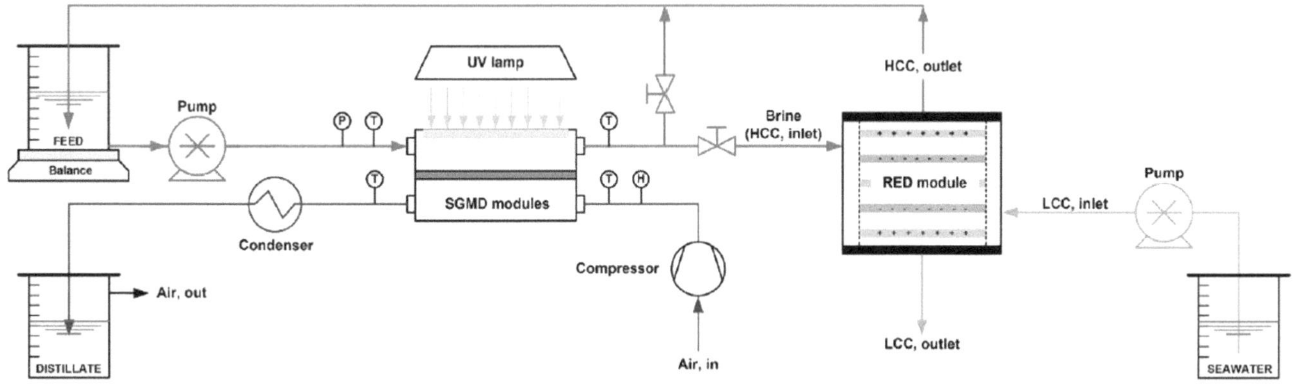

FIGURE 17.5 Illustration of the photothermal SGMD-RED process. Reproduced with permission from Ref. [44].

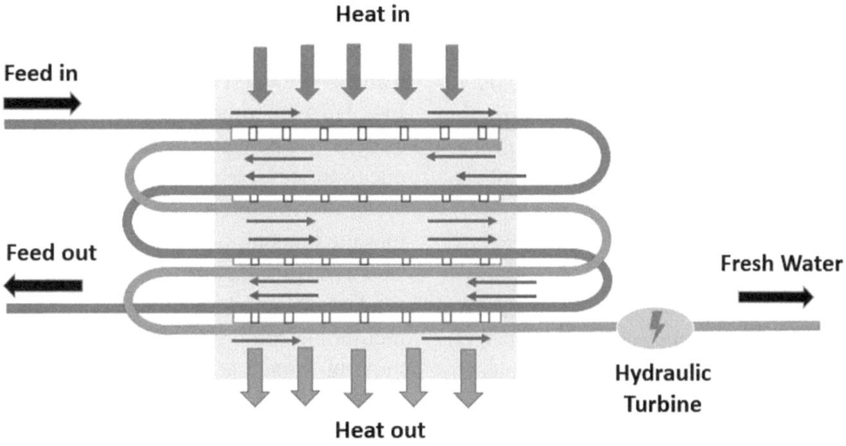

FIGURE 17.6 Schematic of the PRO-MD process. Redrawn from Ref. [47].

water recovery from brine and energy production from the waste steam [37].

17.3.4 Pressure-Retarded Osmosis–Membrane Distillation (PRO- MD)

Membrane distillation and pressure-retarded osmosis integration enable water–energy nexus hastening the transition towards sustainability and a circular economy. MD is preferred for DS regeneration in an osmotic heat engine as it is compatible with low-grade or waste heat. The energy demand minimization for regeneration necessitates the integration of MD-PRO [46].

A multi-stage PRO-MD hybrid unit (Figure 17.6) with a commercial membrane (0.1 μm, Xuanda, China) with a sandwich structure to balance vapour permeability and resistance to hydraulic pressure was used [47]. Two active layers of PTFE nanofibres on either side of the composite membrane with a nominal pore size of 100 nm and a PP support layer (insulating layer) to reduce the heat conduction loss and larger pores to facilitate vapour permeation. The module possesses five working stages with six working plates, each with an effective area of 18 cm^2. Each stage has an evaporation chamber with 0.5 M NaCl solution, and a condensation chamber with fresh water, and the temperature of the first working plate was higher than the last working plate where the freshwater is condensed. The high-pressure freshwater is used to drive the hydraulic turbines for electricity generation. Upon 24 hours of operation, the multi-stage PRMD module produced 188 L of freshwater from a highly saline solution and 27.8 kJ power with a heating area of 1.0 m^2 and temperature varying between 40 and 80°C. It has been reported that a higher working pressure increased the power density production while decreasing the desalination rate due to the increased resistance, membrane wetting, and membrane compaction.

The concentration gradient must be minimized in the PRO system for maximum power density [48]. The product of water permeability and pressure difference represents the power density in the PRO subsystem. The critical operating parameters include feed salinity, salt rejection of membrane, water flux in the MD subsystem, and membrane fouling. The PRO-MD hybrid exhibits advantages of high water recovery, massive osmotic power generation, controlled membrane fouling, less environmental implications [36], and no carbon emission [34].

17.4 INTEGRATED MEMBRANE DISTILLATION FOR RESOURCE RECOVERY

17.4.1 Membrane Bioreactor–Membrane Distillation (MBR-MD)

Membrane bioreactors (MBRs) are advanced wastewater treatment units that integrate membrane filtration and conventional activated sludge process. They are versatile and can be used to produce energy, recover nutrients or resources, and remove trace organic compounds [49].

A DCMD–anaerobic membrane bioreactor (AnMBR) was investigated for simultaneous energy recovery and water reuse from synthetic domestic wastewater [49]. AnMBR demonstrated 0.3–0.5 L/g COD biogas production with a stable methane content of approximately 65% with complete removal of bulk organic matter, phosphate, and 26 trace organic contaminants (removal efficiency: 70%). The digested sludge was applied to the external membrane module (ceramic membrane NGK, Japan, with 1 μm and 0.09 m^2 area) and circulated to the bioreactor with suitable pumping systems. The permeate was then sent to the DCMD process to completely recover the organic matter and phosphates. A commercial PTFE membrane on PP support (Porous Membrane Technology, China) was used for the MD process, and the hybrid system was operated for 30 days continuously with hydraulic backwashing every two days. The removal of NH_4 was 90%, with near-complete removal of COD and phosphate in the MD process. The MD membrane was replaced to maintain the desirable efficiency once in 10 days. However, organic and inorganic matter accumulated in the MD feed solution resulting in fouling of the MD subsystem. MD flux decreased from ≃13 LMH to ≃ 3 LMH at the end of 10 days. The feed solution concentration increased from 3 mS/cm to 58 mS/cm at the end of 30 days.

A similar hybrid system was evaluated to dewater human urine [16]. The human urine was treated with ultrafiltration (UF) to remove harmful bacteria and viruses and then biologically oxidized in the membrane bioreactor (MBR) equipped with braided polyvinylidene fluoride (PVDF) hollow-fibre membrane module (Lotte Chemical, Daejeon, China). A total dissolved organic carbon (DOC) reduction of 96.1% was attained, and the overall organic fraction was reduced by 80–90% in the UF-MBR reactor. The permeate from the UF-MBR reactor was dewatered using a DCMD module housing Durapore®-GVHP PVDF flat-sheet membranes with a nominal pore size of 0.22 μm, water contact angle of 131°, and active layer porosity of 75%. The MD flux diminished from 15 LMH to 1 LMH over time due to membrane fouling which could be recovered by alkaline membrane cleaning. The hybrid system could recover > 80% of the water with a TDS of 280 g/l.

A thermophilic anaerobic membrane distillation bioreactor configuration (Figure 17.7) was used to recover volatile fatty acids (VFAs) and biogas from synthetic wastewater (COD:N:P ratio of 250:5:1, and micronutrients) [50]. A submerged PVDF hollow-fibre membrane module was used in a VMD mode (50 mbar) to minimize losses in the system's thermal efficiency. The thermophilic anaerobic biomass was added to the bioreactor and acclimatised before the experiments started. This configuration exhibited an average flux of 3 LMH and 99.99 % of salts for feed temperatures of 45–65°C for up to 7 days. A 16–50% decline could be explained due to the formation of a compact, non-porous cake layer at this temperature. Membrane wetting was prevented as the permeate was devoid of ions such as Na, K, Fe, etc. Among all VFAs, acetic acid had the highest concentration (58.6 mg/l) in the permeate, while propanoic acid, butyric acid, and isovaleric acid concentrate were also observed in the permeate.

A novel moving sponge–anaerobic osmotic membrane bioreactor–membrane distillation (AnOMBR-MD) method was investigated to treat municipal wastewater [13]. The moving sponge bioreactor consists of polyurethane sponge carriers that house the microorganisms. After sufficient hydraulic retention (anaerobic process), the water is recovered using an FO (CTA-ES membrane, Hydration Technologies, Inc., USA) membrane with 0.2 M Na_3PO_4/0.25 M EDTA-2Na as the DS. The water flux of the FO process declined over time due to the reverse diffusion of these draw solutes resulting in a lower driving force. The diluted DS is sent to an external MD module housing a microporous PTFE commercial membrane (E Creative Co., Ltd., Taipei, Taiwan). The AnOMBR-MD module obtained a stable water flux of 4 $L.m^{-2}.h^{-1}$ with less membrane fouling for 454 days, attributed to the continuous moving sponge around the FO module. The system obtained a 100% nutrient recovery rate with 17.4% of phosphorus as struvite, 99.99% salt rejection, and methane production of 0.16 L/COD removed. MBR coupled with MD could facilitate zero liquid discharge as water can be reclaimed in the MD process [50].

17.4.2 Membrane Distillation–Membrane Crystallization (MD-MCr)

Membrane distillation–membrane crystallization (MD-MCr) is an emerging technology for resource recovery and freshwater production. Membrane crystallization enables the recovery of salts such as sodium, magnesium, barium, strontium. and lithium [51], whereas membrane distillation facilitates freshwater production from brine.

A fractional submerged membrane distillation crystallization (F-SMDC) for sodium sulphate (Na_2SO_4) from SWRO brine was studied using the configuration shown in Figure 17.8 [18]. The module consisted of MD and crystallizer in a single unit with a submerged hollow-fibre membrane (Econity, Republic of Korea). The temperature at the top section was maintained at 50.0 ± 0.2°C by using a heated water circulator and at the bottom section and in the permeate stream the temperatures were 19.8 ± 1.3°C and 20.1 ± 0.3°C, respectively, using a cooling water circulator. The membrane module at the top creates a concentration gradient (MD part) with solution density difference that the feed solution settles down, which is also enhanced by the cold permeate

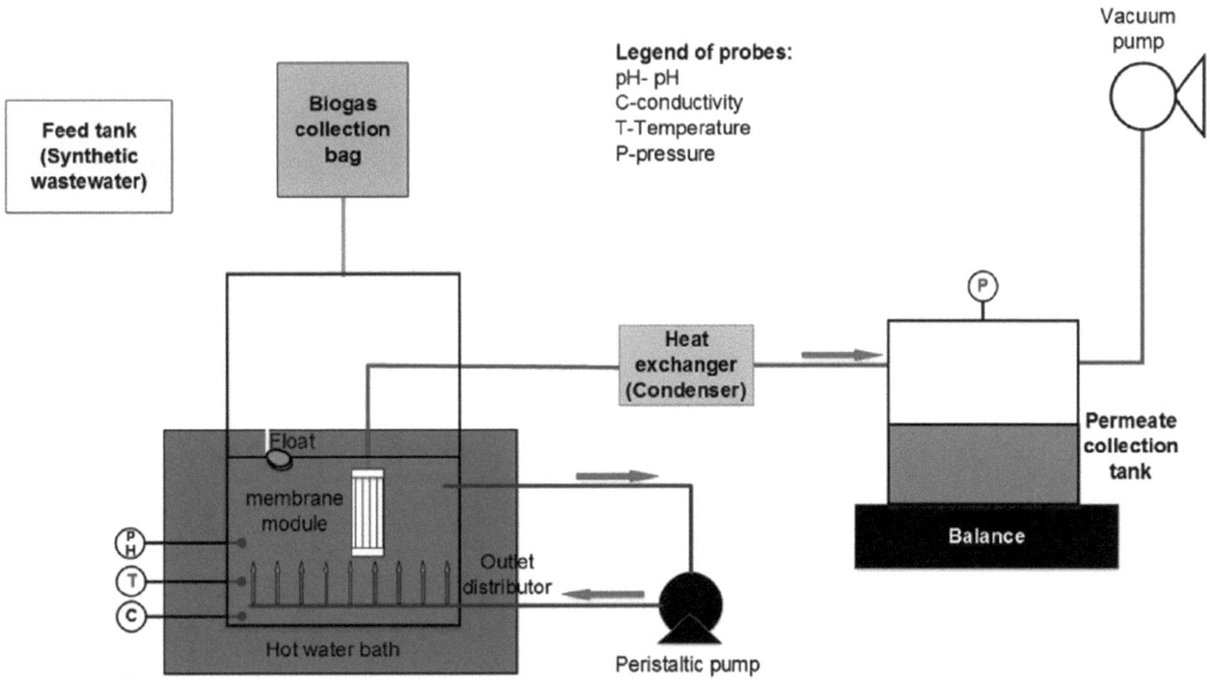

FIGURE 17.7 Schematic of the anaerobic membrane distillation bioreactor. Reproduced with permission from Ref. [50].

stream. The so-produced concentration gradient influences the MD and crystallization efficiency. It was inferred that the concentration and temperature gradient were significant for F-SMDC; also, the presence of calcium sulphate affects the system's performance while a sodium-rich solution enhances the recovery of sodium sulphate. About 72% water and 223.73 g of Na_2SO_4 were recovered with an average flux of 3.3 LMH.

A novel ZLD system that consists of freeze desalination (FD) and membrane distillation–crystallization (MD-C) was investigated through mathematical modelling [52]. The system was operated as follows: seawater (3.5 wt.% of NaCl) is fed into the freeze desalination (FD) unit, where pure ice crystals are generated, and the concentrated brine is fed to the feed tank of MD. The brine is further concentrated by MD and sent to the solid hollow-fibre cooling crystallizer (SHFCC) and cooled for salt crystallization. The brine is recycled after filtering out the salt crystals. For a daily seawater processing capacity of 72 kg with 50% of its heating energy supported by a 50.5 m^2 solar panel, an output of 2.52 kg of salt and 69.48 kg of water per day could be achieved. About 30% of water recovery is achieved through the FD subsystem, while the DCMD unit fulfils the remaining recovery. The transmembrane flux and mass transfer coefficient are the two significant parameters for the MD-MCr hybrid module performance [51]. Laminar conditions govern the crystallization process integrated with membrane distillation b y minimizing shear stress to form well-faceted crystals. Membranes with higher surface area, porosity, and roughness facilitate heterogeneous nucleation with minimum induction time (time between the attainment of supersaturation and formation of the first crystal).

17.5 INTEGRATION OF MEMBRANE DISTILLATION WITH RENEWABLE ENERGY

Renewable energy technologies are emerging technologies to address the clean energy demand and reduce the carbon footprint in energy production [53]. Among renewable energy-powered membrane distillation systems, solar and waste heat or low-grade heat have been widely investigated compared to those driven by wind or geothermal energy. The United States Department of Energy defines waste heat as heat energy with temperatures varying between 25 and 150°C [54].

17.5.1 Solar Energy

Solar energy is used as a heat source for desalination, as thermal energy can be transmitted directly to saltwater by its flow through a field of solar collectors or indirectly by using a heat exchanger with appropriate thermal fluids to transfer the energy in a closed loop [55]. Solar energy can generate working temperatures of 80–85°C, which is in the operating range for MD and works at near atmospheric pressures [55]. A schematic of solar thermal photovoltaic vacuum membrane distillation (STPVMD) is represented in Figure 17.9 [56]. The system employed a PVDF hollow-fibre membrane module in the VMD mode. Polycrystalline silicon solar panels with a BSRN3000 radiation observation system with an effective area of 1.62 m^2, rated power of 235 W, rated voltage of 30 V, and rated current of 7.84 A were used. Also, the system was equipped with a vacuum tube solar collector with a collection area of 1.82 m^2, a storage capacity of 2.5 L, and it could withstand a temperature of 100°C. The solar energy harnessed through solar photovoltaic panels was used to

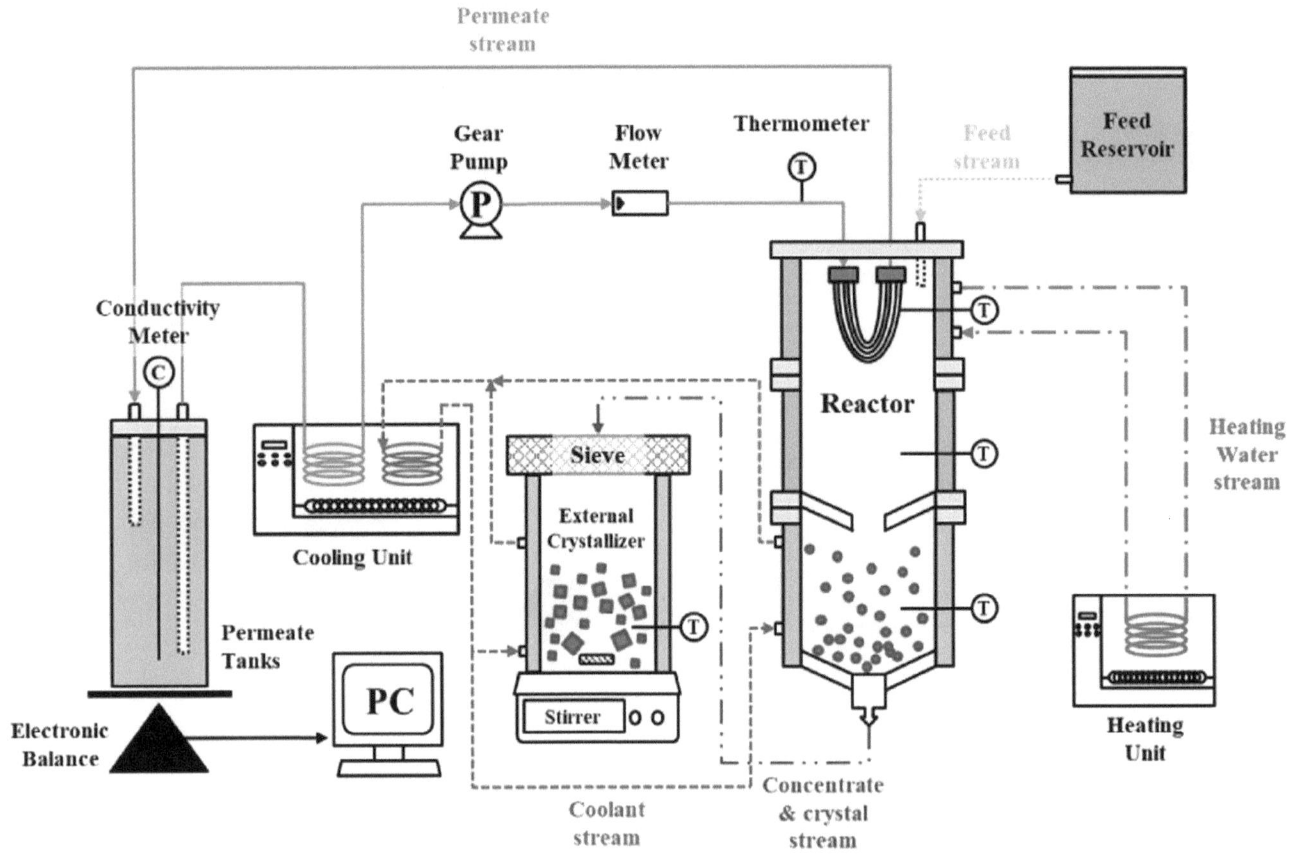

FIGURE 17.8 Illustration of the F-SMDC process. Reproduced with permission from Ref. [18].

power the circulation pumps and the heater, and the excess energy was stored in batteries. The vapour produced from the VMD module was condensed inside the copper tubes inside the water-cooling tank. The response surface methodology (RSM) studies concluded that, at the optimum conditions, the system demonstrated a permeate flux of 6.26 L/m^{2}h and the corresponding electricity consumption of 0.5 kWh, with an energy performance of 12.52 L/(kWm^2).

A nanophotonic enhanced solar energy-powered membrane distillation (NESMD) for seawater and hypersaline brine desalination was investigated [57]. The NESMD membrane used in this study was a bilayer with the base layer consisting of a solar-absorbing, hydrophilic, carbon black nanoparticle-infused polypropylene (PP) layer and the active layer was a microporous hydrophobic polytetrafluoroethylene (PTFE) layer. The surface of the membrane housed in a transparent plate and frame configuration was exposed to sunlight, performing a dual function as a solar collector and hydrophobic barrier between the feed and permeate solutions. The photothermal nanomaterials and coating heat the feed solution in the MD module without the requirement for heat exchangers. A TDS removal rate of 99.5% and average flux of 0.75 LMH were achieved with a solar intensity of 1 kW/m^2 and 0.2 m^2 active membrane area.

A vacuum multi-effect membrane distillation method powered by solar energy for seawater desalination has been investigated [55]. A solar field plate collector with an overall capacity of 17 kW was used for energy harvesting. The heat buffer tank maintained the feed temperature in the range of 60–70°C irrespective of changes in the solar irradiation. The simulation of the VMD system showed that total production of 41.7, 68.4, and 70.5 m^3 could be obtained at feed temperatures of 60, 70, and 80°C, respectively, for the system. A solar-driven membrane distillation (SDMD) with Fe_3O_4/polyvinylidene fluoride-co-hexafluoropropylene (Fe_3O_4/PVDF-HFP) membrane was investigated [58]. The photothermal membrane exhibited a permeate flux of 0.97 kg.m^{-2}.h^{-1} with a salt rejection rate of >99.99% under 1 kW.m^{-2} of solar irradiation and photothermal conversion efficiency of 53%. The Fe_3O_4 nanoparticles on the active surface of the photothermal membranes scatter and absorb light, and scattered light is concentrated near the liquid-irradiated surface leading to intense localized heating and increased membrane surface temperature. Many other studies have investigated solar-driven membrane distillation processes for cogeneration of water and energy under different modes of operation [59–61].

17.5.2 Waste Heat

Waste heat is low-grade heat derived from various thermal energy sources such as marine engines [62], gas-fired power

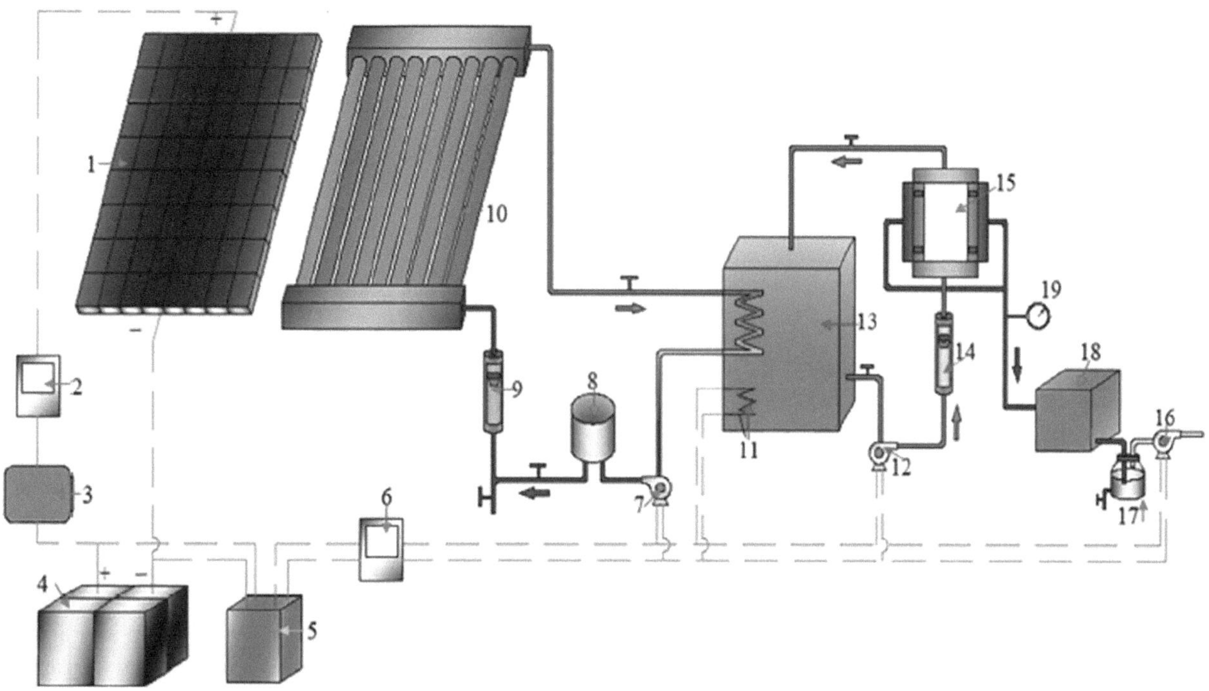

FIGURE 17.9 Schematic diagram of a STPVMD system: (1) solar photovoltaic panel, (2) electric energy measurement meter, (3) charge controller, (4) batteries, (5) inverters, (6) power meter, (7) pressurized circulation pump, (8) buffer water tank, (9) glass rotor flowmeter, (10) vacuum tube collector, (11) electric heating wire, (12) pressurised circulation pump (13) insulated water tank, (14) glass rotor flowmeter, (15) vacuum fibre membrane distillation module, (16) vacuum pump, (17) collection bottle of condensate water, (18) cooling water tank, (19) pressure gauge. Reproduced with permission from Ref. [56].

stations [63], and vent gases [64]. The recovery of waste heat or low-grade heat from absorber vent gases of a CO_2 capture unit and its utilization for freshwater recovery using membrane distillation were modelled using ASPEN Plus [64]. The base conditions assumed were as follows: flue gas flowrate: 3180 t/h; flow rate of CO_2 in the flue gas stream: 127.181 t/h; MD membrane: PTFE membrane; permeate temperature: 20°C; permeate concentration: 0 g/l; and feed concentration: 35 g/l. It was assumed that the cold seawater was warmed by passing through a heat exchanger with purged gas used as feed for the DCMD unit to produce fresh water. The effects of permeate temperature, flow rate, and feed temperature on freshwater production were studied. It was concluded that the freshwater production was 88,599 L/day with a feed temperature of 52.2°C which reduced to 85,050 L/day when the feed temperature decreased to 50.3°C. The freshwater production was 88,950 L/day at the permeate temperature of 10°C, which declined to 78,450 L/day when the permeate temperature increased to 20°C. The freshwater production was increased from 47,850 L/day to 78,450 L/day by increasing the permeate flow rate from 10 L/min to 20 L/min.

Energy-efficiency enhancement of the DCMD process using waste heat for heating and wastewater for cooling was investigated [65]. The liquid effluent from a gold mine containing high concentrations of sulphate, calcium, magnesium, arsenic, and acid pH at a temperature of 60°C was used as the feed solution. The treated surface water and the gold effluent dam supernatant were used as a permeate solution at 25°C. Sterlitech flat-sheet polytetrafluoroethylene (PTFE) membrane produced a flux of 14 kg $m^{-2}{\cdot}h^{-1}$ and removal efficiency rate of > 99% for all contaminants. The permeate conductivity was maintained well below 5 μS/cm with excellent resistance to the membrane wetting for the operation up to 210 days. However, the SEM and electron dispersive (EDS) analysis revealed the membrane scaling caused by calcium sulphate, which was attributed to the supersaturation condition due to the feed solution temperature (60°C).

The performance of the DCMD system in response to the waste heat intermittency in direct and indirect system arrangements, as shown in Figure 17.10, was evaluated [54]. The MD system used a commercial flat-sheet PTFE membrane (Parker Performance Materials, Lee's Summit, MO) with a feed concentration of 10 g/l NaCl and deionized water as permeate solution. The direct arrangement comprises a heat exchanger through which feed solution is fed, whereas, in the indirect arrangement, there are two loops – the heat loop and the membrane loop. The indirect arrangement offers more heat extraction and storage as the high flow rate in the heat loop facilitates heat extraction and storage at a faster rate. In the indirect arrangement, the heat loop was isolated from the system. The direct and indirect systems exhibited fluxes of 22.9 ± 6.0 MH and 18.7 ± 4.6 LMH, respectively. It was reported that the direct arrangement exhibited greater water flux when the heat source was on, while the indirect arrangement exhibited a greater flux when the heat source was

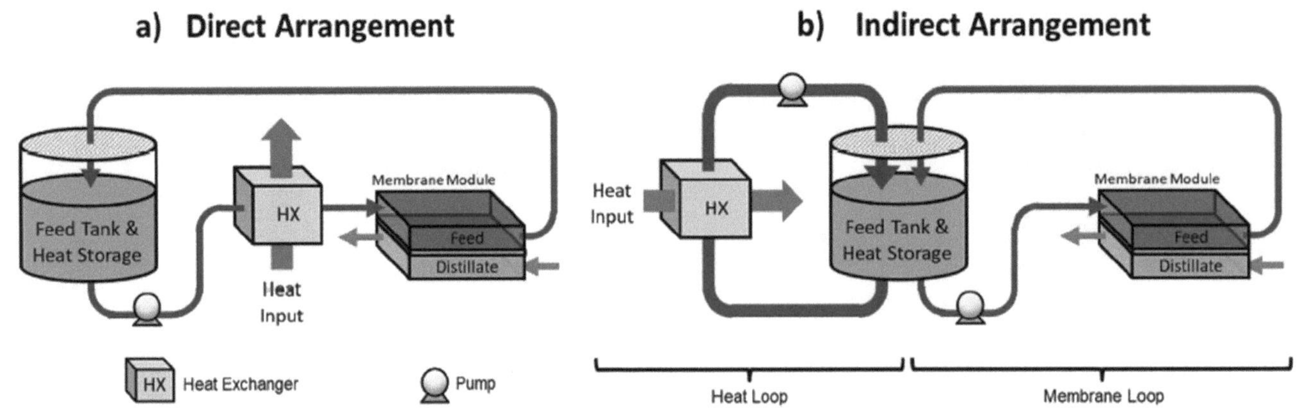

FIGURE 17.10 (a) Direct and (b) indirect arrangement of DCMD integrated with waste heat source. Reproduced with permission from Ref. [54].

off as it stored a greater quantity of heat at a faster rate while the heat source was on. The indirect arrangement produced more water when the degree of intermittency was high (87.5%) than the direct arrangement, where performance was better at a lower level of intermittency (12.5%). The better system control of the indirect arrangement was enhanced with a lower variation in water flux resulting from waste heat intermittency or variability of waste heat source. Therefore, it is evident that an ideal integrated MD system powered by waste heat depends on waste heat availability and variability.

17.5.3 Geothermal Energy

Integrating geothermal energy with membrane distillation is significantly less explored and should be analysed in detail to understand the hybrid system better. In a reported study [66], the hot geothermal water with a TDS concentration of 893 ppm was fed directly into the VMD system. The VMD system consisted of a commercial flat-sheet PVDF membrane (Whatman GmbH, Germany), and the feed temperature was constantly maintained at 60°C. A flux of 9.28 kg/m^2 was achieved with a specific energy consumption of 66.03 KW/kg.h^{-1}. Upon direct supply of geothermal water from the hot reservoir, the specific energy consumption can be reduced to 4.2 kW/kg h$^{-1,}$ which is approximately a 95% reduction. The distillate was produced with a TDS concentration of ≃ 119 ppm, thus enabling the feasibility of water recovery from geothermal water.

17.6 TECHNO-ECONOMIC ANALYSIS OF INTEGRATED MEMBRANE DISTILLATION APPROACHES

Techno-economic assessment is necessary to evaluate the integrated membrane distillation approaches for their economic feasibility for large-scale applications. Membrane distillation-based desalination incurs high energy consumption with costs varying from 1.17$ to 1.60$/m^3, dependent on different parameters, and is generally higher than conventional thermal processes such as multi-effect evaporators [67]. The variables such as costs of membrane unit, membrane replacement, chemical cleaning agents, energy consumption, and system maintenance were considered for estimating capital and operating expenditures [2]. The membrane systems' design life and membrane life span (replacement) were considered to be 15 years and 5 years, respectively. Membrane cleaning and maintenance costs were considered to be 2% and 5% of the initial investment cost, respectively. Capital cost was influenced by permeate recovery rate and operating expenditure (OPEX) was estimated to be 0.50, 0.43, and 1.96 US$/m^3 of water recovered through NF, RO, and MD, respectively. A power density of 2.7 W/m^{-2} was achieved through RED integrated with MD costs up to 4€/m^3, making RED competitive technology for integration [44]. A payback period of 6 years is required for a large stack RED (membrane area 24,000m^2) producing a power density of 0.3 W/m^{2,} incurring a cost of 1,440 € [45]. Therefore, integrated approaches are preferred to minimize the specific energy consumption and cost of water production. For instance, Ali et al. [38] have analysed the costs of an RO-MD hybrid model, which could recover an additional 40% water at the cost of 0.9 $/m^3. However, by configuring the hybrid models in series or multi-effect stages, the specific energy consumption reduces to 0.9 MJ/kg, and consequently, the cost of production also decreases.

The cost of water (COW) for various hybrid membrane-based modules such as RO + MD + antiscalant, RO + CD + UF +MD, RO +NF + MD, and RO + NF + antiscalant was estimated to be 0.633, 0.692, 0.7, and 0.698 USD/m^3, respectively, at recovery rates of 84.59%, 66.93%, 72.73%, and 73.38%, respectively [41]. For a solar-powered V-MEMD unit, the cost of water was estimated to be 5.2 $/m^3 for a system producing 10 m^3/day with a specific energy consumption half of that powered by a conventional heating source. The cost of installing solar thermal photovoltaic vacuum membrane distillation was estimated at 7,713 yuan [56]. However, if the system were operated without photovoltaic power, the system would cost 756.25 yuan only, with 1893.65 L of water production, which produces a value of 1893.65 yuan. Thereby, the cost of return can be attained after 4.24 years. However, from an environmental perspective, by using photovoltaic power, 0.04 kg of coal is saved for each 1 kWh of energy saved, with a significant reduction in emissions [0.272 kg of carbon dust, 0.997 kg of carbon dioxide (CO_2), 0.03 kg of (SO_2), and 0.015 kg of nitrogen oxides (NO_X)]. Also, a geothermal energy-powered VMD module producing 20,000 m^3/d can save US$ 0.72/m^3 of water produced and about 95% of conventional energy is conserved [66].

17.7 CONCLUSIONS AND FUTURE PERSPECTIVES

This chapter comprehensively composes the recent MD integrated membrane technologies for wastewater treatment, sea/brackish water desalination, and resource recovery. With its high potential for high-quality water recovery, membrane distillation can be scaled up for commercialization and integration with existing systems. This chapter discusses how integration with the RO process can lead to higher water recovery (> 80%), leading to zero liquid discharge, making the hybrid process sustainable. The regeneration of DS from the FO process is generally a challenge, which can be addressed by the MD process, which can handle high concentrations of salt in the feed solutions. The integrated system could be more sustainable if renewable energy sources provided heat for heating the feed solution.

The RED and PRO processes require a high-concentration solution to convert the osmotic energy/pressure to electrical energy or mechanical work. The MD process can generate a concentrated stream or regenerate the DS. In turn, the energy produced in the RED or PRO process can be used for the MD process (pumping or heating of feed stream), thus complementing each other. MD can also be explored for resource recovery if coupled with a membrane bioreactor. In addition to resource recovery, it can act as a secondary barrier for removing trace organic compounds to generate ultrapure distillate. The MD process can also concentrate liquid streams, thus providing sites for nucleation and crystallization of inorganic compounds using an integrated MD-crystallizer.

In short, integrating MD with other membrane-based technologies can recover water, enabling zero liquid discharge, reducing the quantity of brine discharged, and mining valuable resources or nutrients from wastewater with the simultaneous production of high-quality permeate. Certain shortcomings, such as a lack of robust commercial membranes and manufacturers, higher specific energy consumptions, and higher operational costs, limit the MD systems' use. Membrane fouling and scaling are another critical challenge that has to be addressed.

Renewable energy sources such as solar, geothermal, and low-grade heat effectively reduce the conventional energy consumption by MD, thus approaching sustainability and a circular economy. The future perspectives of integrated membrane distillation approaches include the development of suitable membranes for long-term operation; mandatory integration with renewable energy sources for its thermal requirements (water–energy nexus); innovative design approaches with multi-stage subsystems, heat exchangers; development of self-heating membranes; and alternative less energy-intensive approaches for permeate condensation.

REFERENCES

1. Chang, H., Liu, B., Zhang, Z., Pawar, R., Yan, Z., Crittenden, J.C., Vidic, R.D. (2021) A critical review of membrane wettability in membrane distillation from the perspective of interfacial interactions. *Environmental Science & Technology* 55: 1395–1418.
2. Fonseca, C.C., Vitória, S.A., Cristina, M.S.A., Lange, L.C., de Andrade, L.H., Foureaux, A.F.S., Fernandes, B.S. (2020) Assessing potential of nano fi ltration, reverse osmosis and membrane distillation drinking water treatment for pharmaceutically active compounds (PhACs) removal. *Journal of Water Process Engineering* 33: 101029.
3. Anderson, W.V., Cheng, C.M., Butalia, T.S., Weavers, L.K. (2021) Forward osmosis–membrane distillation process for zero liquid discharge of flue gas desulfurization wastewater. *Energy and Fuels* 35: 5130–5140.
4. Ravi, J., Othman, M.H.D., Matsuura, T., Bilad, M.R., El-badawy, T.H., Aziz, F., Ismail, A., Rahman, M.A., Jaafar, J. (2020) Polymeric membranes for desalination using membrane distillation: A review. *Desalination* 490: 114530.
5. Tagliabue, M, Tonziello J, Bottino A, Capanelli, G., Comite, A., Pagliero, M., Boero, F., Cattaneo, C. (2021) Laboratory scale evaluation of fertiliser factory wastewater treatment through membrane distillation and reverse osmosis. *Membranes (Basel)* 11: 1–15.
6. Biniaz, P., Ardekani, N.T., Makarem, M.A., Rahimpour, M.R. (2019) Water and wastewater treatment systems by novel integrated membrane distillation (MD). *ChemEngineering* 3: 1–36.
7. Son, H.S., Soukane, S., Lee, J., Kim, Y., Kim, Y.D., Ghaffour, N. (2021) Towards sustainable circular brine reclamation using seawater reverse osmosis, membrane distillation and forward osmosis hybrids: An experimental investigation. *Journal of Environmental Management* 293: 112836.
8. Tan, G., Xue, X., Zhu, Z., Li, J. (2021) Ultrahigh and stable water recovery of reverse osmosis-concentrated seawater with membrane distillation by synchronously optimizing membrane interfaces and seawater ingredients. *ACS ES&T Water* 1: 1577–1586.
9. Chamani, H., Woloszyn, J., Matsuura, T., Rana, D., Lan, C.Q. (2021) Pore wetting in membrane distillation: A comprehensive review. *Progress in Materials Science* 122: 100843.
10. Wang, Z., Lin, S. (2017) Membrane fouling and wetting in membrane distillation and their mitigation by novel membranes with special wettability. *Water Research* 112: 38–47.
11. Anvari, A., Azimi Yancheshme, A., Kekre, K.M., Ronen, A. (2020) State-of-the-art methods for overcoming temperature polarization in membrane distillation process: A review. *Journal of Membrane Science* 616: 118413.
12. Lokare, O.R., Tavakkoli, S., Rodriguez, G., Khanna, V., Vidic, R.D. (2017) Integrating membrane distillation with waste heat from natural gas compressor stations for

produced water treatment in Pennsylvania. *Desalination* 413: 144–153.
13. Nguyen, N.C., Duong, H.C., Nguyen, H.T., Chen, S.S., Le, H.Q., Ngo, H.H., Guo, W., Duong, C.C., Le, N.C., Bui, X.T. (20200 Forward osmosis–membrane distillation hybrid system for desalination using mixed trivalent draw solution. *Journal of Membrane Science* 603: 118029.
14. Khayet, M. (2011) Membranes and theoretical modeling of membrane distillation: A review. *Advances in Colloid and Interface Science* 164: 56–88.
15. Song, X., Luo, W., Hai, F.I., Price, W.E., Guo, W., Ngo, H.H., Nghien, L.D. (2018) Resource recovery from wastewater by anaerobic membrane bioreactors: Opportunities and challenges. *Bioresource Technology* 270: 669–677.
16. Volpin. F., Jiang, J., El Saliby, I., Preire, M., Lim, S., Johir, M.A.H., Cho, J., Han, D.S., Phuntsho, S., Shon, H.K. (2020) Sanitation and dewatering of human urine via membrane bioreactor and membrane distillation and its reuse for fertigation. *Journal of Cleaner Products* 270: 122390.
17. Caroline Ricci, B., Santos Arcanjo, G., Rezende Moreira, V., Rocha Lebron, Y.A., Koch, K., Rodrigues Costa, F.C., Ferreira, B.P., Costa Lisboa, F.L., Miranda, L.D., de Faria, C.V., Celina Lange, L., Santos Amaral, M.C. (2021) A novel submerged anaerobic osmotic membrane bioreactor coupled to membrane distillation for water reclamation from municipal wastewater. *Chemical Engineering Journal* 414: 1–13.
18. Choi, Y., Naidu, G., Nghiem, L.D., Lee, S., Vigneswaran, S. (2019) Membrane distillation crystallization for brine mining and zero liquid discharge: Opportunities, challenges, and recent progress. *Environmental Science: Water Research and Technology* 5: 1202–1221.
19. Cerda, A., Quilaqueo, M., Barros, L., Seriche, S., Gim-Krumm, M., Santoro, S., Avci, A.H., Romero, J., Curcio, E., Estay, H. (2021) Recovering water from lithium-rich brines by a fractionation process based on membrane distillation-crystallization. *Journal of Water Process Engineering* 41 : 102063.
20. Arcanjo, G.S., Costa, F.C.R., Ricci, B.C., Mounteer, A.H., de Melo, E.N.M.L., Cavalcante, B.F., Araujo, A.V., Farioa, C.V., Amaral, M.C.S. (2020) Draw solution solute selection for a hybrid forward osmosis-membrane distillation module: Effects on trace organic compound rejection, water flux and polarization. *Chemical Engineering Journal* 400: 125857.
21. Nawi, N.I.M., Bilad, M.R., Anath, G., Nordin, N.A.H., Kurnia, J.C., Wibisono, Y., Arahman, N. (2020) The water flux dynamic in a hybrid forward osmosis-membrane distillation for produced water treatment. *Membranes (Basel)* 10: 1–13.
22. Sardari, K., Fyfe, P., Ranil Wickramasinghe, S. (2019) Integrated electrocoagulation – Forward osmosis–membrane distillation for sustainable water recovery from hydraulic fracturing produced water. *Journal of Membrane Science* 574: 325–337.
23. Nawaz, M.S., Son, H.S., Jin, Y., Kim, Y., Soukane, S., Al-Hajji, M.A., Abu-Ghdaib, M., Ghaffour, N. (2021) Investigation of flux stability and fouling mechanism during simultaneous treatment of different produced water streams using forward osmosis and membrane distillation. *Water Research* 198: 117157.
24. Alrehaili, O., Perreaul,t F., Sinha, S., Westerhoff, P. (2020) Increasing net water recovery of reverse osmosis with membrane distillation using natural thermal differentials between brine and co-located water sources: Impacts at large reclamation facilities. *Water Research* 184: 116134.
25. Suwaileh, W., Johnson, D., Jones, D., Hilal, N. (2019) An integrated fertilizer driven forward osmosis- renewables powered membrane distillation system for brackish water desalination: A combined experimental and theoretical approach. *Desalination* 471: 114126.
26. Qasim, M., Badrelzaman, M., Darwish, N.N., Darwish, N.A., Hilal, N. (2019) Reverse osmosis desalination: A state-of-the-art review. *Desalination* 459: 59–104.
27. Okampo, E.J., Nwulu, N. (2021) Optimisation of renewable energy powered reverse osmosis desalination systems: A state-of-the-art review. *Renewable & Sustainable Energy Reviews* 140: 110712.
28. Conidi, C., Macedonio, F., Ali, A., Cassona, A., Criscuoli, A., Argurio, P., Drioli, E. (2018) Treatment of flue gas desulfurization wastewater by an integrated membrane-based process for approaching zero liquid discharge. *Membranes (Basel)* 8: 117–129.
29. Li, M., Li, K., Wang, L., Zhang, X. (2020) Feasibility of concentrating textile wastewater using a hybrid forward osmosis-membrane distillation (FO-MD) process: Performance and economic evaluation. *Water Research* 172: 115488.
30. Li, J., Hou, D., Li, K., Zhang, Y., Wang, J., Zhang, X. (2018) Domestic wastewater treatment by forward osmosis-membrane distillation (FO-MD) integrated system. *Water Science and Technology* 77: 1514–1523.
31. Zhang, J., Wang, D., Chen, Y., Gao, B., Wang, Z. (2021) Scaling control of forward osmosis-membrane distillation (FO-MD) integrated process for pre-treated landfill leachate treatment. *Desalination* 520: 115342.
32. Ricci, B.C., Skibinski, B., Koch, K., Mancel, C., Celestino, C.Q., Cunha, I.L.C., Silva, M.R., Alvim, C.B., Faria, C.V., Andrade, L.H., Lange, L.C., Amaral, M.C.S. 2019. Critical performance assessment of a submerged hybrid forward osmosis – membrane distillation system. *Desalination* 468: 114082.
33. Mercer, E., Davey, C.J., Azzini, D., Eusebi, A.L., Tierney, R., Williams, L., Juang, Y., Parker, A., Kolios, A., Tyrrel, S., Cartmell, E., Pidou, M., McAdam, E.J. (2019) Hybrid membrane distillation reverse electrodialysis configuration for water and energy recovery from human urine: An opportunity for off-grid decentralised sanitation. *Journal of Membrane Science* 584: 343–352.
34. Park, K., Kim, D.Y., Yang, D.R. (2017) Theoretical analysis of pressure retarded membrane distillation (PRMD) process for simultaneous production of water and electricity. *Industrial & Engineering Chemistry Research* 56: 14888–14901.
35. Lee, J.G., Kim, Y.D., Shim, S.M., Im, B.K., Kim, W.S. (2015) Numerical study of a hybrid multi-stage vacuum membrane distillation and pressure-retarded osmosis system. *Desalination* 363: 82–91.
36. Han, G., Zuo, J., Wan, C., Chung, T.S. (2015) Hybrid pressure retarded osmosis-membrane distillation (PRO-MD) process for osmotic power and clean water generation. *Environmental Science: Water Research and Technology* 1: 507–515.

37. Tufa, R.A., Di Profio, G., Fontananova, E., Avci, A.H., Curcio, E. (2018) *Current Trends and Future Developments on (Bio-) Membranes: Renewable Energy Integrated with Membrane Operations.* Elsevier Inc.: USA.
38. Ali, E., Orfi, J., Najib, A., Saleh, J. (2018) Enhancement of brackish water desalination using hybrid membrane distillation and reverse osmosis systems. *PLoS One* 13: 1–18.
39. Zhu, Z., Tan, G., Lei, D., Yang, Q., Tan, X., Liang, N., Ma, D. (2021) Omniphobic membrane with process optimization for advancing flux and durability toward concentrating reverse-osmosis concentrated seawater with membrane distillation. *Journal of Membrane Science* 639:119763.
40. Zhang, Z., Lokoare, O.R., Gusa, A.V., Vidic, R.D. (2021) Pretreatment of brackish water reverse osmosis (BWRO) concentrate to enhance water recovery in inland desalination plants by direct contact membrane distillation (DCMD). *Desalination* 508: 115050.
41. Bindels, M., Carvalho, J., Gonzalez, C.B., Brand, N., Nelemans, B. (2020) Techno-economic assessment of seawater reverse osmosis (SWRO) brine treatment with air gap membrane distillation (AGMD). *Desalination* 489: 114532.
42. Rioyo, J., Aravinthan, V., Bundschuh, J. (2019) The effect of 'high-pH pretreatment' on RO concentrate minimization in a groundwater desalination facility using batch air gap membrane distillation. *Separation and Purification Technology* 227: 115699.
43. Parveen, F., Hankins, N. (2021) Integration of forward osmosis membrane bioreactor (FO-MBR) and membrane distillation (MD) units for water reclamation and regeneration of draw solutions. *Journal of Water Process Engineering* 41: 102045.
44. Avci, A.H., Santoro, S., Politano, A., Propato, M., Miceli, M., Aquino, M., Wenjuan, Z., Curcio, E. (2021) Photothermal sweeping gas membrane distillation and reverse electrodialysis for light-to-heat-to-power conversion. *Chemical Engineering and Processing – Process Intensification* 164: 108382.
45. Tufa, R.A., Noviello, Y., Di Profio, G., Macedonio, F., Ali, A., Driolo, E., Fontananova, E., Bouzek, K., Curcio E. (2019) Integrated membrane distillation-reverse electrodialysis system for energy-efficient seawater desalination. *Applied Energy* 253: 1–13.
46. Giwa, A., Hasan, S.W. (2018) *Current Trends and Future Developments on (Bio-) Membranes: Renewable Energy Integrated with Membrane Operations.* Elsevier Inc.: USA.
47. Runze, Z., Ji, L.I., Zikang, Z., Rui, L., Wei, L., ZhiChun, L. (2021) Harvesting net power and desalinating water by pressure-retarded membrane distillation. *Science China Technological Sciences* 64: 1–7.
48. Rahimpour, M.R., Mohsenpour, S. (2018) *Current Trends and Future Developments on (Bio-) Membranes: Renewable Energy Integrated with Membrane Operations.* Elsevier Inc.: USA.
49. Song, X., Luo, W., Mcdonald, J., Khan, S.J., Hai, F.I., Price, W.E., Nghiem, L.D. (2018) An anaerobic membrane bioreactor – membrane distillation hybrid system for energy recovery and water reuse: Removal performance of organic carbon, nutrients, and trace organic contaminants. *Science of the Total Environment* 628–629: 358–365.
50. Yao, M., Woo, Y.C., Ren, J., Tijing, L.D., Choi, J.S., Kim, S.H., Shin, H.K. (2019) Volatile fatty acids and biogas recovery using thermophilic anaerobic membrane distillation bioreactor for wastewater reclamation. *Journal of Environmental Management* 231: 833–842.
51. Salmón, I.R., Simon, K., Clérin, C., Luis, P. (2018) Salt recovery from wastewater using membrane distillation-crystallization. *Crystal Growth & Design* 18: 7275–7285.
52. Lu, K.J., Cheng, Z.L., Chang, J., Luo, L., Chung, T.-S. (2019) Design of zero liquid discharge desalination (ZLDD) systems consisting of freeze desalination, membrane distillation, and crystallization powered by green energies. *Desalination* 458: 66–75.
53. Jones, L.E., Olsson, G. (2017) Solar photovoltaic and wind energy providing water. *Global Challenges* 1: 1600022.
54. Gustafson, R.D., Hiibel, S.R., Childress, A.E. (2018) Membrane distillation driven by intermittent and variable-temperature waste heat: System arrangements for water production and heat storage. *Desalination* 448: 49–59.
55. Andrés-Mañas, J.A., Roca, L., Ruiz-Aguirre, A., Acién, F.G., Gil, J.D., Zaragoza, G. (2020) Application of solar energy to seawater desalination in a pilot system based on vacuum multi-effect membrane distillation. *Applied Energy* 258: 114068.
56. Deng, H., Yang, X., Tian, R., Hu, J., Zhang, B., Cui, F., Guo, G. (2020) Modeling and optimization of solar thermal-photovoltaic vacuum membrane distillation system by response surface methodology. *Solar Energy* 195: 230–238.
57. Said, I.A., Chomiak, T.R., He, Z., Li, Q. (2020) Low-cost high-efficiency solar membrane distillation for treatment of oil produced waters. *Separation and Purification Technology* 250: 1–10.
58. Li, W., Chen, Y., Yao, L., Ren, X., Li, Y., Deng, L. (2020) Fe_3O_4/PVDF-HFP photothermal membrane with in-situ heating for sustainable, stable and efficient pilot-scale solar-driven membrane distillation. *Desalination* 478: 114288.
59. Ma, Q., Xu, Z., Wang, R., Poredoš, P. (2022) Distributed vacuum membrane distillation driven by direct-solar heating at ultra-low temperature. *Energy* 239: 1–13.
60. Lu, H., Shi, W., Zhao, F., Zhang, W., Zhang, P., Zhao, C., Yu, G. (2021) High-yield and low-cost solar water purification via hydrogel-based membrane distillation. *Advanced Functional Materials* 31: 1–7.
61. Miladi, R., Frikha, N., Kheiri, A., Gabsi, S. (2019) Energetic performance analysis of seawater desalination with a solar membrane distillation. *Energy Conversion and Management* 185: 143–154.
62. Bahar, R., Ng, K.C. (2020) Fresh water production by membrane distillation (MD) using marine engine's waste heat. *Sustainable Energy Technologies and Assessments* 42: 100860.
63. Dow, N., Gray, S., Li, J. de., Zhang, J., Ostarcevic, E., Liubinas, A., Athertin, P., Roeszler, G., Gibbs, A., Duke, M. (2016) Pilot trial of membrane distillation driven by low grade waste heat: Membrane fouling and energy assessment. *Desalination* 391: 30–42.
64. Ullah, A., Soomro, M.I., Kim, W.S., Saleem, M.W. (2020) The recovery of waste heat from the absorber vent gases of a CO_2 capture unit by using membrane distillation technology for freshwater production. *International Journal of Greenhouse Gas Control* 95: 102957.
65. Silva, M.R., Reis, B.G., Grossi, L.B., Amaral, M.C.S. (2020) Improving the energetic efficiency of direct-contact membrane distillation in mining effluent by using the

waste-heat-and-water process as the cooling fluid. *Journal of Cleaner Products* 260: 121035.

66. Sarbatly, R., Chiam, C.K. (2013) Evaluation of geothermal energy in desalination by vacuum membrane distillation. *Applied Energy* 112: 737–746.

67. Yan, Z., Yang, H., Yu, Qu, F., Liang, H., der Bruggen, B.V., Li, G. (2018) Reverse osmosis brine treatment using direct contact membrane distillation (DCMD): Effect of membrane characteristics on desalination performance and the wetting phenomenon. *Environmental Science: Water Research & Technology* 4: 428–437.

18 Advancements in Membrane Methodology for Liquid Radioactive Waste Processing

Current Opportunities, Challenges, and the Global World Scenario

Grażyna Zakrzewska-Kołtuniewicz
Institute of Nuclear Chemistry and Technology, Warsaw, Poland
E-mail: g.zakrzewska@ichtj.waw.pl

18.1 INTRODUCTION: MEMBRANE METHODS FOR THE TREATMENT OF LIQUID RADIOACTIVE WASTE

Conventional technologies used for liquid low- and medium-level radioactive waste processing as precipitation coupled with sedimentation, ion exchange, and evaporation are energy-consuming or introduce a third phase that results in the production of secondary wastes (sludge from sedimentation tanks, spent sorbent from ion-exchange columns, or effluents from resin regeneration). These wastes need additional treatment and decontamination. All these disadvantages may be avoided by membrane methods that have already found application in the field of liquid radioactive wastes processing [1–4]. The most advanced are technologies based on pressure-driven membrane processes: microfiltration (MF), ultrafiltration (UF), and reverse osmosis (RO). The choice of process depends on the waste composition or parameters that have to be reached during processing, e.g., decontamination factors (DFs) in relation to the limits defined in national or international regulations, volume reduction coefficients, or the necessity of recycling some components of the solution. Different installations have to be used for treating the wastes from nuclear power plants or reprocessing plants, and for processing the wastes from production of radiopharmaceuticals and medical diagnostics. Radioactive wastes from production of radioisotopes and medicine are usually classified as low- and medium-level waste; they contain mainly β and γ emitters, while the wastes from the waste processing plants contain alpha-bearing elements that can destroy polymeric membrane material. Very often, the organic solvents and complexing agents or acids are presents in this type of waste. The wastes coming from nuclear reactor operations contain the fission (e.g., ^{89}Sr, ^{90}Sr, ^{124}Sb, ^{132}Te, ^{134}Cs, ^{136}Cs, ^{137}Cs, ^{140}Ba, ^{141}Ce), as well as corrosion (^{60}Co, ^{58}Co, ^{51}Cr, ^{54}Mn, ^{59}Fe, 65Zn, ^{95}Zr) products.

Applying the membrane processes it is possible to achieve the following goals:

1. Purification of the effluents to concentration levels enabling safe discharge to the environment;
2. Concentration of the radioactive compounds with volume reduction that is sufficient for solidification;
3. Separation and recycling of the valuable components, e.g., boric acid from cooling waters.

Reverse osmosis enables complete retention of all dissolved compounds, even small monovalent ions. To avoid membrane blocking and scaling before reverse osmosis, microfiltration or ultrafiltration pretreatment can be applied. Apart from preliminary treatment, ultrafiltration can be used for the separation of suspensions or colloids, which are often formed by actinides or ions such as ^{54}Mg, ^{55}Fe, ^{60}Co, and ^{125}Sb. Microfiltration has found application for waste dewatering after precipitation. Nanofiltration (NF) that uses lower pressures than reverse osmosis is applied for the separation of bivalent from monovalent ions. The most common application of the NF process in the nuclear industry is boric acid separation from the reactor coolant.

18.1.1 Selection of Membranes for Nuclear Applications

The membrane is the most important part in separation processes, as its performance controls the efficiency and selectivity of the process. Both flux and selectivity expressed in terms of separation or retention factors determine the process economics and usability, and finally the costs of the installation.

The radiation resistance of the material used as well as the commercial availability of the membrane units and auxiliary equipment, process competitiveness in comparison with conventional methods, its economics and feasibility are important criteria for the application of membranes for radioactive waste processing.

Different membrane materials have found application in nuclear technology; for radioactive waste treatment both

DOI: 10.1201/9781003285656-18

polymer and inorganic membranes are used. The main advantages of polymers are the wide spectrum of different types of material, easy formation of membranes, and easy modification for specific applications. The relatively low cost (less than $1 per square meter) is also beneficial, as well as broad commercial availability. The advantages of inorganic membranes come from their extremely good resistance to temperature, strong chemical environment, and ionizing radiation. Membranes manufactured in some primary processes can be modified for some purposes by regulation of pore size or change of the membrane chemical properties [5]. In the process of modification two kinds of material can be combined, organic and inorganic (hybrid membranes), giving tailored-made structures for special applications. Recent efforts in membrane and membrane systems development are focused on:

- Development of new materials for manufacturing membranes;
- Modification of membrane surface;
- Change of membrane morphologies;
- Membrane systems design and process optimization.

Surface modification plays a vital role governing the performance and separation properties of the membranes. The most common goal of surface modification is reduction of membrane fouling; however separation properties are also of fundamental importance. Various techniques of surface modification are applied such as grafting by photograft copolymerization, UV-assisted photochemical graft polymerization, low-temperature plasma treatment, or γ-ray-induced graft copolymerization [6]. Lately much attention has been paid to the preparation of hydrophobic/hydrophilic composite membranes and blending host polymers with surface-modifying macromolecules (SMMs) [7–9]. Porous composite hydrophobic/hydrophilic membranes prepared by blending fluorinated SMMs with host PS and PES have been proposed for processing low and intermediate radioactive liquid wastes [10].

New developments have been made concerning gas separation membranes and membranes for gaseous diffusion. The tailorable pore sizes and pore chemistry of metal-organic frameworks (MOFs) make them attractive for molecular separations [11–13]. MOF can be used not only for gas storage, but also for the separation of gases like xenon or krypton. Organic–inorganic hybrid membranes for gaseous diffusion were developed by modification of mesoporous Vycor glass with organosilane heptadecafluoro-1,1,2,2-tetrahydrodecyltrichlorosilane (HDFS) [14]. Vycor glass can be also employed as a support for palladium thin foils used for the separation of hydrogen [15]. The membranes with nanopores, in which Knudsen flow dominates, can be applied for separation of isotopes in gaseous phase. There are great hopes for new carbon materials including carbon nanotubes (CNTs) and 2D graphene and graphene-like carbon allotropes [16,17]. Evenly distributed small pore sizes and tubular shape of CNTs make them promising for gas separation membranes. Single-atomic-layer carbon membranes from graphene may be the key to manufacturing separation barriers of the possible lowest thickness. Carbon network structure, as well as high selectivity predicted from the mechanisms of separation, namely size sieving, quantum sieving, and chemical affinity sieving, allow them to be considered as prospective candidates for isotope separation. Amyloid–carbon hybrid membranes can be used for the purification of hospital radioactive wastewater and nuclear wastewater [18]. Different solutions simulated medicine-relevant waste containing isotopes such as technetium (Tc-99m), iodine (I-123), gallium (Ga-68) as well as radioactive compounds having various half-life periods such as technetium (Tc-99m), iodine (I-123), and gallium (Ga-68), as well as real clinical waste containing iodine (I-131) and lutetium (Lu-177) were tested. The obtained results demonstrated the capability of the removal of clinically relevant radioactive substances from hospital wastewater by adsorption mechanism.

A very useful and recently more widely used tool for the prediction of such membrane performance is molecular modeling, which allows describing diffusion of gas molecules through the nanopores of a separation barrier [19].

Much attention has been given to metallic membranes for hydrogen separation, based not only on Pd alloys, but also on alloys of metals of Groups IV and V. Metals such as W, V, Cr, Rb, Nb, Mo, or Ta form body-centered-cubic (BCC) structure and can be used for manufacturing alloys for high-temperature hydrogen separations [20]. Cellulose acetate membranes (CA) modified using acrylamide (AAm) and polyethylene glycol (PEG) have been considered as a sorbent for Cs^+ and Eu^{3+} removal from aqueous waste solutions [21].

Polymeric membranes are manufactured as flat foils, hollow fibers, or tubular shaped. Flat films can be also arranged as a spiral wound configuration, or pleated filter cartridges. The module design permits easy connection in any configuration and up-scaling of the installation. Such an arrangement allows gaining a high membrane area to volume ratio, which relatively lowers the cost of the apparatus and leads to higher unit capacities. The appropriate selection of membrane configuration has to take into account energy expenditure (pressure drop) per product volume and appropriate hydrodynamic conditions in the apparatus promoting turbulence and mass transfer, and reducing boundary-layer phenomena, as well as the easiness of cleaning. In hollow-fiber modules the feed flow in relatively narrow channels is laminar and thus the membrane is particularly susceptible to fouling. The channels in plate-and-frame modules are also small, and membranes are sensitive to fouling, however they are easy to clean. On the contrary, in tubular modules cross-flow velocities are usually large, giving turbulent flow and a large pressure drop. In spiral-wound modules the spacer may act as a turbulence promotor and thus reduce concentration polarization. In order to successfully avoid membrane fouling, it is necessary to carefully understand its causes and nature. Various methods have been used to study this phenomenon, including those based on the use of labeled molecules [22–24].

The problems with fouling are particularly important for nuclear applications, because frequent cleaning produces secondary wastes, which are radioactive and need additional processing. The best practice for membrane installation operation is avoiding the fouling and scaling by all available means (antiscalant injection, feedwater pretreatment to remove oxidizing materials, iron, calcium and magnesium salts, particulates and oils, greases, etc.). Proper selection of the module, which can be easily cleaned by back-flushing or washing with cleaning solution, is very important. In tubular modules direct mechanical cleaning by foam balls is also practiced, however it is limited to small units.

18.1.2 The Effect of Ionizing Radiation

One of the crucial parameters of the membranes employed for radioactive waste treatment is their radiation stability and durability. During long-time operation the membrane is exposed to the action of ionizing radiation. This may cause some structural changes in the membrane that affect its permeability and separation characteristics. The variety of effects in the polymer structure results from ionizing radiation, such as cross-linking and the formation of a new structure, decrease of molecular weight as a result of breaking of the main chains of macromolecules, change of the character and number of double bonds, oxidation of the polymer in the presence of oxygen, etc.

The influence of ionizing radiation on polymer membranes has been studied since attempts were made to use them in nuclear technologies [25–31]. All studies showed that polymeric membranes exhibit limited resistance to ionizing radiation, however the threshold values of doses are sufficiently high to use the membranes for low- and intermediate-radioactive solutions treatment for a period of time. Usually a practical lifetime for most membranes is of the order of 4–5 years, which is sufficient to avoid deterioration of filtration and separation abilities. Ceramic membranes are expected to be more resistant to γ, β, and α emissions, but systematic studies have not been presented.

The influence of ionizing radiation on the performance of polyamide (PA) composite reverse-osmosis (RO) membranes has been thoroughly investigated by French research groups [32–34]. Irradiation was performed with a gamma ^{60}Co source under an oxygen atmosphere to absorb doses of 0.1 and 1 MGy. NaCl rejection and permeability were studied before and after irradiation. Various methods, such as attenuated total reflectance Fourier transform infrared spectroscopy (ATR-FTIR), X-ray photoelectron spectrometry (XPS), atomic force microscopy (AFM), field emission scanning electron microscopy (FE-SEM), and ion chromatography, have been used for characterization. The experiments showed that membranes irradiated at 0.1 MGy did not change their selectivity and permeability properties, while at 1 MGy, NaCl rejection decreased to 65% and permeability increased threefold. The changes in the properties of the RO membranes could be due to both structural relaxation in the polyamide network and modifications to the polysulfone layer. More systematic studies of the degradation of polyamide composite reverse osmosis membranes as a function of the irradiation dose in the range of 0.2–0.5 MGy revealed that the membranes irradiated at 0.2 MGy exhibited a small decrease of permeability, although retention did not change at this dose. Retention began to decrease at 0.5 MGy from 99% to 95%. At this dose permeability exhibited a large two-fold increase. All these changes in selectivity and permeability were attributed to scissions of ester and amide bonds, most precisely the loss of hydrogen bonds between polyamide chains. At the same time, leaching of nitrogen and carbon compounds was identified, as well as oxygen consumption. This was related to degradation of the membranes. Between 0.2 MGy and 0.5 MGy polysulfone layer degradation also occurred. For a dose rate of 0.5 kGy h^{-1}, until a dose of 0.2 MGy, RO composite membranes exhibited irradiation resistance.

The next step in the studies of resistance of these membranes was carried out in aerobic and anaerobic conditions in water [34]. Different behaviors of the RO membranes were observed with presence or absence of oxygen. In anaerobic conditions degradation of the membrane active layer was lower than in aerobic conditions. Confirmation of this was the lower amount of species released in aqueous and liquid phases. Because of less amide bond scissions, the degradation of the membrane active layer was less noticeable. However, the XPS analysis showed that degradation of the membrane top layer was similar in both irradiation cases. The outer PVA layer behaved differently from the rest of the membrane; it was more sensitive towards gamma irradiation. After analysis of the research results it was suggested that the absence of oxygen can influence the concentration of the radicals created by water radiolysis and reduce the degradation depth of the membrane. A different mechanism of degradation in the presence and absence of oxygen was proposed. In anaerobic conditions the degradation is caused by the reductive species and limited to the top surface of the membrane, while in aerobic conditions oxidative species also are involved. This is important information for radioactive waste treatment by RO membranes. The removal of oxygen before the treatment may reduce significantly the degradation of the membranes caused by irradiation.

The impact of irradiation of polymer membranes was studied by other research groups. Cellulose triacetate (CTA) and polyamide thin-film composite (TFC) membranes used in forward osmosis (FO) were investigated by researchers from Tsinghua University, Beijing [35]. The membrane performance was tested using ^{60}Co sources of irradiation at doses of 20 and 200 kGy. Before and after irradiation water flux, nuclide flux, and retention, as well as reverse NaCl flux, were compared. The surface properties of the membranes were examined by FE-SEM, AFM, and XPS. The functional groups were analyzed by FTIR with attenuated total reflectance ATR mode and hydrophilicity of membranes was evaluated by contact angle measurement. Both membranes showed an enhancement of hydrophilicity with increasing dose, resulting in an increase in water flow. However the retention of cobalt and strontium ions and reverse NaCl flux were not affected considerably by irradiation. The surface chemical properties were not changed with irradiation, although the surface morphology and roughness changed with the dose of radiation. TFC membranes synthesized by interfacial polymerization were more resistant to the radiation than CTA membranes prepared by phase separation. Moreover, the Cs ions retention of TFC membranes decreased. For practical application of these membranes in commercial FO for radioactive waste treatment it is necessary to enhance their radiation resistance.

The effect of gamma irradiation on nanofiltration PA membranes used for low- and intermediate-level liquid radioactive wastewaters (LILW) was investigated by the Korea Advanced Institute of Science and Technology (KAIST). It was observed that with an increase of dose to 300 kGy both the salt rejection and the water permeability of NF membranes decreased from 95.6 ± 0.1% to 74.6 ± 0.5% and from 33.7 ± 0.3 L/m^{2}h to 21.4 ± 0.5 L/m^{2}h, respectively [36]. Moreover, a decrease in hydrophilicity and increase in fragility of PA membranes after irradiation were noticed. Membranes were tested by XPS and NMR. It was shown that irradiation caused the formation of new bonds between the unbound carboxyl groups and the amino groups, which increased the cross-linked portion of the amide bond from 28% to 45%. The reduction in salt rejection and water permeability has been attributed to aging of PA membranes by irradiation and should therefore be monitored during LILW treatment using nanofiltration processes.

The stability of nanofiltration membranes was investigated to evaluate their potential in the treatment of low-level liquid radioactive waste and to recover ^{235}U and ^{238}U radionuclides [37]. The NF membranes were immersed in natural waste containing uranium radionuclides for 24–5000 h. The waste was provided by Nuclear Fuel Factory which produces nuclear fuel for Brazilian pressurized water reactors. In experiments, flat-sheet desalination membranes were used with selective polyamide layer on a poly(ether sulfone) support. Static and dynamic tests were performed. Permeate flux and rejection of chloride and sulfate ions were compared before and after the tests, as well as uranium rejection. The morphological and chemical structure of the surface layer was evaluated by such methods such as FE-SEM, ATR-FTIR, AFM, X-ray fluorescence (XRF), and thermogravimetric analysis (TGA). The results of experiments showed that irradiation changed the transport properties of NF membranes. Exposure to radiation increased the permeate flux, but rejection of chlorides decreased after immersion of the membranes in the waste for a short time (24–48 h). Rejection of sulfate ions was reduced only after a long period of exposure, i.e. 5000 h, when the PVA layer was destroyed. In dynamic tests, with a stirring cell 73% of uranium present in the waste was rejected even though the top PVA layer was lost.

The stability of cross-linked microporous polyarylate (PAR) membranes with high Kr/Xe separation performance under irradiation was studied [38]. Gamma irradiation was done using ^{60}Co source with 1.96 and 0.09 kGy/h dose rates. Total doses of 20 kGy and 47 kGy were selected according to the range of total dose absorbed from ^{85}Kr in off-gas for one year of exposure, which is about 86.4 kGy.

For analysis of the chemical structure of the membranes attenuated total reflection-Fourier transform infrared spectroscopy (ATR-FTIR) and X-ray photoelectron spectroscopy were used. The morphology of membranes was characterized by scanning electronic microscopy. The thermal stability of polyarylate and poly(trimethylsilyl)propyne (PTMSP) powders was checked by thermogravimetric analysis (TGA) from 30°C to 900°C at 10°C/min in nitrogen. PTMS was used as a protective layer of PAR/PSf membranes. The experiments showed that γ irradiation does not affects the chemical structure of the membranes, however it could cause accelerated aging of PAR membranes and damage of protective PTMS layer. The dose of 20 kGy caused a decrease in Kr/Xe selectivity by 25% and Kr performance by 44%.

18.1.3 Design of Membrane Systems

There are numerous ways in which the membrane modules can be arranged in filtration systems. The factors influencing the choice of option include the type of process, the expected concentration in the streams, the volume of processed wastes, the acceptable dose to the membrane, and desired costs of the plant. The capital costs play an important role in low- and medium-capacity plants, while the components of operation costs, such as energy consumption, are of less importance. For the design of large-capacity installations the operation costs are key elements. High investment costs for additional equipment, such as control system, washing installation, or measurement apparatuses are justified when they result in a decrease in the operational costs of the plant. Nuclear installations need special control equipment and security systems and their capital costs are relatively high.

The plant can operate in batch or continuous mode. The simplest design is dead-end operation, frequently used in microfiltration, where all the feed is forced through the membrane which results in a continuous increase of feed concentration and worsening of permeate quality. This design is used when complete filtration of the feed is necessary. A more often used arrangement is cross-flow, which decreases the membrane fouling, and can operate as co-current, countercurrent, cross-flow with perfect permeate mixing. The flow in the module is adjusted as a single-pass or flow with recirculation. Batch systems are also employed in small-scale applications when the waste arises discontinuously. A single-pass configuration can be arranged in series arrays, parallel, or tapered design. The loss of volume in a single-pass system is compensated by a tapered arrangement. The tapered array is generally not considered for ultrafiltration and microfiltration, due to the high cross-flow velocities. This results in a high pressure drop which causes low-permeate flux and small volume reduction, as well as a negligible increase in the concentration of the retentate. A very frequent arrangement is a feed and bleed system that consists of a number of stages each fed with a circulation pump. The feed pump generates the applied pressure, while the circulation pump maintains

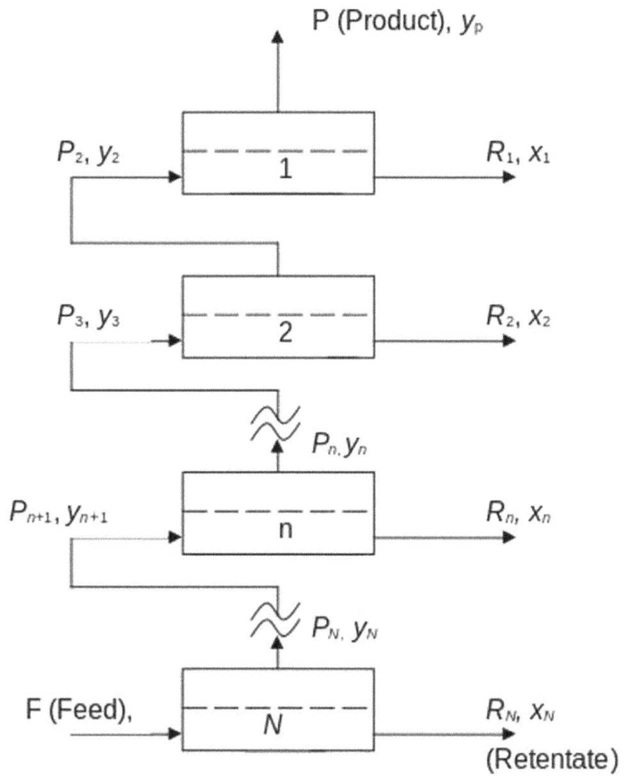

FIGURE 18.1 Cascade without reflux.

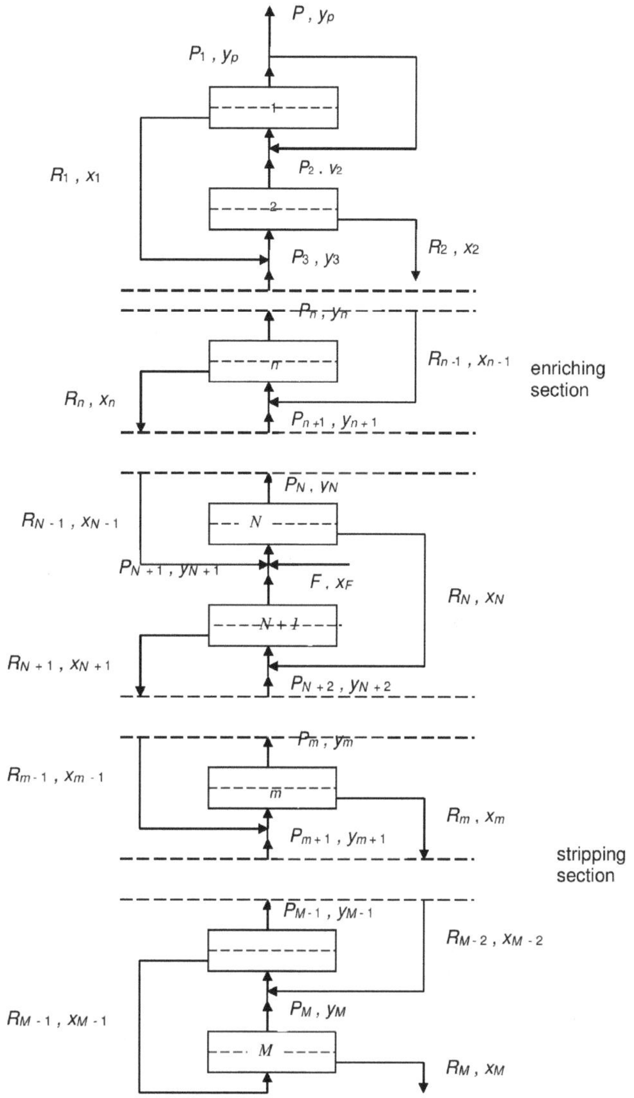

FIGURE 18.2 Recycling cascade with enriching and stripping sections.

the cross-flow velocity. All arrangements of the modules are described in various other books on membrane processes [39–41].

When the product from the single stage does not have the desired quality the cascade system can be employed. The cascade design is frequently used in nuclear industry, especially for isotope separation. The first uranium-235 enrichment as a uranium hexafluoride was conducted in a cascade composed of many stages with porous, metallic membranes. Since the molecular weight ratio of UF_6 isotopomers is about 1.008, the gas diffusion enrichment factor was low (~1.0043) and the unit separation effect had to be multiplied in a large number of stages to achieve high purity of the product. The stage is an element of the separation cascade containing several separation units, connected in series or in parallel, and arranged as a tapered or squared-off design. A permeate from one stage feeds the next one, however permeate is not always a product of the cascade. The simplest scheme is a cascade without reflux (Figure 18.1), which is reasonable when the retentate in the enriching section is not a valuable material. An example of such an arrangement is a plant for production of deuterium by water electrolysis operated by Norsk Hydro [39].

To avoid product losses, recycling cascades are employed (Figure 18.2). The cascades with reflux are more complicated and the capital costs and costs of operation are higher; these are sensible to use only in the case of expensive raw materials. Technical and economic considerations on such systems have been performed in basic books for isotope separation [42,43], as well as by Hwang and Kammermeyer [44].

18.1.4 Process Operation—Safety Aspects

18.1.4.1 Process Control

All installations designed for nuclear industry have to fulfill very strict requirements of safety and reliability over the long term. Control and monitoring systems, automation, and washing installations are considered essential parts of the design.

18.1.4.1.1 Automation

It is common for membrane plants to be automated. Automation is used for starting up and shutting down the installation, controlling the cleaning cycle, and the process parameters. However, some interventions by an operator in the case of an emergency should be foreseen. All valves and remote-controlled devices ought to have the possibility of manual start-up and putting in motion. Membrane installation may be integrated into the central control system in the plant where it is used. However, in the case of the nuclear industry it is rather recommended to design self-contained units, fully automated by simple ladder logic controller or microprocessor control. All operating parameters, such as pressure, temperature, flow, tank level, pH, conductivity, and specific activity are monitored and recorded. Continuous visualization of the main parameters' trends is very important and helpful. The control system includes also automatic start-up and shutdown of pressure pumps, dosing pumps, as well as shutdown of the whole installation in the case of an unexpected event or emergency. Exceeding of the allowable operating parameters' limits should result in the start-up of appropriate blocks and alarms.

In view of radiological safety, personal contact with operating installations should be minimized. The location of data acquisition and control systems has to be planned in a separate room, away from places of potential contamination. However, the necessity for eventual periodic sampling has to be taken into account also. The sampling points have to be located in locations with easy operator access and arranged in a way that eliminates the risks of leaks and contamination.

18.1.4.1.2 *Precautions against Uncontrolled Pressure Excess*

Membrane installations operated in the nuclear industry are pressure-driven systems, most are reverse-osmosis plants. Uncontrolled growth of operation pressure may result in module damage and valve leaks resulting in a contamination hazard. The selection of appropriate pumps and security devices can avoid the danger of pressure overgrowth and its detrimental implications. The security valve outlets have to be connected with existing waste distribution systems to direct any eventual leaks to the waste-collecting tanks.

18.1.4.1.3 *Precautions against Uncontrolled Temperature Excess*

All polymeric membranes and modules have admissible temperatures of operation. For most of them this temperature is 40°C, only for inorganic materials is it much higher. In the case of reverse osmosis there is a possibility of retentate and permeate temperature growth caused by the circulation under high pressure. Continuous temperature control of the streams with an upper alarm has to be enabled. Systems collecting the excess heat (radiators, ribbed pipelines) and allowing intensive cooling have to be designed.

18.1.4.1.4 *Precautions against Uncontrolled Overflow of the Tanks*

All the tanks connected with membrane installation have to be equipped with tank level meters with devices and blockades of the pumps secured against overflow. Special collectors in the floor for contaminated solutions in the case of unexpected overflow have to be designed.

18.1.4.2 Radiological Protection Requirements

Two systems for radiological control are required in the case of membrane plants for radioactive waste processing: the first for process control, and the second for securing the staff and avoiding the spread of contamination. The latter usually exists if the installation is located in a nuclear power plant or other nuclear center. Some security devices have to be applied in the rooms where membrane plant will be placed. The lead shielding in places where increased radioactivity is expected should be installed in accordance with precise preliminary measurements and dose estimations.

For process control purposes the radioactivity measurements of feed, permeate, and retentate should be carried out. A new batch of waste has to be controlled, and precise characteristics (radioactivity, conductivity, and pH) of every new portion of the material should be measured. The radioactivities of pure permeate before discharge have to be controlled, as well as periodic radioactivity of retentate before solidification or other processing stages.

18.1.4.3 Membrane Control

18.1.4.3.1 *Membrane Fouling in Radioactive Waste Treatment*

Fouling is a one of the technical problems of membrane systems operation. This phenomenon occurs mainly in pressure-driven processes. There are four types of fouling: precipitation of dissolved solids, sediment of suspended solids, biological fouling, and organic non-biological fouling caused by carbon-based molecules. The symptoms of fouling include decreased permeate flow, increased pressure drop across the module, and increased total pressure in the unit. Therefore, to keep the flux stable, the pressure needs to be continuously increased. Usually, a 10–15% reduction of the permeate flow should be a cause for immediate membrane cleaning. Otherwise, an increase of pressure in the system and crystallization of salts or silica in the pores may result in irreversible membrane structure change or damage.

Radioactive waste processing is usually a complex process often combining microfiltration and ultrafiltration with some other processes, such as precipitation or binding with macromolecular ligands. During filtration of radioactive wastes, especially when suspensions or macromolecular species are separated, a deposit can be formed on the membrane surface or inside the pores after several hours of operation. Particles deposited on the membrane surface can form a filter cake, which acts as a secondary membrane. The deposit causes a flux decline; however, sometimes separation with a secondary membrane may be better. This secondary membrane usually has smaller pores than the original one and so it retains molecules that pass through the original membrane. As a result, the accumulation of radioactive compounds takes place on the membrane surface and cleaning the membrane is necessary. Membrane cleaning is usually done with a small volume of acidic, basic solutions and water, alternately. For polymeric membranes, the temperature and chemical resistance of the polymers limit the choice of cleaning solutions. There is no such limitation in the case of inorganic membranes—ceramic or metallic. After washing, a high concentration of radioisotopes is obtained in a small volume of cleaning solution. Sometimes, the concentration of radioisotopes in the secondary membrane can reach 90%. During the removal of radioactive strontium by ultrafiltration combined with precipitation with colloid $Ti(OH)_4$ after 2 h of operation, a 50% reduction of permeate stream was observed but a 20-fold volume reduction and 140–150-fold decrease of ^{90}Sr in the permeate were achieved [45]. The strontium balance showed that 94% of ^{90}Sr was found in the secondary membrane. After washing, the UF membrane recovered its initial permeation properties and over 99.2% of strontium was transferred to the mixture of rinsing solution and retentate, the volume of which was 18 times lower than for the feed solution. A similar effect, but not as intensive, was observed during ultrafiltration enhanced by complexation with polyethyleneimine—in the secondary membrane 7% of ^{60}Co was retained, but it was hardly removable with rinsing [46]. Radiometric measurements of the membrane sample showed that some part of ^{60}Co was persistently bound into the membrane. This was disadvantageous, especially for polymeric membranes, as the material of the membranes is permanently exposed to ionizing radiation emitted by bound radioisotopes that can result in the loss of the permeation abilities.

The most common methods of fouling prevention are wastewater pretreatment, the control of hydrodynamic conditions in the apparatus (low flux, high feed flow rates, turbulence promoting, dynamic filtration), or special construction of the module (cross-flow filtration) [47–51].

A thorough understanding of the fouling phenomenon and its nature is paramount to its prevention and further operation of membrane devices [22–24,52–54].

18.1.4.3.2 Scaling Control

During the concentration of radioactive waste by pressure-driven membrane processes, such as reverse osmosis, the solubility limits of some dissolved salts can be exceeded. These salts precipitate on the membrane surface or other parts of the plant, causing damage to the membrane or severe corrosion of the elements of the installation. The most common compounds that precipitate easily causing scaling are calcium salts ($CaCO_3$, $CaSO_4$, CaF_2, $CaHPO_4$), barium and strontium sulfates, and silica and magnesium carbonates. The methods of scale minimization include: proper pretreatment, acid injection to reduce carbonates, water softening by lime or lime soda, ion exchange, removal of bivalent ions with nanofiltration, and antiscalant addition. The scale-inhibiting compounds are injected into the parts of the installation with a high risk of scale formation. locations places have to be specified in the project design of the plant. Antiscalants delay precipitation through the formation of microcrystals with low agglomeration abilities that interfere with crystallization of the other salts. The most common scale-inhibiting chemicals are polyphosphates, polycarboxylates, polymalonates, and polyacrylates. In Chalk River Laboratories, the antiscalant agent effectively preventing scaling in the RO plant was Pretreat Plus (King Lee Technologies, San Diego, California) [55–59], at the Radioactive Waste Management Plant (RWMP) in Poland in a three-stage RO plant, sodium hexametaphosphate (Calgon) was chosen. To prevent scaling in an RO pilot plant designed and operated by ANSTO, the antiscalants AS-1000 (phosphino carboxylic acid) and AS-1300 (polycarboxylic acid) were used [60].

A novel method for scale control is magnetic and electrostatic technology, as suggested by U.S. Department of Energy [61]. This nonchemical technology is recommended for scale and hardness control as a reliable energy saver in certain applications and it can be used as a replacement for most water-softening equipment.

18.1.4.3.3 Membrane Cleaning

Periodic cleaning of the membranes exposed to the action of foulants and scalants present in the wastes is necessary. The frequency of the cleaning depends on the composition of the wastes. Its necessity is recognized based on a pressure drop or rapid flux decrease. If the flux drops by several percent, the membrane has to be restored. The volume of the cleaning solutions should be as small as possible, to minimize the amount of secondary wastes.

The selection of cleaning agents and their concentration depends on the membrane and the kind of foulants present in the waste. Manufacturers of the membranes usually recommend some cleaning solutions, acidic or basic, and detergents, which have to be tested during pilot plant experiments. The most popular agents are citric acid, sulfuric acid, sodium hydroxide, sodium tripolyphosphate, sodium ethylendiaminetetraacetate, and ethylendiaminetetraacetic acid (EDTA). These compounds are relatively inexpensive and widely applied, however sometimes specially prepared formulations of membrane manufacturers are superior and worthy of consideration. Such reagents, however, are very good centers of nucleation and initialization of fouling and scaling. Therefore, subsequent fouling can be even more severe after restoration of the membrane. Various cleaning agents are available in the market. These can minimize the production of secondary wastes because they can clean the membranes very effectively. For polymeric RO membranes applied in Chalk River Laboratories, for example, good results were obtained using MEMCLEAN, an alkaline detergent containing EDTA [55–59,62]. For cleaning ceramic membranes used at the Institute of Nuclear Chemistry and Technology (INCT), Warsaw, the P3-ultrasil, Henkel EKOLAB, was effective. On the other hand, laboratory preparations (e.g. alkali- and acid-based solutions) also have proved very useful [63].

Apart from chemical cleaning, mechanical or compressed air restoration of the membranes can be possible. Mechanical cleaning is restricted only for some configurations, namely tubular modules, with easy access to the membranes. The foam balls introduced into the tubular membrane lumen can scrape the substances deposited on the surface; however, they cannot remove the foulants from the membrane pores. The pores can be cleaned by backflushing with pressurized air or water shock in the opposite direction. This method is limited to those membranes which are strong enough to withstand this force, or to membranes with sufficient support. Some configurations (e.g. spiral wound) are not recommended for this method of cleaning.

A recent approach [64] is a method of direct membrane cleaning (DMC), applied for conductive membranes, such as stainless steel, graphite, and conductive ceramics. By short current pulses (1–5 s at 50–200 mA/cm^2), electrolytic generation of microscopic gas bubbles takes place, which removes the solid deposit from the membrane surface without interrupting the filtration process. Such cleaning enables the cross-flow velocities and a transmembrane pressure to be reduced, which implicates reduced plant wear and lower energy consumption while minimizing the size of the pressure pumps. The process was demonstrated on microfiltration and ultrafiltration membranes of different geometries. DMC was used during the treatment of low-level radioactive waste by microfiltration combined with inorganic sorbents: 44 ppm of nickel hexacyanoferrate for Cs removal, 33 ppm of zirconium phosphate for Sr removal, and 45 ppm hydrous titania for Ru, Ce, and actinides removal. Additionally, $Fe(OH)_3$ was introduced to enhance actinides and Cs retention and to trap fine particles on precipitated floc. When the mixed sorbents were concentrated to 5% by microfiltration, rinsing with 0.1 mM NaOH, which made the slurry conductive and ready for subsequent electrolytic dewatering up to 30% of solids, reduced the salt content. Finally, the concentrate was immobilized with cement powder.

18.1.4.3.4 Secondary Wastes

Membrane installations generate secondary wastes that have to be taken into consideration before plant design. Reverse osmosis produces permeate, which can be discharged after radioactivity control, and retentate that can undergo further processing. Usually the retentate is not suitable for solidification and further volume reduction is necessary.

Secondary wastes are also generated during regular cleaning procedures, because fouled membranes have to be washed or cleaned using cleaning solutions. The concentration of radioisotopes in these solutions is sometimes high and they need treatment by recycling to the feed at the inlet of membrane installation. The additional volume of waste has to be taken into account in the plant design, as well as the influence of extra load on the membrane performance.

The plant itself is a source of secondary solid waste; membranes, spent filter cartridges, and related small parts have to be processed by common methods like compaction or incineration to reduce their volume or they can to be left to decay if the adsorbed radioisotopes have a short half-life.

18.1.4.3.5 Membrane Exchange

For the treatment of conventional, nonradioactive liquid waste the predicted lifetime of membranes is 4–5 years. The effective lifetime depends on the conditions in which the membrane is used: the characteristics of solutions treated, pressure, and temperature. While selecting the membrane for radioactive waste processing, one has to bear in mind its resistance and stability under ionizing radiation exposure. The membranes used in the installation have to be tested before a final decision about their application. In spite of that, the expected lifetime for polymeric membranes will not be longer than 5 years. After this period, the membranes have to be exchanged. Depending on the membrane type and configuration the whole membrane modules or the cartridges where the membranes are assembled can be exchanged. The necessity of membrane exchange has to be taken into account in the design stage. Security considerations, easiness of operations, and easiness of decontamination are very important. Sufficient space has to be reserved for work with membrane modules, which have to be exchanged and means of minimization of contamination have to be foreseen.

Spent membranes and membrane cartridges do not undergo regeneration; they are treated as a solid radioactive waste. They can be stored for decay or processed as with other solid waste materials.

18.1.4.4 Decontamination of the Space and Equipment

All potential points of leaks and other hot spots have to be identified in the design stage. The decontamination procedures in case of accidents have to be elaborated. The specification of the equipment for eventual decontamination and the means of decontamination are defined in laboratory instructions and manuals. All necessary safeguards and security systems are elements of the project design. In the case of an emergency all means limiting the spread of decontamination should be used.

18.2 PRESSURE-DRIVEN MEMBRANE PROCESSES EMPLOYED FOR LIQUID RADIOACTIVE WASTE TREATMENT

18.2.1 Reverse Osmosis

Reverse osmosis has been employed in full scale in many nuclear centers around the world. Permeate after reverse osmosis can be directly discharged to the environment or recycled as service water within the nuclear power plant. There are a number of industrial RO applications and facilities in the pilot-plant stage of operating for radioactive waste processing. Some examples of these installations are presented in Table 18.1.

18.2.1.1 Three-Stage Reverse-Osmosis Plant for Low- and Intermediate-Level Radioactive Waste Processing

Laboratory and pilot-plant experiments carried out at INCT showed that reverse osmosis is very useful for the treatment of low-level liquid radioactive wastes from Polish nuclear laboratories. However, to reach a high decontamination level the process should be arranged as a multistage operation with microfiltration or ultrafiltration pretreatment [71,72].

The RO process was implemented at RWMP in Świerk. The wastes collected there, from all users of nuclear materials in

TABLE 18.1
Examples of nuclear facilities applying RO for liquid radioactive wastes processing

Facility	Type of waste	Process and type of membranes used	Capacity	Results	References
Chalk River Laboratories (Canada)	Mixed aqueous waste	Tubular (TRO) and spiral-wound (SWRO) reverse osmosis (Filmtec SW30HR) with MF pretreatment	3 m^3/h	2200 m^3 per year processed wastes VRC = 200–400 Retention: 99.9% (α); 99% (β and γ); 99.5–99.8% solids	[55–58,65]
ANSTO (Australia)	Low-level radioactive wastes from laboratories and radioisotope production	Two passes for permeate purification and two stages of concentration by RO spiral wound modules (4" FILMTEC XLE-4040 and 2.5" FILMTEC BW30-2540) with UF pretreatment (PCI FPA10 modules)	1.2 m^3/h	High-purity effluent was obtained. Flux decline in UF tubular unit was observed, resulting from irreversible fouling of membranes by surfactants coming from radiopharmaceutical production. The additional pretreatment step was recommended	[60,66]
Nine Mile Point nuclear power plant (United States)	BWR floor drains and other wastes from NPP	RO systems based on Thermex	36,300 m^3/y	11 m^3/y of solid waste generated; total organic carbon in the effluent <50 ppb, conductivity of 0.058 mS/cm. The plant operates without secondary wastes	[67]
Pilgrim nuclear power plant (United States)	BWR floor drains and various other wastes	RO systems based on Thermex		The waste volume decreased in time of the system operation, personnel exposure associated with radioactive waste operation was reduced	[67]
Wolf Creek nuclear power plant (United States)	Floor drains and other wastes	Spiral-wound RO with tubular UF modules for pretreatment	Processing rate of DD: 4620 L/d	The installation equipped with drum dryer (DD) unit and demineralizer system; high-salinity solutions are filtered; VRC: 10–20	[67,68]
Bruce nuclear power plant (Canada)	Wastes from chemical cleaning of steam generator	Ultrafiltration (Zenon ZPF-12 tubular membrane modules) and two-stage RO (Filmtec SW30HR) coupled with wet oxidation technology (WAO)		Permeate meets the criteria for sewer usage; the RO concentrate is returned for further processing (WAO, solidification)	[69]
Comanche Peak nuclear power plant (United States)	Floor drains, resin sluice water, and boron recycling water	UF with RO and NF (RWE NUKEM)	2.28 m^3/h	The suspended matter is rejected by UF unit, dissolved compounds by RO. NF membranes were used for passage of boron. Membrane technology followed by IX performed better than existing demineralizer	[67]
Dresden nuclear power plant (United States)	Wastes contaminated by transuranic elements	UF with RO	Two tanks of total volume 1440 m^3 for processing	Permeate from the installation passed deep-bed demineralizers; after radioactivity control was discharged from the plant. Volume reduction of tank wastes by factor of 10	[67]

(Continued)

TABLE 18.1 (Continued)
Examples of nuclear facilities applying RO for liquid radioactive wastes processing

Facility	Type of waste	Process and type of membranes used	Capacity	Results	References
Savannah River site (United States)	Reprocessing wastes with high concentration of sodium nitrate	MF (cross-flow filters, 0.2 μm) and RO (high salt rejection spiral-wound elements)		Plant met the discharge criteria; one of the major problems was biofouling of the MF membranes	[67,70]
Radioactive Waste Management Plant, Warsaw, Poland	Wastes from nuclear laboratories and application of radioisotopes	Two passes for permeate purification, two stages of concentration by RO spiral wound modules (SU-720R, SU-810 (TORAY))	1 m^3/h	Complete purification of effluent (β- and γ-emitters is lower than 10 kBq/m^3, α-emitters lower than 1 kBq/m^3, TDS<0.1 g/dm^3), concentrated retentate for solidification	[71,72]
DTRO, Tsinghua University	Radioactive wastewater from HTGR (decontamination, floor flush, laboratory draining)	Disc tubular RO, three stages	400–600 L/h	The system can reach CF of 50 and DF >30. The concentration of wastewater did not affect the DF greatly; pH of permeate from the second stage is lower than the raw water	[73]

Poland, have to be processed before safe disposal. Until 1990, the wastes were treated by chemical methods that sometimes did not ensure sufficient decontamination. To reach the discharge standards the system of radioactive waste treatment was modernized. A new evaporator integrated with membrane installation replaced the old technology based on chemical precipitation with sorption on inorganic sorbents. The two installations, EV and 3RO, can operate simultaneously or separately. The membrane plant is applied for the initial concentration of the waste before the evaporator. It may be also used for final cleaning of the distillate, depending on the requirements. The need for additional distillate purification is necessitated due to entrainment of radionuclides with droplets or the presence of volatile radioactive compounds which are carried over.

Evaporation is, however, a very efficient cleaning process, although it is highly energy consuming. The application of the membranes at the beginning of the cleaning cycle significantly reduced the energy consumption.

Membrane installation, with a capacity of ~1 m^3/h pure permeate, was composed of three stages of reverse osmosis (Figure 18.3) preceded by pre-treatment with polypropylene depth filters. The first two stages were used for purification, and the third one for the final concentration. Two types of spiral-wound RO modules were used in the installation: *SU-720R* and *SU-810 (TORAY)*. Both types of modules worked under a pressure of 20 bar and with high salt rejection (higher than 99%). The membrane was manufactured from cross-linked fully aromatic polyamide composite. Two *Model SU-720R* modules connected in series placed in a single housing formed the first stage of reverse osmosis. The second stage was the same as the first one: two modules placed in a single housing. The third stage was composed of two housings in parallel, with two *Model SU-810 (TORAY)* modules in each vessel.

In 2019, the RO plant was modernized. Before the installation, a two-stage pre-treatment filter battery was installed to remove coarse particles and organic substances. This significantly extended the time between subsequent cycles of washing the membranes and prolonged their life. The first stage of pre-treatment is a coarse filter; the second stage is a set of carbon filters. During the modernization, a dosimetric system was added to the computer control unit, allowing for monitoring of the radiation level, archiving measurements, and remote reading at the ZUOP dosimetric center. Additionally, an electronic air flow control and visualization system was installed in the installation during the modernization process. Membrane modules were replaced and control and measurement equipment was extended. This enabled optimization work to be carried out to improve the retention and volume reduction rates as well as enhancing treatment quality. A view of the plant is shown in Figure 18.4.

Liquid radioactive waste was directed from the waste storage tank to the 8 m^3 feed reservoir. After pretreatment with PP depth filters and injection of antiscalant, the wastes were directed to the first stage of RO. The retentate from this stage was concentrated in the third RO unit. The concentrated solution could be directly solidified if the concentration of the total solute was appropriate (<250 g/dm^3). The salt concentration was limited by the conditions of concrete

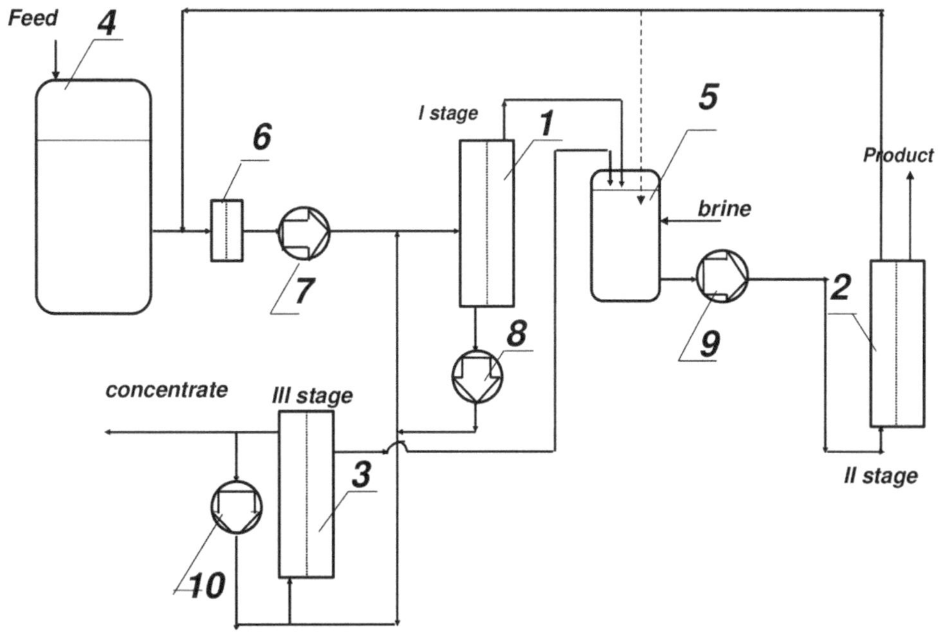

FIGURE 18.3 A scheme of three-stage RO plant. 1, first stage of RO; 2, second stage of RO; 3, third stage of RO; 4, feed tank; 5, intermediate tank; 6, depth filters; 7, 8, 9, 10, pumps.

FIGURE 18.4 Three-stage RO plant for radioactive waste processing at Radioactive Waste Management Plant in Świerk after modernization in 2019.

(Courtesy of Radioactive Waste Management Plant, ZUOP.)

solidification. If the concentration was not sufficient, further concentration took place in the evaporator. Permeate from the first and third stages was directed to the permeate reservoir before the second RO unit. The product from the membrane installation (permeate from the second stage) was of the required radiochemical purity and, after control of the specific activity and salinity, was discharged to the communal sewage.

Laboratory experiments showed the influence of the total concentration of ballast non-active salts on decontamination factors. As the concentrations of total solute in permeate from the first and third stages were very low a decrease in the retention of radionuclides was observed. To improve the efficiency of radioisotope removal, an additional salt injection was performed before the second stage.

The characteristics of permeate and retentate streams in terms of upper limits at the exit of the RO plant are presented in Table 18.2. The concentration of salt in the permeate is lower than 0.1 g/dm^3. The concentration of some specific elements such as heavy metals has to be in conformity with the limits of impurities for wastes discharged to inland waters. The total specific activity for β- and γ-emitters is lower than10 kBq/m^3, while for α-emitters it is lower than 1 kBq/m^3 (the limits for liquid waste). The total salt concentration in retentate is limited by the ability to bind the solution with the concrete, and the specific radioactivity by nuclear safety regulations. On the basis of parameters defined below, the filtration abilities of the membranes to clean radioactive waste and the performance of the installation were assessed. The values calculated from the experimental results are presented in Table 18.3.

$$R = \frac{c_f - c_p}{c_f} * 100\% \quad (18.1)$$

$$DF = \frac{A_f}{A_p} \quad (18.2)$$

$$K_{TDS} = \frac{c_f}{c_p} \quad (18.3)$$

$$CF = \frac{A_R}{A_f} \quad (18.4)$$

The retention of the dissolved salts by the three RO units was higher than 99% in each case; decontamination factors were also high. Most purification took place in the first stage of reverse osmosis; the second stage played the role of final purification. Permeate in the majority of cases was sufficiently pure for discharge. However, when the initial activity of the waste was high (the A feed solution), the radioactivity of the product was too high for discharge and hence the permeate was returned to the installation inlet tank. Injection of the salt solution (NaCl) before the second stage increased the decontamination factor in that stage from 3.09 to 91, and for entire plant to 13,700 (the numbers obtained for A waste sample).

The results of radioactive waste sample C treatment with 3RO are shown in Figure 18.5. The feed radioactivity is 2,200 Bq/dm^3. After processing, the permeate has radioactivity below discharge limits (3,9 Bq/dm^3) and can be discharged; concentrate of radioactivity 16,000 Bq/dm^3 can undergo further processing. The average decontamination factor for the entire plant was 564.

TABLE 18.2
Characteristics of the streams after RO installation

Retentate:	
Total concentration (g/dm^3)	250
Specific activity	10^7 kBq/m^3 (0.3 Ci/m^3)
Permeate:	
Total concentration (g/dm^3)	0.1
Specific activity	0.01 ALIp/m^3 or <10 kBq/m^3 (β and γ) And <1 kBq/m^3 (α)

18.2.1.2 Other Applications of RO

The three-stage RO osmosis system for the treatment of radioactive wastewater of HTGR was constructed at Tsinghua University, China [73]. The wastewater originated from decontamination activities during the repair processes, floor flush, and laboratory drainage. The installation was composed of disc tubular reverse osmosis modules (DTRO) operated under 7 MPa. This design provides high feed rates, and creating turbulence avoids fouling. The influent to the system is 400 L/h and the intake of the first stage was 600 L/h. The time to reach a steady state was 300 minutes. Concentration factors of 50 and decontamination factors greater than 30 were achieved with the system. The pH of the permeate from the second stage was lower than the pH of raw water.

RO preceded by MF or UF is considered an option for the treatment of radioactive wastes from Romanian nuclear centers. Effective studies have been carried out at the Research Center for Macromolecular Materials and Membranes, Bucharest, and at the Institute of Nuclear Research, Pitesti, aiming at employing these pressure-driven techniques for cleaning the wastes from decontamination of nuclear installations and reactor primary circuits [74,75].

There are assessments predicting the use of reverse osmosis for the processing of wastes from medical applications [76,77] and for the removal of cesium-137 from decontamination wastes after an accident in a steel production factory [78]. RO is considered as a method for removal of radioactive pollutants from contaminated water (removal of ^{137}Cs and

TABLE 18.3
Removal of non-radioactive substances (dissolved salts) and radionuclides. (Copyright (2001) from Concentration of Low- and Medium-Level Radioactive Wastes with 3-stage Reverse Osmosis Pilot Plant, by Chmielewski, et al. Reproduced by permission of Taylor & Francis, Ltd., www.informaworld.com)

	Feed		Permeate		*R*	K_{TDS}	*DF*	*CF*
	(ppm)	(Bq/dm^3)	(ppm)	(Bq/dm^3)	(%)			
Stage I	641.3	1.95 × 104	3.1	96	99.52	208	203	
A								
B	320.6	280	1.5	1.4	99.60	214	200	
C	769.5	2200	3.9	10.4	99.49	197	211	
Stage II	4.2	136	1.3	44	68	3.1	3.1	
A								
B	2.3	2.3	1.5	1.6	34	1.5	1.4	
C	4.2	12.3	1.2	3.9	71	3.5	3.2	
Stage III	2344.9	7.1 × 104	8.7	275	99.63	269	258	2.8
A								
B	1265.3	1.1 × 103	5.6	5.4	99.56	226	204	3.9
C	2455.9	7 × 103	7.0	23	99.71	322	304	2.4
Entire Plant	641.3	1.95 × 104	1.3	44	99.79	478	443	10.2
A								
B	320.6	280	1.5	1.6	99.52	209	175	15.4
C	769,5	2200	1.2	3.9	99.84	635	564	7.3

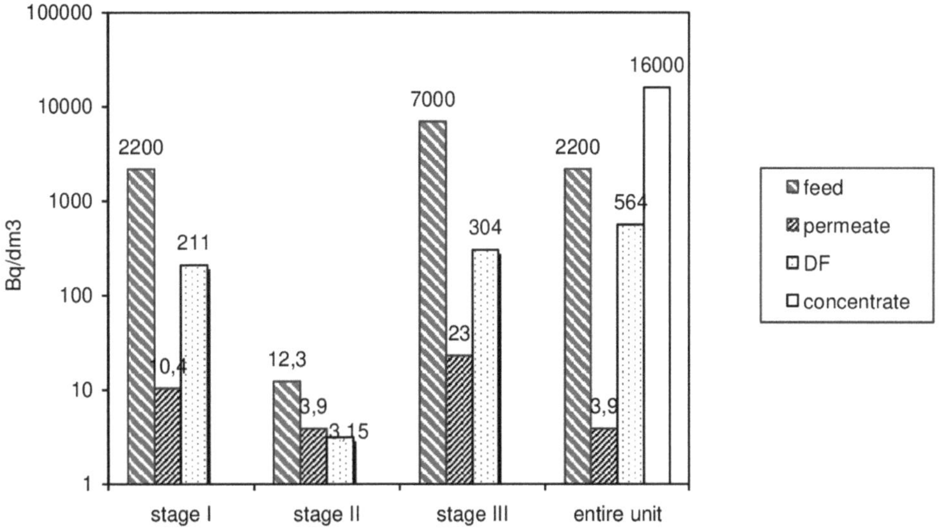

FIGURE 18.5 Removal of radioactive compounds; experiments with three-stage RO plant.

^{90}Sr) in the vicinity of atomic power plants [79], as well as for removal of small quantities of radionuclides (^{222}Rn, ^{234}U, ^{238}U, ^{226}Ra) from drinking water [80,81]. Advanced reverse osmosis membranes allow reducing trace concentrations of Cs, Sr, and iodine in high-salinity water with a rejection rate higher than 99.7% [82].

In the literature, one can find many descriptions of the use of RO membranes for the removal of boron from sources of various origins [83–85]. The radioactive waste treatment with separation of boron from silicon and radionuclides was investigated [86] using a cross-flow filtration apparatus with a flat-sheet RO membrane. A higher temperature, 55°C, resulted in better silica and boron separation efficiency with stable nuclides rejection. Under these conditions 46.58% separation of boron was observed and decontamination factors for other components were as follows: silicon—29.3, Cs—4.54, Sr—58.88, and Co—70.41. These differences in separation factors make it possible to design a high-boron, low-silicone solution-producing system suitable for boron reuse.

RO has found a special place of use after the nuclear accident in Fukushima–Daiichi. A huge amount of contaminated water was released through the cooling of damaged reactors, which had to be cleaned up. After primary cesium and strontium removal, the coolant water was treated by RO to reduce salinity before being recycled to the reactors [87].

Many studies have been carried out in various research centers for the application of RO for decontamination purposes [88,89]. The process of decontamination of two kinds of effluents, contaminated groundwater after nuclear accident and contaminated seawater during a nuclear accident, was studied [88]. Due to the high retention of Cs and Sr, the SW30 HR membranes were selected. The experiments were performed on a laboratory scale with a variety of flat-sheet membranes sized to fit the SEPA CF cell and with two types of semi-industrial spiral-wound RO membranes. For groundwater effluent investigation the results were very optimistic; the Cs and Sr retentions for sea water were slightly lower. Spiral-wound membranes, 2.6 m^2 surface area, were used in the scale-up process. For this module the retention and fluxes were obtained. The aging experiments demonstrated better resistance of SW30 HR membranes to γ-irradiation compared to Osmonics SE membranes, however limited degradation of membranes was observed.

A full cycle of cleaning the cooling water including an Advanced Liquid Processing System (ALPS) to treat the retentates of the RO plant to further purify the water from various radionuclides except tritium was developed by TEPCO Company [90–93].

Reverse osmosis was proposed as a final stage of purification of the flowback fluids obtained after hydraulic fracturing of gas shales [94]. These fluids have complex characteristics; they contain suspended solids and a large spectrum of metal ions, organic pollutants, as well as naturally occurring radionuclides (NORM) and their decay products (^{226}Ra, ^{214}Pb, ^{214}Bi, ^{212}Pb, ^{208}Tl, ^{228}Ac, ^{238}U, ^{234}U). Moreover they have high salinity and so the use of RO to reduce the NaCl content was justified.

18.2.2 Forward Osmosis

In recent years, much attention has been paid to forward osmosis (FO) applications. Unlike reverse osmosis (RO), where the driving force is hydraulic pressure, FO uses the osmotic pressure difference of the solutions on both sides of the membrane as the driving force to transport water through the membrane. Therefore, FO requires the use of two solutions, differing significantly in solute concentration and separated by a semipermeable membrane: the feed and the receiving solution (draw solution) with an osmotic pressure much greater than the osmotic pressure of the feed. The receiving solution is gradually diluted during the process and the feed is concentrated. FO is an alternative to more energy-consuming, evaporative methods of concentration of solutions, but also to membrane distillation or RO. Integrated FO–UF systems are effective for water desalination and reuse, as well as osmotic pumps for the controlled release of drugs and concentration of juices and proteins [95]. To expand commercial application of the process much attention is paid in the current research to designing effective draw solutions, developing new membrane materials, reducing the energy consumption, and elaborating energy-efficient recovery processes [96]. This economical process could replace RO as a suitable method for the concentration of radioactive waste. Despite the lack of applications on an industrial scale, a lot of research has been carried out in the laboratory to assess the suitability of this process for the technology of liquid radioactive waste treatment [97–104]. Commercial FO membranes were applied by researchers from Tsinghua University, namely polyamide-based thin-film composite with embedded polyester screen support (TFC-ES), cellulose triacetate with embedded polyester screen support (CTA-ES), and cellulose triacetate with a cast nonwoven support (CTA-NW), respectively. All these asymmetric membranes consisted of a mechanical support layer and dense active top layer [97–99]. Most of these membranes proved to be effective in rejecting Cs, Co, and Sr ions. The effects of the membrane material, feed solution properties, draw solution concentration, cross-flow rates on the permeate flux, and retention of separated ions were investigated and evaluated. Variations in performance of the membranes, as well as changes to the membrane surface were observed and evaluated. Membrane characterization was made by scanning electron microscopy, atomic force microscopy, and contact angle measurements.

The factor influencing performance of FO membranes was membrane orientation in relation to the separated solutions: active layer facing feed solution (AL-FS) and active layer facing draw solution (AL-DS). Some differences in separation of three ions, Co, Cs, and Sr, were observed in terms of retention and permeability, as well as membrane used. For instance, the rejection of Cs ions with CTA membranes was 90.35–97.15%, while the TFC membrane rejected them with a retention rate lower than 48% [97]. The role of support used is also important. The use of NW support resulted in severe concentration polarization hindering water flux, while a support layer of ES alleviated the concentration polarization.

The transport of radionuclides through the FO membrane can be driven by three mechanisms: convection occurring with water flow, diffusion driven by concentration difference, and transport due to electrostatic interactions. The charge of the FO membrane active layer appeared to be the main factor affecting Co(II) separation [99]. The most important problem of FO applied to radioactive solutions processing is membrane fouling of different origin. All kinds of deposited substances affect the permeate flux and separation, however the deposition of radioactive ions is particularly troublesome as it leads to an increase in the total accumulated radioactivity in the system, especially in the case of long-lived isotopes. Fouling and cleaning protocols for FO membranes were investigated [100]. Studies with simulated liquid radioactive waste containing Co, Sr, and Cs ions revealed that Sr(II) was more prone to foul the membrane than Co(II) and Cs(I). The fouled membrane surface became more hydrophobic and rougher. On-line cleaning with deionized water recovered the flux to 69%, while HCl and ultrasound cleaning were tested also.

Fouling in FO is a difficult problem to address because it is affected by many different parameters such as the properties of the feed, type of draw solution, membrane material type. and operating conditions. The pre-treatment of feed solution is very important, especially when the composition of the wastewater is complex, including substances such as metal ions, radionuclides, macromolecular ligands, and other complexing agents. Organic fouling is the most complicated because it can lead to complete membrane blocking. For membrane cleaning, the most effective treatment is a combination of chemical and physical methods. Innovative techniques such as UV and ultrasound have been proposed [101].

Forward osmosis was also examined for the removal of nuclides and boric acid from the simulated borate-containing radioactive wastewater [102,103]. Boron permeation through the membrane depended on the type of membrane used, membrane orientation, and salt and boron concentration in the feed solution, pH of the feed solution, and osmotic pressure of the draw solution [102]. In the case of the multicomponent solution, none of the elements except boron permeated through the FO CTA membrane. The maximum boron flux was obtained for a pH below 7, at high osmotic pressure, and for an active layer facing a draw solution orientation.

Another study showed the high potential of CTA-ES membranes for retention of Co^{2+} (99.62–99.69%) and Sr^{2+} (96.90–97.85%) when the boric acid concentration was increased from 0 to 2400 mgL^{-1} and the active layer facing feed solution (AL-FS) mode. Cellulose triacetate membranes (CTA-ES) and membranes with a cast nonwoven support (CTA-NW) exhibited high Cs^{+} retention (~95%). The thin-composite membranes supported by a polyester screen had the lowest Cs^{+} retention, at below 40%. It was revealed that due to the highest water flux, nuclide retention, and boric acid flux the CTA-ES membranes had the greatest potential to treat radioactive wastewater containing borate [103].

Medical radioactive liquid waste from radiation therapy rooms in hospitals was treated by FO [104]. The process was assessed for a solution containing both natural iodine ^{125}I and radioactive iodine ^{131}I which is present in real radioactive waste. Different draw solutions were used under various pHs to determine the optimal operation conditions. High rejection rates were obtained of up to 99.85%. The inhibitory effect of reverse solute flux that resulted in higher iodine rejection was noticed. The treatment of real radioactive waste, while allowing high iodine retention, resulted in severe fouling of the membranes and reduction of flux. Hydraulic washing was not successful because of the combined organic and inorganic foulants present on the membrane. The advantage of a medical wastewater treatment process carried out in this way was the expected septic tank volume reduction, which allowed increasing the number of radiation therapy rooms from two to eight at a 75% recovery rate.

FO can be easily combined with other processes into integrated systems. Such combined processes, which are easy to optimize, are very effective in the treatment of radioactive liquid waste. The combination of FO with other membrane processes like electrodialysis or reverse osmosis enhances the separation and reduces energy consumption [105]. With adsorption, FO formed an efficient system for removal of strontium from radioactive waste [106]. The FO stage led to a concentration factor of 10 with a 90% volume reduction. The concentrate from FO was further treated by adsorption on nanostructured layered sodium vanadosilicate which can be a good matrix for strontium solidification.

The integration of a very-high-temperature reactor (VHTR) with a water desalination system employing FO was proposed [107]. The BOTANIC code was developed to analyze the behavior of tritium and migration of T downstream of the process to avoid getting tritium into potable water produced in an FO desalination plant.

18.2.3 Nanofiltration

Membranes having effective pore sizes between 0.001 and 0.01 μm are used in nanofiltration (NF). NF is placed between reverse osmosis and ultrafiltration, and because of that sometimes it is considered as a loose reverse osmosis. Typical operating pressures for NF are 0.3–1.4 MPa. The process allows separating monovalent ions from multivalent ions, which are retained by the NF membrane. The process can be used for the separation of organic compounds of moderate molecular weight from a solution of monovalent salts. The very well-known application in nuclear industry is boric acid recovery from contaminated cooling water in a nuclear reactor. There are some examples of nanofiltration applications and studies done with the aim of implementation in nuclear centers described in literature. Some of these are listed in Table 18.4.

The research into the use of NF, often called "loose reverse osmosis," concerns the development of new membranes and membrane systems, improvement of their performance, e.g. by elevation of temperature, and resistance to radiation [118–120]. Very often it is considered as a component of the installation in which it plays the role of a pre-treatment stage before RO. Optimizing the entire membrane system results in high fluxes and low pressures compared to only RO-based installations [118].

18.2.4 Ultrafiltration

Ultrafiltration (UF) operates at a lower pressure difference (0.2–1 MPa) than reverse osmosis and with higher permeate fluxes. It uses more porous membranes, with a pore size of 0.001–0.1 μm. In such a case, low-molecular-weight dissolved compounds pass through the membrane, while colloid and suspended matters are rejected by the UF membrane.

In the nuclear industry, ultrafiltration has been applied in the pretreatment stage before reverse osmosis that requires the removal of potential foulants from feed streams. Very often, ultrafiltration is combined with precipitation or complexation. Small ions bound by a macromolecular chelating agent form complexes, which are retained by the UF membrane. Such enhanced ultrafiltration becomes an efficient separation process with high decontamination factors, sometimes compared with those obtained by reverse osmosis. Radioactive cations can be removed in the precipitation process, forming less-soluble particles (carbonates, phosphates, oxalates, or hydroxides), which are later filtered with UF membrane. These hybrid methods are effectively used in many plants processing α-bearing radioactive waste streams. A number of installations are under operation in nuclear centers and many efforts are ongoing to implement ultrafiltration for radioactive waste processing. Ultrafiltration installations were tested at nuclear power plants around the United States. These experiments proved the usability of UF systems for the treatment of aqueous wastes from floor drains, water from reactor cooling systems, as well as for some types of waste from reprocessing plants. In some cases, the pilot plant experiments were followed by construction of full-scale installations. Examples of the full-scale plants operated in some nuclear centers, as well as testing facilities applying UF for radioactive waste processing, are provided in Table 18.5.

There was a suggestion to treat the wastes from laundries by cleaning the contaminated cloths through ultrafiltration [123–125]. These wastes contain surfactants, high alkalinity, and high salt concentration. It is difficult to process them by evaporation where considerable foaming takes place with contaminated droplets carryover. The concept of UF use for this kind of waste is advantageous because of the possibility of detergent recycling and reuse. The detergents are recovered in permeate, while all suspended particles, fibers, and radionuclides are rejected in the retentate.

One of the current researches devoted to membrane treatment of radioactive waste is directed toward seeded ultrafiltration and all methods, combined with ultrafiltration, give considerable enhancement of separation (Table 18.6).

Extensive studies have been done within the International Atomic Energy Agency (IAEA) coordinated research program on the use of inorganic absorbers for the treatment of aqueous wastes and backfill of underground repositories [139]. In 1992–1996 the IAEA coordinated a research program on waste treatment and immobilization technologies involving inorganic sorbents that resulted in the elaboration of new technologies for the production of sorbents, which can be applied in the nuclear industry. The sorbents can be used in minced form which can be directly introduced into the feed streams treated by ultrafiltration [140] or in a more coarse form for ion exchange columns followed by the membrane process [141].

The processes of UF and enhanced UF for low- and intermediate-level radioactive waste treatment were studied at INCT, Poland. Liquid radioactive wastes originating mainly from the application of radioisotopes collected from

TABLE 18.4
Examples of application of NF for liquid radioactive wastes processing and isotopes separation

Facility	Type of waste/process stream	Process and type of membranes used	Results	References
Chalk River Laboratories (AECL, Canada)	Reactor coolant (cleaning and boric acid recovery)	Three-stage installation, NF membranes	The products of the plant are concentrated boric acid that after purification can be recycled and the concentrate of radioisotopes for immobilization	[108]
Bugey Nuclear Power Plant, France	Reactor waters (separation of ionized silica and boric acid)	Nanofiltration	Nanofiltration allows the separation with 92% recovery for silica and 16.5% for boron. The corrosive activated cations (Sn^{2+}, Ag^{+}, Co^{2+}) are retained by the NF membranes	[109]
ANSTO, Australia	Uranium mill effluents	NF membranes in cross-flow membrane cell	The rejection for uranium was greater than 75%. Some of the tested membranes showed potential for separation of radium, sulfate, and manganese	[110]
ESWE- Institute for Water Research and Water Technology, Wiesbaden, Germany	Water with dissolved uranium	NF membranes from Osmonic Desal (Desal 5 DK, Desal 5DL, and Desal 51 HL) and Dow (NF 90 and NF 45)	Divalent anion complex UO2(CO3)22 and four-valent anion complex UO2(CO3)34 were rejected between 95% and 98% by four membranes and between 90% and 93% by NF90 membrane	[80]
Kyungpook National University, Korea	Simulated nuclear waste containing strontium	Nanofiltration with complexation with polyacrylic acid, Nitto Denko NTR7410, NTR7250, and NTR729HF membranes	Greater Sr removal at elevated pH was attributed to the formation of $SrCO_3$(s) due to dissolution of atmospheric CO_2. Improvement of Sr removal was achieved with the addition of PAA, an increase of membrane fouling was observed at lower pH	[111]
CEN de Cadarache	Simulated nuclear waste containing strontium	Nanofiltration with complexation with polyacrylic acid, FILMTEC NF 70 membranes	The effect of complexation was diminished in high concentration of non-active sodium nitrate. A 98.2% concentration of strontium was achieved and 70% of sodium nitrate from the waste was eliminated in two-stage process	[112]
Laboratoire de Catalyse et Synthése Organique, Université Claude Bernard, Lyon	Separation of lanthanides and actinides, separation of lanthanide isotopes	NF/complexation with EDTA, DTPA, and new ligands on the basis of DTPA. Sepa MG-17 NF membrane (Osmonics)	Separation factors α(150 Nd/142 Nd) = 1.0021 ± 0.0016 and α(160 Gd/155 Gd) = 1.0028 ± 0.0014 were obtained	[113–116]
Laboratoire de Catalyse et Synthése Organique, Université Claude Bernard, Lyon	High-salinity wastes containing cesium	NF/complexation with resorcinarens and calixarens	The Cs^{+}/Na^{+} selectivity from 3 mol/L aqueous $NaNO_3$ solution was about 90%. With the two-stage process 99% removal of trace quantities of cesium was achieved and sodium retention not higher than 10%	[117]

TABLE 18.5
Examples of nuclear testing facilities applying UF for liquid radioactive wastes processing

Facility	Type of waste	Process and type of membranes used	Results	References
Enhanced Actinide Removal Plant at Sellafield	Wastes containing actinides after precipitation stage	Two-stage UF	First stage produces a concentrate of a few weight per cent solids content that is dewatered in a second stage	[64]
Paks Nuclear Power Plant	Contaminated boric acid solutions	Plate and frame UF modules with polysulfone membranes	The volume reduction of about 45% and decontamination factors in the range 10–100 were obtained.	[121]
Nukem, Hanau	Low-level wastes from a fuel fabrication plant	UF with co-precipitation pretreatment	DFs of about 100–200; VRC achieved by UF 10–15 times greater than that obtained with precipitation	[122]
River Bend nuclear power plant (United States)	Floor drains from BWR	UF with IX	Permeate from UF is additionally polished in ion exchange beds and after that water is recycled in the plant. Feedwater turbidity decreases from 20–150 NTU to <0.1 NTU	[67]
Salem nuclear power plant	Low-level radioactive wastes, which originate from floor drains from PWR, laboratories, sampling points, and auxiliary equipment drains	Tubular UF modules and demineralization unit	UF membranes removes particles smaller than 0.05 μm, oil, grease, colloids, and metal complexes protecting IX beds. The quantity of 58/60Co, 54Mg, and 100Ag, is reduced	[67]
Seabrook nuclear power plant (United States)	Floor drains from PWR and, spent resins tank drain-down	UF with cation resin demineralization	Colloidal 58Co removed below discharge limits and more than 90% of TSS was removed by UF system	[67]
Callaway nuclear power plant (United States)	Floor drains and equipment drains tanks, reactor coolant water	UF with IX	70% of radioactivity and suspended solids were removed from coolant water with UF; 89% of radioactivity and suspended matter from floor drains. The full-scale plant configured in 2001 consisted of UF and ion exchange units, processed the waste for direct disposal	[67]
Diablo Canyon nuclear power plant (United States)	Spent media transfer liquid containing high concentration of radioactive submicron particles	Two mobile, skid-mounted tubular UF plants	Systems produced liquid free from particulate activity that can be introduced into the IX columns without the danger of serious fouling and deterioration of their performance	[67]
Mound Laboratory (United States)	Wastes from a fuel reprocessing plant	4.5 m^3/h installation composed of 32.3 m long tubular UF elements, of total surface area 6.5 m^2	80–99% of α-emitters were removed in the test operation and high retention of transuranic elements in full-scale installation was observed (241Am, 98.9%; 238Pu, 98.6%; 237Np, 69.1%; 233U, 93.7%).	[67]

throughout Poland at RWMP in Świerk. They contain various radioactive substances (total specific activity <10^7 kBq/m^3) and ballast non-active salts (concentration < 5 g/dcm^3) also. In the solution, small radioactive ions, such as $^{51}Cr^{3+}$, $H^{51}CrO_4^-$, $^{60}Co^{2+}$, and $^{137}Cs^+$ were present; most of these can easily pass through the membrane for which the cut-off value is ~2,000 MW. The decontamination factors obtained by using ultrafiltration membranes are low (1.07–1.12). The possibility to improve the removal efficiency and increase decontamination factors is the application of reverse osmosis or ultrafiltration enhanced by complexation. Ultrafiltration membranes that are permeable to small ions retain the macromolecules or particles formed in the process of complexation or sorption.

The selection of an appropriate complexing agent is very important to remove the radioisotopes with high efficiency.

TABLE 18.6
Examples of test facilities using enhanced UF for liquid radioactive wastes processing

Facility	Type of waste/process stream	Process and type of the membranes	Type of sorbent/binding agent	Results	References
Cadarache Nuclear Research Centre	Synthetic solutions and actual liquid radioactive wastes	Ultrafiltration with sorption	Activated charcoal and nickel hexacyanoferrate, soluble polymers	Good removal of plutonium, strontium, cesium isotopes, moderate removal of ruthenium and cobalt	[123]
Harwell Laboratory, UK	PWR wastes (wastes containing actinides—colloidal forms of plutonium and americium—in alkaline solutions, laundry wastes)	Ultrafiltration with sorption (seeded UF)	Manganese dioxide, sodium nickel hexacynaoferrate, hydrous titanium oxide, zirconium phosphate, ferric hydroxide, different commercial sorbents	The process is capable of reducing the amount of radioactivity in aqueous effluents to very low levels	[126–129]
University of Colorado, Department of Chemical Engineering	Environmental waters	Polymer-assisted UF	Hyperbranched chelating polymers (glucoheptonamide, polyamidoamine, and ployethylenimine derivatives)	Removal of boric acid from water solutions, remediation of water	[130]
Los Alamos National Laboratory	Wastes from decommissioning and decontamination of nuclear facilities	UF/complexation hybrid process	Polymers with functional groups—amine, carboxylic, oxime, pyrrolidone, phosphonic, and sulfonic	Purification of cellulose materials contaminated with organic substances, toxic and radioactive metals; separation of 241Am (III) and 238Pu (IV) from high acidic solutions and real radioactive wastes	[131–133]
Bhabha Atomic Research Centre	Radioactive solutions containing cerium	UF/complexation hybrid process	Polyethylenimine	Radioactive cerium removal	[134]
Institute of Colloid Chemistry and Water Chemistry, Ukrainian Academy of Science	Environmental waters	UF/complexation hybrid process	Polyethylenimine, polyethyleneglycol, caroboxymethyl cellulose, polyacrylamide	Enhancing removal of U(VI) from contaminated environmental waters; 99% removal of uranium was achieved	[135,136]
Institute of Nuclear Chemistry and Technology, Poland	Wastes from nuclear laboratories and application of radioisotopes	UF/complexation hybrid process	Chelating polymers, hexacyanoferrates, manganese dioxide, hydrous titanium oxide	Effective removal of radioactive ions and heavy metals	[63,140]

Each ion needs a specific ligand, which has to fulfil special requirements, such as:

1. High molecular weight, selected for each UF membrane cutoff;
2. Good solubility in water;
3. Ability of selective binding the ions and molecules;
4. Stability of complexes in the process conditions;
5. Nontoxic, not causing a potential hazard;
6. Low price and market availability.

Many different complexing agents and adsorbents have been examined. As ligands attaching ions of Cr, Co, and Cs, polyethyleneimine (PEI), microcrystalline chitosan (MCH), polyacrylic acid (PAA) and its derivatives, polyvinylpyrolidone (PVD), and suspension of hexacyanoferrates of transient metals were applied [63,140].

The experiments were carried out in laboratory-scale units and pilot plants. Polymeric (capillary-type UF module AMICON H26P30-43, cutoff = 3×10^4 MW, membrane surface area 2.5 m^2) and ceramic UF membranes (Membralox and CeRam Inside) were employed. A solution of selected radionuclides, simulated sewage, and original liquid low-level radioactive wastes was used in the filtration tests. Chelating complexing agents were added to the feed solution to bind the radioactive ions. After mixing and seasoning the sample was filtered with UF membrane at fixed pH and ambient temperature.

18.2.4.1 Removal of Radionuclides from Water Solutions by Ultrafiltration/Complexation

Membralox tubes, 250 mm long and 7/10 mm diameter, with the membrane placed inside the tube and CeRam Inside three-channel tubes were tested at INCT [63,137–138,140]. The ceramic membranes were from 1 kDa cut-off to 100 nm pore size range. The tests were performed with non-active and with radioactive model solutions. The experiments proved that membranes in the UF range were not sufficient to achieve high decontamination factors and the process has to be combined with chemical complexation or sorption. The experiments showed a significant increase of retention factors and decontamination factors when macromolecular compounds were added. An example of the results for two membranes, Membralox tube with 50 nm filtering layer and CeRam Inside 15 nm, is shown in Figure 18.6. The retention of cesium ions was similar for both membranes. When ultrafiltration was combined with complexation with macromolecular ligands the retention was improved. Retention factors for Membralox 50 nm were higher than for CeRam Inside. An exception was the case when sodium cyanoferrate was added. For both membranes the retention factors were close to 1. For the two membranes the application of macromolecules resulted in increased retention.

The retention and decontamination factors for cobalt ions were low when UF membranes were applied (Figure 18.7). To intensify the effect of separation the complexing agents,

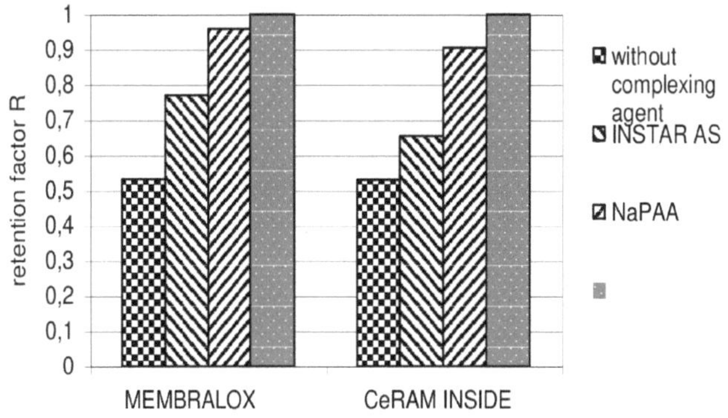

FIGURE 18.6 The influence of complexing agent on retention factor of Cs^+ ions. 1, Membralox 50 nm; 2, CeRam Inside 15 nm.

(Reprinted from Journal of Membrane Science, Vol. 225, Zakrzewska-Trznadel, et al.: Radioactive solutions treatment by hybrid complexation –UF/NF process, pp. 25–39, Copyright with permission from Elsevier.)

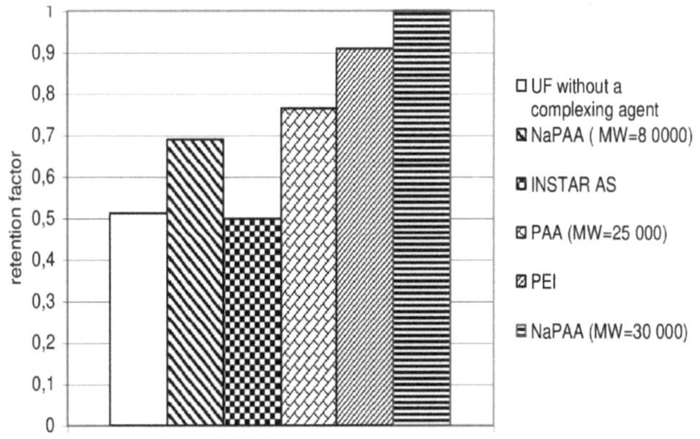

FIGURE 18.7 The influence of complexing agent on retention of Co^{2+} ions, Membralox 50 nm membrane, pH = 6–7.

(Reprinted from Journal of Membrane Science, Vol. 225, Zakrzewska-Trznadel, et.al: Radioactive solutions treatment by hybrid complexation –UF/NF process, pp. 25-39, Copyright with permission from Elsevier.)

polyacrylic acid of different cross-linking, polyacrylic acid salts, polyethylenimine, and Instar AS (complexing agent containing macromolecular acrylamide and sodium acrylate copolymer), were added to the feed solution. The results of ultrafiltration of cobalt ions complexed before filtration by the soluble polymers are presented in Figure 18.7. The experiments showed high influence of the polymer, its chemical form, and average molecular weight on retention factors. The best results were obtained when sodium polyacrylate of high molecular weight or polyethylenimine were employed.

The application of macromolecular chelating polymers allows the catching of small radioactive ions with good efficiency. Soluble polymers efficiently bound radioactive cobalt and chromium, however the complexation of $HCrO_4^-$ ions was difficult. Chromate ions were removed

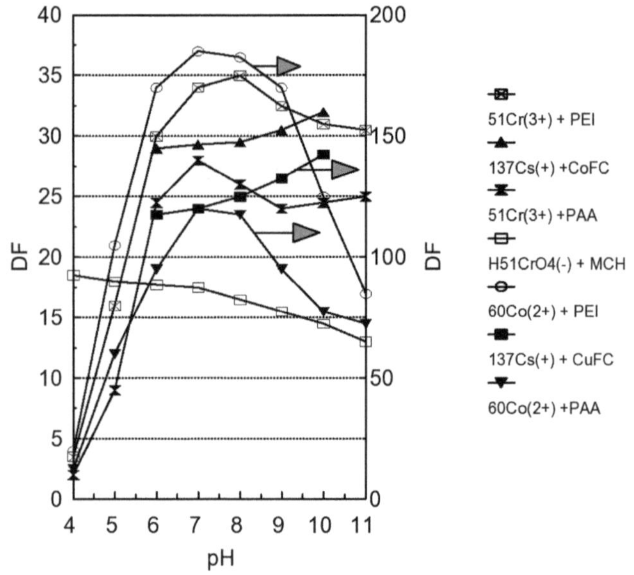

FIGURE 18.8 DF vs. pH for small ions combined with macromolecules.

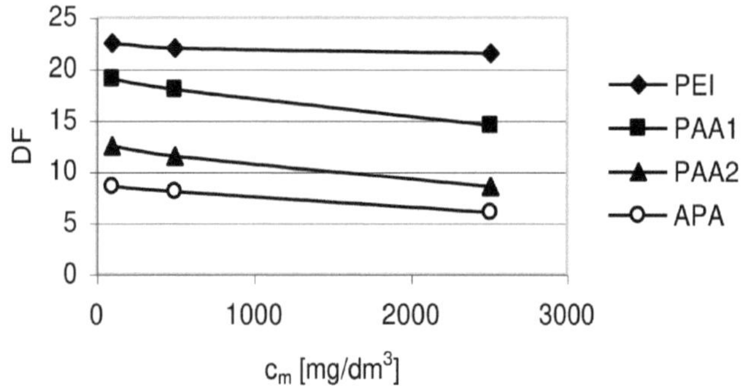

FIGURE 18.9 ^{60}Co removal by UF/complexation method; variation of DF on alkali metals concentration; c_{Co} = 5 mg/l, c_L/c_{Co} = 10, pH = 7, T = 20°C.

with MCH, which is a good sorbent for all metals; however decontamination factors for radioactive chromium in the form of chromate ion $H^{51}CrO_4^-$ were moderate. Cesium isotopes were effectively removed with cyanoferrates of transient metals; copper cyanoferrate more effectively binds cesium than cobalt cyanoferrate. A meaningful dependence of retention and decontamination factors on pH was observed (Figure 18.8). Favorable conditions of binding the cesium with cyanoferrates are an alkali environment, under pH > 8, while the best complexation abilities of soluble polymers are in neutral conditions. It was found that the best conditions of binding $^{51}Cr^{3+}$ and $^{60}Co^{2+}$ ions with PEI macromolecules occur at pH = 7–8. The most effective adsorption of ions $^{137}Cs^+$ by CuFC was found at pH 9.5–10.

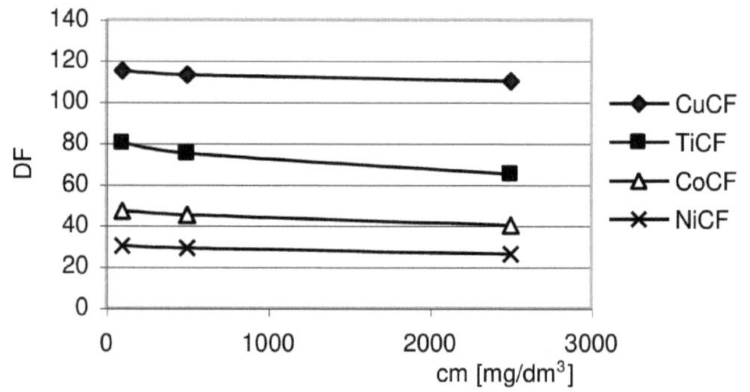

FIGURE 18.10 ^{137}Cs removal by UF/complexation method; variation of DF on alkali metals concentration; c_{Cs} = 5 mg/l, c_L/c_{Co} = 4, pH = 7, T = 20°C.

The binding abilities of the soluble polymers fall slowly with the concentration of alkali metals. In Figure 18.9, the results of UF/complexation with polyethyleneimine (PEI), polyacrylic acid of different cross-linking (PAA1 and PAA2), and polyacrylic acid amide are presented. In each case the decontamination factors for ^{60}Co decrease when the concentration of alkali metals increases from 0.1 to 2.5 g/dm^3. The increase in alkali metals causes a decline of the decontamination factor for radioactive cesium, while cyanoferrates are applied as sorbents (Figure 18.10).

The concentration of the complexing agent should be selected for each radioisotope-complexing agent pair. Usually the concentrations ratio, the ligand to the ion bound by this ligand, is in the range 1–20. Figure 18.11 shows the *DF*'s for europium-152 as a function of concentration of NaPAA of Mw = 8000, which was selected to remove this radioisotope. The best binding conditions were found at a concentration of 8 g/dm^3, which corresponds to the concentration ratio of polymer to europium, such as 18.5:1. For this concentration ratio, *DF* was the highest. A further increase in ligand concentration did not result in an increase of the decontamination factor.

18.2.4.2 Treatment of Original Radioactive Waste

When actual radioactive waste is treated by the UF/complexation method the decontamination factors for each radioactive component of the sample differ from those obtained for single ion-solutions. The ions interfere and compete in a complexation process and the salinity of the wastes influences the removal, which results in lower decontamination factors. The characteristics of radioactive waste used in ultrafiltration experiments are shown in Table 18.7. The sample had relatively low salinity (<1 g/dm^3); its specific radioactivity was in the medium-level liquid waste range (~150 kBq/dm^3). The main portion of radioactivity came from cobalt-60 and cesium-137, but also some quantities of lanthanides and small amount of actinides (^{241}Am) were also present. Decontamination factors obtained for actual waste differed from *DF*s obtained for solutions of single radioisotopes. As was proved before [140], the concentration of alkaline metals influenced the *DF* markedly. The competition between different ions bound by chelating ligands or adsorbents resulted in an increase in the retention and decontamination factors for the original liquid waste. The results of UF/complexation tests are shown in Table 18.8 and decontamination factors calculated for selected radioisotopes in Figure 18.12.

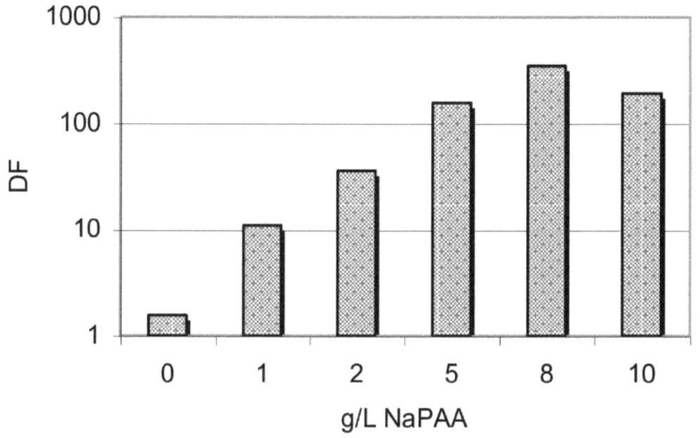

FIGURE 18.11 DFs for europium-152 in function of concentration of NaPAA (MW = 8000), Membralox 5 kDa, 0.5 g/l Eu_2O_3, pH = 8.

(Reprinted from Journal of Membrane Science, Vol. 225, Zakrzewska-Trznadel, et al.: Radioactive solutions treatment by hybrid complexation –UF/NF process, pp. 25–39, Copyright with permission from Elsevier.)

The hybrid UF/complexation process appeared very effective for removal of americium and europium isotopes; in the majority of permeate samples the concentration of these radioisotopes was below detection limits or very low. Good removal of ^{60}Co was observed with sodium polyacrylate of MW>15,000 Da and with polyethyleneimine. The complexation of cesium ions with chelating polymers was rather moderate; good results were obtained with cobalt cyanoferrate, and decontamination factors were higher than 100. Average decontamination factors for total waste radioactivity were low, except for the case when two complexing agents were dosed to the effluent simultaneously: PEI and CoCF (Table 18.8). The results obtained in the experiment when cobalt hexacyanoferrate together with soluble polymer (PEI) were applied showed a significant increase of ^{137}Cs removal, however the decontamination factors for other radioisotopes decreased in comparison with the tests when only a single chelating polymer was applied as a complexing agent.

One of the major limitations of UF/complexation is that removal of isotopes by this method is efficient when the mixture of ligands binds the ions effectively under the same conditions (the same pH, concentration of alkali-metals, salts, etc.). To overcome this problem the process may be run in a multistage system and the complexing agents are dosed in successive stages to remove some specific ions separately. The above process was divided into two steps to enable a high removal efficiency. At first, CoCF slurry was introduced into the feed solution to bind the cesium ions. After some hours the feed was filtered with UF ceramic membranes with up to three-fold volume reduction. Sodium polyacrylate, Mw 30,000, was added to the permeate and after seasoning and

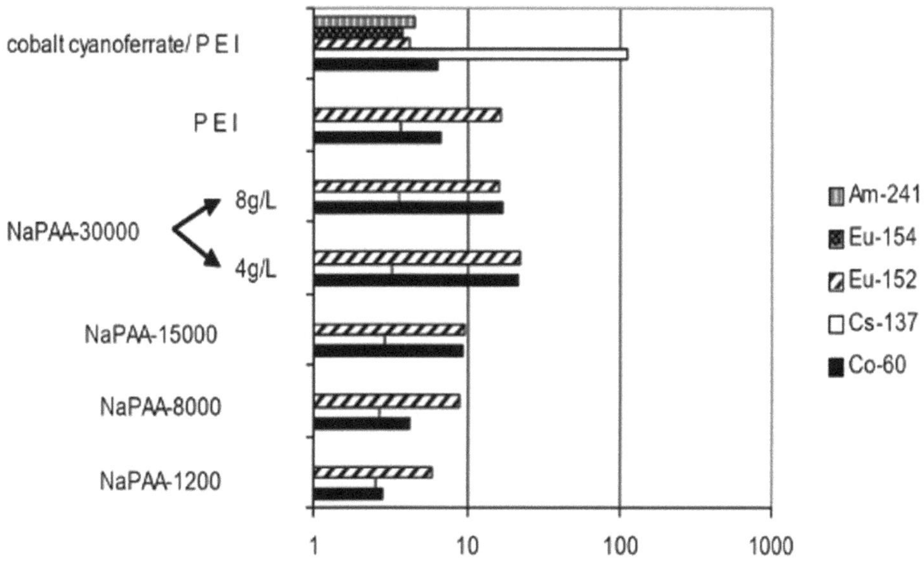

FIGURE 18.12 Decontamination of the sample of liquid radioactive waste in UF/complexation process with a use of different complexing agents, CeRam Inside membrane, 15 kDa.

(Reprinted from Journal of Membrane Science, Vol. 225, Zakrzewska-Trznadel, et al.: Radioactive solutions treatment by hybrid complexation –UF/NF process, pp. 25–39, Copyright with permission from Elsevier.)

pH adjustment the solution was filtered. The *DF*s for such a process arrangement are provided in Figure 18.13. In the first stage of filtration, high removal of Cs-137 was achieved, while other radioisotopes were rejected at a moderate rate. In the second stage, where sodium polyacrylate was employed, decontamination factors were much higher for most of the radioisotopes present in the waste. Only for ^{137}Cs did ultrafiltration give *DF*s smaller than 1, which corresponds with the higher specific activity in permeate. A low $DF<1$ shows that Cs-137 is not bound by NaPAA and passes through the membrane. Additionally, a low concentration of this radioisotope and generally low total salinity of the effluent may cause low decontamination for cesium after the treatment in the first stage. This was earlier observed during operation of a three-stage RO plant, where a decrease in total salinity before the final purification stage caused a significant decrease in decontamination factors [73].

For all radioisotopes present in the waste sample, *DF*s in the case where cyanoferrate and chelating polymer were applied subsequently in two-stage process were higher than those obtained when the complexing agents were added to the feed solution simultaneously. All decontamination factors were higher than *DF*s obtained for a single soluble polymer (NaPAA or PEI; Table 18.9).

Apart from radioactive compounds, radioactive wastes may also contain non-active but chemically toxic substances. The national standards and regulations describe the discharge limits for these substances also. Heavy metals are toxic compounds, most often present in liquid radioactive wastes. Experiments showed that most of these metals are removed by UF/complexation method. The method is inefficient for removal of monovalent ions, bivalent cations, and anions which are retained in 25–50%, but high retention of metals like Mn, Fe, Co, Cu, Pb, and Cr was observed. The retention

TABLE 18.7
Characteristics of radioactive waste sample. (Reprinted from Journal of Membrane Science, Vol. 225, Zakrzewska-Trznadel: Radioactive solutions treatment by hybrid complexation –UF/NF process, pp. 25-39, Copyright (2003) with permission from Elsevier)

Chemical composition* (mg/dm^3)	Na (274), K (28) Mn (< 0.05), Fe (0.60), Cu (< 0.03) Pb (< 0.01), Ca (< 43.35), Mg (16.18) Cr (< 0.05), Au (< 0.1) Cl^- (177.5), F^- (1.58), NO_3^- (3.83), NO_2^- (1.67), SO_4^{2-} (195), PO_4^{3-} (5.37) SiO_2 (dissolved) (25.7) C organic (120)
pH	5.2
Radiochemical composition (kBq/dm^3)	Co-60 (7.1), Cr-51, Sb-124 (0.01), Sb-125 Cs-137 (139.49), Ce-141, Eu-152 (0.44), Eu-154 (0.065), Am-241 (1.68)

* Concentration of metals performed as a total concentration of the element in different chemical forms.

TABLE 18.8
Content of some radioisotopes in the process streams in ultrafiltration/complexation experiments. Membrane CeRam Inside 15 kDa. (Reprinted from Journal of Membrane Science, Vol. 225, Zakrzewska-Trznadel: Radioactive solutions treatment by hybrid complexation–UF/NF process, pp. 25–39, Copyright (2003) with permission from Elsevier)

	Co-60 kBq/dm^3	Cs-137 kBq/dm^3	Eu-152 kBq/dm^3	Eu-154 kBq/dm^3	Am-241 kBq/dm^3	ΣApi kBq/dm^3	DF_{total}
Feed solution	7.12	139.49	0.44	0.06	1.68	148.79	
Permeate							
NaPAA, Mw = 1,200 cL = 4 g/dm^3	2.48	54.92	0.07	–	–	57.47	2.6
NaPAA, Mw = 8,000 cL = 4 g/dm^3	1.71	52.64	0.05	–	–	54.40	2.7
NaPAA, Mw = 15,000 cL = 4 g/dm^3	0.76	48.11	0.05	–	–	48.92	3.0
NaPAA, Mw = 30,000 cL = 4 g/dm^3	0.33	43.56	0.02	–	–	43.91	3.4
NaPAA, Mw = 30,000 cL = 8 g/dm^3	0.42	38.74	0.03	–	–	39.19	3.80
PEI cL = 4 g/dm^3	1.07	38.00	0.03	–	–	39.10	3.9
CoFC, cL = 4 g/dm^3 PEI, cL = 6 g/dm^3	1.13	1.26	0.11	0.02	0.37	2.89	51.5

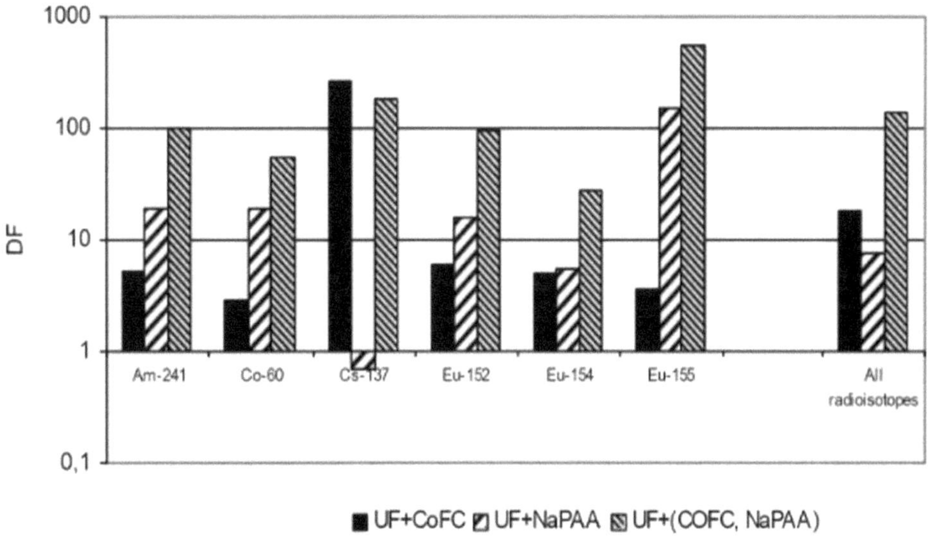

FIGURE 18.13 Decontamination factors for two-stage process, CeRam Inside 15 kDa.

(Reprinted from Journal of Membrane Science, Vol. 225, Zakrzewska-Trznadel, et al.: Radioactive solutions treatment by hybrid complexation –UF/NF process, pp. 25–39, Copyright with permission from Elsevier.)

TABLE 18.9
Decontamination factors in combined process UF/complexation by use of different complexing agents. (Reprinted from Journal of Membrane Science, Vol. 225, Zakrzewska-Trznadel: Radioactive solutions treatment by hybrid complexation–UF/NF process, pp. 25–39, Copyright (2003) with permission from Elsevier)

	Am-241	Co-60	Cs-137	Eu-152	Eu-154
Na PAA, 30,000	$\to \infty$	21.24	3.2	22	$\to \infty$
PEI, 15,000	$\to \infty$	6.62	3.67	16.67	$\to \infty$
CoCF/PEI (simultaneously)	4.54	6.31	110.7	4.14	3.78
CoCF/NaPAA, 30,000 (subsequently)	101.4	55.38	179.5	96.39	26.95

rates for those metals in the two-stage experiment described above were as follows:

Mn 94%
Fe 99%
Cu 99%
Pb 87%
Cr 75%
Co 94%

The advantages of ceramic membranes employed in the nuclear industry were found to be their extremely high chemical and physical stability (full pH range), resistance to oxidants and solvents, and resistance to ionizing radiation. Ceramic materials are advantageous where solutions composed of organic compounds or high radioactive wastes containing α-emitters are treated. High temperature resistance allows washing with warm streams and sterilization by steam. This is very important when macromolecular complexing ligands causing membrane fouling are applied.

18.2.4.3 UF Pilot-Plant Experiments

The apparatus equipped with Sunflower CeRam Inside (23-8-1178), shown in Figure 18.14 and characterized in Table 18.10, was used in pilot plant tests. The plant consisted of a feed tank (1), equipped with a cooler (2), membrane module housing (3), pre-treatment filters (4), pressure pump TONKAFLO (5), circulating pump Grundfos (6), non-return valve (7), and two needle valves and four ball valves.

The filtration experiments were conducted in cross-flow mode at pressure 0.25–0.5 MPa in the system. In that pressure range for filtration of water, permeate flux was 26–85 L/m^{2}h. All process parameters were recorded with a data acquisition system; periodically the samples of permeate and retentate were collected for chemical analysis.

Original radioactive liquid waste from the storage tank in a nuclear center was treated in the UF/complexation process. The main radioactive components of the waste sample were ^{60}Co, ^{65}Zn, ^{133}Ba, ^{134}Cs, ^{137}Cs, ^{152}Eu, and ^{241}Am. When ultrafiltration was applied solely, the rejection of radioactive

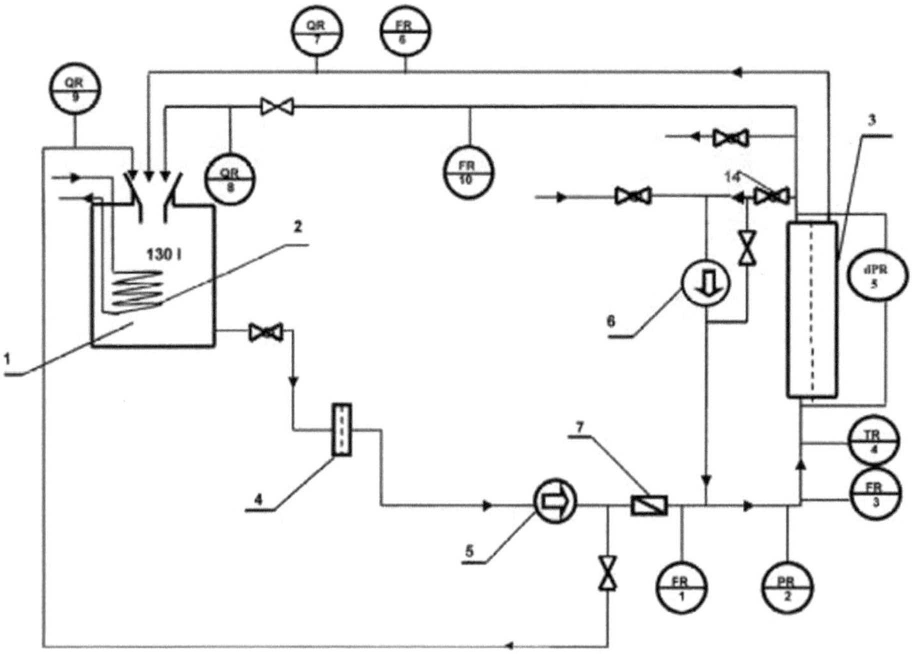

FIGURE 18.14 Pilot plant for treatment of radioactive wastes by UF/complexation.

(Reprinted from Desalination, Vol. 162, Zakrzewska-Trznadel, et al.: Application of ceramic membranes for hazardous wastes processing: pilot plant experiments with radioactive solutions, pp. 191–199, Copyright with permission from Elsevier.)

TABLE 18.10
Characteristics of Sunflower CeRam Inside ceramic membrane

Filtration area (m^2)	0.35
External diameter (mm)	25
Length (mm)	1178
Number of channels	23
Hydraulic channel diameter (mm)	3.6
Flux volume at velocity 1 m/s (m^3/h)	0.86
Pressure drop at velocity 4 m/s (bar)	1.0
Cutoff (kDa)	8
pH	0-14
Temperature (°C)	Up to 150

compounds was very low and the decontamination factors were between 1.02 and 1.05. Therefore, ultrafiltration was combined with complexation by soluble polymers. The application of sodium polyacrylate (NaPAA) with 15,000 and 30,000 molecular weight, in the concentration range 0.4–3 g/L, gave a slight increase of decontamination factors to 1.3–2.8. In addition, dosing of INSTAR AS up to a concentration 3 g/L resulted in an increase in the decontamination factor to 3.1. Since the major portion of total activity came from radioactive isotopes of cesium, the use of cobalt hexacyanoferrate as a binding agent was advisable. The cobalt hexacyanoferrate was introduced as a suspension prepared from 0.02 N solutions of $Co(NO_3)_2{\cdot}6H_2O$ and $K_4[Fe(CN)_6]\ 3H_2O$. As shown (Figure 18.15), the first portion of the suspension caused an increase of DF to 30 and a further increase of complexing agent concentration did not change the decontamination factor radically. The results of the radiochemical analysis are shown in Table 18.11, where the isotopic composition of the samples is performed. The notation of the samples was as follows: S, feed solution composition; P1, permeate treated with 1 g/L of NaPAA; P2, permeate after complexation with 3 g/L NaPAA; and P3, after the treatment with CoCF and INSTAR AS. The calculated decontamination factors for each radioisotope as well as *DF*s for total activity in processed waste are presented in Figure 18.16. The radiochemical analyses of the last sample of permeate showed complete retention of ^{134}Cs, ^{152}Eu, and ^{241}Am when a complexation/UF process was employed. About a 1770-fold decrease in the concentration of ^{137}Cs was observed; the decontamination factors for other radioisotopes were also high. However, due to a high initial ^{65}Zn content, the concentrations of this radioisotope in permeate (P4) were also high, and the radioactive zinc was not removed sufficiently. Therefore, an additional cleaning procedure should be applied to reduce the ^{65}Zn concentration and to meet discharge standards. It is likely that the dosage of the polymer was too small to complex all radioisotopes sufficiently (the increase of *DF* with subsequent dosages of polymer was observed; see Figure 18.16). Further portions of cobalt hexacyanoferrate did not cause an increase in the rate of decontamination.

The membrane performance was good, during an almost 60-hour operation the permeate flux was stable after an initial decline. The biggest flux decrease was observed when macromolecular sodium polyacrylate was introduced: first after dosing 0.4 g/L of NaPAA, then after injection of the next

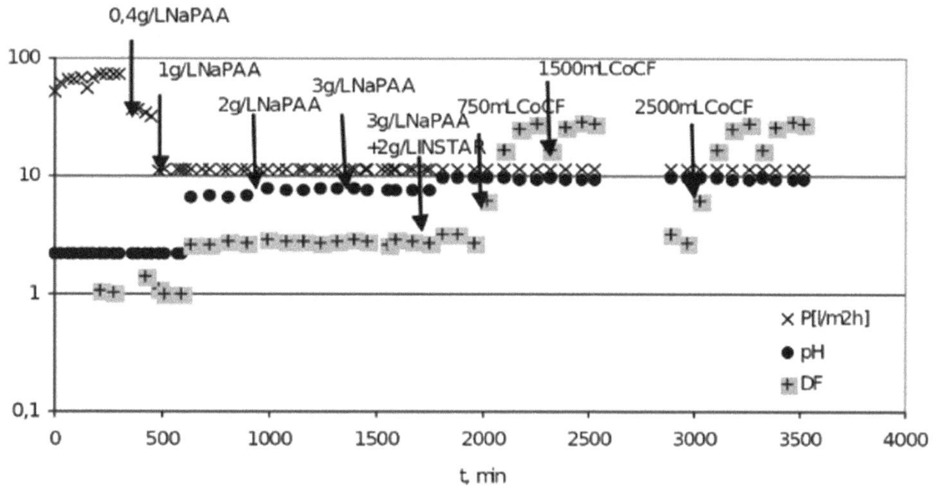

FIGURE 18.15 Radioactive waste treatment by UF/complexation.

(Reprinted from Desalination, Vol. 162, Zakrzewska-Trznadel, et al.: Application of ceramic membranes for hazardous wastes processing: pilot plant experiments with radioactive solutions, pp. 191–199, Copyright with permission from Elsevier.)

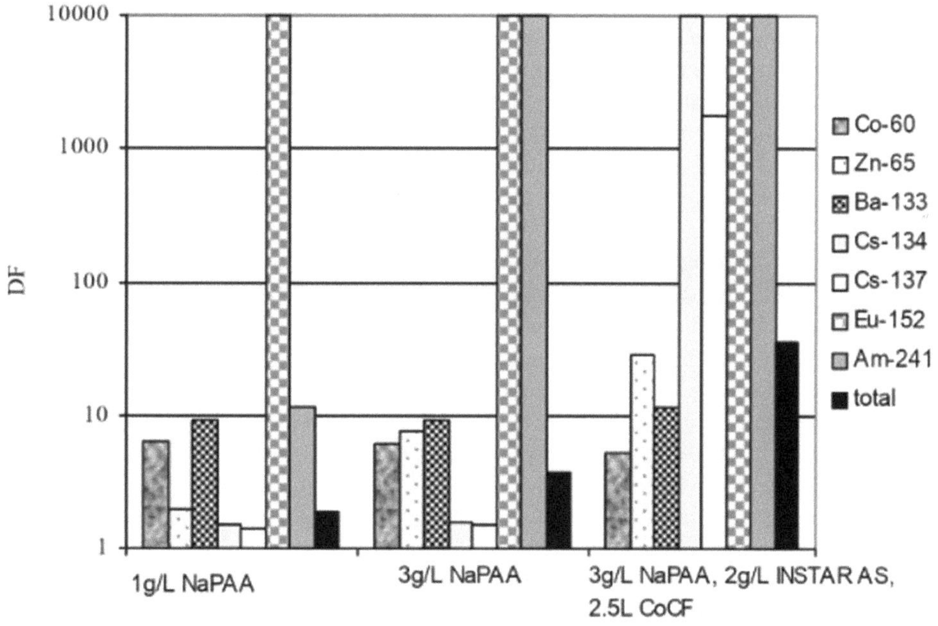

FIGURE 18.16 Decontamination factors for selected components of radioactive wastes treated in UF/complexation process.

(Reprinted from Desalination, Vol. 162, Zakrzewska-Trznadel, et al.: Application of ceramic membranes for hazardous wastes processing: pilot plant experiments with radioactive solutions, pp. 191–199, Copyright with permission from Elsevier.)

portion (1 g/L NaPAA). In that time the permeate flux declined from 52 L/m^{2}h in the beginning (without the complexing agent), to 11 L/m^{2}h after injection of 1 g/L of the polymer. A further increase of polymer concentration did not result in a flux decline, nor did the addition of a CoCF suspension.

The complexation/ultrafiltration process is efficient for the retention of non-active hazardous components of wastes, e.g., heavy metals, the concentration of which has to be reduced according to national regulations. In the process described above, the retention of heavy metals was high, and reached 98% for manganese, 98.6% for iron, 92.4% for copper, 91.5% for nickel, and 94.7% for zinc. Some part of the bivalent ions was also retained, for instance, Mg in 67%, Ca in 58%, and SO_4^{2-} in 22.6%.

The enhanced ultrafiltration is effective but hardly a controlled process, because of its sensitivity to the process conditions, such as pH, temperature, or concentration of alkaline ions. However, it can be easily applied for wastes

TABLE 18.11
Chemical analysis of the samples (CeRam Inside, 8 kDa). (Reprinted from Desalination, Vol. 162, Zakrzewska-Trznadel, et al.: Application of ceramic membranes for hazardous wastes processing: pilot plant experiments with radioactive solutions, pp. 191-199, Copyright (2004) with permission from Elsevier)

	S	P1	P2	P4
Isotope	Bq/L	Bq/L	Bq/L	Bq/L
Co-60	374	59.5	61.4	70.9
Zn-65	22038	11111	2920	784.6
Ba-133	1058	113.7	114.3	93.5
Cs-134	22.7	15.2	14.4	0
Cs-137	8742	6162	5823	4.9
Eu-152	27.2	0	0	0
Am-241	1365	116.8	0	0

containing radionuclides that are effectively attached to the macromolecules of various ligands at the same or similar pH value.

Current development of the hybrid treatment processes is being directed toward searching for new, cheap sorbents and complexing agents, and among them natural polymers like chitin, lignin, huminates, and lignosulfonate are considered [142]. Using huminates in the ratio 1:25–1:50 with a combination of ceramic membranes, selectivity higher than 99% for ions, such as Sr^{2+}, Ni^{2+}, Cu^{2+}, Co^{3+}, and Ca^{2+} was obtained. Chitosan and cellulose which are the most abundant biopolymers are increasingly being used as coagulants and flocculants in the processes of water and wastewater purification [143,144]. Due to the amine groups present in the polymer chain, chitosan is a good sorbent for transient metals [145,146]. It can be formed in granules for fixed beds, as well as a water solution of the acetate [147]. With conjunction of the membrane, chitosan may cause a 6–10-fold increase in the removal of Cu(II), Co(II), Ni(II), and Zn(II) from water solutions [148]. Removal of metals, as well as radionuclides present in liquid radioactive waste, can be attained through the use of yeast biomass with a combination of ultrafiltration and microfiltration [149,150]. The other way to enhance ultrafiltration is through the application of alginic acid or alginates [151–153]. Apart from other natural sorbents, such as biomass, powdered wood bark, nutshell, modified wool, or cotton, the use of cheap, inorganic sorbents like natural zeolites, clays, or fly ash is considered [154–156].

The other approach utilizes the surfactants for separation purposes and the possibility of removal of metals in the process of micellar-enhanced ultrafiltration (MEUF) [157–161]. MEUF can be applied to separate the micellar phase using membranes with a pore size smaller than the micellar diameter. In such a process, surfactant is added to an aqueous solution containing metals or radionuclides, which are too small to be rejected by UF membranes. The surfactant at concentrations greater than the critical micelle concentration (CMC) forms aggregates in which the metallic components of the solution can be dissolved or solubilized. The UF membrane rejects the micelles containing bound solutes or pollutants. As an alternative to extraction, adsorption, or distillation, the MEUF process is not only economical but also an environment-friendly method of removal of metallic pollutants, as well as recovery of metals from water solutions.

New potential applications of sorbents in conjunction with a membrane are expected from the development of molecular-imprinted or ionic-imprinted polymers that are capable of metal-ion recognition. This concept is based on preparation of matrix in the presence of the molecular or ionic template. After removal of the target molecule/ion the prepared solid can react with the solution of the molecules/ions from which the imprinted molecule/ion should thus be preferentially extracted from the mixture [162–164]. Another gate to radioactive waste treatment could be opened by the employment of hybrid processes involving UF membranes, which are efficient and flexible methods of processing [165,166]. Simulation, modeling, and computer-aided design could be very beneficial for further process development [167–170] since hybrid systems need careful optimization to make all components of the process working in synergy.

18.2.5 Microfiltration

In nuclear technology, microfiltration is used either for pretreatment purposes or for the concentration of coarse particles after the precipitation process. For high-level radioactive wastes ceramic filters are used, as some types of effluents have high decontamination and concentration factors. It is also possible to combine MF with other processes, e.g. adsorption, in hybrid systems. An example would be the use of MF membranes combined with a dispersed CuFC adsorbent into a countercurrent two-stage cesium-removal hybrid system [179]. The decontamination factors obtained in such a unit were three times higher than for single-stage adsorption in a jar test.

The MF facilities used in the nuclear industry to treat liquid radioactive wastes are summarized in Table 18.12.

18.3 EMERGING TECHNOLOGIES IN RADIOACTIVE WASTE TREATMENT—MEMBRANE CONTACTORS

The processes involving membrane contactors have become one of the more rapidly developing modern unit operations with the application of membranes. Systems such as membrane emulsifiers, membrane extractors, membrane strippers and scrubbers, and membrane distillation and crystallization systems, may overcome the limitations of conventional units and various more common membrane processes that have been already applied in the nuclear industry [180,181]. Both hydrophobic and hydrophilic membranes are applied for the construction of these systems, in which they play a role of the artificial interface for mass transfer. As opposed to other membrane processes based

TABLE 18.12
Examples of nuclear testing facilities applying MF for liquid radioactive wastes processing

Facility	Type of the waste	Process and type of the membranes used	Results	References
AECL Chalk River Laboratory (Canada)	Mixed low-level radioactive wastes	Microfiltration was used as a pretreatment step in RO installation	MF hollow-fiber membranes remove the suspended solids larger than 0.2 μm. The filtrate from MF forms the feed for the SWRO system, the backwash solution undergo further volume reduction in thin-film evaporator	[56,57]
AECL Chalk River Laboratory (Canada)	Groundwater and soils decontamination	Hollow-fiber MF system consisted of 40 cross-flow filtration modules, ~ 6 cm in diameter, 50 cm long, and 1 m^2 surface area	The system demonstrated the usability of MF for the treatment of soil leachate and removal of radionuclides from ground water. The radioactivity of 90Sr was reduced from 1700–3900 Bq/L to 2 Bq/L	[67,171–173]
Rocky Flats (United States)	Groundwater containing uranium isotopes, organic toxic compounds, and heavy metals	Tubular MF modules, pore size 0.1 μm	The removal efficiency of uranium isotopes by the system was 99.9%	[67]
Idaho National Engineering and Environmental Laboratory (INEEL)	Radioactive wastes from fuel reprocessing	Cross-flow filtration with Mott sintered Hastelloy filter	Good removal of undissolved solids from INEEL radioactive slurries	[174]
Berkeley Nuclear Laboratories	Simulated radioactive wastes	Cross-flow microfiltration; Pall PSS (2.5 μm limit of separation), Fairey Microfiltrex FM4 (1 μm) and APV Ceraver (1.4 μm) ceramic and stainless steel membranes	Crossflow MF was found to be effective dewatering of range of radioactive wastes	[175]
Energy and Environmental Research Center, University of North Dacota	Simulated wastes containing suspended and colloidal solids	SpinTek ST IIL centrifugal membrane filtration technology	Ability to process wastes from decontamination and decommissioning systems within the US DOE; ability to treat hazardous wastewater to a slurry-type level and reduce tank sludge volume	[176]
Los Alamos Nuclear Laboratory (LANL)	Surrogate and real radioactive wastes from LANL	SpinTek ST IIL centrifugal membrane filtration technology	Elaboration of the model for determining the applicability and economics of the system to different DOE waste and process streams	[177]
Oak Ridge National Laboratory, Radiochemical Engineering Development Center (REDC)	Waste after processing irradiated targets containing transuranic nuclides	Cross-flow filtration unit (three Mott Metallurgical Corporation elements), ¼-in ID, 0.5 μm pore size.	Modular design (resorcinol-formaldehyde resign column, cross-flow filtration, solidification unit); radionuclides (transuranium and rare elements) removed as hydroxides, stored as solids, filtrate is going to the LLLW; allows 10–20% concentration by weight of the slurry	[178]

on membrane selectivity, membrane contractors employ porous barriers that do not interact with separated species; they serve only as a barrier between two phases. The most common configuration of the membranes is the hollow-fiber arrangement, which is relatively cheap and provides a large interfacial surface for mass transport. Different types of contacting devices are available in the market with different geometries, enabling the extraction process intensification [182]. Membrane contactors were used for degassing of water (to remove dissolved oxygen), separating hydrocarbons (e.g., olefins from paraffins), removing organic contaminants from water [183–185], concentration of solutions of non-volatile substances (via membrane distillation), or removal of metal compounds [186–188].

For the application of membrane contactors in nuclear science and technology such processes as membrane distillation and membrane extraction/stripping were tested. A very important part of the research in this field was devoted to liquid membranes which also employ membrane contactors to form supported liquid membranes [189].

Liquid–liquid extraction using membrane contactors covers a broad range of applications and different module configurations with a predominance of hollow-fiber and flat-sheet arrangements. In hollow-fiber modules it is worth considering which liquid flows on the shell side and which one is introduced inside the fibers. The danger of bypassing and channeling in the shell side makes it difficult to reach a high extraction efficiency with the solute-containing feed on this side. Fiber-side flow is not advisable when the feed liquid contains particulates. To avoid leakages, the liquid wetting the membrane pores and usually flowing under lower pressure should preferably be placed on the fiber side [190]. Commercially available modules for concentration-driven processes are produced by several manufacturers, although some are designed for pressure-driven processes, such as microfiltration. Some arrangements are equipped with a shell-side baffle that improves the performance by minimizing the bypassing and increasing the mass transfer coefficient by producing a velocity component normal to the membrane surface.

Membrane solvent extraction employing membrane contactors appears to be an efficient technique for the separation of acids, recovery of acids from wastewater, and separation of acids from salts [191]. It is also applied for selective removal of heavy metal ions from industrial effluents. The process can be accomplished using a supported liquid membrane (SLM) route, which has been tested in many research laboratories with moderate success, and simple membrane extraction that only contacts aqueous and organic phases by a porous membrane separating and also contacting these two media. Whereas supported liquid membrane extraction suffers from instability caused by loss or deterioration of the organic phase, membrane solvent extraction avoids this problem with a large portion of the organic medium, which usually consists of extracting agents diluted in common diluents, such as kerosene, n-dodecane, or toluene. However, the risk of membrane solubilization remains, and it should be overcome by all possible means to enable large-scale and widespread applications of membrane extraction in industry.

18.3.1 Membrane Distillation

Membrane distillation (MD) is a separation method, which employs porous lyophobic membrane, which is nonwettable by the liquid. Because of the lyophobicity of the polymer, only vapor is transported through the membrane pores. The condensation takes place on the other side of the membrane in an air gap, cooling liquid, or inert carrier gas. Usually, MD is employed to treat water solutions, therefore hydrophobic membranes, manufactured from polymers, such as polypropylene (PP), polytetrafluoroethylene (PTFE,) or poly(vinylidenefluoride) (PVDF), are used in the process. It is possible to use polyethylene terephthalate (PET) track-etched membranes (TeMs), as well as PET membranes modified by styrene and triethoxyvinylsilane (TEVS) using UV-induced grafting [192,193]. Developed for desalination of sea water, MD works well in the treatment of radioactive wastewater with high salinity [194].

The driving force in the MD process is a gradient of partial pressures of the components of the solution in the gaseous phase. The main advantages of membrane distillation applied for radioactive waste processing are:

1. Moderate process conditions, ambient pressure, and moderate temperature;
2. High retention and decontamination, higher than for conventional evaporators;
3. Low operational costs for medium-size installations, the possibility of utilization of waste heat, e.g., from the cooling system of a nuclear reactor;
4. The possibility to achieve high concentrations, close to saturation;
5. The possibility to use plastics as construction materials for apparatus, because of moderate process conditions—minimization of corrosion and capital costs;
6. Compact installations (one-stage units).

In contrast to other methods, and also to reverse osmosis, MD allows complete purification in a single stage, not involving additional processes for polishing permeate. It is a simple, economic, and environment-friendly method when it is use for radioactive waste processing.

Membrane distillation was developed at INCT for liquid radioactive waste processing [195]. For many years the process was studied as a method for stable isotopes enrichment. Using porous PTFE membranes, stable isotopes of oxygen and hydrogen were enriched in natural water with relatively high separation factors [196–199]. It was proved that MD distillation is a competitive method for enrichment of oxygen-18, the isotope that now finds a big market demand because of its high consumption by positron emission tomography [200]. Membrane distillation installations alone

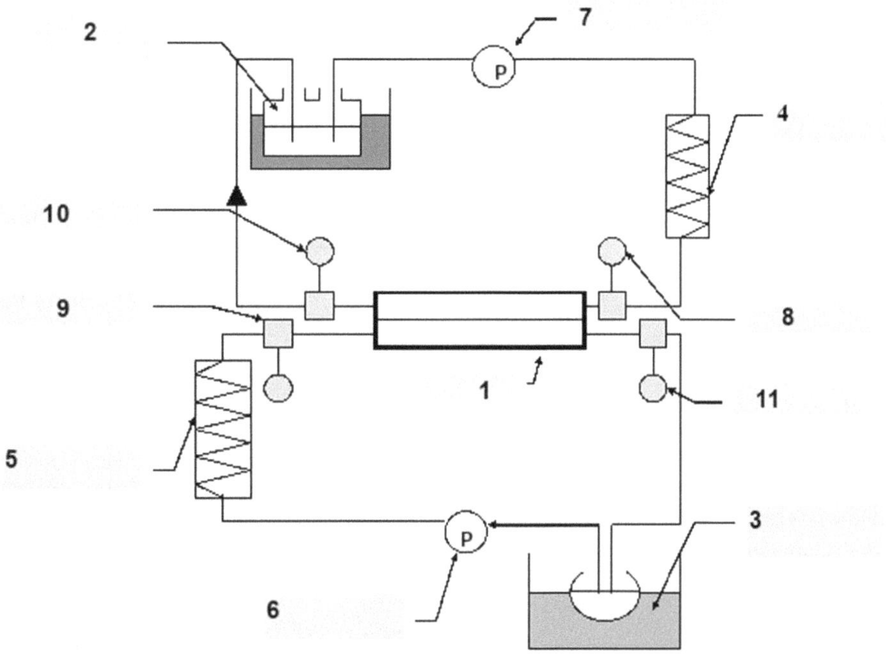

FIGURE 18.17 Laboratory setup for testing MD process. 1, membrane module; 2, feed reservoir; 3, distillate reservoir; 4,5, heat exchangers; 6,7, peristaltic pumps; 8, 9,10,11, thermometers.

or in combination with RO could be an important element of the water management system in a nuclear power plant [201].

Since membrane distillation exhibits a high ability for the concentration of aqueous solutions with high retention of acids, salts, and other low-volatility compounds, and it can be used for the concentration of different radioactive waste streams with high volume reduction and high retention factors [202–204]. Different configurations of MD were studied including direct contact process, air gap MD, and vacuum MD. Vacuum membrane distillation with hollow-fiber membranes made from polypropylene was tested for the removal of strontium from radioactive wastewater [205]. Membrane with a mean pore diameter of 0.18 μm was capable of rejecting 99.6% of strontium ions from an aqueous solution with a concentration of Sr^{2+} 10 mg/L. The membrane flux was 6.71 $Lm^{-2}h^{-1}$.

Vacuum membrane distillation was also proposed for the treatment of low-level radioactive wastewater from the UO_2 fuel element industry [206]. The PTFE hollow-fiber membranes, pore size of 0.22 μm, were used in the experiments. The average permeate flux was 11.3 $Lm^{-2}h^{-1}$ and uranium retention was more than 99.99% at a feed temperature of 75°C. During the experiments, no fouling or membrane wetting were observed. Membrane autopsy confirmed these observations.

At INCT the process was conducted in a batch-type counter-flow apparatus (Figure 18.17) equipped with capillary PP Accurel membranes that showed good effectiveness of membrane distillation for purification of radioactive waste. The permeate obtained was pure water. All solutes together with radioactive compounds were rejected by the hydrophobic membrane. At 10-fold volume reduction of the initial portion of waste and approximately 10-fold concentration of radioactivity in the retentate stream was reached, while the radioactivity of permeate retained was on the level with natural background (Figure 18.18). As was observed in experiments, a small amount of sorption in the system took place. However, permeate was free of radioactive substances and other dissolved compounds, while the concentration and

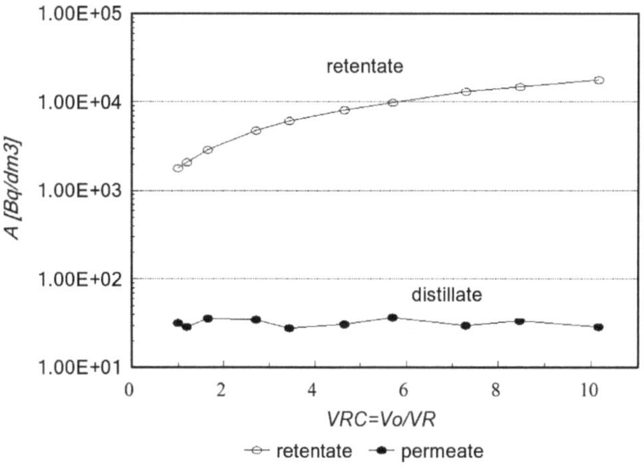

FIGURE 18.18 Concentration of radioactive waste in batch-type MD apparatus.

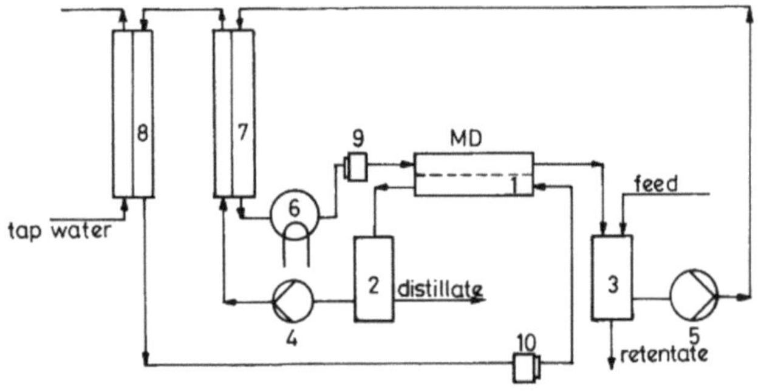

FIGURE 18.19 A scheme of the pilot plant for radioactive waste concentration. 1, MD distillation module; 2, distillate reservoir; 3, feed tank; 4,5, pumps; 6, heater; 7,8, heat exchangers; 9,10, prefilters.

(Reprinted from Journal of Membrane Science, Vol. 163, Zakrzewska-Trznadel, et al.: Concentration of radioactive components in liquid low-level radioactive waste by membrane distillation, pp. 257–264, Copyright (1999) with permission from Elsevier).

radioactivity factors sometimes slightly differed from volume reduction factors.

Membrane distillation for the concentration of radioactive waste was also tested in pilot plant experiments. The facility used in the tests consisted of a spiral-wound module, equipped with PTFE membrane, and an effective surface area of 4 m^2. The installation enabled the recovery of part of the heat by two installed heat exchangers (Figure 18.19).

The experiments were conducted in the temperature range 35°C–80°C at the feed inlet and 5°C–30°C at the distillate inlet, with feed and distillate flow rates up to 1,500 dm^3/h. Under these conditions, the permeate stream was 10–50 dm^3/h (60 300 dm^3/m^2/day). During the experiment run the activity of distillate was stable on the level of natural background radioactivity and the concentrating radioactive compounds took place in retentate. Retention of radioactive ions in retentate was almost complete (decontamination factor $\rightarrow \infty$; Table 18.13). Most of the radionuclides were not detected in distillate; only trace amounts of Co-60 and Cs-137 were present. Also, the retention coefficients of non-active ions were high (Table 18.14).

These experiments proved that membrane distillation can be applied for radioactive wastewater treatment. In one-stage installation the membrane retained all radionuclides and decontamination factors were higher than those obtained by other membrane methods. The distillate obtained in the process was pure water, which could be recycled or safely discharged into the environment. It seems that this process could overcome various problems of evaporation such as corrosion, scaling, or foaming. There is no entrainment of droplets, which cause contamination of the condensate from a thin-film evaporator. Operation at low evaporation temperature can decrease the volatility of some volatile nuclides present in the waste, such as tritium or some forms of iodine and ruthenium. The process is especially economic for the plants, which can utilize waste heat, e.g. plants operating in the power and nuclear industries.

TABLE 18.13
Radiochemical composition of the waste sample used in experiments and effluent after MD plant. (Reprinted from Journal of Membrane Science, Vol. 163, Zakrzewska-Trznadel, et al.: Concentration of radioactive components in liquid low-level radioactive waste by membrane distillation, pp. 257–264, Copyright (1999) with permission from Elsevier)

Radionuclide	Activity of the feed (Bq/dm^3)	Activity of the effluent (Bq/dm^3)	Decontamination factor (DF)
^{60}Co	4510	1.04	4336.5
^{65}Zn	3390	Below detection limit	$\rightarrow \infty$
^{114m}In	86.2	Below detection limit	$\rightarrow \infty$
^{110m}Ag	10.4	Below detection limit	$\rightarrow \infty$
^{133}Ba	2990	Below detection limit	$\rightarrow \infty$
^{134}Cs	7.84	Below detection limit	$\rightarrow \infty$
^{137}Cs	29.5	0.673	43.8
^{140}La	<0.653	Below detection limit	$\rightarrow \infty$
^{170}Tm	526	Below detection limit	$\rightarrow \infty$
^{192}Ir	37.3	Below detection limit	$\rightarrow \infty$

TABLE 18.14
Chemical composition of the waste sample used in experiments and effluent after MD plant. (Reprinted from Journal of Membrane Science, Vol. 163, Zakrzewska-Trznadel, et al.: Concentration of radioactive components in liquid low-level radioactive waste by membrane distillation, pp. 257–264, Copyright (1999) with permission from Elsevier)

Ion	Concentration in the feed (mg/dm^3)	Concentration in the effluent (mg/dm^3)	Retention coefficient (R)
Na^+	1060.6	3.269	0.9969
NH_4^+	207.1	14.584	0.9296
K^+	21	0.212	0.9899
Mg^{2+}	33.7	Not detected	1
Ca^{2+}	87.2	2.375	0.9728
F^-	5.7	0.442	0.9225
Cl^-	744.2	1.485	0.9980
NO_3^-	1832.9	0.065	0.9999
SO_3^{2-}	37.6	0.186	0.9950

In membrane distillation there is a possibility of loss of membrane hydrophobicity during long-term operation. However, during 80-hour tests with a DM pilot plant changes in membrane wettability were not observed. The liquid entry pressure for PTFE membrane used in experiments was about 0.23 MPa, which is sufficiently high in comparison with the pressure in the MD installation. According to the manufacturer's suggestion the pressure applied in the MD module cannot exceed 0.07 MPa. To reduce the risk of liquid entry into membrane pores the pressure on the distillate side was kept slightly higher than the pressure of the retentate side.

However, the presence of compounds lowering the surface tension may cause wetting of the hydrophobic membrane. These compounds have to be removed before entering the membrane system by appropriate pretreatment. At low concentrations these substances can be removed by sorption, oxidation, or phase separation, whereas at high concentrations they can be removed by ultrafiltration. The main non-active components of the wastes treated in the above experiments were inorganic salts such as sulfates, nitrates, or chlorides that increased the surface tension. It was proved that the presence of salts crystallizing in membrane pores may result in wetting of the membrane during long-time operation of MD installations, especially at low distillate temperatures [207]. Such a situation may take place in the module entrance, when the distillate temperature is relatively low. Precipitate deposition, scaling, and pore blocking are problems for all membrane installations. The appropriate pretreatment or dosing antiscaling additives are methods of minimization of these phenomena. The problem of scaling is more serious in evaporation installations that are widely used in the nuclear industry and operating in higher temperatures than membrane distillation.

The experiments with concentrations of inorganic salts showed that 25% concentration of the solute could be achieved by using MD. This is the upper limit for concrete solidification. Usually, such a high concentration is not reached, because the limit of radiation dose is exceeded for solidified waste, originating from concentrated radionuclides. Distillate is pure water and can be discharged to sewage or surface waters or can be recycled.

In Table 18.15 decontamination factors for different processes for low- and medium-level radioactive waste treatment are shown. Membrane distillation with its high-decontamination factors is a competitive method in this field. However, it has to be mentioned that these high decontamination factors are achieved for low-volatile solutes and after adequate pretreatment.

TABLE 18.15
DFs for most common methods of low- and medium-level radioactive wastes processing

Process	DF
Chemical precipitation	$10<DF<10^2$ (β,γ), 10^3 (α)
Organic ion exchange	$10<DF<10^3$
Inorganic ion exchange	$10<DF<10^4$
Evaporation	$104<DF<10^6$
Bioaccumulation	$DF>10^3$
Biosorption	$DF<10^3$
Reverse osmosis	$10<DF<10^3$
Membrane distillation	$DF\rightarrow\infty$

A majority of the methods require multistage systems; even evaporation does not always yield a product of radiochemical purity. In addition, it is an energy-consuming method and has the process barriers described above. The other treatment methods also have some constraints. Ion exchange consumes large volumes of expensive resins that need regeneration and in consequence large quantities of secondary waste are created. Biological processes produce large quantities of sludge with high concentrations of radionuclides and intermediate products.

Many years of operation of membrane processing facility in Chalk River Laboratory have proved the reliability of reverse osmosis. However, reverse osmosis requires high pressures, cannot avoid membrane fouling, and necessitates frequent cleaning operations resulting in the production of secondary wastes. It requires a multistage system to achieve sufficient purity of the product and high-volume reduction. Thin-film composites are less resistant to the radiation and strong chemical environment than PTFE or PP membranes. New possibilities may create the commercialization of inorganic RO membranes.

Opinions on the advantages of membrane distillation are diverse. Low investment costs for small- and medium-capacity installations are pointed out on the one hand. For larger facilities both capital and operation costs can be lower than in conventional processes [208,209], while, on the other hand, more moderate opinions on the benefits of MD have also been published [210].

One of the main barriers to the commercial implementation of membrane distillation is relatively high-permeate fluxes from unit membrane area, along with high-energy consumption in the process. The latter can be overcome when MD installation is used in a nuclear power plant, when waste heat from the cooling circuit can be utilized. Another possible way to minimize energy consumption is efficient recovery of heat by appropriate design of the installation or using membrane modules with integrated heat recuperation based on the spiral-wound GORE-TEX principle [208,211]. According to the evaluation of the company, these modules consume 150–280 $kWh_{th}/1\ m^3$ of distillate, while for modules without integrated heat recovery the energy consumed equals ~600 kWh_{th}/m^3. However, as demonstrated in pilot plant experiments performed at INCT, the latter was hardly achieved with the installation described above. In Figure 18.20, energy consumption per unit of the product for various feed inlet temperatures (T_1), estimated from experimental results, is shown. High process temperatures are preferred; at a temperature T1 above 80°C, the consumed energy specified by the manufacturer of the membrane module, close to 600 $kWhth/m^3$, can be obtained. The thermal energy consumption in the process depends not only on temperatures, but also on the intensity of the distillate cooling (to keep the proper temperature gradient across the membrane) and volume flow rates in the apparatus, V_R and V_D (Table 18.16), however the influence of these parameters is moderate. The lower cooling water flow results in a decrease in energy consumption. Between the flow rate of retentate, distillate, and process temperatures are interactions, therefore

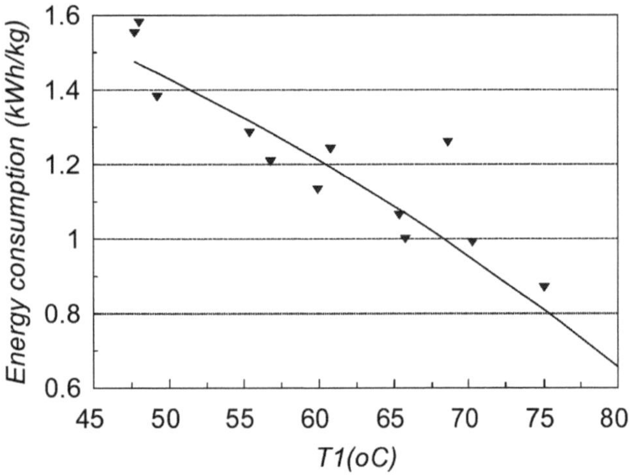

FIGURE 18.20 Energy consumption per unit of the product in MD process.

(Reprinted from Journal of Membrane Science, Vol. 163, Zakrzewska-Trznadel, et al.: Concentration of radioactive components in liquid low-level radioactive waste by membrane distillation, pp. 257–264, Copyright with permission from Elsevier.)

TABLE 18.16
Influence of process parameters on energy consumption per unit of permeate

T1 (°C)	*T2* (°C)	*VR* (L/h)	*VD* (L/h)	*Vcool* (L/h)	*VP* (L/h)	*E/VP* (kWh/L)
69.6	22.8	720	720	1200	23.60	0.96
65.4	21.7	870	870	1200	18.52	1.06
65.5	16.5	870	870	2700	19.36	1.17
49.3	22.1	1380	1380	840	17.28	1.30
49.3	18.6	1380	1380	1260	18.60	1.38
49.2	22.4	1620	1500	840	24.70	1.31
49.2	20.0	1620	1500	1260	24.95	1.35

optimization of process parameters is necessary to get sufficient permeate fluxes and low-energy consumption per unit of the product.

Membrane distillation modules are at present expensive in comparison with RO elements and their costs influence significantly the capital costs of MD installations. The market for MD systems is limited, in spite of its many advantages the MD method is not widely accepted by the industry. The moderate interest of users influences the production capacities and as a consequence reduces the wider implementation of the MD method in different branches of industry. It was proved that the advantages of MD decrease with increased installation capacity; big installations, with productivity comparable with RO, need a large number of modules. A comparison of two processes, RO and MD, proved the technical and economic reasonability of the latter in some cases, such as radioactive waste concentration. The advantages of MD include:

1. Possibility of running the process to high solute concentrations, sufficient for direct solidification;
2. Achieving a high concentration in one stage;
3. Elimination of high pressures, involving RO;
4. Reduction of sorption of some radioactive ions, e.g., $^{60}Co^{2+}$, $^{137}Cs^{-}$, $^{134}Cs^{-}$ inside the pores (the pores are filled with water vapor);
5. Infrequent washing cycles through minimization of fouling and inner sorption, and reduction of quantities of secondary wastes.

Membrane processes are versatile and flexible, and they can be combined with other methods in hybrid processes. Adapted to actual needs they can treat various process streams of different compositions and concentrations. Membrane distillation coupled with evaporation or reverse osmosis may improve the purification efficiency and increase decontamination factors. A flowchart of such hybrid processes is presented in Figure 18.21. In Figure 18.21a the combination of an MD unit with an evaporator is shown. Evaporation is a widely used method for radioactive waste processing. One of the disadvantages of this process is that radionuclides carry over with small droplets. The contaminated condensate needs additional polishing with ion-exchange resins. The installation of an MD module for final cleaning of the condensate can avoid the requirement for ion exchange. The unit that plays the role of demister can be driven with waste heat from a nuclear power plant.

Another possibility of MD use is combination with reverse osmosis which in fact can be loose RO not involving enormously high pressures. The permeate from RO with a reduced amount of multivalent ions is directed to the MD module where final polishing takes place (Figure 18.21b). The preliminary volume reduction takes place in the RO module; the retentate is concentrated and volume-reduced in the MD module. This solution eliminates both the evaporator and ion exchange, replacing these conventional processes with membrane techniques.

The combination of MD with NF has been proposed for the purification and recycling of boric acid from simulated radioactive wastewater [212]. Two types of ceramic membranes were used in a two-step process: NF for purification of boric acid from radionuclides, and a vacuum membrane distillation process for concentration of this acid. Radionuclide removal was achieved through the use of a system of membrane processes (99.99% of Co^{2+} and 95% Ag^{+} in two stages of NF and one VMD) and boric acid was concentrated from 1 to 107 gL^{-1}. Maintaining the permeation flux above 20 L m^{-2} h^{-1}, a >99.9% retention rate of boric was achieved. The hybrid process was found to be a suitable candidate for the treatment and recycling of boric acid from radioactive wastewater.

The use of hybrid methods gives multiple benefits when it combines processes which complement one another or eliminate the drawbacks of a single process. Multistage reverse

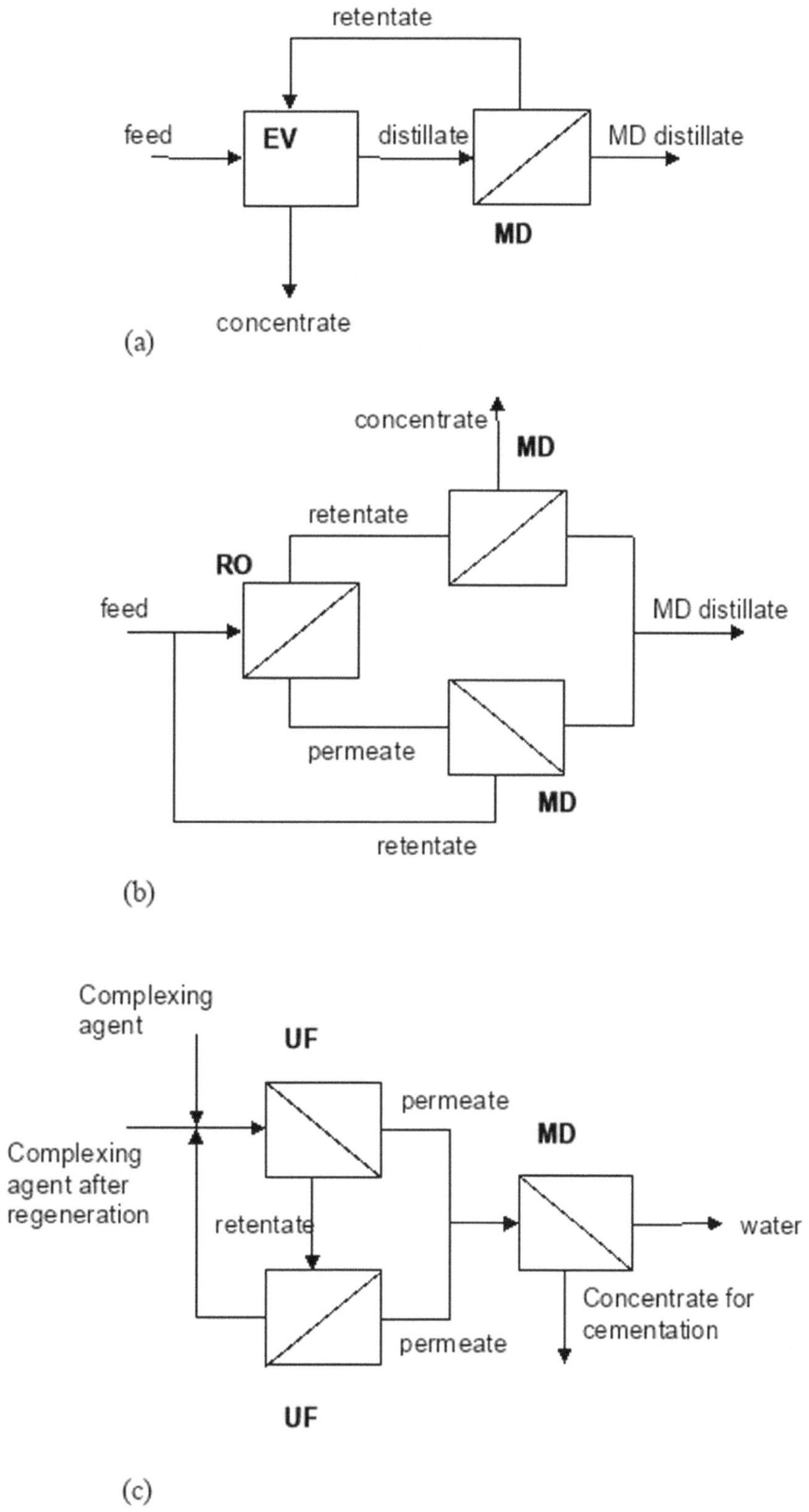

FIGURE 18.21 Hybrid processes for radioactive waste treatment: (a) combination of evaporator with MD module; (b) combination of RO unit with two modules of MD; (c) MD module combined with enhanced UF with regeneration of complexing agent.

osmosis results in high-effluent decontamination, however it cannot be run up to a high retentate concentration. A high concentration is not the limitation for thin-film evaporators, but the condensate needs final polishing by another method. Membrane distillation combines the advantages of both processes, simultaneously avoiding all constraints of RO and EV. In the case of the availability of cheap sources of thermal energy, e.g., waste heat, the complementation of reverse osmosis or evaporation by membrane distillation seems to be justifiable. The application of MD for nuclear desalination seems to be reasonable because nuclear reactors can supply the necessary heat both for radioactive waste treatment and water desalination [201,213,214].

18.3.2 Membrane Solvent Extraction

18.3.2.1 Membrane Contactors for Separation of Radionuclides

In membrane extraction of metals, the mass transport of solute from one phase to another occurs by diffusion. This is controlled by phase equilibrium and the resistances of boundary layers in two phases and the membrane material. Both types of materials are used for membrane extraction and stripping, hydrophilic and hydrophobic, as well as composite hydrophilic/hydrophobic barriers that also have been developed to avoid membrane solubilization [181,182]. To enhance separation, the reactive liquids that induce a chemical reaction with one of the separated species can be used. In membrane solvent extraction of metal-extracting agents, such as tri-n-octylphosphine oxide (TOPO), di(2-ethylhexyl)phosphoric acid (D2EHPA), n-octyl(phenyl)-N,N-diisobutylcarbamoylmethylphosphine oxide (CMPO), and commercial reagents like CYANEX 301, CYANEX 923, LIX622, and LIX622N, are applied.

The microporous hollow-fiber membrane modules were employed for liquid–liquid extraction of neodymium as a surrogate for americium from 2 M nitric acid using DHDECMP (dihexyl N,N-diethylcarbamoylmethylphoshonate) and CMPO extractants in diisopropylbenzene. These modules are applied for back-extraction of neodymium from organic phase into 0.01 M nitric acid [215].

Tests with hollow-fiber modules to select trivalent actinides from lanthanides in acidic media were performed. It was proved that using 2,6-di(5,6-dipropyl-1,2,4-triazin-3-yl)pyridine as extracting agent, up to 94% americium can be recovered from 1 kmol/m^3 HNO_3 with low co-extraction of some lanthanides [216]. The next experiments with extraction followed by lanthanide scrubbing and stripping were performed. They demonstrated that up to 99% americium could be extracted from 0.5 kmol/m^3 HNO_3 with one-third of the fission lanthanides co-extracted when a mixture of bis(chlorophenyl)dithiophosphinic acid and tri-n-octyl phosphine oxide was applied. The tests revealed the potential of such a process for partitioning of minor actinides from spent nuclear fuel in the SANEX process. The application of miniature hollow-fiber modules in experiments allowed using feed volumes as low as several tens of milliliters to run the tests [217]. Separation of trivalent actinides with lanthanides from spent nuclear fuel solutions by the DIAMEX process (i.e. An(III)-Ln co-extraction) was tested with a small hollow-fiber module [218] as a contacting device. The experiments included extraction and back-extraction in which americium was extracted at 99.99% and lanthanides were co-extracted quantitatively. The proposed arrangement exhibited several advantages over common mixer–settler solutions, centrifugal contactors, or pulsed columns, mostly because of the absence of flooding, entrainment, and emulsion formation.

In Bhabha Atomic Research Centre, a porous hydrophobic polypropylene hollow-fiber contactor was proposed to extract Pu(IV) from nitric acid medium [219]. Aliquat-336/Solvesso-100 at various concentrations (5–50%v/v) was used as an extracting agent. The aqueous solutions were composed of Pu(IV) at concentrations of 10^{-5}–10^{-6} mol/L in 1–5 M HNO_3. They were pumped through the tube side, and the organic phase through the shell side of the membrane contactor in a counter-current mode. The optimal flow rates of aqueous and organic phases were adjusted at optimal values: 5.83 cm^{3}s^{-1} and 1.53 cm^{3}s^{-1}, respectively. The stripping of Pu(IV) from loaded Aliquat-336 extractant was accomplished with 0.5 M $NH_2OH{\cdot}HCl$ in 0.3 M HNO_3, and 1 M CH_3COOH in 0.3 M HNO_3 by introducing pregnant with Pu Aliquat-336 organic solution in the shell side, and stripping solution in the tube side of the module.

18.3.2.2 Case Study: Membrane Contactors for Uranium Liquor Processing

At the Institute of Nuclear Chemistry and Technology, membrane contactors were tested in uranium processing technology, at the stage of cleaning the solutions after leaching of uranium ores [220–222].

The uranium liquors contained a variety of metals, both in anionic and cationic forms. Depending on the mineralogy of the ores, alkaline or acidic leaching was applied. If alkaline leaching is applied, the solution may be relatively pure, because of high selectivity of this procedure against uranium. In the case of acidic leaching, apart from uranium, the solution may contain significant amount of metals like zinc, copper, nickel, vanadium, molybdenum, iron, manganese, and lanthanides. Besides, no matter what leaching procedure is applied, the content of uranium is quite low, and the amount of impurities may be significantly high. Such post-leaching solutions should be concentrated and purified. This can be done either by ion exchange (IX) or solvent extraction (SX). Solvent extraction consists of two stages: extraction and stripping. At the extraction stage the leach solution is brought into contact with the organic phase to extract uranium ions. At the stripping step, the organic phase is contacted with the aqueous phase and uranium is transferred back from the organic extracting agent to the aqueous phase. At the same time, the concentration of uranium takes place, since the volumes of applied solutions are appropriately selected. The resulting product is further processed for uranium recovery. SX

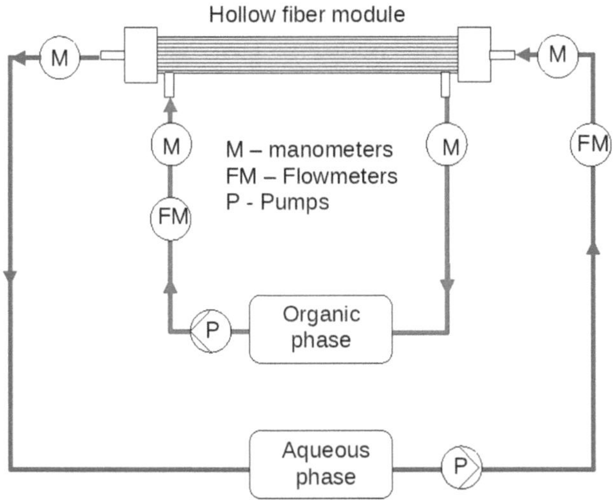

FIGURE 18.22 The scheme of experimental setup for membrane extraction experiments.

TABLE 18.17
Characteristics of the membrane used in X50 Liqui-Cel Extra-Flow, Celgard module

Membrane material	Polypropylene
Average pore size	0.04 µm
Porosity	40%
Internal diameter	220 µm
Outer diameter	300 µm
Inner membrane surface area	1.9 m^2
Outer membrane surface area	2.6 m^2
Number of capillaries	11,000

process involving only liquid–liquid contact is widely applied in uranium processing technology since it is very selective and versatile. Moreover, it can be very easily designed as a continuous operation, while IX needs elution with acids producing the secondary waste. However, SX cannot sustain an appreciable amount of solid material and generally is more sensitive to the grade of the leach solutions treated than IX. The other constrains of SX are the formation of emulsions and losses of organic solvent that is usually expensive. These losses are costly and result in environmental problems since uranium tailings need treatment.

As a standard appliance in the extraction of uranium, a multistage mixer–settler arrangement with concurrent flow of two phases, water and organic, is used. This can be replaced by a novel design with membrane contactors that avoid the constraints of conventional systems. Membrane extraction used for uranium recovery has many advantages over conventional methods, such as no fluid/fluid dispersion, no emulsion formation, no flooding at high flow rates, low solvent holdup, and well-defined and constant interfacial area. Membranes for contractors can be made of ceramics, as well as other chemically resistant materials like PTFE. The system does not need high pressure and elevated temperature. The only problem is membrane stability in contact with two phases that can be achieved by appropriate control of liquid flow rates in the module. Both flow rates in the membrane module should be properly adjusted to avoid a pressure difference across the membrane that may cause membrane solubilization and solvent loss [221].

In Figure 18.22 the scheme of the experimental setup used for membrane extraction experiments is presented. The setup consisted of a membrane contactor with aqueous and organic circuits, two pumps, and the control equipment: flowmeters, pressure gauges, and valves. Two fluids, aqueous and organic, circulated counter-currently. The membrane contactor X50 2.5x8 Liqui-Cel Extra-Flow, Celgard, was used in the system. The characteristics of the membrane are shown in Table 18.17. The small-volume module houses 11,000 capillaries with a 1.9 m^2 inner surface area. The module possesses a central baffle, which enables uniform flow inside the shell.

The process conditions were tested with model solutions of U(VI) in 5% H_2SO_4. The aqueous phase was introduced into the shell side of the device. The extracting agent, 0.2 M D2EHPA in toluene or kerosene diluents, flowed in the fiber side of the membrane contactor. To avoid membrane wettability and entrainment, the flow rates of both fluids in the module were adjusted according to prior model calculations assuming no pressure difference across the membrane. According to this assumption, the linear velocities ratio of the liquids u_w/u_o should not be less than 13. Under these conditions, even the organic phase fills the membrane pores, the mixing of liquids does not occur, and the membrane contactor worked stably. Uranyl ions from the aqueous phase passed through the membrane in its complexed form into the organic phase. The kinetic studies showed that after 4 h, the U extraction efficiency reached almost 100% (Figure 18.23). However, even after 1 h the extraction efficiency was high ~ 95%. In comparison, the extraction experiments done in a separation funnel revealed the extraction equilibrium time to be significantly shorter, at 5–15 minutes.

The same experiments were done with real post-leaching solutions. The liquor after leaching uranium ores, pregnant in metals, was pretreated and introduced to the shell side of the membrane contactor. With application of the same system, uranium was extracted with 99.9% efficiency (Figure 18.24). Together with uranium thorium, ytterbium and molybdenum were extracted with high, over 99%, efficiencies (Table 18.18). The inverse arrangement with aqueous feed solutions introduced into the fibers while the organic phase passed in the shell side of the membrane module was also tested. The first configuration was proved to be better when taking into account the process rate and efficiency of uranium extraction.

The recovery of uranium from loaded organic extractant solutions in kerosene was tested with different stripping agents: sulfuric acid, ammonia carbonate, and sodium carbonate. The highest efficiencies, close to 100%, were achieved with carbonate solutions.

The application of membrane contactors makes possible easy arrangement of the integrated extraction-stripping

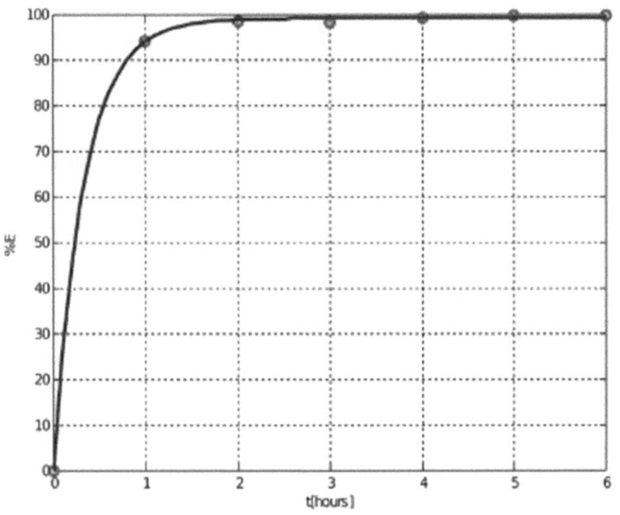

FIGURE 18.23 Extraction of uranium from 0.2 g/L UO_2^{2+} solution in 5% H_2SO_4 into D2EHPA in kerosene using X50 Liqui-Cel membrane contactor.

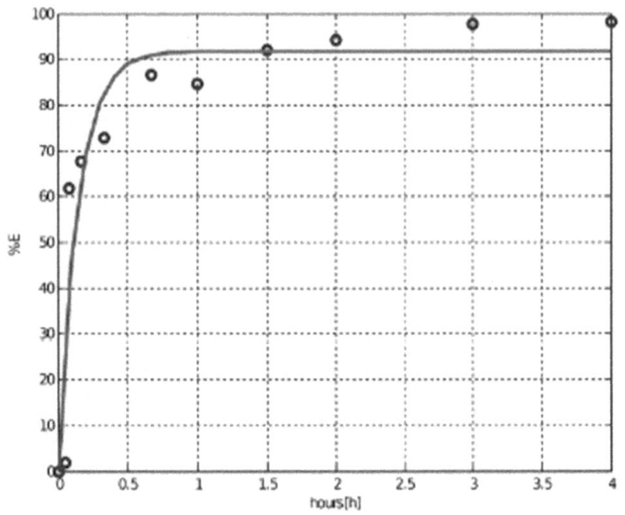

FIGURE 18.24 Extraction of uranium from post-leaching liquor into 0.2 M D2EHPA in toluene using X50 Liqui-Cel membrane contactor.

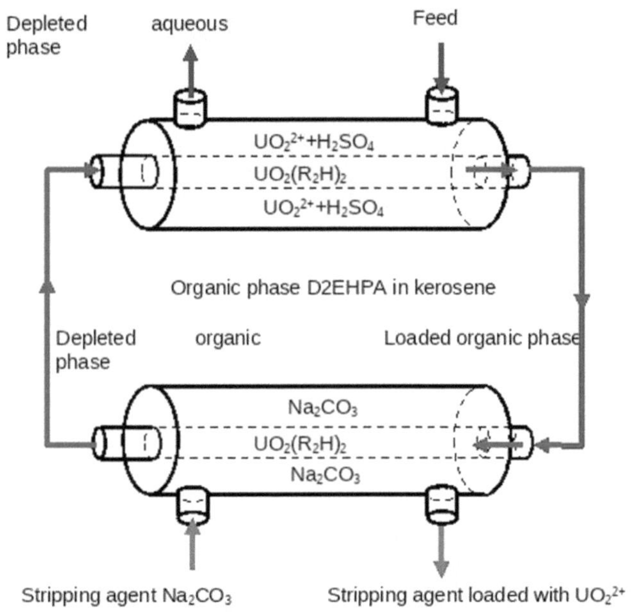

FIGURE 18.25 UO_2^{2+} extracting-stripping scheme with two membrane contactors.

TABLE 18.18
Extraction efficiency of uranium in X50 Liqui-Cel Extra-Flow, Celgard module

Metal	Extraction	Back-extraction		
	% 0,2 M D2EHPA	% 6M H2SO4	% 0.5M Na2CO3	% 0.5M NH4CO3
U	99.9	77.4	99.1	100
Th	100.0	100	36.1	26
Cu	6.8	0	0	0
Co	1.1	0	0	0
Mn	0	0	0	0
Zn	31.7	69.2	6	45.7
La	26.6	100	14.9	6.6
V	14.8	100	58.4	48.5
Yb	99.6	100	100	84.1
Mo	99.4	4.3	93.8	100
Ni	3.6	0	0	0
Fe	76.4	17.9	8.4	7.9

system. Such an arrangement is very beneficial not only because of its simplicity, but also from an economic point of view. The integration of two processes allows continuous regeneration of the organic phase with efficient use of extracting agents, which are sometimes expensive or occasionally synthesized in the laboratory for a specific purpose. The extracting-stripping mechanism is shown schematically in Figure 18.25.

The advantages of extraction in membrane contactors allow them to be considered in the uranium production cycle for low-grade uranium ores [223–225], as well as from secondary sources and waste [225–227].

Another type of contactor included in the uranium production flowchart is a membrane module with the Taylor-Couette helical flow [228]. This device was used in the uranium ore leaching process; thanks to this solution, it was possible to separate the leachate from the remaining solid phase (parent rock) at the same time. The use of this system with circular Couette flow and toroidal Taylor vortex flow created turbulence, thus avoiding filter cake formation and membrane fouling. The same system can be employed in liquid radioactive waste processing with recovery of valuable metals [229].

18.3.3 Supported Liquid Membranes

In supported liquid membranes (SLMs) the transfer of the species occurs by permeation through the organic liquid immobilized in micropores of the membrane matrix used as a support, or by facilitated transport, when the species are subjected to a reversible chemical reaction with the carrier. High selectivity is achieved in facilitated transport if the carrier possesses good affinity to one of separated components. The carrier-mediated process results also in fast mass transport rates that are very important for future industrial applications.

There are many examples of the studies on SLMs for nuclear applications in the literature. SLMs were tested for high-level radioactive waste treatment combined with removal of actinides and other fission products from the effluents from nuclear fuel reprocessing plants. The recovery of the species, such as uranium, plutonium, thorium, americium, cerium, europium, strontium, and cesium, was investigated in various extracting-stripping systems. Selective permeation of Pu(IV) through the SLM containing 2-ethylhexyl phosphonic acid as an ion carrier diluted in n-dodecane was tested in Bhabha Atomic Research Centre [230]. Hollow-fiber SLMs were studied for quantitative separation of plutonium with tri-n-butyl phosphate (TBP) in dodecane, from acidic waste generated in nuclear reprocessing plants [231]. As a stripping liquid, 0.1M $NH_2OH{\cdot}HCl$ in 3 M HNO_3 was applied.

Various carriers like alamine 336 in toluene, N,N-di(2-ethylhexyl)isobutyramide (D2EHIBA) in n-dodecane, N,N,N',N'-tetra-2-ethylhexyl-3-pentane-diamide (T2EHDGA) in n-dodecane, and N,N,N',N'-tetraoctyldiglycolamide (TODGA) in n-dodecane, were tested for U(VI) removal in Bhabha Atomic Research Centre [232–236]. Aliquat 336 (N-methyl-N,N-dioctyloctan-1-ammonium chloride) was used as an extracting agent for uranyl ions from an HCl environment [237]. Extraction of uranium from 6 M HCl was achieved with Alamine 336 [238]. From phosphoric acid, U(VI) was separated by using tri-n-octylphosphine oxide (TOPO) in n-dodecane with ammonium carbonate as a receiving phase [239]. Selective separation of uranium from thorium with hollow-fiber supported liquid membranes was done with application of TBP [240]. For actinides partitioning, extractants, such as TBP, D2EHPA, tri-n-octylamine (TnOA), bis(2,4,4-trimethylpentyl)dithiophosphinic acid (Cyanex 301), and diglycolamides, were applied [241–246]. Am (III) and Eu(III) separation was attained with 2,6-bis(5,6-dipropyl-1,2,4-triazin-3-yl)pyridine as the extractant [247]. Selective separation of cerium(III) from low-level radioactive waste was done with CMPO [248]. From fission products in the PUREX process, separation of Ce(III) was achieved with TBP in kerosene after previous oxidation to Ce(IV) [249].

Selective separation of cesium from fission products of irradiated uranium target was studied with the application of di-t-butyl benzo 18 crown 6 as a carrier employing polypropylene matrix [250]. Synthesized in a laboratory, calix[4]arene crown ethers in a fixed 1,3-alternate conformation gave excellent Cs(I) extraction [251,252]. Hollow-fiber-supported liquid membrane with calix-[4]-bis(2,3-naptho)-crown-6 as carrier in (2-nitrophenyloctyl) ether and n-dodecane was employed for the study of Cs transport in 3M HNO_3 [253]. Separation of Cs from other radionuclides in multi-component nitric acid solution was tested with SLM using chlorinated cobalt dicarbollide (CCD) in phenyltrifluoromethyl sulfone (PTMS) [254]. In Bhabha Atomic Research Centre, SLM systems on the basis of 4,4'(5')di-tert-butyl-dicyclohexano-18-crown-6 (DTBCH18C6) for radioactive strontium removal were investigated [255–257].

Technetium, one of the key elements in the treatment of radioactive liquid waste, because of its long half-life and geochemical mobility, was separated from concentrated alkaline solutions by SLMs containing 2-nitrophenyloctyl ether (NPOE) [258]. The hollow-fiber-supported liquid membrane containing bis(2-ethylhexyl)phosphonic acid, was tested for carrier-free separation of ^{90}Y from ^{90}Sr [259].

SLMs have many advantages such as high selectivity in carrier-mediated transport and low consumption of extracting agents, which is very important when expensive solvents are used. The main drawback of immobilized liquid membranes is their low stability associated with the loss of extracting agent during long-time operation. This instability is a limitation that hinders the commercial application of SLMs in industry. One of the most interesting approaches to avoid this problem is the preparation of plasticized membranes called polymer inclusion membranes (PIMs) [260–262].

18.4 ELECTRIC MEMBRANE PROCESSES

Electric membrane processes are those processes in which the electric potential is the driving force. They proceed with the use of ion exchange membranes, allowing the selective transport of ions and charged molecules through the membrane. These methods have many advantages, such as minimization of secondary waste, moderate working conditions, and low capital and operating costs. The main processes in this group, tested in terms of their application in nuclear technologies, are electroosmosis, electrodialysis, and membrane electrolysis.

Among the electric processes, electrodialysis (ED) is the most mature method, with many industrial applications such as sea water desalination, brackish water concentration, and potable water production. Electrodialysis is a process of membrane separation of the components of aqueous solutions that pass through electrically charged membranes under the driving force of the electrical potential difference. If the ion transport numbers in the solution and in the membrane material are different, then the electric current differentiates the compositions of the solutions separated by the membrane. Among various designs and modifications of electrodialytic systems, electrodialysis reversal (EDR) occupies an important place [263]. This is a process put in place to avoid membrane fouling problems and is economically viable and capable of treating surface water and reusing water from industrial applications. The electrodialysis reversal plant for drinking water treatment, built in Washington, Iowa, is well known as the most cost-effective and practical way to remove radium [264].

Other electric processes have already found some applications in nuclear technologies. Electroosmosis was used to dewater filter sludge or sludge from gravity sedimentation [265]. The high, 99.99%, retention factor meant a significant reduction of membrane fouling was achieved. The only drawback was the low-volume reduction factors due to the decrease in the transport velocity with the increase in the liquid conductivity.

Radioactive waste of various origins was treated by the Chilean Commission for Nuclear Energy with electrodialysis and electroosmosis [266]. The inorganic ED membranes showed high resistance to radiation and performed well over long operation times.

Electrodialysis was studied for removing radioactive ions from low- and intermediate-level radioactive wastes at the Japan Atomic Energy Research Institute (JAERI) [265]. The inactive salts, NaCl, Na_2SO_4, $Ca(NO_3)_2$, as carriers of radioactive ions, such as ^{24}Na, ^{42}K, ^{90}Y and ^{137}Cs, present in the solution in small quantities were used in these experiments. Such a procedure caused a significant reduction of radionuclides, however the removal efficiency decreased over time.

In Bhabha Atomic Research Centre, electrodialysis combined with diffusion dialysis was employed for the separation of fission products, ^{137}Cs and ^{90}Sr, from 3 M nitric acid solutions [267]. Diffusion dialysis as a pretreatment step was carried out in a two-chamber system with the anion exchange membrane. It was the first step to decrease the acidity of the feed solution for improving the performance of the electrodialysis cell at the second stage of processing. In these conditions the transfer of Cs and Sr was greater than 90%.

Electrodialysis combined with electrochemical ion concentration was tested at the same institute for the treatment of low-level waste containing fission products separated from PUREX waste [268].

The separation of strontium and cadmium ions in a three-compartment electrolytic cell was carried out. The influence of ethylenediaminetetraacetic acid (EDTA) as a complexing agent, very often used in nuclear waste processing, was examined at the pH range 2–4 and at 100 mA (12.4 mA cm^{-2}) electric current. The experiments allowed observation of the exclusive movement of strontium ions to the cathode as uncomplexed cations while cadmium migrated to the anode as negatively charged complexes and was removed from the middle compartment of the electrolytic cell [269].

The ED process was tested for separation of molybdenum ions and CO_3^{2-} with HCO^{3-} from liquid radioactive waste containing uranium created in hydrometallurgical uranium ore processing [270]. The anion exchange membranes (AMV Selemion, A229 Morgane, RAI 5035, and AR 204 412 UZL Ionics) were applied. All tested membranes exhibited high retention of carbonate ions. The selectivity of molybdenum to uranium separation was higher than 3.

Another example of the use of electrodialysis for the separation of uranium (VI) from radioactive wastewater is the work carried out by scientists in Iran [271]. In this research, the ED stack was composed of three compartments packed with a pair of ion-exchange membranes and a pair of platinum-coated electrodes (anode and cathode). Simulated radioactive waste was used in these experiments. It has been shown that ED can be used successfully as the main step in a treatment plant to pretreat wastewater of high concentrations. At low concentrations, use of the ED process is limited because of low conductivity of dilute solutions.

Electrodialysis was also employed for salt splitting, e.g. for Na_2SO_4 decomposition, in highly saline waste with the recovery of acid and base [265]. For each pair of anion and cation exchange membranes one pair of electrodes was used. The acid was recovered in the anodic compartment of the ED cell, in the cathode—the base. There was a salt-depleted stream between the membranes. It is possible to simplify the apparatus using a single pair of electrodes and a single cation exchange membrane dividing the chamber into two parts. The process gave 20% NaOH and 20% H_2SO_4, a 12–25% solution of sodium sulfate was the third product of such treatment.

Another way to remove sodium from highly saline radioactive waste is to use sodium super ion conducting ceramic membranes (NaSICON). The wastes produced by the US Department of Energy [272] were used in the studies. Comparative tests of Nafion and NaSICON membranes [273] were performed. Ceramic membranes showed negligible electroosmotic transport of water, high resistance to radiation, and a lack of ^{137}Cs transport. When Nafion membranes were used, about 60% of ^{137}Cs passed from the anolyte to the catholyte compartment.

Another example of salt splitting to regenerate acid and base from various potassium salts by electrodialysis (ED) was studied using a two-compartment membrane electrolysis cell equipped with three types of commercial cation-exchange membranes [274]. These experiments confirmed the usefulness of the method in recovering sulfuric acid, nitric acid, hydrochloric acid, and potassium hydroxide by splitting sulfate, nitrate, and potassium chloride.

The separation experiments on the recovery of boric acid from wastewater [275] were performed in South Korea. AFN ion-exchange membranes allowed for an 87% reduction of the volume of the concentrate.

The ability to remove boron from aqueous solutions by electrodialysis has been proven in numerous studies [276,277]. The results were applied to design a cascade system, composed of three units, for ED boron removal and its concentration. The cost of the removal of boron from 75 mg/L to 0.8 mg/L with simultaneous concentration of this element to 5000 mg/L was estimated at $0.22/$m^3$.

The ion-exchange membranes obtained by radiation grafting of polyacrylic acid on a matrix of polyethylene and Teflon were tested in Egypt to remove zirconium from solutions containing uranium [278].

Elecrodialysis was used to separate ^{125}I and ^{36}Cl ions [279,280]. The novel anion exchange membrane showed high selectivity for iodine ions over chlorine ions; the ratio of electroconductive membrane permeability towards ^{125}I and

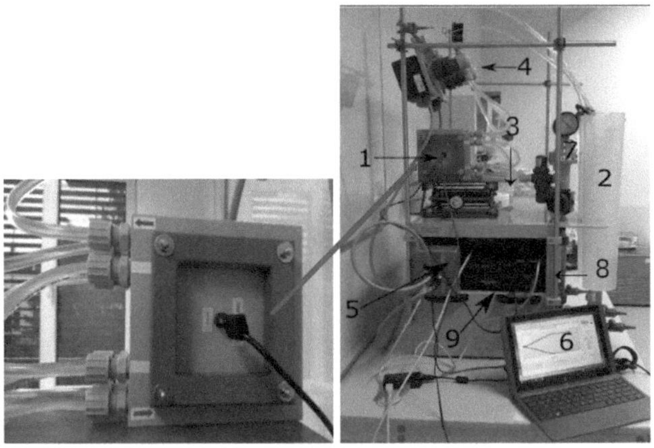

FIGURE 18.26 ED experimental stand. 1, ED cell; 2, cylinders for process solutions; 3, reservoir of the electrode solution; 4, temperature and conductivity meters; 5, data-collecting device; 6, computer; 7, manometer; 8, rotameter; 9, power supply.

TABLE 18.19
Characteristics of the ED cell

Membrane size	110 × 110 mm
Active membrane area	64 cm^2 per membrane
Processing length	80 mm
Cell thickness	0.5 mm
Number of cell pairs	Up to 20
Anode	Pt/Ir-coated with titanium
Cathode	V4A steel
Electrode housing material	Polypropylene

^{36}Cl was 6.2, while the diffusion membrane permeability of the two components was almost the same.

18.4.1 Case Study: Application of Electrodialysis for the Treatment of Liquid Radioactive Waste at the Institute of Nuclear Chemistry and Technology

At the Institute of Nuclear Chemistry and Technology (INCT), Warsaw, the application of ED to treat low-level liquid radioactive waste was considered [281]. The experiments were carried out on the experimental stand shown in Figure 18.26.

The stand was equipped with an ED module composed of ion-exchange membrane pairs: cation-exchange PC-SK membranes and anion-exchange PC-SA membranes. The active membrane area of a single film was 64 cm^2; pairs of membranes could be stacked up to 20 cells (Table 18.19).

The idea to test electrolysis for the processing of liquid radioactive waste arose along with the search for a method of treating waste containing non-ionic organic compounds, which burden further processing steps, such as their solidification or long-term storage in radioactive waste repositories.

The experiments aimed at checking the possibility of separating ions from organic substances present in the

TABLE 18.20
The composition of the radioactive waste sample in ED experiments

Ion	c (mg/dm^3)	Radionuclides	A (Bq/dm^3)
Na^+	635.5	^{60}Co	73
K^+	33.0	^{125}Sb	227
Mg^{2+}	102.7	^{134}Cs	1700
Ca^{2+}	172.2	^{137}Cs	35,000
Cl^-	26.6	Total γ	37,000
NO_3^-	1844.0		
SO_4^{2-}	451.3		
Other ions	16.4		
TDS	3282		

aqueous wastes were carried out with non-radioactive model solutions containing organic compounds, which are used very often to treat liquid radioactive waste: citric acid used for decontamination and octylphenol ethoxylate (Triton X-102) surfactant. The process conditions were selected in the preliminary tests as follows: voltage 20 V, current density 31.3 A/m^2; the velocities of the diluate and the concentrate streams were maintained at the same level, 0.16 cm/s. Although the citrate ions were transported through the membrane less effectively in comparison with other ions, a significant amount passed to the concentrate stream and the ion/citric acid separation effect was negligible. A different effect was observed in relation to the non-ionic compound, which was Triton X-102 surfactant. Almost all surfactant retained in the diluate, only a small amount was transported to the concentrate side [281]. It was shown that the ED process is capable of removing non-ionic organic compounds from radioactive waste before the subsequent stages of processing. Further research carried out both with the model aqueous radioactive solutions and with real liquid radioactive waste showed the possibility of using the ED process for their concentration. The obtained results confirmed that ED can be used in the development of technologies for the treatment of liquid radioactive waste of various compositions: to separate ionic substances from non-ionic constituents.

The concentration of the real low-level radioactive waste sample from the RWMP tanks containing isotopes of ^{60}Co, ^{134}Cs, ^{137}Cs, and ^{125}Sb, is shown in Figure 18.27. The TDS of the sample was 3282 mg/dm^3 and specific γ radioactivity approximately 37 kBq/dm^3 (Table 18.20). The electrodialysis process was carried out for 1 hour. Over 45 minutes, the concentration of ions in the diluate dropped practically to zero, and the radioactivity of the isotopes contained in the sample was reduced to the natural background level. Concentrations of ions and specific radioactivity in the concentrate increased to the stable level: approximately 8 mS/cm and 12 cps.

This research confirmed the possibility of using electrodialysis for concentration of liquid, low-level radioactive waste. The obtained results could be used in the development of technologies for the treatment of liquid radioactive waste

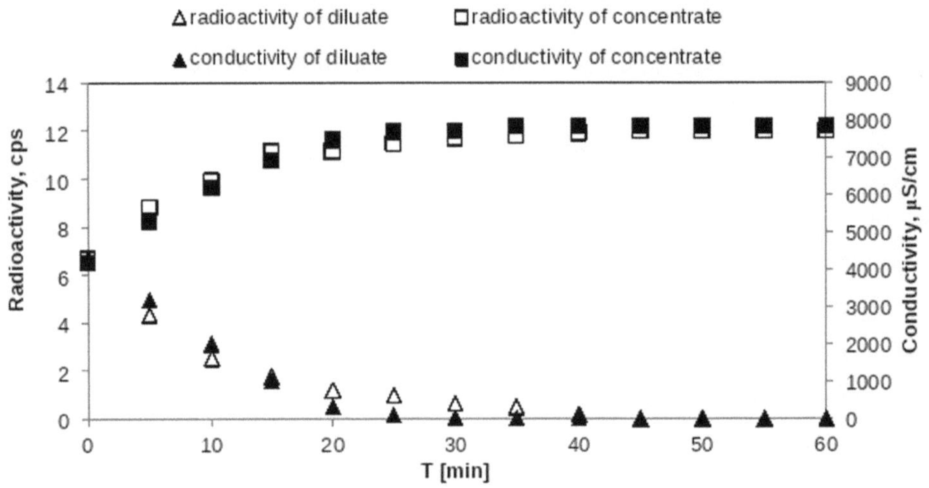

FIGURE 18.27 Variation of the radioactivity and conductivity of the concentrate and diluate in time in ED experiment.

of various compositions: to separate ionic compounds from non-ionic substances. The method could also be used for the treatment of different industrial wastewater contaminated with organic, non-ionic substances.

The advantages of ED, such as high recovery (water/brine, high-volume reduction ratio), the possibility of composition adjustment (separation of monovalent and multivalent ions, separation of ions of the same sign but with different transport numbers in the membrane, separation of electrolyte from non-electrolyte), make the process attractive for the processing of some types of liquid radioactive waste.

18.5 FUTURE OF MEMBRANE PROCESSES IN NUCLEAR TECHNOLOGY

18.5.1 Liquid Radioactive Waste Processing by Membrane Processes—Advantages and Limitations

The continuous development of membrane processes applied in nuclear technologies is of considerable interest [1-4]. Implementation of new membrane materials with high chemical and radiation resistance and new module designs allows spreading applications of membrane processes into different fields of the nuclear industry. However, the industrial applications are still limited to pressure-driven membrane processes.

The main barriers to the use of membrane methods are the following:

1. Necessity of periodical interruption to processing resulting from membrane fouling or blocking and successive cleaning of the installation. It is estimated that about 1% of secondary wastes come from cleaning the apparatus and membranes [64];
2. Need of pretreatment of the streams before membrane installations;
3. The installations operating under high pressure require sophisticated apparatus and pressure pumps.

The elaboration of proper pretreatment methods, application of antiscalants, and minimization of secondary wastes created during cleaning cycles are of great importance. At present the research work on the use of pressure-driven processes for radioactive waste treatment is focused on the following issues:

1. Application of ultrafiltration coupled with other processes such as precipitation, adsorption, complexation, or finely divided ion exchangers;
2. Combining membrane processes with other methods into integrated processes, such as ultrafiltration–evaporation, reverse osmosis–ion exchange;
3. Elaboration of effective methods of fouling control and membrane cleaning, avoiding membrane blockage and flux decline;
4. Use of nanofiltration with pores at the boundary of reverse osmosis and ultrafiltration that allows the separation of multivalent ions like Ca^{2+}, Al^{3+}, and SO_4^{2+} from monovalent. It is possible with nanofiltration to separate multivalent radioactive ions from non-active salts, such as sodium nitrate;
5. Replacing polymeric membranes by inorganic ones, more resistant to ionizing radiation and aggressive chemical environment;
6. Testing the influence of process conditions on the stability and lifetime of particular membrane installation elements—the influence of salts that cause scaling (Na_2SO_4, Na_2CO_3, $NaHCO_3$, $Ca_3(PO_4)_2$), the influence of radiation on membrane material.

Despite some technical and process limitations, membrane techniques are very useful methods for the treatment of different types of effluents. They can be applied in nuclear centers processing low- and intermediate-level liquid radioactive wastes or in fuel reprocessing plants. New techniques, such as liquid membranes or membrane solvent extraction, especially with novel extracting agents, are intensely investigated as potential methods for separation of actinides or lanthanides. Progress in this field may result in the development of effective techniques for plutonium and other actinides partitioning. This enables not only reducing the volume and radioactivity of reprocessing waste, but also minimization of secondary waste volumes generated.

All the methods reported in the chapter have many advantages and can be easily adapted for actual, specific needs. Some of them are good pretreatment methods, while others can be used separately as final cleaning steps, or can be integrated with other processes. Membrane methods can supplement or replace techniques of distillation, extraction, adsorption, ion exchange, etc. An evaluation of membrane processes employed for liquid radioactive wastes treatment is presented in Table 18.21.

18.5.2 New Fields of Application of Membrane Processes in the Nuclear Industry

18.5.2.1 Removal of Tritium from Nuclear Waste (Liquid and Gaseous Effluents)

The Nuclear Energy Agency of the Organisation for Economic Co-operation and Development (OECD/NEA) considers tritium as one of the four volatile, hazardous radionuclides (^{3}H, ^{14}C, ^{85}Kr, and ^{129}I) created in the nuclear cycle that pose a long-term risk [282]. Tritium, a radioactive isotope of hydrogen, emitting mild beta radiation, is created in nuclear reactors and in fuel reprocessing plants by neutron activation of isotopes of ^{2}H, ^{3}He, ^{6}Li, and ^{10}B. The largest quantities of tritium are produced in heavy water reactors (~ 5.42×10^7 GBq/GW$_{el}$ per annum). Smaller amounts are emitted by fast breeder reactors and high-temperature graphite reactors; the smallest quantities are created in light water reactors (~ 5.55×10^5 GBq/GW$_{el}$/annum). It has been evaluated that reprocessing plants with a capacity of 1,400 t/a, processing fuel from LWR can emit $(1.85–3.70)\cdot 10^7$ GBq per year of radioactive tritium [283], which is commensurate with emissions from natural sources.

One of the technical problems of the removal of tritium is its low concentration in technological and waste streams. In heavy water reactors moderator (PHWR) the DTO:D_2O ratio is 10^{-5}, and in water streams from fuel reprocessing plants, HTO:H_2O is 10^{-7} [284]. In this way, great volumes are contaminated with tritium: hundreds and thousands of cubic meters per year need processing. The majority of the treatment methods are based on technologies used for deuterium enrichment, such as water distillation, distillation of hydrogen, electrolysis, ion exchange, and some combined methods such as electrolysis-catalytic exchange, the so-called CECE process [285,286].

Among the eventual technologies for separation of tritium, membrane methods are considered. The application of membranes for detritiation always brings to mind palladium membranes used for hydrogen separation. Extensive research has been done concerning the application of Pd and Pd alloys for gaseous tritium separation, which showed improvements in membrane durability and hydrogen embrittlement, and a decrease in the costs of palladium [287–293]. For HTO/DTO separation by electrolysis, polymeric membranes are more beneficial because of the fact that membranes actually play the role of the electrolyte and perform hydrogen isotope separation at the same time [294].

High hopes for applications in the separation of isotopes field are associated with carbon and graphene membranes [295–298]. The In Pacific Northwest National Laboratory (PNNL) work demonstrating the difference in transport of H_2O, D_2O, and HTO through graphene oxide membrane [295] led to the evaluation of the feasibility of Gox membrane-based isotopic water separation. Proof-of-principle tests proved the capability of these membranes to separate low-concentration tritium from large volumes of tritium-contaminated water. The removed tritium could be concentrated in a small volume to be recycled or stabilized for disposal. The process is low-energy consuming compared to the electrolysis of water in the catalyzed electrochemical exchange process that is currently used in many countries.

Nafion Du Pont membranes were applied for separation of hydrogen isotopes in installations for tritium removal from reactor heavy water and from nuclear reactor atmosphere. The monitor for HTO and HT separate control was constructed in Chalk River Laboratories in Canada [299]. At the High-Energy Accelerators Research Organization (KEK), Japan, work on the apparatus for measurement of tritium emitted from high-energy accelerators, equipped with hollow-filament membranes for gas separation, is carried out [300].

The studies on the separation of water isotopomers (HTO, HDO, H_2O) with membranes from poly[bis(phenoxy) phosphazene and carboxylated derivatives were carried out in the Pacific Northwest Laboratory [301,302]. Using these membranes, tritiated water was extracted from fuel storing pool water with a 3 μCi/L concentration of tritium, and from facilities under the supervision of the US DOE in Hanford and Savannah River. Separation with 10,800 pCi/L tritiated water obtained by the membrane method was not higher than 33% (depletion in the permeate fraction). Water containing 3 μCi/L of HTO was depleted by 22% in a similar system [302].

The membrane permeation was tested for separation of HTO from water at INCT [303]. Membranes made from regenerated cellulose, polysulfone, and polytetrafluoroethylene were used. Separation factors $\alpha_{H2O/HTO}$ obtained through the membrane process were not higher than 1.04 for cellulose and 1.02 for polysulfone, while for PTFE membranes they were as high as 1.06–1.22.

A number of works published recently concern the application of polymeric gas separation (GS) membranes for the separation of hydrogen isotopes, with particular consideration of tritium compounds (HT, HTO) [304–314]. For this purpose, membranes manufactured from glassy and amorphous polymers are applied, mainly polyimide and polycarbonate membranes, as well as polyphenylene oxide membranes assembled in modules of different configuration (e.g., Mc Generon Inc., type B210, Ube Industries, Ltd., type NM-B05A, Parker, type 2112-NX-1-300 [305].

TABLE 18.21
Membrane methods for liquid low- and medium-level radioactive waste processing

Process	Advantages	Disadvantages	Abilities and applications
Reverse osmosis	• Removes dissolved salts • DF 100–1000 • VRF 100–1000 • Economical • Established for large-scale operation	• High-pressure system, limited by osmotic pressure • Subject to fouling, non-backwashable	• Very well developed technology • Ability of rejection of all contaminants • Number of industrial applications for treatment of mixed waste, floor drains, waste from reprocessing, from medical applications
Nanofiltration	• Pressures lower than in RO • Separate multivalent ions and non-ionic organic molecules of MW> 300 Da from monovalent ions and non-active salts • Inorganic membranes of good radiation and chemical resistance available	• Not usable where absolutely purification is necessary; monovalent radioactive ions pass through the membrane • Membranes subjected to fouling	• Ability of separation of monovalent ions from multivalent ions • Separation of organic compounds from monovalent salts • Application in nuclear industry for boric acid recovery from contaminated cooling water
Ultrafiltration	• Separation of dissolved salts from particulate and colloidal material • Inorganic membranes available with good chemical and radiation stability can also operate at elevated temperatures • Pressure <1 MPa	• Fouling—need for chemical cleaning or back flushing • Organic membranes subject to radiation damage	• Good pre-treatment stage for RO • Removes colloids (some actinide compounds) and suspensions • Many industrial applications and pilot tested installation
Enhanced ultrafiltration (UF/complexation, UF/sorption)	• Possibility to treat wide range of waste streams • Removal of small ions without high pressures involved	• Strongly dependent on process conditions (temperature, pH) • The presence of complexing agents may entail further cleaning • Fouling phenomena	• Removes small ions with selected ligands • Enables recycling of some compounds after break-up of the complexes • Pilot plant tests
Microfiltration	• Low pressure operation (100–150 kPa) • Excellent pre-treatment stage for RO • Inorganic membranes available • Low fouling when air backwash is employed	• Backwash frequency can be quite high; depends on solids content of waste stream	• Very well developed technology • Number of implementation in nuclear centers • For pre-treatment purposes or for concentration of coarse particles after precipitation process
Membrane distillation	• Removes all dissolved salts, very high DF • Non-pressure-driven operation • Good chemical and radiation resistance of membranes (Teflon) • Economical for nuclear industry	• High thermal energy consumption comparing with pressure-driven processes unless cheap energy source or waste heat is utilized • Can not be applied for wastes with volatile radioactive compounds	• Tested in laboratory and pilot plant; • Until now no industrial installations in nuclear technology
Electrodialysis	• Well established technology for desalination • Separates ionic substances from non-ionic, concentrates ionic compounds • Minimizes the volume of hazardous wastes and secondary wastes	• The necessity of pretreatment and periodical scale removal • Process limited by concentration (low conductivity for diluted solutions) • Fouling is a problem in higher concentrations	• Intensive laboratory and pilot plant studies, small industrial applications • For separation of ions from radioactive waste, • For recovery of acids and alkali from radioactive wastes and salt solutions

TABLE 18.21 (Continued)
Membrane methods for liquid low- and medium-level radioactive waste processing

Process	Advantages	Disadvantages	Abilities and applications
Liquid membranes	• High selectivity in carrier-mediated transport, high transport rates • Small amount of extracting agent is needed, expensive extracting agents can be used • High interface area in hollow-fiber SLM • Extraction and stripping in one apparatus; • Does not need high pressure and elevated temperature	• Pure VRFs • Sensitive to particulates in aqueous phase • Lack of stability due to the loss of solvent during operation • Suitable combination of solvent and strip phases is necessary • Too low fluxes for industrial applications	• Very promising for fuel reprocessing and partitioning of actinides, • Intensive research in world laboratories on the process and new extracting agents • For recovery of rare earth elements • Needs further development and improvements
Membrane solvent extraction	• Selectivity, as high as in traditional extraction • No fluid–fluid dispersion, no emulsion formation • Well-defined interfacial area • Does not need high pressure and elevated temperature	• Sensitive to particulates in aqueous phase • Risk of membrane wetting; • Poor VRFs	• For partitioning of actinides and rare earth elements recovery • Perspective for fuel reprocessing • Laboratory stage of development, not yet developed at industrial level

Super-high-permeation modules (SHP) produced by UBE Ltd., which have 10 times higher gas permeability than those used at present (high-permeable HP modules), seem to be very promising. This allows reducing the space occupied for membrane installation, eliminating the costs of expensive compressors that can be replaced by blowers [306]. Conventional methods of tritium removal from the atmosphere are limited to catalytic oxidation, followed by adsorption of water on molecular sieves. This method is effective, however, disadvantages result from the size of the apparatus and management of the produced water. This method can be used for small glove boxes, in the case of large compartments the volumes of installations increase; the adsorption system requires large desiccators and high-volume tanks for tritium storage. The advantage of the membrane method is the possibility of a reduction of the processed gas volume to one-tenth of the volume related to conventional methods, as well as simultaneous water removal.

18.5.2.2 Isotope Separation

The history of separation of isotopes by membrane permeation may be derived from the early work of Graham in the mid-19th century on molecular effusion. The most spectacular application of molecular effusion was uranium-235 enrichment for military purposes during the Second World War. Large-scale uranium separation was developed in Oak Ridge employing a molecular effusion process, also called gaseous diffusion. Ni-sponge membranes were applied for the plant K-25 composed of a series of separation columns forming a separation cascade [315]. Despite the competitiveness of the centrifugal method, gaseous diffusion remains an important method for the production of fuel based on natural uranium in countries such as the United States and France, including EURODIF [316–318], supervised by France, Spain, Italy, and Belgium, which have at their disposal the installations in Tricastin and Pierrelatte, as well as in Russia, China, Great Britain, and Argentina [317]. There is no broad information available on membranes used in gaseous diffusion; however it is known that French technology used sintered and anodic alumina, gold-silver alloys, nickel, Teflon, zirconia, and porcelain [319].

Ceramic and metallic porous membranes are still used for the separation of isotopes other than uranium, such as argon and neon [320] or hydrogen [288–293]. Polymeric membranes for the separation of isotopes of hydrogen were applied at the National Institute of Research and Development for Isotopic and Molecular Technologies in Cluj-Napoca, Romania [321,322]. Polymers, such as polyethylene terephthalate (PET), polyethylene (PE), polytetrafluoroethylene (PTFE), cellulose acetate (CA), and polyvinyl chloride (PCV), were used. The selectivity of these membranes was lower than for palladium membranes and microporous *Vycor* glass, however a number of advantages are expected including low costs of polymeric membranes, the possibility of formation in capillaries with a large surface area, and avoiding high pressures and temperatures. The studies on permeation selectivity of seven different isotope pairs, D/H, $^{13}C/^{12}C$, $^{15}N/^{14}N$, $^{18}O/^{16}O$, $^{17}O/^{16}O$, and $^{36}Ar/^{40}Ar$, cover a broad range of molecular weights, through polymer membranes were carried out in France [323]. Two dense polymeric barriers (PDMS and PTFE) were applied in research.

The developments in material science, especially in nanomaterials, open new opportunities for the separation of stable isotopes. Nanoporous metal-organic framework (MOF) membranes have been assessed for the separation of carbon isotopes. They showed high selectivity for $^{12}CH_4/^{13}CH_4$ under ambient conditions (300 K) [324]. Helium isotopes were separated by carbon and graphene membranes [325,326].

Apart from the separation of stable isotopes performed in the gas phase, attempts have been made to carry out this process in the liquid phase. Graphene oxide membranes coated with fluorinated silica nanoparticles were applied for the separation of hydrogen isotopic water in the air gap membrane distillation (AGMD) process [327]. Light and heavy waters were used as model solutions. A superhydrophobic coating prevented liquid water penetration into the membrane. The mean separation factor in the process was 1.151 with a permeation flux of – 0.036 kg m^{-2} h^{-1}.

The enrichment of D and ^{18}O in natural water was carried out using graphene oxide (G-O) and UV-reduced graphene oxide (UV-rG-O) membrane [328]. Pressure-driven dead-end filtration in the liquid phase was applied. The UV-reduced graphene oxide membrane showed better performance than simple graphene oxide membrane (G-O). In a single experiment the enrichment of D was 0.5% for D/H and of ^{18}O was 0.08% for $^{18}O/^{16}O$, without the contribution of the vapor pressure isotope effect.

Polymeric membranes for the separation of hydrogen and oxygen isotopes were studied at INCT, Warsaw [196–199,329,330]. Both hydrophilic barriers, regenerated cellulose and hydrophobic PTFE membranes, were tested. The regenerated cellulose appeared to be a very good system to obtain high separation factors and to consider membrane permeation as a possible and competitive method for the enrichment of deuterium and ^{18}O in natural water. Four different membranes, PVDF, PEI, Psf, and PTFE, were tested by the Korean Atomic Energy Research Institute with a six-stage pressure-driven air-gap membrane distillation system to separate oxygen isotopes [331]. A six-stage system with 3% ^{18}O enrichment appeared to be more efficient from an operational point of view than the arrangements previously tested, which were direct contact and vacuum-driven membrane distillation [332].

The new design of the MD module for isotopic separation of oxygen isotopes was developed on the basis of three-dimensional numerical simulations [333]. An improvement of vapor flux by 15–25% was achieved through the design of triangular baffles which provided uniform temperature distribution inside the module over the membrane surface. The enrichment of ^{18}O isotope by 29 ppm was achieved after processing 20 L of water.

Research on the separation of hydrogen isotopes is focused on aspects related to the safe operation of nuclear reactors and the separation of tritium. Apart from separators based on palladium alloys [334–337], one can find catalytic units with different metallic membranes and various types of integrated systems with catalytic ceramic reactors [338–346].

Using different membranes and various membrane techniques, isotopes of chlorine [347], carbon [348], and uranium in CH_4 [349,350] were separated. Isotopes of gadolinium and neodymium were separated in a hybrid nanofiltration/complexation system [116]. The research on the development of fusion reactors and required tritium breeding has drawn attention to the separation of lithium isotopes [351]. Tritium for fusion reactors is produced from ^6Li by neutron capture reaction. Since the natural concentration of lithium is 7.5% ^6Li and 92.5% ^7Li, the breeding material requires up to 30–90% ^6Li enrichment. The new technique for lithium isotope separation using ionic liquid impregnated organic membranes in electrolysis process was proposed. The ^6Li isotope separation coefficient using this new method was 1.1–1.4 and the process seemed to be competitive with the mercury amalgam method ($\alpha_{6Li/7Li}$=1.06) [352,353].

18.5.2.3 Gaseous Radioactive Wastes—Separation of Noble Gases

Apart from radioactive tritium separation from reactor atmosphere or off-gas, polymeric membranes can be applied for the separation of noble gases produced by nuclear power plants and fuel reprocessing plants as an alternative to the commonly used adsorption or low-temperature distillation methods.

Research into the removal of noble gases by the permeation method with dimethyl silicon membranes was carried out in Oak Ridge National Laboratory [354]. On the basis of experimental work, the calculations for different industrial cascades separating krypton and xenon from the space of molten salt and sodium-cooled breeder reactor or from the off-gas from a plant processing spent reactor fuel were performed.

In the process of separation of noble gases, silicon rubber membranes [355,356] and high-resistance siloxane rubbers (polyvinyltrimethylsilane [PVTMS] and block-copolymer, composed of polyarylate and polydimethylsiloxane of different weigh ratios, so-called sylar) [357,358] were employed.

For this purpose, polymer membranes were formed in tubes [359] placed in modules, operating in a cascade configuration [360,361]. Preliminary storage of treated gases for decay to prolong the lifetime of membranes was practiced. After 20 days' storage of radioactive gases containing ^{85}Kr-0.004 mol and $^{133+135}Xe$ –0.2 mol, at a flow rate of 1 cm^3/s, silicon membranes at the end of the cascade were found to be stable after a number of months. The life time can be prolonged to several years by 30-day storing [362]. During separation of mixtures of noble gases, a decrease in the effective separation factors for particular components was observed. However, the presence of each of them caused a plasticizing effect of the polymer, enhancing the transport through the membrane [363].

In recent years, there has been increasing interest on noble gas separation by membranes [364–366]. Hollow-fiber membranes from polyimide [364,365], flat-sheet membranes from PET, or oriented polypropylene [366] have been applied. The separation of noble gases is a subject of patent applications submitted to European, American, and Japanese patent offices [364,367,368].

Among the noble gases, the separation of krypton and xenon is given the most attention. Separation of these radioisotopes is especially important in fuel reprocessing plants. During spent nuclear fuel treatment, fission products trapped in the fuel are released. The off-gas contains different radioactive species, among which is long-lived ^{85}Kr which must be stored as waste. It is important to separate this gas and purify it to minimize its volume in the storage facility.

The material of the membranes for Kr/Xe separation is not only based on polymers [38,369], but also inorganic materials, including synthetic zeolites [370–375].

The reason so much attention has been devoted to the separation of noble gases stems not only from nuclear uses, but also from a variety of other applications. Noble gases can be used in lighting and photography. The narcotic effect of xenon has been involved in the application in anesthesia. Xenon is also one of the candidate source materials for extreme ultraviolet (EUV) lithography light production; supply of high-purity xenon is required for efficient light production, namely, in laser-produced (LPP) and gas-discharge-produced plasmas (GDPP).

LIST OF SYMBOLS

A_f	specific activity of feed
A_p	specific activity of permeate
A_R	specific activity of retentate
CF	concentration factor
c_p	concentration in permeate
c_f	concentration in feed
DF	decontamination factor
K_{TDS}	TDS reduction coefficient
R,	retention coefficient
T_1	feed (warm stream) inlet temperature
T_2,	cold stream (distillate) inlet temperature
$V_{R,}$	warm stream (retentate) flow rate
V_D	cold stream (distillate) flow rate
V_C	cooling water flow rate
E	energy consumption in MD process
V_P	permeate flow rate
VRC	volume reduction coefficient
x_A, x_B	*A* and *B* components concentration in retentate
y_A, y_B	*A* and *B* components concentration in permeate
$\alpha_{A/B}$	separation factor for *A* and *B* components mixture

LIST OF ABBREVIATIONS

AAm – acrylamide
AECL – Atomic Energy of Canada Limited
ANSTO – Australian Nuclear Science and Technology Organisation
BWR – boiling water reactor
CA – cellulose acetate
CCD – chlorinated cobalt dicarbollide
CECE – combined electrolysis-catalytic exchange
CMPO – n-octyl(phenyl)-N,N-diisobutylcarbamoyl methylphosphine oxide
CNT – carbon nanotubes
DD – drum dryer
DHDECMP – dihexyl N,N-diethylcarbamoy lmethylphoshonate
DIAMEX – diamide An(III)-Ln co-extraction
DMC – direct membrane cleaning
DSC – differential scanning calorimetry
DTPA – diethylenetriaminepentaacetic acid
DTBCH18C6 – 4,4’(5’)di-tert-butyl-dicyclohexano-18-crown-6
D2EHIBA – N,N-di(2-ethylhexyl)isobutyramide
D2EHPA – di(2-ethylhexyl)phosphoric acid
EDTA – ethylendiaminetetraacetic acid
EERC – Energy and Environmental Research Center
EUV – extreme ultraviolet
EV – evaporation
GDPP – gas-discharge-produced plasma
HF – hyperfiltration
HDFS – heptadecafluoro-1,1,2,2-tetrahydrodecyltrichloro silane
IAEA – International Atomic Energy Agency
IX – ion exchange
INCT – Institute of Nuclear Chemistry and Technology
INEEL – Idaho National Engineering and Environmental Laboratory
LANL – Los Alamos Nuclear Laboratory
LPP – laser-produced plasma
MCH – microcrystalline chitosan
MD – membrane distillation
MEUF – micelle-enhanced ultrafiltration
MF – microfiltration
MOF – metal-organic framework
MWCO – molecular weight cut-off
NF – nanofiltration
NPOE – 2-nitrophenyloctyl ether
OECD/NEA – Organisation for Economic Co-operation and Development/Nuclear Energy Agency
PA – polyamide
PAA – polyacrylic acid
PDMS – polydimethylsiloxane
PEG – polyether glycol
PEI- polyethylenimine, polyether imide
PET – polyethylene terephthalate
PHWR – pressurized heavy water rector
PIM – polymer inclusion membranes
PLC – programmable logic controller
PP- polypropylene
PS, Psf – polysulfone
PSA – polysulfonic acid
PTFE – polytetrafluoroethylene
PU – polyurethane
PUREX – plutonium and uranium recovery by extraction
PVA – polyvinyl amide
PVC – polyvinylchloride
PVD – polyvinylpyrolidone
PVDF – poly(vinylidenefluoride)
PVTMS – polyvinyltrimethylsilane
PWR – pressurized water rector
RO – reverse osmosis

RWMP – radioactive waste management plant
SANEX – selective actinide extraction
SLM – supported liquid membrane
SSMs – surface-modifying molecules
SWRO – spiral-wound reverse osmosis
SX – solvent extraction
TBP – tri-n-butyl phosphate
TDS –total dissolved solids
TnOA – tri-n-octylamine
TODGA – N,N,N',N'-tetraoctyldiglycolamide
TOPO – tri-n-octylphosphine oxide
TRO – tubular reverse osmosis
TRU – transuranic elements
TSS – total suspended solids
TUF – tubular ultrafiltration
T2EHDGA – N,N,N',N'-tetra-2-ethylhexyl-3-pentane-diamide
UF – ultrafiltration
WAO – wet air oxidation

REFERENCES

1. Zakrzewska-Trznadel, G. (2013) Advances in membrane technologies for the treatment of liquid radioactive waste. *Desalination* 321: 119.
2. Rana, D., et al. (2013) Radioactive decontamination of water by membrane processes — A review. *Desalination* 321: 77.
3. Pabby, A.K. et al. (2022) Radioactive waste processing using membranes: State of the art technology, challenges and perspectives. *Separation &Purification Reviews* 51(2): 143–173.
4. Zhang, X.; Gu, P.; Liu, Y. (2019) Decontamination of radioactive wastewater: State of the art and challenges forward. *Chemosphere* 215: 543.
5. Zakrzewska-Trznadel, G., Khayet, M. (2012) Membranes in nuclear science and technology: Membrane modification as a tool for performance improvement. In: *Membrane Modification. Technology and Applications*, N. Hilal, M. Khayet, C. J. Wright (Eds.), pp. 2–19. CRC Press Taylor & Francis Group, Boca Raton, FL.
6. Khulbe, K.Ch., Feng, Ch.Y., Matsuura, T. (2011) Separation membrane development. In: *Encyclopedia of Chemical Processing*, pp. 1–19. Taylor & Francis Group, New York.
7. Khayet, M., Mengual, J.I., Matsuura T. (2005) Porous hydrophobic/hydrophilic composite membranes: application in desalination using direct contact membrane distillation. *Journal of Membrane Science* 252: 101.
8. Qtaishat, M., Khayet, M., Matsuura, T. (2009) Novel porous composite hydrophobic/hydrophilic polysulfone membranes for desalination by direct contact membrane distillation. *Journal of Membrane Science* 341: 139.
9. Essalhi, M., Khayet, M. (2012) Surface segregation of fluorinated modifying macromolecule for hydrophobic/hydrophilic membrane preparation and application in air gap and direct contact membrane distillation. *Journal of Membrane Science* 417–418: 163.
10. Khayet, M. (2013) Treatment of radioactive wastewater solutions by direct contact membrane distillation using surface modified membranes. *Desalination* 321: 60.
11. Greathouse, J.A., Kinnibrugh, T.L., Allendorf, M.D. (2009) Adsorption and separation of noble gases by IRMOF-1: Grand canonical Monte Carlo simulations. *Industrial & Engineering Chemistry Research* 48: 3425.
12. Yuying Deng, et al. (2021 Metal-organic framework membranes: Recent development in the synthesis strategies and their application in oil-water separation. *Chemical Engineering Journal* 405: 127004.
13. Jia, Z., Wu, G. (2016) Metal-organic frameworks based mixed matrix membranes for pervaporation. *Microporous and Mesoporous Materials* 235: 151.
14. Singh, R. P., Way, J. D., and McCarley K. C., Development of a Model Surface Flow Membrane by Modification of Porous Vycor Glass with a Fluorosilane, *Industrial & Engineering Chemistry Research* 43, 3033, 2004.
15. Yuna, S., Oyama, S.T. (2011) Correlations in palladium membranes for hydrogen separation: A review. *Journal of Membrane Science* 375: 28.
16. Tong Wu, et al. (2021) Graphene oxide membranes for tunable ion sieving in acidic radioactive waste. *Advanced Science* 8: 2002717.
17. Zhao, H. et al. (2020) Potential application of graphene oxide membranes in high-level liquid waste treatment. *Journal of Cleaner Products* 266: 121884.
18. Bolisetty, S., et al. (2020) Amyloid hybrid membranes for removal of clinical and nuclear radioactive wastewater. *Environmental Science: Water Research & Technology* 6(12): 3249.
19. Jiao, Y., et al. (2013) Modelling carbon membranes for gas and isotope separation. *Physical Chemistry Chemical Physics* 15: 4832.
20. Dolan, M.D. (2010) Non-Pd BCC alloy membranes for industrial hydrogen separation. *Journal of Membrane Science* 362: 12.
21. Zaki, A.A., El-Zakla, T., Abed El Geleel, M.A. (2012) Modeling kinetics and thermodynamics of Cs^+ and Eu^{3+} removal from waste solutions using modified cellulose acetate membranes. *Journal of Membrane Science* 401–402: 1.
22. Miskiewicz, A., Zakrzewska-Koltuniewicz, G. (2018) An application of radiotracer method for investigation of the cake layer formation on the membrane surface in cross-flow flat-sheet membrane module. *Desalination and Water Treatment* 128: 228.
23. Miśkiewicz, A., Dobrowolski, A., Zakrzewska-Trznadel, G. (2013) Using isotopes produced from radionuclide generators as tracers for membrane installations investigation. *European Physical Journal – Web of Conferences* 50: 1–7.
24. Miśkiewicz, A., Zakrzewska-Kołtuniewicz, G., Pasieczna-Patkowska, S. (2019) Photoacoustic spectroscopy as a potential method for studying fouling of flat-sheet ultrafiltration membranes. *Journal of. Membrane Science* 583: 59.
25. Dytnierskij, J.I., Puszkow, A.A., Switcow, A.A. (1973) Purification, concentration of liquid wastes with low level of activity by reverse osmosis. *Atomic Energy* 35(6): 405.
26. Ramachandhran, V., Misra, B.M. (1985) Studies on the radiolytic degradation of cellulose acetate membranes. *Journal of Applied Polymer Science* 30: 35.
27. Ramachandhran, V., Misra, B.M. (1982) Studies on effect of irradiation on semi-permeable membranes. *Journal of Applied Polymer Science* 27: 3427.

28. Ramachandhran, V., Misra, B.M. (1983) Concentration of radioactive liquid streams by membrane processes. *Journal of Applied Polymer Science* 28: 1641.
29. Ramachandhran, V., Misra, B.M. (1986) Studies on the radiation stability of ion exchange membranes. *Journal of Applied Polymer Science* 32: 5743.
30. Chmielewski, A.G., Harasimowicz, M. Influence of gamma and electron irradiation on transport properties of ultrafiltration membranes. *Nukleonika* 37(4): 61.
31. Chmielewski, A.G., Harasimowicz, M., Influence of gamma and electron irradiation on transport properties of nanofiltration and hyperfiltration membranes. *Nukleonika* 42(4): 857.
32. Combernoux, N. et al. (2015) Study of polyamide composite reverse osmosis membrane degradation in water under gamma rays. *Journal of Membrane Science* 480: 64.
33. Combernoux, N. et al. (2016) Effect of gamma irradiation at intermediate doses on the performance of reverse osmosis membranes. *Radiation Physics and Chemistry* 124: 241.
34. Combernoux, N. et al. (2016) Irradiation effects on RO membranes: Comparison of aerobic and anaerobic conditions. *Polymer Degradation and Stability* 134: 126.
35. Xiaojing, L., Jinling, W.., Hou, L., Wang, J. (2020) Performance and deterioration of forward osmosis membrane exposed to various dose of gamma-ray irradiation. *Annals of Nuclear Energy* 135: 106950.
36. Chung, Y., et al. (2021) The impact of gamma-irradiation from radioactive liquid wastewater on polymeric structures of nanofiltration (NF) membranes. *Journal of Hazardous Materials* 403: 123578.
37. de Mello Oliveira, E.E.,Ribeiro Barbosa, C.C., Afonso, J.C. (2013) Stability of a nanofiltration membrane after contact with low-level liquid radioactive waste. *Quimica Nova* 36(9): 1434.
38. Shuwen Yu, et al. (2020) Crosslinked microporous polyarylate membranes with high Kr/Xe separation performance and high stability under irradiation. *Journal of Membrane* Science 611: 118280.
39. Rautenbach, R., Albrecht, R. (1989) *Membrane Processes*. JohnWiley & Sons: Chichester, UK.
40. Mulder, M. (1991) *Basic Principles of Membrane Technology*. Kluwer Academic Publishers: Dordrecht, the Netherlands.
41. Judd, S., Jefferson, B. (Eds.) (2003) *Membranes for Industrial Wastewater Recovery and Re-use*. Elsevier Ltd.: New York.
42. Cohen, K. (1951) *Theory of Isotope Separation as Applied to the Large-scale Production of ^{235}U*. National Nuclear Energy Series III-1B. McGraw-Hill: New York.
43. Benedict, M., Pigford, T.H. (1957) *Nuclear Chemical Engineering*. McGraw-Hill: New York.
44. Hwang, S.-T., Kammermeyer, K. (1975) *Membranes in Separations, Techniques of Chemistry,* Vol. VII. John Wiley & Sons: New York.
45. Fabiani, C. (1986) Strontium separation with ultrafiltration membrane from dilute aqueous solution *Separation Science and Technology* 21(4): 353.
46. Harasimowicz, M. (1995) *Application of ultrafiltration and hyperfiltration for concentration of liquid radioactive wastes*, Ph.D. thesis, University of Mining and Metallurgy, Krakow, Poland.
47. Wu, G., Cui, L., Xu, Y. (2008) A novel submerged rotating membrane bioreactor and reversible membrane fouling control. *Desalination* 228: 255.
48. Ali, S.M. et al. (2021) Dynamic feed spacer for fouling minimization in forward osmosis process. *Desalination* 515; 115198.
49. Wen, X., Li, F., Jiang, B., Zhang, X., Zhao, X. (2018) Effect of surfactants on the treatment of radioactive laundry wastewater by direct contact membrane distillation. *Journal of Chemical Technology & Biotechnology* 93: 2252.
50. Kumar, R., Ismail, A.F. (2015) Fouling control on microfiltration/ultrafiltration membranes: Effects of morphology, hydrophilicity, and charge. *Journal of Applied Polymer Science* 132: 42042.
51. Zhiwei Lv, et al. (2016) Antifouling and high flux sulfonated polyamide thin-film composite membrane for nanofiltration. *Industrial & Engineering Chemistry Research* 55(16): 4726.
52. Rahman, M.M., Al-Sulaimi, S., Farooque, A.M. (2018) Characterization of new and fouled SWRO membranes by ATR/FTIR spectroscopy. *Applied Water Science* 8: 183.
53. Bristow, N.W. et al. (2020) Flow field in fouling spiral wound reverse osmosis membrane modules using MRI velocimetry. *Desalination* 491: 114508.
54. Ruiz-García, A., Melián-Martel, N., Mena, V. (2018) Fouling characterization of RO membranes after 11 years of operation in a brackish water desalination plant. *Desalination* 430: 180.
55. Buckley, L.P., et al. (1992) *Demonstration of Spiral Wound Reverse Osmosis for Liquid Waste Processing*. Rep. AECL Research Pub. RC-796, COG-92-52.
56. Sen Gupta, S.K., Slade, J.A., Tulk, W.S. (1995) *Liquid radwaste processing with crossflow microfiltration and spiral wound reverse osmosis*, 16 pp. AECL-11270, Chalk River, Ontario, Canada.
57. Sen Gupta, S.K. et al. (1996) Liquid radwaste processing with spiral wound reverse osmosis. In: *WM '96 Conference*. Tuscon, AZ, Feb. 25–29.
58. Sen Gupta, S.K., Rimpelainen, S. (1997) *Liquid Radwaste Processing with Spiral-Wound Reverse Osmosis, Ultrapure Water®, Knowledgebase* UP140132: 1.
59. Rimpelainen, S., Tremblay, A.Y. (1999) *Effect of Solute Precipitation on RO Elements During Mixed Wastewater Treatment, Ultrapure Water®, Knowledgebase*, UP160429: 4.
60. Tan, L., Tapsell, G. (2004) Operation of a membrane pilot plant for ANSTO effluent In: *WM'04 Conference*. Tuscon, AZ, February 29–March 4.
61. *Federal Technology Alerts, Non-Chemical Technologies for Scale and Hardness Control. Produced for U.S.* DOE by Battelle Columbus Operations, January 1998, http://www.space-age.com/magwater/fta/fta_nonchem.pdf, accessed on August 11, 2022.
62. Rimpelainen, S., Tremblay, A.Y. (2000) *Verification of Species Responsible for the Fouling of RO Membranes during Mixed Wastewater Treatment, Ultrapure Water®, Knowledgebase*, UP170136: 1.
63. Zakrzewska-Trznadel, G. (2003) Radioactive solutions treatment by hybrid complexation –UF/NF process. *Journal of Membrane Science* 225: 25.
64. International Atomic Energy Agency (1994) *Advances in Technologies for Treatment of Low and Intermediate Level*

Radioactive Liquid Wastes. Technical Report Series No. 370, IAEA, Vienna, Austria.
65. Bourns, W.T., Le, V.T. (1984) *The Reverse Osmosis Plant in CRNL Waste Treatment Centre-Description Design and Operation Principles*. Rep. CRNL-2352, Atomic Energy of Canada Ltd., Chalk River. IAEA Technical Reports Series No. 370.
66. Harries, J., Dimitrovski, L., Hart, K., et al. (2001) Radioactive waste management at ANSTO – managing current and historic wastes. In: *Book of Extended Synopses*: 5-6. IAEA, Vienna. International Conference on Management of Radioactive Waste from Non-power Applications – Sharing the Experience, St. Paul's Bay (Malta) 5–9 Nov. 2001.
67. International Atomic Energy Agency (2004) *Application of Membrane Technologies for Liquid Radioactive Waste Processing*. Technical Report Series No. 431. IAEA: Vienna.
68. Freeman, J. (2001) Wolf Creek's liquid waste processing system improvements. In: *Proceedings of the EPRI International Low-Level Waste Conference 2000*, San Antonio, TX., Electric Power Research Institute, Palo Alto, CA.
69. Evans, D.W., et al. (1995) Treatment of steam generator chemical cleaning wastes: Development and operation of the Bruce spent solvent treatment facility. In: *International Oxidation Application: Proceedings of the Fifth International Symposium*. Nashville, TN.
70. McCabe, et al. (1992) Biofouling of microfilters at the Savannah River F/H area effluent treatment facility. In: *WM'92 Conference*. WM Symposia, Inc.: Tuscon, AZ.
71. Chmielewski, A.G., Harasimowicz, M., Zakrzewska-Trznadel, G. (1999) Membrane technologies for liquid radioactive waste treatment. *Czechoslovak Journal of Physics* 49: 979.
72. Chmielewski, A. G. et al. (2001) Concentration of low- and medium-level radioactive wastes with 3-stage reverse osmosis pilot plant, *Separation Science and Technology* 36(5 6): 1117.
73. Junfeng, L., et al. (2014) Advances in HTGR Wastewater Treatment System Design. *Proceedings of the HTR 2014*, Weihai, China, October 27–31, 2014. Paper HTR2014-3-1399.
74. Dulama, M., Deneanu, N., Popescu, I.V. (2001) Liquid radwaste treatment by microfiltration, ultrafiltration and reverse osmosis. In: *Book of Extended Synopses, International Conference on Management of Radioactive Waste from Non-Power Applications – Sharing the Experience,* pp. 168–169St. Paul's Bay (Malta) 5–9 November, IAEA, Vienna.
75. Roman, G., et al. (2000) Removal of some ions from the radioactive liquid wastes by means of membrane techniques. *National Physics Conference*, Constanta, Romania, September 21–23, 2000. Accessed in: Documentation and Publishing Office, Horia Hulubei National Institute for Physics and Nuclear Engineering, Magurele, Romania.
76. Arnal, J.M., et al. (2000) Declassification of radioactive waste solutions of iodine (I^{125}) from radioimmune analysis (RIA) using membrane techniques. *Desalination* 129: 101.
77. Arnal, J.M., et al. (1998) Concentration of radioactive waste solutions of iodine (I^{125}) from immune analysis (RIA) using membrane techniques. *Desalination* 119: 185.
78. Arnal, J.M., et al. (2003) Treatment of ^{137}Cs liquid wastes by reverse osmosis. Part I. Preliminary testes. *Desalination* 154: 27.
79. Kryvoruchko, A.P., Kornilovich, B.Yu (2003) Water deactivation by reverse osmosis. *Desalination* 157: 403.
80. Raff, O., Wilken, R.-D. (1999) Removal of dissolved uranium by nanofiltration. *Desalination* 122; 147.
81. Huikuri, P., Salonen, L., Raff, O. (1998) Removal of natural radionuclides from drinking water by point of entry reverse osmosis. *Desalination* 119; 235.
82. Sasaki T., et al. (2013) Cesium (Cs) and strontium (Sr) removal as model materials in radioactive water by advanced reverse osmosis membrane. *Desalination and Water Treatment* 51: 1672.
83. Ali, Z. et al. (2019) Defect-free highly selective polyamide thin-film composite membranes for desalination and boron removal. *Journal of Membrane Science* 578: 85.
84. Wolska, J., Bryjak, M. (2013) Methods for boron removal from aqueous solutions — A review. *Desalination* 310: 18.
85. Ding Chen, X.Z., Li, F., Zhang, X. (2016) Rejection of nuclides and silicon from boron-containing radioactive waste water using reverse osmosis. *Separation and Purification Technology* 163: 92.
86. Chen, D., Li, F., Zhao, X., Sun, Y. (2018) The influence of salts on the reverse osmosis performance treating simulated boron-containing low level radioactive wastewater. *Journal of Chemical Technology & Biotechnology* 93; 3607.
87. Sylvester, P., Milner, T., Jensen, J. (2013) Radioactive liquid waste treatment at Fukushima Daiichi. *Journal of Chemical Technology & Biotechnology* 88(9): 1592.
88. Combernoux, N., et al. (2017) Treatment of radioactive liquid effluents by reverse osmosis membranes: From lab-scale to pilot-scale. *Water Research* 123: 311.
89. Lehto, J. et al. (2019) Removal of radionuclides from Fukushima Daiichi waste effluents. *Separation & Purification Reviews* 48(2): 1.
90. TEPCO (2020) *TEPCO Draft Study Responding to the Subcommittee Report on Handling ALPS Treated Water*, March 24, 2020 Tokyo Electric Power Company Holdings, Inc. TEPCO Draft Study Responding to the Subcommittee Report on Handling ALPS Treated Water, accessed on 3.4.2022.
91. TEPCO (2012) *Multi-nuclide Removal Equipment*. Feb. 27, 2012 Tokyo Electric Power Company; hd03-02-03-001-m120328_01-e.pdf (tepco.co.jp) accessed on 3.4.2022.
92. Ministry of Economy, Trade and Industry (2021) *Basic policy on handling of the ALPS treated water*, 13 April, 2021 Ministry of Economy, Trade and Industry. 202104_bp_breifing.pdf (meti.go.jp) accessed on 3.4.2022.
93. IAEA (2022) www.iaea.org/topics/response/fukushima-daiichi-nuclear-accident/fukushima-daiichi-treated-water-discharge, accessed on 3.4.2022.
94. Abramowska, A. et al. (2018) Purification of flowback fluids after hydraulic fracturing of Polish gas shales by hybrid methods. *Separation Science and Technology* 53(8): 1207.
95. Chung, T.S., et al. (2012) Forward osmosis processes: Yesterday, today and tomorrow. *Desalination* 287: 78.
96. Wang, J., Liu, X. (2021) Forward osmosis technology for water treatment: Recent advances and future perspectives. *Journal of Cleaner Production* 280(Part 1): 124354.
97. Liu, X., Wu, J., Wang, J. (2018) Removal of Cs(I) from simulated radioactive wastewater by three forward osmosis membranes. *Chemical Engineering Journal* 344: 353.

98. Liu, X., Wu, J., Hou, L.-a., Wang, J. (2019) Removal of Co, Sr and Cs ions from simulated radioactive wastewater by forward osmosis. *Chemosphere* 232: 87.
99. Liu, X., Wu, J., Liu, C., Wang, J. (2017) Removal of cobalt ions from aqueous solution by forward osmosis. *Separation and Purification Technology* 177: 8.
100. Liu, X., Wu, J., Hou, L.-a., Wang, J. (2020) Fouling and cleaning protocols for forward osmosis membrane used for radioactive wastewater treatment. *Nuclear Engineering and Technology* 52(3): 581.
101. Yadav, S., et al. (2020) Organic fouling in forward osmosis: A comprehensive review. *Water* 12: 1505.
102. Seong, H.D., et al. (2013) *Boron removal in radioactive liquid waste by forward osmosis membrane* In: ICEM2013-ASME2013: 15. International Conference on Environmental Remediation and Radioactive Waste Management, Brussels (Belgium), September 8–12, 2013.
103. Liu, X., Wu, J., Wang, J. (2019) Removal of nuclides and boric acid from simulated radioactive wastewater by forward osmosis. *Progress in Nuclear Energy* 114: 155.
104. Lee, S., Kim, Y., Park, J., Shon, H.K., Hong, S. (2018) Treatment of medical radioactive liquid waste using Forward Osmosis (FO) membrane process. *Journal of Membrane Science* 556: 238.
105. Nigatu Bitaw, T., Park, K., Yang, D.R. (2016) Optimization on a new hybrid forward osmosis-electrodialysis-reverse osmosis seawater desalination process. *Desalination* 398: 265.
106. Zhang, X., Liu, Y. (2021) Integrated forward osmosis-adsorption process for strontium-containing water treatment: Pre-concentration and solidification. *Journal of Hazardous Materials* 414: 125518.
107. Young Park, M., Kim, E.S. (2018) Analysis of tritium behaviors on VHTR and forward osmosis integration. *Nuclear Engineering and Design* 338: 43.
108. IAEA (1996) *Processing of Nuclear Power Plant Waste Streams Containing Boric Acid*, IAEA-TECDOC-911. IAEA: Vienna, Austria.
109. Astruc, C., et al. (1999) *Filtration in Nuclear Power Plants, Euromembrane'99*, Leuven, Belgium, September 19–22.
110. Macnaughton, S.J., et al. (2002) Application of Nanofiltration to the treatment of uranium mill effluents. In: *Technologies for the treatment of effluents from uranium mines, mills and tailings*, p. 55. IAEA-TECDOC-1296. IAEA: Vienna, Austria.
111. Hwang, E.-D., et al. (2002) Effect of precipitation and complexation on nanofiltration of strontium-containing nuclear wastewater. *Desalination* 147: 289.
112. Gaubert, E., et al. (1997) Selective strontium removal from a sodium nitrate aqueous medium by nanofiltration-complexation. *Separation Science and Technology* 32(1–4): 585.
113. Chitry, F., et al. (1999) Separation of gadolinum (III) and lanthanum (III) by nanofiltration-complexation in aqueous medium. *Journal of Radioanalytical and Nuclear Chemistry* 249(2): 931.
114. Chitry, F., et al. (2000) Lanthanides(III)/actinides(III) separation by nano-filtration-complexation in aqueous medium. :n: *Proceedings of the International Conference, Nuclear Waste: From Research to Industrial Maturity*. Montpellier, France, October 2–4, 2000.
115. Chitry, F., et al. (2001) Separation of lanthanides(III) by nanofiltration-complexation in aqueous medium. *Separation Science and Technology* 36 (40); 605.
116. Chitry, F., et al. (2001) Nanofiltration-complexation: A new method for isotopic separation of heavy metals. *Chemistry Letters* 30(8): 770.
117. Chitry, F., et al. (2001) Cesium/sodium separation by nanofiltration-complexation in aqueous medium. *Separation Science and Technology* 36 (5–6): 1053.
118. Chen, D., Zhao, X., Li, F. (2014) Treatment of low level radioactive wastewater by means of NF process. *Nuclear Engineering and Design* 278: 249.
119. Chung, Y., et al. (2021) The impact of gamma-irradiation from radioactive liquid wastewater on polymeric structures of nanofiltration (NF) membranes. *Journal of Hazardous Materils* 403: 123578.
120. Kim, H.-J., Kim, S.-J., Hyeon, S., Kang, H.H., Lee, K.-Y. (2020) Application of desalination membranes to nuclide (Cs, Sr, and Co) separation, *ACS Omega* 5: 20261.
121. Viszlay, J., Toth, S. (1992) Operational experience with pilot ultrafiltration plant. In: *First International Seminar PWR Water Chemistry* Balatonfüred: Hungary.
122. Commission of the European Communities (1982) Research and development on radioactive waste management and storage. In*: Third Annual Progress Report 1982 of the European Community Programme 1980–1984*. Harwood Academic Publishers: Harwood.
123. Barnier, R., et al. (1989) Ultrafiltration treatment of laundry liquid wastes from a nuclear research centre. In: *Proceedings of the International Conference on Waste Management*. Kyoto, Japan.
124. Kichik, V.A., Maslova, M.N., Svitzov A.A., et al. (1987) Method for complex treatment of laundry liquid radioactive wastes by ultrafiltration. *Atomic Energy* 63:(2) 130 (in Russian).
125. Karlin, Y. et al. (2001) Advantageous technology treatment of laundry waters. In: *Proceedings of the International Symposium on Technologies for the Management of Radioactive Waste for Nuclear Power Plants and Back End of Nuclear Fuel Cycle Activities*. Daejon, South Korea, August 30–September 3, 1999. IAEA, Vienna, Austria 2001 (C&S Papers CD Series no. 6).
126. Cumming, I.W., et al. (1992) *Development of Combined Precipitation and Ultrafiltration Process and its Application to the Treatment of Low Active Wastes*, p. 26. IAEA: 26 IAEA-SM-303-21P, Vienna, Austria.
127. Hooper, E.W., Sellers, R.M. (1992) Activity removal from liquid streams by seeded ultrafiltration. In: *Proceedings of the Symposium on Waste Management '91,* Vol. 1, p. 749. University of Arizona: Tucson, AZ.
128. Hooper, E.W. (1992) Activity removal from aqueous waste streams by seeded ultrafiltration. In: Use of inorganic sorbents for treatment of liquid radioactive waste and backfill of underground repositories. In: *Proceedings of final Research Co-ordination Meeting*, p. 408. IAEA-TECDOC-675, Řež, Czechoslovakia, November 1991, International Atomic Energy Agency, Vienna, Austria.
129. Hooper, E.W. (1997) Inorganic sorbents for removal of radioactivity from aqueous waste streams. In: *Waste treatment and immobilization technologies involving inorganic sorbents*, IAEA-TECDOC-947, Final report of a

co-ordinated research programme 1992–1996, International Atomic Energy Agency, Vienna, Austria.

130. Smyth, B.M., Todd, P., Bowman, N. (1999) Hyperbranched chelating polymers for the polymer-assisted ultrafiltration of boric acid. *Separation Science and Technology* 34(10): 1925.
131. Vanderberg, L.A., et al. (1999) Treatment of heterogeneous mixed wastes: Enzyme degradation of cellulosic materials contaminated with hazardous organics and toxic and radioactive metals. *Environmental Science & Technology* 33(8): 1256.
132. Smyth, B.F., et al. (1998) Evaluation of synthetic water-soluble metal-binding polymers with ultrafiltration for selective concentration of americium and plutonium. *Journal of Radioanalytical and Nuclear Chemistry* 234(1–2): 219.
133. Smyth, B.F., et al. (1998) Preconcentration of low levels of americium and plutonium from waste waters by synthetic water-soluble metal-binding polymers with ultrafiltration. *Journal of Radioanalytical and Nuclear Chemistry* 234(1–2): 225.
134. Ramachandhran, V., Misra, B.M. (1998) Radiocerium separation behavior of ultrafiltration membranes. *Journal of Radioanalytical and Nuclear Chemistry* 237(1–2): 121.
135. Kryvoruchko, A.P., et al. (2004) Ultrafiltration removal of U(VI) from contaminated water. *Desalination* 162: 229.
136. Kryvoruchko, A.P., Kornilovich, B.Yu. (1999) Baromembrane decontamination of uranium-containing waste water. In: *Proceedings of the International Conference Euromembrane 99,* Vol. 2, p. 468. Leuven, Belgium.
137. Zakrzewska-Trznadel, G., Harasimowicz, M. (2002) Removal of radionuclides by membrane permeation combined with complexation. *Desalination* 144: 207.
138. Zakrzewska-Trznadel, G., Harasimowicz, M. (2004) Application of ceramic membranes for hazardous wastes processing: pilot plant experiments with radioactive solutions. *Desalination* 162: 191.
139. International Atomic Energy Agency (1992) *Use of Inorganic Sorbents for the Treatment of Liquid Radioactive Waste and Backfill of Underground Repositories*, IAEA-TECDOC-675. Vienna, Austria.
140. Chmielewski, A., Harasimowicz, M. (1995) Application of ultrafiltration and complexation to the treatment of low-level radioactive effluents. *Separation Science and Technology* 30(7–9): 1779.
141. Kallonen, I. (1996) Purification of radioactive liquid wastes at Padliski in Estonia. *Kemia-Kemi* 23(7): 550.
142. Mynin, V.N., Terpugov, G.V. (1998) Purification of waste water from heavy metals by using ceramic membranes and natural polyelectrolytes. *Desalination* 119: 361.
143. Coughlin, R.W., Deshaies, M.R., Davis E.M. (1990) Chitosan in crab shell wastes purifies electroplating wastewater. *Environmental Progress & Sustainable Energy* 9: 35.
144. Muzzarelli, R.A.A. (1983) Chitin and its derivatives: new trends of applied research. *Carbohydrate Polymers* 3: 53.
145. Guibal, E., et al. (1994) Uranium and vanadium sorption by chitosan and derivatives, *Water Science and Technology* 30(9): 183.
146. Onsoyen, E., Skaugrud, O. (1990) Metal recovery using chitosan. *Journal of Chemical Technology & Biotechnology* 49: 395.
147. Zarzycki, R., et al. (2003) The effect of chitosan form on copper adsorption. In: *Environmental Engineering Studies*, Pawłowski et al. (Eds.), p. 199. Kluwer Academic/Plenum Publishers: New York.
148. Juang, R.-S., Shiau, R.-C. (2000) Metal removal from aqueous solutions using chitosan-enhanced membrane filtration. *Journal of Membrane Science* 165: 159.
149. Bayhan, Y.K., et al. (2001) Removal of divalent heavy metal mixtures from water by *Saccharomyces cerevisiae* using crossflow microfiltration. *Water Research* 35: 2191.
150. Kamalika R., Sinha P., Susanta L. (2008) Immobilization of long-lived radionuclides $^{152,154}Eu$ by selective bioaccumulation in *Saccharomyces cerevisiae* from a synthetic mixture of $^{152,154}Eu$, ^{137}Cs and ^{60}Co. *Biochemical Engineering Journal* 40(2): 363.
151. Maureira, A., Rivas, B.L. (2009) Metal ions recovery with alginic acid coupled to ultrafiltration membrane. *European Polymer Journal* 45: 573.
152. Wicaksana, F., et al. (2012) Microfiltration of algae (*Chlorella sorokiniana*): Critical flux, fouling and transmission. *Journal of Membrane Science* 83: 387.
153. Miskiewicz A., Zakrzewska-Kołtuniewicz G. (2021) Application of biosorbents in hybrid ultrafiltration/sorption processes to remove radionuclides from low-level radioactive waste. *Desalination and Water Treatment* 242: 47.
154. Katsou, E., Malamis, S., Haralambous, K.J. (2011) Industrial wastewater pre-treatment for heavy metal reduction by employing a sorbent-assisted ultrafiltration system. *Chemosphere* 82: 557.
155. Bailey, S.E., et al. (1999) A review of potentially low-cost sorbents for heavy metals. *Water Research* 33(2): 2469.
156. Dulama, M., et al. (2010) Application of indigenous inorganic sorbents in combination with membrane technology for treatment of radioactive liquid waste from decontamination processes. *Radiochimica Acta* 98: 413.
157. Hebrant, M., et al. (2001) Micellar extraction of europium (III) by a bolaform extractant and parent compounds derived from 5-pyrazolone. *Colloids and Surfaces A* 143: 77.
158. Xiarchos, I., et al. (2003) Polymeric ultrafiltration membranes and surfactants. *Separation & Purification Reviews* 3(2): 215.
159. Landaburu-Aguirre, J. et al. (2009) The removal of zinc from synthetic wastewaters by micellar-enhanced ultrafiltration: statistical design of experiments. *Desalination* 240: 262.
160. Xiarchos, I., Jaworska, A., Zakrzewska-Trznadel, G. (2008) Response surface methodology for the modelling of copper removal from aqueous solutions using micellar-enhanced ultrafiltration. *Journal of Membrane Science* 321; 222.
161. Niescior-Browinska, P., Zakrzewska-Koltuniewicz, G., Chajduk, E. (2014) The recovery of boric acid from PWR reactor cooling water and wastewater by using micellar-enhanced ultrafiltration. In: *Monographs of the Environmental Engineering Committee Polish Academy of Sciences*, Vol. 119, p. 17. Membranes and Membrane Processes in Environmental Protection. Warsaw-Gliwice.
162. Garcia, R., Pinel, C., Lemaire, M. (1998) Ionic imprinting effect in gadolinum/lanthanum separation. *Tetrahedron Letters* 39: 8651.
163. Vigneau, O., Pinel, C., Lemaire, M. (2002) Ionic imprinting resins based on EDTA and DTPA derivatives for lanthanides(III) separation. *Analytica Chimica Acta* 435: 75.
164. Garcia R. et al. (2002) Solid-liquid lanthanide extraction with ionic-imprinted polymers. *Separation Science and Technology* 37(12): 2839.

165. Koltuniewicz, A.B. (2017) Integrated membrane operations in various industrial sectors. In: *Comprehensive Membrane Science and Engineering*, E. Drioli, L. Giorno (Eds.), pp. 109–164. Elsevier: Oxford, UK.
166. Rao, S.V.S., et al. (2000) Effective removal of cesium and strontium from radioactive wastes using chemical treatment followed by ultrafiltration. *Journal of Radioanalytical and Nuclear Chemistry* 246(2): 413.
167. Cojocaru, C., Zakrzewska-Trznadel, G., Jaworska, A. (2009) Removal of cobalt ions from aqueous solutions by polymer assisted ultrafiltration using experimental design approach, Part 1: Optimization of complexation conditions. *Journal of Hazardous Materials* 169: 599.
168. Cojocaru C., Zakrzewska-Trznadel G., Miskiewicz A. (2009) Removal of cobalt ions from aqueous solutions by polymer assisted ultrafiltration using experimental design approach, Part 2: Optimization of hydrodynamic conditions for a cross-flow ultrafiltration module with rotating part. *Journal of Hazardous Materials* 169: 610.
169. Uzal, N., et al. (2011) Optimization of Co^{2+} ions removal from water solutions by application of soluble PVA and sulfonated PVA polymers as complexing agents. *Journal of Colloid and Interface Science* 362: 615.
170. Foust, H., Ghosehajra, M. (2010) Sizing an ultrafiltration process that will treat radioactive waste. *Separation Science and Technology* 45: 1025.
171. Bonnema, B.E., Navratil, J.D., Bloom, R.R. (1995) SOIL*EX™ process design basic for mixed waste treatment. In: *Proceedings Conference on WM'95.* WM Symposia, Inc.: Tuscon, AZ.
172. Suzuki, K., et al. (1997) A study of removal of hazardous metals and radionuclides in ground water. In: *Proceedings of the conference on WM'97.* WM Symposia, Inc.: Tuscon, AZ.
173. Buckley, L.P., Vijayan, S., Wong, C.F. (1993) Remediation process technology for ground water. In: *Proceedings of the International Conference on Nuclear Waste Managfement and Environmental Remediation,* p. 33. Prague, Czech Republic: American Society of Mechanical Engineers, New York.
174. Mann, N.R., Todd, T.A. (2000) Crossflow filtration testing of INEEL radioactive and non-radioactive waste slurries. *Chemical Engineering Journal* 80: 237.
175. Brown, R.G., et al. (1991) A comparative evaluation of cross-flow microfiltration membranes for radwaste dewatering. In: *Effective Industrial Membrane Processes: Benefits and Opportunities*, M.K. Turner (Ed.). Elsevier Applied Science: New York.
176. Stepan, D.J., Moe, T.A., Collings, M.E. (1996) *Task 9 – Centrifugal Membrane Filtration.* Semi-Annual Report, DOE/MC/31388-5500, April 1–September 30, 1996, Energy and Environmental Research Center, University of North Dacota, Grand Forks, ND.
177. A. Greene, W.A. et al. *Centrifugal Membrane Filtration, Final Report Contract No. DE-AC21-96MC3313*. In: www.osti.gov/servlets/purl/859218 accessed on August 10, 2022.
178. Brunson, R.R. et al. (1999) Waste treatment at the radiochemical engineering development center, *Separation Science and Technology* 34(6–7): 1195.
179. Han, F., Zhang, G.-H., Gu, P. (2012) Removal of cesium from simulated liquid waste with countercurrent two-stage adsorption followed by microfiltration. *Journal of Hazardous Materials* 225–226: 107.
180. Pabby, A.K. (2008) Membrane techniques for treatment in nuclear waste processing: Global experience. *Membrane Technology* 11: 9.
181. Drioli E., Criscuoli A., Curcio E. (2005) *Membrane Contactors: Fundamentals, Applications and Potentialities.* Elsevier.
182. Pabby, A.K., Sastre, A.M. (2008) Hollow fiber membrane-based separation technologies. Performance and design perspectives. In: *Solvent Extraction and Liquid membranes. Fundamentals and Applications in New Materials.* M. Aguilar, J.L. Cortina (Eds.), p. 91. CRC Press/Taylor & Francis Group; Boca Raton, FL.
183. Klaassen, R., Jansen, A.E. (2004) The membrane contactor: Environmental applications and possibilities. *Environmental Progress* 20(1); 37.
184. Gonzalez-Munoz, M.J., et al. (2004) Simulation of integrated extraction and stripping processes using membrane contactors. *Desalination* 163: 1.
185. Witek, A., Szafran R., Kołtuniewicz, A.B., (2006) p-Cresol removal using a membrane contactor enhanced by the micellar solubilization. *Desalination* 200: 575.
186. Urtiaga, A., et al. (2005) Membrane contactors for recovery of metallic compounds. Modelling of copper recovery from WPO process. *Journal of Membrane Science* 257: 161.
187. Juang, R.-S., Huang, H.-L. (2003) Mechanistic analysis of solvent extraction of heavy metals in membrane contactors. *Journal of Membrane Science* 213: 125.
188. Kumar, A., et al. (2005) Comparative performance of non-dispersive solvent extraction using a single module and the integrated membrane process with two hollow fiber contactors. *Journal of Membrane Science* 248: 1.
189. Drioli, E., Giorno, L.N (Eds.) (2010) *Comprehensive Membrane Science and Engineering, Membrane Contactors and Integrated Membrane Operations*, Vol. 4, Elsevier: Kidlington, UK.
190. Gabelman, A., Hwang, S.-T. (1999) Hollow fiber membrane contactors. *Journal of Membrane Science* 159: 61.
191. Koltuniewicz, A.B., Drioli, E. (2008) *Membranes in Clean Technologies, Theory and Practice,* Vol. 1. Wiley-VCH Verlag GmbH&Co: Weinheim, Germany.
192. Zdorovets, M.V., et al. (2020) Liquid low-level radioactive wastes treatment by using hydrophobized track-etched membranes. *Progress in Nuclear Energy* 118: 103128.
193. Korolkov, I.V., et al. (2019) Modification of PET ion track membranes for membrane distillation of low-level liquid radioactive wastes and salt solutions. Separation and Purification Technology 227: 115694.
194. Wen, X., Li, F., Zhao, X. (2016) Removal of nuclides and boron from highly saline radioactive wastewater by direct contact membrane distillation. *Desalination* 394: 101.
195. Chmielewski, A.G., et al. (2000) *The method of purification of radioactive wastes.* PL 179430, Polish Patent Office: Warsaw, Poland.
196. Chmielewski, A.G., et al. (1991) $^{16}O/^{18}O$ and H/D separation factors for liquid/vapour permeation of water through an hydrophobic membrane. *Journal of Membrane Science* 60: 319.
197. Chmielewski, A.G., et al. (1993) Cascades for natural water enrichment in deuterium and oxygen-18 using membrane mermeation. *Separation Science and Technology* 28(1–3); 909.

198. Chmielewski, A.G., et al. (1995) Membrane distillation employed for separation of water isotopic compounds. *Separation Science and Technology* 30(7–9): 1653.
199. Zakrzewska-Trznadel, G., Chmielewski, A.G., Miljević, N. (1996) Separation of protium/deuterium and oxygen-16/oxygen18 by membrane distillation process. *Journal of Membrane Science* 113: 337.
200. Chmielewski, A.G., et al. (2002) *Stable isotopes-some new fields of application.* R. Zarzycki, A.G. Chmielewski, G. Zakrzewska-Trznadel (Eds.) Lodz: Polish Academy of Sciences, Commission for Environmental Protection, Lodz, Poland.
201. Zakrzewska–Kołtuniewicz, G. (2017) Water management in nuclear power plant using advanced low-temperature systems. *European Water* 58: 345.
202. Zakrzewska-Trznadel, G., Harasimowicz, M., Chmielewski, A.G. (1999) Concentration of radioactive components in liquid low-level radioactive waste by membrane distillation. *Journal of Membrane Science* 163: 257.
203. Zakrzewska-Trznadel, G., Harasimowicz, M., Chmielewski, A.G. (2001) Membrane processes in nuclear technology – application for liquid radioactive waste treatment. *Separation and Purification Technology* 22–23: 617.
204. Zakrzewska-Trznadel, G. (1998) Membrane distillation for radioactive waste treatment. *Membrane Technology, International Newsletter* 103: 9.
205. Jia, F., et al. (2017) Removal of strontium ions from simulated radioactive wastewater by vacuum membrane distillation. *Annals of Nuclear Energy* 103: 363.
206. Nie, X. (2021) Decontamination of uranium contained low-level radioactive wastewater from UO_2 fuel element industry with vacuum membrane distillation. *Desalination* 516: 115226.
207. Gryta, M., Concentration of NaCl solutions by membrane distillation integrated with crystallization, *Separation Science and Technology* 37 (15): 3635, 2002.
208. Schneider, K., Gassel, T.S. (1984) Membrane distillation. *Chemie Ingenieur Technik* 56(7): 514.
209. Sarti, G.C., Gostoli, C., Matulli, S. (1985) Low energy cost desalination process using hydrophobic membranes. *Desalination* 56: 277.
210. Hanbury, W.T., Hodgkies, T. (1985) Membrane distillation – an assessment. *Desalination* 56: 287.
211. Schnaider, K., et al. (1989) Membranes and modules for transmembrane distillation. *Journal of Membrane Science* 39: 25.
212. Chen, X., et al. (2019) Ceramic nanofiltration and membrane distillation hybrid membrane processes for the purification and recycling of boric acid from simulative radioactive waste water. *Journal of Membrane Science* 579: 294.
213. Khayet, M., Mengual, J.I., Zakrzewska-Trznadel, G. (2005) Direct contact membrane distillation for nuclear desalination: Part I – Review of membranes used in membrane distillation and methods for their characterization. *Journal of Nuclear Desalination* 1(4): 435.
214. Khayet, M., Mengual, J.I, Zakrzewska-Trznadel, G. (2006) Direct contact membrane distillation for nuclear desalination: Part II – Experiments with radioactive solutions. *Journal of Nuclear Desalination* 2(1): 56.
215. Kathios, D.J. et al. (1994) A preliminary evaluation of microporous hollow fiber membrane modules for the liquid-liquid extraction of actinides. *Journal of Membrane Science* 97: 251.
216. Geist, A. et al. (2003) Application of novel extractants for actinide(III)/lanthanide(III) separation in hollow-fibre modules. *Membrane Technology* 5: 5.
217. Geist, A., Weigl, M., Gompper, K. (2005) Small-scale actinide(III) partitioning process in miniature hollow fiber modules. *Radiochimica Acta* 93: 197.
218. Geist, A., Weigl, M., Gompper, K. (2003) DIAMEX Process Development Studies: Actinide(III)-Lanthanide Co-extraction in a Hollow Fiber Module. In: *Proceedings of the International Workshop on P&T and ADS Development*, SCK-CEN, Mol, Belgium, October 6–8.
219. Gupta, S.K. et al. (2002) Hollow fiber membrane contactor: Novel extraction device for plutonium extraction. *BARC Newsletter*, Founder's Day Special Issue, 181.
220. Zakrzewska, G., et al. (2013) Recovery of uranium(VI) from water solutions by membrane extraction. *Advanced Materials Research* 704: 66.
221. Biełuszka, P. et al. (2014) Liquid-liquid extraction of uranium(VI) in the system with a membrane contactor. *Journal of Radioanalytical and Nuclear Chemistry* 299(1): 611.
222. Biełuszka, P., Zakrzewska-Kołtuniewicz, G., Chajduk, E. (2014) Solvent extraction of uranium in the system with membrane contactors. In: *Monographs of the Environmental Engineering Committee Polish Academy of Sciences*, Vol. 118 Membranes and Membrane Processes in Environmental Protection, Warsaw-Gliwice, s. 117–126.
223. Kiegiel, K., et al. (2014) Dictyonema black shale and Triassic sandstones as potential sources of uranium. *Nukleonika* 60(3); 515.
224. Kiegiel, K., et al. (2017) Solvent extraction of uranium from leach solutions obtained in processing of Polish low-grade ores. *Journal of Radioanalytical and Nuclear Chemistry* 311(1): 589.
225. Zakrzewska-Kołtuniewicz, G., Wołkowicz, S., Kiegiel, K. (2020) Uranium from Domestic Resources in Poland. In: *Proceedings of an International Symposium on Uranium Raw Material for the Nuclear Fuel Cycle: Exploration, Mining, Production, Supply and Demand, Economics and Environmental Issues (URAM-2018)*, pp. 395–398. IAEA: Vienna, Austria, June 25–29, 2018.
226. Zakrzewska, K.G., et al. (2014) Recovery of valuable metals from the waste deriving from uranium production and processing of secondary materials. In: *Proceedings of 2014 – Sustainable Industrial Processing Summit/ Shechtman International Symposium,* F. Kongoli (Ed.), Vol. 2, p. 267–277. Flogen.
227. Katarzyna Kiegiel, K. et al. (2018). Uranium in Poland: Resources and Recovery from Low-Grade Ores. In: *Uranium – Safety, Resources, Separation and Thermodynamic Calculation,* N. S. Awwad (Ed.). IntechOpen, available at: www.intechopen.com/books/uranium-safety-resources-separation-and-thermodynamic-calculation/uranium-in-poland-resources-and-recovery-from-low-grade-ores
228. Miskiewicz, A. et al. (2016) Application of membrane contactor with helical flow for processing uranium ores. *Hydrometallurgy* 163: 108.

229. Zakrzewska-Trznadel, G. et al. (2015) *Method of obtaining and separating valuable metallic elements, specifically from low-grade uranium ores and radioactive liquid wastes*. European Patent EP2604713.
230. Kedari, C.S., Pandit, S.S., Ramanujam, A. (1999) Selective permeation of plutonium(VI) through supported liquid membrane containing 2-ethyljexyl 2-ethylhexyl phosphonic acid as ion carrier. *Journal of Membrane Science* 156: 187.
231. Rathore, N.S., et al. (2001) Hollow fiber supported liquid membrane: a novel technique for separation and recovery of plutonium from aqueous acidic wastes. *Journal of Membrane Science* 189: 119.
232. Lakshmi, D.S., et al. (2004) Uranium transport using a PTFE flat-sheet membrane containing alamine 336 in toluene as the carrier. *Desalination* 163: 13.
233. Shailesh, S., et al. (2006) Transport studies of uranium across a supported liquid membrane containing N,N-di(2-ethylhexyl)isobutyramide (D2EHIBA) as the carrier. *Journal of Membrane Science* 272: 143.
234. Panja, S., et al. (2011) Facilitated transport of uranium(VI) across supported liquid membranes containing T2EHDGA as the carrier extractant. *Journal of Hazardous Materials* 188: 281.
235. Panja, S., et al. (2009) Studies on uranium(VI) pertraction across a N,N,N',N'-tetraoctyldiglycolamide (TODGA) supported membrane. *Journal of Membrane Science* 337: 274.
236. Panja, S., et al. (2012) Uranium(VI) pertraction across a supported liquid membrane contacting a branched diglicolamide carrier extractant: Part III Mass transfer modeling. *Desalination* 285: 213.
237. Mohapatra, P.K., et al. (2006) Pertraction across a PTFE flatsheet membrane containing Aliquat 336 as the carrier. *Separation and Purification Technology* 51: 24.
238. Lakshmi, D.S., Mohapatra, P.K., Mohan, D. (2004) Uranium transport using a PTFE flat-sheet membrane containing alamine 336 in toluene as the carrier. *Desalination* 163: 13.
239. Singh, S.K. et al. (2007) Carrier-mediated transport of uranium from phosphoric acid medium across TOPO/*n*-dodecane-supported liquid membrane. *Hydrometallurgy* 87: 190.
240. Ura, P., et al. (2006) Feasibility study on the separation of uranium and thorium by hollow fiber supported liquid membrane. *Journal of Industrial and Engineering Chemistry* 12: 673.
241. Panja, S., et al. (2008) Facilitated transport of Am(III) through a flat-sheet supported liquid membrane (FSSLM) containing tetra(2-ethyl hexyl)diglycolamide (TEHDGA) as carrier. *Journal of Membrane Science* 325: 158.
242. Ansari, S.A., et al. (2008) Separation of Am(III) and trivalent lanthanides from simulated high-level waste using hollow fiber-supported liquid membrane. *Separation and Purification Technology* 63: 239.
243. Hoshi, H., Tsuyoshi A., Akiba K. (2000) Separation of americium from europium using liquid membrane impregnated with organodithiophosphinic acid. *Journal of Radioanalytical and Nuclear Chemistry* 243(3): 621.
244. Sasaki, Y., et al. (2001) The novel extractants, diglycolamides, for the extraction of lanthanides and actinides in HNO_3-n-dodecane system. *Solvent Extraction and Ion Exchange* 19: 91.
245. Bhattacharyya, A., Mohapatra, P.K., Manchanda, V.K. (2006) Separation of trivalent actinides and lanthanides using a flat sheet supported liquid membrane containing Cyanex-301 as the carrier. *Separation Science and Technology* 50: 278.
246. Panja, P.K., et al. (2012) Liquid–liquid extraction and pertraction behavior of Am(III) and Sr(II) with diglycolamide carrier extractants. *Journal of Membrane Science* 399–400: 28.
247. Bhattacharyya, A., et al. (2011) Liquid–liquid extraction and flat sheet supported liquid membrane studies on Am(III) and Eu(III) separation using 2,6-bis(5,6-dipropyl-1,2,4-triazin-3-yl)pyridine as the extractant. *Journal of Hazardous Materials* 195: 238.
248. Teramoto, M., et al. (2000) Treatment of simulated low level radioactive wastewater by supported liquid membranes: uphill transport of Ce(III) using CMPO carrier. *Separation and Purification Technology* 18: 57.
249. El-Said, N., Rahman, N.A., Borai, E.H. (2002) Modification in Purex process using supported liquid membrane separation of cerium(III) via oxidation to cerium(IV) from fission products from nitrate medium by SLM. *Journal of Membrane Science* 198: 23.
250. Mohapatra, P.K., et al. (2004) Selective transport of cesium using a supported liquid membrane containing di-t-butyl benzo 18 crown 6 as the carrier. *Journal of Membrane Science* 232: 133.
251. Kim, J.K., et al. (2001) Selective extraction of cesium ion with calyx[4]arene crown ether through thin sheet supported liquid membrane. *Journal of Membrane Science* 187: 3.
252. Raut, D.R., et al. (2013) Evaluation of two calix-crown-6 ligands for the recovery of radio cesium from nuclear waste solutions: Solvent extraction and liquid membrane studies. *Journal of Membrane Science* 429: 197.
253. Kandwal, P., et al. (2011) Mass transport modeling of Cs(I) through hollow fiber supported liquid membrane containing calix-[4]-bis(2,3-naptho)-crown-6 as mobile carrier. *Chemical Engineering Journal* 174: 110.
254. Mohapatra, P.K., Bhattacharyya, A., Manchanda, V.K. (2010) Selective separation of radio-cesium from acidic solutions using supported liquid membrane containing chlorinated cobalt dicarbollide (CDD) in phenyltrifluoromethyl sulphone (PTMS). *Journal of Hazardous Materials* 181: 679.
255. Rawat, N., et al. (2006) Evaluation of supported liquid membrane containing a macrocyclic ionophore for selective removal of strontium from nuclear waste solution. *Journal of Membrane Science* 275: 82.
256. Raut, D.R., Mohapatra, P. K., Manchanda, V.K. (2012) A highly efficient supported liquid membrane system for selective strontium separation leading to radioactive waste remediation. *Journal of Membrane Science* 390–391: 76.
257. Kandwal, P., Ansari, S.A., Mohapatra, P. K. (2012) A highly efficient supported liquid membrane system for near quantitative recovery of radio-strontium from acidic feeds. Part II: Scale up and mass transfer modeling in hollow fiber configuration. *Journal of Membrane Science* 405–406: 85.
258. Chen, J., Boerrigter, H., Veltkamp, A.C. (2001) Technetium(VII) transport across supported liquid membranes (SLMs) containing 2-nitrophenyl octyl ether (NPOE). *Radiochimica Acta* 89: 523.
259. Kandwal, P., et al. (2011) Separation of carrier free ^{90}Y from ^{90}Sr by hollow fiber supported liquid membrane containing

bis(2-ethylhexyl) phosphonic acid. *Separation Science and Technology* 46: 904.
260. Kusumocahyo, S.P., et al. (2004) Development of polymer inclusion membranes based on cellulose triacetate: carrier-mediated transport of cerium(III). *Journal of Membrane Science* 244: 251.
261. Arous, O. (2010) Selective transport of metal ions using polymer inclusion membranes containing crown ether and cryptands. *The Arabian Journal for Science and Engineering* 35(2A); 79.
262. Ulewicz, M., Radzyminska-Lenarcik, E. (2011) Transport of metal ions across polymer inclusion membrane with 1-alkylimidazole. *Physicochemical Problems of Mineral Processing* 46: 119.
263. Valero, F., Barceló, A., Arbós R. (2011) *Electrodialysis Technology – Theory and Applications, Desalination, Trends and Technologies*, Michael Schorr (Ed.). InTech, available from www.intechopen.com/books/desalination-trends-and-technologies/electrodialysis-technology-theory-and-applications
264. Hays, J. (2000) Iowa's first electrodialysis reversal water treatment plant, *Desalination* 132(1): 161.
265. International Atomic Energy Agency (1994) *Advances in Technologies for Treatment of Low and Intermediate Level Radioactive Liquid Wastes*. Technical Report Series No. 370. IAEA: Vienna.
266. Andalaft, E., Vega, R., Correa, M., Araya, R., Loyola, P. (1997) Zeta potential control in decontamination with inorganic membranes and inorganic adsorbents. In: *Treatment technologies for low and intermediate level waste from nuclear applications*. Final report of a coordinated research programme 1991–1996, pp. 15–32. IAEA-TECDOC-929. IAEA.
267. Mathur, J.N. (1998) Diffusion dialysis aided electrodialysis process for concentration of radionuclides in acid medium. *Journal of Radioanalytical and Nuclear Chemistry* 232(1–2): 237.
268. Singh, R.K., et al. (1996) Novel method for concentration of low level radioactive waste. *Indian Journal of Chemical Technology* 3: 149.
269. Gasser, M.S., Nowier, H.G. (2004) Separation of strontium and cadmium ions from nitrate medium by ion-exchange membrane in an electrodialysis system. *Journal of Chemical Technology & Biotechnology* 79: 97.
270. Lounis, C., Gavach C. (1997) Treatment of uranium leach solutions by electrodialysis for anion impurities removal. *Hydrometallurgy* 44: 83.
271. Zaheri, A., Moheb, A., Keshtkar, A.R., Shirani, A.S. (2010) Uranium separation from wastewater by Eeectrodialysis. *Iranian Journal of Environmental Health Scienceand Engineering* 7(5): 42.
272. Fountain, M.S., et al. (2008) Caustic recycle from Hanford tank waste using NaSICON ceramic membranes. *Separation Science and Technology* 43: 2321.
273. Hobbs, D.T. (1999) Caustic recovery from alkaline nuclear waste by an electrochemical separation process. *Separation and Purification Technology* 15: 239.
274. Yazicigil, Z. (2007) Salt splitting with cation-exchange membranes. *Desalination* 212: 70.
275. Park, J.K., Lee, K.J. (1995) Separation of boric acid in liquid waste with anion exchange membrane contactor. *Waste Management* 15(4): 283.
276. Kijański, M., et al. (2013) The concept of a system for electrodialytic boron removal into alkaline concentrate. *Desalination* 310: 75.
277. Dydo, P., Turek, M. (2014) The concept for an ED–RO integrated system for boron removal with simultaneous boron recovery in the form of boric acid. *Desalination* 342: 35.
278. Hegazy, E.-S., et al. (1999) Characterization and application of radiation grafted membranes in treatment of intermediate active waste., *Nuclear Instruments and Methods in Physics Research B* 151: 393.
279. Inoue, H., et al. (2004) Radioactive iodine waste treatment using electrodialysis with an anion exchange paper membrane. *Applied Radiation and Isotopes* 61: 1189.
280. Inoue, H. (2003) Radioactive iodine and chloride transport across a paper membrane bearing trimethylhydroxypropylammonium anion exchange groups. *Journal of Membrane Science* 222: 53.
281. Miśkiewicz, A., Nowak, A., Pałka, J., Zakrzewska-Kołtuniewicz, G. (2021) Liquid low-level radioactive waste treatment using an electrodialysis process. *Membranes* 11: 324.
282. International Atomic Energy Agency (1991) *Safe Handling of Tritium, Review of Data and Experience*, Technical Reports Series No. 324. IAEA: Vienna, Austria.
283. International Atomic Energy Agency (2004) *Management of Waste Containing Tritium and Carbon-14*, Technical Reports Series No. 421. IAEA: Vienna, Austria.
284. International Atomic Energy Agency (1984) *Management of Tritium at Nuclear Facilities*, Technical Reports Series No. 234. IAEA: Vienna, Austria.
285. Sadhankar, R.R., Miller, A.I. (2003) New heavy water production and processing technologies. In: *Third Conference Isotopic and Molecular Processes PIM2003*, Cluj-Napoca, Romania, September 25–27.
286. Kitamoto, A., Shimizu, M., Masui, T. (1992) The advanced CECE process for enriching tritium by the chemical exchange method with a hydrophobic catalyst. In: *Proceedings of the International Symposium on Isotope Separation and Chemical Exchange Uranium Enrichment*, Y. Fuji, T. Ishida, K. Takeuchi (Eds.), p. 497. October 29–November 1, 1990, Tokyo, Japan, Research Laboratory for Nuclear Reactors, Tokyo Institute of Technology.
287. Trequattrini, F., et al. (2021) Promising isotope effect in Pd77Ag23 for hydrogen separation. *ChemicalEngineering* 5: 51.
288. Rumyantsev, V.V., Shatalov, V.M., Misuna, G.Ya. (2002) Gas separation of hydrogen isotopes by means of multicell metal membrane. *Desalination* 148; 293.
289. Glugla, M., et al. (2006) Hydrogen isotope separation by permeation through palladium membranes. *Journal of Nuclear Materials* 355: 47.
290. Shi, Y., Wei, Y-J., Su, Y-J. (2008) Experimental studies on the hydrogen isotope recovery using low-pressure palladium membrane diffuser. *Journal of Membrane Science* 322: 302.
291. Semidey-Flecha, L., Hao, S., Sholl, D.S. (2009) Predictions of H isotope separation using crystalline and amorphous metal membranes: A computational approach. *Journal of The Taiwan Institute of Chemical Engineers* 40; 246.
292. Bellanger, G. (2009) Optimization for the tritium isotope separation factor and permeation by selecting temperature

and thickness of the diffusion Pd–Ag alloy cathode. *Fusion Engineering and Design* 84: 2197.
293. Tosti, S., et al. (2011) Design of Pd-based membrane reactor for gas detritiation. *Fusion Engineering and Design* 86: 2180.
294. Bornea, A., et al. (2010) Laboratory studies for development of a plant to concentrate the radioactive waste from tritiated water. *Fusion Engineering and Design* 85: 1970.
295. Sevigny, G.J., et al. (2015) *Separation of tritiated water using graphene oxide membrane*, prepared for U.S. Department of Energy, Fuel Cycle Research and Development Material Recovery and Waste Form Development Campaign, June 2015 FCRD- MRWFD-2015-000773 PNNL-24411.
296. Qu, Y., Li, F., Zhao, M. (2017) Efficient hydrogen isotopologues separation through a tunable potential barrier: The case of a C2N membrane. *Scientific Reports* 7: 1483.
297. Lozada-Hidalgo, M., et al. (2017) Scalable and efficient separation of hydrogen isotopes using graphene-based electrochemical pumping. *Nature Communications* 8: 15215.
298. Rehman, F., et al. (2021) Graphene-based composite membranes for isotope separation: challenges and opportunities. *Reviews in Inorganic Chemistry* 42(4): 327–336.
299. Mc Elroy, R.G.C., Osborne, R.V., Surette, R.A. (1982) A monitor for the separate determination of HT and HTO. *IEEE Transactions on Nuclear Science* 29(1): 816.
300. Sasaki, Sh., et al. (2003) Basic characteristics of hollow-filament polyimide membrane in gas separation and application to tritium monitors. *Journal of Radioanalytical and Nuclear Chemistry* 255(1): 91.
301. Nelson, D.A., et al. (1996) Isotopomeric water separations with supported polyphosphazene membranes. *Journal of Membrane Science* 112: 105.
302. Duncan, J.B., Nelson, D.A. (1999) The separation of tritiated water using supported polyphosphazene membranes. *Journal of Membrane Science* 157: 211.
303. Zakrzewska-Trznadel, G. (2006) Tritium removal from water solutions. *Desalination* 1–3: 737.
304. Le Digabel, M., et al. (2002) Application of gas separation membranes to detritiation systems. *Desalination* 148: 297.
305. Le Digabel, M., et al. (2003) Glovebox atmosphere detritiation process using gas separation membranes. *Fusion Engineering and Design* 69: 61.
306. Labrune, D., et al. (1995) Separation of hydrogen isotopes from nitrogen with polyimide membrane. *Fusion Technology* 28: 676.
307. Iwai, Y., Yamanishi, T., Nishi, M. (1999) A steady-state simulation model of gas separation system by hollow fiber type membrane module. *Journal of Nuclear Science and Technology* 36(1): 95.
308. Hayashi, T., et al. (1998) Effective tritium processing using polyimide films. *Fusion Engineering and Design* 39–40: 901.
309. Ishida, T., et al. (1996) R&D of compact detritiation system using a gas separation membrane module for the secondary confinement. *Fusion Technology* 30: 926.
310. Ishida, T., et al. (2000) Design of a membrane atmosphere detritiation system using super high permeation module. *Fusion Engineering and Design* 49–50: 839.
311. Hirata, S., et al. (1995) Experimental and analytical study on membrane detritiation process. *Fusion Technology* 28: 1521.
312. Hayashi, T., et al. (1995) Gas separation performance of hollow-filament type polyimide membrane module for a compact tritium removal system. *Fusion Technology* 28: 1503.
313. Ito, H., et al. (1992) Separation of tritium using polyimide membranes. *Fusion Technology* 21: 993.
314. Nakagawa, T., Yoshida, M., Kidokoro, K. (1990) Development of rubbery materials with excellent barrier properties to H2, D2 and T2. *Journal of Membrane Science* 52: 263.
315. Maier-Komor, P. (2010) Uranium isotope separation from 1941 to the present. *Nuclear Instruments and Methods in Physics Research A* 613: 465.
316. Petit, J.F. (1990) Technological aspects of the historical development of Eurodif. In: *Proceedings of the International Symposium on Isotope Separation and Chemical Exchange Uranium Enrichment*, Yasuhiko Fuji, Takanobu Ishida, and Kazuo Takeuchi (Eds.), pp. 103–109. Tokyo, October 29–November 1.
317. Douglas, M. (Ed.) (1997) *A Kirk-Othmer Encyklopedia of Separation Technology*, Vol. 1. Ruthven: John Wiley & Sons.
318. Schiel, R. (1972) *Bibliography on uranium isotope separation*. Commission of the European Communities, Directorate-General Dissemination of Information Centre for Information and Documentation – CID Luxembourg, 254 pp, No. EUR 4796 e.
319. Hsieh, H.P. (1996) *Inorganic Membranes for Separation and Reaction,* Membrane Science and Technology Series, Vol. 3, Elsevier: Amsterdam, the Netherlands.
320. Fain, D.E., Brown, W.K. (1976) *Neon isotope separation by gaseous diffusion transport in the transition flow regime with regular geometries*. In: International Symposium on Rarefied Gas Dynamics, Aspen, CO, July 18, 1976, https://inis.iaea.org/search/search.aspx?orig_q=source:"CONF-760710-6" accessed on August 11, 2022.
321. Mercea, P. (1983) Permeation of H_2 andD_2 through polymers. *Isotopenpraxis* 19(5): 153.
322. Mercea, P., Tosa, V. (1985) Quantum isotope effect in gas transport through polymers. *Isotopenpraxis* 21(12): 413.
323. Agrinier, P., et al. (2008) Permeation selectivity of gaseous isotopes through dense polymers: Peculiar behavior of the hydrogen isotopes. *Journal of Membrane Science* 318: 373.
324. Wang, J., et al. (2020) Computational screening and design of nanoporous membranes for efficient carbon isotope separation. *Green Energy & Environment* 5(3): 364.
325. Poteryaeva, V.A., et al. (2021) Helium isotope separation by bi-layer membranes of g-C_3N_4. *Advances in Natural Sciences: Nanoscience and Nanotechnology* 12: 045005.
326. Hauser, A.W., Schwerdtfeger, P. (2012) Nanoporous graphene membranes for efficient 3He/4He separation. *The Journal of Physical Chemistry Letters* 3(2): 209.
327. Wen, M., et al. (2021) Superhydrophobic composite graphene oxide membrane coated with fluorinated silica nanoparticles for hydrogen isotopic water separation in membrane distillation. *Journal of Membrane Science* 626: 119136.
328. Ching, K., et al. (2022) Liquid-phase water isotope separation using graphene-oxide membranes. *Carbon* 186; 344.
329. Chmielewski, A.G., et al. (1991) Investigation of the separation factor between light and heavy water in the liquid/vapour membrane permeation process. *Journal of Membrane Science* 55: 257.

330. Chmielewski, A.G., et al. (1997) Multistage process of deuterium and heavy-oxygen enrichment by membrane distillation. *Separation Science and Technology* 32(1–4); 527.
331. Kim, J., Chang, D.-S., Choi, Y.-Y. (2009) Separation of oxygen isotopic water by using a pressure-driven air gap membrane distillation. *Industrial & Engineering Chemistry Research* 48: 5431.
332. Kim, J., et al. (2004) Isotopic water separation using AGMD and VEMD. *Nukleonika* 49(4): 137.
333. Hossein A., et al. (2018) Experimental and numerical evaluation of membrane distillation module for oxygen-18 separation. *Chemical Engineering Research and Design* 132: 492.
334. Yoshida, H., et al. (1983) Preliminary design of fusion reactor fuel clean-up system by the palladium alloy membrane method. *Nuclear Technology Fusion* 3: 471.
335. Chen, S., et al. (2002) A palladium alloy composite membrane for the purification of hydrogen isotopes. *Separation Science and Technology* 37(11): 2701.
336. Aoki, K., et al. (1998) Applicability of palladium membrane for the separation of protium and deuterium. *International Journal of Hydrogen Energy* 23(5): 325.
337. Bridesell, S.A., Willms, R.S. (1998) Tritium recovery from tritiated water with a two-stage palladium membrane reactor. *Fusion Engineering and Design* 39–40: 1041.
338. Basile, A., et al. (1995) Membrane integrated system in the fusion reactor fuel cycle. *Catalysis Today* 25: 321.
339. Violante, V., et al. (1994) *Ceramic catalytic membrane reactor for separation of hydrogen and/or isotopes thereof from fluid feeds*, US Patent No 5366712, November 22, 1994.
340. Drioli, E., Violante, V., Basile, A. (1990) Membrane separation processes in the fusion fuel cycle. in: *Proceedings of ICOM'90*, Vol. II, p. 1040. Chicago, IL.
341. Drioli, E., Violante, V., Basile, A. (1990) Membrane separation processes in fusion separation plants. In: *Proceedings of 5th World Filtration Congress*, Vol. I, p. 449. Nice, France.
342. Violante, V., Basile, A., Drioli, E. (1995) Composite catalytic membrane reactor analysis for water gas shift reaction in the tritium fusion fuel cycle. *Fusion Engineering and Design* 30: 217.
343. Tosti, S., et al. (2000) Catalytic membrane reactors for tritium recovery from tritiated water in the ITER fuel cycle. *Fusion Engineering and Design* 49–50: 953.
344. Konishi, S., et al. (1998) Development of electrolytic reactor for processing of gaseous tritiated compounds. *Fusion Engineering and Design* 39: 1033.
345. Heinze, S., Bussiere, P., Pelletier, T. (2003) French experience in tritiated water management. *Fusion Engineering and Design* 69: 67.
346. Stolz, T., et al. (2003) Self-radiolysis of tritiated water. *Fusion Engineering and Design* 69: 57.
347. Campbell, D.J. (1985) *Fractionation of stable chlorine isotopes during transport through semipermeable membranes*. MS thesis, University of Arizona: Phoenix, AZ.
348. Fritz, S.J., Hinz, D.L., Grossman, E.L. (1987) Hyperfiltration-induced fractionation of carbon isotopes. *Geochimica Cosmochimica Acta* 51: 1121.
349. Fujii, Y., et al. (1979) Separation of isotopes of lithium and uranium by electromigration using cation-exchange membranes. *Isotopenpraxis* 15(7): 203.
350. Okamoto, M., Oi, T., Fujii, Y., et al. (1980) Dependencies of pH and current density on the isotope effect occurred by the electromigration of uranyl ions in a cation exchange membrane. *Isotopenpraxis* 16(2); 58.
351. Whitworth, T.M., Mariñas, B.J., Fritz, S.J. (1994) Isotopic fractionation and overall permeation of lithium by a thin-film composite polyamide reverse osmosis membrane. *Journal of Membrane Science* 88: 231.
352. Hoshino, T., Terai, T. (2011) High-efficiency technology for lithium isotope separation using an ionic-liquid impregnated organic membrane. *Fusion Engineering and Design* 86: 2168.
353. Hoshino, T., Terai, T. (2011) Basic technology for ^{6}Li enrichment using an ionic-liquid impregnated organic membrane. *Journal of Nuclear Materials* 417: 696.
354. Rainey, R.H., et al. (1968) Separation of radioactive xenon and krypton from other gases by use of permselective membranes. In: *Treatment of Airborne radioactive Wastes*, pp. 323–341. Proceedings of a Symposium, New York, August 26–30, 1968.
355. Kimura, S., et al. (1973) Separation of rare gases by membranes. *Radiochemical and Radioanalytical Letters* 13(56): 349.
356. Stern, S.A., Leone, S.M. (1980) Separation of krypton and xenon by selective permeation. *AICHE Journal* 26(6): 881.
357. Bożenko, E.I. et al. (1983) Primienienije polimernych membran dla oczistki gazowych wybrosow atomnych elektrostancji. *Plasticheskie Massy* 2: 51.
358. Bekman, I.N. et al. (1984) Membrannyje metody wydielienija radioaktiwnych błogorodnych gazow, cz. II. *Radiochimija* 3 337.
359. Nikonow, W.N. et al. (1984) Membrannyje metody wydielienija radioaktiwnych błogorodnych gazow, cz. I. *Radiochimija* 3: 332.
360. Ohno, M., et al. (1976) Separation of rare gases by membranes. *Radioanalytical Letters* 27 (5–6): 299.
361. Ohno, M., et al. (1977) Radioactive rare gas separation using a separation cell with two kinds of membrane differing in gas permeability tendency. *Journal of Nuclear Science and Technology* 14(8): 589.
362. Stern, S.A., Wang, S.C. (1980) Permeation cascades for the separation of krypton and xenon from nuclear reactor atmospheres. *AICHE Journal* 26(6): 891.
363. Jonas, Ch. (1983) Zum Einfluss des Plastifizierungseffektes auf den Stofftrennvorgang. *Chem. Techn.* 35(2): 77.
364. Jensvold J.A., Jeanes T.O. (2001) *Membrane for separation of xenon from oxygen and nitrogen and the method of using same*. US 6168649, January 2, 2001, www.google.com/patents, accessed on October 12, 2014.
365. Sasaki, Sh., et al. (2003) Basic characteristics of hollow-filament polyimide membrane in gas separation and application to tritium monitors. *Journal of Radioanalytical and Nuclear Chemistry* 255(1): 91.
366. Nörenberg, H., et al. (2001) Pressure-dependent permeation of noble gases (He, Ne, Ar, Kr, Xe) through thin membranes of oriented polypropylene (OPP) studied by mass spectrometry. *Polymer* 42: 10021.
367. Ravi J., Whitlock W. (1998) *Recovery of noble gases*. EP0826629, March 4, 1998, http://worldwide.espacenet.com, accessed on October 12, 2014.

368. Schmidt, K. (2010) *Retention of noble gases in the exhaled air of ventilated patients by membrane separation*. US 2010/0031961, February 2, 2010.
369. Li, J., et al. (2018) Porous organic materials with ultra-small pores and sulfonic functionality for xenon capture with exceptional selectivity. *Journal of Materials Chemistry* 6: 11163.
370. Wu, T., et al. (2017) Zeolitic imidazolate framework-8 (ZIF-8) membranes for Kr/Xe separation. *Industrial & Engineering Chemistry Research* 56: 1682.
371. Kwon, Y.H., et al. (2017) Krypton-xenon separation properties of SAPO-34 zeolite materials and membranes. *AIChE Journal* 63: 761.
372. Feng, X., et al. (2016) Kr/Xe separation over a chabazite zeolite membrane. *Journal of the American Chemical Society* 138: 9791.
373. Kwon, Y.H., et al. (2018) *Zeolite Membranes for Krypton/Xenon Separation from Spent Nuclear Fuel Reprocessing off-Gas*. Final Project Report (Oct 2014–Dec 2017). Georgia Tech Research Corporation.
374. Wu, T., et al. (2018) Microporous crystalline membranes for Kr/Xe separation: comparison between AlPO-18, SAPO34, and ZIF-8. *ACS Applied Nano Materials* 1: 463.
375. Kwon, Y.H., et al. (2018) Ion-exchanged SAPO34 membranes for krypton-xenon separation: control of permeation properties and fabrication of hollow fiber membranes. *ACS Applied Materials Interfaces* 10: 6368.

19 Polymer Inclusion Membranes

Spas D. Kolev, M. Inês G.S. Almeida, and Robert W. Cattrall*
School of Chemistry, The University of Melbourne, Victoria 3010, Australia
*Email: s.kolev@unimelb.edu.au

19.1 INTRODUCTION

19.1.1 Liquid Membranes

The liquid membrane (LM) concept combines solvent extraction (SX) and membrane-based technologies, enabling both extraction and back-extraction in a single step with reduced consumption of extractants and diluents. For these reasons separation based on LMs can be viewed as a promising alternative to traditional SX. The LM separation approach involves mass transfer of a target chemical species between two solutions (i.e., feed and receiver solutions) separated by an immiscible liquid membrane [1]. The main types of LMs are bulk liquid membranes (BLMs), emulsion liquid membranes (ELMs), supported liquid membranes (SLMs), and polymer inclusion membranes (PIMs).

A BLM consists of an organic liquid phase partitioned between an aqueous feed phase and a receiver phase, and the mass transfer is enhanced by mechanical stirring of one or both phases. A low interfacial surface area and mass transfer rate are the main disadvantages of this type of membrane.

Emulsion globules composed of the receiver aqueous solution and an organic liquid phase dispersed into the feed aqueous phase form a typical ELM. A surfactant is usually used to disperse the emulsion globules in the feed solution. After equilibrium has been reached, the feed aqueous phase is separated from the emulsion globules, which are subsequently demulsified to form an organic liquid layer and an aqueous receiver solution containing the separated and preconcentrated target chemical species. Unlike BLMs, ELMs are characterized by a large interfacial surface area between the emulsion globules and the receiver phase on one hand and between the organic liquid phase and the feed aqueous phase on the other. This configuration enables fast mass transfer between the two aqueous phases. The main difficulties in conducting the separation process in this case are related to the formation and breakdown of the emulsion itself, which limits the commercial applications of ELMs.

SLMs are composed of a porous hydrophobic membrane acting as a solid support for the organic liquid phase. The liquid phase, retained inside the membrane pores by capillary forces, leaches slowly into the receiver and feed aqueous solutions. Thus, insufficient long-term stability is the main disadvantage of SLMs which prevents the widespread adoption of these otherwise easy to use liquid membranes in industry.

PIMs, on the other hand, retain most of the advantages of SLMs while exhibiting greater stability and versatility.

19.1.2 Polymer Inclusion Membranes

Polymer-based liquid membranes have been known for over 50 years and have been used as the sensing membranes of ion-selective electrodes (ISEs) and optodes where they have been usually referred to as plasticized membranes. However, separation based on this type of membrane, mostly known in such applications as PIMs, has been proposed as a possible alternative to conventional SX [2,3]. The interest in PIM-focused research has been growing exponentially in the past few years (Figure 19.1). PIMs are usually composed of an extractant (carrier), a base polymer, and a plasticizer. In some cases, the carrier also acts as a plasticizer and so an additional plasticizer is not necessary. Also, a modifier is occasionally added to the membrane composition to improve the solubility of the extracted species in the membrane liquid phase. It should be pointed out that the membranes used for sensing and separation differ substantially from PIMs with regards to their transport properties. In the first case, fast ion-exchange at the membrane–sample solution interface combined with an extremely low transport rate of the ion of interest across the membrane is required. For separation purposes, both fast ion-exchange and transport rates are required, which is only possible if membranes with high concentrations of extractants (> 30% m/m) are used as opposed to the low concentrations (~ 1–2% m/m) of extractants involved in sensing applications [2–4].

PIMs are commonly prepared using the solvent casting method (i.e., precipitation by solvent evaporation), which involves weighing the membrane components into a beaker followed by dissolution in a small volume of an appropriate solvent. Tetrahydrofuran (THF) is often used as volatile solvent when poly(vinyl chloride) (PVC) or poly(vinylidene fluoride-co-hexafluoropropylene) (PVDF-HFP) is used as the base polymer, while dichloromethane (DCM) is used for cellulose triacetate (CTA)-based PIMs. In the case of PVC or CTA as the base polymer, the mixture is stirred at room temperature using a magnetic stirrer until all the components are dissolved and the solution is then poured into a glass ring positioned on a glass plate. The ring is covered with a filter paper in order to allow slow evaporation of the solvent. The membrane formed after the complete evaporation of the solvent is then peeled from the glass plate. Successful membranes are normally transparent, homogeneous, non-porous, flexible, and mechanically strong (Figure 19.2). When PVDF-HFP is used as the base polymer, the mixture is stirred at ≥ 40°C until complete dissolution of all components. After cooling the

DOI: 10.1201/9781003285656-19

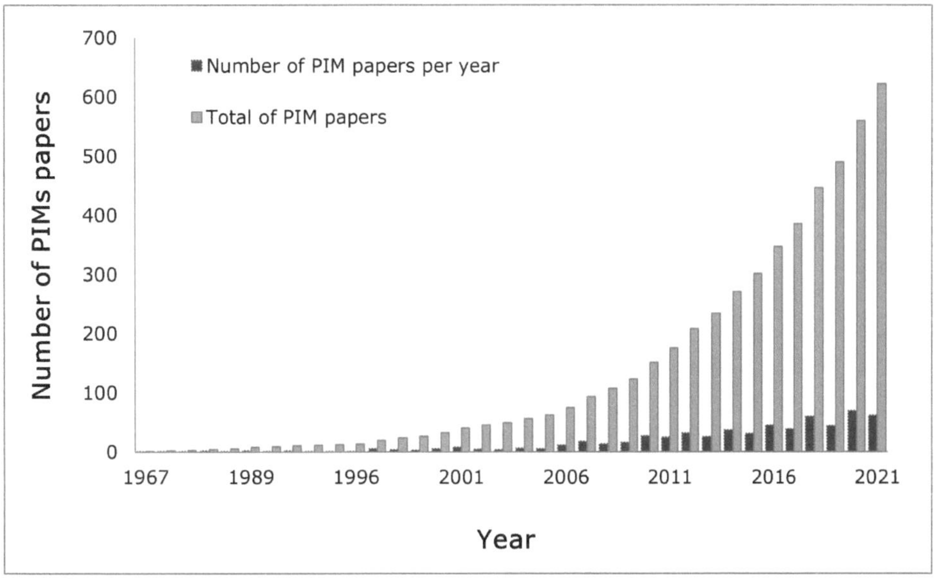

FIGURE 19.1 Evolution of PIM papers (including reviews) from 1967 until 2021 (according to ISI Web of Knowledge).

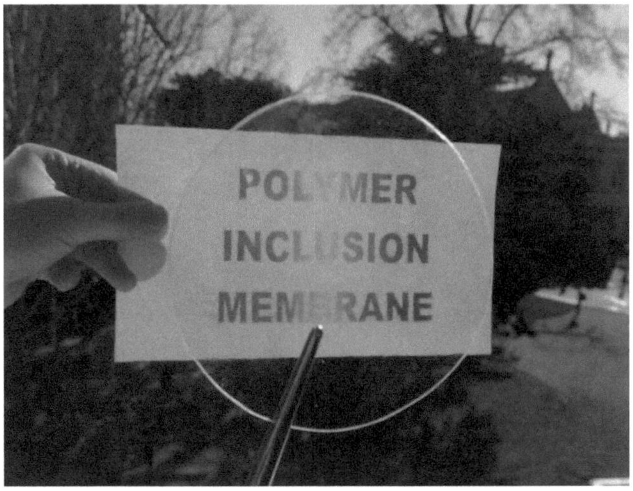

FIGURE 19.2 Photographic image of a typical PIM.

solution to room temperature, it is spread along a glass plate by means of a casting knife made of Teflon and covered using an aluminum tray to slow down the evaporation of THF [5]. After a minimum of 24 h, a thin membrane of rectangular shape is peeled from the glass plate and is considered as successful if it exhibits the same features as those described above.

Greener solvent alternatives to THF have been reported for the fabrication of PIMs [6]. PVC- and PVDF-HFP-based PIMs were successfully fabricated using 2-methylTHF and ethyl acetate, respectively, which are renewably sourced and less hazardous solvents than THF.

19.2 BASE POLYMERS

The base polymer provides the membrane with mechanical strength. PVC, CTA, and PVDF-HFP have been the most widely used base polymers in PIMs since these polymers provide membranes with a high mechanical strength and are compatible with a large range of carriers, plasticizers, and modifiers [7].

CTA is a polar polymer in which most of the cellulose units are fully acetylated. The acetyl groups and residual hydroxy groups are capable of forming highly orientated hydrogen bonding, which gives CTA a semicrystalline structure [2]. The polarity and semicrystalline nature of CTA may render it incompatible with high concentrations of hydrophobic non-polar carriers. Moreover, CTA can be slightly hydrated and, thus, susceptible to acid hydrolysis [8]. Gardner et al. studied different cellulose derivatives [i.e., cellulose acetate propionate (CAP), cellulose acetate butyrate (CAB), and cellulose tributyrate (CTB)] in order to assess the durability of PIMs against hydrolysis under alkaline and acidic conditions [9]. The membranes contained bis-*tert*-butylcyclohexano-18-crown-6 as the carrier. It was observed that the resistance to hydrolysis increased with an increase in the alkyl chain length, although the rate of ion transport across the membrane decreased. CAP and CAB together with cellulose acetate hydrogen phthalate (CAH) were also used in attempts to manufacture PIMs with dinonylnaphthalene sulfonic acid (DNNS) and dinonylnaphthalene disulfonic acid (DNNDS) as carriers and some of these compositions produced membranes which were found to be fragile [10]. Cellulose acetate (CA) has been successfully used without a plasticizer to produce PIMs with macrocyclic carriers [11,12].

PVC is a non-flammable and durable semicrystalline polymer formed from a vinyl chloride monomer. The C-Cl functional group in PVC is relatively polar and non-specific dispersion forces dominate its intermolecular interactions [2]. PVC is used as the base polymer in membranes because of its strength, inertness, and compatibility with a variety of carriers and plasticizers. Unlike CTA, PVC is also resistant to acidic solutions since it is not prone to acid hydrolysis.

Studies involving the use of PVC with different molecular weights (MWs) have been reported. It was observed that the nature of the base polymer had only a small influence on the membrane transport properties, although membranes formed with PVC with low MW (e.g., 80,000 g mol^{-1}) were found to be less fragile [13]. It was noticed that PIMs prepared with PVC with MW of 233,000 g mol^{-1} were more difficult to dissolve in TFH [14].

Polyvinylidene fluoride (PVDF) has also been used as the base polymer of PIMs [15–18], although in some cases a different fabrication method (i.e., phase inversion by immersion precipitation) was used to prepare such membranes which exhibited a porous morphology [16–18]. PVDF is a thermoplastic, semicrystalline fluoropolymer with high hydrophobicity, good chemical resistance, and excellent thermal and mechanical stability which make it attractive as a base polymer for PIMs. The copolymer PVDF-HFP is also highly hydrophobic and exhibits lower crystallinity than PVDF due to the presence of the hexafluoropropylene chain segments. In the past few years, it has been increasingly used as a base polymer in PIMs due to providing higher permeability and extraction/transport efficiency when compared to PIMs containing other base polymers, such as PVC or CTA [19,20]. PVDF-HFP-based PIMs are usually fabricated via the solvent casting method and have a non-porous morphology [19].

19.3 CARRIERS

The carrier (extractant) is usually a complexing agent or an ion-exchanger, responsible for binding the species of interest, and thus facilitating their extraction into the PIM. The transport process across the membrane is driven by the concentration gradient of the extracted species/carrier complex or ion-pair within the membrane. Several classes of SX reagents have been incorporated in PIMs, namely basic, acidic and chelating, neutral or solvating, and macrocyclic and macromolecular. The structures of some of these carriers are presented in Figure 19.3 and examples of carriers according to the classification outlined above are presented in Tables 19.1, 19.2, and 19.3, along with the chemical species they have been applied to.

19.3.1 Basic Carriers

Basic carriers (Table 19.1) consist mainly of amine-based compounds such as quaternary ammonium salts (e.g., Aliquat 336), tertiary amines [e.g., tri-*n*-octylamine (TOA), tri-isooctylamine (TIOA)], weakly basic compounds (e.g., alkyl derivatives of pyridine *N* oxides), and thiadiazine derivatives

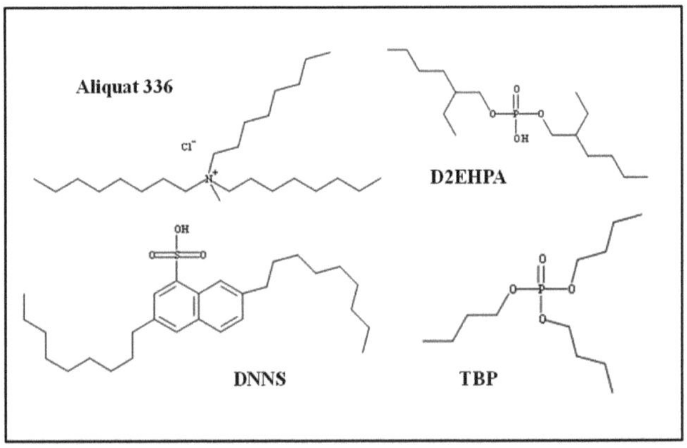

FIGURE 19.3 Chemical structures of examples of basic (Aliquat 336), acidic (D2EHPA, DNNS), and neutral (TBP) carriers.

TABLE 19.1
Examples of basic, acidic, and chelating carriers used in PIMs and chemical species they have been applied to (all abbreviations are defined in the Glossary)

Type of carrier	Examples of carriers	Target species
	Basic	
Quaternary amines	Aliquat 336/Capriquat	Anions [41], AuNPs [26], 1-butanol [15], heavy metals [23, 71], organic compounds [27,72,73]
Tertiary amines	TBA, THA, TIOA, TOA	Heavy metals [74], ReO_4^- [75]
Pyridine and its derivatives	TDPNO	Heavy metals [76]
Thiadiazine derivatives	FFAT	Cr(VI) [77]
	Acidic and chelating	
Alkyl phosphoric acids	D2EHPA, D2EHDTPA, Cyanex 272/301/302/471X	Am(III) [78]), Co-60 [79], gases [80], heavy metals [28], trivalent rare earth ions [81], In(III) [82]
Sulfonic acids	DNNS, DNNDS	Co-60, Cs-137, Sr-90 [83]; H^+ [10]; NH_4^+ [84]; heavy metals [31]
Carboxylic acids	Lauric acid, Lasalocid A, t-DAPA, gluconic acid	Heavy metals [35,85], paracetamol [86]
Hydroxyoximes	LIX® 84-I/984/54-100	Cu(II) [20,87,88]
Hydroxyquinoline	Kelex 100	Au(III) [89]; Cd(II), Pb(II) [90]
β-Diketones	HBA, HBM, HBTA, HTTA, HFTA, HPBI	Cu(II) [91], trivalent rare earth ions [92]

[e.g., 5-(4-phenoxyphenyl)-6H-1,3,4-thiadiazin-2-amine (FFAT)]. The reason for this classification is based on the extraction mechanism of the carriers. In the case of fully substituted quaternary ammonium compounds (e.g., Aliquat 336), the carrier acts as an anion-exchanger by forming an ion-pair with another anion from the feed phase which can be a complex metal anion. In the case of amine and other weak bases, mentioned above, the carrier must be protonated first in order to participate in an anion exchange reaction with another anion from the feed phase or with an anionic metal complex. Alternatively, the carrier may react directly with a protonated anionic metal complex [21].

Aliquat 336 is a commercially available extractant, extensively applied as a carrier in PIMs, which also has plasticizing properties. For this reason, it is frequently used in the manufacturing of PIMs without the addition of a separate plasticizer [22,23]. Aliquat 336-based PIMs have been applied to the extraction and/or transport of anionic complexes of mainly heavy metals [e.g., Au(III), Cd(II), Co(II), Cr(VI), Cu(II), Ni(II), Pd(II), Pt(IV), Re(VII)], common anions (e.g., I^-, SCN^-), and organic compounds (e.g., small saccharides, amino acids, lactic acid, thiourea, ciprofloxacin) [2,3]. Moreover, it has been employed in a PVC-based template for the preparation of Au nanoparticles. The nanoparticles were formed either in the bulk of the PIM or as a monolayer on the PIM surface by first extracting $AuCl_4^-$ into the membrane by anion exchange which was followed by in situ reduction of Au(III) on the membrane surface with BH_4^- [24,25] or EDTA [26]. Capriquat, is another liquid anion-exchanger similar in composition to Aliquat 336 with the major component being trioctylmethylammonium (TOMA) chloride [27].

TABLE 19.2
Examples of neutral or solvating carriers used in PIMs and chemical species they have been applied to (all abbreviations are defined in the Glossary)

Type of carrier	Examples of carriers	Target species
Phosphoric acid esters	TBP	Cd(II), Pb(II) [93]; U(VI) [94]
Phosphonic acid esters	DBBP	As(V) [95]
Ionic liquids	Cyphos IL 101/102/104, Bif-ILEs	1-Butanol [15], Cr(VI) [17], Zn(II) [96], V(V) [97]
Others	CMPO, TODGA, TOPO, TETDS, LSI, polyethylene glycol	Mo(VI) [98], Pb(II) [99], Sr(II), lanthanides [100]

19.3.2 Acidic and Chelating Carriers

The compounds frequently applied in PIMs that are classified as acidic carriers include organophosphorus acids, sulfonic acids, and carboxylic acids. Moreover, there is another group of compounds that have not only acidic but also chelating properties, namely hydroxyoximes, quinolines, and β-diketones. Examples of these carriers are given in Table 19.1. Acidic carriers are usually used in PIMs for the extraction and transport of metal cations, which involve the exchange of the metal cation with the hydrogen ions of the carrier, while maintaining a suitable difference in the concentration of the hydrogen ion the feed and receiver solutions [2].

Among the acidic carriers, di-(2-ethylhexyl) phosphoric acid (D2EHPA) has been the most commonly used acidic carrier in PIMs. D2EHPA is a commercially available extractant which also has the ability to act as a plasticizer

TABLE 19.3
Examples of macrocyclic and macromolecular carriers used in PIMs and chemical species they have been applied to (all abbreviations are defined in the Glossary)

Type of carrier	Examples of carriers	Target species
Crown ethers	Dicyclohexano-18-crown-6, di-*tert*-butylcyclohaxano-18-crown-6, dibenzo-18-crown-6, undecyl-aza-18-crown-6, imidazole azothiacrown ethers, imidazole azocrown ethers	Alkaline metals [101,102], Cs-137 [103], heavy metals [104], picrate [105], ReO_4^- [39]
Ionizable lariat ethers	PNP-lariat ethers	Heavy metals [32]
Calixarenes	*p-tert*-Butylcalix[4]arene, calix[4]arene, calix[4]pyrrole, calix[4]resorcinarenes, thiacalix[4]arenes	Heavy metals [106], Na^+, O_2 [11], citric acid [107]
Cryptands	4,7,13,18-Tetra oxa-1,10-diazabicyclo	Ag(I), Cu(II) [108]
Calix crowns	Calix[4]-crown-6	Heavy metals [109]
Cyclodextrins	β-Cyclodextrin	Heavy metals [33]
Others	Bathophenanthroline, Bathocuproine, 1-alkylimidazole, 1-decylimidazole, 1-decyl-2-methylimidazole, ω-Thiocaprolactam, PVP, BPA, thioether donor macrocycles, lipophilic acyclic polyethers	Anions [110], Ba(II) [111], heavy metals [112], I_2 [113], lanthanides [114]

[28,29]. It has been applied mostly for the extraction and transport of heavy metals [e.g., Zn(II), U(VI), Pb(II), Cd(II), Cr(III), Fe(III)] [2,3]. PIMs incorporating PVC and D2EHPA were successfully coated with Ag nanoparticles after reducing the extracted Ag^+ at the membrane surface with L-ascorbic acid [26].

DNNS is another commercially available acidic carrier used in PIM studies, which is normally supplied as a 50–65% solution in *n*-heptane, 2-butoxyethanol, or ethylene glycol dibutyl ether. It is usually used as received to prepare PIMs, although since these diluents are water soluble Ershad et al. have compared the performance of PIMs prepared with commercial DNNS and purified DNNS [30] in terms of stability as well as extraction efficiency [31]. It was demonstrated that the purified DNNS-based PIMs exhibited superior extraction performance and slightly better stability when compared to the PIM counterparts prepared with the commercial DNNS reagent [31]. It has also been shown that the addition of commercial DNNS to PIMs prepared with macrocyclic compounds as carriers (e.g., bis-PNP-lariat ether, β-cyclodextrin) produces a synergistic effect towards the target chemical species [32,33].

N-6-(*t*-dodecylamido)-2-pyridinecarboxylic acid (*t*-DAPA) is a carrier, proposed as an alternative to LIX® reagents [known to be very selective towards Cu(II)] to overcome problems related to their cost and poor physicochemical stability. However, the facts that *t*-DAPA is not commercially available and very little is known about its stability raise the question whether it will become an attractive alternative to LIX® reagents [34,35]. Its high selectivity for Cu(II), Zn(II), and Ni(II) can be explained by the high affinity of the carboxyl group and the pyridine moiety for these metals, although it is more selective for Cu(II) at low pH values.

The extraction mechanisms for PIMs containing various carriers normally mimic those of the corresponding SX systems. A number of these were described in our previous reviews on PIMs [2,3].

19.3.3 Neutral or Solvating Carriers

Neutral or solvating carriers are commercially available phosphorus-based extractants with high selectivity towards actinides and lanthanides. Examples are presented in Table 19.2, together with the species they have been applied to.

Tri-*n*-butylphosphate (TBP) and dibutyl butyl phosphonate (DBBP) are examples of neutral phosphoric and phosphonic acid esters, respectively, that have been applied in PIMs for the extraction of Cd(II), Pb(II), U(VI), and As(V) [2,3]. Another important group of neutral carriers are ionic liquids (ILs), which consist of salts in the liquid state and have the ability to exchange their anion or cation. Aliquat 336, which is categorized as a basic extractant, can also be viewed as an IL. It is thus possible to design different ILs according to particular applications. Trihexyl(tetradecyl)phosphonium chloride (Cyphos IL 101), trihexyl(tetradecyl)phosphonium bromide (Cyphos IL 102), and trihexyl(tetradecyl) phosphonium-(2,4,4-trimethylpentyl)phosphinate (Cyphos IL 104) are commercially available phosphonium-based ILs which have been investigated recently as carriers in PIMs. Cyphos IL 104 is particularly interesting since it acts as a bifunctional extractant that can extract both cations and anions simultaneously and has been used to extract HNO_3 and Y(III) from nitrate solutions [36] and H_2CrO_4 [16]. Ionic liquids that present high specificity to a particular analyte are known as task-specific ionic liquids (TSILs). Examples include Cyphos IL 104 [37] and tricaprylmethylammonium thiosalicylate [38].

19.3.4 Macrocyclic and Macromolecular Carriers

Macrocyclic and macromolecular carriers are compounds such as crown ethers, ionizable lariat ethers, calixarenes, calix crowns, and cyclodextrins. Table 19.3 provides examples of such carriers as well as their target chemical species.

A considerable number of research papers have described PIMs using macrocyclic and macromolecular carriers [2,3]. The main reason for selecting these carriers is related to their high complexing selectivity towards metal ions. Due to the fact that their structures can be tailored to a selected metal ion, and that they exhibit low solubility in aqueous solutions, these carriers have attracted considerable interest, although the majority of them are still not commercially available. Their synthesis is often expensive, and this may affect the economic viability of their utilization in large-scale separation systems. However, only a relatively small amount of a carrier is required for the preparation of a PIM compared to the amount required in traditional SX, which possibly makes their industrial applications more feasible. Moreover, it is possible to re-dissolve the PIM after use and reform the membrane for subsequent use, thus recycling the expensive carrier [39].

19.4 PLASTICIZERS AND MODIFIERS

As mentioned earlier, unless the carrier also has plasticizing properties (e.g., Aliquat 336, D2EHPA, and TBP), a plasticizer or modifier is additionally incorporated in the membrane preparation. The polymer chains of the base polymer are under the influence of weak and non-specific van der Waals forces and stronger polar interactions resulting in the formation of a rigid three-dimensional membrane structure. Such a structure produces very poor diffusive fluxes for species introduced into the polymer. Consequently, a plasticizer may be required to be added to the PIM composition to penetrate between the polymer chains and reduce the intermolecular forces [7]. This leads to a decrease in the polymer glass transition temperature and an increase in the membrane softness and flexibility, consequently improving the diffusion of the extracted species through the membrane. The most common plasticizers used in PIMs are 2-nitrophenyloctyl ether (NPOE) and 2-nitrophenylpentyl ether (NPPE) [2,3]. Others, such as bis(2-ethylhexyl)adipate (DEHA), dibutylphthalate (DBP), or dibutylsebacate (DBS) have also been tested. These plasticizers contain a hydrophobic alkyl backbone and one or more polar groups which neutralize the polar groups of the polymer. A balance between the non-polar and polar groups is necessary. A longer alkyl chain results in higher hydrophobicity and viscosity and lower polarity of the plasticizer, while an increase in the number of polar groups usually decreases viscosity and increases the hydrophilicity, thus leading to loss of the plasticizer to the aqueous phase(s).

The concentration of the plasticizer, if required, is important since if it is too low a rigid and brittle membrane will be formed. This is termed the "anti-plasticization" effect and the minimum plasticizer concentration required is dependent on the plasticizer type and the base polymer used. For example, for PVC-based PIMs, this concentration is around 20% (w/w). If the plasticizer concentration is too high, then excess plasticizer can exude to the PIM surface and the thin film formed can inhibit transport across the membrane. Also, such PIMs are normally mechanically weak and difficult to use.

Higher initial flux values for the extracted species are usually obtained when PIMs incorporate high-polarity and low-viscosity plasticizers, such as NPOE and NPPE, and this has led to the conclusion that the initial flux values increase with increasing the dielectric constant and decreasing the viscosity of the plasticizer. However, it should be noted that most of the plasticizers used in PIMs have similar viscosity values (Table 19.4) and moreover, the dielectric constant of the membrane liquid phase is also dependent on the dielectric constants of the carrier and the base polymer [40]. Thus, a degree of caution is necessary when attempting to correlate initial PIM flux values with the dielectric constant and viscosity of the plasticizer.

Modifiers are common in SX processes to solubilize extracted species in the organic phase and hence overcome third-phase formation. Long-chain alkyl alcohols, namely 1-dodecanol and 1-tetradecanol, are examples of modifiers that have been tested in PIMs by Cho et al. [41]. The same study has shown that membranes prepared with a range of long-chain alkyl alcohols (C_6 to C_{14}) as modifiers have exhibited similar transport rates and degrees of extraction when first used. However, when used again only the PIMs containing 1-dodecanol or 1-tetradecanol have shown no evidence of deterioration in their performance. The fact that these modifiers have the lowest water solubility values (Table 19.4), in comparison to the other modifiers, suggests a direct relationship between membrane stability and water solubility of the modifier.

19.5 STRUCTURE, STABILITY, AND LIFETIME

Successful PIMs should be transparent and homogeneous, and most studies involving this type of membrane make this evaluation through observation with the naked eye or under an optical microscope. Nevertheless, such evaluation is rather subjective. For instance, SLMs also look transparent and homogeneous to the naked eye, although they have a microporous structure. Hence, several advanced and sophisticated techniques have been employed in the study of the morphology and structure of PIMs in order to determine the distribution and interaction of the various membrane components, and ultimately assess how that affects the membrane transport efficiency.

Such techniques can be used either alone or in conjunction with each other and among them are scanning electron microscopy (SEM), atomic force microscopy (AFM), Fourier transform infrared spectroscopy (FTIR), and transmission infrared mapping microscopy (TIMM) [3]. These techniques are often used in surface morphology studies and have been used with PIMs to provide information on how the carrier and the plasticizer are distributed within the base polymer. Moreover, other techniques have been applied. FTIR has been used to study the type of interactions between PIM constituents while TIMM has been used to provide a distribution profile of the membrane components. High-resolution synchrotron-based FTIR spectroscopy was also used to study PIM homogeneity [42]. PVC-based PIMs containing D2EHPA or Aliquat 336 as carriers and CTA-based PIMs with Aliquat 336 as the carrier were shown to be homogeneous on the micrometer scale. This study also included proton-induced X-ray emission microspectrometry (μ-PIXE) measurements which revealed a homogeneous distribution of the membrane

TABLE 19.4
Physicochemical parameters of some PIM plasticizers/modifiers (all abbreviations are defined in the Glossary)

Plasticizer/modifier	Dielectric constant (ε_r)	Water solubility (g/kg H_2O)	Viscosity (cP)
NPOE	24 (25°C) [115]	–	11.1 (25°C) [115]
NPPE	24 (na) [116]	–	7.58 (na) [104, 109]
DEHA	5 (na) [116]	–	13.7 (na) [104, 109]
DBP	6.58 (20°C) [117]	0.0112 (25°C) [117]	16.6 (25°C) [117]
TEHP	4.8 (25°C) [115]	–	13.1 (25°C) [115]
DBS	4.54 (20°C) [117]	0.04 (20°C) [117]	9.5 (na) [118]
TBEP	8.7 (na) [82]	–	–
DOP	5.22 (20°C) [117]	0.00027 (25°C) [117]	40.4 (na) [119]
TBP	8.34 (20°C) [117]	0.39 (25°C) [117]	3.32 (na) [103]
EB	6.20 (20°C) [117]	0.83 (25°C) [117]	–
1-dodecanol	5.82 (30°C) [117]	0.004 (25°C) [117]	–
1-tetradecanol	4.42 (45°C) [117]	0.00031 (25°C) [117]	–

na: temperature not available.

components both on the surface and in the interior of the membrane. On the other hand, PIMs that cannot be produced reproducibly, such as those containing Cyanex 272 and PVC, have been found to lack homogeneity. Best et al. [43] have studied the instability of Cyanex 272-based PIMs used for the extraction of Co(II) by X-ray absorption spectroscopy (XAS). The results indicated that the blue oil formed on the surface of the membrane corresponded to $\{Co(Cyanex)_2\}_n$ oligomeric species, which could not be back-extracted into an acidic receiver phase. Small- and wide-angle X-ray scattering (SAXS/WAXS) have been used recently to characterize PIMs in terms of their nano-scale internal structure [44]. Using PVC, CTA, and PVDF-HFP polymers as examples, the results have indicated that both pure polymer membranes and PIMs contained alternating crystalline and amorphous phases, which is in agreement with previous polymer studies [45]. Moreover, while miscible plasticizers are capable of associating closely with the polymer amorphous phase, the remainder of the PIM liquid phase (carriers/less miscible plasticizers) is distributed throughout nanometer-size domains which exist between—and are phase-separated from—the polymer chains.

The strong interactions between the organic liquid phase and the base polymer in PIMs are considered to be responsible for their better stability in comparison with other liquid membranes. In SLMs, capillary forces or interfacial tension are responsible for the retention of the liquid phase within the membrane pores. However, because of the weakness of these adhesive forces, membrane breakdown can easily occur by lateral shear forces, emulsion formation, or leaching of the membrane liquid phase to the adjacent aqueous phase(s). In contrast, in PIMs, the carrier, plasticizer, and base polymer are most likely bound together by a form of secondary bonding as discussed in Section 19.4 which includes weak and non-specific van der Waals forces and stronger polar interactions such as hydrogen bonding. These interactions are much stronger than interfacial tension or capillary forces and are thought to provide PIMs with considerably more stability than SLMs.

Several authors have studied the stability of PIMs and their reusability by performing repeated transport experiments with the same membrane, which involved renewing both the feed and receiver solutions each time. In general, the stability of PIMs has been proven to be relatively good, with flux or permeability values varying only slightly within the first several cycles and with no signs of structural weakening of the membrane. In some cases, particularly when very low ionic strength aqueous phases are used, there can be considerable leaching of the membrane liquid phase or of some of its components (plasticizer, modifier, and carrier) into the aqueous phase. However, it has been demonstrated that an equilibrium condition is eventually reached and negligible further leaching occurs [46]. Thus, it is suggested that PIMs need to be conditioned in solutions of similar ionic strength to those to be employed in the actual extraction and transport systems. Although there is little doubt that the stability of PIMs is greater than that of SLMs, lower initial flux values or permeability coefficients are frequently reported for PIMs in comparison with those reported for SLMs.

19.6 TRANSPORT MECHANISMS

The overall transport mechanism across a typical PIM can be viewed as consisting of the following three steps [2]:

1. The extracted species diffuses across the stagnant diffusion layer at the membrane–feed solution interface where it reacts with the carrier and the reacted carrier species is replaced by another carrier species from the bulk of the membrane.
2. The product of the reaction between the extracted species and the carrier, which is usually a complex or an ion-pair, is transported towards the membrane–receiver solution interface along the corresponding concentration gradient.
3. At the membrane–receiver solution interface the extracted species is released into the receiver solution. This process can be facilitated by a suitable stripping reagent. The liberated in this process carrier species is transported back to the membrane–feed solution interface.

The PIM mass transfer process is often referred to as "facilitated transport" and it allows transport of the extracted species from a feed solution with a lower concentration to a receiver solution with a higher concentration. Thus, it is possible to transfer quantitatively the extracted species from the feed to the receiver solution without using an external force (e.g., pressure, electrical field). It should be pointed out that despite the uphill character of the overall transport between the feed and receiver solutions, the transport of the complex or ion-pair of the extracted species within the membrane is always downhill, i.e., along the corresponding concentration gradient.

When the extracted species forms a complex with the carrier, the receiver solution usually contains a stripping reagent (e.g., a water-soluble complexing reagent). The role of this reagent is to facilitate the dissociation of the carrier-extracted species complex at the membrane–receiver solution interface and to subsequently form a water-soluble complex with the extracted species. In fact, the extracted species is present in the receiver solution due to the stripping complexation reaction as a different chemical species. The driving force in this case is the difference between the concentration of the carrier-extracted species complex at the membrane–feed solution interface and the practically zero concentration of the complex at the membrane–receiver solution interface. In the case of extracted ionic species, the driving force for the uphill transport can be the potential gradient generated by the coupled transport of another ionic species across the membrane. The extracted ionic species, in this case, is transported to satisfy the electroneutrality condition within the membrane system. This coupled transport process can be counter- (Figure 19.4b and

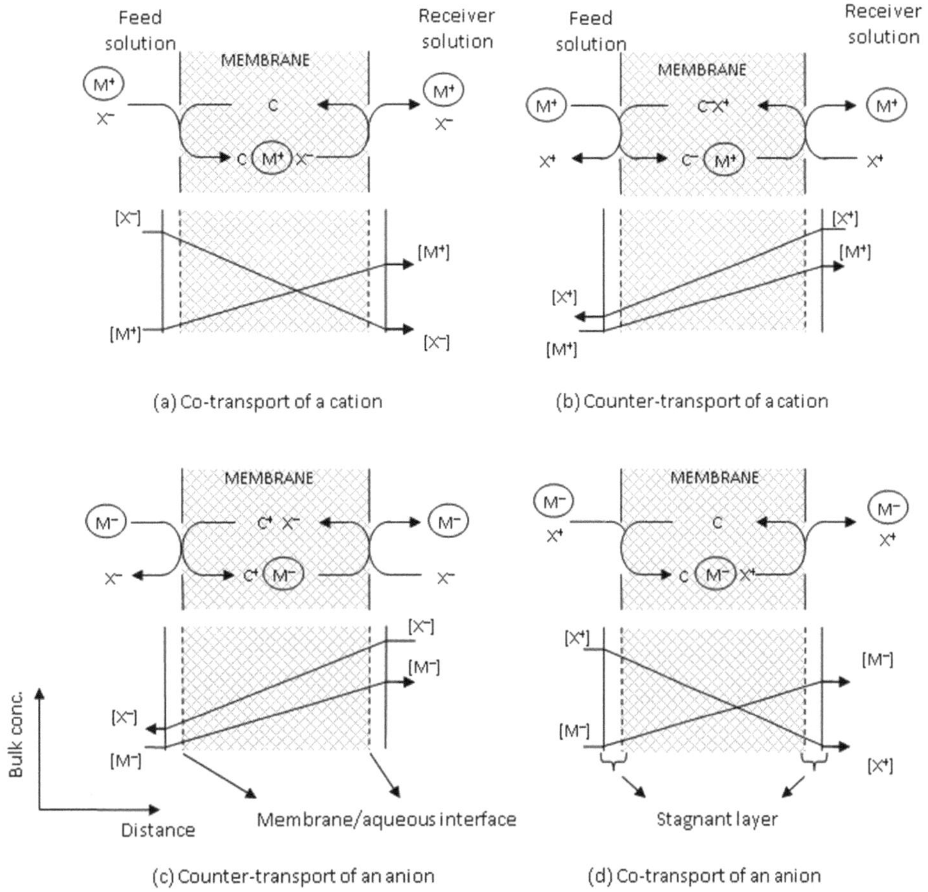

FIGURE 19.4 Schematic description of coupled transport of a positively charged (M^+) or negatively charged (M^-) species through a PIM. C represents the carrier and X is an aqueous soluble coupled transport ion. [M^+], [M^-], [X^-], and [X^+] represent the total analytical concentrations of the respective solutes in the bulk aqueous phases. (a) The target solute is a cation and is concurrently transported with an anion; (b) the target solute is a cation and is counter-currently transported with a cation; (c) the target solute is an anion and is counter-currently transported with an anion; (d) the target solute is an anion and is concurrently transported with an cation. (Reprinted from Reference [2], (Copyright 2006), with permission from Elsevier.)

c) or co- (Figure 19.4a and d) transport with respect to the extracted ionic species.

A possible mechanism for the transport of the extracted species across the membrane can be based on Fickian diffusion of its complex or ion-pair with the carrier through the membrane liquid phase. Another mechanism which can be responsible for the bulk membrane transport of the extracted species is the so-called "chained carrier" mechanism proposed by Cussler et al. [47]. This mechanism applies to ion-exchange membranes with ionic sites covalently bound to the polymeric backbone (e.g., Nafion). Very often the carriers used in PIMs are bulky chemical species and their mobility within the membrane liquid phase is so low that for all practical purposes these carriers can be considered as immobile, and the membrane transport of the extracted species can be viewed as taking place via successive relocations from one carrier to another. Riggs and Smith [48] used this mechanism to explain the transport of sugars across PIMs containing different TOMA salts as carriers. Sugars have a strong ability to form hydrogen bonds and the transport process in this case is most probably due to the formation of heteroconjugate anions between the sugar molecule and the TOMA anion. The "carrier chained" mechanism explains the percolation threshold of PIMs where a certain minimum amount of carrier is required to form a "continuous chain" across the membrane to allow the uninterrupted transport of the extracted species from the feed to the receiver solution [40,48,49]. The percolation threshold, which determines the average distance between the carrier molecules, has been found to depend on the size of the extracted species, i.e., larger extracted species can be transported even when carrier molecules are not in close proximity to each other. For example, White et al. [50] established that the threshold for the disaccharide sucrose was lower than that for the monosaccharide fructose. The same authors suggested that local mobility of the carrier particles could facilitate bulk transport across the membrane. In this case the extracted species can "jump" between two neighboring carrier particles when as a result of their local mobility they come in close proximity.

It can be expected that often the bulk transport mechanism is a combination of the mechanisms outlined above. In both cases the transport rate will be dependent on: (1) the viscosity of the liquid phase, which is determined by the viscosities of the extractant and plasticizer/modifier (if present); (2) the size of the extracted adduct; (3) the size of the nanochannels (discussed in Section 19.5); and (4) the temperature of the system.

19.7 EXTRACTION AND STOICHIOMETRY

SX and ion-exchange are extremely important techniques for the separation of numerous species from aqueous solutions and have been used extensively in separation science and in industrial processes. It has often been said that PIMs mimic the extraction properties of the organic phase in SX systems and the solid phase in ion-exchange systems but have the advantage that both extraction and back-extraction processes can take place in a single step by using the PIM in transport mode (i.e., a PIM separates the feed and receiver solutions). Also, PIM-based separation minimizes and even eliminates the use of often highly flammable, toxic, and volatile diluents.

The majority of PIM studies so far have been concerned with the examination of various carrier/plasticizer/modifier compositions for preparing successful flat sheet membranes to extract and transport the chemical species of interest such as inorganic and organic ions or molecules. Extraction studies can be carried out by simply agitating an aqueous solution containing the species of interest and the PIM studied. The depletion of the extracted species in the aqueous solution is monitored in time until the extraction equilibrium between the membrane and the solution is reached. Alternatively, a two-compartment transport cell can be used where both compartments contain the same feed solution. Once extraction equilibrium has been established, the extraction constant can be calculated provided that the carrier in the membrane is not completely reacted.

The molar concentrations of the membrane components can be calculated if the membrane volume is known. It can be determined by measuring the membrane geometric dimensions or by using the membrane density determined independently and knowing the membrane mass. In PIM extraction and transport studies, emphasis is placed on achieving both high extraction constants and transport rates on one hand and separation from other species present in the aqueous phase on the other. The transport rate is assessed through the determination of the initial flux [J_0, Eq. (19.1)] or the permeability [P, Eq. (19.2)]:

$$J_0 = \left(\frac{V}{A}\right)\cdot\left(\frac{C_t - C_0}{t}\right) \tag{19.1}$$

$$P = \frac{J_0}{C_0} \tag{19.2}$$

where V is the volume of feed or receiver solution, A is the area of the PIM in contact with the feed or receiver solution, and Ct and C_0 correspond to the concentrations of the target chemical species of interest at time t and time 0 (i.e., initial concentration), respectively.

In a number of studies, attempts have been made to determine the stoichiometry of the extracted complex of the species of interest and the carrier and to elucidate the extraction mechanism. In an SX system, a common approach for determining the stoichiometry of the extracted complex is based on the so-called "slope analysis" method [21]. This involves carrying out a series of extraction experiments using different concentrations of the extractant in the organic phase. Let's consider a generic extraction reaction described by the following stoichiometric equation:

$$A_{(aq)} + nE_{(mem)} \leftrightarrows AE_{n\,(mem)} \tag{19.3}$$

where A is the chemical species of interest, E is the extractant, and subscripts aq and mem refer to the aqueous and membrane phases, respectively.

The extraction constant of this process (K_{ex}) is described by Eq. (19.4).

$$K_{ex} = \frac{\left[AE_n\right]_{(mem)}}{\left[A\right]_{(aq)}\left[E\right]_{(mem)}^{n}} \tag{19.4}$$

All concentrations in Eq. (19.4) are the corresponding equilibrium concentrations.

Taking into account that the distribution ratio (D) is defined by Eq. (19.5), Eq. (19.4) can be converted to Eq. (19.6).

$$D = \frac{\left[AE_n\right]_{(mem)}}{\left[A\right]_{(aq)}} \tag{19.5}$$

$$lgD = lgK_{ex} + nlg\left[E\right]_{(mem)} \tag{19.6}$$

If the volumes of the two phases are equal and the initial extractant concentration in the organic phase is selected to be much higher than the initial concentration of the extracted species in the aqueous phase, its equilibrium value can be assumed to be practically equal to its initial value. Under such conditions the slope of the relationship lgD versus $lg[E]_{(mem)}$, where $[E]_{(mem)}$ is the initial extractant concentration in the organic phase, will be equal to the stoichiometric coefficient n of the extracted complex (AE_n).

Another simpler method involves fully reacting the extractant in the organic phase with the extracted species using a high initial aqueous concentration of the extracted species and analyzing the organic phase to determine the (extractant: extracted species) ratio n [Eq. (19.3)].

In the case of a PIM, the latter method can be used by ensuring a sufficiently high initial concentration of the extracted species in the aqueous feed phase. However, the complex stoichiometry obtained may be different from the stoichiometry of the complex formed when the membrane is not saturated with the extracted species.

On the other hand, the "slope analysis" method is not readily applicable to PIMs since it is more difficult to prepare a range of membranes with the same volume but with sufficiently different carrier concentrations. Also, it is not possible to fulfill the requirement that the carrier concentration in the membrane remains essentially constant during the extraction process.

A novel method for determining the stoichiometry in PIM systems has been proposed by St John et al. [29]. This approach has been applied to the extraction of uranium(VI) from sulfate solutions using a PIM containing D2EHPA as the carrier and involves equilibrating segments of the PIM of the same composition (40% w/w D2EHPA) but with varying masses with the aqueous feed solution. This enables different equilibrium concentrations of unreacted D2EHPA in the membrane to be obtained. At constant acidity in the aqueous feed solution, the relationship between the distribution ratio (*D*) and the concentration of unreacted D2EHPA in the PIM can be described by Eq. (19.6) and the unreacted D2EHPA concentration can be determined by the following equation:

$$[E]_{(mem)} = [E]^0_{(mem)} - n'[AE_n] \tag{19.7}$$

where *E* and *A* refer to D2EHPA and the uranyl cation (UO_2^{2+}) in the membrane, while superscript 0 refers to initial concentration.

The parameters *n* and *n′* should have the same value. The method for determining this value involves varying *n′* in Eq. (19.7) in the range where the true value is expected to be, i.e., between 1 and 10, and calculating the unreacted equilibrium D2EHPA concentration ($[E]_{(mem)}$) for each *n′* value. This concentration is then used in determining *n* by using Eq. (19.6). The function $n - n'$ is plotted versus *n′* (Figure 19.5) and the *n′* intercept where $(n - n') = 0$ (i.e., $n = n'$) corresponds to the true value of *n* and gives the number of D2EHPA units in the U(VI)–D2EHPA complex. The value of 4 was obtained, which along with the fact that two hydrogen ions were exchanged for each uranyl cation, suggested that the extraction reaction could be described by Eq. (19.8).

$$UO_2{}^{2+}{}_{(aq)} + 2(HB)_{2\,(mem)}\ \Delta UO_2B_2(HB)_{2\,(mem)} + 2H^+{}_{(aq)} \tag{19.8}$$

where $(HB)_2$ refers to the dimeric form of D2EHPA and B refers to its conjugate base.

19.8 MEMBRANE CONFIGURATION

As mentioned earlier, the common method used in PIM studies involves casting flat sheet membranes, however, the ability to prepare PIMs in other configurations is of considerable importance, particularly if they are to be applied in large-scale industrial separation processes.

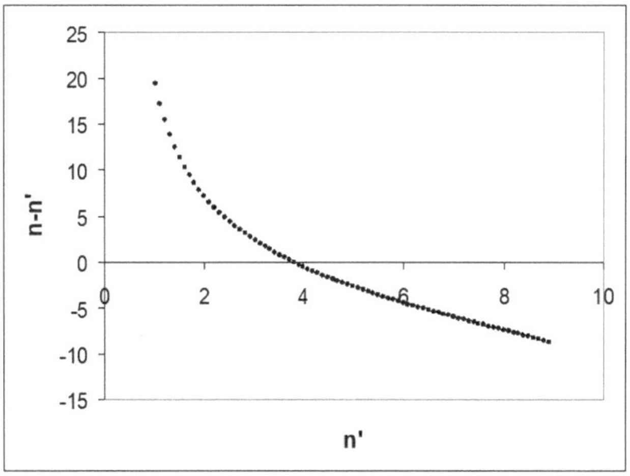

FIGURE 19.5 Method for determining the stoichiometry in PIM systems. Plot of *n – n'* versus *n'*. (Reprinted from Reference [29], (Copyright 2010), with permission from Elsevier.)

Membrane separation on a large scale requires a high membrane surface area to solution volume ratio in order to process large volumes of feed solutions. Separation modules in which the PIM is sandwiched between two meander-shaped flow channels (Figure 19.6) containing the feed and receiver solutions, respectively, provide a longer flow path length and hence a higher contact surface area than is possible with a simple two-compartment cell used in the transport studies outlined earlier [51]. However, the application of such a modular approach is limited by the size of the flat sheet membrane that can be cast and by the number of modules required to achieve the necessary membrane surface area to volume ratio.

The application of membranes to industrial-scale separation problems is a mature technology as demonstrated by the success of processes like reverse osmosis (RO) for desalination and wastewater treatment. Therefore, it can be anticipated that some of the methods for making large surface area membrane modules for these processes could be applicable to PIM-based separation. One such method involves producing spiral-wound membrane modules containing a number of individual flat sheet membranes wound around a central core. A spiral-wound module has a very high membrane surface area to volume ratio and is relatively easy to manufacture and replace. Although PIMs have considerable mechanical strength, it may be necessary to use an inert membrane backing material, similar to those used in RO membranes, to provide additional stability. An example of a spiral-wound module incorporating an SLM is shown in Figure 19.7 [52]. This module could easily accommodate a PIM.

Another common method for making large surface area membrane modules can be based on the hollow-fiber (HF) membrane technology. HFs are manufactured by spinning techniques involving polymer solutions or melts. A polymer solution is pumped through a spinneret consisting of an orifice with a central inlet tube which forms the HF shape followed by immersion in a coagulation bath.

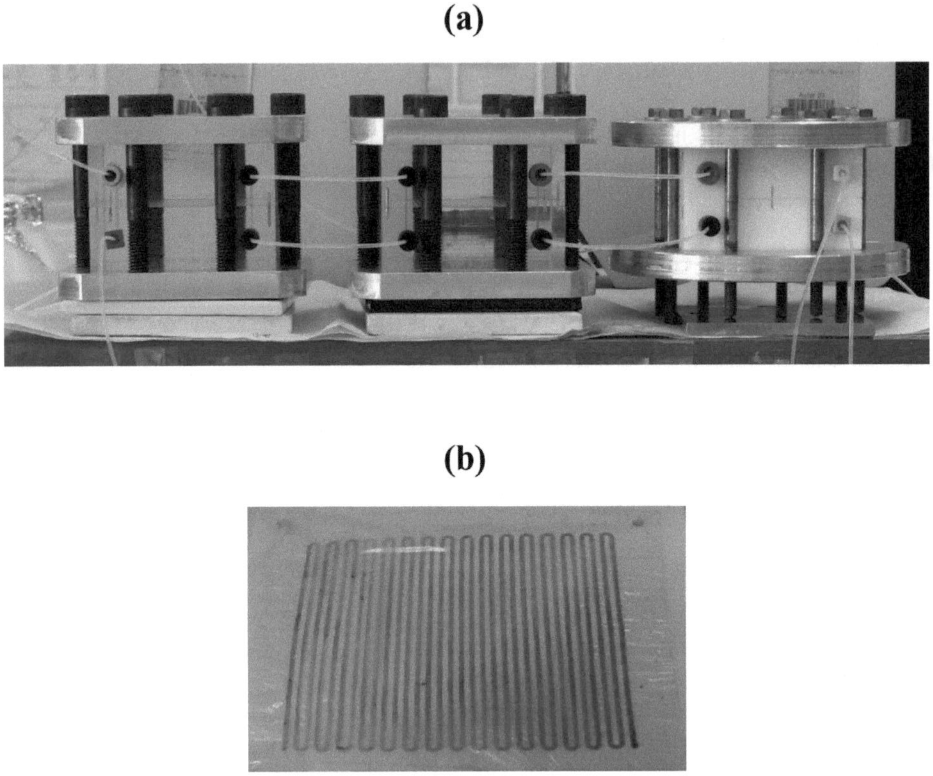

FIGURE 19.6 Laboratory-scale pilot system consisting of three membrane separation modules in series used for the transport of thiocyanate [51]. (a) Photographic image of the three modules in series, each containing a PIM sandwiched between two meander-shaped flow channels. (b) Photographic image of the membrane after exposure to the channels filled with blue ink.

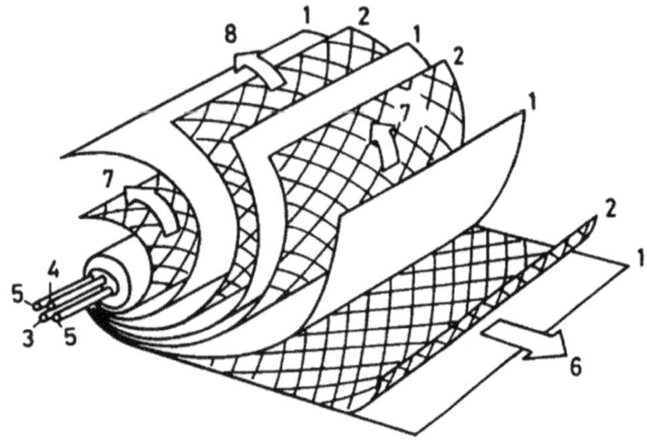

FIGURE 19.7 Schematic diagram of a spiral-type flowing liquid membrane (SLM) module. 1, microporous hydrophobic membrane (support); 2, mesh spacer; 3, inlet pipe of feed solution; 4, inlet pipe of receiver solution; 5, inlet tube of organic membrane solution; 6, feed solution; 7, organic membrane solution; 8, receiver solution. (Reprinted from Reference [52]. Copyright (1989), with permission from Taylor & Francis Ltd.)

In industrial-scale separation, an HF module consists of a bundle of hundreds of individual fibers which provide the required high membrane surface area to volume ratio. In such a module one aqueous solution is flown through the lumen of the fibers while the other solution is flown co- or counter-currently on the outside of the fibers.

HFs have been used successfully for preconcentration in chemical analysis. One technique termed hollow-fiber liquid-phase microextraction (HF-LPME) involves filling the pores of an inert HF material (commonly polypropylene) with an organic phase containing a carrier in a similar way to the manufacturing of SLMs. The fiber is then dipped into an aqueous sample solution and the analyte is transported across the HF to the small volume of the receiver solution in the lumen of the HF which is subsequently analyzed [53,54].

It is possible to use HFs with pores filled with a PIM. However, the real challenge is to prepare HFs made of PIM alone and Kusumocahyo et al. have attempted to produce such HFs for Ce(III) separation using commercially available CTA HFs which have been immersed in a chloroform solution containing *N,N,N′,N′*-tetraoctyl-3-oxapentanediamide (TODGA) as the carrier and NPOE as the plasticizer [55]. The solution penetrates the CTA HFs and dissolves some of the CTA and, after drying, produces what is essentially PIM HFs.

A PIM can be used as a solid-phase extractant, and as such mimics the behavior of an ion-exchange resin. Thus, it is of interest to produce PIMs in the form of beads. A microfluidic system for producing polymer microspheres has been recently reported by Zhang et al. [56] which was subsequently used to fabricate micro-polymer inclusion beads (μPIBs) (Figure 19.8) [57,58]. The formation of μPIBs is based on the

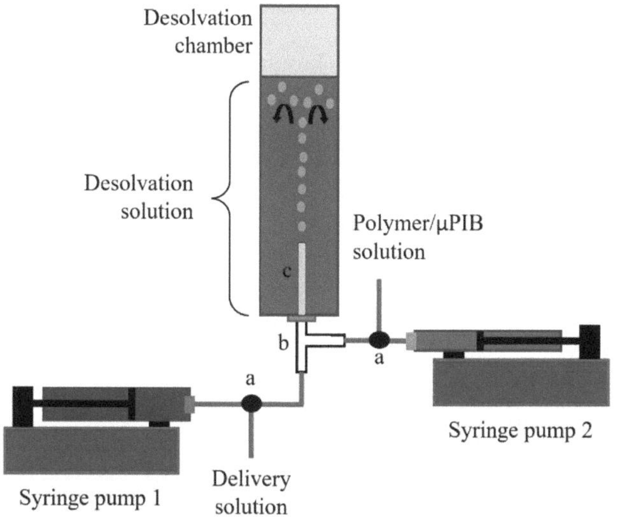

FIGURE 19.8 Schematic representation of the microfluidic system used to fabricate polymer microspheres or micropolymer inclusion beads (μPIBs). a, injection valve; b, tee connector; c, stainless steel capillary tube. (Reprinted from Reference [58], (Copyright 2021), with permission from Elsevier.)

low solubility of THF in a solution of 20% (w/w) NaCl. As such, the polymer and liquid organic phases are dissolved in THF and this μPIB solution is then merged with a 20% (w/w) NaCl solution leading to its segmentation. Once the μPIB droplets reach the desolvation NaCl solution, where the NaCl concentration is lower than 20% (w/w), they rise as the THF is desolvated and μPIBs are formed. The size of the beads can be controlled by manipulating the concentration of the μPIB solution. An alternative approach has also been reported based on coating glass beads with a PIM using a dip coating procedure [59].

19.9 APPLICATIONS, CONCLUSIONS, AND FUTURE TRENDS

The steady increase in research on PIMs will lead to a wide range of laboratory and industry-based separation applications. Their attraction for use in industry is based on the fact that PIMs can be formulated using most of the commercially available SX reagents and they do not require a large inventory of diluents. In addition, losses of PIM components to the corresponding feed and receiver aqueous solutions are small compared to those in SX systems and PIMs have the advantage of allowing extraction and stripping to be conducted in a single step. Most research on PIMs has been focused on flat sheet membranes, however, industrial applications will require very high surface area to solution volume ratios and so other membrane configurations such as HFs or spiral-wound modules will be required to achieve this aim.

In spite of their promise for use in separation science, there are still a few improvements to be made before PIM-based separation is fully accepted as a viable alternative to mature separation techniques such as SX, ion-exchange, and SLM-based separation. These problems are highlighted in previous sections of this chapter and are mostly associated with membrane stability and the relatively low transmembrane transport rate. It has already been demonstrated that PIMs have considerably lower losses of components to aqueous solutions than SLMs and recent research has shown that these losses can be minimized if the membranes are pre-conditioned before use [60]. Also, the incorporation of highly hydrophobic ionic liquids in the membrane composition either as carriers, plasticizers, or modifiers will further minimize such losses.

The common base polymers used in PIMs are PVC, CTA, and PVDF-HFP. PVC and CTA have some limitations such as instability caused by acidity or alkalinity of the adjacent aqueous solutions. CTA suffers from hydrolysis reactions in strongly acid solutions and PVC undergoes dehydrochlorination under alkaline conditions and gradually turns brittle and black. However, alternative base polymers may be of use, and one such example is PVDF and its co-polymers (e.g., PVDF-HFP) which are relatively inert under most conditions [5].

The problem of low diffusion coefficients is more difficult to overcome and both physical and chemical approaches can be used to improve transport rates. Reduction of the membrane thickness will have a dramatic effect on the transport rate, and the manufacturing of membranes as thin as 20 μm has been reported which still retain acceptable mechanical strength [46]. Also, the use of an inert porous backing material should allow even thinner membranes to be produced.

An increase in temperature increases the diffusion coefficients in PIMs and the membranes have been shown to be stable up to around 50°C [46].

The application of sonication to PIM systems increases the transport rate, which has been suggested to be due to a reduction in the thickness of the stagnant diffusion aqueous layer at the aqueous solution–membrane interface [61]. However, it is also possible that sonication provides some internal mixing of the membrane liquid phase.

The chemical approach to increasing transport rates involves manipulation of the PIM composition. It has been observed that some PIM compositions provide much higher transport rates than their SLM counterparts but the reasons for this phenomenon remain unclear. In order to understand the chemical processes occurring in PIMs, a number of researchers have investigated the structure of PIMs with a view to obtaining information regarding the way the carrier and other membrane components interact and the mechanisms for mass transport within the membrane. It is anticipated that once there is a better understanding of the structure of PIMs, it will be possible to better formulate PIM compositions, providing optimum transport rates.

Another area in which PIMs show considerable promise is chemical analysis [62]. The use of PIMs in the construction of ISEs and optodes is well established but their potential use in analytical separation is only starting to be explored. PIMs are particularly useful in solid-phase extraction (SPE) for preconcentration of analytes [22,23].

For example, Fontàs et al. have used a PIM containing Aliquat 336 as the carrier for the preconcentration of Cr(VI) prior to its determination in the membrane by energy-dispersive X-ray fluorescence spectrometry [63]. Flat sheet PIMs can also be conveniently incorporated into separation modules for use in on-line analytical techniques such as flow injection analysis. For example, a D2EHPA-based PIM has been used in a separation module incorporated in a flow injection analysis system for the determination of Zn(II) [46]. More recently PIM-based separation in flow injection analysis has been utilized by Yaftian et al. [64] for the selective determination of V(V) in sulfate solutions with Cyphos IL 101 as the PIM carrier, and by Vera et al. [65] for the determination of arsenate using Aliquat 336 as the carrier. This new approach has interesting implications for use in automated analysis, particularly in portable instruments for the continuous monitoring of pollutants in the field.

PIMs show considerable potential for passive sampling of specific contaminants in water [66–68]. In this application, low membrane diffusion coefficients can be an advantage. Passive samplers are generally left immersed in a river, lake, or contaminated water site for several days and a PIM can slowly and selectively accumulate and transport the analyte to the receiver solution within the passive sampler.

The use of HF-LPME is expected to make an important contribution to analytical separation.

Recent research has demonstrated that PIMs can be used as templates for the production of metallic nanoparticles either imbedded within the PIM [24,25] or as a monolayer on the surface [26]. Such nanoparticles may find use in catalysis and in chemical sensing.

In addition, an electric field can be applied to the PIM-based transport set up to improve the rate of transport of the extracted species across the PIM. This approach is usually termed electromembrane or electro-driven extraction [69]. PIMs can also be used in combination with ion-exchange membranes to perform electrodialysis [70].

GLOSSARY

AFM: Atomic force microscopy
AuNPs: Au nanoparticles
Bif-ILEs: Bifunctional ionic liquid extractants
BLM: Bulk liquid membrane
BPA: Bis(pyridylmethyl)amine
CA: Cellulose acetate
CAB: Cellulose acetate butyrate
CAH: Cellulose acetate hydrogen phthalate
Calix[4]pyrrole: Meso-octamethyl-porphyrinogen
CAP: Cellulose acetate propionate
CMPO: Octyl(phenyl)-*N*,*N*-diisobutyl carbamoylmethyl phosphine oxide
CTA: Cellulose triacetate
CTB: Cellulose tributyrate
Cyanex 272: Bis(2,4,4-trimethylpentyl)phosphinic acid
Cyanex 301: Bis(2,4,4-trimethylpentyl)dithiophosphinic acid
Cyanex 302: Bis(2,4,4-trimethylpentyl)monothio phosphinic acid
Cyanex 471X: Tri-isobutylphosphine sulfide
Cyphos IL 101: Trihexyl(tetradecyl)phosphonium chloride
Cyphos IL 102: Trihexyl(tetradecyl)phosphonium bromide
Cyphos IL 104: Trihexyl(tetradecyl)phosphonium-(2,4,4-trimethylpentyl)phosphinate
***t*-DAPA**: *N*-6-(*t*-dodecylamido)-2-pyridinecarboxylic acid
DBBP: Dibutyl butyl phosphonate
DBP: Dibutylphthalate
DBS: Dibutylsebacate
DCM: Dichloromethane
DEHA: Bis(2-ethylhexyl) adipate
D2EHDTPA: Di(2-ethylhexyl) dithiophosphoric acid
D2EHPA: Di(2-ethylhexyl) phosphoric acid
DNNDS: Dinonylnaphthalene disulfonic acid
DNNS: Dinonylnaphthalene sulfonic acid
DOP: Dioctylphthalate
EB: Ethyl benzoate
ELM: Emulsion liquid membrane
FFAT: 5-(4-Phenoxyphenyl)-6H-1,3,4-thiadiazin-2-amine
FTIR: Fourier transform infrared spectroscopy
HBA: Benzoylacetone
HBM: Dibenzoylmethane
HBTA: Benzoyltrifluoroacetone
HF: Hollow fiber
HF-LPME: HF liquid phase microextraction
HFTA: Furoyltrifluoroacetone
HPBI: 3-Phenyl-4-benzoylisoxazol-5-one
HTTA: Thenoyltrifluoroacetone
IL: Ionic liquid
ISE: Ion-selective electrode
LM: Liquid membrane
LSI: Pb(II) selective ionophore
2-MethylTHF: 2-Methyltetrahydrofuran
MW: Molecular weight
NPOE: 2-Nitrophenyloctyl ether
NPPE: 2-Nitrophenylpentyl ether
μPIBs: Micropolymer inclusion beads
PIM: Polymer inclusion membrane
μ-PIXE: Proton-induced X-ray emission microspectrometry
PNP-lariat ether: Cyclic phosphaza-16-crown-6 ether
PVC: Poly(vinyl) chloride
PVDF: Polyvinyldene fluoride
PVDF-HFP: Poly(vinylidene fluoride-co-hexafluoropropylene)
PVP: Poly(vinylpyrrolidone)
RO: Reverse osmosis
SAXS: Small angle X-ray scattering
SEM: Scanning electron microscopy
SLM: Supported liquid membrane
SPE: Solid-phase extraction
SX: Solvent extraction
TBA: Tri-*n*-butylamine
TBEP: Tris(2-butoxyethyl) phosphate
TBP: Tri-*n*-butylphosphate
TDPNO: 4-(1'-n-Tridecyl)pyridine *N*-oxide

TEHP: Tris-(2-ethylhexyl) phosphate
TETDS: Tetraethylthiuram disulfide
THA: Tri-*n*-hexylamine
THF: Tetrahydrofuran
TIMM: Transmission infrared mapping microscopy
TIOA: Tri-isooctylamine
TOA: Tri-*n*-octylamine
TODGA: *N,N,N′,N′*-Tetraoctyl-3-oxapentanediamide
TOMA: Trioctylmethylammonium
TOPO: Tri-*n*-octylphosphine oxide
TSILs: Task-specific ionic liquids
WAXS: Wide angle X-rat scattering
XAS: X-ray absorption spectroscopy

REFERENCES

1. Kolev, S.D. (2019) Membrane techniques – Liquid membranes. In: *Encyclopedia of Analytical Science*, Worsfold, P., Poole, C., Townshend, A., Miró, M. (Eds.), pp. 1–9. Elsevier: Amsterdam.
2. Nghiem, L.D., Mornane, P., Potter, I.D., Perera, J.M., Cattrall, R.W., Kolev, S.D., (2006) Extraction and transport of metal ions and small organic compounds using polymer inclusion membranes (PIMs). *Journal of Membrane Science* 281: 7–41.
3. Almeida, M.I.G.S., Cattrall, R.W., Kolev, S.D. (2012) Recent trends in extraction and transport of metal ions using polymer inclusion membranes (PIMs). *Journal of Membrane Science* 415: 9–23.
4. Cattrall, R.W. (1997) *Chemical Sensors*. Oxford: Oxford Science Publications.
5. Bonggotgetsakul, Y.Y.N., Cattrall, R.W., Kolev, S.D. (2016) Recovery of gold from aqua regia digested electronic scrap using a poly (vinylidene fluoride-co-hexafluoropropene) (PVDF-HFP) based polymer inclusion membrane (PIM) containing Cyphos (R) IL 104. *Journal of Membrane Science* 514: 274–281.
6. Carner, C.A., Croft, C.F., Kolev, S.D., Almeida, M.I.G.S. (2020) Green solvents for the fabrication of polymer inclusion membranes (PIMs). *Separation and Purification Technology* 239: 116486.
7. Pereira, N., St John, A., Cattrall, R.W., Perera, J.M., Kolev, S.D. (2009) Influence of the composition of polymer inclusion membranes on their homogeneity and flexibility. *Desalination* 236: 327–333.
8. Nowak, Ł., Regel-Rosocka, M., Marszałkowska, B., Wiśniewski, M. (2010) Removal of Zn(II) from chloride acidic solutions with hydrophobic quaternary salts. *Polish Journal of Chemical Technology* 12: 24–28.
9. Gardner, J.S., Walker, J.O., Lamb, J.D. (2004) Permeability and durability effects of cellulose polymer variation in polymer inclusion membranes. *Journal of Membrane Science* 229: 87–93.
10. Ocampo, A.L., Aguilar, J.C., Miguel, E.R.D., Monroy, M., Roquero, P., de Gyves, J. (2009) Novel proton-conducting polymer inclusion membranes. *Journal of Membrane Science* 326: 382–387.
11. Valente, A.J.M., Jimenez, A., Simoes, A.C., Burrows, H.D., Polishchuk, A.Y., Lobo, V.M.M. (2007) Transport of solutes through calix[4]pyrrole-containing cellulose acetate films. *European Polymer Journal* 43: 2433–2442.
12. Benosmane, N., Guedioura, B., Hamdi, S.M., Hamdi, M., Boutemeur, B. (2010) Preparation, characterization and thermal studies of polymer inclusion cellulose acetate membrane with calix 4 resorcinarenes as carriers. *Materials Science & Engineering C-Materials for Biological Applications* 30: 860–867.
13. Kebiche-Senhadji, O., Tingry, S., Seta, P., Benamor, M. (2010) Selective extraction of Cr(VI) over metallic species by polymer inclusion membrane (PIM) using anion (Aliquat 336) as carrier. *Desalination* 258: 59–65.
14. Upitis, A., Peterson, J., Lukey, C., Nghiem, L.D. (2009) Metallic ion extraction using polymer inclusion membranes (PIMs): Optimising physical strength and extraction rate. *Desalination and Water Treatment* 6: 41–47.
15. Matsumoto, M., Murakami, Y., Kondo, K. (2011) Separation of 1-butanol by pervaporation using polymer inclusion membranes containing ionic liquids. *Solvent Extraction Research and Development—Japan* 18: 75–83.
16. Guo, L., Liu, Y.H., Zhang, C., Chen, J. (2011) Preparation of PVDF-based polymer inclusion membrane using ionic liquid plasticizer and Cyphos IL 104 carrier for Cr(VI) transport. *Journal of Membrane Science* 372: 314–321.
17. Guo, L., Zhang, J., Zhang, D., Liu, Y., Deng, Y., Chen, J. (2012) Preparation of poly(vinylidene fluoride-co-tetrafluoroethylene)-based polymer inclusion membrane using bifunctional ionic liquid extractant for Cr(VI) transport. *Industrial & Engineering Chemistry Research* 51: 2714–2722.
18. Huang, S.Y., Chen, J., Chen, L., Zou, D., Liu, C.Y. (2020) A polymer inclusion membrane functionalized by di(2-ethylhexyl) phosphinic acid with hierarchically ordered porous structure for Lutetium (III) transport. *Journal of Membrane Science* 593: 117458.
19. O'Bryan, Y., Cattrall, R.W., Truong, Y.B., Kyratzis, I.L., Kolev, S.D. (2016) The use of poly(vinylidenefluoride-co-hexafluoropropylene) for the preparation of polymer inclusion membranes. Application to the extraction of thiocyanate. *Journal of Membrane Science* 510: 481–488.
20. Wang, D., Cattrall, R.W., Li, J., Almeida, M.I.G.S., Stevens, G.W., Kolev, S.D. (2017) A poly(vinylidene fluoride-co-hexafluoropropylene) (PVDF-HFP)-based polymer inclusion membrane (PIM) containing LIX84I for the extraction and transport of Cu(II) from its ammonium sulfate/ammonia solutions. *Journal of Membrane Science* 542: 272–279.
21. Rydberg, J., Sekine, T. (1992) Solvent extraction equilibria. In: *Principles and Practices of Solvent Extraction*, Rydberg, J., Musikas, C., Choppin, G.R. (Eds.), pp. 144–147. Marcel Dekker, Inc.: New York.
22. Blitz-Raith, A.H., Paimin, R., Cattrall, R.W., Kolev, S.D. (2007) Separation of cobalt(II) from nickel(II) by solid-phase extraction into Aliquat 336 chloride immobilized in poly(vinyl chloride). *Talanta* 71: 419–423.
23. Kagaya, S., Cattrall, R.W., Kolev, S.D. (2011) Solid-phase extraction of cobalt(II) from lithium chloride solutions using a poly(vinyl chloride)-based polymer inclusion membrane with Aliquat 336 as the carrier. *Analytical Sciences* 27: 653–657.
24. Kumar, R., Pandey, A.K., Tyagi, A.K., Dey, G.K., Ramagiri, S.V., Bellare, J.R., Goswami, A. (2009) In situ formation of

stable gold nanoparticles in polymer inclusion membranes. *Journal of Colloid and Interface Science* 337: 523–530.
25. Kumar, R., Pandey, A.K., Goswami, A., Shukla, R., Ramagiri, S.V., Bellare, J.R. (2010) Plasticised polymer inclusion membrane as tunable host for stable gold nanoparticles. *International Journal of Nanotechnology* 7: 953–966.
26. Bonggotgetsakul, Y.Y.N., Cattrall, R.W., Kolev, S.D. (2011) The preparation of a gold nanoparticle monolayer on the surface of a polymer inclusion membrane using EDTA as the reducing agent. *Journal of Membrane Science* 379: 322–329.
27. Sakai, Y., Kadota, K., Hayashita, T., Cattrall, R.W., Kolev, S.D. (2010) The effect of the counter anion on the transport of thiourea in a PVC-based polymer inclusion membrane using Capriquat as carrier. *Journal of Membrane Science* 346: 250–255.
28. Kolev, S.D., Baba, Y., Cattrall, R.W., Tasaki, T., Pereira, N., Perera, J.M., Stevens, G.W. (2009) Solid phase extraction of zinc(II) using a PVC-based polymer inclusion membrane with di(2-ethylhexyl)phosphoric acid (D2EHPA) as the carrier. *Talanta* 78: 795–799.
29. St John, A.M., Cattrall, R.W., Kolev, S.D. (2010) Extraction of uranium(VI) from sulfate solutions using a polymer inclusion membrane containing di-(2-ethylhexyl) phosphoric acid. *Journal of Membrane Science* 364: 354–361.
30. Danesi, P.R., Chiarizia, R., Scibona, G. (1973) A simple purification method for liquid cation exchanger dinonylnaphathalene sulphonic acid (DNNSA). *Journal of Inorganic and Nuclear Chemistry* 35: 3926–3928.
31. Ershad, M., Almeida, M.I.G.S., Spassov, T.G., Cattrall, R.W., Kolev, S.D. (2018) Polymer inclusion membranes (PIMs) containing purified dinonylnaphthalene sulfonic acid (DNNS): Performance and selectivity. *Separation and Purification Technology* 195: 446–452.
32. Kozlowski, C.A., Kozlowska, J. (2009) PNP-16-crown-6 derivatives as ion carriers for Zn(II), Cd(II) and Pb(II) transport across polymer inclusion membranes. *Journal of Membrane Science* 326: 215–221.
33. Kozlowski, C.A., Girek, T., Walkowiak, W., Koziol, J.J. (2005) Application of hydrophobic beta-cyclodextrin polymer in separation of metal ions by plasticized membranes. *Separation and Purification Technology* 46: 136–144.
34. Tasaki, T., Oshima, T., Baba, Y. (2007) Extraction equilibrium and membrane transport of copper(II) with new N-6-(t-dodecylamido)-2-pyridinecarboxylic acid in polymer inclusion membrane. *Industrial & Engineering Chemistry Research* 46: 5715–5722.
35. Tasaki, T., Oshima, T., Baba, Y. (2007) Selective extraction and transport of copper(II) with new alkylated pyridinecarboxylic acid derivatives. *Talanta* 73: 387–393.
36. Liu, Y., Zhu, L., Sun, X., Chen, J. (2010) Toward greener separations of rare earths: bifunctional ionic liquid extractants in biodiesel. *AIChE Journal* 56: 2338–2346.
37. Wang, Y.T., Chen, L., Yan, Y.S., Chen, J., Dai, J.D., Dai, X.H. (2020) Separation of adjacent heavy rare earth Lutetium (III) and Ytterbium (III) by task-specific ionic liquid Cyphos IL 104 embedded polymer inclusion membrane. *Journal of Membrane Science* 610: 118263.
38. Elias, G., Margui, E., Diez, S., Fontas, C. (2018) Polymer inclusion membrane as an effective sorbent to facilitate mercury storage and detection by X-ray fluorescence in natural waters. *Analytical Chemistry* 90: 4756–4763.
39. Lamb, J.D., West, J.N., Shaha, D.P., Johnson, J.C. (2010) An evaluation of polymer inclusion membrane performance in facilitated transport with sequential membrane reconstitution. *Journal of Membrane Science* 365: 256–259.
40. O'Rourke, M., Duffy, N., Marco, R.D., Potter, I. (2011) Electrochemical impedance spectroscopy—A simple method for the characterization of polymer inclusion membranes containing Aliquat 336. *Membranes* 1: 132–148.
41. Cho, Y., Xu, C., Cattrall, R.W., Kolev, S.D. (2011) A polymer inclusion membrane for extracting thiocyanate from weakly alkaline solutions. *Journal of Membrane Science* 367: 85–90.
42. St John, A.M., Best, S.P., Wang, Y., Tobin, M.J., Puskar, L., Siegele, R., Cattrall, R.W., Kolev, S.D. (2011) Micrometer-scale 2D mapping of the composition and homogeneity of polymer inclusion membranes. *Australian Journal of Chemistry* 64: 930–938.
43. Best, S.P., Kolev, S.D., Gabriel, J.R.P., Cattrall, R.W. (2016) Polymerisation effects in the extraction of Co(II) into polymer inclusion membranes containing Cyanex 272. Structural studies of the Cyanex 272-Co(II) complex. *Journal of Membrane Science* 497: 377–386.
44. Nagul, E.A., Croft, C.F., Cattrall, R.W., Kolev, S.D. (2019) Nanostructural characterisation of polymer inclusion membranes using X-ray scattering. *Journal of Membrane Science* 588: 117208.
45. Strobl, G.R. (1997) *The Physics of Polymers*. Vol. 2. Springer.
46. Zhang, L.J.L., Cattrall, R.W., Kolev, S.D. (2011) The use of a polymer inclusion membrane in flow injection analysis for the on-line separation and determination of zinc. *Talanta* 84: 1278–1283.
47. Cussler, E.L., Aris, R., Bhown, A. (1989) On the limits of facilitated diffusion. *Journal of Membrane Science* 43: 149–164.
48. Riggs, J.A., Smith, B.D. (1997) Facilitated transport of small carbohydrates through plasticized cellulose triacetate membranes. Evidence for fixed site jumping transport mechanism. *Journal of the American Chemical Society* 119: 2765–2766.
49. O'Rourke, M., Cattrall, R.W., Kolev, S.D., Potter, I.D. (2009) The extraction and transport of organic molecules using polymer inclusion membranes. *Solvent Extraction Research and Development—Japan* 16: 1–12.
50. White, K.M., Smith, B.D., Duggan, P.J., Sheahan, S.L., Tyndall, E.M. (2001) Mechanism of facilitated saccharide transport through plasticized cellulose triacetate membranes. *Journal of Membrane Science* 194: 165–175.
51. Cho, Y., Cattrall, R.W., Kolev, S.D. (2018) A novel polymer inclusion membrane based method for continuous clean-up of thiocyanate from gold mine tailings water. *Journal of Hazardous Materials* 341: 297–303.
52. Teramoto, M., Tohno, N., Ohnishi, N., Matsuyama, H. (1989) Development of a spiral-type flowing liquid membrane module with high-stability and its application to the recovery of chromium and zinc. *Separation Science and Technology* 24: 981–999.
53. Ghambarian, M., Yamini, Y., Esrafili, A. (2012) Developments in hollow fiber based liquid-phase microextraction: principles and applications. *Mikrochimica Acta* 177: 271–294.
54. Bello-Lopez, M.A., Ramos-Payan, M., Ocana-Gonzalez, J.A., Fernandez-Torres, R., Callejon-Mochon, M. (2012)

Analytical applications of hollow fiber liquid phase microextraction (HF-LPME): A review. *Analytical Letters* 45: 804–830.
55. Kusumocahyo, S.P., Kanamori, T., Iwatsubo, T., Sumaru, K., Shinbo, T., Matsuyama, H., Teramoto, M. (2006) Modification of preparation method for polymer inclusion membrane (PIM) to produce hollow fiber PIM. *Journal of Applied Polymer Science* 102: 4372–4377.
56. Zhang, Y.L., Cattrall, R.W., Kolev, S.D. (2017) Fast and environmentally friendly microfluidic technique for the fabrication of polymer microspheres. *Langmuir* 33: 14691–14698.
57. Zhang, Y.L., Croft, C.F., Cattrall, R.W., Kolev, S.D. (2021) Microfluidic fabrication of micropolymer inclusion beads for the recovery of gold from electronic scrap. *ACS Applied Materials & Interfaces* 13: 61661–61668.
58. Croft, C.F., Almeida, M.I.G.S., Kolev, S.D. (2022) Development of micro polymer inclusion beads (μPIBs) for the extraction of lanthanum. *Separation and Purification Technology* 285: 120342.
59. Ohshima, T., Kagaya, S., Gemmei-Ide, M., Cattrall, R.W., Kolev, S.D. (2014) The use of a polymer inclusion membrane as a sorbent for online preconcentration in the flow injection determination of thiocyanate impurity in ammonium sulfate fertilizer. *Talanta* 129: 560–564.
60. Zhang, L.J.L., Cattrall, R.W., Ashokkumar, M., Kolev, S.D. (2012) On-line extractive separation in flow injection analysis based on polymer inclusion membranes: A study on membrane stability and approaches for improving membrane permeability. *Talanta* 97: 382–387.
61. Bonggotgetsakul, Y.Y.N., Ashokkumar, M., Cattrall, R.W., Kolev, S.D. (2010) The use of sonication to increase extraction rate in polymer inclusion membranes. An application to the extraction of gold(III). *Journal of Membrane Science* 365: 242–247.
62. Almeida, M.I.G.S., Cattrall, R.W., Kolev, S.D. (2017) Polymer inclusion membranes (PIMs) in chemical analysis – A review. *Analytica Chimica Acta* 987: 1–14.
63. Fontàs, C., Queralt, I., Hidalgo, M. (2006) Novel and selective procedure for Cr(VI) determination by X-ray fluorescence analysis after membrane concentration. *Spectrochimica Acta Part B-Atomic Spectroscopy* 61: 407–413.
64. Yaftian, M.R., Almeida, M.I.G.S., Cattrall, R.W., Kolev, S.D. (2018) Flow injection spectrophotometric determination of V(V) involving on-line separation using a poly(vinylidene fluoride-co-hexafluoropropylene)-based polymer inclusion membrane. *Talanta* 181: 385–391.
65. Vera, R., Zhang, Y.L., Fontas, C., Almeida, M.I.G.S., Antico, E., Cattrall, R.W., Kolev, S.D. (2019) Automatic determination of arsenate in drinking water by flow analysis with dual membrane-based separation. *Food Chemistry* 283: 232–238.
66. Almeida, M.I.G.S., Chan, C., Pettigrove, V.J., Cattrall, R.W., Kolev, S.D. (2014) Development of a passive sampler for Zinc(II) in urban pond waters using a polymer inclusion membrane. *Environmental Pollution* 193: 233–239.
67. Almeida, M.I.G.S., Silva, A.M.L., Coleman, R.A., Pettigrove, V.J., Cattrall, R.W., Kolev, S.D. (2016) Development of a passive sampler based on a polymer inclusion membrane for total ammonia monitoring in freshwaters. *Analytical and Bioanalytical Chemistry* 408: 3213–3222.
68. Garcia-Rodriguez, A., Fontàs, C., Matamoros, V., Almeida, M.I.G.S., Cattrall, R.W., Kolev, S.D. (2016) Development of a polymer inclusion membrane-based passive sampler for monitoring of sulfamethoxazole in natural waters. Minimizing the effect of the flow pattern of the aquatic system. *Microchemical Journal* 124: 175–180.
69. See, H.H., Hauser, P.C. (2014) Electro-driven extraction of low levels of lipophilic organic anions and cations across plasticized cellulose triacetate membranes: Effect of the membrane composition. *Journal of Membrane Science* 450: 147–152.
70. Qin, Z.H., Wang, Y.Z., Sun, L., Gu, Y.X., Zhao, Y., Xia, L., Liu, Y., Bruggen, B.V., Zhang, Y. (2022) Vanadium recovery by electrodialysis using polymer inclusion membranes. *Journal of Hazardous Materials* 436: 129315.
71. Sellami, F., Kebiche-Senhadji, O., Marais, S., Colasse, L., Fatyeyeva, K. (2020) Enhanced removal of Cr(VI) by polymer inclusion membrane based on poly (vinylidene fluoride) and Aliquat 336. *Separation and Purification Technology* 248: 117038.
72. Garcia-Rodriguez, A., Matamoros, V., Kolev, S.D., Fontàs, C. (2015) Development of a polymer inclusion membrane (PIM) for the preconcentration of antibiotics in environmental water samples. *Journal of Membrane Science* 492: 32–39.
73. Olasupo, A., Sadiq, A.C., Suah, F.B.M. (2022) A novel approach in the removal of ciprofloxacin antibiotic in an aquatic system using polymer inclusion membrane. *Environmental Technology & Innovation* 27: 102523.
74. Pośpiech, B., Walkowiak, W. (2007) Separation of copper(II), cobalt(II) and nickel(II) from chloride solutions by polymer inclusion membranes. *Separation and Purification Technology* 57: 461–465.
75. Nowik-Zajac, A., Kozlowski, C., Walkowiak, W. (2010) Transport of perrhenate anions across plasticizer membranes with basic ion carriers. *Physicochemical Problems of Mineral Processing*: 179–186.
76. Wionczyk, B., Apostoluk, W.A., Prochaska, K., Kozowski, C. (2001) Properties of 4-(1 '-n-tridecyl)pyridine N-oxide in the extraction and polymer inclusion membrane transport of Cr(VI). *Analytica Chimica Acta* 428: 89–101.
77. Saf, A.O., Alpaydin, S., Coskun, A., Ersoz, M. (2011) Selective transport and removal of Cr(VI) through polymer inclusion membrane containing 5-(4-phenoxyphenyl)-6H-1,3,4-thiadiazin-2-amine as a carrier. *Journal of Membrane Science* 377: 241–248.
78. Bhattacharyya, A., Mohapatra, P.K., Ghanty, T.K., Manchanda, V.K. (2008) A pH dependent transport and back transport of americium(III) through the cellulose triacetate composite polymer membrane of cyanex-301 and TBP: role of H-bonding interactions. *Physical Chemistry Chemical Physics* 10: 6274–6280.
79. Kozlowski, C.A., Kozlowska, J., Pellowski, W., Walkowiak, W. (2006) Separation of cobalt-60, strontium-90, and cesium-137 radioisotopes by competitive transport across polymer inclusion membranes with organophosphorous acids. *Desalination* 198: 141–148.
80. Kebiche-Senhadji, O., Bey, S., Clarizia, G., Mansouri, L., Benamor, M. (2011) Gas permeation behavior of CTA polymer inclusion membrane (PIM) containing an acidic carrier for metal recovery (DEHPA). *Separation and Purification Technology* 80: 38–44.

81. Croft, C.F., Almeida, M.I.G.S., Cattrall, R.W., Kolev, S.D. (2018) Separation of lanthanum(III), gadolinium(III) and ytterbium(III) from sulfuric acid solutions by using a polymer inclusion membrane. *Journal of Membrane Science* 545: 259–265.
82. de San Miguel, E.R., Aguilar, J.C., de Gyves, J. (2008) Structural effects on metal ion migration across polymer inclusion membranes: Dependence of transport profiles on nature of active plasticizer. *Journal of Membrane Science* 307: 105–116.
83. Kozlowski, C.A., Walkowiak, W., Pellowski, W. (2009) Sorption and transport of Cs-137, Sr-90 and Co-60 radionuclides by polymer inclusion membranes. *Desalination* 242: 29–37.
84. Almeida, M.I.G.S., Silva, A.M.L., Cattrall, R.W., Kolev, S.D. (2015) A study of the ammonium ion extraction properties of polymer inclusion membranes containing commercial dinonylnaphthalene sulfonic acid. *Journal of Membrane Science* 478: 155–162.
85. Tayeb, R., Fontàs, C., Dhahbi, M., Tingry, S., Seta, P. (2005) Cd(II) transport across supported liquid membranes (SLM) and polymeric plasticized membranes (PPM) mediated by Lasalocid A. *Separation and Purification Technology* 42: 189–193.
86. Tarhouchi, S., Louafy, R., El Atmani, E., Hlaibi, M. (2022) Kinetic control concept for the diffusion processes of paracetamol active molecules across affinity polymer membranes from acidic solutions. *BMC Chemistry* 16: 2.
87. de San Miguel, E.R., Hernandez-Andaluz, A.M., Banuelos, J.G., Saniger, J.M., Aguilar, J.C., de Gyves, J. (2006) LIX (R)-loaded polymer inclusion membrane for copper(II) transport – 1. Composition-performance relationships through membrane characterization and solubility diagrams. *Materials Science and Engineering A—Structural Materials Properties Microstructure and Processing* 434: 30–38.
88. Wang, D., Cattrall, R.W., Li, J., Almeida, M.I.G.S., Stevens, G.W., Kolev, S.D. (2018) A comparison of the use of commercial and diluent free LIX84I in poly (vinylidene fluoride-co-hexafluoropropylene) (PVDF-HFP)-based polymer inclusion membranes for the extraction and transport of Cu(II). *Separation and Purification Technology* 202: 59–66.
89. de San Miguel, E.R., Garduno-Garcia, A.V., Aguilar, J.C., de Gyves, J. (2007) Gold(III) transport through polymer inclusion membranes: Efficiency factors and pertraction mechanism using Kelex 100 as carrier. *Industrial & Engineering Chemistry Research* 46: 2861–2869.
90. Aguilar, J.C., Sanchez-Castellanos, M., de San Miguel, E.R., de Gyves, J. (2001) Cd(II) and Pb(II) extraction and transport modeling in SLM and PIM systems using Kelex 100 as carrier. *Journal of Membrane Science* 190: 107–118.
91. Mitiche, L., Tingry, S., Seta, P., Sahmoune, A. (2008) Facilitated transport of copper(II) across supported liquid membrane and polymeric plasticized membrane containing 3-phenyl-4-benzoylisoxazol-5-one as carrier. *Journal of Membrane Science* 325: 605–611.
92. Sugiura, M., Kikkawa, M. (1989) Carrier-mediated transport of rare-earth ions through cellulose triacetate membranes. *Journal of Membrane Science* 42: 47–55.
93. Arous, O., Amara, M., Trari, M., Bouguelia, A., Kerdjoudj, H. (2010) Cadmium (II) and lead (II) transport in a polymer inclusion membrane using tributyl phosphate as mobile carrier and $CuFeO_2$ as a polarized photo electrode. *Journal of Hazardous Materials* 180: 493–498.
94. Matsuoka, H., Aizawa, M., Suzuki, S. (1980) Uphill transport of uranium across a liquid membrane. *Journal of Membrane Science* 7: 11–19.
95. Ballinas, M.D., De San Miguel, E.R., Rodriguez, M.T.D., Silva, O., Munoz, M., De Gyves, J. (2004) Arsenic(V) removal with polymer inclusion membranes from sulfuric acid media using DBBP as carrier. *Environmental Science & Technology* 38: 886–891.
96. Kogelnig, D., Regelsberger, A., Stojanovic, A., Jirsa, F., Krachler, R., Keppler, B.K. (2011) A polymer inclusion membrane based on the ionic liquid trihexyl(tetradecyl) phosphonium chloride and PVC for solid-liquid extraction of Zn(II) from hydrochloric acid solution. *Monatshefte Fur Chemie* 142: 769–772.
97. Yaftian, M.R., Almeida, M.I.G.S., Cattrall, R.W., Kolev, S.D. (2018) Selective extraction of vanadium(V) from sulfate solutions into a polymer inclusion membrane composed of poly(vinylidenefluoride-co-hexafluoropropylene) and Cyphos (R) IL 101. *Journal of Membrane Science* 545: 57–65.
98. Bayou, N., Arous, O., Amara, M., Kerdjoudj, H. (2010) Elaboration and characterisation of a plasticized cellulose triacetate membrane containing trioctylphosphine oxyde (TOPO): Application to the transport of uranium and molybdenum ions. *Comptes Rendus Chimie* 13: 1370–1376.
99. Oberta, A., Wasilewski, J., Wodzki, R. (2011). Structure and transport properties of polymer inclusion membranes for Pb(II) separation. *Desalination* 271: 132–138.
100. Sugiura, M. (1993) Effect of quaternary ammonium-salts on carrier-mediated transport of lanthanide ions through cellulose triacetate membranes. *Separation Science and Technology* 28: 1453–1463.
101. Lacan, P., Guizard, C., Legall, P., Wettling, D., Cot, L. (1995) Facilitated transport of ions through fixed-site carrier membranes derived from hybrid organic-inorganic materials. *Journal of Membrane Science* 100: 99–109.
102. Casadella, A., Schaetzle, O., Nijmeijer, K., Loos, K. (2016) Polymer inclusion membranes (PIM) for the recovery of potassium in the presence of competitive cations. *Polymers* 8: 76.
103. Mohapatra, P.K., Lakshmi, D.S., Bhattacharyya, A., Manchanda, V.K. (2009) Evaluation of polymer inclusion membranes containing crown ethers for selective cesium separation from nuclear waste solution. *Journal of Hazardous Materials* 169: 472–479.
104. Ulewicz, M., Sadowska, K., Biernat, J.F. (2007) Facilitated transport of Zn(II), Cd(II) and Pb(II) across polymer inclusion membranes doped with imidazole azocrown ethers. *Desalination* 214: 352–364.
105. Sugiura, M. (1981) Coupled-ion transport through a solvent polymeric membrane. *Journal of Colloid and Interface Science* 81: 385–389.
106. Ulewicz, M., Lesinska, U., Bochenska, M. (2010) Transport of lead across polymer inclusion membrane with *p-tert*-butylcalix[4]arene derivative. *Physicochemical Problems of Mineral Processing* 245–256.
107. Benosmane, N., Boutemeur, B., Hamdi, S.M., Hamdi, M. (2018) Citric acid removal from aqueous solutions using a polymer inclusion membrane based on a mixture of CTA and CA. *Desalination and Water Treatment* 114: 163–168.

108. Arous, O., Amara, M., Kerdjoudj, H. (2010) Selective transport of metal ions using polymer inclusion membranes containing crown ethers and cryptands. *Arabian journal for science and engineering. Section B: Engineering* 35: 79–93.
109. Ulewicz, M., Lesinska, U., Bochenska, M., Walkowiak, W. (2007) Facilitated transport of Zn(II), Cd(II) and Pb(II) ions through polymer inclusion membranes with calix 4 -crown-6 derivatives. *Separation and Purification Technology* 54: 299–305.
110. Gardner, J.S., Peterson, Q.P., Walker, J.O., Jensen, B.D., Adhikary, B., Harrison, R.G., Lamb, J.D. (2006) Anion transport through polymer inclusion membranes facilitated by transition metal containing carriers. *Journal of Membrane Science* 277: 165–176.
111. Elshani, S., Chun, S., Amiri-Eliasi, B., Bartsch, R.A. (2005) Highly selective Ba^{2+} separations with acyclic, lipophilic di- *N*-(X)sulfonyl carbamoyl polyethers. *Chemical Communications* 279–281.
112. Gajda, B., Skrzypczak, A., Bogacki, M.B. (2011) Separation of cobalt(II), nickel(II), zinc(II) and cadmium(II) ions from chloride solution. *Physicochemical Problems of Mineral Processing* 289–294.
113. Bhagat, P.R., Pandey, A.K., Acharya, R., Nair, A.G.C., Rajurkar, N.S., Reddy, A.V.R. (2008) Molecular iodine preconcentration and determination in aqueous samples using poly(vinylpyrrolidone) containing membranes. *Talanta* 74: 1313–1320.
114. Sugiura, M. (1992) Effect of polyoxyethylene normal-alkyl ethers on carrier-mediated transport of lanthanide ions through cellulose triacetate membranes. *Separation Science and Technology* 27: 269–276.
115. Scindia, Y.M., Pandey, A.K., Reddy, A.V.R. (2005) Coupled-diffusion transport of Cr(VI) across anion-exchange membranes prepared by physical and chemical immobilization methods. *Journal of Membrane Science* 249: 143–152.
116. Kozlowski, C.A., Walkowiak, W. (2005) Applicability of liquid membranes in chromium(VI) transport with amines as ion carriers. *Journal of Membrane Science* 266: 143–150.
117. Haynes, W.M. (2011-2012) *CRC Handbook of Chemistry and Physics*, 92nd ed. Taylor & Francis.
118. Pont, N., Salvadó, V., Fontàs, C. (2008) Selective transport and removal of Cd from chloride solutions by polymer inclusion membranes. *Journal of Membrane Science* 318: 340–345.
119. Kumar, R., Pandey, A.K., Sharma, M.K., Panicker, L.V., Sodaye, S., Suresh, G., Ramagiri, S.V., Bellare, J.R., Goswami, A. (2011) Diffusional transport of ions in plasticized anion-exchange membranes. *Journal of Physical Chemistry B* 115: 5856–5867.

20 Electromembrane Processes
Recent Advances, Applications, and Future Perspectives

*Madupathi Madhumala, Tallam Aarti, and Sundergopal Sridhar**
Membrane Separations Group, Chemical Engineering and Process Technology Department,
CSIR-Indian Institute of Chemical Technology,
Telangana, Hyderabad, India
*Email: sridhar11in@yahoo.com, s_sridhar@iict.res.in

20.1 INTRODUCTION TO ELECTRO-MEMBRANE PROCESSES

The speedy evolution of chemical and biochemical industries in various regions has increased the need for consistent eco-friendly processes for downstream operations [1]. Electro-membrane processes are toxic-free and have been widely used in purification and separation processes, and their application has broadened toward energy storage and conversion devices, sensors, etc. The term electro-membrane process originated from the principle involved in their basic model, i.e., the linkage of mass transfer of ions through ion-exchange membrane on the application of an electric field [2]. Nowadays, there is high scope for these processes comprising ion-selective membranes with low electrical resistance, good permselectivity, high strength, and durability. In the early 1950s, the electro-membrane process was initiated with the introduction of an ion-selective barrier, also known as an ion-exchange membrane [3]. Over three decades, an in-depth investigation was carried out in developing different ion-selective membranes with enhancement in quality [4]. These membranes currently have suitable additional features such as high strength, low electric resistance, and high selectivity, which accomplish the bare requirements of electro-membrane processes.

An electro-membrane process involves technologies in which the electric potential difference drives the ion transference. These processes are an essential group of separation techniques, especially for the deduction of charged constituents [5]. Moreover, the separation of charged components in the electro-membrane processes primarily depends on the selectivity of the ion-selective membrane. These membranes are comprised of electrically charged ions attached to the base polymer, which allows penetration of oppositely charged ions through the membrane in the presence of an electric field. Furthermore, ion-exchange membranes have become a vibrant element for many energy conversion and storage systems. Broadly, the electro-membrane processes with ion-exchange membranes are mainly applied in three areas: (1) separation of salt/acid/base components from aqueous solutions, for example, desalination; (2) electrochemical production of organic and inorganic substances; and (3) energy conversion [6].

20.1.1 Evolution of Electro-Membrane Processes

The electro-membrane or ion-exchange membrane-centered processes have been industrially scaled up over the last 45 years, while their principal application has been recognized for more than 130 years. First, in the 1890s, Ostwald brought to light the presence of "membrane potential" at the periphery of a semi-permeable membrane, which is in contact with an aqueous solution due to the variance in concentration [7]. Then, Donnan in 1911 described the actuality of such a borderline and derived a mathematical expression relating the concentration equilibrium that led to "Donnan exclusion potential" [8]. In 1903, the principal study on electrodialysis (ED) was implemented by Morse and Pierce to separate electrolytes using a dialysis membrane in the presence of an electric field [9]. Later, work on ion-exchange membranes was initiated by Michaelis in 1925 [10] by carrying out diffusion experiments between concentrated apple juice and pure water. These studies revealed that apple skin is impermeable for the electrolytes, as the transport of cations arose when diffusion happened against the electrolyte solution. The arrival of cation-selective membranes in the form of water-insoluble resins holding sulfonic ions, prepared by the polycondensation method, led to advancements in ion-selective membrane-based processes for industrial applications [11].

Concurrently, Meyer and Straus projected a multi-stack positioning of cation- and anion-selective membranes set alternately to form parallel chambers between the two electrodes [12]. Using such multi-chamber electrodialyzers, the concentration or demineralization of solutions in various chambers can be accomplished with one pair of electrodes, resulting in permanent energy losses. However, it was difficult to commercialize the ion-exchange membranes due to their high electrical resistance. Then, in 1953, this obstacle was diminished with the development of a highly selective, stable, and low electric-resistance ion-exchange membrane by Winger et al. (1953). The ion-exchange membrane-based

DOI: 10.1201/9781003285656-20

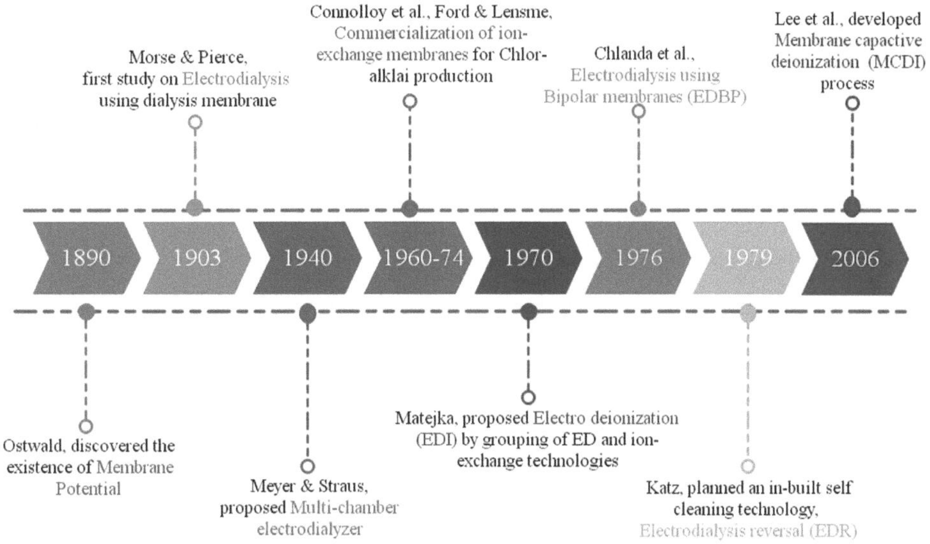

FIGURE 20.1 Time line image on the evolution of electro-membrane processes.

ED technique swiftly turned into an industrial process [13]. Eventually, the ED process was given emphasis on the desalination of seawater. The highly chemical-resistant cation-exchange membrane using sulfonated polytetrafluoroethylene was first developed by E. L. duPont de Nemours and Company in the late 1960s and early 1970s [14,15].

Furthermore, these stable cation-exchange membranes were industrially used for the chloro-alkali manufacturing process and energy conversion or storage application (fuel cells). As an advancement of the ED process, the electrodeionization (EDI) technique was proposed by Matejka et al. in the 1970s, which involves the combination of electrodialysis and ion-exchange technologies [16]. This method was extensively used to attain highly deionized water with excellent resistivity. Moreover, EDI technology has been commercially efficacious in producing ultrapure deionized water, which has become a boundless requisite in medical, analytical, and biochemical industries. Simultaneously, Chlanda et al. tried the combination of anion and cation exchange laminas into a single bipolar membrane that is applied in numerous innovations in the field of electrodialysis [17]. However, the origination of the electrodialysis reversal (EDR) technique amplified the reliability of the membrane process because of its in-built self-cleaning nature. In this process, the mode of operation in terms of stream flow and the polarity in the ED cell are reversed after a period of time, resulting in a reduction in scaling and fouling of the membrane [18]. Later, Lee et al. (2006), suggested a separation method for selective transfer of anions/cations across the membrane in the presence of an electric field with the introduction of the membrane capacitive deionization process (MCDI) [19]. This technology considerably confines the co-ion adsorption, escalates the salt removal rate, and acts like a protective layer against the electrodes [20]. These benefits were the root cause of the MCDI process being extensively applied for the desalination of brackish water [21]. Figure 20.1 illustrates an outline of the progression of the electro-membrane processes. Given the evolution of electro-membrane processes, this chapter describes the current advancements in this area and their importance. Furthermore, the chapter provides a complete overlook of the electro-membrane process applications and potential in different fields.

20.2 ION-SELECTIVE/EXCHANGE MEMBRANE: THE KEY COMPONENT

The ion-exchange membrane plays a crucial role in electro-membrane processes for improving the overall performance in a given practice. Their exclusive transportation and selective separation of ionic components make them remarkable tools in energy and separation processes. Generally, these membranes are in film form, containing extremely swollen gels with negative or positive charges. In an idyllic state, these membranes will transport the ions in opposite directions and oppose the ions of the same charge through the membrane.

20.2.1 Classification and Types of Ion-Exchange Membranes

The ion-exchange membranes are classified based on their charge and structure.

Based on the functionality, ion-exchange membranes are categorized into:

a. *Anion-exchange membranes*: These have positively charged ions attached to the base membrane for selective permeability of anions. The groups that are fixed to the polymer backbone of these membranes are $–NH_3^+$, $–N^+HR_2$, $–N^+R_3$, $–S^+R_2$, etc.
b. *Cation-exchange membranes*: These have negatively charged ions attached to the base membrane and selective permeability of cations. The groups that are

fixed to the polymer backbone of these membranes are $-COO^-$, $-SO_3^-$, $-PO_3H^-$, $-PO_3^{-2}$, $-SeO_3^-$ etc.

c. *Bipolar membranes*: These have positively and negatively charged ions by lamination of cation and anion exchange membranes for water splitting and selective transport of protons and hydroxyl ions.

As far as their structure and preparation technique are concerned, ion-exchange membranes are divided into:

a. *Homogeneous ion-exchange membranes*: These contain fixed charged ions that are uniformly spread over the polymeric matrix of the membrane. Usually, these membranes are achieved by functionalizing a polymeric membrane or polymerizing monomers with functional groups.
b. *Heterogeneous ion-exchange membranes*: These have ion-exchange resins of a distinctive domain in the uncharged polymer matrix. Either dispersion or melting and pressing ion-exchange resin beads in the polymer produce these membranes.

For any electro-membrane process, the selection of ion-exchange membranes is made based on the application of the process, i.e., the type of ions to be permeated through the membrane.

20.2.2 Desired Features of Ion-Exchange Membrane

The primary preferred characteristics of ion-exchange membranes are:

- Selectivity—the ion-exchange membranes should have high permeability to oppositely charged ions and preclude ions of the same charge.
- Mechanical strength and stability—the membranes should have high mechanical strength and withstand heavy loads.
- Electric resistance—the permeation rate of these membranes in the presence of the electric field should probably be high.
- Stability—the ion-exchange membrane should be greatly stable and less prone to swelling or shrinking.
- Chemical resistance—membranes should sustain any oxidizing agents and chemical solutions, having a pH ranging between 0–14.

Considering the desired characteristics and type, the selection of ion-exchange membrane will be accomplished, which plays an integral part in the operation of any electro-membrane process.

20.3 ELECTRO-MEMBRANE PROCESSES: BASIC ASPECTS AND PRINCIPLES

The technologies involving mass transport in the presence of an electric field across the semi-permeable membrane are called electro-membrane processes. These processes belong to the essential group of separation techniques for the selective elimination of charged species from aqueous solutions. In this section, brief descriptions of each electro-membrane process/principle and recent advancements are provided.

The most important electro-membrane processes that are industrially used are as follows:

a. Electrodialysis (ED);
b. Bipolar membrane electrodialysis (EDBP);
c. Electrodialysis reversal (EDR);
d. Membrane capacitive deionization (MCDI);
e. Electro-deionization (EDI).

20.3.1 Electrodialysis (ED)

The theoretically and economically vital electro-membrane process used for ionic separation from electrolyte solutions is traditional electrodialysis. It is an established process with well-engineered equipment and extensive commercial applications [22]. The main principle involved in this technology is illustrated in Figure 20.2, which displays the basic arrangement of the classical ED cell [2].

The cell comprises cation and anion exchange membranes organized alternately in the middle of the cathode and anode to form distinct compartments. When an aqueous/electrolyte solution is pumped into these compartments in the presence of electric potential, the negative charge ions (anions) drift in the direction of the anode, while the positive charge ions (cations) move toward the cathode. The anions are permitted through the anion-selective membrane and forbidden by a cation-selective membrane. Similarly, the cations are permitted through the cation-selective membrane but are forbidden by anion-selective membranes. Altogether, it results in an increase of ion concentration in alternate chambers, whereas the supplementary chambers concurrently become depleted. This solution is termed diluate, and the solution that passes through the semi-permeable membrane is named concentrate [23]. Overall, the energy required for ion separation from the feed solution is expressed as a function of the current applied and utilized and the electric resistance offered by the stack. The energy required for the electrodialysis separation process can be calculated using Equation (20.1) [22].

$$E_{ED} = \frac{InR_e tzFQ\Delta\acute{c}}{\xi} \tag{20.1}$$

where, E is the energy requirement, I is the current applied, R_e is the resistance offered by the cell, n is the number of cells in a stack, t is the time duration, z is the electrochemical valence, F is Faraday's constant, Q is the volumetric flow rate of feed, $\Delta\acute{c}$ is the concentration difference between the feed and diluate solutions, and ξ is the current utilized.

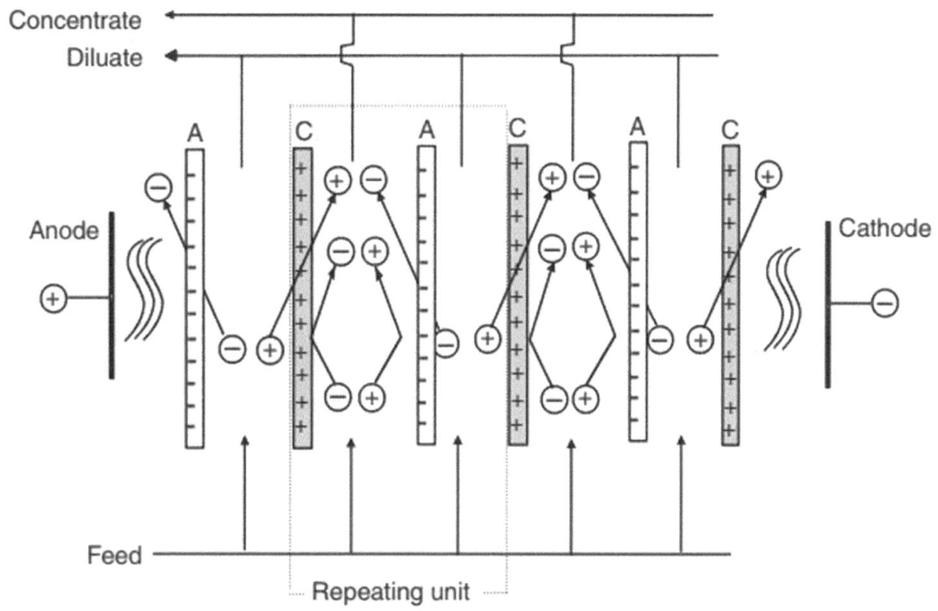

FIGURE 20.2 Schematic representation of ion-exchange in electrodialysis cell. (Adapted with permission from Ref. [2].)

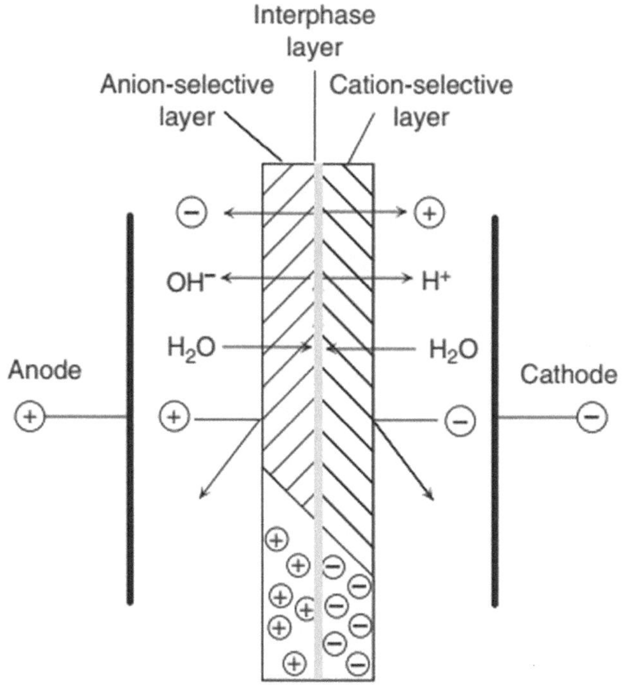

FIGURE 20.3 Schematic of the principles of bipolar membrane electrodialysis process. (Adapted with permission from Ref. [2].)

20.3.2 Bipolar Membrane Electrodialysis (EDBP)

To increase the performance of ED, a number of researchers have integrated the conventional ED process with bipolar membranes. Bipolar membranes are ion-exchange membranes comprising two oppositely charged ion-exchange membranes contiguous to each other. It has distinct characteristics and properties from that of mono-polar membranes. Moreover, the EDBP process has gained significant attention, especially for the electro-synthesis of acids and bases by an electrically enforced water-splitting mechanism. The principle involved in electrodialysis using the bipolar membrane process is demonstrated in Figure 20.3 [2], which shows a bipolar membrane (with an aqueous interphase region) placed in parallel between the cathode and anode.

In the presence of a reverse potential, the bipolar membrane can dissociate the solution molecules [25]. When the electrical potential is developed among the electrodes, the entire charged species will be detached from the interphase layer. At the same time, the water is left in the middle of the cation- and anion-exchange membranes. Additional electric currents will be achieved merely by hydroxyl and hydrogen ions dissociated from the water [2]. Theoretically, the EDBP method can dissociate water into protons and hydroxyl ions with a thermodynamic efficiency of 80% [26]. However, the energy consumption for water dissociation in the electrodialysis process using bipolar membrane can be estimated using the Nernst equation for solutions of different pH by Equation (20.2) [2].

$$\Delta G = F\Delta\varphi = 2.3RT\Delta pH \tag{20.2}$$

where, ΔG is the Gibbs free energy, R is the universal gas constant, T is the absolute temperature, $\Delta\varphi$ is the voltage difference, and ΔpH is the pH difference between the two solutions.

20.3.3 Electrodialysis Reversal (EDR)

An important route for the proficient use of ED is an innovative functioning mode denoted as ED reversal. Like ED, the EDR

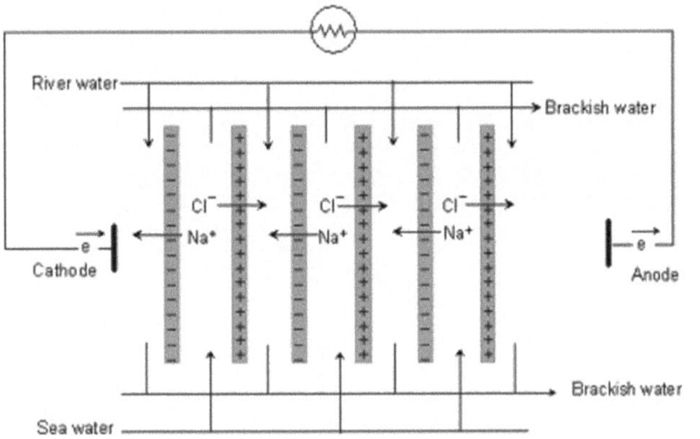

FIGURE 20.4 Schematic representation of energy production in EDR process. (Adapted with permission from Ref. [28].)

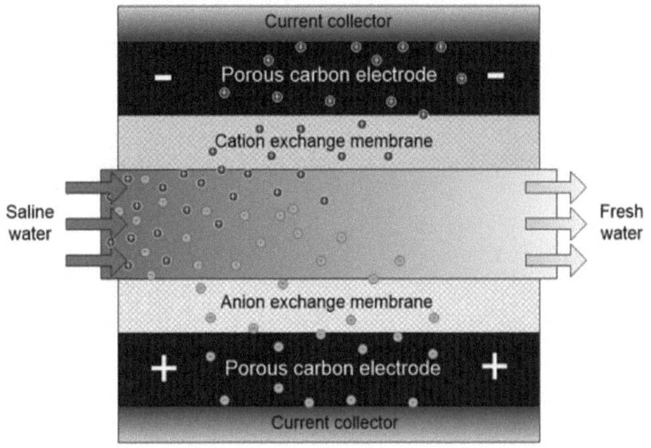

FIGURE 20.5 Schematic representation of ion transport in an MCDI cell.

process is an electrochemical charge-driven technology in which the charged components are separated through a semi-permeable ion exchange membrane in the electric field. However, the only change in the reverse electrodialysis mode of operation is that the polarity of the applied electric current will be periodically (minutes/hours) reversed. Simultaneously, the stream flows are switched such that the concentrate chamber becomes the diluate chamber and vice versa. The prime benefit of the reverse-polarity functional mode is the prevention of the formation of scales (membrane fouling) to produce high water recovery [27]. Even if there is any precipitation on the surface of the membrane, it gets re-dissolved when the polarity of the current is reversed. Typically, the EDR process is used for the selective separation of ions with high recoveries, handling large-scale flow operations, and treating extremely contaminated solutions. The principle involved in energy production by combining river water and seawater in the EDR process is illustrated in Figure 20.4 [28].

Moreover, a simple equation (20.3) derived by Forgacs can be used to estimate the energy flow through the river and seawater channels [83].

$$E_{EDR} = C_r Q_r ln\frac{C_r}{C_m} + C_s Q_s ln\frac{C_s}{C_m} \quad here, C_m = \frac{C_r Q_r + C_s Q_s}{Q_r + Q_s} \tag{20.3}$$

where, E is the Gibbs energy flow, C_r is the concentration of river water, C_s is the concentration of seawater, C_m is the concentration of mixture of both streams, Q_r is the volumetric flow rate of river water, and Q_s is the volumetric flow rate of seawater.

20.3.4 Membrane Capacitive Deionization (MCDI)

The MCDI process combines a capacitive deionization (CDI) system and an ion-exchange membrane. In the conventional CDI process, voltage is applied to the carbon electrodes, instigating the ions to be adsorbed into the pores of the electrodes [29]. Once the electrodes are rich in charges, the ions get free from the electrodes by changing the polarity of the current. During this process, there is a chance of co-ion adsorption, limiting the CDI process's application [30]. To minimize this effect significantly, Lee et al. proposed the concept of introducing the ion-selective membranes in front of the electrodes [19]. Figure 20.5 represents the scheme of ion transport in the MCDI cell [31]. In the deionization phase of the MCDI process, the cation-selective and anion-selective membranes on their respective electrodes restrict the diffusion of counter-ions. Meanwhile, the regeneration step, i.e., under the reverse polarity mode of operation, the anion-selective membrane inhibits the passage of cations on the way to the cathode and the cation-selective membrane inhibits the passage of anions on the way to the anode [28].

During the MCDI process, the energy consumption per molar mass of transferred salt can be estimated using Equation (20.4) [84].

$$E_{MCDI} = \frac{U\left(\int_{t=0}^{t} I_{ads}(t)dt + \int_{t=0}^{t} I_{des}(t)dt\right)}{nM} \tag{20.4}$$

where, E is the energy consumption, U is the applied voltage, $I_{ads}(t)$ is the applied current during the sorption step, $I_{des}(t)$ is the applied current during the desorption step, n is the number of ions absorbed on the electrode, and M is the molar mass of salt to be separated.

20.3.5 Electro-Deionization (EDI)

Conventional ED is an unprofitable process for treating dilute aqueous solutions due to energy losses and high electrical resistance. To overcome these issues, ion-exchange resins are engaged in series in the compartments of the ED cell. This type of ED with resins occupying the diluate compartments is termed electro-deionization. The ion-exchange resin is made from a similar polymer backbone but differs in functional groups. These highly porous beads have a large surface area from which ions are adsorbed and/or desorbed. The existence of the ion-exchange beads in the ED cell augments the ionic conductivity of an aqueous solution [32]. The resulting hybrid process (EDI) does not need any regeneration chemicals as the process aids in situ regeneration of the ion-exchange beads. The fundamental principle involved in the EDI technology is shown in Figure 20.6 [1]. First, the aqueous feed enters the diluate chamber packed with mixed-bed ion-exchange resins which scavenge out the charged components in the feed. In the presence of electrical potential, charged species are drawn off the resin and allowed to migrate toward their respective electrodes. In this manner, ions are continuously removed and shifted to adjacent concentrate chambers, and the ultra-pure solvent is achieved as a product.

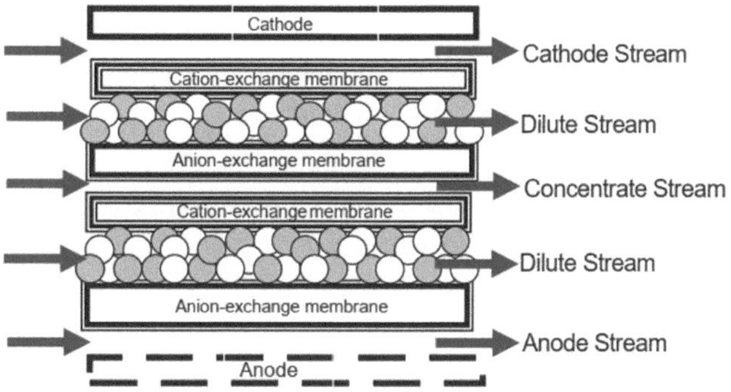

FIGURE 20.6 Schematic arrangement of an EDI cell. (Adapted with permission from Ref. [1].)

Moreover, the energy utilized by the EDI cell for the separation of ions is calculated using Equation (20.5) [85].

$$E_{EDI} = \frac{V\,I\,t}{Q} \tag{20.5}$$

where, E is the energy consumption, V is the applied voltage, I is the applied current, t is the operating time, and Q is the volumetric flow rate of water.

Additionally, an outline of the electro-membrane processes based on its features, limitations, and economic comparison is provided in Table 20.1.

20.4 RECENT ADVANCES AND APPLICATIONS

20.4.1 Water Purification

The requirement for fresh water is increasing tremendously due to the rapid growth of the population and industry all over the world. Desalination has been rapidly developed as a water treatment technology to safeguard water security [33]. Among developed technologies, electro-membrane desalination has been considered one of the promising technologies for water

TABLE 20.1
Comparison of electro-membrane processes

Electro-membrane process	Key features	Limitations	Economic viability
ED	• Simple construction with a continuous mode of operation • Commercially used in the desalination of saline or milk whey solutions	Particles > 1 μm can obstruct the membrane	• Compact system • Commercially viable due to its low operational costs
EDBP	• Produces acids and bases from respective salts • No oxidation–reduction reaction takes place • Maintains constant pH and produces fewer byproducts	Complex manufacturing and low stability	• High price • Bipolar membranes made of special materials can be scaled-up and beneficial
EDR	• Robust technology • Highly resistive to fouling and scaling • The best solution for the treatment of municipal and waste waters	Should always operate under low current density, and may require post-treatment	Requires high capital investments for heavily concentrated feed streams
MCDI	• Non-polluting scalable system • High recovery rate with high purity levels	• Requires prevention against fouling • Works efficiently for low TDS solutions	• Requires high CAPEX, and has less OPEX compared to other processes • Lab-scale trials led to high prices
EDI	• Simple and energy-efficient process • Replaced the conventional ion exchange resin process that requires a regeneration step	Cannot be applicable treating water with hardness > 1	• Highly reliable and cost-effective process • Low energy consumption, i.e., low operating costs

purification. Electrically charged ion-exchange membranes (IEMs) facilitate the removal of ionizable species from water. Numerous studies have attempted to synthesize IEMs with improved properties by introducing functional groups, blending polymers, introducing additives, and surface modification. Chehayeb et al. studied the effect of liquid flow rate, current density, and system size on the design of ED systems for water desalination [34]. The smaller size of the system facilitates improved current density for salt removal, resulting in reduced fixed costs. Xu et al. studied the performance of the ED process at the bench and pilot scales for desalination under varying operating parameters such as current density, linear velocity, hydraulic retention time (HRT), and stack staging [35]. The selectivity toward Na ions remained unchanged even at a high current density. The results obtained from the bench-scale were used to simulate the pilot-scale results. Pellegrino et al. studied the performance of the reverse osmosis and electrodialysis hybrid process for water desalination [36]. RO membranes were used as spacer channels between cation and anion exchange membranes. The system was operated at high pressure to generate two streams, one being RO quality permeate while the other was a concentrated brine solution. The process was found to be economical, with a high salt rejection rate. Reig et al. designed an ED pilot plant to concentrate seawater brine from the RO process [37]. The ion-exchange membranes could concentrate brine from 70 to 245 g/L NaCl at an energy consumption of nearly 0.12 kWh/kg of NaCl. Nowadays, special attention has been paid to heavily fertilized agricultural regions where the nitrate content exceeds the permissible limit of 45 mg/L in groundwater [38]. The consumption of nitrate-contaminated water results in methemoglobinemia disease in humans. Hence, there is a need to develop an efficient technology for the separation of nitrate from water or wastewater. Several investigations have been carried out on the removal of nitrate ions from drinking water [39,40]. Bi et al. in their study investigated the performance of the ED process for the removal of nitrate from contaminated groundwater [41]. The results showed a 99% of nitrate removal rate with a current efficiency of 17–34% and energy consumption of 0–1.7 Wh/L. Other emerging contaminants in drinking water include fluoride (F^-) and perchlorate (ClO_4^-) ions. Arar et al. studied the influence of F^- ion removal from synthetic wastewater in the presence of interfering sulfate (SO_4^{2-}) and chloride (Cl^-) ions [42]. The increase in the ratio of chloride and sulfate ions to fluoride ions results in an enhancement of the ED performance. The process enabled 96% removal of F^- ions from synthetic wastewater. A study on the utilization of Nernst-Planck-based models for determining energy consumption in both ED and MCDI processes was carried out by Patel et al. for the desalination of brackish water [43]. The ED process exhibited a reduction in salinity from 3 g/L in feed to 0.5 g/L in product water with 80% water recovery and flux of 15 L/m^{2}h. Figure 20.7 shows the variation in energy consumption for different feed concentrations at constant water recovery and flux. The specific energy consumption (SEC) was observed to be lower for ED in comparison to MCDI.

20.4.2 Extraction of Charged Analytes from Aqueous Solution

The complexity of isolation of desired compounds from pharmaceutical and environmental samples for product enrichment with respect to the analyte is important for its use in analytical systems [44]. Selective extraction of target drug analytes from their aqueous solution using electro-membrane extraction facilitated by a supported liquid membrane (SLM) has gained huge attention in recent years. The mass transport of target analytes from the sample through SLM into an acceptor solution is illustrated in Figure 20.8. Mass transfer is eased by the application of an external electric field across the ED system comprising solvent-immobilized SLM [45].

Studies on the identification of new SLMs and additives for improved mass transfer of pharmaceutical products are ongoing. So far, the literature describes the use of 2-nitrophenyl octyl ether (NPOE) as an ideal solvent for extraction of non-polar (log $P > 2$) basic pharma drugs [46]. The extraction of polar (log $P < 1$) basic drugs is facilitated by the use of a

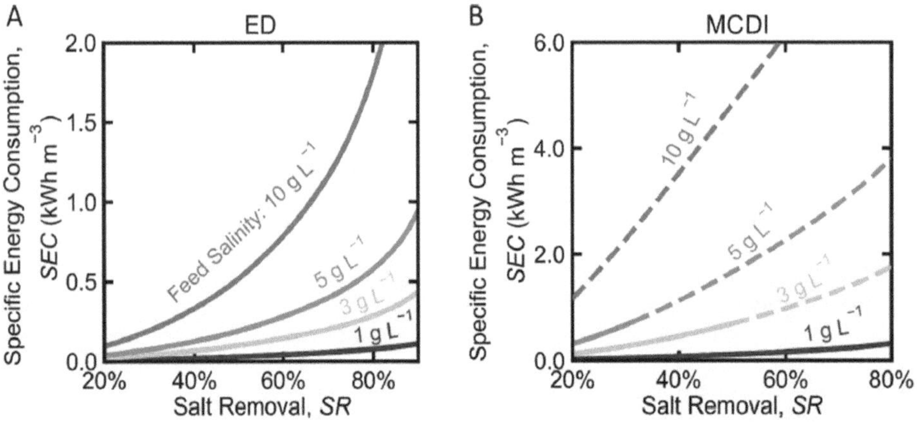

FIGURE 20.7 Specific energy consumption (SEC) for ED and MCDI for varying extent of salt removal. (Adapted with permission from Ref. [43].)

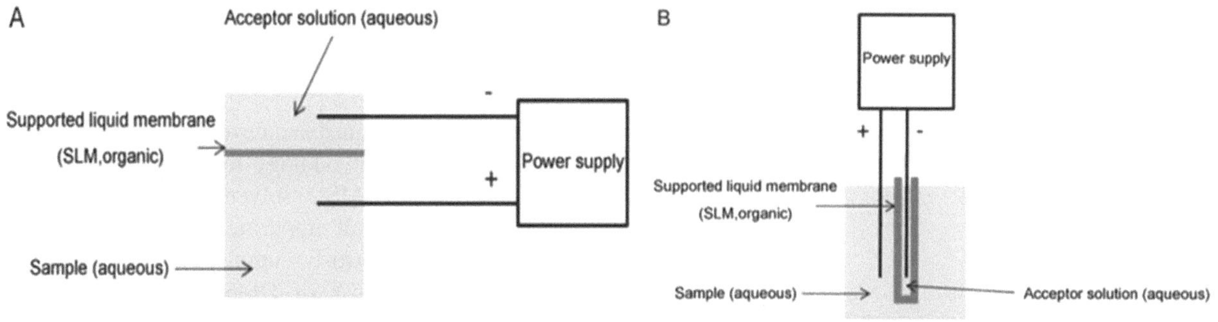

FIGURE 20.8 Principle of EME (A) SLM immobilized in flat sheet porous polymeric membrane. (B) SLM immobilized in hollow-fiber porous polymeric membrane. (Adapted with permission from Ref. [45].)

mixed-solvent system comprising di-(2-ethylhexyl) phosphate or tris-(2-ethylhexyl) phosphate in NPOE solvent to form ion pairs with the analytes [44]. The studies identified that NPOE solvent exhibited properties of high product recovery, no solvent intermixing with the sample or acceptor solution, and low current discharge. Extraction of acidic drugs is done by the use of aliphatic alcohols [47]. The other factors which influence the extraction performance are the viscosity of the organic solvent, applied voltage, pH of the acceptor and donor phases, stirring speed, presence of ionic substances, and temperature. There have been many innovations in EME by researchers to extract drugs, amino acids/peptides, metals, and other ions from their standard solutions. Pedersen et al. studied the performance of SLM for the extraction of drugs such as Methadone, Diclofenac, and Buserelin from plasma and/or urine samples. NPOE and 5% di-(2-ethylhexyl) phosphate in 1-octanol are used as SLM media [45]. A study on the extraction of thebaine from aqueous, biological, and pharmaceutical samples was performed by Seidi et al. [48]. The NPOE solvent enabled the migration of 45–55% of thebaine from the sample solution into an acceptor solution. Middelthon et al. reported the extraction of 35 basic drugs from their standard solutions using 2-nitrophenyl pentyl ether (NPPE) solvent with acidic HCl in standard and acceptor solutions [49]. The study showed that the variation in HCl concentration from 10 mM to 0.1 mM facilitated basic drug extraction. The addition of TEHP (10% by volume) to NPPE solvent enabled the extraction of medium polar analytes ($1 < \log P < 2$), whereas the more polar basic drugs were extracted using a mixture of DEHP and NPPE solvents as the SLM to improve the distribution ratio. Eskandari et al. worked on the extraction of mebendazole from blood and urine samples using NPOE solvent in SLM media [50]. Diluate HCl of 100 mM concentration was added to the sample and acceptor solutions. A kinetic study on the extraction of basic drugs was facilitated by hollow-fiber membrane liquid-phase micro-extraction (HF-LPME) and EME processes [51]. The alkaline feed sample was extracted into the acidic acceptor solution in the HF-LPME process, whereas the feed and alkaline samples were acidic in the EME process. The mass transfer of basic drugs across SLM was found to be lower in HF-LPME in comparison with EME. A study on antidepressant extraction from human blood and urine samples was reported by Davarani et al. using NPOE as SLM at 200 V [52]. A pH of 4.0 was considered for the donor solution, while that of the acceptor solution was pH 2.0. A tricyclic recovery rate of 90–95% was obtained within 20 min. Extraction of amino acids from whole blood including serum and plasma was reported by Strieglerova et al. using 1–85% (v/v) ethyl-2-nitrobenzene (ENB)–15% (v/v) DEHP as SLM [53]. The experiments were repeated to study the migration time and peak area. At operating conditions of 50 V and 500 rpm, the repeatability with respect to migration time was better than 0.3% and the peak area of amino acids was observed to be > 13%. Recovery of peptides was reported by Balchen et al. using 1-octanol (55% w/w)/diisobutyl ketone (35% w/w)/DEHP (10% w/w) as SLM at 50 V using HCl as acceptor and donor solution [54]. An extraction recovery rate of 36–56% was obtained at an agitation speed of 1050 rpm and extraction time of 5 min. Extraction of heavy metal ions from tap water and powdered milk samples was carried out using SLM containing a mixture of 1-octanol in 0.5% (v/v) bis(2-ethyl hexyl) phosphoric acid [55]. The donor solution contained water, while the acceptor contained 100 mM acetic acid. A sample recovery rate of 15–42% was obtained at an operating voltage of 75 V within an operation time of 5 min. Hu et al. studied the extraction of inorganic anions such as chloride, bromide, and sulfate from ethyl acetate using a polypropylene hollow-fiber membrane [56]. The experiment was carried out at 600 V with deionized water as an acceptor. The membrane facilitated the ion recovery across the range of 76% to 110% within 10 min of operating time. Extraction of analytes from wastewater was performed using a compartmentalized membrane envelope fabricated of porous polypropylene membrane [57]. The SLM was formed by impregnating the membrane envelope with toluene (Figure 20.9). The application of 300 V across the electrode facilitated the migration of basic drugs toward the negative electrode and acidic drugs toward the positively charged electrode. The deionized drugs from the middle compartment of the cell were further extracted into an organic acceptor solvent.

A study on the extraction of environmental pollutants such as 4-chlorophenol, 2,4-dichlorophenol, 2,4,6-trichlorophenol, and pentachlorophenol from sea water was carried out using

1-octanol composed SLM at 10 V and an agitation speed of 1,250 rpm [58]. The study revealed a recovery rate of 74% of pentachlorophenol into alkaline acceptor solution within 10 min of extraction time.

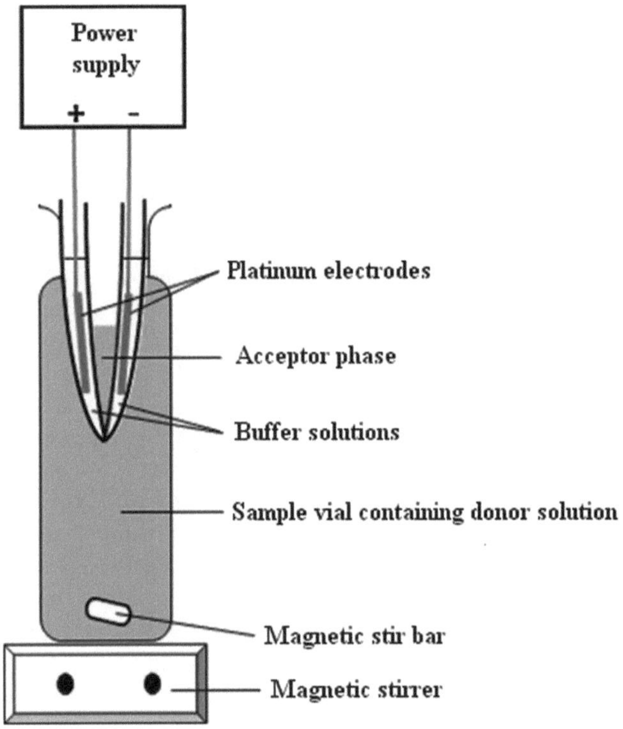

FIGURE 20.9 Experimental setup of EME containing three compartment envelopes. (Adapted with permission from Ref. [57].)

20.4.3 Separation and Recovery of Organic Chemicals and Biomolecules from Process Stream

Ravikumar et al. designed and developed an efficient ED and distillation hybrid methodology for economical recovery of dimethyl sulfoxide (DMSO) solvent from bulk drug industrial effluent containing salt impurities such as the hazardous sodium azide and corrosive ammonium chloride [59]. The feed contained 12–15% of DMSO solvent and 2–3% of impurities. The ED technique effectively separated corrosive and hazardous salts, while the distillation of desalted liquor was carried out at 20–30 mmHg at a reboiler temperature below 150°C to enable recovery and recycling of DMSO solvent for antiretroviral drug manufacture and simultaneously produced water for use in cooling towers. Figure 20.10(a) depicts a schematic of the ED and distillation hybrid process for recovery of 99.5% DMSO solvent and water from pharmaceutical effluents. A commercial ED stack capable of treating a total capacity of 7500 L/batch was commissioned by the pharmaceutical industry in 2010 to recover 30 metric tonnes of DMSO solvent (Figure 20.10b).

Bazinet et al. first reported a study on the separation of polyphenols from tobacco extract using the ED process [60]. A separation rate of 90.8% of chlorogenic acid with 86.5% of scopoletin and 81.3% rutin was obtained using a three-fold membrane stack within 4 h of operation time. Electrodialysis with filtration membrane was employed by Labbe et al. for the separation of epigallocatechin-3-gallate and epigallocatechin from green tea [61]. A product migration rate of 50% within 1 h was obtained by the ED process. Several investigations have been done on the treatment of black liquor using an electrochemical process. Jin et al. used ED for the separation of lignin [62]. On application of an electric field, lignin got precipitated at the anode, while NaOH was recovered at the cathode. Electromembrane processes including plain ED, electro-ions substitution altered ED scheme, ED with bipolar

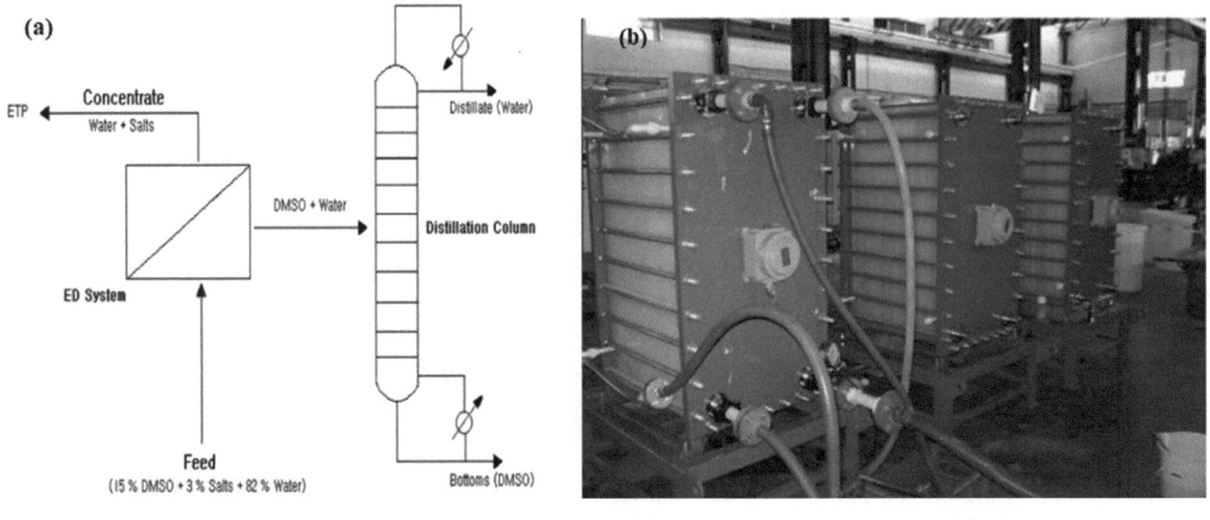

FIGURE 20.10 (a) Proposed scheme of ED–distillation hybrid process for recovery of DMSO solvent. (b) Photograph of commercial electrodialysis stacks installed for sodium azide separation from DMSO effluent. (Adapted with permission from Ref. [59].)

membranes, and electrodeionization have been used for the recovery of various organic and amino acids from different chemical feedstocks. A study on the recovery of amino acids was reported by Readi et al. [63]. The feed contained acid concentrations varying from 25 to 100 mM with pH 12.5. The experiments performed using an ED stack of 36 cm^2 effective membrane area could recover 63% of amino acid with current efficiency and energy consumption being 83% and 3 kWh/kg, respectively. Recovery of tartaric acid from a feed containing 10 kg/m^3 tartaric acid and 60 kg/m^3 glucose using the ED process resulted in a final acid concentration of 170–300 kg/m^3 with an energy consumption of 5.103–12.103 kJ/kg [64]. ED enabled butyric acid separation to 85% acid purity at a current efficiency of 52% [65], whereas > 90% of glyceric and lactic acids were recovered with an energy consumption of 0.25 kWh/kg [66,67].

20.4.4 Wastewater Treatment

Heavy metals are identified as one of the major emerging pollutants in wastewater. The high toxicity, carcinogenic nature, and non-biodegradability of these ions prove to be very harmful to the environment and living beings. Electrodialysis has been identified as a suitable process for the treatment of industrial wastewater. Electroplating of metal involves the use of heavy metal ions including Ni, Cu, Zn, Cr, Cd, and Pb. A pilot ED study was performed by Itoi et al. for the recovery of Ni during metal finishing from wastewater [68]. A metal ion recovery rate of 90% was obtained when using 5 g/L rinse wastewater. An ED trial for the treatment of an artificial solution of electroplating Watts' bath consisting of 65 g/L of $NiCl_2$, 275 g/L of $NiSO_4$, 45g/L of H_3BO_3, and organic additives resulted in Ni ion recovery of 95–99% [69]. Benvenuti et al. scaled up the ED plant for the treatment of 480 L/day of effluent from the industrial plating process (Figure 20.11) [70].

A techno-economic feasibility study reported a total process saving of 3800 US$ per year with 2.8 kWh/m^3 of E_{spec}. Peng et al. used an electrolysis-ED-EDI combined system containing EDI equipped with 13-cell-pairs with mixed IXRs in all channels for Ni recovery from a synthetic solution [71]. The process yielded ~99.8% of Ni^{2+} recovery with 93.9% purity. The use of an ED-electrolysis integrated process successfully recovered over 99% of Cu^{2+} from a synthetic solution with E_{spec} of ~2 kWh/m^3. ED integrated with microbial desalination cells followed by precipitation was developed for the treatment of Cu-containing wastewater and seawater desalination [72]. The ED process is applied for different electroplating processes to recover Zn ion. Studies on the recovery of Zn from phosphate plating baths and Zn cyanide electroplating solutions have been done by researchers. A two-stage ED scheme was developed with a monovalent selective anion exchange membrane (MVA) for the recovery of 418 mg/L Cr(VI) ion from real wastewater (Figure 20.12) [73].

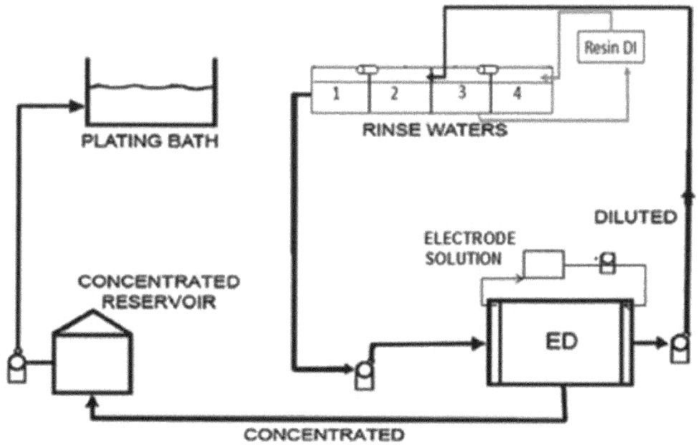

FIGURE 20.11 Flow-chart of the ED treatment for Ni electroplating rinse wastewater (the IXR extended the water cycle within the third and fourth tanks). (Adapted with permission from Ref. [70].)

The wastewater with a pH of 2.2 contained $HCrO_4^-$ ions. Cr ions were concentrated in stage 1. The pH adjustment to 8.5 resulted in stable chromate ion form, i.e. divalent CrO_4^{2-} that are retained in the diluate with the use of MVAs in stage 2. The two-stage ED process enabled Cl^- removal by ~45%. Marder et al. studied the removal of Cd ions from simulated wastewater through a five-compartment ED module [74]. The feed (diluate) contained 0.0089 mol/L of CdO, 0.081 mol/L of NaCN, and 0.018 mol/L of NaOH. The process resulted in $CdCN_4^{2-}$ removal by 86%, and 95% separation of CN^-. Abou-Shady et al. reported recovery of Pb(II) ions from the ED pilot stack integrated with electrolysis and adsorption processes [75]. The ED process resulted in the reduction of Pb^{2+} from the initial concentration of 600 mg/L to a final value of ~16 mg/L in the diluate. Adsorption helped further reduce Pb^{2+} ions to ~1 mg/L, whereas ~90% of Pb was recovered by electrolysis via cathode deposition from the ED concentrate. Several studies have also been reported on the treatment of actual wastewater streams from industries with various metal ions. Zuo et al. carried out the removal of Cr ions (~1 mg/L) from a mixture of heavy metal ions using integrated membrane processes [76]. Microfiltration and ultrafiltration have been used for removing organics and suspended solids, while ED was employed for desalination followed by nanofiltration or reverse osmosis to concentrate the solutes in the reject and recover water in permeate. The ED process enabled 97% removal of Cr (III), Cu (II), and Zn (II) ions and 95% separation of sulfate and chloride ions with an 85% reduction in COD from the feed solution (concentration of 1000 mg/L of total ion content and 300 mg/L COD). A study on the use of electrodialysis with a monovalent selective cation exchange membrane was carried out by Reig et al. for the recovery of Cu and Zn metal ions from a synthetic acidic metallurgical solution containing As(V) metal ions (Figure 20.13) [77].

Experimental trials were carried out using metallurgical wastewater containing mixtures of salts such as $CuSO_4$, $ZnSO_4$, and Na_2HAsO_4. The process enabled recovery of 80% and 87% of Cu^{2+} and Zn^{2+} ions, respectively, while As (V) recovery of 95% was obtained in another solution. The

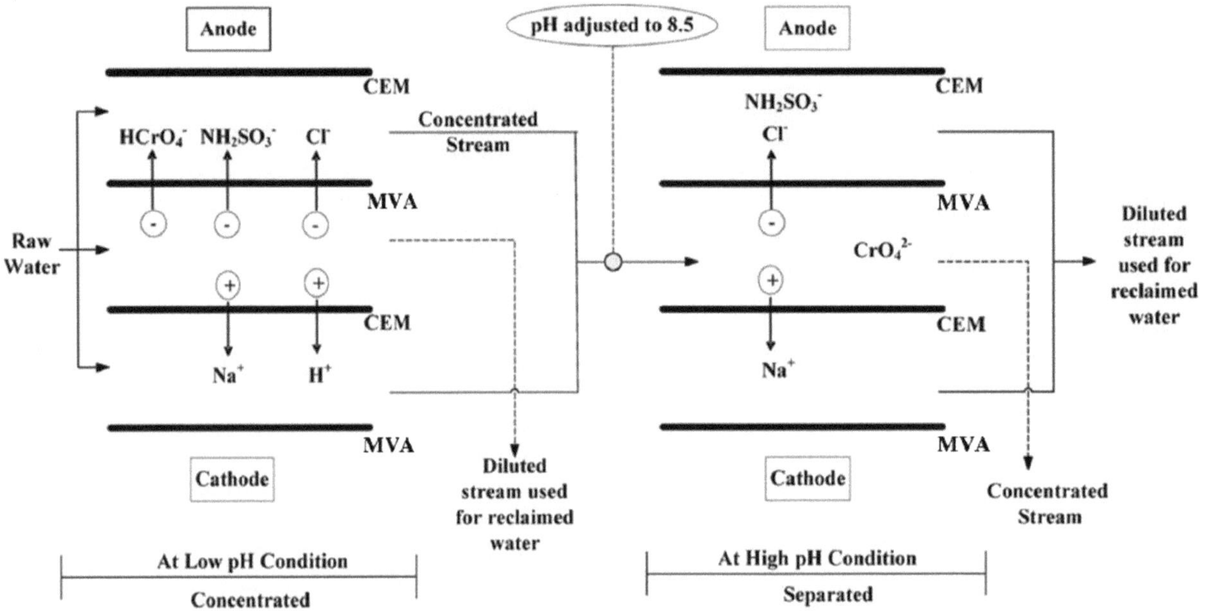

FIGURE 20.12 Two-stage selective ED with MVAs for recovering water and concentrate solution from Cr(VI) electroplating wastewater. (Adapted with permission from Ref. [73].)

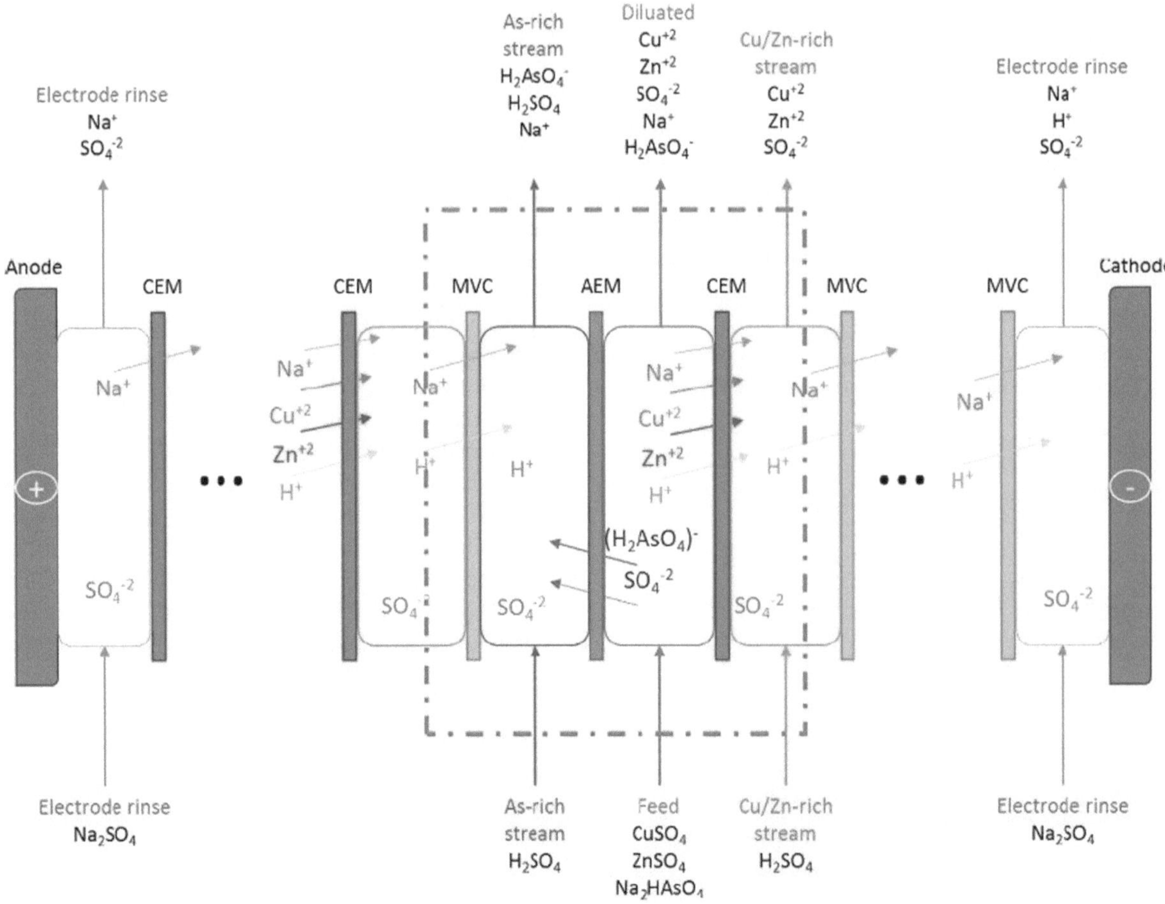

FIGURE 20.13 SED configuration for recovering Cu^{2+} and Zn^{2+} from acidic metallurgical wastewater containing As (VI). (Adapted with permission from Ref. [77].)

TABLE 20.2
Overview of applications of electro-membrane processes

Category	Research field	Scale of operation	Process characteristics	References
Water purification	Concentration of brine solution	Pilot-scale ED in continuous mode	• NaCl concentration 70–245 g/L • Energy consumption: 0.12 kWh/kg NaCl • End-use: Chlor-alkali industry	[37]
	Nitrate removal from groundwater	Lab-scale ED process	• % Nitrate removal: 99% • Current efficiency: 17–34% • Energy consumption: 0–1.7 Wh/L	41
	Desalination of brackish water	50-cell stack ED and MCDI	• Salinity reduction: 3 g/L to 0.5 g/L • Water recovery: 80% • ED has 30% higher energy efficiency than MCDI for the same rate of salt removal	[43]
Extraction of charged analytes from aqueous solution	Recovery of peptides	Lab-scale electro-membrane extraction	• EME-supported liquid membrane with the electrical potential of 50 V • Electric current: 350 μA • Peptide recovery: 55–65%	[54]
	Extraction of antidepressants from human body fluids	Lab-scale electro-membrane extraction	• EME with NPOE impregnated hollow-fiber membrane • Electrical potential difference: 200 V • 95% recovery of tricyclic antidepressants	[52]
Extraction and recovery of organic chemicals and biomolecules	Recovery of DMSO from pharmaceutical effluent	Lab-scale ED with the distillation process	• Feed composition: 12–15% of DMSO solvent and 2–3% impurities • Distillation column maintained at 20–30 mm Hg vacuum • Reboiler temperature: 150°C • DMSO Recovery: 99.5 % from pharmaceutical effluents.	[59]
	Separation of polyphenols from tobacco extract	Lab-scale ED	• 3-fold membrane stack • Operating time: 4 h • Separation rate: 90.8% chlorogenic acid, 86.5% scopoletin, and 81.3% rutin	[60]
Extraction of value-added products from wastewater	Recovery of Ni from wastewater	Pilot-scale ED	• Treatment of an artificial solution of electroplating Watts' bath consisting of 65 g/L of $NiCl_2$, 275 g/L of $NiSO_4$, 45g/L of H_3BO_3, and organic additives. • Ni recovery: 95–99%	[69]
	Recovery of metals from its synthetic solutions	Combination of Electrolysis-ED-EDI systems	• Cell-pairs with mixed IXRs • Yield: 99.8% of Ni^{2+} with 93.9% purity • Recovery of Cu^{2+}: 99% through ED-EDI process	[71]
	Recovery of Cu, Zn, As(V) ions from metallurgical wastewater	Lab-scale ED	• Membrane: cation-selective membrane • The feed contains $CuSO_4$, $ZnSO_4$, and Na_2HAsO_4 salts • Recoveries of Cu^{2+}, Zn^{2+}, and As (V) ions are 80%, 87%, and 95%, respectively	[77]

product with 99.8% purity contained As (V) with a comparable amount of Zn^{2+} ion concentration. The recirculation of the product to the feed can reduce the unwanted divalent cations. As an overview, significant separations from the above-stated processes are mentioned in Table 20.2.

20.5 CHALLENGES AND PROBABLE REMEDIES

Electro-driven membrane processes exhibit high potential to replace the traditional tools for desalination, purification of organic solutions, and downstream operations. Electro-membrane processes such as ED, EDI, EDR, EDBP, and MCDI are highly efficient techniques for selective separation of components without phase transitions and require low maintenance, foot area, and energy. However, membrane fouling and concentration polarization are still the leading challenges in these processes [78]. Fouling is the result of the accumulation of poorly soluble substances over the surface of the membrane or within the pore structure. Fouling results in a drop in output, but also entails a surplus energy source to maintain consistent membrane performance [79]. In addition, concentration polarization also undeniably ensues as a consequence of the difference concerning the number of transport ions in the membrane and the feed solution. This leads to the deposition of ionic components over the surface of the membrane, which thereby restricts the performance of the electro-membrane process. Additionally, the problems resulting from fouling and concentration polarization issues are inevitable but can be overcome by choosing a suitable pretreatment methodology, modified membrane morphology and hydrophilicity, and optimized process conditions [80]. The use of appropriate approaches permits an elongated membrane life with low working costs. Screening and selection of the appropriate pretreatment process can hinder these obstacles by the reduction of impurities in the feed [81]. Optimization of the process operating parameters can reduce the contact between contaminations/foulant and the membrane surface. Hydrophobic membranes are comparatively more prone to fouling than hydrophilic membranes. Therefore, it is necessary to conduct hydrophilic modification either by mixing the base polymer with hydrophilic compounds or adding a hydrophilic layer to the active surface of the membrane [82]. Hence, electro-membrane processes have huge scope for up-gradation to enhance the feasibility of industrial applications.

20.6 CONCLUSIONS AND FUTURE PERSPECTIVES

Ever since the development of ion-exchange membranes, the electro-membrane processes have been renowned for their prime applications. The present study summarized the evolution of ion-exchange membrane-based processes and their uses in various fields. The chapter provided a detailed explanation of the characteristics and principal aspects of ion-exchange membranes along with their classification. Among all the processes, ED has been found to possess great potential and has been implemented in various applications including water desalination, wastewater treatment, and recovery of value-added chemicals from their process streams. Moreover, desalination using ED will be cost-effective in enabling energy recovery besides recycling and reuse of water for various applications. In the current scenario, selective extraction of target analytes such as drugs and active pharmaceutical intermediates from aqueous solutions is being accomplished by electro-membrane extraction through supported liquid membranes. In addition, these processes have shown a great impact on the separation and enrichment of the desired product from the complex pharmaceutical, biological, and ecological samples. Considering the prime advantages of electro-membrane processes, several studies on the identification of new SLMs and additives for improved mass transfer of pharmaceutical products are rapidly intensified. Likewise, these processes have high scope for separation or recovery of heavy metal ions from industrial process streams that can offer solutions for electroplating and tanning industries. Furthermore, the integration of any of these processes with the membrane process can improve the purity with reduced energy consumption when compared to conventional techniques. To overcome these two challenges in electro-membrane processes, advanced research pertaining to membrane fouling and concentration polarization is being carried out. Finally, studies on the commencement of troubles relating to fouling and concentration polarization can be undertaken by choosing suitable pretreatment, and membrane modification to attain greater hydrophilicity and optimization of operating conditions.

REFERENCES

[1] Nagarale, R.K., Gohil, G.S., Shahi, V.K. (2006) Recent developments on ion-exchange membranes and electro-membrane processes. *Advances in Colloid and Interface Science* 119(2–3): 97–130.

[2] Strathmann, H. (2010) Electromembrane processes: Basic aspects and applications. *Comprehensive Membrane Science and Engineering* 2: 391–429.

[3] Juda, W., McRae, W.A. (1950) Coherent ion-exchange gels and membranes. *Journal of the American Chemical Society* 72(2): 1044–1044.

[4] Paidar, M., Fateev, V., Bouzek, K. (2016) Membrane electrolysis—History, current status and perspective. *Electrochimica Acta* 209: 737–756.

[5] Barragan, N., Bedi, D., Sivaraman, M., Loya, J.D., Babaguchi, K., Findlater, M., Yan, W. (2022) Selective removal of barium and hardness ions from brackish water with chemically enhanced electrodialysis. *ACS ES&T Water* 2(2): 288–298.

[6] Pourcelly. G. (2015) Electromembrane processes. *Encyclopedia of Membranes*, 1–3.

[7] Ostwald, W. (1890) Elektrische eigenschaften halbdurchlässiger scheidewände. *Zeitschrift für physikalische Chemie* 6(1): 71–82.

[8] Donnan, F.G. (1995) Theory of membrane equilibria and membrane potentials in the presence of non-dialysing electrolytes. A contribution to physical-chemical physiology. *Journal of Membrane Science* 100(1): 45–55.

[9] Morse, H.W., Pierce, G.W. (1903) Diffusion und Übersättigung in gelatine. *Zeitschrift für Physikalische Chemie* 45(1): 589–607.

[10] Michaelis, L. (1925) Contribution to the theory of permeability of membranes for electrolytes. *The Journal of General Physiology* 8(2): 33–59.

[11] Hans, W., Karl, J. (1940) *U.S. Patent No. 2,204,539*. Washington, DC: U.S. Patent and Trademark Office.

[12] Meyer, K.H., Straus, W. (1940) La perméabilité des membranes VI. Sur le passage du courant electrique a travers des membranes sélectives. *Helvetica Chimica Acta* 23(1): 795–800.

[13] Winger, A.G., Bodamer, G.W., Kunin, R. (1953) Some electrochemical properties of new synthetic ion exchange membranes. *Journal of The Electrochemical Society* 100(4): 178.

[14] Grot, W. (1974) *U.S. Patent No. 3,849,243*. Washington, DC: U.S. Patent and Trademark Office.

[15] James, C.D., Franklin, G.W. (1966) *U.S. Patent No. 3,282,875*. Washington, DC: U.S. Patent and Trademark Office.

[16] Matějka, Z. (1971) Continuous production of high purity water by electro-deionisation. *Journal of Applied Chemistry and Biotechnology* 21(4): 117–120.

[17] Chlanda, F.P., Lee, L.T., Liu, K.J. (1978) *U.S. Patent No. 4,116,889*. Washington, DC: U.S. Patent and Trademark Office.

[18] Katz, W.E. (1979) The electrodialysis reversal (EDR) process. *Desalination* 28(1); 31–40.

[19] Lee, J.B., Park, K.K., Eum, H.M., Lee, C.W. (2006) Desalination of a thermal power plant wastewater by membrane capacitive deionization. *Desalination* 196(1–3): 125–134.

[20] McNair, R., Szekely, G., Dryfe, R.A. (2020) Ion-exchange materials for membrane capacitive deionization. *ACS ES&T Water* 1(2): 217–239.

[21] Hassanvand, A., Wei, K., Talebi, S., Chen, G.Q., Kentish, S.E. (2017) The role of ion exchange membranes in membrane capacitive deionisation. *Membranes* 7(3): 54.

[22] Strathmann, H. (1994) Electrodialytic membrane processes and their practical application. In: *Studies in Environmental Science*, Vol. 59, pp. 495–533. Elsevier.

[23] Strathmann, H.W.S.K.H. (1986) Electrodialysis. In: *Synthetic Membranes: Science, Engineering and Applications*, pp. 197–223. Springer, Dordrecht.

[24] Strathmann, H. (2017) Electromembrane processes: Basic aspects and applications. In: *Comprehensive Membrane Science and Engineering*, pp. 355–392. Elsevier.

[25] Huang, C., Xu, T., Zhang, Y., Xue, Y., Chen, G. (2007) Application of electrodialysis to the production of organic acids: State-of-the-art and recent developments. *Journal of Membrane Science* 288(1–2): 1–12.

[26] Hanada, F., Hirayama, K., Ohmura, N., Tanaka, S. (1993) *U.S. Patent No. 5,221,455*. Washington, DC: U.S. Patent and Trademark Office..

[27] American Membrane Technology Association. *Electrodialysis Reversal Desalination*. www.amtaorg.com/electrodialysis-reversal-desalination (accessed Feb. 4, 2022).

[28] Strathmann, H., Grabowski, A., Eigenberger, G. (2013) Ion-exchange membranes in the chemical process industry. *Industrial & Engineering Chemistry Research* 52(31): 10364–10379.

[29] Porada, S., Zhao, R., Van Der Wal, A., Presser, V., Biesheuvel, P.M. (2013) Review on the science and technology of water desalination by capacitive deionization. *Progress in Materials Science* 58(8): 1388–1442.

[30] Han, L., Karthikeyan, K.G., Anderson, M.A., Wouters, J.J., Gregory, K.B. (2013) Mechanistic insights into the use of oxide nanoparticles coated asymmetric electrodes for capacitive deionization. *Electrochimica Acta* 90: 573–581.

[31] Folaranmi, G., Bechelany, M., Sistat, P., Cretin, M., Zaviska, F. (2020) Towards electrochemical water desalination techniques: a review on capacitive deionization, membrane capacitive deionization and flow capacitive deionization. *Membranes* 10(5): 96.

[32] Widiasa, I.N., Sutrisna, P.D., Wenten, I.G. (2004) Performance of a novel electrodeionization technique during citric acid recovery. *Separation and Purification Technology* 39(1–2): 89–97.

[33] Alabi, A., AlHajaj, A., Cseri, L., Szekely, G., Budd, P., Zou, L. (2018) Review of nanomaterials-assisted ion exchange membranes for electromembrane desalination. *npj Clean Water* 1(1): 1–22.

[34] Chehayeb, K.M., Farhat, D.M., Nayar, K.G. (2017) Optimal design and operation of electrodialysis for brackish-water desalination and for high-salinity brine concentration. *Desalination* 420: 167–182.

[35] Xu, X., He, Q., Ma, G., Wang, H., Nirmalakhandan, N., Xu, P. (2018) Selective separation of mono- and di-valent cations in electrodialysis during brackish water desalination: Bench and pilot-scale studies. *Desalination* 428: 146–160.

[36] Pellegrino, J., Gorman, C., Richards, L. (2007) A speculative hybrid reverse osmosis/electrodialysis unit operation. *Desalination* 214(1–3): 11–30.

[37] Reig, M., Casas, S., Aladjem, C., Valderrama, C., Gibert, O., Valero, F., Cortina, J. L. (2014) Concentration of NaCl from seawater reverse osmosis brines for the chlor-alkali industry by electrodialysis. *Desalination* 342: 107–117.

[38] Elmidaoui, A., Sahli, M.M., Tahaikt, M., Chay, L., Taky, M., Elmghari, M., & Hafsi, M. (2003) Selective nitrate removal by coupling electrodialysis and a bioreactor. *Desalination* 153(1–3): 389–397.

[39] Kabay, N.A.L.A.N., Arda, M., Kurucaovali, I., Ersoz, E.R.K.A.N., Kahveci, H., Can, M., Yuksel, M. (2003) Effect of feed characteristics on the separation performances of monovalent and divalent salts by electrodialysis. *Desalination* 158(1–3): 95–100.

[40] Kabay, N.A.L.A.N., Yüksel, M.Ü.M.İ.N.E., Samatya, S., Arar, Ö.Z.G.Ü.R., Yüksel, Ü. (2007) Removal of nitrate from ground water by a hybrid process combining electrodialysis and ion exchange processes. *Separation Science and Technology* 42(12): 2615–2627.

[41] Bi, J., Peng, C., Xu, H., Ahmed, A. S. (2011) Removal of nitrate from groundwater using the technology of electrodialysis and electrodeionization. *Desalination and Water Treatment* 34(1–3): 394–401.

[42] Arar, O., Yavuz, E., Yuksel, U., Kabay, N. (2009) Separation of low concentration of fluoride from water by electrodialysis (ED) in the presence of chloride and sulfate ions. *Separation Science and Technology* 44(7): 1562–1573.

[43] Patel, S.K., Qin, M., Walker, W.S., Elimelech, M. (2020) Energy efficiency of electro-driven brackish water desalination: Electrodialysis significantly outperforms membrane capacitive deionization. *Environmental Science & Technology* 54(6): 3663–3677.

[44] Krishna Marothu, V., Gorrepati, M., Vusa, R. (2013) Electromembrane extraction—a novel extraction technique for pharmaceutical, chemical, clinical and environmental analysis. *Journal of Chromatographic Science* 51(7): 619–631.

[45] Pedersen-Bjergaard, S., Huang, C., Gjelstad, A. (2017) Electromembrane extraction–Recent trends and where to go. *Journal of Pharmaceutical Analysis* 7(3): 141–147.

[46] Huang, C., Gjelstad, A., Pedersen-Bjergaard, S. (2016) Organic solvents in electromembrane extraction: recent insights. *Reviews in Analytical Chemistry* 35(4): 169–183.

[47] Gjelstad, A., Jensen, H., Rasmussen, K.E., Pedersen-Bjergaard, S. (2012) Kinetic aspects of hollow fiber liquid-phase microextraction and electromembrane extraction. *Analytica Chimica Acta* 742: 10–16.

[48] Seidi, S., Yamini, Y., Heydari, A., Moradi, M., Esrafili, A., Rezazadeh, M. (2011) Determination of thebaine in water samples, biological fluids, poppy capsule, and narcotic drugs, using electromembrane extraction followed by high-performance liquid chromatography analysis. *Analytica Chimica Acta* 701(2): 181–188.

[49] MiddelthonBruer, T.M., Gjelstad, A., Rasmussen, K.E., Pedersen Bjergaard, S. (2008) Parameters affecting electro membrane extraction of basic drugs. *Journal of Separation Science* 31(4): 753–759.

[50] Eskandari, M., Yamini, Y., Fotouhi, L., Seidi, S. (2011) Microextraction of mebendazole across supported liquid membrane forced by pH gradient and electrical field. *Journal of Pharmaceutical and Biomedical Analysis* 54(5): 1173–1179.

[51] Gjelstad, A. (2010) Electromembrane extraction: the use of electrical potential for isolation of charged substances from biological matrices. *LCGC North America* 28(2): 92–112.

[52] Davarani, S.S.H., Najarian, A.M., Nojavan, S., Tabatabaei, M.A. (2012) Electromembrane extraction combined with gas chromatography for quantification of tricyclic antidepressants in human body fluids. *Analytica Chimica Acta* 725: 51–56.

[53] Strieglerová, L., Kubáň, P., Boček, P. (2011) Electromembrane extraction of amino acids from body fluids followed by capillary electrophoresis with capacitively coupled contactless conductivity detection. *Journal of Chromatography A* 1218(37): 6248–6255.

[54] Balchen, M., Jensen, H., Reubsaet, L., Pedersen Bjergaard, S. (2010) Potential driven peptide extractions across supported liquid membranes: Investigation of principal operational parameters. *Journal of Separation Science* 33(11): 1665–1672.

[55] Kubáň, P., Strieglerová, L., Gebauer, P., Boček, P. (2011) Electromembrane extraction of heavy metal cations followed by capillary electrophoresis with capacitively coupled contactless conductivity detection. *Electrophoresis* 32(9): 1025–1032.

[56] Hu, Z., Chen, H., Yao, C., Zhu, Y. (2011) Determination of inorganic anions in ethyl acetate by ion chromatography with an electromembrane extraction method. *Journal of Chromatographic Science* 49(8): 617–621.

[57] Basheer, C., Lee, J., Pedersen-Bjergaard, S., Rasmussen, K.E., Lee, H.K. (2010) Simultaneous extraction of acidic and basic drugs at neutral sample pH: a novel electro-mediated microextraction approach. *Journal of Chromatography A* 1217(43): 6661–6667.

[58] Lee, J., Khalilian, F., Bagheri, H., Lee, H.K. (2009) Optimization of some experimental parameters in the electro membrane extraction of chlorophenols from seawater. *Journal of Chromatography A* 1216(45): 7687–7693.

[59] Ravikumar, Y.V.L., Sridhar, S., Satyanarayana, S.V. (2013) Development of an electrodialysis–distillation integrated process for separation of hazardous sodium azide to recover valuable DMSO solvent from pharmaceutical effluent. *Separation and Purification Technology* 110(7): 20–30.

[60] Bazinet, L., DeGrandpré, Y., Porter, A. (2005) Enhanced tobacco polyphenol electromigration and impact on membrane integrity. *Journal of Membrane Science* 254(1–2): 111–118.

[61] Labbé, D., Araya-Farias, M., Tremblay, A., Bazinet, L. (2005) Electromigration feasibility of green tea catechins. *Journal of Membrane Science* 254(1–2): 101–109.

[62] Jin, W., Tolba, R., Wen, J., Li, K., Chen, A. (2013) Efficient extraction of lignin from black liquor via a novel membrane-assisted electrochemical approach. *Electrochimica Acta* 107: 611–618.

[63] Readi, O.K., Gironès, M., Wiratha, W., Nijmeijer, K. (2013) On the isolation of single basic amino acids with electrodialysis for the production of biobased chemicals. *Industrial & Engineering Chemistry Research* 52(3): 1069–1078.

[64] Andrés, L.J., Riera, F.A., Alvarez, R. (1997) Recovery and concentration by electrodialysis of tartaric acid from fruit juice industries waste waters. *Journal of Chemical Technology & Biotechnology: International Research in Process, Environmental and Clean Technology* 70(3): 247–252.

[65] Du, J., Lorenz, N., Beitle, R.R., Hestekin, J.A. (2012) Application of wafer-enhanced electrodeionization in a continuous fermentation process to produce butyric acid with *Clostridium tyrobutyricum. Separation Science and Technology* 47(1): 43–51.

[66] Habe, H., Shimada, Y., Fukuoka, T., Kitamoto, D., Itagaki, M., Watanabe, K., Sakaki, K. (2010) Two-stage electrodialytic concentration of glyceric acid from fermentation broth. *Journal of Bioscience and Bioengineering* 110(6): 690–695.

[67] Ryu, H.W., Kim, Y.M., Wee, Y.J. (2012) Influence of operating parameters on concentration and purification of L-lactic acid using electrodialysis. *Biotechnology and Bioprocess Engineering* 17(6): 1261–1269.

[68] Itoi, S., Nakamura, I., Kawahara, T. (1980).Electrodialytic recovery process of metal finishing waste water. *Desalination* 32: 383–389.

[69] Benvenuti, T., Krapf, R.S., Rodrigues, M.A.S., Bernardes, A.M., Zoppas-Ferreira, J. (2014) Recovery of nickel and water from nickel electroplating wastewater by electrodialysis. *Separation and Purification Technology* 129: 106–112.

[70] Benvenuti, T., Rodrigues, M.A.S., Bernardes, A.M., Zoppas-Ferreira, J. (2017) Closing the loop in the electroplating industry by electrodialysis. *Journal of Cleaner Production* 155: 130–138.

[71] Peng, C., Jin, R., Li, G., Li, F., Gu, Q. (2014). Recovery of nickel and water from wastewater with electrochemical combination process. *Separation and Purification Technology* 136: 42–49.

[72] Dong, Y., Liu, J., Sui, M., Qu, Y., Ambuchi, J.J., Wang, H., Feng, Y. (2017) A combined microbial desalination cell and electrodialysis system for copper-containing wastewater treatment and high-salinity-water desalination. *Journal of Hazardous Materials* 321: 307–315.

[73] Chen, S.S., Li, C.W., Hsu, H.D., Lee, P.C., Chang, Y.M., Yang, C.H. (2009) Concentration and purification of chromate from electroplating wastewater by two-stage electrodialysis processes. *Journal of Hazardous Materials* 161(2–3): 1075–1080.

[74] Marder, L., Sulzbach, G.O., Bernardes, A.M., Ferreira, J.Z. (2003) Removal of cadmium and cyanide from aqueous solutions through electrodialysis. *Journal of the Brazilian Chemical Society* 14(4): 610–615.

[75] Abou-Shady, A., Peng, C., Bi, J., Xu, H. (2012) Recovery of Pb (II) and removal of NO3− from aqueous solutions using integrated electrodialysis, electrolysis, and adsorption process. *Desalination* 286: 304–315.
[76] Zuo, W., Zhang, G., Meng, Q., Zhang, H. (2008) Characteristics and application of multiple membrane process in plating wastewater reutilization. *Desalination* 222(1–3): 187–196.
[77] Reig, M., Vecino, X., Valderrama, C., Gibert, O., Cortina, J. L. (2018) Application of electrodialysis for the removal of As from metallurgical process waters: Recovery of Cu and Zn. *Separation and Purification Technology* 195: 404–412.
[78] Kim, S., Lee, S., Lee, E., Sarper, S., Kim, C.H., Cho, J. (2009) Enhanced or reduced concentration polarization by membrane fouling in seawater reverse osmosis (SWRO) processes. *Desalination* 247(1–3): 162–168.
[79] Du, J.R., Peldszus, S., Huck, P.M., Feng, X. (2009) Modification of poly (vinylidene fluoride) ultrafiltration membranes with poly (vinyl alcohol) for fouling control in drinking water treatment. *Water Research* 43(18): 4559–4568.
[80] Abdelrasoul, A., Doan, H., & Lohi, A. (2013). Fouling in membrane filtration and remediation methods. *Mass Transfer – Advances in Sustainable Energy and Environment Oriented Numerical Modeling* 195.
[81] Ang, W.L., Mohammad, A.W., Hilal, N., Leo, C.P. (2015) A review on the applicability of integrated/hybrid membrane processes in water treatment and desalination plants. *Desalination* 363; 2–18.
[82] Handojo, L., Wardani, A.K., Regina, D., Bella, C., Kresnowati, M.T.A.P., Wenten, I.G. (2019) Electro-membrane processes for organic acid recovery. *RSC Advances* 9(14): 7854–7869.
[83] Veerman, J. (2020) Reverse electrodialysis: co-and counterflow optimization of multistage configurations for maximum energy efficiency. *Membranes* 10(9); 206.
[84] Mitko, K., Rosiński, A., Turek, M. (2021) Energy consumption in membrane capacitive deionization and electrodialysis of low salinity water. *Desalination and Water Treatment* 214: 294–301.
[85] Wardani, A.K., Hakim, A.N., Wenten, I.G. (2017) Combined ultrafiltration-electrodeionization technique for production of high purity water. *Water Science and Technology* 75(12): 2891–2899.

21 Membrane Applications for Valorization Routes of Industrial Brines and Mining Waters

Examples of Resource Recovery Schemes

J. López[1,2], *M. Reig*[1,2], *and J.L. Cortina*[1,2,3,*]
[1]Chemical Engineering Department, Escola d'Enginyeria de Barcelona Est (EEBE), Universitat Politècnica de Catalunya (UPC)-BarcelonaTECH, Barcelona, Spain
[2]Barcelona Research Center for Multiscale Science and Engineering, Barcelona, Spain.
[3]CETaqua, Carretera d'Esplugues, Cornellà de Llobregat, Spain
*Corresponding author: jose.luis.cortina@upc.edu

21.1 INTRODUCTION

The European Green Deal [1] is a promising change for the European Union (EU) society in general, especially for the European process industries. Europe aims to reach climate neutrality and a circular economy by 2050, and the main actors to achieve this goal are the process industries (e.g. cement, ceramics, steel, mineral non-ferrous metallurgy, chemicals and water industry). They convert primary raw materials and secondary resources into materials that are used in the manufacturing industry to make products or that are used directly. Process industries are a key part of many value chains that develop EU industries and society. As large energy and resource consumers, they are the key to enabling a climate-neutral energy system and to contribute to a circular economy. One of their main objectives is "closing the energy and feedstock loops" to achieve ambitious near-zero landfilling and water discharge by 2050. This "near zero-water-discharge-plant" will help minimize the use of fresh water in processes and close the water loop in the process industries. It will also reduce the environmental footprint notably. However, these streams typically contain multiple impurities (such as salts and organic residues), which make the separation more complex. Achieving this objective should be supported by new technologies to remove low-concentration chemical species from high-flow-rate aqueous streams [2].

21.1.1 Management of Desalination and Industrial Brines

Treatment of wastewater containing high dissolved solids is complex and cost-intensive, and often leads to disposal without appropriate treatment [3]. Wastewater from oil and gas extraction, tanneries, chemical industries, pulp, minerals, and metallurgical industries contains a very large amount of dissolved solids.

Desalination processes of seawater or brackish water produce a large amount of rejected brine (45–55% and 15–25% of inlet seawater and brackish water, respectively). As a consequence, to properly manage such a large amount of brine, several methods have been implemented. The selection of the best option mainly depends on the location of the desalination site. In the case of seawater desalination plants, direct disposal into the sea following engineered dispersion systems is the most used option to reduce the environmental effect generated by the higher density values [4]. In the case of in-land industrial desalination facilities, the options include direct discharge to natural surface water bodies or industrial wastewater treatment facilities and injection in deep wells [5]. Thus, the initial attempts over the last decade were centred on zero liquid discharge (ZLD) or later minimum liquid discharge (MLD) techniques. These aim to reduce or avoid the amount of brine discharged to the environment to promote resource recovery. Both ZLD and MLD approaches include the use of thermal evaporative technologies, which can provide water at large cost due to the high energy consumption [6]. However, apart from water recovery, some residues are still present after applying ZLD and MLD technologies. In fact, a pasty material with 5–10% water content is also obtained, containing a mixture of solids that are originally present as dissolved species in the brine. Thus, ZLD and MLD systems are limited sustainability solutions.

Currently, emerging alternative ZLD technologies such as membrane-based options could be integrated with thermal processes after the reverse osmosis (RO) steps or before that for selectively removing specific salts. Among the membrane options, those that hold more potential to be sustainable solutions for high-salinity streams are conventional electrodialysis (ED), selectrodialysis (SED), bipolar electrodialysis (EDBM) [7–9], membrane distillation (MD) [10], and forward osmosis (FO) [11]. Thermal methods can be subsequently employed to reduce the volumes of brine

DOI: 10.1201/9781003285656-21

disposal via the formation of a slurry, solid-like conventional thermal crystallization, eutectic freeze crystallization, which separates ice and salt by differences in the density [12], as well as membrane crystallization (MCr) [3].

21.1.2 Acid Metal Influenced Mining Waters

Currently, the mining and mineral processing industries face increasing pressure to embody principles of sustainable mining and mineral processing. This need has resulted in a paradigm shift towards modifying actual processing operation routes to reduce the volume of wastes (e.g., solid and liquid wastes) generated, the operating cost, and to improve the treatment efficiency. From the different options postulated, a key component of sustainable mining and mineral processing is the reuse of mining and mineral-processing waste [13]. When focusing on waste processing, tailing management, especially dewatering and safe disposal after cementation in the drilled galleries for mineral extraction, is one of the most sustainable solutions. For example, several new projects are under construction, such as the case of Minera lo Frailes in Spain in the old Aznalcollar mine (https://mineralosfrailes.es/). When focusing on acidic liquid wastes, the state of the art reflects strong bases describing different approaches to reduce their main pollution load (e.g., acidity and metallic and non-metallic content) [14,15]. Only recently, have proposals on metal recovery [16,17], acid recovery [18,19], and water reuse [20] been postulated. When reviewing all the postulated proposals, there is a common and main key driver for enabling a holistic approach to incorporating resource recovery and water reuse by innovative treatment technologies, such as membrane-based processes. However, any solution seems to be based on one membrane-based stage integrated with different global solutions (conventional and alternative processes). This chapter focuses on (i) identifying the main problems associated with acid metal influenced mining waters (AMIMWs), (ii) demonstrating a sustainable proposal based on the use of a membrane technology and the proof of concept, and (iii) describing the main transport mechanisms involved. The main effort is centred on both proposing innovative solutions to recover high added-value by-products [e.g., zinc, copper, and rare earth elements (REE)] and the production of water with sufficient quality to be reused on-site or discharged to natural bodies.

21.2 BRINE RECOVERY BY ELECTROMEMBRANE TECHNOLOGIES

21.2.1 Seawater (SW) and Brackish Water (BW) Reverse Osmosis Brine Recovery

RO is a pressure-driven membrane technique used worldwide for seawater desalination and is highly used for brackish water desalination. The RO process is widely used because it is simple, compact, efficient, and low-cost [21,22]. However, its main drawback is the generation of brines [from seawater reverse osmosis (SWRO) or brackish water reverse osmosis (BWRO)], which should be properly managed to protect the environment and to recover some added-value elements contained in it, promoting circular economy approaches [23].

Currently, coastal countries discharge the seawater reverse osmosis (SWRO) brines generated in their desalination plants into the sea, while inland countries reduce/dilute and dispose of them (for example, the leading method for inland BWRO brine management is the use of evaporation ponds). In both cases, high costs are associated with this management, especially for inland desalination plants [24,25]. Furthermore, apart from the economic impact, both cases result in negative environmental issues [26–29]. Thus, BWRO and SWRO brine reuse and recovery is an important issue to be considered. However, conventional SWRO treatments have some disadvantages, such as low productivity, extensive land use, and generation of solid and/or liquid wastes that require special handling, among others. Thus, several authors have studied new treatments to improve the BWRO and SWRO brine management [26,30–32]. In this sense, ZLD and MLD schemes appeared. The main idea of these to avoid (ZLD) or minimize (MLD) the liquid waste (e.g., brine) disposal with maximal water recovery. For instance, ZLD schemes allow dry salt to be recovered, whereas MLD schemes permit the recovery of wet salt. Thus, both schemes combine concentration and separation treatments and technologies to change the linear economy schemes (based on element removal) to circular ones (based on added-value element recovery) [33,34].

For this reason, BWRO and SWRO brine recovery is an opportunity to apply ZLD/MLD schemes on a large scale and to follow circular economy strategies to promote the "waste to resources" concept, which is encouraged in the SPIRE (Sustainable Process Industry through Resource and Energy Efficiency) programme of the European Union (EU) Commission [23,35,36]. All in all, BWRO and SWRO brine recovery and reuse is a sustainable, environmentally friendly, and promising alternative to its conventional management.

21.2.1.1 Proposed Schemes for SWRO Brine Recovery

Three schemes were proposed for SWRO brine recovery by electromembrane techniques. EDBM was the main technology proposed for SWRO brine treatment and recovery, whereas nanofiltration (NF) and precipitation were considered as pretreatment techniques. Moreover, ED was either applied for brine valorization or pretreatment. Thus, (i) ED was studied for the concentration of SWRO brine, (ii) an integrated train of NF, precipitation, and EDBM was tested to produce HCl and NaOH from NaCl, after purifying the brines by divalent ion removal, and (iii) an integrated train of two electro-membrane processes (ED followed by EDBM) was proposed to produce highly concentrated HCl and NaOH from NaCl-rich brines.

21.2.1.1.1 Electrodialysis for NaCl Concentration

ED is an electrically driven membrane process that separates ions from aqueous solutions to concentrate and dilute them. Thus, ED was historically used for desalination purposes, although a concentrated stream was also obtained. For this reason, another application appeared: table salt recovery from seawater. This application was developed in Japan because they have no other domestic salt supply [37]. Following the same idea, ED was used for concentration purposes in this work. Indeed, this study aimed to recover the SWRO brines for use in the chlor-alkali industry, considering that NaCl is their major component (around 39 g Cl^-/L and 21 g Na^+/L ≈ 60 g NaCl/L ≈ 1 M NaCl) [38–40].

In this case [41], SWRO brines from the "El Prat de Llobregat" seawater desalination plant (Barcelona, Spain) were used as the feed stream of the ED unit. Several experiments were carried out for two years, testing temperatures from 16 to 27°C (depending on the season of the year) and applying different current densities through the ED electrodes (from 0.3 to 0.6 kA/m²). Thus, the effect of these operational parameters on ED performance was studied to achieve the maximum NaCl concentration with the minimum energy consumption, by using the ED pilot plant set-up scheme shown in Figure 21.1.

An SV10-50 Eurodia Aqualizer stack (Eurodia, France) was used to treat 1 m³ of SWRO brine (feed) by ED. Anionic and cationic monovalent selective Neosepta membranes (ACS and CIMS, respectively) with an active membrane area of 1000 cm² each, were located in 50 cell pairs between the electrodes. As shown in Figure 21.1, diluate and electrolyte rinse streams were designed in a single pass, whereas the concentrate stream (0.25 m³) was recirculated to obtain the maximum NaCl concentration value. Diluate and concentrate flow rates were around 0.5 m³/h, whereas the electrolyte rinse stream flow rate was at 0.15 m³/h.

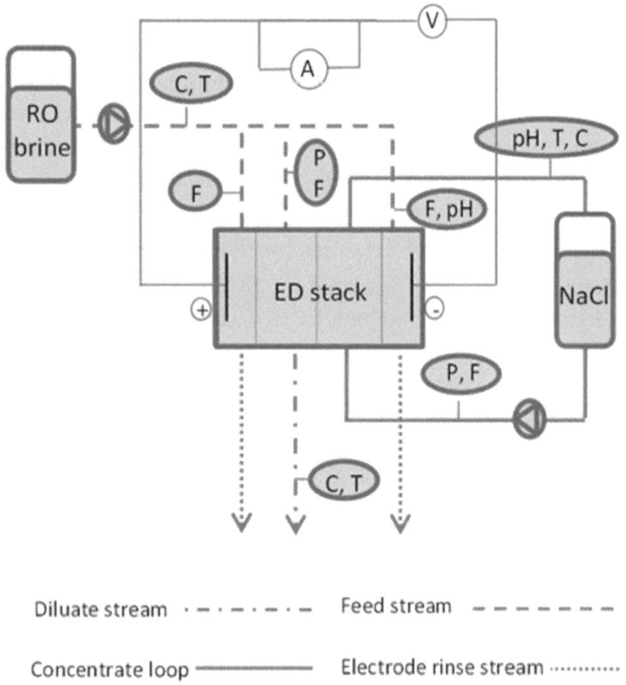

FIGURE 21.1 ED pilot-plant scheme for NaCl concentration from SWRO brine. Pressure (P), temperature (T), conductivity (C), flow (F), pH, electrical current (A), potential (V).

Several operational parameters were monitored over time in all three streams, such as temperature, conductivity, and flow. In addition, pressure and pH were measured in the concentrate and diluate streams. Electrical current and potential measurements were also monitored over time. On the other hand, HCl was added to maintain an acidic pH in all streams (below 7, below 5.5, and below 3 in the diluate, concentrate, and electrode rinse compartments, respectively) and to avoid precipitation issues.

As mentioned above, current densities and temperature were the main studied parameters, and they were varied from 0.30 to 0.60 kA/m² and from 16 to 27°C, respectively. Experiments were stopped when the NaCl concentration in the concentrate stream was constant (around 20 h).

The results indicated that it was possible to concentrate NaCl up to 270 g/L depending on the temperature and the current density applied to the ED system (Figure 21.2). Indeed, Figure 21.2a shows the NaCl concentration evolution during the summer season (T > 25°C), working at different current densities (from 0.3 to 0.6 kA/m²), whereas its evolution at different temperatures, working at a constant current density of 0.3 kA/m², is plotted in Figure 21.2b.

The results plotted in Figure 21.2a demonstrate that higher NaCl concentration values (> 250 g/L) were achieved when working at higher current densities (0.6 kA/m²). This occurred because the high intensity increased the effect of water and ion migration. On the other hand, as shown in Figure 21.2b, much lower NaCl concentration values (around 180 g/L) were achieved at a high temperature (28.3°C) and working at the minimum current density tested (0.3 kA/m²). In this case, it was concluded that higher temperatures implied large osmosis flux through the membranes, diluting the concentrate compartment.

Moreover, the results (data not shown) demonstrated that the monovalent selective Neosepta membranes hindered the transport of divalent ions (e.g., Mg^{2+}, Ca^{2+}, or SO_4^{2-}), lowering their migration through the selective CIMS and ACS membranes, they were diluted by the water transport. Thus, their concentrations decreased over time in the concentrate compartment. However, some divalent ions were also concentrated due to ion-complexation reactions that formed mono-charged species, such as $NiCl^+$ or $CuCl^+$ [41].

Finally, energy consumption (kWh/kg NaCl) during the ED trials was also calculated. In this case, two operation modes were considered: (i) batch – Equation (21.1) and (ii) continuous process – Equation (21.2).

$$Ec_b = \frac{V \cdot I \cdot t}{v \cdot C} \tag{21.1}$$

$$Ec_c = \frac{V \cdot I}{P_{flowrate}} \tag{21.2}$$

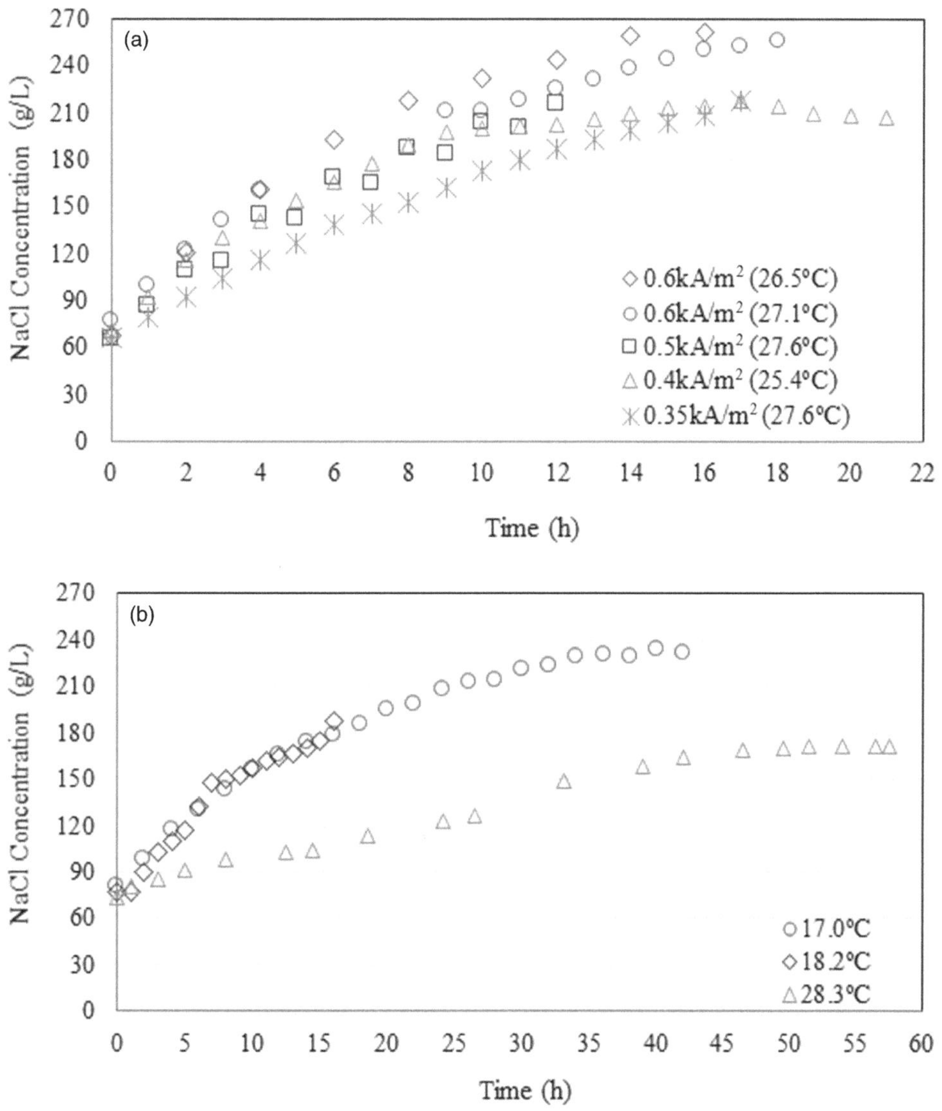

FIGURE 21.2 NaCl concentration evolution at (a) different current densities (T > 25°C) and (b) different temperatures (i = 0.3 kA/m²).

where Ec_b and Ec_c are the energy consumption (kWh/kg NaCl) in the batch and continuous mode, respectively, V and I are the applied voltage (V) and current intensity (A), respectively, t is the operational time (h), v is the volume of the concentrate tank (L), C is the NaCl concentration achieved in the concentrate compartment (kg/L), and $P_{flowrate}$ is the concentrated brine production flow rate (kg NaCl/h). In the case of the continuous process, V is defined as:

$$V = (r_{mem} + r_{dc} + r_{cc}) \cdot I + V_{mem} \tag{21.3}$$

where r_{mem}, r_{dc}, and <u>rcc</u> are the electrical resistance of the ion exchange membrane, the diluate, and the concentrate compartment, respectively, and V_{mem} is the average membrane potential. In this case, the voltage (V) was experimentally measured.

Table 21.1 shows the results obtained in the work and a comparison with an industrial cell.

The main target of Japanese electrodialysers for salt production from seawater is to obtain higher concentrations than 200 g NaCl/L with an energy consumption lower than 0.12 kWh/kg NaCl. As can be seen in Table 21.1, this objective was not accomplished either by the industrial cell or by the proposed ED cell. In this work, with an energy consumption of 0.12 kWh/kg NaCl, it was possible to concentrate the NaCl up to 185 g/L, working at 0.35 kA/m² and 27°C. On the other hand, to achieve a concentrate solution of 203 g NaCl/L, 0.19 kWh/kg NaCl was required, working at 0.50 kA/m² and at the same temperature (27°C).

As a summary, it can be concluded that ED with monovalent selective membranes allowed the SWRO brine to be concentrated up to approximately 200 g NaCl/L, whereas major divalent ions purification was achieved with an energy consumption of around 0.15 kWh/kg NaCl.

TABLE 21.1
Energy consumption of the ED cell treating the SWRO brine of Barcelona, and a comparison of the performance of an industrial electrodialyser using seawater in Japan

	Current density (kA/m²)	Temperature (°C)	Energy consumption (kWh/kg NaCl)	NaCl concentration (g NaCl/L)
Industrial cell [42]	0.27	23.5	0.160	174
Batch mode	0.30	14.0	0.20	199
	0.40	10.0	0.30	209
	0.35	28.0	0.29	208
	0.50	28.0	0.35	186
	0.60	28.0	0.38	208
Continuous mode	0.30	16.0	0.16	176
	0.30	18.0	0.15	176
	0.35	27.0	0.12	185
	0.40	17.0	0.20	178
	0.45	18.0	0.26	219
	0.50	20.0	0.26	246
	0.50	27.0	0.19	203
	0.60	20.0	0.30	245
	0.60	27.0	0.24	244

21.2.1.1.2 Integration of Nanofiltration, Precipitation, and Electrodialysis with Bipolar Membranes for HCl and NaOH Production

In this work [43], NF was used as the SWRO brine pretreatment step before an EDBM stage. The main purpose of the NF was to remove the major divalent ions (e.g., calcium and magnesium) before entering the EDBM and to be able to produce acid and base products. However, because the allowed concentration levels of calcium and magnesium in the EDBM unit are so low (< 10–30 mg/L, depending on the manufacturer), a precipitation unit was installed between the two main stages (NF and EDBM) to precipitate calcium carbonate and magnesium hydroxide (see Figure 21.3).

In this case, the pretreatment stage was composed of five NF elements in series placed inside two pressure vessels. Spiral-wound NF270 membranes (Dow Chemical, USA) with an active membrane area of 38 m² were used, working in open mode configuration at a maximum of 20 bar. Several parameters were monitored, including pressure (P), temperature (T), flow (F), conductivity (C), and pH (Figure 21.4).

Table 21.2 shows the NF results in terms of permeate composition, and rejection and recovery rate at different tested pressures (from osmotic pressure to 20 bar) when treating a feed flow rate of 1250 L/h of brine.

As shown in Table 21.2, multivalent ions were more rejected than monovalent ones, increasing their rejection at higher pressure. At 20 bar, the rejection order was as follows: Al(III) > S (VI) > Mg (II) > Sr (II) > Cu (II) > Ca (II) > Ni (II) > NaCl > K (I). Thus, after the NF step, the concentrate stream (divalent-rich) could be used for phosphorus recovery in wastewater treatment plants (Figure 21.3). On the other hand, NF permitted the purification of the monovalent-rich stream (98%) with a NaCl concentration of around 52 g/L, containing divalent trace elements such as calcium (415 mg/L) and magnesium (760 mg/L). Thus, the permeate NF stream (monovalent-rich) was sent to the precipitation unit. Then, different pH values were tested by chemical precipitation, adding sodium carbonate and sodium hydroxide for calcium and magnesium recovery, respectively (Table 21.3).

As shown in Table 21.3, the maximum precipitation values (removal rate higher than 95%) were achieved at the more alkaline pH conditions, i.e., 10.6 for $CaCO_3$ precipitation by adding Na_2CO_3 and 11.6 for $Mg(OH)_2$ precipitation by adding NaOH. Finally, a NaCl-rich stream (99.9% purity) was obtained, containing around 50 g NaCl/L, 17 mg Ca^{2+}/L, and 9 mg Mg^{2+}/L. After this process, this stream could be used as a feed for the EDBM system.

An ED64-4 lab-scale stack from PCell (Germany) was used (Figure 21.5) to carry out the EDBM experiments. For that, three cell triplets with cationic (PC-SK), anionic (PC-Acid 60), and bipolar membranes (PC-BP) with an active membrane area of 64 cm² were employed. A batch mode configuration was tested by circulating four streams: electrode rinse, salt (diluate), acid, and base. Operational parameters such as pressure, temperature, conductivity, flow, salt pH, electrical current, and potential were monitored over time.

For the EDBM tests, several parameters were kept constant, such as the initial volume of each stream (1 L), initial electrode rinse solution (45 g Na_2SO_4/L), and initial salt concentration (50 g NaCl/L). On the other hand, other experimental parameters were tested, such as the applied voltage (6 and 9 V) to the EDBM stack and the initial acid and base concentrations (0.05, 0.1, and 0.5 M). Besides the concentration evolution in each compartment, energy consumption and current efficiency (η) for the EDBM set-up

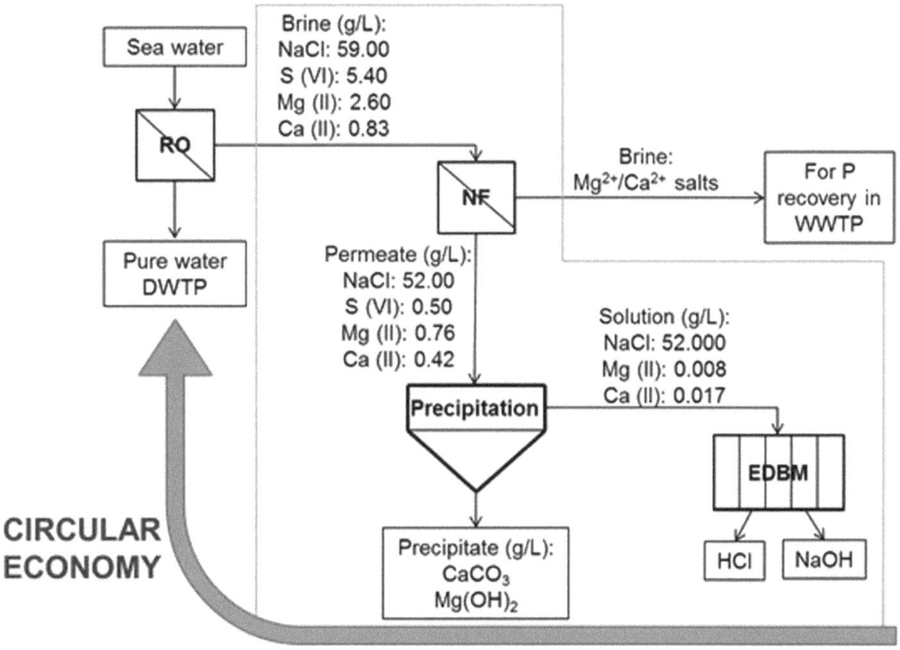

FIGURE 21.3 Nanofiltration, precipitation, and electrodialysis with bipolar membranes scheme for SWRO brine recovery [43].

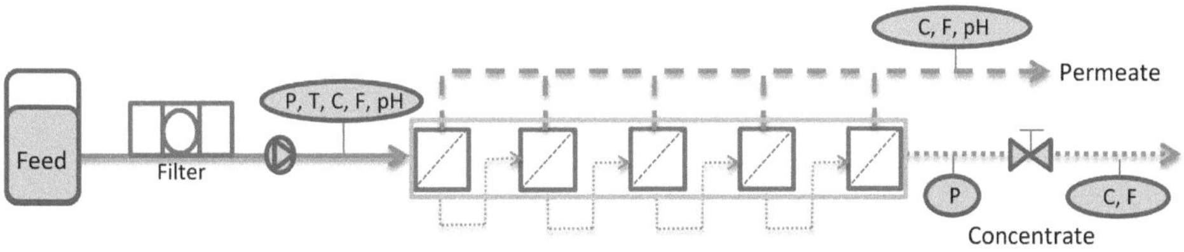

FIGURE 21.4 Nanofiltration pilot-scale scheme [43].

TABLE 21.2
NF results in terms of permeate composition, rejection, and recovery rates

Component	SWRO brine composition (mg/L)	NF permeate composition (mg/L)					% Rejection (at 24 ± 2°C)				
		8 bar	10 bar	12 bar	18 bar	20 bar	8 bar	10 bar	12 bar	18 bar	20 bar
NaCl	59000	57000	55000	54000	52000	52000	3	7	9	12	12
K (I)	700	670	666	670	670	664	4	5	4	4	5
Ca (II)	830	598	498	465	415	415	28	40	44	50	50
Mg (II)	2600	1200	1160	1000	920	760	54	55	62	70	71
S (VI)	5400	1200	1100	1100	600	500	78	79	79	89	91
Al (III)	0.3	<DL	<DL	<DL	<DL	<DL	>93	>93	>93	>93	>93
Ni (II)	0.07	0.069	0.069	0.063	0.054	0.045	0	0	10	23	35
Sr (II)	16	10.7	9.4	8.5	4.8	4.7	33	41	47	70	70
Cu (II)	0.03	0.025	0.023	0.020	0.010	0.010	17	23	33	67	67
Recovery (%)		25	42	46	74	76					

were also calculated, following Equations (21.1) and (21.4), respectively.

$$\eta = \frac{F \cdot \left(v_t \cdot C_t - v_i \cdot C_i\right)}{I_d \cdot S \cdot t \cdot n} \quad (21.4)$$

where η is the current efficiency (%), F is the Faraday constant (96,500 C/mol), v (L) and C (M) are the volume and concentration, respectively, at time "t" or at initial time "i", I_d is the current density applied (A/cm^2), S is the membrane surface (64 cm^2), t is the operational time (s), and n is the number of cell triplets of the EDBM stack (three in this study).

TABLE 21.3
NF results in terms of permeate composition and rejection

Precipitation	NF permeate concentration (mg/L)	pH	Removal (%)	Concentration after precipitation (mg/L)
Calcium with Na_2CO_3	415.0	8.6	0	415
		9.8	67 ± 3	137.0
		10.2	93 ± 3	29.1
		10.6	96 ± 2	16.6
Magnesium with NaOH	760.0	9.1	0	760
		10.3	41 ± 3	448.4
		10.6	76 ± 4	182.4
		11.5	99 ± 1	9.3

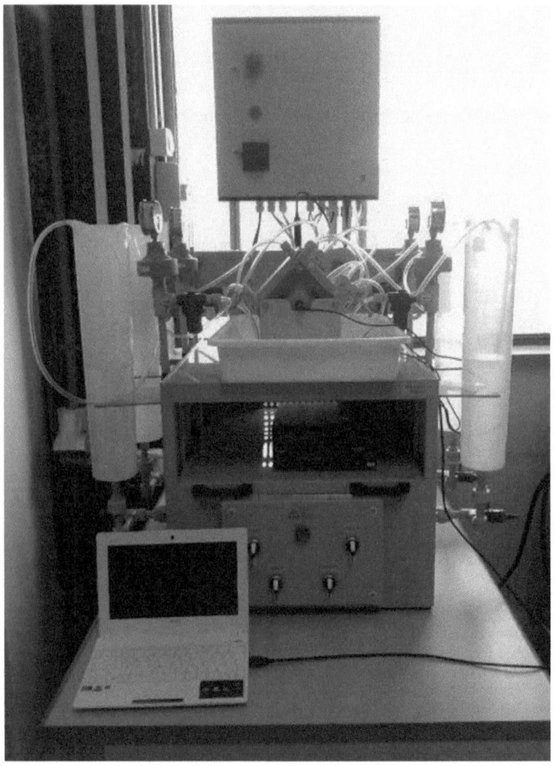

FIGURE 21.5 Electrodialysis with bipolar membrane set-up used at lab-scale.

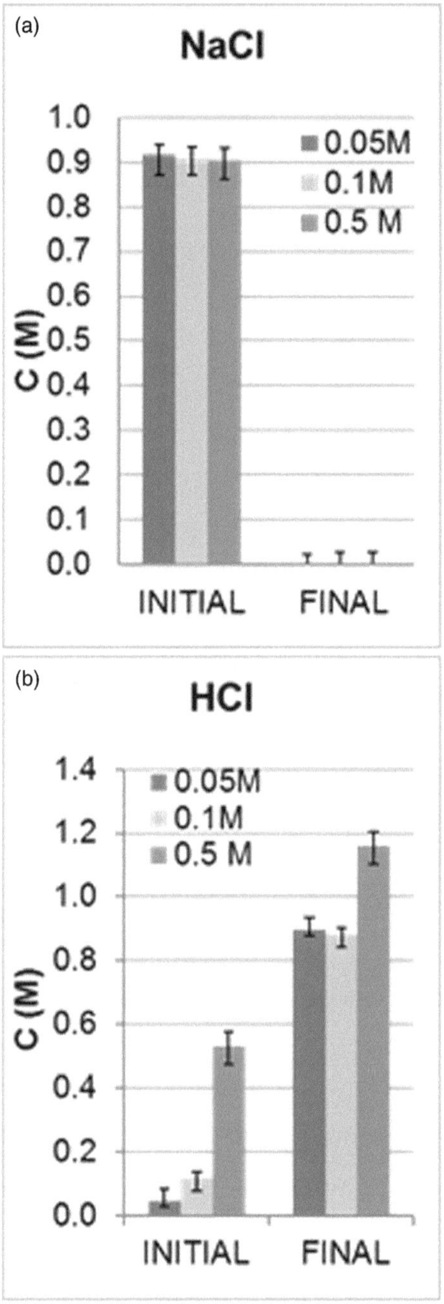

FIGURE 21.6 (a) Salt (NaCl), (b) acid (HCl), and (c) base (NaOH), initial and final concentrations, at different initial acid and base concentrations (0.05, 0.1, and 0.5 M).

Thus, it was possible to obtain HCl and NaOH streams by treating the divalent-free stream from the NF and the precipitation unit using the EDBM system. Regarding the effect of the applied voltage, the results demonstrated that the efficiency values (NaCl percentage that was converted to HCl and NaOH) were around 20% when applying 6 V. In contrast, efficiency values higher than 70% were achieved by applying the high value of the current density tested (9 V). Thus, it was concluded that higher NaOH and HCl concentrated products (almost 1 M) were obtained by applying the maximum voltage. On the other hand, the initial acid and base concentrations were tested to determine their effects on the final HCl and NaOH production (Figure 21.6).

As seen in Figure 21.6, high conversion efficiencies were obtained in all cases, and it seemed that the initial acid and base concentrations had no substantial effect on the overall EDBM performance. Thus, it was concluded that the best initial concentrations to use could be the minimum (0.05 M HCl and 0.05 M NaOH) in order to not require initial concentrated acids and bases. Finally, regarding energy consumption and

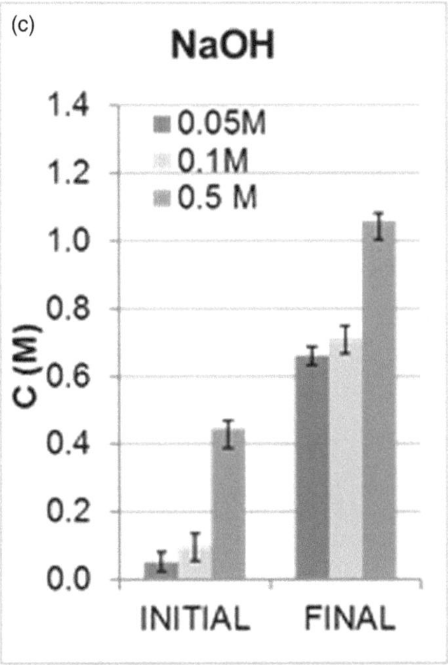

FIGURE 21.6 (Continued)

current efficiency, it was determined that the latter decreased (from 86 to 69%) as the initial NaOH concentration increased (from 0.05 to 0.5 M). The concentration gradient could explain this behaviour, i.e., more OH^- ions migrated from the base compartment to the salt compartment. All in all, an energy consumption value of around 2.5 kWh/kg NaOH was required to achieve a current efficiency of around 77% when treating SWRO brine for HCl and NaOH production.

21.2.1.1.3 Integration of Electrodialysis and Electrodialysis with Bipolar Membranes for HCl and NaOH Production

This work [8] was designed in order to be able to achieve a higher acid and base final concentration after the EDBM process and also to accomplish a circular economy scheme. For that, ED was studied as a SWRO pretreatment before the HCl and NaOH production instead of the NF technique. Thus, in this case, two electrical-driven membrane technologies were used for the recovery of added-value elements from SWRO brine: (i) ED with monovalent selective membranes was used as a pretreatment for NaCl concentration and divalent ion purification and (ii) EDBM was tested for HCl and NaOH production from the concentrate stream of the pretreatment, mainly containing NaCl (Figure 21.7).

An ED pilot plant set-up was installed in the wastewater treatment plant in El Prat de Llobregat, Barcelona (Figure 21.8). The ED stack (Aqualizer SV-10) was purchased from Eurodia (France) and was equipped with 50 cell pairs of Neosepta monovalent selective membranes [cationic (CIMS) and anionic (ACS)] with a membrane area of 0.1 m^2. As shown in Figure 21.8, several filters were installed in the circuit to retain suspended solids, and several indicators were also used to monitor different operational parameters (e.g., pressure, flow, temperature, conductivity). The concentrate stream (0.5 m^3/h) worked in a recirculation mode to concentrate the SWRO brine as much as possible, whereas the diluate (0.5 m^3/h) and electrode rinse (0.15 m^3/h) streams operated in an open-loop design to control the increase in the temperature inside the stack and to achieve higher current density values. HCl was used to maintain a low pH in all streams and to avoid precipitation during the ED process. Thus, the produced HCl after the EDBM process could be recirculated back and be used as a pH controller for the ED step.

Several experiments were performed using the ED set-up at current densities between 0.3 and 0.4 kA/m^2 and at two temperature ranges depending on the season of the year: summer (22–28°C) and winter (15–18°C). The main objective was to concentrate the NaCl of the SWRO brines, purifying it of divalent ions and maintaining a low energy consumption (< 0.3 kWh/kg NaCl). For this reason, experiments were stopped when the NaCl concentration in the concentrate stream was constant (around 20 h). Table 21.4 shows the NaCl concentration and energy consumption results after the ED trials.

As summarized in Table 21.4, NaCl-rich brines of around 200 and 100 g/L were obtained when working at low and high temperatures, respectively. As previously mentioned, summer temperatures did not allow NaCl concentrations as high as those in winter to be obtained, due to the larger osmosis transport phenomenon (water flux transported from the diluate to the concentrate). On the other hand, energy consumption values lower than 0.2 kWh/kg NaCl were achieved at low temperatures, whereas even lower energy consumption values (around 0.1 kWh/kg NaCl) were calculated when working at higher temperatures. All in all, it was concluded that ED permitted the NaCl from SWRO brine to be concentrated up to 100 or 200 g NaCl/L with low energy consumption values (< 0.2 kWh/kg NaCl).

Subsequently, both NaCl-rich brines (at 100 and 200 g/L) were introduced into the EDBM stack to determine the maximum acid and base concentration that this technique could achieve. In this case, a lab-scale EDBM stack from PCCell (Germany), named ED64-4, was used. Anionic, cationic, and bipolar PCCell membranes with an active area of 64 cm^2 were also used, following the membrane scheme shown in Figure 21.9. Three triplets were used in a close loop mode configuration. Flow rates of each stream were kept constant (15–20 L/h for the salt, acid, and base streams and 100 L/h for the electrode rinse stream) during the EDBM experiments, and a constant voltage of 9 V was applied. The electrode rinse solution used was 45 g Na_2SO_4/L, while the initial acid and base solutions were required (0.05 M HCl and NaOH, respectively). Again, operational parameters (e.g., pressure, temperature, conductivity, flow rate) were monitored to follow the experimental performances.

Figure 21.10 shows the initial and final concentrations of each stream (salt, acid, and base). The EDBM tests were performed with initial NaCl solutions of 50 g/L (SWRO brines without pretreatment), 100 g/L (SWRO after ED

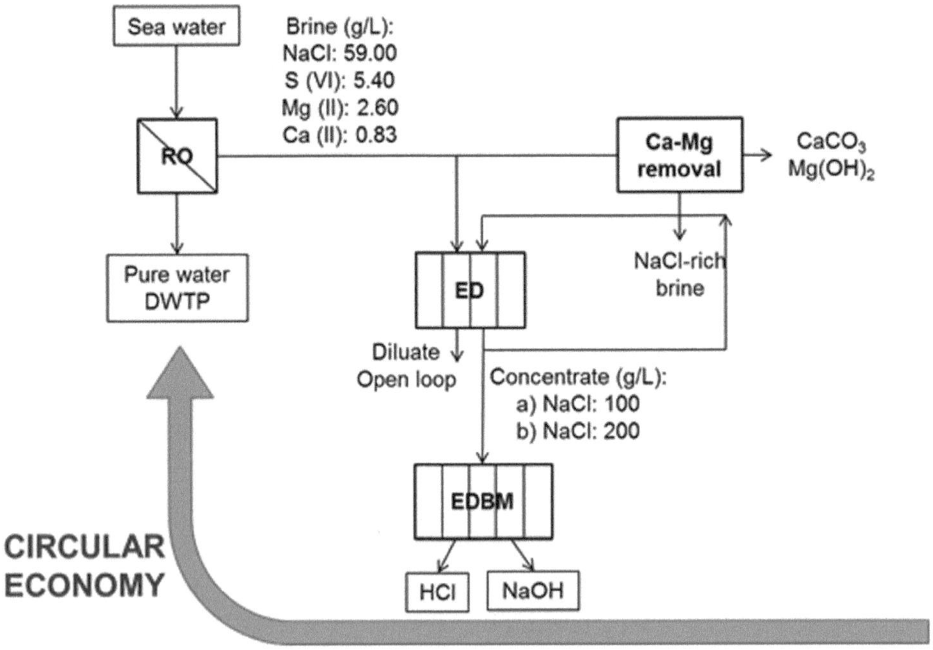

FIGURE 21.7 Electrodialysis and electrodialysis with bipolar membranes scheme for SWRO brines recovery [8].

FIGURE 21.8 Electrodialysis pilot plant, located in the El Prat de Llobregat wastewater treatment plant.

pretreatment, at high temperature), and 200 g/L (SWRO after ED pretreatment, at low temperature).

As plotted in Figure 21.10, higher HCl and NaOH concentrations up to 2 M were obtained by treating the initial NaCl solution at 200 g/L. Thus, the ED pretreatment step allowed more concentrated acid and bases to be obtained. Finally, a factorial design was defined to obtain a linear model using the response surface methodology (RSM), taking into account the initial salt, acid, and base concentrations and fixing the energy consumption as the response (Figure 21.11).

As can be seen in Figure 21.11, the minimum energy consumption of 1.7 kWh/kg NaOH was calculated when

TABLE 21.4
NaCl concentration and energy consumption obtained during the ED experiments, depending on the temperature and current density applied

Temperature (°C)	Current density applied (kA/m²)	Final NaCl concentration (g/L)	Energy consumption (kWh/kg NaCl)
16.6	0.40	204.8	0.217
17.9	0.35	185.4	0.154
18.2	0.30	187.7	0.148
21.7	0.40	135.1	0.096
22.0	0.40	133.9	0.055
28.3	0.30	104.0	0.161

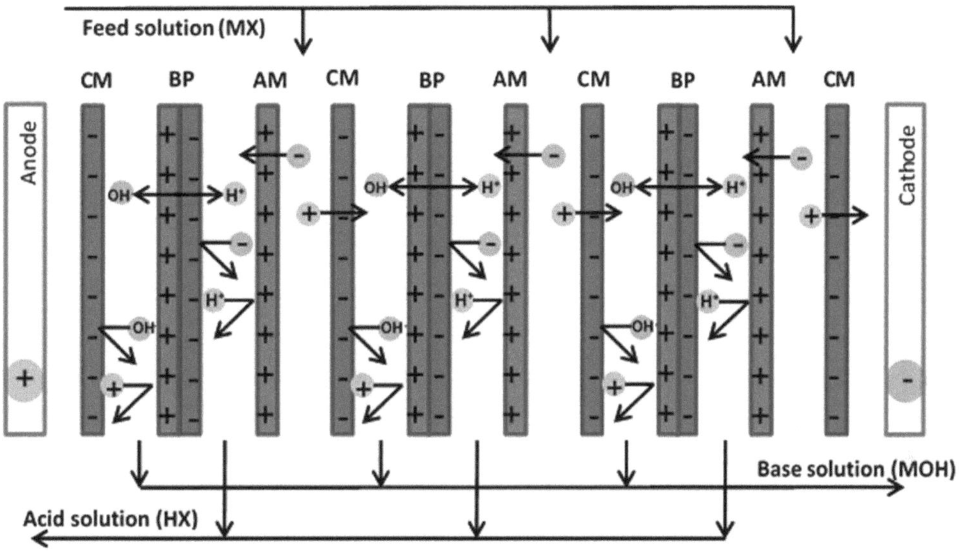

FIGURE 21.9 Membrane disposition in the EDBM set-up.

working by EDBM with initial concentrations of 104 g NaCl/L and 0.24 M HCl and NaOH.

21.2.2 Industrial Brine Recovery

Most industrial companies, especially chemical industries, generate large amounts of wastewater. These liquid wastes are usually rich in different elements and could be valorised, promoting circular economy schemes. In fact, most of these wastewaters are made of several highly concentrated elements, generating salty streams (such as brines), whose recovery is a major concern. For instance, common salts, such as sodium, ammonium, or calcium chloride, as well as sodium, magnesium sulphate, or ammonium nitrate, are some of the potential salts to be recovered from industrial brines. Indeed, according to the Water Supply and Sanitation Technology Platform Report of Brines Management, the reuse of these salts in the same industrial site has potential interest [44]. Nowadays, the main focus is to ensure industrial brine recovery applicability at full-scale because the composition of these wastewater streams is very complex. Thus, selective separation techniques, combined with other wastewater treatments, should be proposed to valorise these industrial brines and be able to recover salts with maximum purity [45].

21.2.2.1 Proposed Scheme for Industrial Brine Recovery: Selectrodialysis and Electrodialysis with Bipolar Membranes

As mentioned above, several valorisation routes could be applied for industrial brines. However, if quality requirements are desired, the main focus should be the separation of the main elements before their concentration/purification. Wastewaters with mixtures of salts, such as NaCl and Na_2SO_4, could be one example of industrial brine. Thus, the main focus for their recovery is (i) to separate both salts (NaCl and Na_2SO_4) in two different streams, minimizing the presence of other minor elements and (ii) to concentrate or purify these streams in different ways, such as producing dry salt or producing acids and bases.

Two electrical-driven membrane processes were proposed to valorise an industrial brine mainly composed of sodium chloride and sodium sulphate. Two electrodialysis-based

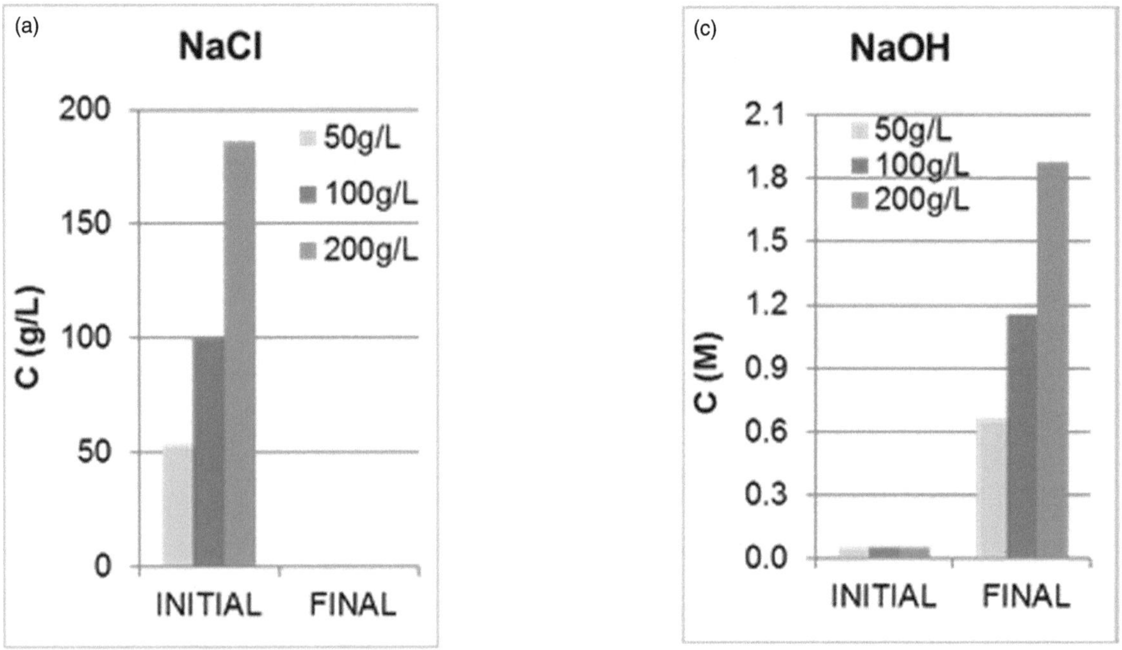

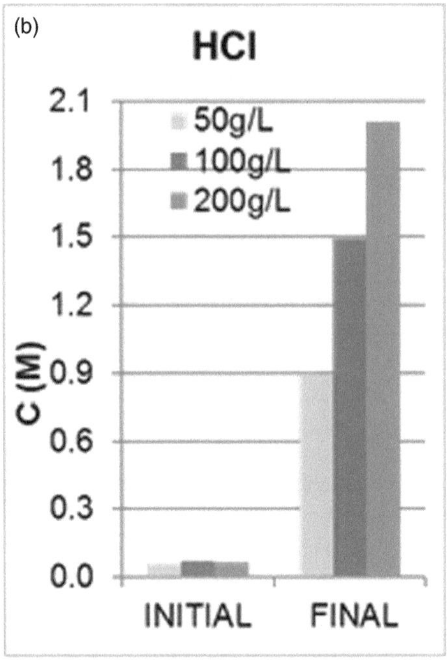

FIGURE 21.10 (a) Salt (NaCl), (b) acid (HCl), and (c) base (NaOH), initial and final concentrations, at different initial salt concentrations (50, 100, and 200 g NaCl/L).

technologies were studied in this case: selectrodialysis (SED) and EDBM [46].

It is worth mentioning that ED with monovalent selective membranes differs from SED. The former only uses monovalent selective membranes (anionic and cationic) placed alternatively between the electrodes to concentrate monovalent ions as much as possible, purifying the concentrate stream from divalent ones. The latter employs standard anionic and cationic membranes, as well as one monovalent selective membrane (anionic or cationic, depending on which ions are required to be separated), in order to produce two concentrated streams (one monovalent-rich and the other divalent-rich).

Thus, as can be seen in Figure 21.12, the idea was to separate the chloride and sulphate ions by means of SED and then to produce the respective acid (HCl and H_2SO_4) and base (NaOH) using EDBM. Finally, these products could be used in the same industrial sites, promoting a circular economy scheme [23].

For the SED tests, a lab-scale setup was used with an ED64-4 PCCell stack working in the batch mode. In this case, standard anionic (PC-SA) and cationic (PC-SK) exchange membranes, as well as monovalent selective anionic (PC-MVA) membranes (all from PCCell), were used in a three-cell triplet configuration (Figure 21.13). All membranes employed had an active membrane area of 64 cm^2. Four streams were employed for the SED trials: electrode rinse solution (0.4 M Na_2SO_4), feed solution (diluate stream), brine solution (monovalent-rich stream), and product solution (divalent-rich stream). Initially, all tanks were filled with 1 L of each solution. In this case, the initial feed composition was composed of 0.5 M Cl^- and 0.8 M SO_4^{2-}, whereas the initial brine and product streams were only composed of NaCl, according to optimal preliminary experimental results [47]. Again, several parameters were monitored during the operation, such as pressure, temperature, conductivity, flow rate, current, and voltage. The electrode rinse flow rate was kept constant at 100 L/h, whereas the other streams were set at around 15–20 L/h. Moreover, a constant voltage of 9 V was applied for the SED experiments, which lasted around 15 h.

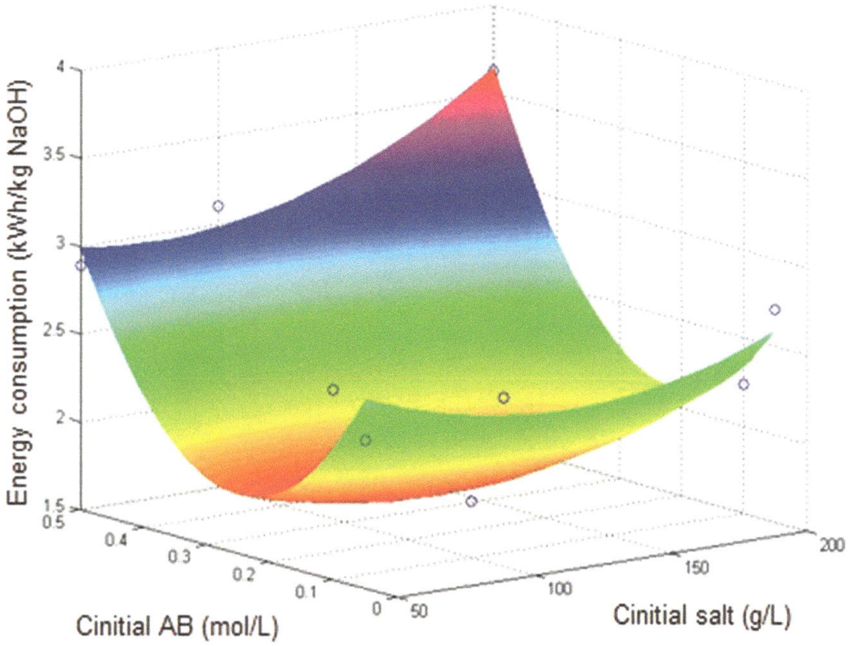

FIGURE 21.11 Optimal energy consumption values depending on the initial salt, acid, and base concentrations, represented with a surface plot.

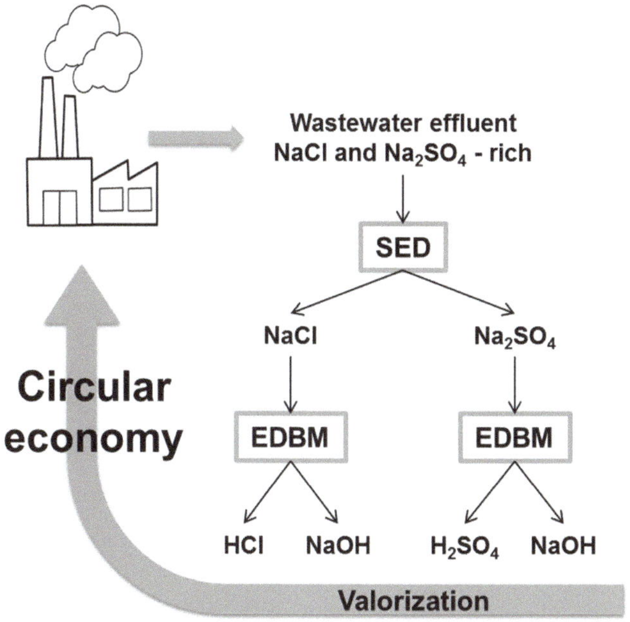

FIGURE 21.12 Selectrodialysis and electrodialysis with bipolar membranes scheme for industrial brines recovery [47].

The results demonstrated that chloride and sulphate ions could be separated by SED (Figure 21.14), obtaining a chloride-rich stream (brine) and a sulphate-rich stream (product).

As can be seen in Figure 21.14a, it was possible to obtain a monovalent-rich stream (composed of 72% Cl^- and 28% SO_4^{2-}). In this case, the final brine composition was 1.5 M chloride and 0.2 M sulphate. On the other hand, Figure 21.14b shows that a divalent-rich stream, containing 0.5 M SO_4^{2-} (97%) and 0.01 M Cl^- (3%) was also obtained.

Once both streams were separated, the EDBM experiments were conducted on each stream separately. In this case, the same PCCell stack was used (ED64-4) with standard anionic and cationic membranes combined with bipolar ones in three-cell triplets. Four streams were also used, although in this case they were: the electrode rinse, the salt (diluate), the acid, and the base. The same operational conditions were tested as for the SED tests: a constant voltage of 9 V and the same flow rates for each stream.

Table 21.5 summarizes the EDBM final compositions for the acid and base compartments.

As can be seen in Table 21.5, it was possible to produce 2.3 M NaOH and 1.3 M HCl (with traces of H_2SO_4, 0.2 M) from the SED monovalent-rich stream, whereas a 1.1 M NaOH base solution and a 0.4 M H_2SO_4 acid solution with some traces of HCl (0.03 M) were obtained from the SED divalent-rich stream.

Thus, it was possible to valorise a high-salinity industrial brine (containing NaCl and Na_2SO_4), obtaining pure NaOH in the base stream and almost pure HCl (87%) and H_2SO_4 (93%) in the acid compartment, depending on the initial solution (monovalent-rich or divalent-rich, respectively). These products could be used as chemicals in the same industrial processes, favouring the valorisation of wastewater and industrial symbiosis.

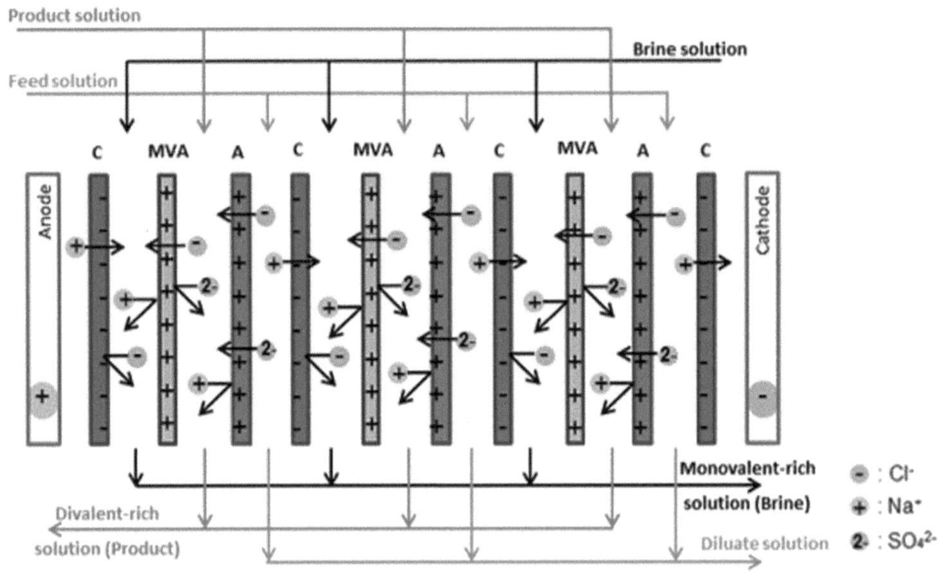

FIGURE 21.13 SED membranes disposition inside the stack.

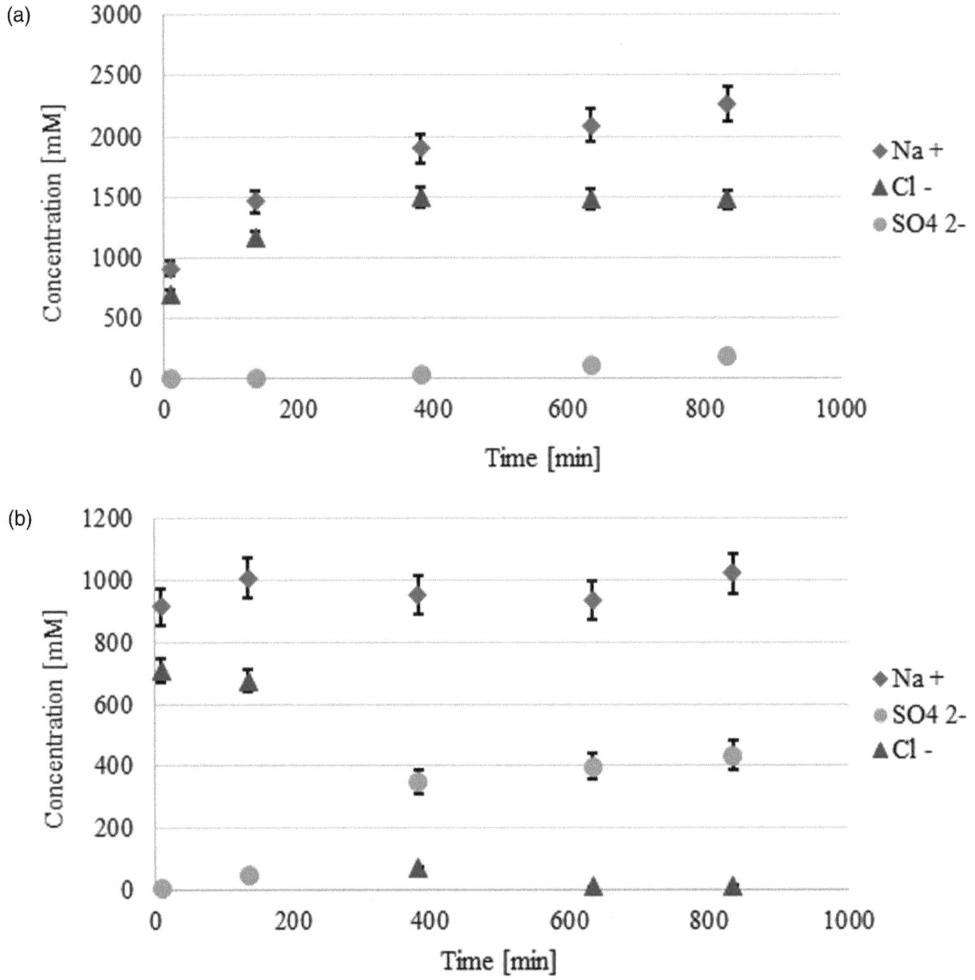

FIGURE 21.14 Ions concentration evolution during the SED experiments in the (a) brine stream (monovalent-rich) and (b) product stream (divalent-rich).

TABLE 21.5
EDBM final base and acid compositions for both treated streams

		Stream from SED	
Compartment	**Element**	**Cl-rich stream**	**SO_4^{2-}-rich stream**
Base	NaOH	2.3 M	1.1. M
Acid	HCl	1.3 M (87%)	0.03 M (7%)
	H_2SO_4	0.2 M (13%)	0.4 M (93%)

21.3 MINING WATER RECOVERY BY MEMBRANE TECHNOLOGIES

21.3.1 Acidic Mine Waters

Membrane technologies have been successfully applied for the treatment of acidic mine waters by their integration or with other technologies (solvent extraction, sorption, etc.) to promote the recovery of valuable metals and/or acid (see Figure 21.15).

Pressure-driven membrane processes have been evaluated for the treatment of acid mine waters. However, the high concentration of metals and their high acidity, which implies a high osmotic pressure of the feed solution, make the use of RO technically and economically unviable. As an alternative, NF is the preferable option, as due to the high permeation of acid, the osmotic pressure difference across the membrane is reduced, allowing work at lower pressures compared to RO. For instance, Ricci et al. [52] studied the integration of NF (MPF-34, Koch) and RO (TFG-HR, Koch) for treating an effluent from a pressure-oxidation process coming from a gold mining company (Brazil). The water was characterized as having a pH of 1.46 and 2.56 g/L Mg, 487 mg/L Ca, 436 mg/L Fe, and 21.48 g/L SO_4, among others. The membrane exhibited metal rejection rates ranging from 76% for Cu to 93% for Al, while acid permeated freely across the membrane after recovering 40% of the permeate in the NF unit. Nevertheless, scaling related to the precipitation of calcium sulphate was observed. The NF permeate was further treated in an RO unit in order to concentrate the acid recovered. In this case, the high rejection rates of the RO (>92%) allowed a 0.1 M sulphuric acid stream in the concentrate to be achieved. In a later study, Amaral et al. [54] showed that a long-term operation inside a mining company allowed the best performance to be achieved, working at recovery ratios of 90%, 40–50%, and 50% for UF (PVDF membrane, ZeeWeed), NF (DK, GE), and RO (SG, GE), respectively. Finally, obtaining a rich-metal stream from the NF unit (concentration factor of 2) was possible, while acid was finally concentrated by a factor of 2.7 in the RO retentate. López et al. [51,55] evaluated the performance of different NF membranes (NF270, Desal DL, HydraCoRe 70pHT, and a 1 nm TiO_2 membrane) for treating synthetic solutions, mimicking acidic mine water from the Iberian Pyrite Belt, aiming to concentrate rare earth elements (REEs) for further recovery. The authors studied the effects of both the Al and Fe concentration on the membrane performance (see Figure 21.16). Regarding the effect of the Al concentration (from 600 to 2200 mg/L) at pH 1, its increase resulted in higher sulphate rejection rates [e.g., from 60% to 80% for the Desal DL (positively charged) and from 38% to 46% for HydraCoRe 70pHT (negatively charged), at 10 bar] due to a shift in the sulphate equilibrium towards a higher presence of Al-SO_4 complexes [$AlSO_4^+$, and $Al(SO_4)_2^-$]. In terms of metal rejection rates, no variations were observed, but the transport of H^+ was favoured across the membrane, showing even lower rejections at higher Al concentrations (e.g., from 35% to 5% for the Desal DL and from 60% to 25% for the HydraCoRe 70pHT). The same conclusions were observed when the levels of Fe(III) were increased. Among the different membranes tested, the TiO_2 membrane showed the worst performance, exhibiting rejection rates below 40% for all the elements in the solution. Recently, Pino et al. [48] proposed the integration of NF and solvent extraction for the recovery of copper from mine water (Andina Division, CODELCO, Chile). The water presented as slightly acidic pH (3.5), with the presence of 531 mg/L Cu, 382 mg/L Al, 186 mg/L Ca, and 4671 mg/L SO_4, among others. The authors evaluated the performance of NF270 (Dow-Filmtec) and observed an overall rejection rate ranging between 70% (10 bar) and 83% (20 bar). Moreover, it was reported that after a recovery rate of 65%, the performance of the membrane decreased due to the precipitation of gypsum [$CaSO_4{:}2H_2O(s)$], which limited the possibility of concentrating Cu to more than 1 g/L. Experiments performed with solvent extraction using LIX 84-IC showed that it would be possible to recover up to 97% Cu from the retentate using two counter-current extraction stages and one stripping stage.

As an alternative to pressure-driven processes, forward osmosis (FO) is emerging as an alternative to concentrate elements in a solution without needing external hydraulic pressure. Choi et al. [56] evaluated the integration of forward osmosis and nanofiltration to treat a synthetic solution mimicking acid mine drainage (AMD) (pH 3.1, 20 mg/L Fe, 5.6 mg/L Cu, 4.2 mg/L Mn, 0.24 mg/L As, and 0.05 mg/L Cd, among others). In relation to the performance of the FO unit, two draw solutions were evaluated (EDTA-Na_4 and PSS-Na). When evaluating the water flux across the membrane at the same osmotic pressure of both draw solutions, the authors reported higher fluxes for the EDTA-Na_4 than for the PSS-Na, which was related to a higher concentration polarization effect with the former one. Moreover, some metals (e.g., Mn, As, Ca, and Pb) were completely rejected by the membrane, while rejection rates varied between 80–85% for Fe, Cu, and Zn. In a recent work, Baena-Moreno et al. [50] studied the performance of FO when treating synthetic effluents simulating the composition of Almanegra (Iberian Pyrite Belt), which is characterized by pH 2.7, 11.7 g/L SO_4, 1.10 g/L Mg, 0.97 g/L Zn, and 0.74 g/L Fe, among others. The authors evaluated the effects of different draw solutions (NaCl, KCl, $CaCl_2$, and $MgCl_2$) and their concentrations (from 1 to 5 mol/L) in water recovery and metal

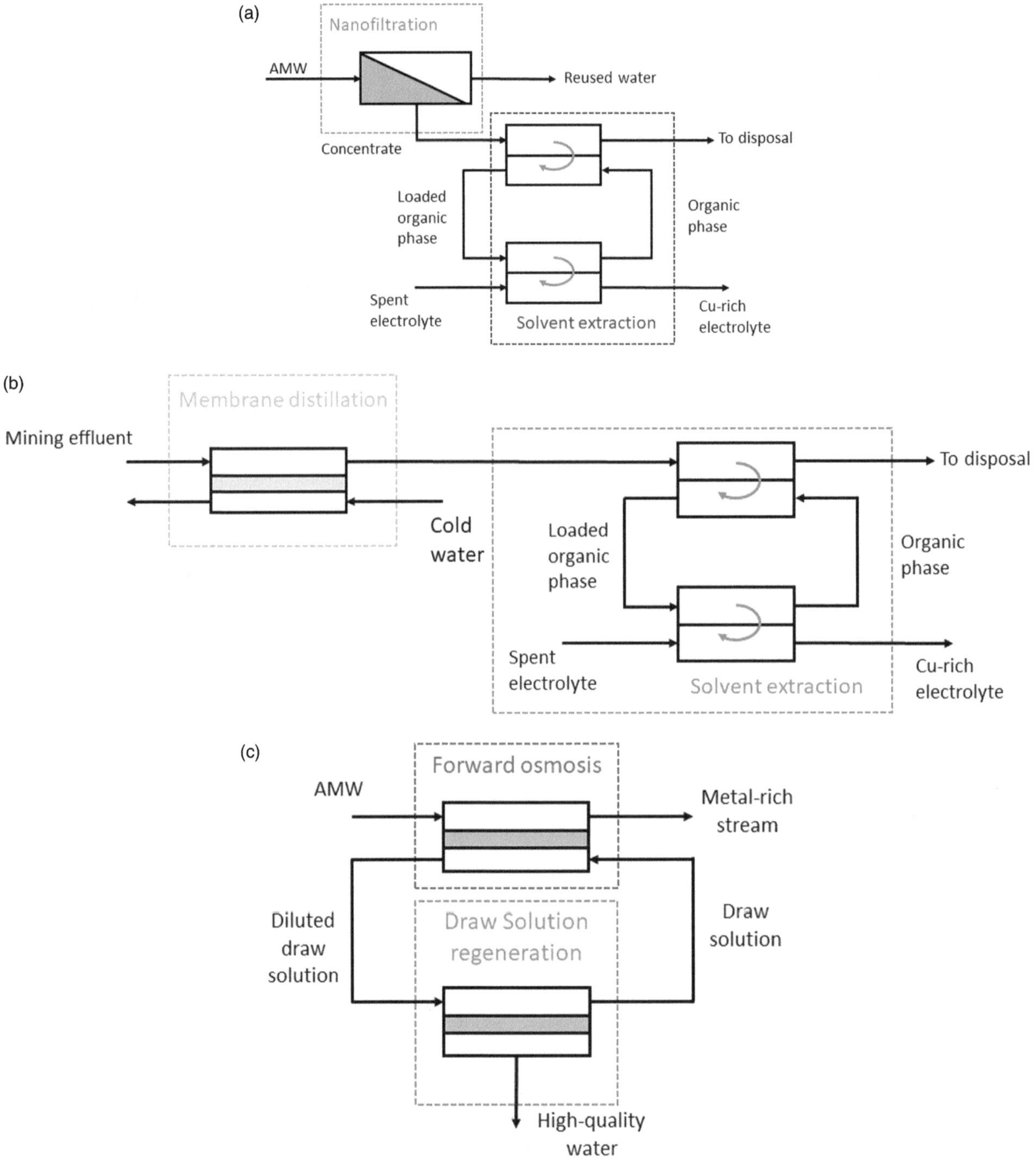

FIGURE 21.15 Examples of proposed valorisation of mining waters by (a) integration of nanofiltration and solvent extraction for Cu recovery (adapted from Ref. [48]), (b) membrane distillation and solvent extraction for Cu recovery (adapted from Ref. [49]), (c) forward osmosis for concentrating metals and water recovery (adapted from Ref. [50]), (d) selective precipitation, ion-exchange, and nanofiltration for rare earth elements (REEs) recovery (adapted from Ref. [51]), (e) integration of microfiltration, nanofiltration, and reverse osmosis for sulphuric acid recovery and metal concentration (adapted from Ref. [52]), and (f) electrodialysis for acid recovery from acid mine drainage (AMD) (adapted from Ref. [53]).

rejection. Independent of the draw solution, rejections were higher than 99.5%, but the flux obtained followed the following order: $MgCl_2$ > $CaCl_2$ > NaCl > KCl at the same concentration. For instance, fluxes across the membrane were 52.5 LMH, 51 LMH, 26.5 LMH, and 25 LMH using 5 mol/L $MgCl_2$, $CaCl_2$, NaCl, and KCl as draw solutions, respectively. In long-term experiments, the permeate flux decreased from 52 to 47.4 LMH after recovering 90% of water using 5 mol/L $MgCl_2$, keeping metal rejection rates higher than 99.5%. However, scaling

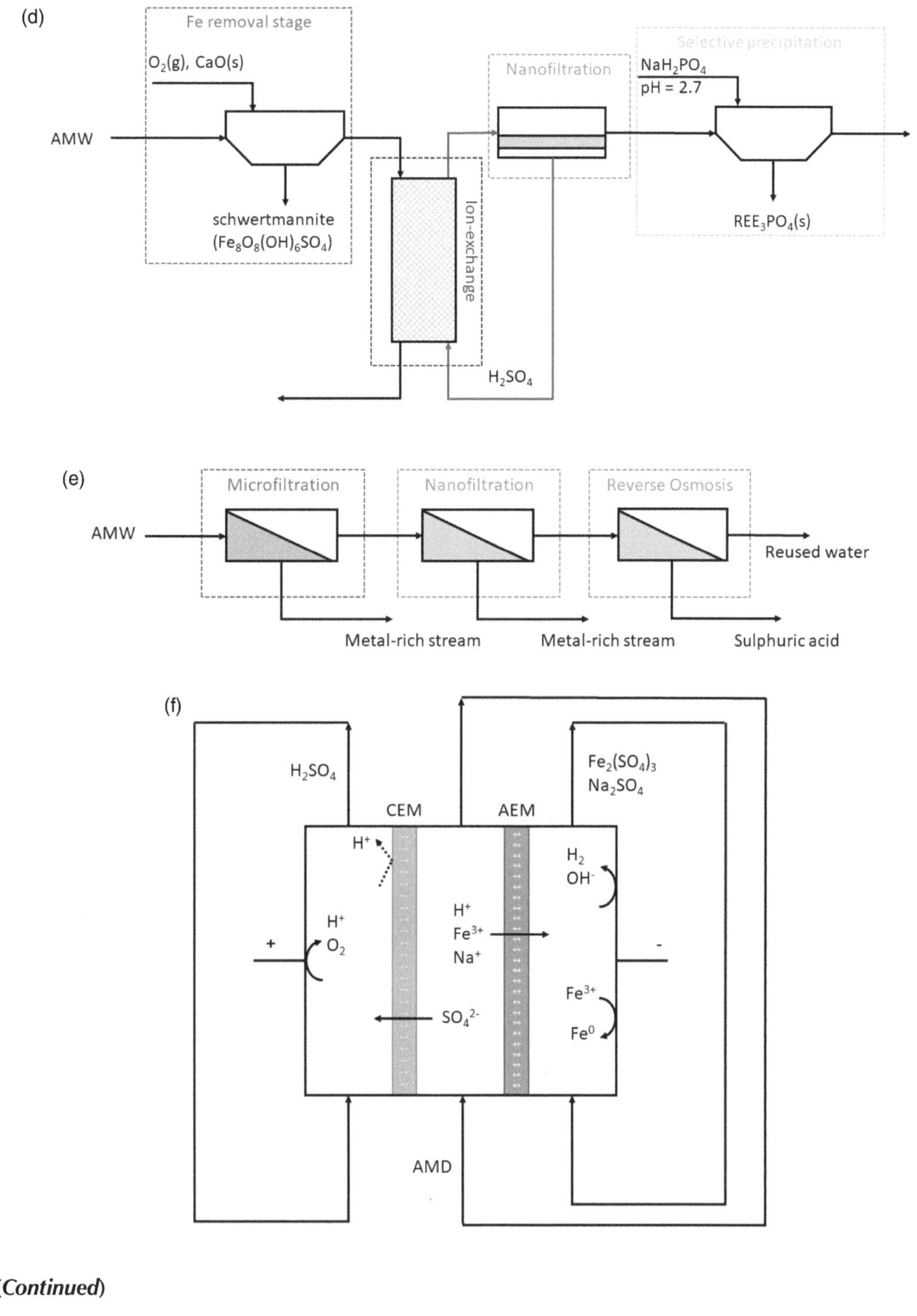

FIGURE 21.15 (*Continued*)

was observed due to the precipitation of a mixture of metallic sulphates and carbonates.

Membrane distillation (MD) has also been studied for the valorisation of acidic mine waters, allowing water recovery due to differences in vapour pressure across both membrane sides. For example, Amaya-Vías et al. [57] tested water and air gap membrane distillation (WGMD and AGMD, respectively) with polytetrafluoroethylene (PTFE) membranes with a sample from the Tinto River (pH 2.1 and composed mainly of 10.5 g/L SO_4, 0.81 g/L Mg, and 0.73 g/L Fe). Both WGMD and AGMD allowed permeate fluxes of 16.8 and 10.8 LMH, respectively, to be achieved with metal rejection rates higher than 99%. Ryu et al. [58] combined the zeolites for metal removal with direct contact membrane distillation (DCMD) for water recovery. In this case, a synthetic solution was used, which was characterized by a pH of 2 and a metal content of 340 mg/L Fe, 220 mg/L Mg, 170 mg/L Ca, 150 mg/L Al, and 120 mg/L Cu. The zeolite removed between 26–30% of the metals in the solution after being thermally treated at 500°C. Moreover, combining sorption and selective precipitation at

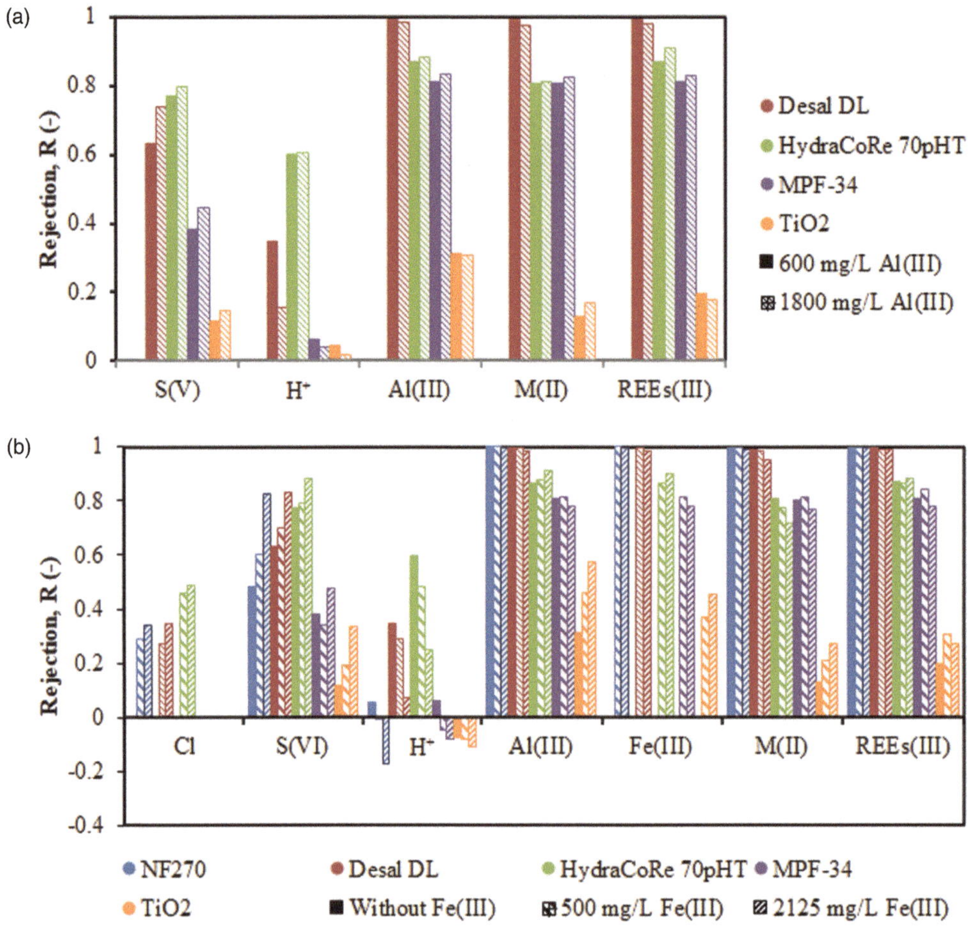

FIGURE 21.16 Effect of (a) Al and (b) Fe concentration on rejections of sulphate, chloride, H+, Al(III), Fe(III), REEs(III), and divalent metals (M(II)), at TMP of 10 bar when treating an acidic mine water (AMW).

pH 4 allowed the removal of Fe and Al completely. DCMD was evaluated at 50% water recovery, with an average flux of 2.5 ± 0.2 LMH. Jimenez and Ulbricht [49] studied the performance of DCMD with synthetic mining effluents (pH 2, 300 mg/L Fe, 50 mg/L Cu) to recover water and concentrate the metals in solution. The authors evaluated D845 and 3M membranes, showing permeate fluxes of 5.9 and 6.2 LMH, respectively, at 60°C, with rejection rates higher than 99.9%. The authors also proposed integrating three stages of DCMD (each of them recovering 80% of water) to reach the concentration required for solvent extraction for copper recovery.

Conventional ED has also been applied for treating acidic mine waters (AMWs), aiming to recover sulphuric acid. For example, Buzzi et al. [59] treated three different types of AMD from a carboniferous area from Brazil using ED, after selectively removing the Fe and following an oxidation/precipitation step to recover sulphuric acid. The authors observed removal rates (cations and anions) of higher than 97% in the three samples, allowing for water recovery. Martí-Calatayud et al. [53] studied acid recovery from synthetic AMD [pH 1.68, 8 g/L $Fe_2(SO_4)_3$, 1.5 g/L Na_2SO_4] in a three-compartment cell under different current densities. The authors were able to concentrate sulphuric acid by a factor of 4.0 at 15 mA/cm^2. Nevertheless, the concentration of Fe(III) did not vary in the central compartment at current densities higher than 10 mA/cm^2, which was related to precipitates of $Fe(OH)_3$ at the cationic exchange membrane. This precipitation of $Fe(OH)_3$ at higher current densities caused an increase in the specific energy consumption from 6 kWh/kg at 5 mA/cm^2 to 20 kWh/kg at 15 mA/cm^2.

21.3.2 Effluents from the Mineral and Metal Processing Industry

Effluents from the mineral and metal processing industry are usually characterized by a higher acidity, as they are created during pickling, electrowinning, gas cleaning, and mineral leaching, among others. Examples of the applicability of membrane technologies for the mineral and metal processing industry are shown in Figure 21.17.

Due to the high acidity of these streams, diffusion dialysis (DD) has been widely applied for acid recovery and purification. The presence of an anion exchange membrane favours the transport of acids across the membrane, while metals are rejected due to electrostatic repulsion. Jeong et al. [62] studied the performance of Selemion DSV membranes for

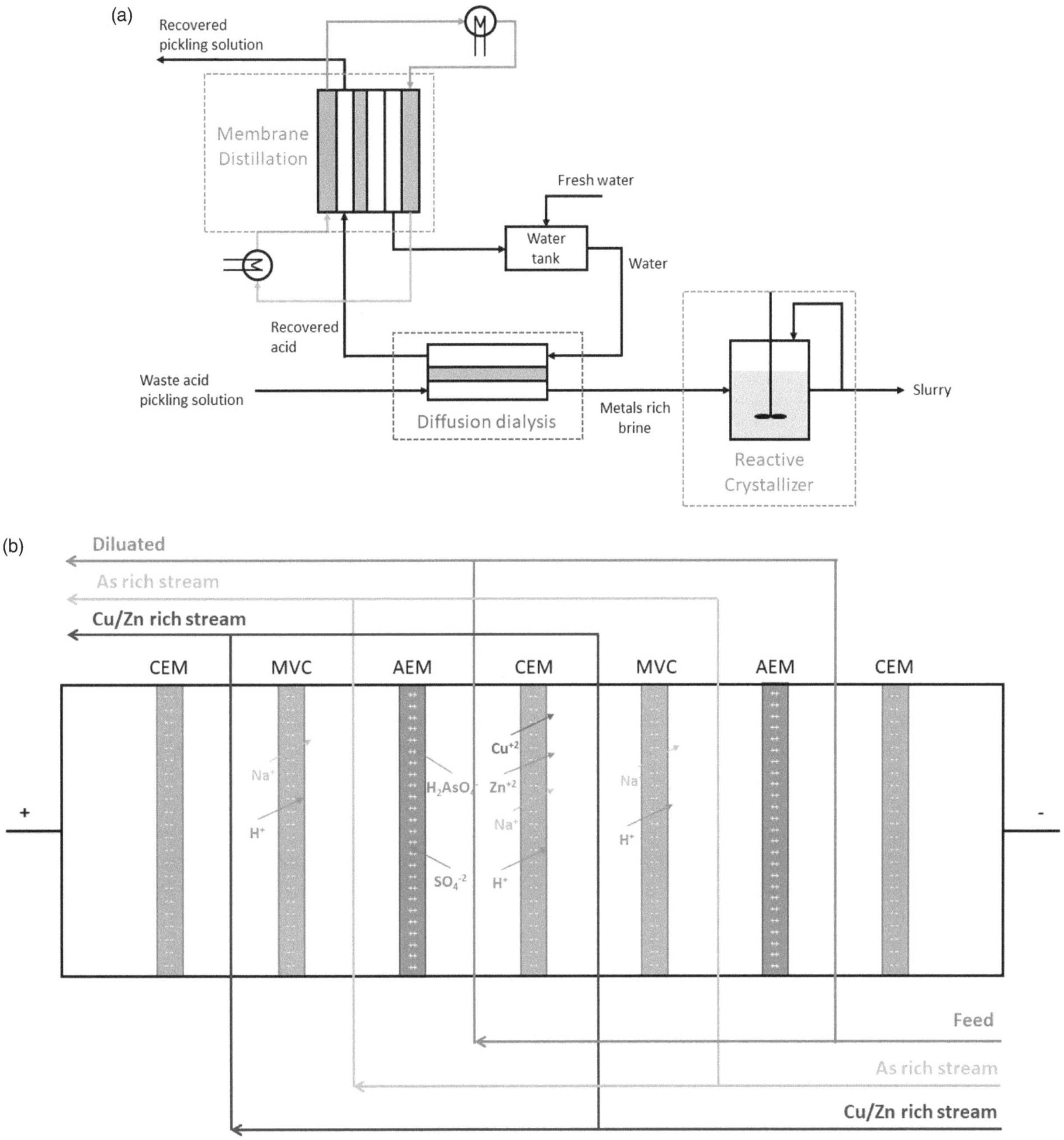

FIGURE 21.17 Examples of valorisation of mineral and metal processing water. (a) Integration of diffusion dialysis and membrane distillation for treating waste acid pickling solution, aiming to recover hydrochloric acid (adapted from Ref. [60]), and (b) selectrodialysis for the selective recovery and separation of Cu and Zn from As in metallurgical process waters (adapted from Ref. [61]).

treating an acidic solution from the diamond manufacturing process (440 g/L H_2SO_4, 52 g/L Fe, and 18 g/L Ni). The authors observed a dependence of the acid recovery on the operational parameters (flow rates and temperature) and initial acid concentration. For instance, lower flow rates (same for acid and water) promoted acid recovery and a higher concentration of acid in the diffusate. On the other hand, by keeping the feed flow-rate constant and varying the flow rate of water, the authors reported an increase in acid recovery at high water flow rates, but obtained more diluted sulphuric acid in the diffusate. In addition, higher temperatures enhanced the transport of sulphuric acid. By studying the effect of metals in the feed solution (Ni, Fe), the authors achieved high rejection rates (> 95%) without significant changes in acid recovery. Under the optimal conditions (0.26 $L/m^2{\cdot}h$), 80% of acid was recovered with impurities below 2 g/L. Li et al. [63] treated a leaching solution for vanadium extraction [2.4 mol/L H^+, 4.2 g/L V(V), 13.8 g/L Al(III)] using DD (DF120 membrane) for acid recovery. The authors evaluated the effect of flow rates on acid recovery and metal rejections. The results demonstrated that it was possible to achieve acid recoveries of 84% (2.2 mol/L H^+ in the diffusate) and metal rejection rates higher than

TABLE 21.6
Composition of the copper smelter effluent, waste stream (dialysate) and recovered acid (diffusate) after treatment with DD at 0.86 LMH (water to acid flow-rate ratio equal to 1)

	Copper smelter effluent (g/L)	Waste stream (g/L)	Recovered acid (g/L)	Recovery or passage rates (%)
H_2SO_4	217.5 ± 8.7	66.7 ± 8.9	146.3 ± 8.6	68.9 ± 2.3
As	3.33 ± 0.06	2.0 ± 0.3	1.26 ± 0.06	38.7 ± 1.3
Zn	0.46 ± 0.02	0.45 ± 0.07	< 0.04	
Fe	0.13 ± 0.01	0.12 ± 0.02	< 0.04	
Pb	$4.9\cdot10^{-3} \pm 4.9\cdot10^{-4}$	$4.4\cdot10^{-3} \pm 6.1\cdot10^{-4}$	$6.9\cdot10^{-4} \pm 2.3\cdot10^{-4}$	13.5 ± 1.5
Cd	0.10± 0.01	0.10 ± 0.01	$7.5\cdot10^{-3} \pm 4.0\cdot10^{-4}$	6.9 ± 0.3
Ni	$6.3\cdot10^{-3} \pm 1.4\cdot10^{-4}$	$6.2\cdot10^{-3} \pm 8.9\cdot10^{-4}$	$< 4\cdot10^{-4}$	
Cu	0.06 ± 0.01	$0.06\pm 7.7\cdot10^{-3}$	$3.2\cdot10^{-3} \pm 4.3\cdot10^{-5}$	5.2 ± 0.1

90% under the optimum conditions (0.21 $L/m^2{\cdot}h$ feed flow rate and water to feed flow rate ratio of 1.0–1.3). The authors also studied the effect of the metal concentration on its rejection and acid recovery, observing that both parameters increased with the metal concentration. Recently, Gueccia et al. [64] evaluated acid recovery from pickling solutions (5–20 g/L Zn, 50–150 g/L Fe, and 70–100 g/L HCl) using batch and continuous configurations with Fumasep membranes (FAD-PET-75). Under the batch configuration, the authors reported the possibility of recovering HCl, whereas the presence of Fe enhanced the possibility of achieving higher acid concentrations. Nevertheless, the presence of Zn limited the transport of HCl due to its passage as anionic species ($ZnCl_3^-$, $ZnCl_4^{2-}$). In the continuous configuration, the authors reported the possibility of recovering 80% of HCl, with leakages of 30% of Fe and 60% of Zn, working with a solution containing 100 g/L HCl, 117 g/L Fe, and 8 g/L Zn at 48 mL/min. Luo et al. [65] evaluated the DF120 membrane in a spiral-wound module for H_2SO_4 recovery. The authors tested the acid recovery and its concentration in the dialysate as a function of the flow rate by connecting different modules (one, two, or three modules), where fluids circulated in a counter-current regime. The results indicated the possibility of achieving the same acid recoveries under different configurations. For example, 65% of acid could be recovered working at 6.5 mL/min in a single module or at 50.5 mL/min with three modules in series. López et al. [66] evaluated the performance of Neosepta-AFX membranes for recovering acid from a gas cleaning effluent from a copper smelter (220 ± 10 g/L H_2SO_4, 3.4 ± 0.2 g/L As, and 0.5 ± 0.1 g/L Zn, among others). At the optimum operating conditions (0.86 LMH flow rates for both acid and water; see Table 21.6), it was possible to recover around 67% of acid, while metal impurities were rejected by more than 85%. Nevertheless, 39% of As permeated easily across the membrane due to its presence as neutral species (H_3AsO_4/ H_3AsO_3).

Studies related to NF have been published from the mineral and metal processing industry. For example, López et al. [67] evaluated the performance of NF270 for treating solutions coming from a gas scrubber from a copper smelter (up to 40 g/L H_2SO_4, 10 g/L HCl, and 1 g/L As, while metals were at mg/L levels). High metal rejection rates (> 80%) were observed, while acid permeated easily across the membrane (sulphate and chloride rejections below 40% and 5%, respectively). Due to the presence of As as a non-charged species [H_3AsO_4(aq)], it also permeated across the membrane (rejections below 30%). In a recent work, López et al. [68] evaluated the Duracid membrane for treating effluents from copper smelters exiting the solvent extraction unit in order to recover sulphuric acid (see Figure 21.18). The solutions were characterized by an acidic pH (0.6 < pH < 1.6), 27–57 g/L S(VI), 8–15 g/L Fe, and 0.7–1.5 g/L Zn, among others. High metal rejection rates (> 80%) were reported, showing the possibility of recovering H_2SO_4 in the permeate, with the presence of some Na and As.

The performance of electro-membrane processes has also been widely studied for the mineral and metal processing industry. Cifuentes et al. [69] tested the treatment of copper electrorefining electrolytes [50 g/L H_2SO_4, 3–9 g/L Cu, 3 g/L As (As(III) or As(V)), 0.025 g/L Sb] using ED (Ionac AEM MA-3745 and CEM MC-3470). Due to the speciation of As, it was observed that As(III) (in solution as H_3AsO_3 and $H_4AsO_3^+$) migrated towards the anode, while As(V) (as H_3AsO_4 and $H_2AsO_4^-$) was transported to the cathode. The authors reported the possibility of separating and concentrating Cu from As, obtaining separation factors of 12.5 and 20.0 for Cu/As(III) and Cu/As(V), respectively. Chekioua and Delimi [70] evaluated the treatment of acid pickling bath [150 g/L H_2SO_4 and 26 g/L Fe(II)] with ED using the anionic exchange membrane (AEM), named AMX AEM, and different cationic exchange membranes (CEMs; CMX, Nafion 117, and CMV). The best results were achieved for the CMX in terms of Fe(II) rejection (66%) and energy consumption (1.85 kWh/kg). Moreover, Fe(II) rejection increased from 7% to 66% when the current density was increased from 1 to 20 mA/cm^2, but at even higher values (30 mA/cm^2), scaling was observed in the membranes. The addition of Fe(II) to the solution enhanced the recovery of H_2SO_4, as its concentration increased from 14

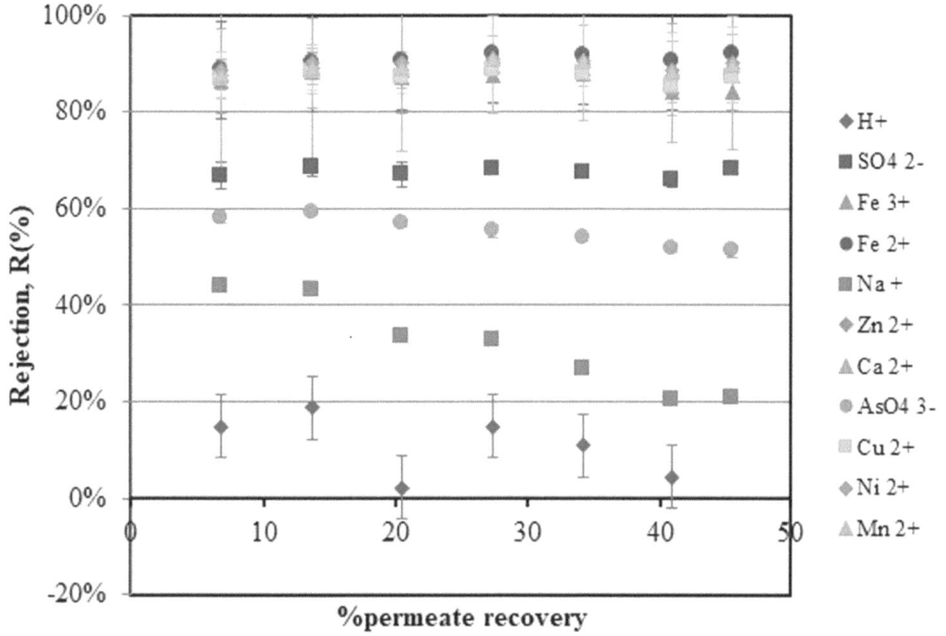

FIGURE 21.18 Rejection curves as a function of permeate recovery at 32 bar using Duracid membrane when treating an effluent from a copper smelter.

to 25 g/L when Fe levels were shifted from 1 to 52 g/L. Liu et al. [71] evaluated the possibility of using EDBM for treating a raffinate from the hydrometallurgical copper industry (pH 1.4, 45 g/L SO_4^{2-}, 11.8 g/L Fe, 336 mg/L Zn, and 135 mg/L Cu, among others). In this work, Liu et al. [71] studied the effect of the current density and the solvent ratios of raffinate and heavy metal chamber (or compartments). Moreover, the effect of the number of triplets in the performance results was also evaluated, and a decrease in energy consumption from 0.160 to 0.089 kWh/L of raffinate was observed. After operating the set-up for 40 h at 3 mA/cm^2 and a volume ratio of 1:15 (raffinate to metal chamber), more than 70% of metals were rejected, with a total presence of 100 mg/L of metals in the acid recovered (39 g/L H_2SO_4), which could be reused as a leaching agent. Another possible configuration is the inclusion of mono-selective membranes (SED). Within this context, Reig et al. [61] studied the performance of SED to selectively separate As(V) from Cu and Zn from a metallurgical process stream (pH 2.3, 2 g/L Cu, 9.6 g/L Zn, and 2.4 g/L As). This configuration resulted in the possibility of obtaining a Cu- and Zn-rich stream, containing 80% and 87% of the initial amounts of Cu and Zn, respectively, with As impurities of around 0.02%.

Thermal-driven membrane processes have also been evaluated for the treatment of processing waters, as they can be used as a pre-concentration unit and to recover water. For instance, Cai and Guo [72] evaluated the performance of PTFE membranes for treating effluents from the hot-dip galvanizing industry (HCl-based, 0.5 < pH < 1.5, 50–300 g/L $FeCl_2$ and 4–35 g/L $FeCl_3$). The authors evaluated the effect of acidity in the feed solution, observing a decrease in the permeate flux from 8 to 6 kg/m^{2}h at 75°C when the HCl concentration was increased from 0 to 200 g/L. In addition, a higher HCl concentration in the permeate was achieved when working at high acidities. For example, an initial concentration of 7 wt.% HCl resulted in a permeate containing around 3 wt.% HCl, while working at 15 wt.% HCl led to a concentration of 7.5 wt.%. The authors also evaluated the salt-effect by adding $FeCl_2$, reaching higher acidities in the permeate with high metal rejections (> 99.99%). Recently, Chen et al. [73] treated refining wastewater (pH 0.03, 48 g/L Cl, 11.4 g/L Na, 10 g/L K, among others) at 60°C using DCMD for recovering precious group metals. By increasing the pH of the solution to 7, membrane scaling was observed due to the precipitation of silica (pH 7) and Cr(III) at pH 3. During operation, the permeate was mainly composed of HCl with low amounts of metals (< 5 mg/L). Tomaszewska et al. [74] treated pickling solutions [86–135 g/L HCl, 361 g/L metals (Cu, Fe, Zn, and Mg)] with capillary polypropylene (PP) membranes at 70°C using MD. The authors reported the possibility of recovering HCl in the permeate with concentrations higher than the feed side (165 g/L). Nevertheless, the total concentration of metals in the feed stream rose to 600 g/L, causing the precipitation of $CuCl_2{\cdot}2H_2O$. The authors suggested the need to perform a rinse with water on the feed side to prevent the formation of salt crystals in the membrane pores.

The integration of DD and MD for the treatment of pickling solutions (100 g/L HCl, 40–120 g/L Fe, 12 g/L Zn) at the pilot plant scale has been assessed by Gueccia et al. [60]. In the DD unit, the authors were able to recover 82% of the HCl, while leakages of 50% and 65% of Fe and Zn, respectively, were observed due to the formation of metal–chloride complexes. The recovered acid exiting the DD unit was fed to the MD unit, aiming to concentrate it to be recycled back to the pickling

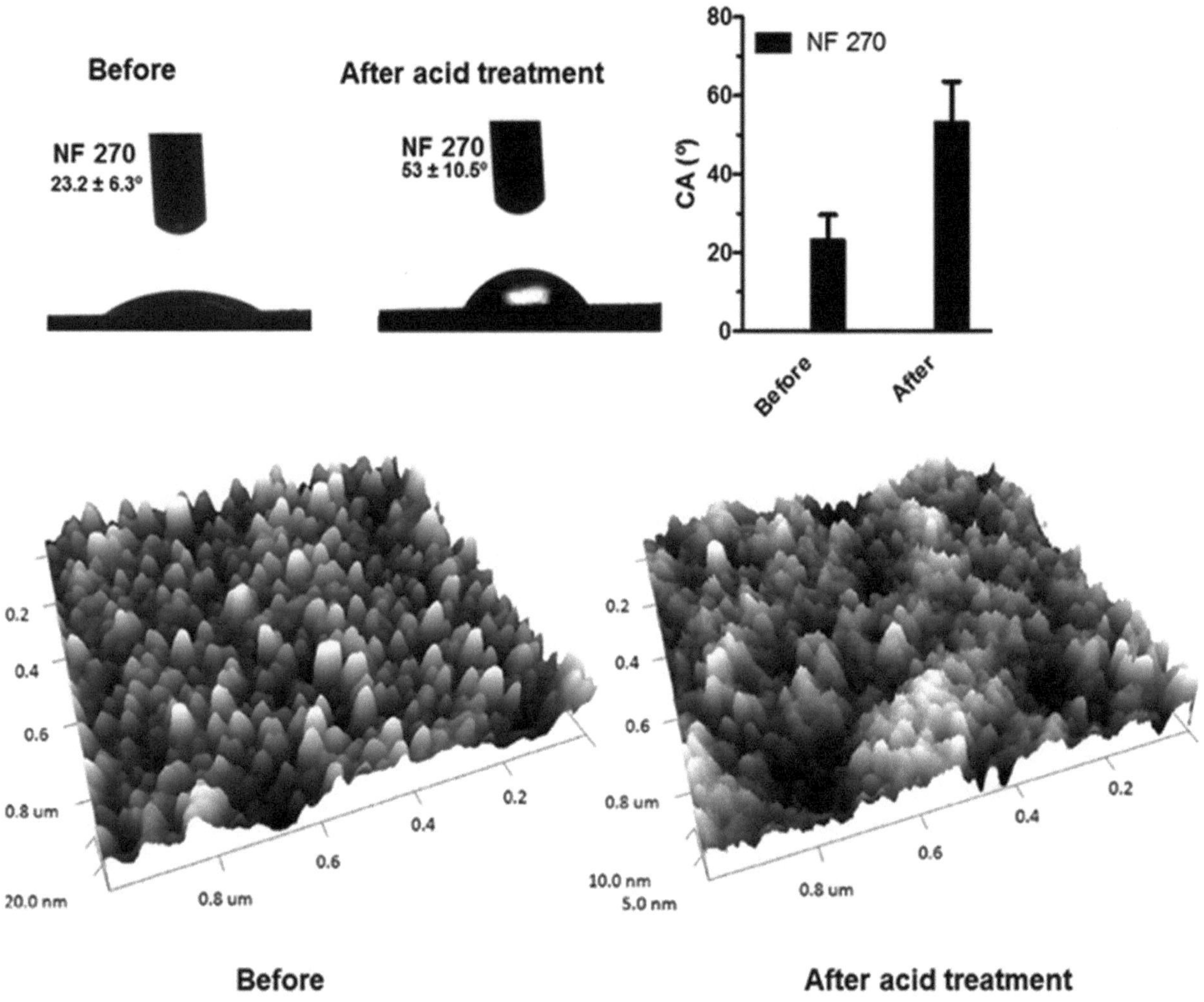

FIGURE 21.19 Contact angle and atomic force microscopy results of virgin and aged NF270 after acidic treatment.

unit. In this case, the salt effect transported around 20% of HCl across the membrane.

21.3.3 Issues to Consider When Treating Mining Effluents

Some issues should be considered when working with acidic effluents using membrane technologies. One of the main concerns is the stability of the membrane in such acidic media. For example, typical NF, RO, and FO membranes are made of an active layer of polyamide, which is susceptible to suffering hydrolysis at long-term exposure [75–77]. Unfortunately, most of the published studies are focused on immersing the membranes in acids for weeks rather than long-term experiments. For example, when studying the stability of commercial polyamide NF membranes, Plat et al. [75] reported membrane degradation in acidic media, which was severe at high temperatures. Manis et al. [76] observed a change in the membrane properties of polyamide membranes after immersing them into H_3PO_4 solutions, affecting the membrane charge and selectivity. López et al. [77] immersed the NF270 in 98 g/L H_2SO_4 for 30 days. Figure 21.19 shows the contact angle and atomic force microscopy (AFM) results of both virgin and treated NF270. It can be observed that the membrane became more hydrophilic after the acid treatment, as an increase in contact angle was noticed. Moreover, the AFM results showed that the virgin membrane surface appeared homogeneous, while for the acid treatment, scattered areas emerged in the selective layer. Roughness measurements from AFM showed an increase from 1.9 ± 0.2 nm to 3.4 ± 0.2 nm.

A further analysis by X-ray photo-electron spectroscopy (XPS) (see Table 21.7) revealed the hydrolysis of the amide groups, promoting the formation of free amine and carboxylic groups. For example, the contribution of C-N, C=N, and O=C-N groups decreased from 95% in the virgin to 52% in the aged membrane, while the fraction of the groups related to -NH_3^+, -NH_2R^+ increased noticeably.

These results confirmed hydrolysis of the amide groups that increased the permanent ionized groups and the size of the free volume, according to the following reaction:

TABLE 21.7
Binding energies and relatively amount of functional groups for virgin and aged membrane, given by XPS

	Virgin membrane		Aged membrane	
	E_B (eV)	%	E_B (eV)	%
C (1s)	**72.5 %**		**66.0 %**	
C-C,C-H, C=C	284.6	60.1	284.6	65.0
C-O, C-N	285.8	24.9	285.9	28.2
C=O-N, N=C, C=O	287.6	15	288.6	5.0
Saturated, π bond			292.7	0.7
Saturated, π bond			291.4	1.2
N (1s)	**16.8 %**		**23.0 %**	
C-N, C=N, O=C-N	399.5	95.3	399.9	51.6
$-NH_3^+$, $-NH_2R^+$	401.3	4.7	401.5	48.4
O (1s)	**10.7 %**		**4.9 %**	
O=C-N, C=O, C-O	531	75.6	531.2	8.7
H···O=C-N, O=C-O	532.4	24.4	532.8	63.7
SO_4^{2-}			531.6	27.6
S (2p)	**-**		**6.1 %**	

In order to overcome the degradation of membranes, a new generation of polymeric membranes devoted to working in high acidic media has been developed [52,78,79], that is formulated on sulphamide or a sulphonated polyether-sulphone. As an alternative, ceramic membranes based on zirconia (ZrO_2) or titania (TiO_2) are also promising options, but their high cost, low active surface area per volume, and low selectivity limit their applicability.

Another important aspect that must be considered is the scaling (i.e., precipitation of inorganic mineral phases) that may occur during operation. For instance, when treating AMWs with low pH, the precipitation of Fe(III) as hydroxide [$Fe(OH)_3$(s)] or oxyhydroxide [FeOOH(s)] and gypsum [$CaSO_4 \cdot 2H_2O$(s)] can occur at the membrane surface. For example, Rieger et al. [80] observed the precipitation of gypsum and metal hydroxides when filtering AMW, while Al-Zoubi et al. [81] also reported gypsum formation using NF. Scaling can also have some implications in electro-membrane processes. As previously stated, Chekioua and Delimi [70] reported Fe(II) scaling at 30 mA/cm^2 in the treatment of acid pickling bath using ED. When treating AMW using ED, Martí-Calatayud et al. [53] reported the formation of $Fe(OH)_3$ precipitates in the anodic side of the CEM as a result of surpassing the limiting current density (16 mA/cm^2). This caused an increase in the specific energy consumption from 6 to 20 kWh/kg due to Fe scaling. Finally, in treating acidic waters using MD, Feng et al. [82] observed supersaturation of $FeSO_4$ leading to pore blocking, while Tomaszewska et al. [74] reported the formation of $CuCl_2 \cdot 2H_2O$ crystals when working with MD. Most of the commercially available anti-scalants cannot work under acidic conditions ($pH < 1$); thus, precipitation of mineral phases and incrustations cannot be avoided. Then, high cross-flow velocities must be used to minimize mineral deposition onto the membrane surface.

21.4 CONCLUSIONS

In the case of industrial brines, the objective of increasing the dependence on secondary resources (i.e., recycled materials that are transferred back to the production of new materials) that today contribute 10–12% of the total material demand in the EU will clearly need new technologies and processes to promote internal reuse. Recycling all materials and valorising all the waste streams by 2050 implies a fundamental re-design of value chains. Industries need to establish new collaborations (e.g., industrial and urban symbiosis) to make a paradigm shift to enhance circularity and up-cycling feasible. The concentrated brines generated in both the industrial and urban cycles are a new target resource that needs to be put in value.

In the case of AMIMWs, the principal challenge under consideration is whether alternative treatment methods that have been postulated can compete with their conventional counterparts. Although several alternative treatment approaches have been presented for AMIMW management and explored in general at the laboratory scale (Section 21.3), none of these have attained full-scale implementation. However, it should be highlighted that some of them are under pilot evaluation at a relevant industrial scale [e.g., with a technology readiness level (TRL) of 5 to 7]. Accordingly, the integration of processes by combining conventional and alternative ones (e.g., membrane-based) may increase the efficiency of the former while overcoming limitations of the latter. As described in Section 21.2, integrated processes have been widely explored for treating industrial brines, concentrates, and wastewater with promising results.

The generation of AMIMWs, metal-rich acid aqueous solution, poses major environmental and economic challenges for the mining and mineral processing industry. The selection of a suitable AMIMW management option (or a few options combined together) relies on geographical site-specific settings, such as resource availability, aqueous stream composition (e.g., acidity level and content of valuable metals and hazardous species), treatment targets, associated risk, and the desired end-use of the site that accumulates the AMIMWs. The varying effectiveness of current AMIMW treatments applied at mining and mineral processing sites is specifically attributed to aspects such as acidity, metal content loads, and flow rates to be gathered. Due to the toxicity of the sludge generated when treating AMIMWs using traditional methods, alternatives are being studied to avoid their generation. The search for safer processes has led to much efforts on exploring more sustainable treatment options for reuse and recovery, given that AMIMW are now considered a new source for raw materials recovery or even for obtaining added-value critical raw materials such as REEs. Moreover, membrane and alternative treatment options can produce high-quality water for reuse from AMIMWs. Nevertheless, the cost associated with wastewater management, efforts in pretreatments to reduce membrane fouling, and increase in membrane lifetime remain challenges for achieving effective full-scale AMIMW treatment. Thus, sustainable solutions only will be achieved on management routes based on integrated processes by increasing the efficiency of conventional processes while overcoming the limitations of alternative processes.

LIST OF ACRONYMS

AEM	Anionic exchange membrane
AFM	Atomic force microscopy
AGMD	Air gap membrane distillation
AMD	Acid mine drainage
AMIMW	Acidic metal influenced mine waters
AMW	Acidic mine water
BW	Brackish water
BWRO	Brackish water reverse osmosis
CEM	Cationic exchange membrane
DCMD	Direct contact membrane distillation
DD	Diffusion dialysis
ED	Electrodialysis
EDBM	Bipolar electrodialysis
EU	European Union
FO	Forward osmosis
MCr	Membrane crystallization
MD	Membrane distillation
MLD	Minimum liquid discharge
NF	Nanofiltration
PP	Polypropylene
PTFE	Polytetrafluoroethylene
REE	Rare earth elements
RO	Reverse osmosis
RSM	Response surface methodology
SED	Selectrodialysis
SPIRE	Sustainable Process Industry through Resource and Energy Efficiency
SW	Seawater
SWRO	Seawater reverse osmosis
TRL	Technology readiness level
WGMD	Water gap membrane distillation
XPS	X-ray photo-electron spectroscopy
ZLD	Zero liquid discharge

ACKNOWLEDGEMENTS

This research was supported by the W4V project (ref. PID2020-114401RB-C21) financed by the Spanish Ministry of Science and Innovation and also by the Catalan Government (ref. 2017-SGR-312), Spain. Additionally, this research has received funding from the European Union's Horizon 2020 research and innovation programme, within the OpenInnoTrain project under the Marie Sklodowska-Curie grant agreement n°823971. The content of this publication does not reflect the official opinion of the European Union. Responsibility for the information and views expressed in the publication lies entirely with the author(s). The work of J. López was supported within the scope of Margarita Salas fellowship, financed by the Ministerio de Universidades (Spain) and European Union – NextGenerationEU.

REFERENCES

[1] European Commission (2019) *The European Green Deal – COM(2019) 640 final*. Brussels.

[2] Kotsanopoulos, K.V., Arvanitoyannis, I.S. (2015) Membrane processing technology in the food industry: Food processing, wastewater treatment, and effects on physical, microbiological, organoleptic, and nutritional properties of foods. *Critical Reviews in Food Science and Nutrition* 55: 1147–1175.

[3] Quist-Jensen, C.A., Macedonio, F., Drioli, E. (2016) Membrane crystallization for salts recovery from brine—an experimental and theoretical analysis. *Desalination and Water Treatment* 57: 7593–7603.

[4] Ghaffour, N., Missimer, T.M., Amy, G.L. (2013) Technical review and evaluation of the economics of water desalination: Current and future challenges for better water supply sustainability. *Desalination* 309: 197–207.

[5] Mohamed, A.M.O., Maraqa, M., Al Handhaly, J. (2005) Impact of land disposal of reject brine from desalination plants on soil and groundwater. *Desalination* 182: 411–433.

[6] Tsai, J.H., Macedonio, F., Drioli, E., Giorno, L., Chou, C.Y., Hu, F.C., Li, C.L., Chuang, C.J., Tung, K.L. (2017) Membrane-based zero liquid discharge: Myth or reality? *Journal of the Taiwan Institute of Chemical Engineers* 80: 192–202.

[7] Reig, M., Casas, S., Gibert, O., Valderrama, C., Cortina, J.L. (2016) Integration of nanofiltration and bipolar electrodialysis for valorization of seawater desalination brines: Production of drinking and waste water treatment chemicals. *Desalination* 382: 13–20.

[8] Reig, M., Casas, S., Valderrama, C., Gibert, O., Cortina, J.L. (2016) Integration of monopolar and bipolar electrodialysis for valorization of seawater reverse osmosis desalination brines: Production of strong acid and base. *Desalination* 398: 87–97.

[9] Reig, M., Valderrama, C., Gibert, O., Cortina, J.L. (2016) Selectrodialysis and bipolar membrane electrodialysis combination for industrial process brines treatment: Monovalent-divalent ions separation and acid and base production. *Desalination* 399: 88–95.

[10] Seraj, S., Mohammadi, T., ahmadzadeh tofighy, M. (2022) Graphene-based membranes for membrane distillation applications: A review. *Journal of Environmental Chemical Engineering* 10: 107974.

[11] Chekli, L., Phuntsho, S., Kim, J.E., Kim, J., Choi, J.Y., Choi, J.S., Kim, J., Kim, J.H., Hong, S., Sohn, J., Shon, H.K. (2016) A comprehensive review of hybrid forward osmosis systems: Performance, applications and future prospects. *Journal of Membrane Science* 497: 430–449.

[12] Fernández-Torres, M.J., Randall, D.G., Melamu, R., von Blottnitz, H. (2012) A comparative life cycle assessment of eutectic freeze crystallisation and evaporative crystallisation for the treatment of saline wastewater. *Desalination* 306: 17–23.

[13] Macías, F., Pérez-López, R., Caraballo, M.A., Cánovas, C.R., Nieto, J.M. (2017) Management strategies and valorization for waste sludge from active treatment of extremely metal-polluted acid mine drainage: A contribution for sustainable mining. *Journal of Cleaner Products* 141: 1057–1066.

[14] RoyChowdhury, A., Sarkar, D., Datta, R. (2015) Remediation of acid mine drainage-impacted water. *Current Pollution Reports* 1: 131–141.

[15] Johnson, D.B., Hallberg, K.B. (2005) Acid mine drainage remediation options: A review. *Science of the Total Environment* 338: 3–14.
[16] Kefeni, K.K., Msagati, T.A.M., Mamba, B.B. (2017) Acid mine drainage: Prevention, treatment options, and resource recovery: A review. *Journal of Cleaner Products* 151: 475–493.
[17] Vecino, X., Reig, M., López, J., Valderrama, C., Cortina, J.L. (2021) Valorisation options for Zn and Cu recovery from metal influenced acid mine waters through selective precipitation and ion-exchange processes: promotion of on-site/off-site management options. *Journal of Environmental Management* 283: 112004.
[18] Nleya, Y., Simate, G.S., Ndlovu, S. (2016) Sustainability assessment of the recovery and utilisation of acid from acid mine drainage. *Journal of Cleaner Products* 113: 17–27.
[19] López, J., Reig, M., Gibert, O., Torres, E., Ayora, C., Cortina, J.L. (2018) Application of nanofiltration for acidic waters containing rare earth elements: Influence of transition elements, acidity and membrane stability. *Desalination* 430: 33–44.
[20] Masindi, V. (2017) Recovery of drinking water and valuable minerals from acid mine drainage using an integration of magnesite, lime, soda ash, CO2 and reverse osmosis treatment processes. *Journal of Environmental Chemical Engineering* 5: 3136–3142.
[21] Patel, C.G., Barad, D., Swaminathan, J. (2022) Desalination using pressure or electric field? A fundamental comparison of RO and electrodialysis. *Desalination* 530: 115620.
[22] Shimura, H. (2022) Development of an advanced reverse osmosis membrane based on detailed nanostructure analysis. *Polymer Journal* 54: 767–773.
[23] European Commission (2020) *Circular Economy Action Plan.*
[24] Malaeb, L., Ayoub, G.M. (2011) Reverse osmosis technology for water treatment: State of the art review. *Desalination* 267: 1–8.
[25] Stanford, B.D., Leising, J.F., Bond, R.G., Snyder, S.A. (2010) Chapter 11 Inland desalination: Current practices, environmental implications, and case studies in Las Vegas, NV. *Sustainability Science and Engineering* 2: 327–350.
[26] Rioyo, J., Aravinthan, V., Bundschuh, J., Lynch, M. (2017) A review of strategies for RO brine minimization in inland desalination plants. *Desalination and Water Treatment* 90: 110–123.
[27] Lee, S., Choi, J., Park, Y.G., Shon, H., Ahn, C.H., Kim, S.h. (2019) Hybrid desalination processes for beneficial use of reverse osmosis brine: Current status and future prospects. *Desalination*. 454: 104–111.
[28] Xu, P., Cath, T.Y., Robertson, A.P., Reinhard, M., Leckie, J.O., Drewes, J.E. (2013) Critical review of desalination concentrate management, treatment and beneficial use. *Environmental Engineering Science* 30: 502–514.
[29] Roberts, D.A., Johnston, E.L., Knott, N.A. (2010) Impacts of desalination plant discharges on the marine environment: A critical review of published studies. *Water Research* 44: 5117–5128.
[30] Kim, D.H. (2011) A review of desalting process techniques and economic analysis of the recovery of salts from retentates. *Desalination* 270: 1–8.
[31] Fernandez-Gonzalez, C., Dominguez-Ramos, A., Ibañez, R., Irabien, A. (2016) Electrodialysis with bipolar membranes for valorization of brines. *Separation & Purification Reviews* 45: 275–287.
[32] Zhang, X., Zhao, W., Zhang, Y., Jegatheesan, V. (2021) A review of resource recovery from seawater desalination brine. *Reviews in Environmental Science and Bio/Technology* 20: 333–361.
[33] Panagopoulos, A. (2022)Brine management (saline water & wastewater effluents): Sustainable utilization and resource recovery strategy through minimal and zero liquid discharge (MLD & ZLD) desalination systems. *Chemical Engineering and Processing – Process Intensification* 176: 108944.
[34] Cipolletta, G., Lancioni, N., Akyol, Ç., Eusebi, A.L., Fatone, F. (2021) Brine treatment technologies towards minimum/zero liquid discharge and resource recovery: State of the art and techno-economic assessment. *Journal of Environmental Management* 300: 113681.
[35] European Commission (2022) *SPIRE – Sustainable Process Industry through Resource and Energy Efficiency*, Roapmap 2030.
[36] Mavukkandy, M.O, Chabib, C.M., Mustafa, I., Al Ghaferi, A., AlMarzooqi, F. (2019) Brine management in desalination industry: From waste to resources generation. *Desalination* 472: 114187.
[37] Fijita, T. (2009) Current and challenges of salt production technology. *Bulletin of the Society of Sea Water Science, Japan* 63:) 15–20.
[38] Baker, R.W. (2012) *Membrane Technology and Applications*, 3rd ed. Wiley editions, Chichester.
[39] Du, F., Warsinger, D.M., Urmi, T.I., Thiel, G.P., Kumar, A., Lienhard, J.H. (2018) Sodium hydroxide production from seawater desalination brine: Process design and energy efficiency. *Environmental Science & Technology* 52: 5949–5958.
[40] Sharkh, B.A., Al-Amoudi, A.A., Farooque, M., Fellows, C.M., Ihm, S., Lee, S., Li, S., Voutchkov, N. (2022) Seawater desalination concentrate—a new frontier for sustainable mining of valuable minerals. *npj Clean Water* 5: 9.
[41] Reig, M., Casas, S., Aladjem, C., Valderrama, C., Gibert, O., Valero, F., Centeno, C.M., Larrotcha, E., Cortina, J.L. (2014) Concentration of NaCl from seawater reverse osmosis brines for the chlor-alkali industry by electrodialysis. *Desalination* 342: 107–117.
[42] Tanaka, Y. (2013) Development of a computer simulation program of batch ion-exchange membrane electrodialysis for saline water desalination. *Desalination* 320: 118–133.
[43] Reig, M., Casas, S., Gibert, O., Valderrama, C., Cortina, J.L. (2016) Integration of nanofiltration and bipolar electrodialysis for valorization of seawater desalination brines: Production of drinking and waste water treatment chemicals. *Desalination* 382: 13–20.
[44] WSSTP (n.d.) *Brines management report, 2012.* http://wsstp.eu/wp-content/uploads/sites/102/2013/11/ExS-Brines.pdf (accessed March 23, 2016).
[45] TechSci Research (2022) *Brine Management Technology Market Size, Share, Trend & Forecast 2027*, www.techsciresearch.com/report/global-brine-management-technology-market/1513.html (accessed May 19, 2022).
[46] Reig, M., Valderrama, C., Gibert, O., Cortina, J.L. (2016) Selectrodialysis and bipolar membrane electrodialysis combination for industrial process brines treatment: Monovalent-divalent ions separation and acid and base production. *Desalination* 399: 88–95.

[47] Reig, M., Valderrama, C., Gibert, O., Cortina, J.L. (2016) Selectrodialysis and bipolar membrane electrodialysis combination for industrial process brines treatment: Monovalent-divalent ions separation and acid and base production. *Desalination* 399: 88–95.

[48] Pino, L., Beltran, E., Schwarz, A., Ruiz, M.C., Borquez, R. (2020) Optimization of nanofiltration for treatment of acid mine drainage and copper recovery by solvent extraction. *Hydrometallurgy* 195: 105361.

[49] Jimenez, Y.P., Ulbricht, M. (2019) Recovery of water from concentration of copper mining effluents using direct contact membrane distillation. Industrial & Engineering Chemical Research 58: 19599–19610.

[50] Baena-Moreno, F.M., Rodríguez-Galán, M., Arroyo-Torralvo, F., Vilches, L.F. (2020) Low-energy method for water-mineral recovery from acid mine drainage based on membrane technology: Evaluation of inorganic salts as draw solutions. *Environmental Science & Technology* 54: 10936–10943.

[51] López, J., Reig, M., Gibert, O., Cortina, J.L.L. (2019) Integration of nanofiltration membranes in recovery options of rare earth elements from acidic mine waters. *Journal of Cleaner Products* 210: 1249–1260.

[52] Ricci, B.C., Ferreira, C.D., Aguiar, A.O., Amaral, M.C.S. (2015) Integration of nanofiltration and reverse osmosis for metal separation and sulfuric acid recovery from gold mining effluent. *Separation and Purification Technology* 154: 11–21.

[53] Martí-Calatayud, M.C., Buzzi, D.C., García-Gabaldón, M., Ortega, E., Bernardes, A.M., Tenório, J.A.S., Pérez-Herranz, V. (2014) Sulfuric acid recovery from acid mine drainage by means of electrodialysis. *Desalination* 343: 120–127.

[54] Amaral, M.C.S., Grossi, L.B., Ramos, R.L., Ricci, B.C., Andrade, L.H. (2018) Integrated UF–NF–RO route for gold mining effluent treatment: From bench-scale to pilot-scale. *Desalination* 440: 111–121.

[55] López, J., Reig, M., Gibert, O., Cortina, J.L. (2019) Recovery of sulphuric acid and added value metals (Zn, Cu and rare earths) from acidic mine waters using nanofiltration membranes. Separation and Purification Technology 212: 180–190.

[56] Choi, J., Im, S.J., Jang, A. (2019) Application of volume retarded osmosis – Low pressure membrane hybrid process for recovery of heavy metals in acid mine drainage. *Chemosphere* 232: 264–272.

[57] Amaya-Vías, D., Tataru, L., Herce-Sesa, B., López-López, J.A., López-Ramírez, J.A. (2019) Metals removal from acid mine drainage (Tinto River, SW Spain) by water gap and air gap membrane distillation. *Journal of Membrane Science* 582: 20–29.

[58] Ryu, S., Naidu, G., Hasan Johir, M.A., Choi, Y., Jeong, S., Vigneswaran, S. (2019) Acid mine drainage treatment by integrated submerged membrane distillation–sorption system. *Chemosphere* 218: 955–965.

[59] Buzzi, D.C., Viegas, L.S., Rodrigues, M.A.S., Bernardes, A.M., Tenório, J.A.S. (2013) Water recovery from acid mine drainage by electrodialysis. *MineralsEngineering* 40: 82–89.

[60] Gueccia, R., Winter, D., Randazzo, S., Cipollina, A., Koschikowski, J., Micale, G.D.M. (2021) An integrated approach for the HCl and metals recovery from waste pickling solutions: pilot plant and design operations. *Chemical Engineering Research and Design* 168: 383–396.

[61] Reig, M., Vecino, X., Valderrama, C., Gibert, O., Cortina, J.L. (2018) Application of selectrodialysis for the removal of As from metallurgical process waters: Recovery of Cu and Zn. *Separation and Purification Technology* 195: 404–412.

[62] Jeong, J., Kim, M.S., Kim, B.S., Kim, S.K., Kim, W.B., Lee, J.C. (2005) Recovery of H_2SO_4 from waste acid solution by a diffusion dialysis method. *Journal of Hazardous Materials* 124: 230–235.

[63] Li, W., Zhang, Y., Jing, H., Zhu, X., Wang, Y. (2016) Separation and recovery of sulfuric acid from acidic vanadium leaching solution by diffusion dialysis. *Journal of Environmental Chemical Engineering* 4: 1399–1405.

[64] Gueccia, R., Aguirre, A.R., Randazzo, S., Cipollina, A., Micale, G. (2020) Diffusion dialysis for separation of hydrochloric acid, iron and zinc ions from highly concentrated pickling solutions. *Membranes (Basel)* 10: 1–17.

[65] Luo, F., Zhang, X., Pan, J., Mondal, A.N., Feng, H., Xu, T. (2015) Diffusion dialysis of sulfuric acid in spiral wound membrane modules: Effect of module number and connection mode. *Separation and Purification Technology* 148: 25–31.

[66] López, J., de Oliveira, R.R., Reig, M., Vecino, X., Gibert, O., de Juan, A., Cortina, J.L. (2020) Acid recovery from copper metallurgical process streams polluted with arsenic by diffusion dialysis. *Journal of Environmental Chemical Engineering* 9(1): 104692.

[67] López, J., Reig, M., Gibert, O., Cortina, J.L. (2019) Increasing sustainability on the metallurgical industry by integration of membrane nanofiltration processes: Acid recovery. *Separation and Purification Technology* 226: 267–277.

[68] López, J., Gibert, O., Cortina, J.L. (2020) Evaluation of an extreme acid-resistant sulphonamide based nanofiltration membrane for the valorisation of copper acidic effluents. *Chemical Engineering Journal* 405: 127015.

[69] Cifuentes, L., Crisóstomo, G., Ibez, J.P., Casas, J.M., Alvarez, F., Cifuentes, G. (2002) On the electrodialysis of aqueous H_2SO_4-$CuSO_4$ electrolytes with metallic impurities. *Journal of Membrane Science* 207: 1–16.

[70] Chekioua, A., Delimi, R. (2015) Purification of H_2SO_4 of pickling bath contaminated by Fe(II) ions using electrodialysis process. *Energy Procedia* 74: 1418–1433.

[71] Liu, Y., Ke, X., Zhu, H., Chen, R., Chen, X., Zheng, X., Jin, Y., Van der Bruggen, B. (2020) Treatment of raffinate generated via copper ore hydrometallurgical processing using a bipolar membrane electrodialysis system. *Chemical Engineering Journal* 382: 122956.

[72] Cai, J., Guo, F. (2020) Mass transfer during membrane distillation treatment of wastewater from hot-dip galvanization. *Separation and Purification Technology* 235: 116164.

[73] Chen, G., Tan, L., Xie, M., Liu, Y., Lin, Y., Tan, W., Huang, M. (2020) Direct contact membrane distillation of refining waste stream from precious metal recovery: Chemistry of silica and chromium (III) in membrane scaling. *Journal of Membrane Science* 598: 117803.

[74] Tomaszewska, M., Gryta, M., Morawski, A.W. (2001) Recovery of hydrochloric acid from metal pickling solutions

by membrane distillation. *Separation and Purification Technology* 22–23: 591–600.
[75] Platt, S., Nyström, M., Bottino, A., Capannelli, G. (2004) Stability of NF membranes under extreme acidic conditions. *Journal of Membrane Science* 239: 91–103.
[76] Manis, A., Soldenhoff, K., Jusuf, E., Lucien, F. (2003) Separation of copper from sulfuric acid by nanofiltration. In: *Fifth International Membrane Science & Technology Conference*.
[77] López, J., Reig, M., Gibert, O., Torres, E., Ayora, C., Cortina, J.L. (2018) Application of nanofiltration for acidic waters containing rare earth elements: Influence of transition elements, acidity and membrane stability. *Desalination* 430: 33–44.
[78] Schütte, T., Niewersch, C., Wintgens, T., Yüce, S. (2015) Phosphorus recovery from sewage sludge by nanofiltration in diafiltration mode. *Journal of Membrane Science* 480: 74–82.
[79] Guastalli, A.R., Labanda, J., Llorens, J. (2009) Separation of phosphoric acid from an industrial rinsing water by means of nanofiltration. *Desalination* 243: 218–228.
[80] Rieger, A., Steinberger, P., Pelz, W., Haseneder, R., Härtel, G. (2009) Mine water treatment by membrane filtration processes – Experimental investigations on applicability. *Desalination and Water Treatment* 6: 54–60.
[81] Al-Zoubi, H., Rieger, A., Steinberger, P., Pelz, W., Haseneder, R., Härtel, G. (2010) Nanofiltration of acid mine drainage. *Desalination and Water Treatment* 21: 148–161.
[82] Feng, X., Jiang, L.Y., Song, Y. (2016) Titanium white sulfuric acid concentration by direct contact membrane distillation. *Chemical Engineering Journal* 285: 101–111.

22 Concentration-Driven Membrane Processes for the Recovery of Valuable Compounds from Industrial Wastes

Eugenio Bringas, Ma.-Fresnedo San-Román, Ana M. Urtiaga, and Inmaculada Ortiz*

Departamento de Ingenierías Química y Biomolecular, ETSIIyT, Universidad de Cantabria, Avda. de los Castros, 39005, Santander, Spain.

*Email: eugenio.bringas@unican.es

22.1 INTRODUCTION

Currently, the generation of industrial wastes and wastewaters containing high concentrations of organic and inorganic pollutants is a worldwide concern due to the high impact that the presence of these compounds can have on the environment, even in low concentrations. Efficient treatment of industrial waste is therefore mandatory before disposal, with alternatives for the recovery of valuable compounds playing a prominent role in the approach to the circular economy paradigm [1].

Industrial wastewaters containing heavy metals have received considerable attention because of the toxicity and high solubility of these compounds in aquatic environments and their ability to be absorbed by living organisms causing their incorporation into the food chain [2]. In many industrial processes, heavy metals are dissolved together with mineral acids or alkalis that provoke extremely low or high pH values, respectively [3]. Before these effluents are discharged it is mandatory for the application of efficient treatments that reduce the concentration of the pollutants and comply with legal standards worldwide. However, this situation presents a good opportunity for the development and implementation of advanced technologies that not only remove existing pollutants but also allow their selective separation, facilitating the recovery of valuable components [4].

Conventional physico-chemical processes (neutralization, coagulation, flocculation, precipitation, and filtration) have traditionally been used for treatment before disposal of waste acid solutions containing heavy metals. However, they often fail to reduce the concentration of these pollutants below the permissible limits. Furthermore, these technologies usually do not provide the necessary selectivity to create valuable product streams suitable for recycling or re-use and, as a consequence, by-product sludge can become a disposal problem [5,6].

Different technologies such as solvent extraction [7], ion exchange [8,9], crystallization [10], and adsorption [11] have been well-demonstrated to overcome the limitations of conventional processes in the treatment of industrial wastewater containing heavy metals and mineral acids. In this context, membrane operations have the potential to replace conventional technologies by accomplishing selective and efficient transport of specific components, improving the performance of reactive processes and, eventually, providing reliable options for sustainable growth [12,13]. However, membrane technologies are commonly limited by the trade-off between permeability and selectivity, a situation that has drawn the attention of the scientific community and companies in the sector and is the subject of continuous research. Membranes that incorporate selective agents, e.g. selective extractants, ionic liquids, etc. that react reversibly with target compounds operate under the facilitated transport mechanism offering a very good alternative to the efficient separation of solutes [14]. The selective separation stage can suitably be combined with the recovery of the solutes in a receiving phase at the same time as membrane regeneration takes place. Furthermore, if the conditions of the receiving phase are well designed, the solute can increase its concentration several times that of the feed phase [15]. In addition, the stability and long-term performance of this technology are being improved by the development of new process configurations based on the use of membrane contactors that allow non-dispersive contact and the independent flow of the fluid phases involved in the process [14,16]. On the other hand, diffusion dialysis using ion exchange membranes has been reported as an effective alternative to recover several mineral acids such as HCl [3], H_2SO_4 [17,18], HNO_3 [19,20], and HF [19,21] based on several advantages such as low energy consumption, low installation and operating costs, and process stability and ease of operation [22]. However, the stand-alone application of individual separation technologies often fails to recover all the possible valuable compounds present in complex wastewaters, and thus the development of integrated processes by a combination of single membrane separation stages is a potential solution to overcome this limitation due to the benefits derived from the modular design and flexibility of membrane processes.

22.2 MEMBRANE PROCESSES FOR THE RECOVERY OF MINERAL ACIDS AND METALS FROM WASTEWATERS

In this section, the fundamentals of liquid membrane and membrane dialysis technologies are initially reported

DOI: 10.1201/9781003285656-22

to further propose a useful methodology for designing membrane separation processes capable of recovering metal compounds and mineral acids from industrial wastes. Two different aqueous wastes generated in the context of the surface treatment activity have been selected as case studies to illustrate the benefits of combining membrane-based separation steps to achieve the recovery of valuable materials: (i) the spent Cr^{3+} passivation baths in nitrate media containing Zn^{2+} and Fe^{3+} derived from the electroplating process and (ii) the spent pickling solutions generated in the hot-dip galvanizing process with high concentrations of Fe^{2+} and Zn^{2+} in HCl media. The first step aims at the achievement of the treatment objectives as well as the simultaneous recovery of zinc from the selected wastewaters by application of the emulsion pertraction technology, which combines the advantages of selective liquid membranes and membrane contactors. Afterwards, the recovery of hydrochloric acid by membrane dialysis using anion-exchange membranes will be analyzed as a potential recovery alternative.

22.2.1 Recovery of Heavy Metals by Selective Liquid Membranes

Separation processes based on facilitated transport membranes offer a number of advantages such as the ability of liquid membranes to promote the uphill transport of target species by the coupling of mass transfer and chemical reaction. As depicted in Figure 22.1, the facilitated transport of a solute through a liquid membranes is promoted by its reaction with a selective carrier to form a solute–carrier complex, which then diffuses through the membrane to finally release the solute at the permeate side. However, if the feed solution contains ionic species as in the case of heavy metals removal, other species are co-transported or counter-transported (coupled facilitated transport) in order to maintain the solution electro-neutrality

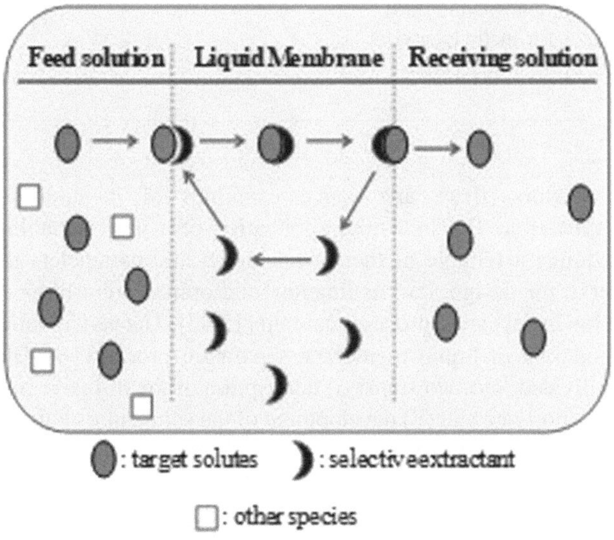

FIGURE 22.1 Facilitated transport mechanism through liquid membranes.

condition. The use of appropriate reacting carriers in the liquid membrane composition promotes the uphill transport of target species and the enhancement of the extraction and stripping stages since the driving forces are always higher than the concentration gradients achieved in non-reacting processes [15,23].

A new time in the development of the liquid membrane technology began in the 1960s when Li and coworkers patented the use of emulsion liquid membrane systems (ELM) for industrial-scale desalination and the separation of hydrocarbons [13]. Nowadays, the development of new separation processes based on the combination of liquid membranes and membrane contactors is becoming a suitable strategy to increase the process competitiveness in terms of process intensification for the following reasons [24]:

i. Higher interfacial mass transfer area/equipment size.
ii. The interfacial contact area is known, constant and independent of the operation conditions thus facilitating the mathematical modeling of mass transfer phenomena.
iii. The fluid phases flow independently and are non-dispersively contacted, avoiding additional separation stages.
iv. The modular design simplifies the process scale-up and allows operation over a wide range of capacities.

Hydrophobic hollow-fiber membrane contactors (HFC) are the most extensive membrane modules used in liquid membrane separation processes. HFCs have been employed under different process configurations and applications, such as hollow-fiber-supported liquid membranes (HFSLMs), hollow-fiber-contained liquid membranes (HFCLMs), non-dispersive solvent extraction (NDSX), and more recently, hollow-fiber renewal liquid membranes (HFRLMs) and emulsion pertraction technology (EPT) also called pseudo-emulsion-based hollow fiber with strip dispersion. The main differences among the different process configurations are in the methods of contacting the fluid phases and the number of HFCs used in the separation process [13,16]. Figure 22.2 shows a flow diagram of the EPT process that combines the advantages of ELM, i.e. high efficiency due to large surface area for mass transfer, and supported liquid membranes (SLMs), i.e. extraction and stripping being performed in one step without dispersion of the organic phase into the aqueous feed phase. After the initial works reported by Ho and co-workers [25–27], several applications have focused on the treatment of waste effluents containing metallic pollutants that have been developed, as reported in Table 22.1.

The flow diagram depicted in Figure 22.2 comprises two essential process units: a mesoporous hollow-fiber membrane contactor (hollow fibers Celgard® X50) and the emulsion vessel with a pseudo-emulsion containing the organic phase formulated with a selective organic extractant and the dispersed stripping solution. Additional process units can be added (i.e. feed tank for homogenization of the solution to be treated) depending on the characteristics of the specific application

TABLE 22.1
Removal of heavy metals from industrial wastes by EPT

References	Feed solution	Liquid membrane composition	Stripping solution
[28]	Co^{2+} (1.7–170 mol m^{-3}) sulfate media; pH: 3–7	DP-8R (bis(2-ethylhexyl) phosphoric acid) (160–1280 mol m^{-3}); Exxsol D100	H_2SO_4 (100 mol m^{-3})
[29,30]	Zn^{2+} (38–180 mol m^{-3}); Fe^{3+} (0.4–1.7 mol m^{-3}) nitrate media; pH: 2–3	Cyanex272 (bis(2,4,4- trimethylpenthyl) phosphinic acid) (200–1000 mol m^{-3}); isodecanol (5% v/v); kerosene	HCl H_2SO_4 (1000–2000 mol m^{-3})
[31]	Fe^{3+} (0.18–18 mol m^{-3}) sulfate media; pH: acid	$(PJMTH^+)_2(SO_4^{2-})$ ionic liquid (1–30% v/v); n-decanol (2.5% v/v); n-decane	H_2SO_4 (1000–3000 mol m^{-3})
[32]	Ni^{2+} (0.17–61 mol m^{-3}) sulfate media; pH: 2–5	Acorga M5640 (1–20% v/v) + DP8R (1–20% v/v); Exxsol D100	H_2SO_4 (1000 mol m^{-3})
[33]	Cr^{6+} (0.4–2.3 mol m^{-3}) chloride media; pH: acid	Cyanex-923 (mixture of four trialkylphosphine oxides) (10% v/v); kerosene	Hydrazine sulfate (77 mol m^{-3})
[34]	Zn^{2+} (1254 mol m^{-3}); Fe^{2+} (1718 mol m^{-3}) chloride media; pH ≈ 0	TBP (tributylphosphate) (100% v/v)	Service water
[35]	Cr^{3+} (0.2–10 mol m^{-3}) sodium hydroxide media; pH: basic	TOMACl (trialkylmethylammonium chloride) (10% v/v); n-decanol (5% v/v); n-decane	H_2SO_4 (500 mol m^{-3})
[36]	U^{6+} (62–312 mol m^{-3}) nitrate media; pH: acid	TBP (tributylphosphate) (5–40% v/v); NPH	HNO_3 (10 mol m^{-3})
[37,38]	Au^+ (0.05 mol m^{-3}) cyanide media; pH = 9–11	LIX-79 (N,N-bis(2-ethyl hexyl)guanidine) (2–18% v/v); n-heptane	NaOH (50–600 mol m^{-3})
[39]	Cr^{6+} (9 mol m^{-3}) sulfate/chloride media; pH = 1.5	Alamine 336 (tri-octyl/decyl amine) (10 % v/v); Pluronic PE3100 (3% v/v); n-dodecanol (1% v/v); Isopar-L	NaOH (3000–6000 mol m^{-3})
[40]	Cu^{2+} (1.6–16 mol m^{-3}) sulfate media; pH = 3	LIX 622N; kerosene	H_2SO_4 (1000 mol m^{-3})
[41]	Zn^{2+}, Fe^{2+} chloride media; pH: acid	Not specified	Not specified
[42]	Cr^{6+} (40–50 mol m^{-3}) sulfate/chloride media; pH = 1.5	Alamine 336 (tri-octyl/decyl amine) (10% v/v); Pluronic PE3100 (5% v/v); n-dodecanol (10% v/v); Isopar-L	NaOH (3000 mol m^{-3})
[25,26]	Cr^{6+} (0.4–115 mol m^{-3}) sulfate; pH = 1.5	Amberlite LA-2 (N-lauryl-N-trialkylmethylamine) (10% v/v); Pluronic PE3100 (3% v/v); n-dodecanol (1% v/v); Isopar-L	NaOH (3000 mol m^{-3})
[27]	Cu^{2+}, Zn^{2+}, Ni^{2+} pH: acid	LIX 973N; Cyanex 301 (bis(2,4,4-trimethylpentyl)dithiophosphinic acid); C12 MTPA (di (2,4,4-trimethylpentyl) dithiophosphinic acid); n-dodecanol; n-dodecane	H_2SO_4 (3000 mol m^{-3})

and operation mode (i.e. batch, semi-batch, or continuous). The aqueous feed solution and the emulsion flow co-currently or counter-currently either through the inner side of the mesoporous hollow-fiber membranes or through the shell side of the membrane contactor without mixing. The target solute is chemically transferred from the aqueous feed to the organic phase that is embedded in the pores of the hollow fibers due to their hydrophobic character. Next, the solute–carrier complex diffuses to the interface at the surface of the droplet of the stripping phase where the back-extraction reaction occurs. The solute is recovered from the internal aqueous phase after breakage of the emulsion [29,34,39,40,42].

However, in spite of the known advantages and applications of liquid membrane separation processes in hollow-fiber contactors, there are scarce examples of its industrial application. The industrial application of a new technology requires a reliable mathematical model and parameters that serve for design, cost estimation, and optimization purposes allowing accurate process scale-up [15,43]. The mathematical modeling of liquid membrane separation processes in HFC is divided into two steps: (i) description of the diffusive mass transport rate and (ii) development of the solute mass balances to the flowing phases.

As depicted in Figure 22.3 which shows an enlarged view of the hollow fibers, several in-series steps are considered to describe the mass transfer pathway of a metallic compound (A) from the feed solution to the stripping solution: (i) diffusion of the solute through the aqueous phase boundary

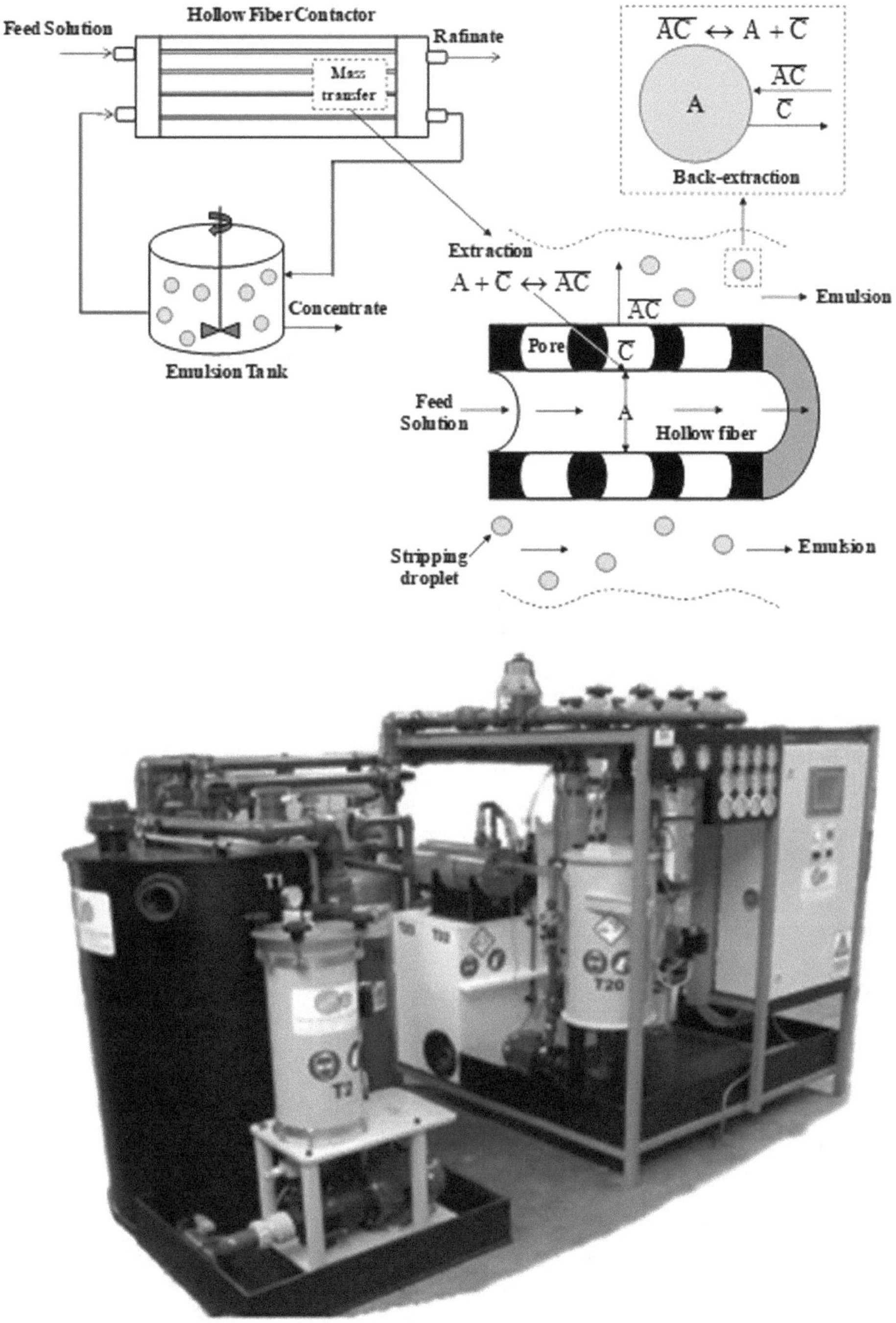

FIGURE 22.2 The EPT process (Courtesy of Ondeo Industrial Solutions). (A) Target solute to be removed from the feed solution; $\overline{C}$: selective extractant in the organic phase; $\overline{AC}$: organometallic complex in the organic phase.

layer [Eq. (22.1)], (ii) instantaneous interfacial reaction between the solute and the selective carrier ($\overline{C}$) to form the organometallic complex species ($\overline{AC}$) [Eqs. (22.7)–(22.8)], (iii) diffusion of the latter species through the organic phase embedded in the pores of the hollow fibers [Eq. (22.3)], (iv) diffusion of the complex species through the boundary layer of the organic phase [Eq. (22.4)], (v) interfacial reaction of the organometallic complex species with the back-extraction agent [Eqs. (22.7)–(22.8)], and (vi) diffusion of the solute through the boundary layer in the stripping solution [Eq. (22.5)]. Depending on the specific application, the counter-transport and/or co-transport of the other species participating in the process should be described in a similar way [i.e. diffusion of the selective carrier trough the membrane (Eq. (22.2))]. The transfer flux through each transport step defined above can be evaluated according to Fick's first law as follows:

$$J_A^a = k_{L,A}^a \cdot (C_A^a - C_A^{a,i_1}) \qquad (22.1)$$

$$J^m_{\bar{C}} = k_{m,\bar{C}} \cdot (C^o_{\bar{C}} - C^{o,i_1}_{\bar{C}}) \tag{22.2}$$

$$J^m_{\overline{AC}} = k_{m,\overline{AC}} \cdot (C^{o,i_1}_{\overline{AC}} - C^{o,i_2}_{\overline{AC}}) \tag{22.3}$$

$$J^o_{\overline{AC}} = k_{o,\overline{AC}} \cdot (C^{o,i_2}_{\overline{AC}} - C^o_{\overline{AC}}) \tag{22.4}$$

$$J^s_A = k^s_{L,A} \cdot (C^{s,i_3}_A - C^s_A) \tag{22.5}$$

where k_L, k_{m}. and k_o are the individual mass transport coefficients. Assuming pseudo-steady-state the overall diffusive flux of solute A, J_A (mol h^{-1} m^{-2}), can be calculated as follows:

$$J_A = J^a_A = J^m_{\bar{C}} = J^m_{\overline{AC}} = J^o_{\overline{AC}} = J^s_A \tag{22.6}$$

The interfacial chemical reactions are considered fast enough to reach equilibrium instantaneously [44]. Thus, the species concentration at the organic–aqueous interfaces is obtained from the expressions of the equilibrium parameters or distribution coefficients of the extraction and back-extraction reactions (included in Figure 22.3) described by a simple mass action law as follows:

$$K^k_{eq} = \prod_i (C^j_i)^{\alpha^k_i} \quad \forall i = A, \bar{C}, \overline{AC}; \quad \forall j = a, o, s; \ \forall k = i_1, i_3 \tag{22.7}$$

$$H^k_{eq} = \prod_i (C^j_i)^{\alpha^k_i} \quad \forall i = A, \overline{AC}; \quad \forall j = a, o, s; \ \forall k = i_1, i_3 \tag{22.8}$$

where α^k_i is the stoichiometric coefficient of species "i" in the reaction "k" which occurs at interfaces i_1 or i_3.

In conclusion, the mathematical description of the overall diffusive flux in liquid membrane separation processes needs the calculation of the corresponding mass transport parameters ($k^a_{L,A}$, $k_{m,\bar{C}}$, $k_{m,\overline{AC}}$, $k_{o,\overline{AC}}$, and $k^s_{L,A}$) and equilibrium parameters of the interfacial reactions.($K^{i_1}_{eq}/H^{i_1}_{eq}$ and $K^{i_3}_{eq}/H^{i_3}_{eq}$)

In a second stage, the solute mass balances in the different process units are developed according to a specific flow model. The most accurate approximation for developing the solute mass balances to the fluid phases flowing through a membrane contactor takes into account the axial and radial variation in the fluid properties by means of two-dimensional models [45]. Nevertheless, the most extensive approach to the mathematical modeling of hollow-fiber contactors is the assumption of ideal plug flow that neglects the concentration variations perpendicular to the bulk flow direction. As a result,

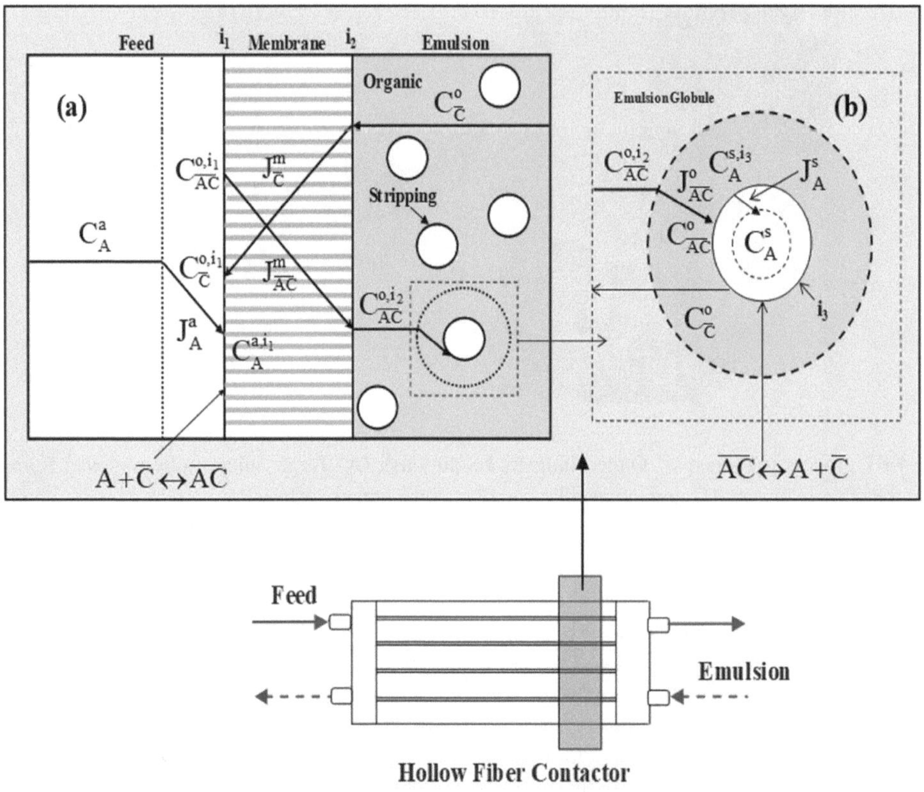

FIGURE 22.3 Enlarged view of a differential volume element of the hollow-fiber membrane (a) and the emulsion globules (b) (a, feed solution; m, liquid membrane; s, stripping solution; i, interface).

only the concentration axial variation is taken into account obtaining a one-dimensional (1D) flow model [12]. The details of the specific considerations to mathematically model the EPT process can be found elsewhere [34,39,40,44].

In the forthcoming sections the potential of the EPT technology to develop separation processes with metal recovery is evaluated.

22.2.1.1 Regeneration of Chromium-Based Passivation Baths with Zinc Recovery

Zinc electroplating is one of the most common galvanic operations, used for corrosion protection of metallic and plastic components. After zinc deposition, usually a chemical passivation layer is applied with the aim of enhancing the corrosion protection, obtaining at the same time the required appearance [8,9]. However, the progressive implementation of conversion coatings based on trivalent chromium introduce new challenges, as formulations free of hexavalent chromium are more sensible to the presence of zinc and iron impurities, and thus their lifetime is significantly reduced with respect to the previously employed hexavalent chromium chromatizing baths, now banned due to their environmentally hazardous properties [46]. From the information provided by different plating workshops, the lifetime of the passivation baths depends on the zinc and iron intakes entering the passivation baths which usually range between 0.06–0.79 kg Zn day^{-1} and 0.01–0.06 kg Fe day^{-1} under normal operation conditions [47]. The long-term accumulation of zinc and iron produces an efficiency loss in the passivation process which finally causes the generation of spent passivation baths with the following typical composition and concentration ranges: pH (1.8–2.5), Cr^{3+} (4.5–9.4 kg m^{-3}), Zn^{2+} (2.5–11.8 kg m^{-3}), and Fe^{3+} (0.02–0.09 kg m^{-3}) [48]. With nitrate being the anion most commonly employed in the formulation of trivalent chromium passivation baths, iron and zinc remain as cationic species at the working pH of the passivation bath. Figure 22.4 illustrates the integration of the EPT process, previously described, into a typical surface treatment line allowing the in situ regeneration of Cr^{3+}-based passivation baths by the selective removal of iron and zinc up to the limits required for bath reuse, with a second objective being the generation of a zinc concentrate with the lowest possible concentration of iron allowing the recovery of metallic zinc.

The information provided by the equilibrium isotherms depicted in Figure 22.5 confirms the commercial selective carrier bis(2,4,4-trimethylpenthyl)phosphinic acid (Cyanex272), as a suitable reagent to formulate the liquid membrane due to its capacity to selectively separate iron and zinc from chromium under the typical pH conditions of the passivation baths. Therefore, the typical composition of the pseudo-emulsion employed in the EPT regeneration process contains Cyanex272 and sulfuric acid as stripping agent and the extraction and back-extraction of zinc and iron are described, respectively, by the following reversible chemical reactions [29,49]:

$$M^{n+}_{(a)} + n\overline{RH}_{(o)} \overset{K^{i_1}_{eq,M^{n+}}}{\longleftrightarrow} \overline{MR}_{n\,(o)} + nH^{+}_{(a)} \tag{22.9}$$

$$\overline{MR}_{n\,(o)} + nH^{+}_{(s)} \overset{K^{i_3}_{eq,M^{n+}}}{\longleftrightarrow} M^{n+}_{(s)} + n\overline{RH}_{(o)} \tag{22.10}$$

where M^{n+} represents the metallic impurities in the passivation bath (a), Zn^{2+} and Fe^{3+}, $\overline{RH}$ and $\overline{MR}_n$ are the free extractant and

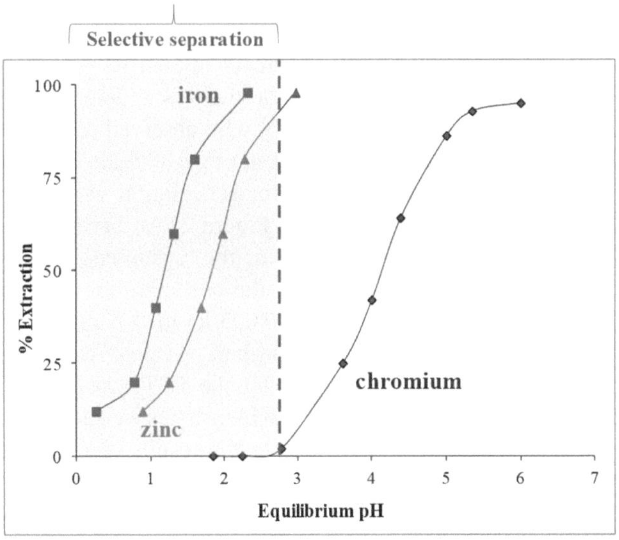

FIGURE 22.5 Extraction isotherms of iron, zinc, and chromium with Cyanex 272 in sulfate media [29].

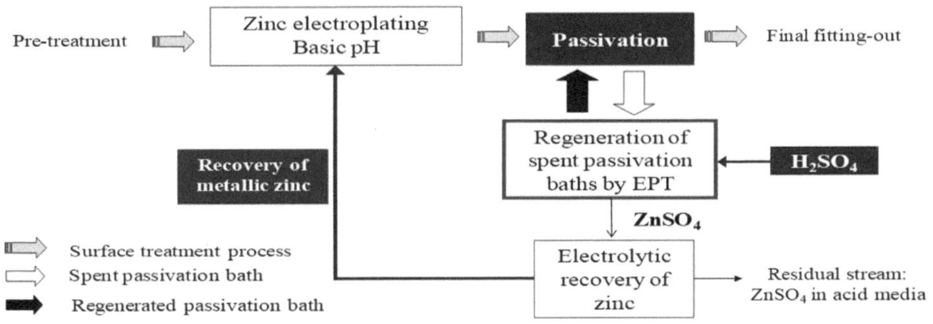

FIGURE 22.4 In situ regeneration of Cr^{3+} passivation baths by the EPT process.

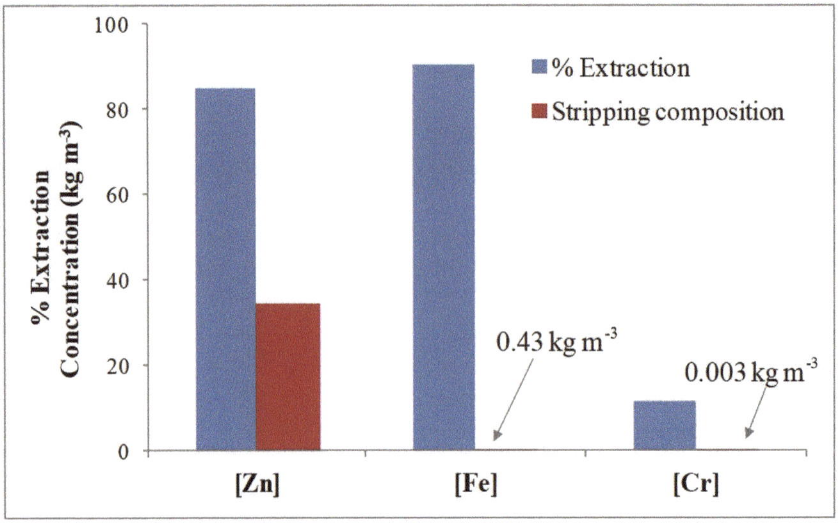

FIGURE 22.6 Extraction percentages and concentrations in the stripping solution after 3 hours of EPT regeneration of spent passivation baths using Cyanex272 as selective extractant [29,30]. $C_{Zn^{2+},initial} = 7$ kg m^{-3}; $C_{Fe^{3+},initial} = 0.04$ kg m^{-3}; $C_{Cr^{3+},initial} = 4.5$ kg m^{-3}

the organometallic complexes formed by reaction dissolved in the liquid membrane (o), and H^+ is the free protons in the stripping solution (s).

As a proof of concept, the efficiency of the EPT process in the regeneration of real passivation baths was demonstrated by Urtiaga and co-workers [29,30] who observed reductions in the zinc and iron contents of greater than 80% after 3 hours of experimental running with an almost negligible variation of the chromium concentration (see Figure 22.6). Furthermore, zinc is selectively recovered in the stripping solution (concentration >35 kg m^{-3}) with the concentrations of iron (<0.45 kg m^{-3}) and chromium (<0.003 kg m^{-3}) being almost negligible (see Figure 22.6). Further, trivalent chromium passivation baths regenerated using the EPT process were compared to the bath used in a local industry as well as to fresh and spent baths [50]. According to the results, the samples passivated in the EPT regenerated bath showed a significant improvement in their electrochemical behavior compared to the samples passivated in the spent baths. Therefore, these results confirm the EPT process as a suitable alternative to perform both the regeneration of the spent passivation baths and the zinc recovery for further reuse.

The design and scale-up of liquid membrane separation processes needs separation and concentration mathematical models as reported in Section 22.2.1. When complex solutions such as wastewaters are treated, several simplifications according to the specific characteristics of the system are usually assumed in order to reduce the number of parameters and mathematical complexity of the EPT model. From a kinetic point of view the transport through the membrane is often the controlling step which is described by the corresponding membrane mass transport coefficients calculated by empirical correlations. On the other hand, the kinetic resistances associated with the transport of the different species involved in the process through the different boundary layers are considered negligible due to the typical high concentrations of those species in both the feed and stripping solutions. The values of the equilibrium parameters of the interfacial extraction and back-extraction reactions are either experimentally obtained or determined by thermodynamic calculations or simulation techniques such as parameter estimation and molecular simulation.

Under the aforementioned assumptions, Bringas et al. [44] developed a suitable EPT mathematical model able to describe the separation and concentration kinetics of zinc and iron initially present in real spent passivation baths. Using the kinetic and equilibrium parameters reported in Table 22.2 and experimental data, the EPT mathematical model was successfully validated leading to the conclusion that 88% of the simulated concentration values of zinc and iron in the spent passivation bath fell within the range $C^a_{T,M^{n+},EXP} \pm 20\% \cdot C^a_{T,M^{n+},EXP}$. On the other hand, 91% of simulated data of the stripping phase fell within the range $C^s_{T,M^{n+},EXP} \pm 20\% \cdot C^s_{T,M^{n+},EXP}$.

This mathematical model can be applied for the design and scale-up of the EPT process to perform the in situ regeneration of passivation baths by integration of the EPT process in the normal operation of the electroplating process at the plating workshop facilities (see Figure 22.4). Figure 22.7 shows the configuration of the regeneration process when a passivation bath with a total volume of 1.5 m^3 and a working pH of 2.1 is treated by EPT. In the final designed process, 20 m^2 of membrane area are required to provide enough flux of zinc to avoid its accumulation in the passivation bath ($C_{Zn^{2+}} = 2$ kg m^{-3} = constant) when the intake flux of zinc ($I_{Zn^{2+}}$) varies from 0.15 kg day^{-1} to 1.2 kg day^{-1}. On the other hand, between 20 and 60 m^2 of membrane are needed to keep constant the initial concentration of iron ($C_{Fe^{3+}} = 0.05$ kg m^{-3}) in the passivation bath allowing the iron removal kinetics to be almost equal to its intake flux ($I_{Fe^{3+}}$) ranging from 0.01 kg day^{-1} to 0.06 kg day^{-1}, respectively.

TABLE 22.2
Kinetic and equilibrium parameters of the EPT mathematical model applied to the regeneration of spent passivation baths [44]

Parameter		Value	Comments
Membrane mass transport coefficients	$k_{m,\overline{ZnR_2}}$ (zinc) $k_{m,\overline{FeR_3}}$ (iron) $k_{m,\overline{RH}}$ (free carrier)	1.4 10^{-3} m h^{-1} 1.1 10^{-3} m h^{-1} 2.1 10^{-3} m h^{-1}	Calculated by empirical correlations which depend on both the species diffusivity and the membrane support characteristics (thickness, tortuosity, and porosity)
Equilibrium parameter of the extraction reaction	$K^{i_1}_{eq,Zn^{2+}}$ (zinc)	3.1 10^{-4}	Experimental equilibrium parameter defined by the mass action law
	$H^{i_2}_{eq,Fe^{3+}}$ (iron)	1.2	Constant distribution coefficient obtained by means of parameter estimation techniques using AspenTech simulation software
Equilibrium parameter of the back-extraction reaction	$K^{i_3}_{eq,M^{n+}}$	∞	The back-extraction reactions are assumed to be completely shifted to the right due to the stoichiometric excess of acid in the stripping solution. The driving force for mass transfer is maximized and thus, all the metallic impurities extracted from the feed solution are totally back-extracted

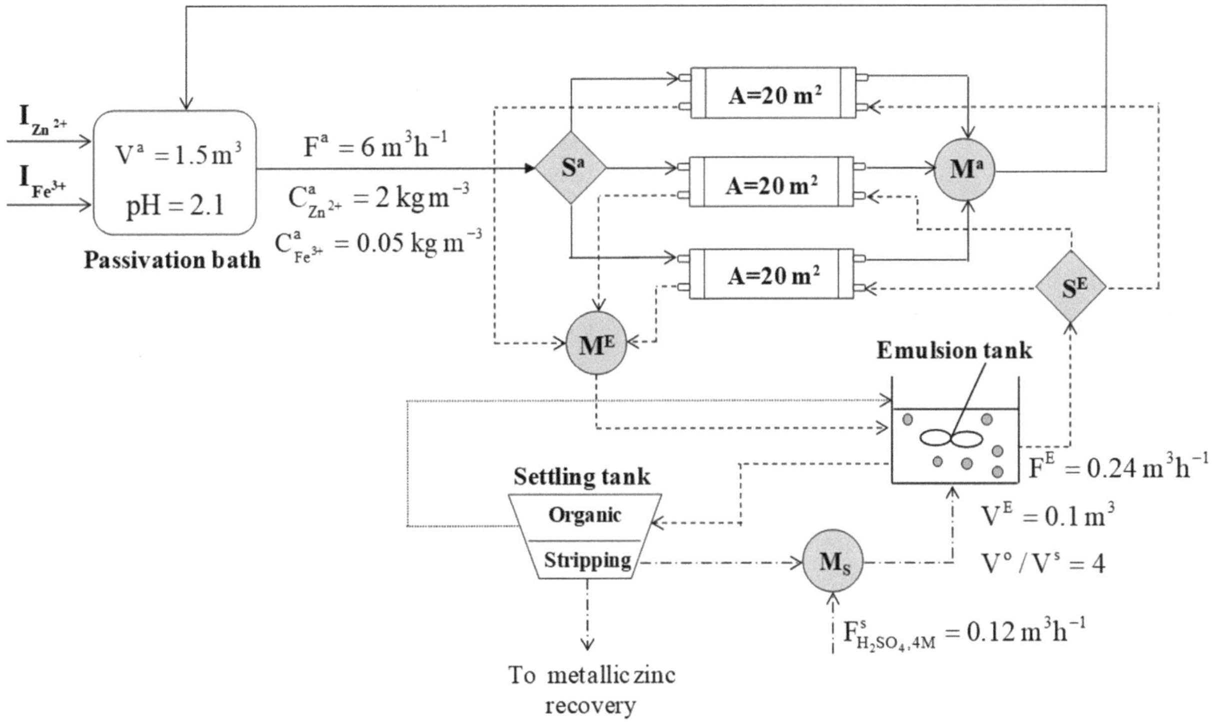

FIGURE 22.7 Flow diagram of the EPT-based in situ regeneration process (S, splitter; M, mixer; V, volume; F, flowrate; a, feed solution; o, organic phase; s, stripping solution).

The viability of employing the stripping solution produced during the EPT process as electrolytic solution to obtain metallic zinc by electrodeposition is an issue of interest since the recovery of zinc can compensate for the costs of the regeneration process. A typical composition of electrolyte solutions used in the literature for zinc recovery is 50–90 kg Zn^{2+} m^{-3} and 120–200 kg H_2SO_4 m^{-3} [51]. Barakat et al. [52] employed an electrolyte solution for zinc electrowinning containing 153 kg Zn^{2+} m^{-3}, 0.014 kg total Fe m^{-3}, and approximately 35 kg m^{-3} of free acid concentration. Diban and co-workers[30] performed a long-term analysis of the EPT process applied to the regeneration of passivation baths obtaining a stripping solution with a suitable composition for zinc recovery by electrodeposition: pH = 1.9, 159 kg Zn^{2+} m^{-3}, 0.08 kg total Fe m^{-3}, and 0.12 kg Cr^{3+} m^{-3}. The application of the electrowinning process to this solution over 30 min at 7.5 V and current density of 700 A m^{-2} produced a solid deposit with a zinc content of approximately 98.5% with this purity grade being valid for several applications such as hot-dip galvanizing, production of brasses and bronzes, pigments for the formulation of paint coatings, in primary batteries, or as sacrifice anode [53]. Concerning the impurities, no traces of chromium were found in the deposit and the contents of iron (0.00054 wt.%) and copper (0.00064 wt.%) accomplished the maximum limits for the special high-grade quality (99.995%) according to the European Standard for Primary Zinc [54].

The environmental sustainability of EPT for regenerating trivalent chromium passivation baths was analyzed by means of the life cycle analysis methodology. The integration of EPT in the zinc electroplating line caused a considerable extension of the passivation bath lifetime, a factor that decreased the generated waste (by 92%) during the manufacture cycle of the passivated product. A reduction of the total environmental burden to air and water and the resource usage during the whole manufacture cycle of the product were stated [55].

In conclusion, the promising results discussed above confirm the effectiveness of EPT as an eco-efficient separation process capable of being incorporated into the operation of existing surface treatment processes to perform the in situ regeneration of chromium-based passivation baths with zinc recovery.

22.2.1.2 Recovery of Zinc from Spent Pickling Effluents

Hot-dip galvanizing is a metallurgic industry process consisting of covering the surface of iron or steel pieces with a layer of molten zinc that forms a very resistant surface so that the metallic piece is protected from oxidation. This process consists of the following steps: (i) degreasing with a hot alkaline solution, (ii) rinsing with water, (iii) pickling with HCl 20% (v/v), (iv) rinsing, (v) fluxing with zinc and ammonium chlorides, and finally, (vi) hot-dip coating with molten zinc at temperatures about 450ºC. In the pickling step, hydrochloric acid is consumed during the process, but the concentration of chloride ions does not change. As a result, spent pickling solutions (SPS) containing among other substances high concentrations of Zn, Fe, and HCl are generated. These waste effluents have a strong hazardous character and need to be treated before disposal [34]. In addition, the high concentrations of valuable materials such as zinc and hydrochloric acid motivate the development of treatment technologies that enable the selective recovery of those valuable compounds according to the principles of a circular economy.

Spent pickling effluents are often neutralized with lime and then released into the environment, but the discharge of such wastes is highly undesirable because of their high acidity and ecotoxicity. Another method that has been considered appropriate for the treatment of spent pickling effluents is the Ruthner process, in which hydrochloric acid is evaporated and iron oxide granules are formed in a fluidized bed at temperatures of around 700°C. However, the presence of zinc in concentrations higher than 0.5 kg m^{-3} disturbs the process [56].

The difficulty underlying the development of valuable compound recovery processes is associated with the physicochemical complexity usually encountered in the hydrochloric acid effluents, where the target species are present in a heterogeneous mixture with different amounts of non-desirable compounds (oils, residual acid, impurities, etc.) [34]. The development of selective processes to treat hazardous streams, allows the recovery of reduction of the effluent toxicity and the recovery as the same time, of those components with higher added value [57]. In this sense, progress in knowledge about membrane-processes has allowed the application of EPT to the recovery of the metallic zinc contained in acidic SPS typically containing high concentrations of zinc (80–145 kg m^{-3}), iron (80–90 kg m^{-3}), and hydrochloric acid (219 kg m^{-3}) and the presence of other metals including manganese, lead, aluminum, cadmium, nickel, cobalt, etc., that can be considered as impurities.

In aqueous effluents with a high concentration of hydrochloric acid, different iron and zinc chlorocomplex species have been reported in the open literature. Fe^{2+} and Fe^{3+} form only cationic or neutral compounds such as $FeCl^+$ and $FeCl_2^+$, $FeCl^{2+}$ and $FeCl_3$, while Zn^{2+} in the presence of chloride usually forms anionic chlorocomplexes which are selectively extracted with the solvating reagents typically employed in the extraction of anionic compounds [58].

Under the usual composition of SPS the ionic strength reaches high values (μ>5 *m*) so that more than 90% of zinc is in the form of $ZnCl_4^{2-}$ (HCl>98.6 kg m^{-3}) and only a few percent exists as $ZnCl_3^-$ (29.2<HCl<98.6 kg m^{-3}). According to Cierpiszewski et al. [59], tributyl phosphate (TBP) is reported as the most suitable reagent enabling the extraction (EX) of zinc chloride from HCl solutions, and the subsequent back-extraction (BEX) with service water. The extraction of zinc from hydrochloric acid solutions using TBP and zinc stripping from TBP with service water are described by the following chemical reactions [60,61]:

$$2H^+ + ZnCl_4^{2-} + TBP \leftrightarrow \overline{H_2ZnCl_4TBP} \quad (22.11)$$

$$\overline{H_2ZnCl_4TBP} \leftrightarrow ZnCl_4^{2-} + 2H^+ + \overline{TBP} \quad (22.12)$$

Working with the system SPS-TBP-water, Ortiz and co-workers [57] reported zinc selectivity values, defined as the ratio between the zinc and iron fluxes entering the stripping solution, higher than 146 kg zinc/kg iron, which confirmed the viability of the separation and recovery process. The performance of the EPT process in the treatment of SPS with zinc recovery was analyzed for the first time by Carrera et al. [34] who evaluated the influence of the variability in the initial composition of SPS on the rate of zinc chloride separation. These authors concluded that the initial diffusive flux of zinc from the SPS to the organic liquid membrane decreased from 0.1 kg m^{-2} h^{-1} to 0.03 kg m^{-2} h^{-1} when the initial concentration of zinc in the SPS varied from 80 to 20 kg m^{-3} (see Figure 22.8) obtaining average extraction percentages of zinc of 56% when steady-state conditions are reached (t ≈ 1 hour). On the other hand, the recovery of zinc in the stripping solution

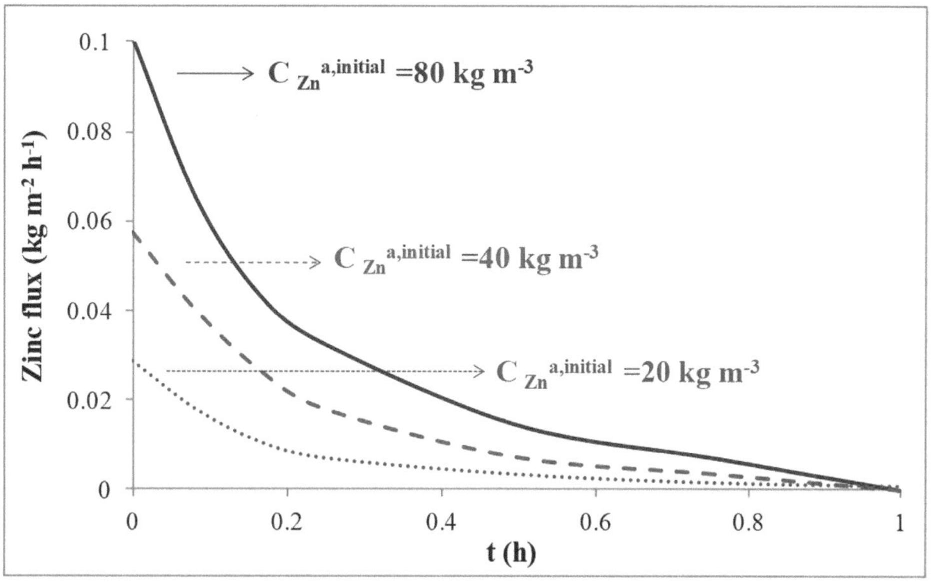

FIGURE 22.8 Change with time of the zinc diffusive flux from the SPS to the organic liquid membrane [34].

TABLE 22.3
Recovery ratios of zinc at steady-state conditions [34]

$C_{Zn}^{a,initial}$ (kg m^{-3})	% (v/v) TBP /% (v/v) water (emulsion)	R_{Zn}
80	80/20	16.2
40	80/20	12.9
20	80/20	11.2
80	60/40	21.5
80	50/50	28.5

was quantified by the recovery ratio of zinc (R_{Zn}) defined as follows (see Table 22.3):

$$R_{Zn} = \frac{V^s \cdot C^s_{T,Zn}}{V^a \cdot C^{a,initial}_{Zn}} \cdot 100 \quad (22.13)$$

where V^S is the volume of the stripping solution, V^a is the volume of SPS, $C^s_{T,Zn}$ is the time-dependent zinc concentration in the stripping solution, and $C^{a,initial}_{Zn}$ is the initial concentrations of zinc in the SPS.

Table 22.3 shows that the higher the initial concentration of zinc in the SPS, the higher the flux, and thus, the higher the recovery ratio. Nevertheless, the obtained values of the recovery ratio are low because of the equilibrium limitation imposed by the solubility of zinc chloride in the stripping solution (water). This hypothesis was confirmed by the higher recoveries of zinc obtained when higher volumes of water are employed in the emulsion formulation. In addition, the characteristics of the stripping solution ($C_{Zn^{2+}} \approx 50$ kg m^{-3}, $C_{Fetotal} \approx 0.01$ kg m^{-3}, high HCl concentration, and low level of impurities) obtained by EPT fulfill the requirements needed to employ this phase as an electrolytic solution ($C_{Zn^{2+}} > 31$ kg m^{-3}, $C_{Fetotal} < 0.01$ kg m^{-3}, [HCl] = 4.7–7.3 kg m^{-3}, and low level of impurities) [62].

Samaniego and co-workers [56] studied the interfacial extraction and back-extraction chemical equilibria described by the reactions given by Eq. (22.11) and Eq. (22.12) concluding that these reactions can be accurately described by constant distribution coefficients (see Table 22.4). Carrera et al. [34] carried out mathematical modeling of the EPT process applied to the treatment of SPS, concluding that the mass transfer resistance to zinc transport was shared between the liquid membrane embedded in the pores of the hollow fibers, and the organic phase boundary layer. Table 22.4 summarizes the equilibrium parameters and mass transport coefficients needed to describe the separation and concentration kinetics of zinc from SPS by means of EPT. Considering the complex nature of this kind of wastewater, it is thought that the EPT mathematical model permits a satisfactory description (95% of the simulated data fall within the range $C^a_{T,Zn^{2+},EXP} \pm 15\% \cdot C^a_{T,Zn^{2+},EXP}$) of the zinc separation and recovery process from SPS using TBP as extractant and service water as the back-extraction agent [34].

Laso and co-workers [63] compared the performance of EPT and non-dispersive solvent extraction (NDSX), which unlike EPT is a process configuration that employs two membrane contactors to carry out the simultaneous EX and BEX stages. In addition, the influence of TBP concentration (20–100% v/v) and the stripping phase/feed phase volume ratio in the range V_s/V_a 0.2–2 on both the extraction kinetics and the selectivity of the recovery of zinc over iron ($\alpha_{Zn/Fe}$) in the stripping phase was analyzed. This study shows that EPT and NDSX are effective when removing zinc versus iron from SPS using TBP as an extractant and water as a stripping agent. The kinetics of the extraction and back-extraction of

TABLE 22.4
Kinetic and equilibrium parameters of the EPT mathematical model applied to the treatment of SPS [34,56]

Parameter		Value	Comments
Membrane mass transport coefficients	$k_{m,\overline{H_2ZnCl_4 \cdot TBP}}$	$9.7\ 10^{-4}$ m h^{-1}	Determined by means of parameter estimation techniques using AspenTech simulation software and experimental data
Mass transport coefficient in the organic phase boundary layer	$k_{o,\overline{H_2ZnCl_4 \cdot TBP}} \cdot A_{i_3}$	45 h^{-1}	
Equilibrium parameter of the extraction reaction	$H^{i_1}_{eq,Zn^{2+}}$	1.16	Constant distribution coefficient experimentally obtained
Equilibrium parameter of the back-extraction reaction	$H^{i_3}_{eq,Zn^{2+}}$	1.16	

zinc were demonstrated to be promoted by EPT, increasing the TBP concentration in the range between 20% (v/v) and 50% (v/v) and increasing the stripping volume from 0.2 to 1 L. On the other hand, the transport of iron was enhanced by EPT, high TBP concentration, and increasing stripping volume. To optimize the further electrochemical recovery of zinc from the stripping solution, the transport of iron should be limited by the selection of the process configuration and operation conditions. Carrillo-Abad et al. [64] evaluated the electrochemical recovery of zinc from stripping solutions obtained by the treatment of SPS with NDSX and containing different Zn/Fe molar ratios. For all the tested experimental conditions, approximately 80% of the zinc is deposited previous to the iron co-deposition. Moreover, an increase in the applied current provides a higher zinc recovery but at a higher energy costs.

The viability of membrane-based solvent extraction technologies to treat SPS allowing the selective recovery of zinc has been extensively confirmed at bench-scale, however the implementation of the technology in real scenarios needs confirmation of its performance on a larger scale. Recently, Arguillarena and co-workers [65] evaluated the scale-up of membrane-based solvent extraction technology aimed at the selective separation of zinc from industrial SPAs as a purification step prior to zinc electrowinning. The results reported in this work demonstrate the NDSX technology in an industrially relevant environment equivalent to TRL 6. The experiments were carried out at a pilot scale treating SPAs batches of 57–91 L in NDSX configuration. The pilot plant was equipped with four hollow-fiber contactors and 80 m^2 of total membrane area, which was approximately 30 times higher than previous bench-scale studies reported in the literature. Starting with SPAs with high Zn (71.7 ± 4.3 kg·m^{-3}) and Fe (82.9 ± 5.0 kg·m^{-3}) content, the NDSX process achieved a stripping phase with 55.7 kg Zn·m^{-3} and only 3.2 kg Fe·m^{-3} which could be reused either by galvanizers or as supply for other secondary zinc markets. The reproducibility of results confirmed the stability of the organic extractant and its adequate regeneration in the NDSX operation.

22.2.2 Recovery of Mineral Acids by Membrane Diffusion Dialysis

Diffusion dialysis (DD) is an ion-exchange membrane (IEM) separation process driven by the concentration gradient of chemical species present on either side of the membrane. The process is based on the fact that ion-exchange membranes generally show a high permeability for counter-ions through the membrane while co-ions are rejected. Exceptions are protons and hydroxide ions, which can easily permeate through both cation- and anion-exchange membranes since they exhibit high mobility within the membrane due to their small hydration radius and little charge. The transport mechanism in diffusion dialysis is more complex than in conventional dialysis due to the electrostatic interaction between positive and negative charges and the electroneutrality requirement. DD is a slow and spontaneous process that does not require external gradients to promote separation. This fact makes it impossible to control the global rate of the process. Like any spontaneous process, it leads to an increase in entropy and a decrease in Gibbs free energy, being, therefore, thermodynamically favorable. The only external energy required in the system is that necessary to circulate the solutions to their corresponding compartments [66].

Environmental pollution is continually increasing by rapid industrial development. Electroplating, paper milling, metal refining, steel processing, hydrometallurgy, alkaloid production, and many more industries are involved in the discharge of huge amounts of toxic metal-ion-containing acidic waste, which gives rise to severe environmental pollution. In the recent past, various processes have been employed to treat industrial acidic effluent, namely solvent extraction, thermal decomposition, crystallization, neutralization, and distillation, and are effective in addressing the above issue but regrettably high-energy consumption and high capital cost limit their implementation. Nowadays, DD has drawn immense attention in industrial effluent treatment and has been considered an attractive technology for the separation and concentration of inorganic acids such as HCl, H_2SO_4, HNO_3, and HF, from wastewater that contains, in addition to acid, salts. These effluents are, among others, those generated in the titanium dioxide industry, aluminum etching, steel production, metal

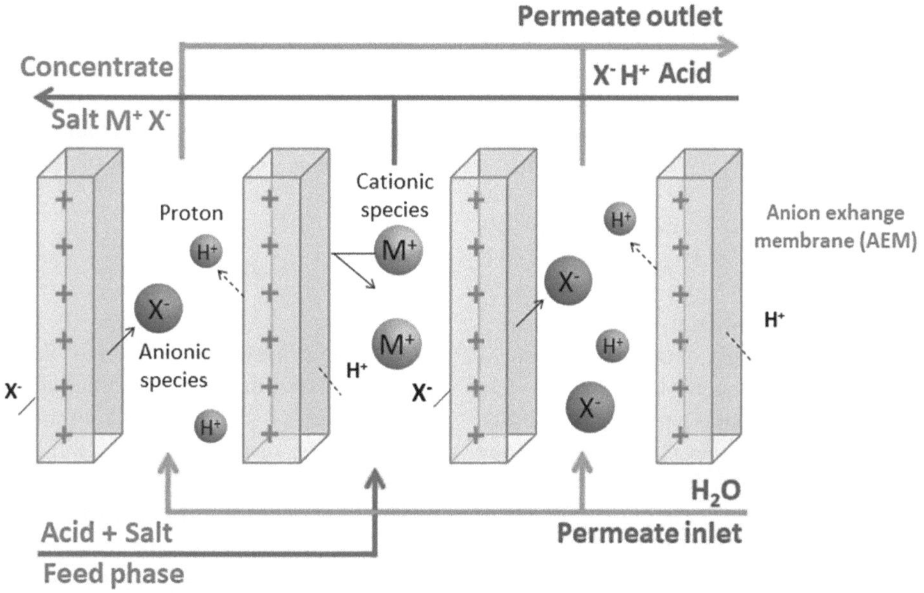

FIGURE 22.9 Principle of dialysis technology with ion exchange membranes in a system containing salts in acid medium.

refinery, non-ferrous metal smelting industry, foil industries, steel pickling treatment industries, industries of electrolytic coatings, etc. [67–69]. DD using anion exchange membranes has proven to be an energy-efficient and eco-friendly process for these applications [70].

The separation of the acid from the salt is achieved only through the arrangement of multiple anion exchange membranes. Figure 22.9 describes the principle of the DD process for the specific case dealing with the recovery of acid from mixtures with salt using anion exchange membranes [71]. As shown in Figure 22.9, both the anions and metal salts in feed solution tend to be transported into the permeate compartment because of the concentration difference on both sides of the membrane. However, due to the presence of the anion exchange membrane (AEM), anions (X^-) can pass easily through it, while cations (M^+), except for protons, will be more or less retained due to their positive charge and large size. H^+ ions, despite being positively charged, are more competitive in diffusion than metal ions because of their small size, low valence state, and higher mobility [72]. Hence, they can be spread through the mechanism of "tunneling" suggested by Grotthus, together with the anions, to meet the requirement of electrical neutrality. H^+ transport is the key to the DD process for the recovery of acids from complex solutions [73].

Sulfuric acid (H_2SO_4), hydrochloric acid (HCl), or a combination of hydrofluoric and nitric acids (HF + HNO_3) are often used as pickling agents in industries, such as steel production, metal refining, and non-ferrous metal smelting, where large quantities of spent liquor are produced. Though the acid liquor can be reused, the accumulation of metal ions in the solution will result in decreased efficiency of the pickling agent [74]. The separation of acids from salt–acid mixtures by DD allows the simultaneous recovery of the acid and rejection of the salts, thus enhancing the industrial and environmental benefits. The recovery and rejection rates during the DD process are determined by different factors, such as the membrane properties, the feed composition, and the operation parameters [75]. Accordingly, significant differences have been reported for diverse systems of DD due to the "salt effect," which is obvious in the systems HCl + chlorides, such as HCl + $FeCl_2$, HCl + $FeCl_3$, HCl + $NiCl_2$, and HCl + $ZnCl_2$, while in the systems H_2SO_4 + sulfates, such as H_2SO_4 + $FeSO_4$, H_2SO_4 + $NiSO_4$, H_2SO_4 + $CuSO_4$, or H_2SO_4 + $ZnSO_4$, it is not observed. Low absorption and high diffusivity in the membrane are always observed in the strong acid solutions, while in the weak acid solution, the trends are just the opposite [66].

Since 1964, a number of steel pickling industries started to select HCl as pickling agent instead of H_2SO_4, inducing faster and cleaner pickling, lower acid consumption, fewer quantities of waste pickle liquor, and more uniform products. Consequently, the research on the HCl recovery by diffusion dialysis has emerged. When metallic compounds and chloride are dissolved together, anionic or cationic chlorocomplexes of metal ions are formed, and thus the permeability of chloride anion through anionic membranes is influenced. Currently, pickling baths are stressed until the acid concentration decreases by 75–85% and the concentration of metals reaches the highest solubility threshold; Zn^{2+} concentration increases even up to 130 kg m^{-3} (≈ 271 kg m^{-3} $ZnCl_2$, 2.0 kmol m^{-3} $ZnCl_2$) and HCl concentration is lower than 80 kg m^{-3} (≈ 2.2 kmol m^{-3}) [76]. For systems such as HCl + $NiCl_2$ and HCl + $FeCl_2$ where cationic chlorocomplexes of metals are formed, the mass transport fluxes of the metallic species through the anionic membrane are low, and thus the selectivity and recovery of HCl are high. However, in the system HCl + $ZnCl_2$, the separation efficiency of HCl over $ZnCl_2$ is affected because of the formation of anionic complexes ($ZnCl_3^-$ and

$ZnCl_4^{2-}$) which can go through the anionic membrane. Recently, San Román and co-workers [71,77] analyzed the recovery of hydrochloric acid by DD from synthetic solutions containing a mixture of HCl and zinc chlorides that simulate the composition of the stripping solutions obtained by the treatment of real SPS effluents by EPT as described in Section 22.2.1.2.

In this work, the influence of the variability of the SPS initial composition, (0.5 kmol m^{-3} (68.2 kg m^{-3}) < [$ZnCl_2$] < 1.0 kmol m^{-3} (136.4 kg m^{-3}) and 0.5 kmol m^{-3} (18.3 kg m^{-3}) < [HCl] < 3.0 kmol m^{-3} (109.5 kg m^{-3})) in the HCl and zinc diffusive fluxes and in the HCl recovery of the permeate phase at steady-state conditions, was evaluated. Experiments were performed in continuous mode working with 10^{-3} m^3 of each feed and recovery solution. Eight experiments were carried out under the following conditions: 0.5 $ZnCl_2$ kmol m^{-3} + (0.5/1.0/2.0/3.0) HCl kmol m^{-3} and 1.0 $ZnCl_2$ kmol m^{-3} + (0.5/1.0/2.0/3.0) HCl kmol m^{-3} that allow to calculate eight values of J_{HCl} and J_{Zn} in both the feed and the permeate compartments. Commercial anion exchange membranes (FUMASEP FAD) were used in a two-compartment arrangement. Ten repeating units were placed in the stack in order to obtain 0.1 m^2 of membrane area (Figure 22.10).

In the first stage, the composition of the SPS residue was characterized, determining the distribution of cationic and anionic species present. Figure 22.11 shows the percentage of cationic zinc (and neutral species) chlorocomplexes and anionic zinc chlorocomplexes in the initial composition of SPS when different initial concentrations of $ZnCl_2$ and HCl are used during the experiments [78].

It can be observed in Figure 22.11 that the percentage of zinc chlorocomplexes cationic (and neutral species) is greater than 50% when the concentration of HCl is less than 1.0 kmol m^{-3} (Figure 22.11a). Conversely, the percentage of zinc chlorocomplex anionic species is higher than 70% for HCl concentrations greater than 2.0 kmol m^{-3} (Figure 22.11b). The trend is very similar for both $ZnCl_2$ concentrations. Next, in order to evaluate the behavior of the DD technology, the fluxes of HCl and zinc (J_i) in the feed and permeate compartment to each experiment were determined using the following equation under steady-state conditions:

$$J_i = \frac{Q \cdot \left(C_{\text{initial},i} - C_{\text{final},i}\right)}{A} \tag{22.14}$$

where Q (m^3 s^{-1}) is the flow-rate of the flowing phase and $C_{\text{initial},i}$ and $C_{\text{final},i}$ (kmol m^{-3}) are the initial and final concentrations (steady-state) of zinc and HCl in the feed and permeate compartments, respectively. A is the area of the membrane (0.1 m^2). The results obtained are shown in Figure 22.12; each pair of Zn and HCl dots represents the flow of an experiment in a given phase. For example, in Figure 22.12a, the pink triangles represent the flux of Zn at two initial concentrations of Zn [0.5 (light) and 1.0 (dark) kmol m^{-3}] and different initial concentrations of HCl (0.5–1.0–2.0–3.0 kmol m^{-3}), and the blue circles correspond to the flux of HCl at two initial concentrations of Zn 0.5 (light) and 1.0 (dark) kmol m^{-3} and different HCl (0.5–1.0–2.0–3.0 kmol m^{-3}), in the feed phase.

As depicted in Figure 22.12, the authors concluded that under the selected operation conditions, the flux of HCl is not dependent on the initial concentration of $ZnCl_2$ and also increases as the initial concentration of HCl in the SPS increases. Additionally, it is observed that the values of the HCl flux steady-state conditions are approximately four and five times higher than the flux of zinc, in the feed phase (Figure 22.12a) and in the permeate phase (Figure 22.12b), respectively, reaching a maximum value HCl of 1.4 10^{-3} kmol m^{-2} s for an initial concentration of HCl of 3.0 kmol m^{-3}. It should be noted that the flow of HCl is similar in

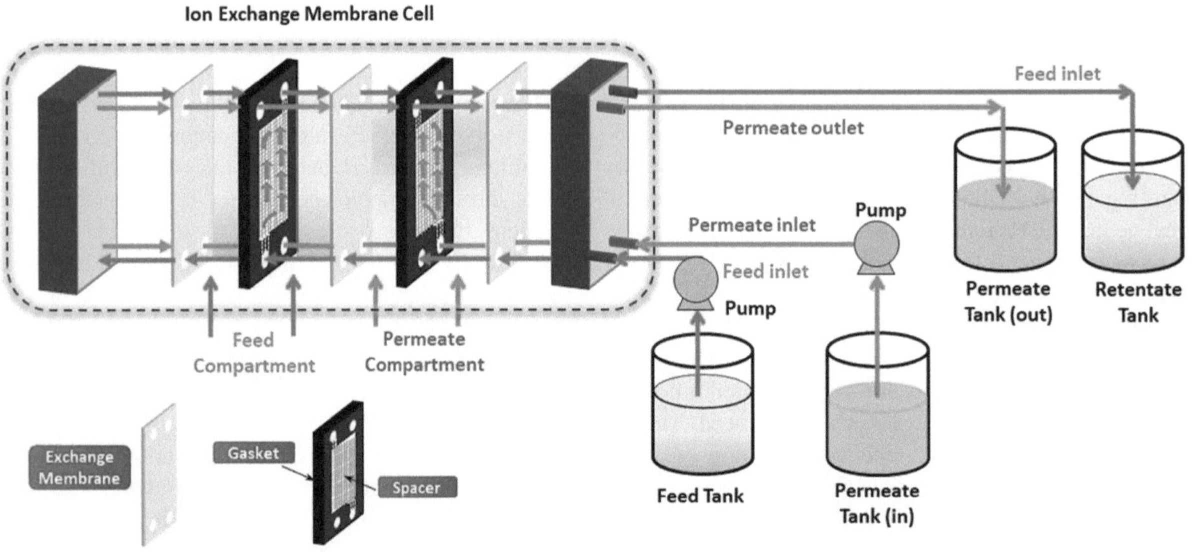

FIGURE 22.10 Experimental set up of DD [78].

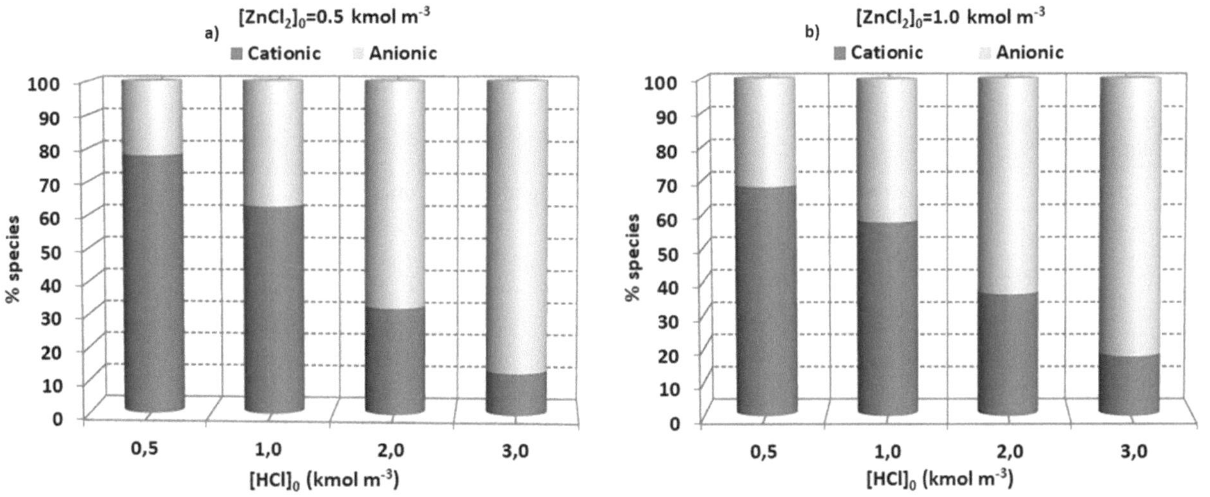

FIGURE 22.11 Percentage of cationic (and neutral species) and anionic zinc chlorocomplexes in the initial composition of SPS for different initial concentrations of $ZnCl_2$ versus the initial concentration of HCl [78].

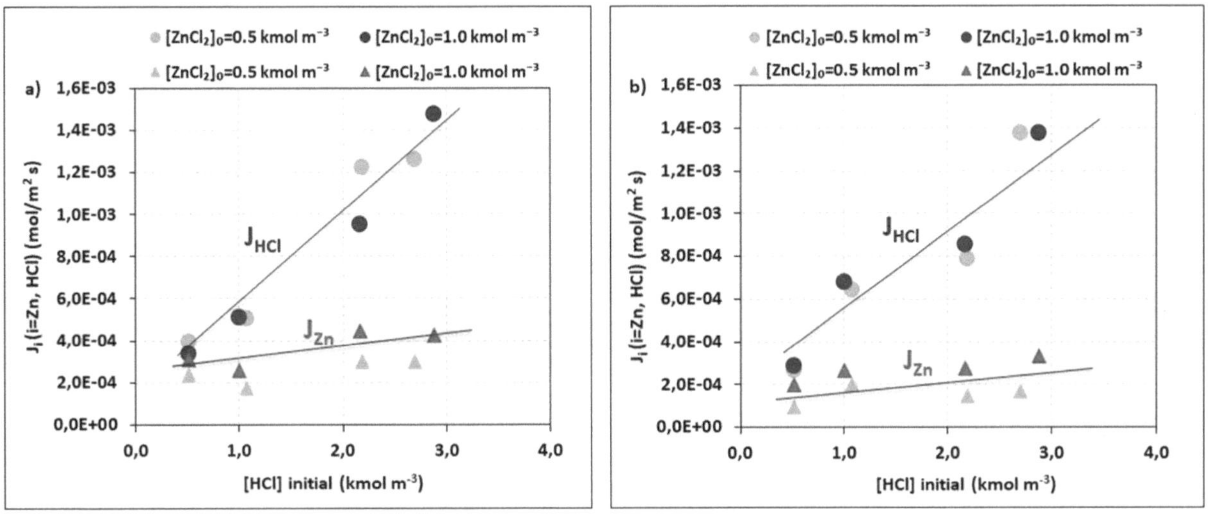

FIGURE 22.12 Steady-state HCl and zinc flux values in the (a) feed compartment and (b) permeate compartment.

both phases. Finally, in order to quantify the influence of the process variables, the separation selectivity and HCl recovery percentage were evaluated. The separation selectivity was defined according to the ratio between HCl flux and zinc flux in the permeate phase. The percentage of HCl recovered (R_{HCl} %) was calculated from the final HCl concentration in the permeate phase referred to the initial HCl concentration in the feed phase. These values are shown in Table 22.5.

In Table 22.5, the selectivity of HCl towards Zn permeation increases as the initial HCl concentration increases and as $ZnCl_2$ concentration decreases. This behavior is due to two different contributions: (i) the increasing trend of the concentration of free chloride with increasing initial HCl and decreasing initial $ZnCl_2$ concentrations and (ii) the predominance of the zinc cationic species at low total $ZnCl_2$ concentration (yellow color in Table 22.5), thus minimizing the co-transport of anionic zinc chlorides and improving the flux of HCl.

Table 22.5 shows that the lower the initial concentration of HCl in the SPS, the higher the concentration of HCl recovered in the permeate phase, around 40% when the initial concentration of HCl is 0.5 kmol m^{-3} (18.3 kg m^{-3}). Once again, the performance of the recovery process is not dependent on the initial concentration of zinc in the SPS. This behavior can be attributed to the fact that for initial $ZnCl_2$ concentrations higher than 1.0 kmol m^{-3} (136.4 kg m^{-3}) the predominant species are $ZnCl_3^-$ and $ZnCl_4^{2-}$, which penetrate into the anion-exchange membrane competing with HCl and therefore the recovery of HCl falls to 20%. However, as the negatively charged complex proceeds to the permeate side within the membrane, a portion of these complexes is converted to $ZnCl_2$ and $ZnCl^+$ due to the low HCl concentration in the permeate

TABLE 22.5
Fluxes, selectivity, and recovery of HCl obtained by DD

Feed phase conditions initial (kmol m^{-3})					
$[HCl]_0$	$[ZnCl_2]$	Feed phase J_{HCl} (mol m^{-2} s^{-1})	Permeate phase J_{HCl} (mol m^{-2} s^{-1})	Selectivity	R_{HCl} %
0.5	**0.5**	3.9 10^{-4}	2.7 10^{-4}	2.9	38
1.0		5.0 10^{-4}	6.4 10^{-4}	3.2	35
2.0		1.2 10^{-3}	7.9 10^{-4}	5.3	20
3.0		1.3 10^{-3}	1.1 10^{-3}	6.5	20
0.5	**1.0**	3.4 10^{-4}	2.9 10^{-4}	1.4	38
1.0		5.1 10^{-4}	6.8 10^{-4}	2.5	38
2.0		9.5 10^{-4}	8.6 10^{-4}	3.1	20
3.0		1.5 10^{-3}	1.4 10^{-3}	4.2	20

side of the membrane, and thus compensating for the transport of chloride. At lower zinc concentrations, neutral or cationic complexes unable to penetrate into the membrane are formed, and thus the permeation of HCl is not affected. On the other hand, the permeation of HCl through the membrane is favored at lower acid concentrations in the feed phase because under these conditions the permeation of zinc species is hindered, whereas the opposite trend is observed with increasing of the concentration of HCl in the feed solution.

In conclusion, the optimization of the HCl recovery process depends on: (i) the initial concentration of HCl and zinc in the SPS that determines the distribution of ionic species and (ii) the required concentration and purity of the recovered acid for reuse in the pickling step of the hot-dip galvanizing process or in another industrial process. Therefore, the results described above confirm the viability of the diffusion dialysis technology as a promising alternative to carry out the recovery of hydrochloric acid from the spent pickling solutions produced during the hot-dip galvanizing process. However, in spite of the promising results obtained through this work, there remain challenges that must be overcome before this process can be scaled up. Among others, the research and development of new membranes able to guarantee lower permeation of anionic species in order to completely achieve competitive conditions in terms of fluxes and selectivity are required.

22.3 CONCLUDING REMARKS AND FUTURE DIRECTIONS

The number of applications of membrane technology dealing with the recovery of valuable compounds from wastewaters and industrial effluents is increasing due to its potential to replace conventional processes by accomplishing selective and efficient transport of specific components and providing reliable options for sustainable growth. In particular, liquid membrane separation processes based on facilitated transport offer additional advantages due to their ability to promote the uphill and selective transport of the target species by the coupling between mass transfer and chemical reaction. However, the stand-alone application of individual membrane processes often fails to recover all the possible valuable materials present in complex wastewaters. Therefore, the development of integrated processes consisting of single membrane separation stages is a promising strategy to design eco-innovative processes.

As a proof of concept the initial hypothesis has been illustrated in the text through two different applications focused on the recovery of heavy metals and mineral acids from industrial effluents: (i) the recovery of zinc from spent chromium-based passivation baths and spent pickling solutions using the emulsion pertraction technology and (ii) the recovery of hydrochloric acid from spent pickling solutions by means of membrane diffusion dialysis. From the analysis previously undertaken, it is concluded that the process design needs the kinetic analysis of the separation and recovery stages followed by the development of accurate mathematical models as a previous step in the process design and scale-up.

According to the previous explanations and promising results, the proposed technologies are suitable to be managed as more common chemical engineering processes allowing their application in new scenarios where valuable compounds could be recovered.

ACKNOWLEDGMENT

This research was developed in the framework of the projects RTI2018-093310-B-I00 and PID2020-115409RB-I00 financed by the Ministry of Science, Innovation and Universities (Spain).

REFERENCES

[1] Ortiz, I., Bringas, E., Samaniego, H., San Román, M.F., Urtiaga, A. (2006) Membrane processes for the efficient recovery of anionic pollutants. *Desalination* 193: 375.

[2] Babel, S., Kurniawan, T.A. (2004) Cr(VI) removal from synthetic wastewater using coconut shell charcoal and commercial activated carbon modified with oxidizing agents and/or chitosan. *Chemosphere* 54: 951.

[3] Xu, J., Lu, S., Fu, D. (2009) Recovery of hydrochloric acid from the waste acid solution by diffusion dialysis. *Journal of Hazardous Materials* 165: 832.

[4] Regel-Rosocka, M. (2010) A review on methods of regeneration of spent pickling solutions from steel processing. *Journal of Hazardous Materials* 177: 57.

[5] Kentish, S., Stevens, G. (2001) Innovation in separations technology for the recycling and re-use of liquid waste streams, *Chemical Engineering Journal* 84: 149.

[6] Ortiz Uribe, I., Mosquera-Corral, A., Lema, J., Esplugas, S. (2015) Advanced technologies for water treatment and reuse. *AIChE Journal* 61(10): 3146.

[7] Silva, J.E., Paiva, A.P, Soares, D., Labrincha, A., Castro, F. (2005) Solvent extraction applied to the recovery of heavy metals from galvanic sludge. *Journal of Hazardous Materials* 120: 113.

[8] Fernández-Olmo, I., Ortiz, A., Urtiaga, A., Ortiz, I. (2008) Selective iron removal from spent passivation baths by ion exchange. *Journal of Chemical Technology & Biotechnology* 83: 1616.

[9] Ortiz, A., Fernández-Olmo, I., Urtiaga, A., Ortiz, I. (2009) Modeling of iron removal from spent passivation baths by ion exchange in fixed-bed operation. *Industrial & Engineering Chemistry Research* 48, 7448, 2009.

[10] Ozdemir, T., Oztin, C., Kincal, N.S. (2006) Treatment of waste pickling liquors: process synthesis and economic analysis. *Chemical Engineering Communications* 193: 548.

[11] Hua, M., Zhang, S., Pan, B., Zhang, W., Lv, L., Zhang, Q. (2012) Heavy metal removal from water/wastewater by nanosized metal oxides: A review. *Journal of Hazardous Materials* 211–212: 317.

[12] Drioli, E., Curcio, E., Di Profio, G. (2005) State of the art and recent progresses in membrane contactors. *Transactions of the Institution of Chemical Engineers, Part A. Chemical Engineering Research and Design* 83(A3): 223.

[13] San Román, M.F., Bringas, E., Ibáñez, R., Ortiz I. (2010) Liquid membrane technology: Fundamentals and review of its applications. *Journal of Chemical Technology and Biotechnology* 85: 2.

[14] Zarca, R., Ortiz, A., Gorri, D., Ortiz, I. (2017) Generalized predictive modeling for facilitated transport membranes accounting for fixed and mobile carriers. *Journal of Membrane Science* 542: 168.

[15] Bringas, E., San Román, M.F., Irabien, J.A., Ortiz, I. (2009) An overview of the mathematical modeling of liquid membrane separation processes in hollow fiber contactors. *Journal of Chemical Technology & Biotechnology* 84: 1583.

[16] de Agreda, D., Garcia-Diaz, I., López, F.A., Alguacil, F.J. (2011) Supported liquid membranes technologies in metals removal from liquid effluents. *Revista de Metalurgia* 47: 146.

[17] Palaty, Z., Zakova, A. (2004) Separation of H_2SO_4 + $ZnSO_4$ mixture by diffusion dialysis. *Desalination* 169: 277.

[18] Jeong, J., Kim, M.S., Kim, B.S., Kim, S.K., Kim, W.B., Lee, J.C. (2005) Recovery of H_2SO_4 from waste acid solution by a diffusion dialysis method. *Journal of Hazardous Materials* B124; 230.

[19] Xu, T.W., Yang, W.H. (2003) Industrial recovery of mixed acid ($HF+HNO_3$) from the titanium spent leaching solutions by diffusion dialysis with a new series of anion exchange membranes. *Journal of Membrane Science* 220: 89.

[20] Palaty, Z., Zakova, A. (2004) Transport of nitric acid through the anion-exchange membrane NEOSEPTA-AFN. *Desalination* 160: 51.

[21] Hichour, M., Persin, F., Molenat, J., Sandeaux, J., Gavach, C. (1999) Fluoride removal from diluted solutions by donnan dialysis with anion-exchange membranes. *Desalination* 122: 53.

[22] Luo, J., Wu, C., Xu, T., Wu, Y. (2011) Diffusion dialysis-concept, principle and applications. *Journal of Membrane Science* 366: 1.

[23] Gabelman, A., Hwang, S.H. (1999) Hollow fiber membrane contactors. *Journal of Membrane Science* 159; 61.

[24] Ferraz, H.C., Duarte, L.T., Di Luccio, M., Alves, T.L.M., Habert, A.C., Borges, C.P. (2007) Recent achievements in facilitated transport membranes for separation processes. *Brazilian Journal of Chemical Engineering* 24: 101.

[25] Ho, W.S.W., Poddar, T.K. (2001) New membrane technology for removal and recovery of chromium from wastewaters. *Environmental Progress* 20: 44.

[26] Ho, W.S. (2001) Supported liquid membrane process for chromium removal and recovery. *U.S. Patent, 6171563.*

[27] Ho, W.S.W., Wang, B., Neumuller, T.E., Roller, J. (2001) Supported liquid membranes for removal and recovery of metals from waste waters and process streams. *Environmental Progress* 20: 117.

[28] Alguacil, F.J., García-Díaz, I., López, F., Sastre, A.M. (2011) Cobalt(II) membrane-extraction by DP-8R/Exxsol D100 using pseudo-emulsion based hollow fiber strip dispersion (PEHFSD) processing. *Separation and Purification Technology* 80: 467.

[29] Urtiaga, A., Bringas, E., Mediavilla, R., Ortiz, I. (2010) The role of liquid membranes in the selective separation and recovery of zinc for the regeneration of Cr(III) passivation baths. *Journal of Membrane Science* 356: 88.

[30] Dibán, N., Mediavilla, R., Urtiaga, A., Ortiz, I. (2011) Zinc recovery and waste sludge minimization from chromium passivation baths. *Journal of Hazardous Materials* 192: 801.

[31] Alguacil, F.J., Alonso, M., López, F.A., López-Delgado, A., Padilla, I., Tayibi, H. (2010) Pseudo-emulsion based hollow fiber with strip dispersión pertraction of iron(III) using $(PJMTH^+)_2(SO_4^{2-})$ ionic liquid as carrier. *Chemical Engineering Journal* 157: 366.

[32] González, R., Cerpa, A., Alguacil, F.J. (2010) Nickel(II) removal by mixtures of Acorga M5640 and DP8R in pseudo-emulsion based hollow fiber with strip dispersion technology. *Chemosphere* 81: 1164.

[33] Sonawane, J.V., Pabby, A.K., Sastre, A.M. (2010) Pseudo-emulsion based hollow fiber strip dispersion (PEHFSD) technique for permeation of Cr(VI) using Cyanex-923 as carrier. *Journal of Hazardous Materials* 174: 541.

[34] Carrera, J.A., Bringas, E., San Román, M.F., Ortiz, I. (2009) Selective membrane alternative to the recovery of zinc from hot-dip galvanizing effluents. *Journal of Membrane Science* 326: 672.

[35] Alguacil, F.J., Alonso, M., López, F.A., López-Delgado, A. (2009) Application of pseudo-emulsion based hollow fiber strip dispersion (PEHFSD) for recovery of Cr(III) from alkaline solutions. *Separation and Purification Technology* 66: 586.

[36] Roy, S.C., Sonawane, J.V., Rathore, N.S., Pabby, A.K., Janardan, P., Changrani, R.D., Dey, P.K., Bharadwaj, S.R. (2008) Pseudo-emulsion based hollow fiber strip dispersion

technique (PEHFSD): Optimization, modelling and application of PEHFSD for recovery of U(VI) from process effluent. *Separation Science and Technology* 43: 3305.

[37] Sonawane, J.V., Pabby, A.K., Sastre, A.M. (2008) Pseudo-emulsion based hollow fiber strip dispersion: a novel methodology for gold recovery. *AIChE Journal* 54: 453.

[38] Sonawane, J.V., Pabby, A.K., Sastre, A.M. (2007) Au(I) extraction by LIX-79/n-heptane using the pseudo-emulsion-based hollow-fiber strip dispersion (PEHFSD) technique. *Journal of Membrane Science* 300: 147.

[39] Bringas, E., San Román, M.F., Ortiz, I. (2006) Separation and recovery of anionic pollutants by the emulsion pertraction technology. Remediation of polluted groundwaters with Cr(VI). *Industrial & Engineering Chemistry Research* 45: 4295.

[40] Urtiaga, A., Abellán, M.J., Irabien, J.A., Ortiz I. (2005) Membrane contactors for the recovery of metallic compounds. Modelling of copper recovery from WPO processes. *Journal of Membrane Science* 257: 161.

[41] Klaassen, R., Jansen, A.E. (2001) The membrane contactor: Environmental applications and possibilities. *Environmental Progress* 20: 37.

[42] Ortiz, I., San Román, M.F., Corvalán, S.M, Eliceche, A.M. (2003) Modeling and optimization of an emulsion pertraction process for removal and concentration of Cr(VI). *Industrial & Engineering Chemistry Research* 42: 5891.

[43] Klaassen, R., Feron, P.H.M., Jansen, A.E. (2005) Membrane contactors in industrial applications. *Chemical Engineering Research and Design* 83: 234.

[44] Bringas, E., Mediavilla, R., Urtiaga, A., Ortiz, I. (2011) Development and validation of a dynamic model for regeneration of passivating baths using membrane contactors. *Computers & Chemical Engineering* 35: 918.

[45] Alonso, A.I., Pantelides, C.C. (1996) Modelling and simulation of integrated processes for recovery of Cr(VI) with Aliquat 336. *Journal of Membrane Science* 110: 151.

[46] European Union Law. *Directive 2000/53/EC of the European Parliament and of the Council of 18 September 2000 on End-of Life Vehicles*. http://eur-lex.europa.eu (accessed 1 April 2022).

[47] García, V., Steeghs, W., Bouten, M., Ortiz, I., Urtiaga, A. (2013) Implementation of an eco-innovative separation process for a cleaner chromium passivation in the galvanic industry. *Journal of Cleaner Products* 59: 274.

[48] Bringas, E., San Roman, M.F., Urtiaga, A., Ortiz, I. (2013) Integrated use of liquid membranes and membrane contactors: Enhancing the efficiency of L-L reactive separations. *Chemical Engineering and Processing: Process Intensification* 67: 120.

[49] Kanungo, S.B., Mohapatra, R. (1995) Coupled transport of Zn(II) through a supported liquid membrane containing bis(2,4,4-trimethylpentyl) phosphinic acid in kerosene. I A model for the rate process involving binary and ternary complex species. *Journal of Membrane Science* 105: 217.

[50] García-Antón, J., Fernández-Domene, R.M., Sánchez-Tovar, R., Escrivà-Cerdán, C., Leiva-García, R., García, V., Urtiaga, A. (2014) Improvement of the electrochemical behaviour of Zn-electroplated steel using regenerated Cr (III) passivation baths. *Chemical Engineering Science* 111: 402.

[51] Tsarikidis, P.E., Oustadakis, P., Katsiapi, A., Agatzini-Leonardou, S. (2010) Hydrometallurgical process for zinc recovery from electric arc furnace dust (EAFD). Part II: downstream processing and zinc recovery by electrowinning. *Journal of Hazardous Materials* 179: 8.

[52] Barakat, M.A., Mahmoud, M.H.H., Shehata, M. (2006) Hydrometallurgical recovery of zinc from fine blend of galvanization processes. *Separation Science and Technology* 41: 1757.

[53] Gordon, R.B., Graedel, T.E., Bertram, M., Fuse, K., Lifset, R., Rechberger, H., Spatari, S. (2003) The characterization of technological zinc cycles. *Resources, Conservation and Recycling* 39: 107.

[54] European Standard EN 1179 (2003) *Zinc and Zinc Alloys: Primary Zinc*. Technical Committee CEN/TC 209, 2003.

[55] Garcia, V., Margallo, M., Aldaco, R., Urtiaga, A., Irabien, A. (2013) Environmental sustainability assessment of an innovative Cr (III) passivation process. *ACS Sustainable Chemistry & Engineering* 1: 481.

[56] Samaniego, H., San Román, M.F., Ortiz I. (2006) Modelling of the extraction and back-extraction equilibria of zinc from spent pickling solutions. *Separation Science and Technology* 41: 757.

[57] Ortiz, I., Bringas, E., San Román, M.F., Urtiaga A.M. (2004) Selective separation of zinc and iron from spent pickling solutions by membrane-based solvent extraction, *Separation Science and Technology* 39: 1.

[58] Regel, M., Sastre, A.M., Szymanowski, J. (2001) Recovery of zinc(II) from HCl spent pickling solutions by solvent extraction. *Environmental Science and Technology* 35: 630.

[59] Cierpezewski, R., Miesiac, I., Regel-Rosocka, M., Sastre, A.M., Szymanowski, J. (2002) Removal of zinc(II) from spent hydrochloric acid solutions from zinc hot galvanizing plants, *Industrial & Engineering Chemistry Research* 41: 598.

[60] Forrest, V.M.P., Scargill, D., Spickernell, D.R. (1969) The extraction of zinc and cadmium by tri-n-butyl phosphate from aqueous chloride solutions. *Journal of Inorganic Nuclear Chemistry* 31: 187.

[61] Morris, D.F.C., Short, E.L. (1962) Zinc chloride and zinc bromide complexes Part II. Solvent-extraction studies with zinc-65 as tracer. *Journal of the Chemical Society Abstracts* 2662.

[62] Samaniego, H., San Román, M.F., Ortiz I. (2007) Kinetics of zinc recovery from spent pickling effluents. *Industrial & Engineering Chemistry Research* 46: 907.

[63] Laso, J., García, V., Bringas, E., Urtiaga, A.M., Ortiz, I. (2015) Selective recovery of zinc over iron from spent pickling wastes by different membrane-based solvent extraction process. *Industrial & Engineering Chemistry Research* 54: 3218.

[64] Carrillo-Abad, J., Garcia-Gabaldon, M., Ortiz-Gandara, I., Bringas, E., Urtiaga, A.M., Ortiz, I., Perez-Herranz, V. (2015) Selective recovery of zinc from spent pickling baths by the combination of membrane-based solvent extraction and electrowinning technologies. *Separation and Purification Technology* 151: 232.

[65] Arguillarena, A., Margallo, M., Arruti-Fernández, A., Pinedo, J., Gómez, P., Urtiaga, A. (2020) Scale-up of membrane-based zinc recovery from spent pickling acids of hot-dip galvanizing. *Membranes* 10(12): 1.

[66] Luo, J., Wu, C., Xu, T., Wu, Y. (2011) Diffusion dialysis: concept, principle and applications. *Journal of Membrane Science* 366: 1.

[67] Zhuang, J.X., Chen, Q., Wang, S., Zhang, W.M., Song, W.G., Wan, L.J., Ma, K.S., Zhang, C.N. (2013) Zero discharge process for foil industry waste acid reclamation: coupling of diffusion dialysis and electrodialysis with bipolar membranes. *Journal of Membrane Science* 432: 90.
[68] Cheng, C., Yang, Z., Pan, J., Tong, B., Xu, T. (2014) Facile and cost effective PVA based hybrid membrane fabrication for acid recovery. *Separation and Purification Technology* 136: 250.
[69] Ruiz-Aguirre, A., Lopez, J., Gueccia, R., Randazzo, S., Cipollina, A., Cortina, J.L., Micale, G. (2021) Diffusion dialysis for the treatment of H_2SO_4-$CuSO_4$ solutions from electroplating plants: Ions membrane transport characterization and modelling. *Separation and Purification Technology* 2661: 118215.
[70] Yadav, V., Raj, S.K., Rathod, N.H., Kulshrestha, V. (2020) Polysulfone/graphene quantum dots composite anion exchange membrane for acid recovery by diffusion dialysis. *Journal of Membrane Science* 611: 118331.
[71] San Román, M.F., Ortiz Gándara, I., Ibañez, R., Ortiz, I. (2012) Hybrid membrane process for the recovery of major components (zinc, iron and HCl) from spent pickling effluents. *Journal of Membrane Science* 415–416: 616.
[72] Zhang, C., Zhang, W., Wang, Y. (2020) Review diffusion dialysis for acid recovery from acidic waste solutions: Anion exchange membranes and technology integration. *Membranes* 10: 169.
[73] Gueccia, R., Ruiz Aguirre, A., Randazzo, S., Cipollina, A., Micale, G. (2020) Diffusion dialysis for separation of hydrochloric acid, iron and zinc ions from highly concentrated pickling solutions. *Membranes* 10: 129.
[74] Kavitha, E., Oonguzhali, E., Nanditha, D., Kapoor, A., Arthanareeswaran, G., Prabhakar, S. (2022) Current status and future prospects of membrane separation processes for value recovery from wastewater. *Chemosphere* 291: 132690.
[75] Lin, J., Huang, J., Wang, J., Yu, J., You, X., Lin, X., Van der Bruggen, B., Zhao, S. High-performance porous anion exchange membranes for efficient acid recovery from acidic wastewater by diffusion dialysis. *Journal of Membrane Science* 624: 119116.
[76] Regel-Rosocka, M., Cieszyñska, A., Wioeniewski, M. (2011) Methods of regeneration of spent pickling solutions from steel treatment plants. *Polish Journal of Chemical Technology* 9(2).
[77] San Román, M.F., Ortiz Gándara, I., Bringas, E., Ibañez, R., Ortiz, I. (2018) Membrane selective recovery of HCl, zinc and iron from simulated mining effluents. *Desalination* 440: 78–87.
[78] Ortiz, I. (2017) *Recovery of HCl from galvanizing effluents by ion exchange membranes.* Doctoral Thesis. University of Cantabria.

23 Salts Recovery from Brines through Membrane Crystallization Processes

*Mirko Frappa, Francesca Alessandro, Enrico Drioli, and Francesca Macedonio**
Institute on Membrane Technology (CNR-ITM), Rende, Italy
*Email: f.macedonio@itm.cnr.it

23.1 INTRODUCTION

Global water stress, raw material depletion, environmental pollution, and energy production and consumption are already severe problems that our modern society has to solve and overcome to maintaining and/or increase quality of life. According to the latest report of the World Health Organization and UNICEF in 2020, 74% of the global population uses safely managed drinking water services [1]. However, three out of 10 people, or 2.3 billion people around the world, lack basic hygiene services. In the field of water treatment for recovery and purification, membrane technology is increasingly emerging as an alternative (or complementary) approach to traditional techniques. This is because membrane processes are more cost-effective in terms of energy consumption, separation efficiency, environmental sustainability, and process management. Today, membranes are used to make production processes increasingly innovative, in the context of a circular economy that responds to environmental and human needs. Based on the size and physical state of the substances present in the mixtures to be processed, we can distinguish different types of membrane processes. The main membrane operations used in water treatment and recovery are microfiltration (MF), ultrafiltration (UF), nanofiltration (NF), and reverse osmosis (RO). Emerging membrane processes involve membrane distillation (MD) and membrane-assisted crystallization (MCr). Water stress, which is the pressure on the quantity and quality of water resources, exists in many places throughout the world. Desalination is a viable solution to freshwater scarcity and several technologies have already been successfully applied (such as reverse osmosis) to produce around 100 million m^3 of freshwater daily. An estimated 6 billion people will suffer from water shortages by 2050 [2]. Desalination represents a valid solution to the scarcity of fresh water [3]. Over time, different desalination technologies have been adopted: multi-stage flash distillation (MSF), multiple effect distillation (MED), mechanical vapor compression (MV), reverse osmosis (RO), and membrane distillation (MD) [4]. In the past few decades, membrane technologies (particularly RO and NF) have replaced heat-based technologies (such as MSF and MED) with high energy consumption. About 90% of today's desalinated water is produced by RO [5]. MD is a cutting-edge technology for water desalination that is attracting interest around the world. During MD operations, water evaporates at the feed–membrane interface and diffuses through non-wetted pores to the permeate side under the influence of a vapor pressure gradient. The membrane placed between the liquid and the vapor phase prevents the mixing of liquid–vapor phases. The separation mechanism of MD retains all non-volatile solutes on the feed side so that fresh water is collected on the permeate size. MD is suitable for separating highly concentrated salt solutions, such as brine from RO installations, produced water, and industrial wastewaters. It can be considered as a hybrid membrane and thermal desalination method [6]. Historically, different thermal processes have been used in desalination: MSF, MED, and TVC (thermal vapor compression). However, since the 1950s, they have been replaced by more efficient membrane processes such as RO [7]. The advantages of MD include the ability to operate at low temperatures and pressures, a small plant footprint, low capital costs compared to conventional distillation processes, and the near complete absence of flux limitations due to concentration polarization. Unfortunately, MD permeation fluxes are several times lower than those obtained with RO [8]. With the aim of achieving a zero-liquid discharge in desalination, the concept of membrane crystallization (MCr) has been introduced by Drioli and his colleagues in recent times [9–12]. The raw materials industry needs water in the right quantity, at the right time, and with the required quality. To this end, mining and processing plants often rely on water self-sufficiency from local bodies of water, groundwater, or the sea (Figure 23.1). Water can also be supplied via public networks. In the specific case of mining operations, there are usually additional water resources available from mining effluents, which often meet the needs of the facility and may also provide additional resources for other water users. As a drawback, drainage can lower the water table under certain circumstances.

Depending on the water quality requirements for the mining and industrial processes, water supplies may require pretreatment. After the water is used, the wastewater is purified and released into the environment. Facilities can set up various types of dedicated ponds and dams for wastewater treatment. Often, wastewater is reused and/or recycled and loops back

DOI: 10.1201/9781003285656-23

into the facility for processes that are less demanding in terms of water quality (cascading use).

The EU is very advanced in terms of water management in mining and other industrial activities. Recently, the industry also has been able to actively learn from other sectors through pre-applications and regulations for desalinated seawater. In addition to the direct use of water on-site, the extraction of raw materials depends on auxiliary materials (such as chemicals) for on-site production, technical equipment, and often off-site generated energy sources. The production of all these items requires additional water sources.

23.2 DESALINATION

Water desalination is a practical route to recover clean and reusable seawater. In this field membrane technology accounts for a large number of separation processes, including nanofiltration, reverse osmosis, membrane distillation, and membrane crystallization.

The major sources of clean waters are the oceans and seas. However, in order to recover clean water from the sea it is necessary to apply processes to remove the salts from seawater. Desalination is the process of removing the saline fraction from waters containing salt, generally from marine waters, in order to obtain water with a low saline content; the water is then often used for food use, but also for industrial use, as cooling water. Drinking water must not be free of salts: both for health reasons and because the addition of certain salts is recommended (however, this is a practice generally carried out downstream of the distillation itself, to allow the addition of the correct salts) and while fully deionized water would be completely tasteless, it is not palatable. A small amount of saline entrainment is therefore left in the treated water, of the order of 25–50 mg/L. Technologies currently used for water desalination can be classified according to the following main criteria:

- Phase change of treated water;
- Type of energy involved;
- Process employed.

According to these criteria, the most important, relevant technologies can be categorized as shown in Figure 23.1.

As shown in Figure 23.1, the first distinction that can be done is between processes based on phase change and processes based on no phase change. A second major distinction that can be made between processes developed for water purification is between processes based on a membrane system and processes that do not use membrane systems. Table 23.1 reports the main desalination processes not based on membranes utilized

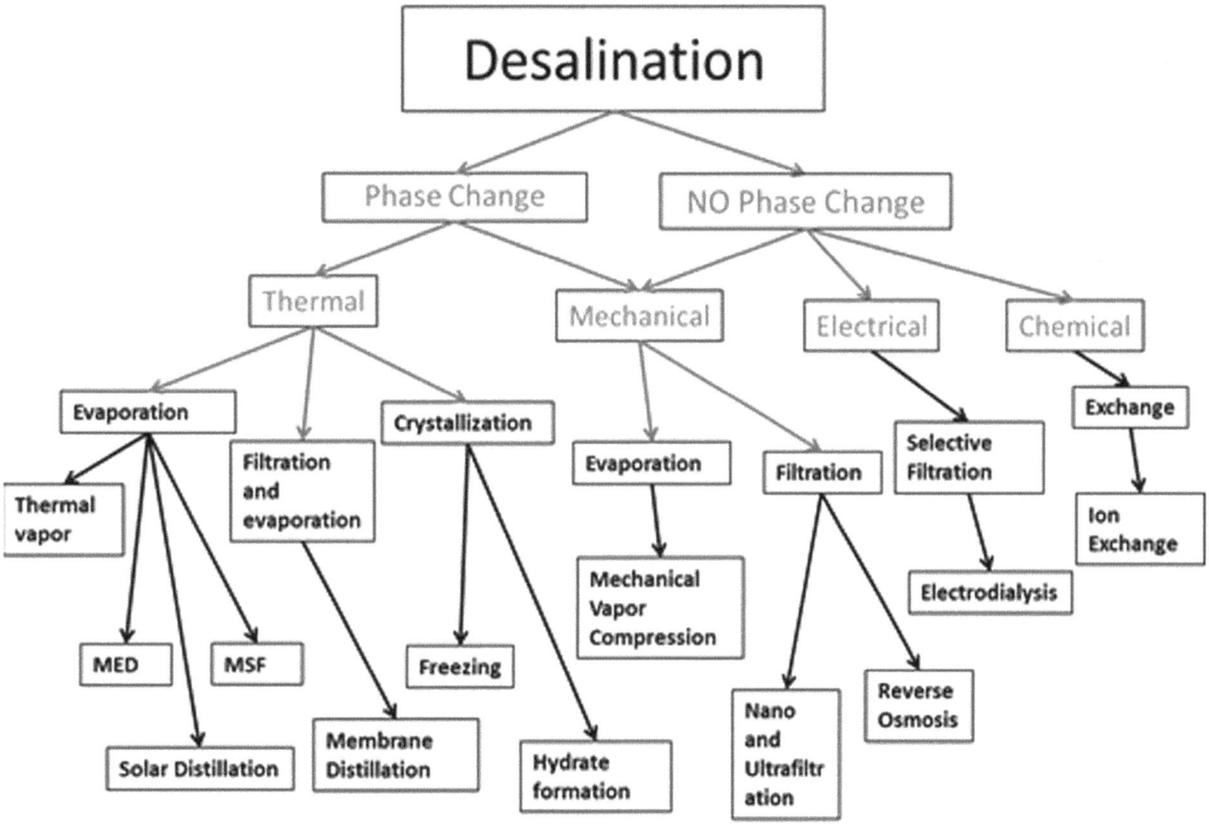

FIGURE 23.1 Schematic processes for desalination.

TABLE 23.1
NO-membrane-based processes for water recovery

Process	Setup	Applications	Information	References
Multi-effect distillation (MED)	MED unit consists of maintained low pressure where saline water is sprayed	Seawater	from 91,000 m^3/day to 320,000 m^3/day	[13,14]
Multi-stage flash (MSF)	The process is composed of a series of elements, called *stages*, where condensing steam is used to preheat the seawater feed. By splitting the overall temperature difference between the hot source and seawater into a large number of stages, the system approaches ideal total latent heat recovery	Seawater or recycled brine	heat input and electric power required of 3.5–4.5 kWh/m^3	[15–17]
Mechanical vapor compression (MVC)	The vapors separated in the evaporation chamber are brought to higher pressure by the compressor, driven by the engine	Production of fresh water for small- to medium-scale purposes such as resorts, industries, and petroleum drilling sites	Capacity of 10,240 day/m^3	[18–22]
Solar desalination	The sun provides the energy to evaporate the saline water	Generally used for small-scale operations	The salt and un-evaporated water left behind in the still basin forms the brine solution that must be discarded appropriately	[23,24]
Thermal vapor compression (TCV)	A technology based on multiple effect distillation but by using thermal compressors (or thermocompressors) as a source of thermal energy	Used on an industrial scale for seawater desalination	Desalination capacity can be much greater by allowing greater adaptability for the input from steam production plants	[25–28]
Desalination by ion exchange	Obtained by removing Na^+ and Cl^- ions from resins, respectively, in the H^+ and OH^- cycle	Used for small and very small plants with very high purities of water produced	1 m^3/h maximum	[29–32]

for desalination of seawater. In general, these processes are characterized by high management costs.

On the contrary, membrane technologies represent interesting solutions to the production of freshwater and at the same time reduce costs and have a low environmental impact. In fact, the growing global demand for water makes membrane processes the principal source of water from desalination and wastewater treatment. Moreover, thermal desalination is generally more cost-intensive than RO desalination, although the cost of desalination has been declining over time. In Table 23.2, the capacity and energy consumption of the main processes utilized in water recovery are reported.

The results achieved to date confirm the potential of micro- and ultrafiltration in the removal of suspended solids and colloidal species. Nanofiltration is used to reduce water hardness and limit fouling in subsequent distillation processes [43–45].

Figure 23.2 displays the percentage of the different membrane-based technologies used in desalination processes. Reverse osmosis, over the last few decades, has become successful because it has the highest water recovery factor with respect to any other conventional distillation process [46,47]. Obviously, the membrane characteristics influence the performance of the process and also the temperature and concentration of feed.

Membrane techniques are often evolutions or hybridizations of the main techniques. This category includes direct osmosis (FO), nanofiltration (NF), freeze desalination (FD), and hybrid approaches.

In accordance with the "process intensification" strategy, the future will focus on greater production capacity, energy and raw material savings, an increase in plant safety, an improvement in automation and control devices, and a reduction in the overall dimensions of equipment and costs, to mitigate the environmental impact. Potentially, membrane operations have all the characteristics mentioned. This potentially makes them the technologies of the future by replacing conventional energy-intensive techniques, enabling selective transport, and increasing the efficiency of numerous processes. Furthermore, the creation of compact membrane systems, capable of carrying out the operations of traditional process units, today represents a real prospect.

Another challenge is membrane performance improvements in various membrane-based applications. For example, numerous studies have been conducted to improve membrane performance and stability in FO and PRO applications. However, progress is still not good enough for commercialization. In other words, a suitable membrane should be manufactured and developed for its commercial applications. Furthermore, the world is also facing challenges in terms of environmental protection, water demand, and energy regeneration. Desalination technology will enable fresh water production and energy production in one application within decades. At present, there are some limitations, but these issues can be resolved as science and engineering evolve. The first step is to realize an increase in the total supply of freshwater for future generations, harvest valuable resources, and generate energy with a simply designed desalination application. The requirements of process intensification are well satisfied even when membrane engineering is applied to the agro-food industry (fruit juice treatment) and a similar concept is now also being investigated in the petrochemical industry. In the production of ethylene via thermal cracking, membranes are proposed for the separation of gases, the production of oxygen-enriched air, the removal of hydrocarbons and acid gases from wastewater and furnace effluents, and the elimination of coke from the water for microfiltration [12,48,49].

23.2.1 Reverse Osmosis (RO)

Growing global demand for water makes membrane filtration the prominent technology in desalination and wastewater treatment; the global cumulative contracted capacity is dominated by seawater reverse osmosis (SWRO) [50], with an increase of 6.8% per year in the last decade, equivalent to an annual addition in fresh water production of 4.6 million m^3/day. Membrane desalination technologies account for more than 90% of all desalination plants [51]. RO uses semipermeable membranes with excellent separation performance and good chemical stability [52]. The water to be treated is pushed into the membrane module by a pump, which exerts a pressure higher than the osmotic pressure of the feed water in order to allow pure water to pass through the membrane, while the remaining part comes out with a high salt concentration, due to the retention of all the components that do not pass the membrane. The separation takes place thanks to diffusion and dissolution mechanisms, which intervene in varying degrees and allow action up to ionic level. The performance of the membrane depends on the membrane structure, membrane material, as well as on the temperature and concentration of the feed. As a matter of fact, feed osmotic pressure increases with increasing feed temperature and concentration parameters as indicated by the following van‘t Hoff’s law (valid for dilute solutions):

$$\pi_s = \frac{n_s}{V} RTi$$

Where π_s is the osmotic pressure, n_s is the total amount of moles of solutes in solution, R is the ideal gas constant, V is the volume of solvent and i is van‘t Hoff’s coefficient.

Current SWRO plants consume 3–4 kWh/m^3 and emit 1.4–3.6 kg of CO_2 per cubic meter of water produced [53,54]. This is highly dependent on the fuel used to generate electricity. Less efficient desalination technology (as the thermal processes) typically emits 8–20 kg CO_2/m^3, with the exception of stand-alone MEDs at 3.4 kg CO_2/m^3. These numbers may be small when viewed through a global lens, but they can be large in local networks and ecosystems. In terms of cost, energy consumption is one of the major cost components of RO desalination [55,56], while concentration polarization and fouling are the main problems of this process. Concentration polarization is the result of selective transport of some species across membranes. Retained species accumulate in front of the membrane and can cause the creation of a concentration gradient between the solution at the membrane surface and the bulk. This leads to back-transport of the material accumulated at the membrane surface by diffusion. A direct consequence of concentration polarization is the reduction of both water flux and rejection. Membrane fouling is due to the dissolved, colloidal, or biological matter that can accumulate at the membrane surface, building a continuous layer that reduces or inhibits mass transfer across the membrane. For efficient RO desalination, adequate pretreatment, supplying high-quality feed water is essential. Notable examples of very productive and large seawater reverse osmosis (SWRO) desalination plants are those in Israel (such as the Sorek SWRO desalination plant), United States (such as the Carlsbad Desalination SWRO Plant in San Diego County), Oman (such as the Al Ghubrah plant or the Barka IWPP expansion—both SWRO), and the United Arab Emirates (for example, the Al Fujairah IWPP expansion). Although membrane desalination technology has been very successful, there remain significant challenges in terms of desalinated water cost, increased productivity (i.e., increased water recovery), improved water quality, and improved environmental sustainability of the desalination process. The further improvement of SWRO desalination processes requires high-permeability and/or antifouling membranes. Recently, the application of nanotechnology and biotechnology to membrane fabrication has heralded a new generation of RO membranes, of which the water permeabilities potentially surpass conventional polymeric membranes by several orders of magnitude. Examples can be found in carbon nanotubes (CNTs) and other carbon-based membranes (like graphene and graphene oxide), as well as

TABLE 23.2
Operating conditions, capacity, and energy consumption of the main process for water recovery

Process	Energy consumption (kWh/m³)	Operating condition	Feed	Separation process	Membrane type	Capacity	References
MSF	14–25	90–110°C	Seawater	Thermal process	–	< 20 million (m³/day) (cumulative in the world)	[33,34]
MED	7–25	60–75°C	Seawater (limit 70 g/L)	Thermal process	–	< 10 million (m³/day) (cumulative in the world)	[33,35]
BWRO SWRO	0.5–3 3–6	17–20 bar 52–69 bar	Brackish water Seawater (limit 70 g/L)	Pressure driven	Polyamide, cellulose acetate	> 40 million (m³/day) (cumulative in the world)	[33,36,37]
MD	671–699	Feed temperature = 40–80°C; possibility to work with solar energy	Seawater; brackish water. reclamation of wastewater and the treatment of various industrial wastewaters	Difference of vapor pressure	Hydrophobic porous membranes made of polymeric or ceramic materials	120 (kg/day)	[38]
ED	0.49–1.75 (brackish) 10–25 (seawater)	18–22°C	Brackish water with a salinity of 1000–5000 mg/L; seawater	Non-pressure driven	Electricity and specialized membranes to separate ionic substances	350 (kg/day)	[39,40]
NF		Feed pH > 1–3, feed pH < 9–11, and a maximum operating temperature commonly between 40–50°C	River water; brackish water or inland water; seawater; brine or concentrated seawater	Pressure-driven separation processes to remove most of the suspended or undissolved ingredients like suspended solid, inorganic, and organic compounds	Hydrophilic porous membrane made of polymeric or ceramic materials	10 (kg/day)	[41]
FO	0.11	high temperature not necessary	Seawater, MBR, and conventional activated sludge (CAS) effluent, sewage, and landfill leachate	Difference in chemical potential of water between two solutions separated by a semipermeable membrane	Dense polymeric membranes	n.a.	[42]

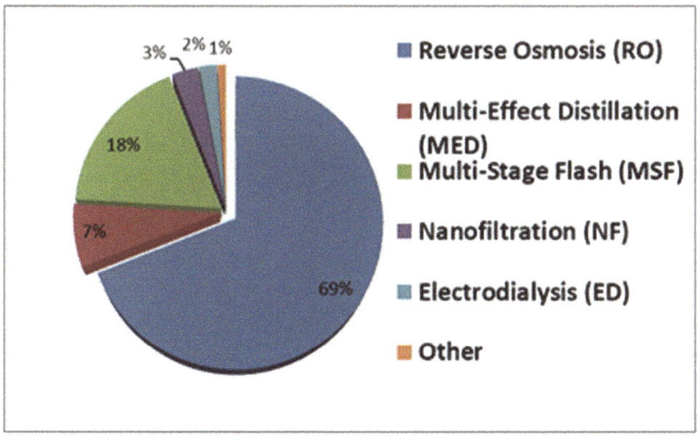

FIGURE 23.2 Percentage of the most desalination techniques used based on the productivity of fresh water in 2019. Reprinted with permission from Ref. [5].

in inorganic membranes, mixed-matrix membranes, and biomimetic membranes. These are emerging as membranes with superior permeability, durability, and selectivity, in particular for water purification.

23.2.2 Nanofiltration (NF)

Nanofiltration is a membrane process for the removal of divalent ions. This implies that the separation process is based on the charge of the solute and the molecular size of the unfilled solute. The selectivity of the NF membrane for solute rejection depends on the charge and size of the solute. This aspect differentiates NF from RO in which all solutes are rejected, also requiring a lower working pressure than RO. However, NF is not suitable for seawater desalination, but can preferably be used as a pretreatment unit in a hybrid method. NF membrane devices are specifically designed to eliminate viruses from contaminated sources. In most cases, the working parameters such as flow rate, temperature, membrane nature, virus load, and filter area are optimized for efficient and reproducible elimination of the virus in question. Furthermore, the exclusion of viruses from contaminated water based on existing techniques other than NF membrane techniques can be costly and time-consuming. The main applications of NF are in water treatment for the production of drinking water, as well as in wastewater treatment and also in reuse. NF can be used to treat all types of water including groundwater, surface water, and wastewater, or as a pretreatment for desalination.

23.2.3 Electrodialysis (ED)

Electrodialysis (ED) and electrodialysis reversal (EDR) are driven by a direct current (DC) in which ions flow through an ion-selective membrane to the oppositely charged electrode. Similarly to the electrolysis principle, EDs have different chambers for positive and negative electrodes [40]. This is the opposite compared to water in pressure-driven processes, as outlined above. The EDR system periodically reverses the polarity of the electrodes. Ion-transfer (perm-selective) anion and cation membranes separate the ions in the feed water. These systems are used primarily in waters with low total dissolved solids (TDS) content. Desalination through ED occurs via a direct current from an external source applied through the electrodes in a brine solution containing ion-selective membranes, which are connected in parallel to form channels. When the electrodes are charged, the negative salt ions move through the anion-permeable membrane towards the anode and vice versa the positive salt ions move the other way through the cation-permeable membrane to the cathode. These movements of ions create the separation of concentrated brine, allowing the formation of fresh water. In general ED is used for desalination of brackish water with low concentration of salt (with TDS <5000 mg/L) operating at 85–90% recovery. The largest EDR plant is in Barcelona, Spain, which produces 200,000 m^3/day of drinking water.

To calculate the energy consumption for an ED plant, it is fundamental to consider two components: the energy for separation and the energy for pumping. The first can be approximated as requiring 3.59 kWh/m^3 per 1000 ppm of salt removed. This energy requirement varies significantly with temperature and is stated reflecting typical ambient conditions of 18–22°C. The energy for pumping is approximately 7.58 kWh/m^3 of water produced. The energy consumption of EDR at a 75% recovery rate and 25°C has been shown to vary between 0.49 kWh/m^3 at 1000 mg/L TDS and 1.75 kWh/m^3 at 5000 mg/L TDS [57].

23.2.4 Forward Osmosis (FO)

Forward osmosis (FO) is the opposite process to reverse osmosis, which uses the natural osmotic pressure existing across the membrane to draw water from the less concentrated side of the membrane into the more concentrated solutes. Compared to RO, which forces brine through a membrane, forward osmosis (FO) instead uses the osmotic pressure generated by the natural salt gradient as the driving force across the membrane. In this method, one side has the feed water (often seawater) and the other has the membrane and drawing solution. The diluted solution is then processed to separate the product from the reusable draw solution. The desalination process when using FO is completed simply by adding a desalination unit. Therefore, FO is optimally used as a pretreatment device for other desalination methods such as RO. Hybrid FO systems are best suited for desalination of salty feed water because there is an RO difference that makes them unsuitable for desalination of salty water. Moreover, the hybrid FO process consumes less energy than RO. Other advantages of FO over RO are high water recovery and minimal fouling [58]. The main limitations of FO are its high energy requirements when used as a stand-alone desalination method and its limited range of solutes.

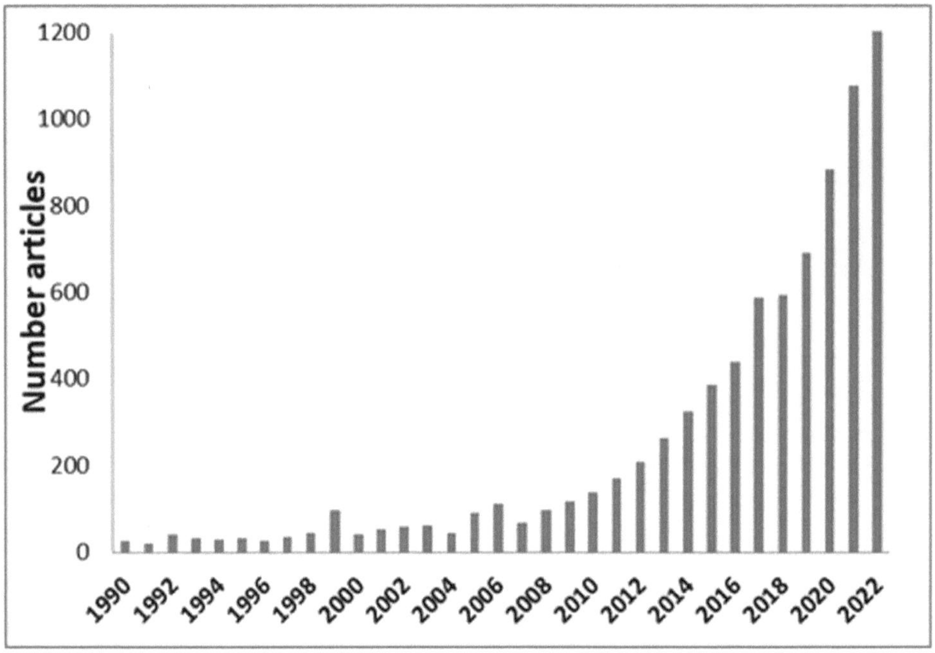

FIGURE 23.3 Number of membrane process articles in the last 30 years through September 2022 (ScienceDirect).

23.2.5 Membrane Distillation (MD)

Membrane distillation (MD) is a membrane contactor-based process dedicated to water treatment, including water desalination. In comparison with the other technologies previously described, MD is a relatively new process. However, over the years, MD has received constantly growing interest across the world. As shown in Figure 23.3, the number of publications in the early 1990s was only a few tens, while in 2020 the number of publications exceeded 800, confirming the growing interest in this new eco-sustainable process for water desalination.

Freshwater can be recovered from saline streams with a theoretical salts and non-volatile components rejection rate of 100%. MD requires low operating temperatures with respect to conventional distillation columns, reduced concentration polarization with respect to pressure-driven membrane processes, and low operating pressure with respect to reverse osmosis (RO). This process mainly involves two-phase changes of water: evaporation and condensation. The first occurs at the liquid–vapor interface (i.e., feed side); successively, the water vapor passes through non-wetted pores of hydrophobic membrane and then condenses om the opposite side of the membrane (i.e. the permeate side).

The success of the separation depends on the capacity of the membrane to keep a stable and durable interface between two phases and subsequently constant mass transfer. The transmembrane flux is measured according to the following equation:

$$J = B * (P_F - P_p) = B\Delta P$$

$$\mathrm{B} \propto (r^a \varepsilon)/(\delta\tau)$$

where *B* is a membrane distillation coefficient, which depends on the membrane morphology, *r* is the nominal pore size of the membrane, *a* is the exponent of *r* in the range of 1–2 (a = 1 for Knudsen diffusion, a = 2 for viscous flux), τ is the pore tortuosity, and δ is the membrane thickness.

The understanding of heat and mass transfer phenomena in MD allows identifying membrane characteristics for enhancing the MD performance.

23.3 RECOVERY OF RAW MATERIALS

Various minerals have been mined since ancient times. Historical sources have shown that sodium chloride was extracted from seawater even before 2000 BCE [59]. Magnesium was extracted from seawater by precipitation of magnesium as magnesium hydroxide through the addition of lime or dolime [60]. The recovery of precious metals [e.g., gold (Au), uranium (U), lithium (Li), bromine (Br), etc.] from seawater and brine were widespread practices at that time [5]. Nowadays, salts present in higher concentrations such as sodium (Na), magnesium (Mg), calcium (Ca), and potassium (K) are commercially extracted from seawater. On the contrary, less attention has been paid to the recovery of elements that are present in low concentration due to economic and operational challenges. One study estimated that the successful recovery of Na, Mg, Ca, and K from desalination of brine can lead to the generation of approximately US$ 18 billion/year [61]. The main factors to take into consideration are the concentration of salts contained and the potential market value they may have. The operating cost of extracting salts from seawater can depend on their concentration. The recovery or extraction of minerals with relatively higher concentrations (> 1 mg/L) in seawater will be more economically feasible. In fact, most of Li, U, and Rb in seawater range from 0.003 to 0.17 mg/L, making it difficult to selectively remove them in the presence of other dominant ions present in higher concentrations, such as Ca, Mg, and K [62]. Furthermore, although the concentrations of salts in seawater are relatively low, their total quantity in the oceans is enormous, which makes them interesting from a point of view of potential applications. Elements potentially attractive for extraction include Na, Ca, Mg, K, Li, Sr, Br, B, I, and U. However, the main challenge is to develop more efficient, easier, and cheaper extraction techniques [6]. Deposits of pegmatite granites, brine, seawater, and hot springs are the main lithium resources in the world. Since the use of lithium and its compounds continues to increase in various fields, lithium mining takes on an increasingly relevant and broader concept in the field of research [63]. Among the main sources of lithium extraction, the work of literature concerning the extraction from the salt lake of Cha'erhan (Qinghai, China) has been reported. The ratio of lithium to magnesium is a crucial factor in the recovery of lithium from brines. The main methods for lithium extraction from the low mass ratio of Li^{+}/Mg^{2+} brine are solvent extraction, ion exchange, precipitation, and calcination-leaching. In the case of rare earth elements (REEs), different techniques have been developed for more efficient extraction of materials from the environment [64]. The principal methods are precipitation, filtration, adsorption, ion exchange, and solvent extraction, which in general shows a higher efficiency and lower cost for large-scale production [65]. Solvent extraction could be introduced to treat in situ leaching solutions of ion-adsorption type rare earth ores; it may exhibit many potential advantages over traditional precipitation methods. The recovery of salts and minerals has shifted the interest of researchers and scientists toward a mentality of sustainability and reduction of environmental impacts. Hence the ZLD (zero liquid discharge) systems were introduced. Originally, ZLD systems were designed to recover as much fresh water as possible and eliminate waste [66]. The technique achieves three key objectives: (1) to reduce the expense and need for brine disposal; (2) allows the marketing of salts produced according to a circular economy policy; and (3) allows the solid salts to be used internally by the company. In the MEDINA project, the improvement and design of membrane-based desalination plants was already developed [67]. The project team sought to solve, or at least mitigate, several critical problems in seawater and brackish water desalination systems, including: to solve and/or mitigate these problems, approaches based on the integration of various operations in the reverse osmosis pretreatment stage have been proposed and investigated. In particular, MCr has been investigated as a technique to improve the productivity of desalination systems, recover some of the valuable ions present in high-concentration streams of desalination plants, and reduce the environmental impact. In fact, as reported above, it was estimated that an integrated membrane-based system with MCr units working on NF and RO retentate streams could increase the plant recovery rate to 92.4% [11]. From an energy point of view, MCr introduces, with respect to an RO desalination process, a thermal energy requirement. However, the possibility to work at relatively low temperature (30–90°C) provides the opportunity to utilize waste heat or other sustainable energy resources (such as geothermal or solar energy), thus lowering energy consumption to 1.61 kWh/m^3 [68]. Other examples can be found in the Megaton project in Japan and the SEAHERO project in South Korea [67,69]. In the first part of both projects the emphasis was mainly on increasing the desalination capacity. However, the second part of the project addressed the issue of brine treatment. Hybrid systems with MD and PRO units are proposed for the extraction of valuable resources from brine, minimization of the environmental impact of brine, and the recovery of energy. Moreover, the SEAHERO project has suggested a hybrid forward osmosis/reverse osmosis system for increasing the recovery factor by 30% and thereby reducing the brine volume. Another interesting example of the rapidly growing interest toward minerals/water/energy production from the sea can be found in the 5-year global MVP research program (2013–2018) in Korea. Hybrid RO technology processes with emerging technologies such as FO–RO, RO–PRO, and RO–MD developed in the project can not only solve brine problems, but also create new desalination markets [70].

23.4 MEMBRANE CRYSTALLIZATION PROCESS

Membrane crystallization (MCr) is emerging as an interesting technique for the recovery of raw materials from high-concentrate brine. MCr is an extension of MD where the continuous evaporation of volatile components from the feed (with a high concentration of solute) generates supersaturation of the solute. In Figure 23.4, a schematic representation of membrane crystallization set-up with the separation process is shown. The technical feasibility of the MCr process for recovery of salts from seawater brine is well acknowledged in the literature [71–73]. MCr also has the potential to reach the industry's goal of zero liquid discharge.

In MCr, control of the crystallization process is critical to minimize crystal deposition and growth on the membrane and to maximize crystal formation and removal in the crystallizer. Depending on the chemical-physical properties of the membrane and the process parameters (temperature, concentration, flow rate, etc.), the evaporation rate of the solvent and therefore the degree of supersaturation and supersaturation rate can be controlled very precisely. The effect of control of the nucleation and growth rates is by selecting a wide range of kinetic trajectories available in the thermodynamic phase diagram that lead to the formation of specific crystals (that are not readily achievable by conventional crystallization methods). The surface structure of the membrane yields heterogeneous nucleation because it may trap solute molecules in its cavities, thus leading to localized supersaturation, which assists nucleation and crystal formation at supersaturation conditions [75–77]. Traditionally,

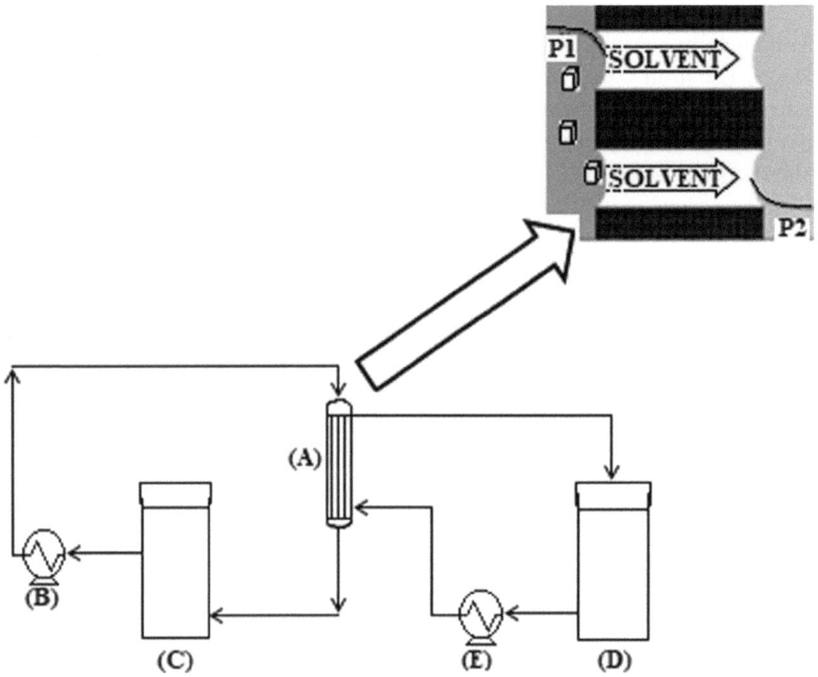

FIGURE 23.4 Schematic representation of a DCMCr unit comprising: (A) membrane module, (B) pump, (C) crystallizer tank and (D) permeate tank, [74] (open access).

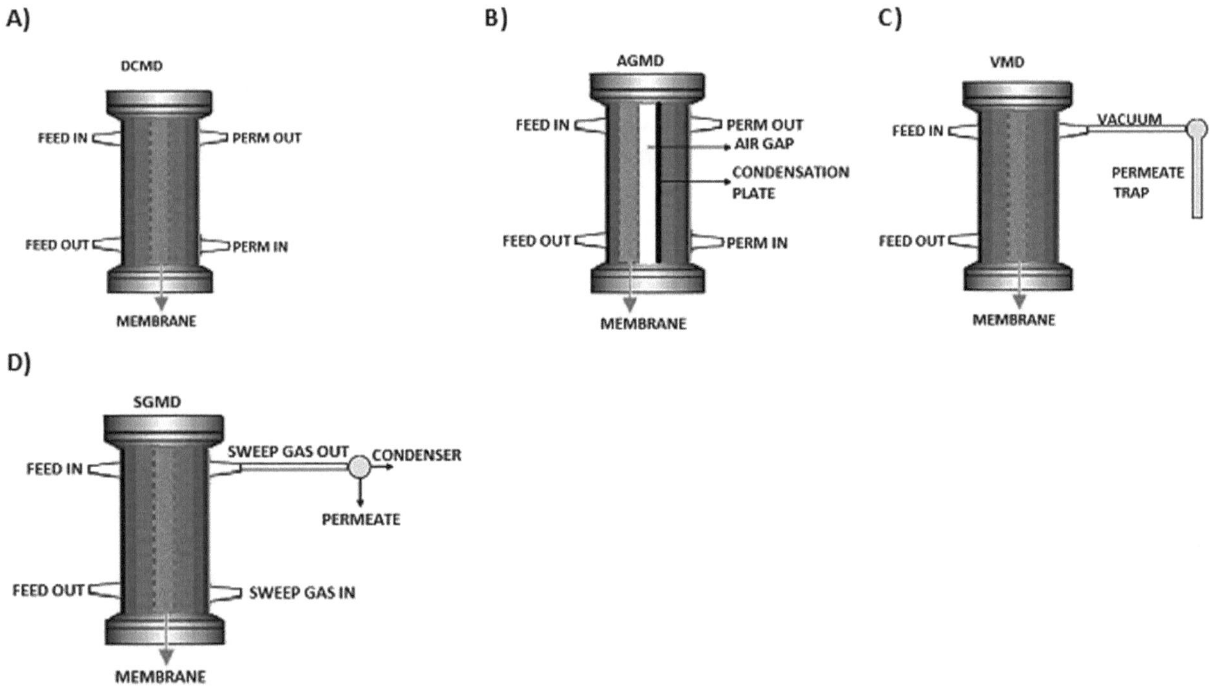

FIGURE 23.5 MD configurations schematic setup: (A) direct contact membrane distillation (DCMD), (B) air gap membrane distillation (AGMD), (C) vacuum membrane distillation (VMD), and (D) sweep gas membrane distillation (SGMD).

crystallization is a separation process, based on the limited solubility of a compound in a solvent at a certain temperature, pressure, etc. A change of these conditions to a state where the solubility is lower will lead to the formation of a crystalline solid. Although crystallization has been applied for thousands of years in the production of salt and sugar, many phenomena that occur during crystallization are still poorly understood. Above all, the nucleation and growth mechanisms of crystals and the complex behavior of industrial crystallizers remain elusive. One reason for this is the lack of adequate tools to

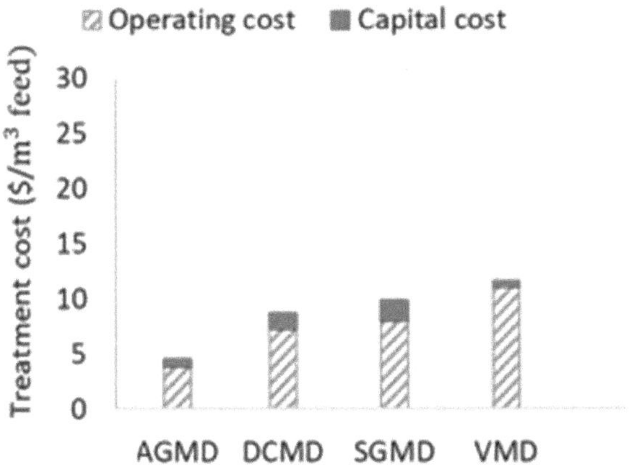

FIGURE 23.6 Treatment cost, for the main MD configurations, [81] (pen access).

measure and monitor crystallization processes. On the other hand, the demands for constant product quality (purity, crystal size, etc.) are continuously increasing, thus creating great interest in crystallization research. Membrane-assisted crystallization has been evaluated to provide important advantages against traditional crystallization due to easy scalability and good control of crystal nucleation growth [78]. This process arises, in fact, from the need to produce substances in the solid crystalline state required in various sectors of industry, technology, and scientific research. The solid crystalline state makes these products more stable for storage and more functional to manage by users, where morphology is the dominant feature with consequent safe interest in countless applications [46,79].

23.4.1 Membrane Crystallization Configuration

Principally, MCr can be operated in all the MD configurations. The main MD configurations utilized, as shown in Figure 23.5, are: direct contact, vacuum, sweep gas, and air gap MD [80]. Depending on the approach used to induce water evaporation and condensation, different MD basic configurations can be assembled. A brief description of the most traditional and recent configuration types is given below.

A thorough analysis based on optimization and comparison of the main MD configurations has been carried out by Khanna et al. [81] in continuous recirculation mode showing that all configurations had low recoveries. Among the configurations, AGMD showed the highest recovery rate of 6.5% when operating near saturation. As a result, the salinity of the feed stream entering the membrane module approaches the salinity of the waste brine stream. Except for VMD, the sensitivity analysis of the waste brine salinity revealed that operating at a higher waste brine salinity (corresponding to higher recoveries) increases the energy intensity of all configurations. This is due to increased vapor pressure lowering at higher salinity, which has a direct impact on permeate production and heat recovery within the system. This leads to a higher energy intensity regardless of the salinity of the brine. However, regardless of the approach, membrane crystallization results in lower energy consumption than conventional evaporator assembly [82]. Figure 23.6 shows a comparison of the produced water treatment cost obtained from the optimization model for the main MD configurations. AGMD outperforms all other configurations with $4.57/m^3 feed treatment cost followed by DCMD, SGMD, and VMD, with 8.7, 9.8, and 11.63 $/m^3 feed, respectively. The operating cost accounts for the majority of the total treatment cost and is the primary reason for treatment cost differences across configurations [81].

Membrane-assisted crystallization, being an extension of MD, can be performed utilizing the same MD configurations:

1. *Direct contact membrane crystallization* (DCMCr) is obtained when the saline solution faces the feed side, while a cold, pure water solution is circulated tangentially to the permeate side of the membrane. A difference of temperature or concentration across the membrane induces the necessary vapor pressure difference to promote transport through the membrane [83].
2. In *air gap membrane crystallization* (AGMCr), an air gap is interposed between the membrane and a cold condensing surface placed inside the membrane module. The air gap helps in increasing the conductive heat transfer resistance, thus decreasing the amount of heat lost by conduction through the membrane [84,85]. The heat loss is one of the most critical issues of membrane distillation, as detailed hereafter.
3. *Vacuum membrane crystallization* (VMCr) requires the application of a vacuum or low pressure at the permeate side and an external condenser in order to condense the vapor water and collect the permeate [86–88].
4. In *sweep gas membrane crystallization* (SGMCr), a gas, such as air or nitrogen, sweeps the permeate side of the membrane carrying the evaporated molecules outside the membrane module for condensation in place of the vacuum used in VMCr [43,87,89,90].

The membranes used in MD/MCr applications can be made from a combination of both polymeric or inorganic materials, or hybrid or composite configurations. Hollow-fiber and flat membranes can be used equally well. When the membrane is prevented from being wetted by adjacent solutions, no direct mass transfer is observed in the liquid phase through its porous structure, but the two subsystems in contact undergo mass transfer in the gas phase. A vapor pressure gradient between the two subsystems causes evaporation of volatile components from the feed solution, migration across the porous membrane, and finally re-condensation on the distillate side (Figure 23.7a). Continuous removal of solvent from the feed solution in a film crystallizer increases the concentration of solutes and causes supersaturation. Given the process design, the membrane can be used for membrane-assisted operation (that is, mixing recirculation loops) or directly for in situ crystallization purposes. The first case can be seen as a typical

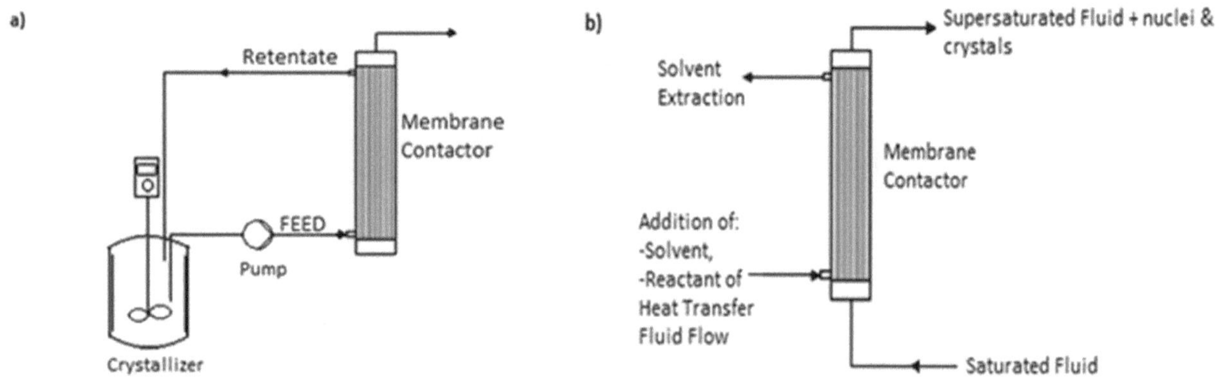

FIGURE 23.7 Schematic representation of the two process designs: (a) hybrid membrane crystallization process; (b) the crystallization takes place directly in the membrane module. Reprinted with permission from Ref. [12].

hybrid process approach, and is shown in Figure 23.7a. The membrane module is here used to generate supersaturation, or simply to concentrate the solid phase, but nucleation and crystal growth take place in the crystallizer (e.g. [9,91]). In the second case (Figure 23.7b), the crystallization takes place directly in the membrane module where the supersaturation is generated [92–94].

Recently, Di Profio et al. [95] proposed a new MCr design in which crystallization is induced by using antisolvent. This new approach operates in two configurations: first, solvent/antisolvent demixing (Figure 23.8b) and second, antisolvent addition (Figure 23.8b and c). In both cases, according to the general concept of membrane crystallization, solvent/antisolvent transport occurs in the vapor phase and is not driven through the membrane in the liquid phase, in contrast to the configuration described above. Selective and precise dosing of antisolvent controlled by the porous membrane allows finer control of solution composition during the process and at the nucleation point, resulting in improved final crystal properties. According to the above considerations, crystallization using membranes can be classified depending on the different working principles, such as [96]:

1. A process based on membrane distillation/osmotic distillation where the diffusion of solvent molecules in the gas phases through the porous membrane driven by the chemical potential gradient results in supersaturation in the crystallization solution.
2. Membrane-assisted crystallization in which pressure-driven membrane operations (MF, NF, and RO) are used to concentrate the solution by liquid-phase solvent removal and collect the crystals in a separate tank. It is often operated with seeds at low temperatures;
3. Solid (non-porous) hollow fibers used as heat exchangers to create supersaturation by cooling;
4. Antisolvent (or crystallization solution) extruded in liquid state directly through the pores of the membrane under a pressure gradient onto the crystallization solution (or antisolvent);

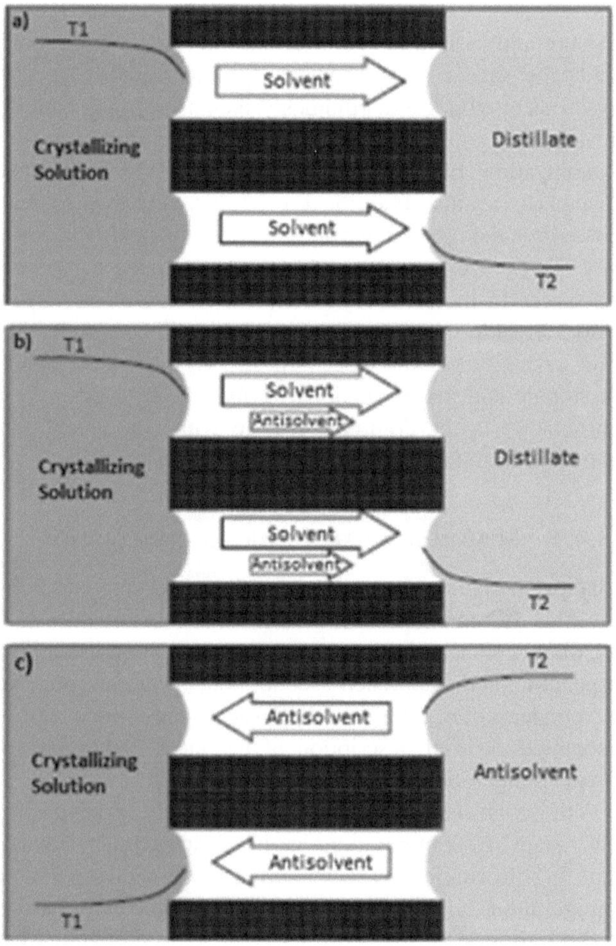

FIGURE 23.8 Basic principle of a membrane crystallizer: (a) solvent removal MCr, where solvent is removed from the crystallizing solution under a temperature gradient (T1 > T2); (b) solvent/antisolvent demixing MCr, in which the preferential evaporation of the solvent induces the increase of the antisolvent volume fraction thus reducing solubility (T1 > T2); (c) antisolvent addition MCr, where an antisolvent is evaporated into the crystallizing solution in vapor phase from the other side of the membrane (T1 < T2). Reprinted with permission from Ref. [97].

5. Antisolvent film crystallization, where the dosing of the antisolvent to the crystallization solution is performed by a membrane according to the operating principle of point 1, in two configurations of solvent/antisolvent separation and addition of anti-solvent.

Energy consumption of crystallization with antisolvent addition was studied by Das et al. [82]. They found that the addition of antisolvent reduces energy consumption by 36% in comparison to conventional evaporative crystallization.

23.4.2 The Nucleation and Crystal Growth Process in the Crystallizer

As mentioned, crystal nucleation and growth in the feed solution are induced by the supersaturation of salts. In general, during crystallization various phenomena such as nucleation, growth, agglomeration, breaking, and even dissolution of crystals can be observed [98]. Detailed relations and models used to explain the heat and mass transport through the membrane in MCr can be described by using the same concepts developed for membrane distillation [99–103]. As a general description, heat and mass transport through membranes occurs only if the overall system is not in thermodynamic equilibrium. For what concerns the mass transport, it can be separated into three steps: mass transfer in the feed boundary layer, mass transfer *through the membrane pores*, and mass transfer in permeate boundary layers. The mass transfer in the permeate boundary layer is not considered since the mole fraction of the transporting species in the permeate stream is approximately equal to one. The mass transfer in boundary layers is analyzed by film theory, whereas the dusty gas model (DGM) is usually employed to describe mass transfer across the membrane. The dusty gas model elucidates mass transfer in porous media by four possible mechanisms: viscous flow, Knudsen diffusion, molecular diffusion, and surface diffusion. It is typical for membrane crystallization in direct contact configuration to neglect surface diffusion and viscous flow, and to employ a Knudsen-molecular diffusion transition model [99,104]:

$$N' = \frac{\varepsilon P D_{ij}}{\tau \delta R T} ln \left(\frac{p_a^{2} \frac{2r}{3}\left(\frac{8RT}{\pi M_i}\right)^{1/2} + PD_{ij}}{p_a^{1} \frac{2r}{3}\left(\frac{8RT}{\pi M_i}\right)^{1/2} + PD_{ij}} \right)$$

where N' is the molar transmembrane flux, ε is the porosity, P is the total pressure, D_{ij} is the diffusivity, τ is membrane tortuosity, δ is membrane thickness, R is ideal gas constant, T is temperature, r is pore size, M_i is molecular weight, and p_a^1 and p_a^2 are the partial pressure of air at the feed and membrane surface, respectively. The model can be further simplified in specific cases. The Knudsen diffusion model is suitable for the system in which the collision between molecule and pore wall dominates the mass transport. On the other hand, the molecular diffusion model is preferred when the collision between the molecules plays the main role in the mass transfer across the membrane. Nevertheless, if both molecule–pore wall and molecule–molecule collisions occur frequently, the Knudsen-molecular diffusion transition model must be employed. Both the molecular diffusion limit and Knudsen-molecular diffusion transition model were successfully applied to describe the flux in a DCMD system [100,105–107]. In both cases the transmembrane flux is proportional to membrane porosity ε, whereas it is inversely proportional to membrane thickness, δ. Therefore, membrane structural properties will greatly affect the membrane crystallization performance both in terms of solvent evaporation rate and crystal nucleation and growth. As a matter of fact, a crystallizing solution can be imagined as a certain number of solute molecules moving among the molecules of solvent and colliding with each other, so that a number of them merge, forming a cluster. The critical size n^* that an assembly of molecules must have in order to be stabilized by further growth is given by [108],

$$n^* = \frac{32\pi v_0 \gamma^3}{3\left(k_B T\right)^2 ln^3 S}$$

where v_0 is the molecular volume, γ is the interfacial energy, k_B is Boltzmann's constant, and S is the supersaturation. The previous equation correlates the critical size n^* with the supersaturation S. When the supersaturation level is high, the crystal size tends to decrease (typically a few tens of molecules). Therefore, the proper choice of membrane chemical-physical properties and process parameters (temperature, concentration, etc.), allow regulating flowrate, supersaturation degree, supersaturation rate, nucleation, and growth.

Concerning crystal nucleation, most of the crystallization models in MDCr have been developed on the basis of the classical theory of nucleation and growth. The equation for classical homogeneous nucleation (HON) is [109]:

$$B = zf^* c_0 e^{-\frac{W}{K_b T}}$$

where B is the stationary nucleation rate ($m^{-3}s^{-1}$), z is the Zeldovich factor, f^* is the attachment frequency (s^{-1}), W is the work to form a cluster of size n (J), and c_0 is the nucleation site concentration (m^{-3}) [98]. In the specific example, W is defined as:

$$W = -nK_b T lnS + \gamma A_c$$

where A_c is the surface area of the cluster (m^2), γ is the interfacial energy (Jm^{-2}), and S is the supersaturation.

For MCr, not homogeneous but heterogeneous crystallization occurs. In this case, the Randolph-Larson model can be used for estimation of the crystal growth rate (G) and nucleation (D^0) can be estimated according to the following equations [71,110]:

$$\ln(n) = \frac{-L}{Gt} + \ln(n^0)$$

$$D^0 = n^0 G$$

where n is the crystal population density, L is the crystal size, t is retention time, and n^0 is population density at L equal to zero. A plot of $\ln(n)$ versus L is a straight line whose intercept is $\ln(n^0)$ and whose slope is $-1/Gt$. Thus, from a given product sample of known slurry density and retention time, it is possible to obtain the nucleation rate and growth rate for the conditions tested when the sample satisfies the assumptions of the derivation and yields a straight line [111].

23.4.3 Application and Future Directions: New Strategies to Overcome Current Limitations

Membrane crystallization arises from the need to produce substances in the solid-crystalline state required in various sectors of industry, technology, and scientific research. The principal advantages of MCr are control of the crystallization process in order to minimize deposition and growth of crystals on the membrane, and maximizing crystal production by choosing a broad set of available kinetic trajectories in the thermodynamic phase diagram [96]. This can be done through the choice of membranes with specific chemical-physical properties and through the choice of the process parameters, such as temperature, concentration, flow rate, etc., which enable the solvent evaporation rate to be controlled, thereby leading to the production of crystalline morphologies and structures.

In general, the different polymorphic forms of the same substance are considered different materials, with their characteristic physical, chemical, and biological properties, making each of them a different and patentable drug. In the case of proteins, large crystals (100 mm at least in two dimensions) are required with high order in their crystal lattice for determination of the structure at the atomic level by X-ray diffraction analysis [112]. Other, but not less important, applications have been developed for the treatment of wastewater for the recovery of high-purity silver [113] or sodium sulfate [114], CO_2 capture [115,116], nanotechnology, such as the synthesis of $BaSO_4$ and $CaCO_3$ particles [93,117], or the recovery of antibiotics [118] or polystyrene microparticles [119].

The solid-crystalline state is useful for its greater stability for storage and increased functionality for users. For this reason, the solid-crystalline phase is utilized in several fields of industry, technology, and scientific research, or as additives for cosmetics, hygiene and personal care products, pharmaceuticals, fine chemicals, pigments, and foods [95,120]. In the organic semiconductor field, the crystalline phase is utilized for the preparation of single-crystal-based devices and photonic crystals due to experience of wide diffusion in heterogeneous catalysis. MCr could be applied for the preparation of new structure-based drugs in medical devices. These crystal forms are required to obtain the highest surface–volume ratio and greater catalytic efficiency [121,122]. The different polymorphic forms of the same substance, in some cases, are considered different materials, with their characteristic physical, chemical, and biological properties. This aspect is fundamental in the pharmaceutical industry where it is possible to make different and patentable drugs with different polymorphic shapes of a single crystal [123,124].

For proteins, atomic-level structure determination by X-ray diffraction analysis requires large crystals (at least 100 mm in two dimensions) with high crystal lattice order [123,124]. In principle, MCr could not only overcome the limitations of thermal systems, but also those of conventional membrane systems such as RO. In fact, it is possible to achieve high recoveries and high concentrations, as concentration polarization does not significantly affect the driving force of the process. Thanks to its properties, the integration of MC into RO brine offers the potential to produce high-quality solids and controlled properties with significant added value, reducing the traditional problem of brine disposal with the potential to turn it into a new profitable market [75]. In recent literature, MCr was mainly applied in the recovery of a single salt with the main focus on the recovery of Na, Ca, and Mg, with some studies on Rb and Li. Some works have been conducted on mixtures of salts recovered using the MCr process [42]. Li et al. [132] proposed the recovery of Na_2CO_3 and Na_2SO_4 with a combination of membrane crystallization and an intermediate stage for the $CaCO_3$ precipitation. $CaCO_3$ precipitation was possible by adding $Ca(OH)_2$ to the residual mixture solution of Na_2SO_4 and remaining Na_2CO_3.

Since the discovery of MCr, the development of specific membranes for MCr has been limited [133]. The membrane plays a dominant role in MCr technology in comparison to MD as it also provides an interface for nucleation. The physicochemical properties of the membrane material significantly influence crystal formation. Solute molecules interact with chemical functional groups on the membrane polymer chain, resulting in increased heterogeneous nucleation and supersaturation of the solute at the membrane boundary. For example, Edwie et al. [134] developed PVDF hollow-fiber membranes in various configurations and found that a thinner mixed-matrix cell membrane with a small pore size increased the wetting resistance compared to a thicker membrane with large pores and spherical structure. By depositing a coarse layer of PP on the surface of hollow-fiber membranes, Meng et al. [135] fabricated superhydrophobic membranes using fluorosilane functionalization. Despite the increase in the number of studies done, membrane advancement is still lagging behind, hampering the applicability of MCr technology. Advancements in membrane materials and their modifications could significantly improve the overall performance of the process by suppressing scaling and wetting phenomena. As previously discussed, MCr technology has relatively high operating costs associated with it. Such costs could be reduced by supplementing the heating modules with alternative energy sources such as geothermal energy, solar energy, or other low-quality waste heat. Several factors, such as flow rate, inlet temperature, etc., significantly influence energy consumption. However, the energy consumption of MCr is comparable to that of MD, as is shown in Guan et al. [136]. They analyzed the total energy consumption in MCr technology by varying the feed stream and permeate temperatures. They stated that the crystallization subunit uses negligible energy compared to MD technology. In contrast, 97.8% of the energy required in MCr technology is consumed by the heater itself. This implies that the crystallizer has a negligible contribution to the overall MCr energy consumption [133].

In conclusion, the development and application on an industrial scale of membrane crystallization technology requires facing and overcoming the following problems:

- *Development of proper membranes for MCr application*: as for MD, the development of new material/membrane preparation methods with extended life, with stable hydrophobic character and properties tailored to this technology (in terms of pore size, porosity, thickness, thermal conductivity, surface roughness, etc.).
- *Technology transfer to industry*: application and testing of the technology on an industrial scale and valuation of the benefits/advantages/drawback with respect to conventional processes.
- Development of water treatment systems coupled with renewable energy sources in order to allow a significant reduction in energy consumption.
- Recourse to membrane pretreatment in order to (1) decrease costs associated with plant footprint, RO membrane replacement costs, and chemical costs, and (2) increase recovery and permeate flux related to lower fouling rates.
- Enhancement of transport mechanisms and improvement of module design (e.g., further research in the transport properties of polymer-, carbon- and carbon-nanotube-based, zeolite, and mixed-matrix membranes).

REFERENCES

[1] WHO and UNICEF (n.d.) *Progress on household drinking water, sanitation and hygiene, 2000-2020: Five years into the SDGs*. https://data.unicef.org/resources/progress-on-household-drinking-water-sanitation-and-hygiene-2000-2020/.

[2] Boretti, R.L. (2019) Reassessing the projections of the world water development report. *Npj Clean Water* 2: 1–6.

[3] Shannon, M.A., Bohn, P.W., Elimelech, M., Georgiadis, J.G., Marĩas, B.J., Mayes, A.M. (2008) Science and technology for water purification in the coming decades. *Nature* 452: 301–310.

[4] Pinto, F.S., Marques, R.C. (2017) Desalination projects economic feasibility: A standardization of cost determinants. *Renewable and Sustainable Energy Reviews* 78: 904–915.

[5] Jones, E., Qadir, M., van Vliet, M.T.H., Smakhtin, V., mu Kang, S. (2019) The state of desalination and brine production: A global outlook. *Science of the Total Environment* 657: 1343–1356.

[6] Kesieme, U.K., Milne, N., Aral, H., Cheng, C.Y., Duke, M. (2013) Economic analysis of desalination technologies in the context of carbon pricing, and opportunities for membrane distillation. *Desalination* 323: 66–74.

[7] Ali, A., Tufa, R.A., Macedonio, F., Curcio, E., Drioli, E. (2018) Membrane technology in renewable-energy-driven desalination. *Renewable and Sustainable Energy Reviews* 81: 1–21.

[8] Subramani, A., Jacangelo, J.G. (2015) Emerging desalination technologies for water treatment: A critical review. *Water Research* 75: 164–187.

[9] Curcio, E., Criscuoli, A., Drioli, E. (2001) Membrane crystallizers. *Industrial & Engineering Chemical Research* 40: 2679–2684.

[10] Gugliuzza, A., Basile, A. Membrane contactors: Fundamentals, membrane materials and key operations. In: *Handbook of Membrane Reactions*, pp. 54–106 Elsevier.

[11] Quist-Jensen, C., Macedonio, F., Drioli, E. (2016) Integrated membrane desalination systems with membrane crystallization units for resource recovery: A new approach for mining from the sea. *Crystals* 6: 36.

[12] Chabanon, E., Mangin, D., Charcosset, C. (2016) Membranes and crystallization processes: State of the art and prospects. *Journal of Membrane Science* 509: 57–67.

[13] Garg, M.C. (2018) Renewable energy-powered membrane technology: Cost analysis and energy consumption. In: *Current Trends and Future Developments on (Bio-) Membranes: Renewable Energy Integrated with Membrane Operations*, pp. 85–110. Elsevier.

[14] Sharon, H., Reddy, K.S. (2015) A review of solar energy driven desalination technologies. *Renewable and Sustainable Energy Reviews* 41: 1080–1118.

[15] Al-Othman, A., Tawalbeh, M., El Haj Assad, M., Alkayyali, T., Eisa, A. (2018) Novel multi-stage flash (MSF) desalination plant driven by parabolic trough collectors and a solar pond: A simulation study in UAE. *Desalination* 443: 237–244.

[16] Bandi, C.S., Uppaluri, R., Kumar, A. (2016) Global optimization of MSF seawater desalination processes. *Desalination* 394: 30–43.

[17] Mezher, T., Fath, H., Abbas, Z., Khaled, A. (2011) Techno-economic assessment and environmental impacts of desalination technologies. *Desalination* 266: 263–273.

[18] Farahat, M.A., Fath, H.E.S., El-Sharkawy, I.I., Ookawara, S., Ahmed, M. (2021) Energy/exergy analysis of solar driven mechanical vapor compression desalination system with nano-filtration pretreatment. *Desalination* 509: 115078.

[19] Chen, Q., Kum Ja, M., Burhan, M., Akhtar, F.H., Shahzad, M.W., Ybyraiymkul, D., Ng, K.C. (2021) A hybrid indirect evaporative cooling-mechanical vapor compression process for energy-efficient air conditioning. *Energy Conversion and Management* 248: 114798.

[20] Wen, C., Gong, L., Ding, H., Yang, Y. (2020) Steam ejector performance considering phase transition for multi-effect

distillation with thermal vapour compression (MED-TVC) desalination system. *Applied Energy* 279: 115831.

[21] Fan, S., Li, J., Liu, Y., Xiao, Z. (2019) Bioethanol production in membrane distillation bioreactor with permeate fractional condensation and mechanical vapor compression. *Energy Procedia* 158: 21–25.

[22] Wen, C., Ding, H., Yang, Y. (2020) Performance of steam ejector with nonequilibrium condensation for multi-effect distillation with thermal vapour compression (MED-TVC) seawater desalination system. *Desalination* 489: 114531.

[23] Dahayat, A.K., Somwanshi, A., Patel, B. (2022) Economic analysis of conventional single slope solar distillation plant for different climates in India. *Materials Today: Proceedings* 63: 23–34.

[24] Li, C., Zhou, X., Xiao, Y., Zhang, T., Ye, M. (2022) Inhibition of typical phenolic compounds entering into condensed freshwater by surfactants during solar-driven seawater distillation. *Science of the Total Environment* 815: 152694.

[25] Farahat, M.A., Fath, H.E.S., Ahmed, M. (2022) A new standalone single effect thermal vapor compression desalination plant with nano-filtration pretreatment. *Energy Conversion and Management* 252: 115095.

[26] Zhou, S., Liu, X., Zhang, K., Shen, S. (2022) Investigation and optimization for multi-effect evaporation with thermal vapor compression (MEE-TVC) desalination system with various feed preheater arrangements. *Desalination* 521: 115379.

[27] Shahzamanian, B., Varga, S., Soares, J., Palmero-Marrero, A.I., Oliveira, A.C. (2021) Performance evaluation of a variable geometry ejector applied in a multi-effect thermal vapor compression desalination system. *Applied Thermal Engineering* 195: 117177.

[28] Ettouney, H., El-Dessouky, H. (1999) Single-sffect thermal vapor-compression desalination process: Thermal analysis. *Heat Transfer Engineering* 20: 52–68.

[29] Smitha, B., Sridhar, S., Khan, A.A. (2005) Solid polymer electrolyte membranes for fuel cell applications – A review. *Journal of Membrane Science* 256: 10–26.

[30] Mei, Y., Tang, C.Y. (2018) Recent developments and future perspectives of reverse electrodialysis technology: A review. *Desalination* 425: 156–174.

[31] Zhang, C., Song, J., Huang, T., Zheng, H., He, T. (2020) Increasing lithium extraction performance by adding sulfonated poly (ether ether ketone) into block-copolymer ethylene vinyl alcohol membrane. *Journal of Chemical Technology & Biotechnology* 95: 1559–1568.

[32] Xu, T., Huang, C. (2008) Electrodialysis-based separation technologies: A critical review. *AIChE Journal* 54: 3147–3159.

[33] Takabatake, H., Taniguchi, M., Kurihara, M. (2021) Advanced technologies for stabilization and high performance of seawater ro membrane desalination plants. *Membranes (Basel)* 11: 1–20.

[34] Al-Karaghouli, A., Kazmerski, L.L. (2013) Energy consumption and water production cost of conventional and renewable-energy-powered desalination processes. *Renewable and Sustainable Energy Reviews* 24: 343–356.

[35] Panagopoulos, A. (2020) Process simulation and techno-economic assessment of a zero liquid discharge/multi-effect desalination/thermal vapor compression (ZLD/MED/TVC) system. *International Journal of Energy Research* 44: 473–495.

[36] Zhao, D., Lee, L.Y., Ong, S.L., Chowdhury, P., Siah, K.B., Ng, H.y. (2019) Electrodialysis reversal for industrial reverse osmosis brine treatment. *Separation and Purification Technology* 213: 339–347.

[37] Roy, S., Ragunath, S. (2018) Emerging membrane technologies for water and energy sustainability: Future prospects, constraints and challenges. *Energies* 11: 2997.

[38] Miladi, R., Frikha, N., Gabsi, S. (2021) Modeling and energy analysis of a solar thermal vacuum membrane distillation coupled with a liquid ring vacuum pump. *Renewable Energy* 164: 1395–1407.

[39] Mir, N., Bicer, Y. (2021) Integration of electrodialysis with renewable energy sources for sustainable freshwater production: A review. *Journal of Environmental Management* 289: 112496.

[40] Sun, L., Chen, Q., Lu, H., Wang, J., Zhao, J., Li, P. (2020) Electrodialysis with porous membrane for bioproduct separation: Technology, features, and progress. *Food Research International* 137: 109343.

[41] Paul, M., Jons, S.D. (2016) Chemistry and fabrication of polymeric nanofiltration membranes: A review. *Polymer* 103: 417–456.

[42] Mazlan, N.M., Peshev, D., Livingston, A.G. (2016) Energy consumption for desalination — A comparison of forward osmosis with reverse osmosis, and the potential for perfect membranes. *Desalination* 377: 138–151.

[43] Bernardo, P., Drioli, E., Golemme, G. (2009) Membrane gas separation: A review/state of the art. *Industrial & Engineering Chemical Research* 48: 4638–4663.

[44] Castro-Muñoz, R., Barragán-Huerta, B.E., Fíla, V., Denis, P.C., Ruby-Figueroa, R. (2018) Current role of membrane technology: From the treatment of agro-industrial by-products up to the valorization of valuable compounds. *Waste and Biomass Valorization* 9: 513–529.

[45] Wu, J., Fan, Q., Xia, Y., Ma, G. (2015) Uniform-sized particles in biomedical field prepared by membrane emulsification technique. *Chemical Engineering Science* (2015). doi:10.1016/j.ces.2014.08.016.

[46] Macedonio, M.F., Drioli, F.E. (2019) Progress of membrane engineering for water treatment. *Journal of Membrane Science Results* 6: 269–279.

[47] Abdelhamid, A.E., Elawady, M.M., El-Ghaffar, M.A.A., Rabie, A.M., Larsen, P., Christensen, M.L. (2015) Surface modification of reverse osmosis membranes with zwitterionic polymer to reduce biofouling. *Water Science and Technology: Water Supply* 15: 999–1010.

[48] Giacalone, F., Catrini, P., Tamburini, A., Cipollina, A., Piacentino, A., Micale, G. (2018) Exergy analysis of reverse electrodialysis. *Energy Conversion and Management* 164: 588–602.

[49] Favre, E. (2011) Membrane processes and postcombustion carbon dioxide capture: Challenges and prospects. *Chemical Engineering Journal* 171: 782–793.

[50] Feria-Díaz, J.J., Correa-Mahecha, F., López-Méndez, M.C., Rodríguez-Miranda, J.P., Barrera-Rojas, J. (2021) Recent desalination technologies by hybridization and integration with reverse osmosis: A review. *Water* 13: 1369.

[51] Loganathan, P., Naidu, G., Vigneswaran, S. (2017) Mining valuable minerals from seawater: A critical review. *Environmental Science: Water Research & Technology* 3: 37–53.

[52] Naidu, G., Tijing, L., Johir, M.A.H., Shon, H., Vigneswaran, S. (2020) Hybrid membrane distillation: Resource, nutrient and energy recovery. *Journal of Membrane Science* 599: 117832.
[53] Elimelech, M., Phillip, W.A. (2011) The future of seawater desalination: Energy, technology, and the environment. *Science* 333: 712–717.
[54] Lienhard, L.D., Thiel, J.H., Warsinger, G.P., Banchik, D.M. (2016) *Low Carbon Desalination: Status and Research, Development, and Demonstration Needs*. Report of a workshop conducted at the Massachusetts Institute of Technology in association with the Global Clean Water Desalination Alliance. http://hdl.handle.net/1721.1/105755.
[55] Zhao, R., Porada, S., Biesheuvel, P.M., van der Wal, A. (2013) Energy consumption in membrane capacitive deionization for different water recoveries and flow rates, and comparison with reverse osmosis. *Desalination* 330: 35–41.
[56] Thamaraiselvan, C., Arnusch, C.J. (2021) Integrated nanofiltration membrane process for water and wastewater treatment. In: *Handbook of Nanotechnology Applications*, pp. 147–168. Elsevier.
[57] Zarzo, D. (2018) Beneficial uses and valorization of reverse osmosis brines. In: *Emerging Technologies for Sustainable Desalination Handbook*, pp. 365–397. Elsevier.
[58] Teow, Y.H., Mohammad, A.W. (2019) New generation nanomaterials for water desalination: A review. *Desalination* 451: 2–17.
[59] Ihsanullah, I., Mustafa, J., Zafar, A.M., Obaid, M., Atieh, M.A., Ghaffour, N. (2022) Waste to wealth: A critical analysis of resource recovery from desalination brine. *Desalination* 543: 116093.
[60] Sharkh, B.A., Al-Amoudi, A.A., Farooque, M., Fellows, C.M., Ihm, S., Lee, S., Li, S., Voutchkov, N. (2022) Seawater desalination concentrate—a new frontier for sustainable mining of valuable minerals. *Npj Clean Water* 5: 9.
[61] Abdulsalam, A., Idris, A., Mohamed, T.A., Ahsan, A. (2017) An integrated technique using solar and evaporation ponds for effective brine disposal management. *International Journal of Sustainable Energy* 36: 914–925.
[62] Kumar, A., Naidu, G., Fukuda, H., Du, F., Vigneswaran, S., Drioli, E., Lienhard, J.H. (2021) Metals recovery from seawater desalination brines: Technologies, opportunities, and challenges. *ACS Sustainable Chemistry & Engineering* 9: 7704–7712.
[63] Liu, W., Xu, H., Shi, X., Yang, X. (2017) Fractional crystallization for extracting lithium from Cha'erhan tail brine. *Hydrometallurgy* 167: 124–128.
[64] Chen, Q., Ma, X., Zhang, X., Liu, Y., Yu, M. (2018) Extraction of rare earth ions from phosphate leach solution using emulsion liquid membrane in concentrated nitric acid medium. *Journal of Rare Earths* 36: 1190–1197.
[65] Abdel-Aal, E.A., Mahmoud, M.H.H., Sanad, M.M.S., Criscuoli, A., Figoli, A., Drioli, E. (2010) Membrane contactor as a novel technique for separation of iron ions from ilmenite leachant. *International Journal of Mineral Processing* 96: 62–69.
[66] Semblante, G.U., Lee, J.Z., Lee, L.Y., Ong, S.L., Ng, H.Y. (2018) Brine pre-treatment technologies for zero liquid discharge systems. *Desalination* 441: 96–111.
[67] Kurihara, M., Hanakawa, M. (2013) Mega-ton water system: Japanese national research and development project on seawater desalination and wastewater reclamation. *Desalination* 308: 131–137.
[68] Tufa, R.A., Noviello, Y., Di Profio, G., Macedonio, F., Ali, A., Drioli, E., Fontananova, E., Bouzek, K., Curcio, E. (2019) Integrated membrane distillation-reverse electrodialysis system for energy-efficient seawater desalination. *Applied Energy* 253: 113551.
[69] Kim, S., Cho, D., Lee, M.-S., Oh, B.S., Kim, J.H., Kim, I.S. (20090 SEAHERO R&D program and key strategies for the scale-up of a seawater reverse osmosis (SWRO) system. *Desalination* 238: 1–9.
[70] Park, J., Lee, S. (2022) Desalination technology in South Korea: A comprehensive review of technology trends and future outlook. *Membranes (Basel)* 12: 204.
[71] Macedonio, F., Drioli, E. (2010) Hydrophobic membranes for salts recovery from desalination plants. *Desalination and Water Treatment* 18: 224–234.
[72] Macedonio, F., Katzir, L., Geisma, N., Simone, S., Drioli, E., Gilron, J. (2011) Wind-Aided Intensified eVaporation (WAIV) and Membrane Crystallizer (MCr) integrated brackish water desalination process: Advantages and drawbacks. *Desalination* 273: 127–135.
[73] Cui, Z., Li, X., Zhang, Y., Wang, Z., Gugliuzza, A., Militano, F., Drioli, E., Macedonio, F. (2018) Testing of three different PVDF membranes in membrane assisted-crystallization process: Influence of membrane structural-properties on process performance. *Desalination* 440: 68–77.
[74] Tsai, J.-H., Perrotta, M.L., Gugliuzza, A., Macedonio, F., Giorno, L., Drioli, E., Tung, K.-L., Tocci, E. (2015) Membrane-assisted crystallization: A molecular view of NaCl nucleation and growth. *Applied Science* 8: 2145.
[75] Macedonio, F., Quist-Jensen, C.A., Al-Harbi, O., Alromaih, H., Al-Jlil, S.A., Al Shabouna, F., Drioli, E. (2013) Thermodynamic modeling of brine and its use in membrane crystallizer. *Desalination* 323: 83–92.
[76] Quist-Jensen, C.A., Ali, A., Mondal, S., Macedonio, F., Drioli, E. (2016) A study of membrane distillation and crystallization for lithium recovery from high-concentrated aqueous solutions. *Journal of Membrane Science* 505: 167–173.
[77] Quist-Jensen, C.A., Ali, A., Drioli, E., Macedonio, F. (2019) Perspectives on mining from sea and other alternative strategies for minerals and water recovery – The development of novel membrane operations. *Journal of the Taiwan Institution of Chemical Engineering* 94: 129–134.
[78] Ji, X., Curcio, E., Al Obaidani, S., Di Profio, G., Fontananova, E., Drioli, E. (2010) Membrane distillation-crystallization of seawater reverse osmosis brines. *Separation and Purification Technology* 71: 76–82.
[79] Curcio, E., Di Profio, G., Drioli, E. (2003) A new membrane-based crystallization technique: tests on lysozyme. *Journal of Crystal Growth* 247: 166–176.
[80] Ruiz Salmón, I., Luis, P. (2018) Membrane crystallization via membrane distillation. *Chemical Engineering and Processing: Process Intensification* 123: 258–271.
[81] Mohammadi Shamlou, E., Vidic, R., Khanna, V. (2022) Optimization-based modeling and economic comparison of membrane distillation configurations for application in shale gas produced water treatment. *Desalination* 526: 115513.
[82] Das, P., Dutta, S., Singh, K.K.K., Maity, S. (2019) Energy saving integrated membrane crystallization: A sustainable

technology solution. *Separation and Purification Technology* 228: 115722.
[83] Baghbanzadeh, M., Lan, C.Q., Rana, D., Matsuura, T. (2016) Membrane distillation. In: *Nanostructured Polymer Membranes*, pp. 419–455. John Wiley & Sons, Inc.: Hoboken, NJ, USA.
[84] Khayet, M., Cojocaru, C. (2012) Artificial neural network modeling and optimization of desalination by air gap membrane distillation. *Separation and Purification Technology* 86: 171–182.
[85] Meindersma, G.W., Guijt, C.M., de Haan, A.B. (2006) Desalination and water recycling by air gap membrane distillation. *Desalination* 187: 291–301.
[86] Kiefer, F., Spinnler, M., Sattelmayer, T. (2018) Multi-effect vacuum membrane distillation systems: Model derivation and calibration. *Desalination* 438: 97–111.
[87] Ko, C.C., Ali, A., Drioli, E., Tung, K.L., Chen, C.H., Chen, Y.R., Macedonio, F. (2018) Performance of ceramic membrane in vacuum membrane distillation and in vacuum membrane crystallization. *Desalination* 440: 48–58.
[88] Ma, A., Ahmadi, A., Cabassud, C. (2018) Direct integration of a vacuum membrane distillation module within a solar collector for small-scale units adapted to seawater desalination in remote places: Design, modeling & evaluation of a flat-plate equipment. *Journal of Membrane Science* 564: 617–633.
[89] Xie, Z., Duong, T., Hoang, M., Nguyen, C., Bolto, B. (2009) Ammonia removal by sweep gas membrane distillation. *Water Research* 43: 1693–1699.
[90] Gasconsviladomat, F., Souchon, I., Athes, V., Marin, M. (2006) Membrane air-stripping of aroma compounds. *Journal of Membrane Science* 277: 129–136.
[91] Edwie, F., Chung, T.-S. (2013) Development of simultaneous membrane distillation–crystallization (SMDC) technology for treatment of saturated brine. *Chemical Engineering Science* 98: 160–172.
[92] Kieffer, R., Mangin, D., Puel, F., Charcosset, C. (2009) Precipitation of barium sulphate in a hollow fiber membrane contactor, Part I: Investigation of particulate fouling. *Chemical Engineering Science* 64: 1759–1767.
[93] Kieffer, R., Mangin, D., Puel, F., Charcosset, C. (2009) Precipitation of barium sulphate in a hollow fiber membrane contactor: Part II: The influence of process parameters. *Chemical Engineering Science* 64: 1885–1891.
[94] Curcio, E., Simone, S., Di Profio, G., Drioli, E., Cassetta, A., Lamba, D. (2005) Membrane crystallization of lysozyme under forced solution flow. *Journal of Membrane Science* 257: 134–143.
[95] Di Profio, G., Stabile, C., Caridi, A., Curcio, E., Drioli, E. (2009) Antisolvent membrane crystallization of pharmaceutical compounds. *Journal of Pharmceutical Sciences* 98: 4902–4913.
[96] Di Profio, G., Curcio, E., Drioli, E. (2010) Membrane crystallization technology. In: *Comprehensive Membrane Science Engineering*, pp. 21–46. Elsevier.
[97] Drioli, E., Di Profio, G., Curcio, E. (2012) Progress in membrane crystallization. *Current Opinion in Chemical Engineering* 1: 178–182.
[98] Jiang, X.,. Tuo, L., Lu, D., Hou, B., Chen, W., He, G. (2017) Progress in membrane distillation crystallization: Process models, crystallization control and innovative applications. *Frontiers of Chemical Science and Engineering* 11: 647–662.
[99] Srisurichan, S., Jiraratananon, R., Fane, A. (2006) Mass transfer mechanisms and transport resistances in direct contact membrane distillation process. *Journal of Membrane Science* 277: 186–194.
[100] Schofield, R.W., Fane, A.G., Fell, C.J.D. (1987) Heat and mass transfer in membrane distillation. *Journal of Membrane Science* 33: 299–313.
[101] Laganà, F., Barbieri, G., Drioli, E. (2000) Direct contact membrane distillation: modelling and concentration experiments. *Journal of Membrane Science* 166: 1–11.
[102] Phattaranawik, J. (2003) Effect of pore size distribution and air flux on mass transport in direct contact membrane distillation. *Journal of Membrane Science* 215: 75–85.
[103] Martínez, L., Rodríguez-Maroto, J.M. (2008) Membrane thickness reduction effects on direct contact membrane distillation performance. *Journal of Membrane Science* 312: 143–156.
[104] Lawson, K.W., Lloyd, D.R. (1997) Membrane distillation. *Journal of Membrane Science* 124: 1–25.
[105] Phattaranawik, J., Jiraratananon, R., Fane, A. (2003) Heat transport and membrane distillation coefficients in direct contact membrane distillation. *Journal of Membrane Science* 212: 177–193.
[106] Bandini, S., Gostoli, C., Sarti, G.C. (1991) Role of heat and mass transfer in membrane distillation process. *Desalination* 81: 91–106.
[107] Macedonio, F., Ali, A., Drioli, E. (2017) 3.10 Membrane distillation and osmotic distillation. In: Comprehensive. Membrane Science Engineering, pp. 282–296. Elsevier.
[108] Wagner, C. (1939) Kinetik der Phasenbildung. Von Prof. Dr. M. Volmer. (Bd. IV der Sammlung "Die chemische Reaktion", herausgegeben von K. F. Bonhoeffer.) XII und 220 S. Verlag Th. Steinkopff, Dresden und Leipzig 1939. *Angewandte Chemie* 52: 503–504.
[109] Rubbo, M. (2013) Basic concepts in crystal growth. *Crystal Research and Technology* 48: 676–705.
[110] Macedonio, F., Politano, A., Drioli, E., Gugliuzza, A. (2018) Bi_2Se_3-assisted membrane crystallization. *Materials Horizons* 5: 912–919.
[111] Frappa, M., Macedonio, F., Gugliuzza, A., Jin, W., Drioli, E. (2021) Performance of PVDF based membranes with 2D materials for membrane-assisted crystallization process. *Membranes (Basel)* 11: 302.
[112] Bensaadi, S., Nasrallah, N., Amrane, A., Trari, M., Kerdjoudj, H., Arous, O., Amara, M. (2017) Dialysis and photo-electrodialysis processes using new synthesized polymeric membranes for the selective removal of bivalent cations. *Journal of Environmental Chemical Engineering* 5: 1037–1047.
[113] Tang, B., Yu, G., Fang, T., Shi, T. (2010) Recovery of high-purity silver directly from dilute effluents by an emulsion liquid membrane-crystallization process. *Journal of Hazardous Materials* 177: 377–383.
[114] Quist-Jensen, C.A., Macedonio, F., Horbez, D., Drioli, E. (2017) Reclamation of sodium sulfate from industrial wastewater by using membrane distillation and membrane crystallization. *Desalination* 401: 112–119.
[115] Frappa, M., Xue, L., Enrico, D., Francesca, M. (2020) Membrane crystallization and membrane condenser: Two membrane contactor applications. *Journal of Chemical Science and Chemical Engineering* 1: 7–17.

[116] Ruiz Salmón, I., Janssens, R., Luis, P. (2017) Mass and heat transfer study in osmotic membrane distillation-crystallization for CO_2 valorization as sodium carbonate. Separation and Purification Technology 176: 173–183.

[117] Drioli, E., Curcio, E., Criscuoli, A., Di Profio, G. (2004) Integrated system for recovery of $CaCO_3$, NaCl and $MgSO_4{\cdot}7H_2O$ from nanofiltration retentate. *Journal of Membrane Science* 239: 27–38.

[118] Li, S., Li, X., Wang, D. (2004) Membrane (RO-UF) filtration for antibiotic wastewater treatment and recovery of antibiotics. *Separation and Purification Technology* 34: 109–114.

[119] Zheng, R., Chen, Y., Wang, J., Song, J., Li, X.M., He, T. (2018) Preparation of omniphobic PVDF membrane with hierarchical structure for treating saline oily wastewater using direct contact membrane distillation. Journal of Membrane Science 555: 197–205.

[120] Curcio, E., Drioli, E. (2005) Membrane distillation and related operations—A review. *Separation and Purification Review* 34: 35–86.

[121] Margolin, A.L., Navia, M.A. (2001) Protein crystals as novel catalytic materials. *Angewandte Chemie International Edition* 40: 2204–2222.

[122] Falkner, J.C., Al-Somali, A.M., Jamison, J.A., Zhang, J., Adrianse, S.L., Simpson, R.L., Calabretta, M.K., Radding, W., Phillips, G.N., Colvin, V.L. (2005) Generation of size-controlled, submicrometer protein crystals. *Chemistry of Materials* 17: 2679–2686.

[123] McPherson, A., Gavira, J.A. (2014) Introduction to protein crystallization. *Acta Crystallographica Section F: Structural Biology Communications* 70: 2–20.

[124] Smatanová, I.K. (2002) Crystallization of biological macromolecules. *Materials and Structures* 9: 14–15.

[125] Li, W., Van der Bruggen, B., Luis, P. (2014) Integration of reverse osmosis and membrane crystallization for sodium sulphate recovery. *Chemical Engineering and Processing: Process Intensification* 85: 57–68.

[126] Ye, W., Lin, J., Shen, J., Luis, P., Van der Bruggen, B. (2013) Membrane crystallization of sodium carbonate for carbon dioxide recovery: Effect of impurities on the crystal morphology. *Crystal Growth & Design* 13: 2362–2372.

[127] Luis, P., Van der Bruggen, B. (2013) The role of membranes in post-combustion CO_2 capture. *Greenhouse Gases: Science and Technology* 3: 318–337.

[128] Jia, Z., Liu, Z., He, F. (2003) Synthesis of nanosized $BaSO_4$ and $CaCO_3$ particles with a membrane reactor: effects of additives on particles. *Journal of Colloid and Interface Science* 266: 322–327.

[129] Zhou, J., Cao, X., Yong, X., Wang, S., Liu, X., Chen, Y., Zheng, T., Ouyang, P. (2014) Effects of various factors on biogas purification and nano-$CaCO_3$ synthesis in a membrane reactor. *Industrial & Engineering Chemical Research* 53: 1702–1706.

[130] Drioli, E., Carnevale, M.C., Figoli, A., Criscuoli, A. (2014) Vacuum membrane dryer (VMDr) for the recovery of solid microparticles from aqueous solutions. *Journal of Membrane Science* 472: 67–76.

[131] Ali, A., Quist-Jensen, C.A., Jørgensen, M.K., Siekierka, A., Christensen, M.L., Bryjak, M., Hélix-Nielsen, C., Drioli, E. (2021) A review of membrane crystallization, forward osmosis and membrane capacitive deionization for liquid mining. *Resources, Conservation and Recycling* 168: 105273.

[132] Li, W., Van der Bruggen, B., Luis, P. (2016) Recovery of Na_2CO_3 and Na_2SO_4 from mixed solutions by membrane crystallization. *Chemical Engineering Research and Design* 106: 315–326.

[133] Yadav, A., Labhasetwar, P.K., Shahi, V.K. (2022) Membrane distillation crystallization technology for zero liquid discharge and resource recovery: Opportunities, challenges and futuristic perspectives. *Science of the Total Environment* 806: 150692.

[134] Edwie, F., Chung, T.-S. (2012) Development of hollow fiber membranes for water and salt recovery from highly concentrated brine via direct contact membrane distillation and crystallization. *Journal of Membrane Science* 421–422: 111–123.

[135] Meng, S., Ye, Y., Mansouri, J., Chen, V. (2015) Crystallization behavior of salts during membrane distillation with hydrophobic and superhydrophobic capillary membranes. *Journal of Membrane Science* 473: 165–176.

[136] Guan, G., Wang, R., Wicaksana, F., Yang, X., Fane, A.G. (2012) Analysis of membrane distillation crystallization system for high salinity brine treatment with zero discharge using Aspen flowsheet simulation. *Industrial & Engineering Chemical Research* 51: 13405–13413.

24 A Sustainable Approach for Dye Removal from Industrial Effluent Using Membrane-Based Techniques

Recent Advances and Future Perspectives

Pallavi Mahajan-Tatpate[1], Vikrant Gaikwad[1], and Anil Kumar Pabby[2]*

[1]School of Chemical Engineering, Dr. Vishwanath Karad MIT World Peace University, Pune, Maharashtra, India

[2]Formerly associated with BARC Complex, Nuclear Recycle Board, BARC, Tarapur, Maharashtra, India

* E-mail: (pallavi.tatpate@mitwpu.edu.in)

24.1 INTRODUCTION

All living beings need water for their survival. However, with population growth, industrialization, and water over-utilization the per capita water requirement is not being satisfied [1,2]. Apart from this, the failure in managing the climate, basin level water resources, and poor rain water harvesting systems have influenced water scarcity [3]. Recently it was recorded that around 800 million people globally do not have access to drinking water of desired quality. It is predicted that 47% of the world's population will face clean water scarcity by 2030 [4]. One good approach to overcoming this issue of water scarcity is the treatment of industrial effluent. Thus, considering industrial effluent treatment as a burden, it can be optimistically considered as a water resource. Industrial effluent is contaminated with different pollutants, such as organic pollutants (pesticides, dyes, etc.), inorganic pollutants (heavy metals, acid, etc.), and biological pollutants (bacteria, viruses, worms, etc.) [5,6]. Among these different pollutants dyes are abundantly present in wastewater and have major health and environmental concerns [4,7,8].

24.1.1 Sources of Dyes in Wastewater

The world looks more beautiful with attractive colors of the surroundings, resulting in increased use of dyes. The various industries, including the textile, paint, leather, cosmetics, tannery, printing, paper, pulp, food, and medical industries, introduce dyes in wastewater [7,8]. Worldwide, the dye effluent discharge from textile industry is 54%, dyeing industry is 21%, paper and pulp industry is 10%, tannery and paint industry is 8%, and dye manufacturing industry is 7% [9]. These industries require dyes and produce approximately 1,000,000 tons of dyes globally [10]. Currently, 100,000 commercial dyes are in use with approximately 7×10^7 tons of dyes being annually introduced [10,11]. The textile sector is one of the major sources of dyes in wastewater. In the textile industry, 0.15 m^3 of water is consumed to process 1 kg of fabric, with 3000 m^3 water discharged to process 20 tons of fabric. The dye industry annually contributes to 7.5 metric tons of dyes discharged in wastewater, whereas the textile industry consumes 700,000 tons of dyes, of which even if only 10% gets discharged, it creates a huge negative impact on living beings and the environment [10].

24.1.2 Objectives of Dye Removal

The dye effluent stream assessment before its discharge into the environment is essential as a minute quantity (1 mg/L) can produce color in water that is unacceptable for consumption [10]. Apart from this, the chemical structure of dyes is comprised of aromatic rings which gives them good stability against light, oxidants, and biological degradation [12] and also makes them calcitrant and persistent in nature. Also, it decreases the aesthetic value of water bodies and its contamination can have adverse impacts on living beings and the environment which necessitates dye removal [8].

24.1.2.1 Effect on the Environment

The presence of dyes in wastewater alters the physical and chemical properties of soil. It disturbs the quality of water bodies on the Earth's surface and adversely impacts flora and fauna [13]. Dye contamination disturbs aquatic habitats, and through the food chain it affects humans and disturbs the natural food cycle [4]. The presence of azo type dyes in water in very small amounts (<1 mg/L) affects water transparency and gas-water solubility parameters. Also, it reduces the penetration of light through water and hence affects the photosynthesis process, and depletes the oxygen level which disrupts the aquatic biocycle [12–14]. Dye contamination causes the nutrient imbalance of soil, kills soil microorganisms, and deteriorates seed germination, plant growth, and agricultural productivity [13,14].

DOI: 10.1201/9781003285656-24

24.1.2.2 Effects on Humans

The consumption of dye-contaminated water can cause allergy, nausea, hemorrhage, dermatitis, skin ulceration, mucous membrane ulceration, mutagenesis, defects in bone marrow, anemia, kidney damage, etc. [14,15]. Azo dyes are carcinogenic, and disturb respiratory, urinary, and intestinal activities. Also, this contamination can result in hypertension, porphyria, sensitivity, etc. [10]. The infection of 1,4-diamino benzene dye causes diseases such as dermatitis, skin irritation, gastritis, vomiting, lacrimation, chemosis, exophthalmos, hypertension, and respiratory issues [13]. Congo red dye contains hydroxyl, sulfonic, and chromophore groups which combine to form highly toxic chemical substances resulting in death [10]. Overall, in general, these ionic and non-ionic dyes can cause severe impacts on human health, including increasing the heart rate. They can also cause vomiting, cyanosis, jaundice, tissue necrosis, Heinz body formation, etc. [16].

24.2 TYPES OF DYES USED IN INDUSTRY

Dyes are basically classified based by their source of production as natural or synthetic dyes. Natural dyes are derived from plants, animals, and minerals, whereas manmade dyes are called synthetic dyes [11]. Most synthetic dyes are based on their chemical nature or surface charge [11,14], based on water solubility [9], or ionic charge [17], and are classified as ionic and non-ionic dyes, etc., as shown in Figure 24.1.

24.2.1 Ionic Dyes

The dyes which ionize upon dissolving or dispersing in an aqueous medium are called ionic dyes [11,14]. Dyes can also be classified based on solubility; water-soluble dyes are of cationic and anionic types [9].

24.2.1.1 Cationic Dyes

Water-soluble cationic dyes are the basic type of dyes [9]. These dyes generate colored cations in solution, and are electrostatically attracted by negative-charged substrate [18]. The chromophore in basic dye contains $-NR_{3+}$ or $-NR_{2+}$ functional groups. These have good tinctorial strength and are economical but do not have good color stability, and are carcinogenic and toxic to aquatic environments. These dyes are applied to silk, wool, cotton, nylon, acrylic fibers, and paper for brightness. They are less suitable to dye natural fiber, but show excellent performance for acrylics or synthetic fibers [9,18]

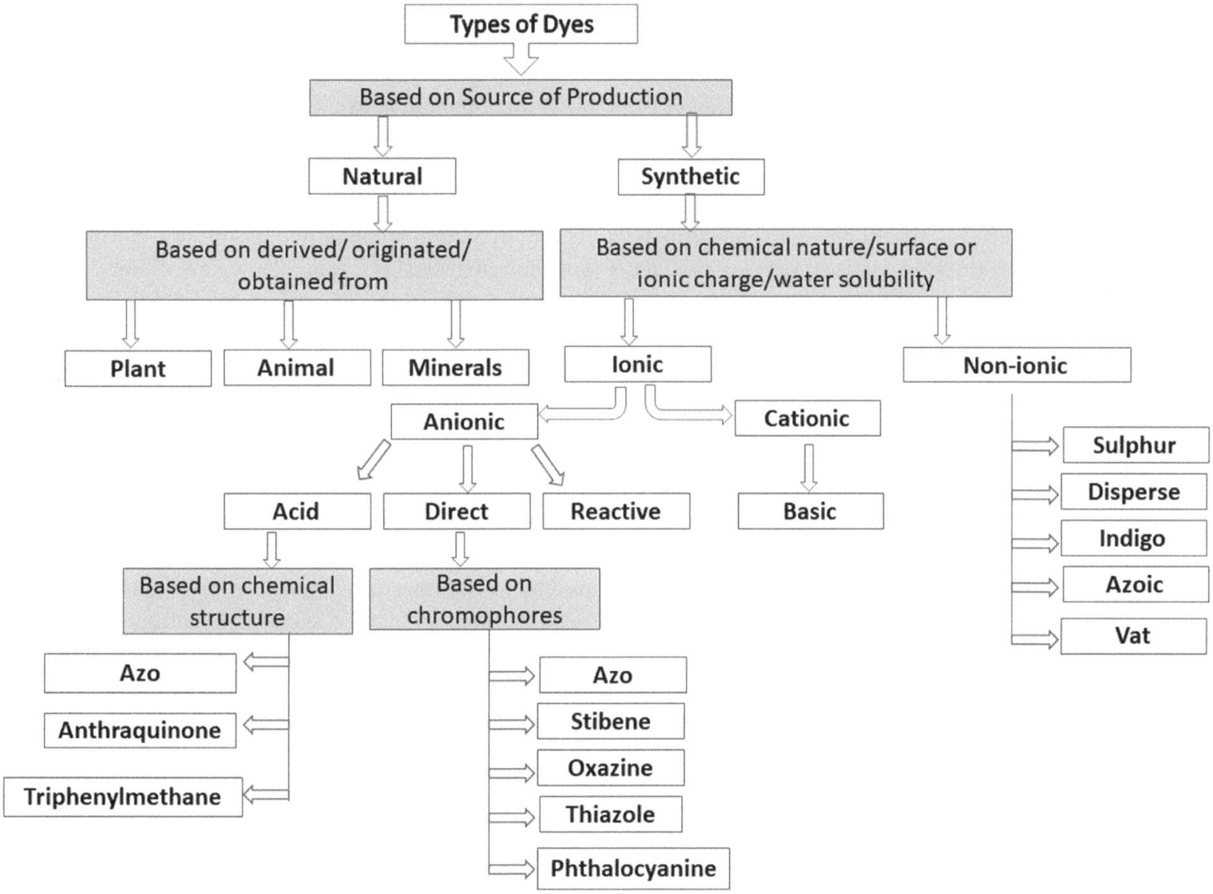

FIGURE 24.1 Classification of dyes.

2.4.2.1.2 Anionic Dyes

Water-soluble anionic dyes are the acid type of dyes. Anionic dyes are further classified as acid, direct, and reactive types [9].

a. Acid Dyes

As their name implies, acid dyes are applied in acidic conditions and contain acidic functional groups (SO_3H, COOH). These acidic functional groups have affinity for basic functions such as polyamides. They comprise 30–40% of total dye consumption. Acid dyes, mostly sulfonic dyes, are used in natural fibers, such as wool, silk, angora, mohair, and alpaca, and synthetic polyamide-nylon fibers [9], and are used in the leather, paper, printing, pharmaceutical, cosmetics industries, etc. due to their high solubility and bright colors [18]. Based on their chemical structure dyes are divided into azo, anthraquinone, and triphenylmethane. Chromophores include nitro, indigoid, quinolone, azine, xanthane, phthalocyanine, etc. They are inexpensive and have excellent fastness properties. Example of acid dyes are C.I. acid red, acid orange IV and 7, acid blue 349, 147, and 83, C.I. acid blue 7, C.I. acid blue 25 (AB25), C.I. acid blue 25 (anthraquinone), acid red 337, acid yellow 36, acid violet 17, acid black 234, etc. Dyeing auxiliaries and organic acids are possible pollutants and are carcinogenic; C.I. acid orange 150 and 165 are toxic; and C.I. acid violet 17 is allergic [9].

b. Direct Dyes

These dyes can be directly applied on fabric from an aqueous solution, bind to fabric by nonionic, noncovalent force [19], and do not need any binder or mordant for dyeing [9,18]. They are classified based on chromophore, fastness property, and applications. According to the chromophore they are classified as azo dyes, such as mono-, di-, tri-, or polyazo dyes, stilbene, oxazine, phthalocyanine, thiazole, copper complex azo, etc. These dyes are substitutes for carcinogenic benzidine (congo red) dyes [9]. They are used for dyeing cellulose, cotton, viscose fiber, leather, paper, etc. [9,18].

c. Reactive dyes

These dyes form a covalent bond with fibers and do not degrade upon exposure to sunlight [9,18]. They require temperatures above 60°C for dyeing cellulosic, wool, silk, and cotton fibers. Some of these dyes bind with heavy metals, such as Cu, Cr, and Ni, and decompose, resulting in the formation of toxic heavy metal ions in the environment [11].

24.2.2 Non-Ionic Dyes

The dyes which do not ionize upon dissolving or dispersing in aqueous medium are called non-ionic dyes [11,14,20,21]. These dyes are mostly insoluble or sparingly soluble in water, and so also classified as water-insoluble dyes. Some examples of these dyes are disperse, vat, sulfur, azoic, indigo, etc.

24.2.2.1 Disperse Dyes

These dyes are often insoluble or sparingly soluble in water [11]. Temperature of 100°C and acidic pH (4.5 to 5.5) are required to dye. A total of 85% of these dyes are composed of mono-, diazo-, and anthraquinone, and the remaining 15% is composed of methylene, styryl, acetylene benzimidazole, quinonaphthalon, nitro, etc. These dyes are persistent and non-biodegradable. They are used to dye hydrophobic fabrics, such as cellulose acetate, acrylic, polyamide, etc. These dyes are carcinogenic and allergic [9].

A. Vat Dyes

These dyes are soluble in water when the water alkalinity is reduced and by oxidation reaction they can be converted into insoluble form. They are used to dye cotton, cellulose, wool, nylon, polyester, acrylic, etc. fibers. They contain Mn, Fe, Cu, etc. heavy metal impurities and hence can cause serious health and environmental issues [9].

B. Sulfur Dyes

Sulfur dyes are available in powder, paste, and liquid forms. The liquid form is easy to apply and requires less time. These dyes are eco-friendly, and are used to dye nylon, polyester, cellulose, etc. fibers [9].

C. Azoic Dyes

These are also known as naphthalene dyes and contain the -N=N- azo group. They are soluble in alkaline solution, and are cheap and used for brighter shades as compared with other dyes and applied to dye cellulose, cotton, rayon, silk fibers, however they are carcinogenic [9].

D. Indigo Dyes

These dyes are soluble in chloroform, nitrobenzene, and concentrated sulfuric acid. They are divided into natural (water-insoluble) and leuco (water-soluble) forms. They are used to dye cotton fibers and are used mainly for denim jeans [9,22].

24.3 CONVENTIONAL METHODS FOR DYE REMOVAL FROM WASTEWATER

There are various physical, chemical, and biological methods for the separation of dyes from wastewater. Depending on the wastewater contamination level and properties, the most suitable method of separation can be selected [4,7–9]. The conventional and membrane-based methods and their advantages and disadvantages are described below. Various conventional methods, such as coagulation, flocculation, adsorption, ion-exchange, electrochemical, etc. are used for removing dyes from wastewater.

24.3.1 Coagulation-Flocculation

This is a chemical method of separation, in which initially coagulants (aluminum or ferrous sulfate, ferric chloride, poly aluminum chloride, magnesium carbonate, etc.) are added to wastewater and mixed vigorously [4,23,24]. It forms hydrophobic colloids from suspended dye particles [25]. Then the flocculants (polyacryl amide, colfloc RD-Ciba, etc.) are added to the solution, which forms flocs of larger size which precipitate and with further filtration clean water is obtained. This is cost-effective, as sludge settles fast and it has good settling and dewatering characteristics. However, it generates a large volume of concentrated sludge, which has further disposal issues, which is pH-dependent [4,26]. This limits its practical applicability.

24.3.2 Adsorption

This is a physical method of separation, in which adsorbent is added to the solution, and pollutant is adsorbed on the surface of the adsorbent. The different adsorbents include agricultural waste, such as sawdust, natural material, such as zeolite and kaolinite, activated carbon/carbon nanotubes, or bio-adsorbents such as non-living/algal/microbial biomass, etc. Depending on the dye molecule's (adsorbate) affinity, it is bound to adsorbent by physical or chemical interaction and is further removed by filtration [4,8,19,25]. In this process, low-cost adsorbents have been utilized for separation, and it is an easy process, less sludge is generated, it works over a wide pH range, and it can be used to separate all types of dyes. However, for reusing the adsorbents for treating the next batch of wastewater, adsorbent is required to be regenerated or in continuous operation as its surface gets saturated after a certain amount of time. The regeneration or desorption process is expensive and dyes are not obtained in their original form, which prevents their further reuse. If regeneration is avoided, it generates issues of toxic sludge dumping and secondary pollution, which limit its application [4,25].

24.3.3 Ion-Exchange

This is one of the physical methods of separation. In this method, ion-exchange resin beds are used. These ion-exchangers are of three types, cation, anion, and amphoteric, which exchange positive, negative, and both positive and negative ions, respectively. Ion-exchangers are long-chain organic polymers, such as amberlite IRA 400, or acid resins with -SO_3H or -COOH groups, etc. These processes show good separation efficiency and fast kinetics. However, after a certain amount of use, the resin bed gets saturated which requires it to be regenerated, which is done by an acid or alkaline wash, causing secondary pollution. Also, the process is expensive, a large amount of resins are required to treat a large amount of water, and it cannot be used for continuous operations [4,8,9].

24.3.4 Electrochemical Treatment

This is one of the chemical separation methods in which an anode and cathode are placed in a water bath and an electric current is passed through it. This generates hydrogen and oxygen on cells and results in oxidation and reduction reactions in the solution. Electrocoagulation, electro-Fenton, and anodic oxidation are three types of electrochemical treatments in combination with coagulation, Fenton reaction, and oxidation reaction, respectively. In the electrocoagulation method an iron anode acts as a catalyst and coagulant agent and in situ coagulant is generated. In the anodic oxidation process, at the anode an organic substance is adsorbed which is degraded by the anodic electron transfer process. Here oxidant O_3 and H_2O_2 are electrochemically produced. In the electro-Fenton process, H_2O_2 is used as a Fenton reagent that is generated continuously at the cathode by a dissolved oxygen reduction. It interacts with pollutants and starts oxidation. These processes are simple, fast, and ecofriendly but need a heavy electric supply, and are expensive processes needing periodic maintenance and with recovery/regeneration and disposal issues [7,8].

24.3.5 Advanced Oxidation Process (AOP)

In this process in situ hydroxyl radicals (OH^*), strong oxidizing agents, are generated which attack the pollutant (dye molecule) and produce organic peroxide radicals. This radical is finally converted into CO_2, H_2O, and inorganic salt [8]. Ozonation, photocatalysis, Fenton, photo-Fenton, and electrochemical oxidation are different advanced oxidation processes. In photocatalysis process, ZnO, TiO_2 etc. nanoparticles are used as photocatalytic agents which generate free radicals for dye degradation. In Fenton and photo Fenton processes, H_2O_2 is used as a Fenton reagent to form OH^* radicals. In electrochemical oxidation, O_3 and H_2O_2 oxidants are electrochemically generated. These are rapid processes that can remove dye molecules in harsh conditions without the generation of sludge but they are costly methods and separation is based on the pH of the solution [7,21,27].

24.3.6 Biological Processes

In this process, microorganisms, fungi, yeast, bacteria, algae, and enzymes are used to convert dye molecules into non-toxic products. The process is carried out either in aerobic or anaerobic conditions. In aerobic conditions, the products are CO_2, water, and/or biomass and in anaerobic conditions the product is methane. Bacteria have been proven to be more efficient than microorganisms, due to their easy cultivation and high growth rate but their efficiency depends on their adaptation to the environmental conditions. Fungi can accelerate their metabolism with help of intra- and extracellular enzymes (e.g. lignin or manganese peroxide) to change the environmental conditions. Yeast has the potential to sustain itself in adverse conditions such as low pH or at high concentrations due to its rapid growth rate. Algae do not need the addition of carbon or other components (unlike bacteria and fungi) for dye degradation. Enzymes have the highest separation efficiency, but they are costly in pure form, and hence less preferred for separation. Instead of pure enzymes, industrial enzymes are preferred due to their low cost and availability in liquid form. Biological processes can even be combined with AOP which has been proved to be more efficient than a single process [7]. These processes are cheap, produce less sludge, and are energy saving and ecofriendly [4,7,8,27]. Although this method has many advantages, enzyme production limitations, pH, temperature, or environmental conditions, and the sensitivity of algae, fungi, yeast, and bacteria limit their practical application [27].

24.4 SIGNIFICANCE OF MEMBRANE-BASED METHODS

As discussed earlier, the conventional processes have their own limitations of chemical utilization during the process, separation is pH-dependent, secondary pollutant generation, and disposal or discard issues [4,26,27]. Meanwhile, membrane processes are attracting increased attention due to the easy process, no additional chemicals utilization during the process, less space requirement, low operating pressure, high separation efficiency, etc. Also, in membrane processes the pollutants (dye molecules) are not only removed from wastewater but also recovered, and so they can be used as raw material in other processes [20]. Figure 24.2 shows the benefits of membrane technology.

24.5 MEMBRANE-BASED METHODS FOR DYE REMOVAL FROM WASTEWATER

Different membrane-based methods, namely ultrafiltration, nanofiltration, reverse osmosis, and electrodialysis, can be used for dye removal from wastewater.

24.5.1 Microfiltration (MF)

The pore size of MF membrane ranges from 100 to 10,000 nm and the operating pressure is less than 2 bar.

Lee et al. studied the separation of black 5 and orange 16 dye using a hybrid coagulation–adsorption–membrane process. In this hybrid process initially coagulation and sedimentation process were carried out and then in a single tank the adsorption and membrane processes were carried out. The authors observed that when the MF process alone is used, membrane fouling was observed to be almost 50% within 5 hours. The water flux was reduced from 325 to approximately 100–200 LMH with 2–3% dye removal efficiency for wastewater containing orange 16 and red 5 dye, respectively. However, when the hybrid coagulation–adsorption–membrane process was used, the water flux for 5 hours was observed to be almost constant (330–324 LMH) with 100% dye removal efficiency. This hybrid process reduced the amount of sludge produced because a lower amount of adsorbent and coagulant was required to carry out the process [20,28].

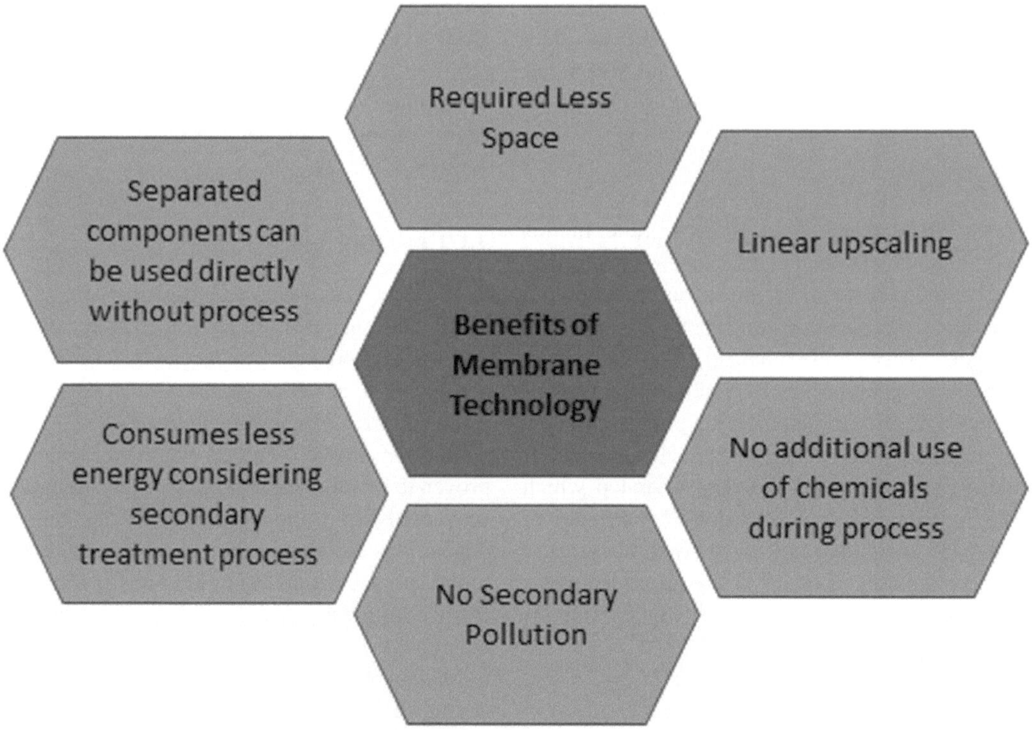

FIGURE 24.2 Benefits of membrane technology.

Daraei et al. examined the performance of polyvinylidene fluoride (PVDF) MF membranes coated with organoclay (grade cloisite 15A and 30B) and chitosan for removal of methylene blue and acid orange 7 from wastewater. The coated membranes showed better rejection than membranes without coating. The 15A shows maximum flux as compared to 30B at neutral pH at 0.5%.

It is observed that 2% 30B membranes are homogeneous and dense with the methylene blue separation around 80% and the separation was found to be 72% for 1% 30B. In an acidic environment, cloisite 30B provided more positive charge and an efficient coating to the used membrane. The increase in amount from 0.5% to 2% had no effect on dye (acid orange 7) removal, but permeate flux decreases with an increase in the amount of organoclay. Also, there is no decrease in separation efficiency or water flux with time (up to 3 hours) in normal and acidic environments, which indicates that the membranes are quite stable. The 15A cloisite shows more water flux than 30B. Considering water flux and separation efficiency, 2% 15A and 1% 30B showed better performance for acid orange 7. Overall, the separation efficiency of PVDF microfiltration membrane was increased from 30% to 85% when the coating was applied on the membrane [20,29].

Shi et al. evaluated the rhodamine B (RhB) rejection efficiency of polyether sulfone MF membranes decorated with polyethylene glycol (PEG) and tannic acid (TA). PEG of various molecular weights (600, 800, 1000, 2000, 4000, 6000, and 10,000) were used to decorate the polysulfone membrane. For this, 1 wt.% PEG and TA were taken in a 1:1 volume ratio and dissolved in water. Its suspension was formed, and then it is vacuum filtered using polyether sulfone membrane. This suspension adhered to membrane as per the procedure and these membranes were used for RhB separation. The maximum separation efficiency of these membranes were more than 98.9% with 4777 LMH water flux, when PEG of 10,000 molecular weight was used [30].

Homem et al. observed blue corazol (BC) dye rejection using polyether sulfone MF membrane modified with polyethyleneimine (PEI) and graphene oxide (GO). Layer by layer PEI and GO is deposited on polyether sulfone using a pressurized filtration system. The first layer of PEI is formed on MF membrane followed by GO and then again a PEI layer was formed by permeating PEI, GO solution through the membrane at defined pH. The concentration of the first layer of PEI was taken as 0, 1.5, and 3%, whereas the GO concentration was taken as 0.025% and the third layer of PEI was applied with concentrations of 0, 1.5, and 3%. All membranes with different concentrations showed more than 92% BC dye rejection. On the other hand, pristine polyether sulfone shows a much lower rate of BC dye rejection (11.8%). The best membrane performance was observed for the membrane formed with 3% PEI, 0.025% GO, and 1.5% PEI concentrations. It showed a 97.8% BC rejection rate with 99.4 LMH/bar water permeability. Also, the 3% PEI, 0.025% GO, and 0% PEI in the third layer showed maximum water permeability of 258.5 LMH/bar with 96.2% BC dye rejection. Based on the desired separation efficiency and water permeability, the membrane was chosen [31]. Table 24.1 shows the separation efficiencies of MF membrane for different types of dye removal.

Table 24.1 shows the MF membrane used along with the coating or in combination with other processes. Different types of coating options are used because the pore size of MF membrane ranges from 100 to 10,000 nm, which is not

TABLE 24.1
Dye separation from wastewater using MF membrane

MF membrane material	Modifier/concentration (working parameters)	Process of MF modification	Type of dye separated	Rejection efficiency	References
Polyethylene	Coagulant-alum, adsorbent-activated carbon	Hybrid coagulation-adsorption-membrane	Black 5 and Orange 16	100%	[28]
Polyvinylidene fluoride	Chitosan and organoclay: grade cloisite 30B/2% (pH 7)	Coating	Methylene blue	80%	[29]
Polyvinylidene fluoride	Chitosan and organoclay: grade cloisite 30B/1% (pH 7)	Coating	Methylene blue	72%	[29]
Polyvinylidene fluoride	Chitosan and organoclay: grade cloisite 15A/2% (pH 3.5)	Coating	Acid orange 7	≈70%	[29]
Polyvinylidene fluoride	Chitosan and organoclay: grade cloisite 15A/1% (pH 3.5)	Coating	Acid orange 7	≈50%	[29]
Polyether sulfone	Polyethylene glycol (MW 10,000) and tannic acid	Decorated/coated using vacuum filtration	Rhodamine B	98.9	[30]
Polyether sulfone	Polyethylene glycol (MW 600, 800, 1000, 2000, 4000, 6000, 10,000) and tannic acid	Decorated/coated using vacuum filtration	Rhodamine B	60-80	[30]
Polyether sulfone	Polyethyleneimine (3%), graphene oxide (0.025%), polyethyleneimine (1.5%)	Coating: layer-by-layer deposition	Blue Corazol	97.8	[31]

suitable to separate dissolved dye molecules in water. The MF membrane is applicable to separate only suspended particles and colloid dye [20]. Also, use of MF with conventional processes needs secondary treatment to recover dye particles in nascent form. In addition, the coating, with layer-by-layer deposition, has fouling issues. Membrane modification is required to avoid concentration polarization or fouling issue of membrane. The membrane used should be hydrophilic and smooth, and periodic cleaning is required to overcome issues related to membrane fouling.

24.5.2 Ultrafiltration (UF)

The pore size of UF membrane ranges from 10 to 100 nm and the operating pressure ranges between 1–10 bar [20,25].

Baek et al. studied optimization of the micellar-enhanced UF membrane for separation of ferricyanide and nitrate dye from wastewater. The UF membrane is made of regenerated cellulose with a molecular weight cut off (MWCO) between 3000 and 10,000. Cetylpyridinium chloride (CPC) is used as a cationic surfactant. The separation efficiency for 1 mM nitrate dye was observed to be 40, 62, 81, and 89% for CPC concentrations of 1, 2, 4, and 6, respectively, for membrane with 10,000 MWCO. On the other hand, for membrane with 3000 MWCO, nitrate dye rejection has increased to 64, 78, 87, 91, and 95% for CPC concentrations of 1, 2, 4, 6, and 10, respectively. As expected, the rejection rate was increased with a decrease in MWCO. Also, with an increase in CPC concentration, the rejection rate increased due to the formation of large micelles. Similarly, for 10,000 MWCO membrane, ferricyanide dye rejection was observed to be 65, 81, 98, 99, and >99% for 1, 2, 3, 5, and 10 CPC concentrations, respectively, and for 3000 MWCO membrane it was increased to 89, 93, 99, 99.5, and >99.9% [20,32].

Petrov and Stoychev reported the removal of dyes using polyacrylonitrile (PAN) membrane. They used two PAN membranes in series, the first membrane was prepared using copolymer polyacrylonitrile-methyl acrylate-sodium vinyl sulfonate and the second membrane from acrylonitrile-methylacrylate-2-acrylamide-2-methylpropane sulfonic acid. After the first membrane, precipitant $FeSO_4.7H_2O$ was used and then the second membrane was used. In the first stage at 10 PPM dye concentration, the rejection of dyes was observed to be 63, 87, 89, and 93%, which was increased to 94, 96, 97, and 97% for blau RNS, gelb 3RS, ROT 4BS, and Schwarz GRS, respectively. In the second stage, rejection for all type of dyes reached 99%. Also, it was observed that with an increase in concentration the rejection rate was increased, and water flux was decreased. This could be because of complex formation and an increase in the van der Waal's interaction at higher concentrations [20,33].

Huang et al. analyzed the rejection efficiency of polysulfone hollow-fiber membrane (MWCO: 10 kDa) using sodium dodecylsulfate (SDS) anionic surfactant for cationic dye-methylene blue (MB) dye of molecular weight 373.90. The authors determined the effect of pressure and concentration on water flux, resistance, rejection, and other factors. The MB rejection was observed to be ≥ 99% for 0.01–0.09 MPa operating pressure at 6 mg/L MB concentration and 8 mM SDS concentration. With an increase in pressure, there was a slight decrease in rejection efficiency from 99.6% (at 0.01 MPa) to 98.9% (at 0.09 MPa). This suggests that at high pressure the micelles become compact, resulting in low micelle solubility and hence a lower quantity of MB dye is solubilized in micelles. This leads to diffuse, permeate solute molecules (dye molecules) through the membrane. Similarly, with increasing SDS concentration from 8 to 72 mM, the MB rejection decreased from 99 to 95%. This could be due to a higher SDS concentration although there was an increase in the number of micelles, the shape of micelles changed from spherical to cylindrical. This allows MB to easily diffuse through the pores of membranes [20,34].

Rambabu et al. investigated the rejection of Congo red and orange II dye using polyethersulfone-$BaCl_2$ blend UF membrane for 18% polyether sulfone (PES). The authors studied the optimization of $BaCl_2$ concentration from 0 to 4% using dimethyl formamide (DMF) solvent. The surface roughness and porosity were observed to increase with $BaCl_2$ concentration, whereas the contact angle was reduced with an increase in concentration. With an increase in porosity, water flux was increased. The dye rejection rate was observed to increase till 2% $BaCl_2$ concentration and thereafter it decreased [36].

Yang et al. examined the rejection of direct red 23 and congo red (CR) dye using polyimide (PI) UF membrane. The authors observed 98.65% direct red 23 dye rejection at 0.1 MPa with 345.10 LMH water flux. For CR dye the rejection rate was observed to decrease from 99.09 to 91.82% with an increase in temperature from 20 to 95°C, respectively, with 305.58–539.85 LMH water flux. This is because with an increase in temperature, the viscosity of the solution decreases resulting from the formation of comparatively larger pores [37].

Fradj et al. observed the separation of methyl orange and direct blue dye using polymer-enhanced ultrafiltration (PEUF) membrane using chitosan. It was observed that the methyl orange rejection rate was 86% and the direct blue dye rejection rate was 89% at 0.1 mM chitosan concentration using PEUF membrane. The increase in polymer concentration increased the dye retention rate [39]. Table 24.2 shows the separation efficiencies of different UF membranes for various dyes.

From Table 24.2, it can be observed that the separation efficiency of dyes using micellar-enhanced ultrafiltration (MEUF) or PEUF membrane is higher as compared to UF membrane. MEUF or PEUF is preferred because the molecular weight cut-off (MWCO) of UF membrane is higher than the molecular weight of dye molecules and hence there is lower separation efficiency.

24.5.3 Nanofiltration (NF)

The pore size of the NF membrane ranges from 1 to 10 nm and the operating pressure ranges between 5–20 bar [20,25].

TABLE 24.2
Dye separation from wastewater using UF membrane

Membrane type	Membrane material	Surfactant	Surfactant conc/operating parameter	Type of dye	% Dye removal	References
MEUF	Regenerated cellulose (MWCO: 10 kDa)	CPC	1 mM	Nitrate	40	[38]
		CPC	2 mM		62	
			4 mM		81	
			6 mM		89	
MEUF	Regenerated cellulose (MWCO: 3 kDa)	CPC	2 mM	Nitrate	78	[38]
			4 mM		87	
			6 mM		91	
			10 mM		95	
MEUF	Regenerated cellulose (MWCO: 10 kDa)	CPC	1 mM	Ferricyanide	65	[38]
			2 mM		81	
			3 mM		98	
			5 mm		99	
			10 mM		>99	
MEUF	Regenerated cellulose (MWCO: 3 kDa)	CPC	1 mM	Ferricyanide	89	[38]
			2 mM		93	
			3 mM		99	
			5 mM		99.5	
			10 mM		>99.5	
MEUF	Polysulfone (MWCO: 10 kDa)	SDS	8 mM/0.01 MPa	Methylene blue (6mg/L)	99.6	[34]
			8 mM/0.09 MPa		98.9	
			72 mM/0.0 1MPa		95	
UF	Polyimide	TMPDA, DDM, PMDA	0.1 MPa	Direct red 23	98.65	[37]
UF	PAN	NA	10 PPM	BLAU RNS	63	[33]
				GELB 3RS	87	
				ROT 4BS	89	
				SCHWARZ GRS	93	
UF	Polyimide	TMPDA, DDM, PMDA	20°C, 0.1 MPa	CR	99.09	[37]
			95°C, 0.1 MPa		91.82	
PEUF	NA	Chitosan	0.1 mM	Methyl orange	86	[39]
				Direct Blue	89	
PEUF	PES (30 kDa)	PEI	50 PPM	RR120	99	[14]
		Chitosan			88	
PEUF	Cellulose (10 kDa)	SA	1250 PPM	MB	98	[14]
		PAA	28.8 PPM		99	
		$PANH_4$	35.6 PPM		98	

Abid et al. optimized PES membrane using polyvinylpyrrolidone (PVP) as a pore former and magnetic graphene-based composite (MMGO) using dimethylacetamide (DMAc) solvent for direct red 16 dye rejection. MMGO incorporates hydrophilicity and smoothness to membrane and hence helps to enhance water flux and importantly membrane life span by reducing membrane fouling. It was observed that virgin PES can show a dye rejection rate of up to 91%. It reached 99% at different MMGO concentrations (0.1, 0.5, 1 wt.%). Also, the water flux was increased through the use of MMGO and it was observed to be maximum at 0.5 wt.% MMGO. The use of MMGO is beneficial over GO, as the use of graphene sheets aggregates and affects membrane structure, which affects membrane efficiency and reusability [6].

Kebria et al. studied the rejection rate of crystal violet dye using polysulfone membranes. The membrane was formed using 18% PSF, 1.5% PVP as pore former, and DMF as solvent. It was modified using 2 wt.% PEI, triphthaloyldechloride (TPC) (0.1–0.5 wt.%), and SiO_2 nanoparticles (0.03, 0.05, and 0.1 wt.%) using an interfacial polymerization process. The dye was dissolved in water and organic solution (2-propanol), and the membrane performance was observed. The organic solvent is used to determine membrane performance in a harsh environment which can cause destruction of membrane. In an organic environment, the dye rejection rate of membrane decreased from 91–99% to 50–85% at 0.1 and 0.5 TPC concentrations. However, at 0.1 wt.% concentration, the rejection rate was approximately 99–100% in an aqueous

TABLE 24.3
Dye separation from wastewater using NF membrane

Membrane material	Membrane material composition/ operating parameter	Type of dye	Rejection efficiency	References
PES/MMGO	20/0 (wt.%)	Direct Red 16	91	[6]
	20/0.1 (wt.%)		99	
	20/0.5 (wt.%)		99	
	20/1 (wt.%)		99	
PSF/PVP/PEI/TPC/SiO_2	18/1.5/2/0.1/0.03 (wt.%)	Crystal Violet (aqueous)	94	[40]
	18/1.5/2/0.1/0.05 (wt.%)		96	
	18/1.5/2/0.1/0.1 (wt.%)		≈100	
PSF/PVP/PEI/TPC/SiO_2	18/1.5/2/0.5/0.03 (wt.%)	Crystal Violet (aqueous)	≈100	[40]
	18/1.5/2/0.5/0.05 (wt.%)			
	18/1.5/2/0.5/0.1 (wt.%)			
PSF/PVP/PEI/TPC/SiO_2	18/1.5/2/0.1/0.03 (wt.%)	Crystal Violet (organic)	50	[40]
	18/1.5/2/0.1/0.05 (wt.%)		65	
	18/1.5/2/0.1/0.1 (wt.%)		75	
PSF/PVP/PEI/TPC/SiO_2	18/1.5/2/0.5/0.03 (wt.%)	Crystal Violet (organic)	55	[40]
	18/1.5/2/0.5/0.05 (wt.%)		85	
	18/1.5/2/0.5/0.1 (wt.%)		≈99	
PA-TFC and PA-TFC/DEA		Congo Red	99.6	[41]
		Methyl Blue	99.8	
		Sunset Yellow	97.6	
		Neutral Red	80.6	
Polyamide	50 PPM feed, 8.3 pH	Acid Red	91	[42]
	65 PPM feed, 8.3 pH		93	
Polyamide	50 PPM, 8.3pH	Reactive black	94	[42]
	65 PPM, 8.3pH		95	
Polyamide	50 PPM, 8.3 pH	Reactive blue	95.5	[42]
	65 PPM, 8.3 pH		96.5	

and organic environment. Also, the flux was observed to be maximum at 0.1% SiO_2 concentration [40].

Liu et al. investigated the rejection of congo red, methyl blue, sunset yellow, and neutral red dyes using polyamide (PA) thin-film composite (TFC) NF membrane with and without modifying it with diethanolamine (DEA). By interfacial polymerization PA membrane was prepared using trimesoyl chloride and piperazine, the DEA was poured on this membrane which formed a covalent bond with the membrane and made it more hydrophilic. It also changed the morphology, compactness of the active layer, and charge-carrying capacity of the membrane. The dye rejection rate was observed to be more than 99% for congo red and methyl blue, and exceed 97% for sunset yellow, and > 80% for neutral red dyes for modified and unmodified membranes. The water flux was observed to be 57.3, 42.5, and 52.1 LMH for congo red, methyl blue, and sunset yellow, respectively, and increased by approximately 22 LMH for membrane modified with DEA. The deposited dyes after filtration can be easily removed by simple hydraulic washing for membrane modified with DEA [41]. Table 24.3 shows the separation efficiencies of different NF membranes for dye molecules.

Table 24.3 shows that the NF membrane rejection efficiency is suitable for the removal of low-molecular-weight dyes (reactive dyes) but it has concentration polarization and membrane fouling issues. The deposition of retentate on the surface layer causes the blocking of pores and hence results in a reduction in water flux. It necessitates the feed to be preprocessed before it passes for NF treatment [20].

24.5.4 Reverse Osmosis (RO)

The pore size of the RO membrane ranges from 0.1 to 1 nm and operating pressure ranges between 20–100 bar [20,25].

Mustafa et al. studied the removal of direct blue 6 and direct yellow from wastewater using polyamide thin-film composite (TFC). The authors studied the effect of dye feed concentration, feed temperature, and operating time on the dye rejection efficiency of polyamide composite membranes. It was observed that with an increase in dye feed concentration, permeate concentration increases, and there is a clear decrease in the rejection efficiency of the membrane. Also, the water flux was observed to be reduced with increasing feed concentration. A similar observation was noticed with increased operating time. This is attributed to fouling of membrane at high feed concentration and an increase in operating time. Meanwhile, with an increase in temperature, the density and viscosity of the solution decrease, and hence water flux is increased, which results in an increase in solute concentration in the

TABLE 24.4
Dye separation from wastewater using RO membrane

Membrane material	Operating parameters	Type of dye	Rejection efficiency	References
Polyamide	100 PPM feed	Direct Blue 6	98.6	[43]
	450 PPM feed		97.5	
Polyamide	100 PPM feed	Direct Yellow	98.25	[43]
	450 PPM feed		97	
Polyamide	30°C	Direct Blue 6	98.4	[43]
	50°C		97.4	
Polyamide	30°C	Direct Yellow	97.75	[43]
	50°C		96.75	
Polyamide	50 PPM feed, 8.3 pH,8 bar, 39°C	Acid Red	96	[42]
	65 PPM feed, 8.3 pH, 8 bar, 39°C		97	
Polyamide	50 PPM feed, 8.3 pH, 8 bar, 39°C	Reactive black	98.5	[42]
	65 PPM feed, 8.3 pH, 8 bar, 39°C		99.5	
Polyamide	50 PPM feed, 8.3 pH, 8 bar, 39°C	Reactive blue	99.25	[42]
	65 PPM feed, 8.3 pH, 8 bar, 39°C		99.75	
Polyamide	50 PPM feed, 4.5 pH, 8 bar, 39°C	Acid Red	93	[42]
	50 PPM feed, 7 pH, 8 bar, 39°C		97	
Polyamide	50 PPM feed, 8.3 pH, 10 bar, 39°C	Acid Red	96	[42]
	65 PPM feed, 8.3 pH, 10 bar, 39°C		97	
Polyamide	50 PPM feed, 8.3 pH, 8 bar, 39°C	Acid Red	96.5	[42]
	50 PPM feed, 8.3 pH, 8 bar, 26°C		97.5	
Polyamide	50 PPM feed, 8.3 pH, 8 bar, 39°C	Reactive Black	98.5	[42]
	50 PPM feed, 8.3 pH, 8 bar, 26°C		99	
Polyamide	50 PPM feed, 8.3 pH, 8 bar, 39°C	Reactive Blue	98.5	[42]
	50 PPM feed, 8.3 pH, 8 bar, 26°C		99	

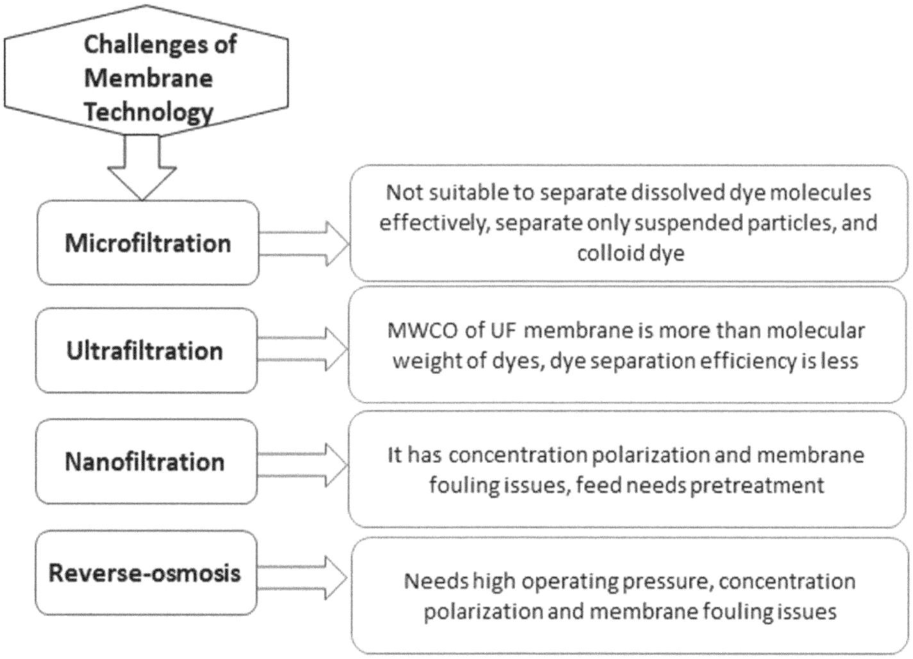

FIGURE 24.3 Challenges of membrane technology.

product and thus a decrease in the rejection efficiency of the membrane [43].

Abid et al. analyzed the effect of dye concentration, feed temperature, operating pressure, and pH on acid red, reactive black, and reactive blue dye using a polyamide membrane [42]. The rejection efficiency was observed to increase with these parameters, as shown in Table 24.4.

RO is a well-known technology for desalination. The high salt concentration reduces the permeate flux and enhances the membrane fouling, which has a negative impact on membrane performance. The dense membrane structure and membrane fouling have high operating and maintenance costs [20].

24.6 CONCLUDING REMARKS, CHALLENGES, AND FUTURE PERSPECTIVES

Membrane technology is a sustainable, eco-friendly, and green technology for the separation of dyes from water considering its advantages of no use of chemicals during the process, and no secondary pollution. It is a simple process in which the pollutant is separated in its nascent form and can be reused as a raw material in processing industries. Also, the membrane processes can be easily upscaled. Though there are many advantages, each membrane process has its limitations, as shown in Figure 24.3. However, the limitations of conventional separation methods, such as byproduct generation, treatment for recovery, or dumping issues, are overcome by membrane technology, and hence it is a promising approach for the separation of dyes from wastewater. Also, it is energy efficient as compared to conventional processes considering the cost involved in overcoming the limitations of conventional processes. The separation efficiency for conventional and membrane processes are almost the same. However, after use of adsorbents, surfactants, coagulants, etc. in conventional processes for a few cycles, it must be replaced. If it is not replaced, the separation efficiency is reduced. Also, for separation or recovery of dyes it needs further treatment which requires the use of chemicals. If the dyes are not recovered there are issues related to dumping of toxic chemicals or pollutants in soil and hence the generation of secondary pollution which badly affects the environment. Hence, considering the aspect of green technology in terms of a physical separation method without the use of chemicals during the process and thus no secondary treatment, no dumping of toxic pollutants (dyes) in soil, and dyes can be reused membrane technology is a sustainable approach toward dye separation from water. Also, the cost of chemicals for single use or in fixed intervals during processing and secondary treatment is cost-comparable with membrane technology. Though the use of membranes is a sustainable approach, it has some challenges, as shown in Figure 24.3.

The main challenges of membrane technology include concentration polarization which leads to membrane fouling. In order to resolve this problem, thermally, chemically, and mechanically stable hydrophilic material should be developed which has less tendency to foul. Researchers are modifying the membrane surface with a hydrophilic material. Also, the use of nanomaterials in the membrane matrix will help to improve membrane performance. The surface modification post membrane synthesis such as coating, grafting, etc. is a short-term solution and is prone to fouling issues. Instead, with the incorporation of hydrophilic material and nanoparticles by physical blending, the membrane pre-synthesis process will improve membrane stability. The UF membrane is preferred over NF and RO as it consumes comparatively less energy without compromising separation efficiency. MEUF and PEUF are not preferred as they need retentate post-treatment to get the product in its purified form, thus simple UF with a modified surface will help to resolve many concerns.

REFERENCES

[1] Biswas, A.K. (2003) Water availability and use. *Water Resources of North America* November: 163–174.

[2] Kummu, M., et al. (2016) The world's road to water scarcity: Shortage and stress in the 20th century and pathways towards sustainability. *Scientific Reports* 6(May): 1–16.

[3] Dolan, F., Lamontagne, J., Link, R., Hejazi, M., Reed, P., Edmonds, J. (2021) Evaluating the economic impact of water scarcity in a changing world. *Nature Communications* 12(1): 1–10.

[4] Al-Tohamy R., et al. (2022) A critical review on the treatment of dye-containing wastewater: Ecotoxicological and health concerns of textile dyes and possible remediation approaches for environmental safety. *Ecotoxicology and Environmental Safety* 231: 113160.

[5] Singh, J., Yadav, P., Pal, A.K., Mishra, V. (2020) *Water Pollutants: Origin and Status*. Springer.

[6] Abdi, G., Alizadeh, A., Zinadini, S., Moradi, G. (2018) Removal of dye and heavy metal ion using a novel synthetic polyethersulfone nanofiltration membrane modified by magnetic graphene oxide/metformin hybrid. *Journal of Membrane Science* 552: 326–335.

[7] Shindhal, T., et al. (2021) A critical review on advances in the practices and perspectives for the treatment of dye industry wastewater. *Bioengineered* 12(1): 70–87.

[8] Sharma, S., Kaur, A. (2018) Various methods for removal of dyes from industrial effluents – a review. *Indian Journal of Science and Technology* 11(12): 1–21.

[9] Velusamy, S., Roy, A., Sundaram, S., Kumar Mallick, T. (2021) A review on heavy metal ions and containing dyes removal through graphene oxide-based adsorption strategies for textile wastewater treatment. *Chemical Record* 21(7): 1570–1610.

[10] Subramanian, S.M., Ali, K. (2021) *Novel Materials for Dye-containing Wastewater Treatment*. Springer.

[11] Said, B., Souad M.R., Ahmed, E.H. (2020) A review on classifications, recent synthesis and applications of textile dyes. *Inorganic Chemistry Communications* 3(March): 107891.

[12] Dalvand, A., Gholami, M., Joneidi, A., Mahmoodi, N.M. (2011) Dye removal, energy consumption and operating cost of electrocoagulation of textile wastewater as a clean process. *Clean – Soil, Air, Water* 39(7): 665–672.

[13] Hassaan, M.A., El Nemr, E. (2017) Health and environmental impacts of dyes: Mini review. *American Journal of Environmental Science and Engineering* 1(3): 64.

[14] Oyarce, E., et al. Removal of dyes by polymer-enhanced ultrafiltration: An overview. *Polymers* 13(19): 2021.
[15] Pandit, P., Basu, S. (2004) Dye and solvent recovery in solvent extraction using reverse micelles for the removal of ionic dyes. *Industrial and Engineering Chemistry Research* 43(24): 7861–7864.
[16] Ruan, W., Hu, J., Qi, J., Hou, Y., Zhou, C., Wei, X. (2019) Removal of dyes from wastewater by nanomaterials: A review. *Advanced Materials Letters* 10(1): 9–20.
[17] Yazdani, M.R. (2018) *Engineering adsorptive materials for water remediation – Development, characterization, and application.* Aalto University Publication Series, June.
[18] Benkhaya, S., Rabet, M., El Harfi, A. (2020) A review on classifications, recent synthesis and applications of textile dyes. *Inorganic Chemistry Communications* 115: 107891.
[19] Kiernan, J.A. (2001) Classification and naming of dyes, stains and fluorochromes. *Biotechnic and Histochemistry* 76(5–6): 261–278.
[20] Moradihamedani, P. (2022) Recent advances in dye removal from wastewater by membrane technology: a review. *Polymer Bulletin* 79(4): 2603–2631.
[21] Li, Q., Yue, Q.Y., Sun, H.J., Su, Y., Gao, B.Y. (2010) A comparative study on the properties, mechanisms and process designs for the adsorption of non-ionic or anionic dyes onto cationic-polymer/bentonite. *Journal of Environmental Management* 91(7): 1601–1611.
[22] Chakraborty, J.N., Chavan, R.B. (2004) Dyeing of denim with indigo. *Indian Journal of Fibre and Textile Research* 29(1): 100–109.
[23] Kasperchik, V.P., Yaskevich, A.L., Bil'Dyukevich, A.V. (2012) Wastewater treatment for removal of dyes by coagulation and membrane processes. *Petroleum Chemistry* 52(7): 545–556.
[24] Golob, V., Vinder, A., Simonič, M. (2005) Efficiency of the coagulation/flocculation method for the treatment of dyebath effluents. *Dyes and Pigments* 67(2): 93–97.
[25] Mahajan-Tatpate, P., Dhume, S., Chendake, Y. (2021) Removal of heavy metals from water: Technological advances and today's lookout through membrane applications. *International Journal of Membrane Science and Technology* 8: 1–21.
[26] Mahajan-Tatpate, P., Dhume, S., Chendake, Y. (2021) Removal of heavy metals from water: Technological advances and today's lookout through membrane applications. *International Journal of Membrane Science and Technology* 8(1).
[27] Samsami, S., Mohamadi, M., Sarrafzadeh, M.H., Rene, E.R., Firoozbahr, M. (2020) Recent advances in the treatment of dye-containing wastewater from textile industries: Overview and perspectives. *Process Safety and Environmental Protection* 143(May): 138–163.
[28] Lee, J.W., Choi, S.P., Thiruvenkatachari, R., Shim, W.G., Moon, H. (2006) Submerged microfiltration membrane coupled with alum coagulation/powdered activated carbon adsorption for complete decolorization of reactive dyes. *Water Research* 40(3): 435–444.
[29] Daraei, P., et al. (2013) Novel thin film composite membrane fabricated by mixed matrix nanoclay/chitosan on PVDF microfiltration support: Preparation, characterization and performance in dye removal. *Journal of Membrane Science* 436: 97–108.
[30] Shi, P., Hu, X., Wang, Y., Duan, M., Fang, S., Chen, W. (2018) A PEG-tannic acid decorated microfiltration membrane for the fast removal of Rhodamine B from water. *Separation and Purification Technology* 207(June): 443–450.
[31] Homem, N.C., et al. (2019) Surface modification of a polyethersulfone microfiltration membrane with graphene oxide for reactive dyes removal. *Applied Surface Science* 486: 499–507.
[32] Baek, K., Lee, H.H., Yang, J.W. (2003) Micellar-enhanced ultrafiltration for simultaneous removal of ferricyanide and nitrate. *Desalination* 158(1–3): 157–166.
[33] Petrov, S.P., Stoychev, P.A. (2003) Ultrafiltration purification of waters contaminated with bifunctional reactive dyes. *Desalination* 154(3): 247–252.
[34] Huang, J.H., et al. (2010) Micellar-enhanced ultrafiltration of methylene blue from dye wastewater via a polysulfone hollow fiber membrane. *Journal of Membrane Science* 365(1–2): 138–144.
[35] Marcucci, M., Nosenzo, G., Capannelli, G., Ciabatti, I., Corrieri, D., Ciardelli, G. (2001) Treatment and reuse of textile effluents based on new ultrafiltration and other membrane technologies. *Desalination* 138(1–3): 75–82.
[36] Rambabu, K., Srivatsan, N., Gurumoorthy, A.V.P. (2017) Polyethersulfone-barium chloride blend ultrafiltration membranes for dye removal studies. *IOP Conference Series: Materials Science and Engineering* 263(3).
[37] Yang, C., et al. (2020) Fabrication and characterization of a high performance polyimide ultrafiltration membrane for dye removal. *Journal of Colloid and Interface Science* 562: 589–597.
[38] Baek, K., Kim, B.K., Cho, H.J., Yang, J.W. (2003) Removal characteristics of anionic metals by micellar-enhanced ultrafiltration. *Journal of Hazardous Materials* 99(3): 303–311.
[39] Ben Fradj, A., Boubakri, A., Hafiane, A., Ben Hamouda, S. (2020) Removal of azoic dyes from aqueous solutions by chitosan enhanced ultrafiltration. *Results in Chemistry* 2: 100017.
[40] Kebria, M.R.S., Jahanshahi, M., Rahimpour, A. (2015) SiO_2 modified polyethyleneimine-based nanofiltration membranes for dye removal from aqueous and organic solutions. *Desalination* 367: 255–264.
[41] Liu, M., et al. (2017) High efficient removal of dyes from aqueous solution through nanofiltration using diethanolamine-modified polyamide thin-film composite membrane. *Separation and Purification Technology* 173: 135–143.
[42] Abid, M.F., Zablouk, M.A., Abid-Alameer, A.M. (2012) Experimental study of dye removal from industrial wastewater by membrane technologies of reverse osmosis and nanofiltration. *Iranian Journal of Environmental Health Science and Engineering* 9(17): 0–9.
[43] Mustafa, N., Nakib, H.A.L. (2013) Reverse osmosis polyamide membrane for the removal of blue and yellow dye from waste water. *Iraqi Journal of Chemical and Petroleum Engineering* 14(2): 49–55.

25 Membrane Bioreactors for Wastewater Treatment

Recent Advances, Challenges, and Future Perspectives

Nethravathi[1]*, Arun M. Isloor*[1,*]*, and Ahmad Fauzi Ismail*[2]

[1]Membrane and Separation Technology Laboratory, Department of Chemistry, National Institute of Technology Karnataka, Surathkal, Mangalore, Karnataka, India

[2]Advanced Membrane Technology Research Centre, Universiti Teknologi Malaysia, Skudai, Malaysia

* Author for the correspondence

E-mail: isloor@yahoo.com

25.1 INTRODUCTION

Wastewater is indeed a water resource, provided there exists a trustworthy process for its effective conversion. To deal with the limited water resources available, yet mostly being contaminated, today the world seeks the exploration of promising technologies among which membrane technology is dominant. A membrane is typically a homogeneous or heterogeneous structure acting as a selective barrier separating different phases. Based on different viewpoints membranes have various classifications. By nature, they can be synthetic or biological. Further, synthetic membranes are subdivided into inorganic and organic (liquid and polymeric) ones. Based on the morphology, there are symmetric and asymmetric membranes. Transport across the membrane happens in multiple ways, i.e., active, passive, carrier-mediated, and diffusion transport.

The driving force for the transport may be temperature (ΔT), pressure (ΔP), or electrical potential ($\Delta\mu$), of which $\Delta\mu$ at constant T, P, and E are carried out for most applications, as illustrated in Figure 25.1.

$$\mu_i = \mu_i^O + RTlna_i + V_iP \tag{25.1}$$

where R is the universal gas constant.

Pressure-driven membrane processes are further classified into four types, namely reverse osmosis (RO), nanofiltration (NF) [1,2] ultrafiltration (UF) [3–5], and microfiltration (MF). Table 25.1 presents some characteristics of typical pressure-driven membrane separation

Membrane science has several advantages compared to conventional techniques (namely, flocculation, skimming, centrifugation, electrocoagulation) including its easy integrability with other techniques. Moreover, it is an energy-efficient and cost-competitive process [7–10].

Nowadays, hybrid membrane processes are employed to obtain better performances compared to the standalone processes where a drawback of one technique is complimented by the other. The main objective of a hybrid process is to segregate wastewater into water of different purities that are applicable in different fields such as janitorial services, irrigation, industrial processes, etc.

One such hybrid process is the technology of membrane bioreactors which simply combines advanced membrane technology and the conventional activated sludge process [11]. This is a promising combination as it complements some of the disadvantages of the sludge process such as the inability to cope with the effluent flow-rate and composition fluctuation [12]. Additionally, MBRs offer outstanding disinfection capacity as they completely retain bacteria and suspended solids at the earlier stages of operation [13]. Other advantages include lower footprint of MBRs with obviation in the secondary sedimentary tanks, and fine control of the solid retention time (SRT) [14].

The basic process of MBR follows either submerged or sidestream configurations with two processing steps, as shown in Figure 25.2. First, a bioreactor that contains aerobic bacteria has to digest the organic impurities. As the name aerobic indicates, this process happens in the presence of DO (dissolved oxygen) [15]. Then comes a module-containing membrane separating relatively purified water from the bacteria and organic matter [16]. These two operations can be run successively, which is referred to as a sidestream/external mode. If the membranes are immersed with reasonable partition inside the bioreactor slurry it is referred to as submerged MBR [17].

DOI: 10.1201/9781003285656-25

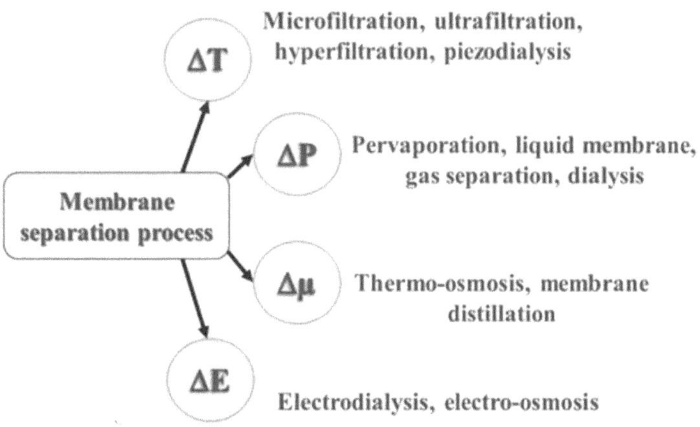

FIGURE 25.1 Classification of the membrane processes based on the driving forces.

TABLE 25.1
Some characteristics of the pressure-driven membrane processes. Reproduced from Ref. [6]

Process	Diameters retained (μm)	MWCO (kDa)	Average permeability (L/ m^2 h bar)	Pressure range	Membrane type	Retained solutes
MF	10^{-1}–10	100–500	500	1–3	Porous, symmetric, or asymmetric	Bacteria, oil, fat, colloids, grease, microparticles, organics
UF	10^{-3}–1	20–150	150	2–5	Asymmetric, microporous	Pigments, proteins, oils, organics, sugar, microplastics
NF	10^{-3}–10^{-2}	2–20	10–20	5–15	Asymmetric, thin-film composites (TFC), tight porous	Sulfates, divalent cations, lactose, divalent anions, NaCl, pigments
RO	10^{-4}–10^{-3}	0.2–2	5–10	15–75	Asymmetric, TCF, semi-porous	Monovalent ions

* MWCO, Molecular weight-cut-off.

The building blocks of MBR have been presented in Figure 25.3. The contaminated water is taken in the feed water tank which is connected to the sedimentation tank which is capable of removing gross settleable solid impurities. The flow from the outlet of this tank is equalized in the next step in the equalization tank. The anoxic tank is used for denitrification, whereas the aeration tank carries out nitrification and biological fixation.

It is absolutely necessary to optimize the various operational parameters while designing a MBR system, as presented in Table 25.2.

Minimizing the cost of MBR maintenance is a great challenge that has to be taken up. Recently, functional machine learning (FML) and functional profile monitoring (FPM) have been used to assess the MBR life-time triggers. Table 25.3 gives an idea about the ways of maintaining MBRs.

Further polymer selection plays an important role in deciding the application area, as described in Table 25.4.

Ceramic membranes are also employed in MBR applications due to their added advantage over the polymeric ones. Resistance to extreme pH in the range 1–14 and temperature (100°C), chemical inertness, reliable life time, better control over pore size distribution, higher regeneration capacity, geometric simplicity, etc. are a few added advantages. However, maintaining all of these magnificent properties, while at the same time optimizing the manufacturing cost and minimizing energy consumption is a great challenge that future research needs to focus on [19–21].

MBR has tremendous applications in various domains. Sirus et al. studied MBR designed using PES/GO membranes for the treatment of wastewater from the dairy industry [22]. Enzymatic MBRs are used in the synthesis of fine chemicals, raw material transport, and in the food industry. MBRs are also employed for the removal of aromatic pharmaceuticals, estrogenic compounds, and micropollutant degradation [23].

Since the BOD/COD ratio of municipal water is 0.4–0.6, under steady-state conditions most of the installations of MBRs have a BOD level less than 2 mg/L. Total suspended solids (TSS) and volatile suspended solids (VSS) are other important effluent characterization tools. Both these tests are cost-effective and require a typical GFC filter paper, desiccator, and drying oven (105°C). A high value of the TSS/VSS ratio indicates cake layer formation. In one work, the effluent quality from MBR and CAS was compared by measuring the TDS, TSS, TP, and TN. MBR was shown to improve the TDS and COD by 5% and 3%, respectively [24]. A glimpse of the characterization tools available for MBR is presented in Table 25.5

MBR is useful for the treatment of both municipal and industrial wastewater, and the required process design is as shown in Figure 25.4.

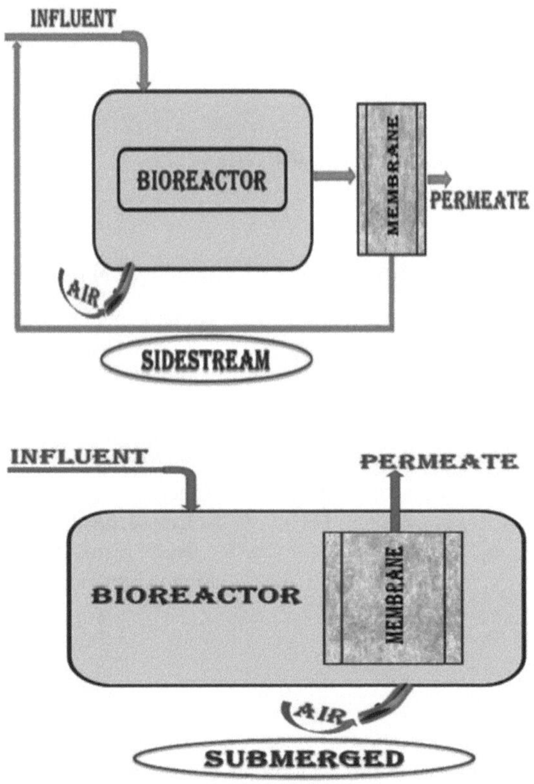

FIGURE 25.2 Basic MBR processes.

MBR works greatly for the removal of micropollutants, organic pollutants, nutrients, phosphorus, and organic contaminants both from municipal and industrial wastewater. The chemical energy contribution of the carbonaceous matter is up to 1.66 kWh/m^3 and nitrogenous matter up to 0.3 kWh/m^3. Rui Li et al. designed a pilot-scale ceramic membrane for the removal of micropollutants such as endocrine-disrupting compounds, personal care products, pharmaceuticals, etc. They investigated an osmosis membrane system with the intention of achieving high water quality [25].

Ying Wang et al. used anoxic/aerobic MBR for the simultaneous removal of carbon and nitrogen from wastewater from food processing. They obtained TN, NH_4^+-N, and COD removal rates of 74%, 91%, and 94%, respectively [26]. The effect of SRT on phosphorus and nitrogen removal from municipal wastewater was studied using sequencing batch MBR (SBMBR). SRT reduction helped to increase the TP removal from 40% to 49% and TN removal from 80% to 86%. However implementation of a pre-aeration phase in SBMBR helped to enhanced TP removal up to 76% [27]. Hadi et al. studied simultaneous nutrient and organic removal from Shiraz municipal wastewater using anaerobic–anoxic MBR. On operating at a constant SRT of 60 days, the results revealed a removal ratio of TP:TN:COD equal to 3.4:21.4:100 [28].

TABLE 25.2
MBR process design parameters

Feed water/waste water (WW) parameters	Membrane parameters	
Flow rate	Membrane module	Process design
Influent BOD		
Influent COD	Module packing density	Membrane flux
Influent BOD		Membrane area
Influent TSS	Specific Aeration demand	Membrane module volume
Influent VSS		
Influent NH_4-N		Scouring air flow
Influent alkalinity		
Aeration WW temperature		

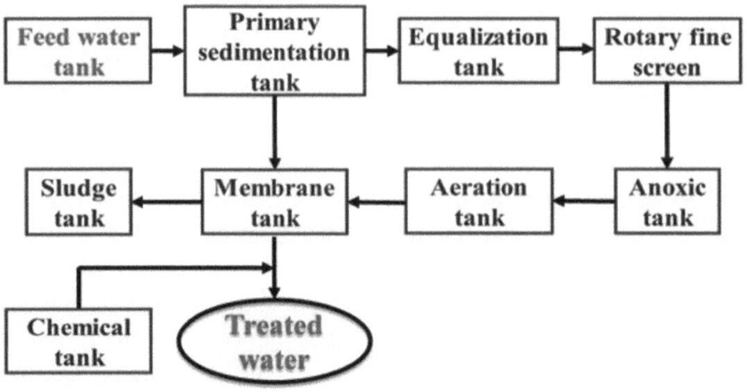

FIGURE 25.3 Building blocks of MBR.

TABLE 25.3
MBR maintenance systems. Reproduced from Ref. [18], copyright Elsevier

	Manual maintenance	FPM-based maintenance		Incompetent maintenance	
	Performance	Performance	Improvement	Performance	Improvement
Operation period	4.22	4.17	18 days	4.30	–31 days
Permeability (LMH/kPa)	1.100	1.108	0.66%	1.1086	–1.33%
Pumping energy (kWh)	22,380	22,278	0.45%	22,271	–1.84%
NaOCl cost ($)	15110.94	14404.25	3.55%	15,586.07	–1.11%

TABLE 25.4
Selection of polymeric and ceramic membrane materials in MBR

Polymer	T_g (°C)	Advantages	References
Polysulfone (PSf)	180–250	Good mechanical strength, chemical resistance	[143]
Polyether sulfone (PES)	220	Compaction resistant, rigid, highly permeable, narrow pore size distribution, oxidant resistant	[22]
Polyacrylonitrile (PAN)	95	High hydrophilicity and mechanical strength	[144]
Polyvinylidene fluoride (PVDF)	–38	Highly oxidant tolerant, chlorine resistant	[145]
Polyethylene (PE)	100	Highly resistant to organic solvents, oxidants, low cost	[21]
Poly propylene (PP)	–25	Decent mechanical strength, good organic solvent resistance	[146]
Cellulose acetate (CA)	198	Renewable source	[145]
Nylon	47	Excellent rejection and selectivity, small pores	[147]

25.2 OPERATIONAL PARAMETERS OF MEMBRANE BIOREACTORS

Proper optimization of the operational parameters is a key factor determining the performance of membrane bioreactors. There are several parameters, namely hydraulic retention time (HRT), solid retention time (SRT), recirculation ratio (α), transmembrane pressure (TMP), temperature, critical flux, etc.

25.2.1 Hydraulic Retention Time (HRT) and Solid/Sludge Retention Time (SRT)

This is the average time duration a substrate spends inside the bioreactor's digester. HRT determines the extent of fouling of the membrane used in the bioreactor. A few researchers have reported that the sludge settleability would be promoted by reducing the HRT value from around 10 to 4, which in turn enhances the resistance of the membrane [29]. A higher HRT value is generally expected for the degradation of complex compounds present in industrial wastewater. This increases the biomass adaptation period. Many authors have studied the effects of HRT on the chemical oxygen demand. It is essential to note that a balance between the impurity removal efficiency and the operational cost must be balanced by HRT. A very long HRT is not preferable [30].

SRT is usually kept longer than HRT, which increases the bacterial adaptability for the desired biodegradation. Qirong Dong et al. studied the impact of SRT of 40–100 days and HRT of 2.5–8.5 hours on municipal wastewater treatment [31]. The sludge and methane production were found to increase with the extension of SRT [32]. The investigation concluded that the increased SRT and decreased HRT are capable of reducing the footprint, thereby bringing down the overall operational cost. However, a reduction in HRT and increase in SRT can promote fouling of the membranes [33].

TABLE 25.5
MBR sample characterization tools

Analysis	Tools	General purpose
Ion analysis	Colorimetry	Phosphate and nitrogen determination
Sum total analysis	BOD_5	To examine the consumption of oxygen by microbes for 5 days
	COD	To quantify the oxidizable organic compound concentration
	TOC	To quantify total bound nitrogen

Meng Xu et al. studied the uncoupling between HRT and SRT and also the ratio of SRT/HRT on the nutrient removal

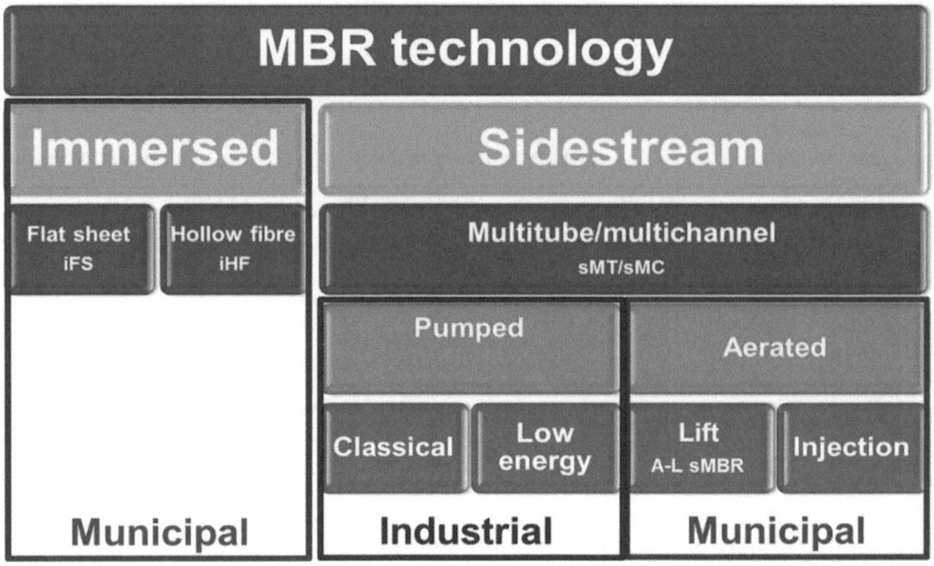

FIGURE 25.4 MBR process configuration for industrial and municipal wastewater treatment.

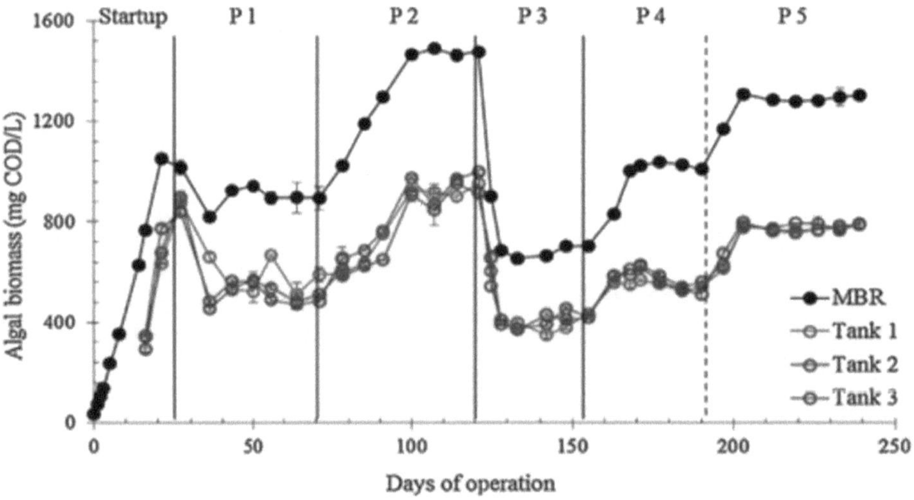

FIGURE 25.5 Concentration of algal biomass at different HRT and SRT. Reproduced from Ref. [34], copyright Elsevier.

and algal growth in an MBR, which is presented in Figure 25.5. They fixed the SRT to 5 days and varied HRT from 12 to 6 hours in first batch. Later, they fixed HRT to 12 hours and varied SRT from 10 days to 5 days. In this research, an SRT of 10 days and HRT of 24 hours showed maximum removal efficiency (total N, 73.4 ± 6.3% and total P, 91.3 ± 3.8%) [34].

25.2.2 Recirculation Ratio (α)

In many studies, the concentrate stream itself is used as a feed stream. The recirculation ratio is calculated as follows (Figure 25.6 gives the recirculation concentration of a pilot MBR)

$$\textit{Recirculation ratio, } \alpha = \frac{\textit{Return activated sludge}}{\textit{Average daily influent flow rate}} \tag{25.2}$$

Gundogdu and co-workers used MBR effluent with the concentrate stream from NF membrane. The mixture was used as a feed solution for another NF membrane. As expected, the flux gradually decreased with a rise in concentration which was attributed to fouling [35]. Miyoshi et al. investigated how recirculation durations affect the removal of nitrogen, organic matter, and phosphorus in an MBR set up. With a rise in the fractions of recirculation time, the efficiency of nitrogen and phosphorus removal was found to deteriorate by a small extent [36]. However, total nitrogen and BOD were lowered below 5 mg/L and 3 mg/L, respectively, irrespective of the operating parameters [37]. They concluded that it is necessary to select appropriate intensity of recirculation in order to achieve efficient removal [38]. Kappel et al. showed continuously produced reusable water by combining NF and MBR and concluded that the recirculation of NF concentrate was

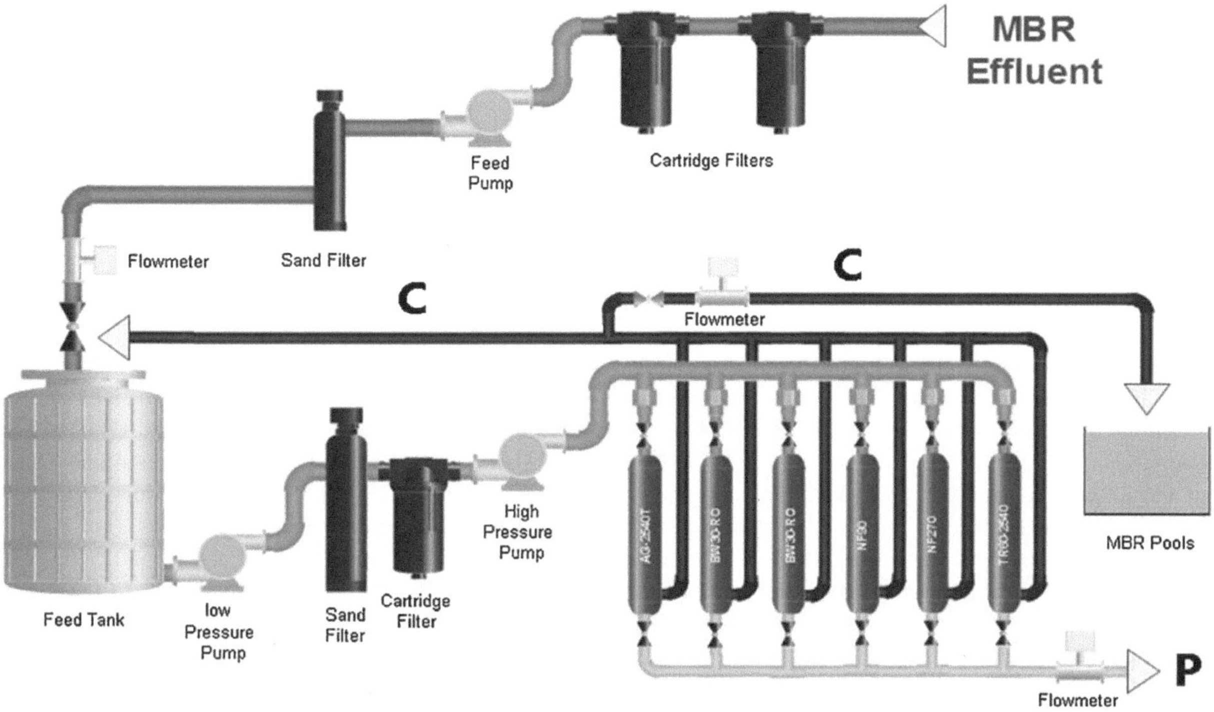

FIGURE 25.6 Concentrate recirculation in mini-pilot membrane system. Reproduced from Ref. [35], copyright Elsevier.

TABLE 25.6
Different definitions of critical flux. Reproduced from Ref. [40], copyright Elsevier

Definition	Restriction	Method of determination
Long-term stable operation	Initial decline in flux is not taken into account	Observation of flux and TMP
Transition between pressure-independent and pressure-dependent flux	Short-term experiment	Hydraulic tests
Inertial lift-velocity (V_{IL})	Theoretical based	Hydraulic tests
Limiting flux	Difficulty in fouling control	Stepwise rise in TMP
No deposition of material	Invisibility of soluble deposition	Mass balance

feasible without compromising on the permeate quality. Moreover, recirculation increased the NF flux provided regular cleaning was carried out [39].

25.2.3 Critical Flux (J_c)

This concept is generally applied to microfiltration. This is a flux point above which membrane gets fouled and below which it does not show any flux decline; other definitions are given in Table 25.6

In order to reduce fouling in MBR, Nguyen et al. compared the fouling behavior of different hydrophilic/hydrophobic membranes at various sub-critical fluxes and investigated the vertical distribution of foulants in the cake layer on the membrane. The results showed that the polysaccharides and proteins foul the membrane at a higher value of sub-critical flux [41].

25.2.4 Temperature

The temperature at which wastewater treatment is carried out is one of the significant factors to be considered during MBR operation. Pretel and co-workers assessed the temperature dependence of the treatment efficiency of a submerged anaerobic MBR. Four factors were considered (namely, energy consumption, sludge disposal, nutrient recovery, and energy obtained from biogas capture) at two ambient temperatures (20 and 33°C). An overall energy balance (OEB) of 0.19 Kw h m^{-3} was obtained [39]. Sarioglu et al. presented dynamic modeling for analyzing the performance of MBR. With the increase in temperature from 24 to 38°C, 40–95% phosphorus removal was observed in the study. The competition substrate storage between GAOs and PAOs (glycogen accumulating and phosphorus organisms) was found to be greatly dependent on temperature. GAOs had lower storage rates than PAOs.

Further, the response from GAOs to the rise in temperature was greater compared to PAOs [42].

25.3 CONFIGURATIONS OF MEMBRANE BIOREACTORS (MBRS)

Typically used configurations often elaborated in the literature are external (sidestream) and internal (submerged) membrane bioreactors.

Submerged MBR systems are preferred for wastewater treatments over sidestream ones due to their hydraulic efficiency, simpler design, and lower consumption of energy. However, the sidestream configuration is comparatively free from fouling that could otherwise reduce flux during the process.

Liu et al. worked on "computational fluid dynamics" (CFD) to theoretically optimize the membrane module and operational parameters of a submerged MBR. Through the investigation of membrane stress and water velocity, a handful of key indexes, including aeration design, membrane distance (d), freeboard height, and gas–liquid dispersion height (h_m) were optimized [43]. Li et al. analyzed the fouling behavior of a submerged MBR considering the sub-critical flux. A three-stage fouling process was observed in their analysis: stage I, linear slow increase in transmembrane pressure (TMP); stage II, an exponential rise with an increase in TMP; and stage III, a rapid linear rise with TMP. At the third stage, a cake-like layer of flocks on the membrane was observed [44].

Ersahin and co-workers studied the impact of a dynamic AnMBRs configuration on its filtration performance. They found that external MBR requires a longer duration than submerged MBR in order to get a high-quality permeate. Also, pyrosequencing analysis performed as a part of the investigation distinguished the structure of the microbial community in the two configurations. Further results showed that the external configuration provides a longer start-up period and a slightly poorer permeate quality in terms of biogas production and COD concentration [45].

Chen and co-workers studied the effect of configurations on the efficiency of a granular anaerobic MBR for the treatment of municipal wastewater. In this research, both submerged and external granular AnMBRs were tested for their wastewater treatment efficacy. Submerged-G-AnMBRs showed a negligible difference in terms of methane yield and organic removal. Also, the fouling propensity of the membrane in submerged configuration was higher [46].

In one attempt, Clouzot et al. investigated the impact of MBR configuration design on the microbial activity and sludge structure via the measurement of rheology. In the external MBR, deflocculation of the activated sludge induced a higher shear stress due to the reduced viscosity. Moreover, SMP (soluble microbial product) was higher in the external configuration. However, the growth of autotropic microbes was found to be easier in submerged MBR [47].

Andrade et al. compared external and submerged MBR efficiency in dairy wastewater treatment. External polymeric substances (EPS) and SMP removal was the focus of the study. EPS are the aggregates of organic matter forming a hydrated matrix of gel thus prompting the growth of flocs and biofilms on the MBR. Here, both configurations exhibited excellent color (98%), COD (98%), and nutrient (86%) removal efficacy [48].

There are several other MBR configurations based on geometry, of which the currently employed are spiral-wound, hollow-fiber, plate-and-frame, tubular, and pleated filter cartridge. Of these the first three are housed in a module after properly sealing the bundle of membranes with the help of epoxy resin [17].

Recently, configurations such as vertical-submerged MBR (VSMBR), hybrid biofilm MBR (HBMBR), submerged-rotating MBR (SRMBR), osmotic MBR (OMBR), reverse osmosis integrated MBR (MBR-RO), membrane-distillation bioreactor (MDBR), jet loop MBR, air-sparging MBR (ASMBR), ammonium oxidation MBR, etc. have been explored to a greater extent. Of these, MBR-RO, MDBR, NFMBR, and OMBR are designed for wastewater reuse. ASMBR, JLMBR, and membrane adsorption/coagulation bioreactors are specially designed for fouling control.

Gholami and co-workers presented an innovative jet-loop airlift type bioreactor for nutrient removal from soft-drink wastewater (COD:N:P = 100:16:1). JLMBRs can provide a system with regions having different reduction–oxidation potentials. Two independent variables, air-flow rate and HRT, were chosen for the modeling, trend, and optimization. Eleven hours of HRT and an air-flow rate of 41/min gave 83.6% and 98% of total nitrogen removal and total COD removal, respectively. μ_{max} calculated from the kinetic studies revealed that JLMBR is a cost-effective treatment strategy for wastewater with high nitrogen and carbon [49]. Parveen et al. integrated FO-MBR and MDBR units together for the regeneration and reclamation of draw solution. Inorganic NaCl, tetraethylammonium bromide (TEAB), and polyelectrolytes were used as draw solutes for the forward osmosis process with different operating temperatures [50]. Tomasini et al. used integrated MBR-RO for wastewater treatment. They observed that the solute concentration did not significantly affect the permeate quality and rejection. However, a very high solute concentration in the feed has the potential to damage the membrane surface [51]. In one attempt, a pilot-scale study of an osmotic MBR was done by a group of researchers. Compared to conventional MBR, OMBR has certain benefits such as lower consumption of energy, improved flux, and lower fouling propensity [52]. Figure 25.7 shows the retrofitting of MBR into OMBR.

Wang et al. tried to clean municipal wastewater by removing the pharmaceutical and personal care products from it with the help of an MBR-RO/NF integrated system. A total of 27 PPCPs were found in real wastewater, with caffeine being in the highest concentration (18.4 ng/L) and trimethoprim being present in the lowest concentration (7.12 ng/L). The designed integrated MBR was able to remove them at rates of 95.41% and 41.08%, respectively [53].

Vertical MBR was implemented by Chae and co-workers for the removal of nutrients and organic matter from municipal

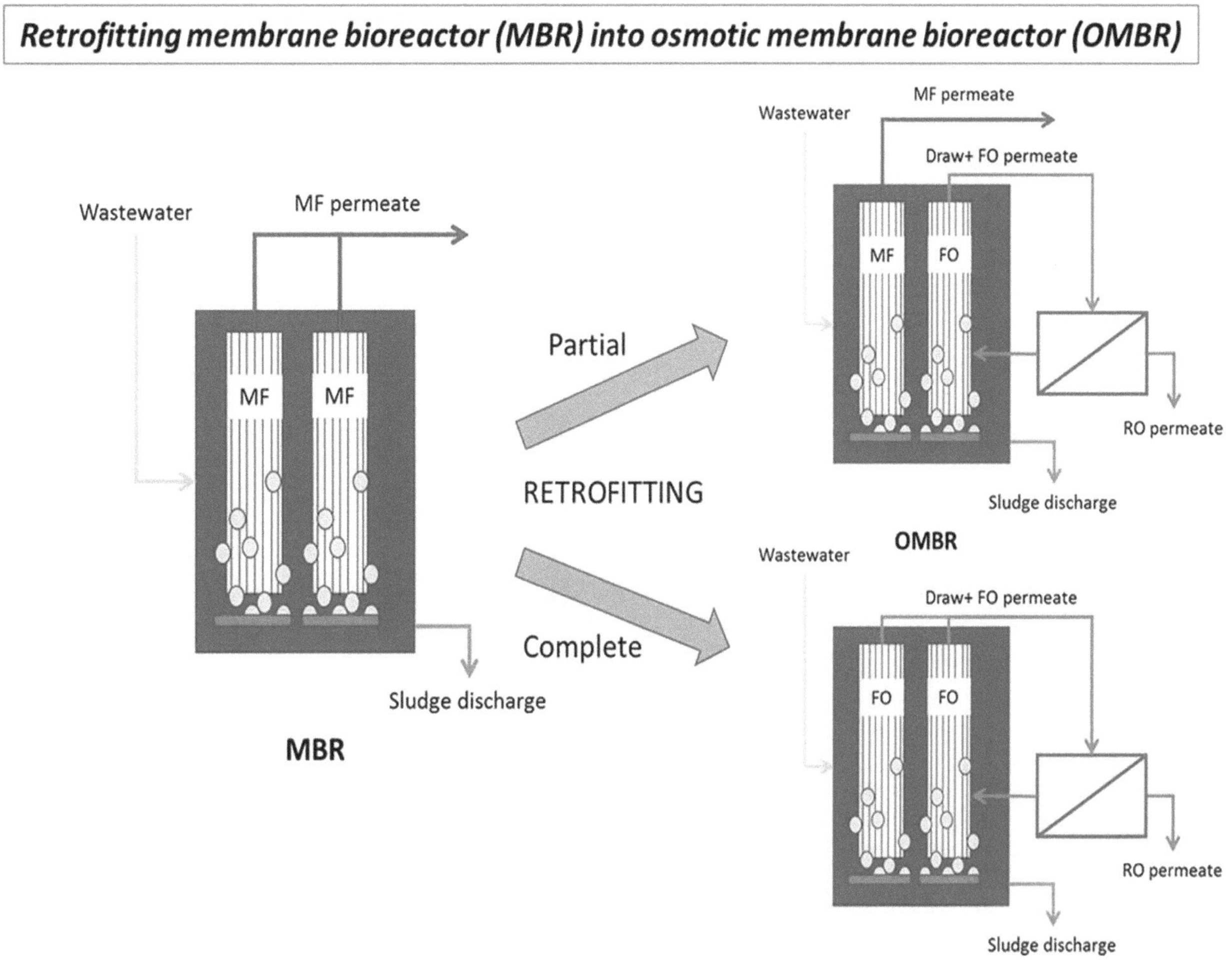

FIGURE 25.7 Retrofitting MBR into OMBR. Reproduced from Ref. [52], copyright Elsevier.

drain out. TMP < 40 kPa was maintained in the study and a flux of 18 L/m^2/h was obtained with an energy consumption of 0.94 kWh/m^3 for a full-scale set up [54]. Jiang et al. studied the effect of HRT on the performance of an HBMBR in the removal of micropollutants from wastewater. Biodegradation was the primary removal mechanism here, and 18 hours of HRT was found to be the ideal condition to minimize fouling propensity [55].

The membrane aerated biofilm reactor (MABR) is another configuration of great interest. At its core there is a spirally wound, self-respiring membrane capable of passively delivering oxygen into wastewater to support the growth of useful bacteria. The setup enables simultaneous nitrification and denitrification. This technique combines gas separation with the biofilm process. In comparison to conventional techniques, MABR has heterogeneous mass transfer, high oxygen transfer efficiency, and biofilm stratification. Xiang Mei et al. achieved 97.15% formaldehyde removal from wastewater using MABR [56]. Corsino et al. monitored a hybrid MABR for real wastewater treatment for 304 days. A biofilm nitrification rate of 2.40 g NH_4 m^{-2} d^{-1} was obtained in 150 days [57].

The hydrogen-based membrane biofilm reactor (MBfR) is another recently explored configuration that can be employed in the treatment of ground water contaminated with scarce organic sources. The configuration uses HF film for biofilm growth, and a hydrogen gas delivery system as an electron donor which provides bubbleless hydrogen for autotropic microorganisms. The electron acceptors are the microbes that respire the oxidized pollutants [58]. Pang et al. studied the effect of carbon fixing on denitrification using MBFR. They found that higher salinity (7%) inhibited nitrate reduction completely [59]. Many researches have proved that both selenium and nitrate present in groundwater can be remove successfully with the help of MBFR. Xia and co-workers used hydrogenophaga and Proteobacteria for selenium and nitrate removal through MBFR [60].

Extractive MBR is yet another interesting configuration which uses highly selective membranes for specific pollutant removal. Thermodynamic affinity and polymer accessibility are the major two aspects governing selectivity. Recently, volatile organic compounds were extracted from waste gas. Mullins et al. used EMBR for biodegradation and selective transport of hydrophobic and hydrophilic contaminants. Excellent contaminant removal was observed in both continuous and batch configurations [62]. Figure 25.8 presents the design of an extractive membrane bioreactor.

Recently, hybrid MBRs have been used in many applications. Media-based MBRs, dense membrane MBRs,

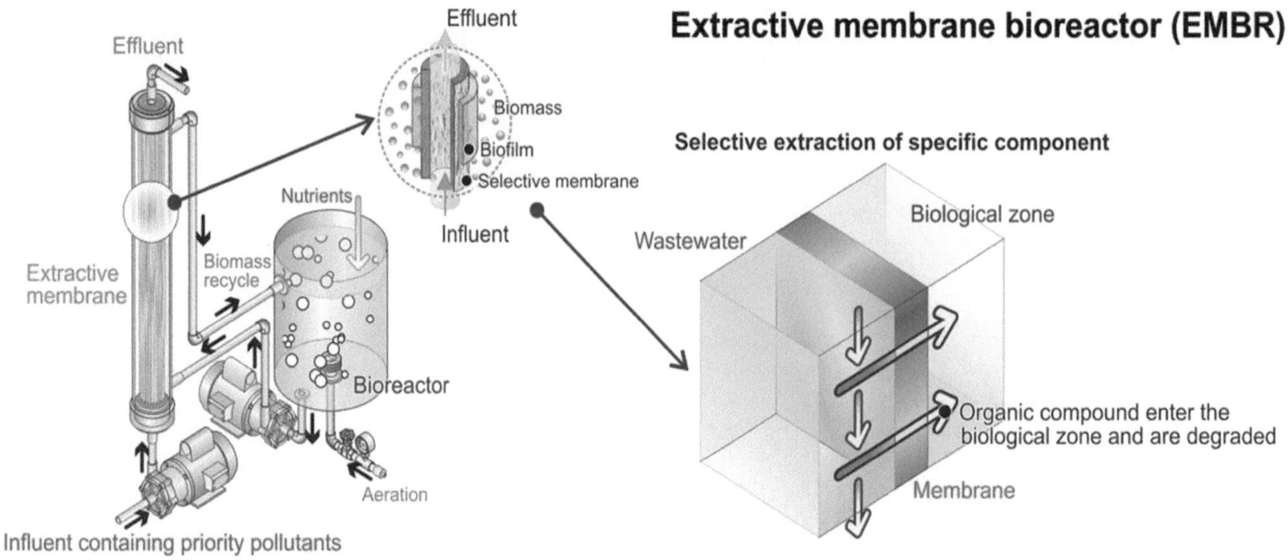

FIGURE 25.8 EMBR for wastewater treatment. Reproduced from Ref. [61], copyright Elsevier.

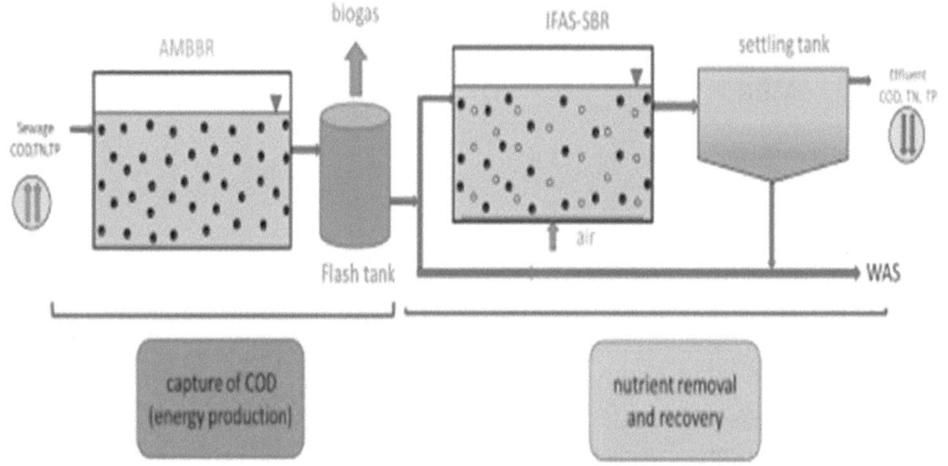

FIGURE 25.9 Media-based MBR for biomass generation and nutrient recovery. Reproduced from Ref. [64], copyright Elsevier.

and osmotic MBRs are the broad domains. Integrated fixed-film activated sludge (IFAS) and moving-bed bioreactors (MBBRs; shown in Figure 25.9) are the two main commercial media-based MBRs. In IFAS, recirculation of the sludge is performed, whereas MBBR has no sludge return line. In IFAS, the resistance from organic loading and hydraulic fluctuations is enhanced by lowering the carrier filling ratio, whereas MBBR has a typical carrier filling ratio of 60–70% [63].

The challenge here is to understand the composition and dynamics of the biofilm. The seeding phenomenon affects the bacterial dynamics in the suspension which also has the possibility of enhancing the nutrient removal. This needs to be explored in greater depth. Since the industrial application of IFAS coupled membrane has been less explored, a detailed study on the IFAS-MBR foiling control and microbial community could help in this regard [65].

Dense membrane MBRs typically adopt reverse osmosis (RO), electrodialysis (ED), forward osmosis (FO), and nanofiltration (NF) technologies which are capable of rejecting even the dissolved charged species [66]. Of these, NF and ED are more of an extractive process than just filtration. Recent research is focusing more on using hybrid membranes of MBR/NF and MBR/FO to achieve desalination and biological treatment in a single step. However, reduced permeability is the challenge concerned with dense membranes [67].

Shi Zhang et al. studied the effect of periodic replenishment of powdered activated carbon on the stability and performance of hybrid MBRs. Effective mitigation in the MBR membrane fouling was achieved at a 1.67% PAC replenishment ratio [68].

25.4 ADVANTAGES AND LIMITATIONS OF MEMBRANE BIOREACTORS

Excellent quality of treated water, smaller reactor and footprint requirements, higher biodegradable efficiency, ease of operation, and great biomass retention are a few commonly observed advantages of MBR over CASP ("conventional activated sludge process"). MBRs are also capable of displaying flexible operation and can handle higher volume loading of up to 20 kg COD/m^3/day, good disinfection capability (< 0.5 NTU), compactness, and low production of sludge. The biological as well as chemical oxygen demand can be removed effectively through MBRs.

Alongside the advantages, there are a few limitations that MBR possesses that hamper its application in various domains. Fouling is one such serious issue to worry about that reduces the output of the treatment process, declines flux, increases the TMP rapidly, and increases consumption of energy. Maintenance and operational costs are added disadvantages.

25.5 SOME KEY COMPANIES MANUFACTURING MEMBRANES FOR MBR APPLICATIONS

Mitsubishi Rayan Co. Ltd. came up with a new energy-efficient hollow fiber named sterapore featuring a PVDF HF with high chlorine resistance for MBR application. More than 4000 municipal as well as industrial MBRs are equipped with sterapore. Asahi Kasei company manufactures PVDF HF membranes and their module is popular worldwide as MICROZA. These membranes are capable of withstanding up to 5000 ppm chlorine, 4% NaOH, 1% H_2O_2, and up to 10% HCl, H_2SO_4, and citric acid cleaning conditions [69]. Evoqua is another company standing as company of the year (2018) in the water technology solution market; MemPulse is the MBR system supplied by this company which provides reduced biosolids, energy, and footprint. Toray's NHP and TMR09 MBR series are comprised of membranes coated with a frame of stainless steel, pervaded water manifold, and aeration feature demanding minimum maintenance. Tianjin MOTIMO Membrane Technology Ltd. (MOTIMO) provides HF column membranes, curtain style UF for continuous membrane filtration, submerged membrane filter, and MBR applications. Kubota Membrane Europe supplies membranes for municipal, industrial, and commercial MBR applications helping users achieve TN removal of 3 mg/L and TP removal of 2 mg/L [70]. Ovivo/GLV group is another global force with leading MBR innovations targeting energy neutrality, treated effluent reuse, and complete nutrient recovery all at an affordable cost, and with market experience of 200 years. Koch Membrane Company manufactures MBR membranes under the trade name PURON which is a reinforced PVDF UF (0.03 micron). The single header design provides minimal clogging, high flux, and reduced footprint. The complete MBR package can provide flow rates of up to 250,000 GPD (950 m^3/day). In addition, these systems can reduce nitrogen concentration and BOD to 10 mg/L and 5 mg/L, respectively [71]. Beijing Origin Water Technology Co., Ltd. supplies membranes (pore size = 0.3 micron) with high tensile strength (≥ 200 N) and high chemical stability producing high-quality permeate (turbidity close to 0) [72]. GE Water Process and Technology supplies MBRs with ZeeWeed HF-UF membranes. Nearly 1000 water treatment plants use this technology [73].

25.6 NANOMATERIALS USED IN MEMBRANE BIOREACTORS (NMS-MBR TECHNOLOGY)

With the intention of overcoming the limitations of MBR, the membranes used here are modified by incorporating several types of nanomaterials, which provides a great leap toward improved wastewater treatment [74]. Nanomaterial-based membranes are capable of elevating the MBR performance in terms of thermal stability, fouling control, hydraulic stability, water permeability, and better selectivity [75]. To date, numerous nanomaterials, namely nanofibers (NFs), nanotubes (NTs), nanoparticles (NPs), nanocrystals (NCs), nanowires (NWs), and nanosheets (NSs) have been employed to improve membrane properties in MBR [76], as shown in Figure 25.10

Despite the several advantages, NM incorporation in the membrane matrix of MBR also poses certain challenges, such as leaching. If leached into aquatic regions, NMs can get transformed and be taken up by aquatic creatures, threatening their life.

NF-MBR contains membranes incorporated with fibers of diameter typically about 100 nm providing a super high aspect ratio which enables them to interlock in a subtle form [77]. Electrospinning is commonly employed for the fabrication of nanofibers and provides several advantages including material selectivity, operational ease, and low cost. The chemical as well as physical properties of electrospun NFs can easily be tuned as per the requirements. Arslan and co-workers studied pressure-driven MBR having fiber-incorporated membrane [78]. Polyamide-coated electrospun nanofiber was used here. The system showed >99% removal of organic carbon, 70.8% removal of total nitrogen, and 99.8% removal of total phosphorus [79]. Membrane hydrophilicity is a prominent factor deciding the membrane performance, especially when it comes to water purification applications. In one attempt, PVA-based nanofibers were fabricated as PVA has abundant hydroxyl groups on its side chain [80].

NP-MBR embeds nanoparticles in the membrane matrix due to the availability of a larger surface area and flexible shape and size-dependent properties [81–84]. Further, they possess functionalizable chemical groups. However, it is absolutely necessary to understand the fate of the nanoparticles used in wastewater purification systems [85–88]. Yuan et al. assessed the fate of Ag-nanoparticles in an anaerobic-anoxic-oxic MBR. They observed the aggregation of Ag NPs and formation of Ag-sulfur complex in the sludge. Only < 0.5% Ag could pass across the HF membranes with a 0.1 μm pore size [89]. Wang et al. studied the long-term effects of zinc oxide (ZnO) nanoparticles. They concluded that these ZnO nanoparticles have no adverse environmental effects [90].

NTs are another class of nanomaterials attracting significant attention due to their large surface area, higher aspect ratio, tunable physical and chemical properties, light weight, and easier functionalizability. Carbon nanotubes, unlike activated carbon, can be self-assembled and immobilized on the supporting material through techniques like chemical vapor deposition. In one study, Almusawy and co-workers designed a nanocomposite membrane made of ZnO/PPSU

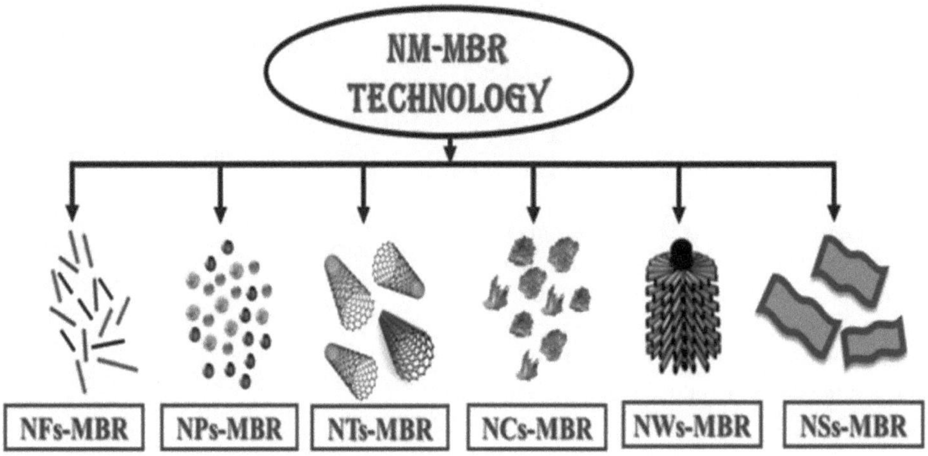

FIGURE 25.10 Types of nanomaterials used in MBR.

and employed it in an MBR along with polyurethane coated with multi-walled carbon nanotubes (MWCNTs). The results showed efficient adsorption of activated sludge by spongy-CNTs, thereby minimizing fouling of the membrane. Further, 81% of phosphorus and >90 % of nitrogen were eliminated by 0.3 mg/CNT [91]. Generally, in MBR, bacteria survive by consuming the organic wastes thereby producing clean water. Therefore, the fabricated membranes should be non-toxic to bacteria while still possessing antifouling character. CNTs via their oxidative-stress stimuli act as resilient antibacterial nanomaterials exerting a toxic effect through the release of reactive oxygen and ions. Ayyaru et al. fabricated a PVDF membrane blended with CNTs and sulfonated CNTs (SCNTs) with a mean pore size of 50 and 60 nm, respectively. SCNTs showed 90% bovine serum albumin (BSA) rejection. On the other hand, both PVDF-CNTs and PVDF-SCNTs were non-toxic to the useful bacteria [92].

Nanocrystals (NCs) are yet another form of nanoparticles used by researchers for improving membrane properties. Cellulose acetate NCs are commonly used filler due to their superior thermal property, biodegradability, higher tensile strength (~7 GPa), and higher Young's modulus (~130 GPa). It has been reported in several literatures that the incorporation of CNCs increases the membrane hydrophilicity, surface roughness, and charge density. Jinling Lv et al. synthesized graphene oxide-cellulose NC composites and used them as a hydrophilic agent on PVDF. The fabricated membrane showed higher wettability, and lower polysaccharide and protein adsorption compared to the virgin membrane. The MBR exhibited long-term antifouling behavior, higher porosity, and a high negative zeta potential value. The hydrogen bond formed hydroxyl groups of cellulose nanocrystals and the oxygen-possessing groups of GO facilitate its better dispersion in solvent [93].

NWs provide an aspect ratio of over 1000 in comparison to other 1D nanomaterials. Nanowires offer uniform, flexible, and multifunctional activity. Wang et al. hydrothermally synthesized nanowires of sodium niobate with a strong piezo-catalytic property for wastewater dye (Rhodamine B) degradation. The piezoelectric coefficient (d_{33}) was ~12 p.m./V. When vibration is applied externally, the nanowires of sodium nitride bend, which in turn accumulates large negative and positive charges on its surface. Nearly 80% dye degradation was achieved using this approach. Parameters including initial dye concentration, vibration time, and amount of nanowire additive decide the degradation efficiency [94]. Yin et al. reported the microfiltration conductive membrane made of Cu-nanowires for fouling mitigation. A spontaneous electric field of over 0.073 V/cm was responsible for the mitigation of fouling. The fouling cake layer on the nanowire-incorporated membrane was about 80 µm, which was much thinner compared to neat PVDF. First-class quality of discharge was achieved with 94.5% COD removal, 78.5% total nitrogen removal, 86.6% removal of total phosphorus, and 99.8% removal of $(NH_4)^+$- N without much power consumption [95].

NSs are the next-generation membrane materials which are two-dimensional nanomaterials with molecular or atomic ratios. Graphene oxide (GO), carbon nitride (C_3N_4), boron nitride (BN), and molybdenum disulfide (MoS_2) are a few widely used nanosheets in water purification. Among them, GO-derived nanosheets were found to be more efficient due to their different oxidation state, higher mechanical strength, and higher charge density. Zhibo Ma and co-workers fabricated a super-hydrophilic yet relatively oleophobic membrane via surface functionalization. In this study, layered double hydroxide (M-LDH) and modified SiO_2 tethered PVDF membranes were used. The lamellar LDH serves as an oleophobic barrier giving 1.5 times better permeability compared to the virgin membrane. Moreover, the surface functionalization imparted superior antifouling capability [96]. Liang et al. reported zwitterionic functionalized nanosheets of MoS_2 for dye/salt separation. Polyether sulfone, which is highly hydrophobic, was the base polymer used in this research. The hydrophobicity was reduced by the incorporation of MoS_2/PSBMA prepared using the RATRP technique ("Reverse Atom Transfer Radical Polymerization"). Rejection rates of 98.2% for reactive black 5 and 99.3% for reactive green 19 dye were achieved with this technique [97].

25.7 ANAEROBIC MEMBRANE BIOREACTORS

MBR embedded with a membrane that operates in the absence of oxygen is called as an anaerobic MBR. This is a cost-effective technology compared to an aerobic one as the sludge handling and aeration costs are negligibly lower; other comparisons are provided in Table 25.7.

The energy recovery in AnMBR is governed by the transformation of organic carbon to methane gas. Compared to CAS, one of the key operating challenges of AnMBR is to optimize the membrane performance whilst minimizing fouling. Open-loop and closed-loop control strategies have been developed with this in mind. The main challenge here is the lack of reliable and robust online instrumentation to implement these strategies. Prolonged SRT and clarified effluent are a few added advantages of AnMBR. Sidestream is a well-established configuration of AnMBR which can provide high shear operational stability. Hence they are suitable to be operated under extreme conditions such as high salinity, higher concentration of suspended solids, or poor granulation of biomass. There are many operational advantages also. AnMBR does not require oxygen for biotransformation of the organic matter, which reduces the total energy consumption. Further, biogas is produced as a useful end product from the organic matter treatment which can then be combusted to generate heat and electricity. In turn, the heat generated is used to maintain appropriate temperature for the digestion process in the reactor.

Ultrafiltration (UF) and microfiltration (MF) membranes are commonly used in AnMBR. With the objective of evaluating the temperature effect on the performance of industrial HF membranes in a submerged AnMBR, Robles et al. operated two commercial HF-UF membrane modules at temperatures of 20°C, 25°C, and 33°C. Higher permeability was observed under psychrophilic conditions than mesophilic ones due to lowering of the biomass activity in the former condition. Also, at the psychrophilic condition, EPS (extracellular polymeric substance) and SMP (soluble microbial products) levels were observed to be lower, which helped to reduce the fouling propensity. The system did not demand any chemical cleaning as the fouling observed was reversible [99]. In another attempt, the same group of researchers studied the effects of other operating parameters such as intensity of gas sparging and back-flush (BF) frequency on the performance of HF AnMBR. They observed a need for low gas sparging intensity and low BF frequency at the sub-critical conditions and significantly higher BF frequency and higher gas-sparging intensity was demanded by supra-critical conditions. Even after the longer operation (> 2 years), significant irreversible fouling was not observed, which made the cleaning easy [99].

TABLE 25.7
Comparison of aerobic and anaerobic MBR. Reproduced from Ref. [98], copyright Elsevier

Feature	Anaerobic	Conventional Aerobic
Bioenergy recovery	Yes	No
Sludge production	Low	High to moderate
Footprint	Low	Low
Alkalinity requirement	High to moderate	Low

Several configurations of AnMBR have been designed recently. Broadly, (1) "completely stirred tank reactor" (CSTR) and (2) UASB reactor are the two types. Svojitka et al. studied the performance of an AnMBR in the treatment of pharmaceutical wastewater, operating it for 580 days. The main objective of the study was to recognize the inhibitory factors, thereby testing its long-term application. Around 97% removal of COD was observed here. Optimum efficiency was procured with the addition of methanol into the influent. Further, poor production of methane, acidification, and methanogenic population shift were witnessed with the addition of co-substrate to the influent [100].

Chen and co-workers investigated the treatment of brewery wastewater with the help of an AnMBR. One litre of beer is capable of generating 3–10 L of wastewater, having soluble starch, sugars, ethanol, and some suspended solids. A total of 98% of COD removal was observed with 0.53 ± 0.015 m^3 biogas/kg COD biogas yield. The result showed that 0.04 μm ultrafiltration membrane was the best performing membrane with a critical flux in the range 8.64 ± 0.69 L/m^2/h [101]. Gu et al. used an energy-efficient innovative RO-ion exchange MBR process for the reclamation of wastewater. A total of 76.8% of COD of the influent was converted into methane, thereby treating wastewater equivalent to 0.41 kWh/m^3. Moreover, around 95% of phosphate, organic carbon, ammonium, and some other major anions and cations were successfully removed [102].

Yurtsever and co-workers investigated the performance of an AnMBR in the treatment of real textile water. In addition, the fouling rate was also investigated. COD and color of the permeate were decreased to 70 mg/L and 150 Pt-Co, respectively, with a weekly cleaning. Gel permeation chromatography showed that the membrane present in the system cuts off all the molecules with a molecular weight greater than 15 kDa with 4.1 ± 0.7 LMH [103]. Figure 25.11 shows other creative configurations of AnMBR.

A study of a pilot plant of a newly designed HF membrane module in a submerged AnMBR was conducted by a group of researchers. Assessment of the experiment showed 17–20% improvement in the performance of the membrane with rotation at sub-critical conditions. Further, the technique was capable of removing the fouling cake from the membrane surface, which provides cleaning-free filtration. At the supra-critical conditions, flux of 16 L/hm^2 was observed with the enhancement of the efficiency of backwashing [105].

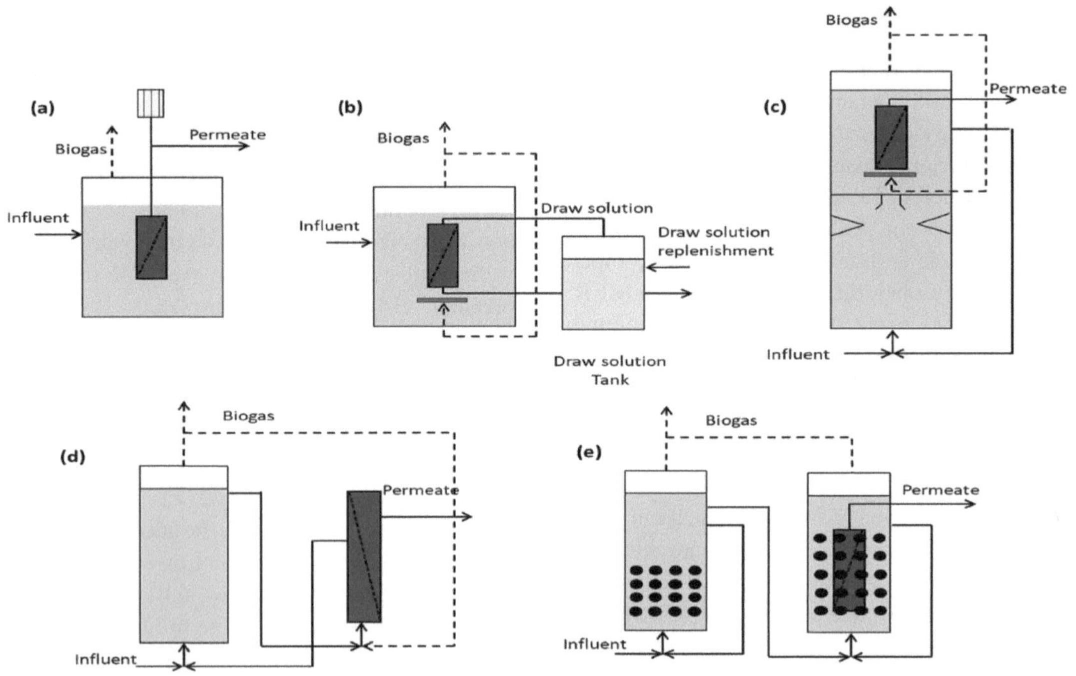

FIGURE 25.11 AnMBR configurations: (a) rotating membranes; (b) osmotic AnMBR; (c) coupled AnMBR-UASB; (d) gas-lift AnMBR; (e) two-staged fluidized AnMBR. Reproduced from Ref. [104], copyright Elsevier.

25.8 FOULING OF MEMBRANES

Fouling is a serious issue which results in frequent replacement of the membrane, and lowers the quality of effluent adversely affecting the overall operational cost [106]. Fouling can be classified in various ways. Broad categorization includes external, internal, and concentration fouling. Reversible and irreversible fouling are the types of fouling based on the degree of removal. Further, based on the materials causing fouling there are inorganic fouling, organic fouling, and biofouling, of which biofouling is comparatively more serious. Biofouling is generally caused by the accumulation of "soluble microbial products" (SMP) and "extracellular polymeric substance" (EPS) that are secreted during the fouling development by the microorganisms finally creating a biofilm [107]. Four stages govern biofilm formation, as shown in Figure 25.12: (1) microbial cell attachment; (2) secretion of the microbial products on the surface of the membrane; (3) EPS production on the membrane surface; and (4) microbial cell multiplication and detachment.

There are several factors which decide the extent of fouling in MBR. Predominantly, (i) the material from which the membrane is made of; polysulfone (PS), polyethylene (PE), polyether sulfone (PES), polyvinylidene fluoride (PVDF), and polyacrylonitrile (PAN) are a few organic polymers popularly used as base materials, of which PVDF has a greater antifouling nature. (ii) The membrane's intrinsic properties; hydrophobicity/hydrophilicity, pore size, roughness, surface charge, membrane module structure, and porosity of the membrane [109].

Operating conditions have a great influence on the fouling phenomenon. First, the membrane flux and the transmembrane pressure (TMP), which are interrelated, affect the fouling in such a way that operation above the critical TMP results in the formation of a thick cake of fouling. Aeration is another parameter that not only washes the membrane surface but also provides oxygen required for activated sludge metabolism. Cross-flow velocity (CFV) is another minor factor which affects the thickness of the fouling layer. As the CFV increases, deposition of large particles on the membrane surface is reduced. However, a higher value of CFV will break the larger sludge particle, thereby making the cake layer denser. Also, a rise in CFV would stimulate EPS release, thereby increasing fouling. Next, the SRT and HRT values both have direct effects on the flux of the membrane and also on fouling. It was observed that a shorter HRT value produces a higher organic loading rate, which in turn influences the metabolic activities of the microbes and increases the sludge particle size, finally causing fouling. Temperature alters the rate of mass transfer, enzymatic activity, particle solubility, microbial activity, and also the viscosity of the medium. It was found that, at higher temperatures, anaerobic microbes are capable of exhibiting higher activity and high COD removal, while at lower temperatures microbes can release more EPS.

The major microbes present in the MBR are bacteria (e.g., Proteobacteria), Metazoa (e.g., Rotifers, Nematodes), Protozoa (e.g., Flagellates, Amoebae), algae, filamentous bacteria, and fungi. However, the majority of the microbes are bacteria. Chains, paired forms, and clusters are a few biomass morphologies. The metabolisms include numerous energy

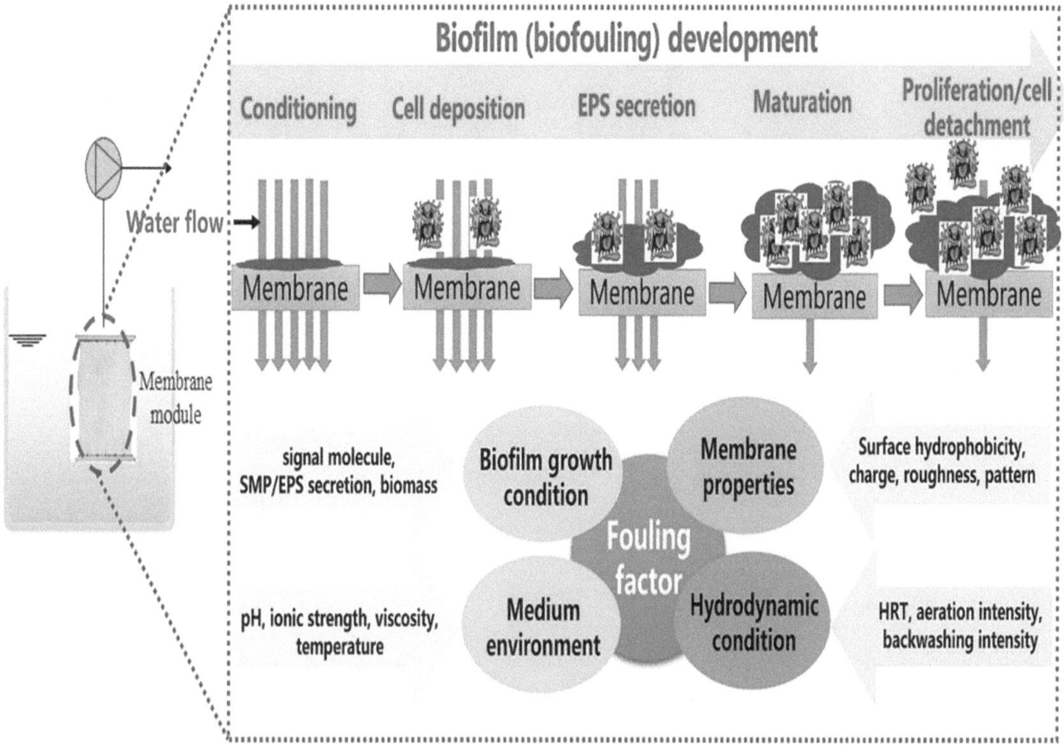

FIGURE 25.12 Stages of biofouling and the governing factors. Reproduced from Ref. [108], copyright Elsevier.

sources, electron acceptors, electron donors, and carbon sources. Biomass yield is the ratio of generated biomass to the substrate consumed. Figure 25.13 depicts the factors determining the performance of MBR.

25.8.1 Fouling Control

Different biological and physicochemical strategies have been adopted to regulate fouling of membranes in MBRs.

25.8.1.1 Biological Approach

With the motive of preventing biofilm formation, techniques such as quorum quenching (QQ) [111], energy uncoupling (EU), and enzymatic disruption (ED) have been employed. Of these, QQ is an in situ control of fouling method [112]. This technique inhibits the synthesis of a signal molecule responsible for fouling [113]. However, this approach has limitations of high cost and instability of enzymes in the sludge [108]. Bio-stimulation quorum quenching is another innovative approach where researchers use some synthetic or natural analogs of signal molecules [114]. In a few studies, metabolic uncouplers such as nitrated and chlorinated phenols and tetrachlorosalicylanilide (TCS) were used for the inhibition of fouling. These special molecules were used because of their capacity to dissipate proton-motive force and disrupt ATP synthesis of the biofilm microorganism [115].

QQ is an enzymatic strategy which governs the mechanism of degradation of AHL structures, as shown in Figure 25.14 and Table 25.8. Investigations have been carried out to immobilize the quenchers onto the carrier to reduce the loss of enzymatic activity.

The addition of bio-stimulants is an efficient way to promote the growth of QQ bacteria. Gamma-caprolactone (GCL) is one of the most widely studied biostimulants which enhances the growth of *Rhodococcus* that can degrade AHLs, resulting in the reduction of biofouling.

Enzymatic disruption is another smart way of preventing biofouling. Soluble microbial products (SMP) and extracellular polymeric substances (EPS) are the major components responsible for bio-cake formation. Since proteins and polysaccharides are the key ingredients of these pollutants, proteolytic (proteinase K, trypsin, subtilisin, etc.) and polysaccharide-degrading enzymes (glucanase, cellulase) can help inhibit EPS. However, cost0effective enzyme extraction maintaining the stability of these enzymes at various temperature, pH, and salt concentration is a future research challenge.

Energy uncoupling comes under enzymatic disruption. The main source of energy for microbial metabolism comes from ATP (adenosine triphosphate). The coupling between oxidative phosphorylation and electron transport produces proton motive force (PMF) on which EPS synthesis depends. The addition of uncoupling agents would inhibit the coupling process. A few uncoupling agents have been listed in Table 25.9

25.8.1.2 Physicochemical Approach

Several in situ chemical and physical approaches have been adopted widely to clean fouled membrane and enhance

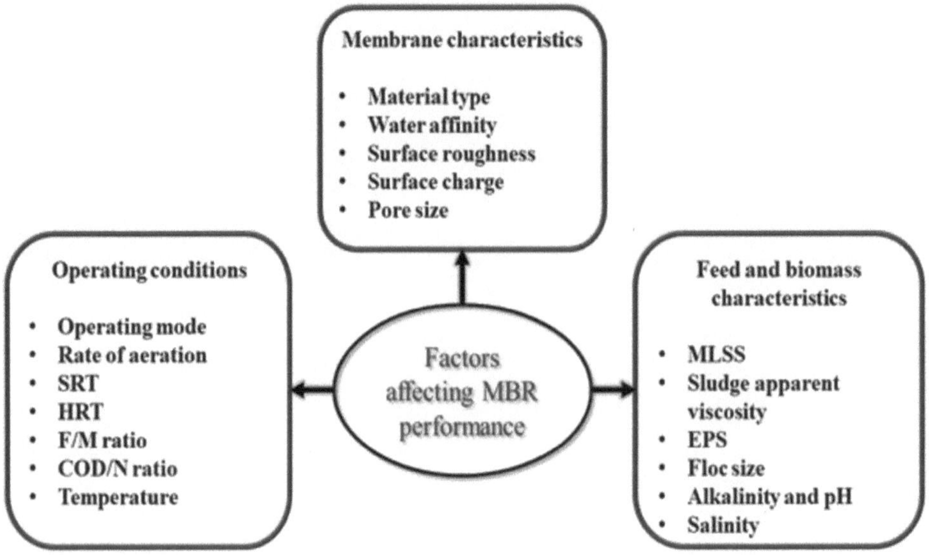

FIGURE 25.13 Factors affecting MBR performance.

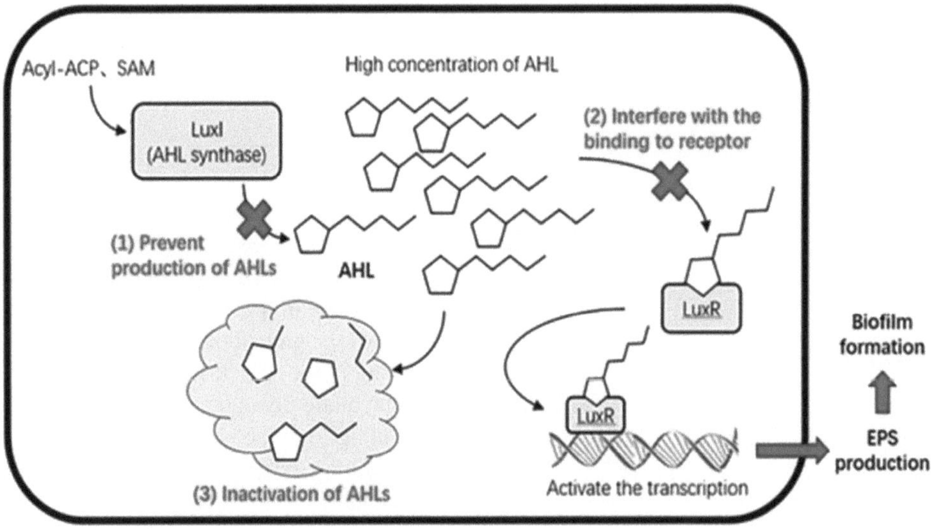

FIGURE 25.14 Quorum quenching strategy for preventing biofilm formation in MBR. Reproduced from Ref. [116], copyright Elsevier.

its reusability [113]. Physical cleaning techniques involve the application of forces such as hydraulic backflushing, mechanical/air scouring, etc. Implementation of dynamic membrane technologies (DMTs) like rotating, oscillating, and vibrating membranes also belongs to this category [117]. A few membrane cleansing strategies have been presented in Figure 25.15.

The chemical method uses cleaning agents like oxidants (H_2O_2, NaOCl, etc.), acids (HCl), bases (NaOH), surfactants, chelates, and some disinfectants which are preinjected into the backflushing solution. This process is called chemical enhanced backflushing (CEB) [118].

Recently, hybrid electrothermal methods are employed where an electrical field is applied to suppress fouling. This approach includes electrophoresis, electroosmosis, electrochemical QQ, and electrocoagulation that are capable of degrading the foulants and can control sludge mobility. Electrophoresis can drift the charged foulants away from the membrane, thereby reducing the chances of their deposition on the membrane [119].

Bansi et al. investigated the parameters responsible for fouling and tried to find a promising solution in a pilot step-aerating MBR taking filamentous bacteria as a model. With a F/M loading of ≤ 0.65 ± 0.2 g, COD/g MLSS/d at the temperature of 20 ± 3°C, HRT = 1.6 h, DO = 2.5 ± 0.1 mg/L, the bacteria were under control at an index of 1.5–3.0 [120].

Yang et al. tried to optimize the MBR hydrodynamics for the formation of fouling cake layer control using computational fluid dynamics (CFD) and response surface methodology (RSM). The effects of different factors on the fouling rate were

TABLE 25.8
QQ strategies

Categories	Names	Active component	Mode of action	Targeted molecule	Biofouling control capability
Enzymes	AHL-lactonase	*Halomonas* sp. strain 33	AHLs degradation		
	AHL-acylase	*Tenacibaculum discolor*		C_4-HSL, C_{12}-hsl and C_8-HSL	Fouling was reduced from 58.0 to 85.8%
	AHL-oxidoreductase	*Burkholderia* strain GG4		3-oxo-C_6-HSL	
	AHL- oxidase	*Bacillus megaterium*		C_4-HSL and C_{12}-HSL	
QQ bacteria	*Rhodococcus* sp. BH4	Lactonase	Signal molecule degradation	(C_6, C_7, C_8, C_{10}, 3-oxo-C_{12}, 3-oxo-C_6)-HSL	Fouling was reduced by 75.0–89.0%
	Bacillus methylotrophicus sp. WY	Lactonase		(C_6, C_7, C_8, C_{10}, 3-oxo-C_{12}, 3-oxo-C_6)-HSL	Targeted molecule degradation exceeds 90% and flux was increased 3–4 times
	Enterococcus sp. HEMM-1 *Serratia* sp. Z4	Lactonase		(C_4, C_6)-HSL, BHL	Reduction in biofilm by 15–44%
	Candida albicans	Farnesol		AI-2 (DPD)	Reduction in biofilm by about 70%
	Pseudomonas nitroreducens JYQ3	Acylase		C_6-HSL	Membrane flux was increased by 19%
	Pseudomonas JYQ4	Acylase		C_6-HSL	Membrane flux was increased by 22%
	Pseudomonas sp. 1A1	Acylase		(C_6, C_8, C_{10}, C_{12})-HSL	Fouling was reduced by about 63.6%
	Delftia sp. T6	Acylase		C_8-HSL	Reduction in biofilm by 76%
	Bacillus sp. T5	Acylase		C_8-HSL	Reduction in biofilm by 85%

TABLE 25.9
Uncoupling agents

Name	Effect	References
TCS (3,3′,4′,5-tetrachlorosalicylic acid)	ATP synthesis reduced by 75–90% resulting in the slowdown of fouling by 2 times	[116]
NP (Nitrophenols)	ATP synthesis reduced by 75–81.8%, EPS release was reduced from 26.98 mg/g VSS to 20.52 mg/g VSS	[148]
DNP (2,4-dinitrophenol)	High dosage retarded membrane fouling	[149]
OCP (O-chlorophenol)	Reduced fouling	[150]
CCCP (Carbonyl cyanide m-chlorophenylhydrazone)	Significant inhibition in the biofilm formation	[151]

analyzed and these were found in the order, bubble diameter in the presence of shear stress < aeration intensity and MLSS < bubble diameter < MLSS, respectively. Optimum fouling control was achieved with MLSS = 8820 mg L^{-1}, airflow rate = 2.0 m^3 h^{-1}, and bubble diameter = 4.88 mm [121].

Use of coagulants is one the most widely adopted techniques in fouling control. Gkotsis et al. compared the efficiency of different commercially available coagulants/flocculants. $Fe_2(SO_4)_3.5H_2O$, $FeCl_3.6H_2O$, $PSF_{0.3}$, $FeClSO_4$, PAC, A9-M, $Al_2SO_4.18H_2O$, $NaAlO_2$, PAC-A16, and FO4350SSH were the selected commercial coagulants. After selection of the most efficient flocculants, major fouling indices, such as TTF, TMP, and SMP concentrations, were correlated with the intention of improving the overall processability [122]. In one attempt, Suh et al. investigated different physical and biological fouling control models. The model was allowed to undergo some biological processes such as nitrification, biomass growth, denitrification, formation, and degradation of SMP and EPS. In this study, with the increase in idle-cleaning time, fouling was found to converge [123].

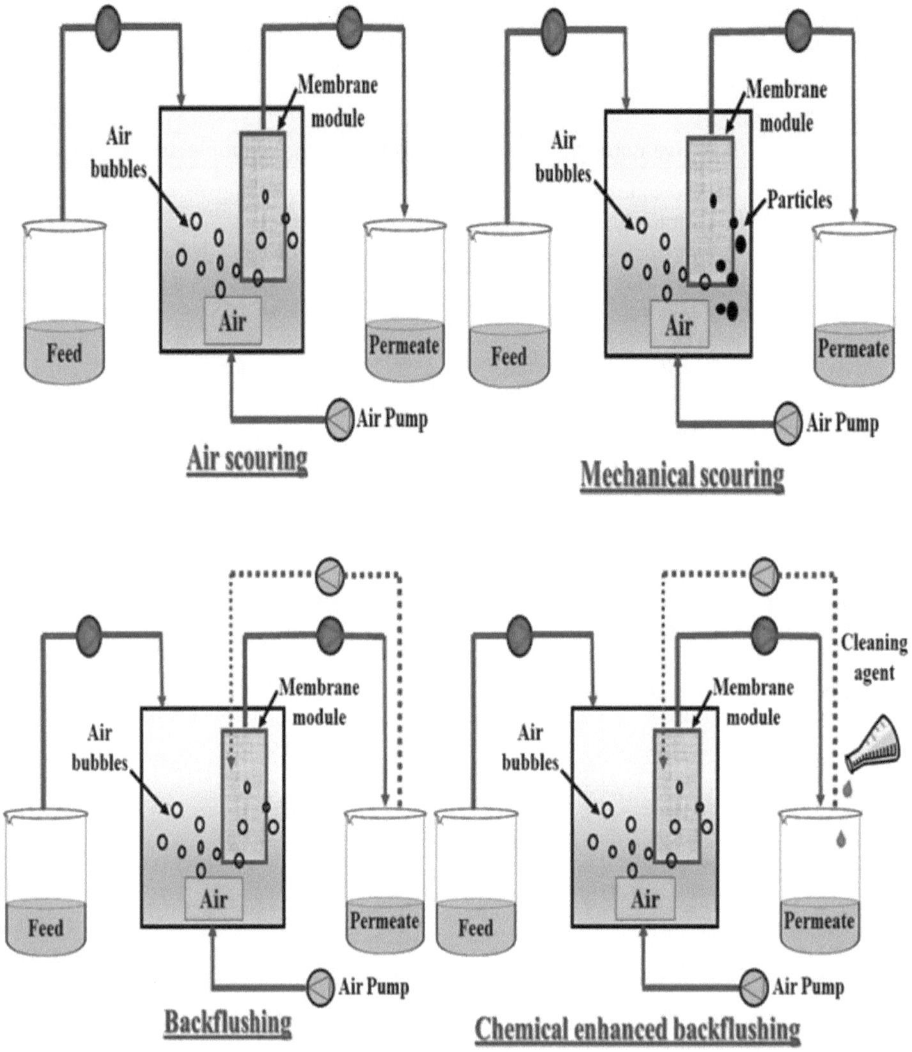

FIGURE 25.15 Common membrane cleaning strategies. Reproduced from Ref. [108], copyright Elsevier.

Recently, several innovative cures have been explored to mitigate fouling. With the use of carriers, additives, sludge granulation, surface modification of the membrane, biomass pre-setting, etc., Bohm et al. investigated fluid dynamics to mitigate MBR fouling. The rheology of the fluid influences the fluid dynamics, which in turn affects the fouling of the membrane [124].

There are several additives used exclusively to prevent the fouling of membranes [125–127]. Panchami et al. synthesized a zwitterionic nanoparticle to reduce the fouling of polyphenyl sulfone (PPSU) ultrafiltration membranes and achieved effective protein separation [4]. Hebbar et al. incorporated functionalized halloysite nanotubes into polyetherimide membranes to treat hazardous dye effluents [128].

In situ shear-enhanced fouling control methods have been proven to be significant in past research studies. Traditionally practiced air-sparging is a costly technique. Wang et al. established a novel vibrating flat-sheet membrane made up of ceramic material for the comparison of shear stress rates with the conventional air-sparging method. The shear stress experienced with a vibration speed of 40, 80, and 120 RPM, aeration rate of 0.5, 1.0, and1.5 LPM has been examined as low, medium, and high shear phases. Thus, the established VMBR was capable of removing 89.89% COD, 78.35% TOC, and 99.9% NH_4-N in the three phases [129].

25.9 MASS TRANSFER PHENOMENON IN MBR

The mass transfer in MBR is governed by two factors: (a) hydraulic resistance posed by the membrane for which flux equation can be given as

$$J = \frac{\Delta P}{\mu R_m} \tag{25.3}$$

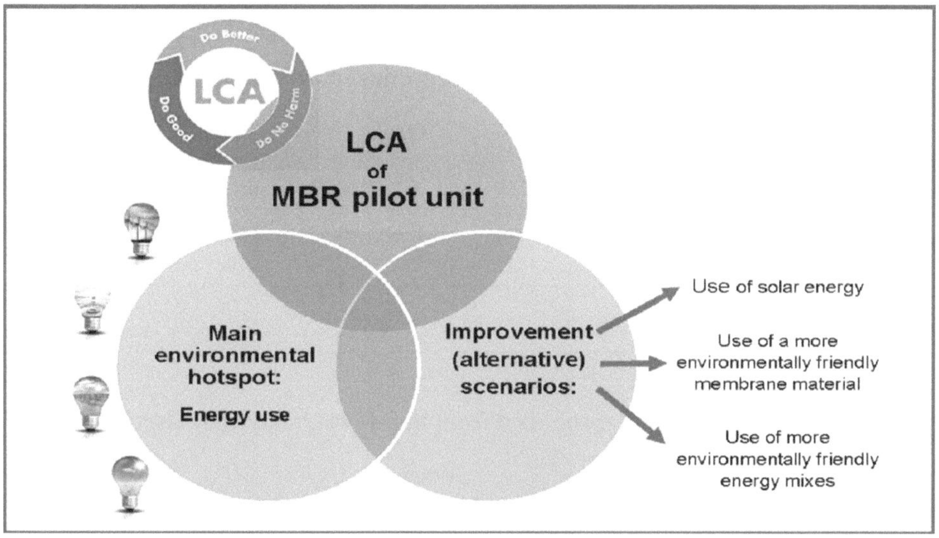

FIGURE 25.16 LCA of MBR. Reproduced from Ref. [133], copyright Elsevier.

where J = flux (ms^{-1}), ΔP = transmembrane pressure, μ = fluid viscosity, R_m = membrane hydraulic resistance. Further, R_m can be calculated from porosity through the following equation

$$R_m = \frac{K\left(1-{}_m\right)^2 S_m^2 I_m}{{}^3_m} \quad (25.4)$$

ε = porosity, S_m = ratio of pore surface area to volume, and I_m = thickness of the membrane.

(b) Resistance offered by fouling/cake layer (R_c):

$$R_c = \frac{K^I\left(1-{}_c\right)S_c^2 I_c}{\varepsilon_c^3} \quad (25.5)$$

Usually, cake layer resistance increases with time. However, in cross-flow operation as the adhesive forces help to retain the cake layer it is balanced by shearing forces acting near the boundary of hydrodynamic layer, and resistance is expected to remain steady [130].

The overall mass transfer in MBR is governed by adsorption, diffusion, and desorption. For tubular membranes, the inverse of K_0, gives the resistance to mass transfer. The pollutant removal and mass transfer efficiency are affected by the contact time as well.

$$\frac{1}{K_0} = \frac{1}{K_t} + \frac{r_i \ln\left(r_0 r_i\right)}{PC * D_{mem}} + \frac{r_i}{r_0 * K_s} \quad (25.6)$$

where K_t is the coefficient for mass transfer of the tube liquid film, K_s is for shell liquid, PC is the partition coefficient, D_{mem} is the diffusion coefficient of the membrane, and r_i and r_o are the inner and outer radii.

Some researchers have also included the effect of biofilm layer on mass transfer.

$$\frac{1}{K_o} = \frac{1}{K_t} + \frac{r_i ln\left(r_o r_i\right)}{PC_{bio} * D_{bio}} + \frac{r_i ln\left(r_f r_i\right)}{PC_{bio} * D_{bio}} + \frac{r_i}{r_f K_s} \quad (25.7)$$

where D_{bio} is the biofilm diffusion coefficient and PC_{bio} is the biofilm–solute partition coefficient [131].

25.10 LIFE CYCLE ASSESSMENT (LCA)

LCA is the study of the environmental effect of any product/process/services with the consideration of environmental load produced by that particular process, as shown in Figure 25.16. The study of vectors such as hazardous substance emission, toxicological data, emission sources, etc. provides the complete LCA of any process. Material acquirement, manufacture, large-scale production, use/maintenance, and finally waste management are the five important stages of LCA of any process/product [132].

Table 25.10 gives a comparison of some past MBR LCA analyses.

Banti et al. carried out a comparison study of MBR with the conventional activated sludge process examining their life cycle impact on the environment [135]. They used an influent having BOD = 400 mg/L and were able to get an effluent of high quality with BOD < 5 mg/L. The MBR under study contained a membrane tank, an aeration tank, and a denitrification unit, whereas the CAS under study contained nitrification, denitrification, sedimentation, equalization, mixing and flocculation tanks, and a drum with a filter. The study finalized the environmental superiority of MBR over CAS [134].

Holloway et al. assessed the lifecycle of a hybrid MBR system consisting of ultrafiltration osmotic MBR (UFO-MBR) and a full advanced treatment (FAT) that combined RO, low-pressure filtration, and ultraviolet advanced oxidation

TABLE 25.10
LCA; past research results; comparison of MBR with CAS. Reproduced from Ref. [134]

Treatment unit	Eutrophication potential	Acidification potential	Ozone depletion potential	Global warming potential	Photochemical ozone creation potential
CAS	4.77×10^{-3}	1.15×10^{-2}	1.70×10^{-7}	2.68	1.50×10^{-4}
MBR	8.60×10^{-4}	1.98×10^{-3}	3.13×10^{-8}	4.96×10^{-1}	8.40×10^{-4}

TABLE 25.11
Analysis of LCA studies on different MBRs. Reproduced from Ref. [141], copyright Elsevier

Goal	Source of water	Functional unit	Software	System boundary	Approach	Findings	References
To compare three decentralized treatment systems	Decentralized wastewater	1 m^3 of treated wastewater for irrigation	Sima Pro	Gate-to-gate	Midpoint	AnMBR—best configuration due to reduced energy consumption and biogas output	[153]
To compare three biological treatment systems for non-potable reuse	Graywater	Delivery of non-potable reuse water for the whole building	Not specified	Cradle-to-grave	Midpoint	Graywater AeMBR meeting onsite non-potable demands has lowest human health and environmental implications	[154]
To evaluate two treatment plants	Municipal wastewater	1 m^3 treated wastewater	OpenLCA	Gate-to gate	Midpoint	Compared to CAS, this had considerable environmental benefit excluding excess sludge production	[155]
To evaluate community-based sewage water use	Sewage	Annual provision of space and water heating, water treatment per person, irrigation	OpenLCA	Gate-to-gate	Midpoint	Compared to CAS, to MBR showed lower impact contribution	[156]
To compare AnMBR with CAS	Domestic	Treatment of 5 million gallon medium-strength wastewater having same effluent characteristics	Not specified	Gate-to-gate	Midpoint	AnMBR was found to be more efficient due to the integration of sedimentation with anaerobic digestion, alternate procedure for methane dissolution, and removal of biological sulfide	[157]
To evaluate full-scale AnMBR	Urban wastewater	Volume of treated water in (m^3)	Sima Pro	Cradle-to-grave	Midpoint	Medium to high organic loading rate, helped to enhance the AnMBR performance, increasing dissolved methane recovery	
To evaluate management systems	Graywater	Annual treatment of graywater generated per person	OpenLCA	Cradle-to-gate	Midpoint	MBR is preferred when large quantity of graywater is used	[158]

TABLE 25.12
Major input parameters governing MBR cost. Reproduced from Ref. [142], copyright Elsevier

Parameter Process	Units	Value	Parameter Labor Rates	Units	Value	Parameter Financial	Units	Value
Building cost	$/m^2$	380	Construction labor rate	$/h	4	Interest rate	%	8.0
Excavation	$/m^3$	28	Operator labor rate	$/h	5.33	Construction period	Years	3.0
Wall concrete	$/m^3$	720	Administration labor rate	$/h	8	Operating life plant	Years	39
Slab concrete	$/m^3$	380	Laboratory labor rate	$/h	8	Other costs	%	
Crane rental	$/m^2$	60	Chemical costs	$/kg		Engineering design fee	%	15.0
Canopy roof	$/h	172	Hydrated lime [$Ca(OH)_2$]	$/kg	0.4	Miscellaneous	%	5.0
Electricity	$/kWh	0.096	$Al_2(SO_4)_3$ *14 H_2O	$/kg	0.6	Administration/legal	%	2.0
Hand rail	$/m	200	Ferric chloride	$/kg	0.8	Inspection	%	2.0
Land costs	$/m^2$	15	Polymer	$/kg	2.86	Contingency	%	10.0
			Citric acid 50%	$/kg	1.146	Technical	%	2.0
			NaOCl 14%	$/m^3$	344.68	Profit and overhead	%	15.0

process (UV-AOP). The FAT showed lower environmental impact than UF-MBR (which has a larger surface area) [136].

Ttofa et al. examined the environmental sustainability of a pilot-scale MBR which indicated that the majority of the impacts were attributed to indirect emissions. The system emitted only 1% greenhouse gas from 1 m^3 of wastewater which signified that the MBR was environmentally friendly [133].

Kobayashi and co-workers examined the LCA of an MBR designed for graywater treatment. They focused on global-warming potential, carcinogenic potential, and eutrophication potential of the system on natural as well as engineered graywater [137]. In one study, the LCA of an integrated-fixed film activated-sludge MBR was analyzed taking cost in consideration. Pollutant removal was of the order: BOD (99.86%), COD (94.54%), TSS (99.99%), TN (97%), and total carbon (93.45%). Emission of greenhouse gases was about 27.5% during the entire process [138].

Dominguez et al. reported the LCA of three cases: solar-driven photovoltaic photocatalysis, a simple photocatalysis, and an MBR process. Removal of a surfactant, sodium-dodecylbenzenesulfonate, was the target of the process. Two LCA indicators, environmental burden and natural resources exploited, were examined for both, of which energy consumption was a critical aspect [139]. In one attempt at assessing the life cycle of a community wastewater treatment system, Cabling et al. concluded that the integration of MBR with heating schemes could provide more environmental savings when operated on perchlorinated water [140]. A well-established collaboration between the LCA experts and researchers is an absolutely necessary to avoid data selection difficulty. Table 25.11 analyses the LCA of different MBRs.

25.11 COST ANALYSIS

Several factors affect the cost of water treatment by MBR. Load and size of the plant, type of technology and construction materials used, nature of wastewater (whether industrial or municipal), discharge control (nutrient control), sludge treatment type and disposal method (incineration, composting), energy recycling and energy supply, degree of automation are just a few. The total unit cost (TUC) of MBR WWTP is calculated as follows:

$$TUC = \frac{ADCAPEX + OPEX}{Q_{inf}} / m^3 \qquad (25.8)$$

where Q_{inf} is the flow of the influent (m^3/year), ADCAPEX is the annual depreciation charge of capital cost, and OPEX is operation and maintenance cost. A few other parameters impacting the cost of MBR are depicted in Table 25.12.

Developments in the configuration of MBR also helped greatly to reduce the cost.

25.12 CHALLENGES AND CONTROVERSIAL AREAS IN MBR RESEARCH AND FUTURE PERSPECTIVES

Many numerical and mathematical strategies have been developed for the prediction of membrane resistance, while considering just some simple operational parameters. However, the complexity of fouling and the associated non-linear equations are mostly based on hypotheses and assumptions. Focusing more on strategies like artificial neural networks (ANNs) would help in predicting membrane fouling in MBR plants. For applying ANN to large-scale plants, ANN architecture and optimization of experimental databases are needed along with carefully examining the operational duration.

Although MBRs are being furnished to an appreciable extent, there is a great demand for technical advancements. Generally, MBRs use a higher concentration of MLSS compared to CAS, which helps in reducing the size and cost of the process. However, this increases the trade-off between the surface area of the membrane and the expense of oxygen transport.

Equipment availability is another issue being faced by developing countries. Also, most of these countries lack local resources such as financial capital, trained human resources, manufacturing capability, and technical support.

Improving the cost-effectiveness is one of the main challenges of MBR. As one cannot predict the market price of the membrane, constant research is required to enhance the membranes used in MBR. Further, operative energy consumption needs to be reduced, and specific flux has to be enhanced to ensure affordable cost. Much work needs to be done with regard to process optimization in order to improve pollutant removal efficiency. In AnMBR, much research needs to be done with respect to the recovery of methane, control of process inhibition, facilitating biohydrogen logistics, etc.

Employing lower MLSS concentration provides higher water fluxes and is also cost-effective. However, several years of operating experience are necessary to establish constant higher flux with good lifetime of the membrane. It seems that more focus needs to be placed on pretreatment strategies such as removing fibrous materials (e.g., hair). Currently, screens with openings of ≤ 2 mm are being used. However, the quantity and noxiousness of the material removed may pose problems for further operation.

To control fouling, economically hydrodynamic conditions of the module/tank configuration also need to be optimal. With this in mind, aeration and chemical cleaning have to be thoroughly implemented to tackle fouling complexity. Proper plans to establish MBRs with limited use of land, especially in densely populated areas, is necessary. The construction of underground MBR treatment plants also should be investigated.

Recently adopted techniques like osmotic MBRs can further be improved by integrating forward osmosis membranes with different biological processes such as moving-bed biofilm reactors, anaerobic digesters, attached growth biofilm reactors, sequential batch reactors, etc.

25.13 CONCLUSIONS

The technology behind membrane bioreactors has been exclusively studied and developed to mitigate the problem of water scarcity. The combination of membrane technology with biological processes such as activated sludge is a great way to reclaim potable water from municipal or industrial wastewater. This chapter presents a thorough discussion on membrane bioreactor technology starting with its history, operational parameters required for efficient pollutant removal, and latest configurations. The use of advanced nanomaterials has the potential to enhance the properties of the membranes used in MBR for which the different categories of nanomaterials available were discussed in the next part of the chapter. Of the different types of MBRs, anaerobic MBRs have been given great importance because of their cost-competitiveness as is also reviewed thoroughly. Every fascinating technology, no matter how useful it is, has its own pitfalls. Fouling is one such difficulty which researchers work tirelessly to resolve. The chapter discusses different techniques available to clean fouled membrane and also provides brief precautionary measures to avoid fouling. Finally, the life cycle assessment of MBR technology has been analyzed to present an awareness of the detrimental impacts of the technology on the environment, if any. However, many more investigations need to be carried out in the future to maximize MBR performance both from technical and economic points of view. The main objective of this chapter has been to focus researchers' attention on integrated MBR systems and the optimization of operational parameters that results in the best MBR designs.

REFERENCES

[1] Padaki, M., Isloor, A.M., Nagaraja, K.K., Nagaraja, H.S., Pattabi, M. (2011) Conversion of microfiltration membrane into nanofiltration membrane by vapour phase deposition of aluminium for desalination application. *Desalination* 274(1–3): 177–181.

[2] Ibrahim, S., Mohammadi Ghaleni, M., Isloor, A.M., Bavarian, M., Nejati, S. (2020) Poly(homopiperazine-amide) thin-film composite membrane for nanofiltration of heavy metal ions. *ACS Omega* 5(44): 28749–28759.

[3] Moideen, I.K., Isloor, A.M., Garudachari, B., Ismail, A.F. (2016) The effect of glycine betaine additive on the PPSU/PSF ultrafiltration membrane performance. *Desalination and Water Treatment* 57(52): 24788–24798.

[4] Panchami, H.R., Isloor, A.M., Ismail, A.F. (2021) Improved hydrophilic and antifouling performance of nanocomposite ultrafiltration zwitterionic polyphenylsulfone membrane for protein rejection applications. *Journal of Nanostructure in Chemistry* 1–22.

[5] Nair, A.K., Isloor, A.M., Kumar, R., Ismail, A.F. (2013) Antifouling and performance enhancement of polysulfone ultrafiltration membranes using $CaCO_3$ nanoparticles. *Desalination* 322: 69–75.

[6] Ezugbe, E.O., Rathilal, S. (2020) Membrane technologies in wastewater treatment: A review. *Membranes (Basel)* 10(5).

[7] Kolangare, I.M., Isloor, A.M., Inamuddin, Asiri, A.M., Ismail, A.F. (2019) Improved desalination by polyamide membranes containing hydrophilic glutamine and glycine. *Environmental Chemistry Letters* 17(2): 1053–1059.

[8] Balakrishna Prabhu, K., Saidutta, M.B., Isloor, A.M., Hebbar, R. (2017) Improvement in performance of polysulfone membranes through the incorporation of chitosan-(3-phenyl-1h-pyrazole-4-carbaldehyde). *Cogent Engineering* 4(1).

[9] Chandrashekhar Nayak, M., Isloor, A.M., Inamuddin, Lakshmi, B., Marwani, H.M.,. Khan, I. (2020) Polyphenylsulfone/multiwalled carbon nanotubes mixed ultrafiltration membranes: Fabrication, characterization and removal of heavy metals Pb^{2+}, Hg^{2+}, and Cd^{2+} from aqueous solutions. *Arabian Journal of Chemistry* 13(3): 4661–4672.

[10] Heggabailu Rajashekhara, P., Isloor, A.M., Fauzi Ismail, A. (2022) One-step synthesis and characterization of hydrophilic polymer microspheres immobilized with polyphenylsulfone ultrafiltration membranes for protein rejection application. Research square preprint article.

[11] Zhang, D., et al. (2016) Characterization of soluble microbial products (SMPs) in a membrane bioreactor (MBR) treating

synthetic wastewater containing pharmaceutical compounds. *Water Research* 102: 594–606.

[12] Lares, M., Ncibi, M.C., Sillanpää, M., Sillanpää, M. (2018) Occurrence, identification and removal of microplastic particles and fibers in conventional activated sludge process and advanced MBR technology. *Water Research* 133: 236–246.

[13] Tadkaew, N., Hai, F.I., McDonald, J.A., Khan, S.J., Nghiem, L.D. (2011) Removal of trace organics by MBR treatment: The role of molecular properties. *Water Research* 45(8): 2439–2451.

[14] Alturki, A.A., Tadkaew, N., McDonald, J.A., Khan, S.J., Price, W.E., Nghiem, L.D. (2010) Combining MBR and NF/RO membrane filtration for the removal of trace organics in indirect potable water reuse applications. *Journal of Membrane Science* 365(1–2): 206–215.

[15] Fenu, A., et al. (2010) Energy audit of a full scale MBR system. *Desalination* 262(1–3): 121–128.

[16] Sipma, J., et al. (2010) Comparison of removal of pharmaceuticals in MBR and activated sludge systems. *Desalination* 250(2): 653–659.

[17] Tan, X., Li, K. (2015) *Fundamentals of Membrane Reactors*. Wiley Online Library.

[18] Woo, T.Y., Nam, K.J., Heo, S.K., Lim, J.Y., Kim, S.Y., Yoo, C.K. (2022) Predictive maintenance system for membrane replacement time detection using AI-based functional profile monitoring: Application to a full-scale MBR plant. *Journal of Membrane Science* 649: 120400.

[19] Meabe, E., Lopetegui, J., Ollo, J., Lardies, S. (2011) Ceramic membrane bioreactor: potential applications and challenges for the future," *MBR Asia International Conference*, April 2011, pp. 1–9. [Online]. Available: www.yumpu.com/en/document/view/9234763/ceramicmembrane-bioreactor-potential-applications-and-likuid

[20] Koyuncu, I.. et al. (2020) Applications of ceramic membrane bioreactors in water treatment. *Current Trends and Future Developments on (Bio-)Membranes: Ceramics Membranes Bioreactions* 141–176.

[21] Amini, M., Etemadi, H., Akbarzadeh, A., Yegani, R. (2017) Preparation and performance evaluation of high-density polyethylene/silica nanocomposite membranes in membrane bioreactor system. *Biochemical Engineering Journal* 127: 196–205.

[22] Zinadini, S., Vatanpour, V., Zinatizadeh, A.A., Rahimi, M., Rahimi, Z., Kian, M. (2015) Preparation and characterization of antifouling graphene oxide/polyethersulfone ultrafiltration membrane: Application in MBR for dairy wastewater treatment. *Journal of Water Process Engineering* 7: 280–294.

[23] Neoh, C.H., Noor, Z.Z., Mutamim, N.S.A., Lim, C.K. (2016) Green technology in wastewater treatment technologies: Integration of membrane bioreactor with various wastewater treatment systems. *Chemical Engineering Journal* 283: 582–594.

[24] Scholes, E., Verheyen, V., Brook-Carter, P. (2016) A review of practical tools for rapid monitoring of membrane bioreactors. *Water Research* 102: 252–262.

[25] Li, R., et al. (2022) Removal of micropollutants in a ceramic membrane bioreactor for the post-treatment of municipal wastewater. *Chemical Engineering Journal* 427.

[26] Wang, Y., Huang, X., Yuan, Q. (2005) Nitrogen and carbon removals from food processing wastewater by an anoxic/aerobic membrane bioreactor. *Process Biochemistry* 40(5): 1733–1739.

[27] Belli, T.J., Bernardelli, J.K.B., da Costa, R.E., Bassin, J.P., Amaral, M.C.S., Lapolli, F.R. (2017) Effect of solids retention time on nitrogen and phosphorus removal from municipal wastewater in a sequencing batch membrane bioreactor. *Environmental Technology (United Kingdom)* 38(7): 806–815.

[28] Falahti-Marvast, H., Karimi-Jashni, A. (2015) Performance of simultaneous organic and nutrient removal in a pilot scale anaerobic-anoxic-oxic membrane bioreactor system treating municipal wastewater with a high nutrient mass ratio. *International Biodeterioration and Biodegradation* 104: 363–370.

[29] Mannina, G., Capodici, M., Cosenza, A., Di Trapani, D., Ekama, G.A. (2018) The effect of the solids and hydraulic retention time on moving bed membrane bioreactor performance. *Journal of Cleaner Products* 170: 1305–1315.

[30] Viero, A.F., Sant'Anna, G.L. (2008) Is hydraulic retention time an essential parameter for MBR performance? *Journal of Hazardous Materials* 150(1): 185–186.

[31] Cheng, C., et al. (2017) Effects of side-stream ratio on sludge reduction and microbial structures of anaerobic side-stream reactor coupled membrane bioreactors. *Bioresource Technology* 234: 380–388.

[32] Jiang, M., Westerholm, M., Qiao, W., Wandera, S.M., Dong, R. (2020) High rate anaerobic digestion of swine wastewater in an anaerobic membrane bioreactor. *Energy* 193: 116783.

[33] Dong, Q., Parker, W., Dagnew, M. (2016) Influence of SRT and HRT on bioprocess performance in anaerobic membrane bioreactors treating municipal wastewater. *Water Environment Research* 88(2): 158–167.

[34] Xu, M., Li, P., Tang, T., Hu, Z. (2015) Roles of SRT and HRT of an algal membrane bioreactor system with a tanks-in-series configuration for secondary wastewater effluent polishing. *Ecological Engineering* 85: 257–264.

[35] Gündoğdu, M., Kabay, N., Yiğit, N., Kitiş, M., Pek, T., Yüksel, M. (2019) Effect of concentrate recirculation on the product water quality of integrated MBR – NF process for wastewater reclamation and industrial reuse. *Journal of Water Process Engineering* 29: 100485.

[36] Chan, R., Chiemchaisri, C., Chiemchaisri, W. (2020) Effect of sludge recirculation on removal of antibiotics in two-stage membrane bioreactor (MBR) treating livestock wastewater. *Journal of Environmental Health Science and Engineering* 18(2): 1541–1553.

[37] Odriozola, J. et al. (2017) Model-based methodology for the design of optimal control strategies in MBR plants. *Water Science and Technology* 75(11): 2546–2553.

[38] Miyoshi, T., Tsumuraya, T., Nguyen, T.P., Kimura, K., Watanabe, Y. (2018) Effects of recirculation and separation times on nitrogen removal in baffled membrane bioreactor (B-MBR). *Water Science and Technology* 77(12): 2803–2811.

[39] Kappel, C., Kemperman, A.J.B., Temmink, H., Zwijnenburg, A., Rijnaarts, H.H.M., Nijmeijer, K. (2014) Impacts of NF concentrate recirculation on membrane performance in an integrated MBR and NF membrane process for wastewater treatment. *Journal of Membrane Science* 453: 359–368.

[40] Le Clech, P., Jefferson, B., Chang, I.S., Judd, S.J. (2003) Critical flux determination by the flux-step method in a

submerged membrane bioreactor. *Journal of Membrane Science* 227(1–2): 81–93.
[41] Nguyen, T.N.P., Su, Y.C., Pan, J.R., Huang, C. (2014) Comparison of membrane foulants occurred under different sub-critical flux conditions in a membrane bioreactor (MBR). *Bioresource Technology* 166: 389–394.
[42] Sarioglu, M., Sayi-Ucar, N., Cokgor, E., Orhon, D., van Loosdrecht, M.C.M., Insel, G. (2017) Dynamic modeling of nutrient removal by a MBR operated at elevated temperatures. *Water Research* 123: 420–428.
[43] Liu, M., et al. (2018) Numerical optimization of membrane module design and operation for a full-scale submerged MBR by computational fluid dynamics. *Bioresource Technology* 269: 300–308.
[44] Li, J., Zhang, X., Cheng, F., Liu, Y. (2013) New insights into membrane fouling in submerged MBR under sub-critical flux condition. *Bioresource Technology* 137: 404–408.
[45] Ersahin, M.E., Tao, Y., Ozgun, H., Gimenez, J.B., Spanjers, H., van Lier, J.B. (2017) Impact of anaerobic dynamic membrane bioreactor configuration on treatment and filterability performance. *Journal of Membrane Science* 526: 387–394.
[46] Chen, C., et al. (2017) Impact of reactor configurations on the performance of a granular anaerobic membrane bioreactor for municipal wastewater treatment. *International Biodeterioration and Biodegradation* 121: 131–138.
[47] Clouzot, L., Roche, N., Marrot, B. (2011) Effect of membrane bioreactor configurations on sludge structure and microbial activity. *Bioresource Technology* 102(2): 975–981.
[48] de Andrade, L.H., dos S. Mendes, F.D., Espindola, J.C., Amaral, M.C.S. (2014) Internal versus external submerged membrane bioreactor configurations for dairy wastewater treatment. *Desalination and Water Treatment* 52(16–18): 2920–2932.
[49] Gholami, F., Zinatizadeh, A.A., Zinadini, S., McKay, T., Sibali, L. (2020) An innovative jet loop-airlift bioreactor for simultaneous removal of carbon and nitrogen from soft drink industrial wastewater: Process performance and kinetic evaluation. *Environmental Technology & Innovation* 19: 100772.
[50] Parveen, F., Hankins, N. (2021) Integration of forward osmosis membrane bioreactor (FO-MBR) and membrane distillation (MD) units for water reclamation and regeneration of draw solutions. *Journal of Water Process Engineering* 41: 102045.
[51] Tomasini, H.R., Hacifazlioglu, M.C., Kabay, N., Bertin, L., Pek, T.O., Yuksel, M. (2019) Concentrate management for integrated MBR-RO process for wastewater reclamation and reuse-preliminary tests. *Journal of Water Process Engineering* 29: 100455.
[52] Blandin, G., Gautier, C., Sauchelli Toran, M., Monclús, H., Rodriguez-Roda, I., Comas, J. (2018) Retrofitting membrane bioreactor (MBR) into osmotic membrane bioreactor (OMBR): A pilot scale study. *Chemical Engineering Journal* 339: 268–277.
[53] Wang, Y., Wang, X., Li, M., Dong, J., Sun, C., Chen, G. (2018) Removal of pharmaceutical and personal care products (PPCPs) from municipal waste water with integrated membrane systems, MBR-RO/NF. *International Journal of Environmental Research and Public Health* 15(2): 269.
[54] Chae, S.R., Chung, J.H., Heo, Y.R., Kang, S.T., Lee, S.M., Shin, H.S. (2015) Full-scale implementation of a vertical membrane bioreactor for simultaneous removal of organic matter and nutrients from municipal wastewater. *Water* 7(3): 1164–1172.
[55] Jiang, Q., et al. (2018) Effect of hydraulic retention time on the performance of a hybrid moving bed biofilm reactor-membrane bioreactor system for micropollutants removal from municipal wastewater. *Bioresource Technology* 247: 1228–1232.
[56] Mei, X., et al. (2019) Treatment of formaldehyde wastewater by a membrane-aerated biofilm reactor (MABR): The degradation of formaldehyde in the presence of the cosubstrate methanol. *Chemical Engineering Journal* 372: 673–683.
[57] Corsino, S.F., Torregrossa, M. (2022) Achieving complete nitrification below the washout SRT with hybrid membrane aerated biofilm reactor (MABR) treating municipal wastewater. *Journal of Environmental Chemical Engineering* 10(1): 106983.
[58] Long, M., Zeng, C., Wang, Z., Xia, S., Zhou, C. (2020) Complete dechlorination and mineralization of para-chlorophenol (4-CP) in a hydrogen-based membrane biofilm reactor (MBfR). *Journal of Cleaner Products* 276: 123257.
[59] Pang, S., et al. (2022) Changed carbon fixing affects denitrification in a hydrogen-based membrane biofilm reactor under high salinity conditions. *Journal of Water Process Engineering* 49: 103097.
[60] Xia, S., Xu, X., Zhou, L. (2019) Insights into selenate removal mechanism of hydrogen-based membrane biofilm reactor for nitrate-polluted groundwater treatment based on anaerobic biofilm analysis. *Ecotoxicology and Environmental Safety* 178: 123–129.
[61] Gede Wenten, I., Friatnasary, D.L., Khoiruddin, K., Setiadi, T., Boopathy, R. (2020) Extractive membrane bioreactor (EMBR): Recent advances and applications. *Bioresource Technology* 297: 122424.
[62] Mullins, N.R., Daugulis, A.J. (2019) The biological treatment of synthetic fracking fluid in an extractive membrane bioreactor: Selective transport and biodegradation of hydrophobic and hydrophilic contaminants. *Journal of Hazardous Materials* 371: 734–742.
[63] Tuluk, B., et al. (2022) High-speed treatment of low strength domestic wastewater for irrigation water production in pilot-scale classical, moving bed and fixed bed hybrid MBRs. *Journal of Cleaner Products* 375: 134084.
[64] Waqas, S., et al. (2020) Recent progress in integrated fixed-film activated sludge process for wastewater treatment: A review. *Journal of Environmental Management* 268: 110718.
[65] Vergine, P., Salerno, C., Berardi, G., Pollice, A. (2018) Sludge cake and biofilm formation as valuable tools in wastewater treatment by coupling Integrated Fixed-film Activated Sludge (IFAS) with Self Forming Dynamic Membrane BioReactors (SFD-MBR). *Bioresource Technology* 268: 121–127.
[66] Ribera-Pi, J., Badia-Fabregat, M., Espí, J., Clarens, F., Jubany, I., Martínez-Lladó, X. (2020) Decreasing environmental impact of landfill leachate treatment by MBR, RO and EDR hybrid treatment. *Environmental Technology* 42(22): 3508–3522.
[67] Wang, H., et al. (2020) Molecular insight into variations of dissolved organic matters in leachates along China's largest

A/O-MBR-NF process to improve the removal efficiency. *Chemosphere* 243: 125354.
[68] Zhang, S., et al. (2019) Characteristics of the sludge filterability and microbial composition in PAC hybrid MBR: Effect of PAC replenishment ratio. *Biochemical Engineering Journal* 145: 10–17.
[69] A. K. Corporation, "Asahi Kasei Overview," 1971.
[70] "KUBOTA Membrane USA Corporation | MBR Information." www.kubota-membrane.com/page/mbr-information (accessed Nov. 12, 2022).
[71] C. Submerged, "Puron ® mbr."
[72] "The MBR Site | Origin Water MBRU." www.thembrsite.com/directories/membrane-products/mbru-beijing-origin-water/ (accessed Nov. 12, 2022).
[73] "GE's Next-Generation MBR Wastewater Treatment System Slashes Energy Use, Boosts Productivity | GE News." www.ge.com/news/press-releases/ges-next-generation-mbr-wastewater-treatment-system-slashes-energy-use-boosts-0 (accessed Nov. 12, 2022).
[74] Shaw, S., Patra, A., Misra, A., Nayak, M.K., Chamkha, A.J. (2021) A numerical approach to the modeling of Thomson and Troian slip on nonlinear radiative microrotation of Casson Carreau nanomaterials in magnetohydrodynamics. *Journal of Nanofluids* 10(3): 305–315.
[75] Ahsani, M., Hazrati, H., Javadi, M., Ulbricht, M., Yegani, R. (2020) Preparation of antibiofouling nanocomposite PVDF/Ag-SiO2 membrane and long-term performance evaluation in the MBR system fed by real pharmaceutical wastewater. *Separation and Purification Technology* 249: 116938.
[76] Pervez, M.N., et al. (2020) A critical review on nanomaterials membrane bioreactor (NMs-MBR) for wastewater treatment. *npj Clean Water* 3(1): 1–21.
[77] Kajekar, A.J., et al. (2015) Preparation and characterization of novel PSf/PVP/PANI-nanofiber nanocomposite hollow fiber ultrafiltration membranes and their possible applications for hazardous dye rejection. *Desalination* 365: 117–125.
[78] Pereira, V.R., Isloor, A.M., Bhat, U.K., Ismail, A.F. (2014) Preparation and antifouling properties of PVDF ultrafiltration membranes with polyaniline (PANI) nanofibers and hydrolysed PSMA (H-PSMA) as additives. *Desalination* 351: 220–227.
[79] Arslan, S., Eyvaz, M., Güçlü, S., Yüksekdağ, A., Koyuncu, İ., Yüksel, E. (2021) Pressure assisted application of tubular nanofiber forward osmosis membrane in membrane bioreactor coupled with reverse osmosis system. *Journal of Water Chemistry and Technology* 43(1): 68–76.
[80] Pervez, M.N., Stylios, G.K., Liang, Y., Ouyang, F., Cai, Y. (2020) Low-temperature synthesis of novel polyvinylalcohol (PVA) nanofibrous membranes for catalytic dye degradation. *Journal of Cleaner Products* 262: 121301.
[81] Ghalamchi, L., Aber, S., Vatanpour, V., Kian, M. (2019) Comparison of NLDH and g-C_3N_4 nanoplates and formative Ag_3PO_4 nanoparticles in PES microfiltration membrane fouling: Applications in MBR. *Chemical Engineering Research and Design* 147: 443–457.
[82] Noormohamadi, A., Homayoonfal, M., Mehrnia, R., Davar, F. (2019) Employing magnetism of Fe_3O_4 and hydrophilicity of ZrO_2 to mitigate biofouling in magnetic MBR by Fe_3O_4-coated ZrO_2/PAN nanocomposite membrane. *Environmental Technology* 41(20): 2683–2704.
[83] Shahzad, A., et al. (2021) Advances in the synthesis and application of anti-fouling membranes using two-dimensional nanomaterials. *Membranes* 11(8): 605.
[84] Balakrishnan, M., Yadav, S., Singh, N., Batra, V.S. (2022) Membrane-based hybrid systems incorporating nanomaterials for wastewater treatment. *Nano-Enabled Technologies for Water Remediation* 71–144.
[85] Pereira, V.R., Isloor, A.M., Al Ahmed, A., Ismail, A.F. (2014) Preparation, characterization and the effect of PANI coated TiO_2 nanocomposites on the performance of polysulfone ultrafiltration membranes. *New Journal of Chemistry* 39(1): 703–712.
[86] Syed Ibrahim, G.P., Isloor, A.M., Bavarian, M., Nejati, S. (2020) Integration of zwitterionic polymer nanoparticles in interfacial polymerization for ion separation. *ACS Applied Polymer Materials* 2(4): 1508–1517.
[87] Pereira, V.R., Isloor, A.M., Bhat, U.K., Ismail, A.F., Obaid, A., Fun, H.K. (2015) Preparation and performance studies of polysulfone-sulfated nano-titania (S-TiO_2) nanofiltration membranes for dye removal. *RSC Advances* 5(66): 53874–53885.
[88] Hebbar, R.S., Isloor, A.M., Ismail, A.F. (2014) Preparation and evaluation of heavy metal rejection properties of polyetherimide/porous activated bentonite clay nanocomposite membrane. *RSC Advances* 4(88): 47240–47248.
[89] Yuan, Z.H., Yang, X., Hu, A., Zheng, Y.M., Yu, C.P. (2016) Assessment of the fate of silver nanoparticles in the A2O-MBR system. *Science of the Total Environment* 544: 901–907.
[90] Wang, Z., et al. (2014) Long-term operation of an MBR in the presence of zinc oxide nanoparticles reveals no significant adverse effects on its performance. *Journal of Membrane Science* 471: 258–264.
[91] Almusawy, A.M., Al-Anbari, R.H., Alsalhy, Q.F., Al-Najar, A.I. (2020) Carbon nanotubes-sponge modified electro membrane bioreactor (EMBR) and their prospects for wastewater treatment applications. *Membranes* 10(12): 433.
[92] Ayyaru, S., Pandiyan, R., Ahn, Y.H. (2019) Fabrication and characterization of anti-fouling and non-toxic polyvinylidene fluoride-sulphonated carbon nanotube ultrafiltration membranes for membrane bioreactors applications. *Chemical Engineering Research and Design* 142: 176–188.
[93] Lv, J., Zhang, G., Zhang, H., Yang, F. (2018) Graphene oxide-cellulose nanocrystal (GO-CNC) composite functionalized PVDF membrane with improved antifouling performance in MBR: Behavior and mechanism. *Chemical Engineering Journal* 352: 765–773.
[94] Wang, S., et al. (2019) Lead-free sodium niobate nanowires with strong piezo-catalysis for dye wastewater degradation. *Ceramics International* 45(9): 11703–11708.
[95] Yin, X., Li, X., Wang, X., Ren, Y., Hua, Z. (2019) A spontaneous electric field membrane bioreactor with the innovative Cu - nanowires conductive microfiltration membrane for membrane fouling mitigation and pollutant removal. *Water Environment Research* 91(8) 780–787.
[96] Ma, Z., et al. (2020) Surface functionalization via synergistic grafting of surface-modified silica nanoparticles and layered double hydroxide nanosheets for fabrication of superhydrophilic but relatively oleophobic antifouling membranes. *Separation and Purification Technology* 247: 116955.

[97] Liang, X., et al. (2019) Zwitterionic functionalized MoS2 nanosheets for a novel composite membrane with effective salt/dye separation performance. *Journal of Membrane Science* 573: 270–279.

[98] Lin, H., Peng, W., Zhang, M., Chen, J., Hong, H., Zhang, Y. (2013) A review on anaerobic membrane bioreactors: Applications, membrane fouling and future perspectives. *Desalination* 314: 169–188.

[99] Robles, A., Ruano, M.V., Ribes, J., Ferrer, J. (2013) Factors that affect the permeability of commercial hollow-fibre membranes in a submerged anaerobic MBR (HF-SAnMBR) system. *Water Research* 47(3): 1277–1288.

[100] Svojitka, J., Dvořák, L., Studer, M., Straub, J.O., Frömelt, H., Wintgens, T. (2017) Performance of an anaerobic membrane bioreactor for pharmaceutical wastewater treatment. *Bioresource Technology* 229: 180–189.

[101] Chen, H., Chang, S., Guo, Q., Hong, Y., Wu, P. (2016) Brewery wastewater treatment using an anaerobic membrane bioreactor. *Biochemical Engineering Journal* 105: 321–331.

[102] Gu, J., Liu, H., Wang, S., Zhang, M., Liu, Y. (2019) An innovative anaerobic MBR-reverse osmosis-ion exchange process for energy-efficient reclamation of municipal wastewater to NEWater-like product water. *Journal of Cleaner Products* 230: 1287–1293.

[103] Yurtsever, A., Sahinkaya, E., Çınar, Ö. (2020) Performance and foulant characteristics of an anaerobic membrane bioreactor treating real textile wastewater. *Journal of Water Process Engineering* 33: 101088.

[104] Vinardell, S., et al. (2020) Advances in anaerobic membrane bioreactor technology for municipal wastewater treatment: A 2020 updated review. *Renewable and Sustainable Energy Review* 130: 109936.

[105] Ruigómez, I., Vera, L., González, E., Rodríguez-Sevilla, J. (2016) Pilot plant study of a new rotating hollow fibre membrane module for improved performance of an anaerobic submerged MBR. *Journal of Membrane Science* 514: 105–113.

[106] Christensen, M.L., Niessen, W., Sørensen, N.B., Hansen, S.H., Jørgensen, M.K., Nielsen, P.H. (2018) Sludge fractionation as a method to study and predict fouling in MBR systems. *Separation and Purification Technology* 194: 329–337.

[107] Tian, Y., Su, X. (2012) Relation between the stability of activated sludge flocs and membrane fouling in MBR: Under different SRTs. *Bioresource Technology* 118: 477–482.

[108] Liu, Q., et al. (2021) A review of the current in-situ fouling control strategies in MBR: Biological versus physicochemical. *Journal of Industrial and Engineering Chemistry* 98: 42–59.

[109] Du, X., Shi, Y., Jegatheesan, V., Ul Haq, I. (2020) A review on the mechanism, impacts and control methods of membrane fouling in MBR system. *Membranes* 10(2): 24.

[110] Al-Asheh, S., Bagheri, M., Aidan, A. (2021) Membrane bioreactor for wastewater treatment: A review. *Case Studies in Chemical and Environmental Engineering* 4: 100109.

[111] Yu, H., Du, C., Qu, F., He, J., Rong, H. (2022) Efficient biostimulants for bacterial quorum quenching to control fouling in MBR. *Chemosphere*. 286: 131689.

[112] Mohamadi, S., Hazrati, H., Shayegan, J. (2020) Influence of a new method of applying adsorbents on membrane fouling in MBR systems. *Water and Environment Journal* 34(S1): 355–366.

[113] Gao, D.W., Wen, Z.D., Li, B., Liang, H. (2014) Microbial community structure characteristics associated membrane fouling in A/O-MBR system. *Bioresource Technology* 154: 87–93.

[114] Yu, H., et al. (2016) Biofouling control by biostimulation of quorum-quenching bacteria in a membrane bioreactor for wastewater treatment. *Biotechnology and Bioengineering* 113(12): 2624–2632.

[115] Qiong, T., Song, L., Cheng, Y., Xiong, J.R. (2013) Improving the performance of membrane bioreactors by adding a metabolic uncoupler, 4-nitrophenol. *Applied Biochemistry and Biotechnology* 169(7): 2126–2137.

[116] Cui, Y., Gao, H., Yu, R., Gao, L., Zhan, M. (2021) Biological-based control strategies for MBR membrane biofouling: a review. *Water Science and Technology* 83(11): 2597–2614.

[117] Wang, Z., Ma, J., Tang, C.Y., Kimura, K., Wang, Q., Han, X. (2014) Membrane cleaning in membrane bioreactors: A review. *Journal of Membrane Science* 468: 276–307.

[118] Jiang, C.K., et al. (2019) Effect of scrubbing by NaClO backwashing on membrane fouling in anammox MBR. *Science of the Total Environment* 670: 149–157.

[119] Ensano, B.M.B. et al. (2019) Applicability of the electrocoagulation process in treating real municipal wastewater containing pharmaceutical active compounds. *Journal of Hazardous Materials* 361: 367–373.

[120] Banti, D.C., Mitrakas, M., Samaras, P. (2021) Membrane fouling controlled by adjustment of biological treatment parameters in step-aerating MBR. *Membranes* 11(8): 553.

[121] Yang, M., et al. (2017) Optimization of MBR hydrodynamics for cake layer fouling control through CFD simulation and RSM design. *Bioresource Technology* 227: 102–111.

[122] Gkotsis, P.K., Batsari, E.L., Peleka, E.N., Tolkou, A.K., Zouboulis, A.I. (2017) Fouling control in a lab-scale MBR system: Comparison of several commercially applied coagulants. *Journal of Environmental Management* 203: 838–846.

[123] Suh, C., Lee, S., Cho, J. (2013) Investigation of the effects of membrane fouling control strategies with the integrated membrane bioreactor model. *Journal of Membrane Science* 429: 268–281.

[124] Böhm, L., Drews, A., Prieske, H., Bérubé, P.R., Kraume, M. (2012) The importance of fluid dynamics for MBR fouling mitigation. *Bioresource Technology* 122: 50–61.

[125] Hebbar, R.S., Isloor, A.M., Ismail, A.F. (2014) Preparation of antifouling polyetherimide/hydrolysed PIAM blend nanofiltration membranes for salt rejection applications. *RSC Advances* 4(99): 55773–55780.

[126] Syed Ibrahim, G.P., Isloor, A.M., Ismail, A.F., Farnood, R. (2020) One-step synthesis of zwitterionic graphene oxide nanohybrid: Application to polysulfone tight ultrafiltration hollow fiber membrane. *Scientific Reports* 10(1): 1–13.

[127] Kumar, M., Isloor, A.M., Somasekhara Rao, T., Ismail, A.F., Farnood, R., Nambissan, P.M.G. (2020) Removal of toxic arsenic from aqueous media using polyphenylsulfone/cellulose acetate hollow fiber membranes containing

zirconium oxide. *Chemical Engineering Journal* 393: 124367.
[128] Hebbar, R.S., Isloor, A.M., Zulhairun, A.K., Sohaimi Abdullah, M., Ismail, A.F. (2017) Efficient treatment of hazardous reactive dye effluents through antifouling polyetherimide hollow fiber membrane embedded with functionalized halloysite nanotubes. *Journal of the Taiwan Institute of Chemical Engineers* 72: 244–252.
[129] Wang, C., Ng, T.C.A., Ng, H.Y. (2021) Comparison between novel vibrating ceramic MBR and conventional air-sparging MBR for domestic wastewater treatment: Performance, fouling control and energy consumption. *Water Research* 203: 117521.
[130] Pabby, A.K., Sonawane, J.V., Gupta, S.K., Sawant, S.R., Rathore, N.S., Kulkarni, Y. (2015) Overview and the current status of membrane-based processing of radioactive nuclear plant waste: Evaluation of some case studies. In: *Handbook of Membrane Separation: Chemical, Pharmaceutical, Food, Biotechnology Applications, Second Ed.*, pp. 709–722.
[131] Manetti, M., Tomei, M.C. (2022) Extractive polymeric membrane bioreactors for industrial wastewater treatment: Theory and practice. *Process Safety and Environmental Protection* 162: 169–186.
[132] Krzeminski, P., Leverette, L., Malamis, S., Katsou, E. (2017) Membrane bioreactors – A review on recent developments in energy reduction, fouling control, novel configurations, LCA and market prospects. *Journal of Membrane Science* 527: 207–227.
[133] Ioannou-Ttofa, L., Foteinis, S., Chatzisymeon, E., Fatta-Kassinos, D. (2016) The environmental footprint of a membrane bioreactor treatment process through life cycle analysis. *Science of the Total Environment* 568: 306–318.
[134] Banti, D.C., Tsangas, M., Samaras, P., Zorpas, A. (2020) LCA of a membrane bioreactor compared to activated sludge system for municipal wastewater treatment. *Membranes (Basel)* 10(12): 1–15.
[135] Chu, T., Abbassi, B.E., Zytner, R.G. (2022) Life-cycle assessment of full-scale membrane bioreactor and tertiary treatment technologies in the fruit processing industry. *Water Environment Research* 94(1): e1661.
[136] Holloway, R.W., et al. Life-cycle assessment of two potable water reuse technologies: MF/RO/UV-AOP treatment and hybrid osmotic membrane bioreactors. *Journal of Membrane Science* 507: 165–178.
[137] Kobayashi, Y., Ashbolt, N.J., Davies, E.G.R., Liu, Y. (2020) Life cycle assessment of decentralized greywater treatment systems with reuse at different scales in cold regions. *Environment. International* 134: 105215.
[138] Nowrouzi, M., Abyar, H. (2021) A framework for the design and optimization of integrated fixed-film activated sludge-membrane bioreactor configuration by focusing on cost-coupled life cycle assessment. *Journal of Cleaner Products* 296: 126557.
[139] Dominguez, S., et al. (2018) LCA of greywater management within a water circular economy restorative thinking framework. *Science of the Total Environment* 621: 1047–1056.
[140] Cabling, L.P.B., Kobayashi, Y., Davies, E.G.R., Ashbolt, N.J., Liu, Y. (2020) Life cycle assessment of community-based sewer mining: Integrated heat recovery and fit-for-purpose water reuse. *Environments – MDPI* 7(5).
[141] Razman, K.K., Hanafiah, M.M., Mohammad, A.W. (2022) An overview of LCA applied to various membrane technologies: Progress, challenges, and harmonization. *Environmental Technology & Innovation* 27: 102803.
[142] Arif, A.U.A., Sorour, M.T., Aly, S.A. (2020) Cost analysis of activated sludge and membrane bioreactor WWTPs using CapdetWorks simulation program: Case study of Tikrit WWTP (middle Iraq). *Alexandria Engineering Journal* 59(6): 4659–4667.
[143] Tizchang, A., Jafarzadeh, Y., Yegani, R., Khakpour, S. (2019) The effects of pristine and silanized nanodiamond on the performance of polysulfone membranes for wastewater treatment by MBR system. *Journal of Environmental Chemical Engineering* 7(6): 103447.
[144] Hashemi, T., Mehrnia, M.R., Ghezelgheshlaghi, S. (2022) Influence of alumina nanoparticles on the performance of polyacrylonitrile membranes in MBR. *Journal of Environmental Health Science and Engineering* 20(1): 375–384.
[145] Razzaghi, M.H., Tavakolmoghadam, M., Rekabdar, F., Oveisi, F. (2018) Investigation of the effect of coagulation bath composition on PVDF/CA membrane by evaluating critical flux and antifouling properties in lab-scale submerged MBR. *Water and Environment Journal* 32(3): 366–376.
[146] Hosseinpour, S., Azimian-Kivi, M., Jafarzadeh, Y., Yegani, R. (2021) Pharmaceutical wastewater treatment using polypropylene membranes incorporated with carboxylated and PEG-grafted nanodiamond in membrane bioreactor (MBR). *Water and Environment Journal* 35(4): 1249–1259.
[147] Zhou, M., Shi, Q., Wang, Y. (2021) Application of hydrophilic modified nylon fabric membrane in an anammox-membrane bioreactor: performance and fouling characteristics. *Environmental Science and Pollution Research* 29(4): 5330–5344.
[148] Liang, Z., Hu, Z. (2012) Biodegradation of nitrophenol compounds and the membrane fouling trends in different submerged membrane bioreactors. *Journal of Membrane Science* 415–416: 93–100.
[149] Ding, A., et al. (2019) Effect of metabolic uncoupler, 2,4-dinitrophenol (DNP) on sludge properties and fouling potential in ultrafiltration membrane process. *Science of the Total Environment* 650: 1882–1888.
[150] Fang, F., et al. (2019) Characterization of interactions between a metabolic uncoupler O-chlorophenol and extracellular polymeric substances of activated sludge. *Environmental Pollution* 247: 1020–1027.
[151] Baugh, S., Ekanayaka, A.S., Piddock, L.J.V., Webber, M.A. (2012) Loss of or inhibition of all multidrug resistance efflux pumps of Salmonella enterica serovar Typhimurium results in impaired ability to form a biofilm. *Journal of Antimicrobial Chemotherapy* 67(10): 2409–2417.
[152] Razman, K.K., Hanafiah, M.M., Mohammad, A.W. (2022) An overview of LCA applied to various membrane technologies: Progress, challenges, and harmonization. *Environmental Technology & Innovation* 27: 102803.
[153] Arias, A., Feijoo, G., Moreira, M.T. (2020) Environmental profile of decentralized wastewater treatment strategies based on membrane technologies. Chapter 11. In: *Curr. Dev. Biotechnol. Bioeng. Adv. Membr. Sep. Process. Sustain. Water Wastewater Manag. – Case Stud.*

Sustain. Anal., pp. 259–287, Jan. 2020, doi: 10.1016/B978-0-12-819854-4.00011-3.

[154] Arden, S., et al. (2020) Human health, economic and environmental assessment of onsite non-potable water reuse systems for a large, mixed-use urban building. *Sustainability* 12(13): 5459.

[155] Banti, D.C., Tsangas, M., Samaras, P., Zorpas, A. (2020) LCA of a membrane bioreactor compared to activated sludge system for municipal wastewater treatment. *Membranes* 10(12): 421.

[156] Cabling, L.B.P., Kobayashi, Y., Davies, E.G.R., Ashbolt, N.J., Liu, Y. (2020) Life cycle assessment of community-based sewer mining: Integrated heat recovery and fit-for-purpose water reuse. *Environments* 7(5): 36.

[157] Jiménez-Benítez, A., Ferrer, J., Rogalla, F., Vázquez, J.R., Seco, A., Robles, Á. (2020) Energy and environmental impact of an anaerobic membrane bioreactor (AnMBR) demonstration plant treating urban wastewater. Chapter 12. In: *Curr. Dev. Biotechnol. Bioeng. Adv. Membr. Sep. Process. Sustain. Water Wastewater Manag. – Case Stud. Sustain. Anal.*, pp. 289–310, Jan. 2020, doi: 10.1016/B978-0-12-819854-4.00012-5.

[158] Kobayashi, Y., Ashbolt, N.J., Davies, E.G.R., Liu, Y. (2020) Life cycle assessment of decentralized greywater treatment systems with reuse at different scales in cold regions. *Environment International* 134: 105215.

26 Process Intensification in Water Treatment Using Membrane Technology

Enrico Drioli, Elena Tocci, and Francesca Macedonio*
Institute on Membrane Technology of the Italian National Research Council (CNR-ITM), Rende (CS), Italy
*Email: e.tocci@itm.cnr.it

26.1 INTRODUCTION

The world faces critical global challenges, including energy shortages, warming, material resources, water, and food shortages, and environmental pollution.

One of the key elements that can overcome to these challenges and realize sustainable development is process intensification (PI). This is defined as a set of radically innovative principles ("paradigm shifts") in process and equipment design. Such principles drastically improve processes and products, reduce capital investments, environmental impact, energy, and material, and yield substantial benefits in quality, waste, process safety, and more [1–4]. PI principles and explicit goals cover a wide range of processing equipment types and methodologies (Figure 26.1). They must maximize the synergistic effects from partial processes at all possible scales and the effectiveness of intra- and intermolecular events.

The four approaches to PI involve space (structure), thermodynamics (energy), function (synergy), and time [6]. These principles look for the engineering methods to control better, improve and, therefore, change the inherent kinetics of chemical reactions—i.e., number/frequency of collisions, the mutual orientation of molecules, and their energy [6,7]. In the logic of PI, optimizing the driving forces at every scale and maximizing the specific surface area to which these forces apply is essential. The principle that lies behind it refers to maximization of the driving force effects and not simply to the driving forces. Moreover, to give each molecule the same processing experience (considering that processes deliver uniform products with minimum waste when all molecules undergo the same history). In the last few years, since new developments were introduced, i.e., the Internet of Things (IoT), cloud computing, artificial intelligence (AI), cyber-security, big data, and analytics, a new definition of Process Intensification 4.0 has been coined. These new technological tools modernized different product designs and process systems controlled in an interconnected framework [8]. Consequently, a fifth approach has been defined to transform PI into the next generation of process engineering. It is the knowledge domain denoted as Data (Figure 26.2). Today data and their fast processing time, data-driven algorithms, and data flow from diverse, interconnected sensors are easily available. The fifth approach enhances the current intensified equipment or designs new ones and better controls the process by employing data-driven decisions [5].

Modern membrane engineering is an important case of process intensification since it satisfies the requirements of innovative processes and equipment design, i.e., smaller, cleaner, more energy-efficient technologies. It can replace conventional energy-intensive separation techniques, achieving the selective and efficient transport of specific components, and improving the performance of reactive processes. Moreover, exploiting the compartmentalization of two regions on two sides of the membranes, membrane devices can combine two processes carried out on two sides of the membrane in one membrane device [9,10].

Membrane engineering has already shifted toward sustainable production of strategic processes in all industrial sectors, such as potable water production, energy generation, sustainable separation processes for chemicals, regenerative medicine, food and beverages, pharmaceutical production, electronics, and packaging [10]. All the traditional and novel membrane operations and their integrated schemes fit zero-liquid-discharge logic, low energy consumption, and total raw materials utilization. For these reasons, they have outstanding potential to reach several sustainable development goals and can promote a resource-conserving and environmental-friendly society [11,12], such as sustainable agriculture by recycling water and nutrients from wastewater (SDG 2), clean water and sanitation (SDG 6), and sustainable energy production (SDG 7). Various membrane technologies are recognized as Best Available Technologies (BATs) in various processes and in waste management [13].

Water demand is increasing worldwide, while water availability in many regions is likely to decrease due to climate change and socioeconomic development patterns. Recent accelerated climate change has exacerbated existing environmental problems around the world and in particular in the Mediterranean Basin where observed precipitation trends are characterized by high variability in space and time, but climate models clearly indicate a trend toward reduced rainfall in coming decades. The combination of reduced rainfall and warming generates strong trends towards drier conditions. A recent study by the Potsdam Institute for Climate Impact Research published in Nature [14] revealed that major heatwaves in Western Europe have increased three to four times faster than in other mid-latitude boreal zones

DOI: 10.1201/9781003285656-26

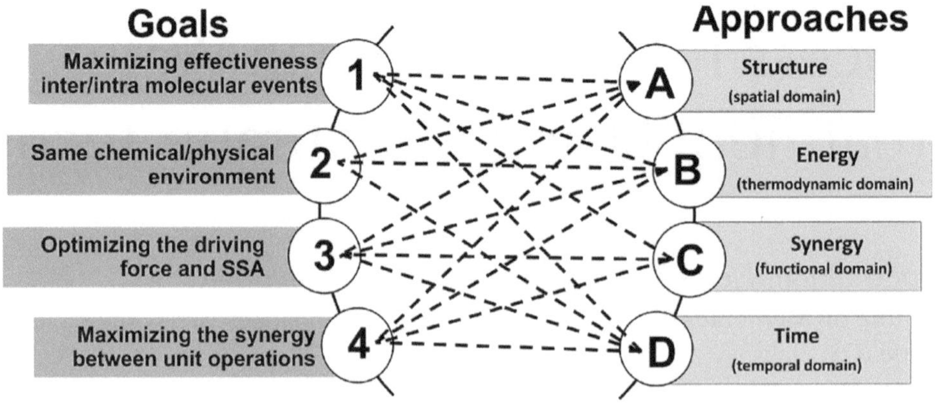

FIGURE 26.1 Goals and approaches in the process intensification classical framework. Reprinted from Ref. [5] (Open Access Publication). SSA, specific surface area.

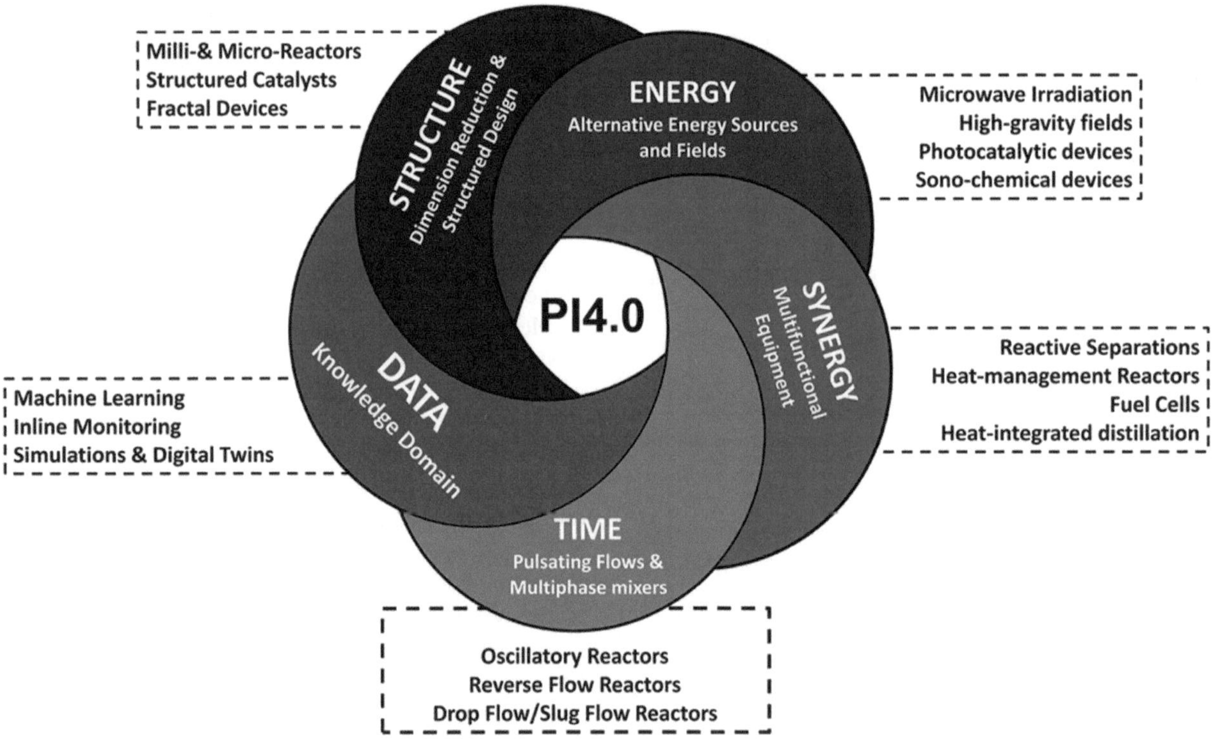

FIGURE 26.2 Introduction of data approach (knowledge domain) in PI4.0. Reprinted from Ref. [5] (Open Access Publication).

over the last 40 years. A global atmospheric temperature increase of 2°C is expected to be accompanied by a reduction in summer precipitation of about 10–15% in southern France, northwestern Spain, and the Balkans and up to a 30% reduction in Turkey. Scenarios with 2–4°C temperature increases by the 2080s for southern Europe would imply stronger and more widespread decreases in precipitation of up to 30%. The coupled effects of warming and drought are expected to lead to a general increase in aridity and subsequent desertification of many Mediterranean land ecosystems. Deserts would expand in southern Spain and Portugal, northern parts of Morocco, Algeria, Tunisia, Sicily, southern Turkey, and parts of Syria [15].

With the aim to address water scarcity within the European Union (EU) and moving the EU toward a water-efficient and water-saving economy, the EU Commission identified an initial set of policy options to be taken at European, national, and regional levels. The Mediterranean Strategy for Sustainable Development (MSSD) 2016–2025 was adopted as a strategic guiding document for all stakeholders and partners in order to translate the 2030 Agenda for Sustainable Development at the regional, sub-regional, and national levels [16]. The MSSD provided an integrative policy framework for securing a sustainable future for the Mediterranean region consistent with the Sustainable Development Goals. One of its targets is "Target 6.4: By 2030, substantially increase water-use efficiency across all sectors and ensure sustainable withdrawals and supply of freshwater to address water scarcity and substantially reduce the number of people suffering from water scarcity." Ongoing climate and environmental changes, increasing water scarcity, and desertification affect living standards, and industrial and agricultural development.

Furthermore, mineral deficiency is becoming a threat to future sustainable development. Supply risk and supply disruption of certain minerals and metals might negatively affect sustainable industrial growth. Lithium and phosphorus are two main examples in this context. Lithium is particularly interesting for its increasing use in lithium-ion batteries. Similarly, phosphorus is a fundamental raw material for fertilizer production and has several other uses in domestic products. The availability of both lithium and phosphorus in required quantities in the near future is under hot debate [17,18]. In this perspective, the supply of other strategic minerals, such as barium, strontium, uranium, cesium, etc., from their traditional sources may not be enough to fulfill their demand [19].

At the same time, the environmental regulations are becoming increasingly stringent. To improve sustainability, waste has to be turned into sustainable resources of minerals and water. These facts clearly indicate the need for revolutionary changes in state-of-the-art norms and practices in the process industry. Consequently, efficient technologies able either to minimize the utilization of water and minerals or to recover valuable components from wastes have to be developed.

Membrane engineering is a powerful discipline able to significantly contribute to solving the water- and raw materials-related issues [20]. Regarding water shortage, more than 80 million m^3/d of fresh water is obtained via membrane processes (mostly through reverse osmosis). From 2010 to today, around 90% of desalination capacity employs membrane technologies, leaving just under 10% to the more traditional distillation/evaporation systems (with the use of thermal technologies for large-scale projects—often outdated plants—remaining concentrated in the Middle East) [21].

The key to widespread interest and implementation of seawater RO plants has been a significant reduction in capital and operation/maintenance costs over the past 40 years. Several factors have helped in reducing RO energy consumption and costs, including improvements in membrane materials and technology (higher flux, higher salt rejection, lower hydrostatic pressure required, lower materials cost) and the use of pressure recovery devices. RO has also become less expensive than thermal processes.

The reliability of RO has increased over the years also as a consequence of the development of various other membrane operations (such as MF and UF) that can be introduced upstream of RO for reducing fouling phenomena and extending membrane life-time: MF allows removing suspended solids and large bacteria, and lowering the silt density index (SDI) and the ratio COD/BOD; in addition to these, UF also retains dissolved macromolecules, colloids, and the smallest bacteria. The membrane bioreactor (MBR) was also investigated as a pretreatment to RO because it allows achieving high organic removal [biopolymers (94–97%) and humic (71–76%)] [22]. A submerged membrane adsorption bioreactor was used as a pretreatment in seawater desalination for biofouling control [22].

This chapter aims to update on the contributions of membrane engineering to sustainable processes in water treatment. We present an overview of the membrane engineering approaches successfully employed in water treatment on Earth and in space, in producing raw materials, and in membrane reactors.

26.2 SUCCESS CASES OF PROCESS INTENSIFICATION IN WATER TREATMENTS

26.2.1 Membrane Bioreactors

Membrane bioreactors (MBRs) combine membrane filtration with biological treatment [23–27].

They use the cellular activity of microorganisms to convert the organic matter in the wastewater water into a sludge, while fresh water is recovered by permeating through membranes [28,29]. MBR systems have several advantages in comparison with the traditionally used conventional activated sludge (CAS). CAS employs an aeration tank, using aeration and a biological floc composed of bacteria and protozoa, for biological oxidization of organic pollutants. It contains also a secondary clarifier (sedimentation tank), where the sludge is separated from the treated wastewater, producing a waste sludge (or floc) containing the oxidized material. The conventional activated sludge and membrane bioreactor process is shown in Figure 26.3.

MBR uses advanced oxidation processes (AOPs) since it generates highly reactive free radicals, for organic pollutant degradation, which is a quick and non-selective reaction, instead of the traditional activated sludge process [31–33]. AOP involves highly concentrated microorganisms to degrade recalcitrant organics (such as pesticides, endocrine-disrupting chemicals, pharmaceuticals) and remove nutrients from wastewater. It produces quality effluent of low biological oxygen demand (BOD) (Table 26.1).

The compact nature of the MBR delivers better performances. In CASP, the separation of biomass from water occurs based on gravity, while in a membrane bioreactor, the activated sludge removal process is led by means of UF or MF membranes, and higher qualities of effluent water can be achieved. When MF membranes are used, having pore sizes usually in the range 0.1–0.4 μm, complete retention of suspended matter and reduction of bacteria in the outflow are obtained.

Moreover, several key benefits of membrane bioreactors in wastewater treatment are offered, such as high biomass concentration, short hydraulic retention time and less sludge (< 40%) production compared to conventional activated sludge, low hydrostatic pressure with low flux, small footprint (< 50%) compared to traditional activated sludge (purifying bacteria) plants, and easy combination in existing plants with relatively simple retrofitting [12].

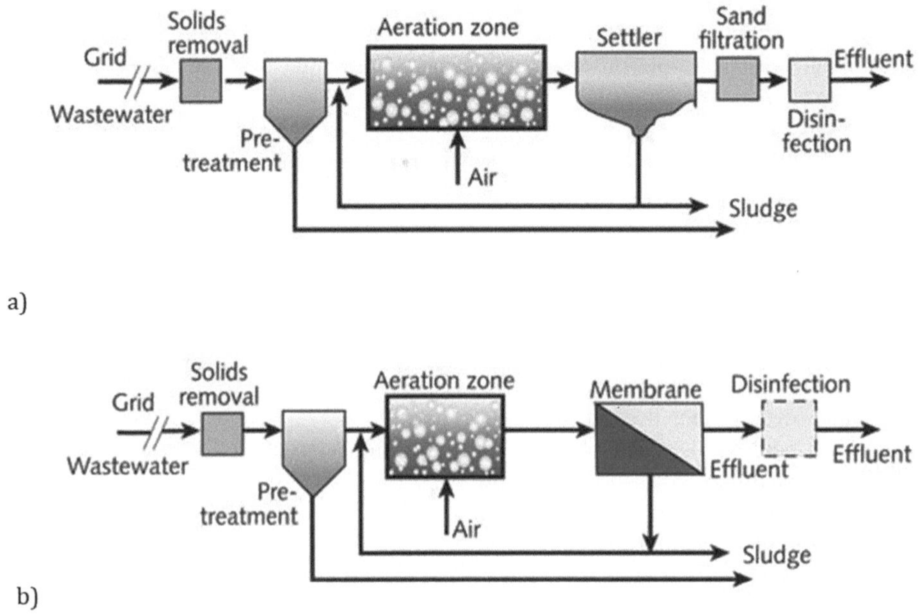

FIGURE 26.3 (a) Activated sludge process; (b) membrane bioreactor (MBR) process. Reprinted from Ref. [30] (Open Access Publication).

TABLE 26.1
Comparison of key process conditions and degrees of removal in CASP and MBR

	CASP	MBR
MLSS (kg/m^3)[a]	5	12-20
BOD Loading rate (kg/m^3d)[a]	0.25	0.3–2.5
BOD removals (%)[a]	85–95	98–99
TSS removal (%)[a]	~60	>99
NH_4^+ removal (%)[a]	99	>99
P_{total} removal (%)[a]	89	97–99
Footprint (relative)[b]	100	30–50
Energy (kWh/m^3)[b]	<0.5	0.4–0.7
Solid waste (relative)[b]	100	<100

[a] From table 3 in Ref. [34]. [b]From table 1 in Ref. [9]. TSS, total suspended solids; MLSS, mixed liquor suspended solids.

MBRs can be classified based on different categories depending on the membrane positioning (outside or submerged), membrane configurations (flat sheet or cylindrical shape), and bio-treatment process (aerobic, anoxic, and anaerobic). In outside (side-stream) MBRs, the sludge is pumped into the membrane module, which generates cross-flow at the membrane surface, permeating through them. The concentrated sludge rejected by the membrane is recycled to the MBR. This pressure-driven membrane filtration process generally employs micro- (MF) and ultrafiltration (UF) membranes. MF membranes are preferred for their capability to remove colloids and viruses and for their lower fouling tendency.

An example of an industrial application is NEOSEP MBR which uses enhanced activated sludge and immersed membrane filtration in producing water with quality appropriate for reuse for industrial and municipal uses. The system consists of separate aeration tanks followed by membrane tanks containing submerged membrane modules. The membrane elements separate the mixed liquor solids from the treated effluent, which is drawn through the membrane by either gravity or pumps. The permeability of the elements is maintained through air scouring of the membrane surface, a membrane relaxation phase, and periodic in situ chemical cleaning. Excess mixed liquor solids are wasted directly from the aeration tanks.

The MBR development worldwide (Figure 26.4) is evident since the number of scientific publications and patents, starting from the late 1990s, has increased exponentially [35]. Recent Business Communication Company reports indicated that the global market size of MBR was 3.0 billion USD in 2019 and is projected to reach USD 4.9 billion by 2026 at a compound annual growth rate (CAGR) of 8.3% between 2021 and 2026.

Worldwide, the emerging economies of Asia Pacific (APAC) and South America are considered to have the highest MBR market growth in the world due to the increasing water scarcity and the growing demand for clean water (www.marketsandmarkets.com/Market-Reports/membrane-bioreactor-market-484.html. Accessed 25th August 2022).

MBR is a classic example of a hybrid membrane process and process intensification and has experienced rapid growth for over three decades in the treatment of industrial and municipal wastewater and is still in rapid growth today.

The reasons for this are the extensive advantages over conventional processes, such as the high-quality effluent with enhanced nutrient removal, optimal microbial separation, complete control of sludge retention time (SRT) and hydraulic retention time, and small footprint requirement but with high volumetric loadings [35]. It is considered by the European

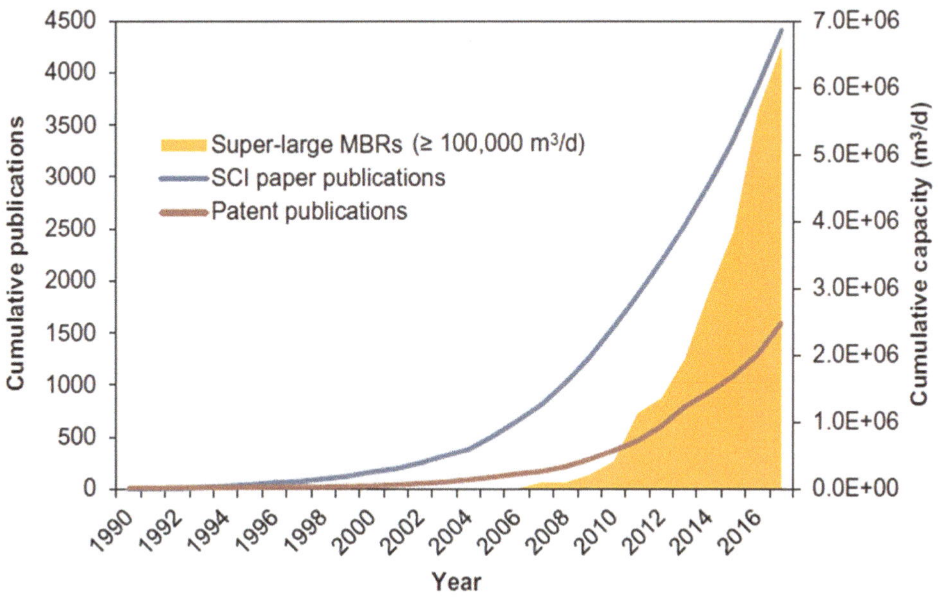

FIGURE 26.4 Worldwide development of MBR paper publication, patent publication, and engineering application. Reprinted with permission from Ref. [35].

Union as one of the Best Available Technologies (BATs) for municipal and industrial wastewater treatment [36].

Fouling phenomena are the main challenges encountered by an efficient MBR process since they cause high energy consumption and low flows. Fouling depends on many factors. The main responsibilities are the membrane characteristics, the operating conditions, and the properties of the wastewater influents. Further improvements in MBR technology require a reduction of the fouling problem, such as the optimization of hydrodynamic conditions [25]. Moreover, MBR requires higher energy than CASP to work, although it has a smaller footprint and higher cost, both in terms of operational cost, per unit volume of treated wastewater, and of capital cost per unit capacity. Notwithstanding this, MBRs are applied for their easy maintenance, high flux, and simple module replacement. Various researchers have continued to explore this configuration, promoting advances that may operate the process with an energy need of about 0.3 kWh/m^3 of purified water.

26.2.2 Membrane Distillation and Crystallization

In the case of sea- and brackish-water desalination, the integration of various membrane units in the RO post-treatment offers a reliable solution not only to reduce the brine disposal problem but also to recover the valuable minerals contained in seawater. Membrane contactors and, in particular, membrane distillation (MD) and membrane crystallization (MCr) are well suited to this logic. MD and MCr are processes that, when utilized for the treatment of RO brine, allow increasing water recovery factor and reducing brine disposal problem [37]. These processes utilize microporous hydrophobic membranes to promote the mass transfer between phases. In particular, a heated, aqueous feed solution is brought into contact with one side (feed side) of a hydrophobic, microporous membrane (Figure 26.5). The hydrophobic nature of the membrane prevents penetration of the aqueous solution into the pores, resulting in a vapor–liquid interface at each pore entrance. When the feed is water containing salts, the water will be vaporized close to the pores and will then pass as a vapor through the membrane pores.

Since MD and MCr operate on the principles of vapor–liquid equilibrium, 100% (theoretical) of ions, macromolecules, colloids, and other non-volatile components (including pollutants difficult to remove with other operations, such as boron and arsenic) are rejected, thus producing high-purity waters. Moreover, MD and MCr are not limited by concentration polarization phenomena as is the case in pressure-driven processes. Therefore, fresh water can be obtained also from high-salinity waters (such as oil-produced water, RO brine, etc.).

In MCr, the evaporation of the feed solution and its consequent concentration are extended until supersaturation is reached, thus attaining a supersaturated environment where crystals may nucleate and grow. When an MCr follows an NF or RO stage, the highly concentrated brine does not represent waste but the mother liquor in which salt crystals could nucleate and grow.

In an MCr, the membrane matrix acts as a selective gate for solvent evaporation, modulating the final degree and rate for the generation of the supersaturation. Hence, the possibility to act on the transmembrane flow rate, by changing the driving force of the process, allows to tailor the final properties of the crystals produced both in terms of structure (polymorphism) and morphology (habit, shape, size, and size distribution). The experimental evidence that can be found in the literature [39], mainly at lab-scale, validate the effectiveness of MCr as an advanced method for performing well-behaved crystallization

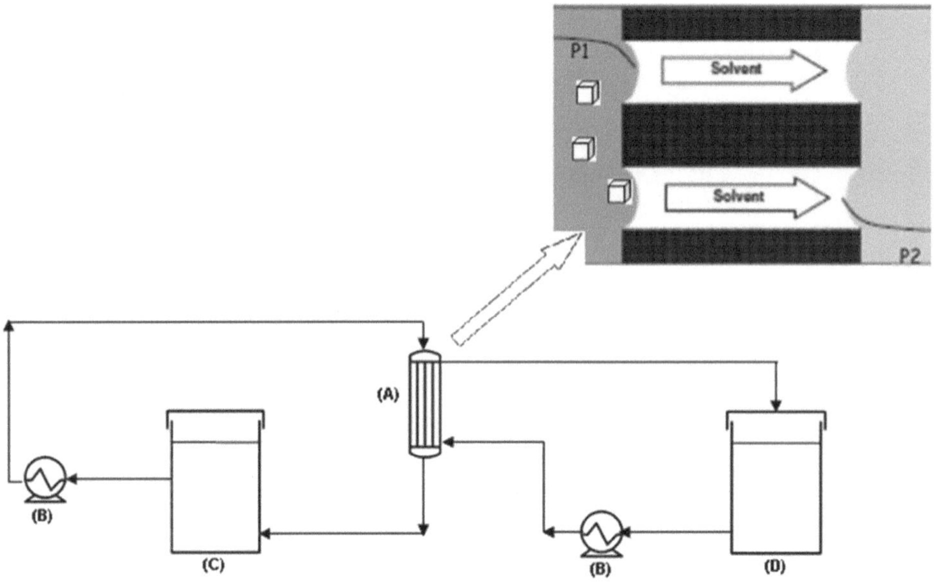

FIGURE 26.5 Schematic representation of a membrane distillation/crystallization: (A) membrane module; (B) pump; (C) crystallizer tank; (D) permeate tank. In the box at the top right, the scheme of the basic principle of the phenomena occurring inside the membrane module: solvent is removed from the crystallizing solution under a vapor pressure gradient (P1 > P2, where 1 refers to the feed/crystallizing side, 2 refers to permeate side). Reprinted from Ref. [38] (Open Access Publication).

processes. The studies carried out by Drioli and co-workers [40–45] showed that the introduction of a MCr unit on NF and RO retentate streams of an integrated membrane-based desalination system constituted by MF/NF/RO increases plant recovery factor so much as to reach 92.8%, higher than that of an RO unit (about 45%). Interesting examples of the fast growing of the integrated membrane-based desalination strategy described above can be found in various international projects, such as MEDINA (funded by the European Union in the framework of the FP6 program), SEAHERO and Global MVP research program (2013–2018) (funded and developed in South Korea). The emphasis of the projects was, on one side to increase the desalination capacity of RO desalination plants and, on the other side, to extract valuable resources from RO brine, thus minimizing the environmental impact and improving the sustainability of desalination plants.

26.2.3 Membrane Engineering in Space: Water Treatment

Water treatment processes are needed for space missions since it is impossible to carry massive reserves of water and food in shuttles and space stations due to limitations in weight and in-room space. All the resources necessary to sustain life need to be, as much as possible, recovered and reused to reduce waste disposal. Membrane engineering applied to space is a case in point of intensified processes because space requires minimizing volume, weight, energy consumption, costs, and safe procedures, particularly in long-duration crewed missions. Challenges in space stations are the need for oxygen and atmosphere regeneration with the elimination of CO_2, fresh water, the energy supply, and liquid and solid waste management.

Earlier research activities on a feasibility study of membrane separators for spatial applications were carried out in the 1990s in the framework of the Project "Liquid Gas Phase Separation," Contract ESTeC n. 11523/95/NL/SB (European Space Agency) [46–48].

As reported by the European Space Agency, membrane operations involve various sectors for both water and gas [49,50]. Both traditional separations and membrane contactors are used, separating H_2O from CH_4, and H_2 from CH_4. Membrane bioreactors are used for CO_2 capture and oxygen production using a photosynthetic process; a microgravity condenser is employed for collecting water; free gas extractor from a liquid stream for microgravity water system (hollow-fiber polymeric membrane contactor). In summary, membrane systems are used for filtration and recovery of water for portable use, without contamination by microbes, and for hygiene, gas separations, and solid and liquid waste management.

However, the challenge to handle for membrane systems to work correctly is the absence of gravity. These effects can influence the membrane lifetime, transport processes, and fouling phenomena. Moreover, specific attention to radiation resistance of materials and shielding is required in space missions due to the extreme conditions of cosmic radiation, heat, and debris impact.

Water recycling is vital for maintaining the International Space Station (ISS). On board the ISS, a water bottle can cost ca. USD 10,000. Several sources of wastewater can be reused in a space station: urine, fecal, food waste, hygiene, cabin condensate, laundry, and Sabatier water streams. The

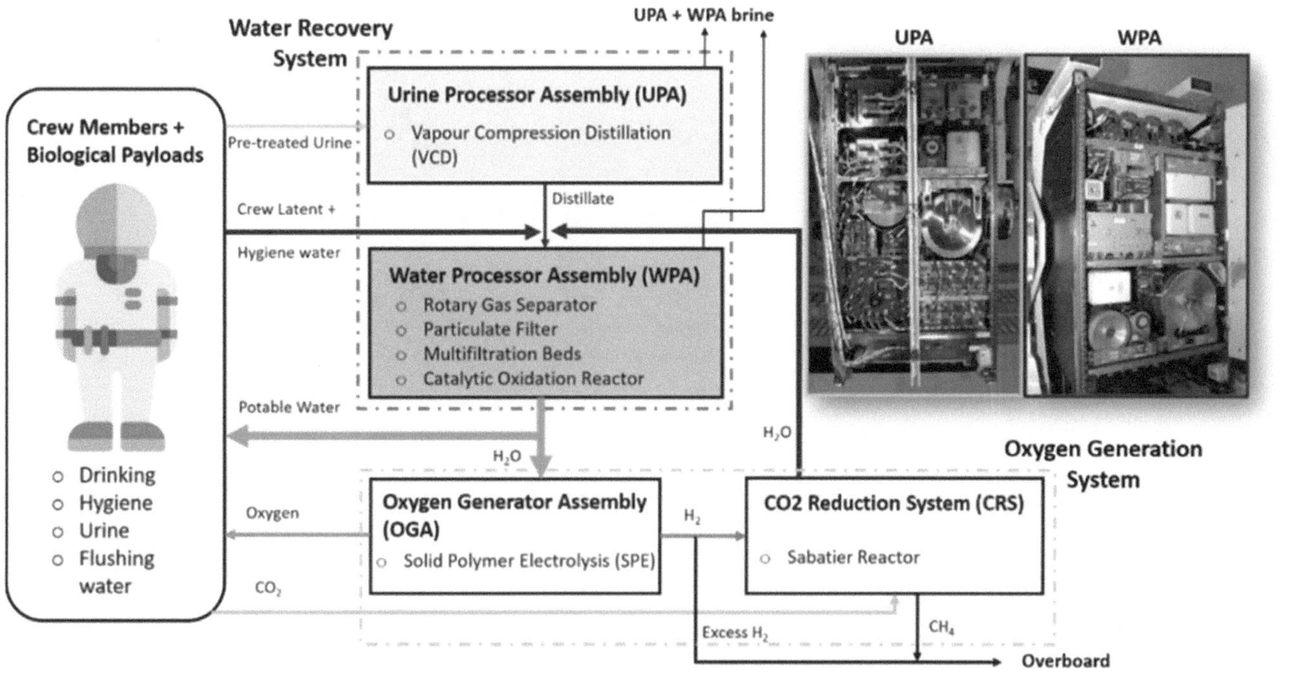

FIGURE 26.6 The International Space Station (ISS) water purification process. Reprinted from Ref. [51] (Open Access Publication).

water purification process (so-called "Water Recovery and Management") ensures potable water for crew drinking and hygiene, oxygen production, and urinal flush water. Figure 26.6 shows the Water Recovery and Management schematic diagram, split into Urine Processor Assembly (UPA) and Water Processor Assembly (WPA).

The wastewater from crew urine and hygiene and the condensed water vapor from crew humidity condensate (both perspiration and respiration) are treated for recovering clean water [51]. In addition to these sources, the Sabatier product water (produced from the carbon dioxide reduction system) is processed to potable standards [52].

The water treatment technologies on board employ only physical/chemical systems for their reliable performance. Comparing the energy inputs, the consumables, and the waste products among chemical (both pressure-driven like FO, MF, UF, NF, RO, and thermally driven like MD and MCr) membrane processes and the biological processes (including MBr), it is observed that they are always lower than traditional chemical methods and the opportunities for resource recovery are particularly high in hybrid processes. Membrane bioreactors (MBr) combining chemical and physical processes can degrade and convert microbes and separate solid/liquid with membrane filtration [50]. One example is given by the Urea Bioreactor Electrochemical (UBE) system, employing forward osmosis for resource recovery from water recycling systems.

The urea bioreactor (GAC-urease) was used to recover urea from the wastewater stream and convert it to ammonia (Figure 26.7). The latter was fed to an electrochemical cell to generate electrical energy. The UBE systems removed >80% of organic carbons and converted approximately 86% of the urea to ammonia. This system provided a method of targeting urea and removing it as N_2 while generating electrical current [53].

Another wastewater source is the water produced from the Sabatier reaction [54]. The Sabatier reaction produces water and methane from hydrogen and exhaled carbon dioxide ($CO_2 + 4H_2 \rightleftarrows CH_4 + 2H_2O$). The Sabatier reaction is almost a closed loop since the water generated can be decomposed again, through water electrolysis, into hydrogen and oxygen, and collected separately. Then, reusing hydrogen gas with the carbon dioxide in the atmosphere, the Sabatier reaction produces drinkable water and methane for heating and rocket propellant.

Water can be saved also from air dehumidification of the ISS originated by the crew's metabolic moisture. The dehumidification of aircraft improves thermal comfort, preventing bacteria growth and preserving the onboard electronics. Currently, the dehumidification devices are based on condensing heat exchangers. The biggest challenge is separating water from the air because of the microgravity. In many spacecraft, the water/air separator is usually used in conjunction with the hygroscopic coating applied on the fins of the condensing heat exchanger.

However, there are some limitations, such as dissolution or peeling of the hygroscopic coating, causing the separator to clog.

Membrane condensers could be applied for the dehumidification of cabins. Microporous hydrophobic membranes are used to promote water vapor condensation and recovery from a humid gaseous stream avoiding water droplet dragging (Figure 26.8) [55–58]. Water recovery is obtained on the retentate side with the simultaneous transfer of the other gases across the membrane.

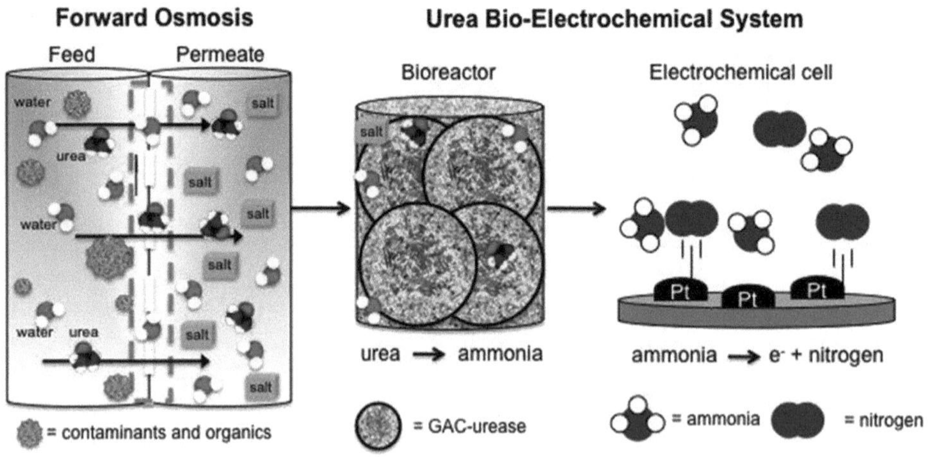

FIGURE 26.7 Urea Bioreactor Electrochemical (UBE) system, with forward osmosis for resource recovery from water recycling system. Reprinted with permission from Ref. [53] Copyright (2021) American Chemical Society.

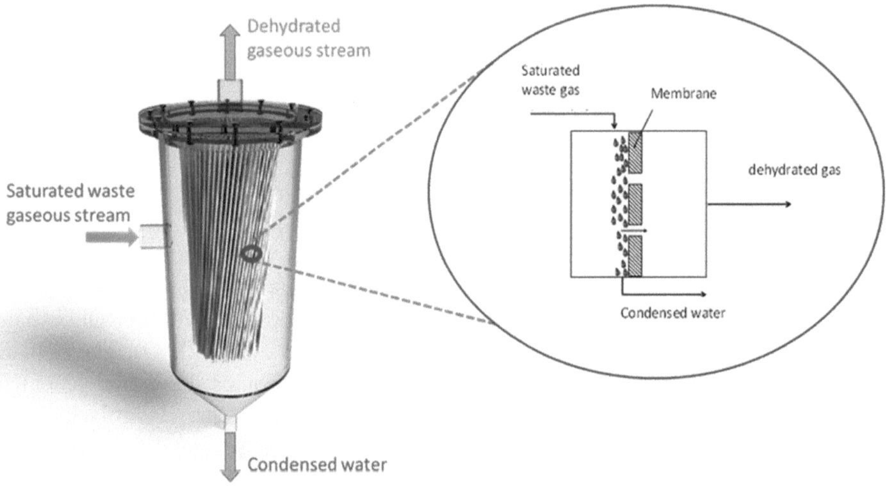

FIGURE 26.8 Membrane condenser module and process for recovering evaporated "waste" water from a gaseous stream. Reprinted from Ref. [57] (Open Access Publication).

The advantages of the process include the lack of corrosion since the membranes can be fabricated in highly resistant material, low-temperature problems, and the ability to retain the contaminants contained in the fed gaseous streams [59,60].

26.3 CONCLUSION AND OUTLOOK

Membrane systems working in separation, contacting, and reaction processes have proved to contribute to sustainable processes. Membrane engineering has demonstrated a shift toward process intensification, of strategic sustainable processes and products, in all industrial sectors.

This has been achieved since membrane engineering can satisfy the requirements of minimization of volume, weight, energy consumption, and costs, attaining multistep processes in a single device with functioning flexibility. In this chapter, we have shown through some positive examples the success of the process intensification strategy in membrane engineering, indicating how membrane systems can help find the solutions for wastewater treatments on Earth and in space. However, many other cases not considered here demonstrate that membrane engineering could introduce extraordinary levels of process intensification in the coming years. For example, in biomedical applications (e.g., artificial brains) and innovative medicine, including tissue engineering, microplastic removal, blue energy production from salinity gradients, etc.

Some challenges remain to be overcome. For example, more efficient fouling control and the reduction of energy consumption. Fouling represents the key challenge in long-term operations, for which new solutions must be proposed, such as developing new membranes, more resistant to chlorine and extreme pH conditions, and new approaches for efficient cleaning.

REFERENCES

1. Stankiewicz, A.I., Moulijn, J.A. (2000) Process intensification: Transforming chemical engineering. *Chemical Engineering and Processing* 96: 22–34.
2. Moulijn, J.A., Stankiewicz, A., Grievink, J., Górak, A. (2008) Process intensification and process systems engineering: A friendly symbiosis. *Computers & Chemical Engineering* 32: 3–11.
3. Kuo, W.-H.K., Chase, H.A. (2010) Process intensification for the removal of poly-histidine fusion tags from recombinant proteins by an exopeptidase. *Biotechnology Progress* 26: 142–149.
4. Górak, A., Stankiewicz, A. (2011) Research agenda for process intensification towards a sustainable world of 2050. *Research Agenda for Process Intensification towards a Sustainable World of 2050* 90.
5. López-Guajardo, E.A., Delgado-Licona, F., Álvarez, A.J., Nigam, K.D.P., Montesinos-Castellanos, A., Morales-Menendez, R. (2021) Process intensification 4.0: A new approach for attaining new, sustainable and circular processes enabled by machine learning. *Chemical Engineering and Processing – Process Intensification* 108671.
6. Van Gerven, T., Stankiewicz, A. (2009) Structure, energy, synergy, time—The fundamentals of process intensification. *Industrial & Engineering Chemistry Research* 48: 2465–2474.
7. Charpentier, J.-C. (2007) Modern chemical engineering in the framework of globalization, sustainability, and technical innovation. *Industrial & Engineering Chemistry Research* 46: 3465–3485.
8. Sharma, A.K., Bhandari, R., Pinca-Bretotean, C., Sharma, C., Dhakad, S.K., Mathur, A. (2021) A study of trends and industrial prospects of Industry 4.0. *Materials Today: Proceedings* 47: 2364–2369.
9. Sirkar, K.K., Fane, A.G., Wang, R., Wickramasinghe, S.R. (2015) Process intensification with selected membrane processes. *Chemical Engineering and Processing: Process Intensification* 87: 16–25.
10. Drioli, E., Giorno, L. (2009) *Membrane Operations: Innovative Separations and Transformations*, Drioli, E., Giorno, L. (Eds.). Wiley.
11. Macedonio, F., Drioli, E. (2017) Membrane engineering for green process engineering. *Engineering* 3: 290–298.
12. Drioli, E., Macedonio, F., Tocci, E. (2021) Membrane science and membrane engineering for a sustainable industrial development. *Separation and Purification Technology* 275: 119196.
13. JRC (2015) *Best Available Techniques (BAT) Reference Document for Waste Treatment Industries (Draft).*
14. Rousi, E., Kornhuber, K., Beobide-Arsuaga, G., Fei, L., Coumou, D. (2022) Accelerated western European heatwave trends linked to more-persistent double jets over Eurasia. *Nature Communications.*
15. Guiot, J., Cramer, W., Guiot, J., Cramer, W., Paris, T. (2017) *Climate change: The 2015 Paris Agreement thresholds and Mediterranean basin ecosystems.*
16. MedECC Network Science-policy interface (2019) *Risks Associated to Climate and Environmental Changes in the Mediterranean Region. A preliminary assessment by the MedECC Network Science-policy interface* (2019). (from https://ufmsecretariat.org/wp-content/uploads/2019/10/MedECC-Booklet_EN_WEB.pdf; last accessed 14 September 2023).
17. Wanger, T.C. (2011) The lithium future — resources, recycling, and the environment. *Conservation Letters* 4: 202–206.
18. Ulrich, A.E., Frossard, E. (2014) Science of the total environment on the history of a reoccurring concept: Phosphorus scarcity. *Science of the Total Environment* 490: 694–707.
19. Wouters, H., Bol, D. (2009) *Material scarcity. An M2i study.* Delft: Stichting Materials Innovation Institute.
20. Issaoui, M., Jellali, S., Zorpas, A.A., Dutournie, P. (2022) Membrane technology for sustainable water resources management: Challenges and future projections. *Sustainable Chemistry and Pharmacy* 25: 100590, doi:10.1016/j.scp.2021.100590.
21. IDA (2018–2019) *Water Security Handbook 2018-2019.* IDA.
22. Naidu, G., Jeong, S., Vigneswaran, S., Rice, S.A. (2013) Microbial activity in bio fi lter used as a pretreatment for seawater desalination. *DES* 309: 254–260.
23. Fontananova, E., Drioli, E. (2014) Membrane reactors: Advanced systems for intensified chemical processes. *Chemie-Ingenieur-Technik* 86: 2039–2050.
24. Amiri, T.Y., Ghasemzageh, K., Iulianelli, A. (2020) Membrane reactors for sustainable hydrogen production through steam reforming of hydrocarbons: A review. *Chemical Engineering and Processing – Process Intensification* 157: 108148.
25. Al-Asheh, S., Bagheri, M., Aidan, A. (2021) Membrane bioreactor for wastewater treatment: A review. *Case Studies in Chemical and Environmental Engineering* 4.
26. Olajire, A.A. (2020) Recent advances on the treatment technology of oil and gas produced water for sustainable energy industry-mechanistic aspects and process chemistry perspectives. *Chemical Engineering Journal Advances* 4: 100049.
27. Sutherland, K. (2010) The rise of membrane bioreactors. *Filtration and Separation* 47: 14–16.
28. Giorno, L., Drioli, E. (2000) Biocatalytic membrane reactors: Applications and perspectives. *Trends in Biotechnology* 18: 339–349.
29. Mazzei, R., Drioli, E., Giorno, L. (2010) Biocatalytic membranes and membrane bioreactors. In: *Comprehensive Membrane Science and Engineering*, Drioli, E., Giorno, L., (Eds.), pp. 195–212. Elsevier Science.
30. Esfahani, B.A., Koupaei, M.S., Ghasemi, S.Z. (2014) Industrial waste water treatment by membrane systems. *Indian Journal of Fundamental and Applied Life Sciences* 4: 1168–1177.
31. Stasinakis A.S. (2008) Use of selected advanced oxidation processes ({AOPs}) for wastewater treatment – a mini review. *Global NEST Journal* 2008: 376–385.
32. Argurio, P., Fontananova, E., Molinari, R., Drioli, E. (2018) Photocatalytic membranes in photocatalytic membrane reactors. *Processes* 6.
33. Besha, A.T., Gebreyohannes, A.Y., Tufa, R.A., Bekele, D.N., Curcio, E., Giorno, L. (2017) Removal of emerging micropollutants by activated sludge process and membrane bioreactors and the effects of micropollutants on membrane fouling: A review. *Journal of Environmental Chemical Engineering* 5: 2395–2414.

34. Kraume, M., Bracklow, U., Vocks, M., Drews, A. (2005) Nutrients removal in MBRs for municipal wastewater treatment. *Water Science and Technology* 51: 391–402.
35. Xiao, K., Liang, S., Wang, X., Chen, C., Huang, X. (2019) Current state and challenges of full-scale membrane bioreactor applications: A critical review. *Bioresource Technology* 271: 473–481.
36. Le-Clech, P., Chen, V., Fane, T.A.G. (2006) Fouling in membrane bioreactors used in wastewater treatment. *Journal of Membrane Science* 284: 17–53.
37. Drioli, E., Ali, A., Macedonio, F. (2015) Membrane distillation: Recent developments and perspectives. *Desalination* 356: 56–84.
38. Ali, A., Quist-Jensen, C.A., Macedonio, F., Drioli, E. (2015) Application of membrane crystallization for minerals' recovery from produced water. *Membranes* 5: 772–792.
39. Drioli, E., Di Profio, G., Curcio, E. (2015) *Membrane-Assisted Crystallization Technology*. World Scientific.
40. Macedonio, F.. Drioli, E. (2008) Pressure-driven membrane operations and membrane distillation technology integration for water purification. *Desalination* 223: 396–409.
41. Macedonio, F., Curcio, E., Drioli, E. (2007) Integrated membrane systems for seawater desalination: energetic and exergetic analysis, economic evaluation, experimental study. *Desalination 203*: 260–276.
42. Macedonio, F., Drioli, E., Curcio, E., Di Profio, G. (2009) Experimental and economical evaluation of a membrane crystallizer plant. *Desalination and Water Treatment* 9: 49–53.
43. Macedonio, F., Drioli, E. (2010) Hydrophobic membranes for salts recovery from desalination plants. *Desalination and Water Treatment* 18: 224–234.
44. Gugliuzza, A., Macedonio, F., Politano, A., Drioli, E. (2019) Prospects of 2D materials-based membranes in water desalination. *Chemical Engineering Transactions* 73: 265–270.
45. Macedonio, F., Criscuoli, A., Gzara, L., Albeirutty, M., Drioli, E. (2021) Water and salts recovery from desalination brines: An exergy evaluation. *Journal of Environmental Chemical Engineering* 9: 105884.
46. European Space Agency (1996) *Viability study for a membrane separatorfor spatial applications: Project "Liquid Gas Phase Separation,"* Contract ESTeC n. 11523/95/NL/SB (European Space Agency) – Subcontract DO 1410093 I00222 (DASA Dornier GmbH) (1.7.1995–30.6.1996).
47. Drioli, E., Golemme, G., Santella, F. (1996) *Liquid-Gas Phase Separation: List of Processes. Report in the Framework of the Contract ESTEC n. 11523/95/NL/SB.* European Space Agency.
48. Drioli, E., Golemme, G. (1996) *Liquid-Gas Phase Separation: Quantitative Evaluations. Report in the Framework of the Contract ESTEC n. 11523/95/NL/SB.* European Space Agency.
49. Report ESA/PB-HME (2019) European Space Agency Human Spaceflight, Microgravity and Exploration Programme Board, 30.
50. Bernardo, P., Iulianelli, A., Macedonio, F., Drioli, E. (2021) Membrane technologies for space engineering. *Journal of Membrane Science* 626: 119177.
51. Volpin, F., Badeti, U., Wang, C., Jiang, J., Vogel, J., Freguia, S., Fam, D., Cho, J., Phuntsho, S., Shon, H.K. (2010) Urine treatment on the International Space Station: Current practice and novel approaches. *Membranes* 10.
52. Pruitt, J.M., Carter, D.L., Bagdigian, R.M., Kayatin, M.J. (2016) Upgrades to the ISS Water Recovery System. *45th International Conference on Environmental Systems* 10–14.
53. Nicolau, E., Fonseca, J., Rodríguez-Martínez, J., Richardson, J.T.-M., Flynn, M., Griebenow, K., Cabrera, R.C. (2014) Evaluation of a urea bioelectrochemical system for wastewater treatment processes. *ACS Sustainable Chemistry & Engineering* 2: 749–754.
54. Sabatier, P., Senderens, J. (1902) Hydrogénation directe des oxydes du carbone en présence de divers métaux divisés. *Comptes Rendus.*
55. Macedonio, F., Brunetti, A., Barbieri, G., Drioli, E. (2012) Membrane condenser as a new technology for water recovery from humidified "waste" gaseous streams. *Industrial & Engineering Chemistry Research* 52: 1160–1167.
56. Macedonio, F., Brunetti, A., Barbieri, G., Drioli, E. (2017) Membrane condenser configurations for water recovery from waste gases. *Separation and Purification Technology* 181: 60–68.
57. Macedonio, F., Brunetti, A., Barbieri, G., Drioli, E. (2019) Membrane condenser as emerging technology for water recovery and gas pre-treatment: Current status and perspectives. *BMC Chemical Engineering* 1: 1–15.
58. Macedonio, F., Frappa, M., Bamaga, O., Abulkhair, H., Almatrafi, E., Albeirutty, M., Tocci, E., Drioli, E. (2022) Application of a membrane condenser system for ammonia recovery from humid waste gaseous streams at a minimum energy consumption. *Applied Water Science* 12: 90.
59. Macedonio, F., Cersosimo, M., Brunetti, A., Barbieri, G., Drioli, E. (2014) Water recovery from humidified waste gas streams: Quality control using membrane condenser technology. *Chemical Engineering and Processing: Process Intensification* 86: 196–203.
60. Macedonio, F., Frappa, M., Brunetti, A., Barbieri, G., Drioli, E. (2020) Recovery of water and contaminants from cooling tower plume. *Environmental Engineering Research* 25: 222–229.

Index

E

N

O

P

Q

R

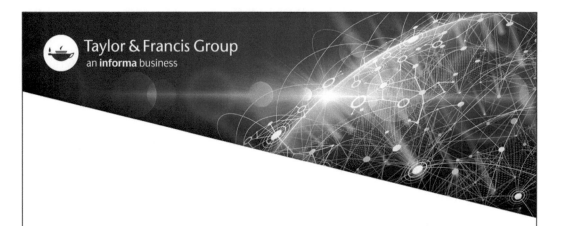